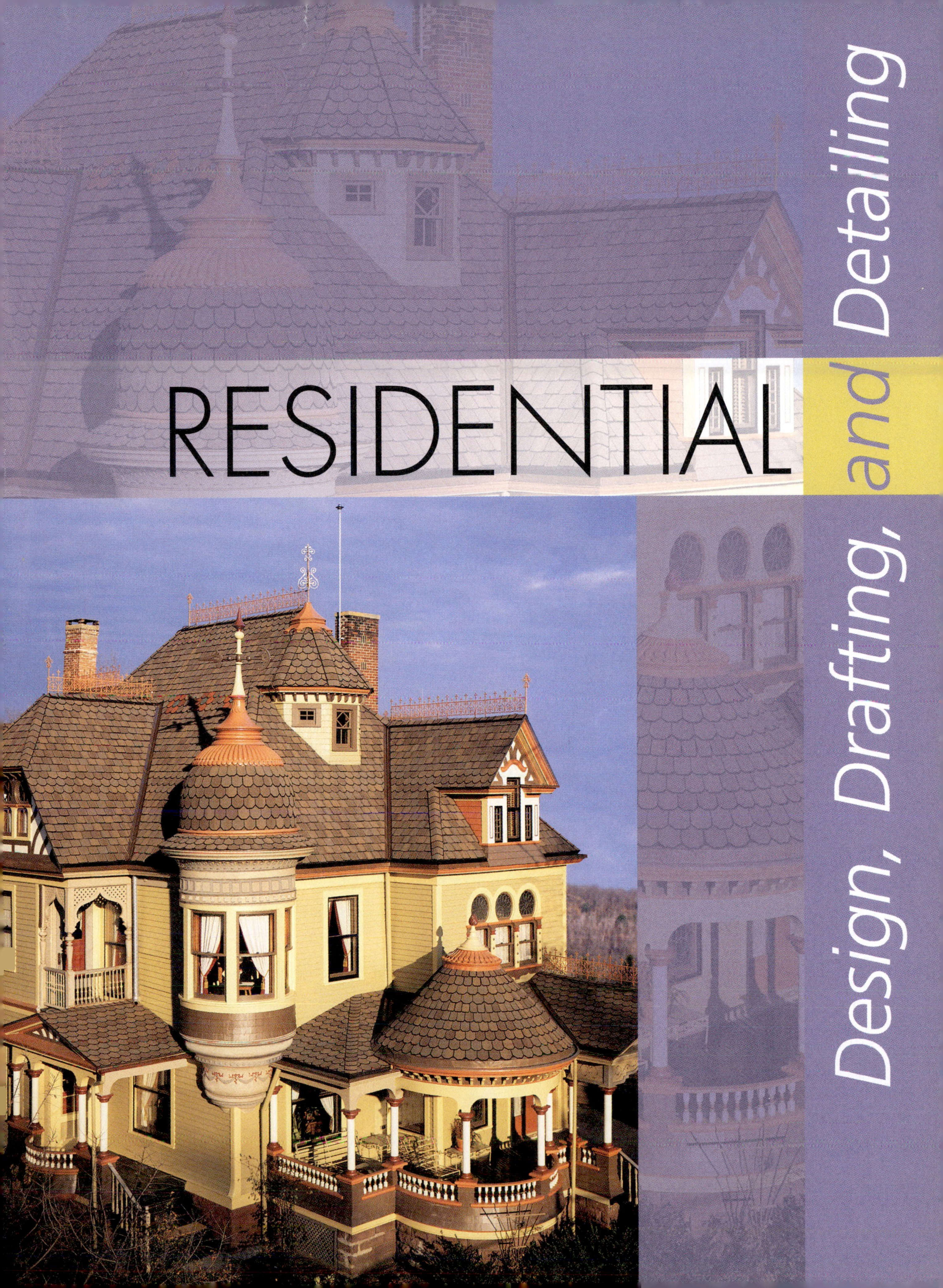
RESIDENTIAL
Design, Drafting, and Detailing

RESIDENTIAL Design, Drafting, and Detailing

ALAN JEFFERIS, JANICE JEFFERIS

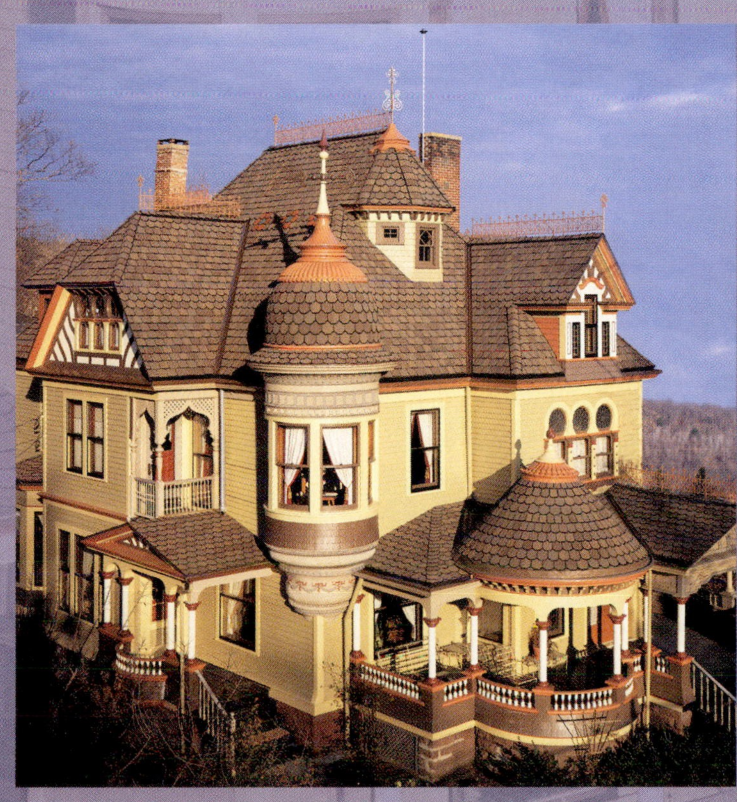

THOMSON
DELMAR LEARNING

Africa • Australia • Canada • Denmark • Japan • Mexico • New Zealand
Philippines • Puerto Rico • Singapore • Spain • United Kingdom • United States

Residential Design, Drafting, and Detailing
Alan Jefferis, Janice Jefferis

Vice President, Technology and Trades ABU:
David Garza

Director of Learning Solutions:
Sandy Clark

Managing Editor:
Larry Main

Acquisitions Editor:
James Devoe

Product Manager:
John Fisher

Marketing Director:
Deborah Yarnell

Marketing Manager:
Kevin Rivenburg

Marketing Coordinator:
Mark Pierro

Director of Production:
Patty Stephan

Production Manager:
Stacey Masucci

Content Project Manager:
Andrea Majot

Technology Project Manager:
Kevin Smith

Editorial Assistant:
Tom Best

Printed in United States of America
1 2 3 4 5 XX 09 08 07

For more information contact
Thomson Delmar Learning,
Executive Woods
5 Maxwell Drive, PO Box 8007,
Clifton Park, NY 12065-2919
or find us on the World Wide Web at
www.delmarlearning.com

Library of Congress Cataloging-in-Publication Data
Jefferis, Alan.
Residential design, drafting, and detailing / Alan Jefferis.
p. cm.
ISBN 1-4180-1275-0 (alk. paper)
1. Architectural drawing—Technique.
2. Architectural design—Technique.
3. Architecture, Domestic—Designs and plans.
I. Title.
NA2708.J447 2007
728.028--dc22 2007015295

ISBN-10: 1-4180-1275-0
ISBN-13: 978-1-4180-1275-5

NOTICE TO THE READER

CONTENTS

29—Section Layout 745

30—Stair Construction and Layout 785

31—Fireplace Construction and Layout 797

SECTION IX

SUPPLEMENTAL DRAWINGS 815

32—Renovations, Remodeling, and Additions 816

Abbreviations 849

Glossary 853

Index 867

PREFACE

Residential Design, Drafting, and Detailing is a practical, comprehensive textbook that is easy to use and understand. The content may be used as presented, by following a logical sequence of learning activities for design and drafting, or the chapters may be rearranged to accommodate alternative formats for traditional or individualized instruction.

APPROACH

Practical

Residential Design, Drafting, and Detailing provides a practical approach to architectural drafting as it relates to current common professional practices. The emphasis on standardization of CAD techniques will provide an excellent and necessary foundation of drafting training and implement a common approach to drafting. After students become professional CAD drafters, this text will serve as a valuable desk reference.

Realistic

Chapters contain professional examples, illustrations, step-by-step layout techniques, drawing problems, and related tests. The examples demonstrate recommended drafting presentation with actual architectural drawings used for reinforcement. The correlated text explains CAD techniques and provides useful information for skill development. Step-by-step layout methods provide a logical approach to beginning and finishing complete sets of working drawings.

Practical Approach to Problem Solving

The professional CAD technician's responsibility is to work as part of a design team and convert the sketches and ideas of other team members into formal drawings. The text explains how to prepare construction documents from design sketches by providing the learner with the basic guidelines for drawing layout, common design guidelines, and the minimum code requirements related to each drawing. The concepts and skills learned from one chapter to the next allow students to prepare complete sets of residential working drawings. Problem assignments are presented on the student CD in order of difficulty and in a manner that provides students with a wide variety of residential drafting and design experiences.

The problems are presented as preliminary designs or design sketches in a manner that is consistent with actual architectural office practices. It is not enough for students to duplicate drawings from given assignments; they must be able to think through the process of drawing development with a foundation of how drawing and construction components are implemented. The goals and objectives of each problem assignment are consistent with recommended evaluation criteria based on the progression of learning activities. The drafting problems assume that you will be using AutoCAD 2004 or newer.

FEATURES OF THE TEXT

Applications

Realistic problems are presented as preliminary drawings in a manner that is consistent with industry practices. The problems have been supplied by architectural designers and architects and have each been built to meet a specific need. Each problem solution is based on the step-by-step layout procedures provided in the chapter discussions. Problems are given in order of complexity so students may be exposed to a variety of drafting experiences. Problems require students to go through the same thought and decision-making processes that a professional CAD technician faces daily, including view layout, dimension placement, section placement, and many other activities. Chapter tests provide complete coverage of each chapter and may be used for student evaluation or as study questions.

DIRECTIONS

CHAPTER 9

Drawing Problems 

Unless your instructor assigns a specific site, select one of the following site plans using the following guidelines.

Visit the website for the municipality that governs your area and use the setbacks for the given size of the lot you have chosen, or use the following setbacks:

Minimum front setback, 25'-0''
Minimum rear yard setback, 20'-0''
Minimum side yard setback: one level, 5'-0''; two level, 6'-0''

Use the specified paper size for the municipality that governs your area. If your zoning department does not specify a paper size, draw your drawing to fit on an 8.5'' × 14'' sheet of paper.

Select the appropriate scale to draw the required site drawing.

PROBLEMS

Problems 9.1 through 9.7. Use the subdivision map in Figure 8.7 and draw the site plan for one of the parcels of land. Begin the selected site plan problem by representing the given information in preparation to complete a preliminary design study for one of the homes in Chapter 11. Once the floor plan for your preliminary floor plan is approved, complete the required site drawings.

9.8. Draw a vicinity map of your school showing major access routes and important landmarks within a 5-mile radius.

9.9. Use the following legal description and draw the site plan for a tract of land situated in the SW 1/4 of the NE 1/4 of section 17, T2S, R1W of Mount Diablo Meridian, Rancho Santa Barbara:

Beginning at a 5/8'' iron rod marking the true point of beginning of the southwest property corner that lies 32' directly north of the center of Rancho Santa Barbara Place, proceed N 3° 15' W 115.0', thence N8° 30' E 140.0', thence proceed N 90° 00'E a distance of 92.10; thence proceed S38° 30' E 91.5'; thence proceed due south a distance of 181.5'; and thence proceed due West 163.25' back to the true point of beginning.

9.10. Use the following legal description and draw the site plan for the following tract of land:

Parcel 12, tax lot 215682 of the Miller Addition, situated in section 12, T3N, R1E, 6th Principal Meridian, Cloud County, Kansas. Beginning at a 5/8'' iron rod marking the southerly corner and the true point of beginning of the southeast property corner, go a distance of 185.0' N49° 50' W to the far westerly corner. Said property line lies 32.00' north of the center of Miller Drive. Thence N 65° 15' E a distance of 104.50'; thence N49° 00' W a distance of 147.0'; thence N43° 30' E a distance of 93.0' back to the true point of beginning.

9.11. Use the following description to draw and label a site plan that can be plotted at a scale of 1/8'' = 1'–0'':

Beginning at a point that is the NE corner of the G.M. Smith D.L.C., which lies 250 feet north of the centerline of N.E.122 street which is in Section 1AB, Township 6 north, Range 1 west of Los Angles county, California thence south 225.00 feet to a point which is the southwest corner

PROBLEMS

Computer-Aided Design Drafting (CADD)

CADD is a valuable tool used in most professional offices. This book is based on the use of AutoCAD 2000 or newer. Other CAD programs can be used, but the drawings on the student CD may not be suitable. The CD contains skeleton drawings for many of the drawing projects. These drawings can serve as a base for a specific drawing project, but students will need to be responsible to set all drawing variables to meet the demands of their program. It is important to remember that professionals have donated these drawings with the understanding that students will use them as a learning experience. Any reproduction of these drawings for personal use is illegal. Although laws vary for each state, unless more than 50% of the plans are altered, you can be prosecuted for copyright infringement.

Note:

Because professional architects and designers have prepared many of the drawings in this text, some of the drawings may vary from the techniques that have been presented. Keep in mind that the architectural design world does not conform to one standard. The company that signs your check is correct; the rest of the world is out of step.

Every office has a library of stock notes, symbols, and details. The CD also contains symbols, notes, and details that can be inserted into your drawings to save time. Many of the symbols should be converted to blocks to save additional time. As a professional CAD technician, you will be expected to verify the accuracy of the information that you insert into a drawing. Notes that are placed on one floor plan may not be suitable for another floor plan. Before you insert any notes or details into your drawing, verify that they meet the needs of the project.

Code and Construction Techniques

The 2006 International Residential Code is introduced throughout the text as it relates to specific instructions and applications. Construction techniques differ through-

STAIR TYPE	IRC
Straight stair	
Max. rise	7 3/4″ (195 mm)
Min. run	10″ (250 mm)
Min. headroom	6′–8″ (2000 mm)
Min. tread width	3′–0″ (900 mm)
Handrail height	34″ (850 mm) min. 38″ (950 mm) max.
Guardrail height	36″ (900 mm) min.
Winders	
Min. tread depth	6″ (150 mm) min. 10″ (250 mm) @ 12″ (300 mm)
Spiral	
Min. width	26″ (650 mm)
Min. tread depth	7 1/2″ (190 mm) @ 12″ (300 mm)
Max. rise	9 1/2″ (240 mm)
Min. headroom	6′–6″ (1950 mm)

CONSTRUCTION TECHNIQUES

out the country. This text clearly acknowledges the difference in construction methods and introduces the student to the format used to make complete sets of working drawings for each method of construction. Students may learn to prepare drawings from each construction method or, more commonly, for the specific construction techniques that are used in their locality. The problem assignments are designed to provide drawings that involve a variety of construction alternatives.

Additional Readings

An Additional Readings section is offered at the end of selected chapters. This is a list of websites for companies or organizations that offer services or materials related to the chapter content. This list is by no means complete. Use a search engine such as Google to increase your resource base.

Organizing Your Course

Architectural drafting is the primary emphasis of many technical drafting curricula, whereas other programs offer only an exploratory course in this field. This text is appropriate for either application because its content reflects the common elements found in an architectural drafting curriculum.

Prerequisites

An interest in architectural drafting plus basic computer skills, basic arithmetic, written communication, and reading skills are the only prerequisites required. Basic drafting skills and layout techniques are presented as appropriate. Students should have an understanding of basic AutoCAD drawing and editing skills. Students with an interest in architectural drafting who begin using this text will end with the knowledge and skills required to prepare complete sets of working drawings for a residence.

Fundamental through Advanced Coverage

This text may be used in an architectural drafting curriculum that covers the basics of residential architecture in a one-, two-, or three-semester sequence. In this application, students use the chapters directly associated with the preparation of a complete set of working drawings for a residence, in which the emphasis is on the use of fundamental skills and techniques. The balance of the text may remain as a reference for future study or as a valuable desk reference.

ADDITIONAL READINGS

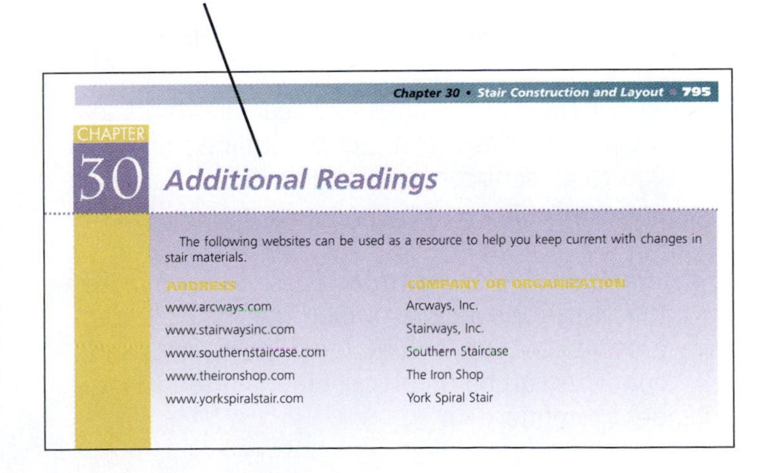

Chapter 30 • Stair Construction and Layout ■ 795

CHAPTER 30 Additional Readings

The following websites can be used as a resource to help you keep current with changes in stair materials.

ADDRESS	COMPANY OR ORGANIZATION
www.arcways.com	Arcways, Inc.
www.stairwaysinc.com	Stairways, Inc.
www.southernstaircase.com	Southern Staircase
www.theironshop.com	The Iron Shop
www.yorkspiralstair.com	York Spiral Stair

The text may also be used in the comprehensive architectural drafting program in which a four- to six-semester sequence of architectural drafting and design is required. In this application, students may expand on the primary objective of preparing a complete set of working drawings for the design of residential projects with the coverage of any one or all of the following areas: residential drafting, energy-efficient construction techniques, solar and site orientation design applications, planning for heating and cooling needs, and structural load calculations.

Back of Book CD

The back of book CD contains AutoCAD drawing files for use in many of the drawing projects in the text. This is explained throughout the preface. Open the Read Me file on the CD to thoroughly understand the file structure of the CD. The CD also contains eight appendices, drawing blocks, and a folder containing extra photographs that depict architectural styles and concepts.

SUPPLEMENTS

E.Resource

This is an educational resource that creates a truly electronic classroom. It is a CD-ROM containing tools and instructional resources that enrich your classroom and make the instructor's preparation time shorter. The elements of e.resource link directly to the text and tie together to provide a unified instructional system. With e.resource you can spend your time teaching, not preparing to teach. ISBN: 1-4180-1276-9.

Features contained in e.resource include:

Syllabus: Lesson plans created by chapter. You have the option of using these lesson plans with your own course information. www.worldclasslearning.com.

Chapter Hints: Objectives and teaching hints that provide the basis for a lecture outline that helps you to present concepts and material. Key points and concepts can be graphically highlighted for student retention.

PowerPoint® Presentation: These slides provide the basis for a lecture outline that helps you to present concepts and material. Key points and concepts can be graphically highlighted for student retention.

ExamView Computerized Test Bank: Over 800 questions of varying levels of difficulty are provided in true/false and multiple choice formats so you can assess student comprehension.

Video and Animation Resources: These AVI files graphically depict the execution of key concepts and commands in drafting, design, and AutoCAD and let you bring multi-media presentations into the classroom.

Photos: Additional photos are included to provide additional resources for lectures.

Solutions Manual

A solutions manual is available with answers to end of chapter review questions and solutions to end of chapter problems. ISBN: 1-4180-1277-7.

Videos

Two video sets, containing four 20-minute tapes each, are available. The videos correspond to the topics addressed in the text:

Set #1, ISBN 0–7668–3094–2
Set #2, ISBN 0–7668–3095–0

Video sets are also available on Interactive Video CD-ROM.

Set #1, ISBN 0–7668–3116–7
Set #2, ISBN 0–7668–3117–5

ACKNOWLEDGMENTS

We would like to thank and acknowledge the many professionals who reviewed the manuscript to help us publish our architectural drafting text. A special acknowledgment is due the instructors who reviewed the chapters in detail.

Reviewers

Tom Bledsaw
ITT Technical Institute, Indianapolis, IN

Walter B. Cheever
South Central Technical College, N. Mankato, MN

LeRoy Cook
Clackamas Community College, Oregon City, OR

Margaret Ann Jeffries
Pellissippi State Technical Community College, Knoxville, TN

Tom Kane
Pueblo Community College, Florence, CO

David LaRue
ITT Technical Institute, Strongsville, OH

Jeff Plant
Salt Lake Community College, Salt Lake City, UT

Joe Ramos
Mt. San Antonio College, Walnut, CA

Joey Spillyards
Red Rocks Community College, Lakewood, CO

Roy Trouerbach
University of Advancing Computer Technology, Tempe, AZ

Tony Whitus
Tennessee Technology Center, Crossville, TN

Contributing Companies

The quality of this text is also enhanced by the support and contributions from architects, designers, engineers, and vendors. The list of contributors is extensive and acknowledgment is given at each illustration. The following individuals and companies gave an extraordinary amount of support with technical information and art for this edition:

Glen Becker
ABTCO Inc.

Alan Mascord
Alan Mascord Design Associates, Inc.

W.A. Erdos
Alside, Inc.

Pamela Russom
American PolySteel

Ray Clark
APA—The Engineered Wood Association

Dennis Dinser
Arcadian Designs, Inc.

Kathy Kelly and Larry Bowa
BOWA Builders

J. Gregg Borchelt, P.E.
Brick Industry Association

Tracy Lacht
Broan-NuTone LLC

Pamela Allsebrook
California Redwood Association

LeRoy Cook
Custom Home Design

David P. Schulze
DPS-Labs

Meredith Brandes
Elk Roofing

Eric S. Brown
Eric S. Brown Design Group Inc.

Dr. Carol Ventura
ICF home photos

Peggy Nila and Cheryl Melendez
International Conference of Building Officials

Ken Smith
Ken Smith Architect and Associates, Inc.

Laine M. Jones
Laine M. Jones Design

Steve Webb
Lennox Industries, Inc.

Pamela Winikoff
Leviton

Kathleen M. Arnelt
Louisiana-Pacific Corporation

Debbie Teague
McElroy Metal

Leah West
Mississippi Concrete Industries Association

Richard Branham
National Concrete Masonry Association

Richard R. Chapman and Robert C. Guzikowski
Simpson Strong-Tie Company, Inc.

Richard Wallace
Souther Pine Council

Ray Clark
Southern Forest Production Association

Kate Cammack and Kerry Vanden Heuvel
Stephen Fuller, Incorporated

Robin Kelley
Trus Joist, A Weyerhaeuser Business

Lisa Kotasek
Uponor Wirsbo

W. Lee Roland
W. Lee Roland, Builders

W.D. Farmer FAIBD
W.D. Farmer Residence Designer, Inc.

Eric C. Wilson
Western Wood Products Association

William E. Poole
William E. Poole Designs, Inc.

Rascoe Clark
Wood Basements

Special thanks goes to the International Code Council (ICC) for granting permission to use text from the International Residential Code/2006:

Thanks also to many of my excellent former students who have contributed drawing material to this text and who are now pulling down the big bucks as professionals designers: Kimi Barnham, Lisa Echols, Teresa Jefferis, Eli Jefferson, Benigno Molina-Manriquez, Aarron Phillips, Katja Poschwatta, Tracy Reif, Pavel Sandu, and Stanley Tiffany.

And finally, thanks to my many family members who have provided photos of homes from across the country, including David, Sara, Michael, Tereasa, Aaron, Jordan, Zachary, Megan, and Matthew Jefferis. Thanks also for patiently waiting for me to finish this book. And most of all, thanks to Him who is able to keep you from falling and is able to present you before His glorious presence without fault and with great joy.

Artwork for title pages and section openers in the text were provided courtesy of the following companies:

Half-title page: CertainTeed

Title page: CertainTeed

Section I: Alan Mascord, Photography by Bob Greenspan, Alan Mascord Design Associates, Inc.

Section III: BOWA Builders, Photography by Greg Hadley

Section IV: Velux

Section V: Laine M. Jones, Laine M. Jones Design

Section VI: David Jefferis

Section VIII: Michael Jefferis

TO THE STUDENT

Residential Design, Drafting, and Detailing is designed for you, the student. The development and format of the presentations have been tested in classroom instruction. The information presented is based on architectural standards, common drafting practice, and trends in the drafting industry. Use the text as a learning tool while in school, and take it along as a desk reference when you enter the profession. Examples and illustrations are used extensively. Drafting is a graphic language, and most drafting students learn best by observation of examples. Here are a few helpful hints.

1. *Read the text.* (Duh! Why else would you buy the book, other than to keep my wife and kids off the streets?) The text content is intentionally designed for easy reading. Sure, it doesn't read the same as an exciting, sexy short story, but it does give construction facts in as few, easy-to-understand words as possible. Don't pass up the reading, because the content will help you to understand the drawings clearly.
2. *Look carefully at the examples.* Look at the examples carefully in an attempt to understand the intent of specific applications. If you are able to understand why something is done a certain way, it will be easier for you to apply those concepts to the drawing problems and, later, to the job. Preparing construction documents is a precise technology based on rules and guidelines. These drawings not only guide construction but also serve as legal documents in a court of law if misunderstandings occur. To avoid misunderstandings, the CAD drafter must prepare drawings that are well organized, with information clearly specified in a method that is easy to interpret. There will always be situations when rules must be altered to handle a unique situation. Then you will have to rely on judgment based on your knowledge of accepted standards. Drafting is often like a puzzle; there may be more than one way to solve a problem.
3. *Use the text as a reference.* Few drafters, with the exception of Jesus, Monk, or John Doe, remember everything they read. Trying to remember everything about drafting standards, techniques, and concepts is impossible. It's imperative that after you

have read a chapter, you highlight key information so you'll be able to use the book as a reference. A very wise employer once told me, "Don't remember facts, remember where to find them." Become familiar with the definitions and use of technical terms. It would be difficult to memorize everything noted in this text, but after considerable use of the concepts, architectural drafting applications should become second nature.

4. *Learn each concept and skill before you continue to the next.* The text is presented in a logical learning sequence. You need to understand key information to be placed on one drawing, before you can advance to another. Failure to master each chapter in sequential order may cause your drawings to be cluttered with information that may belong on a different drawing. Each chapter is designed for learning development, and chapters are sequenced so that drafting knowledge grows from one chapter to the next. Problem assignments are presented in the same learning sequence as the chapter content and also reflect progressive levels of difficulty.
5. *Practice.* Development of good computer skills depends to a large extent on practice. Some individuals have an inherent talent for CAD drafting, and some people are readily compatible with computers. If you fit into either group, great! If you don't, then practice. Practice your CAD skills to help improve the quality of your drafting presentation, and practice communicating and working with a computer.
6. *Use sketches or preliminary drawings.* When you are drawing with a computer, the proper use of a sketch or preliminary drawing can save a lot of time in the long run. Prepare a preliminary layout for each problem. Even with the powerful MOVE command, planning will give you a chance to organize thoughts about drawing scale, view selection, dimension and note placement, and paper size. After you become a veteran, you may be able to design a sheet layout in your head, but until then, you will be sorry if you don't use sketches.

Alan Jefferis
Residential designer,
CAD instructor, drafting geek

Janice Jefferis
Residential designer, interior
decorator, drafting goddess

SECTION 1

Introduction to Architectural Drafting

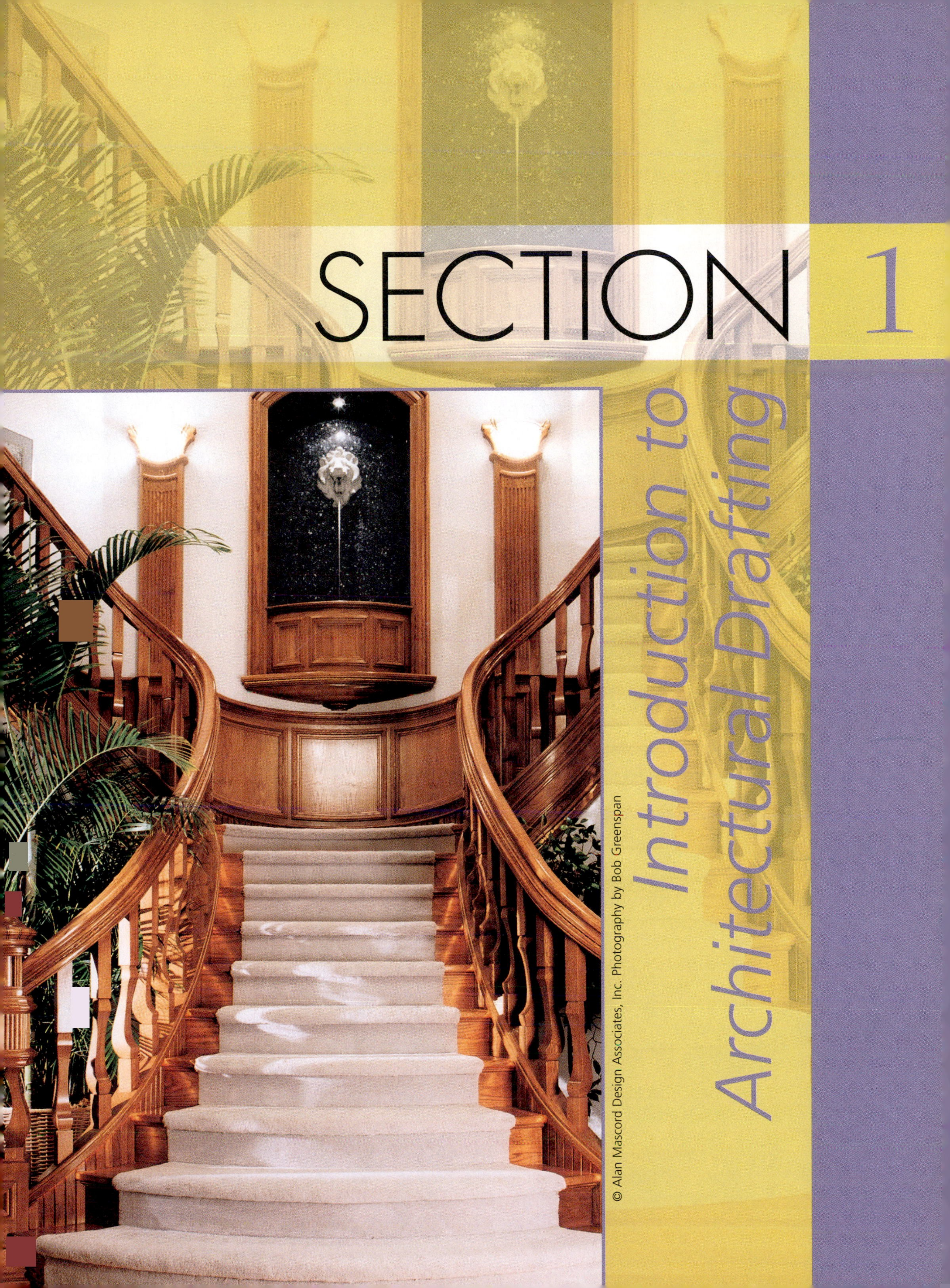

© Alan Mascord Design Associates, Inc. Photography by Bob Greenspan

1

CHAPTER

Professional Architectural Careers, Office Practice, and Opportunities

As you begin working with this text, you are opening the door to many exciting careers. Each career, in turn, has many different opportunities within it. Whether your interest lies in theoretical problem solving, artistic creations, or working with your hands creating something practical, a course in residential architecture will help prepare you to satisfy that interest. An architectural drafting class can lead to a career as a computer-aided design (CAD) technician, designer, interior decorator, interior designer, architect, engineer, model maker, specification writer, or illustrator. Once you have mastered the information and skills presented in this text, you will be prepared for each of these fields as well as many others.

DRAFTER/CAD TECHNICIAN

The term ***drafter*** has been used for years to describe a person who creates the drawings and details for another person's creations. Now that computers dominate design offices, the term has been replaced by the title ***CAD technician.*** In addition to the terms *drafter* and *CAD technician,* job listing may also be described by the terms *CAD designer, engineering technician, CAD operator, CAD design technician, CAD engineering design technician,* technician and *design drafter.* The term *drafter* is used for that rare individual who draws manually rather than with a computer. Many training programs require entry-level students to master manual drafting skills before they incorporate computers into their studies. Many people involved with interior design still use manual skills because of the artistic effects that can be obtained. It is the CAD technician's responsibility to use the proper line properties, dimensions, and text styles and to properly lay out the required drawings necessary to complete a project. Such a task requires great attention to detail as the technician draws the sketches created by other team members. Because the technician could possibly be working for several architects or engineers within an office, he or she must be able to get along well with others.

The Entry-Level Technician

Your job as an entry-level technician will generally consist of making corrections to drawings created by others. There may not be a lot of mental stimulation to making changes, but it is a very necessary job. It is also a good introduction to the procedures and quality standards within an office.

As your knowledge of company standards and confidence increase, your responsibilities will be expanded. To advance in an office, you'll need to become proficient in using the firm's computer standards and any special menus and list processing (LISP) routines needed to work efficiently. No matter what tools are used to create the drawings, your supervisor will typically give you a sketch and expect you to draw the required drawing. Figure 1.1a shows a designer's sketch. Figure

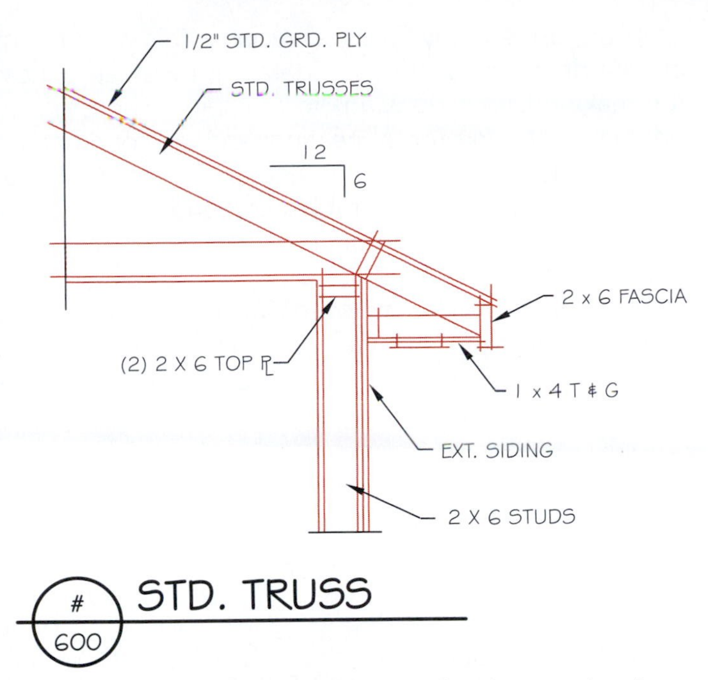

FIGURE 1.1a ■ *A sketch or skeleton drawing may be given to an entry-level technician to follow until experience is gained. Information for completing the drawing can be found by examining similar jobs in the office.*

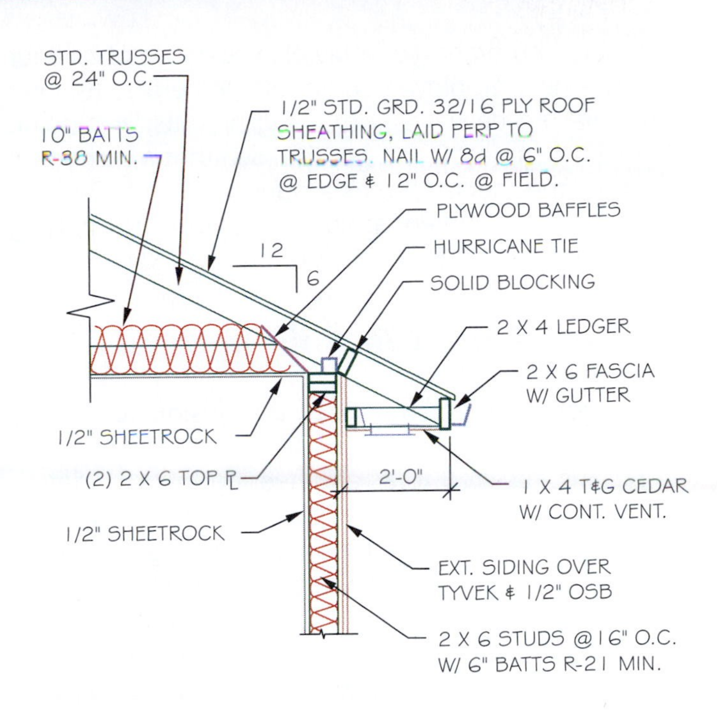

FIGURE 1.1b ■ *A detail drawn by a CAD technician using the sketch shown in Figure 1.1a.*

1.1b shows the drawing created by a CAD technician. As you gain an understanding of the drawings you are making and gain confidence in your ability, the sketches you are given generally will become more simplified. Eventually your supervisor may just refer you to a similar drawing and expect you to be able to make the necessary adjustments to fit it to the new application.

The decisions involved in making drawings without sketches require the technician to have a good understanding of what is being drawn. This understanding does not come just from a textbook. To advance and become a leader on the drawing team will require you to become an effective manager of your time. This would include the ability to determine what drawings will need to be created or selected from a stock library and edited and to estimate the time needed to complete these assignments and meet deadlines established by the team captain, the client, the lending institution, or the building department. An even better way to gain an understanding of what you are drafting is to spend time working at a construction site so that you understand what a craftsperson must do as a result of what you have drawn. Figure 1.1b shows a detail created by a CAD technician, with the resulting construction shown in Figure 1.2.

To become a good team leader you will also need to develop skills that promote a sense of success among your teammates. Although it is against the law to discriminate on the basis of race, color, religion, gender, sexual orientation, age, marital status, or disability, exceeding the minimum requirement of the law to create a friendly and productive work environment is a critical skill for a team leader. Depending on the size of the of-

FIGURE 1.2 ■ *Details created by a CAD technician are used to create every aspect of construction.* Courtesy Benny Molina-Manriquez.

fice where you work, you may also spend a lot of your time as a new employee editing stock details, running prints, making deliveries, obtaining permits, and doing other office chores. Don't get the idea that a technician does only the menial chores around an office. But you do need to be prepared, as you go to your first drafting job, to do things other than drafting.

The Experienced Technician/Detailer

Although other members of the design team may prepare the basic design for a project, experienced technicians, referred to as engineering techs or detailers, are expected to make decisions about construction design. These might include determining structural sizes and connection methods for intersecting beams, drawing renderings or three-dimensional models, visiting job sites, and supervising entry-level technicians. The types of drawings you will be working on will also change as you gain experience. Instead of drawing site plans, electrical plans, cabinet elevations, or roof plans or revising existing drawings, an experienced technician and team leader may be working on the floor and foundation plans, elevations, and sections. Another member of the design team will probably still make the initial design drawings but will pass these drawings on to you as soon as a client approves the preliminary drawings.

In addition to drafting, you may work with the many city and state building departments that govern your work. This will require you to research the codes that govern the building industry. You will also need to become familiar with vendors' catalogs. The most common is the *Sweet's Catalog.* Sweet's, as it is known, is a series of books that contain product information on a wide variety of building products. Sweet's is updated quarterly or on the Internet. Information in Sweet's can be found by listings of the manufacturer, trade name, and type of product.

Employment Opportunities

Engineering technicians are employed in firms of all sizes. Job opportunities for architectural technicians can be found by searching the categories of architectural drafter, architectural design, architectural CAD operator, CAD drafter, CAD technician, and engineering technician in the classified ads of your local newspaper or by looking on the Internet. Designers, architects, and engineers all require entry-level and advanced technicians to help produce their drawings. Architectural equipment suppliers also employ CAD technicians. This work might include drawing construction details for a steel fabricator, making layout drawings for a cabinet shop, or designing ductwork for a heating and air-conditioning installer. Many manufacturing companies hire CAD operators with an architectural background to help draw and sometimes sell a product or draw installation diagrams for instruction booklets or sales catalogs. CAD technicians are also employed by many government agencies. These jobs include working in planning, utility, or building departments; on survey crews; or in other related municipal jobs.

Educational Requirements

In order to get your first CAD drafting job, you will need a solid education, a good understanding of basic computer skills, good CAD drawing skills, and the ability to sell yourself to an employer. CAD skills must include a thorough understanding of drawing and editing commands as well as the ability to quickly decide which option is best for the given situation. This ability will also come with practice. If you work full time editing details, you'll quickly become proficient in deciding on the best commands to use. The education required for a CAD technician can range from one or more years in a high school drafting program to a diploma from a one-year accredited technical school to a degree from a two-year college program and all the way to a master's degree in architecture.

Helpful areas of study for an entry-level CAD technician would include math, writing, drawing, and computer drafting. The math required ranges from simple addition to calculus. Although the CAD technician may spend most of the day adding dimensions expressed in feet and inches, knowledge of advanced math will be helpful for solving many building problems. You'll often be required to use basic math skills to determine quantities, areas, and volume. Many firms provide their clients with a list of materials required to construct the project. Areas will need to be calculated to determine the size of each room, required areas to meet basic building code requirements, the loads on a structural member, or the size of the structure. Writing skills will also be very helpful. As a senior CAD technician, you're often required to complete the paperwork that accompanies any set of plans, such as permits, requests for variances, written specifications, or environmental impact reports. In addition to standard CAD drafting classes, classes in photography, art, surveying, and construction will be helpful.

DESIGNER

The meaning of the term ***designer*** varies from state to state. Many states restrict the use of the term by requiring those calling themselves designers to have had formal training and to have passed a competency test. A designer's responsibilities are very similar to those of an experienced technician and are usually based on both education and experience. A designer is usually the coordinator of a team of CAD technicians. The designer may work under the direct supervision of an architect, an engineer, or both and supervise the work schedule of the drafting team.

In addition to working in a traditional architectural office setting, designers may have their own office practice in which they design residential and multi-family structures. The ***American Institute of Building Design*** is a national association of designers and CAD technicians who are certified to be knowledgeable in the field of residential design. State laws vary regarding the types and sizes of buildings a designer may work on without the stamp of an architect or engineer. Students wishing more information about a career as an architectural designer can obtain information from the:

American Institute of Building Design (AIBD)
991 Post Road East
Westport, CT 06880
1-800-366-2423
Website: www.aibd.org
E-mail: aibdnat@aol.com

Students can also contact the American Design and Drafting Association, the U.S. Department of Labor, or the U.S. Office of Education.

KITCHEN AND BATH DESIGNER

Some CAD technicians and designers choose to specialize in the area of residential kitchen and bath design. Many kitchen and bath designers are hired by a homeowner to do design work. Others are hired by residential contractors to help with projects. Kitchen and bath designers also work directly for the manufacturers of kitchen and bath equipment fixtures. In order to meet the demand for qualified professionals in this area of design, the ***National Kitchen & Bath Association*** (NKBA) has created its own training program that offers the foundation for professional career growth through its course offerings, technical manuals, and multilevel certification programs. The three professional levels for which the NKBA certifies its members, as well as their requirements, include:

- **Associate of Kitchen & Bath Designer** (AKBD): One year of experience related to the kitchen and bath industry and one year in a related field, thirty hours of NKBA professional development training, and successful completion of the AKBD exam.
- **Certified Kitchen Designer** (CKD) or Certified Bathroom Designer (CBD): three years of experience related to the kitchen and bath industry and four years in a related field, sixty hours of NKBA professional development training, and successful completion of the AKBD exam.
- **Certified Master Kitchen & Bathroom Designer** (CMKBD): Ten years of experience related to the kitchen and bath industry, 100 hours of NKBA professional development training, and both the CKD and CBD certification.

For further information about careers in the field of kitchen and bath design, contact:

National Kitchen and Bath Association (NKBA)
687 Willow Grove Street
Hackettstown, NJ 07840
1-800-843-6522
Website: www.nkba.org

INTERIOR DECORATOR

An ***interior decorator*** plans the interiors of buildings with the aim of making rooms more attractive, comfortable, and functional. Most interior decorators are hired to decorate homes, but they may also be hired to decorate interiors of businesses and offices. They may work on the entire interior of a building or a single room. An interior decorator's work may involve a variety of elements, including space planning, determination of color schemes, furniture placement, and the coordination of interior finishes such as paint and wallpaper, window coverings, and flooring. It may also include the arrangement of lighting fixtures, art objects, furnishing accessories, and interior plants. Specific job requirements may include:

- Meeting with clients to determine the scope of a project
- Reviewing and measuring of the space to be decorated
- Preparing proposed room layouts and obtaining cost estimates
- Providing samples and colors of materials to be used

- Arranging and overseeing painting, wallpapering, and flooring
- Selecting and purchasing furnishings and other items

There are no formal educational requirements to enter this career, but many schools offer classes to enhance your design skills. You can start calling yourself an interior decorator as soon as you start doing interior decorating.

INTERIOR DESIGNER

Interior designers work with the structural designer to optimize and harmonize the interior design of structures. In addition to health and safety concerns, interior designers help plan how the space will be accessed, how a space will be used, the amount of light that will be required, acoustics, seating, storage, and work areas. The elements of interior design include the consideration of how the visual, tactile, and auditory senses of the occupants will be affected.

- Visual considerations include the study and application of color, lighting, and form to improve how the occupants function in the space.
- Tactile considerations include the design of surfaces, the shape of individual rooms within a structure, and how the texture of finished surfaces and furnishing will affect the usage of areas within a structure.
- Auditory design considerations include the relationship of how noise and echo will be created and how they can be controlled.

An interior designer must have an aesthetic, practical, and technical appreciation for how people use and respond to these elements and how the elements interact with one another.

Designers must also be knowledgeable about the many types and characteristics of furnishings, accessories, and ornaments used in creating interiors. Furniture, lighting, carpeting and floor covering, paint and wall covering, glass, wrought metal, fixtures, art, and artifacts are just some of the many items and materials designers select from. In addition, they must be familiar with the various styles of design, art, and architecture and their history. Figure 1.3 shows a room that was created with the aid of an interior designer. Interior designers provide various services, including:

- Consulting to help determine project goals and objectives

FIGURE 1.3 ■ *Interior designers help plan how space will be accessed and used, the amount of light that will be required, acoustics, seating, storage and work areas.* Courtesy Interior Design by Daphne Weiss Inc.; Maxwell Mackenzie, photographer.

- Generating ideas for aesthetic possibilities of the space, and arranging space to suit its intended function
- Creating illustrations and renderings of proposals
- Developing documents and specifications related to interior spaces in compliance with applicable building codes
- Specifying and purchasing fixtures, furnishings, products, materials, and colors
- Designing and managing fabrication of custom furnishings and interior details
- Monitoring and managing construction and installation of the design

Although a college degree is currently not a requirement, the trend among employers and in states that have licensing requirements is to require a degree from an accredited institution. This can range from training in a two-year program to earn an associate's degree or certificate to a four- or five-year program leading to a bachelor's (BA, BS, BFA) or master's (MFA, MA, MS) degree. The option chosen may depend on the licensing requirements in your state and whether you have completed a degree in another field.

In the United States, interior designers are registered by title. People can't represent themselves using the title interior designer or registered interior designer unless they have met the requirements for education, experience, and examination as set forth in the statutes established by the **National Council for Interior Design Qualification (NCIDQ).** Candidates who apply to take the NCIDQ examination must demonstrate an acceptable level of professional work experience and completion of related course work. The minimum examination requirements include two years of formal interior design experience and four years of full-time work experience in the practice of interior design. Passage of the examination is required in twenty jurisdictions in the United States and eight provinces in Canada that regulate the profession of interior design. For further information about careers in the field of interior design, contact:

American Society of Interior Designers (ASID)
202-546-3480
Website: www.asid.org
International Interior Design Association
Website: www.iida.org

ARCHITECT

An ***architect*** is a licensed professional who designs commercial and residential structures. Few architects work full time in residential design. Although many architects design some homes, most devote their time to commercial construction projects such as schools, offices, and hospitals. An architect is responsible for the design of a structure and for the way the building relates to the environment. Architects perform the tasks of many professionals including designer, artist, project manager, and construction supervisor. The architect serves as a coordinator on a project to ensure that all aspects of the structure blend together to form a pleasing relationship. This coordination includes working with the client, the contractors, and a multitude of engineering firms that may be working on the project. Figure 1.4 shows a home designed by an architect to blend the needs and wishes of the client with the site, materials, and financial realities.

Use of the term *architect* is legally restricted to individuals who have been licensed by the state where they practice. Obtaining a license typically requires three years to complete a master's program, three years as an intern, and two years to complete the registration exam process. Some states allow a designer to take the licensing test through practical work experience. Although standards vary for each state, five to seven years of experience under the direct supervision of a licensed architect or engineer is usually required.

FIGURE 1.4 ■ *Architects use their training to blend the needs and wishes of the client with the site, materials, and financial realities.* Courtesy Cornerstone Developers, Inc.

Positions in an architectural firm include:

Technical staff—Consulting engineers such as mechanical, electrical, and structural engineers; landscape architects; interior designers; CAD operators; and drafters.
Intern—Unlicensed architectural graduates with less than three years of experience. An intern's responsibilities typically include developing design and technical solutions under the supervision of an architect.
Architect I—Licensed architect with three to five years of experience. An architect I's job description typically includes responsibility for a specific portion of a project within the parameters set by a supervisor.
Architect II—Licensed architect with six to eight years of experience. An architect II's job description typically includes responsibility for the daily design and technical development of a project.
Architect III—Licensed architect with eight to ten years of experience. An architect III's job description typically includes responsibility for the management of major projects.
Manager—Licensed architect with more than ten years of experience. A manager's responsibilities typically include management of several projects, project teams, and client contacts as well as project scheduling and budgeting.
Associate—Senior management architect, but not an owner in the firm. This person is responsible for major departments and their functions.
Principal—Owner/partner in an architectural firm.

Education

High school and two-year college students can prepare for a degree program by taking classes in fine arts, math, science, and social science. Many two-year drafting programs offer drafting classes that can be used for credit in four- or five-year architectural programs. A student planning to transfer to a four-year program should verify with the new college which classes can be transferred. Preparation to enter a degree program in architecture should include:

- Fine arts classes—such as drawing, sketching, design, and art along with architectural history—which will help the future architect develop an understanding of the cultural significance of structures and help transform ideas into reality.
- Math and science—including algebra, geometry, trigonometry, and physics—which will provide a stable base for the advanced structural classes that will be required.
- Sociology, psychology, cultural anthropology, and classes dealing with human environments, which will help develop an understanding of the people who will use the structure.
- Literature and philosophy courses to help prepare you to read, write, and think clearly about abstract concepts.

In addition to formal study, students should discuss with local architects the opportunities and possible disadvantages that may await them in pursuing the study and practice of architecture.

Areas of Study

The study of architecture is not limited to the design of buildings. Although the architectural curriculum typically is highly structured for the first two years of study, students begin to specialize in an area of interest during the third year of the program.

Students may branch from architecture into related fields such as urban planning, landscape architecture, and interior architecture.

- Urban design is the study of the relationship among the components within a city.
- Interior architects work specifically with the interior of a structure to ensure that all aspects of the building will be functional.
- Landscape architects specialize in relating the exterior of a structure to the environment.

Students wishing further information about training or other related topics can contact:

American Institute of Architects (AIA)
1735 New York Avenue, NW
Washington, DC 20006
202-626-7300
Website: www.aia.org

American Society of Landscape Architects (ASLA)
636 Eye Street NW
Washington, DC 20001-3736
202-898-2444
Website: www.asla.org

ENGINEER

An ***engineer*** is a licensed professional who applies mathematical and scientific principles to the design and construction of structures. The term covers a wide variety of professions. In the construction fields, structural engineers are the most common, although many jobs exist for electrical, mechanical, and civil engineers. Although most engineers specialize in the design of structures built of steel or concrete, many also directly supervise CAD technicians and designers in the design of single and multifamily structures. Many municipalities require an engineer or an architect to stamp plans drawn by designers to ensure that the planned structure will resist lateral and wind loads. Figure 1.5 shows an example of a drawing created by a CAD technician based on a sketch provided by an engineer.

Electrical engineers work with architects and structural engineers and are responsible for the design of lighting and communication systems. They supervise the design and installation of specific lighting fixtures, communication services, surround-sound systems, security features, and requirements for computer networking.

Mechanical engineers are also instrumental parts of the design team. They are responsible for the sizing and layout of heating, ventilation, and air-conditioning (HVAC) systems and plan how treated air will be routed throughout the project. They work with the project architect to determine the number of occupants of the completed building and the heating and cooling load that will be generated.

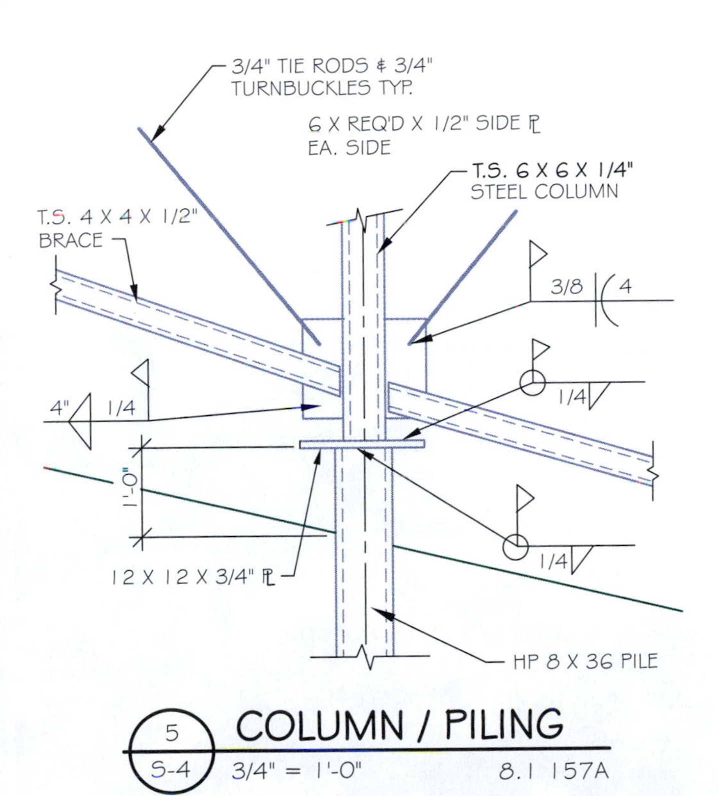

FIGURE 1.5 ■ *The structural engineering team is responsible for determining the types and sizes of materials to resist the loads and stresses imposed on a building. This detail was drawn from calculations provided by the engineer to support a hillside residence. Loads were determined and materials selected that are capable of resisting the anticipated stress.* Courtesy Ron Sellards, structural engineer; David Jefferis, CAD detailer.

Civil engineers are responsible for the design and supervision of a wide variety of construction projects, such as highways, bridges, sanitation facilities, and water treatment plants. They are often directly employed by construction companies to oversee the construction of large projects and to verify that the specifications of the design architects and engineers have been carried out. They often provide the land drawings such as topography, grading plans, street design, and other land-related improvements for residential subdivisions or individual sites.

A license is required to function as an engineer. The license can be applied for after several years of practical experience or after obtaining a bachelor's degree and three years of practical experience. Success in any of the engineering fields requires high proficiency in math and science, including courses in physics, mechanics, print reading and architecture, mathematics, and material sciences. As with the requirements for becoming an architect, an engineer must have five years of education at an accredited college or university, followed by successful completion of a state-administered examination. Certification can also be accomplished by training under a licensed engineer and then successfully completing the examination. Additional information can be obtained about engineering by writing to:

American Society of Civil Engineers (ASCE)
1015 15th Street NW, Suite 600
Washington, DC 20005
1-800-548-2723
Website: www.asce.org

RELATED FIELDS

In addition to the careers that involve drawing and design, there are many related careers that require an understanding of drafting and building principles but do not require the CAD skills. These include model maker, illustrator, specification writer, inspector, and construction-related trades.

Model Maker

In addition to presentation drawings, many architectural offices use models of a project to help convey design concepts. Models such as the one shown in Figure 1.6 are often used as a public display to help gain support for large projects. Model makers need basic

FIGURE 1.6 ■ *Models are used to convey design ideas from the design team to the owner and review boards.* Courtesy James Eismont, photographer.

drafting skills to help interpret the plans required to build the actual project. Model makers may be employed within a large architectural firm or for a company that only makes models for architects. For more information, contact the Association of Model Makers at:

www.modelmakers.org

Illustrator/Renderer

Many drafters, designers, and architects have the basic skills to draw architectural renderings. Very few, though, have the expertise to make this type of drawing rapidly. Most illustrators have a background in art. By combining artistic talent with a basic understanding of architectural principles, the illustrator is able to produce drawings that show a proposed structure realistically. Figure 1.7a shows a drawing that was prepared by an architectural illustrator. The rendered elevation and floor plan in Figure 1.7b were also prepared by an architectural illustrator to advertise a home. Presentations range from traditional renderings created with ink, colored pencils, or watercolors to computer generated renderings. For more information, contact the American Society of Architectural Illustrators at:

www.asai.org

Specification Writer

Specifications are written instructions that consist of the data needed to build a structure; they are used to clearly convey the intentions of the owner, the ar-

FIGURE 1.7a ■ *An architectural illustrator prepares drawings from sketches and preliminary drawings to help the owner visualize how the project will look when completed.* Courtesy Alan Mascord Design Associates, Inc.

FIGURE 1.7b ■ *Many offices sell plans that have been built on multiple sites. These 'stock plans' can be selected from catalogues featuring drawings created by an architectural illustrator.* Courtesy Design Basics, Inc.

chitectural team, and each consultant to be involved with the project. Written specifications provide a method of supplementing the working drawings regarding the quality of materials to be used and labor to be supplied. The construction drawings and written specifications must be compatible so that the project can be accurately bid and built. They must be clear in order to avoid misinterpretation. The drawings visually define the relationships between materials, products, and systems within the structure by showing the location and size of each element. Written specifications express the requirements in words. Specifications describe:

- Quality and type of material
- Required gauge, size, or capacity of material and equipment
- Quality of workmanship
- Methods of fabrication
- Methods of installation
- Test and code requirements

Specifications provide information regarding the quality of materials and workmanship, methods of installation, the desired performance to be achieved at completion, and how performance is to be measured. A specification writer must have a thorough understanding of the construction process and have a good ability to read plans. Generally, a specification writer must have good technical writing skills. These skills can often be acquired by taking classes in technical writing at the two-year-college level.

Inspector

Building departments require that plans and the construction process be inspected to ensure that the required codes for public safety have been met. A plans examiner must be licensed by the state, to certify minimum understanding of the construction process. In most states, there are different levels of examiners. An experienced CAD technician or designer may be able to qualify as a low-level or residential-plans inspector. Generally, a degree in engineering or architecture is required to advance to an upper-level position.

The construction that results from the plans must also be inspected. Depending on the size of the building department, the plans examiner may also serve as the building inspector. In large building departments, one group inspects plans and another inspects construction. To be a construction inspector requires an exceptionally good understanding of codes limitations, print reading, and construction methods. Each of these skills has its roots in a beginning drafting class.

Jobs in Construction

Many CAD technicians are employed directly by construction companies. The benefits of this type of position have already been discussed. These CAD technicians typically not only do drafting but also work part time in the field. Many CAD technicians give up their jobs for one of the high-paying positions in the construction industry. The ease of interpreting plans as a result of a background in drafting is of great benefit to any construction worker.

DESIGN BASICS

Designing a home for a client can be an exciting but difficult process. Rarely can a designer or architect sit down and create a design that meets the needs of the client perfectly in one try. The time required to design and complete the drawings for a home can range from a few days to several months. It is important for you to understand the design process and the role the technician plays in it. This requires an understanding of basic design, financial considerations, and common procedures of design. In the rest of this chapter, the terms *architect* and *designer* can be thought of as synonymous.

Financial Considerations

Both designers and CAD technicians need to be concerned with costs. Finances influence decisions made by the technician about framing methods and other structural considerations. Often the advanced technician must decide between methods that require more materials with less labor and those that require fewer materials but more labor. These are decisions of which the owner may never be aware, but they can make the difference in the affordability of the house. The designer makes the major financial decisions that affect the cost and size of the project. The designer needs to determine the client's budget at the beginning of the design process and must work to keep the project within these limits.

Through past experience and contact with builders, the designer should be able to make an accurate estimate of the cost of a finished house. This estimate is often made on a square footage basis in the initial stages of design. For instance, in some areas, a modest residence can be built for approximately $75 per square foot. In other areas, the same house may cost approximately $150 per square foot. A client wishing to build a 2500-square-foot house could expect to pay between $187,500 and $375,000, depending on where it will be built. Keep in mind that these are estimates. A square footage price tells very little about the home. It's like saying that you can buy a nice car for $20,000. Although this is a true statement, it doesn't define new or used, a VW or BMW. In the design stage, estimates of square footage help set parameters for the design. Such an estimate is based on typical cost for previous clients. The price of materials such as lumber, concrete, and roofing will vary throughout the year, depending on supply and demand. Basic materials will typically account for approximately 50 percent of the project cost (see Table 1.1). Finished materials are the part of the estimate that will bring wide variations in cost. If you have ever been in a home supply store, you know that a toilet can be purchased for between $30 and $300. Every item in the house will have a range of possible prices. The final cost of the project is determined once the house is completely drawn and a list of materials prepared. With a list of materials, contractors are able to make accurate decisions about cost.

The source of finances can also affect the design process. Certain lending institutions may require some drawings that the local building department may not. The Federal Housing Administration (FHA) and Veterans Administration (VA), which provide loans, often require extra forms, drawings, and specifications that need to be taken into account in the initial design stages.

The Client

Most houses are not designed for one specific family. In order to help keep costs down, houses are often built in subdivisions and designed to appeal to a wide variety of people. One basic plan may be flopped or built with several different options, saving the contractor the cost of paying for several different complete plans. Some families may make minor changes to an existing plan. These modified ***stock plans*** give the prospective buyer a chance to have a personalized design at a cost far below that of a custom-drawn plan (see Figure 1.7b). If finances allow, or if a stock plan cannot be found to meet its needs, a family can have a plan custom-designed.

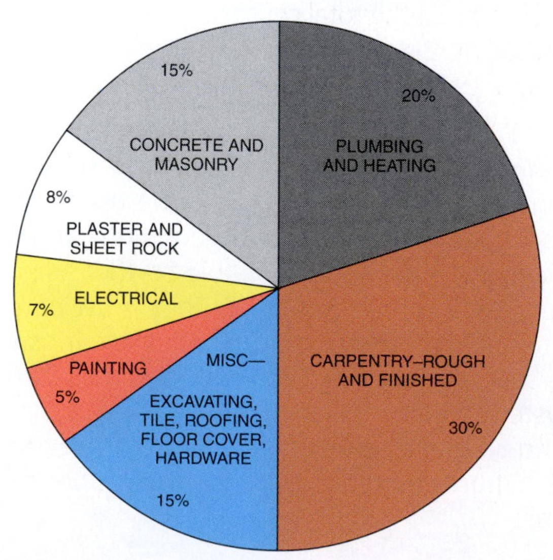

TABLE 1.1 ■ *Approximate breakdown of costs by trade groups. Prices vary based on area of the country and the time of year for construction.*

THE DESIGN PROCESS

The design of a residence can be divided into several stages. These generally include initial contact, preliminary design studies, room planning, initial working drawings, final design considerations, completion of working drawings, permit procedures, and job supervision.

Initial Contact

Clients often approach designers to obtain background information. Design fees, schedules, and the compatibility of personalities are but a few of the basic questions to be answered. This initial contact may take place by telephone or a personal visit. The questions asked are important to both the designer and the client. The client needs to pick a designer who can work within budget and time limitations. Drawing fees are another important consideration in choosing a designer. Design fees vary based on the type of project and the range of services to be provided. Fees are generally based on hourly rates, price per square foot of construction, a percentage of construction cost, or a combination of these methods. Square footage prices vary based on the size of the project and the area of the country, but they must be set to cover design and drafting time, overhead, and profit margin. New CAD technicians are often shocked to learn that their office bills clients at a rate that is often three to ten times their hourly rate of pay for drafting time. It is important to remember that this billing price must include supervision time, overhead, state and federal taxes, and hopefully a profit for the office. Design fees for a custom home range from 5 to 15 percent of the total construction cost. The amount varies based on the services provided, the workload of the office, and the local economy. The designer needs to screen clients to determine if the client's needs fit within the office schedule.

Client Criteria

Once an agreement has been reached, the preliminary design work can begin. The selection process usually begins with the signing of a contract to set guidelines that identify which services are to be provided and when payment is expected. This is also the time when the initial criteria for the project will be determined based on the goals and objectives of the clients. This portion of the design phase will focus on general requirements such as four bedrooms, an office, three baths, a kitchen/family area suitable for entertaining large groups, and a three-car garage. Generally, clients have a basic size and a list of specifics in mind, a sketch of proposed floor plans, and a file full of pictures of items that they would like in their house. Photos of desired features can be very helpful in planning for the family's needs and lifestyle. Many architectural firms have a written questionnaire to provide insight into how the structure will be used. Important design considerations include the following factors:

- The number of inhabitants
- The ages and genders of children
- Future plans to add onto the dwelling
- A list of general family activities to be done in the home
- The minimum number of rooms
- The minimum size of each room
- A listing of how each room is to be furnished.
- The family's entertainment habits
- The desired number of bedrooms and bathrooms
- Kitchen appliances desired
- Planned length of stay in the residence
- Live-in guests or requirements for people with disabilities
- The budget for the home
- The style of the home
- Neighborhood covenants, conditions, and restrictions (CC&R)

Additional design considerations can be found in Appendix A on the CD. The written list is an excellent way of enabling the clients to agree on the design criteria before the design team spends hours pursuing unwanted features. Providing minimum room sizes is also much more helpful than allowing a client to request "three big bedrooms." An understanding of how a room is to be furnished will also help the designer make sure that the owner has realistic expectations for the size of the room. It is also helpful to have clients provide a "wish list" of items for their residence. These items can be eliminated if the budget is exceeded or added if it is possible to provide them.

Site Criteria

During this initial phase of design, it is important to become familiar with the lifestyle of the client as well as the site where the house will be built. This will require a visit to the client's current residence and to the proposed building site. A tour of the existing residence allows the designer to get a firm grasp on the client's likes and dislikes. By visiting the site, the designer can get a first-hand picture of the slope, view, and surrounding structures that will influence the project. Visual inspection of the site will make it possible to:

- Determine whether a soils lab will be required to analyze the site
- Consider the existing slope to determine whether a land survey will be required
- Study drainage patterns
- Consider solar, view, and weather orientation
- Photograph surrounding structures and job site elements, such as landscaping, that will remain, and document existing views

Figure 1.8 shows the results of the preliminary site visit.

Preliminary Design Studies

Once a thorough understanding of the client's lifestyle, design criteria, and financial limits has been developed, the preliminary studies can begin. These include research with the building and zoning departments that govern the site, verification of existing utility locations, and discussions with any review boards or covenants, conditions, and restrictions (CC&Rs) that may be required. Once this initial research has been done, preliminary design studies can be started. The preliminary drawings usually involve two stages: bubble drawings and scaled sketches. Bubble drawings are freehand sketches used to help determine room locations and relationships. Several sketches similar to Figure 1.9 are usually created. It is during this stage of the preliminary design that consideration is given to the site and energy efficiency of the home.

Once a satisfactory layout has been sketched, these shapes are transformed into scaled drawings. Figure 1.10 shows a preliminary floor plan. Usually several alternatives are developed to explore different design possibilities. Consideration is given to building code regulations and room relationships and sizes. In this stage, consideration would shift from general to specific needs. If, in the initial meeting, four bedrooms were requested, the primary users for each area, the intended use, and intended furniture would also have to be considered to help shape the design. After the design options are explored, the designer selects a plan to prepare for the client. This could be the point when a CAD technician first becomes involved in the design process. Depending on the schedule of the designer, a senior technician might prepare the refined preliminary drawings, which can be seen in Figure 1.11. The first preliminary drawings usually include the

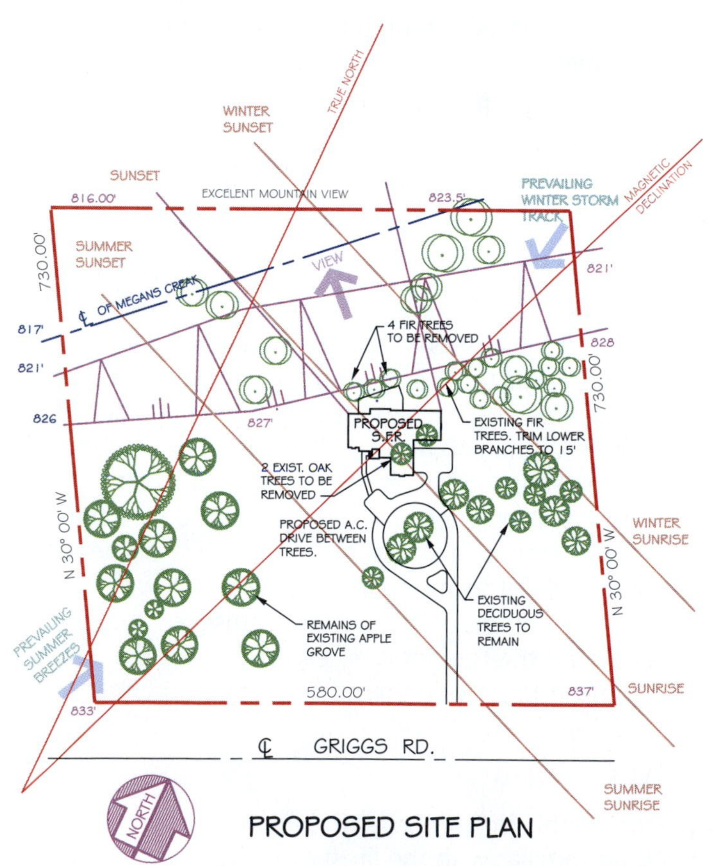

FIGURE 1.8 ■ *Visiting the site allows key features of the site to be identified and a home to be designed that will blend with the site. In a small office, the CAD technician is often involved in locating and drawing the site plan to represent existing features.*

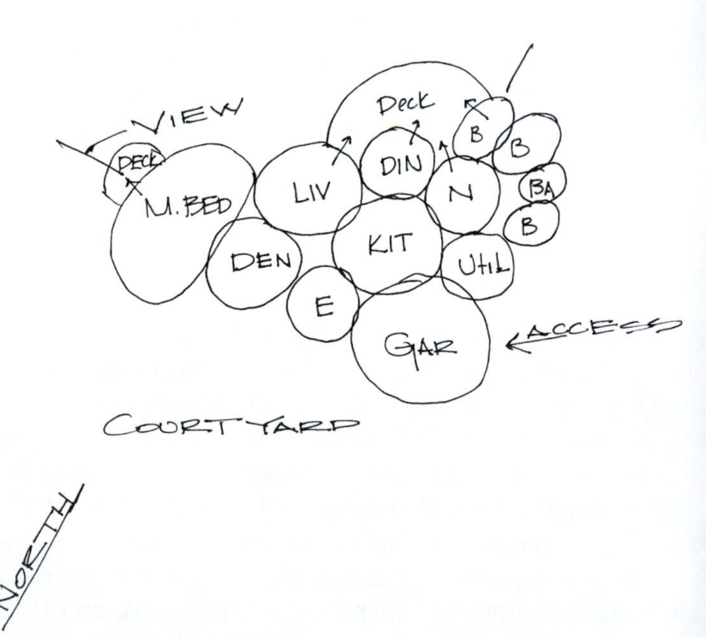

FIGURE 1.9 ■ *Bubble designs are the first drawing in the design process. These drawings are used to explore room and site relationships.*

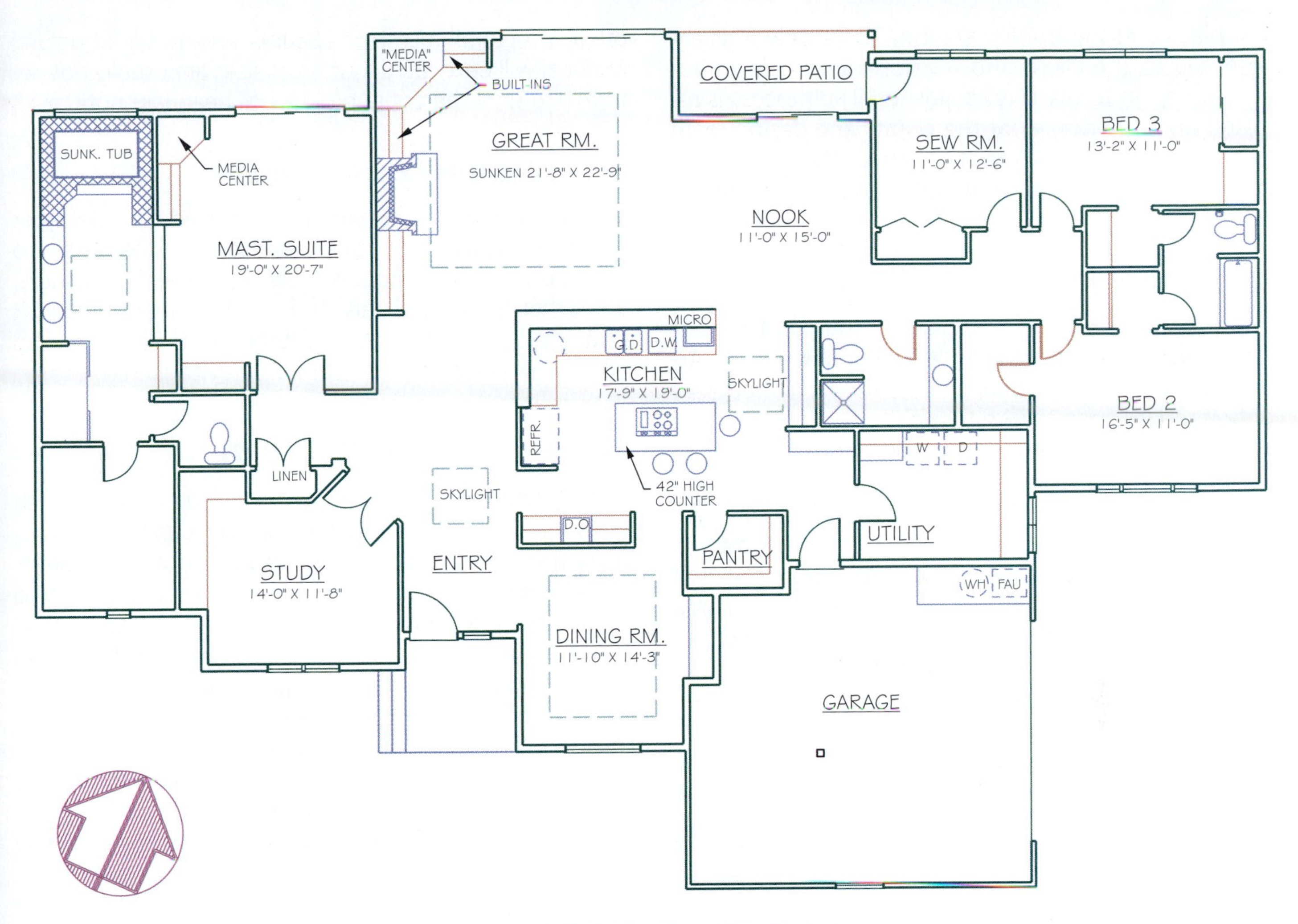

FIGURE 1.10 ■ *Once a satisfactory layout has been determined, the bubble drawings are transformed into scaled drawings.*

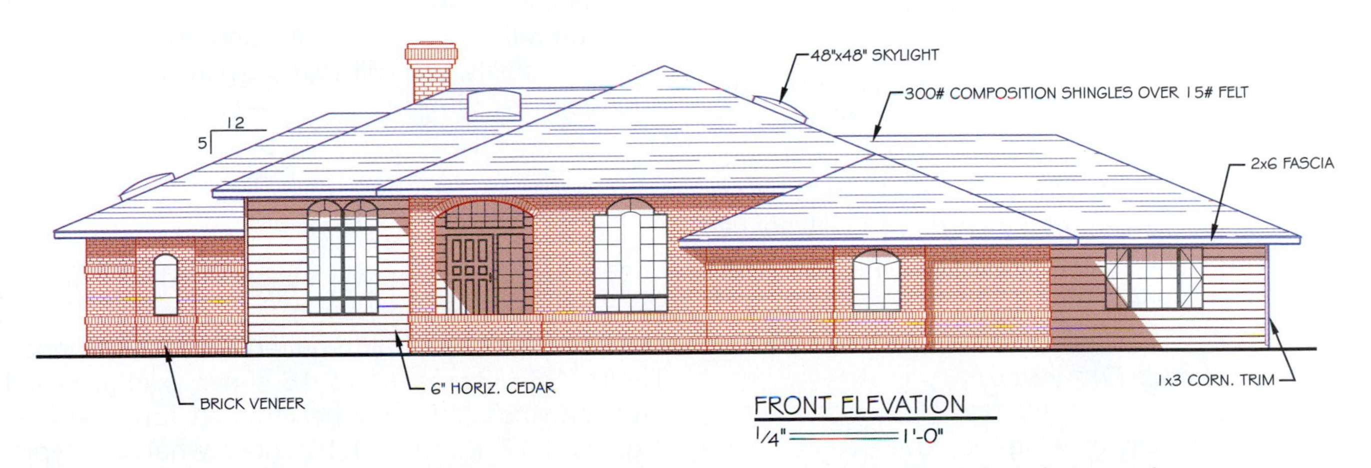

FIGURE 1.11 ■ *Based on the preliminary floor plans, the front elevation is drawn to explore design options.*

floor plans and the front elevation. These drawings are presented to the client and revisions are made to adjust the design to the clients' wishes. The process is repeated until the desires of the clients and design team are satisfied. Once the client approves the preliminary drawings, they are ready to be converted to design drawings.

Room Planning

Room usage must be considered throughout the design process. Many professionals use books such as *Architectural Graphic Standards* and vendor catalogs to verify standard sizes of furniture or equipment. Chapter 5 introduces major design concepts to be considered. Furniture is often drawn on the preliminary floor plan to show how the space within a room can be utilized. Typically the designer will have talked with the owners about how each room will be used and what types of furniture will be included. Occasionally, placement of a family heirloom will dictate the entire layout. When this is the case, the owner should specify the size requirements for a particular piece of furniture. When placing furniture outlines to determine the amount of space in a room, the CAD technician can use blocks from DesignCenter or by third party vendors. Common pieces of furniture can be found on the CD included with the text. See Appendix B for additional furniture sizes.

Figure 1.12 shows a preliminary floor plan with the furniture added. In the final stages of the preliminary process, the designer will often work with the clients or bring in the assistance of interior designers to start planning interior finishes. With the interior space designed, attention can now be focused on interior details. This might include the use of interior elevations, interior perspectives, and presentation boards. Presentation boards are used to present samples of fabric and other materials that can be used for wall finishes and furniture. This is also the time when, with the help of an interior designer, the clients make decisions about key pieces of furniture, lighting fixtures, and equipment.

Initial Working Drawings

With the preliminary drawings approved, a CAD technician can begin to lay out the working drawings. The procedure will vary with each office, but generally each of the drawings required for the project will be started. This would include the foundation, site, roof, electrical, cabinet, and framing plans. At this stage the technician must rely on past experience for drawing the size of beams and other structural members. Beams and other structural material will be located, but exact sizes are usually not determined until the entire project has been laid out.

Final Design Considerations

Once the drawings have been started, the designer will generally meet with the client several times again to get information on flooring, electrical needs, cabinets, and other finish materials. These conferences will result in a set of marked drawings that the CAD technician will use to complete the working drawings.

Completion of Working Drawings

The complexity of the residence will determine which drawings are required. Most building departments require a site plan, a floor plan, a foundation plan, elevations, and one cross section as the minimum drawing to get a building permit. On a complicated plan, a wall framing plan, a roof framing plan, a grading plan, and construction details may be required. Depending on the lending institution, interior elevations, cabinet drawings, a 3D exterior rendering, a scale model, a landscaping plan, mechanical plan, schedules for finish, hardware, electrical and plumbing fixtures, cost analysis, and construction and finish specifications may also be required.

The skills of the CAD technician will determine his or her participation in preparing the working drawings. As an entry-level CAD technician, you will often be given the job of making corrections on existing drawings or, for example, drawing site plans or cabinets. As you gain skill, you will be given more drawing responsibility. With increased ability, you will start to share in the design responsibilities.

Working Drawings

The drawings that will be provided vary for each residence and within each office. Figure 1.13 shows the site plan that was required for the residence in Figure 1.8. Figures 14 to 16 show examples of the civil drawings that may be required for a residence. Figures 1.17 through 1.19 show what are typically considered the architectural drawings. Figures 1.20a and 1.20b show the electrical drawings. These are drawings that show finishing materials. Figures 1.21 through 1.24 show what are typically called the structural drawings. On a large project involving consulting firms, the page number on each page is based on the

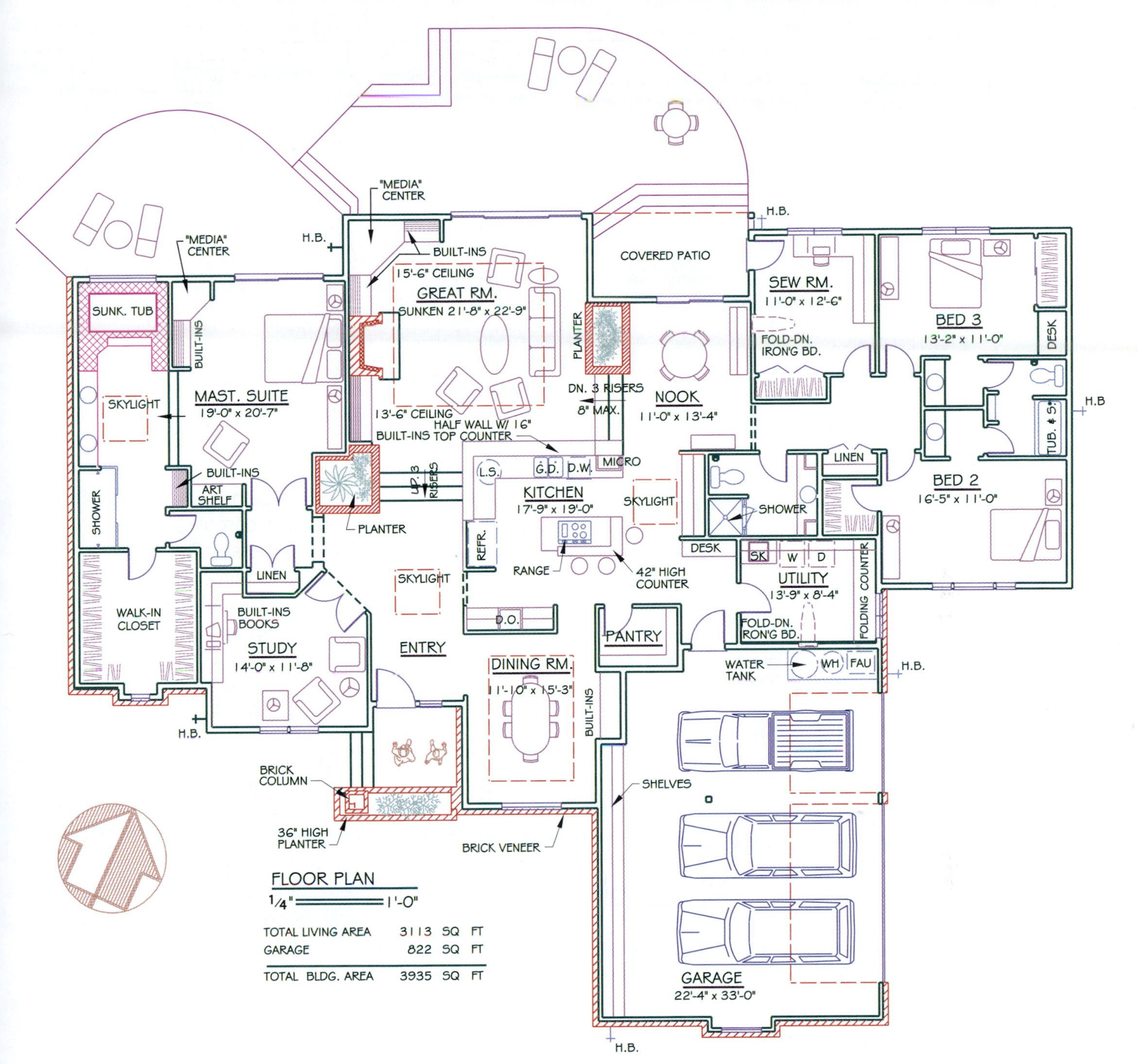

FIGURE 1.12 ■ *Although the drawing in Figure 1.10 met all of the stated goals of the family, changes were made to the original criteria once the owners saw their ideas drawn to scale. The final preliminary floor plan shows the owner's changes, with furniture added.*

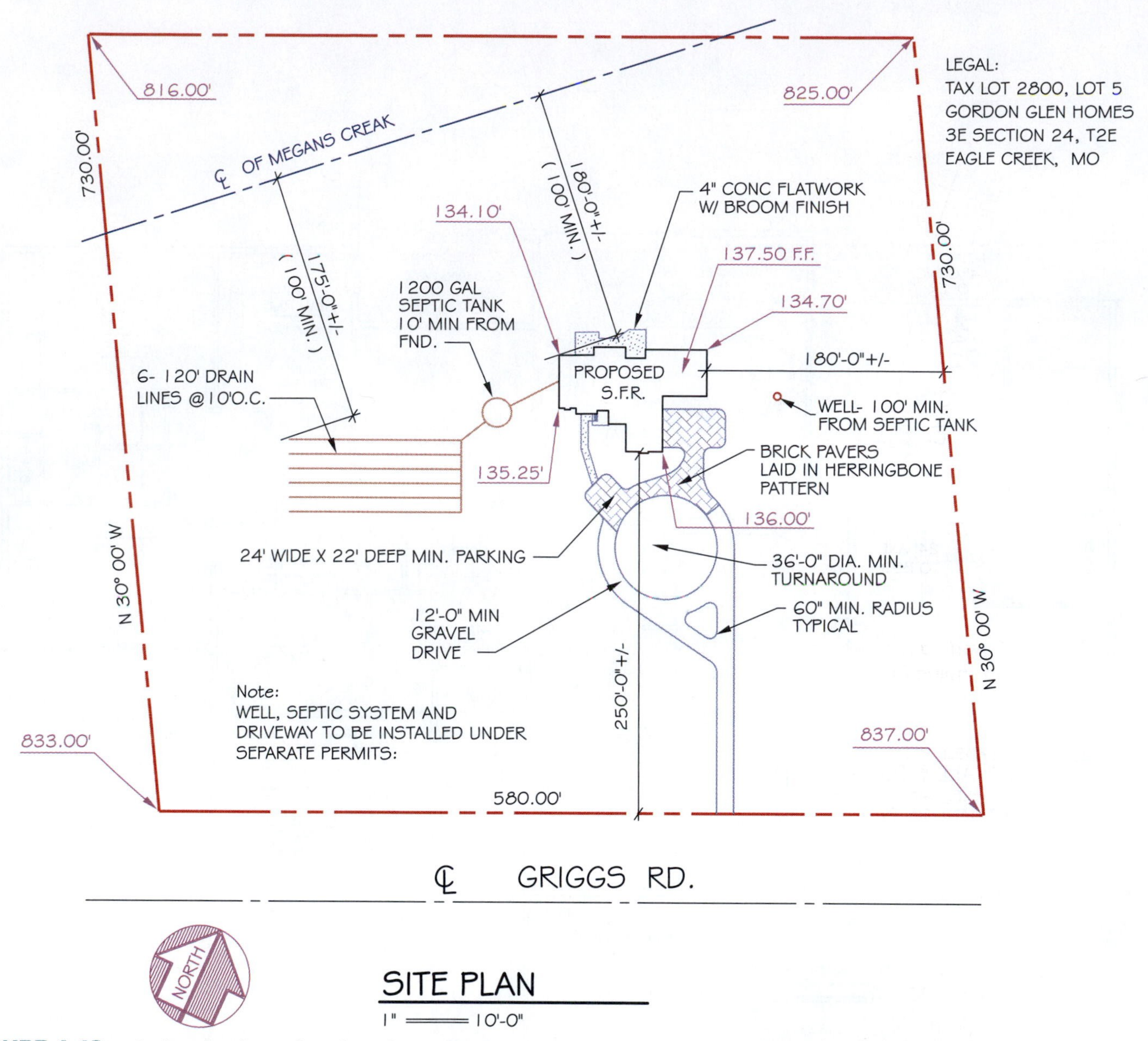

FIGURE 1.13 ■ *A site plan is used to show how the structure relates to the site. When the site is relatively flat, grading data referred to as "spot grades" are placed on the drawing to show drainage patterns.*

type of drawing on that page. Pages would be represented by:

A—Architectural
S—Structural
M—Mechanical
E—Electrical
P—Plumbing

An entry-level CAD technician often completes the site plan using a sketch provided by the senior drafter or designer. A grading plan is not required for every residence. When a home is to be built on a fairly level site, the soil elevations are indicated at each corner of the structure. Figure 1.14 shows a site plan for a hillside home. Because of the large amount of soil to be excavated for the lower floor, the grading plan shown in Figure 1.15 was provided. The junior CAD technician may draw the base drawing, showing the structure and site plan, and the designer or senior CAD technician will usually complete the grading information. Typically the owner is responsible for hiring a surveyor, who will provide the topography and site map, although the designer may coordinate the work between the two offices. Section II provides insight into how these drawings are developed.

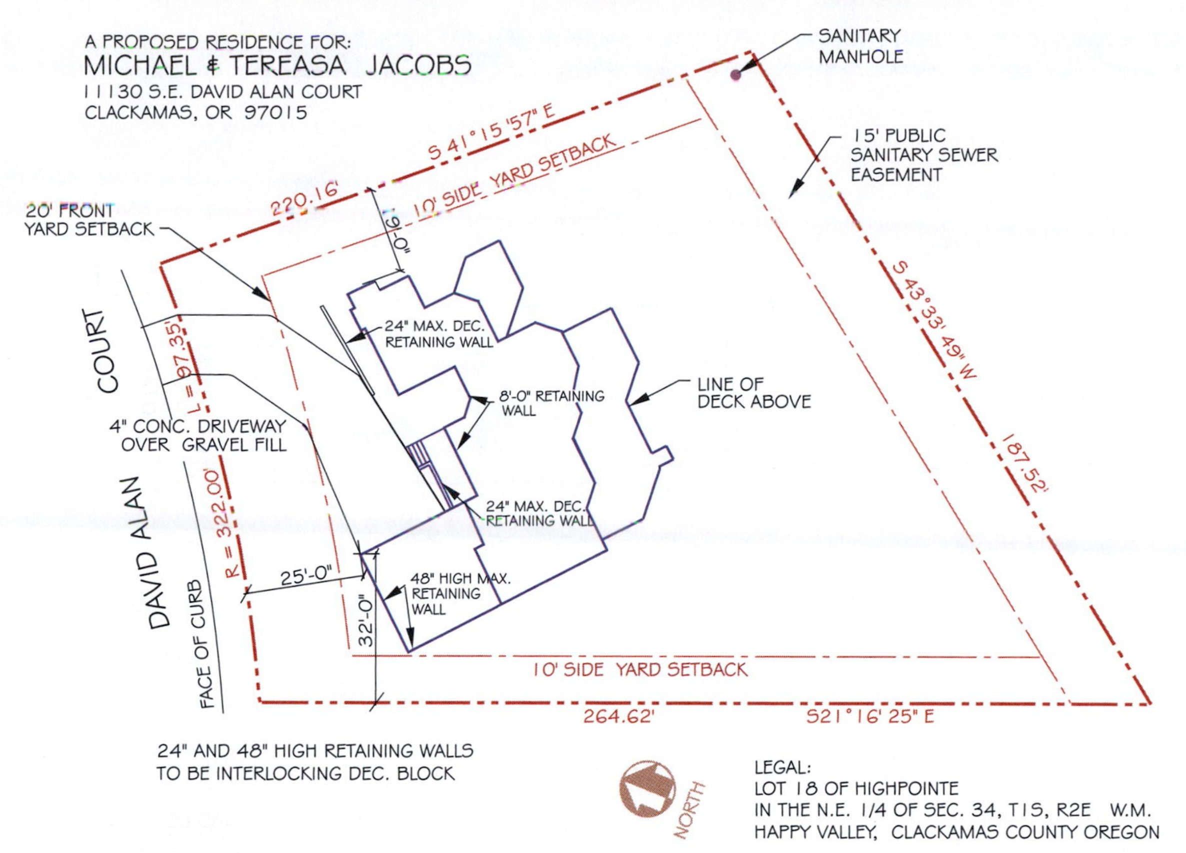

FIGURE 1.14 ■ *If grading is extensive, the site plan is separated from the grading plan. This site plan is used with the drawing shown in Figure 1.15 to describe the site.*

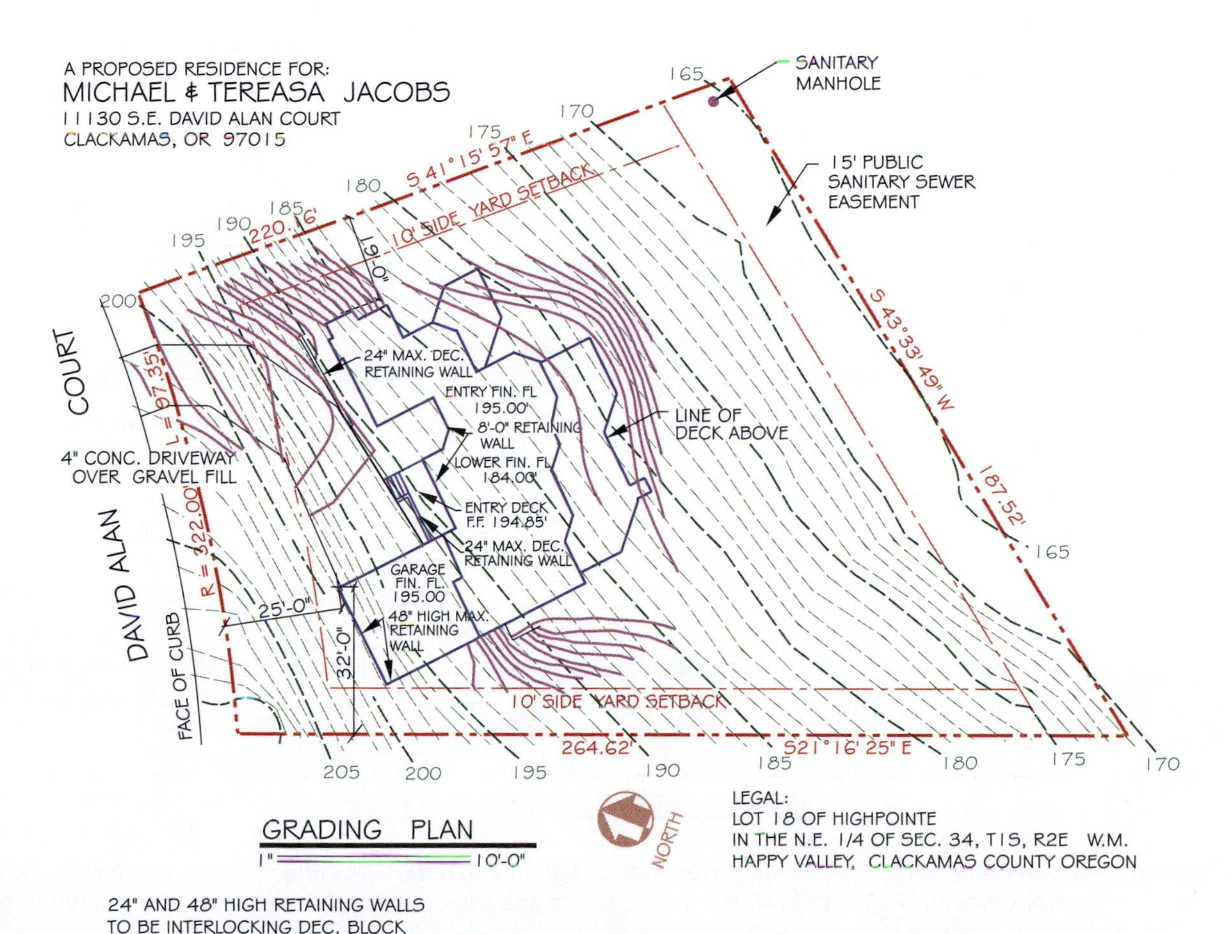

FIGURE 1.15 ■ *A grading plan is used to define new and existing grades and cut-and-fill banks that will be created as soil is relocated. Using layers, the grading plan was created from the site plan shown in Figure 1.14. By careful use of linetypes and lineweights, existing and new contours can be represented. On this site plan, existing soil contours are represented with dashed lines and new grades are represented with continuous lines.*

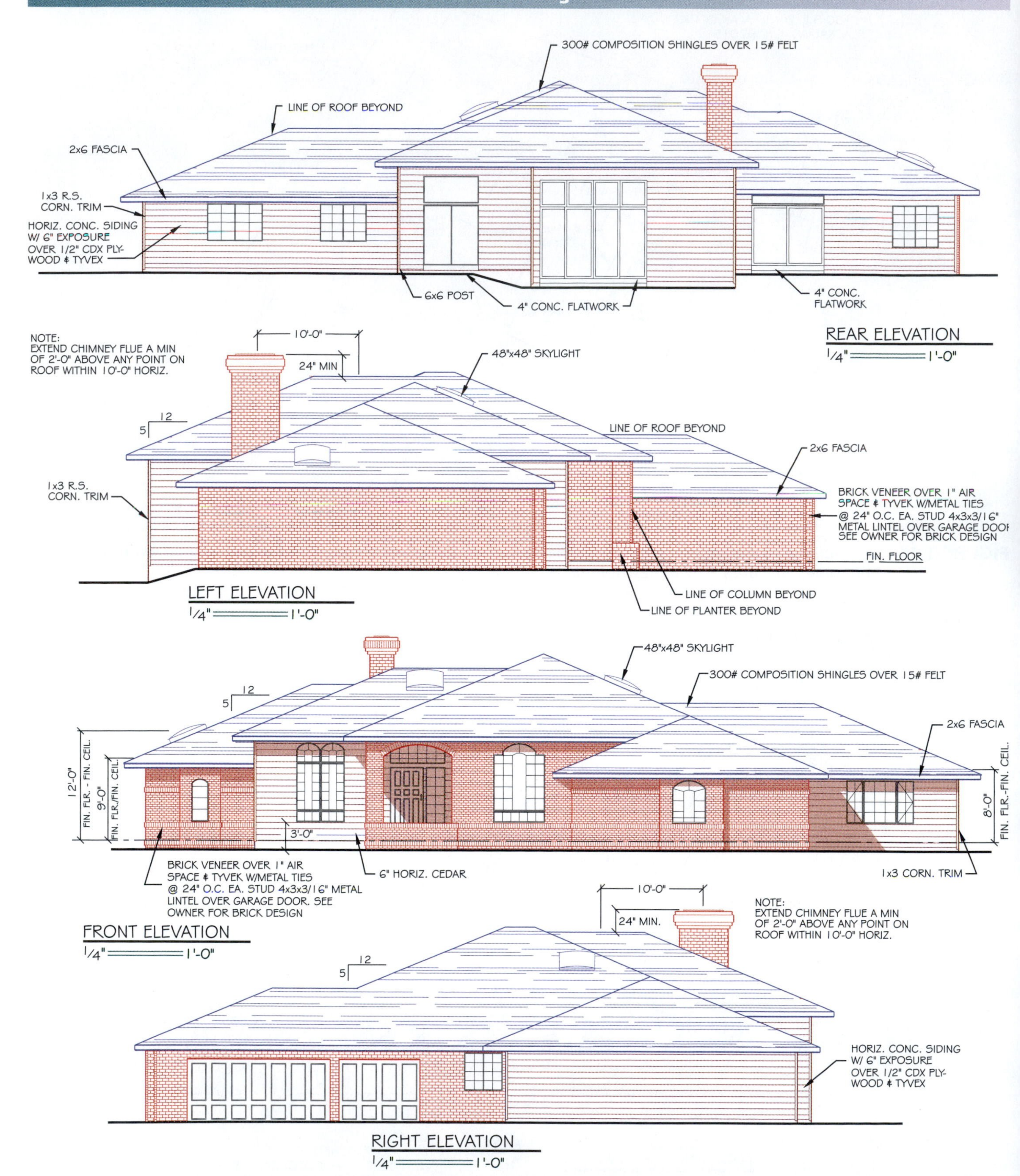

FIGURE 1.16 ■ *Once the client has approved the preliminary design, the exterior elevations can be completed. The working elevations provide a view of each side of the house and show the shape and the exterior materials to be used.*

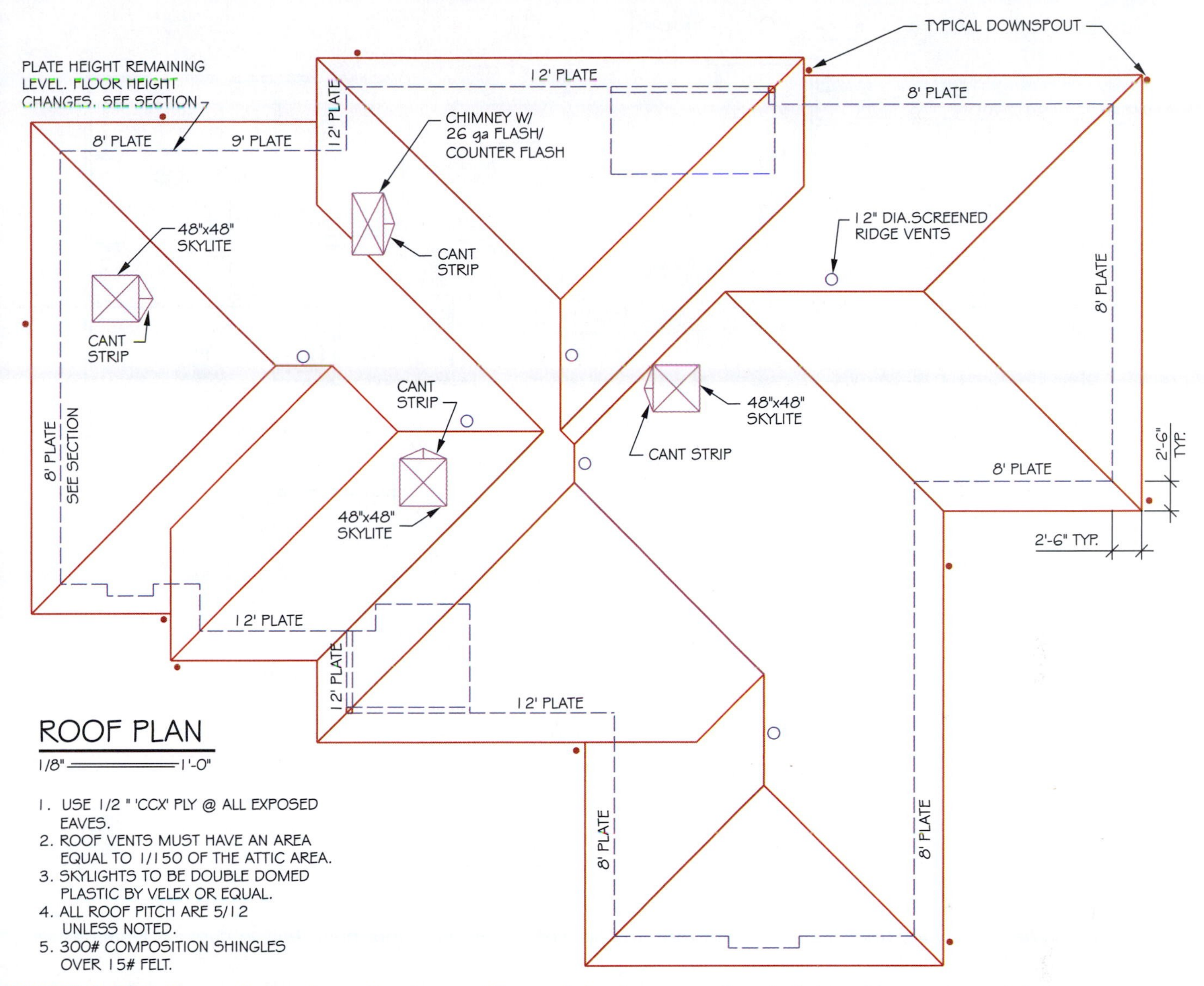

FIGURE 1.17 ■ *The roof plan shows the shape of the roof structure as well as skylights, ridge vents, and drains.*

Using the preliminary elevation shown in Figure 1.11, the working elevations can be completed. Depending on the complexity of the structure, they may be completed by either the junior or the senior CAD technician. Figure 1.16 shows the working elevations for this structure. Section V provides information needed to complete working elevations, and Section IX introduces presentation drawings.

Figure 1.17 shows the roof plan for the structure. The roof plan is often completed by the junior technician using sketches provided by the senior technician. Section IV provides information on roof plans. The plan view represented in Figure 1.18 is usually completed by the senior CAD technician. Figure 1.19 shows some of the schedules used to explain the materials to be used on the interior of the residence. A junior drafter usually completes the schedules based on information compiled by the project leaders. Junior CAD technicians also work on the plan views by making corrections or adding notes, which are typically placed on a marked-up set of plans provided by the designer or senior technician. Section III provides information for completing floor plans.

Figure 1.20 shows the electrical plan. Depending on the complexity of the residence, electrical information may be placed directly on the floor plan. A junior technician typically completes the electrical drawings by working from marked-up prints provided by the designer. Chapter 12 provides information for completing the electrical plans.

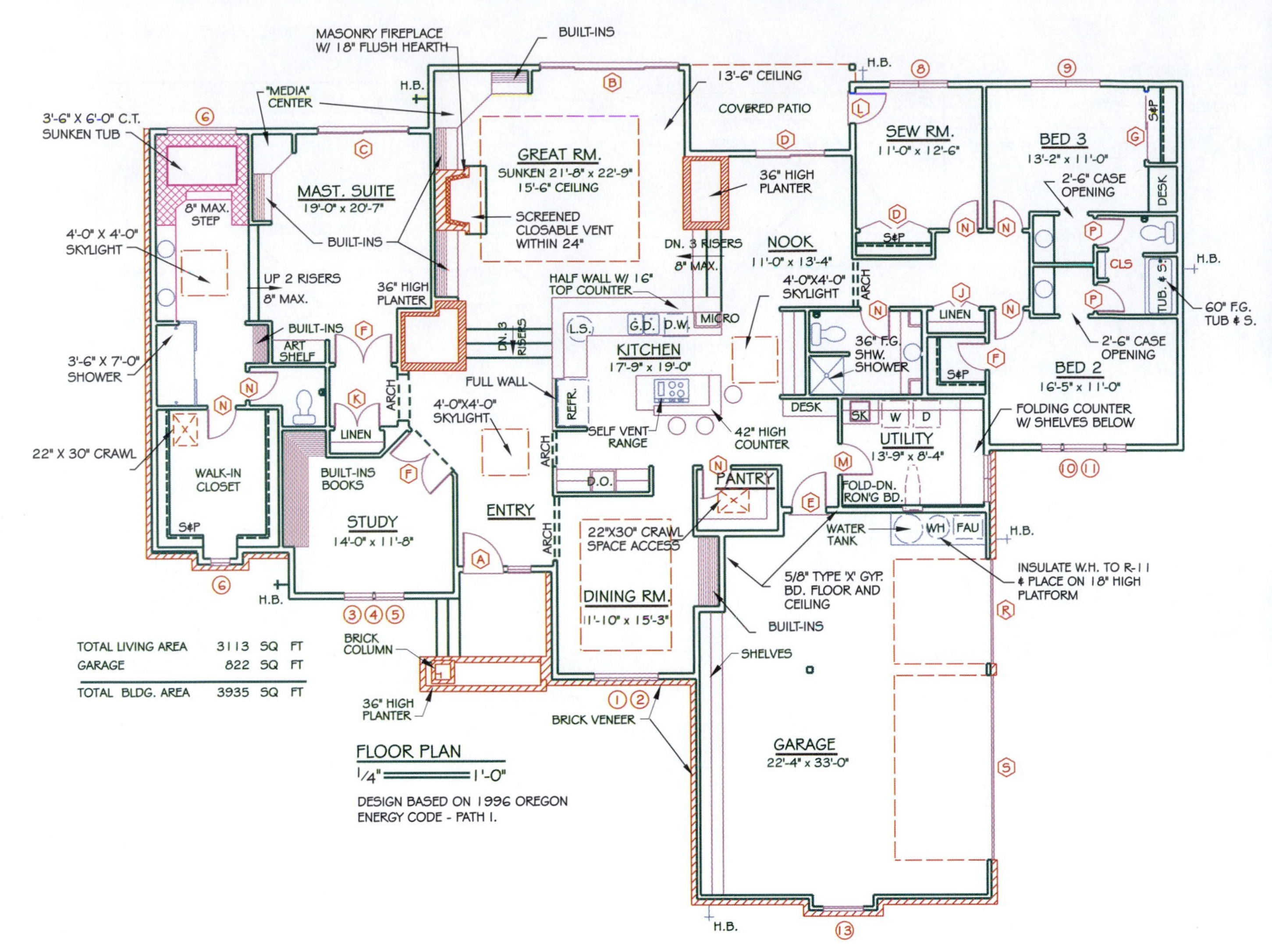

FIGURE 1.18 ■ *The preliminary floor plan is used to form the base of the finished floor plan. Text to describe all materials must be provided.*

DOOR SCHEDULE

SYM.	SIZE	TYPE	QUAN.
A	3'-6" x 8'-0"	S.C.,R.P., W/ 24" SIDELIGHT W/ SQ. TRANSOM ABOVE	1
B	12'-0" x 8'-0"	SL. GLASS W/ 12'-0" x 3'-0" SQ. TRANSOM ABOVE	1
C	8'-0" x 6'-8"	SL. FRENCH DRS. W/ 12" SQ. TRANSOM ABOVE	1
D	6'-0" x 6'-8"	SL. GLASS DRS. W/6'-0" x 3'-0" SQ. TRANSOM ABOVE	1
E	3'-0" x 6'-8"	SELF CLOSING, M.I.	1
F	PR 2'-6" X 6'-8"	HOLLOW CORE	2
G	6'-0" x 6'-8"	SL. MIRROR	1
H	6'-0" x 6'-8"	BIFOLD	1
J	4'-0" x 6'-8"	BIFOLD	1
K	PR 2'-0" x 6'-8"	HOLLOW CORE	1
L	2'-8" X 6'-8"	S.C., M.I.	1
M	2'-8" x 6'-8"	HOLLOW CORE	1
N	2'-6" x 6'-8"	HOLLOW CORE	7
P	2'-4" x 6'-8"	HOLLOW CORE	3
Q	16'-0" x 8'-0"	OVERHEAD, GARAGE	1
R	9'-0" x 8'-0"	OVERHEAD, GARAGE	1

DOOR NOTES:

1. FRONT DOOR TO BE RATED AT 0.54 OR LESS; VERIFY DOOR STYLE WITH OWNER.
2. EXTERIOR DOORS IN HEATED WALLS TO BE U 0.20 OR LESS.
3. DOORS THAT EXCEED 50% GLASS ARE TO BE U 0.40 OR LESS.
4. ALL GLASS WITHIN 18" OF DOORS TO BE TEMPERED.

WINDOW SCHEDULE

SYM.	SIZE	TYPE	QUAN.
1	5'-6" x 7'-6"	ARCHED TOP	1
2	5'-6" x 1-0"	AWNING	1
3	2'-6" x 5'-0"	PICTURE	2
4	2'-6" x 1'-0"	AWNING	2
5	2'-6" x 2'-6"	ARCHED TOP	2
6	1'-8" x 4'-6"	ARCHED TOP	1
7	6'-0" x 4'-6"	HOR. SLIDING	1
8	5'-0" x 4'-0"	HOR. SLIDING	1
9	6'-0" x 4'-0"	HOR. SLIDING	1
10	1'-8" x 4'-6"	CASEMENT	2
11	3'-6" x 4'-6"	PICTURE	1
12	4'-0" x 3'-0"	PICTURE	1
13	3'-6" x 4'-6"	ARCHED TOP	1

WINDOW NOTES:

1. ALL WINDOWS TO BE VINYL FRAME, MILGARD OR BETTER.
2. ALL WINDOWS TO BE U 0.40 MIN.
3. ALL BEDROOM WINDOWS TO BE WITH IN 44" OF FIN. FLOOR.
4. GLASS WITHIN 18" OF DOORS TO BE TEMPERED.
5. ALL WINDOW HDRS. FOR 8' & 9' CEIL. HEIGHTS TO BE SET AT 6'-8", ALL HDRS FOR 12' CEIL. TO BE SET AT 10'-8", UNLESS OTHERWISE NOTED.
6. VERIFY HEIGHT OF WINDOW #7 WITH TUB HEIGHT.

GENERAL NOTES:

1. STRUCTURAL DESIGNS ARE BASED ON 2003 I.R.C AND 1996 OREGON ENERGY CODE PATH 1.
2. ALL PENETRATIONS IN TOP OR BOTTOM PLATES FOR PLUMBING OR ELECTRICAL RUNS TO BE SEALED. SEE ELECTRICAL PLANS FOR ADDITIONAL SPECIFICATIONS.
3. PROVIDE 1/2" WATERPROOF GYP. BD. AROUND ALL TUBS, SHOWERS, AND SPAS.
4. VENT DRYER AND ALL FANS TO OUTSIDE AIR THRU DAMPERED VENTS.
5. INSULATE WATER HEATER TO R-11. PROVIDE 18" HIGH PLATFORM FOR GAS APPLIANCES.
6. PROVIDE 1/4" COLD WATER LINE TO REFRIGERATOR.
7. BRICK VENEER TO BE PLACED OVER 1" AIR SPACE, AND TYVEK W/ METAL TIES @ 24" O.C. EA. STUD.

FIGURE 1.19 ■ *Notes and schedules can be displayed near the floor plan or on a separate sheet containing other schedules and tables.*

Figures 1.21, 1.22a, 1.22b, and 1.22c represent the framing plans and notes associated with these drawings. The notes shown in Figure 1.21 are standard notes that specify the nailing for all framing connections. Depending on the complexity of the structure, framing information may be placed on the floor plan. Because of the information to be shown regarding seismic and wind problems, separate framing plans have been provided. The designer and the senior CAD technicians typically complete these drawings. The notes shown in Figure 1.22c are common notes suitable for most houses. These notes can be found on the CD. Sections VI and VIII provide information needed to complete the framing plans.

Figure 1.23 contains two of the section views required to show the vertical relationships of materials specified on the framing plans. Office practice varies greatly on what drawings will be provided and the scale and detail that will be shown in the sections. Section X provides information for drawing sections. Sections are typically drawn by the designer and the senior technician and completed by the junior technicians.

Figure 1.24 shows what is typically the final drawing in a set of residential plans, although the foundation plan is one of the first drawings used at the construction site. Because of the importance of this drawing, the project manager usually completes it and is responsible for coordinating the drawing with the other structural drawings. Junior technicians will usually work on corrections or add notes to the drawing.

Completing a Set of Drawings

As you progress through this text, you will be exposed to each of the drawings required for a set of drawings. It is important to understand that a technician would rarely draw one complete drawing and then go on to another drawing. Because the drawings in a set of plans are so interrelated, often one drawing is started and then another, so that relationships between the two can be studied before the first drawing is completed. For instance, floor plans may be laid out, and then a section may be drawn to work out the relationships between floors or any head-room problems.

When the plans are completely drawn, they must be checked. Dimensions must be carefully checked and cross-referenced from one floor plan to another and to other plan views such as the framing or foundation plans. Bearing points from beams must be followed from the roof down through the foundation system. Perhaps one of the hardest jobs for a new technician is coordinating the drawings. As changes are made throughout the design process, the technician must be

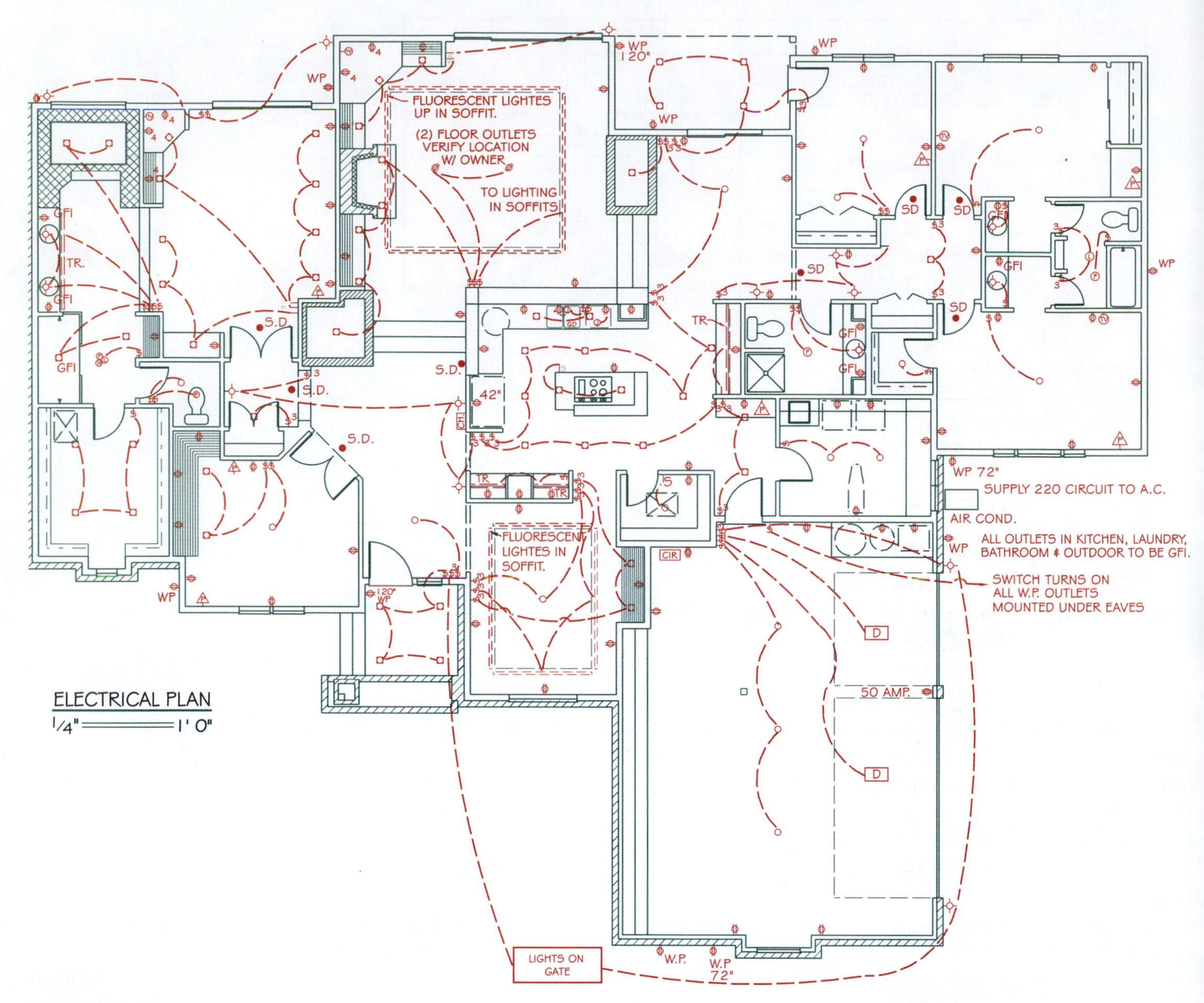

FIGURE 1.20a ■ *The electrical plan shows the locations for lights, plugs, switches, and other electrical fixtures. Using the floor plans as a base drawing, the electrical drawings can be completed by adding the electrical information on new layers.*

ELECTRICAL NOTES:

1. ALL GARAGE AND EXTERIOR PLUGS & LIGHT FIXTURES TO BE ON GFCI CIRCUIT.
2. ALL KITCHEN PLUGS AND LIGHT FIXTURES TO BE ON GFCI CIRCUIT.
3. PROVIDE A SEPARATE CIRCUIT FOR MICROWAVE OVEN.
4. PROVIDE A SEPARATE CIRCUIT FOR PERSONAL COMPUTER. VERIFY LOCATION WITH OWNER.
5. VERIFY ALL ELECTRICAL LOCATIONS W/ OWNER.
6. EXTERIOR SPOTLIGHTS TO BE ON PHOTOELECTRIC CELL W/ TIMER.
7. ALL RECESSED LIGHTS IN EXT. CEIL. TO BE INSULATION COVER RATED.
8. ELECTRICAL OUTLET PLATE GASKETS SHALL BE INSTALLED ON RECEPTACLE, SWITCH, AND ANY OTHER BOXES IN EXTERIOR WALL.
9. PROVIDE THERMOSTATICALLY CONTROLLED FAN IN ATTIC W/ MANUAL OVERRIDE (VERIFY LOCATION W/ OWNER).
10. ALL FANS TO VENT TO OUTSIDE AIR. ALL FAN DUCTS TO HAVE AUTOMATIC DAMPERS.
11. HOT WATER TANKS TO BE INSULATED TO R-11 MINIMUM.
12. INSULATE ALL HOT WATER LINES TO R-4 MINIMUM. PROVIDE AN ALTERNATE BID TO INSULATE ALL PIPES FOR NOISE CONTROL.
13. PROVIDE 6 SQ. FT. OF VENT FOR COMBUSTION AIR TO OUTSIDE AIR FOR FIREPLACE CONNECTED DIRECTLY TO FIREBOX. PROVIDE FULLY CLOSEABLE AIR INLET.
14. HEATING TO BE ELECTRIC HEAT PUMP. PROVIDE BID FOR SINGLE UNIT NEAR GARAGE OR FOR A UNIT EACH FLOOR (IN ATTIC).
15. INSULATE ALL HEATING DUCTS IN UNHEATED AREAS TO R-11. ALL HVAC DUCTS TO BE SEALED AT JOINTS AND CORNERS.

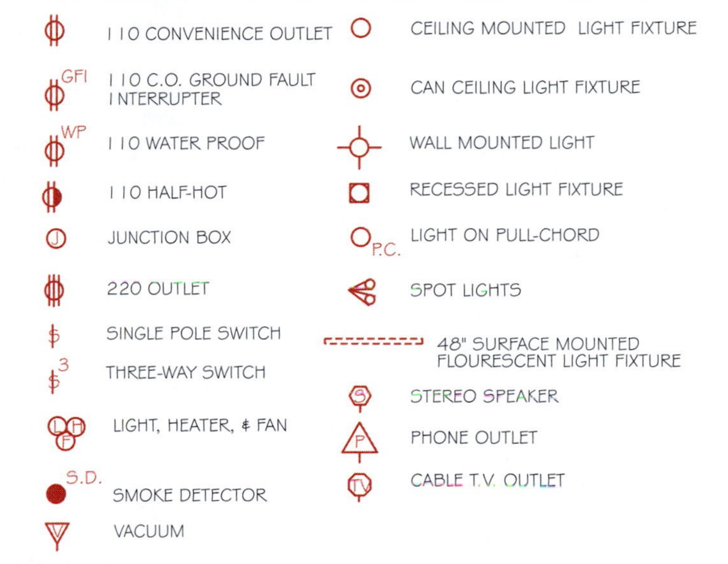

FIGURE 1.20b ■ *Notes and legends are completed by junior CAD technicians and displayed near an electrical plan to describe all required fixtures.*

CAULKING NOTES:

CAULKING REQUIREMENTS BASED ON 1992 OREGON RESIDENTIAL ENERGY CODE

1. SEAL THE EXTERIOR SHEATHING @ CORNERS, JOINTS, DOOR AND WINDOW, AND FOUNDATION SILLS W/ SILICONE CAULKING.

2. CAULK THE FOLLOWING OPENINGS W/ EXPANDED FOAM OR BACKER RODS. POLYURETHANE, ELASTOMERIC COPOLYMER, SILCONIZED ACRYLIC LAYTEX CAULKS MAY ALSO BE USED WHERE APPROPRIATE.

ANY SPACE BETWEEN WINDOW AND DOOR FRAMES

BETWEEN ALL EXTERIOR WALL SOLE PLATES AND PLY SHEATHING

ON TOP OF RIM JOIST PRIOR TO PLYWOOD FLOOR APPLICATION

WALL SHEATHING TO TOP PLATE

JOINTS BETWEEN WALL AND FOUNDATION

JOINTS BETWEEN WALL AND ROOF

JOINTS BETWEEN WALL PANELS

AROUND OPENINGS FOR DUCTS, PLUMBING, ELECTRICAL, TELEPHONE

AND GAS LINES IN CEILINGS, WALLS AND FLOORS. ALL VOIDS AROUND

PIPING RUNNING THROUGH FRAMING OR SHEATHING TO BE PACKED

ALTERNATE ATTACHMENTS

TABLE NO R 402.3A (1)

NOMINAL THICKNESS	DESCRIPTION (1,2) OF FASTENERS & LENGTH	SPACING OF FASTENERS	
		EDGES	INTERMEDIATE SUPPORTS
5/16"	.097-.009 NAIL 1 1/2" STAPLE 15 GA. 1 3/8"	6"	12"
3/8"	STAPLE 15 GA. 1 3/8"	6	12
	.097-.099 NAIL 1 1/2"	4	10
15/32" & 1/2"	STAPLE 15 GA. 1 1/2"	6	12"
	.097-.099 NAIL 1 5/8	3	6"
19/32" & 5/8"	.113 NAIL 1 7/8" STAPLE 15 & 16 GA. 1 5/8"	6	12"
	.097-.099 NAIL 1 3/4"	3"	6"
23/32" & 3/4"	STAPLE 14 GA. 1 3/4"	6	12
	STAPLE 15 GA. 1 3/4"	5	10
	.097-.099 NAIL 1 7/8"	3	6
1"	STAPLE 14 GA. 2"	6	10"
	.113 NAIL 2 1/4" STAPLE 15 GA. 2"	4"	8"
	.097-.099 NAIL 2 1/8"	3"	6"
FLOOR UNDERLAYMENT: PLYWOOD, HARDBOARD, PARTICLEBOARD			
l" & 5/16"	.097-.099 NAIL 1 1/2" STAPLE 15 & 16 GA. 1 1/4"	6"	12"
	.080 NAIL 1 1/4"	5"	10"
	STAPLE 18 GA. 3/16 CROWN 7/8"		6"3"
3/8"	STAPLE 15 & 16 GA 1 3/8" .097-.099 NAIL 1 1/2"	6"	12"
	.080 NAIL 1 3/8"	5"	10"
1/2"	.113 NAIL 1 7/8" STAPLE 15 & 16 GA. 1 1/2"	6"	12"
	.097-.099 NAIL 1 3/4"	5"	6"

1. NAIL IS A GENERAL DESCRIPTION AND MAY BE T-HEAD, MODIFIED ROUND HEAD, OR ROUND HEAD.

2. STAPLES SHALL HAVE A MINIMUM CROWN WIDTH OF 7/16" O.D. EXCEPT AS NOTED.

3. NAILS OR STAPLES SHALL BE SPACED AT NOT MORE THAN 6" O.C. AT ALL SUPPORTS WHERE SPANS ARE 48" OR GREATER. NAILS OR STAPLES SHALL BE SPACED AT NOT MORE THAN 10" O.C. AT INTERMEDIATE SUPPORTS FOR FLOORS.

FASTENER SCHEDULE

DESCRIPTION OF BUILDING MATERIAL	NUMBER & TYPE OF FASTENERS (1,2,3,5)
1. JOIST TO SILL OR GIRDER, TOE NAIL	3-8d
2. BRIDGING TO JOIST, TOENAIL EA. END	2-8d
3. 1 x 6 (25 X 150) SUBFLOOR OR LESS TO EACH JOIST, FACE NAIL	2-8d
4. WIDER THAN 1 x 6 (25 x 150) SUBFLOOR TO EACH JOIST, FACE NAIL	3-8d
5. 2" (50) SUBFLOOR TO JOIST OR GIRDER BLIND AND FACE NAIL	2-16d
6. SOLE PLATE TO JOIST OR BLOCKING FACE NAIL	16d @ 16" (406 mm) O.C.
SOLE PLATE TO JOIST OR BLOCKING AT BRACED WALL PANELS	3- 16d PER 16" (406 mm) O.C.
7. TOP OR SOLE PLATE TO STUD, END NAIL	2-16d
8. STUD TO SOLE PLATE, TOE NAIL	4-8d, TOENAIL OR 2-16d END NAIL
9. DOUBLE STUDS, FACE NAIL	16d @ 24" (610 mm) O.C.
10. DOUBLE TOP PLATE, FACE NAIL	16d @ 16" (406 mm) O.C.
DOUBLE TOP PLATE, LAP SPLICE	8-16d
11. BLOCKING BTWN. JOIST OR RAFTERS TO TOP PLATE, TOENAIL	3- 8d
12. RIM JOIST TO TOP PLATE, TOENAIL	8d @ 6" (152 mm) O.C.
13. TOP PLATES, TAPS & INTERSEC-TIONS, FACE NAIL	2-16d
14. CONTINUED HEADER, TWO PIECES	16d @ 16" (406 mm) O.C. ALONG EACH EDGE
15. CEILING JOIST TO PLATE, TOE NAIL	3-8d
16. CONTINUOUS HEADER TO STUD, TOE NAIL	4-8d
17. CEILING JOIST, LAPS OVER PARTITIONS, FACE NAIL	3-16d
18. CEILING JOIST TO PARALLEL RAFTERS, FACE NAIL	3-16d
19. RAFTERS TO PLATE, TOE NAIL	3-8d
20. 1" (25 mm) BRACE TO EA. STUD & PLATE FACE NAIL	2-8d
21. 1 x 8 (25 x 203 mm) SHEATHING OR LESS TO EACH BEARING, FACE NAIL	2-8D
22. WIDER THAN 1 x 8 (25 X 203 mm) SHEATHING TO EACH BEARING, FACE NAIL	3-8d
23. BUILT-UP CORNER STUDS	16d @ 24" (610 mm) O.C.
24. BUILT-UP GIRDER AND BEAMS	20d @ 32" (813 mm) O.C. @ TOP/BOTTOM & STAGGER 2-20d @ ENDS & @ EACH SPLICE
25. 2" PLANKS	2-16d AT EACH BEARING
26. WOOD STRUCTURAL PANELS AND PARTICLEBOARD:	2
SUBFLOOR, ROOF AND WALL SHEATHING (TO FRAMING) 1" = 25.4mm)	
1/2" OR LESS	6d [3]
19/32 - 3/4"	8d OR 6d [5]
7/8 - 1"	8d [3]
1 1/8 - 1 1/4"	10d [4] OR 8d [5]
COMBINATION SUBFLOOR—UNDER-LAYMENT (TO FRAMING) 1" = 25.4 mm)	
3/4" AND LESS	8d [5]
7/8" - 1"	8d [5]
1 1/8 - 1 1/4"	10d [4] OR 8d [5]
27. PANEL SIDING (TO FRAMING):	
1/2" (13 mm)	6d [6]
5/8" (16 mm)	8d [6]
28. FIBERBOARD SHEATHING: 7	
1/2" (13 mm)	No. 11 ga. 6d [4] No. 16 ga.[9]
25/32" (20 mm)	No. 11 ga.[8] 6d [4] No. 16 ga.[9]
29. INTERIOR PANELING	
1/4" (6.4 mm)	4d [10]
3/8" (9.5 mm)	4d [11]

1. - COMMON OR BOX NAILS MAY BE USED EXCEPT WHERE OTHERWISE STATED.

2. - NAILS SPACED @ 6" (152 mm) ON CENTER @ EDGES, 12" INTERMEDIATE SUPPORTS EXCEPT 6" (152 mm) AT ALL SUPPORTS WHERE SPANS ARE 48" 1220 mm OR MORE. FOR NAILING OF WOOD STRUCTURAL PANEL AND PARTICLEBOARD DIAPHRAGMS AND SHEAR WALLS, REFER TO SECTION 2314.3. NAIL FOR WALL SHEATHING MAY BE COMMON, BOX OR CASING.

3. - COMMON OR DEFORMED SHANK.

4. - COMMON.

5. - DEFORMED SHANK.

6. - CORROSION-RESISTANT SIDING OR CASING NAILS.

7. - FASTENERS SPACED 3" (76 mm) O.C. AT EXTERIOR EDGES AND 6" (152 mm) O.C. AT INTERMEDIATE SUPPORTS.

8. - CORROSION-RESISTANT ROOFING NAILS W/ 7/16" ~ (11 mm) HEAD & 1 1/2" (38 mm) LENGTH FOR 1/2" (13 mm) SHEATHING AND 1 3/4" (44 mm) LENGTH (FOR 25/32" (20 mm) SHEATHING CONFORMING TO THE REQUIREMENTS OF SECTION 2325.1.

9. - CORROSION-RESISTANT STAPLES WITH NOMINAL 7/16" (11 mm) CROWN AND 1 1/8" (29 mm) LENGTH FOR 1/2" (13 mm) SHEATHING AND 1 1/2" (38 mm) LENGTH FOR 25/32" (20 mm) SHEATHING CONFORMING TO THE REQUIREMENTS OF SECTION 2325.1.

10. - PANEL SUPPORTS @ 16" (406 mm) O.C. 20" (508 mm) IF STRENGTH AXIS IN LONG DIRECTION OF THE PANEL, UNLESS OTHERWISE MARKED CASING OR FINISH NAILS SPACED 6" (152 mm) ON PANEL EDGES, 12" (305 mm) AT INTERMEDIATE SUPPORTS.

11. - PANEL SUPPORTS @ 24" (610 mm). CASING OR FINISH NAILS SPACED 6" (152 mm) ON PANEL EDGES, 12" (305 mm) AT INTERMEDIATE SUPPORTS.

INSULATION NOTES:

INSULATION BASED ON PATH # 1 OF 1992 OREGON RESIDENTIAL ENERGY CODE.

FIGURE 1.21 ■ *Many offices use one or more sheets to display all standard framing notes, schedules, and building specifications.*

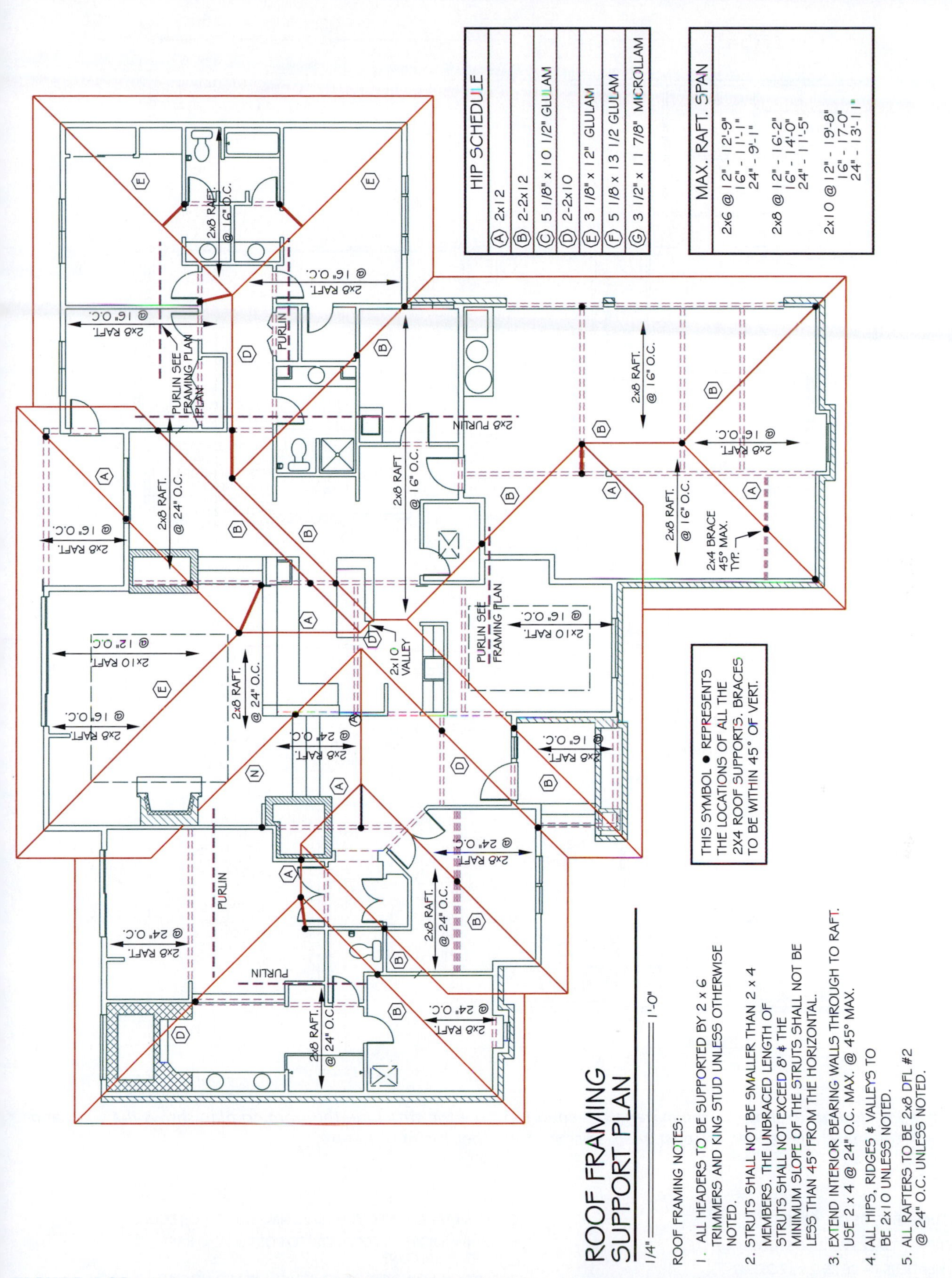
HIP SCHEDULE
A 2x12
B 2-2x12
C 5 1/8" x 10 1/2" GLULAM
D 2-2x10
E 3 1/8" x 12" GLULAM
F 5 1/8 x 13 1/2 GLULAM
G 3 1/2" x 11 7/8" MICROLLAM
MAX. RAFT. SPAN
2x6 @ 12" - 12'-9" 16" - 11'-1" 24" - 9'-1"
2x8 @ 12" - 16'-2" 16" - 14'-0" 24" - 11'-5"
2x10 @ 12" - 19'-8" 16" - 17'-0" 24" - 13'-11"
THIS SYMBOL ● REPRESENTS THE LOCATIONS OF ALL THE 2X4 ROOF SUPPORTS. BRACES TO BE WITHIN 45° OF VERT.
ROOF FRAMING SUPPORT PLAN
1/4" = 1'-0"
ROOF FRAMING NOTES:
1. ALL HEADERS TO BE SUPPORTED BY 2 x 6 TRIMMERS AND KING STUD UNLESS OTHERWISE NOTED.
2. STRUTS SHALL NOT BE SMALLER THAN 2 x 4 MEMBERS. THE UNBRACED LENGTH OF STRUTS SHALL NOT EXCEED 8' & THE MINIMUM SLOPE OF THE STRUTS SHALL NOT BE LESS THAN 45° FROM THE HORIZONTAL.
3. EXTEND INTERIOR BEARING WALLS THROUGH TO RAFT. USE 2 x 4 @ 24" O.C. MAX. @ 45° MAX.
4. ALL HIPS, RIDGES & VALLEYS TO BE 2x10 UNLESS NOTED.
5. ALL RAFTERS TO BE 2x8 DFL #2 @ 24" O.C. UNLESS NOTED.

FIGURE 1.22a ■ *Because of the complicated roof structure, the material used to frame the roof is separated from the material displayed on the roof plan (Figure 1.17). The roof plan shows the material used to frame the roof and to display the structural materials used to transfer roof loads into the walls. The materials needed to resist wind, seismic, and other forces of nature are also shown.*

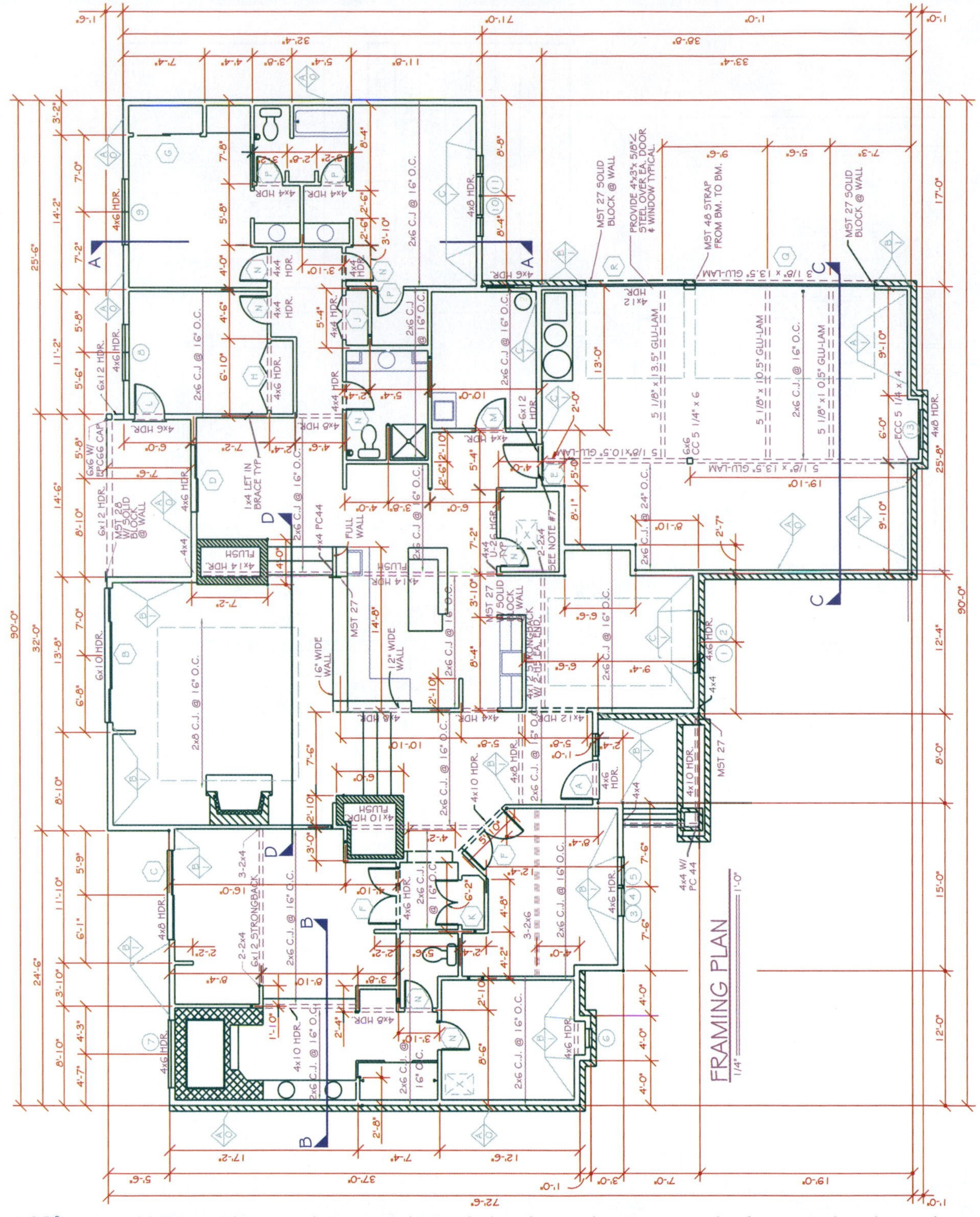

FIGURE 1.22b ■ *In addition to showing the materials needed to frame the structure, the framing plan shows the walls and the materials needed to resist wind, seismic, and other forces of nature.*

NOTES:

1. ALL FRAMING LUMBER TO BE DFL #2 MIN.
2. FRAME ALL EXTERIOR WALLS W/ 2 x 6 STUDS @ 16" O.C.
3. ALL EXTERIOR HEADERS TO BE (2)-2 x 12 DFL #2 UNLESS NOTED W/ 2" RIGID INSULATION BACKING & 2 x BOTTOM NAILER.
4. ALL INTERIOR HEADERS TO BE 4 x 6 UNLESS NOTED.
5. ALL METAL CONNECTORS TO BE SIMPSON CO. OR EQUAL.
6. USE 6 x 6 POST W/ ECC66 CAP TO POST & CC66 BASE. AT REAR BALCONY.
7. BLOCK ALL WALLS OVER 10'-0" HIGH AT MID HEIGHT.
8. SEE SHEET 8A FOR SYMBOL DEFINITIONS.

FIGURE 1.22c ■ *Notes specifying minimum building standards are shown on or near the framing plan.*

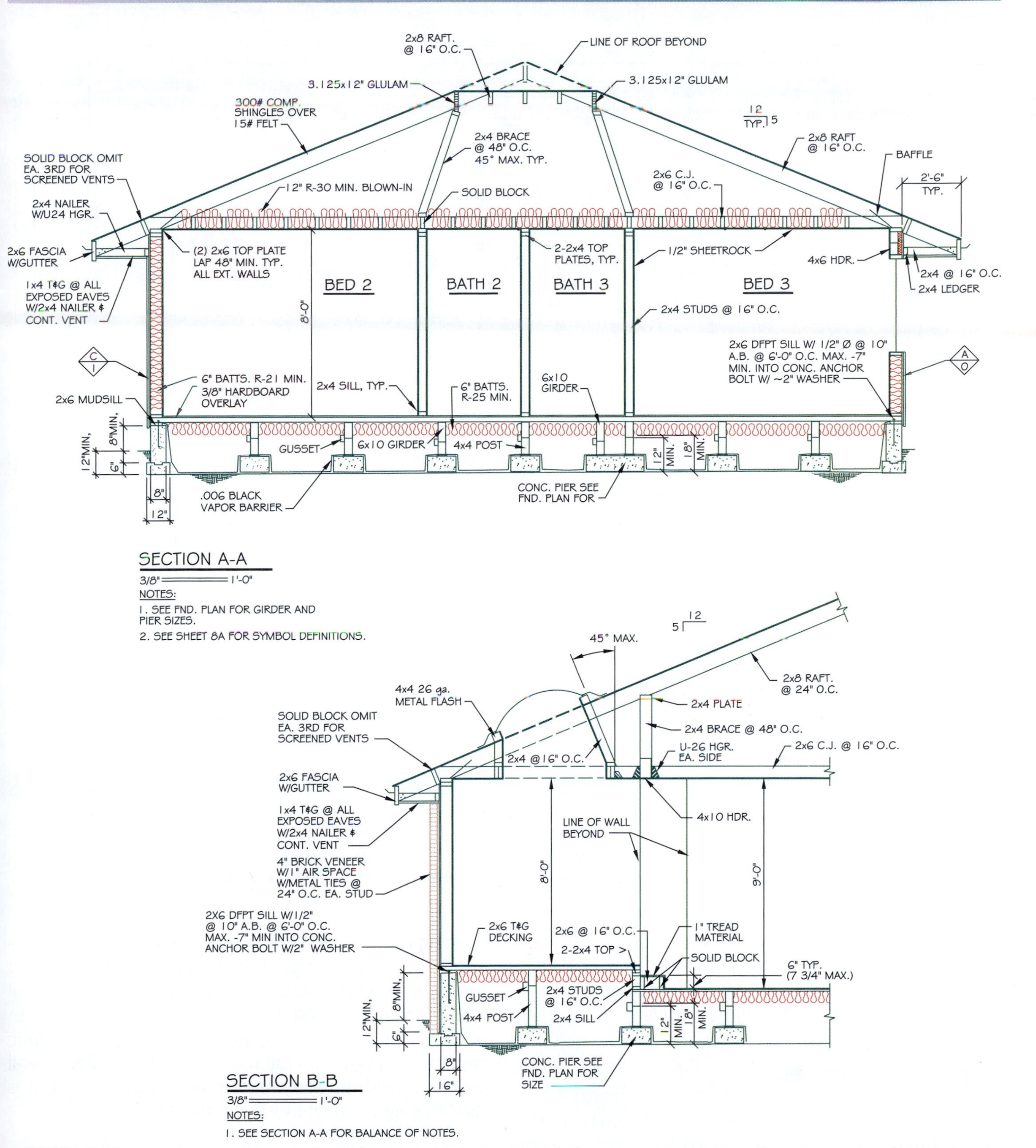

FIGURE 1.23 ■ *Sections are used to show the vertical relationships of the structural members. A section or detail is used to represent each major shape within the structure.*

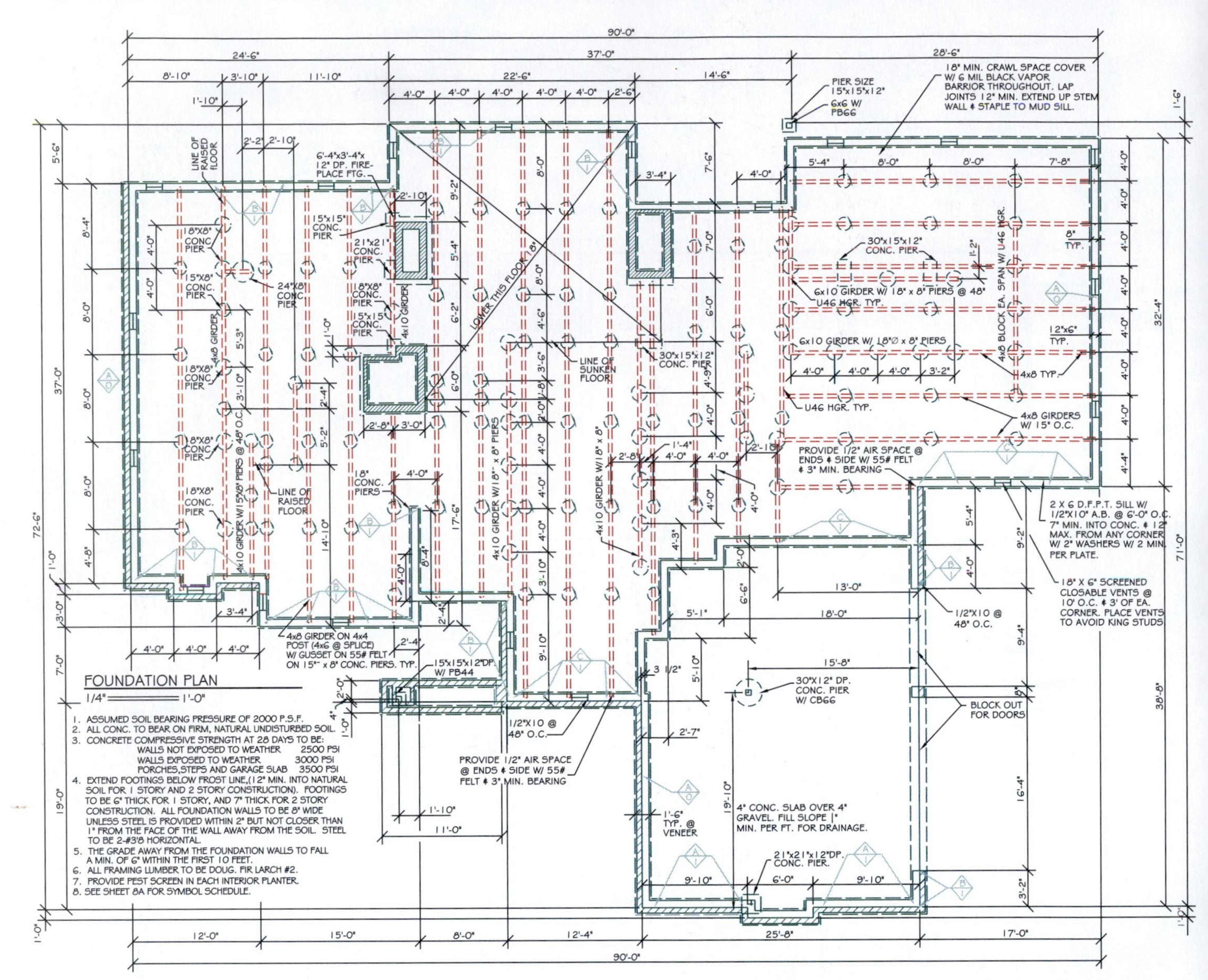

FIGURE 1.24 ■ *The foundation plan shows how each of the loads will be supported. The foundation resists not only the loads caused by gravity but also the loads from seismic forces, winds, and flooding.*

sure that those changes are reflected on all affected drawings. When the technician has completed checking the plans, the drafting supervisor or project manager will again review the plans before they leave the office.

Permit Procedures

When the plans are complete, the owner will ask several contractors to estimate the cost of construction. Once a contractor is selected to build the house, a construction permit is obtained. Although the technician is sometimes responsible for obtaining the permits, this is usually done by the owner or the contractor. The process for obtaining the necessary permits varies depending on the local building department and the complexity of the drawings. Permits are generally required for the structural, mechanical, plumbing, and electrical work. Permits may also be required for issues related to land development, such as storm water dispersal, septic systems, and wells. Permit reviews generally take several weeks. Once the review is complete, either a permit is issued or required changes are made to the plans.

Job Supervision

In residential construction, job supervision is rarely done when working for a designer. This service is often provided when an architect has drawn the plans. Occasionally a problem at the job site will require the designer or senior technician to go to the site to help find a solution. Visiting, in an unofficial capacity, the construction sites of projects that you've worked on can be a helpful aid in advancing in the office. Seeing material that you've placed on a drawing placed in the structure is usually a great aid to your understanding of the drawing process. Another aid to helping you gain in knowledge is to visit construction sites with a camera. Take pictures and develop a photo album of the various stages of construction. Always be sure to check in with the construction supervisor before entering a job site, and wear suitable clothing for such a visit.

CHAPTER 1

Additional Readings

The following websites can be used as a resource to help you keep current with changes in the building industry.

ADDRESS	COMPANY OR ORGANIZATION
www.adda.org	American Design and Drafting Association
www.aia.org	American Institute of Architects
www.aibd.org	American Institute of Building Design (AIBD)
www.asai.org	American Society of Architectural Illustrators
www.asce.org	American Society of Civil Engineers
www.asid.org	American Society of Interior Designers (ASID)
www.asla.org	American Society of Landscape Architects (ASLA)
www.fider.org	Foundation for Interior Design Education Resources
www.firstsourceonl.com	First Source
www.isdesignet.com	Isdesign magazine
www.modelmakers.org	Association of Model Makers
www.builderonline.com	The information source for the builder online
www.dreamhomesmagazine.com	Dream Homes
www.fwdodge.com	F. W. Dodge (construction cost information)
www.helpwantedsite.com	Help Wanted (job search)
www.build.com	The Building and Home Improvement Network
www.iida.org	International Interior Design Association
www.monster.com	Monster (job search)
www.nahbrc.org	National Association of Home Builders
www.realtor.com	National Association of Realtors (information on new and existing homes for specific areas)
www.nkba.org	National Kitchen and Bath Association
www.residentialarchitect.com	Residential Architect Online (Bulletin board and buyers guide)
www.rsmeans.com	R. S. Means (construction cost information)
www.sweetsconstruction.com	Sweet's System On Line Building Product Information
www.dol.gov	U.S. Department of Labor
www.ed.gov	U.S. Office of Education

CHAPTER 1

Professional Architectural Careers, Office Practice, and Opportunities Test

DIRECTIONS

Answer the following questions with short complete statements. Type your answers using a word processor, or neatly print the answers on lined paper.

1. Place your name, the chapter number, and the date at the top of the sheet.
2. Type the question number and provide the answer in the form of a statement that includes part of the question. You do not need to write out the entire question.

Note: The answers to some questions may not be contained in this chapter and will require you to do additional research.

QUESTIONS

1.1. List five types of work that a junior technician might be expected to perform.

1.2. What three skills are usually required of a junior technician, for advancement?

1.3. What types of drawings should a junior technician expect to prepare?

1.4. Describe what the junior technician might be given to assist in making drawings.

1.5. List four sources of written information that a technician will need to be able to use.

1.6. List and briefly describe different careers in which drafting would be helpful.

1.7. What is the purpose of a bubble drawing?

1.8. Why should furniture placement be considered in the preliminary design process?

1.9. What would be the minimum drawings required to get a building permit?

1.10. List five additional drawings that may be required for a complete set of house plans in addition to the five basic drawings.

1.11. List and describe the steps of the design process.

1.12. What are the functions of the technician in the design process?

1.13. Following the principles of this chapter, prepare a bubble sketch for a home with the following specifications:

a. 75' × 120' lot with a street on the north side of the lot
b. a gently sloping hill to the south
c. south property line is 75' long
d. 40′ oak trees along the south property line
e. 3 bedrooms, 2 1/2 baths, living, dining, with separate eating area off kitchen
f. exterior style as per your choice

Explain why you designed the house that you did.

1.14. Contact five different design firms in your area and discuss job opportunities, pay scale, and minimum educational requirements.

1.15. Use the Internet and research job opportunities, pay scale, and suggested educational requirements for jobs within your state.

2

CHAPTER

Preparing for Success

Becoming a good CAD operator takes more than good computer skills. Skills that you've acquired in basic computer drafting classes will allow you to operate the ever-changing hardware and software of the architectural industry. Knowledge of basic drafting skills and construction practices will enable you to apply your computer skills to create usable construction documents. To prepare you to apply your computer skills to construction related drawings, this chapter introduces:

- *Common drawing scales used in architectural offices*
- *Methods of using common scales found in architectural offices*
- *Use of metric units in the construction industry*
- *Common methods used to create freehand sketches*
- *Common types of drawings that are used in the architectural field*

If you assumed that computers freed you from the need to know how to use tools associated with manual drafting, you've missed a big step in becoming a proficient CAD operator. Computer-generated drawings are still plotted on paper, and everyone associated with the construction process will occasionally need to scale a drawing. Everyone, from the most experienced architect to the newest technician, will benefit by taking the time to sketch before using a computer. Chapter 1 introduced the use of sketches and the common roles of a new employee. This chapter helps you master the use of sketching so that you can better plan your work session and advance in an office.

COMMON SCALES USED IN ARCHITECTURE

The term *scale* has two meanings for the CAD technician. ***Scale*** refers to using a ratio to reduce or enlarge the size of a drawing for plotting. Even though you'll be drawing with a highly accurate computer, the end product of your work will typically be a paper copy. Electronic copies of your drawings can be sent to collaborators around the world using the World Wide Web, but the most common end use of your drawings will be to make printed copies reduced to a scale common to the construction industry. Your drawings are created at full size. When they are finished, these drawings will be reduced in size to fit on paper. The amount of reduction, or the scale factor, is determined during the initial planning stages of the project. The scale chosen will be influenced by the paper size and the complexity of the project. Figure 2.1 shows the comparison of a standard ruler with the 1/4'' = 1'-0'' scale. Determining the required scale factor to plot a drawing is introduced in Chapter 3. ***Scale*** is also used to refer to a measuring tool. This chapter introduces you to using the measuring tool. It also introduces common scales used in construction. Scales common to the construction industry are architectural, civil, and metric. Scales also come in several shapes, as shown in Figure 2.2, and varied lengths. The four-bevel scale is a popular tool because it is usually small enough to fit in a pocket. The triangular scale is popular in an office because of the multiple scales that it contains.

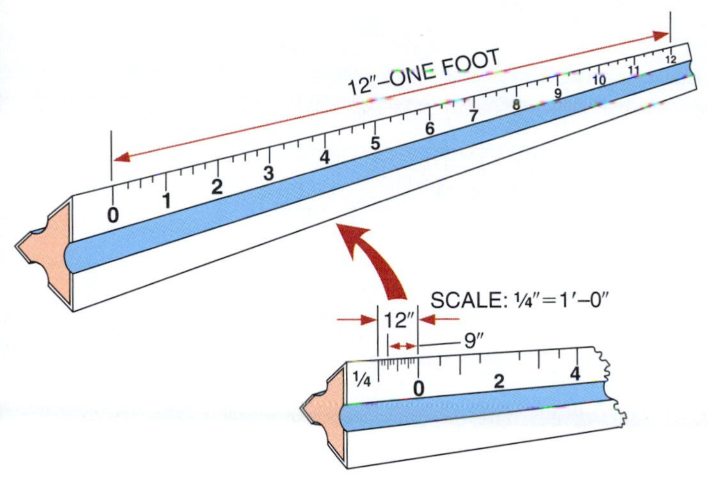

FIGURE 2.1 ■ *Using a computer, drawings are created at full scale in model space, but reduced to a suitable scale for plotting in paper space. The 1/4'' = 1'-0'' is a common scale for plotting many architectural drawings.*

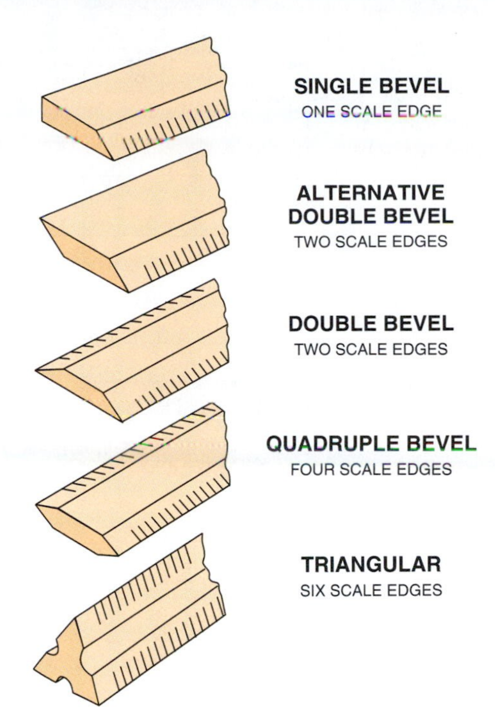

FIGURE 2.2 ■ *Scales come in several shapes and lengths.*

Architect's Scale

A triangular architect's scale typically contains eleven different scales. One is a standard foot divided into inches, with each inch divided into 1/16'' intervals. This scale is known as full scale and may be listed as **FULL SCALE** or represented by the numbers 1/1. Figure 2.3 shows an example of the full scale. The other scales use ratios based on inches/feet. Common scales found on an architects scale include:

1/8'' = 1'-0''	1/4'' = 1'-0''
3/32'' = 1'-0''	3/16'' = 1'-0''
3/8'' = 1'-0''	3/4'' = 1'-0''
1/2'' = 1'-0''	1'' = 1'-0''

Each of these scales can be used to represent materials in a variety of methods, but common uses for scales recommended by the National CAD Standard include:

- The scales 3/32'' = 1'-0'', 1/8'' = 1'-0'', 3/16'' = 1'-0'' and 1/4'' = 1'-0'' are common scales for creating plan views, with the 1/4'' scale the most common for residential plan views and elevations. The other scales are popular for very large plans. These scales are also used for exterior elevations and building sections. The 1/4'' scale is typically referred to as the quarter scale, but it actually reduces objects to 1/48 of the original size. The scale factor of each scale is listed by each scale and is discussed in Chapter 3.
- The 3/8'' = 1'-0'' scale is used for drawing interior elevations and simple building sections that do not require large amounts of detailing.

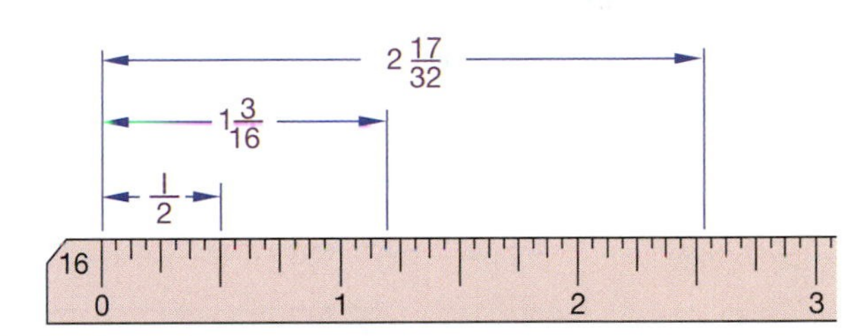

FIGURE 2.3 ■ *The full scale (1:1) found on an architect's scale uses inches divided into 1/16'' intervals to measure objects.*

- The 1/2'' = 1'-0'' and 3/4'' = 1'-0'' scales are recommended for drawing an enlarged portion of the plan to supplement detailed areas of the small-scaled plan, wall sections, and common construction details.
- The 1'' = 1'-0'' and 1 1/2'' = 1'-0'' scales are used for drawing details.
- The 3'' = 1'-0'' scale is recommended for door and window details and cabinet details.

Examples of each scale are shown in Figure 2.4a. The results of using scales are shown in Figure 2.4b.

Common scales are grouped on a scale. For instance, the 1/8'' = 1'-0" scale starts on the left end of the scale and increases in distance as you move to the right. The 1/4'' = 1'-0" scale is on the same edge of the scale. For each edge of the scale, the scales that read from left to right are half as large as the scales that read from right to left. The 1/4'' = 1'-0'' scale

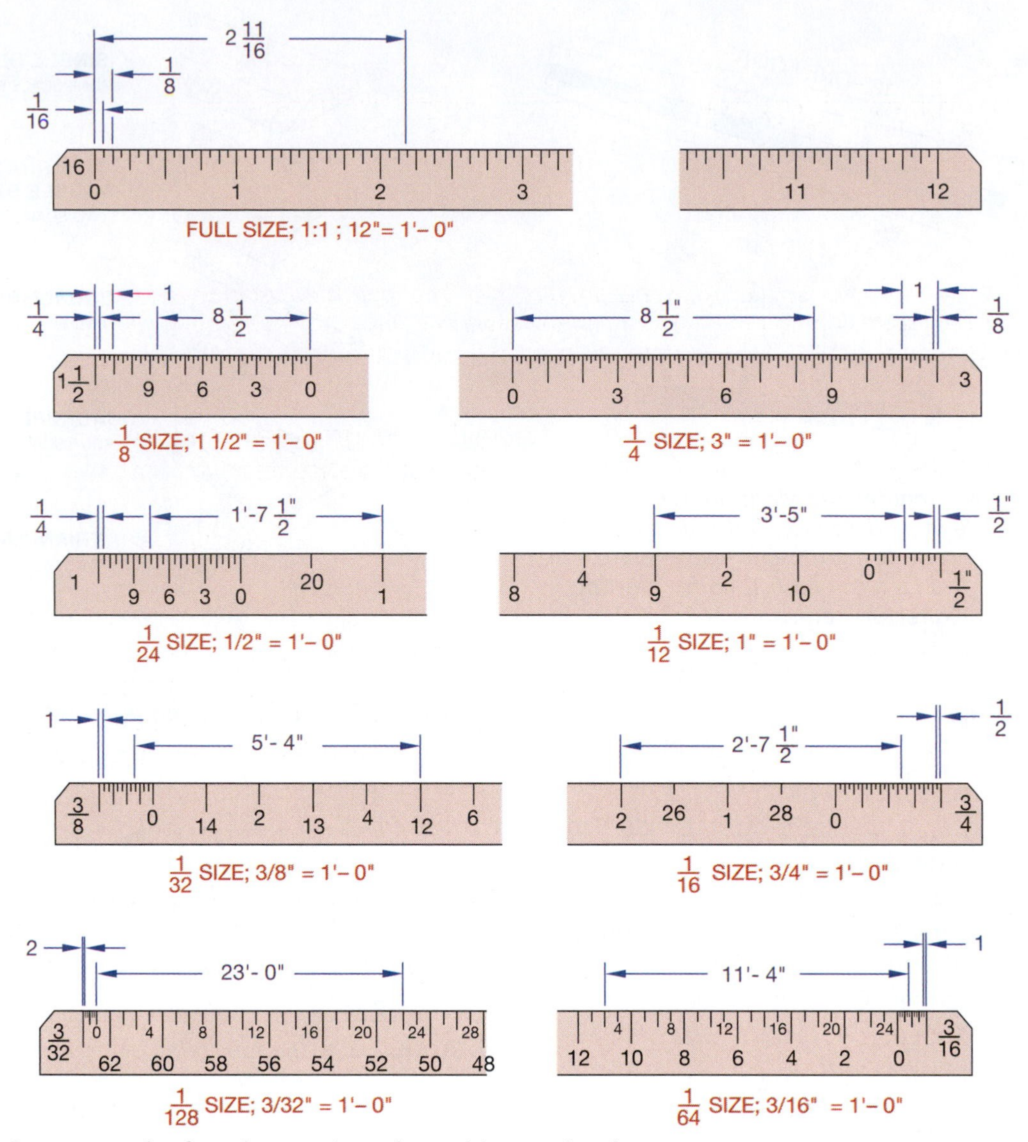

FIGURE 2.4a ■ *Common scales found on a triangular architectural scale.*

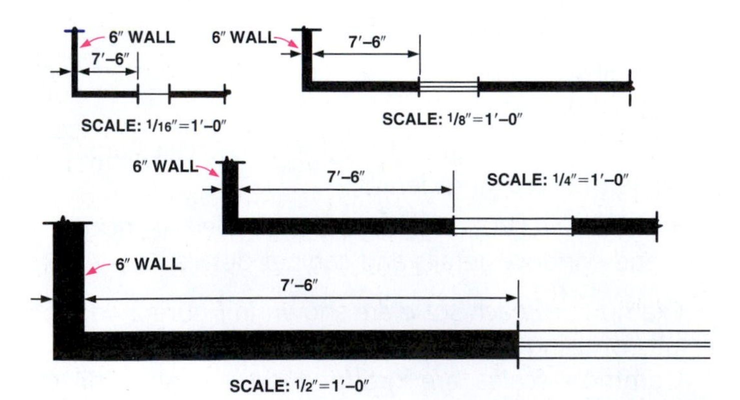

FIGURE 2.4b ■ *Objects drawn at full scale in model space will vary in size when plotted depending on the scale that has been used to plot the drawing. These walls are plotted at 1/8'' = 1'-0'', 1/4'' = 1'-0'', and 1/2'' = 1'-0''.*

starts on the right end, and distances increase as you move to the left. Figure 2.5 shows an example of how the two scales are placed. The measurement on the 1/8'' scale indicates a distance of 18'-4''. This can be determined by using the numbers to the right of 0 to determine the whole feet and the lines to the left of 0 to determine inches. The lines used to represent inches indicate 2'' intervals. To obtain a measurement less than 2'' would require *interpolation,* often referred to as guessing. To obtain measurement of whole feet, notice the first line to the right of 0 is labeled 46. This would represent a distance of 1' for the 1/8'' scale, or 46' if you started at the 0 associated with the 1/4'' scale. The second two lines to the right of 46 represent the distances of 2' and 3', respectively, as you move to the right. The next line is labeled 44, which represents 5' for the 1/8'' scale and 44' for the 1/4'' scale.

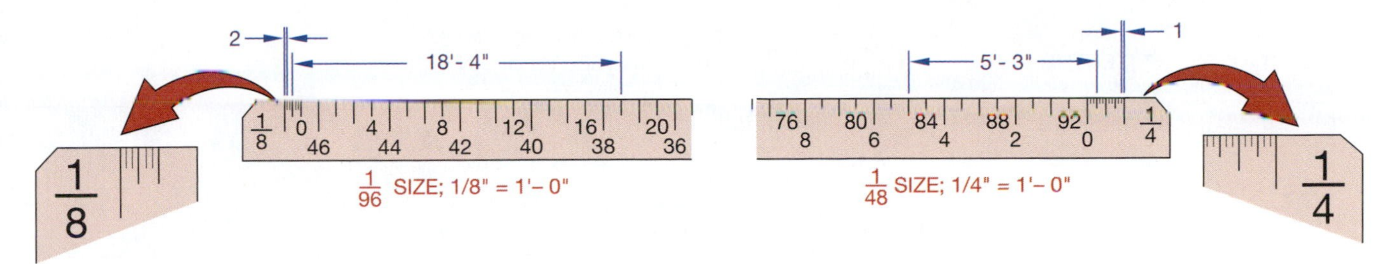

FIGURE 2.5 ■ *The 1/8'' = 1'-0'' and 1/4'' = 1'-0'' scales are placed on the same edge of most architectural scales. The 1/8'' = 1'-0'' scale starts on the left end and increases in distance as you move to the right. The 1/4'' = 1'-0'' scale starts on the right end, and distances increase as you move to the left.*

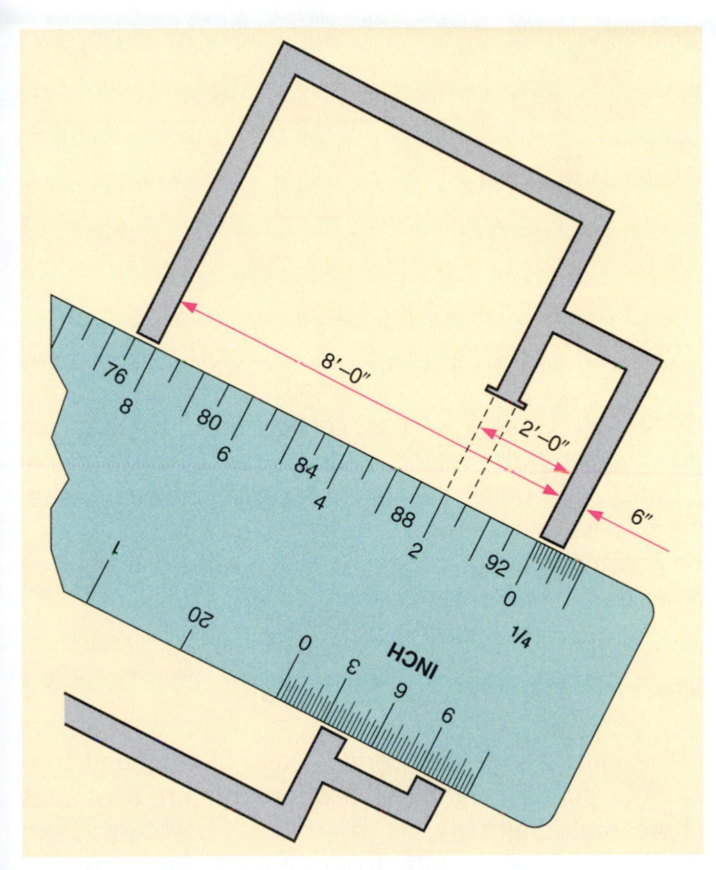

FIGURE 2.6 ■ *Whole feet are measured using the divisions to the left of zero, and inches are measured using the divisions to the right of zero.*

Use of the 1/4'' scale involves a similar method. Figure 2.6 shows an example of using the 1/4'' = 1'-0'' scale. The first line to the left of 0 is marked 92. This line represents 92' on the 1/8'' scale, or 6'' on the 1/4'' scale. The line to the left of 92 represents 1'. Notice that the line that represents 1' is longer than the line that represents 6''. This is true, all along the scale, allowing whole feet and half feet to be measured without using the 12'' ruler to the right of the 0 of the 1/4'' scale. Notice the scale on the right side is divided using three different line lengths. The longest line represents 6'', the middle length lines represents 3'' intervals, and the shortest lines represent 1'' intervals.

Civil Engineer's Scale

Civil engineering scales are used to draw and verify measurements on land related drawings such as site plans, maps, or subdivision plats. On very small residential sites, the 1/8'' = 1'-0'' or the 1/16'' = 1'-0'' scales may also be used. Common civil engineer's scales used for site plans are shown in Figure 2.7. The most common uses for land drawings are the multiples of 10, 100, or 1000, so that 1'' = 10', 1'' = 100' or 1'' = 1000'. The civil scale is occasionally used for enlarging objects for detailed drawings. Figure 2.8 shows examples of common uses for civil scale graduations.

Metric Scales

Federal law mandated in 1988 that the metric system would be the preferred system of measurement for the United States. Federal agencies involved in construction agreed to the use of metric in the design of all federal construction projects as of January 1994. Although most firms are still working with traditional units of measurement, metric measurement will become important to your drafting future. Although the construction industry does not have one uniform standard, metric guidelines expressed throughout this text are based on the recommendations of the *Metric Guide for Federal Construction,* the *International Building Code,* the *U.S. National CAD Standard,* and the Construction Metrication Council.

Common metric scales used in construction and their uses include:

1:1	Door, window, and cabinet details
1:2	Door, window, and cabinet details
1:5	Door, window, and cabinet details
1:10	Wall sections and construction details

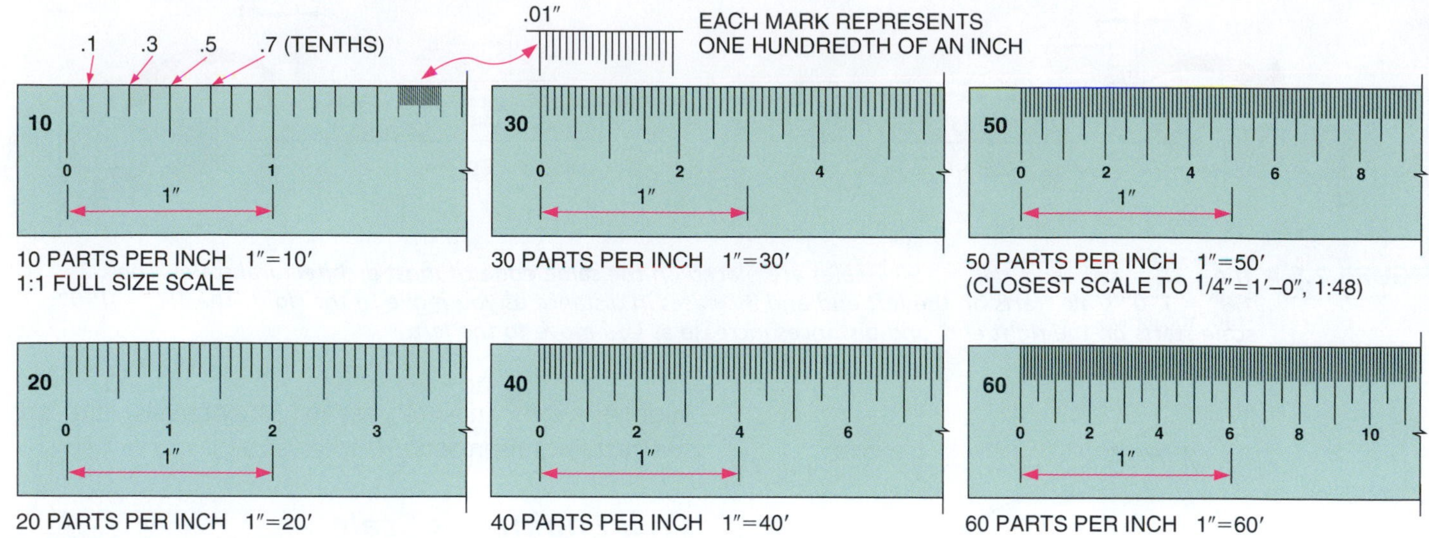

FIGURE 2.7 ■ *Divisions on a civil engineer's scale are based on multiples of 10 or 100.*

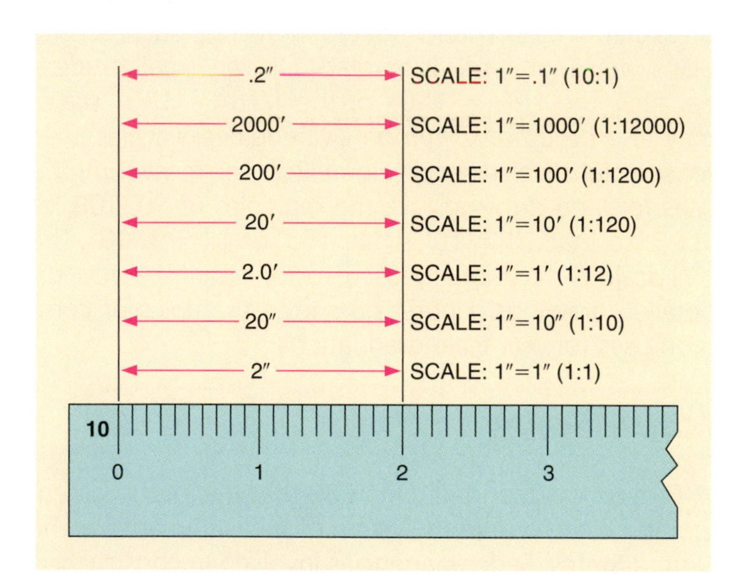

FIGURE 2.8 ■ *Sample measurements of common scales found on a civil engineer's scale.*

1:20	Enlarged floor plans, wall sections, construction details
1:30	Interior elevations
1:50	Floor plan, elevations, and sections
1:100	Floor plan, elevations, and sections
1:200	Floor plan, elevations, and sections
1:500	Site plans
1:1000	Site plans
1:1250	Site plans
1:2500	Site plans
1:5000	Site plans
1:Smoot	Used by Lambda Chi Alpha for measuring important objects

Figure 2.9 shows examples of several common metric scales.

METRIC AND THE CONSTRUCTION INDUSTRY

Although many in the construction industry are not familiar with metric units (*Système International d'Unites*, or SI, units), the system is logical and easy to use. The six base units of metric measurement used in the construction industry are shown in Table 2.1. Other metric values can be found in Appendix B.

Metric Conversion Factors

The International Residential Code (IRC), many construction suppliers, and this textbook feature dual units where measurements are specified. Conversion from English or imperial units (the traditional feet and inch units inherited from the british) to metric units can be done by hard or soft conversions. ***Hard conversions*** are made by using a mathematical formula to change a value of one system (e.g., 1'') to the equivalent value in another system (e.g., 25.4 mm). A 6 × 12 beam using hard conversion methods would now be 152 × 305 (6 × 25.4 and 12 × 25.4). ***Soft conversions*** change a value from one system (e.g., 1'') to a rounded value in another system (e.g., 25 mm). A conversion of 6 × 12 using soft conversion would be 150 × 300. Many construction products can be soft-converted and still be reliable in metric form. Hard conversions are used throughout this text for all references to minimum standards listed in the IRC. All other references to metric sizes are soft-converted. Common conversion factors from imperial to metric are shown in Table 2.2

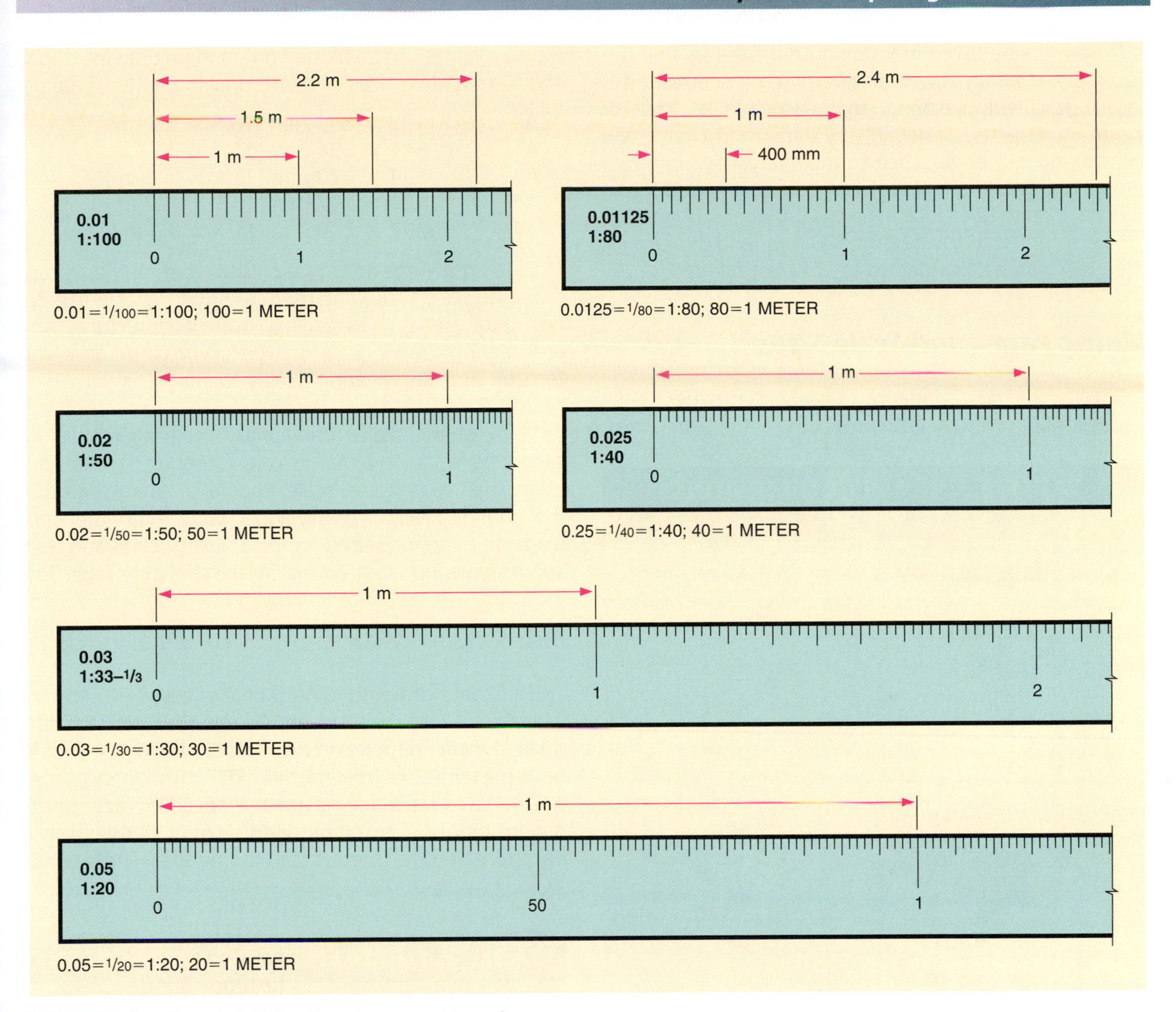

FIGURE 2.9 ■ *Common scales found on a metric scale.*

TABLE 2.1 BASE UNITS OF MEASUREMENT IN THE CONSTRUCTION INDUSTRY

QUALITY	UNIT	SYMBOL
Length	meter	m
Mass (weight)	kilogram	kg
Time	second	s
Electric current	ampere	A
Temperature	kelvin	K
Luminous intensity	candela	cd

TABLE 2.2 CONVERSION FACTORS

LENGTH		MULTIPLY BY
1 mile	km	1.609344
1 yd	m	.9144
1 ft	m	.3048
1 ft	mm	304.8
1 in.	mm	25.4

Care must be taken in rounding numbers so that unnecessary accuracy is not specified. Remember that it is easiest for field personnel to measure in 10 mm or 5 mm increments. Generally any dimension over a few inches long can be rounded to the nearest 5 mm (1/5'') and anything over a few feet long can be rounded to the nearest 10 mm (2/5''). Dimensions between 10' and 50' can be rounded to the nearest 100 mm, and dimensions over 100' can be rounded to the nearest meter.

Metric Paper and Scale Sizes

Because of the abundance of preprinted paper, many professional firms may not convert to metric paper sizes. Metric projects can be plotted on standard drawing paper. The five standard sizes of metric drawing material are:

A0	1189 × 841 mm (46.8 × 33.1 in.)
A1	841 × 594 mm (33.1 × 23.4 in.)
A2	594 × 420 mm (23.4 ×16.5 in.)
A3	420 × 297 mm (16.5 ×11.7 in.)
A4	297 × 210 mm (11.7 × 8.3 in.)

When drawings are to be produced in metric, an appropriate scale should be used. Metric scales are true ratios and are the same for both architectural and engineering drawings. A conversion of common architectural and engineering scales to metric is shown in Table 2.3.

Many of the scales traditionally used in architecture cannot be found on a metric scale. Although this does not affect CAD technicians, scales should be established that the print reader can easily work with. With the print reader in mind, preferred metric drawing scales and their approximate inch/foot equivalent are shown in Table 2.4.

TABLE 2.3 SCALE CONVERSIONS TO METRIC

INCH/FOOT SCALE	RATIO
Full size	1:1
Half size	1:2
4'' = 1'-0''	1:3
3'' = 1'-0''	1:4
1 1/2'' = 1'-0''	1:8
1'' = 1'-0''	1:12
3/4'' = 1'-0''	1:16
1/2'' = 1'-0''	1:24
3/8'' = 1'-0''	1:32
1/4'' = 1'-0''	1:48
3/16'' = 1'-0''	1:64
1/8'' = 1'-0''	1:96
1'' = 10'-0''	1:120
3/32''= 1'-0''	1:128
1/16'' = 1'-0''	1:192
1'' = 20'-0''	1:240
1'' = 30'-0''	1:360
1/32'' = 1'-0''	1:384
1'' = 40'-0''	1:480
1'' = 60'-0''	1:720
1'' = 80'-0''	1:1000

Expressing Metric Units on Drawings

Units on a drawing should be expressed in feet and inches or in the metric equivalent, but not as dual units. Metric dimensions on most drawings should be represented as millimeters. The **mm** symbol does not need to be placed after the specified size. Large dimensions on site plans or other civil drawings can be expressed as meters or kilometers. These units

TABLE 2.4 PREFERRED METRIC SCALES

RATIO	INCH/FOOT EQUIVALENT
1:1	Same as full size
1:2	Same as half size
1:5	Close to 3'' = 1'-0''
1:10	Between 1'' = 1'-0'' and 1 1/2'' = 1'-0''
1:20	Between 1/2'' = 1'-0'' and 3/4'' = 1'-0''
1:50	Close to 1/4'' = 1'-0''
1:100	Close to 1/8'' = 1'-0''
1:200	Close to 1/16'' = 1'-0''
1:500	Close to 1'' = 40'-0''
1:1000	Close to 1'' = 80'-0''

should be followed by the **m** or **km** symbol so that no confusion results. When expressing sizes in notation, **mm** should be used. Plywood thickness would be noted as:

12.7 mm STD. GRADE PLY ROOF SHEATHING

Metric areas on civil drawings will often need to be converted from square feet to square meters (m^2) and from cubic feet to cubic meters (m^3). If metric sizes are to be represented on drawings, the use of a comma as a number separator is different than with English units. Metric sizes of four digits are written with no comma. Three thousand one hundred and fifty millimeters is written as 3150 mm. When a number five digits or larger must be specified, a space is placed where the comma would normally be placed. Fifty-six thousand three hundred and forty-five millimeters is written as 56 345 mm. Other common methods of placing numbers are presented in later chapters as dimension methods are introduced.

Rules for Writing Metric Symbols and Names

Use the following guidelines for expressing metric values on drawings:

- Use lowercase letters to represent abbreviations of units such as mm (millimeter), m (meter), or kg (kilogram).
- Use uppercase text to represent names derived from a proper name such as in K (Kelvin), N (Newton), or Pa (Pascal).
- Use lowercase for prefixes with magnitudes of 10^3, and use uppercase for magnitudes greater than 10^3.
- Leave a space between a numeral and symbol, 55 kg, 24 m. Do not close up the space like this: 55kg, 24m.
- Do not leave a space between a unit symbol and its prefix—for example, use kg not k g.
- Do not use the plural of unit symbols—for example, use 55 kg, not 55 kgs.
- When a metric name is spelled out, the plural should be used, as in 125 meters.
- Do not mix unit names and symbols; use one or the other. Symbols are preferred on drawings where necessary. Millimeters (mm) are assumed on architectural drawings unless otherwise specified.
- Always use decimals and not fractions. Write 0.50 g, not 1/2 g.
- Place a zero before the decimal marker for values less than one. Write 0.45 g, not 0.45 g.

Other guidelines for expressing written metric measurements and the effects of metric conversion on specific building materials are discussed in later chapters. See Appendix B for other common metric conversions.

COMMON TYPES OF DRAWINGS

The two major types of drawings used to show construction information are orthographic projections and isometric drawings. Isometric drawings show three surfaces of an object in one view. Figure 2.10 shows an isometric drawing of a floor system. The drawings gives the impression that the object is three-dimensional. Methods of creating isometric sketches are introduced later in this chapter. Isometric drawings are primarily used in cabinet and plumbing drawings as well as in sales brochures and nonproduction drawings. Their best use for CAD technicians will be as a tool for problem solving. Isometric drawings are used throughout the balance of this text to introduce many of the orthographic drawing projects, but they are not considered as a drawing option.

Orthographic projections show only one side of an object. Several different planes can be represented in one view, but only the portion of the object seen from a specific side is represented. Figure 2.11 shows an orthographic projection of the foundation and floor intersections shown in Figure 2.10. The balance of this

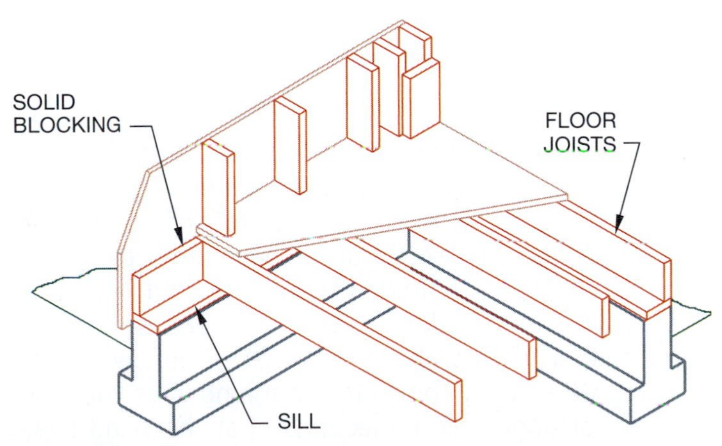

FIGURE 2.10 ■ *Isometric drawings are a key design tool for helping to visualize construction details.*

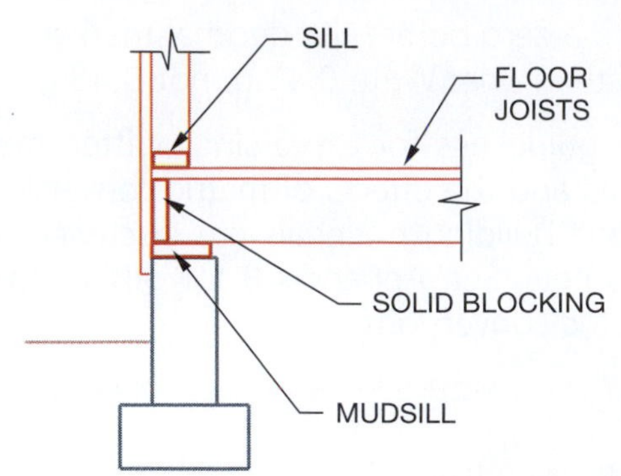

FIGURE 2.11 ■ *Orthographic drawings are used to convey information from the design to the construction crew.*

chapter introduces methods for creating orthographic projections.

THE THEORY OF ORTHOGRAPHIC PROJECTION

Orthographic projection is a drawing method that projects the features of an object onto an imaginary plane. In theory, an object is placed in a glass cube similar to that in Figure 2.12a, and each surface of the object is projected to a parallel surface of the glass cube, as seen in Figure 2.12b. Any portion of the object that is not parallel to the surface of the cube will not be seen in true size and shape. Methods of viewing nonparallel surfaces are discussed later in this chapter. Once each surface of the object is projected to the surface of the glass cube, the cube is unfolded to provide six views of the object. The six views are the front, top, right side, rear, left side, and bottom. Figure 2.13 shows the unfolding of the glass cube. Figure 2.14 shows the box completely unfolded.

In theory, the structure in Figure 2.12 was projected to the surface of the glass cube and the cube was then unfolded to allow viewing of each plane of the object. The edges of the cube are referred to as fold lines. To accurately view an object that has been projected to the cube, the cube is unfolded along the fold lines and viewed as seen in Figure 2.13. As the cube is unfolded to lie flat, a 90-degree bend is placed between each drawing view. It is important to remember, as you start drawing orthographic projections, that the bend is always 90 degrees between views. If any other angle is used, the projection will be distorted.

The Glass Cube

The key to completing an orthographic projection is the placement of the object in the glass cube so that the sides of the object are parallel to the sides of the cube. When parallel surfaces of the object are projected to the cube, the 3D object will be reduced to a 2D projection. Figure 2.15 shows projection of the right side view of the object from Figure 2.12. Only surfaces of the object that are parallel to a surface of the box are projected in true size and shape. Nonparallel surfaces will be projected to the cube but will appear distorted on the surface of the cube. Given only one projection, it is impossible to determine the true size and shape of this plane. By combining the information of a variety of projections, the print reader can accurately determine the true size and shape of an object. Chapter 18 explores the projection of irregular shaped structures further.

Comparing Surfaces

Notice how the views are aligned in Figures 2.12 and 2.14. The top view is directly above, and the bottom view is directly below the front view. The left side will be directly to the left, while the right side is directly to the right side of the front view. This alignment allows the CAD technician to project features from one view to the next to help establish each view. Remember that a bottom view is not used in residential drawing. Once projected, surfaces on the object can be found on each of the six planes of the glass cube. In two views, the surface will be represented as a plane, and in the other views the surface will be represented as a line.

Freehand Sketching

Sketching is a freehand drawing technique that uses no drafting equipment. Students often assume that because they are using a computer, mastering a skill such as sketching will not be necessary. Nothing could be further from the truth. The computer is just a drawing tool. Sketching is a drawing skill and a tool for organizing your thoughts for drawing layout. A sketch can be a valuable tool for communicating ideas between the engineer and the CAD operator during the initial stages of design. As you progress through the creation of a set of drawings, a sketch will be a valuable tool in helping to solve problems and plan structural components. As you progress through the advanced chapters in this text, a sketch will be an invaluable tool for planning where notations and dimensions should be placed. Even with the use of palm-sized computers, sketching is invaluable in

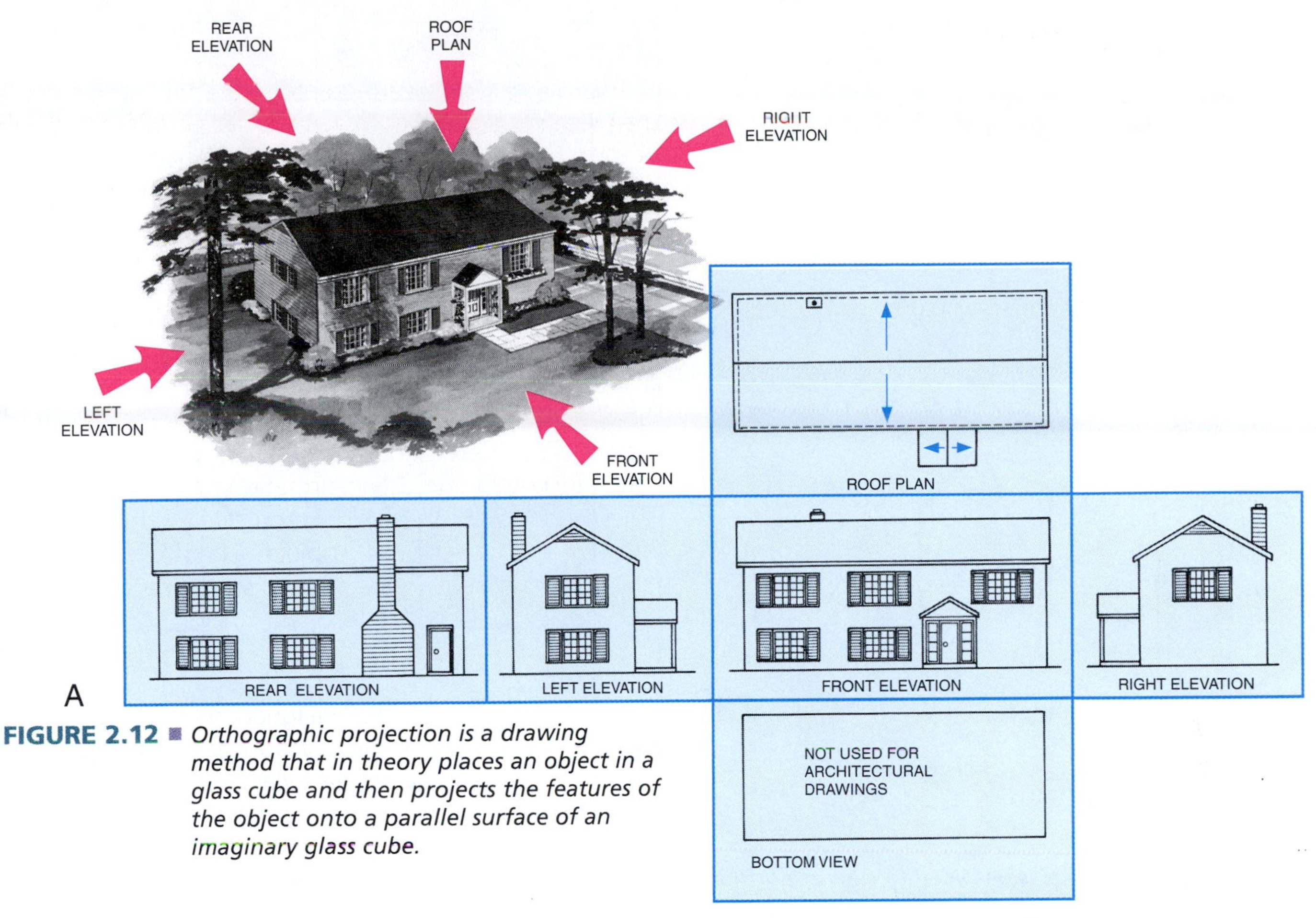

FIGURE 2.12 ■ *Orthographic projection is a drawing method that in theory places an object in a glass cube and then projects the features of the object onto a parallel surface of an imaginary glass cube.*

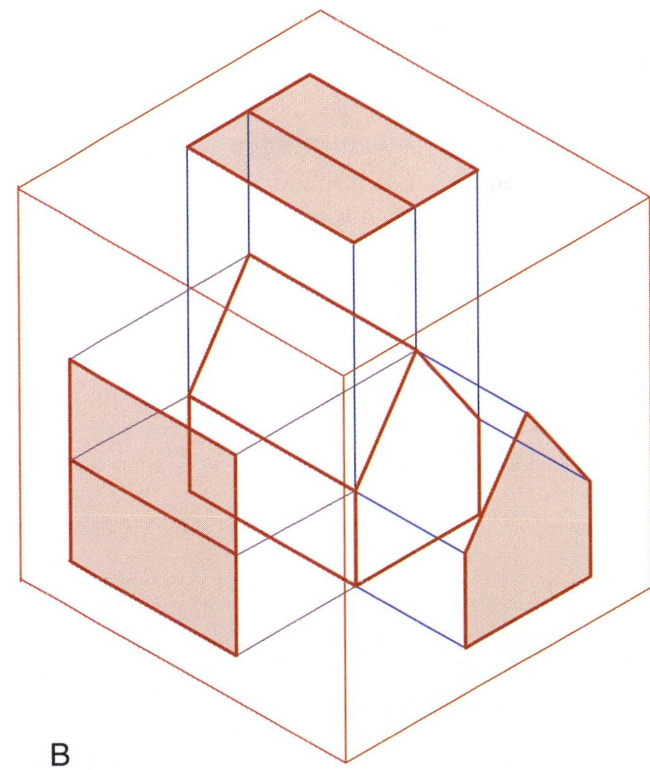

creating drawings in the field for existing parts. Your ability to master the sketching of common geometric shapes in proportion is a tool, which is just as valuable as the CAD commands that you will need to master.

Sketching Tools

Sketches can be created using a variety of materials and tools, including AutoCAD using the SKETCH command. The most common tools for freehand sketching are paper, pencil, and eraser. Sketches for designs worth millions of dollars have been created on a used lunch sack or a paper napkin. When lunch sacks aren't available, paper with a slight matte finish is easier to use than paper with a glossy finish. Graph paper works well for sketching because of its texture and the obvious aid for creating lines using a uniform grid. Pencils with a soft lead, such as a common number 2 lead, or mechanical pencils with F, H, or HB leads are excellent for sketching. If a common pencil is used, the lead should be worn to a smooth, rounded edge before you start a sketch. A sharp lead point can tear the drawing paper. Forming the lead into a wedge and then drawing with the edge of the wedge allows thin lines to be

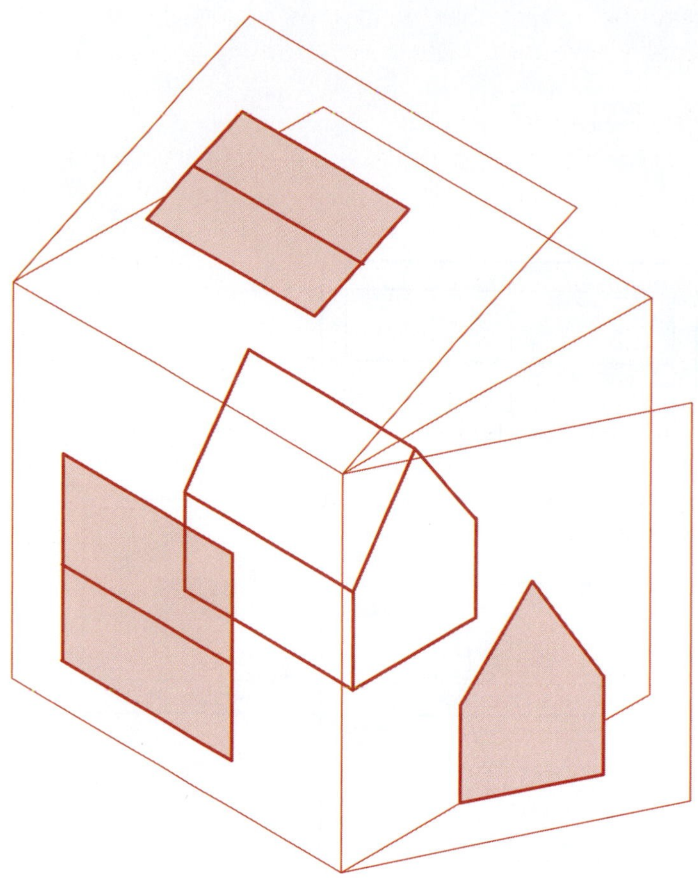

FIGURE 2.13 ■ *The glass box is unfolded to view the orthographic projections. Only three of the possible six surfaces are currently displayed in this drawing.*

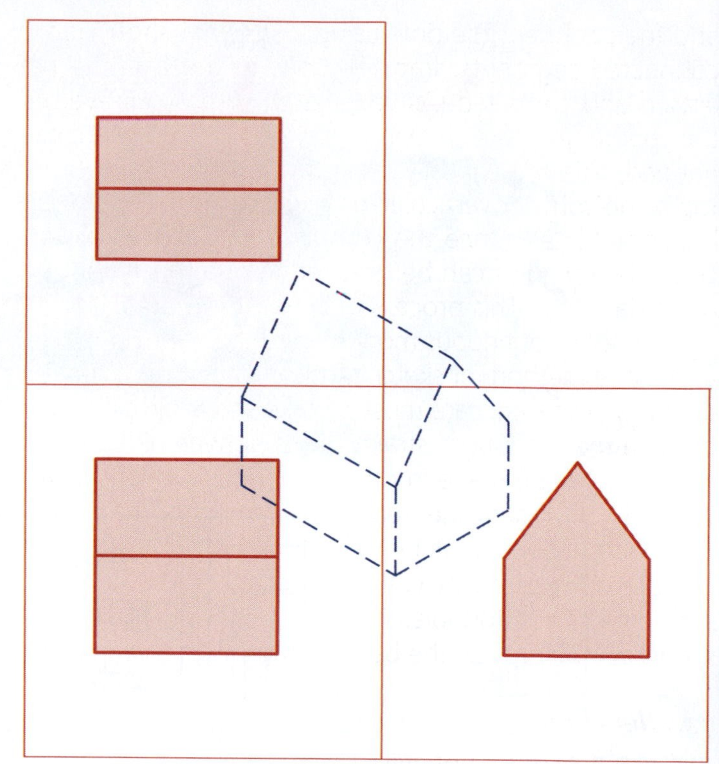

FIGURE 2.14 ■ *Once the glass box is unfolded, the views align with each other to aid the print reader.*

made. Using the full width of the wedge will create wide lines. The pressure applied to the pencil will control the lightness and darkness of the line. For correcting mistakes, a soft eraser will work best. An eraser can be either at the end of the pencil, or it may be a separate rectangular or stick eraser. If you work at keeping construction lines very light, erasing can be kept to a minimum.

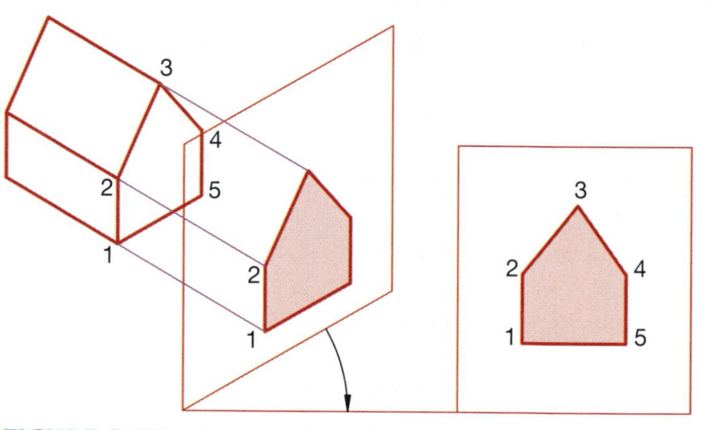

FIGURE 2.15 ■ *Each portion of a structure is projected to a parallel surface of the glass box.*

Sketching Techniques

The following guidelines will aid in the development of sketches:

- Hold the pencil firmly but not so tightly that tension is created. If your fingers are sore after a few minutes of sketching, you're choking the pencil rather than holding it.
- Grip the pencil approximately 1'' to 1.5'' up from the pointed end.
- Draw horizontal lines from left to right using short, light, overlapping strokes.
- Draw vertical lines from top to bottom using short, light, overlapping strokes.
- Create square boxes to frame a circle and then make short, overlapping strokes.

Notice that short, light, overlapping strokes are a key to good sketching.

Sketching Straight Features

Straight features can be made easily if graph paper is being used. If a straight line is to be made on plain paper, place two points to represent the ends of the line and then connect the dots using a series of short, light, connected segments. Imperfections in the line can be erased and corrected. As you sketch, keep your eye on the point you are moving toward. If you're right-handed, this will typically mean pulling the pencil from top to bottom for vertical lines and from left to right for horizontal lines. Once the complete line has been successfully placed, it can be retraced in one movement to get a dark line. This process is shown in Figure 2.16.

Using one continuous motion to connect the dots is an alternative method for sketching short lines. As the length of line is increased, care must be taken to avoid placing a curve in the line. Short, straight lines between 2'' and 3'' in length can be made in one continuous motion of the wrist. Slightly longer lines require movement of the elbow. Lines longer than about 12'' will require movement of the entire arm. Experiment with each method to see which seems most comfortable. For most users, placing short segments will provide the best chance of success.

Parallel Lines

If graph paper is not available, placing a series of dots parallel to a line will help to form the second line. To draw two parallel lines, draw a line in the desired position and then place dots parallel to the original line. The distance from the original line to the string of dots can be kept uniform by measuring with the end of a pencil to another point on the pencil marked by a finger or by marking the space between two dots with a second piece of paper. Once the original distance is determined, slide the makeshift ruler along the original line and mark a new location. This process can be repeated several more times until a series of dots is parallel to the original line. Short, light lines can then be sketched between the set of dots. The process is shown in Figure 2.17.

Once the art of making parallel lines has been mastered, rectangles and squares can be created to aid in the creation of sketches. Most objects can be sketched using a series of rectangles to form the outline of major components. To sketch an object, sketch a box to contain the entire object and then divide the box into smaller areas to outline major items.

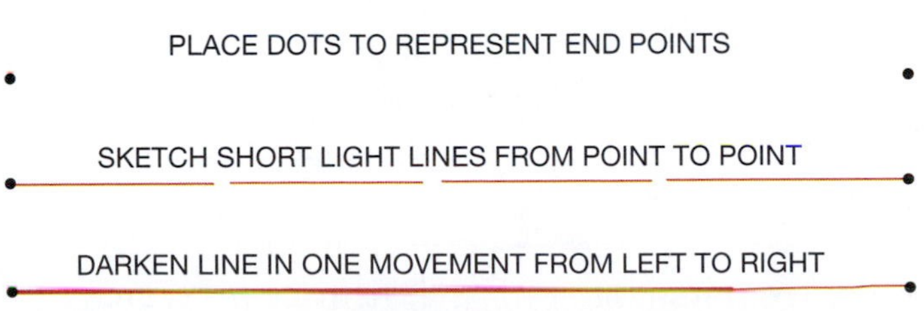

FIGURE 2.16 ■ *To sketch a straight line, place two points to represent the ends of the line, and then connect the dots using a series of short, light, connected segments. Imperfections in the line can be erased and corrected.*

Angular Lines

Whatever the angle to be drawn, lightly drawing a 90° angle is an excellent way to start the process of sketching angled lines. Be sure to keep each leg of the perpendicular lines of equal length. Next draw a line between the end points of the perpendicular lines, similar to Figure 2.18. By placing a dot at the middle of the inclined line and then drawing a line from the intersection to this dot, you create a 45° angle. By dividing the original inclined line into thirds, you can draw lines at 30° and 60°. By dividing the inclined line into thirds and then dividing each of these spaces in half, you can determine 15° increments. Angles smaller than 15° can be determined by breaking the orig-

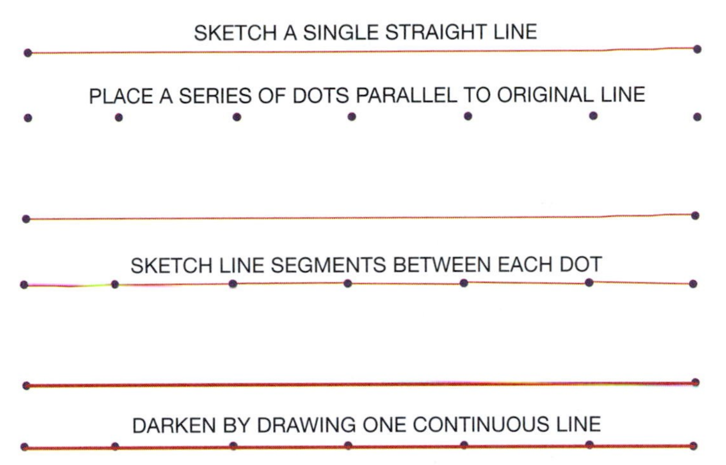

FIGURE 2.17 ■ *To draw two parallel lines, draw a line in the desired position and then place dots parallel to the original line.*

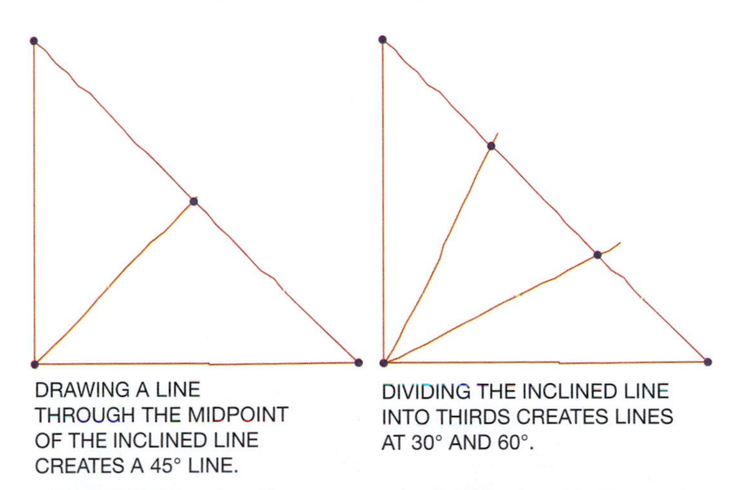

DRAWING A LINE THROUGH THE MIDPOINT OF THE INCLINED LINE CREATES A 45° LINE.
DIVIDING THE INCLINED LINE INTO THIRDS CREATES LINES AT 30° AND 60°.

FIGURE 2.18 ■ *To draw an angle, lightly draw a 90° angle to start the process of sketching angled lines. Next draw a line between end points of the perpendicular lines. Placing a dot at the middle of the inclined line and then drawing a line from the intersection to this dot will create a 45° angle.*

inal 90° angle into 15° increments, then breaking a 15° segment into 5° increments, and then breaking this space into 1° units. Figure 2.19 shows an example of a line sketched at approximately 42°. Once the angle has been determined, the easiest method of drawing the inclined line is to rotate the paper so that the line to be drawn will be horizontal or vertical.

Projection of Rectangular Features on an Inclined Plane

When a rectangular feature such as a skylight projects out of a sloped roof, the intersection of the skylight with the roof appears as a line when the roof also appears as a line. This intersection may then be projected onto adjacent views, as shown in Figure 2.20.

Sketching Circular Features

Common circular features that occur throughout construction drawings include arcs, circles, ellipses, and cylinders.

Sketching Arcs

An arc can be completed by drawing a box equal to the size of the radius and then forming the arc around the center point. As the process is started, the closer the original box is to square, the better the resulting arc. Be sure to use marks on a second sheet of paper to ensure equal lengths for the square. An arc can be completed using the following steps:

1. Draw a square to represent the desired radius.
2. Draw a diagonal line from one corner of the square to mark the center point of the arc.
3. Using a second sheet of paper as a ruler, mark the radius on the diagonal line.
4. Lightly sketch an arc from one corner of the box to the centerline.
5. Complete the layout by sketching the remaining portion of the arc.
6. Darken the rough draft by completing the arc in one movement.

This process is shown in Figure 2.21

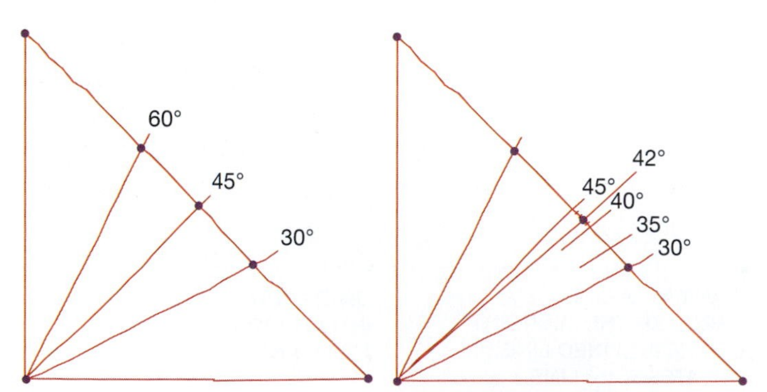

FIGURE 2.19 ■ *By dividing the original inclined line into thirds, you can draw lines at 30° and 60°. By dividing each of these segments in half, you can determine 15° increments. Angles smaller than 15° can be determined by breaking a 15° segment into 5° increments, and then breaking this space into 1° units.*

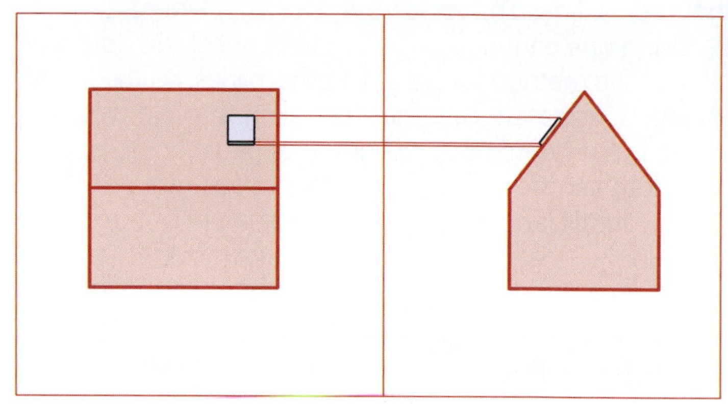

FIGURE 2.20 ■ *When a rectangular feature such as a skylight projects onto a sloped surface, the object will be reduced in height from its true size. The width will not be affected.*

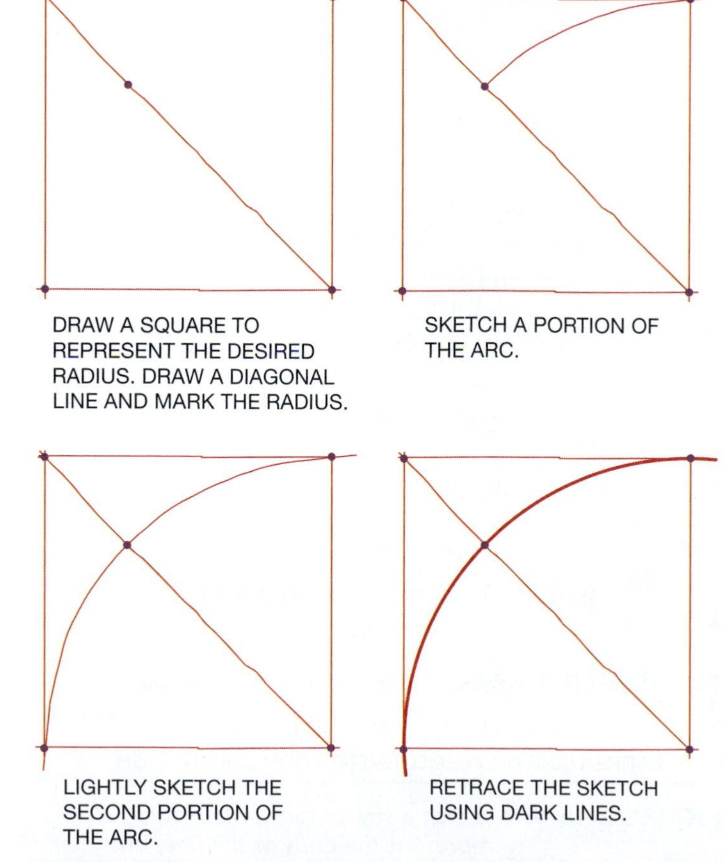

DRAW A SQUARE TO REPRESENT THE DESIRED RADIUS. DRAW A DIAGONAL LINE AND MARK THE RADIUS.
SKETCH A PORTION OF THE ARC.
LIGHTLY SKETCH THE SECOND PORTION OF THE ARC.
RETRACE THE SKETCH USING DARK LINES.

FIGURE 2.21 ■ *An arc can be completed by drawing a box equal to the size of the arc radius and then forming the arc around the center point.*

Sketching Circles

Circular features can be completed by drawing a box, drawing its center points, and then forming the circle around these center points. As the process is started, the closer the original box is to square, the better the resulting circle. Be sure to use marks on a second sheet of paper to ensure equal lengths for the square. A circle can be completed using the following steps:

1. Draw a square to represent the desired diameter.
2. Mark the center of each edge of the square.
3. Draw diagonal lines from each corner of the square to mark the center point of the circle.
4. Using a second sheet of paper as a paper ruler, mark a distance on the diagonal line equal to the distance from the center point to an edge.
5. Lightly sketch an arc from one of the centerlines to a point on the nearest diagonal line.
6. Complete the layout by sketching the seven remaining arcs.
7. Darken the rough draft by completing one-half of the circle in one movement, and then darken the other side.

This process is shown in Figure 2.22

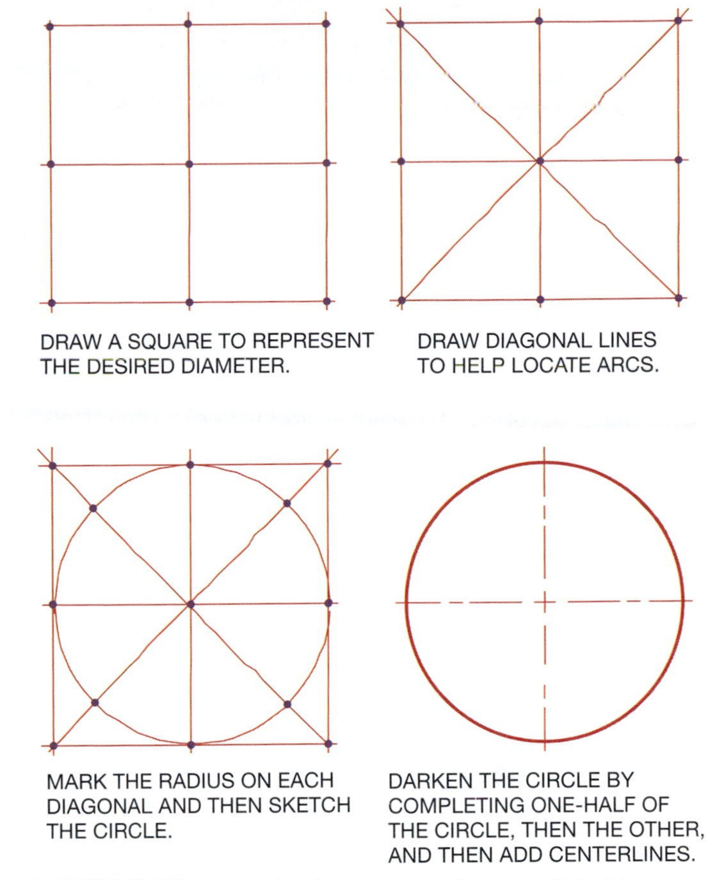

DRAW A SQUARE TO REPRESENT THE DESIRED DIAMETER.
DRAW DIAGONAL LINES TO HELP LOCATE ARCS.
MARK THE RADIUS ON EACH DIAGONAL AND THEN SKETCH THE CIRCLE.
DARKEN THE CIRCLE BY COMPLETING ONE-HALF OF THE CIRCLE, THEN THE OTHER, AND THEN ADD CENTERLINES.

FIGURE 2.22 ■ *Circular features can be completed by drawing a box, drawing its center points, and then forming the circle around these center points.*

Sketching Ellipses

An ellipse is completed with the use of steps similar to those used to draw a circle. Start by drawing a rectangle that represents the lengths of the major and minor diameters. An ellipse can be completed using the following steps:

1. Draw a rectangle that represents the lengths of the major and minor diameters.
2. Find the center point of the rectangle by drawing diagonal lines through the rectangle.
3. Draw an arc to represent one quadrant of the ellipse.
4. Measure the point where the arc passes through the diagonal, then mark this distance on the remaining diagonal lines.
5. Sketch arcs in the remaining grids.
6. Darken the ellipse using methods similar to those that were used to darken a circle.

This process is shown in Figure 2.23.

Sketching Cylinders

Sketching the rectangle that will form the outline of the cylinder will help ensure the accuracy of the sketch. To create the top of the cylinder, follow the steps to draw an ellipse. Repeat the process for drawing the ellipse at the other end of the rectangle. The portion of the bottom ellipse that lies beyond the edges of the rectangle can be erased or left if it is sketched lightly. The lines forming the cylinder can be darkened using the methods that were used to darken arcs, ellipses, and straight lines. This process is shown in Figure 2.24.

Sketching Circles on an Inclined Plane

When the line of sight in a view is perpendicular to an object such as a round window, the window appears round. When a circle is projected onto an inclined surface, such as a round skylight projected onto a sloped roof, the view of the inclined circle is elliptical, as shown in Figure 2.25.

Keeping Proportions in Freehand Sketches

Even though sketches are not drawn to scale, if they are to be a useful production tool, it is important that they be proportionately accurate. It has been mentioned that a pencil or a piece of paper can be used as a measuring tool, but using these tools is time-consuming. With practice, accurate measurements can often be

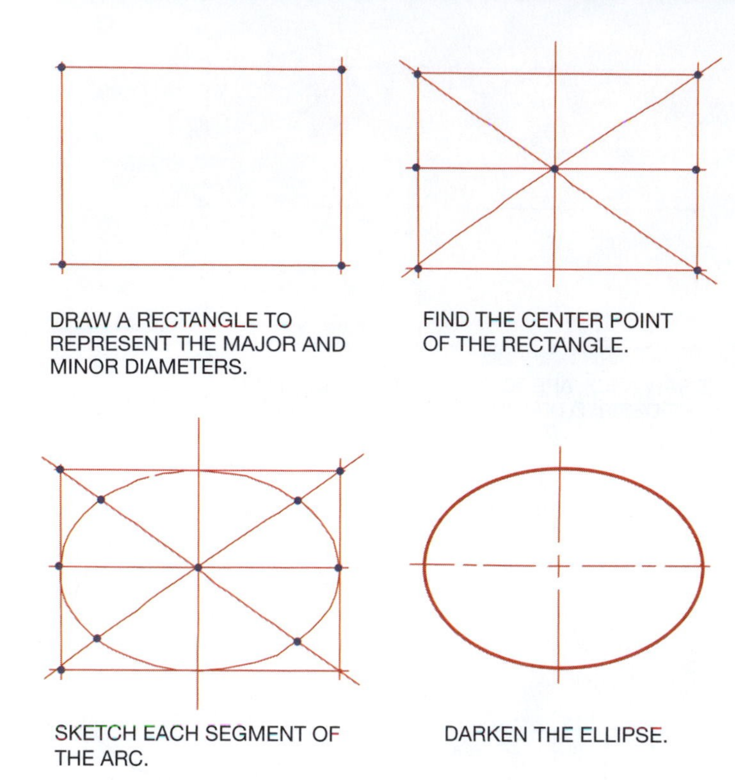

DRAW A RECTANGLE TO REPRESENT THE MAJOR AND MINOR DIAMETERS.
FIND THE CENTER POINT OF THE RECTANGLE.
SKETCH EACH SEGMENT OF THE ARC.
DARKEN THE ELLIPSE.

FIGURE 2.23 ■ *An ellipse is completed by using steps similar to those used to draw a circle. Start by drawing a rectangle that represents the lengths of the major and minor diameters.*

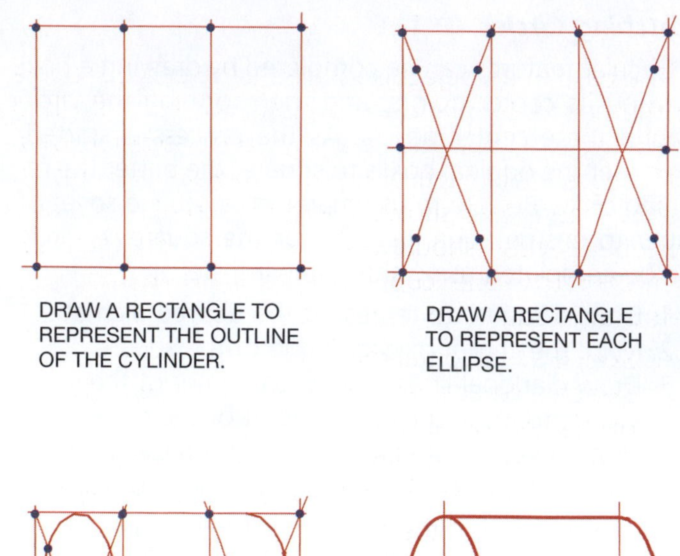

DRAW A RECTANGLE TO REPRESENT THE OUTLINE OF THE CYLINDER.
DRAW A RECTANGLE TO REPRESENT EACH ELLIPSE.

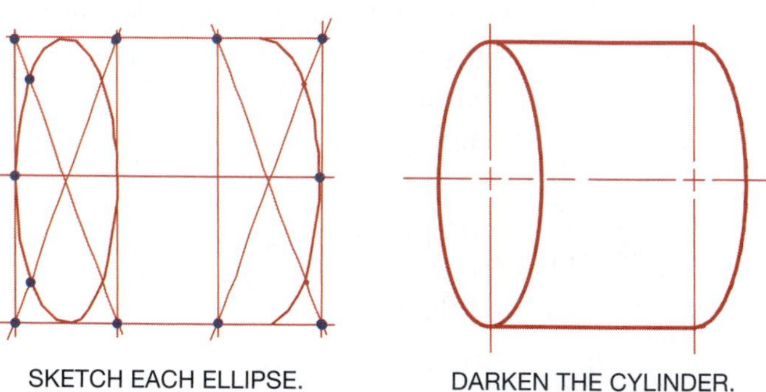

FIGURE 2.24 ■ *Sketch the rectangle that will form the outline of the cylinder to ensure the accuracy of the sketch. Follow the steps to draw an ellipse to create the top of the cylinder. Repeat the process for drawing the ellipse at the other end of the rectangle.*

done by eye. As you sketch, remember that the lines that make up the object are related to each other by size and direction. As you add a line to the sketch, visually compare the new line to previous lines. Taking time to verify that lengths and direction are kept in proportion will enable you to communicate clearly through your sketch. The following guidelines will aid in keeping sketches accurate:

- Before starting to sketch, mentally divide the part into its major components.
- Start the sketch by drawing the largest component and work to the smallest details.
- Use squares and rectangles to form a base for irregular shapes.
- The first line drawn determines the size of other lines. Keep all parts of the object in proportion.

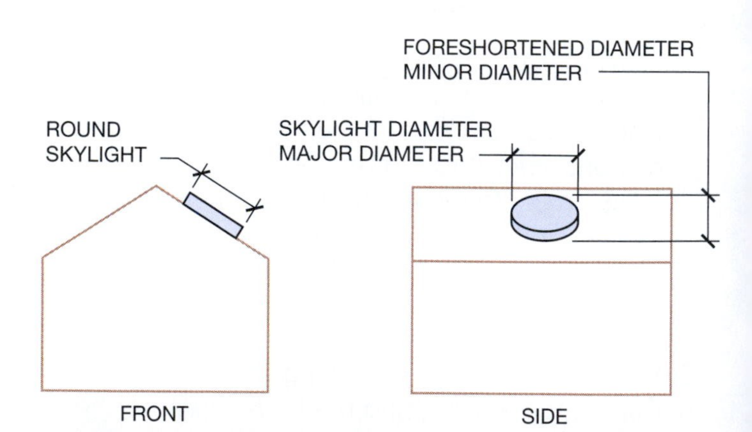

FIGURE 2.25 ■ *Circles projected onto an inclined surface will appear as ellipses.*

Isometric Sketches

Isometric sketches are three-dimensional representations that display the objects much as they appear to the eye. Isometric drawings are an ideal tool to help in planning multiview drawings. An isometric sketch can be used to visualize how each feature of an object will affect other features of the part and to visualize what should be represented in each view of a drawing.

Forming an Isometric Drawing

The base of an object is rotated to be at a 30° angle to the horizontal plane to form an isometric drawing or sketch. All horizontal planes are represented by a 30° angle, and vertical lines represent all vertical planes. Figure 2.26 shows the formation of an isometric cube. A cube should be created to start each isometric sketch. The cube can be created using the following steps:

1. Sketch a horizontal reference line.
2. Sketch a vertical reference line from the midpoint of the horizontal line.
3. Sketch an inclined line at 30° from horizontal on each side of the vertical reference line. To represent the top of the cube, sketch lines that are parallel to the two lines just drawn.
4. Sketch the remaining two lines to close the cube.

Once the cube is completed, an isometric sketch can be completed. The footing shown in Figure 2.27 can be completed using the following steps:

1. Draw an isometric cube to represent the objects to be drawn.
2. Divide the cube into smaller cubes to represent features that must be represented. This would include boxes for the footing, stem wall, and floor system (see Figure 2.28).
3. Lay out the shapes to provide details such as individual floor joists (see Figure 2.29).

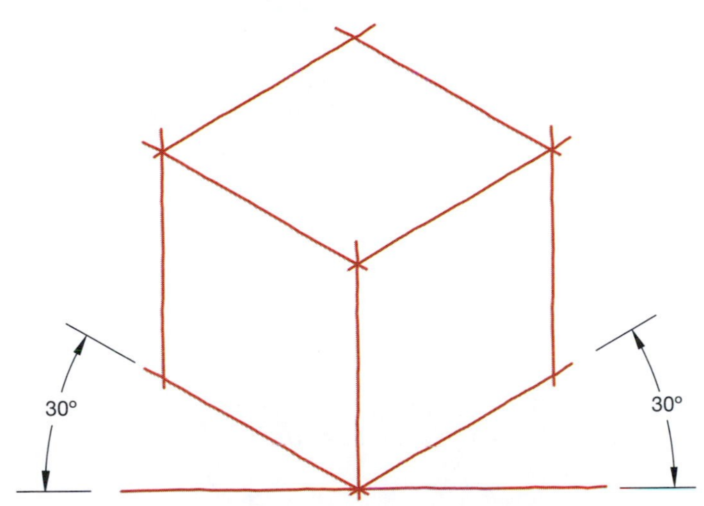

FIGURE 2.26 ■ *The base of an object is rotated to be at a 30-degree angle to the horizontal plane to form an isometric drawing or sketch. All horizontal planes are represented by 30-degree angles, and vertical lines represent all vertical planes.*

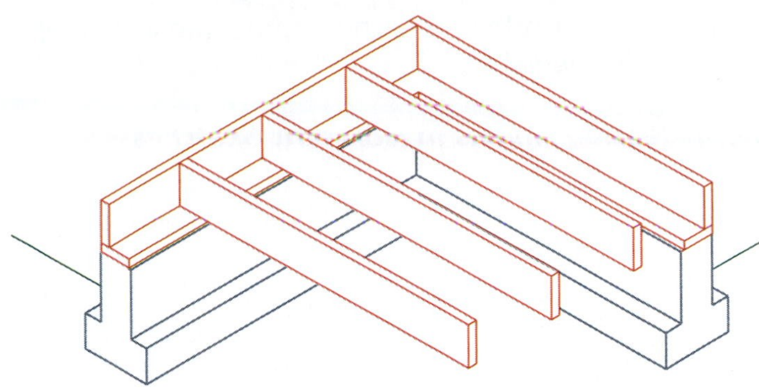

FIGURE 2.27 ■ *An isometric drawing of a joist floor system.*

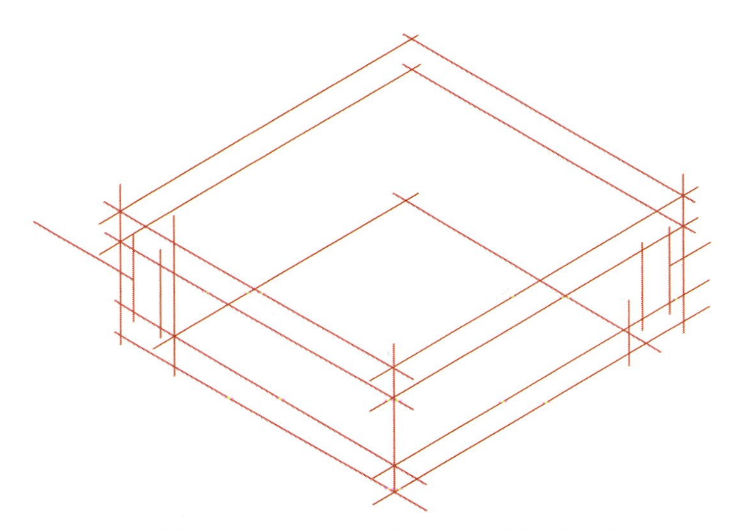

FIGURE 2.28 ■ *A drawing can be created by drawing a cube to contain the entire object. Next, divide the cube into smaller cubes to represent features that must be represented. This would include boxes for the total drawing, the footing, stem wall, and floor system.*

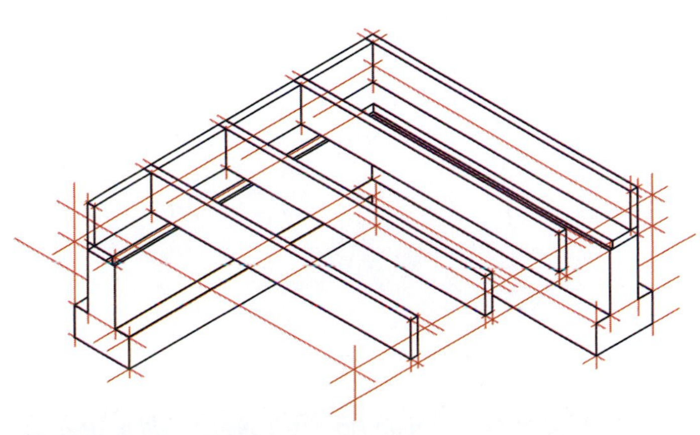

FIGURE 2.29 ■ *Lay out additional boxes to provide details of individual objects such as floor joists.*

4. Darken each feature using appropriate line types and line weights.

Representing Angles in Isometric Sketches

Angled features of an object cannot be accurately measured in isometric sketches, but they can be accurately represented. When a surface is not parallel to one of the surfaces of the glass cube, it can't be measured directly. Each end of the inclined plane must be located and then connected by an object line.

Isometric Circles

A circle in an isometric sketch will appear as an ellipse. The position of the ellipse will vary depending on the surface of the cube on which the circle is located. Figure 2.30 shows the three common orientations for isometric circles. Each circle can be sketched by dividing the circle into four smaller arcs and drawing these arcs first. Figure 2.31 shows how the three planes can be divided to locate the center point for each of the four arcs that will form the ellipse. Figure 2.32 shows how the arcs can be created using the following steps:

1. Draw the cube that will contain the circle.
2. Draw lines from the corner of the cube to the midpoint of the opposite lines.
3. Repeat the last step using the opposite corner.
4. Use points 1 and 2 to form the centers for the large arcs.
5. Use points 3 and 4 to draw the smaller arcs that form the end of the ellipse.
6. Notice that the minor diameter of the ellipse is always lined up on the centerline parallel to one of the edges of the cube.

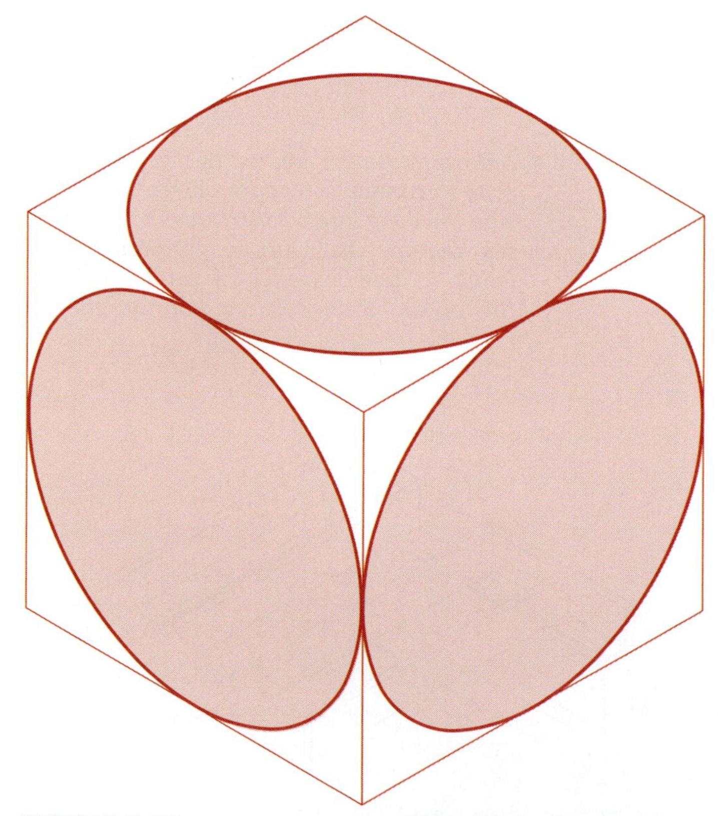

FIGURE 2.30 ■ *Circles in an isometric sketch will appear as ellipses. The position of the ellipse will vary depending on the surface of the cube on which the circle is located.*

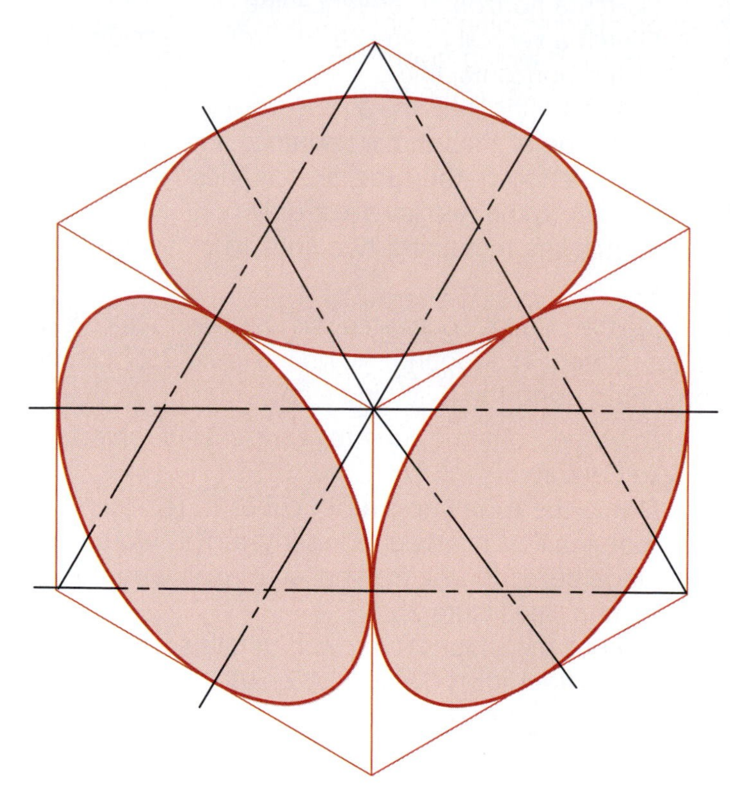

FIGURE 2.31 ■ *The three planes can be divided to locate the center point for each of the four arcs that will form each ellipse.*

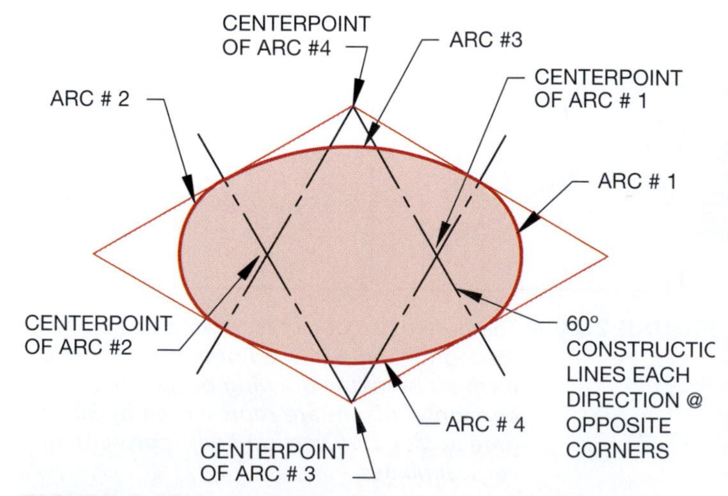

FIGURE 2.32 *The steps for sketching an isometric circle.*

DRAWING LAYOUT USING A COMPUTER

AutoCAD and other computer programs can be used to create sketches using commands such as the SKETCH command of AutoCAD. More useful, however, is the use of common commands such as LINE, OFFSET, TRIM, and FILLET to create scaled drawings using a similar mindset that is used in sketching. The foundation detail shown in Figure 2.11 is used here to illustrate the method of combining common AutoCAD commands to create quick drawing layouts. To create the drawing, it is necessary to know the following sizes:

- The footing (the bottom portion of the foundation) is 12'' wide and 6'' high.
- The footing must extend a minimum of 12'' into the grade.
- The stem wall (the vertical portion of the foundation) is 6'' wide and centered on the footing.
- The stem wall must extend 8'' above the grade.
- A 2 × 6 mudsill will sit on top of the stemwall.
- 2 × 10-floor joists will sit on the mudsill.

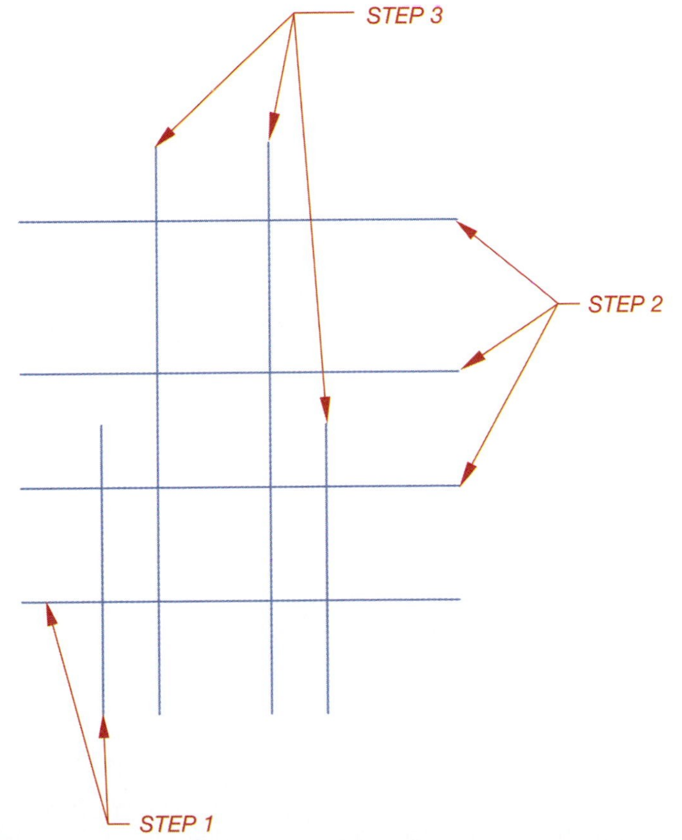

FIGURE 2.33 ■ *A drawing can be started by drawing two lines perpendicular to each other and then using the OFFSET, TRIM, and FILLET commands.*

- 3/4'' plywood will cover the floor joists.
- A 2 × 4 stud wall will be framed on top of the floor.

This information is based on standard construction practices. Once the basic sizes to be represented are known, the drawing or any object can be quickly sketched by drawing two lines perpendicular to each other. To lay out the footing shown in Figure 2.11 would require the following steps.

1. Draw two lines that are perpendicular to each other. Although the exact length is not important, make them longer than the object you will be drawing. For this drawing, the horizontal line is 15'' long and the vertical line is 8'' long.
2. Use the OFFSET command and offset the horizontal line up 6'', 12'', and 20''.
3. Offset the vertical line over at distances of 3'', 6'', and 12'', so that the drawing resembles that shown in Figure 2.33.
4. Use the FILLET command with a radius of 0 to extend the stem wall to its upper limits.
5. Using the TRIM command, use the left edge of the stem wall to trim the line that represents the grade.
6. Use the FILLET command to clean up the corners of the foundation.The foundation portion of the layout has now been completed. The drawing should now resemble Figure 2.34. The floor portion of the drawing can be created using the following steps.
7. Offset the line representing the top of the stem wall up 2''.
8. Highlight the left edge of the stem wall to activate the grip. Stretch this line to be approximately 18'' taller.
9. Offset the line from step 7 a distance of 6'' to represent the mudsill. Fillet the vertical line as required.
10. Use the grip feature, highlight the line representing the top of the mudsill and stretch this line approximately 15'' to represent the bottom of the floor joists (see Figure 2.35).
11. Offset the bottom of the floor joist up 10'', then 1'' to represent the top for the plywood, and then 2'' to represent the top of the sill.
12. Offset the vertical line to the left 3/4'' to represent the OSB overlay on the exterior edge of the wall. Offset the line 4'' to the right to represent the interior edge of the wall.
13. Use the TRIM and FILLET commands to clean up the layout.
14. Use the PROPERTIES command to assign the desired line width and line types.

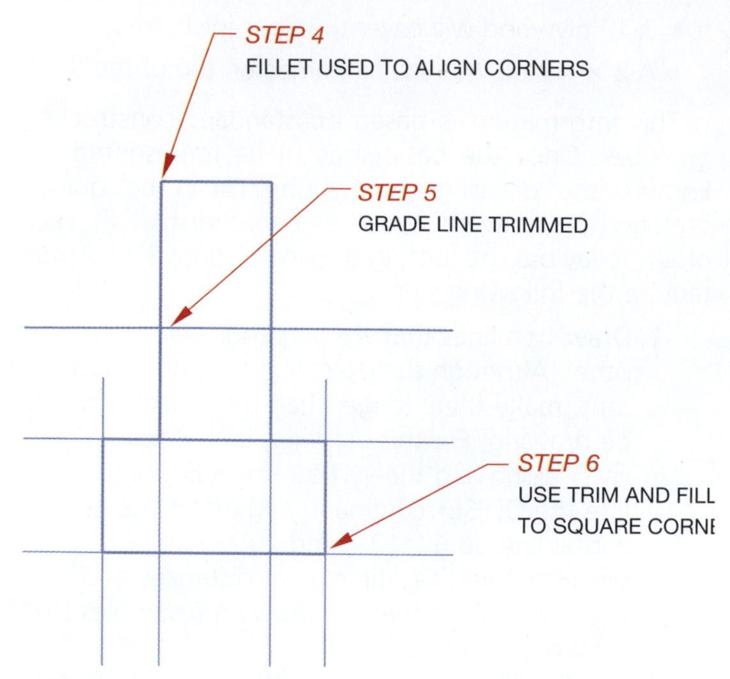

FIGURE 2.34 ■ *The TRIM and FILLET commands were used to clean up the corners of the lines used to create the foundation detail.*

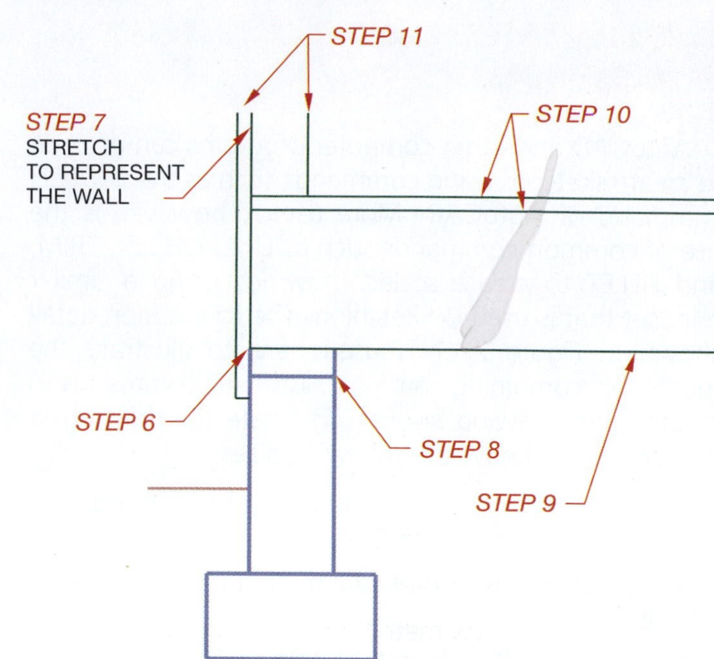

FIGURE 2.35 ■ *The floor portion of the drawing was completed using the OFFSET, TRIM, and FILLET commands and the STRETCH function of GRIPS.*

CHAPTER 2

Additional Readings

The following websites can be used as a resource to help you keep current with changes in CAD drafting–related fields.

ADDRESS	COMPANY OR ORGANIZATION
www.engr.psu.edu	Metric Design and Construction
www.nationalcadstandard.org	U.S. National Cad Standard
www.nibs.org	National Institute of Building Sciences
www.suppliesnet.com	Professional drafting supplies
www.metric.org	U.S. Metric Association

CHAPTER 2

Preparing for Success Test

DIRECTIONS

Answer the following questions with short, complete statements. Type your answers using a word processor or neatly print the answers on lined paper.

1. Type your name, the chapter number, and the date at the top of the sheet.
2. Type the question number and provide the answer in the form of a statement that includes part of the question. You do not need to write out the entire question.

QUESTIONS

2.1. List two definitions of the word *scale*.
2.2. Why should a sketch be used when the drawing will be created using AutoCAD?
2.3. List the common scale and two alternatives for drawing a floor plan.
2.4. What value would you use to convert inches to millimeters?
2.5. List the common divisions found on a full architectural scale.
2.6. List the common divisions found on the 1/4'' = 1'-0'' scale.
2.7. Define hard metric conversion.
2.8. Describe how to create parallel lines when sketching.
2.9. List the three common tools of sketching and the guidelines for selecting each.
2.10. What scale should be used to draw a floor plan that will be in metric units?
2.11. List how the divisions can be used on the 10 scale on a civil scale.
2.12. What is the basic unit used in metric measurement?
2.13. Show how the number two hundred and fifty thousand millimeters would be listed on a drawing.
2.14. The artist in you wants to place the top, front, and side views of a drawing so that they are neatly aligned in a side-by-side arrangement. What are the problems with this layout?
2.15. A surface is at an incline to the glass cube. How will it appear in an orthographic projection?
2.16. An object contains an inclined surface that must be represented at 37° from vertical. Explain how to represent the surface in a sketch.
2.17. What angle forms the base of an isometric drawing or sketch?
2.18. You will be completing drawings using a computer. Why do you need to be able to read a scale?
2.19. Define soft metric conversion.
2.20. What do proportions have to do with sketching techniques?
2.21. On a metric drawing, how would the distance of 1000 millimeters be written?
2.22. Define sketching.
2.23. Describe the proper sketching tools.
2.24. Define an isometric sketch.
2.25. Define orthographic projection.

2.26. Use the indicated scale and draw lines that are 20' long.
a. Scale: 1/8'' = 1'-0'' b. 1'' = 60' c. 1:100
d. Scale: 1/4'' = 1'-0'' e. 1'' = 50' f. 1:50

2.27. Use the indicated scale and draw lines that are 10' long.
a. Scale: 3/8'' = 1'-0'' b. 1'' = 40'-0'' c. 1:20
d. Scale: 1/2'' = 1'-0'' e. 1'' = 30'-0'' f. 1:50

2.28. Use the indicated scale and draw lines that are 5'-6'' long.
a. Scale: 3/4'' = 1'-0'' b. 1'' = 10'-0'' c. 1:20

2.29. Use the indicated scale and draw lines that are 5'-6'' long.
a. Scale: 1'' = 1'-0'' b. 1'' = 60'-0'' c. 1:10

2.30. Use the indicated scale and draw lines that are 3 1/2'' long
a. Scale: 1 1/2'' = 1'-0'' b. 10 c. 1:10
d. Scale: 3'' = 1'-0'' e. 1:5

DIRECTIONS

Use proper sketching materials and techniques to solve the following sketching problems, on 8 1/2 × 11 bond paper or newsprint. Use very lightly sketched construction lines for all layout work. Darken the lines of the object but do not erase the layout lines.

PROBLEMS

2.1. Using proper proportions, sketch two orthographic views of your home or any other local single-family residence.

2.2. Sketch circles with approximate diameters of 1/2'', 2'', and 4''.

2.3. Sketch lines at 13°, 30°, 45°, and 60° inclines from a vertical line.

2.4. Draw orthographic sketches of each surface of the following structure.

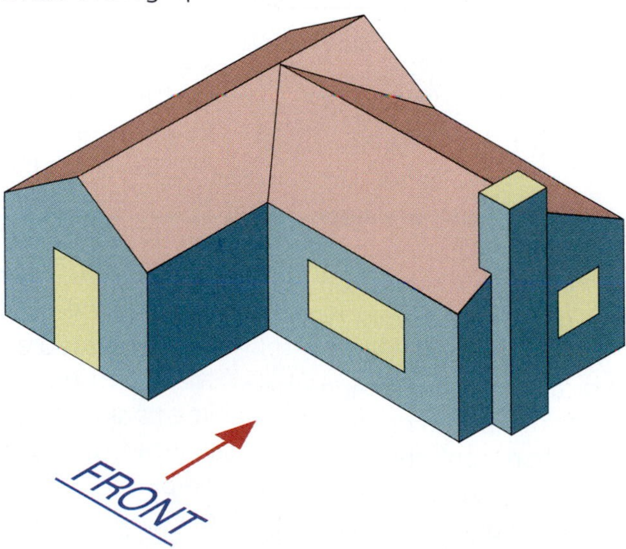

2.5. Use Figure 6.7 to draw the front elevation for this structure.

2.6. Draw a front view of three of the homes shown in Chapter 6. Identify by figure number each house that you sketch.

2.7. Draw isometric sketches of three of the home styles from Chapter 6. Identify by figure number each house that you sketch.

2.8. Sketch a plan view of the kitchen where you live.

2.9. Make a sketch of an elevation showing one wall of cabinets found in the kitchen from problem 8.

2.10. Make an isometric sketch of the cabinets shown in Problem 9.

3 CHAPTER

Computer-Aided Design in Architecture

This chapter provides an overview and practical application of the computer skills necessary to produce a set of residential drawings. It is assumed that you have already completed an entry-level class in CAD drafting prior to working through this text and that you are working with AutoCAD 2000 or newer. This chapter will help you utilize your CAD skills in the preparation of a set of residential drawings. Consideration will be given to:

- *Managing your work environment*
- *Managing the drawing environment*
- *Managing drawing properties with a template*
- *Managing drawing templates, assembling drawings for plotting at multiple scales, working with multiple documents, using DesignCenter*
- *Management of drawing information*

MANAGING YOUR WORK ENVIRONMENT

Although becoming a proficient architectural designer will offer many exciting opportunities, spending long hours at a workstation also has its pitfalls. You might start to notice problems with your eyes, back, and wrist. Fortunately, most physical problems arising from computer use can be minimized or eliminated by proper workstation configuration, good work habits, and good posture. Figure 3.1 shows qualities of a well-equipped workstation.

Equipment

You might have little control over your workstation at school, but if you're setting up your own workstation, it should meet the following requirements:

- Your chair should be adjusted to a height that allows your feet to be supported on an inclined footrest or to rest on the floor. The chair should also provide a tilt adjustment so that your back is firmly supported as you work. It should also have an armrest to provide support for your wrists. The height of the chair should allow your wrists to remain level as you type.
- The monitor should be placed so that the top of the monitor is level with your line of sight and between 18 and 30 inches from your eyes. Lighting should be placed so that it will not produce a glare on the display screen.
- Use a desk with a keyboard drawer that allows your wrists to remain level.
- Use an ergonomic keyboard, a foam wrist support, or forearm supports.

Work Habits

You can have a state-of-the-art workstation, but if you don't maintain good work habits, your body will wear out from fatigue.

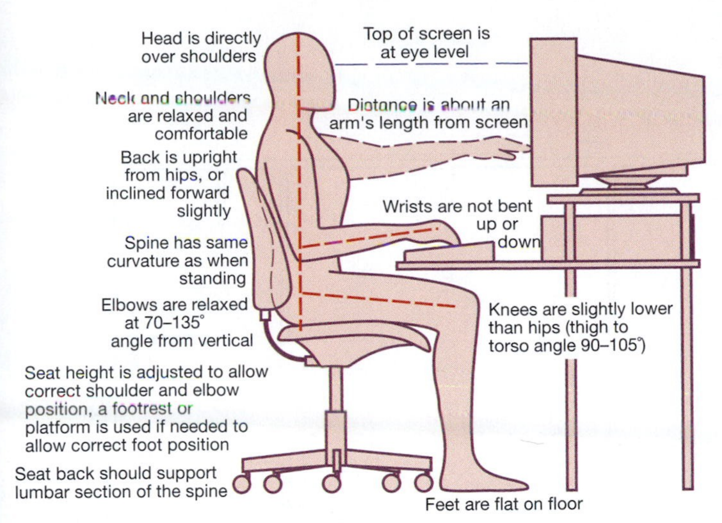
Head is directly over shoulders
Top of screen is at eye level
Neck and shoulders are relaxed and comfortable
Distance is about an arm's length from screen
Back is upright from hips, or inclined forward slightly
Wrists are not bent up or down
Spine has same curvature as when standing
Elbows are relaxed at 70–135° angle from vertical
Knees are slightly lower than hips (thigh to torso angle 90–105°)
Seat height is adjusted to allow correct shoulder and elbow position, a footrest or platform is used if needed to allow correct foot position
Seat back should support lumbar section of the spine
Feet are flat on floor

FIGURE 3.1 ■ *Adjusting equipment to meet the needs of your body will aid in maintaining the health of your wrists, back, and eyes.*

To keep your productivity high and maintain your sanity, you'll need to control your work habits and posture.

- Avoid looking at a monitor for long periods of time. To ease eyestrain, look away from the monitor several times an hour and force your eyes to focus on an object several feet away from your workstation.
- Avoid resting your hand on the mouse when you're not actually using it. When you are using the mouse, avoid resting your hand on the desk. Get in the habit of keeping your wrist level, or adjust the armrest of your chair to provide support.
- Take time to stretch your legs, neck, shoulders, back, and wrist several times each hour. In addition to reducing tension throughout your body, a stretch provides an excellent opportunity to get your eyes off the monitor.
- Take time to walk away from your workstation at least once an hour. It's easy to get caught up with a deadline and convince yourself that you just don't have time for a break, but short breaks will help to keep you productive and to eliminate drawing errors.

MANAGING THE DRAWING ENVIRONMENT

A well-drawn set of plans doesn't just happen because you've worked really hard. Long hours and hard work may be required, but they will not assure that effective communication can take place. To facilitate effective communication between the owner, the design team, and the construction team requires plans that accurately present information in a clear, concise format similar to the drawing in Figure 3.2. This can be accomplished by complying with industry standards for presenting information, by organizing material in an orderly fashion within the project, and by using a uniform method of organizing information within each drawing of the project.

NATIONAL DRAWING STANDARDS

Unlike the field of mechanical drafting, there is no one standard way of doing things in the field of architecture. Mechanical drafters have an ASME standard that governs almost every aspect of drawing creation. In the field of architecture, each office has its own standard. The office standard can even vary based on time constraints or the fees that have been set for the project. The need for a drawing standard was also hindered by the mindset that an architect was part artist. As computers, the Internet, and the World Wide Web have come to dominate the office setting, most professionals have found that a standard method for creating drawings is a must. A uniform drawing standard offers the following advantages as offices share information:

- Consistent display of information for all projects, regardless of the project type or client
- Seamless transfer of information between team members and consulting architects, engineers, and design professionals
- Reduced preparation time for translation of electronic data files between different proprietary software file formats; predictable file translation results
- Reduced data file formatting and setup time
- Reduced staff training time to teach "office standards"

Several organizations have worked to provide the industry with a uniform drawing standard. Among the leaders in developing a CAD standard are:

- The **National Institute of Building Sciences,** which publishes the ***National CAD Standard.*** More information can be obtained at www.nibs.org.
- The **Construction Specifications Institute** (CSI), which publishes the ***Uniform Drawing Standard*** (UDS)—a standard consisting of eight modules that cover drawing set organization, sheet organization, drafting conventions, terms and abbrevia-

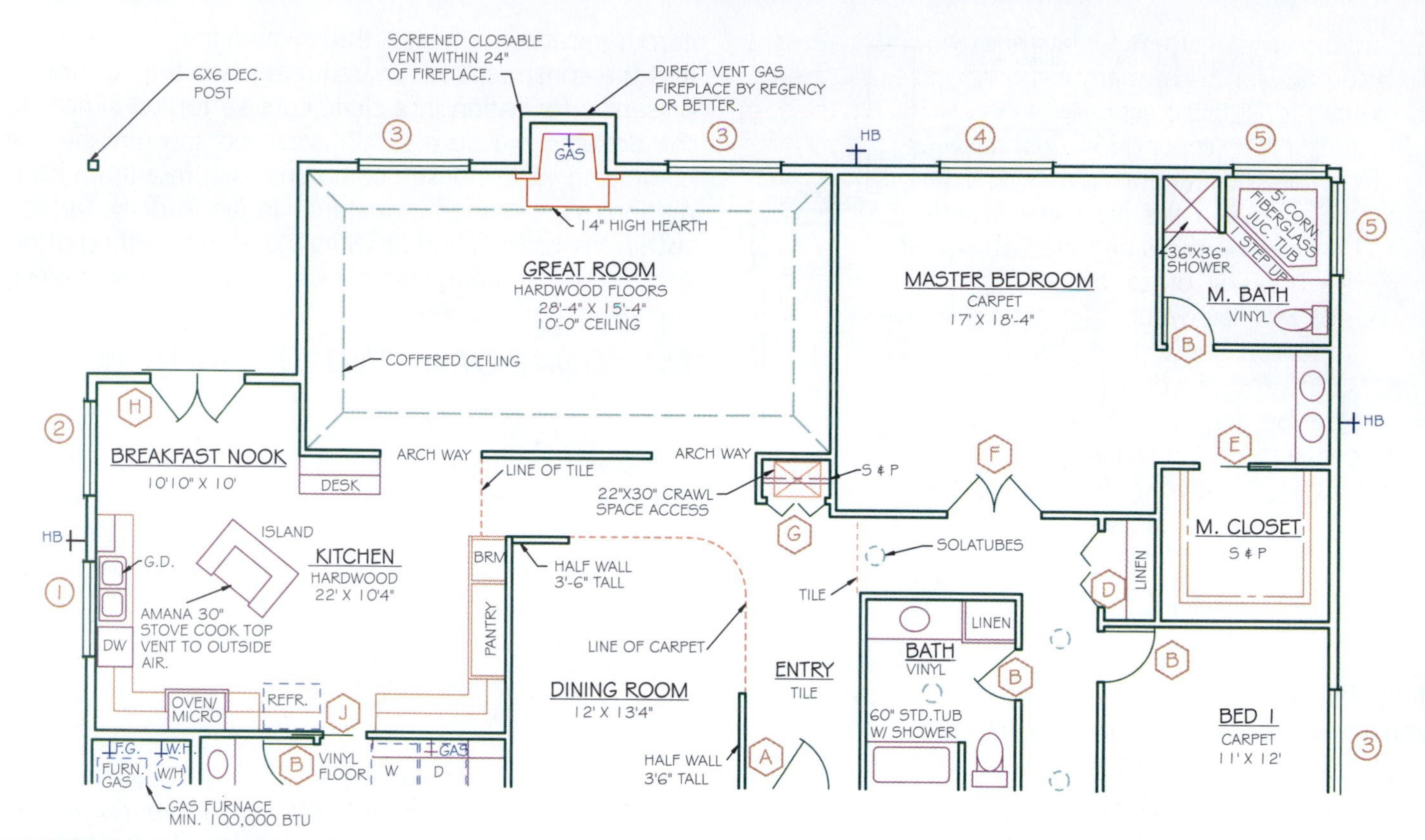

FIGURE 3.2 ■ *Drawings must be created with a uniform set of standards to help team members at consulting firms work with the drawings efficiently.*

tions, symbols, notations, and general regulatory information. More information can be obtained at www.csinet.org.

- **The American Institute of Architects,** which publishes the ***CAD Layer Guidelines.*** More information can be obtained at www.aia.org.

These groups have come together and developed the ***National CAD Standard***. This is a set of standards based on the AIA ***CAD Layer Guidelines***; ***CSI Uniform Drawing System,*** modules 1-8; and Tri-Service (and U.S. Coast Guard) Plotting Guidelines. The standard is published by the National Institute of Building Sciences. The most recent edition of this standard is the ***U.S. National CAD Standard, version 3.1.*** This text will present key features of the standard. Verify with your instructor what standards will govern your projects. If you are allowed to modify one of the base models or the standards presented in this text, just remember to be consistent.

Organizing Folders and Files

The method used to organize drawing files will vary depending on existing office practice and the size of the structure. This chapter is intended to provide you with a method of organizing your drawing files while you are in school and to give you an understanding of some of the common methods that offices use to organize their computer files. Consideration is given to common drawing storage methods, the naming of drawing folders and files, methods of storing drawing files, and methods of maintaining drawing files. It's important to remember that when you leave school and enter an office, it's rarely the CAD technician's role to come into an office and develop a new filing system. It will be your job as a CAD technician to learn the existing system and make your work conform to your employer's standards.

Storage Locations for Drawing Files

If you are a network user, your instructor or network administrator must provide you with a user name or account number before you can access or store information on a network. Once you have access, you will be given a folder located on the network server. Network servers should not be used for long-term storage, but they do provide an excellent storage location for active drawing projects. Many design firms place active drawing files in folders that can be accessed by anyone on the network, ensuring that each member of the design team is working on the most current file without having to pass a diskette between team members. Networks can also be configured to allow access over the Internet to consulting firms. A designer working for the interior designer can be allowed access to the floor plan created by the architectural firm, ensuring the electronic transfer of the most up-to-date drawing files between offices.

Your school or office administrator will determine where you will save your drawing files. Most users save their projects at about 15-minute intervals on the hard drive. At the end of a drawing session, the file is saved in your folder on the network and to a portable USB drive. USB drives and fixed drives are often used for the day-to-day storage of drawing files because of their access speed. Once the drawing session or project is completed, it can be stored on a portable USB memory stick, CD, Zip, or tape cartridge where it is less subject to damage.

Naming Folders

You were exposed to creating folders in an introductory computer class, but their importance can't be overlooked as you work on large architectural projects. An efficient CAD technician will create folders using Windows Explorer to aid filing. It is also wise to divide your folder into subfolders. This might include subfolders based on specific classes you're enrolled in, different types of drawing, or different projects. Many small offices keep work for each client in separate folders or disks with labels based on the client name. Drawings are then saved by contents such as floor, foundation, elevation, sections, specs, or site. Some offices assign a combination of numbers and letters to name each project. Numbers are usually assigned to represent the year the project is started, as well as a job number with letters representing the type of drawing. For example, 0753fl would represent the floor plan for the fifty third project started in 2007. You're honing your CAD and drawing skills at a time when the construction industry is becoming increasingly connected by the Internet. Plans that were transported between offices by messenger are now shared between offices electronically. Because of the need by so many firms to work with a set of drawings, an efficient layer system is essential. Most architectural and engineering offices use file names based on the National CAD Standard. These standards are based on the standards developed by the American Institute of Architects (AIA).

The NCS guidelines provide a uniform file naming system that can be recognized by the various consultants that work on each project. This system covers both model and sheet files. NCS considers model files as those that contain the individual drawing components that will make up the finished drawings. Sheet files are the files that contain the drawing elements that will be plotted, including the model files and the border, title block, and all of the notation contained in a title block.

NCS Guidelines for File Names

File names based on NCS standards consist of the following four components:

- A project code
- A discipline designator
- A model file type
- A definable code

Project code The project code can contain up to 20 characters, which are determined by the project coordinator to meet the demands of the office. The code is optional, but usually it is used to designate the year and the project number or name. A code of 07MILLER could be used to represent a home done for the Miller family in 2007.

Discipline designator This is a 2-character code that represents the discipline that originates the drawings. This portion of the code consists of one letter followed by a hyphen. Common letters used include:

DESIGNATOR	DISCIPLINE	DESIGNATOR	DISCIPLINE
Architectural	A	Contractor/Shop	Z
Civil	C	Electrical	E
Equipment	Q	Other disciplines	X
Fire protection	F	Plumbing	P
General	G	Process	D
Interiors	I	Resources	R
Landscape	L	Structural	S
Mechanical	M	Survey/Mapping	V
Operations	O	Telecommunications	T

A small project that will be completed in office will not require a discipline designator. On drawings that require collaboration, the letter designator clarifies the drawing originator. Designators A, C, E, Q, F, I, L, M, P, S, and T are the most common designators used on large residential projects.

Model file type This code represents the type of drawing contained in a file. Within the structural drawings, you might expect to find details, elevations, schedules, and sections. Each would be considered an S drawing if it were supplied by a structural engineering firm working with the architectural team. The NCS divides model file type modifiers into several groups. Modifiers common to residential design include:

DEFINITION	CODE	DEFINITION	CODE
Area calculations	AC	Floor plan	FP
Border sheet	BS	Isometric/3D	3D
Details	DT	Key plan	KP
Elevations	EL	Roof plan	RP
Equipment plan	QP	Schedule	SH
Existing/demolition plan	XD	Section	SC

Keep in mind that drawings such as the details, elevations, and plans may be found in several designator groups. The architectural team will have a floor plan to show all of the material specified in Chapter 1. The mechanical contractor and the fire protection subcontractors will also have their own floor plans to specify their materials. Drawings for a complicated residence could include drawings with codes such as A-FP, E-FP, F-FP, M-FP, and T-FP.

Definable code This is a user-defined code consisting of four characters. It would typically be used to define multiple levels for floor and framing plans or multiple sheets of elevations, sections, and details. A home with four sheets of sections created by the architectural team could be represented by the code A-SC03XX. The 03 represents the third sheet of sections, and the XX is a space holder suggested by the NCS when four letters are not required in the definable code.

NCS Guidelines for Sheet Names and Numbers

Sheet names for drawings that are assembled for plotting are named using the same guidelines used to name model drawings. A second letter can be added to the discipline designator to define specific applications, such as AD FP to represent the demolition floor plan developed by the architectural team.

Sheet identifiers are placed in the title block. NCS standards recommend the use of the discipline designator and a sheet number so that a page number would resemble A156. The "A" specifies that the drawing originated with the architectural team, the drawing sheet is number 6 of a set of 15 sheets.

Alternative Methods

The NCS is an excellent step in moving the architectural world toward a uniform method of assembling and naming drawings. Years of varied standards, however, mean that you may work in a firm that still uses a more traditional method of naming files. One common practice is to name drawing files by the page number it will occupy in the drawing set. These numbers include:

Common Page Numbers and File Names

A0.01—Index, symbols, abbreviations, notes, and location maps
A1.01—Demolition, site plans, and temporary work
A2.01—Plans, and schedules such as room material, door, or windows and keyed drawings
A3.01—Sections and exterior elevations
A4.01—Detailed large-scale floor plans
A5.01—Interior elevations
A6.01—Reflected ceiling plans
A7.01—Vertical circulation drawings such as stair, elevator, and escalator plans and sections
A8.01—Exterior details
A9.01—Interior details

When this system is followed, a drawing file name of A4.01 would represent a detailed floor plan. If plans for a three-level building were to be drawn, each floor could be saved as a separate drawing file using drawing names such as:

A2.01—level one
A2.02—level two
A2.03—level three

Project Storage Methods

Before you begin a project, you also need to develop a method of saving drawing files. Common methods of saving drawing projects, in their order of usefulness and power, include single-drawing files, layered files, referenced files, and layouts.

Single Drawing Files

In previous basic computer classes, you probably saved each drawing project as a separate drawing (.dwg) file. This method has advantages as you work on larger drawing projects, but it also has its limitations. The main advantage of storing one drawing per file is ease of plotting. Individual drawings can be saved using the project name and contents such as \ARTIMIZ\SITE and be easily retrieved and plotted. The drawback with this storage system is the difficulty in making revisions. If the client changes the overall size of the project, the correction would need to be reflected in the \ARTIMIZ\SITE, \ARTIMIZ\GRADE, and \ARTIMIZ\LANDSCAP drawing files. This would triple drafting time, cutting into company profits and your break time.

Layered Drawing Files

One way to reduce the problems created with single-drawing files is to store related drawings in one drawing file. For plan views, this might include:

Site plan	Floor plan	Framing plan
Electrical plan	Heating plan	Plumbing plan

To be effective, these drawings must be stacked one above the other, as shown in Figure 3.3a. Common elements such as walls will be drawn only once rather than repeated for each drawing, effectively reducing the file size. Drawings can be separated using the LAYER command. Prefixes for layer names can be assigned to easily distinguish the drawing contents and make drawing assembly for plotting easier. Figures 3.3b and 3.3c show portions of floor and framing plans, which were stored in one drawing file and separated by the use of layers.

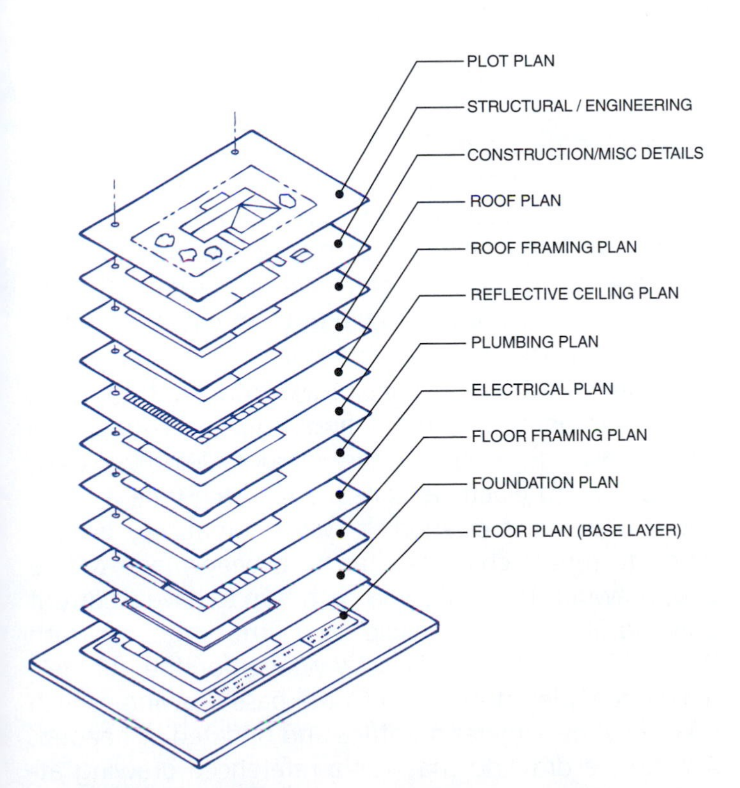

FIGURE 3.3a ■ *Effective use of layers requires drawings to be stored one above the other.*

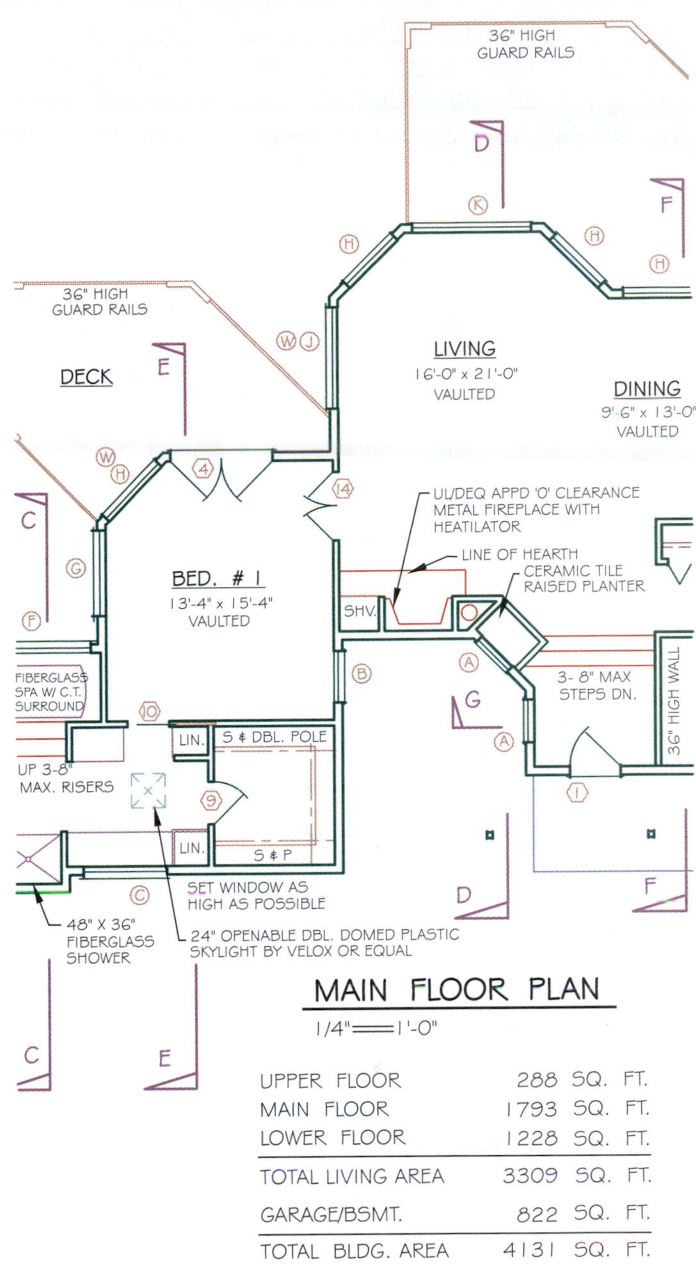

FIGURE 3.3b ■ *Information to be displayed on the floor plan can be displayed separately from information related to other plan views. Layer information can easily be controlled using the State Manager option of the Layer Properties Manager of AutoCAD.*

Naming Layers

Although every drawing file should make extensive use of layering, storing related drawings in one file can be done only if each major group of components is placed on a separate layer. Layers should be given names that describe both the base drawing and the contents of the layer to make plotting easier. Prefixes such as *FLOR, HVAC, PLMB,* and *ELCT* will help identify the floor, mechanical, plumbing, and electrical drawings

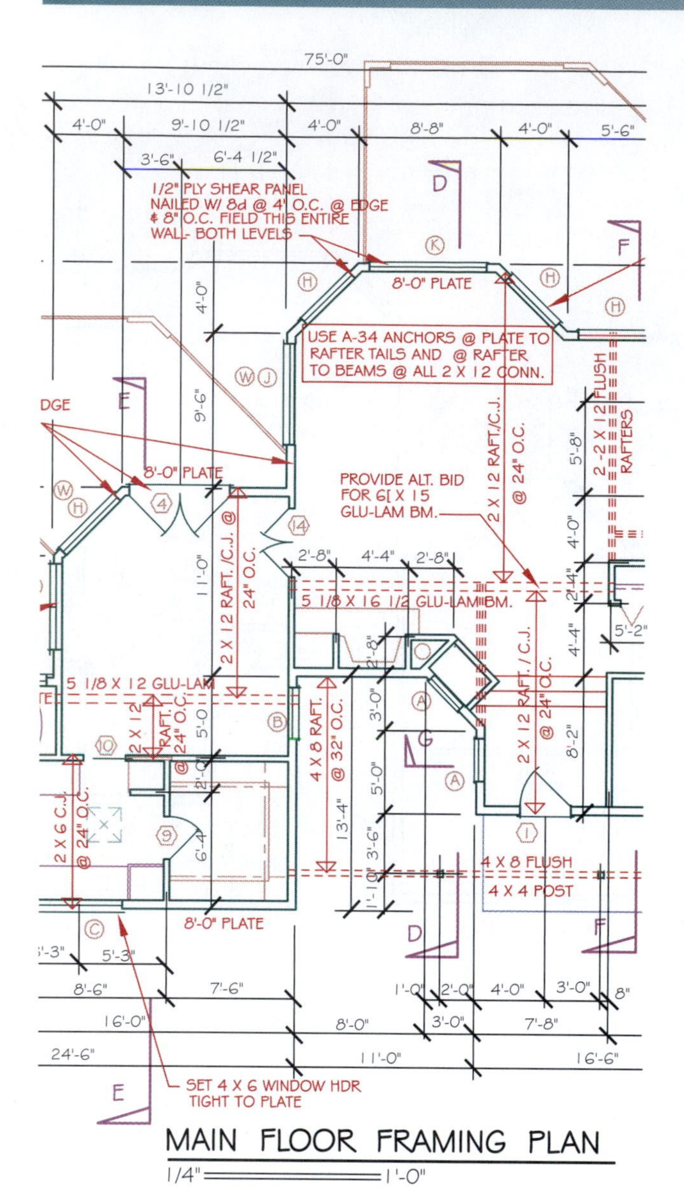

•• SEE SHEET 9 FOR FRAMING NOTES
SHEET 12 FOR INSULATION & CAULKING NOTES
SHEET 12 - 15 FOR SECTIONS
SHEET 16 FOR FOUNDATION NOTES
SHEET 4 FOR WINDOW SCHEDULE
SHEET 6 FOR DOOR SCHEDULE
ALL EXTERIOR HEADERS TO BE 2-2 X 12 W/ RIGID INSULATION BACKING UNLESS NOTED.
USE A-34 ANCHORS AT ALL 2 X 12 RAFTER TAIL/ PLATE INTERSECTIONS AND RAFT./ BEAM CONNECTIONS

DECK FLOOR JOIST /BEAMS:
LIVE LOAD 60# / HEAD LOAD 10
2 X 8 @ 8" O.C. = 16'-0" MAX SPAN
2 X 8 @ 12" O.C. = 13'-0" MAX.
2 X 8 @ 16" O.C. = 11'-0" MAX.

FIGURE 3.3c ■ *This framing plan used base layers from the floor plan in Figure 3.3a as a starting point. Material related to the floor plan is frozen, and material relevant to the framers is displayed.*

and clearly describe the contents within each subgroup. Information about the walls, windows, doors, and appliances that will be required on all drawings can be stored on a layer with a prefix of *BASE*. To plot the framing plan, the base and all framing layers would be set to THAW, with all layers related to other drawings set to FREEZE. Further guidelines on naming layers regardless of the filing method are given later in this chapter.

Viewing Complex Drawings

One of the disadvantages of using files containing multiple files is that because they contain so much information, moving through the drawings using the ZOOM command can be slow. You can overcome this by using the PARTIALOAD or PARTIALOPEN commands or by using the VIEW command. When you work with large files, the PARTIALOAD and PARTIALOPEN commands can be used to allow a specific portion of the drawing to be accessed. If the entire drawing is to be opened, the VIEW command can be used to display specific areas of the drawing. This command allows specific areas to be named, saved, and quickly retrieved and magnified. Another method to control the drawing is to use the Layer State Manager to control what layers will be displayed.

External Referenced Drawings

The XREF command is an effective method for working with related drawings when multiple firms will be collaborating on a project. Floor-related drawings provide an excellent example of drawings that can be referenced. A CAD technician working for the design team can be working on the floor plan while other technicians are working on the electrical or mechanical plan. A technician for the architectural team can draw information in a base drawing that will be reflected on all other plan drawings. Copies of this drawing file can be given electronically to other firms that will develop related drawings, such as the interior design, HVAC, plumbing, and electrical consultants. Each time a referenced drawing is opened, it will automatically be updated to reflect changes that have been made to the base drawing. This will allow each firm to have a current drawing file as a base while each firm progresses with its work. A CAD technician working for the cabinetmaker can add information to the base drawing, which is kept by the originating office and updated as needed. Any time a drawing that has a referenced drawing attached to it is accessed, the most current version of the base drawing is provided.

Layouts

A layout is a paper space tool that consists of one or more viewports to aid plotting. Layouts allow the ease of plotting associated with single-drawing files and the benefits of referenced drawings to be used to arrange multiple drawing components for display and plotting at varied scales. To understand the process of viewing a drawing in a layout, visualize a sheet of 24 × 36'' vellum with a title block and border printed on the sheet. Imagine a hole, the viewport, cut in the vellum that allows you to look through the paper and see the floor plan on another sheet of vellum. Figure 3.4 shows the theory of displaying a drawing in model space in a layout created in paper space. This is a simplified version of what is required to display a drawing for plotting. Now imagine that the floor plan shown is 90'-0'' wide in model space. If you were to hold a sheet of D-size vellum in front of the plan, the paper would be minute. To make the floor plan fit inside the viewport in the paper, you're going to have to hold the paper a great distance away from the floor plan until the drawing is small enough to be seen through the hole. AutoCAD will figure the distance as drawings are placed in the viewport. By entering the desired scale factor for plotting, you can reduce the floor plan to fit inside the viewport and maintain a scale typically used in the construction trade. Multiple viewports can be placed in a single layout, and multiple layouts can be created within a single drawing file. Multiple viewports are discussed later in this chapter.

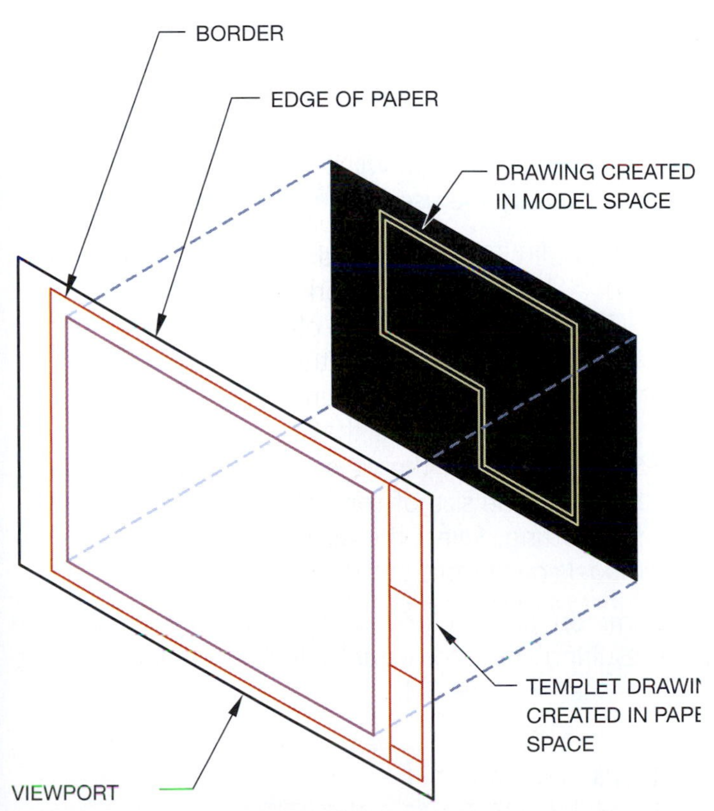

FIGURE 3.4 ■ *A viewport is provided in each drawing template created in paper space to allow the drawing to be viewed at a specific scale when plotted.*

Cleaning Drawing Files

No matter what method is used to store drawing files, care should be taken to keep the files free of unwanted material. As you draw on a computer screen, you need to think of how you work with pencil and paper. Notes, file folders, and sketches tend to accumulate on and around your desk until it is buried. The same can happen to a computer file. AutoCAD is fairly efficient at removing its scratch paper and notes from past files, but you can aid in the cleaning by using the WBLOCK* or the PURGE commands.

*WBLOCK**

The ***WBLOCK**** option will save your drawing as a block, but each object is saved individually rather than as one object. Information about commands that might have been tried and then erased or deleted is removed from the drawing base.

PURGE

The ***PURGE*** command is a useful tool within a drawing file to eliminate unused items and minimize the amount of storage space used. Objects that have been named during a drawing session—such as blocks, dimstyles, layers, linetypes, shapes, and text styles—are examples of items that can be purged from a drawing file. PURGE can be used, for instance, if a template drawing for a multilevel home was used to start a drawing file for a single-level home. The original template may contain layers for *UPPR* and *LWR* levels that are not required for this project. Unused information on the template drawing can be eliminated, reducing the file size.

MANAGING DRAWING PROPERTIES WITH A TEMPLATE

Once the decision has been made about how drawing projects will be organized, consideration should be given to how individual drawing files can be developed. One of the best methods of creating a drawing is by using a drawing template. A drawing template is a base drawing file prepared for a specific class or company that contains all of the typical settings normally required for a specific type of drawing. The template should include:

- A border, and title block with company- and job-specific attributes
- Common settings for units of measurement for lines and angles
- Drawing limits
- Layers
- Scaling information
- Linetypes
- Lineweight
- Text styles and formatting settings
- Text specific to the title block
- Dimension styles and formatting settings

A key element of a template is the border and title block. Key elements of architectural title blocks are introduced later in this chapter. Drawing borders are thick lines, approximately .90'' thick, which surrounds the entire sheet. Top, bottom, and right-side borderlines are usually offset 1/2'' in from the paper edge. The exact size will vary depending on plotting limitations. The left border is approximately 1 1/2'' from the edge of the paper to allow for space to staple the drawing set. Template drawings are usually created for plotting projects on a specific size, such as:

ENGLISH	METRIC
C - 24 × 18''	C - 610 × 460 mm
D - 36 × 24''	D - 915 × 610 mm
E - 48 × 36'' paper	E - 1220 × 915 mm

Later portions of this chapter cover placing information in the title block.

Units of Measurement

The units of measurement should be included in a template drawing. Architectural and engineering units are the units of measurement normally associated with construction drawings. With the exception of large-scale site drawings that are drawn using the engineering units, most architectural drawings use architectural units. This would produce numbers that are expressed as 125'–6 3/4''. Occasionally offices use engineering units that express numbers in decimal fractions, such as 75'–3.5''. The unit of measurement factor can be changed throughout a drawing session, but it should be selected for a template drawing based on the office standard for a particular type of drawing.

Angle Measurement

Decimal degrees expressed as 28.30° and surveyor's units expressed as N30°30'15''E are the two most common methods of angular measurement used on construction drawings. More important than the method of expressing the measurement is adjusting the starting point from which angles are measured. Standard AutoCAD setup allows for 0° to be measured from the 3 o'clock position. This is not appropriate for site-related drawings. Common office practice for adjusting the direction of north is to draw the site plan while assuming north is at the top of a page. This may result in property lines that are not parallel to the border of the paper. Once drawn, however, the site plan and the north arrow can be rotated so that the longest property line is parallel to a border.

Drawing Limits

The limits for a drawing template should be set for both the model and layout views as the template is created. Buildings should be drawn at full scale in model space. Completed drawings are plotted in paper space at a reduced scale. What will be seen in each setting can be controlled by the use of the LIMITS command.

- Set the limits for model space at approximately 150 × 100'. It is important to remember that a drawing can be started with the limits set to any convenient size and that the model space limits can be adjusted at any point throughout the drawing session.
- Set the limits for the viewport in paper space based on the size of the paper that will be used for plotting. Select the desired display scale for the model space from the Viewports toolbar.

Common architectural and engineering scales and the resulting limits for establishing a template drawing can be seen in Table 3.1.

Layers

Although layers are used on all drawings, they are particularly important to a template drawing. Layers can be created, saved, and reused for all similar drawings in

TABLE 3.1 COMMON ARCHITECTURAL SCALES AND LINETYPE SCALES

ARCHITECTURAL VALUES		ENGINEERING VALUES	
DRAWING SCALE	LT SCALE	DRAWING SCALE	LT SCALE
1'' = 1'-0''	6	1'' = 1'-0''	6
3/4'' = 1'-0''	8	1'' = 10'	60
1/2'' = 1'-0''	12	1'' = 100'	600
3/8'' = 1'-0''	16	1'' = 20'	120
1/4'' = 1'-0''	24	1'' = 200'	1200
3/16'' = 1'-0''	32	1'' = 30'	180
1/8'' = 1'-0''	48	1'' = 40'	240
3/32'' = 1'-0''	64	1'' = 50'	300
1/16'' = 1'-0''	96	1'' = 60'	360

a template drawing. Layering can also be especially useful in assigning particular linetypes and colors to specific layers of the template. Because of their importance to a template, layer titles must be easily recognized by each member of the design team. Possible layer prefixes for simple drawings were given earlier in this chapter as drawing setup was considered. On drawing projects involving multiple design firms, a uniform layer and drawing-filing-naming system will aid efficiency.

The U.S. National CAD Standards are incorporated in the AIA CAD Layer Guidelines that are used by many architectural firms. Copies of the standard can be obtained from the NIBS at www.nibs.org. The method for naming layers is similar to naming files. Each naming system can be compared in Figure 3.5. A layer name can be composed of the discipline and major group name. A minor name and status code can be added to the sequence. The discipline code is the same as the one- or two-character code that is used for model and sheet file names. Because name components are common to both file and layer names, care must be taken to avoid creating project-specific references within a layer name. Project designations should not be included in folder names.

Major Group Name

The major group code is a four-character code that identifies a building component specific to the defined layer. Major group layer codes are divided into the major groups of architectural, civil, electrical, fire protection, general, hazardous, interior, landscape, mechanical, plumbing, equipment, resource, structural, and telecommunication. Codes such as *ANNO* (annotation), *EQIP* (equipment), *FLOR* (floor), *GLAZ* (glazing), and *WALL* (walls) are examples of major group codes that are associated with the architectural layers. A complete listing of major codes for each group can be found in the U.S. National CAD Standards.

Minor Group Name

The minor group code is an optional four-letter code that can be used to define subgroups to the major group. The code *A FLOR* (architectural–floor) might include minor group codes for *OTLN* (outline), *LEVL* (level changes), *STRS* (stair treads or escalators), *EVTR* (elevators), or *PFIX* (plumbing fixtures). A layer name of *A FLOR IDEN* would contain room names, numbers, and other related titles or tags. A complete listing of minor group codes specific to each discipline is listed in the national standard. Figure 3.6 shows a listing of common layer names and their contents.

Status Code

The status code is an optional single-character code that can be used to define the status of either a major or minor group. See Layer Name Format in Figure 3.5.

MODEL FILE NAME FORMAT

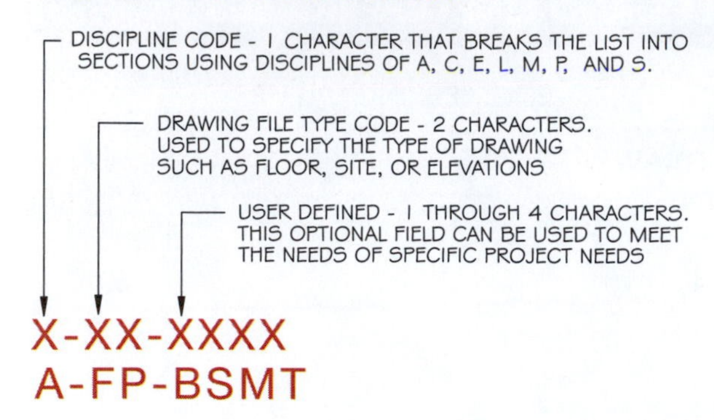

SHEET FILE NAME FORMAT

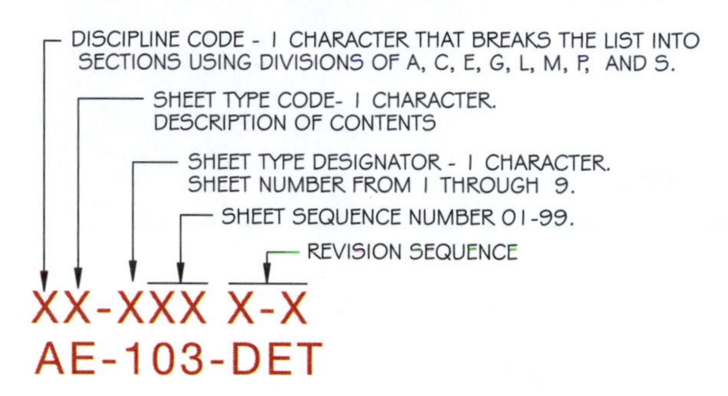

LAYER NAME FORMAT

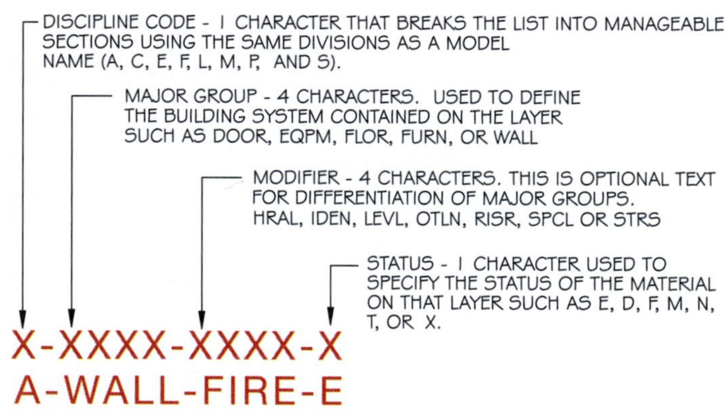

DISCIPLINE CODE - 1 CHARACTER THAT BREAKS THE LIST INTO MANAGEABLE SECTIONS USING THE SAME DIVISIONS AS A MODEL NAME (A, C, E, F, L, M, P, AND S).
MAJOR GROUP - 4 CHARACTERS. USED TO DEFINE THE BUILDING SYSTEM CONTAINED ON THE LAYER SUCH AS DOOR, EQPM, FLOR, FURN, OR WALL
MODIFIER - 4 CHARACTERS. THIS IS OPTIONAL TEXT FOR DIFFERENTIATION OF MAJOR GROUPS. HRAL, IDEN, LEVL, OTLN, RISR, SPCL OR STRS
STATUS - 1 CHARACTER USED TO SPECIFY THE STATUS OF THE MATERIAL ON THAT LAYER SUCH AS E, D, F, M, N, T, OR X.
X-XXXX-XXXX-X
A-WALL-FIRE-E

FIGURE 3.5 ■ *File and layer names use abbreviations to describe four categories of information based on the NCS format.*

ARCHITECTURAL LAYER NAMES

ANNOTATION

A-ANNO-DIMS	DIMENSIONS
A-ANNO-KEYN	KEY NOTES
A-ANNO-LEGN	LEGENDS & SCHEDULES
A-ANNO-NOTE	NOTES
A-ANNO-NPLT	NON-PLOTTING INFORMATION
A-ANNO-NRTH	NORTH ARROW
A-ANNO-REVS	REVISIONS
A-ANNO-REDL	REDLINE
A-ANNO-SYMB	SYMBOLS
A-ANNO-TEXT	TEXT
A-ANNO-TTLB	BORDER & TITLE BLOCK

FLOOR PLAN

A-FLOR-EVTR	ELEVATOR CAR & EQUIPMENT
A-FLOOR-FIXT	FLOOR INFORMATION
A-FLOR-FIXD	FIXED EQUIPMENT
A-FLOR-HRAL	HANDRAILS
A-FLOR-NICN	EQUIPMENT NOT IN CONTRACT
A-FLOR-OVHD	OVERHEAD ITEMS (SKYLIGHTS)
A-FLOR-PFIX	PLUMBING FIXTURES
A-FLOR-STRS	STAIR TREADS
A-FLOR-TPIN	TOILET PARTITIONS

ROOF PLAN

A-ROOF	ROOF
A-ROOF-OTLN	ROOF OUTLINE
A-ROOF-LEVL	ROOF LEVEL CHANGES
A-ROOF-PATT	ROOF PATTERNS

ELEVATIONS

A-ELEV	ELEVATIONS
A-ELEV-FNSH	FINISHES & TRIM
A-ELEV-IDEN	COMPONENT IDENT. NUMBERS
A-ELEV-OTLN	BUILDING OUTLINES
A-ELEV-PATT	TEXTURES & HATCH PATTERNS

SECTIONS

A-SECT	SECTIONS
A-SECT-IDEN	COMPONENT IDENT. NUMBERS
A-SECT-MBND	MATERIAL BEYOND SECTION CUT
A-SECT-MCUT	MATERIAL CUT BY SECTION PLANE
A-SECT-PATT	TEXTURES AND HATCH PATTERNS

FIGURE 3.6 ■ *Common layer names and their contents based on the National CAD Standard Layer Guidelines.*

The code is used to specify the phase of construction. Throughout the balance of this text, it will be assumed that you are working on projects that will not require consulting firms, thus allowing the status code to be omitted. The layer names for the walls of a floor plan *(A WALL FULL)* could be further described using one of the following status codes:

E	Existing to remain
D	Existing to demolish
F	Future work
M	Items to be moved
N	New work
T	Temporary work
X	Not in contract
1–9	Phase numbers

Common Linetypes on Construction Drawings

Although each office may have its own standard, guidelines established by the American National Standards Institute (ANSI) and the National CAD Standards are used throughout much of the construction industry to ensure drawing uniformity. Common linetypes include object, hidden, center, cutting plane, section, break, phantom, extension, dimension, and leader. Examples of most of these types of lines can be seen in Figure 3.7. Specific uses for each linetype are discussed throughout the text.

> **Note:**
>
> *AutoCAD will assign lengths to each line segment based on the LTSCALE settings. Line segment sizes are included in the following discussion so that you'll know how the lines should appear when plotted.*

Object Lines

Object lines are continuous lines used to describe the shape of an object or to show changes in the surface of an object. Object lines can be thick or thin depending on how they are being used. The common line width for thick lines is .6 mm. On a drawing such as a floor plan, thick object lines are used to represent walls, and thin object lines are used to provide contrast when representing doors, windows, cabinets, or appliances. Figure 3.8 shows both uses of object lines in a framing plan.

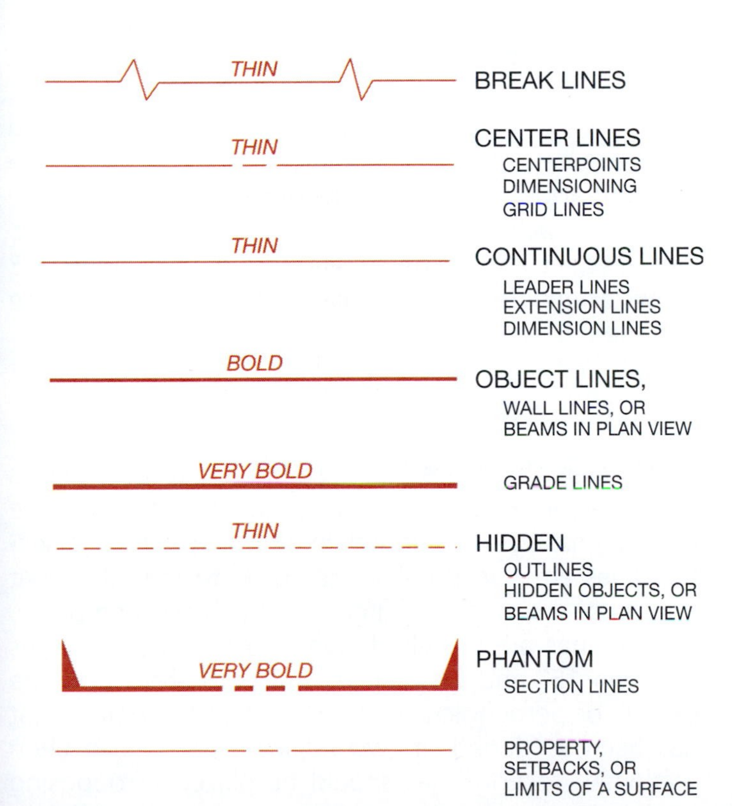

FIGURE 3.7 ■ *Common linetypes found throughout construction drawings.*

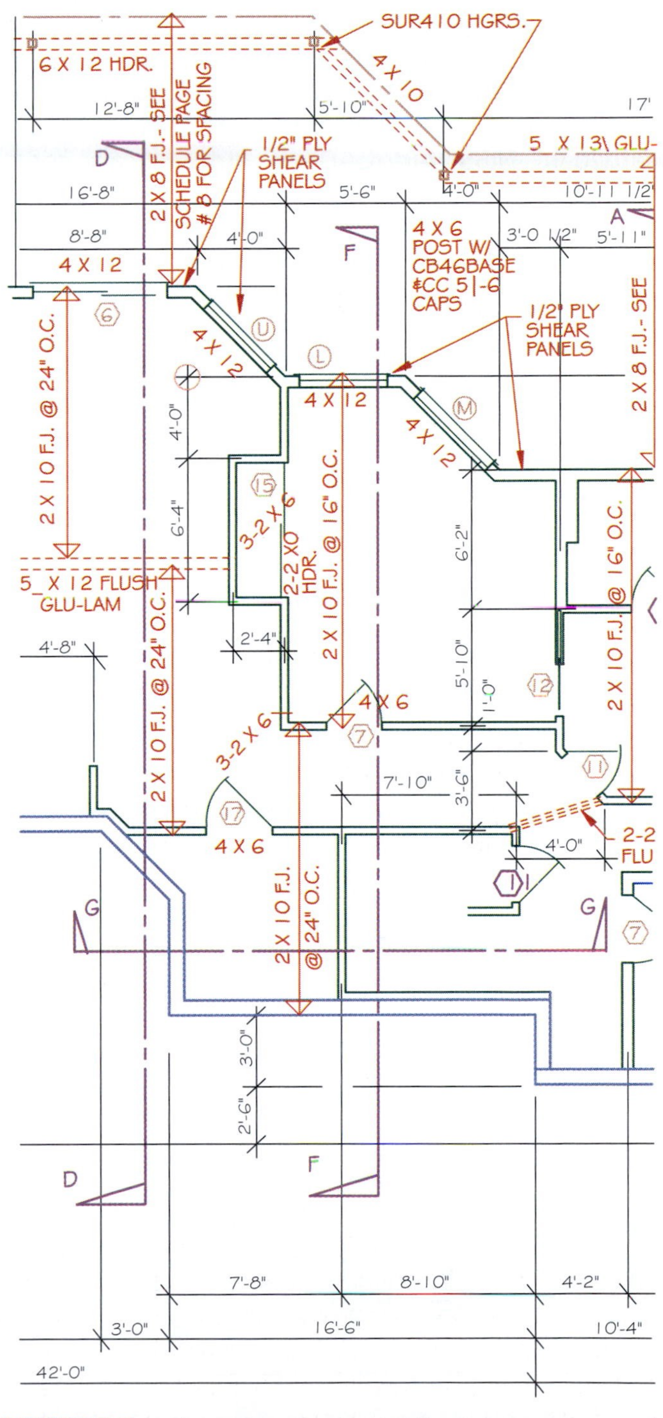

FIGURE 3.8 ■ *Examples of common linetypes used on residential drawings. Assigning varied linetypes to a specific layer on a template drawing can save a great deal of drafting time.*

Several line thicknesses will be required to represent material in details such as the wall section in Figure 3.9. Office standards will vary greatly, but it is important that a detailer use line widths that will allow key features to be easily identified.

Hidden Lines

Dashed or hidden lines are thin lines used to represent a surface or object that is hidden from view. ANSI specifies that the lines should be 1/8'' long with 1/16'' space between lines. A line width of .25 mm is recommended by NIBS. AutoCAD will automatically control the spacing, based on the line length. The size of the line can be altered using the LTSCALE command. Figure 3.8 shows an example of hidden lines used to represent headers that support a deck at the top of the drawing.

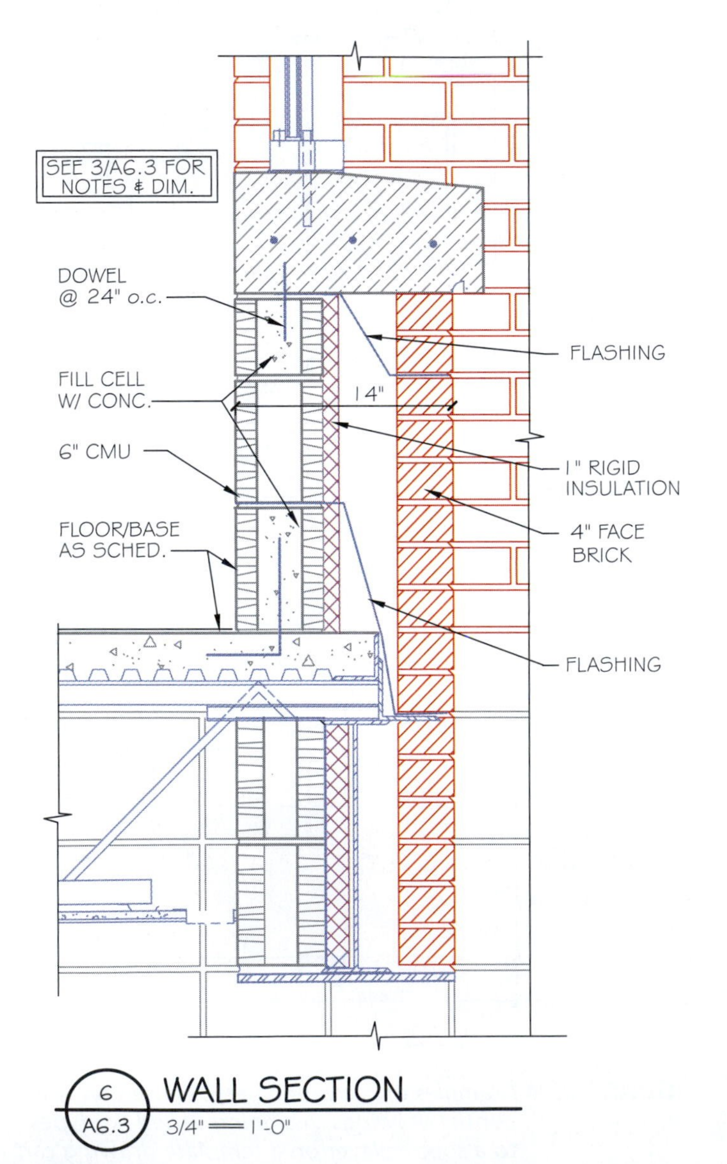

FIGURE 3.9 ■ *Hatch patterns are used to show various materials that have been sectioned.*

Centerlines

A centerline is composed of thin lines that create a long-short-long pattern. The short line should be 1/8'' long, the space should be 1/16'', and the long segment must be between 3/4'' and 1'' long, with a width of .25 mm. AutoCAD will control the length and space of each portion of the line. Centerlines are used to locate the center axis of circular features such as drilled holes, bolts, columns, and piers. Figure 3.8 shows the use of centerlines to represent the outline of the deck at the top of the drawing. Centerlines have also been used to represent dimensions that extend to the center of interior wood walls.

Note:

AutoCAD can reproduce each of the linetypes that contain a line pattern, but the lengths of the lines needs to be adjusted based on the scale that will be used to display the drawing.

Cutting Plane Lines

A cutting plane line is placed on the floor or framing plan to indicate the location of a section and the direction of sight when viewing the section. Cutting planes are represented by a long-short-short-long line pattern. Cutting plane lines should be drawn with a width of .50 mm. The long line segment should be between 3/4'' and 1'' long. The short lines should be 1/8'' long, and the space between lines should be 1/16''. Terminate the line with arrows to indicate the viewing direction. Place a letter at each end of the line using 1/4'' high text to indicate the view or a detail reference bubble. Figure 3.8 shows examples of cutting planes (D, F, and G) and line terminators.

Section Lines

When an object has been sectioned to reveal its cross section, a pattern is placed to indicate the portion of the part that has been sectioned. In their simplest form, section lines are thin parallel lines used to indicate what portion of the object has been cut by the cutting plane. Section lines are usually drawn at a 45° angle. Angles between 15° and 75° can be used but should not be parallel or perpendicular to any part of the object that has been sectioned. If two adjacent parts have been sectioned, section lines should be placed at opposing angles. Spacing between section lines can be altered, depending on the size of the part to be sectioned. Typically, a spacing of 1/8'' should be provided between section lines. Section lines should be placed using the HATCH command. Section lines are generally used only to represent masonry in plan or sectioned views. Most materials have their own special pattern. Figure 3.9 shows common patterns used to indicate that an object has been sectioned.

Break Lines

Break lines are used to remove unimportant portions of an object from a drawing so that it will fit into a specific space. If you consider a 10' steel column that supports a beam and rests on a concrete footing, the column is the same between the beam and the bottom plate. If you need to draw a 10' steel column at full scale in a 5' space, break lines can be used to remove 5' of the column. The three common break lines used on construction drawings are the short, long, and cylindrical break lines. A short break line is a thick jagged line placed where material has been removed (see Figure 3.10). A long break line is a thin line with a zigzag shape or inverted S shape inserted into the line at intervals (see Figure 3.9). A cylindrical break line is a thin line resembling a backward S (see Figure 3.10).

Phantom Lines

Phantom lines are thin lines in a long-short-short-long pattern that are used to show motion or to show an alternative position of a moving part. They can also be used in place of centerlines to represent upper levels or projections of a structure or on site plans to represent easements or utilities. The long line segment should be between 3/4''–1'' long but remain a constant length within each drawing. The short lines should be 1/8'' long. The space between lines should be 1/16''. Examples of phantom lines can be seen in Figure 3.7 and 3.8.

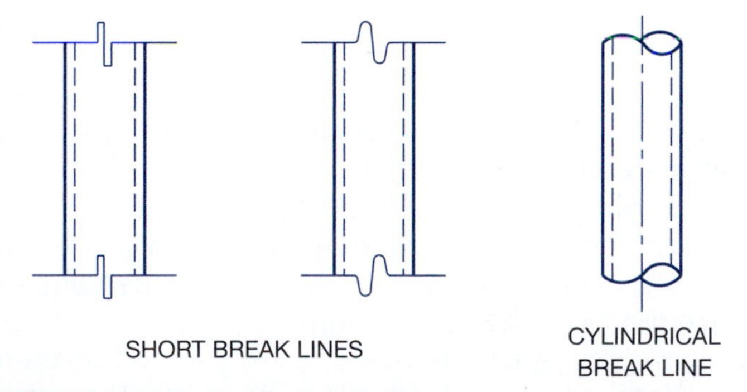

FIGURE 3.10 ■ *Short and cylindrical break lines are used to shorten long objects with a consistent shape.*

Extension Lines

Extension lines are thin lines used to relate dimension text to a specific surface of a part (see Figure 3.8). Provide a 1/16'' to 1/8'' space between the object being described and the start of the extension line. Extend the extension line 1/8'' past the dimension line. An extension line can cross other extension lines, object lines, centerlines, hidden lines, and any other line with one exception. An extension line can never cross a dimension line. Never! Later chapters introduce other guidelines for placing extension lines using the various commands for placing dimensions. Extension lines should be placed using the dimensioning tools of AutoCAD.

Dimension Lines

Dimension lines are thin lines used to relate dimension text to a specific part of an object (see Figure 3.8). Dimension lines extend from one extension line to another. They require a line terminator where the line intersects the extension line. The terminator is usually a tick mark or a solid arrowhead. Dimension text is typically 1/8'' high. It should be centered over the dimension line, although it might need to be placed off center to aid clarity when several lines of text are near each other. The space between the dimension text and the dimension line should be equal to half the height of the dimension text. Dimension lines should be placed using the dimensioning tools of AutoCAD. Later chapters provide specific uses of dimension lines and dimension text as the various commands for dimensioning are introduced.

Leader Lines

Leader lines are thin lines used to relate a dimension or note to a specific portion of a drawing. A solid arrowhead should be used at the object end of the leader line. A horizontal line approximately 1/8'' long should be used at the note end of the leader line. Leader lines can be placed at any angle, but angles between 15° and 75° are most typical. Vertical and horizontal leader lines should never be used. The arrow end of the leader line should touch the edge of the part it describes. The leader line should extend far enough away from the object it is describing so that the attached note is a minimum of 3/4'' from the outer surface of the object (see Figure 3.8 and 3.9). Leader lines should be placed using the QUICK LEADER command.

Linetype Scale

In addition to linetypes being created and set based on a specific layer, the scale factor for linetypes should also be set for a template drawing. The scale can be adjusted using the LTSCALE command. To display linetypes for a drawing in a paper space layout, PSLTSCALE must be set to 0. Common line scales are shown in Table 3.1. The line scale is determined by the size at which the drawing will be plotted. Using an LTSCALE of 2 will make the line pattern twice as big as the line pattern created by a scale of 1. For a drawing plotted at 1/4'' = 1'–0'', a dashed line that should be 1/8'' when plotted would need to be 3" long when drawn at full scale. This was determined by multiplying .125 × 24 (the LTSCALE). The linetype scale factor can be determined by using half of the drawing scale factor. The drawing scale factor is always the reciprocal of the drawing scale. Using a drawing scale of 1/4'' = 1'–0'' would equal .25'' = 12''. Dividing 12 by .25 produces a scale factor of 48. Keep in mind that when the LTSCALE is adjusted, it will affect all linetypes within the drawing. This is typically not a problem unless you are trying to create one unique line and CTLTSCALE can be used.

Lineweight

Varied line widths will help you and the print reader keep track of various components of a complex drawing. As you assign lineweight to template drawings, remember that you're using thick and thin lines to add contrast, not represent thickness of an object. Lineweights should be assigned to a layer by selecting Lineweights from the FORMAT pull-down menu. Common lineweights and their width in millimeters based on NCS recommendations can be seen in the table that follows.

MILLIMETERS	DESCRIPTION	USE
.18	Fine	Material indications, surface marks, hatch lines and patterns
.25	Thin	1/8'' (3mm) annotation, setback, and grid lines.
.35	Medium	5/32'' to 3/8'' (4 to 10 mm) annotation, object lines, property lines
.50	Wide	7/32'' to 3/8'' (5 to 10 mm) annotation, edges of interior and exterior elevations, profiling cut lines
.70	x-wide	1/2 to 1'' (13 to 25 mm) annotation, match lines, borders
1.00	xx-wide	Major titles underlining and separating portions of designs
1.40	xxx-wide	Border sheet outlines and cover sheet line work
2.00	xxxx-wide	Border sheet outlines and cover sheet line work

MANAGING DRAWING TEXT

Text is an important part of every architectural drawing. Before text is placed on a drawing, several important factors should be considered. Each time a drawing is started, consider who will use the drawing, the information the text is to define, and the scale factor of the text. Once these factors have been considered, the text values can be placed in a template and saved for future use, similar to the stock notes seen in Figure 3.11. Architectural offices dealing with construction projects use a style of lettering that features thin vertical strokes, with thicker horizontal strokes to give a more artistic flair to the drawing. Other common variations are compressed and elongated letter shapes. In addition to the basic shape of the letters, some offices use a forward or backward slant to make their office lettering more distinctive. You are unlikely to find all of these variations on one professional drawing. Figure 3.12 shows some of the variations that can be found on drawings.

Text Similarities

With all of the variation in styling, two major areas that are uniform throughout architectural offices are text height and capital letters. Text placed in the drawing area is approximately 1/8'' high. Titles are usually between 1/4'' and 1'' in height, depending on office practice. The other common feature is that text is always made up of capital letters, although there are exceptions based on office practice. The letter "d" is always printed in lowercase when representing

GENERAL FOUNDATION NOTES:

1. EXTEND FOOTINGS (18") MIN. INTO NATURAL SOIL.
2. SLOPE ALL GRADES AWAY FROM FOUNDATION FOR 60" MINIMUM.
3. CONCRETE EXPOSED TO WEATHER TO BE 3000 PSI. CONCRETE NOT EXPOSED TO WEATHER TO BE 2500 PSI. CONCRETE SLAB TO BE 2500 PSI. ALL CONC. COMP. STRENGTH MIN. TO BE @ 28 DAYS.
4. ALL MUDSILLS TO BE DFPT W/ 1/2"Ø x 10" A.B. PROVIDE A MINIMUM OF TWO A.B. FOR EA. PLATE W/ 1 BOLT LOCATED NOT MORE THAN 12" FROM EACH END OF EA, ℙ PIECE. PROVIDE 2" DIA. WAHERS AT EACH BOLT.
5. NOTCHES IN THE ENDS OF JOISTS SHALL NOT EXCEED 1/4 OF THE DEPTH. HOLES DRILLED IN JOISTS SHALL NOT BE IN THE UPPER/LOWER 2" OF THE JOISTS. THE DIA. OF HOLES DRILLED IN JSTS. SHALL NOT EXCEED 1/3 THE DEPTH OF THE JST.
6. BLOCK ALL FLOOR JOIST AT SUPPORTED ENDS AND @ 10'-0" O.C. MAX. ACROSS SPAN.
7. PROVIDE 3/4" STD. GRADE T. & G. PLY. FLOOR PERP. TO FLOOR JOIST. GLUE & NAIL W/ 10 d'S @ 6" O.C. @ EDGES AND BLOCKING AND 12" O.C.

FIGURE 3.11 ■ *Text is a common element of every construction drawing. Settings that control text should be stored in a template and saved for future use.*

ARCHITECTURAL OFFICES USE MANY STYLES OF LETTERING.

S O M E T E X T I S E L O N G A T E D .

SOME OFFICES USE A COMPRESSED STYLE.

NO MATTER THE STYLE THAT IS USED,

IT MUST BE EASY TO READ.

THIS IS A VERY COMMON FONT FOR THE FIELD OF ARCHITECTURE.

FIGURE 3.12 ■ *Common text font variations that can be found on drawings.*

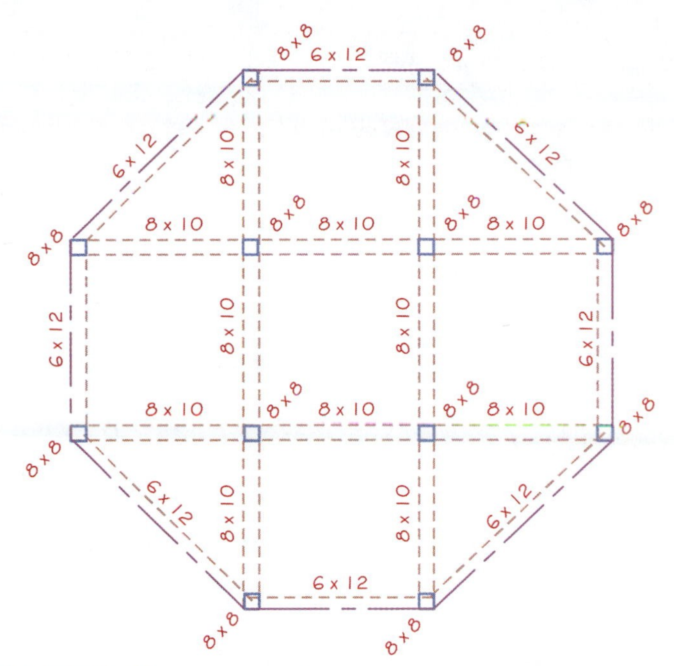

FIGURE 3.13 ■ *Care should be taken with the text orientation so that it is placed parallel to the bottom or the right edge of the page. Text used to describe structural material is placed parallel to the member being described. Specifications for posts or columns are typically placed at a 45° angle.*

the pennyweight of a nail. A typical use would resemble

USE (3) -16d @ EA. JOIST / PLATE CONNECTION

Occasionally manufacturers use lowercase letters to represent the size of special nails or fasteners specific to their products. With Caps Lock activated, all lettering will be capitalized unless Shift is pressed. Pressing Shift will make any letters typed while the key is depressed lowercase letters. Numbers are generally the same height as the text with which they are placed. When a number is used to represent a quantity, it is generally separated from the balance of the note by a dash. Another common method of representing the quantity is to place the number in parentheses ( ) to clarify the quantity and the size. This would resemble

USE (3) -3/4" DIA. M.B. @ 3'' O.C.

Fractions are generally placed side by side (3/4'') rather than one above the other, so that a larger text size can be used. When a distance such as one foot two and one half inches is to be specified, it should be written as 1'-2 1/2''. The ' (foot) and '' (inch) symbols are always used, with the numbers separated with a dash.

Text Placement

Text placement refers to the location of text relative to the drawing and within the drawing file. Care should be taken with the text orientation so that it is placed parallel to the bottom or the right edge of the page. When structural material shown in plan view is described, the text is placed parallel to the member being described. Figure 3.13 shows methods of placing text in plan views. On plan views such as Figure 3.14, text is

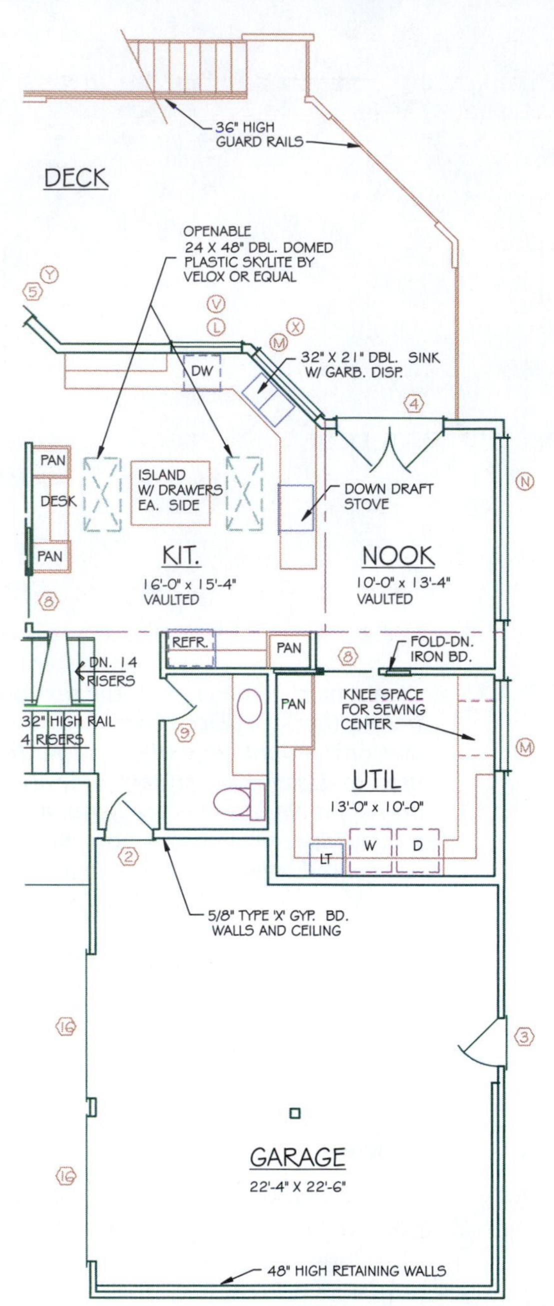

FIGURE 3.14 ■ *Text can be placed within the drawing, but arranged so that it does not interfere with any part of the drawing. Text is generally placed within 2" of the object being described, and a leader line is used to connect the text to the drawing.*

typically placed within the drawing but arranged so that it does not interfere with any part of the drawing. Text is generally placed within 2'' of the object being described. A leader line is used to connect the text to the drawing. The leader line can be either a straight line or an arc, depending on office practice.

On details, text can be placed within the detail if large open spaces are part of the drawing. It is preferable to keep text out of the drawing. Text should be aligned to enhance clarity and can be either aligned left or right. Figure 3.15 shows a detail with good text placement. The final consideration for text placement is to decide on layer placement. Text should be placed on a layer with a major group code of *ANNO*. Minor group codes include:

KEYN (key notes)
LEGN (legends and schedules)
NOTE (general notes)
NPLT (non-plotted text)
REDL (redline)
REVS (revisions)
SYMB (symbols)
TEXT (local notes)
TTLB (border and title block)

Types of Text

Text on a drawing is considered to be either a general or local note. General notes can refer to an entire project or to a specific drawing within a project. Local notes, such as the notes in Figure 3.14, refer to specific

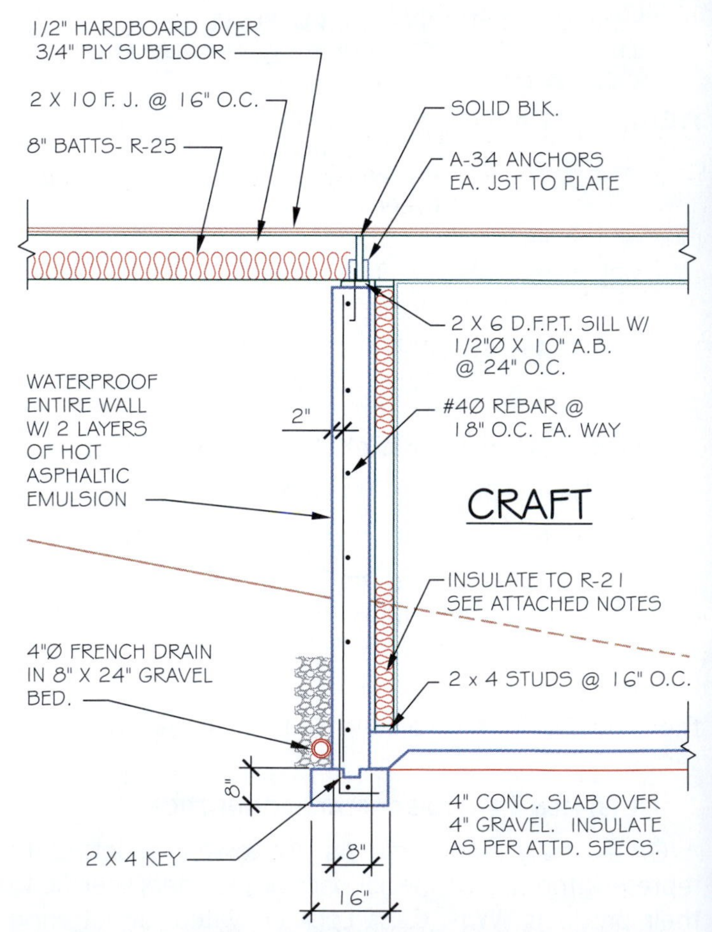

FIGURE 3.15 ■ *Text can be placed within a detail if large open spaces are part of the drawing, but it is preferable to keep text out of the drawing. Text should be aligned to enhance clarity and can be aligned either left or right.*

areas of a project. Small amounts of local notes should be placed using the TEXT command. The notes shown in Figure 3.11 are general notes that specify the materials of a framing plan. The MTEXT command should be used to place large amounts of text. Practical uses for each command in template drawings are discussed later in this chapter.

General Notes

Much of the text used to describe a drawing can be standardized and placed in a wblock or template drawing. Figure 3.11 shows an example of standard notes that an office includes for all foundation plans. These notes can be typed once and saved as a wblock. The wblock can then be nested in the template drawing on the *X XXXX ANNO GENN* layer. The layer can be frozen, thawed when needed, and then edited using the DDEDIT command to make minor changes based on specific requirements of the job. General notes can also be saved as a block and stored with attributes that can be altered for each usage.

Local Notes

Many drawings, such as sections, contain basically the same notes. Local notes can be placed in the template drawing as a block, as seen in Figure 3.16a. These notes can be thawed, moved into the needed position and edited, as seen in Figure 3.16b. This can greatly reduce drafting time and increase drafting efficiency. The MIRRTEXT command should also be adjusted when you work with sections or other drawings that are created using the MIRROR command. A MIRRTEXT variable setting of 0 will flip a drawing while leaving the text readable.

Title Block Text

A title block similar to Figure 3.17 is a key part of every professional drawing template. It is generally placed along the right side of the sheet. Although each company uses a slightly different design, several elements are found in most title blocks including:

- Office logo and office name, address, phone and fax numbers, web site, and e-mail address
- Legal disclaimers
- Professional seal space
- Titles for information boxes including:
  - Sheet number
  - Date of completion
  - Revision date
  - Client name
  - Contractor name and contact information if known
  - Sheet contents
  - Project number
  - Drawn by
  - Checked by

This text is usually created on a layer such as *ANNO TTLB* (title block). It is the black text in Figure 3.17, and should not be altered. Other text in the title block must be altered for each project. This would include the information required for each of the titles that were just listed. Contents would include information such as:

- **Sheet number**—The page identifier based on the position within the drawing set, such as A 201.
- **Date of completion**—The date of completion is placed in this box.
- **Revision date**—A date is not placed in this box unless the project has been revised after the initial printing of the project. Some offices place a date in this box only when the project is revised after the permit is issued.
- **Client name**—Typically a client name or project title is provided.
- **Sheet contents**—A drawing name such as MAIN FLOOR PLAN.
- **Project number**—The number representing this project is placed in this box.
- **Drawn by**—The CAD technician's initials are placed in this box to identify the drawing originator.
- **Checked by**—The initials of the person who approved the drawing for release are placed in this box.

This text is usually created on a layer such as *ANNO TTLB TITL* and must be altered as each sheet is completed.

Revision Text

An alternative to marking the revision date in the title block is to provide a revision column where changes in the drawing are identified and recorded. This method is used for large or complex projects. The date of each revision is listed in the order that each is made next to a letter or number. The letter is then placed next to the drawing revision, which is surrounded by a revision cloud. Figure 3.18 shows a floor plan that includes a revision.

2x RIDGE BLOCK
2 X 10 RIDGE

SCREENED VENTS @ 10'-0" O.C.±

1/2 STD. GRADE 32/16 PLY. ROOF SHEATH. LAID PERP. TO RAFT. NAIL W/ 8d @ 6" O.C. @ EDGE & 12" O.C. FIELD

300# COMPO. SHINGLES OVER 15# FELT

MED. CEDAR SHAKES OVER 15# FELT W/ 30# X 18" WIDE FELT BTWN. EA. COURSE W/ 10 1/2" EXPOSURE.

STD. ROOF TRUSSES @ 24" O.C. INSTALL AS PER MANUF. SPECS. SUBMIT TRUSS DRAWINGS TO BLDG. DEPT. PRIOR TO ERECTION.

2-2 x 6 TOP PLS, LAP 48" MIN.

2 X 6 STUDS @ 16" O.C.

2 X 6 SILL

EXTERIOR SIDING OVER 1/2" WAFERBOARD & TYVEK

5 1/2" BATTS (R-21) FIBERGLASS INSULATION.

4 X 6 HDR.

2-2 X 12 HDR.

LINE OF INTERIOR FINISH

SOLID BLOCK @ MID POINT FOR ALL WALLS OVER 10'-0" HIGH

FIGURE 3.16a ■ *Local notes can be placed in the template drawing as a block and then edited and moved into a drawing as needed.*

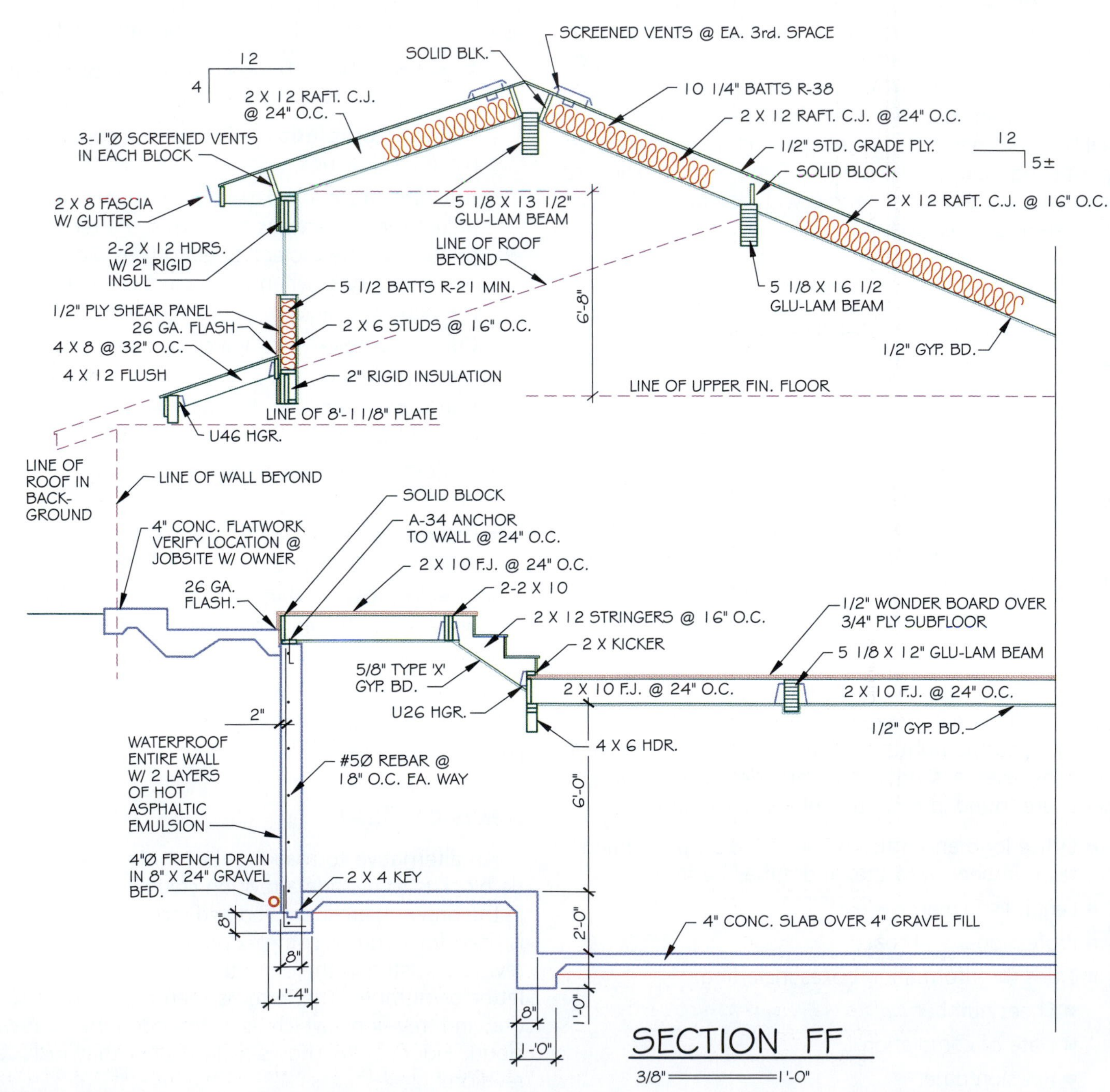

FIGURE 3.16b ■ *The notes from Figure 3.16a were thawed, exploded, edited, and then moved into the needed position.*

FIGURE 3.17 ■ *A title block is a key part of every professional CAD template. Key elements include the company information, key client information, and drawing origination information.*

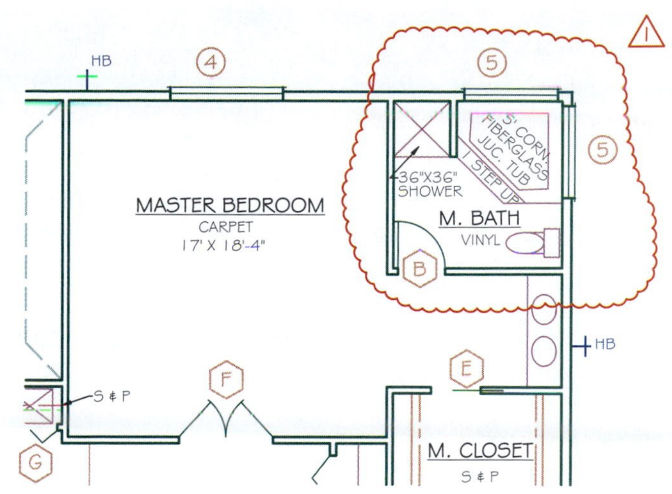

FIGURE 3.18 ■ *Changes can be highlighted on a drawing by using a revision cloud. If more than one series of changes has been made, a schedule containing the date and the revision number should be added to the title block.*

Lettering Height and Scale Factor

Earlier in this chapter, the scale factor required to make line segments the desired size when plotted was considered. Similar factors must be applied to text height to produce the desired 1/8″-high text when plotting is finished. Although text height can be adjusted while preparing to plot, it is best to adjust the text as the text is placed in the drawing. To determine the required text height for a drawing multiply the desired height (1/8'') by the scale factor. The text scale factor is the reciprocal of the drawing scale. For drawings at 1/4'' = 1'-0'', this would be 48. By multiplying the desired height of 1/8'' (.125) × 48 (the scale factor), you see that the text should be 6'' tall. Quarter-inch-high lettering should be 12'' tall. You'll notice that the text scale factor is always twice the line scale factor. Other common text scale heights are shown in Table 3.2.

MANAGING ARCHITECTURAL DIMENSIONING

In addition to the visual representation and the text used to describe a feature, dimensions are needed to describe the size and location of each member of a structure. Figure 3.19 shows a floor plan and the di-

TABLE 3.2 ARCHITECTURAL AND ENGINEERING SCALE FACTORS

ARCHITECTURAL VALUES		ENGINEERING VALUES	
DRAWING SCALE	TEXT SCALE	DRAWING SCALE	TEXT SCALE
1" = 1'-0"	12	1" = 1'-0"	12
3/4" = 1'-0"	16	1" = 10'	120
1/2" = 1'-0"	24	1" = 100'	1200
3/8" = 1'-0"	32	1" = 20'	240
1/4" = 1'-0"	48	1" = 200'	2400
1/16" = 1'-0"	64	1" = 30'	360
1/8" = 1'-0"	96	1" = 40'	480
3/32" = 1'-0"	128	1" = 50'	600
1/16" = 1'-0"	192	1" = 60'	720

Note: Students using AutoCAD 2008 can select the appropriate ANNOTATIVE SCALE factor to display the proper text height.

mensions used to describe the location of structural members. In this chapter you will be introduced to

- Basic principles of dimensioning
- Guidelines for placing dimensions on plan views and on drawings showing vertical relationships
- Future chapters will introduce dimensioning guidelines specific to each type of drawing

Dimensioning Components

Dimensioning features include extension and dimension lines, text, and line terminators. Each type of line was introduced earlier as linetypes were examined. These features should be set as described below using the DIMSTYLE command of AutoCAD. The proper location of extension lines can be seen in Figure 3.20. Two different types of linetypes may be used for extension lines. Solid lines are used to dimension to the exterior face of an object, such as a wall or footing. A centerline is used to dimension to the center of a wood wall or the center of other objects. Figure 3.21 shows examples of each. The exact location of dimension lines will vary with each office, but dimension lines should be placed

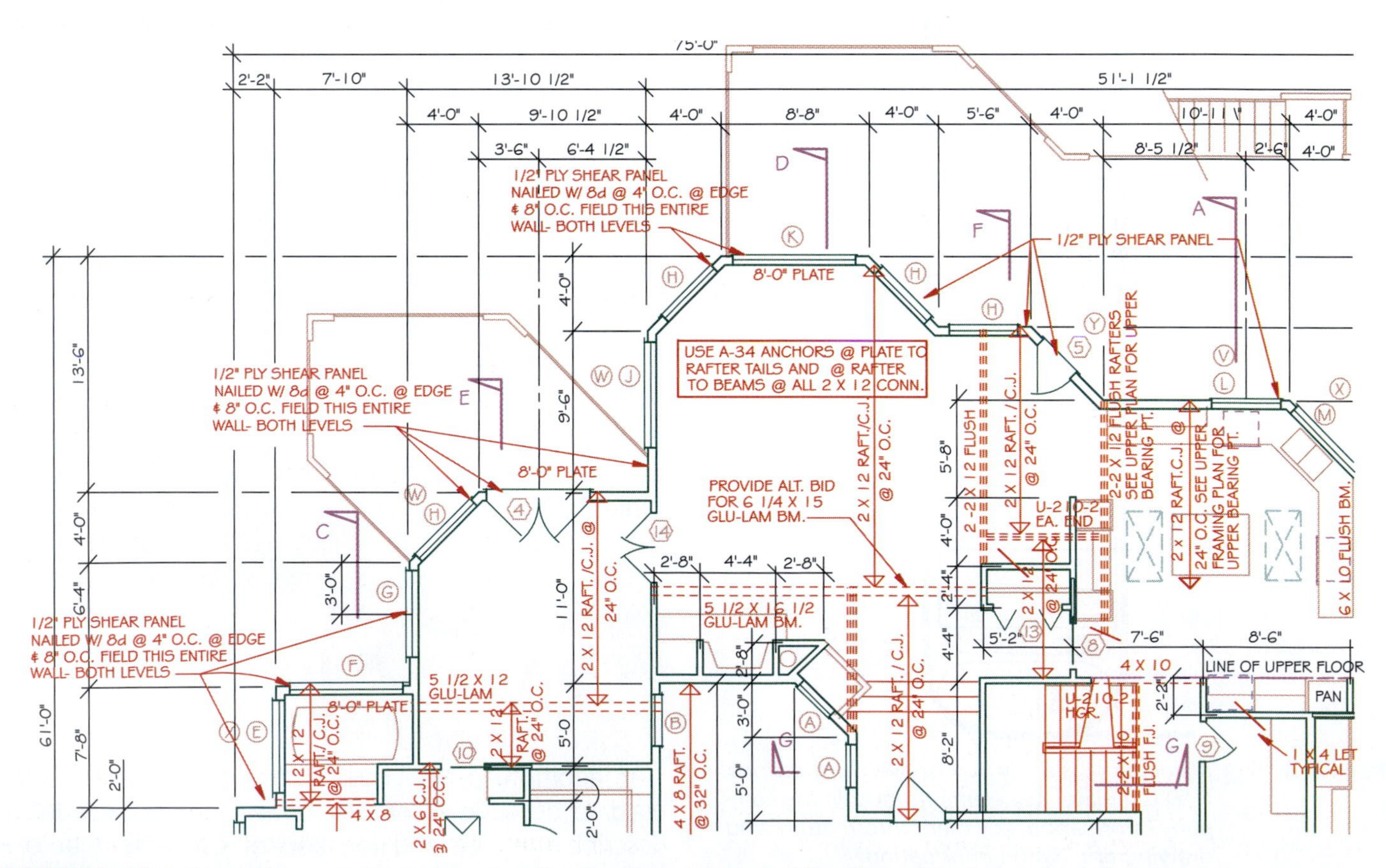

FIGURE 3.19 ■ *Dimensions are needed to describe the size, shape, and location of each member of a structure.*

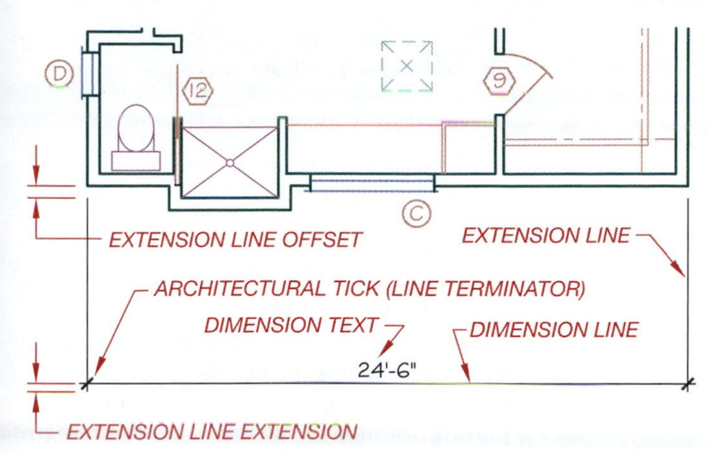

FIGURE 3.20 ■ *Dimensions are composed of the dimension text, a dimension line, extension lines, and line terminators.*

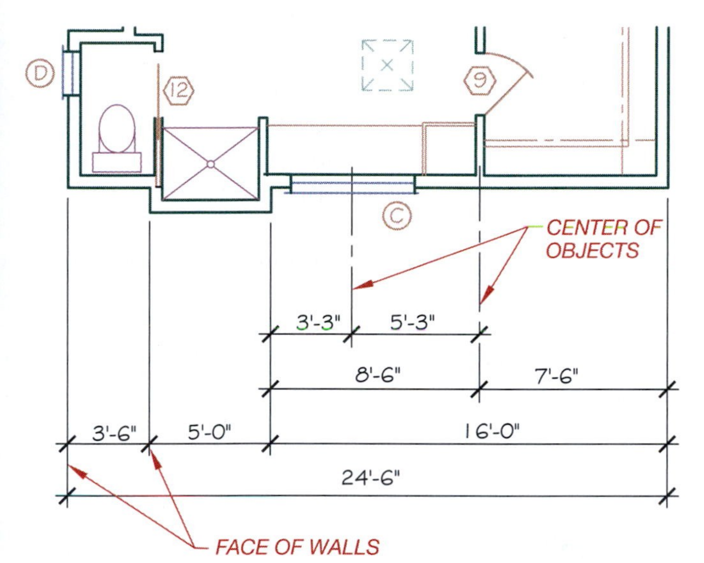

FIGURE 3.21 ■ *Continuous extension lines are used to dimension to the edge of a surface; centerlines are used to dimension to the center of objects.*

in such a way as to leave room for notes, but still close enough to the features being described so that clarity will not be hindered. Guidelines for placement will be discussed later in this chapter.

Dimension Text

Dimension text is expressed as feet and inches using the feet and inch symbols, with a dash placed between each feet and inch numbers. A distance of twelve feet six inches would be expressed as 12'-6''. If a whole number of feet is to be specified such as ten feet, 10'-0'' is preferred over 10'. Most professionals use the zero as a placeholder to keep all dimensions in a similar format. If a dimension less than one foot is to be specified, do not use a zero as a placeholder for feet. Six inches should be expressed as 6'' rather than 0'-6''. When fractions must be represented, side-by-side (1/4'') listings are preferred by most professionals rather than stacked fractions ($\frac{1}{2}$'') because they require less space between lines of text.

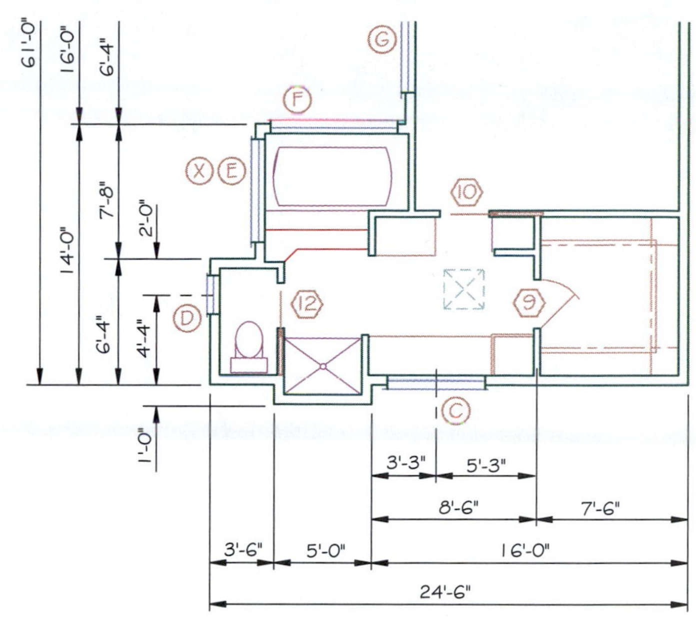

FIGURE 3.22 ■ *Text should be placed above the dimension line and read from the bottom or right side of the drawing.*

The text for dimensions is placed above the dimension line, and centered between the two extension lines. On the left and right sides of the structure, text is placed above the dimension line. Text is rotated so that the text can be read from the right side of the drawing page using what is called ***aligned text.*** Examples of each placement can be seen in Figure 3.22. On objects placed at an angle other than horizontal or vertical, dimension lines and text are placed parallel to the oblique object, as seen in Figure 3.23. Often not enough space is available for the text to be placed between the extension lines when small areas are dimensioned. Although options vary with each office, several alternatives for placing dimensions in small spaces can be seen in Figure 3.24.

Terminators

The default method for terminating dimension lines at an extension line is with a thickened tick mark (architectural tick). Other common options include a dot, a thin tick mark, or an arrow. All four options can be seen in Figure 3.25.

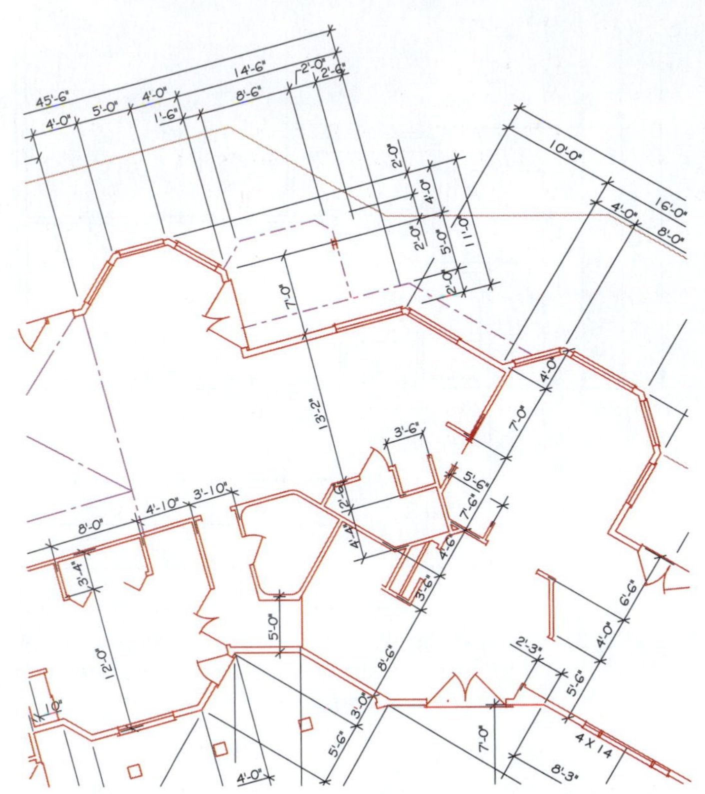

FIGURE 3.23 ■ *When objects drawn on an angle are dimensioned, the dimension lines and text are usually placed parallel to the inclined surface.*

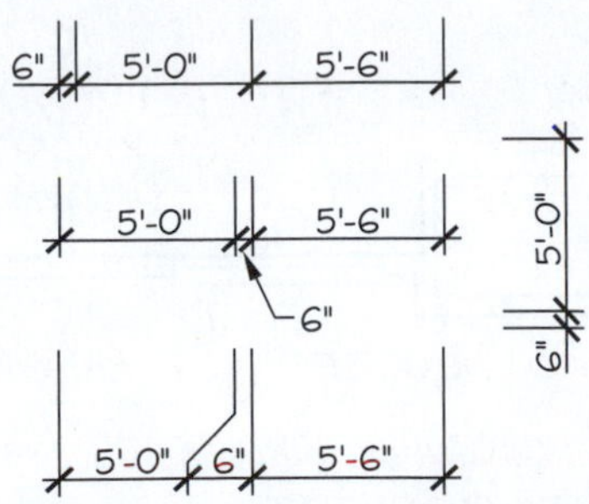

FIGURE 3.24 ■ *Common alternatives for placing dimensions in small spaces include placing the text on the outside of the extension lines, placing the text above or below the dimension line, and "bending" one extension line to provide space for the text.*

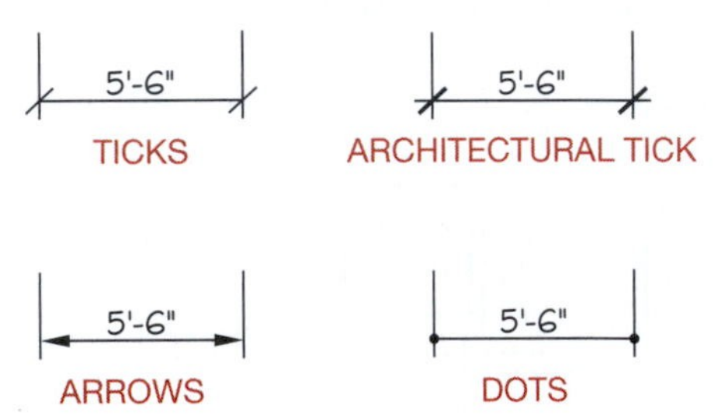

FIGURE 3.25 ■ *Alternatives for terminating dimension lines.*

Dimension Placement

Construction drawings requiring dimensions typically consist of plan views, such as the floor, foundation, and framing plans, and drawings showing vertical relationships, such as exterior and interior elevations, sections, details, and cabinet drawings. Each type of drawing has its own set of challenges to be overcome in placing drawings. No matter the drawing type, dimensions should be placed on a layer titled * *ANNO DIMS*. The * represents the letter of the proper originator such as *A* or *S*, *ANNO* represents annotation, and *DIMS* represents dimensions.

Plan Views

Consideration about dimensions in plan view can be divided into the areas of interior and exterior dimensions. Whenever possible, dimensions should be placed outside the drawing area.

Exterior Dimensions

Exterior dimensions for wood- and steel-frame structures are expressed from the outside edge to the outside edge of another exterior wall. These dimensions are grouped based on what information they provide. Exterior dimensions are usually placed using the following groupings, starting from the outside and working in toward the residence:

- **Overall dimension—**Placed on all sides of the residence from outer edge to outer edge.
- **Major jogs—**Placed from outer edge to outer edge of jogs in the structure so that the dimensions add up to the overall dimension.
- **Wall to wall—**Placed to extend from a known location, such as the edge of a major jog, to one or more unknown locations (interior walls), and then ending at a known location (the opposite end of a major jog). In describing the location of interior walls, the extension line extends to the center of the wall.
- **Wall to openings—**Placed to extend from a known wall location to one or more openings and ending at a known wall location so that the sum of these dimensions adds up to the sum of a wall-to-wall dimension.

Each of these dimensions can be seen in Figure 3.26a. Additional examples will be given in chapters related to specific types of drawings.

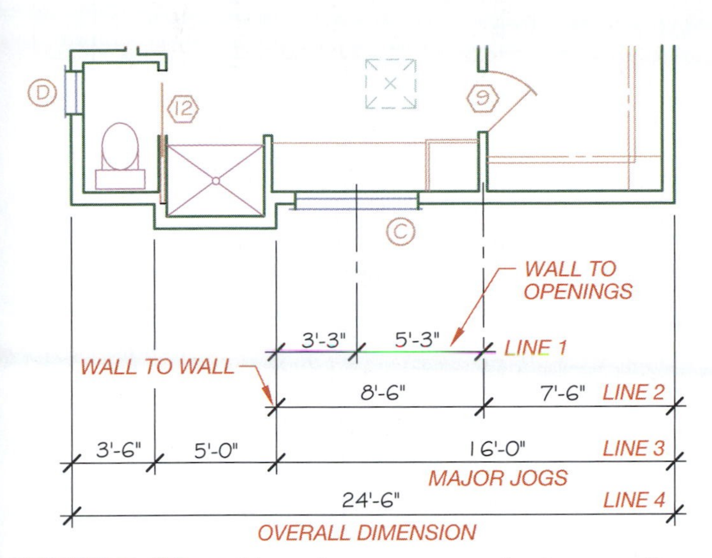

FIGURE 3. 26a ■ *Dimensions are grouped using overall, major jogs, wall to wall, and wall to openings. Dimensions are placed so that dimensions in line 4 will add up to the corresponding sum in line 3. The sum of the dimensions in line 3 will add up to the corresponding sum in line 2, and the sum of line 2 will add up to the dimension in line 1.*

Most offices start by placing an overall dimension on each side of the structure that is approximately 2'' from the exterior wall. Moving inward, with approximately 1/2'' between lines, are dimension lines used to describe major jogs in exterior walls, the distance from wall to wall, and the distance from wall to window or door to wall. Examples of the placement of these four dimension lines can be seen in Figure 3.26b.

Two different systems are used to represent the dimensions between exterior and interior walls. Architects tend to represent the distance from edge to edge of walls as seen in Figure 3.27. Engineering firms tend to represent the distance from exterior edge to center of interior wood walls using methods shown in Figure 3.28. Concrete walls are dimensioned to the edge by both disciplines.

Interior Dimensions

The three main considerations in placing interior dimensions are clarity, grouping, and coordination. Dimension lines and text must be placed so that they can be read easily and that neither interferes with other information that must be placed on a drawing. Information should also be grouped together as seen in Figure 3.29, so that construction workers can find dimensions

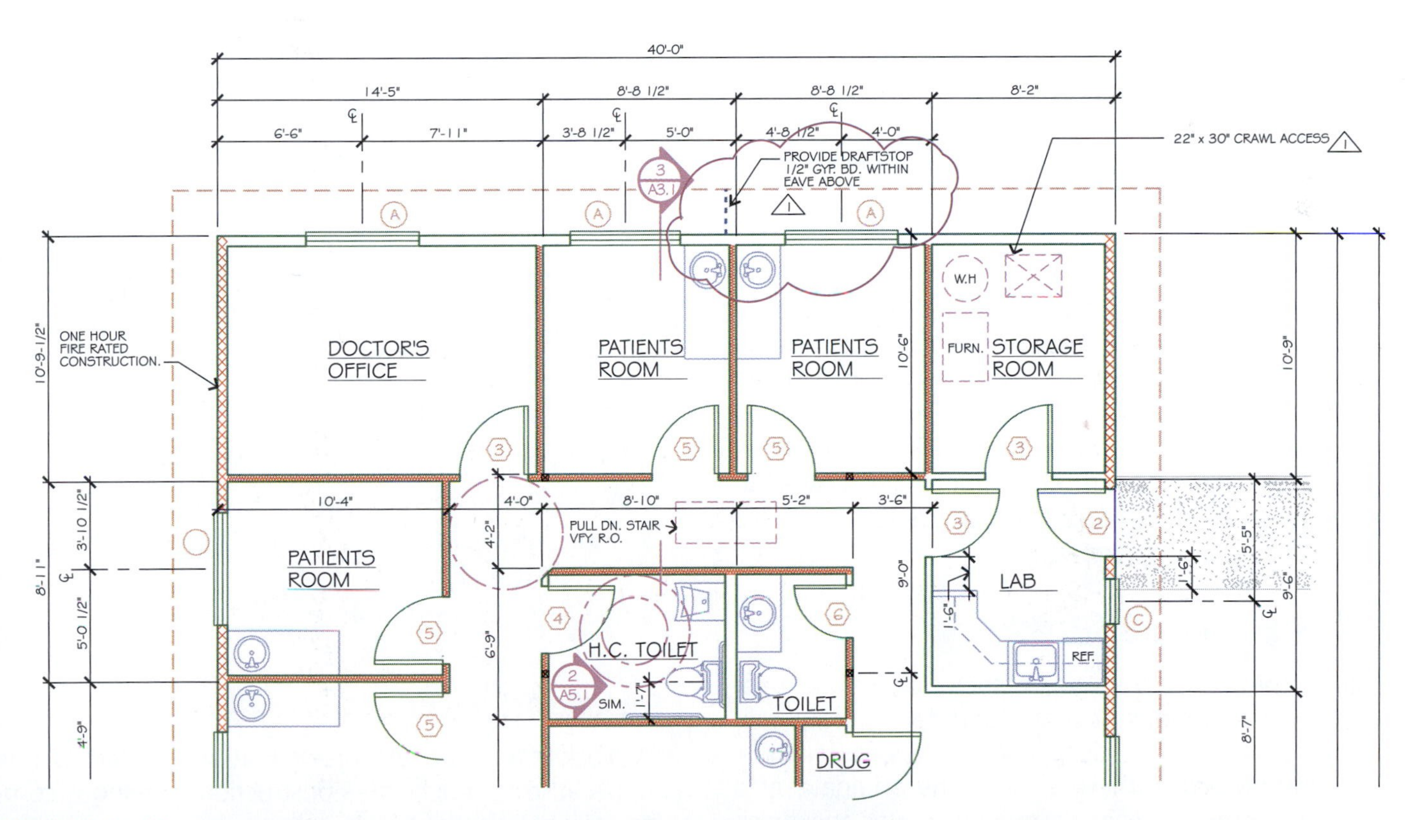

FIGURE 3.26b ■ *Placement of the overall dimension, dimensions to locate major jogs, wall-to-wall dimensions, and wall-to-opening dimensions on a floor plan.* Courtesy Scott R. Beck, architect.

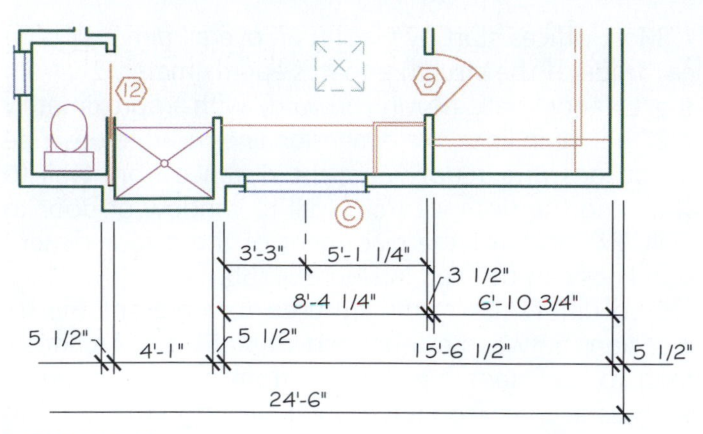

FIGURE 3.27 ■ *Some architectural firms dimension from edge to edge of interior walls.*

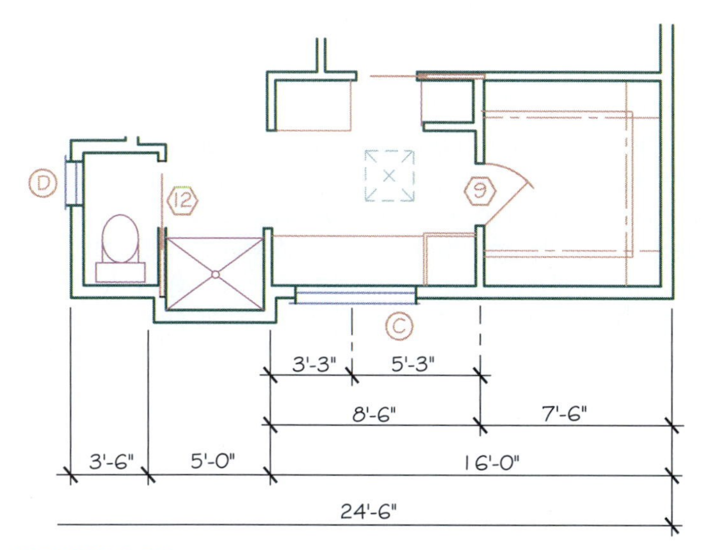

FIGURE 3.28 ■ *Engineering firms often dimension to the center of interior walls.*

easily. More importantly, interior dimensions must be placed so that they match corresponding exterior dimensions. Notice, in Figure 3.29, that the interior dimensions have been placed from center to center of the interior walls. The interior dimensions of 3'-8'' + 5'-2'' + 3'-2'' plus 4'' from the center of the 4'' wall to the exterior edge of the 6'' wall match the total distance of 12'-4'' listed on the exterior.

Vertical Dimensions

Unlike the plan views, which show horizontal relationships, the elevations, sections, and details require dimensions that show vertical relationships, as seen in Figure 3.30. Typically, these dimensions originate at a line that represents a specific point, such as the finish grade, a finish floor elevation, or a plate height.

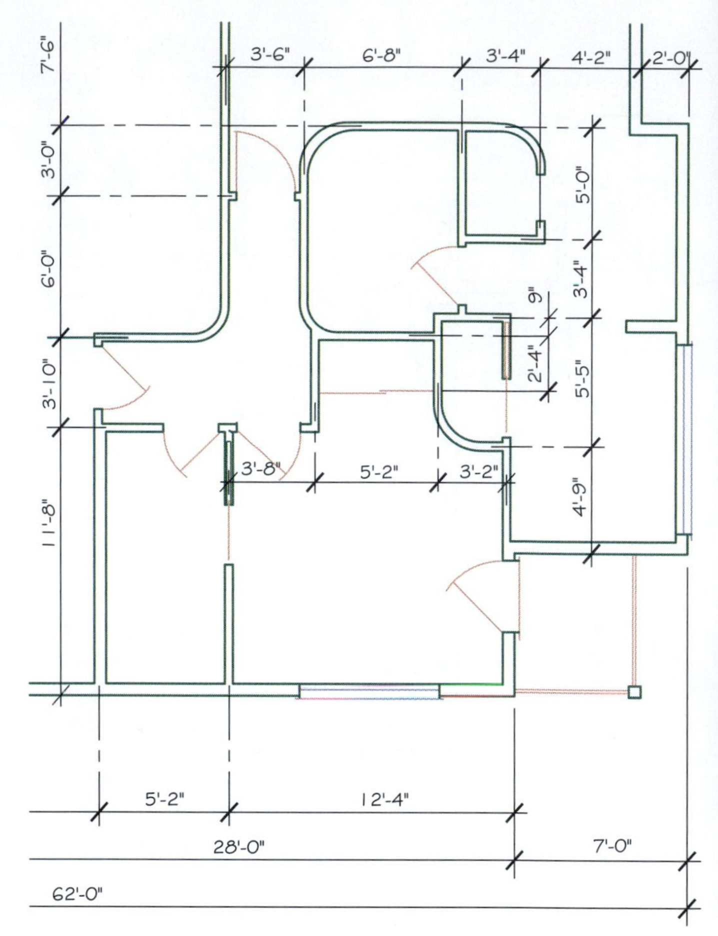

FIGURE 3.29 ■ *Dimensions placed on the inside of a drawing should be grouped together and be clearly visible.*

Managing Drawings

Throughout this chapter you've explored how to control information in one drawing. To advance beyond the jobs typically assigned to a beginning CAD technician, you also need to be able to access and use information from other drawings. This would include:

- Using blocks and wblocks from other drawings
- Managing drawings at multiple scales
- Working in a multidocument environment
- Using the CUT, COPY, and PASTE commands to edit drawings
- Building drawings with DesignCenter

Blocks and Wblocks

A block is a group of objects that are treated as one unit. Because so much of a construction drawing is comprised of symbols, they can be drawn once and stored for future use. Storing blocks in a template drawing places

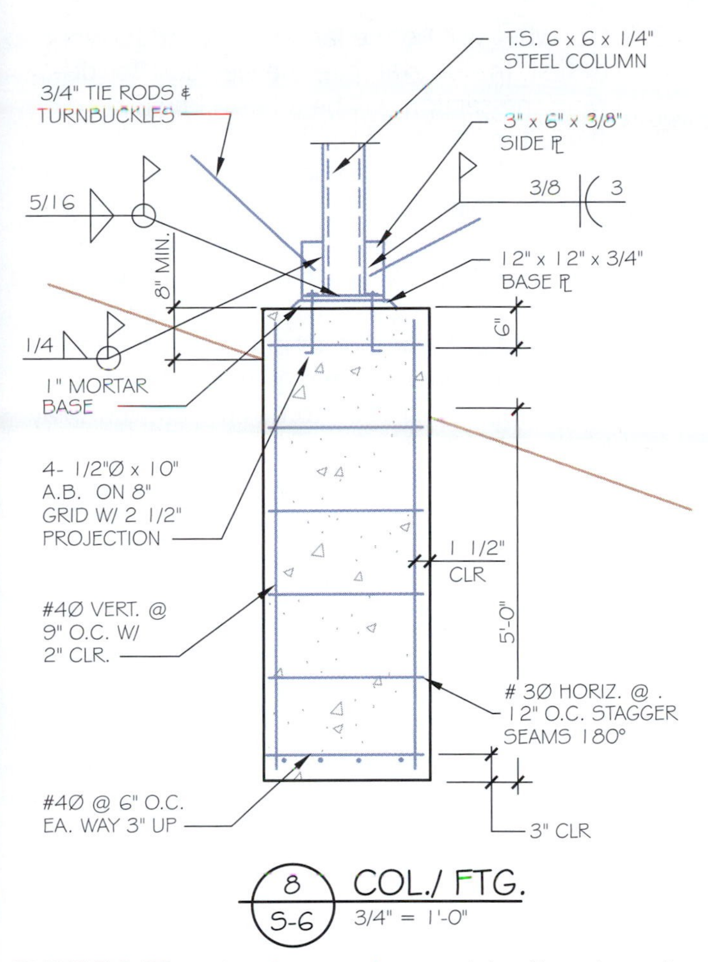

FIGURE 3.30 ■ *Elevations, sections, and details each require dimensions to show vertical and horizontal relationships.* Courtesy Residential Designs.

the symbol where it is frequently used. Common fixtures such as toilets and sinks can be stored on the *A-FLOR-PLMB* layer, frozen, and then thawed as the plan is drawn.

A wblock is similar to a block but it has a wider range of uses. Blocks are stored in a specific drawing but can be moved from drawing to drawing with DesignCenter. Wblocks can be stored as a separate .dwg file and can be dragged into another drawing using the INSERT command. The entire template drawing can be stored as a wblock and inserted into another drawing file, or another wblock can be inserted into the template and then stored with a new file name. Figure 3.31 shows common symbols that are used throughout construction drawings that can be stored as wblocks.

Managing Drawings at Multiple Scales

One of the most valuable computer skills needed to complete drawing projects is the ability to combine drawings to be plotted at multiple scales. Because of its importance, a brief review is provided here. Drawing sheets can be assembled for plotting using layouts and viewports. Each time you enter a layout, you're working with a floating viewport. The viewport allows you to look through the paper and see the drawing created in model space and to display the drawing in the layout for plotting. An unlimited number of viewports can be created in paper space, but only sixty-four viewports can be visible at once. Because floating viewports are considered objects, drawing objects displayed in the viewport cannot be edited. For objects to be edited within the viewport, model space must be restored. You can toggle between model space and a floating viewport by choosing either the MODEL or

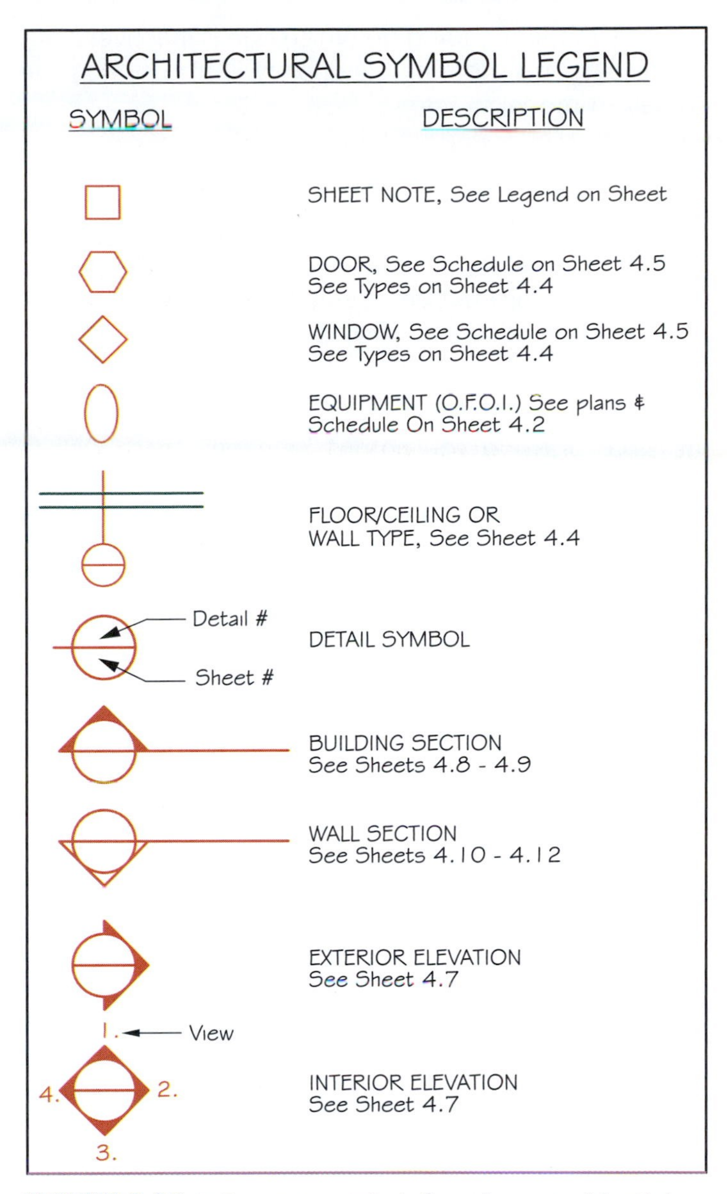

FIGURE 3.31 ■ *Common symbols found on a residential floor plan can be saved as blocks, wblocks, or as part of a drawing template.*

LAYOUT tab, depending on which one is inactive. Double clicking in a floating viewport will also switch the display to model space.

Note:

Remember that while you are editing in model space in a layout, you risk changing the drawing setup. This should be used only by experienced AutoCAD users or when in the initial setup of the view inside of the viewport.

Most professionals work in model space and arrange the final drawing for output in paper space. While working in a layout with the viewport in paper space, any material added to the drawing will be added to the layout but not shown in the model display. This will prove useful as the final plotting layout is constructed.

Creating a Single Floating Viewport

Floating viewports provide an excellent means of assembling multiscaled drawings for plotting. Figure 3.32 shows a sheet of details assembled for plotting using multiple viewports. Construction projects typically comprise drawings that are drawn at a variety of scales. These details range in scale from 1/2'' = 1'-0'' to 1'' = 1'-0''. This sheet of details cannot be easily assembled without the use of multiple viewports. The process for creating multiple viewports will be similar to that for the creation of one viewport. The following example demonstrates the steps used to prepare the section and three details shown in Figure 3.32 for plotting. The section will be displayed at a scale of 1/4'' = 1'-0'', two of the details will be displayed at a scale of 3/4'' = 1'-0'', and the stair section will be displayed at a scale of 1/2'' = 1'-0''. The following sections will walk you through the process of creating multiple viewports in a template and inserting multiple drawings. For this discussion, the INSERT command will be used.

Displaying Model Space Objects in the Viewport

The easiest way to display multiple drawings at multiple scales is by inserting each of the drawings to be plotted into the template in model space. With the details arranged in one drawing, similar to Figure 3.32, switch to the desired layout. Use the following steps to display the drawings in multiple viewports:

1. In the layout, use the PAPER button on the status bar to toggle the drawing to model space, and use the ZOOM ALL option to enlarge the area of model space to be displayed in the existing viewport. Each of the four drawings will be displayed, but each will be at an unknown scale.
2. Select what will be the largest drawing to work with first. In our example, the viewport for displaying the section will be adjusted first.
3. Activate the Viewport toolbar.
4. Set the scale of the viewport to the appropriate value for displaying the section using the viewport scale control menu on the viewport toolbar. For this example a scale of 1/4'' = 1'-0'' was used.
5. Click the MODEL button on the status bar to return back to paper space in your current layout.
6. New viewports will need to be created to display the other details at different scale factors. Refer to the section "Creating Additional Viewports," below. Once all of the drawings to be plotted have been displayed in a viewport with the proper scale factor, you are ready to plot.

Adjusting the Existing Viewport

If the entire drawing can't be seen once the desired drawing is inserted into the viewport, the size of the viewport can be altered. To alter the viewport size, move the cursor to touch the viewport and activate the viewport grips. Select one of the grips to make it hot, and then use the hot grip to drag the window to the desired size. The section has now been inserted into the template and is ready to be plotted at a scale of 1/4'' = 1'-0''. The display would resemble Figure 3.33.

Creating Additional Viewports

Use the following steps to prepare additional viewports:

1. Set the current layer to viewport.
2. Select "single viewport" from the Viewport toolbar.
3. Select the corners for the new viewport. Although you should try to size the viewport accurately, it can be stretched to enlarge or reduce its size once the scale has been set.

The results of this command can be seen in Figure 3.34.

Altering the Second Viewport

As a new viewport is created, the display from the existing viewport will be displayed in the new viewport as well. Use the following steps to alter the display.

1. Click the PAPER button on the status bar to toggle the drawings to model space.

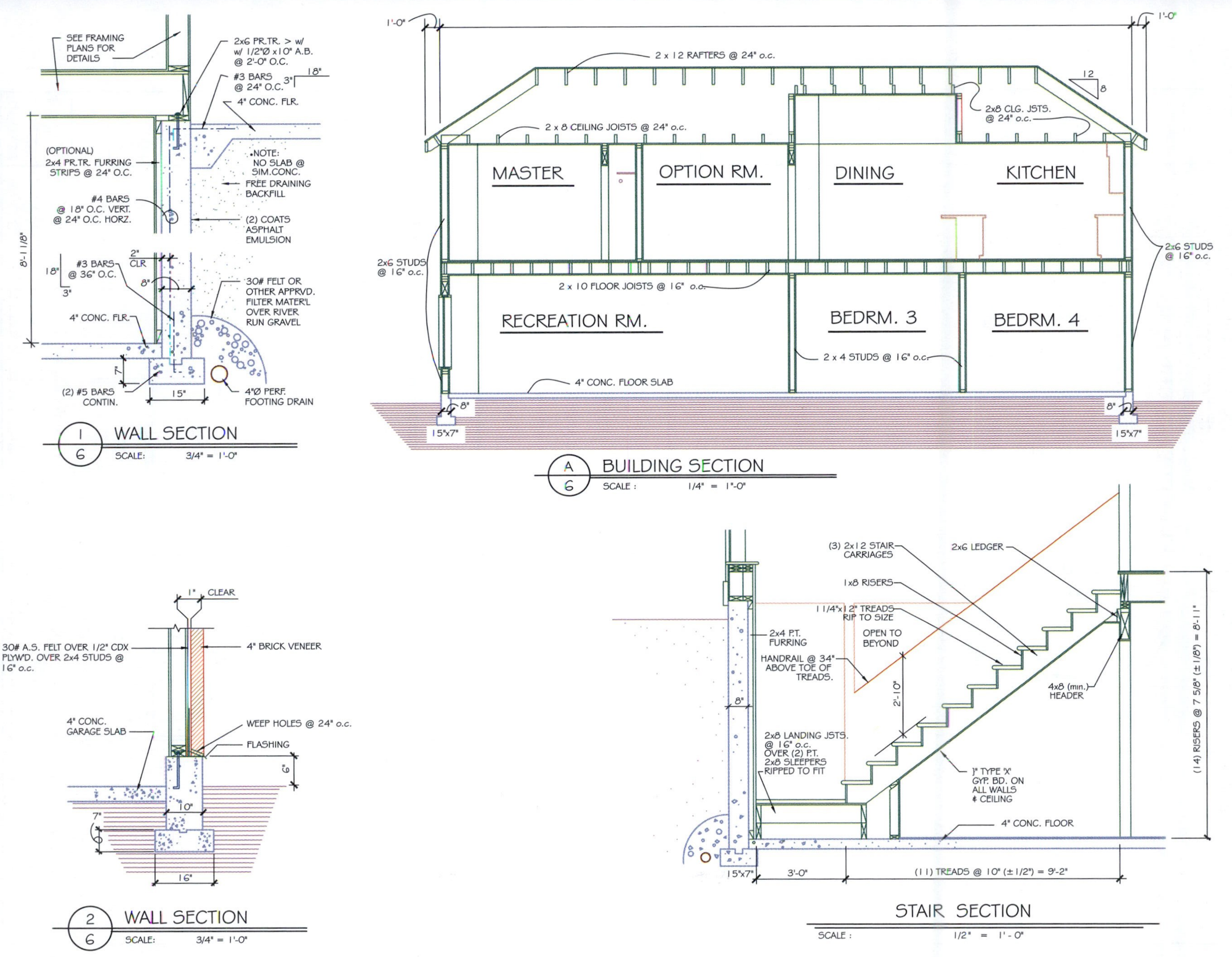

FIGURE 3.32 ■ *The appropriate scale for plotting can be set as the drawing is inserted in the viewport. The size of the viewport can be adjusted if the entire drawing can't be seen.*

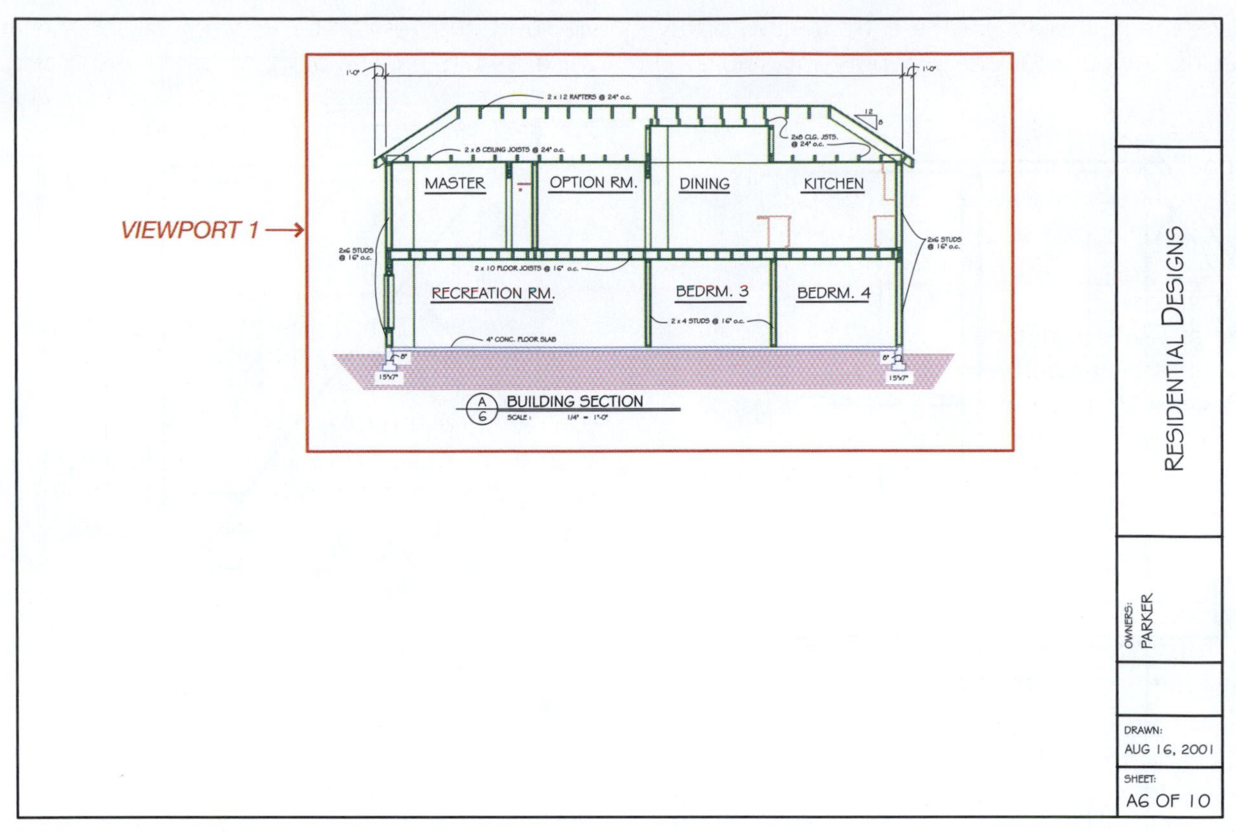

FIGURE 3.33 ■ *The appropriate scale for plotting can be set as the drawing is inserted in the viewport. The size of the viewport can be adjusted if the entire drawing cannot be seen.*

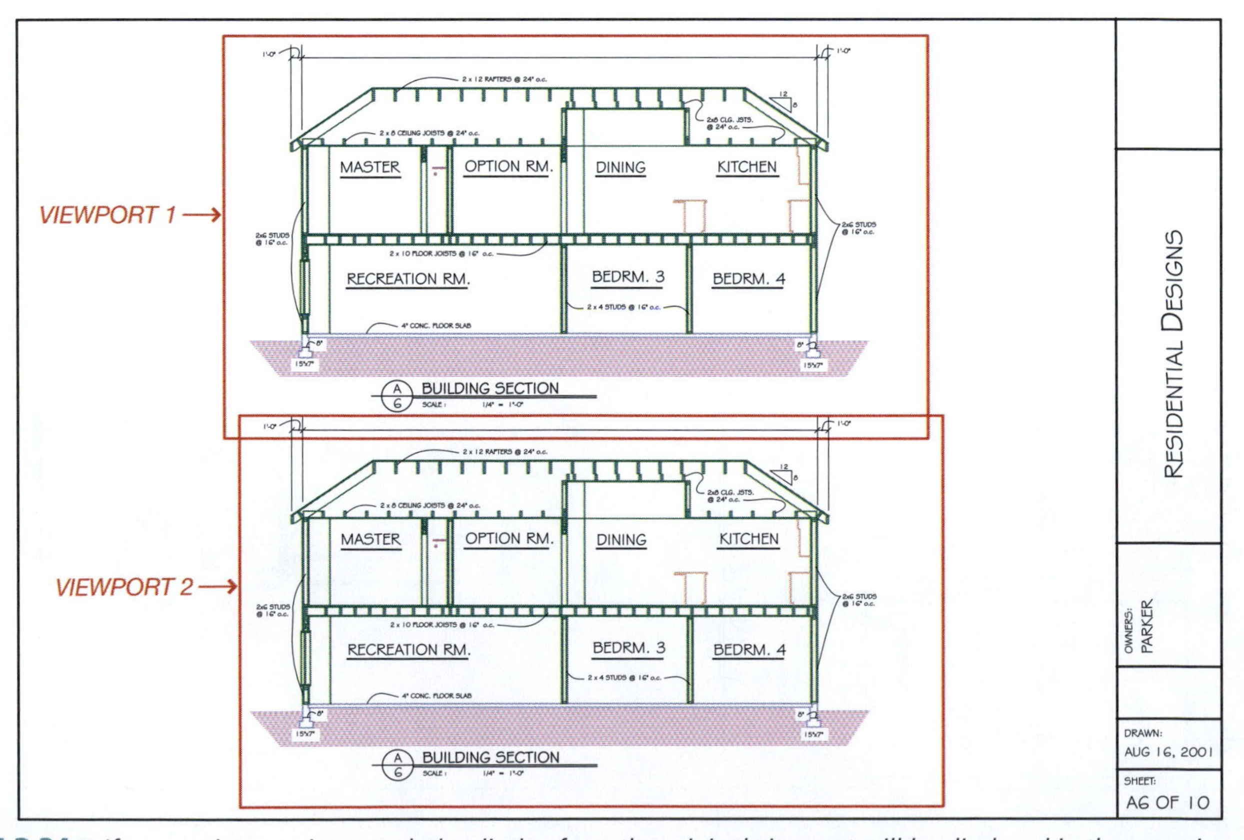

FIGURE 3.34 ■ *If a new viewport is created, the display from the original viewport will be displayed in the new viewport.*

2. Make the new viewport the current viewport by placing the cursor in the viewport and single clicking.
3. Use the ZOOM All option to display all of the model contents in the second viewport.
4. Set the viewport scale to the appropriate value for displaying the section. For this example, a scale of 3/4'' = 1'-0'' will be used to display the wall section.
5. Use the PAN command to center the wall section in the viewport.
6. Click the MODEL button on the status bar to toggle the drawings to paper space.
7. Select the edge of the viewport to display the viewport grips.
8. Make a grip hot and shrink the viewport so that only the wall section is shown.
9. Use the MOVE command to move the viewport to the desired position in the template.
10. Save the drawing for plotting.

The drawing should now resemble Figure 3.35. Once the viewport for the wall section has been adjusted, additional viewports can be created. Once all of the desired details have been added to the layout, the viewport outlines can be frozen. Freezing the viewports will eliminate the line of the viewport being produced as the drawing is reproduced. The finished drawing with the viewports frozen will resemble Figure 3.36.

Working in a Multidocument Environment

An alternative to using a template drawing to store frequently used styles, settings, and objects is to move objects and information between drawing files. Just as you can have several programs open on your desktop at once and rapidly switch from one program to another, AutoCAD allows you to have several drawings open at the same time. Objects or drawing properties can be moved between drawings using CUT, COPY, and PASTE; object drag and drop; the property painter; and concurrent command execution. Important considerations to remember when having multiple drawings open include:

- Any one of the open drawings can be made active.
- Any one of the open drawings can be maximized or minimized to ease viewing, or all can be left

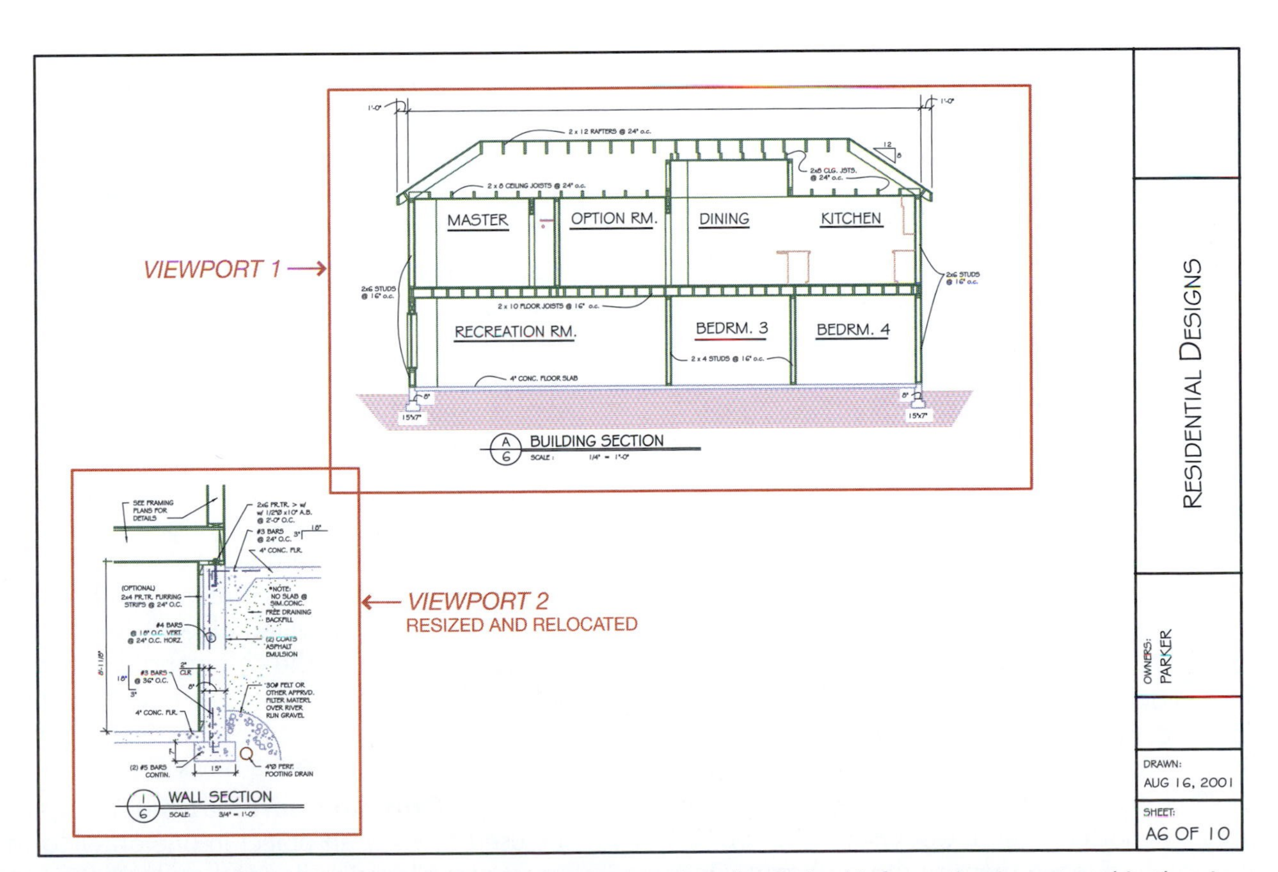

FIGURE 3.35 ■ *Use the PAN command to move the drawings so that the desired information is centered in the viewport. Once it's centered, adjust the size of the viewport so that only the desired material is displayed.*

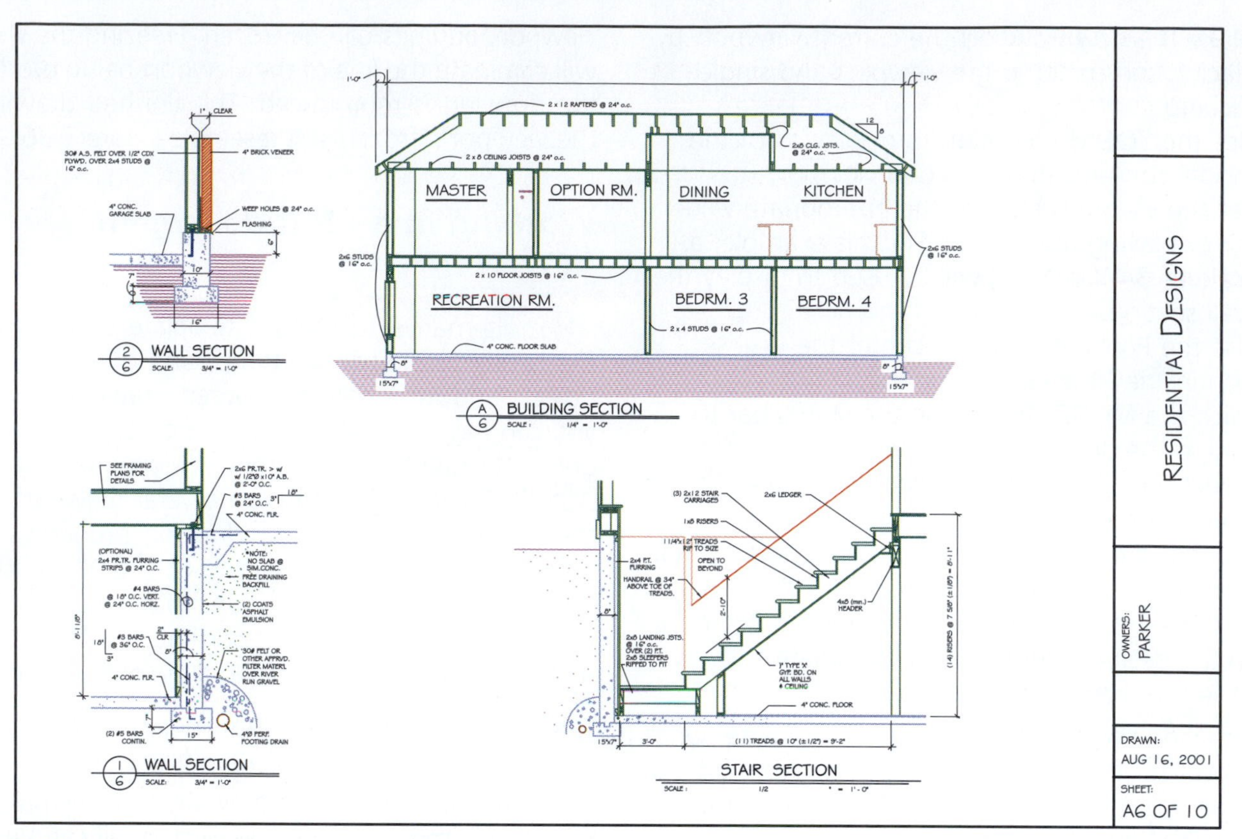

FIGURE 3.36 ■ *The assembled drawing components ready for plotting with each viewport frozen.*

open and you can switch from drawing to drawing as needed.

- You can start a command in one drawing, switch to another drawing and perform a different command, and then return to the original drawing and complete the command in progress. As you return to the original drawing, single-click anywhere in the drawing area and the original command will be continued.
- Each drawing is saved as a separate drawing file, independent of the other open files. You can close a file and not affect the contents of the remaining open files.

Using the Cut, Copy, and Paste Commands

AutoCAD allows drawings, objects, and properties to be transferred directly from one drawing to another using the Cut, Copy, and Paste commands. Each command can be performed using the Edit menu, by keyboard, or by shortcut menu. Cut will remove objects from a drawing and move them to the clipboard. Once cut, the object can be placed in a new drawing with Paste. Copy allows an object to be reproduced in another location. The object to be copied is reproduced in its new location with the Paste command. As the object is copied to the new drawing, so are any new layers, linetypes, or other properties associated the objects.

- **Copy With Base Point—**This option of the shortcut menu is similar to the Copy option. With this option, before you select objects to copy, you'll be prompted to provide a base point. This option works well when objects need to be inserted accurately.
- **Paste as Block—**This option can be used to paste a block into a different drawing. It functions similarly to the Paste command. Select the objects to be copied in the original drawing and select the Copy command. Selected objects do not need to be a block because the command will turn them into a block. Next activate the new drawing and right-click to select the Paste as Block option. The copied objects are now a block.
- **Paste to Original Coordinates—**This option can be used to copy an object in one drawing to the exact same location in another drawing. The command could be useful in copying an object from

one apartment unit to the same location in another unit. The command is active only when the clipboard contains AutoCAD data from a drawing other than the current drawing. Selecting an object to be copied starts the command sequence. The sequence is completed by making another drawing active and then selecting the Paste to Original Coordinates option.

Building Drawings Using DesignCenter

DesignCenter can be used to locate, organize, and customize drawing information. The DesignCenter can be used to transfer data such as blocks, layers, and external references directly from one drawing to another without using the clipboard. Drawing content can be moved from other open files in AutoCAD or from files stored on your machine, on a network drive, or from an Internet site. Access the DesignCenter by selecting the DesignCenter icon from the Standard toolbar.

Adding Content to Drawings Using the DesignCenter

One of the best uses for the DesignCenter is for adding content from the palette into a new or existing drawing. Information can be selected from the palette or from Find and dragged directly into a drawing without opening the drawing containing the original. DesignCenter can also be used to attach external referenced drawings and to copy layers between drawings.

- **Inserting Blocks**—The DesignCenter provides two methods for inserting blocks into a drawing. Blocks can be inserted using the default scale and rotation of the block, or the block parameters can be altered as the block is dragged into the new drawing.
- **Attaching Referenced Drawings**—The DesignCenter can be used to attach a referenced drawing using similar steps used to attach a block and provide the parameters.
- **Working with Layers**—The DesignCenter can be used to copy layers from one drawing to another. Typically, a template drawing containing stock layers can be used in creating new drawing files. This option of the DesignCenter is useful when you're working with drawings created by a consulting firm and the drawing needs to conform to office standards. DesignCenter can be used to drag layers from a template or any other drawing into the new drawing and ensure drawing consistency.
- **Accessing Favorite Contents**—As with most programs, the DesignCenter allows frequently used sources to be stored for easy retrieval.

COMMON ALTERNATIVES TO AUTOCAD

Some professionals in the field of architecture create their drawings using AutoCAD straight out of the box. It's a nice program, but it was really created for general drafting. Although the program keeps adding features to aid the architectural profession, it still has many limitations. Many professionals use AutoCAD with the addition of specialized menus and LISP routines to meet their drawings needs. This might include creating codes that will automatically insert a specific block by one keystroke rather than using several steps with the INSERT command. Although these types of improvements increase productivity, two programs are available to further increase productivity. These tools are AutoCAD Architecture and Revit Architecture.

Exploring AutoCAD® Architecture

AutoCAD® Architecture (formerly Autodesk® Architectural Desktop) allows you not only to use traditional drafting methods used with AutoCAD but also to build a model that can be used to automate tedious coordination tasks. With AutoCAD, if you draw a wall, and insert a window block, the wall and the window are separate, unrelated objects. With AutoCAD® Architecture (AA), the window is associated with the wall, so that if the wall thickness is altered, the thickness of the window is automatically updated. This association saves editing time and eliminates the chance of errors that might occur if you had to make these changes using traditional editing methods. Automated documentation routines in AA allow standard drawings such as elevations, sections, and schedules to be generated automatically from information placed on the floor plan.

In addition to key drawing features that link drawings together, AA increases the accuracy of information in the construction set. Desktop allows common architectural items such as elevation, section markers, and other callout tools to be coordinated throughout the project. Similar to the features of AutoCAD's Sheet Set Manager, AA coordinates the annotation for elevation and section symbols, schedules, sheet numbers, sheet indexes, callouts, and general annotation, greatly increasing your productivity. AA also links information using the Microsoft® Access database, allowing annotation throughout the construction documents to be dynamically updated.

Exploring Revit® Architecture

Revit® Architecture (formerly Autodesk® Revit® Building) is a building design and documentation system produced by Autodesk that works independently of AutoCAD. The program is specifically designed for building information modeling and is similar in many ways to AutoCAD® Architecture. The key difference is that Revit® Architecture is a stand-alone product. Its parametric change technology lets you change anything, anytime, anywhere, while the program coordinates the change everywhere. Information about a feature is entered once in the project database, and is then available for use on other drawings. If the height of a wall is specified on the floor plan, this information is used to generate the same wall in the elevations, sections, and 3D models. All your design information is coordinated, consistent, and complete. This ability to maintain coordination between design changes and construction documents eliminates mistakes of the sort caused by human error with manual editing methods of traditional CAD drafting.

CHAPTER 3

Additional Readings

The following websites can be used as a resource to help you keep current with changes in floor plan–related materials.

ADDRESS	COMPANY OR ORGANIZATION
www.aia.org	American Institute of Architects
www.asme.org	ASME Technical Journal List
www.autodesk.com/archdesktop	Autodesk Architectural Desktop product information
www.autodesk.com/revit	Autodesk Revit product information
www.csinet.org	Construction Specifications Institute
www.metric.org	U.S. Metric Association
www.nationalcadstandard.org	United States National Cad Standards
www.nibs.org	National Institute of Building Sciences

CHAPTER 3

Computer-Aided Design And Drafting Test

DIRECTIONS

Answer the following questions with short complete statements. Type your answers using a word processor or neatly print the answers on lined paper.

1. Type your name, the chapter number, and the date at the top of the sheet.
2. Type the question number and provide the answer in the form of a statement that includes part of the question. You do not need to write out the entire question.

QUESTIONS

3.1. Define aligned dimensioning text.
3.2. Explain how dimension lines should be placed relative to the object being described.
3.3. Show an example of how dimension numerals less than 1' are lettered.
3.4. Show an example of how dimension numerals greater than 12'' are lettered.
3.5. How are the overall dimensions on frame construction placed?
3.6. Describe the dimensional information provided in each line of dimensions when four lines of dimensions are provided on a floor plan. Describe the information associated with each line by starting at the outside and working in.
3.7. Describe and give an example of a specific note.
3.8. Describe and give an example of a general note.
3.9. What is the advantage of using folders when working on the hard disk?
3.10. List a benefit in storing related drawings in a single drawing file.
3.11. List six items typically included in a title block.
3.12. What are the disadvantages in storing related drawings in a single drawing file?
3.13. Define and list five advantages of using a template drawing.
3.14. List the two types of measurement units typically associated with construction drawings.
3.15. How can external referencing be used within a set of construction drawings?
3.16. What are the benefits of the PURGE command, and when can it be used?
3.17. List the major groups of layer names recommended by the AIA.
3.18. Describe guidelines for naming minor layer groups.
3.19. Describe the difference between a block and a wblock.
3.20. List options where drawings can be saved and explain whatever advantages or disadvantages each may have.
3.21. List methods of naming model files.
3.22. Enter AutoCAD and research the difference between PARTIALOAD and PARTIALOPEN.
3.23. Other than plotting, what advantages do layouts provide?
3.24. What is the difference between a layout and a viewport?
3.25. You've created a second viewport, but the contents of the first viewport are displayed. How can this be corrected?

PROBLEMS

3.1 Create a drawing template for plotting a drawing using a scale of 1/4'' = 1'-0'' on D-size material. Create a title block using your school name as the company name, along with other recommended contents. Use a 1/2'' margin on the top, right, and bottom of the page. Use a 1 1/2'' margin on the left side. Establish measurement units and angle measurements suitable for an architectural drawing.

3.2 Create a template drawing for plot plans drawn at a scale of 1'' = 10'-0'' using B-size material. Create a title block using your school name as the company name along with other recommended contents. Use a 1/2'' margin on the top, right, and bottom of the page. Use a 1 1/2'' margin on the left side. Establish measurement units and angle measurements suitable for an architectural drawing.

3.3 Create a template drawing for architectural drawings using metric features. Assume that a scale of 1:50 will be used. Create a title block using your school name as the company name along with other recommended contents. Use a 1/2'' margin on the top, right, and bottom of the page. Use a 1 1/2'' margin on the left side. Establish measurement units and angle measurements suitable for an architectural drawing.

3.4 Using pencil on 8 1/2'' × 11'' paper, sketch an example of a floor plan for a wood-framed structure and show the required dimensions to locate at least one interior partition, one exterior door, and one window. See Chapter 10 if you are unfamiliar with door and window symbols.

3.5 Using pencil on 8 1/2'' × 11'' paper, sketch the portion of the home shown in Figure 3.14. Place the required dimensions that will be needed. Represent the actual dimensions with X's so that no actual distances will be placed.

4 CHAPTER

Building Codes and Interior Design

Building codes are intended to protect the public by establishing minimum standards of safety. Although landowners often assume that they can build whatever they want because they own the property, building codes are intended to protect others from such shortsighted individuals. Consider the stories you've seen in the national news about structures that have been damaged by high winds, raging floodwaters, hurricanes, tornadoes, mudslides, earthquakes, and fire. Hundreds of other structures, which don't make the national news, become uninhabitable because of inadequate foundation design, failed members from poor design, mold, rot, or termite infestation. Building codes are designed to protect consumers by providing minimum guidelines for the construction and inspection of a structure to prevent fire, structural collapse, and general deterioration. Many other aspects of building construction—such as electrical wiring, heating equipment, and sanitary facilities—represent a potential hazard to occupants. Building codes are enforced as a safeguard against these risks.

The regulation of buildings can be traced through recorded history for over 4000 years. In early America, George Washington and Thomas Jefferson encouraged building regulations as minimum standards for health and safety. Building codes are now used throughout most of the United States to regulate issues related to fire, structural stability, health, security, and energy conservation. Architects, engineers, interior designers, and contractors also rely on building codes to help regulate the use of new materials and technology. Each major building code provides testing facilities for new materials intended to develop safe, cost-effective, timely construction methods. In addition to the testing facilities of each major code, several major material suppliers have their own testing facilities.

Although most areas of the country understand the benefits of regulating construction, each city, county, and state has the authority to write its own building codes. Some areas choose to have no building codes. Some areas have codes, but compliance is voluntary. Other areas have building codes but exempt residential construction. This variation in attitudes has produced over 2000 different codes in the United States. Trying to design a structure can be quite challenging when one is faced with so many possibilities.

NATIONAL CODES

Most states have adopted one of four national building codes:

The ***2006 IRC International Residential Code***—written by the International Code Council (ICC)

The ***1999 BOCA***—Building Officials and Code Administrators International, Inc.

The ***1999 SBC***—Standard Building Code written by the SBCCI (Southern Building Code Congress International, Inc.)

The ***1997 UBC***—(Uniform Building Code) written by the ICBO (International Conference of Building Officials)

It is important to remember that each state and municipality has the right to adopt all or a portion of the indicated code. Several states—such as California, Florida, Michigan, New Jersey, and Oregon—have modified the national code to meet problems specific to their locale. In addition to these codes, there are many other national codes that can affect the design of a residence, including:

Fire and Life Safety
International Energy Code
International Fire Code
International Mechanical Code
International Plumbing Code
National Design Specifications for Wood Construction
National Electrical Code
Uniform Mechanical Code
Uniform Plumbing Code

In addition to these national codes, the Department of Housing and Urban Development (HUD), the Federal Housing Authority (FHA), and the Americans with Disabilities Act (ADA) each publish guidelines for minimum property standards for residential construction. Other federal agencies related to construction are listed at the end of this chapter.

INTERNATIONAL BUILDING CODE

In 1972, in order to serve designers better, BOCA, SBCCI, and ICBO joined forces to form CABO (the Council of American Building Officials) to create a national residential code. In 1994, in another attempt to create a national code, the International Code Council (ICC) was formed. Because many regulations affecting construction come from the federal government, in 1997 CABO and ICC merged in an attempt to provide an agency that could channel the many regulations to the state level. The goal of ICC is to develop a single set of comprehensive, coordinated national codes that will eliminate disparities among the three major codes. The ICC has published two codes governing construction. The 2006 International Building Code covers structures with four or more dwelling units as well as commercial and industrial structures. The 2006 edition of the International Residential Code is written to ensure quality residential construction. Each of these codes is referred to as a model code. They each combine features of the three major codes within one base code. Most municipalities are now using one of these codes or their own edited version of one to govern residential construction. Verify the building code that governs your area with your local building department prior to starting construction drawings.

Note:

Information in this text is based on the 2006 International Residential Code (IRC) for One- and Two-Family Dwellings published by ICC. All listings of metric sizes in this chapter are shown as hard conversions in order to be consistent with the IRC listings.

Choosing the Right Code

The architect and engineer are responsible for determining which code will be used during the design process and ensuring that their structure complies with all required codes. Although the drafter is not expected to make decisions regarding the codes as they affect the design, drafters are typically expected to know the content of the codes and how they affect construction.

Each of the major codes is divided into similar sections that specify regulations covering these areas:

- Fire and life safety
- Structural
- Mechanical
- Electrical
- Plumbing

The Fire and life safety and structural codes have the largest effect on the design and construction of a residence. The effects of the codes on residential construction are discussed throughout this text.

BASIC DESIGN CRITERIA FOR BUILDING PLANNING

Building codes have their major influence on construction methods rather than on design. The influence of codes on construction methods is discussed in Sections IV, VI, VII, and VIII. Designers and technicians must be familiar with several areas in order to meet minimum design standards, including basic design criteria. Areas of the code such as climatic and geographical design criteria will affect the materials used to resist forces such as wind, snow, and earthquakes. These forces and how they affect a residence are discussed in Chapter 22. Keep in mind that the following discussion is only an introduction to building codes. As a technician, you will need to become familiar with and constantly use the building code that governs your area.

Occupancy

Building codes divide structures into different categories or occupancies. Houses are considered a Group R occupancy by the IRC. This occupancy includes hotels, apartments, convents, single-family dwellings, and lodging houses with more than ten inhabitants. The R occupancy is further divided into three categories. Single-family dwellings and apartments with fewer than ten inhabitants are defined as R-3. This designation allows the least restrictive type of construction and fire rating group to be used.

In addition to the division of structures into different occupancy ratings, the space within a home or dwelling unit is subdivided into habitable and nonhabitable space. A room is considered habitable space when it is used for sleeping, living, cooking, or dining. Nonhabitable spaces include closets, pantries, bath or toilet rooms, hallways, utility rooms, storage spaces, garages, darkrooms, and other similar spaces.

Location on the Property

For building code purposes, exterior walls of a type R occupancy cannot be located within 5' (1500 mm) of the property lines unless special provisions are made. Keep in mind that zoning regulations may further restrict the location of the structure to the property lines. Typically, an exterior wall that is built within 5' of the property line must be made from materials that will resist a fire for one hour. This is known as a one-hour fire rating. Although only 1/2'' gypsum board is required on each side of the wall, using 5/8'' type X gypsum board on each side of the wall is a common method of achieving a one-hour wall. Many municipalities require any walls built into the minimum side yard to be of one-hour construction. Alternatives based on the siding materials used are available in the building code. Openings such as doors or windows are not allowed in a wall with a fire separation distance less than 5' (1500 mm). Projections such as a roof or chimney cannot extend more than 12'' (305 mm) from the line used to determine the fire separation distance. The one exception to this rule is that detached garages located within 24'' (610 mm) of the property line may have a 4'' (102 mm) eave projection.

Exit Facilities

The major subjects discussed in this section of the code are access doors, emergency ***egress*** (exits), and stairs.

Access Doors

Each dwelling unit, as a residence is referred to in the codes, must have a minimum of one door that is at least 36'' (914 mm) wide and 6''-8' high (2032 mm). Any hallway adjacent to that door must also be a minimum of 36'' (914 mm) clear. All other door sizes may be determined by the designer. Notice that the codes do not mandate that the front door be 36'' (914 mm) wide—only that one exit door must have this dimension. Chapter 10 provides guidelines for determining door sizes throughout a residence.

A floor or landing must be provided on each side of an exterior door. The landing for an exterior door must be within 7 3/4'' (197 mm) from the top of the door threshold provided that the door does not swing over the landing. The IRC requires the landing to be a maximum distance of 1 1/2'' (38 mm) from the top of the door threshold. The minimum landing width must be equal to or greater than the width of the door at the landing, and the landing length must be a minimum of 36'' (914 mm) measured in the direction of travel. The landing is required on each side of the required egress door. Interior landings are to be level. Exterior landings may have a maximum slope of 1/4''/12'' (2% slope). This often becomes a design problem when trying to place a door near a stairway. Figure 4.1 shows some common door and landing problems and their solutions.

Emergency Egress Openings

The term egress is used in the building code to specify areas of access or exits. It is used in reference to doors, windows, and hallways. Windows are a major consideration in designing exits. Emergency egress is required in every sleeping room and in every basement with habitable space. Escape may be made through a door or window that opens directly into a public street, alley, yard, or exit court. The emergency escape must be operable from the inside without the use of any keys or tools. The sill of all emergency escape bedroom windows must be within 44'' (1118 mm) of the floor. Windows used for emergency egress must have a minimum net clear area of 5.7 sq ft (.530 m^2). The net clear opening area must have a minimum width of 20'' (508 mm) and a minimum height of 24'' (610 mm). All net clear openings must be obtainable during normal operation of the window. This opening gives occupants in each sleeping area a method of escape in case of a fire (see Figure 4.2).

If a basement contains one or more sleeping rooms, emergency egress is required in each sleeping room but

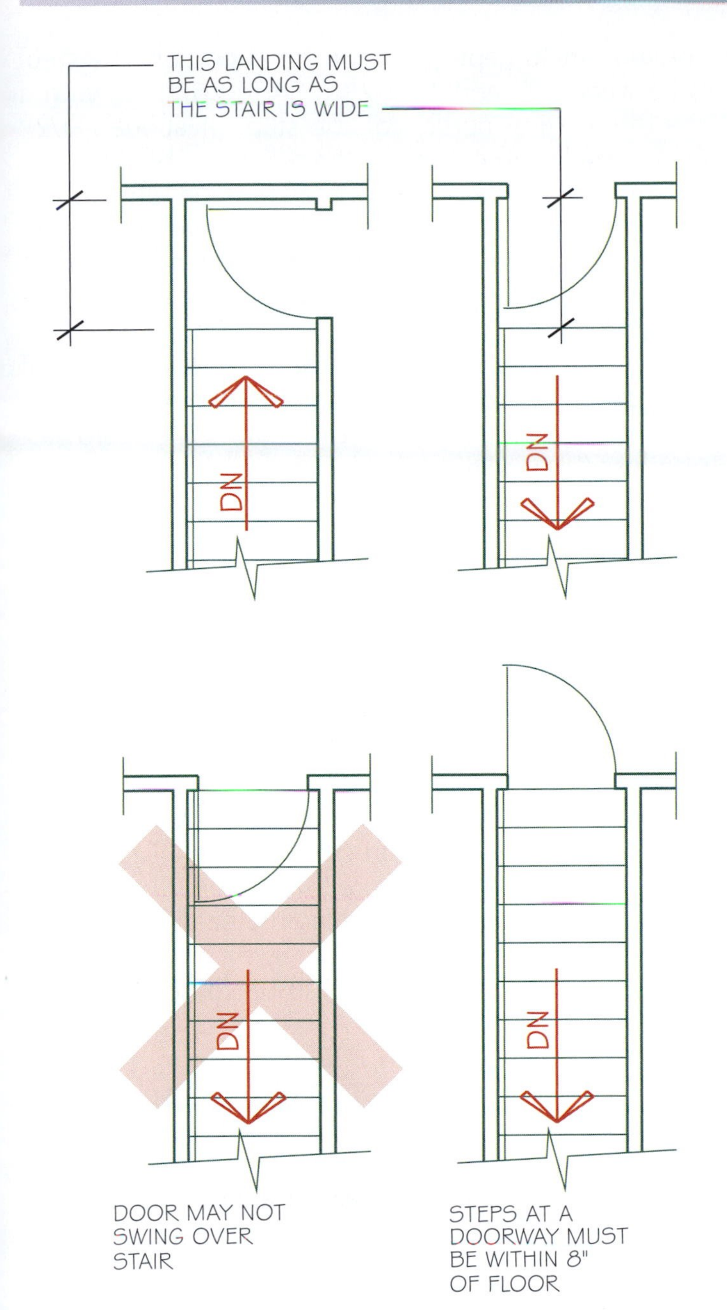

FIGURE 4.1 ■ *Placement of doors near stairs. Landings by doors must be within 1 1/2'' (38 mm) of the door threshold. When the door does not swing over the landing, the landing can be a maximum of 7 3/4'' (197 mm) below the threshold.*

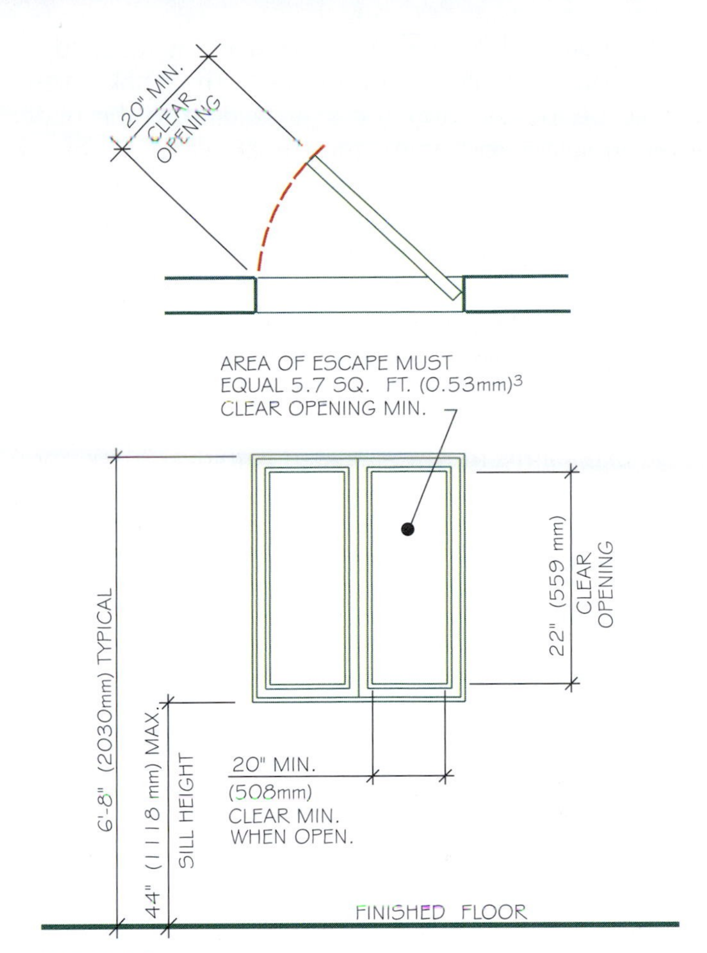

FIGURE 4.2 ■ *Minimum window opening sizes for emergency escape. A means of escape is required for all sleeping units.*

not in adjoining areas of the basement. Emergency escape windows with a finished sill height below the surrounding ground elevation must have a window well that allows the window to be fully opened. The window well is required to have a clear opening of 9 sq ft (.9 m^2) with a minimum horizontal projection and width of 36'' (914 mm) when the window is fully open. If the window well has a depth of more than 44'' (1118 mm), the well must have an approved permanently fixed ladder or stair that can be accessed when the window is fully opened. The ladder or stair cannot encroach more than 6'' (152 mm) into the required well dimensions and must have rungs that have a minimum inside width of 12'' (305 mm). The rungs must be a minimum of 3'' (76 mm) from the wall and must be spaced at a maximum distance of 18'' (457 mm) on center (o.c.) vertically for the full height of the window well.

Smoke Alarms

Closely related to emergency egress is the use of smoke alarm to provide an opportunity for safe exit through the early detection of fire and smoke. A smoke alarm must be installed in each sleeping room as well as at a point centrally located in a corridor that provides access to the bedrooms. For a one-level residence, a smoke detector must be located:

- At the start of every hall that serves a bedroom
- In each sleeping room

For multilevel homes, a smoke alarm is required on every floor, including the basement. The smoke alarm should be located over the stair leading to the upper level. In split-level homes, the smoke alarm is required only on the upper level if the lower level is less than one full story below the upper level. If a door separates the levels of a split-level home, a smoke alarm is required on each level. A smoke alarm is required on the lower level if a sleeping unit is located on that level. Smoke alarms should not be placed in or near kitchens or fireplaces because a small amount of smoke can set off a false alarm.

Smoke alarms must be located within 12'' (305 mm) of the ceiling or mounted on the ceiling. Alarms must be connected to electrical wiring with a battery-powered backup system. Smoke alarms must be interconnected so that if one alarm is activated, all will sound.

Halls

Hallways must be a minimum of 36'' (914 mm) wide. This is very rarely a design consideration because hallways are often laid out to be 42'' (1067 mm) wide or wider to create an open feeling and enhance accessibility from room to room.

Stairs

Stairs can often dictate the layout of an entire structure. Because of their importance in the design process, stairs must be considered at an early design stage. For a complete description of stair construction, see Chapter 30. Minimum code requirements for stairs can be seen in Figure 4.3. Following the minimum standards would result in stairs that are extremely steep and very narrow, with very little room for foot placement. Good design practice provides stairs with a width of between 36'' and 42'' (914 to 1067 mm) for ease of movement. A common tread depth (the horizontal portion of a step) is between 10'' and 10 1/2'' (254 to 267 mm) with a rise (the vertical portion of the step) of about 7 1/2'' (191 mm). Figure 4.4 shows the difference between the minimum stair layout and some common alternatives. Within any flight of stairs, the largest tread depth cannot exceed the smallest tread by 3/8'' (9.5 mm). The difference between the largest and smallest riser cannot exceed 3/8'' (9.5 mm).

Winding stairs may be used if a minimum width of 10'' (254 mm) is provided at a point not more than 12'' (305 mm) from the side of the treads. The narrowest portion of a winding stairway may not be less than 6'' (152 mm) wide at any point.

When a spiral stair is the only stair serving an upper floor area, it can be difficult to move furniture from one floor to another. Spiral stair treads must provide a clear walking width of 26'' (660 mm) measured from the outer edge of the support column to the inner edge of the handrail. A tread depth of 7 1/2'' (191 mm) must be provided within 12'' (305 mm) of the narrowest part of the tread. No riser for a spiral stair may exceed 9 1/2'' (241 mm), and all risers must be equal. Circular stairways must have a minimum tread run of 11'' (279 mm) measured at a point not more than 12'' (305 mm) from the narrow edge. At no point can the tread be less than 6'' (152 mm).

Headroom over stairs must also be considered as the residence is being designed. Straight-run and circular stairs are required to have 6'-8'' (2032 mm) minimum headroom. Headroom of 6'-6'' (1982 mm) is allowed for spiral stairs. This headroom can have a great effect on wall placement on the upper floor over the stairwell. Figure 4.5 shows some alternatives in wall placement over stairs.

INTERNATIONAL RESIDENTIAL CODE MINIMUM STAIR GUIDELINES		
Width (minimum)	*36″	*914 mm
Rise (maximum)	7 3/4″	196 mm
Tread (minimum)	10″	254 mm

*Minimum clear width above the permitted handrail and below the required headroom height. Below the handrail, 31.5″ (787 mm) minimum width with handrail on one side, 27″ (698 mm) minimum width with handrail on each side.

FIGURE 4.3 ■ *Basic design values for stairs.*

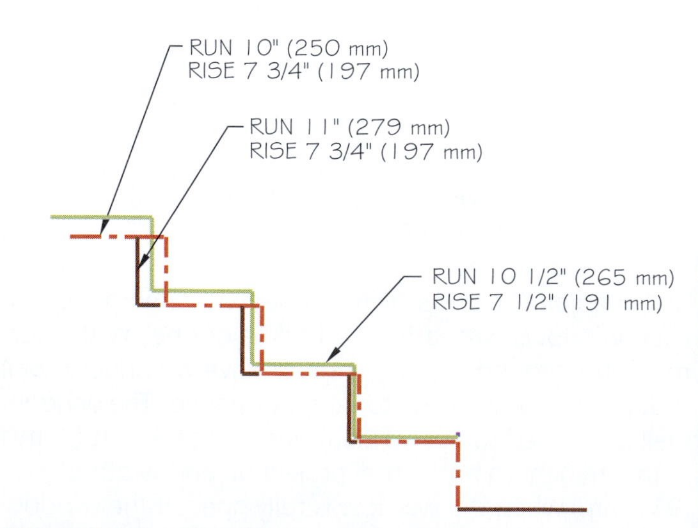

FIGURE 4.4 ■ *Stair layout comparing minimum run and maximum rise based on common practice.*

All stairways with four or more risers are required to have at least one smooth handrail that extends the entire length of the stair. The rail must be placed on the open side of the stairs and must be 34'' to 38'' (864 to 965 mm) above the front edge of the stair. Handrails are required to be 1 1/2'' (38 mm) from the wall but may not extend into the required stair width by more than 4 1/2'' (114 mm).

A guardrail must be provided at changes in floor or ground elevation that exceed 30'' (762 mm). Guardrails are required to be 36'' (914 mm) high. Railings must be constructed so that a 4'' (102 mm) diameter sphere cannot pass through any opening in the rail. The triangular opening formed by the bottom of the rail and the stairs must be constructed so that a 6'' (152 mm) sphere cannot pass through the opening. Horizontal rails or other ornamental designs that form a ladder effect cannot be used.

Room Dimensions

Room dimension requirements affect the size and ceiling height of rooms. Every dwelling unit is required to have at least one room with a minimum of 120 sq ft (11.2 m²) of total floor area. Other habitable rooms except kitchens are required to have a minimum of 70 sq ft (6.5 m²) and shall not be less than 7' (2134 mm) in any horizontal direction. These code requirements rarely affect home design. One major code requirement affecting room size governs the space allowed for a toilet, which is typically referred to as a water closet. A space 30'' (762 mm) wide must be provided for water closets. A distance of 21'' (533 mm) is required in front of a toilet.

Habitable rooms, hallways, corridors, bathrooms, toilet rooms, laundry rooms, and basements must have a minimum ceiling height of 7'-0'' (2134 mm). Exceptions include:

- Bathroom ceilings can be reduced over a counter or plumbing fixture to 6'-8'' (2032 mm). This lowered ceiling can often be used for lighting or for heating ducts.
- Ceilings with beams that are spaced at a maximum distance of 48'' (1219 mm) o.c. are allowed to extend 6'' (152 mm) below the required ceiling height. In rooms that have a sloping ceiling, the minimum ceiling must be maintained in at least half of the room. The balance of the room height may slope to 5' (1524 mm) minimum.
- Any part of a room that has a sloping ceiling less than 5' (1524 mm) high, or
- 7'-0''(2134 mm) for furred ceilings, may not be included as habitable square footage (see Figure 4.6).

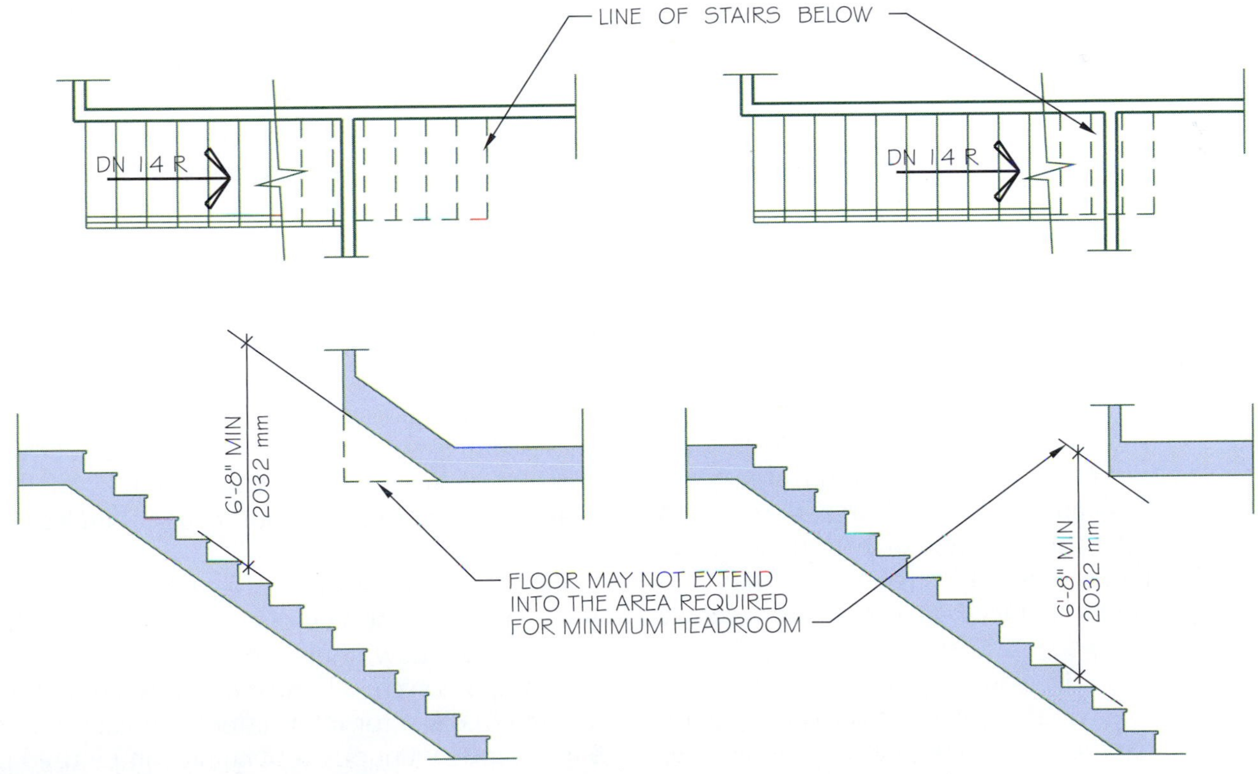

FIGURE 4.5 ■ *Wall placement over stairs.*

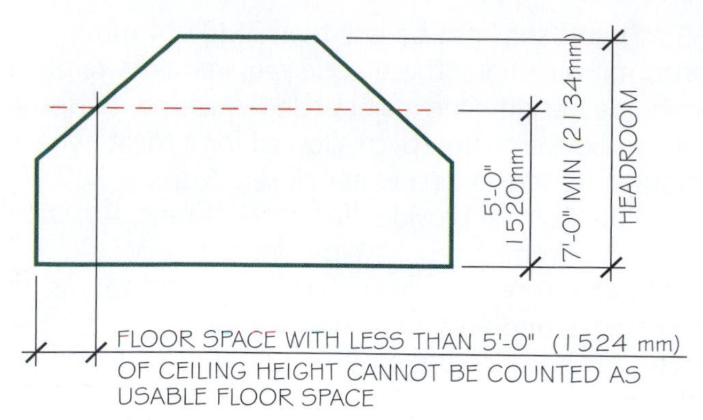

FIGURE 4.6 ■ *Minimum ceiling heights for habitable rooms.*

Light, Ventilation, Heating, and Sanitation Requirements

The light and ventilation requirements of building codes have a major effect on window size and placement. The heating and sanitation requirements are so minimal that they rarely affect the design process. The codes covering light and ventilation have a broad impact on the design of the house. Many preliminary designs show entire walls of glass to take advantage of beautiful surroundings. At the other extreme, some houses have very little glass and thus no view but also very little heat loss. Building codes affect both types of designs.

All habitable rooms must have natural light provided by windows. Kitchens are habitable rooms but are not required to meet the light and ventilation requirements. Windows for other habitable rooms are required to open directly into a street, public alley, or yard on the same building site. Windows may open into an enclosed structure such as a porch as long as the area is at least 65% open. Required areas for light and ventilation may be provided by skylights.

The IRC requires all habitable rooms to have a glazing area equal to 8% of the room's floor area and that half of the area used to provide light also be openable to provide ventilation. A bedroom that is 9 × 10' (2743 × 3050 mm) is required to have a window with a glass area of 7.2 sq ft (.66 m^2) to meet minimum standards. Another limit to the amount of glass area would be in locations subject to strong winds or earthquakes. Because of the lateral movement created by winds and earthquakes, window and wall areas must be carefully proportioned to provide lateral stability. Chapter 24 discusses lateral design. In planning window locations, it must be determined whether an engineer will supervise the design or whether the home will be designed to meet the prescriptive path of the IRC. If the windows are to be based on the IRC, the following design standards must be met:

- Each end of each exterior wall must have wall bracing that is within 12'-6'' (3810 mm) of the end of the wall. This would limit the size of a window to 12'-0'' wide, assuming a 6''-wide wall.
- The wall reinforcing must have a minimum width of 2'-8'' (810 mm).
- Wall reinforcing must occur every 25' (7500 mm) along the wall.

This is rarely a problem, but use of the prescriptive path limits the design of a house with large expanses of glass.

Two alternatives to the light and ventilation requirements are typically allowed in all habitable rooms. Mechanical ventilation and lighting equipment can be used in place of openable windows in all habitable rooms except bedrooms. Glazing can be eliminated except when required for emergency egress when artificial light is provided capable of producing 6 footcandles (6.46 lux) over the area of the room at a height of 30'' (762 mm) above the floor.

Ventilation can be eliminated when an approved mechanical ventilation system capable of producing .35 air changes per hour is provided for the room. Ventilation is also waived if a whole-house mechanical ventilation system capable of producing 15 cfm (7.08 L/s) per occupant is provided. Occupancy is based on two for the first bedroom and one for each additional bedroom.

The IRC requires bedrooms to have a window for emergency egress even if mechanical light and ventilation are provided. The second method allows floor areas of two adjoining rooms to be considered as one if half of the area of the common wall is open and unobstructed. This opening must also be equal to one-tenth of the floor area of the interior room or 25 sq ft (2.3 m^2), whichever is greater. Openings required for light and ventilation may be allowed to open into a thermally isolated sunroom or covered patio. To use this ventilation option, the openable area between the adjoining rooms must be equal to 1/10 of the floor area of the interior room, but it cannot be less than 20 sq ft (1.86 m^2).

Although considered nonhabitable, bathrooms and laundry rooms must be provided with an openable window. The window must be a minimum of 3 sq ft (.279 m^2), of which half must open. A fan that provides 50 cfm (23.6 L/s) for intermittent ventilation or 20 cfm (9.4 L/s) for continuous ventilation can be used instead of a window. These fans must be vented to outside air.

If mechanical ventilation is to be used, careful consideration must be given to the type and placement of the intake and exhaust vents. Exterior vents are required to be protected from local weather conditions. They must also be covered with corrosion-resistant grills, louvers, or screens that have 1/4'' (6.4 mm) minimum and 1/2'' (12.7 mm) maximum openings. Intake openings must be located a minimum of 10' (3048 mm) from any hazardous or noxious contaminants. This minimum distance also includes plumbing vents, chimneys, alleys, parking lots, or loading docks. The minimum distance can be reduced if the intake vent is placed at least 2' (610 mm) below the contaminate source.

Code Alternatives

Many municipalities develop their own code variations to meet the needs of light and code restrictions. The 1996 Oregon Residential Energy Code, for example, allows nine different design alternatives to develop an energy-efficient structure. Rather than strict percentages for windows, variable areas are allowed depending on the quality of insulation, the framing system used, and the quality of windows to be installed. Figure 4.7 shows part of this energy code.

Heating

The heating requirements for a residence are very minimal. IRC requires a heating unit capable of producing and maintaining a room temperature of 68°F (20°C) at a point 3' (914 mm) above the floor and 2' (610 mm) from exterior walls for all habitable rooms. Portable space heaters can no longer be used to comply with the heating requirements of current codes. Chapter 14 provides information on the various types of heating units.

Sanitation

Code requirements for sanitation rarely affect the design of a structure. Each residence is required to have a toilet, a sink, and a tub or shower, but not all are required to have hot and cold water. All plumbing fixtures must be connected to a sanitary sewer or an approved private sewage disposal system. The room containing the toilet must be separated from the food preparation area by a tight-fitting door. Chapter 10 provides further information regarding the layout of each fixture, and Chapter 13 provides information regarding plumbing plans.

CLIMATIC AND GEOGRAPHIC DESIGN CRITERIA

The major influence of the IRC on a structure has to do with structural components. Chapters covering foundations, wall construction, wall coverings, floors, roof-ceiling construction, roof coverings, and chimney and fireplace construction are provided to specify minimum construction standards. To use the information in the codes as it relates to each area of construction requires some basic climatic and geographic information about the area where the structure will be constructed. This includes knowledge of various factors of the building site:

Roof live load
Roof snow load
Wind pressure
Seismic condition by zone
Susceptibility to damage by weathering
Susceptibility to damage by termites
Susceptibility to damage by decay
Frost-line depth
Winter design temperature for heating facilities

Information for each of these categories can be obtained from a table in the applicable building code or from the governing building department. Building codes specify how a building can be constructed to resist the forces of wind pressure, seismic activity, freezing, decay, and insects. Each of these areas is discussed in detail in succeeding chapters. In the initial design stage, an understanding of these factors will determine what types of materials will be required.

ACCESSIBILITY

Although this is not required for most R-3 occupancies, the IRC addresses accessibility that affects type R-1 occupancies (hotels, apartments, and private homes with four or more dwelling units). R-1 units are divided in type A and type B units. Type A units are to be designed and constructed to meet ICC A117.1–1998 standards. This publication is printed by the ***American National Standards Institute, Inc.*** (ANSI) with the cooperation of BOCA, SBCCI, and ICBO. Major sections of this standard include accessible routes, general site and building elements, plumbing elements and fixtures, communication elements and fixtures, special rooms and spaces, and built-in furnishing and equipment.

PRESCRIPTIVE COMPLIANCE PATHS FOR RESIDENTIAL BUILDINGS

PRESCRIPTIVE COMPLIANCE PATHS FOR RESIDENTIAL BUILDINGS									
Building Components	PATH 1	PATH 2 Sun Tempered[4]	PATH 3	PATH 4 Sun Tempered[4]	PATH 5	PATH 6 Sun Tempered[4]	PATH 7 Sun Tempered[4]	PATH 8 House Size Limited[5]	PATH 9 Log Homes/ Solid Timber
Maximum Allowable Window Area[6]	No Limit	No Limit	No Limit	No Limit	No Limit	No Limit	No Limit	12%	No Limit
Window Class[7]	U=0.40	U=0.40	U=0.50	U=0.50	U=0.60	U=0.60	U=0.60	U=0.40	U=0.40
Doors, Other Than Main Entry	U=0.20	U=0.20	U=0.20	U=0.20	U=0.20	U=0.20	U=0.20	U=0.20	U=0.54
Main Entry Door, maximum 24 sq. ft.	U=0.54	U=0.54	U=0.20	U=0.20	U=0.20	U=0.20	U=0.54	U=0.20	U=0.54
Wall Insulation	R-21[9]	R-15	R-21A[8]	R-15A[8]	R-24A[8]	R-21A[8]	R-21A[8]	R-15	___[3]
Underfloor Insulation	R-25	R-21	R-25	R-21	R-30	R-21	R-25	R-21	R-30
Flat Ceilings	R-38	R-49	R-49A[8]	R-38	R-49A[8]	R-49A[8]	R-49A[8]	R-49	R-49
Vaulted Ceilings[10,11]	R-30	R-30	R-30	R-38	R-38	R-38	R-38	R-38	R-38
Skylight Class[7]	U=0.50	U=0.50	U=0.50	U=0.50	U=0.50	U=0.50	U=0.50	U=0.50	U=0.50
Skylight Area[12]	<2%	<2%	<2%	<2%	<2%	<2%	<2%	<2%	<2%
Basement Walls	R-21	R-21	R-21	R-21	R-21	R-21	R-21	R-21	R-21
Slab Floor Edge Insulation	R-15	R-15	R-15	R-15	R-15	R-15	R-15	R-15	R-15
Forced Air Duct Insulation	R-8	R-8	R-8	R-8	R-8	R-8	R-8	R-8	R-8

Notes:

[1] Path 1 is based on cost-effectiveness. Paths 2–7 are based on energy equivalence with Path 1. Cost-effectiveness of Paths 2–9 not evaluated.

[2] As allowed in current Chapter 13, section 1306, thermal performance of a component may be adjusted provided that overall heat loss does not exceed the total resulting from conformance to the required U-value standards. Calculations to document equivalent heat loss shall be performed using the procedure and approved U-values contained in Table 13-B.

[3] R-values used in this table are nominal, for the insulation only and not for the entire assembly. The wall component for Path 9 shall be a minimum solid log or timber wall thickness of 3.5 inches (88.9 mm).

[4] The sun-tempered house shall have one lot line which borders on a street oriented within 30 degrees of true east-west and 50 percent or more of the total glazing area for the heated space on the south elevation. An approved alternate to street orientation based on solar design and access shall be accepted by the building official.

[5] Path 8 applies only to homes with less than 1,500 sq. ft. (139 m^2) heated floor space and glazing area less than 12 percent of heated space floor area.

[6] Reduced window area may not be used as a trade-off criterion for thermal performance of any component, except as noted in Table 13-B.

[7] Window and skylight U-values shall not exceed the number listed. U-values may also be listed as "class" on some windows and skylights (i.e., CL40 is the same as U = 0.40).

[8] A = advanced frame construction as defined in Section 1307.1.4 for walls and Section 1307.1.5 for ceilings.

[9] R-19 Advanced Frame or 2×4 wall with rigid insulation may be substituted if total nominal insulation R-value is 18.5 or greater.

[10] Partially vaulted ceilings and ceilings totaling not more than 150 sq. ft. (139m^2) in area for dormers, bay windows or similar architectural features may be reduced to not less than R-21. When reduced, the cavity shall be filled (except for required ventilation spaces) and a 0.5 perm (dry cup) vapor retarder installed.

[11] Vaulted area, unless insulated to R-38, may not exceed 50 percent of the total heated space floor area.

[12] Skylight area is a percentage of the heated space floor area. Any glazing in the roof/ceiling assembly above the conditioned space shall be considered a skylight.

FIGURE 4.7 ■ *Many municipalities make additions to their building codes to cover specific concerns. This table is part of the 1996 Oregon Residential Energy Code. Although most construction is built to conform to Path 1, designers can use tradeoffs in design such as a higher quality glazing for thinner walls with less insulation using path 4.*

These sections are applicable to R-1 occupancies. Chapter 10 of the ANSI guidelines applies specifically to type A units. Major guidelines apply to bathroom and kitchen design. The main consideration in bathroom design is wheelchair access. See Figure 4.8. A major consideration in kitchen design is access to work areas. Figure 4.9 shows counter requirements.

R-3 structures with four or more dwelling units must be designed to meet the requirements of type B units. Type B units are to be designed in accordance with chapter 11 of the ***International Building Code*** (IBC). Consideration is given in each code to accessible routes; operating controls for electrical, environmental, intercom, and security controls; doorways; kitchens; and bath design.

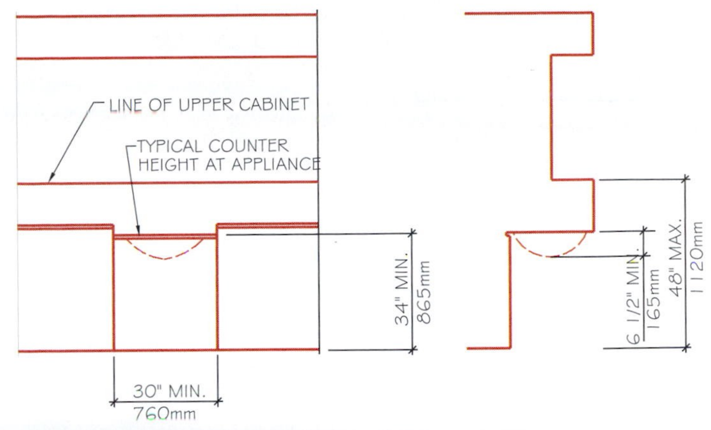

FIGURE 4.9 ■ *A major consideration of the kitchen guidelines is access to the work area.*

Accessible Route

An ***accessible route*** is the walking surface from the exterior access through the residence. The exterior access (the front door) cannot be located in a bedroom. At least one route is required to connect all spaces that are part of the dwelling unit. If only one route is provided, it cannot pass through a bathroom, closet, or similar space.

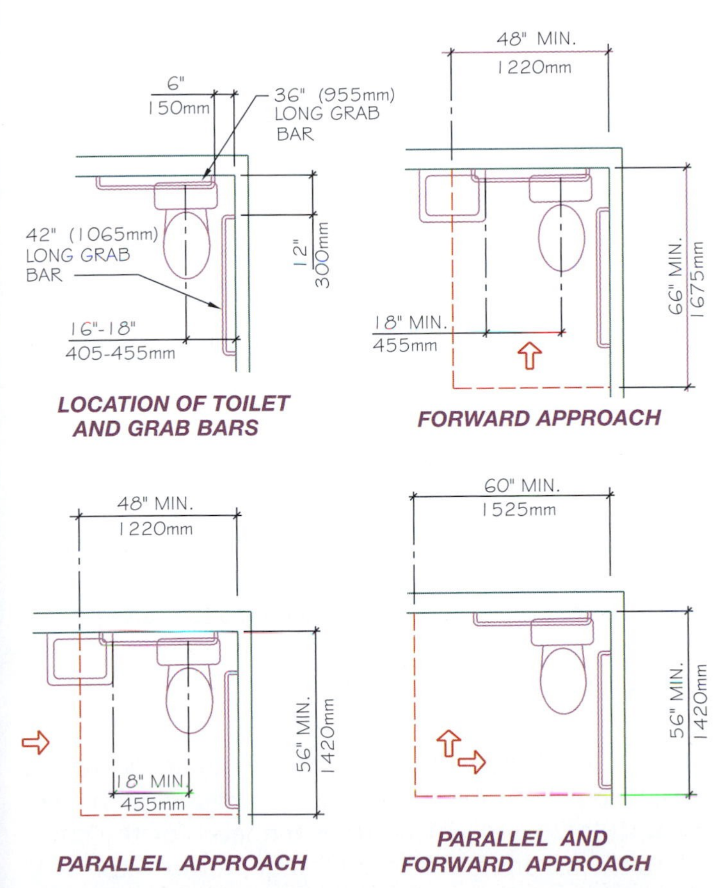

FIGURE 4.8 ■ *Water closet clearance in type A dwelling units.*

The access route is required to have a minimum width of 36'' (914 mm) except at doors. Changes in floor level greater than 1/2'' (12.7 mm) are not allowed unless a ramp, elevator, or wheelchair lift is provided. Ramps cannot have a slope greater than 1:48. The only exception to these elevation requirements is at an exterior door that leads to a deck, patio, or balcony. Here the lower exterior floor surface is allowed to be a maximum of 4'' (102 mm) below the finished interior floor surface.

Operating Controls

This portion of the building code regulates placement of and access to controls for electrical, environmental, security, and intercom controls. Guidelines for placement of controls can be seen in Figure 4.10. The control should be centrally located in a clear floor space measuring a minimum of 30 × 48'' (762 × 1219 mm), and should be no more than 48'' (1219 mm) and no less than 15'' (381 mm) from the finished floor. Exceptions to these locations include:

- Electrical receptacles serving a dedicated use (i.e., a 110 C.O. for a refrigerator)
- Appliance-mounted controls or switches
- A single receptacle located above a portion of a countertop uninterrupted by a sink or appliance (this need not be accessible as long as one receptacle is provided)
- Floor electrical receptacles
- Plumbing fixture controls

Doors

All doors are required to provide a minimum clear opening of 32'' (813 mm), as shown in Figure 4.11. Although maneuvering space is not required on the

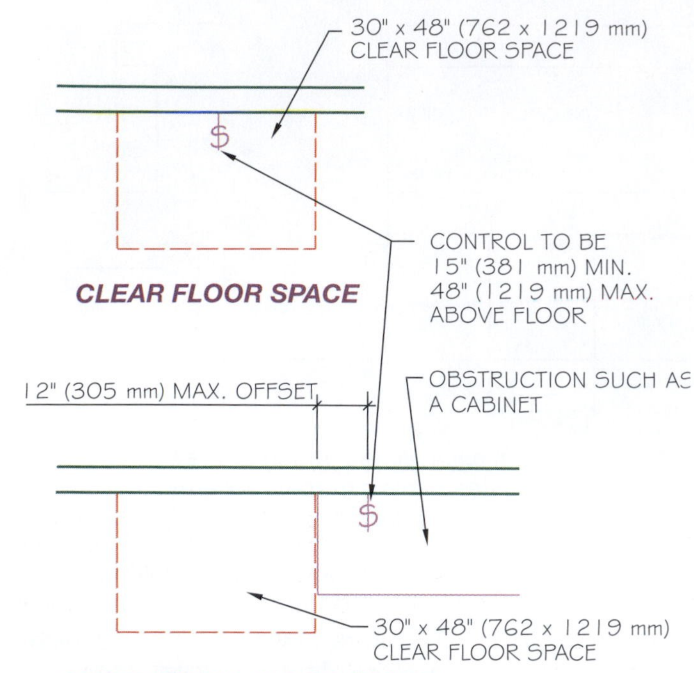

FIGURE 4.10 ■ *A clear floor area of 30 × 48'' (762 × 1219 mm) must be provided to access controls.*

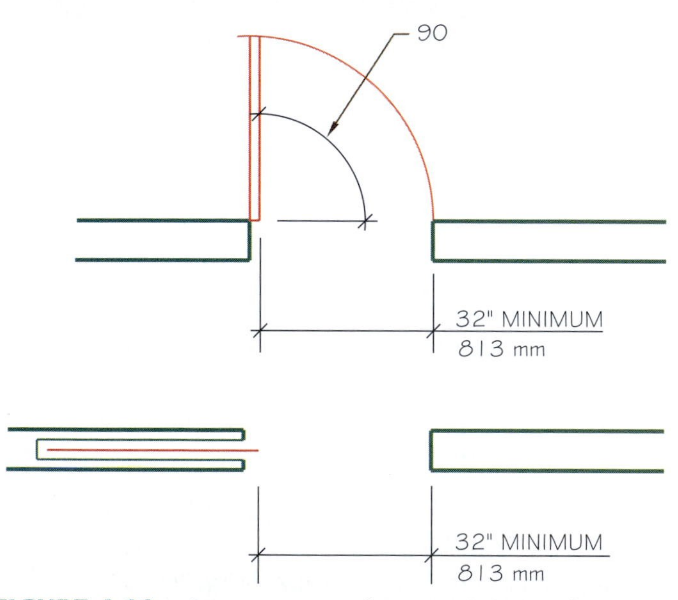

FIGURE 4.11 ■ *Doorways must have a minimum clear opening of 32'' (813 mm) measured between the face of the door when open 90° and the stop. A maximum tolerance of 1/4'' (6.4 mm) is allowed.*

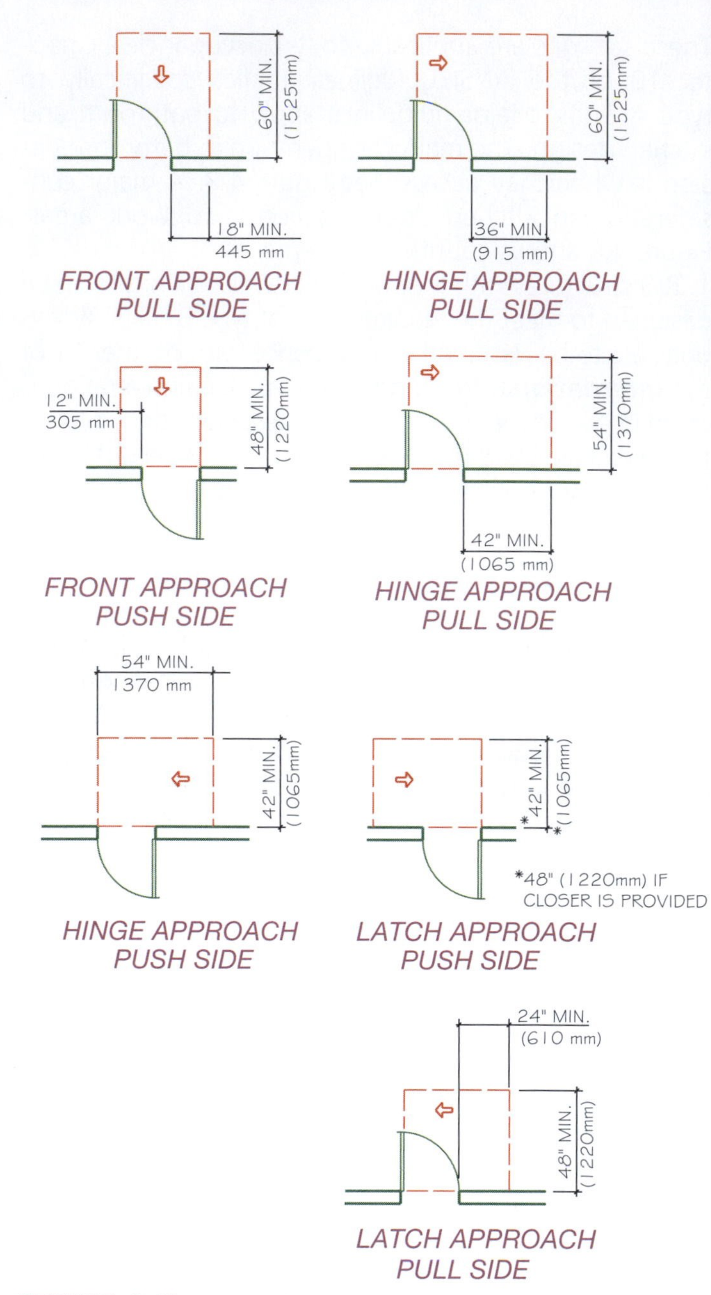

FIGURE 4.12 ■ *Clear floor spaces required for type A units. For exterior swinging doors, a threshold cannot exceed a maximum height of 1/2''.*

dwelling unit side of the door, good design dictates that space be provided for turning a wheelchair. Figure 4.12 shows required clear access for type A units.

For exterior swinging doors, a threshold cannot exceed a maximum height of 1/2'' (12.7 mm). Sliding exterior doors are allowed to have a beveled threshold with a maximum height of 3/4'' (19 mm). The bevel cannot exceed one vertical unit per two horizontal units (50% slope).

Kitchens

Figure 4.13 shows the required minimum clearances for kitchen base cabinet placement. A clear floor area of 30 × 48'' (762 × 1219 mm) is required for the cooktop, dishwasher, freezer, oven, range, refrigerator, sink, and trash compactor when provided.

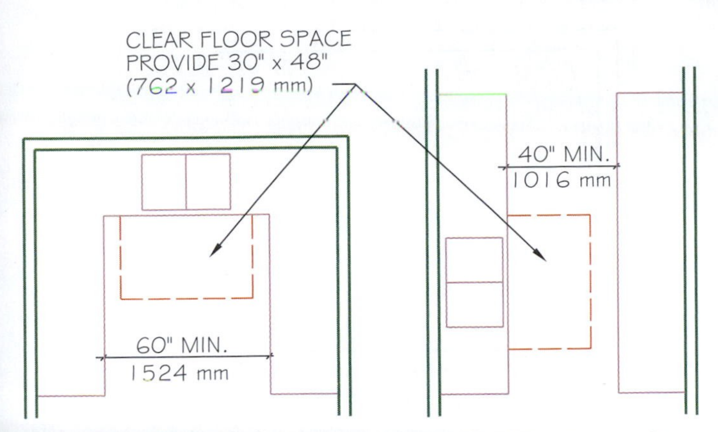

FIGURE 4.13 ■ *Clear floor spaces required between cabinets in a kitchen.*

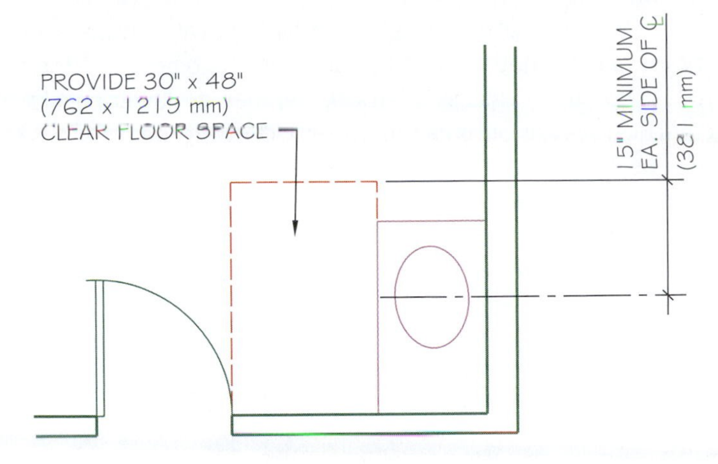

FIGURE 4.15 ■ *Required space for a lavatory. A water closet must be at least 18'' (457 mm) from the centerline of the fixture and 15'' (381 mm) on the other side when located between a bathtub or lavatory. When it is located by a wall, 18'' (457 mm) must be provided from the centerline of the fixture to the wall.*

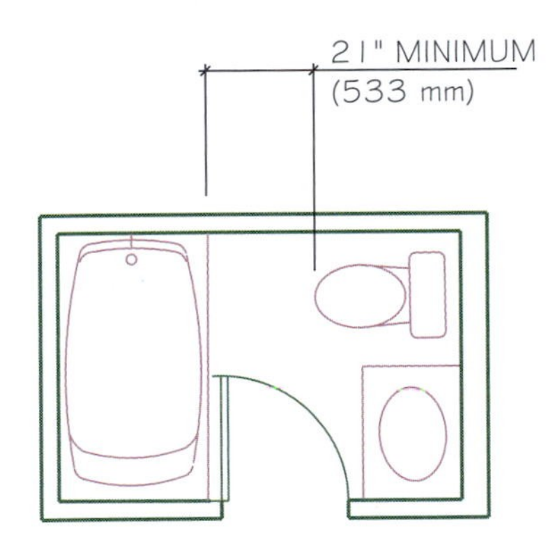

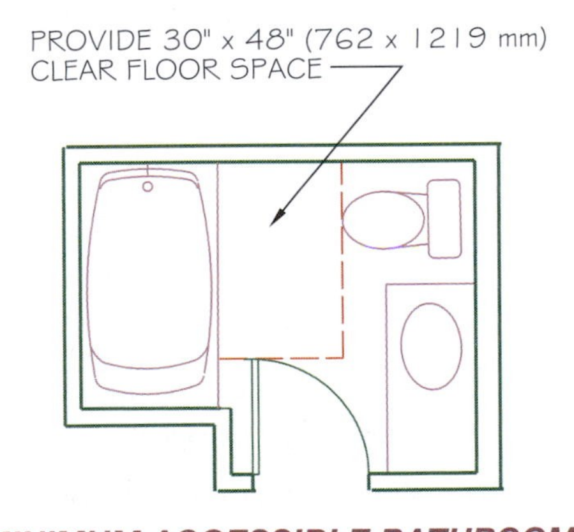

FIGURE 4.14 ■ *A typical bathroom layout (top) is often too small to provide accessibility. Doors are not allowed to swing into the clear floor space required for any fixture unless a 30 × 48'' (762 × 1219 mm) minimum clear floor space is provided beyond the swing of the door.*

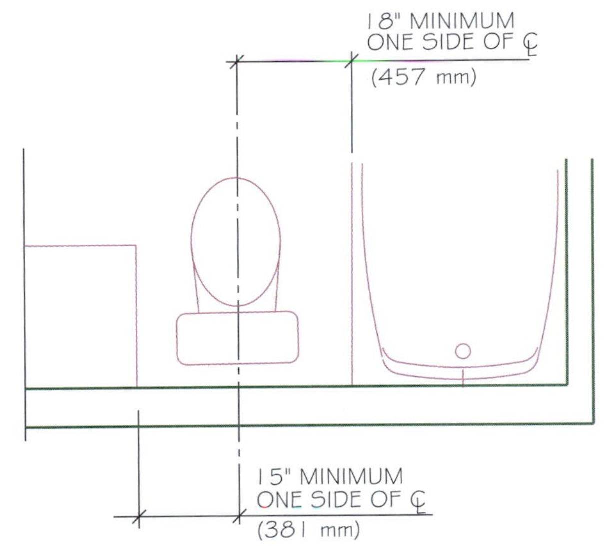

FIGURE 4.16 ■ *A water closet must be 18'' (457 mm) from the centerline of the fixture and 15'' (381 mm) minimum on the other side when located between a bathtub or lavatory. When it is located by a wall, 18'' (457 mm) must be provided from the centerline of the fixture to the wall.*

Toilet and Bathing Facilities

Figure 4.14 shows a typical bathroom layout. Doors are not allowed to swing into the clear floor space required for any fixture unless a clear floor space of at least 30 × 48'' (762 × 1219 mm) is provided beyond the swing of the door. This space allows a person to enter, close the door, and then move to a fixture. If this space is provided, the door may swing into the required space for each bath fixture. Figure 4.15 shows the required space for a lavatory. As shown in Figure 4.16, a water closet must have a minimum of 18'' (457 mm) from the centerline of the fixture and a minimum of 15'' (381 mm) on the other side when located between a bathtub or lavatory. When it is located by a wall, a distance of 18'' (457 mm) must be provided from the centerline of the fixture to the wall. Figure 4.17 shows access methods and clear floor space required for a toilet. For each access method, vanities or lavatories located on a wall behind the water closet are permitted to overlap the clear floor space. Where a tub and/or shower is provided, it must comply with Figure 4.18. If a separate

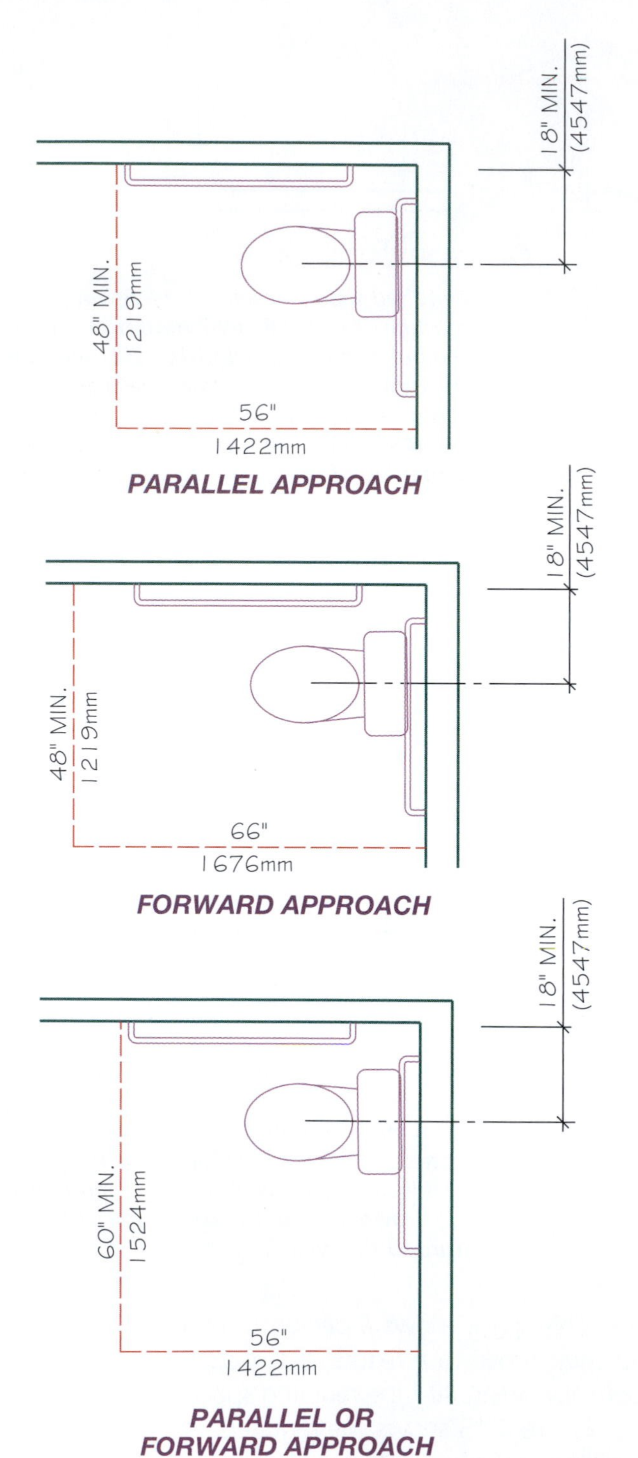

FIGURE 4.17 ■ *Access methods and clear floor space required for a toilet. For each access method, vanities or lavatories located on a wall behind the water closet are permitted to overlap the clear floor space.*

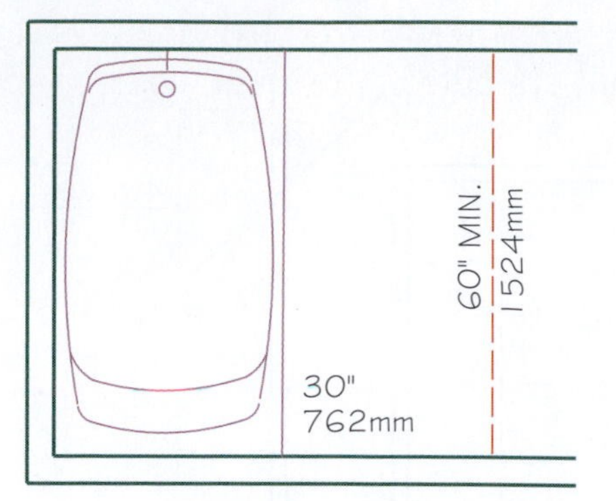

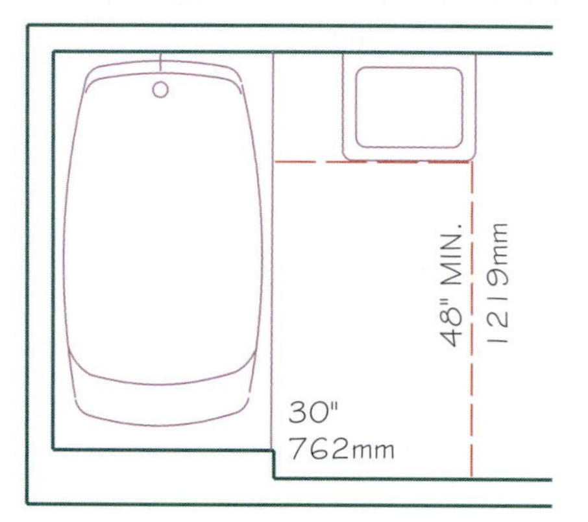

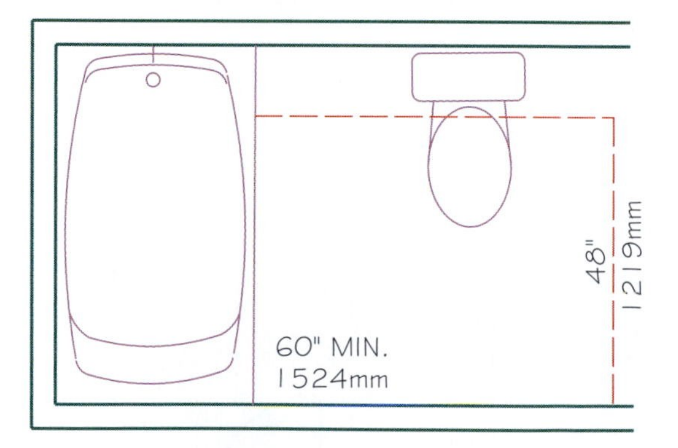

FIGURE 4.18 ■ *Where a tub and/or shower is provided, it must have a clear floor area of 30 × 60'' (762 × 1524 mm). A lavatory placed at the control end of the tub may extend into the clearance if a 30 × 48'' (762 × 1219 mm) clear area is maintained. When the forward approach is used, the floor space must be 48 × 60'' (1219 × 1524 mm). A toilet placed at the control end of the tub may be placed in the clear floor area as long as the minimum requirements for the toilet are still met. Access to the toilet can be achieved using the parallel or the parallel/forward approach.*

shower and tub are provided, the guidelines apply to only one of the fixtures. Figure 4.19 shows the clear floor requirements when only a shower is provided.

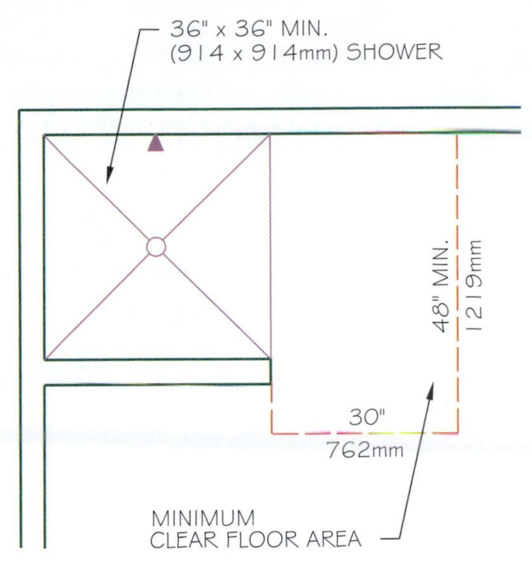

FIGURE 4.19 ■ *If a separate shower and tub are provided, the guidelines apply to only one of the fixtures. If only a shower is provided, minimum clear floor requirements must be met.*

CHAPTER 4

Additional Readings

The following websites can be used as a resource to help you keep current with changes in floor plan–related materials.

MAJOR BUILDING CODE ORGANIZATIONS

ADDRESS	COMPANY OR ORGANIZATION
www.intlcode.org	The International Code Council (ICC). The 2003 IRC is their current code.
www.bocai.org	Building Officials and Code Administrators International, Inc. The 1999 BOCA is their current code.
www.sbcci.org	Southern Building Code Congress International, Inc. The 1999 SBC (Standard Building Code) is their current code.
www.icbo.org	International Conference of Building Officials. The 1997 Uniform Building Code is their current code.

FEDERAL AGENCIES RELATED TO CONSTRUCTION

ADDRESS	COMPANY OR ORGANIZATION
www.access-board.gov	The Access Board—U.S. Architectural and Transportation Barriers Compliance Board
www.eren.doe.gov	DOE—Department of Energy's Energy Efficiency and Renewable Energy Network
www.epa.gov	EPA—U.S. Environmental Protection Agency
www.nara.gov/fedreg	Federal Register—Government Printing Office
www.thomas.voc.gov	Thomas—Federal legislative information, status of bills, Congressional Record, and related information
www.nist.gov	NIST—National Institute of Standards and Technology

MAJOR TESTING LABS, QUALITY ASSURANCE AND INSPECTION AGENCIES, AND THEIR WEBSITES

ADDRESS	COMPANY OR ORGANIZATION
www.apawood.org	APA—The Engineered Wood Association
www.csa.ca	Canadian Standards Association
www.etlsemko.com	ETL Semko
www.osha.gov	Occupational Safety & Health Administration
www.approvals.org	International Approval Services, Inc.
www.nahbrc.org	NAHB Research Center
www.ul.com	Underwriters Laboratories Inc.
www.wwpa.org	Western Wood Products Association

OTHER AGENCIES AND THEIR WEBSITES THAT MAY AFFECT BUILDING CODES IN YOUR AREA

ADDRESS	COMPANY OR ORGANIZATION
www.ada.gov	Americans with Disabilities Act
www.ansi.org	ANSI—American National Standards Institute
www.asce.org	ASCE—American Society of Civil Engineers
www.ashrae.org	ASHRAE—American Society of Heating, Refrigerating, and Air Conditioning Engineers
www.astm.org	ASTM—American Society for Testing and Materials
www.cabo.org	CABO—Council of American Building Officials
www.nfpa.org	NFPA—National Fire Protection Association

ORGANIZATIONS OFTEN USED BY DESIGNERS TO ENSURE THE QUALITY OF MATERIALS

ADDRESS	COMPANY OR ORGANIZATION
www.asfconline.org	IAFC—International Association of Fire Chiefs
www.iec.ch	IEC—International Electrotechnical Commission
www.ieee.org	IEEE—Institute of Electrical and Electronics Engineers
www.iesna.org	IESNA—Illumination Engineering Society of North America
www.iso.ch	ISO—International Organization for Standardization
www.nahb.com	NAHB—National Association of Home Builders
www.ncma.org	NCMA—National Concrete Masonry Association
www.nibs.org	NIBS—National Institute of Building Sciences
www.nmhc.org	NMHC—National Multi-Housing Council
www.nspe.org	NSPE—National Society of Professional Engineers
www.mbinet.org	MBI—Modular Building Institute
www.usgbc.org	USGBC—U.S. Green Building Council

CHAPTER 4

Building Codes and Interior Design Test

DIRECTIONS

Answer the following questions with short, complete statements. Type your answers using a word processor, or neatly print the answers on lined paper.

1. Type your name, the chapter number, and the date at the top of the sheet.
2. Type the question number and provide the answer in the form of a statement that includes part of the question. You do not need to write out the entire question.
3. Answers should be based on the code that governs your area unless otherwise noted.

QUESTIONS

Answers to all questions should be based on the code that governs your area.

4.1. What is the minimum required size for an entry door?

4.2. List the five major building codes used throughout the United States, starting with the code that governs your area.

4.3. All habitable rooms must have a certain percentage of the floor area provided in window glass for natural light. What is the required percentage?

4.4. What is the required width for hallways?

4.5. List the minimum width for residential stairs.

4.6. What is the minimum ceiling height for a kitchen?

4.7. What is the maximum height that bedroom windows can be above the finish floor?

4.8. Toilets must have a space how many inches wide?

4.9. Are bathrooms in a single-family residence required to be handicapped-accessible?

4.10. List three sanitation requirements for a residence.

4.11. What are the area limitations of spiral stairs?

4.12. What is the minimum window area required to meet the ventilation requirements for a bedroom that is 10 × 12'?

4.13. List the minimum square footage required for habitable rooms.

4.14. For a sloping ceiling, what is the lowest height allowed for usable floor area?

4.15. What is the minimum size opening for an emergency egress?

4.16. What is the minimum required size of the main entry door to a single-family residence?

4.17. Visit the website of the building department that governs residential construction in your area. List the site address and determine the following current design criteria:
   a. Roof live loads
   b. Wind pressure
   c. Risk of weathering
   d. Frost-line depth
   e. Snow loads
   f. Seismic zone
   g. Risk of termites

4.18. Visit the website of the agency that regulates your state's building code. List the site address and describe any code changes currently under consideration.

4.19. Visit the website of one of the testing labs and obtain information related to a class drawing project. List the site address and information specific to your project.

4.20. Visit the website of one of the regulatory agencies that influence construction in your area. List the site address and information specific to your project.

CHAPTER 5

Interior Design Considerations

Architectural design probably has more amateur experts than any other field. Wide exposure to houses causes many people to feel that they can design their own house. Exposure to home design programs on HGTV (Home & Garden Television) and TLC (The Learning Channel) and the availability of home improvement retail stores such as Lowe's and Home Depot that offer in-store seminars and product demonstrations, have produced educated consumers with an increased sensitivity to design. Because they know what they like, many people attempt to draw their own plans, only to have them rejected by building departments. Designing can be done by almost anyone. Designing a home to meet a variety of family needs takes more knowledge. This chapter explores the design of the floor plan and interior of a home as it relates to:

- *The living areas of a home*
- *The sleeping areas of a home*
- *The service areas of a home*
- *Traffic flow between these areas*
- *Universal accessibility*

The floor plan is usually the first drawing to be started in designing a home for a family. The floor plan must be designed to harmonize with the site and the inhabitants. Planning for the site is introduced in Chapters 7 and 8. This chapter introduces both the design process and basic requirements for the interior of a home. Once this information has been provided, the design process described in Chapter 1 can be started. The bubble drawings and preliminary sketches will need to integrate the living, sleeping, and service areas as well as movement among these areas.

LIVING AREAS

The living areas of a home include the living, dining, and family rooms, den, and breakfast nook. These are the rooms or areas of a house where family and friends will spend most of their leisure time. The rooms of the living area should be clustered together near the entry to allow easy access for guests.

Entries

Entries serve as a transition point between areas of the home. A well-planned home will have at least two points of entry that include:

- The main entry, to draw guests into the home
- The service entry that is used by the family for access between the garage, yards, and service areas

Main Entry

The main entry in Figure 5.1 provides an outside focal point to draw guests to the front door and serves as a hub for traffic to the living areas. A raised ceiling is a common method of accenting the front door. Figure 5.2 shows an example of an inviting entry created by using an arbor and good landscaping. Both examples clearly define access to the front door. Other key elements to consider when designing the entry are that:

- The entry should be kept proportional to the entire structure and should be made of compatible materials.
- The size of the entry will be influenced by the number of entry doors and their size.
  - Single entry doors are generally 3' (900 mm) wide.
  - Larger homes may have doors that are 42'' (1050 mm) or 48'' (1200 mm) wide.
  - Double doors are typically a total of 5' (1500 mm) or 6' (1800 mm) wide, and one or both panels can be operable. Single and double doors are usually 6'–8'' (2000 mm) high but are also available in 8' (2400 mm) and custom heights.

A second major consideration in planning the entry is providing protection from the weather. Depending on your area of the country, this might include direct sunlight (in arid climates), rain, snow, and wind. A well-planned entry will provide a covered access into the home as well as a windbreak. The windbreak can often

FIGURE 5.1 ■ *The main entry should provide a focal point to draw guests to the front door and to protect them from the weather.* Courtesy Pavel Adi Sandu.

FIGURE 5.2 ■ *An inviting entry created using an arbor and good landscaping to guide guests to the main entry.*

be achieved by recessing the entry area. The depth of the recess should be kept proportional to the width and height of the entry.

Foyer

The ideal entry will open to a foyer rather than directly into one of the living areas. The foyer serves as a place either for welcoming guests or saying good-byes. It also provides a place to put on or remove coats or other weather-related clothing. It should provide access to a closet for seasonal clothing as well as space for guests' coats. The foyer also provides access to each area of the home. This might include a hall leading to the sleeping area, a stair leading to an upper or lower floor, and access to the living room or dining areas. Figure 5.3 shows an example of a foyer for a large residence. The foyer should be proportional to the rest of the home, but it should also be of sufficient size to hold several people. The size will be based on the client's needs and personal preferences, as well as the size and cost of the home. In a small home, a foyer 48'' (1200 mm) wide may be tolerable, although 6 × 6' (1500 × 1500 mm) is still considered small. In a larger home, a room 8' to 12' (2400 to 3600 mm) wide would be more appropriate. The height of the foyer should be considered as the size is determined. A small foyer with tall ceiling will create a feeling of being in an elevator shaft. Remember, the goal of the foyer is to create a warm, inviting feeling and to enhance traffic flow throughout the home. Figure 5.4 shows a floor plan with a foyer that opens to the dining room, great room, study, and master bedroom.

FIGURE 5.3 ■ *The foyer for a home must be proportional to the balance of the home but still create a warm focal point for entering the home. This entry features access to the living room, the outside living areas, and the upper level.* Courtesy Alan Mascord AIBD, Alan Mascord Design Associates, Inc. Bob Greenspan, photographer.

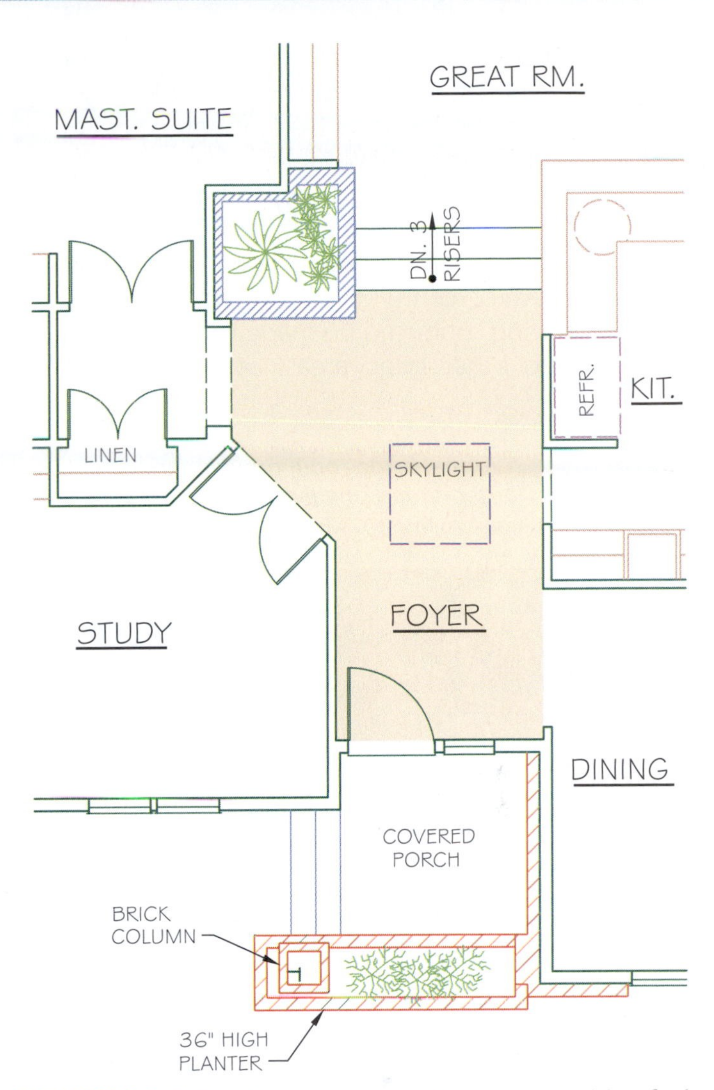

FIGURE 5.4 ■ *The foyer should create a warm, inviting feeling and enhance traffic flow throughout the house.* Courtesy David Jefferis, Residential Designs.

Service Entry

The service entry links the garage, kitchen, utility room, bathrooms, and patios or decks. In many homes, although guests typically use the main entry, the family's main access is through the service entry. Depending on the size and location of the home, the service entry may serve as a mudroom. The service entry is discussed later in this chapter.

Living Room

The purpose of the living room will vary, depending on the size of the home and the preferences of the homeowner. Points to consider in planning the living room include:

- How the room will be used
- How many people will use it
- How often it will be used
- What type and size of furniture that will be placed in it

Some families use the living room as a place for quiet conversation and depend on other living areas for noisier activities. In homes designed to appeal to a wide variety of interests, the formal living room has given way to a large multipurpose room.

If both a living room and a family room are provided, the living room is typically used to entertain guests or for quiet conversations. Figure 5.5 shows a formal living room. If the home has no family room, the living room will also need to provide room for recreation, hobbies, and relaxation.

The size of the living room should be determined by the number of people who will use it, the furniture intended for it, and the budget for the entire construction project. Table 5.1 lists common sizes for living room furniture. See Appendix B on the CD. Rectangular rooms are typically easier to plan for furniture arrangements. The closer a room is to being square, the harder furniture placement becomes. The minimum width needed for common activities is 12' to 14' (3660 to 4270 mm). A room 13 × 18'

(3960 × 5500 mm) serves for most furniture arrangements and allows for moving furniture easily. Furniture blocks can be found in DesignCenter, on the CD, or from third-party vendors.

Within the room, an area approximately 9' (2740 mm) in diameter should be provided for the primary seating area. This area is often arranged to take advantage of a view or centered on a fireplace. In smaller homes, the living room can be made to seem larger if it is attached to other living areas, as in Figure 5.6. In larger residences such as the home shown in Figure 5.7, the living room is often set apart from other living areas to give it a more formal effect.

The living room is usually placed near the entry so that guests may enter it without having to pass through the rest of the house. The living room should also be placed so that the rest of the home can be accessed without having to pass through it. Many designers place the living room one or more steps below the level of the balance of the home, as in Figure 5.8, to enhance a feel-

FIGURE 5.5 ■ *A formal living room provides an area for quiet conversation away from the noise of other living areas. In larger homes, the living room is often set apart from the rest of the living areas to create a formal area.* Courtesy Alan Mascord AIBD, Alan Mascord Design Associates, Inc. Bob Greenspan, photographer.

FIGURE 5.6 ■ *If the living room is open to other living areas, a casual feeling is created.* Courtesy Dennis Dinser Principal, Arcadian Design, LLC.

FIGURE 5.8 ■ *Many designers place the living room one or more steps below the level of the rest of the home to enhance a feeling of separation and to discourage through traffic.* Courtesy Michael Jefferis.

FIGURE 5.7 ■ *In addition to being open, this room is an elegant area in which to entertain.* Courtesy Eric S. Brown Design Group Inc. Oscar Thompson, photographer.

FIGURE 5.9 ■ *A casual living room that draws the outside in. The living areas of this residence are close to each other for easy entertaining but separated for noise control.* Courtesy Eric S. Brown Design Group Inc. Laurence Taylor, photographer.

ing of separation and discourage through traffic. The location of the room should also take energy efficiency into consideration. If the room is used primarily in the evening, it should be placed on the west side of the residence so it will have natural light at that time. In cooler climates, a southwestern orientation will provide the most hours of sunlight; in warmer climates, a northwestern orientation will help the room take advantage of evening light while offering some protection from the low summer sun. The overhang of roofs, wall projections, type and amount of glazing, and landscaping can all be used to control the effects of sunlight on the structure. Figure 5.9 shows a casual living room that is open to the rest of the home and to the yard.

Family Room

The family room is probably the most used area in the house. A multipurpose area, it is used for such activities as watching television, informal entertaining, sorting laundry, playing pool, or eating. With such a wide variety of possible uses, planning the family room can be quite difficult. The room needs to be separated from the living room but still be close enough to have easy access from the living and dining rooms for entertaining. Figure 5.10 shows a possible layout of living areas providing close but separate rooms. The family room needs to be near the kitchen because it is often used as an eating area. Placing the family room adjacent to the

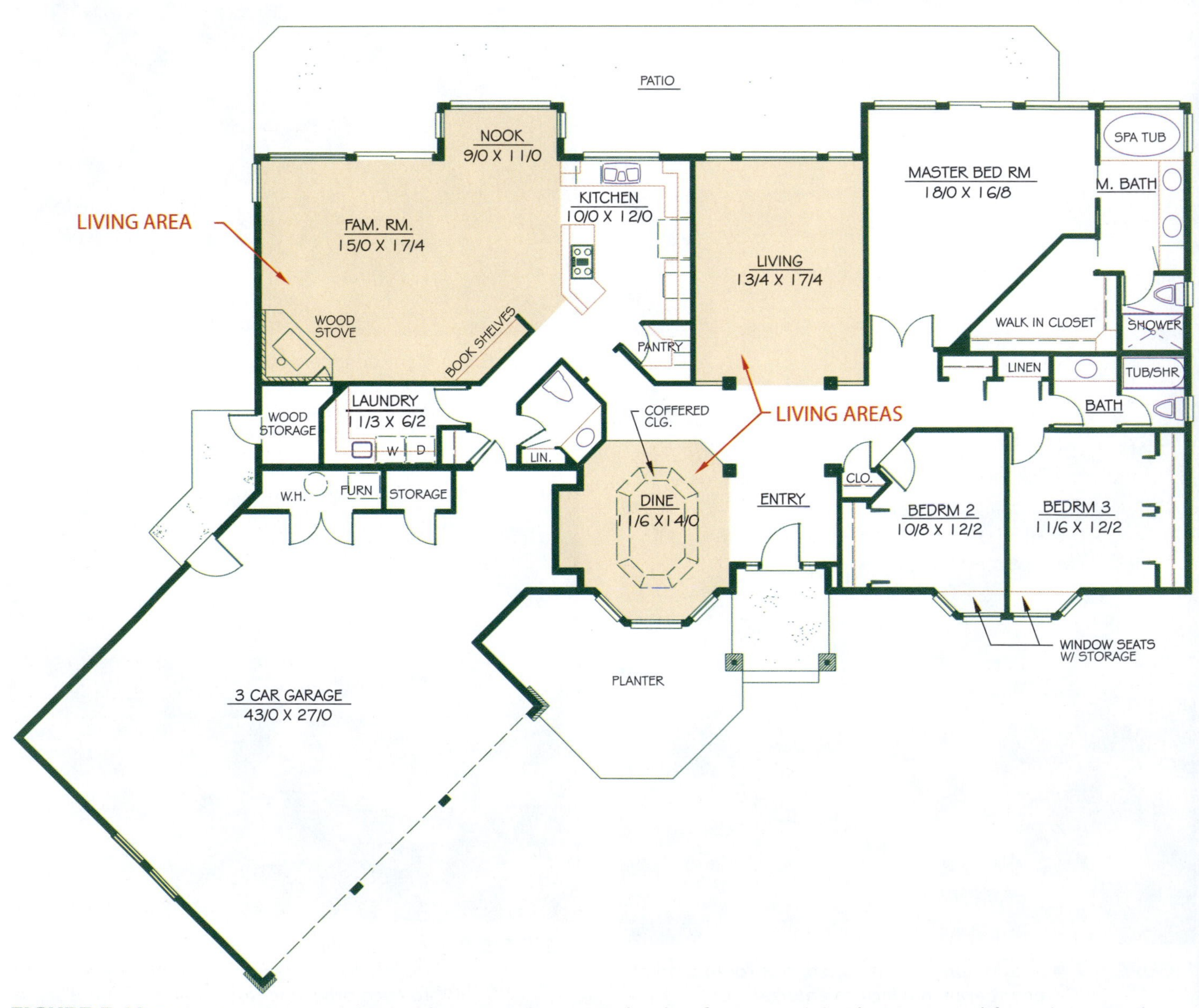

FIGURE 5.10 ■ *The living areas of this residence are close to each other for entertaining but separated for noise control.* Courtesy Wally Griener, Sunridge Designs, CPBD.

FIGURE 5.11a ■ *A pleasing arrangement of fireplace, television, and storage units should be part of every family room.* Courtesy BOWA Builders Inc. Greg Hadley, photographer.

FIGURE 5.11b ■ *This family room is used as a game room with a separate room for family gatherings and informal entertaining.* Courtesy BOWA Builders Inc. Bob Narod, photographer.

kitchen combines the two most used rooms into one area called a country kitchen. It is also common to have the family room near the service areas of the residence.

Sizes can vary greatly depending on the design criteria for the residence. When the family room is designed for a specific family, its size can be planned to meet their needs. When the room is being designed for a house where the owners are unknown, it must be of sufficient size to meet a variety of needs. An area of about 13 × 16' (3900 × 4800 mm) should be the minimum. This is a small room but would provide sufficient area for most activities. If a wood stove or fireplace is in the family room, the room should be large enough to allow for the heat from the stove. In an enclosed room where air does not circulate well, the area around a stove or fireplace is often too warm for comfort.

For most families, space should be provided for an entertainment center storing the television, stereo system, and related equipment. Figure 5.11a shows a pleasing arrangement of a fireplace as well as television and entertainment units. Figure 5.11b shows a family room used as a game room in a large residence.

Bar

A bar or minikitchen similar to Figure 5.11c is a key hub of the entertainment centers of many custom homes. Typically located in the family room or near the dining room, the bar provides a gathering place for casual entertainment. A sink, dishwasher, undercounter refrigerator, microwave, and cooktop are often found in the minikitchen. A raised counter set at a height suitable for bar stools generally surrounds these appliances.

FIGURE 5.11c ■ *A bar or minikitchen is a key hub of the entertainment centers of many custom homes.* Courtesy Dick Schmitke.

Dining Room

Dining areas, depending on the size and atmosphere of the residence, can be treated in several ways. The dining area is often part of the living area or next to it, as in Figure 5.12. For a more formal eating environment, the dining area will be near but separate from the living room area. See Figures 5.13a and 5.13b. The two areas are usually adjoining, so that guests may go easily

FIGURE 5.12 ■ ***The dining area can be combined with other living areas to create a spacious environment.*** Courtesy BOWA Builders Inc. Bob Narod, photographer.

FIGURE 5.13b ■ ***A formal eating area should be near to but separate from the living room area and very close to the kitchen.*** Courtesy Alan Mascord, AIBD. Alan Mascord Design Associates, Inc. Bob Greenspan, photographer.

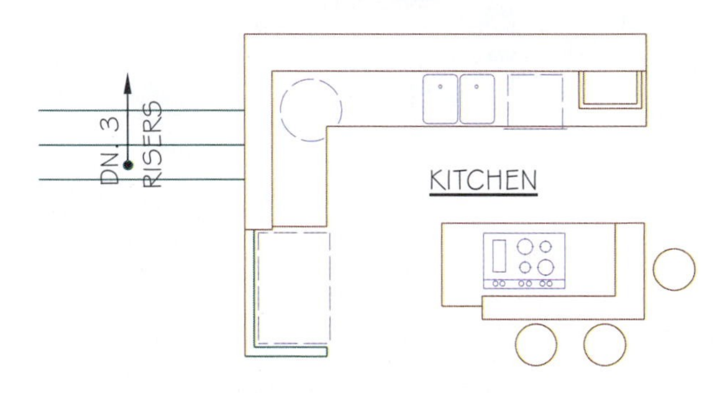

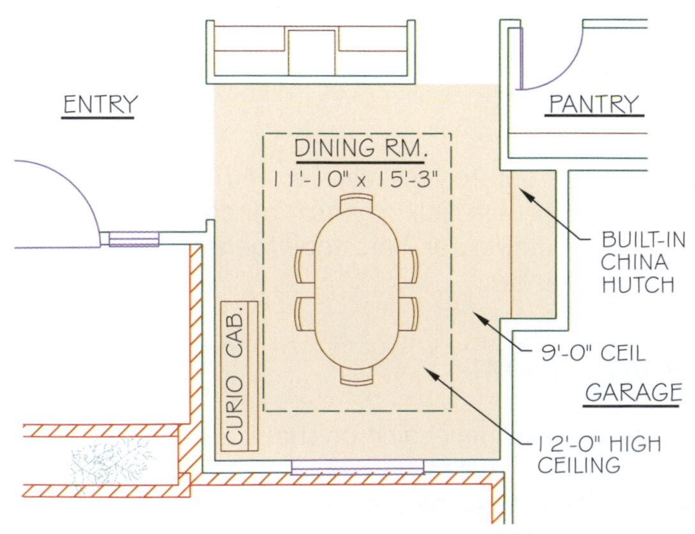

FIGURE 5.13a ■ ***Separating the dining room from the other living areas creates a more formal eating area.*** Courtesy Tereasa Jefferis, Stark Designs.

from one to the other without passing through other areas of the house. The dining room should also be near the kitchen for easy serving of meals but without providing a direct view into the work areas of the kitchen. Because of the need to be near both the living room and kitchen, the dining room is often placed between these two areas or in a corner between them, as in Figure 5.14.

A casual dining room can be as small as 9 × 11' (2700 × 3300 mm) if it is open to another area. This size would allow for a table seating four to six people and a small storage hutch. A formal dining room should be about 11 × 14' (3300 × 4200 mm). This will allow room for a hutch or china cabinet as well as the table and chairs. Allow for a minimum of 32'' (800 mm) from the table edge to any wall or furniture for placement of chairs and a minimal passage area. A space of approximately 42'' (1050 mm) will allow room for walking

FIGURE 5.13c ■ *Although it is open to other rooms, a formal atmosphere is created by enclosing the dining area with columns and arches.* Courtesy Eric S. Brown Design Group Inc. Launce Taylor, photographer.

FIGURE 5.14 ■ *Combining the dining, family, and living areas provides ample space for entertaining.* Courtesy Wally Griener, SunRidge Designs, CPBD.

around a chair when it is occupied. Furniture typically associated with the dining area is shown in Appendix B, Table 5.2, on the CD.

Nook

When space and finances allow, a nook or breakfast area is often included in the design. A nook similar to Figure 5.15 is where the family will eat most meals and snacks. The dining room then becomes an area for formal eating only. Both the dining room and nook need to be near the kitchen. If possible, the nook should also be near the family room. Where space is at a premium, these areas are often placed together in one room called a grand or great room (Figure 5.16). In larger homes, a patio or deck may be enclosed to provide additional informal dining, as shown in Figure 5.17.

FIGURE 5.15 ■ *A nook is a casual area for family dining. This nook is near the kitchen food bar and the family room for convenient access to other activities.* Courtesy Sara Jefferis.

Den/Study/Sitting Room

Although the name may vary, many families plan for a room for quiet conversation, reading, and study. This room is typically located off the entry and near the liv-

FIGURE 5.16 ■ ***A grand room is a combination of a living, dining, and family room.*** Courtesy Alan Mascord, AIBD. Alan Mascord Design Associates, Inc. Bob Greenspan, photographer.

FIGURE 5.17 ■ ***A sunroom or solarium can provide added square footage at a relatively low construction cost.*** Courtesy Velux America, Inc.

FIGURE 5.18a ■ ***Many homes provide a room that can be used for quiet conversation, while louder activities are located in the family room. In smaller homes, a den often serves as a spare bedroom.*** Courtesy William E. Poole, William E. Poole Designs, Inc.

ing room. The den often serves as a buffer between the living and sleeping areas of a residence. In many tract houses, the den is used as a spare bedroom or guest room. Figure 5.18a shows a den used as sitting room. The room usage will determine the furniture required in the room. Built-in shelves and cabinets for storage are often provided. If the room is to be used as a spare bedroom or den, a Murphy bed similar to Figure 5.18b is often used so that the area can double as a sitting and sleeping area. A desk or built-in cabinets for storage and a small work area will also typically be provided when the room is expected to serve as a small office.

Home Office

If a room is to be a true office within the home, its entrance should be directly off the entry. An entry separate from the residence is ideal. Entry to the office can be provided from a covered porch, but it should be distinctive enough that clients will know where to enter. The size of the office depends on the equipment to be used and the number of clients to be accommodated. It should have room for a desk, computer workspace, storage areas for books and files, and a separate area

FIGURE 5.18b ■ *If the study is to be used as a spare bedroom, a Murphy or day bed is often provided.*

for phones, fax, and photocopying machines. Figure 5.18c shows a home office. Chapter 12 examines electrical requirements, and Chapter 19 explores cabinets for a home office.

Home Theater

Although not included in most tract homes, personal theaters or media rooms similar to Figure 5.19 are becoming increasingly popular in many high-end custom homes. What used to be unfinished space or a bonus room can be made into a home theater. Home theaters are custom-designed rooms created, as the name implies, for high-quality viewing of movies. The home designer will work closely with media consultants to meet the needs of the family and help plan the locations of sound equipment. Speakers and electrical control equipment are indicated on the electrical drawings; these drawings are introduced in Chapter 12. Other points to consider are the size and type of television set,

FIGURE 5.18c ■ ***Many homes now include a true office to accommodate professionals who work at home and telecommute.*** Courtesy BOWA Builders Inc. Bob Narod, photographer.

the number of viewers, and the type of seating. A wall-mounted plasma television will require far less room than a projection television or even a standard big-screen TV. Drawings for the entertainment center will be part of the interior elevations. Although most home theaters are designed for eight to twelve people, larger-capacity rooms are occasionally found. Sofas, recliners, and easy chairs make for casual, comfortable seating. Theater-type seating is also available. For larger seating groups, an inclined or stepped floor should be part of the design. The changes in floor elevation must be represented on the floor plan and on interior sections or details. No matter how small the room, space should be provided for snack preparation. Items that may be included are a popcorn machine, a counter with a microwave, and a small refrigerator or wet bar.

SLEEPING AREAS

The age, gender, and number of children will determine how many bedrooms will be required. Preteens can share a bedroom if space is limited. Teens of the same gender have even been known to survive sharing a room together, although this situation often results in distinctive markings of territory during tense times. The ideal situation is to provide a separate bedroom for each child. Each room should have space for sleeping, relaxation, study, storage, and dressing.

The sleeping area consists of bedrooms. These rooms should be placed away from the noise of the living and service areas and out of the normal traffic patterns. Even in a family with no children, a minimum of two

FIGURE 5.19 ■ *Although not included in most tract homes, personal theaters or media rooms are popular in many high-end custom homes.* Courtesy BOWA Builders Inc. by Greg Hadley, photographer.

bedrooms should be provided. A third bedroom greatly increases the value of the home for resale. Most homes have a minimum of three bedrooms; four bedrooms and a den are a common option for many subdivision homes. The extra room provides space for guests, or it can double as a craft, sewing, or hobby room. Many custom homes plan for a space about the size of a bedroom to be used as a multipurpose room. An exercise or equipment room, a game room, and a darkroom are common types of multipurpose rooms located near the bedrooms.

The arrangement of the bedrooms will vary greatly depending on the needs of the family. Common arrangements include placing all the bedrooms together or placing the master bedroom separate from the bedrooms for children. It is also becoming common to plan a bedroom-living unit for long-term care of a live-in relative.

Bedrooms are generally located with access from a hallway for privacy from living areas and to be near bathrooms, as in Figure 5.20. Care must be taken to keep the bathroom plumbing away from bedroom walls. One way to accomplish this is to place a closet between the bedroom and the bath. If plumbing must be placed in a bedroom wall, insulation should be used to help control noise.

Bedrooms

Bedrooms function best on the southeast side of the house. This location brings morning sunlight to the rooms. When a two-level layout is used, bedrooms are often placed on the upper level, away from the living ar-

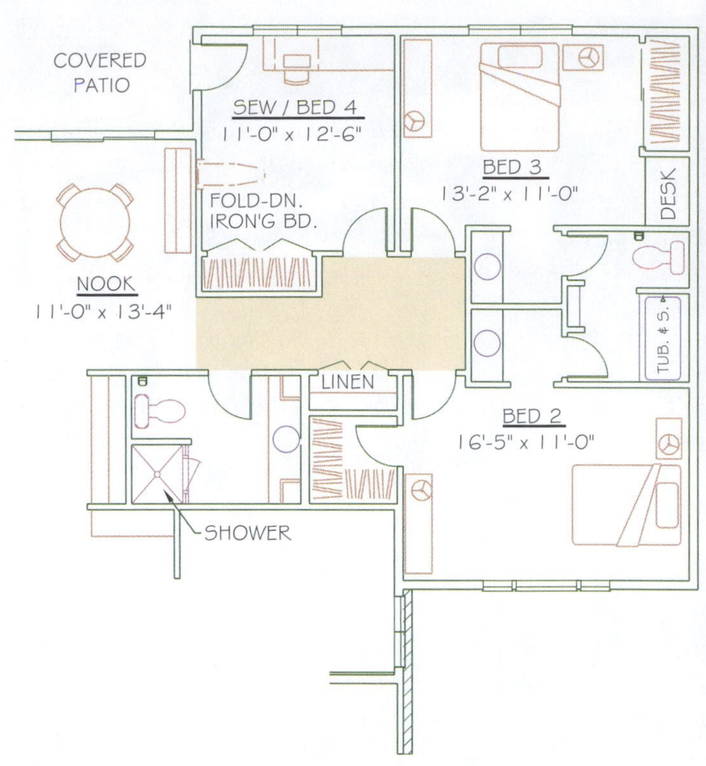

FIGURE 5.20 ■ *Bedrooms should be located with access from a hallway for privacy from the living areas; they should also be near bathrooms.* Courtesy Tereasa Jefferis, Stark Designs.

eas. This arrangement not only provides a quiet sleeping area but often allows the bedrooms to be heated by the natural convection of heat from the living area. Another option is placing the bedrooms in a daylight basement. Care must be taken in basement bedrooms to provide direct emergency exits to the outside. An advantage of a bedroom in a daylight basement is the cool sleeping environment that the basement provides.

Sizes of bedrooms vary greatly depending on the size of the home, the age of the occupants, and the furniture to be placed in them. The IRC requires a minimum of 70 sq ft (6.5 m^2). Homes financed by FHA loans are required to provide a minimum of 100 sq ft (9.3 m^2). Spare or children's bedrooms are often as small as 9 × 10' (2700 × 3000 mm). This might be adequate for a sleeping area, but not if the room is also to be used as a study or play area. Each bedroom should have enough space for a single bed, a small bedside table, and a dresser. Clients often specify enough space for twin beds or a double bed. Plan for a minimum of 24'' (600 mm) on each side of a bed, where space to walk is required. A space of approximately 36'' (900 mm) should be provided between dressers and any obstruction to allow for the opening of drawers or doors. This requires a space of about 12 × 14' (3600 × 4200 mm) plus closet areas. Minimum closet space is covered in the next section.

More than just a room for sleeping, the master bedroom in many custom homes is a retreat from the problems of the day. The master suite serves as a bedroom, sitting area, and bathing area. A well-designed master suite, as in Figure 5.21, provides spacious room for a queen- or king-sized bed plus room for a fireplace, sitting area, direct access to the wardrobe and bathing areas, and access to a private deck or balcony. A master suite also includes its own bathroom, similar to Figure 5.22. Bathroom design considerations are considered as service areas are explored.

The master bedroom should be at least 12 × 14' (3600 × 4200 mm) plus closet areas. An area of 13 × 16' (3900 × 4800 mm) will make a spacious bedroom suite, although additional room is required for a sitting or reading area. A window seat can often be used to provide sitting room without greatly increasing the room's square footage. Try to arrange bedrooms so that at least two walls can be used for bed placement. This is especially important in the master bedroom, to allow for periodic furniture movement. Table 5.3 lists common sizes of bedroom furniture. See Appendix B on the CD.

FIGURE 5.21 ■ *A well-designed master suite provides spacious room for a queen- or king-size bed, a fireplace, a sitting area, direct access to the wardrobe and bathing areas, and access to a private deck or balcony.* Courtesy W. Lee Roland, Builder. Photo by Hayman Studios.

FIGURE 5.22 ■ *A spa or soaking tub is often a part of the master suite.* Courtesy W. Lee Roland, builder. Photo by Hayman Studios.

Window planning is very important in a bedroom. Although morning sun is desirable, space should be planned so that sunlight will not shine directly on the bed. In planning windows for a bedroom, it is especially important to allow for emergency egress. (See Chapter 4 for a review of emergency window requirements.) For upper-level bedrooms used by young children, try to provide a roof, porch, or balcony as an escape route. A flexible escape ladder can be used for older children.

Closets

Building codes do not require bedroom closets. The FHA recommends a length of:

- 4' (1200 mm) of closet space for males.
- 6' (1800 mm) for females.
- 6' (1800 mm) should be considered the practical minimum for resale if a traditional shelf-and-pole storage system is to be used for storage.
- Space can be reduced if closet organizers are used.
- Minimum closet depth is typically considered to be 24'' (600 mm).
- A depth of 30'' (750 mm) is adequate to keep clothing from becoming wrinkled.

FIGURE 5.23 ■ *An area containing shelves over single poles, double poles, adjustable shelves, baskets, and drawers are common methods of providing storage for the master wardrobe.* Courtesy Tom Worcester.

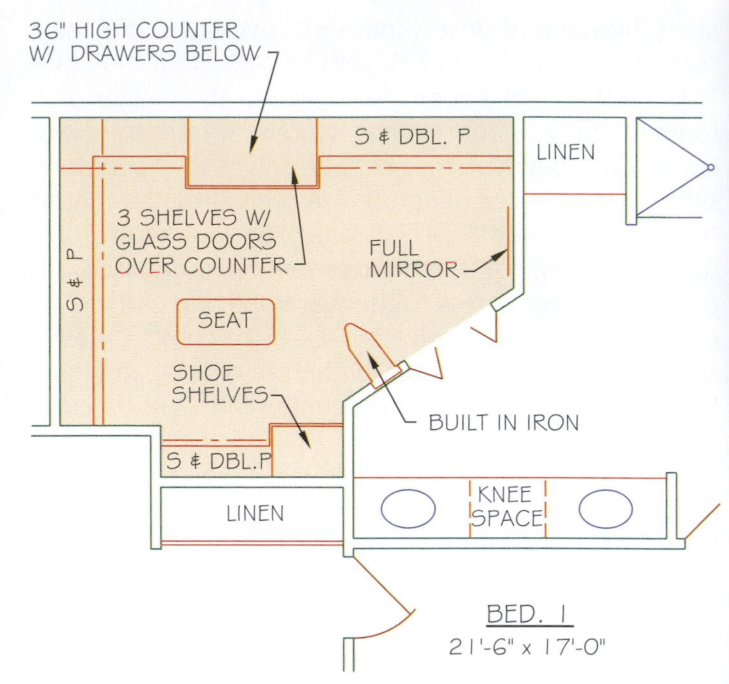

FIGURE 5.24 ■ *A well-planned wardrobe provides ample storage space for hanging and folded clothes and easy access to the bedrooms and bathrooms.*

If a closet must be small, a double-pole system can be used to double the storage space or a premanufactured closet organizer can be used to provide storage. The lower pole should be placed a minimum of 30'' (750 mm) above the floor. The upper pole should be placed 36'' (900 mm) above the upper pole. Keep shelving 2'' (50 mm) minimum above the upper pole. Closets can often be used as a noise buffer between rooms or located in what would be considered wasted space in areas with sloping ceilings. The master bedroom will often have a walk-in closet. It should be a minimum of 6 × 6' (1800 × 1800 mm) to provide adequate space for clothes storage. A closet of 6 × 8' (1800 × 2400 mm) provides much better access to all clothes. Providing multiple rods for hanging pants, shirts, skirts, and blouses can increase storage. A single pole should be provided for hanging dresses and seasonal coats. An area containing shelves, baskets, and drawers, as in Figure 5.23, is also desirable if space permits. Figure 5.24 shows a well-planned wardrobe area for a master suite. Many clients request space for a freestanding armoire to provide additional storage. Space must also be provided near the bedrooms for linen storage. A space 2' (600 mm) wide and 18'' (450 mm) deep would be minimal.

SERVICE AREA

Bath, kitchen, utility rooms, and the garage are each considered a part of the service area. Custom homes also tend to include areas such as bonus rooms, gymnasiums, pet grooming areas, gardening rooms, and greenhouses. Notice that many of these areas require plumbing. Because of the plumbing and services that each provides, an attempt should be made to keep the service areas together. Another consideration in placing the service area is noise. Each of these four areas tends to give rise to noises that will interrupt activities of the living and sleeping areas.

Bathrooms

To provide privacy, bathrooms are often reached by a short hallway, apart from living areas. A house with only one bathroom must have it located for easy access from both the living and sleeping areas. Access to the bathroom should not require having to pass through the living or sleeping areas.

Options for bathrooms include half-bath, three-quarter bath, full bath, and bathroom suite. A half-bath has a lavatory and a water closet (a toilet). A three-quarter bath combines a shower with a toilet and a lavatory. A full bath has a lavatory, a toilet, and a tub or a combination tub-and-shower unit. A bathroom suite typically includes the features of a full bath with a separate tub and shower similar to the bathroom in Figure 5.25. The tub is often an enlarged tub or spa. If the tub is to be raised, skidproof steps should be provided. If windows are placed around a tub or spa, the glass is required to be tempered. The wardrobe, dressing area, and a sitting area are also usually part of the bathroom suite or adjoining it. Figure 5.26 shows a plan view for a master suite. Many custom homes feature a full bath with an adjoining exercise room, sauna, or steam room.

The style of the house affects the number and locations of bathrooms. A single-story residence typically has one bathroom for the master bedroom and a second to serve the living room and the other bedrooms. A two-story house typically has two full bathrooms upstairs with a minimum of a half-bath downstairs. Multilevel homes should have a bath on each level containing bedrooms or a half-bath on each level with any type of living area.

When a residence has two or more baths, they are often placed back to back or above each other to reduce plumbing costs. A full bath is typically provided near the master bedroom, with a separate bathroom to serve the balance of the bedrooms. If space and budget allow, a bathroom for each bedroom is common. For families with young children, a tub is usually a must. Many bathrooms designed to be shared place the sinks in one room with the toilet and tub in an adjoining room (Figure 5.27). For homes designed to appeal to a wide variety of families, a combination of a tub and shower is often used. Homes designed for outside activities often have a three-quarter bath near the utility room or combine the laundry facilities with a bathroom to create what is often called a mudroom.

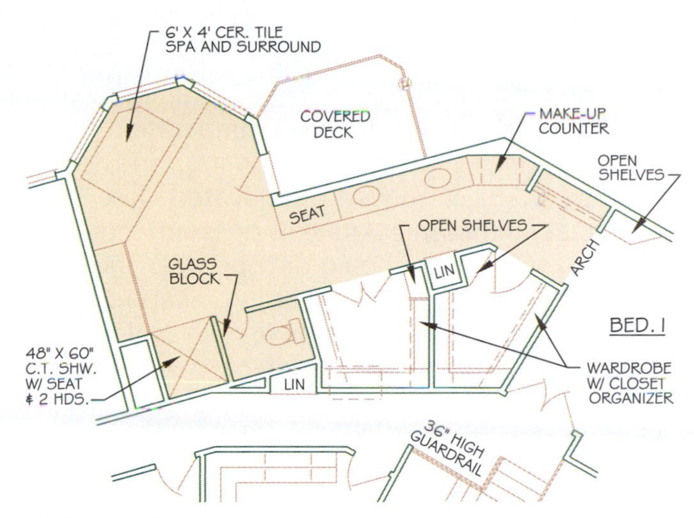

FIGURE 5.26 ■ *Master bath suite includes the sinks, makeup area, bathing area, wardrobe, and sitting area.*

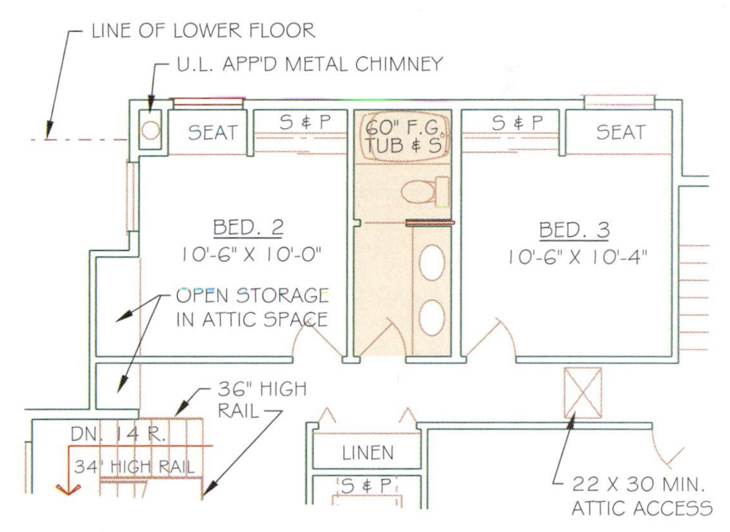

FIGURE 5.27 ■ *Many bedrooms share a bathroom. When the lavatories are separated from the bathing area, two people can use the room at once.*

FIGURE 5.25 ■ *A master bedroom often includes a soaking tub with a separate tub and shower.* Courtesy William E. Poole, William E. Poole Designs, Inc.

Kitchen

A kitchen is used through most of a family's waking hours. It serves not only for meal preparation but often includes areas for eating, working, and laundry (see Figure 5.28). It needs to be located close to the dining areas so that serving meals does not require extra steps. The kitchen should also be near the family room. This will allow those preparing meals to be part of family activities. When possible, avoid placing the kitchen in the southwest corner of the home. This location will receive the greatest amount of natural sunlight, which could easily cause the kitchen to overheat. Because the kitchen creates its own heat, try to place it in a cooler area of the house unless venting and shading precautions are taken. One advantage of a western placement is the natural sunlight available in the late evening.

In a house for a family with young children, a kitchen with a view of indoor and outdoor play areas is a valuable asset. This will allow for the supervision of playtimes and the control of traffic in and out of the house. The kitchen is often closely related to the utility area. Because these are the two major workstations in the home, a kitchen close to the utility room can save valuable time and energy as daily chores are done. It is also helpful to place the kitchen near the garage or carport. This location will allow groceries to be unloaded easily (Figure 5.29).

FIGURE 5.28 ■ *The kitchen is used for meal preparation and is often a place for eating as well as a social hub at family gatherings and parties.* Courtesy Dennis Dinser, Arcadian Design, LLC.

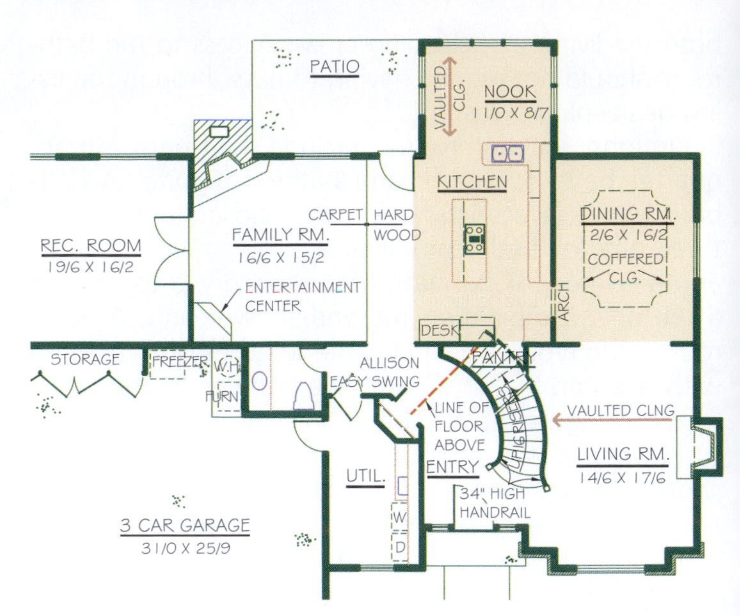

FIGURE 5.29 ■ *For efficient living, the kitchen must be near the dining, living, family, and utility areas as well as the service entry. If traffic must flow through the kitchen, it should not pass through the work triangle.* Courtesy Wally Griener, Sunridge Designs.

Kitchen Work Areas

The space within the kitchen demands great consideration. Perhaps the greatest challenge facing the designer is to create a workable layout in a small kitchen. The ***National Kitchen & Bath Association*** (NKBA) considers any kitchen with less than 150 sq ft (13.9 m^2) as a small kitchen, and kitchens with greater than 150 sq ft (13.9 m^2) as a large kitchen. Layout within the kitchen includes the relationship of the appliances to the work areas. The main work areas of a kitchen are those for storage, preparation, and cleaning. Each can be seen in Figures 5.30a and 5.30b.

Storage Areas

The storage areas consist of the refrigerator, the freezer, and cabinet space for food and utensils. Most families prefer to have a refrigerator in the kitchen, with a separate freezer in the laundry or garage. Storage for cans and dried foods is typically in base cabinets 24'' (600 mm) wide or in a pantry unit, similar to Figures 5.31a and 5.31b. Upper units 12'' (300 mm) deep are also typically used for dishes and nonperishable foods. A minimum counter surface 18'' (450 mm) wide should be provided next to the refrigerator to facilitate access to it (see Figure 5.32). This area can also be useful for preparing snacks if the width is increased to approximately 36'' (900 mm). As the size is in-

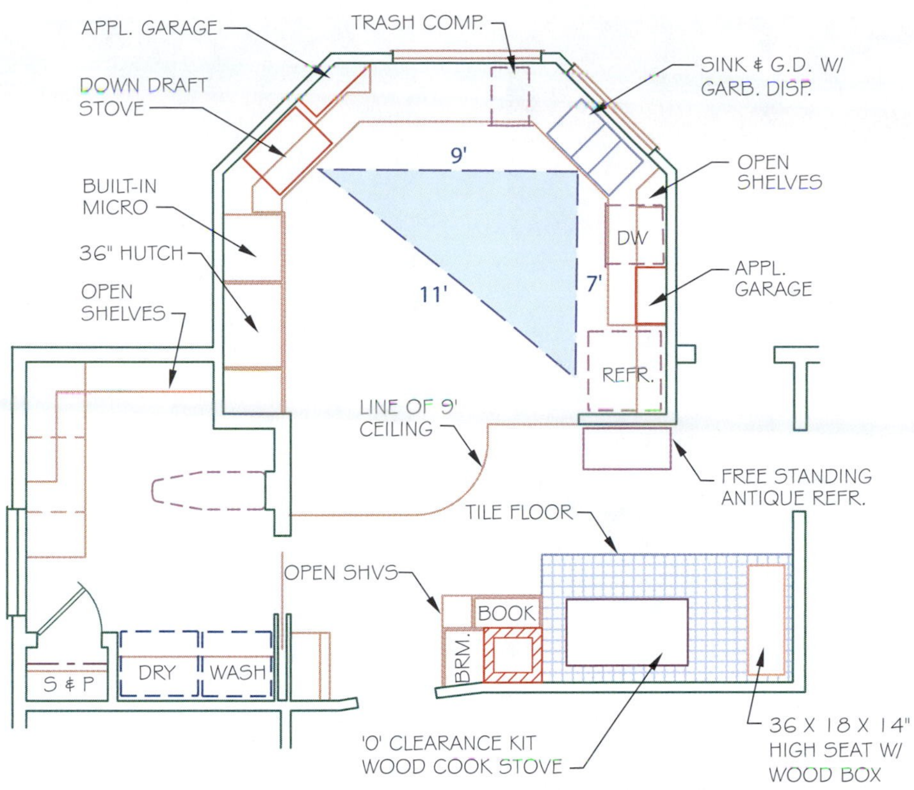

FIGURE 5.30a ■ *A well-planned kitchen will provide preparation areas, storage areas, and cleaning areas separated by adequate counter space. These areas are generally not specified on the floor plan except to denote specific appliances required for each area. Related information is specified on the interior elevations (see Chapter 19).*

FIGURE 5.30b ■ *Minimum preparation areas of approximately 48'' (1200 mm) of work space should be provided for each appliance. This length should be increased to 72'' (1800 mm) if two or more people will be involved in preparing meals.* Courtesy Dennis Dinser, Arcadian Design, LLC.

creased, this counter can also be used to store appliances, such as mixers and blenders in an appliance garage. This is a useful area for a breadboard and a drawer for silverware storage. Additional drawer space could be used for mixing bowls, baking pans, and other cooking utensils similar to those in Figure 5.33. Enclosed cabinet space should also be provided for storage of cookbooks, writing supplies, and telephone books. The NKBA recommends that the main storage area include at least five of the following items located 15'' to 48'' (375 to 450 mm) above the finished floor. The items useful for storage include wall cabinets (see the left side of Figure 5.31b), raised base cabinets, appliance garage, tall cabinets, bins, racks, swing-out pantry closets, interior vertical dividers, specialized drawers, and pull-out shelves.

Preparation Areas

The main components of the preparation areas are the sink, cooking units, and a clear counter workspace. Clear counter workspace should be placed near the storage area, sink, and cooking areas with a minimum of one counter that provides approximately 48'' (1200 mm) of workspace. This length should be increased to 72'' (1800 mm) if two or more people will be involved in preparing a meal. The preparation center can be placed between the primary sink and the cooking unit, between the refrigerator and the primary sink, near a secondary sink, or in an island. Each work area should have a counter at least 18'' (450 mm) wide. Specific requirements will be introduced as each appliance is dis-

FIGURE 5.31a ■ *A pantry should be located in or near the kitchen to provide storage for dry and canned goods. This walk-in pantry is located between the kitchen and garage of the remodel shown in Chapter 32. It uses shelves, bins, and pull-out drawers for long-term storage and provides space for a second refrigerator. Photo* Courtesy Janice Jefferis.

FIGURE 5.31b ■ *Storage cabinet located in the kitchen should provide pull-out shelves and drawers for storage of foods for daily use. Because the cabinet is located just to the left of the baking center, the lower drawers provide convenient storage for large pans and other baking utensils.* Courtesy Cindy Stead.

cussed. Large kitchens often have a small vegetable sink in the preparation area and a larger sink for cleaning utensils. A minimum size for a vegetable sink is 16 × 16'' (400 × 400 mm).

The cooking appliances are usually a range that includes both an oven and surface heating units or separate appliances for baking and cooking. Most kitchens also provide space for a microwave oven that can be part of a built-in oven, be mounted below the upper cabinets, or be placed on the counter. A warming drawer is often located below oven units for additional heating capacity. Larger kitchens also often include built-in coffee or espresso centers. A minimum counter space of 18'' (450 mm) should be placed on each side of a range or oven unit to prevent burns and provide temporary storage while cooking. Figure 5.34 shows a double oven and a freestanding range. A water faucet can be placed in the wall behind the cooktop, about

FIGURE 5.32 ■ *The refrigerator is used for the short-term cold storage of food. A counter 18'' (450 mm) wide should be provided next to the refrigerator. In larger homes, a second refrigerator may be kept in the pantry for additional storage. An undercounter cooler is also common for additional cold storage.* Courtesy Judy Schmitke.

FIGURE 5.33 ■ *Storage can consist of drawers, pull-out shelves behind doors, and bins. The drawers on the left side have false fronts for the display of dry food items such as lentils, beans, and corn.* Courtesy Lynn Worcester.

18'' (450 mm) above the cooking elements to provide easy access for filling large pots with water.

Cleaning Center

The cleaning center typically includes the sink, garbage disposal, and dishwasher, as in the arrangement shown in Figure 5.35. The sink and the surrounding cabinets and counter space are the most heavily used work area of the kitchen, so this area should be centrally located. Most clients prefer a double sink, but a single sink is convenient for washing large items. A typical double sink is 32 × 21'' (800 × 525 mm), but 36'' (900 mm) and 42'' (1050 mm) wide units are also available. Double sinks are also available with the bowls at 90° to each other for use in a corner. If the sink is connected to a public sewer, a garbage disposal can be installed in one side of the sink to eliminate wet garbage. A garbage disposal should not be used on sinks connected to septic tanks because the waste can often overload the system. The NKBA recommends that if a kitchen contains only one sink, the sink should be:

- Located between or across from the cooking surface, preparation center, or the refrigerator.

FIGURE 5.34 ■ *The cooking center of this home includes a freestanding stove, a microwave oven, and preparation areas on each side of the stove. The cooking elements allow for the use of burners, a grill, or a rotisserie. Storage includes open upper cabinets for spices, hanging storage for common utensils, and enclosed upper cabinets for cooking supplies. The counter provides space for the storage of baking supplies. Lower cabinets feature breadboards and drawers and doors for storage.* Courtesy Sam Griggs.

- It should have a clear floor space of 30 × 48'' (750 × 1200 mm) in front of the sink and the dishwasher.
- A minimum area 36'' (900 mm) wide should be provided on one side of the sink for stacking dirty dishes.
- A minimum space of 24 to 30'' (600 to 750 mm) on the other side for clean dishes.

- The dishwasher should be placed on either side of the sink, depending on the client's wishes within 36'' (900 mm) of the edge of the sink.
- A minimum of 21'' (525 mm) of clear floor space should be allowed between the edge of the dishwasher and cabinets or appliances that are placed at right angles to the dishwasher to allow loading and unloading.
- An upper cabinet should be provided near the dishwasher to store dishes.

Many larger kitchens include more than one sink. The second sink can be used near a food bar to provide an additional preparation area.

The Work Triangle

The mere fact that a kitchen is big and has all the latest appliances does not make it efficient. Designing a functional kitchen requires careful consideration of the relationship between the work areas. This relationship, referred to as the work triangle, is a key aspect of kitchen ergonomics. The work triangle is formed by drawing lines from the centers of the storage, preparation, and cleaning areas. This triangle outlines the main traffic area required to prepare a meal. Food will be taken from the refrigerator, cleaned at the sink, and cooked at the microwave or range, and then leftovers will be returned to the refrigerator. Using this work pattern and considering the work triangle, good design places the work centers at approximately equally spaced points of the triangle connected by counters. General rules for efficient kitchen design include:

- Always place workspace between each workstation of the triangle.
- No side of the work triangle should be less than 4' (1200 mm) or greater than 7' (2100 mm).
- The sum of the sides of the work triangle should be at least 15' (4500 mm) but not more than 22' (6600 mm).
- Never arrange rooms so that traffic is required to pass through the work triangle.

Keep in mind that these are only guidelines and that a large kitchen consists of more than just three appliances or workstations. The triangle may need to be modified to accommodate doors or preserve a view. Appliances such as convection ovens, grills, microwave ovens, multiple sinks, and even multiple refrigerators will all affect the arrangement of work areas and appliances.

In addition to traffic within the kitchen, the relationship of the kitchen to other rooms also needs to be considered. If care is not taken in the design process, the kitchen can become a hallway. The kitchen needs to be in a central location, but traffic must flow around the kitchen, not through it. Figure 5.36 shows the traffic pattern in three different kitchens.

FIGURE 5.35 ■ *The kitchen sink is the hub of the cleaning center. Located near windows providing a view of the outside entertainment areas, the sink is flanked by a trash compactor on the left and a dishwasher on the right. Storage for dishes is provided above the dishwasher. The drawers provide storage for eating utensils and other basic kitchen supplies.* Courtesy Marilyn Clifton.

Common Counter Arrangements

Designers typically use one of six common counter arrangements to define the work triangle: straight (one wall), L-shape, corridor, U-shape, peninsula, and island (see Figure 5.37). Each arrangement presents its own design challenges.

Straight

The straight-line or single-wall kitchen layout is ideal when appliances must be placed in a very small space. This type of layout is often used in a small apartment, a mother-in-law unit, or a recreation room that is on a different floor from the main kitchen. Although this arrangement requires few steps to move between appliances, it also provides a very limited amount of cabinet space and work area. Using a movable cart or cutting block to expand the work triangle can improve this arrangement.

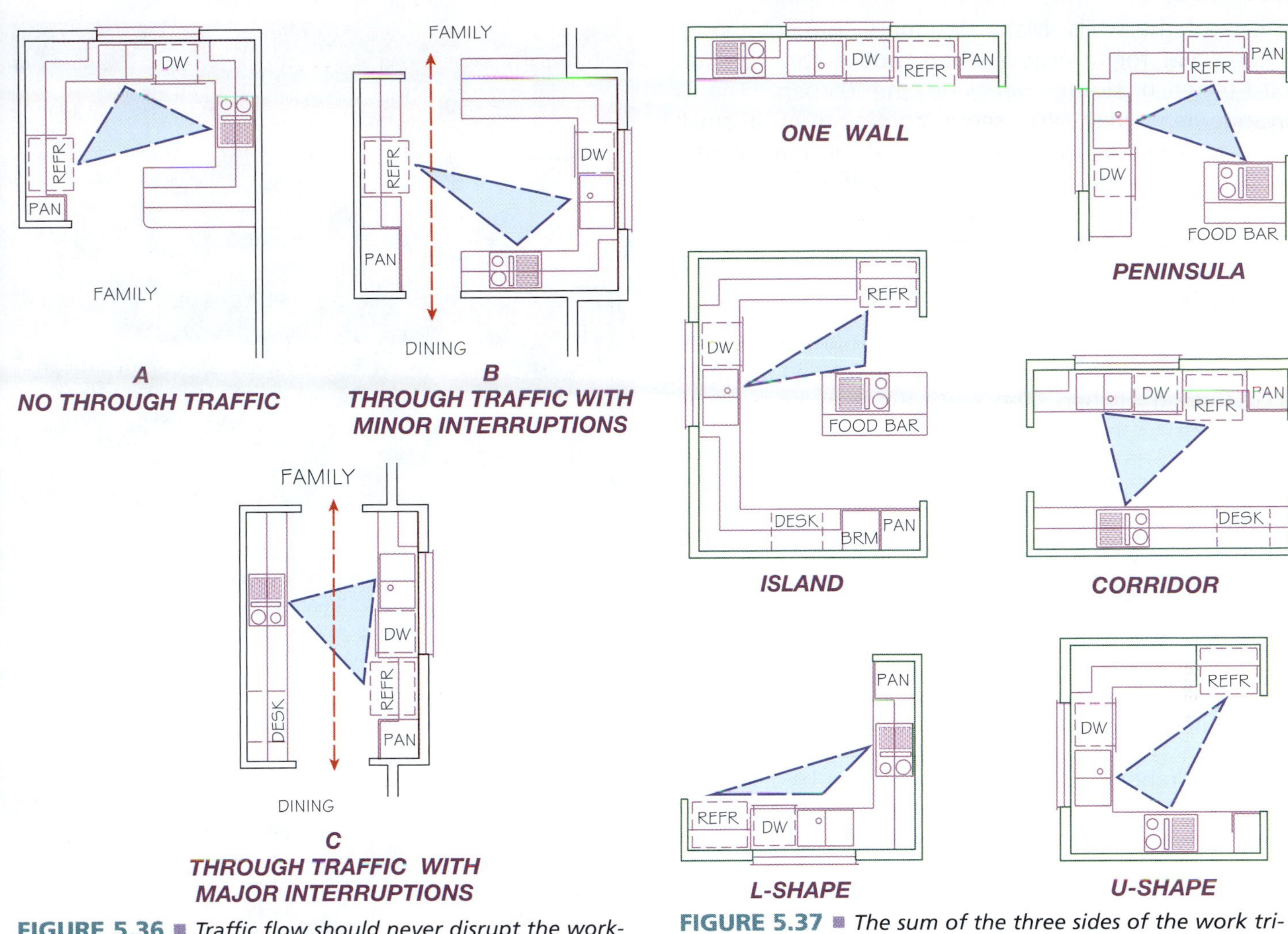

FIGURE 5.36 ■ *Traffic flow should never disrupt the working area of a kitchen; plan A accomplishes this. Plan B allows traffic through the kitchen, but it will not disrupt work. Plan C shows traffic that will be very disruptive.*

FIGURE 5.37 ■ *The sum of the three sides of the work triangle must be between 15 and 22' (4500 and 6600 mm) for an efficient work pattern.*

Corridor or Galley

The corridor or galley arrangement places all of the cabinets on two parallel walls. This arrangement is a great improvement over a straight-line kitchen because it allows convenient storage, provides ample counter space in a small area, and minimizes walking distance in the triangle. The distance between cabinet faces should be a minimum of 48'' (1200 mm). A width of 54'' to 64'' (1350 to 1600 mm) will allow two or more cooks to use the kitchen simultaneously. This arrangement is appropriate for a small home but should not be used in larger homes because of the possibility of poor ***traffic flow.*** A galley kitchen is usually placed between two living areas. Unless an alternative route is provided, the galley kitchen will become a thoroughfare for room-to-room traffic.

L-Shape

An L-shaped cabinet arrangement is suitable for both small and large kitchens, but it loses efficiency as the size of the kitchen increases. Cabinets are placed on two adjacent walls with two workstations placed along one wall and the third workstation placed on the remaining wall. The L-shape enables efficient travel between workstations located close to the right angle, but the arrangement loses its effectiveness as the legs of the L become longer. This type of cabinet arrangement eliminates traffic through the work area and is well suited to great rooms.

U-Shape

The U-shaped cabinet arrangement provides an efficient layout with easy access between workstations. It also eliminates through traffic unless a door is placed in one of the walls. Many designers consider the U-shape ideal for a large kitchen, but it can result in hard-to-reach storage areas in the corners. The U arrangement may also seem confining in a small kitchen. A minimum space of 60'' (1500 mm) should be used between cabinet faces. As the distance increases, an island or peninsula can be added to improve the cabinet arrangement.

Peninsula

A peninsula arrangement provides ample cabinet and workspace by adding one additional leg to an L-shaped or U-shaped kitchen. The peninsula can be used as workspace, as the location for a work center, as a food bar, or as a combination of these features.

Island

An island can be added to any of the other five cabinet arrangements. The island can be used to provide added counter space or work space. A range or cooktop or a small sink is often placed in the island. A minimum of 42'' (1050 mm) should be provided between the island and other counters to allow for traffic flow around the island. If the island contains an appliance such as a range, 48'' (1200 mm) should be the minimum distance provided between counters to allow for traffic flow when the oven door is open.

FIGURE 5.38a ■ *When the food bar is level with the main countertop, allow deep clear space of at least a 15'' (375 mm) for the counter.* Courtesy BOWA builders, Inc. Jim Tetro photographer.

FIGURE 5.38b ■ *For a counter that is 30'' (750 mm) high, allow a space 18'' (450 mm) deep for each person seated at the counter. When the food bar is 42'' (1050 mm) above the floor, allow a deep clear space of at least a 12'' (300 mm) for the food bar.* Courtesy Alan Mascord, AIBD. Alan Mascord Design Associates, Inc. Bob Greenspan, photographer.

Counter and Cabinet Sizes

The standard kitchen counter is 36'' (900 mm) high. The NKBA recommends that at least two different counter heights be provided in the kitchen, with one work area ranging from 28'' to 36'' (700 to 900 mm) and a second area ranging from 36'' to 45'' (900 to 1125 mm) above the finish floor. For custom homes, the height may be adjusted to meet the physical needs of the owners, but it is important to remember that radical changes from the norm may greatly limit resale value. Common cabinet dimensions are listed in Table 5.4 of Appendix B on the CD.

If a food bar is provided in a peninsula or island, consideration must be given to the height and width of the eating counter. Common heights include 30'', 36'', and 42'' (750, 900, and 1050 mm). Widths range from 12'' to 18'' (300 to 450 mm). For a counter 30'' (750 mm) high, allow a space 30'' wide × 18'' deep (750 × 450 mm) for each person seated at the counter. When the food bar is level with the main countertop, as shown in Figure 5.38a, allow a deep clear space of at least a 15'' (375 mm) for the counter. When the food bar is 42'' (1050 mm) above the floor similar to the counter in Figure 5.38b, allow a deep clear space of at least a 12'' (300 mm) for the food bar. If a counter 30'' (750 mm) high is provided, chairs can be used with it. The other heights will require stools for seating. When the food bar is the same height as the other counters, the counter area will seem much larger than when two different heights are used. A width of 12'' (300 mm) will provide sufficient space for eating. If the food bar is

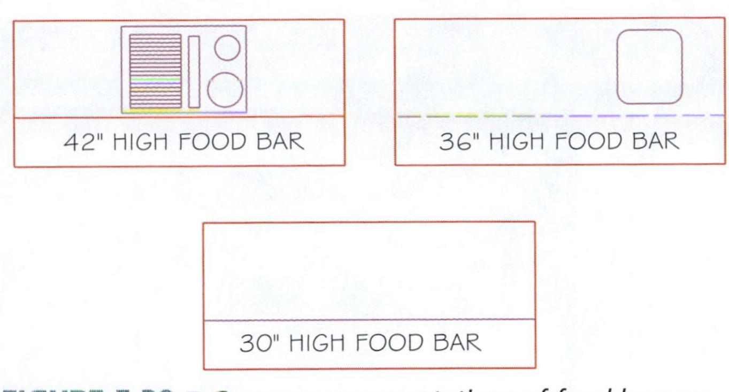

FIGURE 5.39 ■ *Common representations of food bars on floor plans.*

higher than the other kitchen cabinets, it can be used as a visual shield to hide clutter in the kitchen. Figure 5.39 shows how each type of food bar could be represented on the floor plan.

Arrangement of Appliances

Many homeowners prefer the sink to be located in front of a window, but a location with a view into other living areas is also popular. Placement of the sink at a window allows for supervision of outdoor activities and provides a source of light at this workstation. Because of its placement at a window, the sink is often located first as the kitchen is being planned. The sink should also be placed to promote easy movement between work areas. The sink should be near the cooking units and the refrigerator to facilitate the preparation of fresh foods. Provide a minimum of 48'' (1200 mm) between counters to accommodate someone working at the sink and another person walking behind him or her. Try to avoid placing the sink and dishwasher on different counters even if they would be just around the corner from each other. Such a layout often leads to accidents from water dripping onto the floor.

Refrigerator

Typically a space 36'' (900 mm) wide is provided for the refrigerator. The refrigerator should be placed near the service entry to ease unloading of groceries. The refrigerator should be placed at the end of a counter so that it will not divide the counter into small workspaces. If it must be placed near a cabinet corner, allow a minimum of 15'' (375 mm) in the corner for access to the back of the counter. A minimum of 15'' (375 mm) of counter space should be provided on the latch side of a single-door refrigerator and on both sides of a side-by-side unit. If space cannot be provided adjacent to the refrigerator, provide a counter within 48'' (1200 mm) across from the refrigerator. The location must also be convenient for access to the sink for food preparation and to the cooking areas for food storage. The refrigerator should be placed within 60'' to 72'' (1500 to 1800 mm) of both the sink and the cooking center. If possible, it should be placed so that the door will not block traffic through the work triangle. Avoid placing the refrigerator beside an oven.

Range

The cooking units should be located near the daily eating area. A range contains the surface cooking elements and an oven. It can be either freestanding or built-in to the cabinets with a cabinet drawer below the oven unit. The range should be placed so that the person using it will not be standing in the path of traffic flowing through the kitchen. This will reduce the chance of hot utensils being knocked from the range. Ranges should be placed so that there is approximately 18'' (450 mm) of counter space between the range and the end of the counter to prevent burns to people passing by. If a food bar is provided behind and at the same height as the cooking surface, provide a minimum space of 9'' (225 mm) behind the cooking surface. Ranges should not be placed within 18'' (450 mm) of an interior cabinet corner. This precaution will allow the oven door to be open and still permit access to the interior cabinet. A walkway 48'' (1200 mm) wide should be provided to allow someone to walk behind a person working at the range.

Cooktop/Ovens

A cooktop is a cooking unit that is built into the counter and contains only the heating elements. There are generally four of five heating elements, but units with up to eight elements are available. Most units contain interchangeable elements such as simmer plates, a grill, griddles, a rotisserie, and fryers. Cooktops are available in units 30'', 36'', 42'', and 48'' (750, 900, 1050, and 1200 mm) wide. Oven units are typically powered by electricity and can be single or double units. The units may include a conventional oven, a convection oven, or a combination of these units with a microwave oven. Oven units are typically 27'', 30'', or 33'' (675, 750, or 825 mm) wide. The bottom of the oven is typically 30'' (750 mm) above the floor. A cooktop and oven should be within three or four steps of each other if separate units are to be used. Avoid placing a range next to a refrigerator, a trash compactor, or a storage area for produce or breads.

Microwave

A microwave unit may be placed above the oven unit, but the height of each unit above the floor should be considered in regard to the height of the clients. A microwave mounted with the bottom of the appliance between 30'' and 48'' (750 and 1200 mm) above the floor and near the daily eating area often proves to be very practical. Provide a counter 15'' (375 mm) wide beside the microwave for placing hot dishes coming out of the unit.

Breadboards

In addition to the major kitchen appliances, care must be given to other details such as breadboards, counter workspaces, and specialized storage areas. A breadboard can range from the typical lightweight pullout board to a custom, heavy-duty board that pulls out from behind a false drawer front that is suitable for heavy baking needs, similar to Figure 5.40. The breadboard should be placed near the sink and the range but not in a corner. Ideally a minimum of 15'' (375 mm) of counter space can be placed between appliances to allow for food preparation. Specialized storage often needs to be considered to meet clients' needs. These needs are covered as cabinets are explained in Section V.

Communication Center

A central area of most medium and large kitchens is a desk center similar to Figure 5. 41. This area in its simplest form will provide space in the kitchen for a desk area. More than just a desk, the communication center should provide for a telephone, intercom, computer, and security controls.

FIGURE 5.41 ■ *Most midsize and larger homes provide a desk area in the kitchen. This area can be used to pay bills, organize records, post notes, record phone messages, and sort through recipes. As the price of the home increases, the area often includes a computer, the master intercom, and the master security controls.* Courtesy Karen Griggs.

FIGURE 5.40 ■ *A breadboard can range from the typical lightweight pull-out board to a custom, heavy-duty board suitable for heavy baking needs that pulls out from behind a false drawer front.* Courtesy Floyd Miller.

Butler's Pantry

Many upper-end homes supplement the kitchen with a room with additional preparation and storage areas referred to as a butler's pantry or butler's kitchen. This area combines the storage features of a walk-in pantry with preparation features such as a sink, undercounter or standard size refrigerator, and possibly cooking units. These extra features are used as a second kitchen to supplement the activities of the kitchen for large gatherings. Typically a butler's kitchen is located adjacent to

FIGURE 5.42 ■ *Many custom homes supplement the kitchen with a room with additional preparation and storage areas referred to as a butler's pantry or butler's kitchen. This kitchen provides a sink, dishwasher, ice maker, refrigerator, and wine cooler as well as counter space for the preparation and storage of seasonal decorations.* Courtesy Norm Cooper.

FIGURE 5.43 ■ *A wine cellar can be located in an insulated room near the kitchen or family room or in a cool area of the basement.* Courtesy Sherry Benfit.

the kitchen, dining, and family rooms. Figure 5.42 shows a portion of a butler's kitchen.

Wine Cellar

For most kitchens a cabinet or an undercabinet rack or wine cooler is provided for wine. When space allows, a special room may be requested for storing wine. Although this is referred to as a wine cellar, it could be as simple as a small, insulated closet near the kitchen or family room. Figure 5.43 shows a small wine cellar that also provides a quiet sitting area.

Utility Room

The room that is called a utility room or mudroom should be planned to include space for washing, drying, folding, mending, ironing, and storing clothes. Additional space is often used for long-term storage of dry and canned food as well as for a freezer. In upscale homes, multiple front-loading appliances similar to those in Figure 5.44 are often used, or stacking units

FIGURE 5.44 ■ *A small utility room near the kitchen or garage can be used for basic cleaning needs. As the size of the utility room increases, it can be used for additional storage as well as for washing, drying, and folding clothes. Most new laundry appliances are available as front-loading machines, allowing for increased storage above and below the appliance.* Courtesy Judy Cooper.

that require less space for the appliances and more space for storage. In a well-planned clothing center, a means of hanging wet clothing should be provided in addition to the standard clothes dryer. A well-planned utility room must provide for hanging wet clothes over the laundry sink as well as large amounts of storage to supplement kitchen storage. A counter for folding and storing clean clothes as well as a closet for hanging clothes as they come out of the dryer should also be included. A built-in ironing board should also be provided.

Utility rooms are often placed near either the bedrooms or the kitchen area. There are advantages to both locations. Placing the utility room near the bath and sleeping areas puts the washer and dryer near the primary source of laundry. Care must be taken to insulate the sleeping area from the noise of the washer and dryer. Placing the utilities near the kitchen allows for a much better traffic flow between the two major work areas of the home. With the utility room near the kitchen, space can often be provided in the utility room for additional kitchen storage. In smaller homes, the laundry facilities may be enclosed in a closet near the kitchen. In warm climates, laundry equipment may be placed in the garage or carport. In homes with a basement, laundry facilities are often placed near the water heater source. Many homeowners find that laundry rooms in a basement are too remote from the living areas. If bedrooms are on the upper floor, a laundry chute to the utility room can be a nice convenience. The utility room often has a door leading to the exterior. This allows the utility room to function as a mudroom. Entry can be made from the outside directly into the mudroom, where dirty clothes can be removed. This allows for cleanup near the service entry and helps keep the rest of the house clean. Figure 5.45 shows this type of utility layout. Another common use for a utility room, as shown in Figure 5.46, is to provide an area for sewing and ironing.

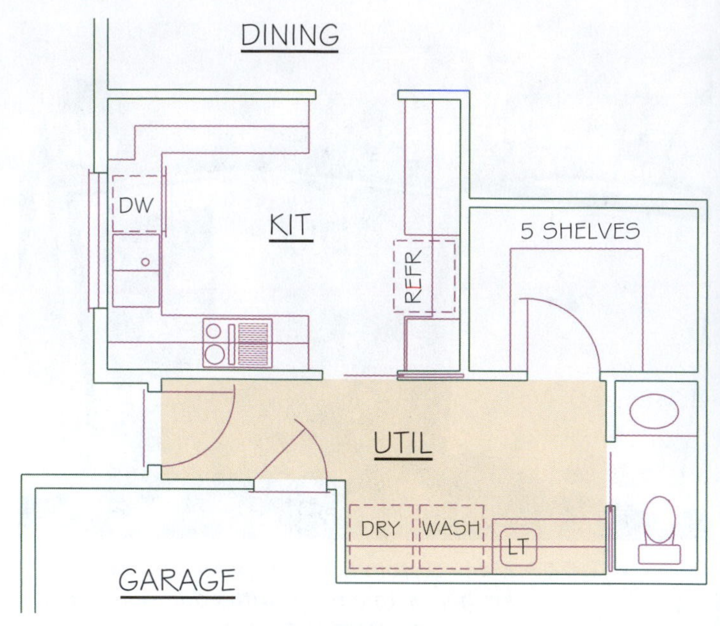

FIGURE 5.45 ■ *The utility room may serve as a service entry where dirty clothes can be removed as the home is entered.*

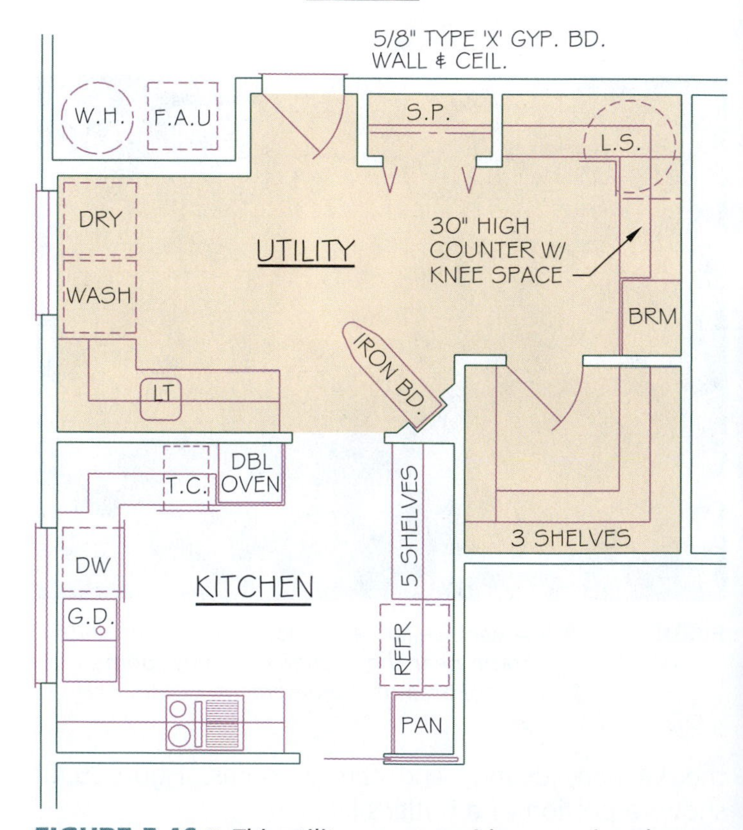

FIGURE 5.46 ■ *This utility room provides a spacious layout for chores.*

Garage and Carport

The garage is used to park vehicles and an assortment of sporting equipment such as a Jet Ski or snowmobile. In addition to traditional uses, the garage often becomes a storage area, a second family room, or a place for the water heater and furnace. Common sizes for parking are as follows:

- The minimum space for a small single car is 11 × 20' (3300 × 6000 mm).
- A space of 13 × 25' (3900 × 7500 mm) allows for a large single parking space.

- The minimum space for two cars should be 21 × 21' (6300 × 6300 mm). These sizes will allow minimal room to open doors and to walk around the car. Additional space should also be included for a workbench.
- A garage 22 × 25' (6600 × 7500 mm) provides for parking, storage, and room to walk around for two cars.

Many custom homes include enclosed parking for three or more cars. Space is often provided for parking a recreational vehicle. Common widths for an RV range from 12' to 15' (3600 to 4500 mm), with lengths ranging upward from 25' (7500 mm). Space should also be planned for the storage of a lawn mower and other yard equipment. Size requirements will vary based on the equipment to be stored and should be verified with the manufacturer. Additional space must be provided if a post is located near the center of the garage to support an upper floor. It will also be helpful if a required post can be located off center from front to back so that it will not interfere with the opening of car doors.

The site often dictates the garage location. Although access and size are important, the garage should be designed to blend with the residence similar to the home in Figure 5.47. Many municipalities require that the garage doors not be seen from the street as the home is being approached. Clients usually have a preference for either one large single door or smaller doors for each parking space. A single door 8' wide × 7' high (2400 × 2100 mm) is common for small and midsize cars. A door 9' (2700 mm) wide is the smallest that should be used for a full-size car, truck with wide side mirrors, or for most boat trailers. A 32'' (800 mm) wide door should be installed to provide access from the garage to the yard. A double door is typically 16' (4800 mm) wide. Aside from the owner's preference, posts required to support upper floors may influence the type of door to be used. The space between multiple doors will be based on the amount of space needed inside the garage to allow car doors to open without hitting anything. Usually 12'' is the minimum space required between doors to allow for framing members. When multiple single doors are to be used, consideration should also be given to lateral bracing for the front of the garage. See Chapter 24 for lateral bracing considerations. Although a good engineer can make any structure stand up, keeping the space between doors to a minimum and increasing the size of the walls at each outer edge of the garage can minimize the cost of keeping the front wall rigid. As a general guideline, do not let structural considerations dictate the initial design.

FIGURE 5.47 ■ *The garage should blend with the design of the residence and not dominate the building site.* Courtesy Laine M. Jones Design, W. Newbury Ma.

In areas where cars do not need to be protected from the weather, a carport can be an inexpensive alternative to a garage. Provide lockable storage space on one side of the carport if possible. This can usually be provided between the supports at the exterior end.

Bonus Rooms

One of the benefits of having a garage is the space that is created above the garage. This space can be used to provide additional storage between the rafters or trusses. The space above the garage can also be turned into living space similar to that in Figure 5.48 if access can be provided. A hallway connected to other upper level living areas typically provides access. The use of bonus rooms is based on the homeowner. Common uses for the area include a family room, additional sleeping areas, or a home office. When used for one of these uses, windows must be provided to meet code requirements for light, ventilation, and emergency egress. Dormers are usually provided to meet building code requirements. Verify code requirements for each of these requirements in Chapter 4 as well as minimum wall heights. See Chapter 15 for a discussion of dormers.

Health Centers

A health center can range from a spare bedroom that will be used to store an exercise bike to an in-house weight room or gymnasium. If an exercise room will be

FIGURE 5.48 ▪ *The attic space above a garage can often be turned into living space if access can be provided.* Courtesy William E. Poole, William E. Poole Designs, Inc.

FIGURE 5.50 ▪ *This custom home provides a full weight room with access to a bathroom provided through the left door and a sauna to the right.* Courtesy BOWA Builders, Inc. Jim Tetro, photographer.

provided, space for the equipment should be provided, as well as hallways and doors that are wide enough to provide access. The room should be near a bathroom, or it may have it's own bath in larger custom homes. Features such as a steam room, sauna, and massage room are also located close to the exercise room. Figure 5.49 shows a room dedicated to health located in a dormer on the upper floor. Figure 5.50 shows a full health center from a custom home. Figure 5.51 shows a basketball court that was provided for a hillside home. Because the home is located on a very steep hill, the cost of providing the gymnasium was minimal. An interior lap pool such as the pool shown in Figure 5.52 is common for many custom homes. When included, consideration must be given to venting the hot, moist air displaced by the pool. Consideration must also be given to the additional structural needs associated with the exercise equipment.

FIGURE 5.49 ▪ *An exercise room is often provided in homes to contain health equipment. This room should be near a bathroom.* Courtesy Margarita Miller.

FIGURE 5.51 ▪ *A sports court was installed in the basement of this home. The balcony is at the basement floor level and the court is excavated below the basement level. The court was financially feasible because of the steep construction site.* Courtesy Judy Word.

FIGURE 5.52 ■ *A lap pool or swimming pool is sometimes included in exclusive custom homes. In the planning stages, it is critical to design a ventilation or dehumidification system.* *Photo* Courtesy BOWA Builders, Inc Bob Narod, photographer.

TRAFFIC PATTERNS

A key aspect of any design is the traffic flow between the interior areas of the residence. It is also important to consider how each area of the home relates to the building site.

Interior Traffic

An important consideration of the design process is the traffic flow between areas of the residence. Traffic flow is the route that people follow as they move from one area to another. Often this means hallways, but areas of a room are also used to aid circulation. Rooms should be of sufficient size to allow circulation through the room without disturbing the use of the room. By careful arrangements of doors, traffic can flow around furniture rather than through conversation or work areas. Codes that cover the width of hallways were covered in Chapter 4. Good design practice will provide a width of 36 to 48'' (900 to 1200 mm) for circulation pathways. A space of 48'' (1200 mm) is appropriate for pathways that are used frequently, such as those that connect main living areas. Pathways used less frequently can be smaller. In addition to hallways, the entries of a residence are important in controlling traffic flow.

The foyer is typically the first access to the residence for guests. In cold climates, many homes have an enclosure that leads to the foyer, to prevent heat loss. The foyer should provide access to the living areas, the sleeping area, and the service areas of the house. The size of the foyer will vary. Provide ample room to open a door completely and allow it to swing out of the way of people entering the home. A closet should be provided near the front door for storage of outdoor coats and sweaters. Provide access to the closet that will not be blocked by opening the front door.

In addition to considering traffic flow between rooms, it is important to determine how the levels of a home will relate to each other. The ideal location for stairs is off a central hallway. Many floor plans locate the stairs in or near the main entry to provide an attractive focal point in the entry as shown in Figure 5.53, as well as convenient access to upper areas. Stairs should be located for convenient flow from each floor level and should not require access through another room. The location will also be influenced by how often the stairs will be used. Stairs serving mechanical equipment in a basement will be used less frequently than stairs connecting the family room to other living areas.

Traffic Flow

Just as important as how the major interior areas relate to each other is the traffic flow between each of these areas and exterior areas. A well planned home will expand the inside living areas to outside areas such as courtyards, patios, decks, and balconies. Sunrooms and solariums are interior areas that combine interior and exterior features. Exterior structures that enhance outdoor living include gazebos and arbors. Each should be planned so that it blends with the overall design of the home.

A sunroom or solarium is a common way to bring the outside into the interior living areas. A sunroom is a multipurpose room, similar to the one shown in Figure 5.54 that can be used as a reading room, a breakfast

FIGURE 5.53 ■ *Stairs are a focal point in the foyer.* Courtesy Alan Mascord, AIBD. Alan Mascord Design Associates, Inc. Bob Greenspan, photographer.

FIGURE 5.54 ■ *A sunroom is a multipurpose room that can be used as a reading room, a breakfast room, a music room, or as extra space for entertaining while providing a transition between interior and exterior spaces.* Courtesy BOWA Builders, Inc. Greg Hadley, photographer.

room, or a music room, or as extra space for entertaining while providing transition between the interior and exterior spaces. A solarium is typically associated with passive solar heating, but the room also provides an excellent location for casual living and may include a lap pool, a spa, or an area for indoor gardening. Each of these uses will require that ventilation, sun orientation, and views be carefully planned. Chapter 6 introduces materials to help in planning a sunroom or solarium.

Most of the terms describing exterior areas are strictly defined and regulated by building codes, but the use of these terms by the general public will vary slightly. IRC defines a court as an exterior space that is at grade level, is enclosed on three or more sides by walls or a building, and is open and unobstructed to the sky. A courtyard provides an excellent screen, allowing privacy for outdoor living in the front yard and a secure play area for young children. Courtyards can be used as an extension of the living, dining, and family rooms by providing a patio or deck and landscaped areas.

A patio is a ground-level exterior entertaining area made of concrete, stone, brick, or treated wood. A patio provides good access to the interior, ample areas of sunny and shaded seating, and room for pool activities. The patio in Figure 5.55 is designed to provide an elegant extension of the living areas to the outside. Balconies and decks are elevated floors. The model codes define a deck as an exterior floor supported on at least two opposing sides by adjoining structures, posts, or piers. Figure 5.56 shows a covered deck supported by columns on the upper floor of a home. A ***balcony*** is an above ground deck that projects from a wall or building with no additional

FIGURE 5.55 ■ *This patio is designed to provide an elegant extension of the living areas to the outside. It provides good access to the interior of the home, ample areas of sunny and shaded seating, and space for pool activities.* Courtesy John B. Scholz, architect.

FIGURE 5.56 ■ *The veranda of this southern-style home is defined as a deck by the IRC. A deck is an exterior floor supported on at least two opposing sides by adjoining structures, posts, or piers.* Courtesy Eric S. Brown Design Group, Inc. Oscar Thompson, photographer.

FIGURE 5.58 ■ *In warm climates, a covered patio is a good way of tying the interior and exterior living areas together.* Courtesy Eric S. Brown Design Group, Inc. Oscar Thompson, photographer.

FIGURE 5.57 ■ *The cantilevered balconies of Fallingwater, designed by Frank Lloyd Wright, might possibly be the most recognized residential balconies in the world. The IRC considers a balcony an above-ground deck that projects from a wall or building with no additional supports.* "Fallingwater" Courtesy Western Pennsylvania Conservancy. Christopher Little, photographer.

FIGURE 5.59 ■ *The exterior kitchen is the hub of the exterior living area.* Courtesy John Earp.

supports. Figure 5.57 shows perhaps the most famous residential balconies in the world. Although a ***porch*** is not defined by the building codes, a common perception is that it is an enclosed patio or deck as seen in Figure 5.58. In some areas of the country, the term *enclosed* would refer to walls, windows and doors, and a roofed area. In warmer areas of the country, an enclosed patio may include only a roof or awning-type covering to protect the areas from sunlight, insects, or other natural elements. The usefulness of an outdoor area may be greatly increased if it is enclosed with screens, plastic, or glass panels. Gazebos and arbors are freestanding structures that provide protection from the elements. Common uses include protected outdoor eating areas, covered pools or spas, and shaded garden living.

Most families will require one or more outdoor entertaining areas. Figure 5.59 shows an outdoor kitchen that is the hub of the exterior living area. To determine which areas are to be included in a home, needs, function, location, orientation, climate, and size must be considered. Key questions to be considered with the client include:

- When will the area be used? Who will use it? What activities are planned for the area?

- Will traffic flow through the area?
- Will play areas need to be provided for children?
- Will seating for dining, quiet activities, or lounging in the sun be required?
- Should there be space for outdoor games such as Ping Pong, shuffleboard, or tennis?

A patio or deck that will be used for outdoor eating will have different size requirements if a pool or spa is to be included. A deck will seem smaller than a patio of the same size because of the addition of a rail. Any deck, patio, or balcony is required to have a guardrail when the finished floor height is 30'' (750 mm) or more above the finish grade or floor below. A balcony as small as 36'' (900 mm) wide can be used to place a lawn chair in a sunny spot for one person. A minimum balcony width of 48'' (1200 mm) should be provided for two or more small chairs, to allow access. The minimum space for patios and decks that will include a seating and traffic area is 10 × 14' (3000 × 4200 mm).

Once the function is determined, the location of the deck or patio must be considered. A patio or deck located near main living areas will need to incorporate traffic flow in addition to the space required for its own functions. Exterior areas located near a bedroom or bathroom suite should have some type of screen or landscaping to provide privacy.

A designer must consider the orientation of the outdoor area and the local climate. A balcony or deck may be placed on the west side of a home in a northern region to take advantage of the sun. In a southern region, designing for shade will be of great importance. Just as important as the movement of the sun will be prevailing wind directions and the view. Chapter 7 discusses methods of designing homes to enhance the relationship of the home to terrain, view, sun, wind, and sound.

PROVIDING UNIVERSAL ACCESSIBILITY

Well-planned, attractive, accessible housing is essential for millions of people with disabilities—those with temporary injuries who walk with the aid of crutches or a walker or those who depend on a wheelchair for mobility. A house can be made accessible with just a few minor design changes based on the Americans with Disabilities Act (ADA), 1997, and the guidelines of the International Code Council.

Exterior Access

An accessible home has a level site for parking with paved walkways from parking areas to the main entry. A parking space 9' (2700 mm) clear should be provided with a minimum walkway of 42'' (1050 mm) between cars. A minimum ceiling height of 9' (2700 mm) is required for raised-top vans. Any changes in elevation between the parking area and the front door can be no more than a 1/12, with a maximum rise of 30'' (760 mm) per run. Inclined ramps should be covered with a nonskid surface to provide sure footing for those walking and to help keep a wheelchair from going out of control. Ramps longer than 30' (9000 mm) or higher than 30'' (750 mm) should have a landing 48'' (1200 mm) wide at the midpoint. Provide handrails that are 29'' to 32'' (740 to 810 mm) high. Guidelines for ramp access are shown in Figure 5.60.

Interior Access

Beyond the scope of the codes presented in Chapter 4, a commonsense approach to design can greatly aid the quality of a home. Doors with a 32'' (800 mm) clear

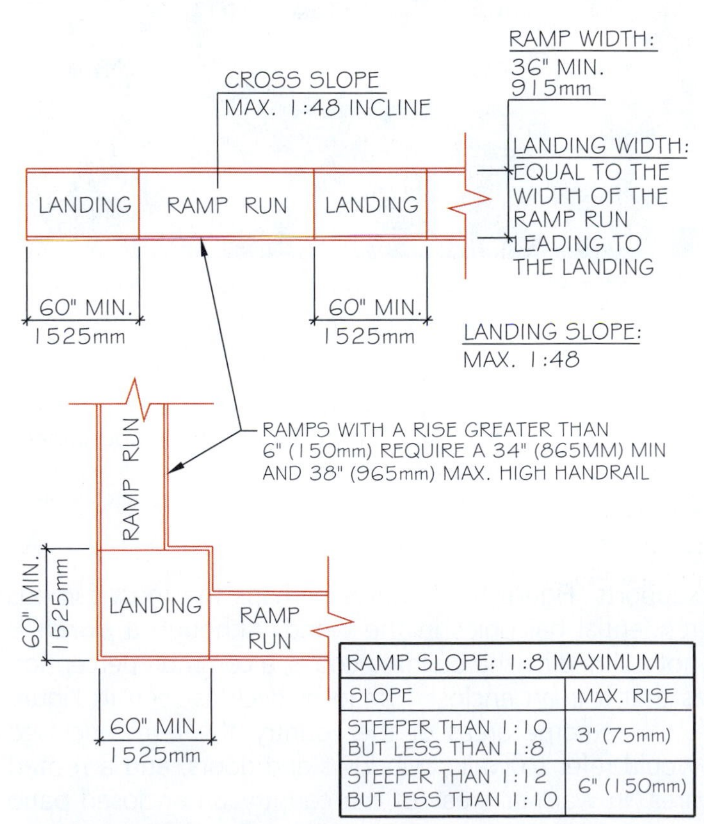

RAMP SLOPE: 1:8 MAXIMUM	
SLOPE	MAX. RISE
STEEPER THAN 1:10 BUT LESS THAN 1:8	3" (75mm)
STEEPER THAN 1:12 BUT LESS THAN 1:10	6" (150mm)

FIGURE 5.60 ■ *Guidelines for designing ramp access based on ICC/ANSI A117.1-1998 standards.*

opening should be provided throughout the residence to ease movement in a wheelchair or with a walker or crutches. Hinges are available that swing the door out of the doorway for maximum clearance. Hallways with a width of 42'' (1050 mm) will provide convenient traffic flow, but 48'' (1200 mm) will make turning from the hall to doorways much easier. Doors with a lever-action handle or latches are typically easier to open than round doorknobs. Windows should be within 24'' (600 mm) of the floor and should not require more than 8 lb of pressure to open.

An area 60'' (1500 mm) square will allow for a 360° turn in a wheelchair. An area of this size should be provided in each room. Floors covered with hardwood or tiles provide excellent traction, although a low-pile carpet is also suitable. Avoid using small area rugs that move easily. Figure 5.61 shows minimum turning alternatives.

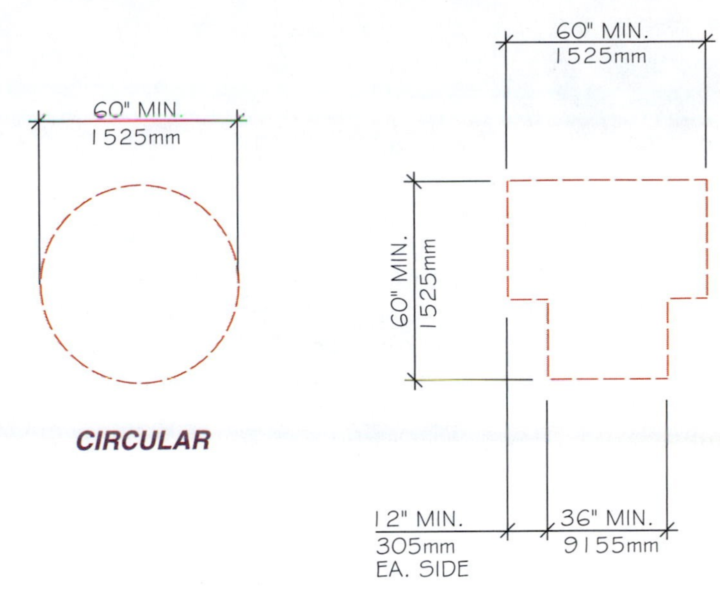

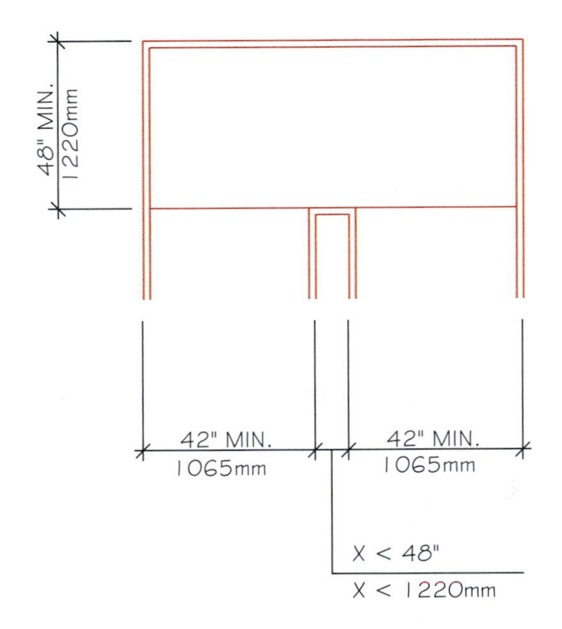

FIGURE 5.61 ■ *Guidelines for turning radii based on ICC/ANSI A117.1-1998 standards.*

Work Areas

To provide a suitable kitchen work area, countertops should be 30 to 32'' (760 to 810 mm) high with pull-out cutting boards. A range should have all controls on a front panel rather than being top-mounted. A wall oven should be mounted between 30 and 42'' (762 to 1067 mm) above the floor. A side-by-side frost-free refrigerator with door-mounted water and ice dispensers provides easy access.

Bathrooms

Doors that swing out or pocket doors will provide the most usable space in a small room such as a bathroom. Vanities should have a roll-under countertop with a lowered or tilted mirror. If possible, reinforce areas of each wall where a grab bar will be used with 3/4'' (20 mm) plywood backing. Grab bars should be provided next to toilets and inside a tub or shower. Each tub and shower should have nonskid floor surfaces. Showers with a seat and a handheld showerhead will provide added safety for people with limited strength. A roll-in shower with no curb should be provided for clients using a wheelchair.

CHAPTER 5

Additional Readings

One of the best ways to improve your ability to design pleasing homes is to look at successful designs. Excellent sources for review include magazines, television, and the Internet. To increase your reading options, use your favorite search engine and research major listings such as architectural magazines, interior design magazines, home magazines. Magazines used by designers include:

Architect's Journal
Architectural Design
Architectural Digest
Better Homes and Gardens
Building Design & Construction
Country Home
Country Living
Design Architecture
DesignLine (AIBD)
Design Times
Home
House Beautiful
Home Design
In Style
Kitchen & Bath Design News
Metropolitan Home

If you have access to cable or satellite television, HGTV and TLC provide an excellent selection of programs that explore architectural design, interior design, and construction. New programs are aired each season, so be sure and check program listings or the websites for your favorite station for available programs. Current popular programs include:

Awesome Interiors
Before & After
Decorating Cents
Decorating with Style
Design Inc.
Designers' Challenge
Designing for the Sexes
Devine Design
Dream Builders
Dream House
Garden Architecture
Homes by Design
Interiors by Design
Kitchen Designs
Landscape Smart
reDesign
reZONED
Room by Room
This Old House Classics
This Small Space
What You Get for the Money
Worlds Most Extreme Houses

Information about each program can be obtained by contacting www.hgtv.com or by using a search engine to search the Internet. Other websites that promote interior design include:

ADDRESS	COMPANY OR ORGANIZATION
www.about.com	Home planning guide to the Internet
www.aibd.org	American Institute of Building Design
www.aham.org	Association of Home Appliance Manufacturers
www.arcadiandesign.net	Arcadian Residential Design
www.decoratorsecrets.com	Decorator Secrets
www.iida.com	International Interior Design Association
www.kcma.org	Kitchen Cabinet Manufacturers
www.maplefloor.org	Maple Flooring Manufacturers Association, Inc.
www.marble-institute.com	Marble Institute of America
www.nspi.org	National Spa and Pool Institute
www.glasswebsite.com/nsa	National Sunroom Association

The best way to improve your skills is to visit open homes. Walking through a wide variety of homes and construction sites will help you visualize the materials you will be drawing. As you walk through homes, it may be helpful to take measurements of each room and develop a list of sizes that you find pleasing or confining. Photographing homes and their decor is an additional method to help you remember fine details of interior design. Photos of each area of a home can be a valuable aid in your future as you help clients define their taste.

CHAPTER 5

Room Relationships and Sizes Test

DIRECTIONS

Answer the following questions with short, complete statements. Type your answers using a word processor or neatly print the answers on lined paper.

1. Type your name, the chapter number, and the date at the top of the sheet.
2. Type the question number and provide the answer in the form of a statement that includes part of the question. You do not need to write out the entire question.

QUESTIONS

5.1. What are the three main areas of a home?
5.2. How much closet area should be provided for each bedroom?
5.3. What rooms should the kitchen be near? Why?
5.4. What are the functions of a utility room?
5.5. Give the standard size of a two-car garage.
5.6. List five functions of a family room.
5.7. What are the advantages of placing bedrooms on an upper floor?
5.8. What are the advantages of placing bedrooms on a lower floor?
5.9. List four design criteria to consider in planning a dining room.
5.10. What are the service areas of the home?
5.11. List and describe two types of entries.
5.12. Using the Internet, research five sources of information about the design of a home theater.
5.13. Describe the use of a foyer in a home.
5.14. Describe the difference between a formal and a casual living room.
5.15. What should be considered in placing a living room?
5.16. You are designing a home for a family with no children. The clients insist on having only one bedroom. How would you counsel them?
5.17. Visit several open homes in your area and take pictures or obtain floor plans of a minimum of five different master bedroom suites.
5.18. Visit several different home supply stores or use the Internet to obtain information on several different closet storage options.
5.19. Trace Figure 5.14 and create a sketch designed to allow access to the dining room without passing through the living room or kitchen.
5.20. Use the Internet to visit builder organizations such as the National Association of Home Builders. Research current issues affecting home building in your region.

CHAPTER 6

Exterior Design Factors

The design of a house does not stop once the room arrangements have been determined. The exterior of the residence must also be considered. Often a client has a certain style in mind that will dictate the layout of the floor plan. In this chapter, ideas are presented to help you better understand the design process. To design a structure properly, consideration must be given to the site, the style and shape of the floor plan, exterior style, and the energy efficiency of the design. Site considerations are explored in Chapter 7. Energy-efficient design is explored in Chapters 7, 14, and 20.

ELEMENTS OF DESIGN

The elements of design are the tools the designer uses to create a structure that will be both functional and pleasing to the eye. These tools are line, form, color, and texture.

Lines

Line provides a sense of direction or movement in the design of a structure and helps relate it to the site and the natural surroundings. Lines may be curved, horizontal, vertical, or diagonal and can accent or disguise features of a structure. Curved lines in a design tend to provide a soft, graceful feeling. Curves are often used in decorative arches, curved walls, and round windows and doorways. Figure 6.1 shows how curved surfaces can be used to accent a structure.

Horizontal lines seen in a long roof or floor, window and siding trim, and balconies and siding patterns can be used to minimize the height and maximize the width of a structure, as seen in Figure 6.2. Horizontal surfaces often create a sense of relaxation and peacefulness. Vertical lines create an illusion of height, lead the eye upward, and tend to provide a sense of strength and stability. Vertical lines are often used in columns, windows, trim, and siding

FIGURE 6.1 ■ *Curved lines are used to provide smooth transitions and accent a structure.* Courtesy California Redwood Association.

FIGURE 6.2 ■ *Horizontal lines can be used to accent the length or hide the height of a structure.* Courtesy Edna Carin, Remax, Danville, CA.

FIGURE 6.3 ■ *Vertical lines formed by the columns and windows can be used to accent the height of a structure.* Courtesy W.D. Farmer FAIBD, W.D. Farmer Residence Designer, Inc.

patterns, as in Figure 6.3. Diagonal lines can often be used to create a sense of transition. Diagonal lines are typically used in rooflines and siding patterns laid parallel to the roof.

Form

Lines are used to produce forms or shapes. Rectangles, squares, circles, ovals, and ellipses are the most common shapes found in structures. These shapes are typically three-dimensional, and the proportions between them are important to design. For the best results, the form of a structure should be dictated by its function. Forms are typically used to accent specific features of a structure. Figure 6.4 shows how varied forms can be used to break up the length of a residence. Form is also used to create a sense of security. For example, large columns provide a greater sense of stability than thinner columns.

FIGURE 6.4 ■ *Forms such as rectangles, circles, and ovals are used to provide interest.* Courtesy California Redwood Association, Robert Corna Architect, Photo by Balthazar Korab.

Color

Color is an integral part of interior design and decorating and helps distinguish exterior materials and accent shapes. A pleasing blend of colors creates a dramatic difference in the final appearance of any structure. Color is described by the terms *hue, value,* and *intensity.*

Hue represents what you typically think of as the color. Colors are categorized as primary, secondary, or tertiary on a color wheel similar to that in Figure 6.5. Primary colors—red, yellow, and blue—cannot be created from any other color. All other colors are made by mixing, darkening, or lightening a primary color. Secondary colors—orange, green, and violet—are made from an equal combination of two primary colors. Mixing a primary color with a secondary color will produce a tertiary or intermediate color—for example, red-orange, yellow-orange, yellow-green, blue-green, blue-violet, and red-violet. Mixing each of the primary colors in equal amounts will produce black.

Mixing black with a color will darken the color, producing a shade of the original color. White has no color pigment and gray is a mixture of black and white. Adding white to a color lightens the original color, producing a tint. ***Value*** is the darkening or lightening of a hue.

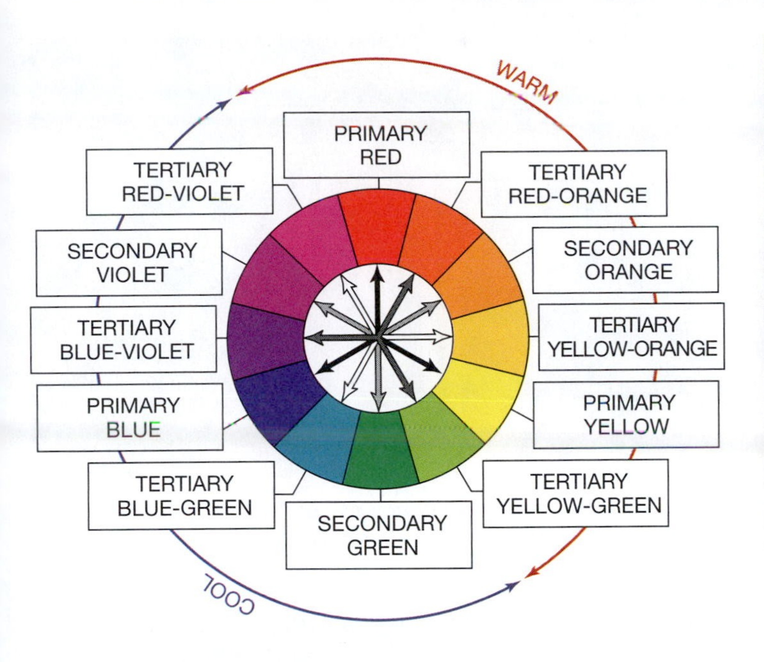

FIGURE 6.5 ■ *Colors are primary, secondary, or tertiary. These categories are then divided into warm and cool colors.*

FIGURE 6.6 ■ *Most designers use neutral colors for major portions of the home and rely on the trim and roofing for accent colors. Color can also be added through landscaping and interior design.* Courtesy Eric S. Brown Design Group, Inc. Photo by Laurence Taylor.

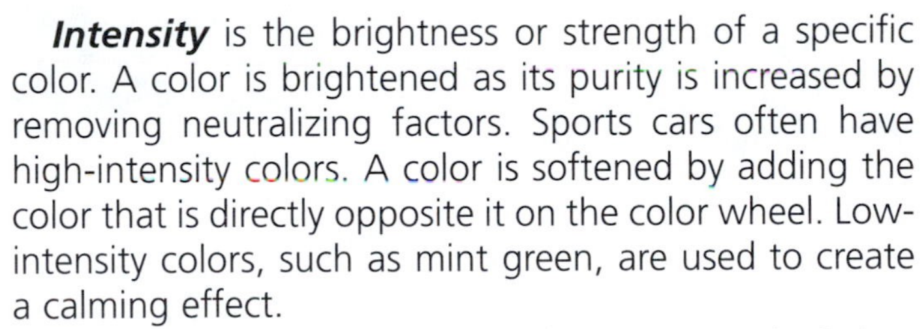

Intensity is the brightness or strength of a specific color. A color is brightened as its purity is increased by removing neutralizing factors. Sports cars often have high-intensity colors. A color is softened by adding the color that is directly opposite it on the color wheel. Low-intensity colors, such as mint green, are used to create a calming effect.

Colors are also classified as warm or cool. Colors seen in warm objects, such as the reds and oranges of burning coals, are warm colors. Warm colors tend to make objects appear larger or closer than they really are. Blues, greens, and violet are cool colors. These colors often make objects appear farther away. Most designers use neutral colors similar to those in Figure 6.6 for the major portions of a home and rely on the trim and roofing for accent colors. Color can also be added through landscaping and interior design features. Figure 6.7 shows a home designed for wide appeal that uses colors common to the architectural style and historic era.

FIGURE 6.7 ■ *Color is used on the exterior of homes to distinguish architectural features and trim, to match a specific architectural time period, or please owners' individual tastes.* Courtesy Abtco, Inc.

Texture

Texture, the roughness or smoothness of an object, is an important factor in selecting materials to complete a structure. Rough surfaces tend to create a feeling of strength and security—examples include concrete masonry and rough-sawn wood. Rough surfaces also give an illusion of reduced height. Resawn wood, plastic, glass, and most metals have a very smooth surface and create a sense of luxury. A smooth surface tends to give an illusion of increased height; also, it reflects more light and makes colors seem brighter.

PRINCIPLES OF DESIGN

Line, form, color, and texture are the tools of design. The principles of design affect how these tools are used to create an aesthetically pleasing structure. The basic principles to be considered are *rhythm, balance, proportion,* and *unity.*

Rhythm

Most people can usually recognize ***rhythm*** in music. The beat of a drum is a repetitive element that sets the foot tapping. In design, a repetitive element provides rhythm and leads the eye through the design from one place to another in an orderly fashion. Rhythm can also be created by a gradual change in materials, shape, and color. Gradation in materials could be from rough to smooth, gradation in shape from large to small, and gradation in color from dark to light. Rhythm can also be created with a pattern that appears to radiate outward from a central point. A consistent pattern of shapes, sizes, or material can create a house that is pleasing to the eye while also providing a sense of ease for the inhabitants and conveying a feeling of equilibrium. Figure 6.8 shows elements of rhythm in a structure.

FIGURE 6.9 ■ *Formal balance places features evenly along an imaginary centerline.* Courtesy Aaron Jefferis.

FIGURE 6.10 ■ *Informal balance is achieved by moving the centerline away from the mathematical center of a structure and altering the sizes and shapes of objects on each side of the centerline.* Courtesy Tom Price, Architect; Phil Eschbach, photographer.

Balance

Balance is the relationship between the various areas of the structure and an imaginary centerline. Balance may be formal or informal. Formal balance is symmetrical; one side of the structure matches the opposite side in size. The residence in Figure 6.9 is an example of ***formal balance.*** Its two sides are similar in mass. ***Informal balance*** is nonsymmetrical; in this case, balance can be achieved by placing shapes of different sizes in various positions around the imaginary centerline. This type of balance can be seen in Figure 6.10.

FIGURE 6.8 ■ *Repetitive features lead the eye from form to form.* Courtesy Monier.

Proportion

Proportion is related to both size and balance. It can be thought of in terms of size, as in Figure 6.11, where one exterior area is compared with another. Many current design standards are based on the designs of the ancient Greeks. Rectangles using the proportions 2:3, 3:5, and 5:8 are generally considered very pleasing. Proportion can also be thought of in terms of how a residence relates to its environment.

FIGURE 6.11 ■ *Shapes that are related by size and shape create a pleasing flow from form to form.* Courtesy Metal Roofing Alliance.

FIGURE 6.12 ■ *Unity is a blend of varied shapes and sizes to create a pleasing appearance.* Courtesy BOWA Builders Inc, Jim Tetro, photographer.

Proportions must also be considered inside the house. A house with a large living and family area needs to have the rest of the structure in proportion. In large rooms, the height must also be considered. A 24' × 34' (7300 × 10 900 mm) family room is too large to have an 8' (2400 mm) flat ceiling. The standard ceiling height of 8' (2400 mm) would be out of proportion to the room size. A 10' (3000 mm) or vaulted ceiling would be much more in keeping with the size of the room.

Unity

Unity relates to rhythm, balance, and proportion. Unity ties a structure together with a common design or decorating pattern. Similar features that relate to each other can give a sense of well-being. Avoid adding features to a building that appear to be just there or tacked on. See Figure 6.12.

As an architectural drafter, you should be looking for these basic elements in residences that you see. Walk through houses at every opportunity to see how other designers have used these basics of design. Many magazines are available that feature interior and exterior home designs. Study the photos for pleasing relationships and develop a scrapbook of styles and layouts that are pleasing to you. This will provide valuable resource material as you advance in your drafting career and become a designer or architect.

COMMON FLOOR PLAN STYLES

Many clients come to a designer with specific ideas about the kind and number of levels they want in a house. Some clients want a home with only one level, so that no stairs will be required. For other clients, the levels in the house are best determined by the topography of the lot. Common floor plan layouts are single, split-level, daylight basement, two-story, dormer, and multilevel.

Single-Level

The single-level house shown in Figure 6.13 represents one of the most common styles. It is a standard of many builders because it provides stair-free access to all rooms, which makes it attractive to people with limited mobility. It is also preferred by many homeowners because it is easy to maintain and can be used with a variety of exterior styles.

Split-Level

The split-level plan (Figure 6.14) is an attempt to combine the features of a one- and a two-story residence. This style is best suited to sloping sites, which allow one area of the house to be two stories and another area to be one story. Many clients like the reduced num-

FIGURE 6.13 ■ *A single-level residence is popular because it allows stair-free access to each area.* Courtesy Debbie Frederick.

FIGURE 6.14 ■ *Homes with split floor levels are ideal for gently sloping sites.* Courtesy Alan Mascord AIBD, Alan Mascord Design Associates, Inc. Bob Greenspan, photographer.

ber of steps from one level to another that is found in the split-level design. The cost of construction is usually greater for a split-level plan than for a single-level structure of the same size because of the increased foundation cost required by the slope.

Split-level plans may be split from side to side or front to back. In side-to-side split levels, the front entrance and the main living areas are typically located on one level, with the sleeping area placed over the garage and service areas. In front-to-rear splits, the front of the residence is typically level with the street, with the rear portion stepping to match the contour of the building site.

Daylight Basement

The style of house called daylight basement could be either a one-story house over a basement or garage or two complete living levels. This style of house is well

FIGURE 6.15 ■ *A home with a daylight basement allows living areas on the lower level to have access to the exterior on one or more sides of the home while maintaining a lowered profile on the other sides.* Courtesy Jordan Jefferis.

FIGURE 6.16 ■ *Two-level homes provide the maximum square footage of living area using the minimum amount of the lot.* Courtesy William E. Poole, William E. Poole Designs, Inc.

suited for a sloping lot. From the high side of the lot, the house will appear to be a one-level structure. From the low side of the lot, both levels of the structure can be seen. See Figure 6.15.

Two-Story

A two-story house (Figure 6.16) provides many options for families that don't mind stairs. Living and sleeping areas can be easily separated, and a minimum of land will be used for the building site. On a sloping lot, depending on the access, the living area can be on either the upper or lower level. The most popular feature of a two-level layout is that it provides the maximum building area at a lower cost per square foot than other styles of houses. This saving results because less material is used on the foundation, exterior walls, and roof.

Dormer

The dormer style provides for two levels, with the upper level usually about half of the square footage of the lower floor. This floor plan is best suited to an exterior style that incorporates a steep roof, similar to that in Figure 6.17. These small dormers, often called doghouse dormers, allow space for windows into the living space in the attic. Another common type of dormer is a full dormer that allows a room with full headroom to be added to the edge of the attic space. The dormer level is formed in what would have been the attic area. The dormer has many of the same economic features as a two-story home. Chapter 15 explores other types of dormers.

Multilevel

With a multilevel layout, the possibilities for floor levels are endless. Site topography and the owners' living habits will dictate the use of this style. Figure 6.18 shows a multilevel layout. The cost of this type of home exceeds that of all other styles because of the problems of excavation, foundation construction, and roof intersections.

FIGURE 6.17 ■ *A small or "doghouse" dormer allows windows to be placed to allow light into the living space in an attic.*

FIGURE 6.18 ■ *Multilevel homes provide endless possibilities for floor-level arrangements.* Courtesy Alan Mascord AIBD, Alan Mascord Design Associates, Inc. Bob Greenspan, photographer.

COMMON RESIDENTIAL STYLES

Exterior home styles are often based on historic periods. New homes usually don't try to replicate the exact layout of past styles of homes, but they do usually try to copy the exterior style and charm associated with a specific historic style. The balance of this chapter examines home styles that were common to the original settlers. Styles will be explored based on the time period of their popularity and the area of the world where the style originated.

Colonial Homes

Some early colonists lived in lean-to shelters called wigwams, formed with poles and covered with twigs woven together and then covered with mud or clay. Log cabins were introduced by Swedish settlers in Delaware and soon were used throughout the colonies by the poorer settlers. Many companies now sell plans and precut kits for assembling log houses. Many of the colonial houses were similar in style to the houses that had been left behind in Europe. Construction methods and materials were varied depending on the local weather and building materials available.

Many colonial homes date back to historic New England. A colonial home can be described as a home where the second floor living area is 100% of the first floor living area. More specifically, it was very symmetrical, with a center door and windows equally spaced on both sides, as well as a gable roof with a ridge that ran parallel to the front side of the home. Bedrooms were usually on the second floor. Newer versions of the Colonial style often have a covered front porch, attached garage, and a family room situated behind the garage. The second floor living area frequently extends over the garage. Figure 6.19 shows an example of a home built using the Colonial style.

Colonial influence can still be seen in houses built to resemble the Georgian, saltbox, garrison, Cape Cod, Federal, Greek Revival, and southern colonial styles of houses. Other house styles from the colonial period include influences of the English, Dutch, French, and Spanish.

Georgian

The Georgian style is a good example of a basic style that was modified throughout the colonies in response to available materials and weather. This style is named for the kings of England who were in power when it flourished in the eighteenth and nineteenth centuries. The Georgian style follows the classic principles of design of ancient Rome. This English style came to America by way of British pattern books and from the masons, carpenters, and joiners who emigrated from England. The style became a favorite of well-to-do colonists who wanted their homes to convey a sense of dignity and prestige.

Georgian house plans were rectangular in shape, with a symmetrical floor layout. Homes were usually two or three stories high with a chimney at one end or side. The roofs had a medium pitch with small overhangs. A square, tooth-like trim, called dentil molding, shown in Figure 6.20, was often placed along the eaves. A central hall flanked by one or two rooms was common.

The principles of form and symmetry are most evident in the front elevation of a Georgian home. The front entry was centered on the wall, and equally spaced double-hung windows surrounded by decorative framework with nine or twelve windowpanes were usually placed on each side. These homes almost always featured an orderly row of five windows just below the eave across the front façade on the second story. The front entry was usually covered with a porch supported by pilasters or columns, and the paneled doorway was trimmed with carved wood detailing and pediments similar to Figure 6.21 or with a transom window above the door. In the South, much of the facade was made of brick. In northern states, wood siding was the major covering. Other common exterior materials were stucco and stone. Modern-day builders often combine features of the refined Georgian style with decorative flourishes from the more formal Federal style. An example of a

FIGURE 6.19 ■ *A colonial home is a very symmetrical structure in which the second-floor living area is the same as the first-floor living area.* Courtesy Leonard Conkling.

FIGURE 6.20 ■ *Square tooth-like trim called dentil molding is often placed along the eaves of traditional homes.* Courtesy Georgia-Pacific.

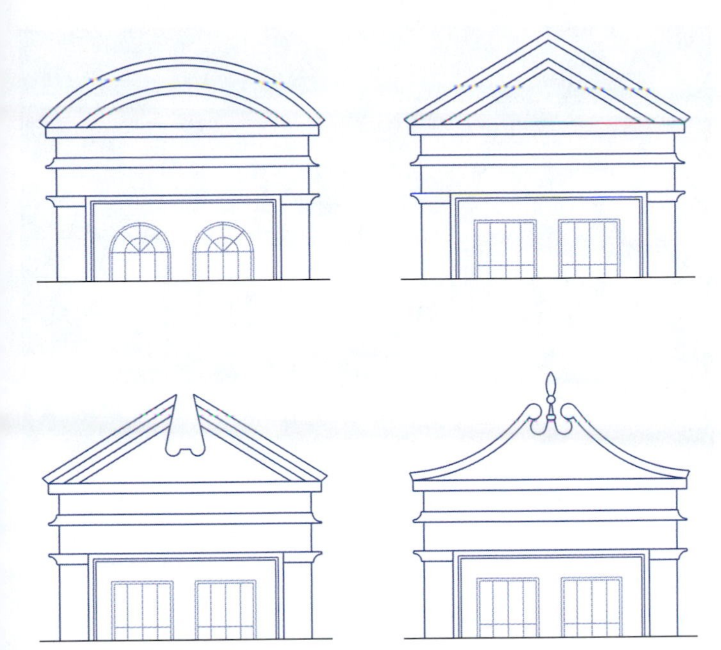

FIGURE 6.21 ■ *Common pediment styles of Colonial-style architecture.*

FIGURE 6.22 ■ *Georgian-style homes have formal balance, with the entry on the centerline. Equally spaced double hung windows with nine or twelve windowpanes are placed on each side with a transom window above the door.* Courtesy William E. Poole, William E. Poole Designs, Inc.

home built in the Georgian style is shown in Figure 6.22.

Saltbox

One of the most common modifications of the Georgian style is the saltbox. The saltbox maintained the symmetry of the Georgian style but omitted much of the detailing. This housing style received its name because its steep roof resembled a colonial-era salt container. In colonial times a saltbox was typically a

FIGURE 6.23 ■ *A traditional saltbox residence has a two-level front and a one-level rear.* Courtesy Janet Rowell.

two-story structure at the front but tapered to one story at the rear. The lower rear portion was often used as a partially enclosed shed, which was oriented north as a windbreak. These square or rectangular homes typically had a large central chimney and large, double-hung windows with shutters. The exterior walls were covered with clapboard or shingles. Figure 6.23 shows the features of saltbox styling in a residence.

Garrison

The garrison style combines saltbox and Georgian style with the construction methods of log buildings. The garrison was originally modeled after the lookout structures of early forts. Many of these homes had steep gabled roofs, small diamond-paned windows, and a second story overhang across the front facade. Originally, heavy carved timbers were used to support the overhang. Garrison-style homes were usually sided in unpainted clapboard or wood shingles. Figure 6.24 shows the features of a garrison-style residence.

Cape Cod

Cape Cod style is seen in a small, one-level home with a steep gabled roof and a central chimney. This style of home was developed in New England during the seventeenth and eighteenth centuries. Early Cape Cod homes were shingle-sided, one-story cottages with no dormers. They had two basic rooms: the great room and the parlor. The great room was used for daily living and the parlor served as the master bedroom. The cen-

FIGURE 6.24 ■ *A Garrison-style home has an upper level that extends over the lower level.*

FIGURE 6.25 ■ *A Cape Cod home features the formal balance of other styles but adds dormers and shutters.*

tral chimney had several connecting fireplaces to warm each area. Over time the kitchen was moved to the back of the house, with small bedrooms at the rear corners. The steep roof originally associated with the style allowed an upper-floor level to be formed throughout the center of the house. Dormers were often placed on the front side of the roof to make the upper floor habitable. The upper bedrooms were either covered with a shed-type dormer covering the entire front width or with small dormers to allow window placement. Windows on the lower floor were placed symmetrically around the door and had shutters. An example of Cape Cod styling is shown in Figure 6.25. An offshoot of Cape Cod style is Cape Ann, which has many of the same features as a Cape Cod house but is covered by a gambrel roof. During the mid-twentieth century, the small, uncomplicated Cape Cod shape became popular in suburban developments. A typical twenty-first century Cape Cod is square or rectangular with 1 or 1 1/2 stories and a steeply pitched, gabled roof. It may have dormers and shutters. The siding is usually clapboard or brick.

Federal

The Federal style, which was popular between 1780 and 1840, combines Georgian architecture with classic Roman and Greek styles; it had a very dignified feeling. Although Georgian homes were square and angular, a Federal-style building was more likely to have curved lines and decorative flourishes, such as swags, garlands, and elliptical windows. Federal homes were built of wood or brick with low-sloped roofs and dentil moldings below the eaves, or with a flat roof surrounded by

FIGURE 6.26 ■ *A main entry door with a fanlight and sidelights is a common feature of the Federal style.* Courtesy Marvin Windows and Doors.

a balustrade. Other key features would include a high, covered entry porch or portico supported on Greek-style columns centered over the front door. A semicircular fanlight window was usually placed above the front door, surrounded by arched trim; narrow side windows usually flank the front door. Other windows on the front elevation were usually arranged symmetrically around a center doorway and trimmed with shutters. Figure 6.26 shows an entry with a fanlight and sidelight windows. These windows are arranged symmetrically around the central doorway and were capped with a projected pediment and shutters. A Palladian window was often used in the center of the second story. This is a large window

FIGURE 6.27 ■ *Federal-style homes combine features of Georgian and classical Roman and Greek architecture. A key feature of the style is a high, covered entry portico supported on Greek-style columns centered over the front door.* Courtesy William E. Poole, William E. Poole Designs, Inc.

FIGURE 6.28 ■ *Greek Revival homes are large and rectangular; they have a very boxlike exterior. The bold, simple exterior lines feature classic Greek proportions and ornamentation.* Courtesy William E. Poole, William E. Poole Designs, Inc.

that is divided into three parts. The center section is larger than the two side sections, and it is usually arched. Circular and elliptical windows were often used to add variety. Figure 6.27 shows an example of a Federal-style residence.

Greek Revival

The Greek Revival style first appeared in public buildings and then quickly became the chief residential motif as settlement occurred in the United States from 1830 through the 1850s. Homes of this style reflect classic proportions and decorations reminiscent of classic Greek architecture and have features that resemble the Parthenon. These homes were large, rectangular, and very boxlike, with bold, simple lines. The exterior was usually covered with clapboard siding or brick. Roofs were usually low-sloped gables or hips with small overhangs with cornices featuring a wide trim band comprising frieze and architrave. A two-story portico with a low-sloped gable roof supported on Greek columns was centered on the front of the residence. The front door was usually topped with a transom window or capped by a prominent cornice (see Figure 6.21) and flanked by narrow sidelights. Figure 6.28 shows an example of Greek Revival architecture.

Southern Colonial

Several home styles emerged from the South as builders sought refuge from the heat, humidity, and generally oppressive weather of the area. One of the most popular southern styles is the Antebellum, which is Latin for "before the war." The term refers to the elegant plantation homes built in the American South in the thirty or so years preceding the Civil War. Antebellum homes used features found in the Greek Revival or Federal style, such as grand, symmetrical facades with evenly spaced windows; boxy layouts with center entrances in the front and rear, balconies; and Greek columns or pillars. Roofs were usually low-sloped gables or hips, with small overhangs and cornices featured a wide trim band with frieze. Figure 6.29 shows an example of a southern plantation home.

English

English-style houses are fashioned after houses that were built in England prior to the early 1800s. Two styles that are often copied based on their English roots are the Tudor and the English cottage.

English Tudor

The traditional Tudor home is a masonry or stucco home modeled after English manor homes but also were influenced by Dutch and Gothic styling. There are several different styles of Tudor homes, but the Elizabethan variation is most often copied in twenty-first

FIGURE 6.29 ■ *Southern Colonial or Plantation-style homes reflect classic symmetry. They typically have a large covered porch to provide protection from the sun.* Courtesy Dixie Pacific.

FIGURE 6.30a ■ *English Tudor styling combines half-timber with brick, stone, and plaster.* Courtesy Metal Roofing Alliance.

FIGURE 6.30b ■ *English cottage styling combines half-timber with brick, stone, and plaster. With a sloping roof and a massive chimney at the front, the cottage-style home served as a model for many storybook homes.*

century architecture. The traditional Tudor home features an asymmetrical layout and walls two or more stories in height. Exterior walls were typically constructed of stone or brick set in intricate patterns or of heavy timber and plaster. The half-timbered exterior featured exposed wood timbers, which formed the structural framework, while the spaces in between were filled with brick or stucco. The roofs typically were comprised of one or more steeply pitched gables. Parapeted gable end walls that extend above the roof and massive chimneys topped with decorative chimney pots were also common. Door tops were often rounded, and windows were usually multipaned casement windows with diamond-shaped glass rather than the more traditional rectangle. Half-timbered bay window projections and projecting oriel windows were other common window features. Figure 6.30a shows common features of English Tudor styling.

English Cottage

A subtype of the Tudor Revival style is the Cotswold Cottage. With a sloping roof and massive chimney at the front, the cottage-style home served as a model for many storybook homes. Cottage homes were patterned after the rustic cottages constructed in southwestern England since medieval times; they often featured steep, half-hipped thatched roofs. The exterior might feature half-timbering, stone, brick, and shingle or stucco siding. Other common features included a sloping, uneven gable roof, asymmetrical shape, a prominent chimney made of brick or stone, small dormer windows, and casement windows of leaded glass. Figure 6.30b shows common features of English cottage styling.

Dutch

As early as the seventeenth century, Dutch settlers in New York, New Jersey, and Pennsylvania often built brick or stone homes with roofs that reflected their Flemish culture. The Dutch Colonial style has many of the features of homes already described. The major difference is a broad gambrel roof with flaring eaves that were slightly rounded into barn-like gambrel shapes. This roof is made of two levels. The lower level was usually very steep and provided the walls for the second floor of the structure.

The upper area of the roof is the more traditional gable roof. Early homes had a single room, which was often expanded by additions at each end, creating a distinctive linear floor plan. End walls were generally of stone, and there was usually a chimney at one or both ends. Double-hung sash windows with outward swinging wood casements, dormers with shed-like overhangs, and a central Dutch double doorway were also common features. The double door, which was divided horizontally, was used to keep livestock out of the home while allowing light and air to pass through the open top. An example of the Dutch Colonial style is shown in Figure 6.31 The gambrel roof is described in Section VI.

French Colonial Homes

Colonial French influence can be seen in several common home styles found throughout the United States. These styles include French Provincial, French Normandy, French ***Mansard,*** and Second Empire. Second Empire homes are explored later in this chapter, as Victorian homes are examined.

French Provincial

French Provincial colonial homes had their origins in the style of the rural manor homes or chateaus built by French nobles during the mid-1600s. These homes are recognized by their balance and symmetry, lending stateliness and formality. These homes were two-level brick or stucco structures with detailing in copper or slate. They generally had steep, hipped roofs; a square balcony; and porch balustrades. Windows and chimneys were usually balanced on each side of the entrance. Other defining features included rectangular doors set in arched openings and double French windows with shutters. Second-story windows usually had an arched top that extended through the cornice to rise above the eaves. Figure 6.32 shows a home with French Provincial features.

FIGURE 6.31 ■ *A common feature of a Dutch Colonial home is a gambrel roof. It is very steep and serves as the walls for the second floor of the structure. The upper area of the roof is the more traditional gable roof.* Courtesy Laine M. Jones Design-W. Newbury, Ma.

French Normandy

French Normandy homes were patterned after homes in Normandy and the Loire Valley of France. Many of the original French homes had farm silos attached to the home instead of to a separate barn. American designers adopted the shape and style of the traditional French farmhouse and created a turret with living space where the silo would traditionally have been. These homes have many of the features of the Tudor style, such as decorative half timbering with stone, stucco, or brick used between the timbers. French Normandy homes are distinguished from Tudor homes by a round stone tower topped by a cone-shaped roof. This tower is usually placed near the center, serving as the entrance to the home. French Normandy and French Provincial details are often combined to create a style simply called French Country or French Rural. Figure 6.33 shows a home with French Normandy features.

French Plantation

In the southern states, the French Normandy style was modified into the French Plantation style. This type of home typically had two full floors with a wraparound porch covered by a hip roof. The porch was an impor-

FIGURE 6.32 ■ *French Provincial homes are recognized by their balance and symmetry, which serves to create a feeling of stately formality.* Courtesy Pamela Russom, American PolySteel, LLC.

FIGURE 6.33 ■ *A French Normandy home originally featured a silo on the front. This feature was altered in America to provide living space.* Courtesy BOWA Builders Inc. Jim Tetro, photographer.

FIGURE 6.34 ■ *This southern home reflects French features in the upper deck, with its iron railing covered by a hip roof.* Courtesy William E. Poole, William E. Poole Designs, Inc.

tant passageway because traditional French Colonial homes did not have interior halls. A third floor is a common component, with dormers added for light and ventilation. Figure 6.34 shows a French Plantation home. This style was modified around New Orleans into what is now known as the Louisiana French style. The size of the balconies is diminished, but the supporting columns and rails have become fancier.

French Mansard

French Mansard styling uses a hipped or mansard roof to hide the upper floor area. A mansard roof is an angled wall; it is discussed in Section IV. Originally found in the northern states, these homes were usually based on a rectangle with a smaller wing on each side. The roof was usually a mansard to hide the upper floor, but hip roofs were also used. Figure 6.35 shows an example of French Mansard styling.

FIGURE 6.35 ■ *A French mansard home hides the upper floor behind an inclined wall referred to as a mansard roof. These homes are usually based on a rectangle with a smaller wing on each side.* Courtesy Pamela Russom, American PolySteel, LLC.

Spanish

In the southwestern United States, Florida, and California, settlers drew upon Hispanic and Moorish building traditions. Spanish colonial buildings were constructed of adobe or plaster and usually had just one story. Deeply shaded porches and dark interiors make these homes particularly suited for warmer climates. Some Spanish Colonial homes featured a Monterey-style second-story porch. Arches and red tiled roofs are two of the most common features of Spanish, or mission-revival style, architecture. Timbers were often used to frame a flat or very low pitched roof. Windows with grills or spindles and balconies supported on square pillars with wrought-iron railings were also common features. Round or quatrefoil windows were often used for accent. A quatrefoil window is a round window composed of four equal lobes resembling a four-leafed clover. Figure 6.36 shows an example of Spanish-style architecture.

Italianate and Italian Villas

Two common Italian home styles that were popular in American architecture include the Italianate and the Italian villas.

FIGURE 6.36 ■ *Spanish-style homes typically have low-sloping tile roofs, arches, and window grills.*

FIGURE 6.37 ■ *A contemporary adaptation of the Italian Villa style. Key features of this style are large, asymmetrical structures that range in height from two to four stories and are covered with a low-pitched, tiled roof.* Courtesy John Henry, architect.

Italianate

Italianate-style homes were popular in the United States during the period of 1840–1885. This style was common throughout most of the country because the wide variety of materials associated with it made it adaptable to varied budgets. Key features of this symmetrical style are the large rectangular wood-framed structure that emphasizes vertical proportions and elaborate decorations. Structures usually ranged in height from two to four stories and were covered with a low-pitched roof with wide eaves featuring decorative brackets, large cornices, and a square cupola. A single-story arcade porch was centered on the front to cover heavily molded double-entry doors. The porch was often topped with ***balustraded*** balconies and flanked by bay windows. The front entry door was usually a double rectangular or arched entry doors with elaborate surrounds. Tall, narrow, double-paned windows with hood moldings were often used with windows, although Roman or segmented arches above windows and doors were also common. Windows were usually placed in pairs.

Italian Villas

The Italian Villa style remains popular in the United States and is often used for large custom homes. The style became popular in America during the nineteenth century and is recognized by its rambling, asymmetrical floor layout, low-pitched tile roofs, and tower. Large overhangs supported on decorative brackets under the eaves are also typical of a villa-style home. Exteriors were usually covered with stucco. Villa homes usually have windows topped with arches and doors crowned with heavy trim. Figure 6.37 shows an example of an Italian villa.

Victorian

Italianate-style homes gave way to the Victorian styles in the late 1800s that featured irregularly shaped floor plans and very ornate detailing. Victorian houses often borrowed elements from many other styles of architecture, including partial mansard roofs, arched windows, and towers. Exterior materials typically included a combination of wood, brick, and wrought iron. Victorian-style homes can be divided into Gothic, Second Empire, Folk, and Queen Anne.

Victorian Gothic

In the 1800s, fashionable houses in England began to resemble medieval Gothic cathedrals, convents, and storybook castles. The style moved to America in the period of 1840–1880. The exterior walls of these homes were typically made of stone or brick with an asymmetrical floor plan. Exterior walls were often topped to resemble the battlement of a castle or fort and used to surround steeply pitched roofs. Several types of windows may be found in Gothic architecture including pointed windows with decorative leaded glass, ***oriel,*** and quatrefoil shaped windows. An oriel window projects from the wall and forms a small turret that does not extend to the ground. Brackets or ***corbels*** usually supported the curved walls that held the windows. Figure 6.38 shows an example of a Gothic Victorian home.

Second Empire

Second Empire homes were inspired by the architecture in Paris during the reign of Napoleon III and reached the height of their popularity in the United States be-

FIGURE 6.38 ■ *Gothic Victorian homes resemble medieval Gothic cathedrals, convents, and storybook castles. Exterior walls are often topped to resemble a battlement and used to surround steeply pitched roofs.*

tween 1855 and 1885. The example of Victorian style shown in Figure 6.39 is easily recognizable because of its high mansard roofs, which cover the upper floors. The roofs were often covered with patterned slate shingles. Although these homes are often seen in horror movies, they were thought to be very elegant at the time they were built. Walls were often covered with ornate wood trim or intricate brick patterns. Wrought-iron crestings were also common above the upper cornice. Other common features of this style of home include dormer windows projecting from the roof, classic pediments above each opening, and tall double-hung windows on the first floor. Decorative brackets supported protrusions such as eaves, window bays, and balconies.

Folk Victorian

Folk Victorian homes developed for less affluent consumers during the period of 1870–1910. These homes were much simpler than the traditional Victorian home but feature trimwork made possible by mass production. These homes were traditionally square or symmetrical with porches with spindle work that wrap around them. They are similar to Queen Anne homes of the same period but do not have the towers, bay windows, or elaborate moldings of that style. Figure 6.40 shows a home representing the Folk Victorian style.

FIGURE 6.39 ■ *Second Empire Victorian homes are easily recognizable because of their high mansard roofs, which covered the upper floors.*

FIGURE 6.40 ■ *Folk Victorian homes are much simpler than the traditional Victorian homes but feature trimwork made possible by mass production. Homes were traditionally square or symmetrical. The porches, adorned with spindlework, wrapped the home.* Courtesy Benigno Molina-Manriquez.

Queen Anne

This style of home was popular during the period of 1880–1910, as the machine age was sweeping America. As a result of mass production in factories, precut architectural components were available for shipping across the country. The style is named after the eigh-

FIGURE 6.41 ■ *The Queen Anne style is characterized by elaborate moldings and trim called gingerbread, a complicated roofline, and a round tower, turret, or a large round bay window.* Courtesy CertainTeed Corporation.

teenth-century English queen and is characterized by elaborate bric-a-brac, a complicated roofline, and expansive porches that wrap around the front and side of the house. These homes usually had a round tower, turret, or large round bay window. Elaborate moldings and trim called gingerbread is also a key feature of the Queen Anne style. Figure 6.41 shows an example of a Victorian Queen Anne-style home.

American Foursquare

The American Foursquare home style was originally popular during the period of 1895–1930. This style is recognizable because of the simplicity of its design. The style features a simple box shape that was typically two stories high with an attic and a full basement. Each floor generally had four rooms. The first floor usually contained an entry foyer, living room, dining room, and kitchen. The second floor typically consisted of three bedrooms and a bathroom.

Walls were usually made of wood, but brick, stone, and stucco sided with clapboards, cedar shingles, or a combination of the two were are also seen in these homes. The structure was usually covered with a low-sloped hip roof. Large overhangs supported by cornice-line brackets or other details copied from Craftsman and Italian Renaissance styles were a common feature. Other features sometimes associated with a Foursquare home are a large central dormer and a front porch that covered most of the front facade of the home. Figure 6.42 shows an example of a home patterned after the Foursquare style.

Surface features and limited decoration helped this style to adapt to a wide range of architectural styles. A Foursquare house with crisp white clapboard siding and black shutters is usually identified as a Colonial Revival. The same Foursquare shape takes on Queen Anne airs when it has bay windows and decorative brackets. Add stucco siding and a porch with stone columns and the house resembles a Craftsman Bungalow. Depending on the details that are added, a Foursquare home may be Spanish Eclectic, French Provincial, Tudor, Neoclassical, or Art Deco.

Farmhouse

The Farmhouse style makes use of two-story construction and is usually surrounded by a covered wrap-around porch. The roof ridge runs parallel to the front surface. A steep roof with dormers usually covers the home with a shallow pitch at the porch. The exterior material was usually clapboard siding. Trim and detail work, common in many other styles of architecture,

FIGURE 6.42 ■ *Homes built to copy the American Foursquare style feature a low-sloped roof covered by large overhangs supported by cornice-like brackets.*

FIGURE 6.43 ■ *The Farmhouse style features a simple structure with a covered porch.* Courtesy Pamela Russom, American PolySteel, LLC.

were rarely found in the Farmhouse style. Figure 6.43 shows an example of a home built to resemble a farmhouse.

Prairie

Prairie-style houses that became popular in the Midwest between 1900 and 1920, are recognizable for their low horizontal lines, one-story projections, low-pitched hipped roofs, and large overhanging eaves designed to help the home blend into the landscape. Roofs were typically tiled or shingled, depending on the area of the country where the home was built. Made popular by Frank Lloyd Wright, the floor plan of a prairie home usually consisted of open interior spaces divided by leaded or stained glass panels or built-in furniture with natural finishes intended to blur the distinction between indoors and outdoors. Rooms were usually based on square, L-shaped, T-shaped, or Y-shaped floor plans. Exterior walls varied between stucco, masonry, or wood-frame construction covered with brick and beveled horizontal siding. Windows were typically arranged in rows of small casement windows framed in horizontal bands that accent the horizontal lines. This style of home can now be found throughout the country and is shown in Figure 6.44.

FIGURE 6.44 ■ *Prairie-style houses are recognizable for their low horizontal lines, with one-story projections, low-pitched hipped roofs, and large overhanging eaves.*

Bungalow

The Bungalow is a popular American style that originated in India, where British colonists adapted the native one-story thatch-roofed houses in the province of Bengal for use as summer homes. This style became popular in America as a 1 1/2–story home between 1900 and 1920; it featured open, balanced, but non-symmetrical floor plans. Roofs were usually steep enough to hide the upper floor or featured low-sloping gables with front and rear dormers. Large extended eaves with decorative, bracketed supports were also typical. A large veranda covered by an extension of the main roof was a common feature at the front of such a home. Common exterior materials included shingles, lap siding, or board-and-batten siding. Windows were typically casement or double-hung with small-paned glass and were usually trimmed with wide simple casings. The front door often featured glass-panes with double sidelights.

FIGURE 6.45 ■ *The Bungalow style is a popular American style featuring a large veranda covered by an extension of the main roof.*

Bungalow homes typically made extensive use of built-in furniture, shelves, seating, and wood trim throughout the house. The dining rooms, bedrooms, kitchens, and bathrooms often were arranged around a central living room on the main floor. Additional bedrooms were placed upstairs in the center of the gable roof. Figure 6.45 shows a home built with bungalow features.

Craftsman

The Craftsman-style home became popular in America in the early twentieth century. Floor plans usually consist of asymmetrical, free-flowing two-story layouts. The upper level was often cantilevered over the lower level. Lower levels were typically made of stone or brick, and upper levels were often covered in shingles, stucco, stone, or brick. Windows were typically double-hung with grids in the upper half and were usually asymmetrically placed. Casement windows placed in series was also a common alternative. Window trim was usually plain wide board casings. Rectangular cantilevered bays with multiple windows were also common.

These homes feature a large wraparound porch elevated several feet above the surrounding grade. Porch roofs were often supported by masonry or large decorative wood columns with wood-framed walls covered with shingles used as a railing between the columns. Open, large-scale balusters were also a popular alternative for porch railings. Tapered wood columns were also common means of supporting the porch roof. The porch was typically covered at the front under a low-sloped roof that extended from the main gable. Roofs, typically steep, and multiple perpendicular gables supported by decorative brackets were common. A large dormer on the front and back of the main gable was also common to the Craftsman style. Figure 6.46 shows a home that includes many features of the Craftsman style.

FIGURE 6.46 ■ *The Craftsman style features a large wraparound porch elevated several feet above the surrounding grade. Porch roofs are often supported by tapered wood columns, with large-scale balusters used for the porch railings.* Courtesy Stephanie Earp.

Ranch

The Ranch style of construction comes from the Southwest, where it has been the dominant style from the mid-1930s to the present. This style is usually defined by a one-story, rambling layout, which is possible because of the mild climate and plentiful land. The major attraction of this style today is the one-level room layout. The roof is typically low-pitched, with a large overhang to block the summer sun. The major exterior material is usually stucco or adobe. Figure 6.47 shows a Ranch-style home. Common features of this style is the long, horizontal, single-level construction covered with a low-sloped roof with large overhangs. The floor plan is usually nonsymmetrical, using either a rectangular, L-, or U-shaped design. No matter what the shape, the floor plans usually feature simple open spaces with very few interior walls.

Contemporary Homes

It is important to remember that although a client may like the exterior look of one of the traditional styles, the traditional floor plan of one of those houses would very rarely be desired. Quite often the floor plans of older houses produced small rooms with poor traffic flow. A designer must take the best characteristics of a particular style of house and work them into a plan that will best suit the needs of the owner. Contemporary, or modern, does not denote any special style of house but implies the use of new materials and features, and may include the use of geometric shapes such as those seen in Figures 6.48, 6.49, and 6.50. Some houses are now being designed to meet a wide variety of needs, and others reflect the lifestyle of the owner. No matter the style that is favored, several recent trends in the construction of new homes are helping buyers realize their

FIGURE 6.47 ■ *Ranch-style homes feature a one-level elongated floor plan covered by a low, sloping roof.*

FIGURE 6.48 ■ *Many contemporary homes use clean lines with little or no trim to provide an attractive structure.* Courtesy Barry Sugerman, AIA.

FIGURE 6.50 ■ *This contemporary home uses regional influences to create a pleasing exterior.*

FIGURE 6.49 ■ *Contemporary homes that are not based on previous styles offer a unique, inviting effect.* Courtesy APA–The Engineered Wood Association.

dreams. The most popular floor plans offer flexibility, adaptability, and plenty of room to grow. A new home can usually incorporate features of any architectural style because decorative details don't dictate the size and placement of the rooms. As you explore the design possibilities throughout this text, give consideration to the following popular trends in homes of the early twenty-first century:

- **Open, Informal Spaces.** Because of high building costs and family lifestyles, interest in formal living rooms is waning. Most home buyers prefer an informal "great room" rather than formal rooms for entertaining. Combining the family room with the dining area gives the home a greater sense of spaciousness. A half wall or a work counter defines the kitchen area while allowing unobstructed views.
- **Fewer Hallways.** Homes with fewer hallways have an open, airy feeling. Instead of long, dark corridors, rooms flow together with doors leading directly to a living area or other shared space. The home may appear larger because less square footage is taken up by passageways.
- **Bonus Rooms.** A spare room near the kitchen, over the garage, or in another area apart from the bedrooms provides extra value when a home is not designed for a specific buyer. These bonus rooms give buyers extra space that can be used in various ways including: as a play area, a multimedia home theater, an art studio, a workshop, an exercise room, a high-tech home office, or a quiet sanctuary.
- **Spacious Laundry Rooms.** No longer relegated to the garage or basement, a cleaning center should be bright, spacious, and conveniently located. A common location would be near the kitchen or bedrooms. The area should provide space for the basic appliances of a utility room in addition to space for storage, for sewing, or for craft supplies. Some designers use this area as a multipurpose room with plentiful cabinets, play space for children, and ample room for crafts and other hobbies.
- **Ample Storage Space.** Walk-in closets, linen closets, dressing rooms, pantries, and easy-to-reach kitchen cabinets add enormous appeal to any home.
- **Accessibility.** Many buyers are seeking homes that can comfortably accommodate family members or guests with mobility problems or visual impairments. Fewer stairs, wider doors and hallways, and larger bathrooms make a home easier to navi-

gate. One-story homes are increasingly popular with baby boomers. Two-story homes with the master bedroom suite on the main level and additional sleeping areas upstairs are also very popular.

- **Spacious Garages.** SUVs, snowmobiles, Jet Skis, and other recreational vehicles make spacious garages a must for many families. A two-car garage is considered the minimum, but many buyers opt for an even wider three-car garage.
- **Sliding Partitions.** Movable partitions, sliding doors, pocket doors, and other types of movable partitions allow for flexibility in living arrangements. A great room can be transformed into a more intimate living area and a secluded dining room.
- **Outdoor Living.** The yard and garden become a part of the floor plan when sliding glass doors lead to decks, patios, and porches. Upscale features include outdoor fireplaces, grills, and wet bars. The home in Figure 6.51 takes advantage of large areas of glass as well as a solarium to overlook the outdoor living areas.

FIGURE 6.51 ■ *This elegant residence reflects no particular style but has pleasing proportions, symmetry, and balance. The well-placed windows and solarium help bring the outside in.* Courtesy BOWA Builders Inc. Greg Hadley, photographer.

CHAPTER 6

Additional Readings

One of the best ways to improve your ability to design pleasing homes is to look at successful designs. Excellent sources for review include magazines, television, and the Internet. To increase your reading options, use your favorite search engine and research major listings such as architectural magazines, interior design magazines, and home magazines. Magazines used by designers include:

Architect's Journal
Architectural Digest
Country Living
Country Home
Design Architecture
Design Times
Better Homes and Gardens
House Beautiful
Home
Metropolitan Home

If you have access to cable or satellite television, HGTV and TLC provide an excellent selection of programs that explore interior design and construction. New programs are aired each season, so be sure and check program listings or the websites for your favorite station for available programs. Current popular programs include:

Before and After
Curb Appeal
Dream Builders
Dream Drives
Dream House
Extreme Home
Homes Across America

Information about each program can be obtained by contacting www.hgtv.com or by using a search engine to search the Internet. Other websites that promote interior design include:

ADDRESS	COMPANY OR ORGANIZATION
www.about.com	Home styles
www.excelhomes.com	Many websites related to design
www.homeportfolio.com	Home design directory
www.greatbuildings.com	Documents 1000's of structures/many related links
www.realtor.org	Realtor magazine online
www.iida.org	International Interior Design Association

CHAPTER 6

Exterior Design Factors Test

DIRECTIONS

Answer the following questions with short, complete statements. Type your answers using a word processor or neatly print the answers on lined paper.

1. Type your name, the chapter number, and the date at the top of the sheet.
2. Type the question number and provide the answer in the form of a statement that includes part of the question.

QUESTIONS

6.1. How are horizontal lines reflected in residential design?

6.2. Describe how lines can be used in the design of a structure.

6.3. Use the Internet to find examples of floor plans showing a simple two-bedroom house in two of the following shapes: L, U, T, or V.

6.4. Define how the terms *formal balance, informal balance, symmetrical,* and *asymmetrical* relate to residential design.

6.5. Describe four floor-plan styles and explain the benefits of each.

6.6. What factors make a two-story house more economical to build than a single-story house of similar size?

6.7. What house shape is the most economical to build?

6.8. Photograph or sketch examples of the following historical styles found in your community:

a. Italianate
b. Garrison
c. Saltbox
d. Queen Anne or Second Empire

6.9. What forms are most typically seen in residential design?

6.10. List and define the three terms used to describe color.

6.11. Identify the primary and secondary colors.

6.12. List the major features of the following styles:

a. Federal
b. English Tudor
c. Italianate
d. Spanish

6.13. How is a tint created?

6.14. Photograph or sketch three contemporary houses that have no apparent historic style.

6.15. What historic period influenced the Georgian style?

6.16. Sketch a Dutch Colonial house.

6.17. Sketch or photograph a house in your community built with a traditional influence and explain which styles this house has copied.

6.18. What are some of the drawbacks of a traditional style?

6.19. What are the common proportions of classic Greek design?

6.20. What style of residence has two full floors with a wraparound porch covered by a hip roof?

6.21. Define the following terms and list the styles with which they are associated.
Portico
Quatrefoil window
Balusters
Corbel
Palladian window

6.22. List differences between the Georgian and Federal styles.

6.23. Explain key features of a Foursquare-style home.

6.24. Explain key features of a Prairie-style home.

6.25. Use the Internet and research a home style that is popular in your area. Explain key features of the style you selected and explain why that style is suited to your area.

CHAPTER 7

Environmental Design Considerations

Previous chapters have considered the home style, the number of floors, and key factors that control the layout of the interior and exterior living spaces. This chapter explores site-related design features that will make the home energy-efficient and environmentally friendly. Design features to be addressed in this chapter include zoning restrictions on design, methods of integrating the structure to the site, and environmentally friendly design.

ZONING CONSIDERATIONS

Several site factors affect the design of a house. Among the most important are the zoning of the property, the neighborhood, and access to the lot.

Zoning Regulations

Each municipality has the right to regulate how property will be used. One of the first jobs of the design team is to verify the governing agency that regulates the building site and then determine how the construction site is zoned. Keep in mind that zoning regulations are separate from building regulations. Zoning regulations control the density of an area by regulating the number of structures that can be built per acre. These regulations for density are often based on a comprehensive plan developed at the county or state level for 5-, 10-, 20-, and 50-year growth patterns. Although regulations will vary greatly for each area, common examples of zoning divisions include:

Urban Low-Density Residential (R-2.5, R-5, R-7, R-10, R-15, R-20, R-30)
Medium-Density Residential (MR-1)
High-Density Residential (HDR)
Special High-Density (Residential) (SHD)
Recreational Residential (RR)
Mountain Recreational Resort (MRR)
Rural (Agricultural) Residential (RA-1)
Rural Residential Farm Forest 5-Acre (RRFF-5)
Farm Forest 10-Acre (FF-10)

Notice after each zone is a listing of subzones. Although they may be called by a different name in your area, the zones in the Urban Low-Density Residential group are similar in most areas. These zones regulate the minimum average lot or parcel area per dwelling of the building site, similar to Table 7.1.

TABLE 7.1 MINIMUM AVERAGE LOT OR PARCEL AREA PER DWELLING

DISTRICT	LOT AREA
R-2.5	2500 sq ft
R-5	5000 sq ft
R-7	7000 sq ft
R-8.5	8500 sq ft
R-10	10,000 sq ft
R-15	15,000 sq ft
R-20	20,000 sq ft
R-30	30,000 sq ft

Once the specific zone for the construction site has been determined, the usable area of the site can be determined. The building area is regulated by setbacks. If a 100 × 100' site is to be developed, it is considered an R-10 zone. Zoning departments establish common minimum setbacks to protect structures on adjoining property from the risk of fire or other structural damage. The setback size will vary greatly, but examples for the minimum design requirements for primary structures in these Urban Low-Density Residential Districts are as follows:

Minimum front yard setback: 20 ft
Minimum rear yard setback: 20 ft
Minimum side yard setback: 5 ft
Maximum building height: 35 ft
Maximum lot coverage of the primary use structures: 40%

Using these guidelines, the 100 × 100' site would now be reduced to a usable area 90' wide × 60' deep. The zoning guidelines for a 10,000-sq-ft site would allow:

- 5400 sq ft of buildable area within the setbacks. (60 × 90')
- A maximum height of 35'
- A maximum of 4000 sq ft of residence (40% of 10,000)
- An area of 6000 sq ft of open space (60% of 10,000 sq ft)

In addition to these guidelines, zoning regulations may vary based on the number of levels or on solar access guidelines. A designer should always verify the site limitations with the zoning department website prior to starting the planning process.

The home in the previous example is unrealistic for many urban settings. A large open space may be available in rural areas, but most urban areas are fairly well developed. Because most buildable urban sites have already been developed, many urban areas allow established areas to be subdivided to create small building sites. Notice that the R2.5 site (25 × 100') would allow a property owner of an R5 or larger site to subdivide and create a new buildable site. Because of the narrow width of the site, many municipalities allow homes to be constructed with one zero side-yard setback. Figure 7.1 shows an example of three narrow infill sites in a high-density residential zone. Even though the zoning department allows high-density construction, different construction methods will be required to construct the wall that abuts the property line. The design team will need to coordinate zoning regulations with building department regulations.

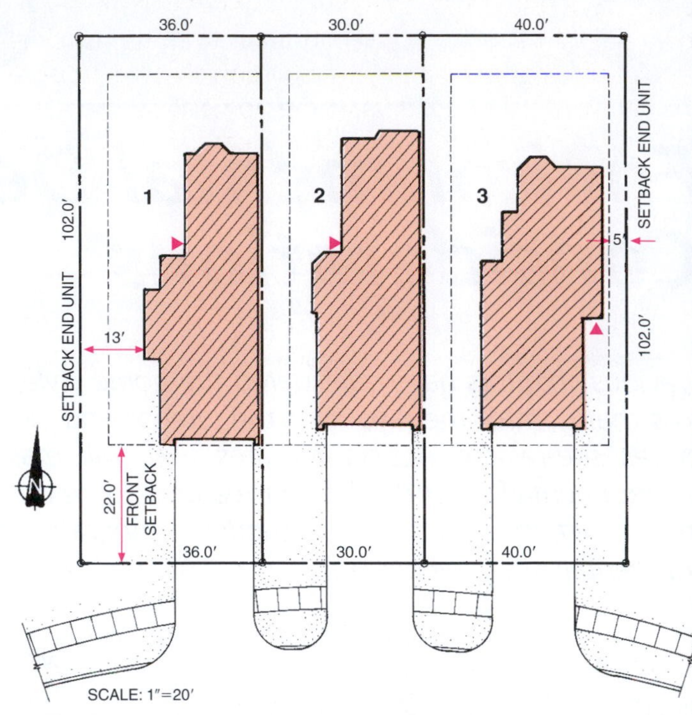

FIGURE 7.1 ■ *Many municipalities allow homes in a high-density residential zone to be constructed with one zero side-yard setback.*

Once the zoning and building regulations that govern the site have been determined, the design team can proceed. If the zoning regulations will hinder the intended project, the design team can search the complete zoning regulations for exceptions or substitutions to allow the project to proceed. Exceptions are often listed in the code based on the specific zone and the size of the proposed structure. Common exceptions allowed to the stated building setbacks include:

- Architectural features may project into the required yard not more than one-third (1/3) the distance of the setback requirement and not exceeding 40'' (1000 mm) into any required yard adjoining a street right-of-way.
- Open unenclosed fire escapes may project a distance not exceeding 48'' (1200 mm).
- An uncovered porch, terrace, patio, or underground structure extending no more than 30'' (750 mm) above the finished elevation may extend within 3' (900 mm) of a side lot line or within 10' (3000 mm) of a front or rear lot line.

If the code restricts the design of the project once the listed exceptions have been applied, the owner can apply for a design review, a ***variance,*** or apply to have the land rezoned to a less restrictive use. A design review allows the owner and the design team to meet with the regulating agency to obtain a solution that will ensure the safety of the occupants of the home. A variance is a legal request to allow a specific project to vary from the general guidelines.

Neighborhood Restrictions

Once the legal restrictions of the zoning department have been considered, the practical aspects of design must be considered. In the initial planning of a residence, the neighborhood must be considered. Given a choice, it is extremely poor judgment to design a $500,000 residence in a neighborhood of $200,000 houses. This is not to say that the occupants would not be able to coexist with their neighbors, but the house will have poor resale value because of the lower value of the houses in the rest of the neighborhood. The style of the houses in the neighborhood should also be considered. Not all houses should look alike, but some unity of design can help keep the value of all the properties in the neighborhood high.

Review Boards

To help keep the values of the neighborhood uniform, many areas have architectural review committees. These are review boards made up of residents who determine what may or may not be built. Although once found only in the most exclusive neighborhoods, review boards are now common in undeveloped subdivisions, recreational areas, and retirement areas. These boards often set standards for minimum square footage, height limitations, and the type and color of siding and roofing materials. A potential homeowner or designer is usually required to submit preliminary designs showing floor plans and exterior elevations to the review board.

Site Access

Site access can have a major effect on the design of the house. Access will be influenced by the size of the building site, the location of streets or alleys, and the site terrain. The narrower the lot, the more access will affect the location of the entry and the garage. Figure 7.2 shows typical access and garage locations for a narrow lot with access from one side. Usually, because of space restrictions, only a straight driveway is used on interior lots.

When a plan is being developed for a corner lot, there is much more flexibility in garage and house placement. To enhance livability, some municipalities are moving away from the layouts shown in Figure 7.2. The traditional layout has produced what is referred to as a snout house, meaning a home that is dominated by a view of the garage. In a design where the home is dominated by the garage, the main entrance is often secondary to the entrance for cars, and the driveway often dominates the front yard. In an attempt to eliminate barriers between homes and enhance visibility, building covenants may call for:

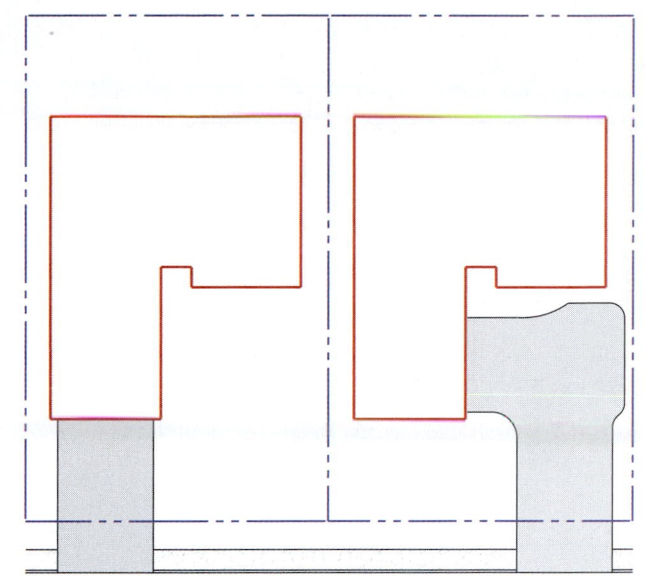

FIGURE 7.2 ■ *Access to an inner lot is limited by the street and garage location.*

- At least one main entrance to the house that meets one of the following requirements:
  - The main entrance can be no further than 6'-0'' (1800 mm) behind the longest wall of the house that faces the street.
  - The main entry must face the street or be at an angle of up to 45°.
  - At least 15% of the area of the street-facing facade of the home must be windows.
  - The length of the garage wall facing the street may not be greater than 50% of the length of the entire facade of the home.
  - A garage wall that faces a street may be no closer to the street property line than the longest street-facing wall of the home (see Figure 7.3).

Figure 7.4 shows an example of a pleasing relationship between the garage and the balance of the home.

Another popular way to make a neighborhood more livable is to remove the garage from the front of the site altogether. Many planned areas in Florida, Maryland, Ore-

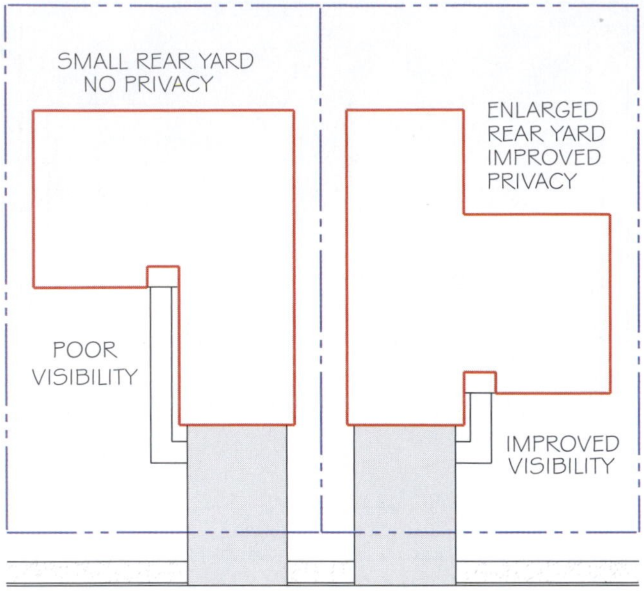

FIGURE 7.3 ■ *Limiting the distance that the garage can project from the balance of the home improves visibility and provides for a friendlier neighborhood atmosphere.*

FIGURE 7.4 ■ *A pleasing relationship between the residence and the garage.* Courtesy Debbie Fredrick.

gon, and Tennessee have moved the garage to the rear of the site so that no driveway or parking is provided on the entry or front side of the home. Automobile access is provided to the rear of the lot by an alley. Figure 7.5 shows a garage carefully blended into a home on a large site.

Driveway Planning

Once the location of the driveway has been determined, the driveway, turnaround, and exterior parking must be considered. The design options include:

- Arrange the driveway at a 90° angle to the access road when possible. A range of 60° to 90° to the access road is acceptable.

FIGURE 7.5 ■ *On larger sites, the entrance to the garage can often be moved away from the main entry so that it is not a part of the front elevation.* Courtesy Barbara Conkling.

- Arrange the driveway so that good visibility of the access road is provided.
- Provide a minimum driveway slope of 1/4'' per foot.
  - Single-car driveway minimum width: 10-0'' (3000 mm)
  - Double-car driveways minimum width:18'-0'' (5400 mm)
  - The minimum turning radius for a driveway: 15'-0'' (4500 mm), with a turning radius of 20' (6000 mm preferred if space permits)
- Provide a minimum space of 10 × 20' (3000 × 6000 mm) for off-street parking.
- If possible, maintain a 5% grade for the driveway.

In addition to providing access to the home for the owners, additional parking space should be planned for guests and future drivers. Figure 7.6 shows a variety of driveway parking and turnaround options. The dimensions are given as commonly recommended minimums for small to standard-size cars. Additional room should be provided if available.

Rural Access

When a residence is being planned for a rural site, weather and terrain can affect access. Studying weather patterns at the site will help reveal areas of the lot that may be inaccessible during parts of the year because of poor water drainage or drifted snow. The shape of the

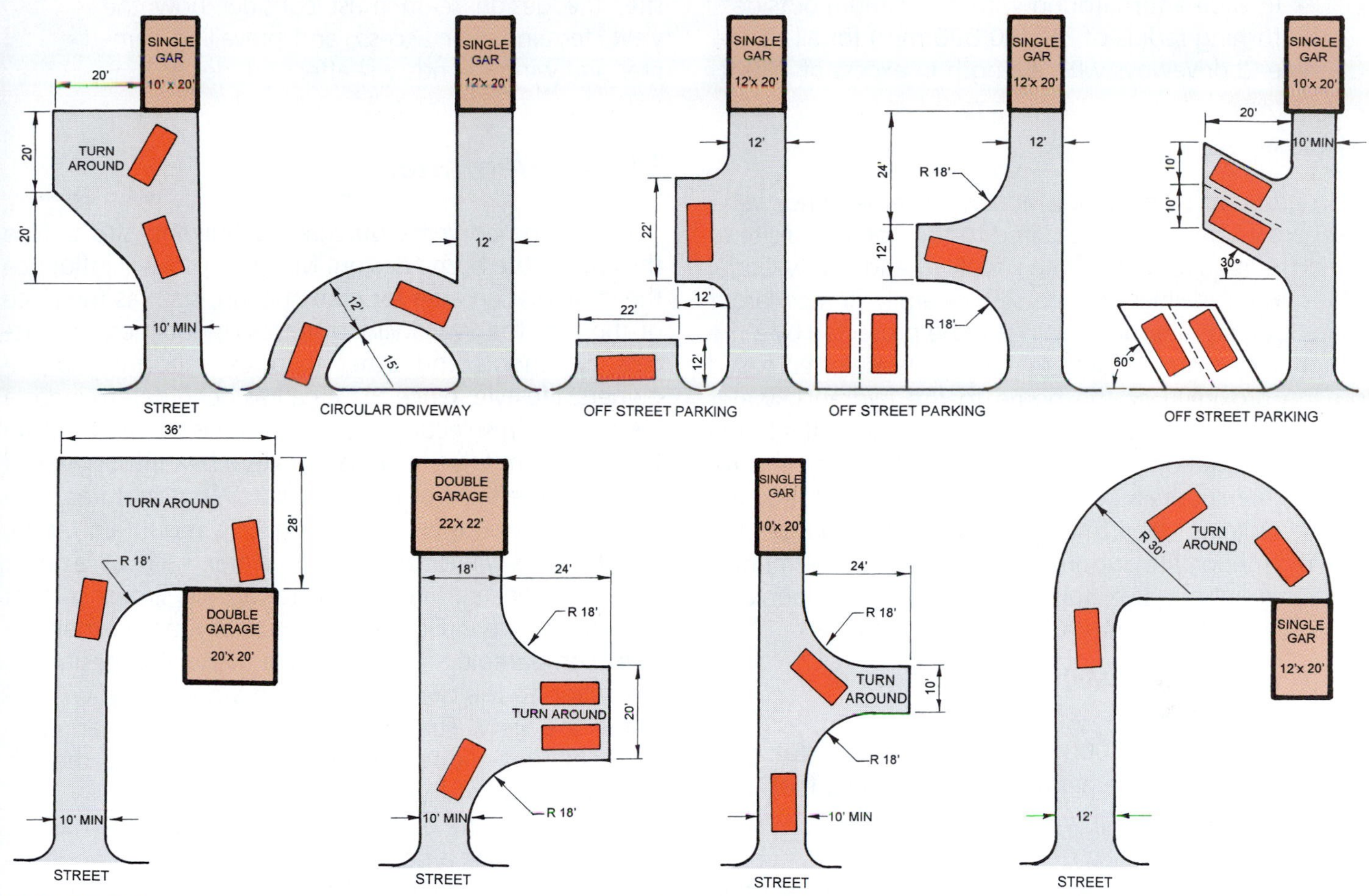

FIGURE 7.6 ■ *Common driveway, parking, and turnaround options and sizes.*

land will determine where the access to the house can be placed. In addition to incorporating the design features for urban drives and turnarounds, access for emergency vehicles must be considered. Standards will vary with each municipality and must be verified with the local building department and fire marshal. Additional requirements for fire safety include:

- Provide an all-weather surface of gravel, concrete, or asphalt paving.
- Provide a minimum driveway width of 15' (4500 mm) with an additional space of 3' (900 mm) of vegetation-free space on each side of the roadway.
- The driveway slope must not exceed 15%.
- All bridges and culverts in the driveway must be able to support 75,000 lb.
- A radius of 30' (9000 mm) or greater should be provided for emergency vehicles, delivery trucks, or vehicles pulling trailers.
- Provide a pull-off equal to twice the road width 40' (12,000 mm) long near the midpoint of the driveway, but at no more than 1/4-mile intervals.
- Maintain tree limbs above the driveway to provide 15' (4500 mm) of clear access.
- For lakefront properties, provide a water access location that is obvious in appearance, cleared, and has safe footing and a footpath that will facilitate the use of portable pumps.
- Provide a turnaround area at or near the end of the drive.
  - For ambulance and smaller firetrucks, access to the turnaround must have a turning radius of 25' (7500 mm) and/or a pull-out of 30 × 12' (9000 × 3600 mm).
  - For large firetrucks you must have a turnaround with a radius of 60' (18,000 mm) and/or a pull-out of 60 × 20'.

- Provide a turnaround with a minimum outside turning radius of 36' (10 800 mm) for all dead-end driveways with a length in excess of 150 ft.

Fire Protection

In addition to providing access for emergency vehicles, provisions may be required to the home and site to meet the requirement of the Life, Fire, and Safety Code. Many municipalities require single-family homes larger than 3500 sq ft in rural settings to be protected by a fire suppression system. These requirements are often based on the proximity of the home to fire hydrants, water pressure at the hydrant, and the travel time to the home from a fire station. Insurance underwriters may also require a fire suppression system even when not required by the building department. Chapter 13 introduces the use of interior fire suppression systems. Some municipalities might waive this requirement for sprinklers when certain guidelines are met, including:

- The use of fire-retardant roofing and siding materials.
- Creating a 30' (9000 mm) minimum firebreak around the entire perimeter of the home that contains no vegetation exceeding 24'' (600 mm) in height.
- Provide a sprinkler system at the top of all exterior banks to control the spread of a possible fire.
- Provide a water storage system containing 20,000 gallons of water for use in fire suppression. This would include a swimming pool, pond, or storage tank.

INTEGRATION OF THE HOME TO THE SITE

How the residence is placed on the site will have a major effect on the livability, energy efficiency, and cost of building and maintaining the home. If a home is to be constructed in a subdivision, the design team may have little control over how the structure will relate to the site. The street location will dictate the front of the home as well as the main entry and the garage location. As the area of the site increases, the options for integrating the site with the environment also increase. Blending the home with the site becomes one of the preliminary factors that the design team must consider as the design process is started. Chapter 1 and Appendix A introduced site features to be evaluated during the initial stages of the design process. To further integrate the home to the site, the design team must consider how the view, terrain, solar access, and prevailing summer and winter winds will affect the design.

View Orientation

Many homeowners purchase a building site before they begin the home design. Many factors will influence the choice of an area for construction, such as the price of the site, the availability of necessary utilities, and access to schools and other features important to the owners' lifestyle. Once the area has been determined, a key feature in selecting the actual site is often based on the view from the site. What makes a beautiful view will vary with the taste of each individual, but features such as a specific mountain peak or even a mountain range, forest, city skyline, or body of water such as a lake, stream, or ocean. Sites that offer a view similar to that from the home in Figure 7.7 are usually more expensive than comparable sites without a view. The designer's obligation to the client is to design a home that will optimize the view. The ideal design will provide an environment that allows the occupants to feel as though they were part of the view.

Rarely will a site be perfect for each aspect of design. A home that has a gorgeous view may have a poor solar or wind orientation. The design team will need to work closely with the owners to evaluate orientation conflicts and evaluate tradeoffs. A home designed to take advantage of a view will generally have large amounts of glass on the view side. The openings for the glass may require the help of a structural engineer to make sure that the home will meet the lateral design

FIGURE 7.7 ■ *An orientation that captures a specific view can greatly increase the value of a home and be a priceless asset to the owner.* Courtesy Endless Pools.

loads caused by wind or earthquakes. If the home is oriented to take advantage of a view but has poor solar or wind orientation, design alternatives or energy-saving building materials can be required to offset possible problems. Construction options to overcome poor solar orientation are explored throughout Section VI.

Even if a home is to be built in a subdivision with no spectacular view, view orientation should be considered. Although the size of the lot and building setbacks may limit placement of the home, consideration should be given to the view from each window. If homes are on the adjacent sites, plan room views to minimize looking directly into the windows of an adjoining home. If the homes have not been constructed on adjoining sites, plan for where new homes could be constructed and add or delete windows accordingly.

Integration of the Home to the Terrain

The land at the building site will greatly influence the type of structure to be built and the drawings needed to complete the construction documents. Land considerations such as natural features, contour, and soil-bearing capacity will play important roles in determining the number of floor levels, where the structure will be placed, and how loads will be transferred into the ground.

Natural Features

Few if any reputable designers would consider bulldozing a site of its major natural features in order to have a clean site at which to begin work. Quite the opposite; one of the first considerations in planning how the home will blend with the site is to map key natural features so that they can be incorporated into the design. This would include locating streambeds, marshlands, major clusters of natural vegetation, and natural rock clusters. Most areas have very strict environmental regulations protecting natural features, even to the point of requiring that no soil leave the construction site, either by erosion or on the wheels of construction equipment. Because great care must be taken in locating each feature, a civil engineering firm is normally hired to locate key natural features to ensure accuracy. In working with bodies of water such as creeks or streams, the design firm must research state or federal regulations that may regulate development within a specified distance. Minimum distances from a stream may be based on a minimum distance such as 100' (30 000 mm) from the average yearly stream edge. Minimum distances may also be based on a minimum height above a 50- or 100-year flood plane. Methods of measurement can be determined by checking with the local building department.

When the location of trees must be determined, this might require knowing the diameter of the trunk, the average diameter of the ground coverage, and the estimated size of the root ball. Once key natural features have been located, the home can be designed so that each feature can safely remain. Figure 7.8 shows Fallingwater, designed by Frank Lloyd Wright. It is what many architects consider to be the best example of a home that blends with its environment. A second major consideration in planning how the home will fit the site is to carefully plan how it will match the terrain.

Terrain

Terrain refers to the shape or contour of a specific parcel of land. Although stock plans are usually drawn as if the construction site were flat, in reality, most of the flat, easy-to-build sites in urban areas have already been developed. Custom home plans are drawn to blend with a specific area at a specific construction site. In working on a site with a gentle slope, someone from the design team may visit the site to determine spot elevations and general trends in drainage. When a large

FIGURE 7.8 ■ *Fallingwater, designed by Frank Lloyd Wright, is considered by many architects to be the best example of a home that blends in with its environment.* Courtesy Western Pennsylvania Conservancy. Christopher Little, photographer. Western Pennsylvania Conservancy, Thomas A. Heinz, photographer.

area of land or a slope is involved, a civil engineering firm will be hired by the owner to develop a map of the topography. Figure 7.9 shows an example of a topographic drawing for a small site. Notice that the elevation for each property corner is given in engineering units. The contours for each foot of elevation change are represented with dashed lines. The process for creating a topographic drawing is presented in Chapter 9. As you consider the effect of terrain on the design of a home, it's important to remember a few basic concepts about reading a topographic plan. Notice that the contour lines on the east side of the site are closer together than those on the west side. The spacing of the lines indicates the slope of the terrain. The closer the lines are to each other, the steeper the slope. For this site, the east side is steeper than the west side of the property.

A second method of describing the slope of land is by the use of slope percentages. Slopes can be specified by percentages, such as 2%, 8%, or 25% slope. The number used represents the rise in inches for 100 in, or the number of feet per 100 ft. A 2% slope would rise 2 ft over a distance of 100 ft. The east side of the property falls 10 ft, or has an approximately 10% slope. The west side slopes approximately 5 ft per 100 ft for a 5% slope. The term approximate is used to describe each slope because the slopes are not uniform throughout the property length. Figure 7.10 shows a representation of various slopes expressed in percentages.

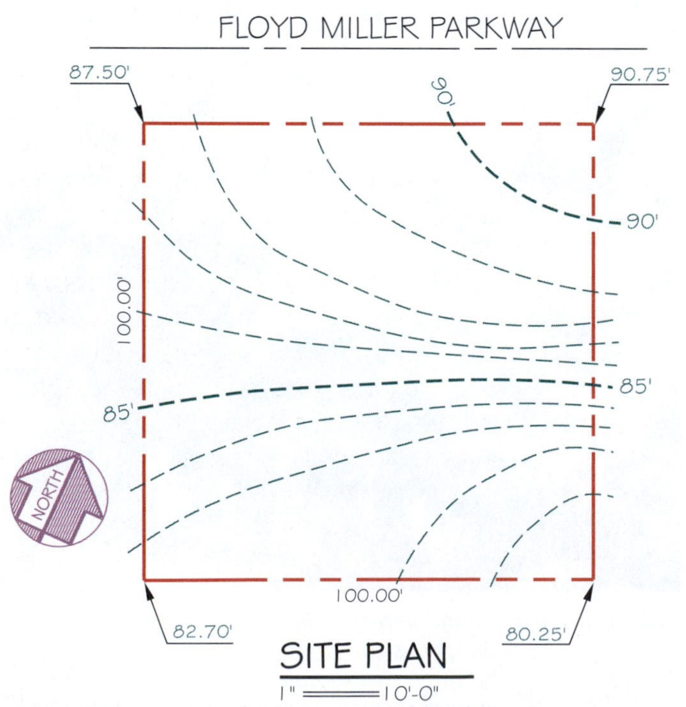

FIGURE 7.9 ■ *A topography plan for a small site with the elevations for each property corner and the contours for each foot of elevation change.*

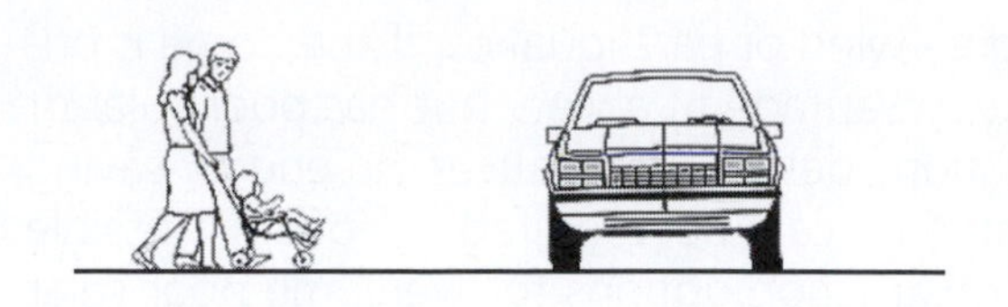

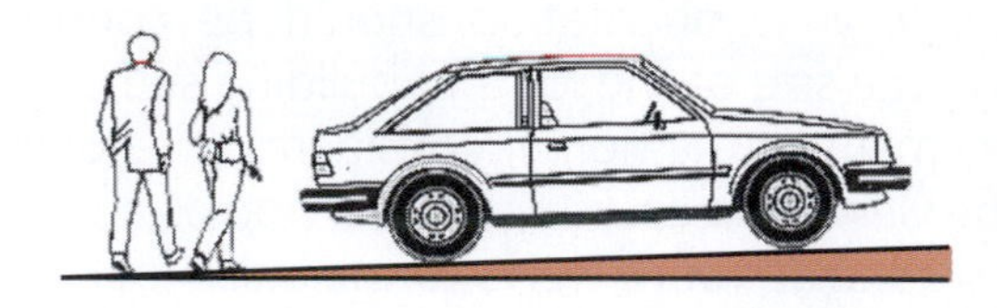

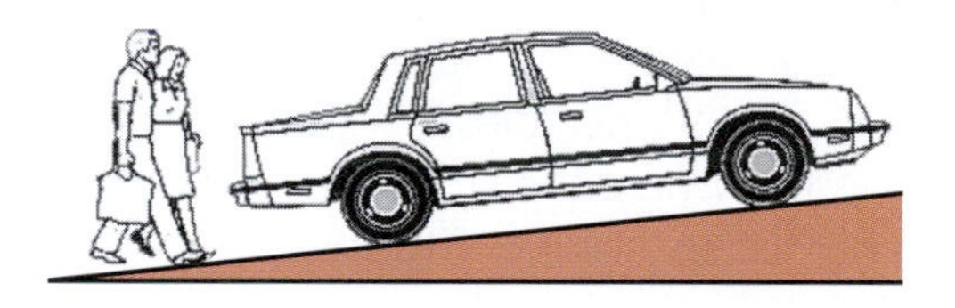

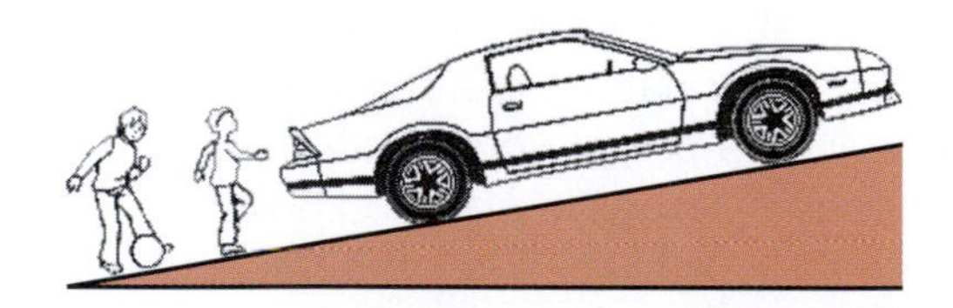

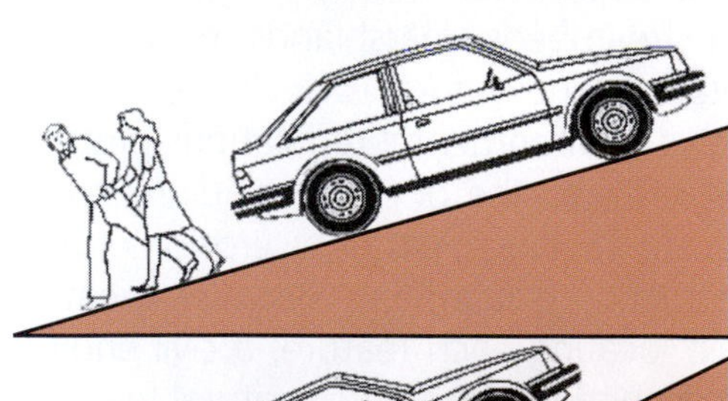

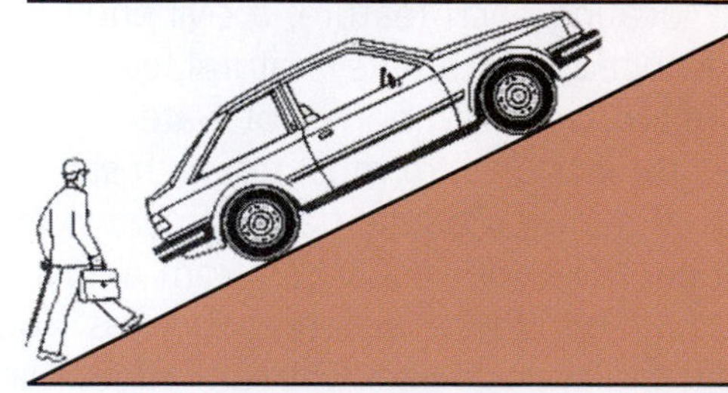

FIGURE 7.10 ■ *Many zoning departments specify the grade for a driveway by a percentage of slope. A slope of 10% is the maximum that many municipalities will allow.*

Modifying Terrain

The contours of the land at a construction site can be altered by adding or removing soil. When soil is removed from a site, it is referred to as a cut. Soil can be cut to form a vertical bank, but this will lead to landslides. An incline of 1.5 horizontal units to 1 vertical unit (1.5/1) is typically used for cut banks. To represent a cut bank on a site plan, lines would need to be placed 1.5' apart. Fill is the process of adding soil over the existing grade. Fill banks are typically created to be at an incline of 2 horizontal units to 1 vertical unit (2/1). To represent a fill bank on a site plan, lines would need to be placed 2' apart. Methods of creating and representing cut and fill banks are introduced in Chapter 9. During the design process, it's important to remember that any portion of the home that is built over fill material will need a special foundation design. (Foundation design is explored in Section 8.) As the house is designed, it's important to remember that the land can be modified to fit the design and style of the home. In designing the portion of the site where the house will be located, two key points must be considered before altering the site:

- The soil must slope away from the residence on all sides for a distance of 5' (1500 mm). The IRC refers to this as positive drainage. On the low side of the house, this is quite easy. On the high side, this will require cutting some of the existing soil to provide drainage, as shown in Figure 7.11. The low point formed between the two slopes must also be inclined to allow water to drain away from the house.

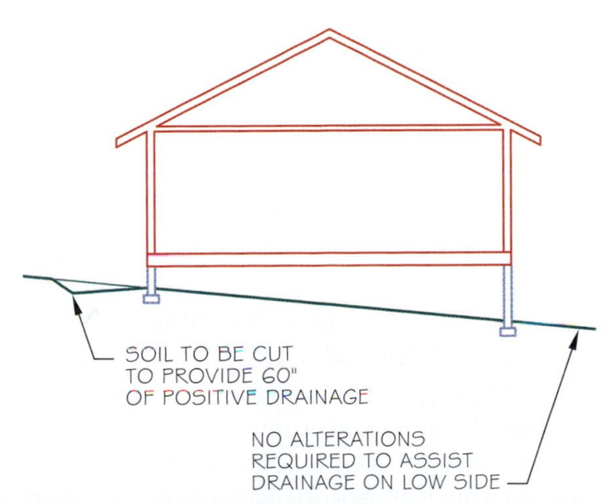

FIGURE 7.11 ■ *Positive drainage must be provided for a distance of 5' (1500 mm) on all sides of the residence. On the high side of the home, the existing soil will need to be cut to provide positive drainage.*

- The practical maximum grade for a driveway is 10%, although some municipalities will allow a slope of 14%. When a car is parked on such a steep grade, it will be awkward to open its doors. Many municipalities have regulations covering the length, grade, width, and turning radius of residential driveways to ensure that service vehicles can navigate the driveway. A slope of 3% to 7% is convenient for access.

The slope of the driveway also plays an important role in determining the finish floor level of the home. Chapter 9 introduces methods of moving soil to provide positive drainage and introduces the process of planning the driveway access to the home.

Soil Considerations

In addition to the slope of the site, the type of soil must also be considered during the design stage. The type of soil will affect the design of the foundation. Although you do not want the foundation to dictate the design of the home, the type of foundation should be considered as the home is designed. The type of soil, its drainage capacity, and its tendency to freeze should all be known prior to starting the design. A soils engineer can provide the design team with this information.

Floor Layout and Terrain Design

The terrain of the site will affect the design of the home. A single-level or two-story home is well suited to a level or a gently sloping site. Land at the upper end of the site can be cut and pushed to the lower portion to form a level pad for a floor (see Figure 7.12a). On sites with a gentle slope, a multilevel floor similar to that in Figure 7.12b, which steps with the site, may be used. A single-level home on a sloped site will require extra construction cost for excavation or building up the foundation. If a single floor is used on a sloping site similar to that in Figure 7.12c, a wood-framed floor will need to be used. A crawl space will be created on the low side of the site. This space can generally be developed into a lower floor formed with a concrete slab, similar to that in Figure 7.12d. The cost of excavation will be increased, but the additional living space will be more economical than using the crawl space for storage. A multilevel home or a home with a daylight basement, similar to that in Figure 7.12e, is well suited to a sloped site. Another alternative for a very steep site is to design a home on stilts. This will require that the design team

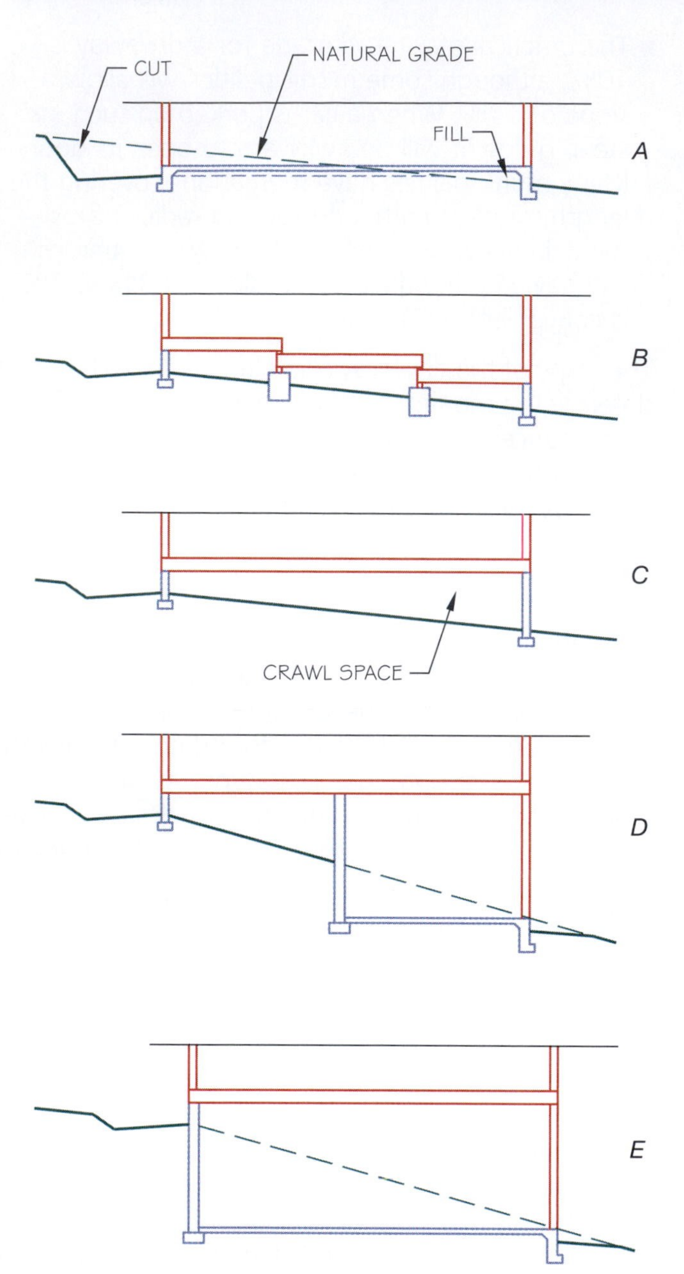

FIGURE 7.12 ■ *The shape of the building will affect the style of the home to be built. A. On mild slopes, land at the upper end of the site can be cut and pushed to the lower portion of the site to form a level pad for a level floor. B. A multilevel floor is well suited to a gentle slope. C. A single-level home on a sloped site will require a floor framed with wood or extra excavation for building a concrete slab. D. A wood-framed floor over a partial basement. E. A multilevel home or a home with a daylight basement is well suited to a steep site.*

coordinate the home design with a structural engineer to design the supports and a soils engineer to study the soil and determine its bearing capacities. The steel frame for a multilevel home constructed on a steep site is shown in Figure 7.13. An alternative to

FIGURE 7.13 ■ *Construction sites with steep slopes will require additional input from structural and soils engineers. The site for this multilevel home required the use of steel pilings driven 27' into the soil to reach solid rock. At the far corner of the home, stilts 20' high were required to span from the top of the pilings to the beams supporting the lower floor.*

building up is to build down. Subterranean construction is used for some high-end homes in frigid regions of the country. Underground homes will have increased excavation and material cost but will have economical advantages in energy consumption over above-ground homes.

Solar Orientation

Solar orientation of a home to the site may be based on heating or cooling requirements. If you're building in one of the southern states, blocking the summer heat will be the goal of your solar design. In northern states, maximizing the winter sun will be the goal of the solar design. In either case, the position of the sun throughout the day and the year is an important factor in home orientation. Even if the home is not designed to take advantage of solar heat, the orientation of the home to the sun should be an important consideration. Chapter 5 considered the relationship of rooms and their usage to the sun during the day. Factors such as placing bedrooms on the east side of the home and living areas on the west side will allow the family to take advantage of natural lighting throughout their daily activities. Figure 7.14 shows considerations of the sun's position throughout the day.

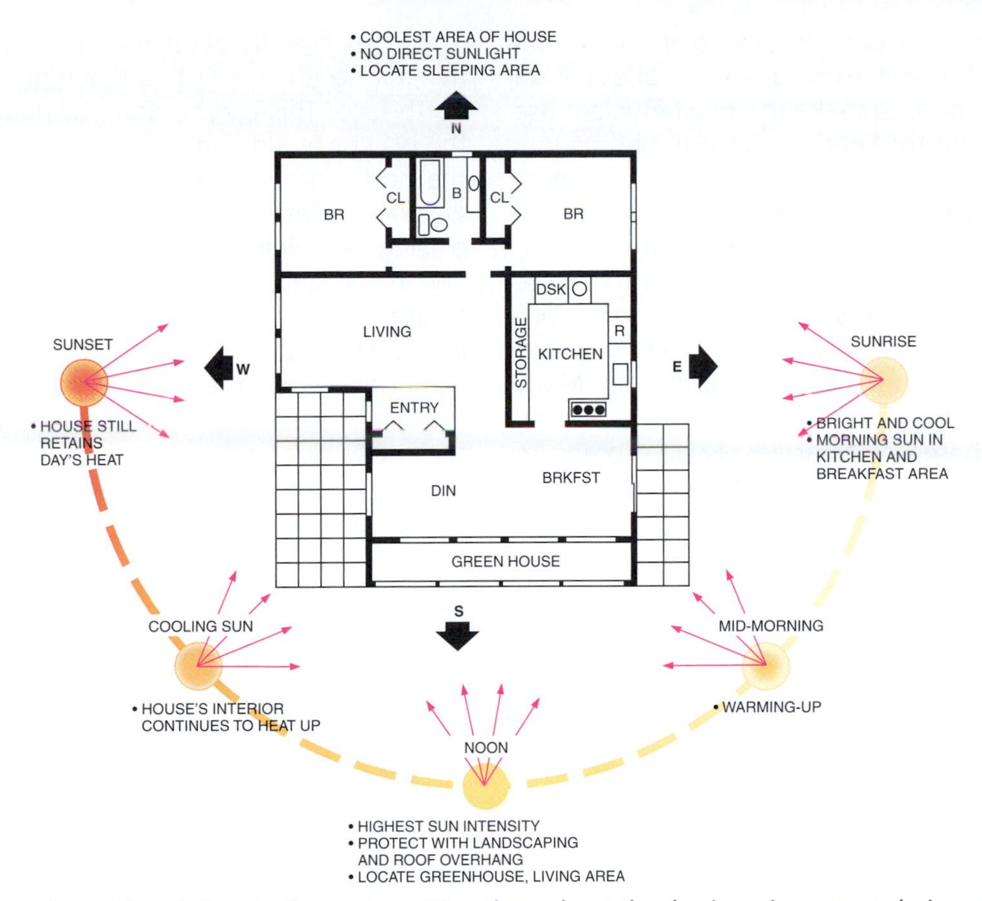

FIGURE 7.14 ■ *Room layout in relation to the sun's position throughout the day is an important design consideration.*

Locating the Home to Maximize Sunlight

The yearly cycle of the sun is as important as its daily effect on room layout. As seen in Figure 7.15, the sun is higher in the sky during the summer, when it rises in the northeast and sets in the northwest. During the winter, the sun is much lower in the sky, rising in the southeast and setting in the southwest. Depending on your location, on December 21 the sun may be as low as 20° above the horizon. This difference in sunrise, sunset, and height above the horizon need to be considered as the windows are placed and the size of the roof overhang is planned. Careful planning of the location and angle of the windows where sunlight can enter the home will affect the heat gained within the residence. Knowing the location of the sun at specific times of the year will allow the designer to size and place glazing to maximize heating and cooling effects. Glazing can be as much as 25° away from perpendicular to the suns rays and still receive 90% of the radiation. Chapter 10 explores window placement and Chapter 15 explores determining the size of the roof overhang based on heating and cooling needs.

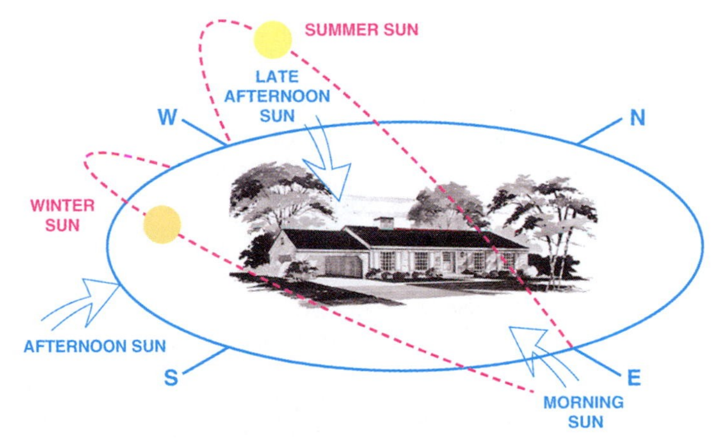

FIGURE 7.15 ■ *The sun is higher in the sky during the summer, rising in the northeast and setting in the northwest. During the winter, the sun is much lower in the sky, rising in the southeast and setting in the southwest. This difference in sunrise, sunset, and height above the horizon will need to be considered as the home is designed.*

Sites that will be developed with a solar orientation should allow for full sun exposure from about 8:00 a.m. through 3:00 p.m. to make the most efficient use of the sun's radiation for heating. Place the home in a northern portion of the sunny area of the site to maximize the exposure of the southern face of the home to the sun. This will limit the chance of future development to cast shade on the residence and maximize southern outdoor living areas. There should not be obstacles such as tall homes, evergreen trees, or other obstructions with the potential to block the sun. Many areas restrict the height and the placement of the home on the site based on solar access laws to ensure that each site will have such access. If there are tall obstructions to solar access, their location and effect on the site should be mapped to determine the extent of the solar blockage. The types of trees on the site will also affect solar heat gain. ***Coniferous*** trees on the south side of the home will hinder solar gain. ***Deciduous*** trees on the south side of the home provide shade from the summer sun. In the winter, when these trees have lost their leaves, winter sun exposure is not substantially reduced.

Building Shape and Solar Planning

In addition to the location of the home on the site, the shape of the home will affect the solar gain admitted into the residence. A rectangular home that is elongated along the east-west axis will maximize solar gain. This orientation is the most efficient shape for both heating and cooling in all climates. Homes with a long south face maximize solar gain during the winter and minimize the solar gain during the summer by having short east and west faces. During the summer, most of the sun will be on the southern roof, with glazing protected by overhangs. Glazing on the east and west walls is not exposed to the hottest sun of the day and can be protected by roof overhangs. By carefully placing rooms in the home based on the need for heat, the heat gained on the south face and use of the cool north face can be maximized.

The depth of the home along the east-west axis is also an important consideration to maximize the solar gain. Studies by the Illuminating Engineering Society of North America have determined that the depth of rooms along the south face of a home should not exceed 2.5 times the window height from the floor. Using this guideline will allow sunlight to penetrate the entire room. Sunlight can be provided to areas that exceed the window height ratio by installing skylights in a south-sloping roof.

A third consideration for the shape of the home is minimizing the height and length of the structure's north face. This can be done by berming earth against the wall or by sloping the roof so that its low side is on the north face of the structure. Providing taller south walls with glazing and shorter north walls with minimal glazing will minimize heat loss. The low north walls will also minimize the shadow cast by the home and increase outdoor living areas on the north side of the home. Using a roof pitch that approximates the sun's winter angle will maximize the exterior living areas on the north face of the home. On a multilevel home, place the upper floor on the south side when possible and protect the north side with attic space over the lower portion of the home.

Establishing South

In order to maximize solar gain, true south must be determined when the solar potential of a site is being evaluated. If view orientation requires that a structure be turned slightly away from south, it is possible that the solar potential will not be significantly reduced. The exact amount will vary based on your location, but the south face of the home can vary up to 45° to the east or west and still maintain efficiency. True south is determined by a line that stretches from the North Pole to the South Pole. When a compass is used to establish north, the compass points to magnetic north, which is different from true north. The difference between true north and magnetic north is referred to as magnetic declination. The amount of magnetic declination differs throughout the country. Figure 7.16 shows a map of the United States that represents the magnetic declination at various locations. A magnetic declination of 15° east, which occurs in central California, means that the compass needle points 15° to the east of true north and 15° to the west of true south. If you were to face magnetic south, true south would be 15° to the left.

Wind Orientation

The term ***prevailing wind*** refers to the direction in which the wind typically blows. The prevailing winds in the United States are from west to east. If you live on the West Coast, the winds generally flow from the southwest off the Pacific Ocean. The prevailing winds are said to be southwesterly. During the late summer and early fall, winds known as Santa Anna winds come from the northeast, producing hot, dry weather. These differences in prevailing winds are caused by differences in atmospheric pressure. These types of patterns

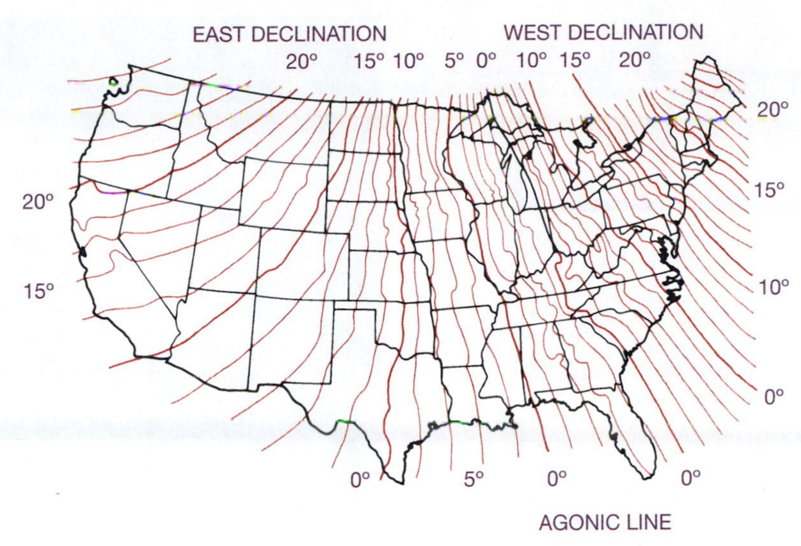

FIGURE 7.16 ■ *Magnetic declination, the variation between magnetic and true north, varies based on your location.*

are large regional patterns. Features such as the Rocky Mountains, the Great Lakes, the Gulf of Mexico, the Atlantic Ocean, and the Great Plains are all examples of natural features that greatly influence the winds in specific regions. Within each region, wind patterns can vary because of the mountains, valleys, canyons, or river basins. A third type of pattern exists within local areas. On hillside sites, heat will cause a breeze to blow uphill during the daylight hours and cooler air will settle during the evening, causing a breeze to flow downhill. The design of the home should be based on the winds expected for the construction site. Wind conditions that influence site location may be found on the Internet, in almanacs, or at the local library by researching subjects such as climate, microclimate, prevailing wind, and wind conditions. The local weather bureau can also provide information regarding prevailing winds in an area.

In planning the home site, select an area with protection from the prevailing winter winds. Landmass, other buildings, and vegetation can all provide protection from the prevailing winds. When the view will not be blocked, place the structure so that existing structures can block winds. Placing a home on the crest of a hill will usually provide the best view, but it will also maximize the force of the wind. Placing a home just below the ridge of a sloping site will allow the land to block the wind. As shown in Figure 7.17, the orientation of the home to the site will also affect wind currents. By altering the shape of the home, protected pockets for outdoor living can be created. The shape of the roof can also be used to deflect wind. A sloping roof can be used to deflect wind currents. Placing the low edge of the roof toward the direction of the wind will cause the

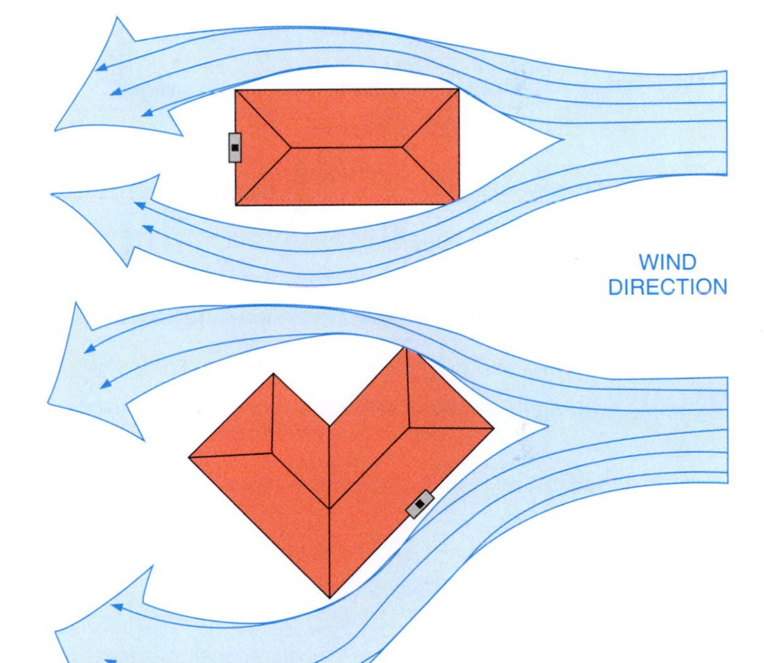

FIGURE 7.17 ■ *The orientation of the home to the prevailing winds and the shape of the home will affect wind flow around it.*

wind to rise up and over the structure. Another method of blocking strong prevailing winds is to sink the home into the grade. This would include such options as daylight basements, earth-bermed, and earth-sheltered homes. Earth-bermed homes have soil on three sides.

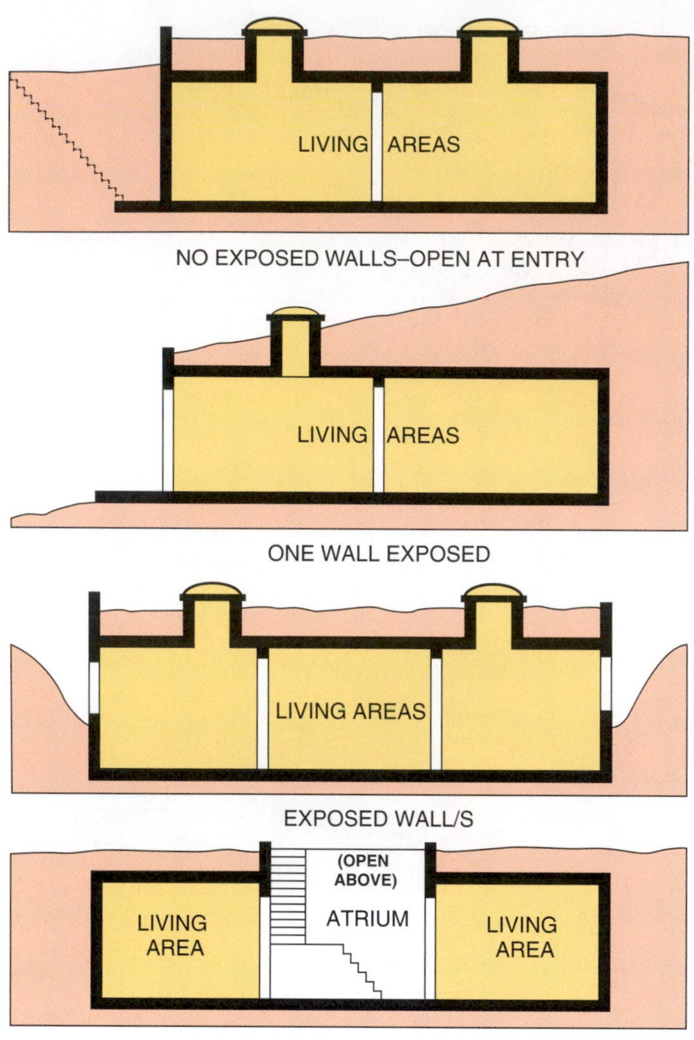

FIGURE 7.18 ■ *Earth-sheltered homes have soil on three sidewalls and the roof. Four options often used for earth-sheltered homes are seen here.*

Earth-sheltered homes have soil on the roof as well as on three sides. Figure 7.18 shows options for earth-sheltered homes. Be sure to verify emergency egress requirements with the local building department if bermed or sheltered options are to be used.

Room location and construction methods can also influence wind protection. Place rooms such as utility rooms, pantries, and bathrooms that do not require large areas of glass for view or solar use on the north side or the side toward severe winter winds. The kitchen produces heat, so it can be placed on the windward side of the house. A garage can also be used to provide a break between cold winter winds and the living areas of the home. Bedrooms with fairly small windows may be placed to provide a barrier between the wind and the balance of the home.

FIGURE 7.19 ■ *Trees and bushes provide an effective barrier to wind.* Courtesy William E. Poole, William E. Poole Designs, Inc.

In addition to natural features or other structures, landscaping is an effective means of blocking wind. Plantings and windbreaks such as fences or decorative barriers can also be used to moderate the effects of winter winds. The decision to plant deciduous or conifer trees for wind protection will need to be balanced with view considerations. Coniferous trees or other evergreen landscaping planted in staggered rows can provide an effective windbreak. A landscape architect or other landscaping professional should be consulted to verify the types of trees to be used. If the wrong types of trees or too few of them are planted, they could become a danger to the home during high winds. Figure 7.19 shows a home built in a wooded area that provides an effective wind buffer.

Cooling Summer Winds

In costal regions and areas affected by large bodies of water, summer winds are usually mild and contribute to a more comfortable living environment. In inland areas, summer breezes may be nonexistent or hot and dry. In areas where cooling is the goal, it is especially important for the home design to take advantage of any cooling available. Comfort can be achieved through design for natural ventilation and by landscape design. The use of high ceilings, ceiling fans, and large south- and west-side eave overhangs can all contribute to keeping a home cool. Dormer windows and openable skylights on the north side of the roof can also be useful in allowing hot air to escape. Windows placed on the windward and leeward sides of the home will allow for cross ventilation. In multiple-level homes, the stairwell can be used to provide for continuous ventilation. The placement of windows on the low and high ends of the stair allows a draft to be created as hot air rises up the stairwell and cooler air is drawn in through the low window.

Landscaping can also be used to provide relief from summer heat. Coniferous trees can be used with deciduous trees to help block the sun. They can also be used to funnel summer winds into the site. Deciduous trees serve a triple purpose. In the summer they provide needed shade and act as a filter to cool heated wind. In the winter they lose their leaves, thus allowing the sun's rays to help warm the house.

Sound Orientation

If the construction site is in a rural setting, sound orientation may not be a concern. Sound considerations are often important in working with an urban setting. The sound of neighbors and traffic 24/7 can make life unbearable. The placement of the home in relation to neighboring structures and to the site will influence how noise can be filtered out. A building site that is level with a road or slightly below one may have less noise than a site that is above and overlooking the sound source.

Good landscaping design can contribute to a quieter living environment. Berms, trees, hedges, and fences can all be helpful. Some landscape materials deflect sounds; others absorb them. The density and thickness of the sound barrier also has an influence on sound reduction. The greater the width of the plantings for sound insulation, the better the control. Trees planted in staggered rows provide the best design.

Construction methods and materials can also reduce the transmission of sound into the building. Construction materials such as masonry are very effective at blocking noise, but wood and sheetrock can also be used to mask sound. Providing breaks in floors or double-wall construction are common methods of stopping sound from carrying from room to room. Sound-deadening board and some types of insulated foams have excellent ability to stop sound. Triple-glazed windows can also help reduce outside sound.

ENVIRONMENTAL DESIGN CONSIDERATIONS

Much of the information presented throughout this chapter is intended to help a home blend with its site. The concept of green building involves considering not only the building site but also how the home integrates with its total environment. A well-designed structure, such as the home in Figure 7.20, may be called environmentally friendly, earth-friendly, or green construction, but increasingly the goal of design is to blend a home that is efficient at all stages: during design, during construction, as it is inhabited, and as it is recycled at the end of its life. A green building design can be defined as any building that is:

- Designed, constructed, and maintained for the health and well-being of the occupants.
- Operated to maximize present and future beneficial impacts on the environment.
- Offer an opportunity to create environmentally sound and resource-efficient buildings by using an integrated approach to design.
- Promote resource conservation by including design features that encourage energy-efficiency, use of renewable energy, and water conservation.

Green buildings are resource-efficient buildings. They are designed to be energy-efficient and to utilize construction materials wisely—including recycled, renewable, and reused resources—to the maximum extent practical. They are designed and constructed to make sure that they will be healthy to live in, are typically more comfortable and easier to live in due to lower operating and owning costs, and are good for the planet. The overall environmental impact of new home and community development and the choices made when

FIGURE 7.20 ■ *A well-planned home will be constructed with the health and well-being of the occupants as a prime consideration. It must also be designed to maximize present and future beneficial impacts on the environment and offer an opportunity to create environmentally sound and resource-efficient buildings by using an integrated approach to design.* Courtesy Dennis Dinser Principal, Arcadian Design, LLC. www.arcadiandesigns.net.

we either reuse or demolish existing structures is very important. Examples of materials that are often featured in green homes include:

- Tankless water heaters
- Low-flow plumbing fixtures
- Gray-water reuse
- Air-admittance vents
- Solar water heaters
- Radiant barriers
- Miniduct air distribution systems
- High-efficiency air conditioners without HCFCs (hydrochlorofluorocarbons)
- Low-impact development techniques
- ENERGY STAR windows, doors, appliances, and insulation levels
- Bamboo flooring
- Low- or no-VOC (volatile organic vapor) paints

Each of these features is explored in the chapters that follow. Members of the building industry have banned together to form the U.S. Green Building Council (USGBC) to help promote the use of these materials and to develop new materials. The council has created a standard called the Leadership in Energy and Environmental Design (LEED™) to measure the effectiveness of green buildings. LEED has become the national standard to measure high-performance sustainable homes. Major areas defined by LEED include:

- Sustainable sites: This portion of the standard evaluates the selected site, the amount of site disturbance, erosion and sedimentation control, urban redevelopment, public transportation, storm water management, landscaping, and light pollution.
- Efficient use of water: Including efficient use of water for living purposes and landscaping as well as wastewater management and dispersal.
- Efficient use of energy and low impact on the atmosphere: This portion of the standard affects the heating and cooling systems, reduction of CFCs (chlorofluorocarbons) in HVAC equipment, ozone depletion, and renewable energy.
- Efficient use of materials: This portion of the standard affects the storage and collection of recyclables, material reuse, management of construction waste, resource reuse, recycled content, use of local and regional materials, use of rapidly renewable materials, and certified wood.
- Indoor environmental quality: This portion of the standard affects indoor air quality, carbon dioxide monitoring, and ventilation effectiveness.

The complete guidelines for green certification can be obtained from the USGBC website.

CHAPTER 7

Additional Readings

The following websites can be used as a resource to help you keep current with environmental issues.

ADDRESS	COMPANY OR ORGANIZATION
www.envirolink.org	Environmental resources
www.greenbuilder.com	Sustainable Building Sourcebook
www.greenerbuilder.com	Resource center for environmentally friendly building
www.iesna.org	Illuminating Engineering Society of North America
www.nesea.org	Northeast Sustainable Energy Association
www.nfpa.org	National Fire Protection Association
www.oikos.com	Green Building Source
www.ourcoolhouse.com	Energy-efficient home construction
www.physicalgeography.net	Fundamentals of physical geography
www.renewableenergyaccess.com/rea/home	Renewable Energy Access
www.usgbc.org	U.S. Green Building Council
www.wbdg.org	Whole Building Design Guide

CHAPTER 7

Environmentally Friendly Design Test

DIRECTIONS

Answer the following questions with short, complete statements. Type your answers using a word processor or neatly print the answers on lined paper.

1. Type your name, the chapter number, and the date at the top of the sheet.
2. Type the question number and provide the answer in the form of a statement that includes part of the question.

QUESTIONS

7.1. Explain how the neighborhood can influence the type of house that will be built.
7.2. List four functions of a review board.
7.3. Define site orientation.
7.4. List and briefly describe five factors that influence site orientation.
7.5. Define magnetic declination.
7.6. Use the Internet to determine the magnetic declination of the area where you live.
7.7. Describe how trees can be an asset in solar orientation.
7.8. Describe what influences the prevailing summer and winter winds in your area.
7.9. Describe features that can be used to protect a structure from wind.
7.10. Describe how landscaping can be used as an effective sound control.
7.11. List three methods of construction common to subterranean construction.
7.12. What is a common goal of the zoning department?
7.13. Why does construction on any site require that a slope be graded away from the structure?
7.14. Define terrain.
7.15. Why might view orientation be more important than other orientation considerations?

SECTION 2

Site Plans

8

CHAPTER

Land Descriptions and Drawings

The drawings required to specify the information for the site plan are usually the first drawings started in a residential project. As described in Chapter 1, a preliminary site plan is drawn to help study and access the design criteria. To complete the working drawings for the permit process, the preliminary site plan is converted to a working site plan. Every construction project requires one or more site-related drawings to describe the work to be done. To complete this process, a drafter must understand the use of legal descriptions to describe land, site-related drawings, and the general process of completing drawings.

LEGAL DESCRIPTIONS

For legal purposes, each piece of land is described by a description of the property known as a ***legal description.*** This description is used by municipalities to specify land parcels as they are bought, sold, and taxed. The legal description of a parcel of land can be obtained from the local zoning department that governs the property or from a title company. The legal description for any parcel of land is a matter of public record and can be obtained if the mailing address is known. The legal description may be given in several forms such as a metes-and-bounds system, a rectangular system, or a lot-and-block system. The type of description to be used depends on the contour of the area to be described or the requirements of the municipality reviewing the plans.

Metes and Bounds

A metes-and-bounds description is also referred to as a long description. A copy of the description can be obtained from a title company, the zoning department, or the tax assessor's office. A complete description can be added to the site plan or attached to the drawings on a separate sheet of paper. This system provides a written description of the property in terms of measurements of distance and angles of direction from a known starting point. The known starting point is referred to as the true point of beginning in a legal description. The true point of beginning is usually marked by a steel rod or a benchmark established by the U.S. Geological Society (USGS). These are referred to as monuments.

Metes

The metes are measured in feet, yards, rods, or surveyor's chains. A rod is equal to 16.5' (5000 mm) or 5.5 yd. A chain is equal to 66' (20 100 mm) or 22 yd. Directions are given from a monument such as a benchmark established by the USGS or an iron rod set from a previous survey. The point of beginning may be several hundred feet away from the property to be described. Directions are given from the point of beginning to a specified point on the perimeter of the property to be described. Directions are then given to outline the property, with all distances set in units of feet expressed in one hundredth of a foot rather than the traditional feet and inches normally associated with construction. Metric units should be expressed in either millimeters or meters. Centimeters are not used on construction drawings.

Bounds

The boundaries of property are described by bearings, which are angles referenced to a quadrant on a compass. Figure 8.1 shows the four compass quadrants and descriptions of lines within each quadrant. Bearings are always described by starting at north or south and turning to the east or west. Bearings are expressed in degrees, minutes, and seconds. Each quadrant of the compass contains 90°, each degree contains 60 minutes, and each minute can be divided into 60 seconds. A degree is represented by the ° symbol, a minute is represented by the ' symbol, and seconds are represented by the '' symbol. Bearings are used to describe the angular location of a property line. Some properties are also defined by their location to the centerline of major streets. When property abuts a body of water such as a creek or river, a boundary in angles may not be given and the property boundary is defined by the centerpoint of the body of water. A metes and bounds legal description would resemble the following description:

> A tract of land situated in the Southeast quarter of the Northeast quarter of Section 17, T3S, R1W of the Willamette Meridian, Clackamas County, Oregon, being more particularly described as follows, to wit: Beginning at the 5/8 inch iron rod at the Southwest corner of the Southeast quarter of the Northeast quarter of said section 17; thence north 0°06'10'' East along the West line of the Southeast quarter of the Northeast quarter, 322.50'; thence leaving said West line North 89°38'15'' East 242.00 feet; thence South 0° 06'10'' West parallel with said West line of the Southeast quarter of the Northeast quarter, 50.00 feet, thence North 89° 38' 15'' East 310.74 feet to the Westerly right of way line of Bell Road No. 113; thence South 3° 31' East along said Westerly right of way line, 272.91 feet to a 5/8 inch iron rod; thence leaving said Westerly right of way line, South 89° 38' 15'' West 569.97 feet to the point of beginning.

This description would be listed on legal documents describing the property as well as the site plan. Figure 8.2 shows the land described by the metes and bounds legal description. A short description of the property should be copied exactly onto the site plan. The short description would read:

> A tract of land situated in the S.E. 1/4, of the N.E. 1/4 of Section 17, T3S, R1W of Willamette Meridian, Clackamas County, Oregon.

Rectangular Systems

Many areas of the country refer to land based on its latitude and longitude. Parallels of latitude and meridians of longitude were used by the U.S. Bureau of Land Management in states that were originally defined as public land states. As the land was divided, large parcels of land were defined by what are known as basic reference lines. There are thirty-one pairs of standard lines in the continental United States and three in Alaska. Principal meridians and base lines can be seen in Figure 8.3. These divisions of land are described as the great land surveys. As the initial division of land was started, the first six principal meridians were numbered. The last-numbered meridian passes through Nebraska, Kansas, and Oklahoma. All subsequent meridians are defined by local names. The great land surveys were further divided by surveys that define land by townships and sections.

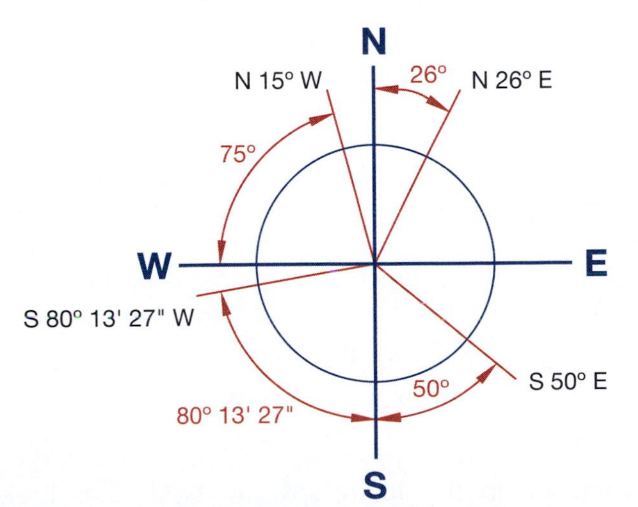

FIGURE 8.1 ■ *Bearings are referenced by quadrants on a compass beginning at either north or south.*

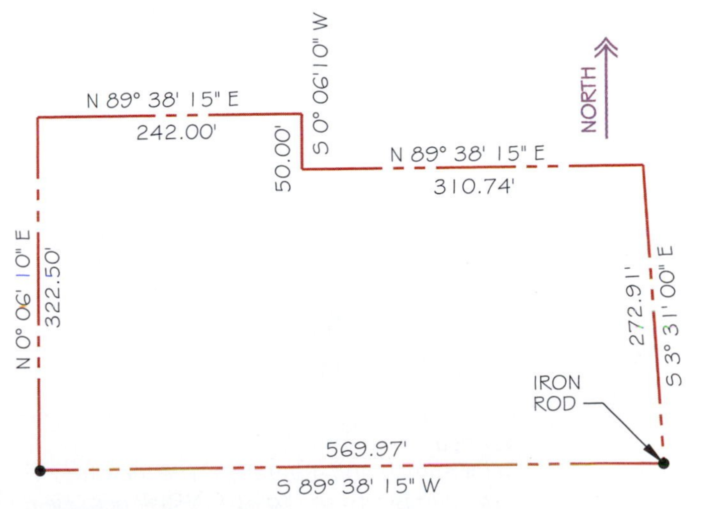

FIGURE 8.2 ■ *The parcel of land described in the metes-and-bounds legal description.*

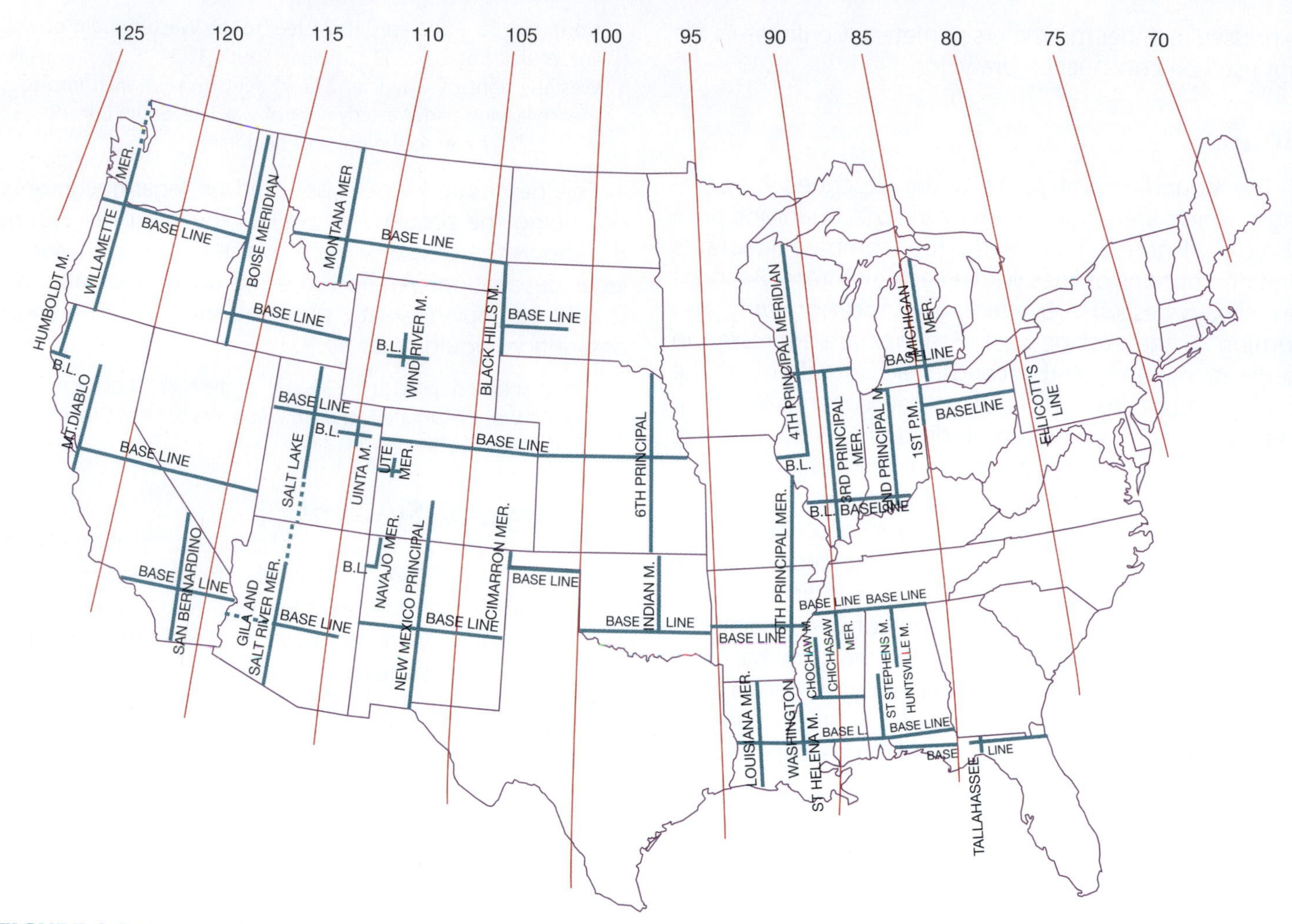

FIGURE 8.3 ■ *Principal meridians and basic reference lines are used to divide land in the continental United States.*

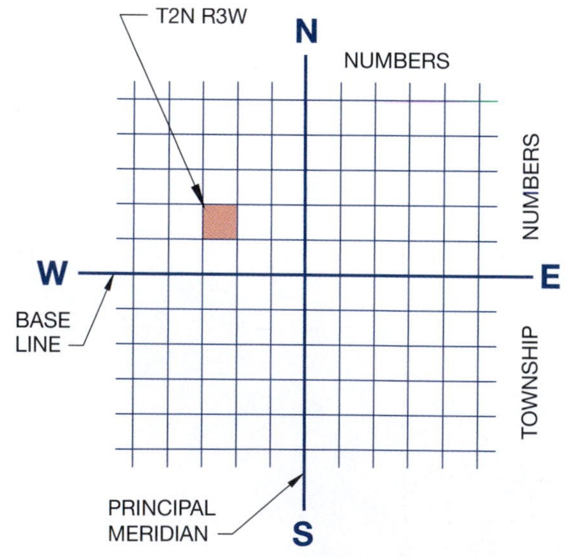

FIGURE 8.4 ■ *A township is a 6-square-mile portion of land defined by its position in reference to a principal meridian and a baseline. The indicated township is referred to as T2NR3W because it is in the second tier north of the baseline and three rows west of the principal meridian.*

Townships

Baselines and meridians are divided into 6-mile-square parcels of land called townships. Each township is numbered based on its location to the principal meridian and baseline. Horizontal tiers are numbered based on their position above or below the baseline. Vertical tiers are defined by their position east or west of the principal meridian. Township positions are shown in Figure 8.4. The township highlighted in Figure 8.4 would be described as Township No. 2 North, Range 3 West because it is in the second tier north of the baseline and in the third row west of the principal meridian. The name of the principal meridian would then be listed. This information is abbreviated as T2R3W on the site plan.

Sections

Land within the townships can be further broken down into 1-mile-square parcels known as sections. Sections are assigned numbers from 1 to 36 as shown in Fig-

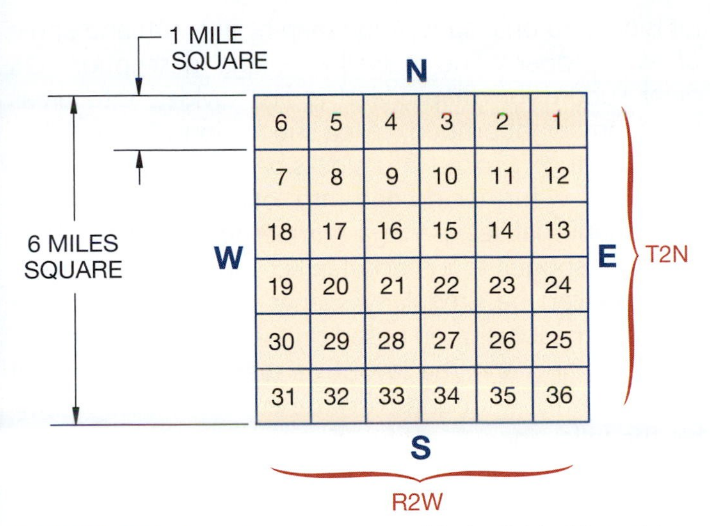

FIGURE 8.5 ■ *Land within a township is divided into thirty-six 1-mile-square portions called sections.*

ure 8.5, beginning in the northeast corner of the township. Each section is further broken down into quarter sections. Each section contains 640 acres or 43,560 square feet. Quarter sections can be further broken down as seen in Figure 8.6. The areas are defined by quarters of quarters or halves of quarters. These small segments are further broken down by quarters or halves again.

Specifying the Legal Description

The legal description, based on the rectangular system, lists the portion of the land to be developed described by its position within the section, the section number, the township, and the principal meridian. A typical legal description resembles the following description:

> The southeast one quarter, of the southwest one quarter of the southwest one quarter of Section 31, Township No. 2 North, Range 3 West of the San Bernardino meridian, City of El Cajon, in the state of California.

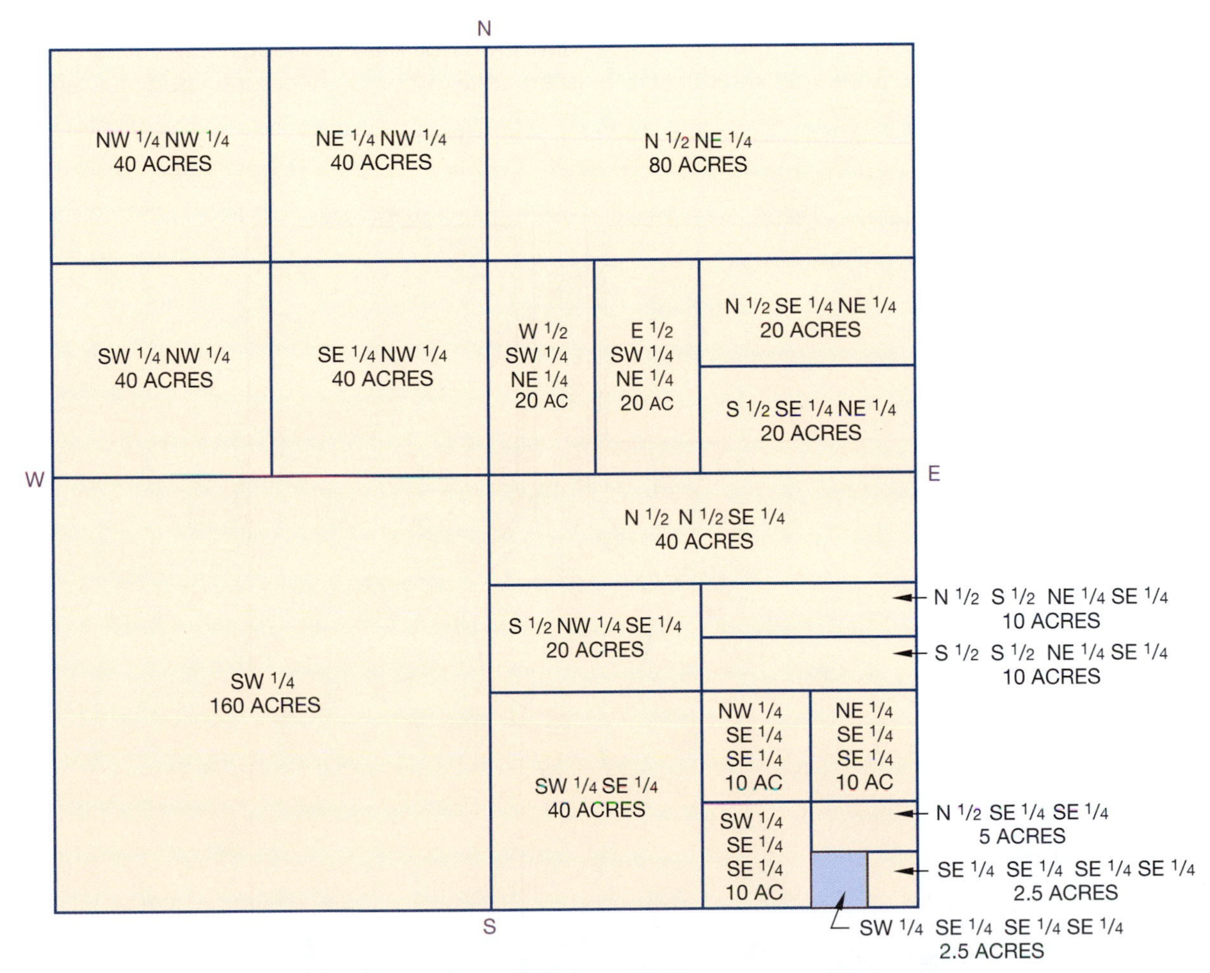

FIGURE 8.6 ■ *A section can be further broken into quarter sections (the NW 1/4 of the NW 1/4) and then divided again into a quarter of a quarter section (the NW 1/4 of the SE 1/4 of the SE 1/4) and then divided one more time into quarters (the SE 1/4 of the SE 1/4 of the SE 1/4 of the SE 1/4).*

On the site plan, this is often abbreviated into a legal description as follows:

The SE 1/4, of the SW 1/4 of the SW1/4, S31, T2N, R3W, San Bernardino meridian, El Cajon, California.

Because the method of describing quarters of quarters of quarters can become confusing, many municipalities have gone to a labeling system using letters. Quarter sections are labeled as A, B, C, or D. Quarter sections are further divided into quarters by the letters A, B, C, or D. The northeast quarter of the northeast quarter would then be listed as Parcel AA. This method of land description works especially well in areas where the land contour is fairly flat. In areas with irregular land shapes, the rectangular land description can be linked with a partial metes and bounds description to accurately locate the property.

Lot-and-Block System

This system of describing land is usually found within incorporated cities. Most states require the filing of a subdivision map, similar to the one in Figure 8.7, that defines individual lot size and shape as land is being divided. Each lot is defined on a subdivision map by a length and angle of each property line as well as a legal description. On older subdivision maps, land is first divided into areas based on neighborhood, called subdivisions. The subdivision is, in turn, broken into blocks based on street layout. The block is further divided into lots. On newer maps, most municipalities assign a number to a parcel of land that corresponds to either a tax account or a map number. Lots can be irregularly shaped or rectangular. The shape of the lot is often based on the contour of the surrounding land and the layout of streets. A typical legal description for a lot and block description would resemble the following:

Lots No. 1, 2, and 3 of Map #17643 of the City of Bonita, County of San Diego, California.

LAND DRAWINGS

The most common plan used to describe property is the site plan. The site plan is started before or in conjunction with the floor plan, although it probably will not be finished until other architectural drawings are completed.

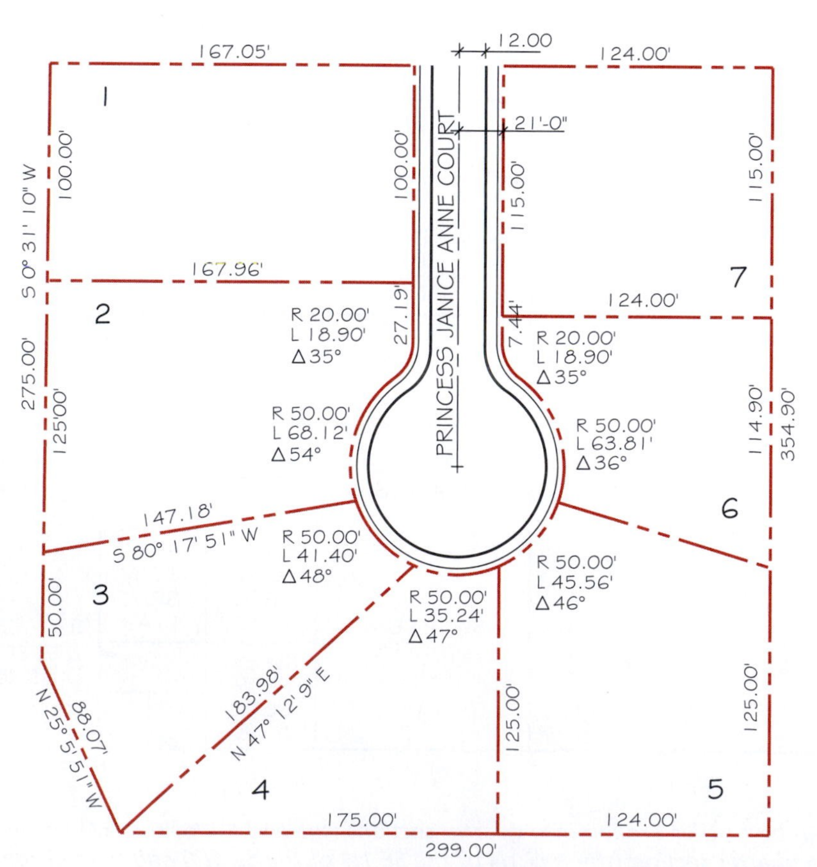

FIGURE 8.7 ■ *A subdivision map shows the size and shape of each parcel of land within a specific area.*

With the property boundaries drawn, the preliminary design for the floor plan can be inserted into the site plan. Preliminary designs for access, walkways, landscaping, and parking can then be determined and adjustments to both the site and floor plans can then be made.

The size of the project and the complexity of the site will dictate the drawings required to describe site-related construction, who will do them, and when they will be done. In addition to the site plan, a ***vicinity map,*** topography, grading plan, landscape, sprinkler, freshwater, and sewer plan may be required to describe the alterations to be made to the site. On simple construction projects, all of these plans can be combined into one site plan. This plan is typically the first sheet of the architectural drawings and labeled A-1. On most plans, the site-related drawings are placed at the start of the architectural drawings and listed as civil drawings. On a complicated custom home, the topography, grading plan, demolition, landscape, sprinkler, freshwater, and sewer plans may comprise the civil drawings (C-1 or L-1) and precede the site plan. Another consideration in the placement of the site plan is municipal regulations controlling the size of the site plan. Many municipalities require a copy of the site plan to be submitted on 8 1/2 × 11'' or 8 1/2 × 14'' paper with specific scale required for submittal. The mandatory uses of small scale for a complex site will often require a simple site plan to locate the residence. Large-scale drawings on a paper size that matches the balance of the drawings may be used to represent landscaping, irrigation, and topography information.

Drawing Origin

The size and complexity of the project will determine who will draw the project. On most projects, the architectural team will draw the site plan. These drawings are usually completed by a drafter working for a civil engineer, surveyor or landscape architect and then given to the architectural team to be incorporated into the working drawings. On simple projects, the architectural team, under the supervision of a landscape architect and a civil engineer, may complete the drawings.

Vicinity Map

A vicinity map is used to show the area surrounding the construction site. It is placed near the site plan and is used to show major access routes to the site. This could include major streets, freeway on and off ramps, suggested routes for large delivery trucks, and rail routes. A drafter working for the architectural team will prepare a vicinity map. It is not drawn to scale, but it should show an area that reflects the size of the project in proportion to the area represented. If building components primarily come from the surrounding area, the map only needs to reflect the immediate construction area. If materials come from several different cities or states, the area of the vicinity map should be expanded to aid drivers who may not be familiar with the area. Figure 8.8 shows an example of a vicinity map.

Site Plan

The site plan for a residential project is the basis for all other site-related drawings. It shows the layout and size of the property, the outline of the structure to be

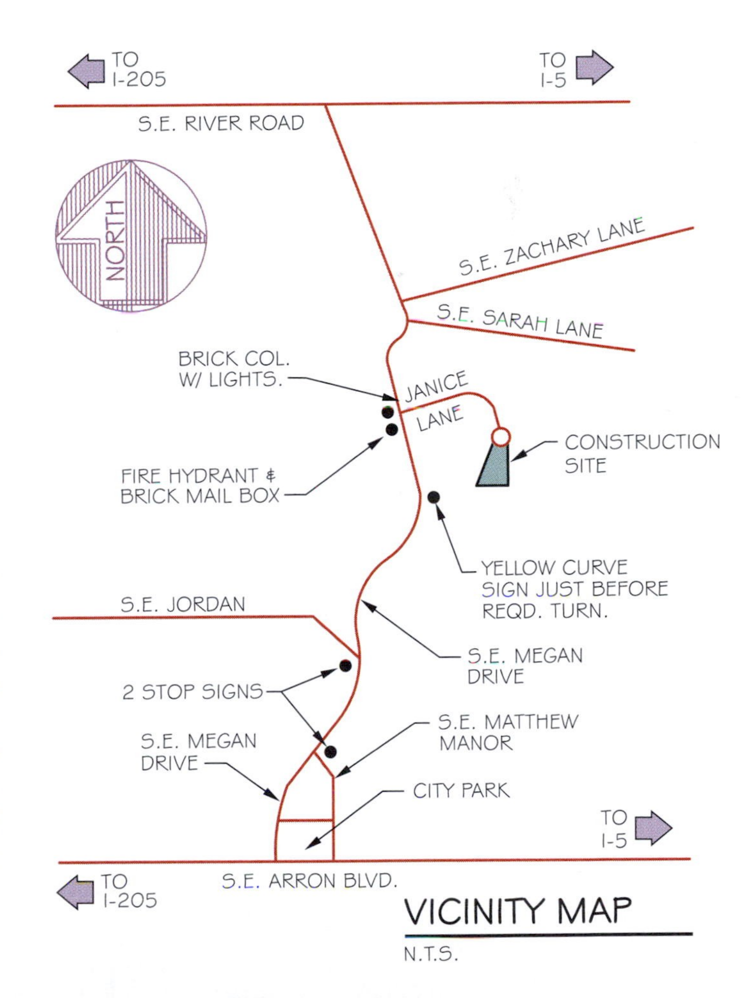

FIGURE 8.8 ■ *A vicinity map is used to show the surrounding area and major access routes to the construction site.*

built, north arrow, ground and finish floor elevations, setbacks, parking and access information, and information about utilities. The results of engineering studies and soils reports related to the site may be summarized on the site plan.

Common Linetypes

The shape of the construction site can be drawn based on information provided by the legal description or subdivision map. Common linetypes used to represent major materials include:

- Property lines, represented by a line with a long-short-short-long pattern. PHANTOM2 or PHANTOMX2 from the AutoCAD line file can be used for representing property lines.
- Centerlines of access roads or easements, represented by a long-short-long line pattern using CENTER, CENTER2, or CENTERX2.
- Sidewalks, patios, stairs, driveway, and exterior parking outlines, typically represented by a continuous linetype.
- Edges of easements and building setbacks, often represented by dashed or hidden lines using DASHED or HIDDEN lines.
- Utility lines, usually represented by thicker lines using either a dashed or hidden pattern. Each line is labeled with a G (gas), S (sewer), W (water), P (power), or T (communications) to denote its usage. See Chapter 13 for requirements and drawing standards when the home is not on public sewers.

Specifications must also be provided to distinguish between existing utilities and those that must be extended within the site. The location of each utility referenced to the property should also be provided. Figure 8.9 shows a site plan for a residence and the appropriate linetypes.

Representing the Structure

The outline of the residence, also known as the building footprint, must be accurately represented and easily distinguishable from the property and utilities. Common

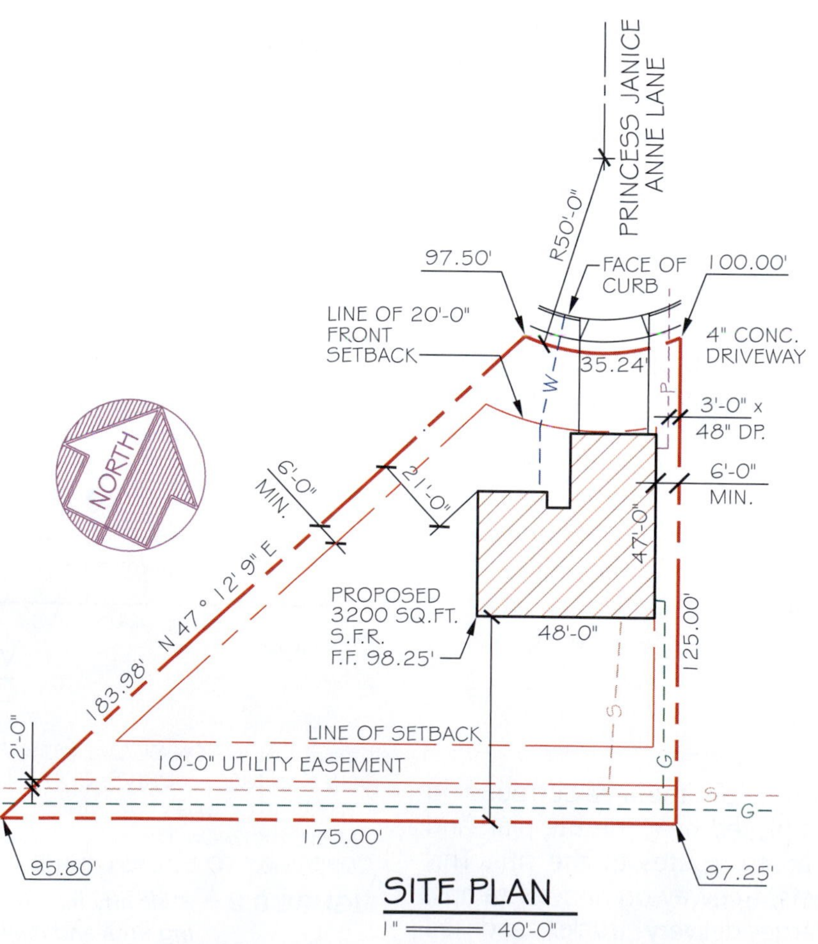

FIGURE 8.9 ■ *Varied linetypes and lineweights are used to represent the materials on a site plan.*

methods to represent a structure include the use of a thick line to define the perimeter and a hatch pattern such as ANSI31 or ANSI37 to further highlight the structure. This method is shown in Figure 8.10. A common alternative is to draw the outline of the residence with a dashed line and bold lines for the outline of the roof. The complexity of the project and building department requirements will determine which method is used. Many firms X-REF the floor plan into the site plan. This offers the advantage of an up-to-date site plan each time the drawing is opened. This can be especially important when the footprint of the structure is altered or door locations are moved.

The use of the structure should also be specified within its outline, with titles such as PROPOSED 2 LEVEL SFR (single family residence). Consideration must also be given by the drafter to accurately distinguish between portions of the structure that are in contact with the ground and those that are supported on columns. Projects often include structures that are to be demolished to make way for the new project. These structures must be accurately located and distinguished from new construction. A separate demolition plan may be used to supplement the site plan. See Chapter 32 for site plan requirements when demolition or alterations are required.

Access, Walks, and Parking Information

Information about access—including driveways and walks, areas to be paved, parking spaces, and ramps—must be shown on the site plan. With the exception of the centerline for access roads, continuous lines are used to represent each. Many offices hatch concrete walkways with a random pattern of dots so they can be easily distinguished from asphalt paving areas (see Figure 8.11).

Off-Street Parking Information

Common arrangements for residential projects were presented in Figure 7.6. Multifamily projects will require off-street parking, but requirements will vary from city to city. The drafter must verify requirements for each project. Off-street parking may be in a garage or open-air parking area. When a parking structure is provided, it can be represented with the same methods used to represent a single-family residence. The size and location of each open air space must be represented. Parking spaces are specified as full, compact, handicapped, or van-handicapped. Common sizes include:

- 9 × 20' (2700 × 6000 mm) for a full space
- 8 ×16' (2400 × 4800 mm) for a compact space
- 14 ×19' (4200 × 5700 mm) for an ADA-approved handicapped space
- 16 × 19' (4800 × 5700 mm) minimum for an ADA-approved handicapped van space

Many municipalities require a parking schedule to be part of the site plan. A parking schedule can be used to

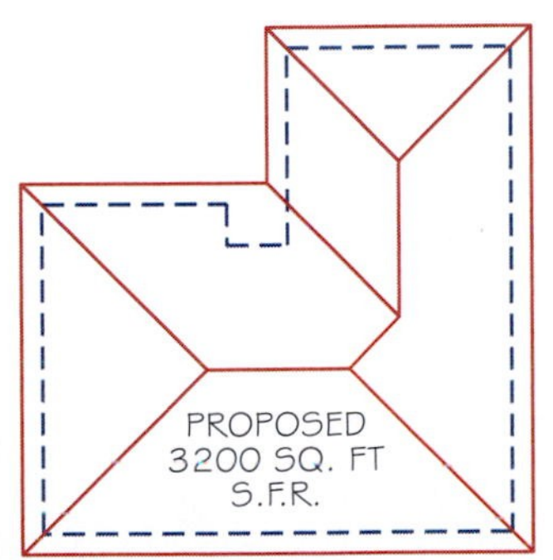

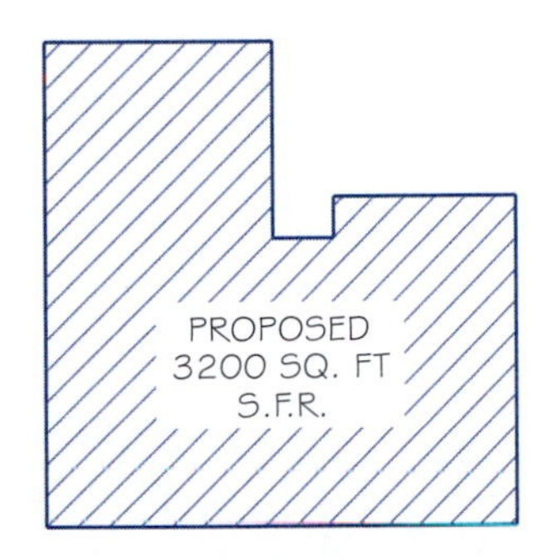

FIGURE 8.10 ■ *The building footprint must be accurately represented and easily distinguishable from the property and utilities. On a simple plan, the structure can be shown with dashed lines and the outline of the roof can be represented. When the plan is more complex, a hatch pattern can be used to distinguish the home from surrounding features.*

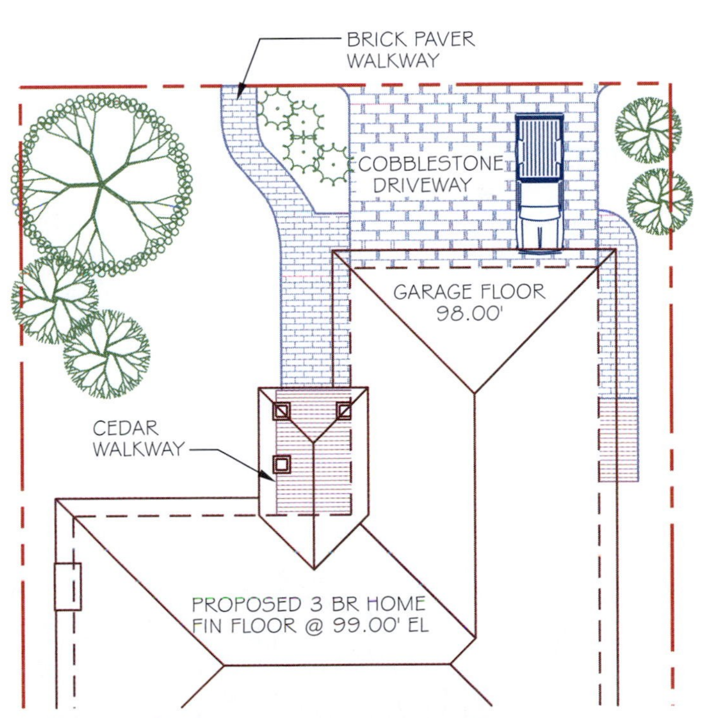

FIGURE 8.11 ■ *Information regarding access—including driveways and walks, areas to be paved, parking spaces, and ramps—is typically represented with continuous lines.*

specify the number, type, and size of each type of parking space.

It is important to remember that there is a wide variance in sizes depending on the municipality and the direction of entry into the space. Figure 8.12 shows an example of common alternatives for parking based on the angle of entry. Perpendicular spaces can be shorter than spaces that require parallel parking. Spaces placed on an angle require different lengths, widths, and a different driveway size as the entry angle is varied. Parking spaces next to obstructions such as raised planters or building supports should have added width to ease access and allow the minimum required width to be maintained. Wheel stops must be drawn and specified for individual spaces. Wheel stops 6 ft (1800 mm) long are typically used for spaces that are 90° to the access drive. Stops are normally placed 24'' (600 mm) from the front end of the stall and straddle the dividing line between spaces so that one stop is shared by two spaces. Building supports often require a permanent protective device to be installed for protection from vehicle damage. A steel column filled with concrete added near each wood column is a common method of protecting the structure from damage caused by careless drivers. These columns must be represented and specified on the site plan. Building supports and wheel stops can be represented as shown in Figure 8.13.

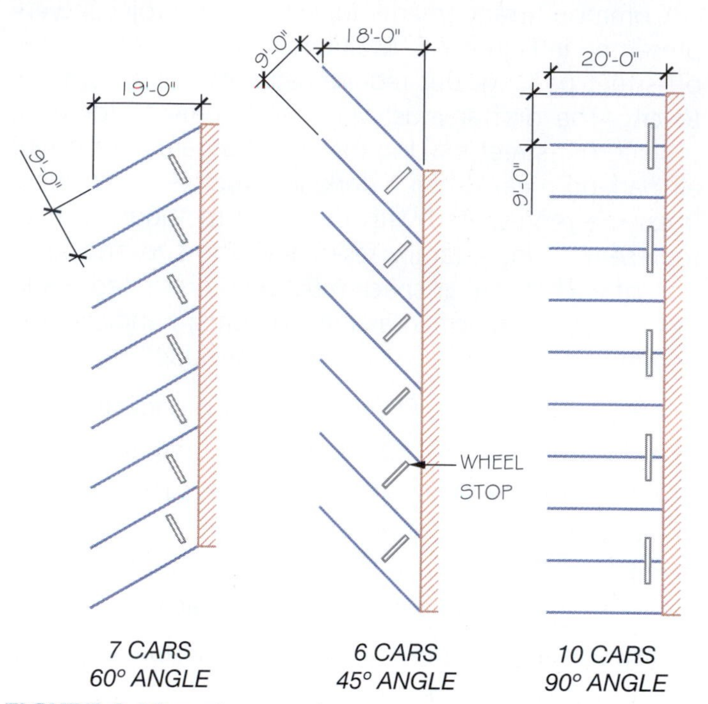

FIGURE 8.12 ■ *The size of each parking space will vary depending on the angle of approach. Each city has it own parking requirements, which the drafter should verify prior to starting the site plan.*

Elevations and Swale

Projects with major changes of contour that require large quantities of soil to be excavated usually require a grading plan. A drafter working for the architectural team or for a civil engineer will help prepare the grading plan. Sites that do not require extensive excavation

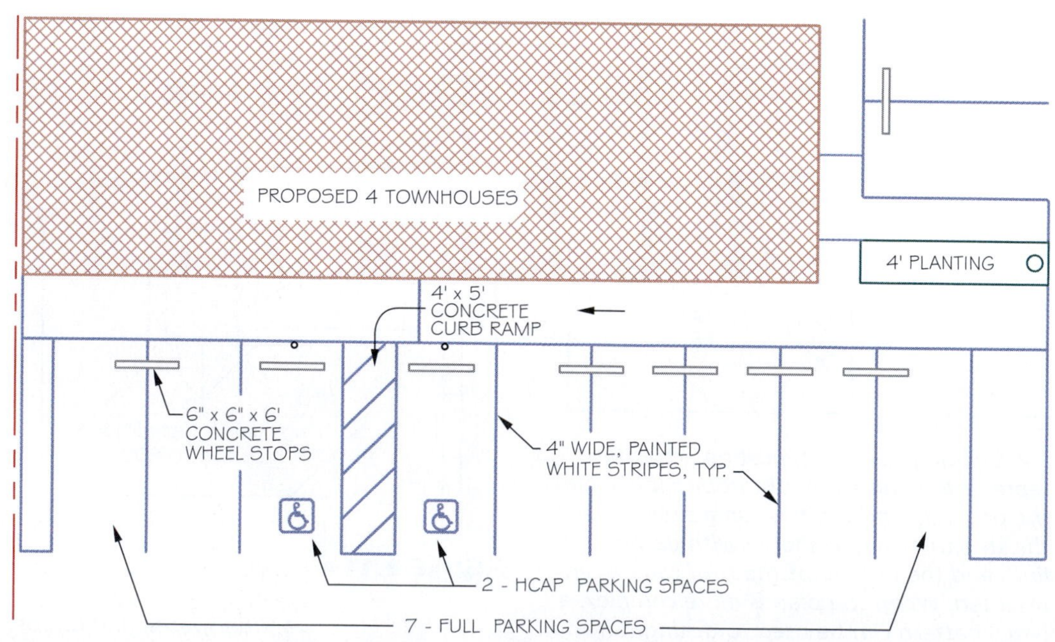

FIGURE 8.13 ■ *Parking spaces must be clearly defined on the site plan either by referencing each space or referencing each group of spaces.*

often reflect finished grade elevations on the site plan. Four common methods of denoting elevation are shown in Figure 8.14. The elevation of ground level is indicated on the site plan with a note similar to:

FINISHED FLOOR ELEVATION 101.25'

A symbol similar to a leader line is used to indicate the elevation of each property corner as well as other important features. The elevation of the specified location is indicated above the leader line. An elevation marker should be placed on each corner of the property. Comparisons of elevations can also be specified by indicators such as TW (top of wall), which are placed above the symbol, or BF (bottom of footing), which is placed below the symbol.

A third method to show minor change of elevation is with a swale indicator. A swale is a small valley used to divert water away from a structure. The slope of the swale is dependent on the surface material being drained. A minimum slope of 2% should be provided for dirt, and a slope of 1% to 2% is used for asphalt or concrete paving areas.

The fourth method to show change of contour is with a bank indicator. A bank indicator is represented by a V placed between lines that indicate the top and toe of the bank. The bottom of the V is placed at the bottom of the bank. Three short lines are normally placed between the V's to indicate the top of the bank. Figure 8.15 shows how each could be used on a site plan.

Drainage

On complex projects, the drainage system may be shown on a separate drainage plan. An underground concrete box called a ***catch basin*** is often placed at the low point of swales to divert water from the site. A catch basin acts as a funnel to collect water and channel it into the storm system. It is covered by a metal cover level with the ground surface called a ***drainage grate.*** The grate allows water to flow into the catch basin without letting anyone fall in. Water flows through the grate, into the catch basin, and then into pipes leading to public storm sewers. The pipe that connects the construction site to the public sewer pipe is referred to as a ***lateral.*** The location of the lateral should be indicated on the site plan and be noted as existing or new.

An engineer designs the system and determines the required change of elevation to ensure proper runoff. Elevation markers referred to as spot grades should be established on the plan to specify the finished grading to ensure runoff. The elevation, size, type, and location of all grates and drainage lines should also be specified on either the site or drainage plan. The grate elevation is shown using a symbol, as in Figure 8.14c. If a

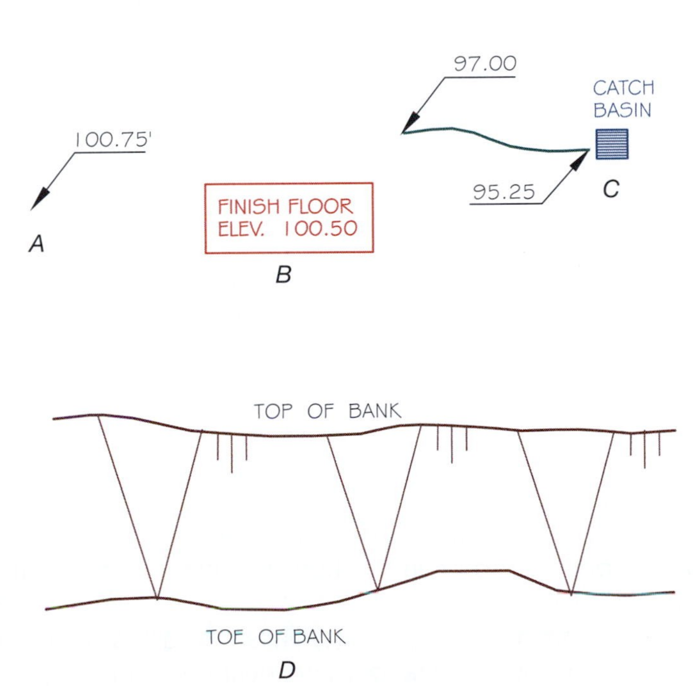

FIGURE 8.14 ▪ *Four common methods of referencing ground elevation include (A) an elevation symbol, (B) a note to describe the elevation, (C) a swale line with elevation symbols, and (D) a bank indicator to represent the top and toe of a bank.*

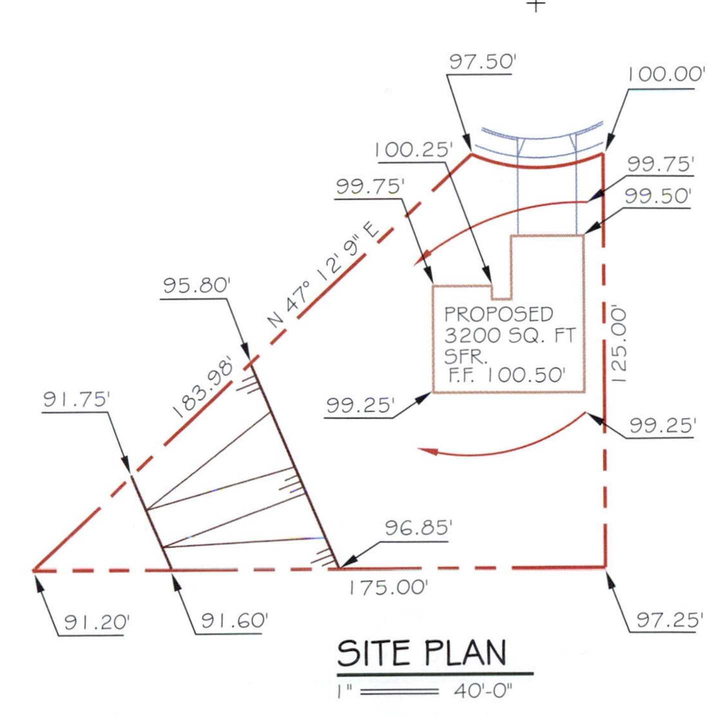

FIGURE 8.15 ▪ *Positive drainage must be provided on each side of the residence. This can be assured by the use of elevation symbols, notes, swales, and bank indicators.*

drainage plan is drawn, the elevation of the inside surface of the bottom of the drainage pipe is also specified. This elevation is known as an invert elevation. Either the minimum slope required or specific elevations along the drainage pipe are specified on a drainage plan.

Common Site Plan Details

The details required for a site plan vary widely from office to office and will vary based on the complexity of the construction. Common areas requiring details on a site plan include sidewalks, curbs, planting details for large plants, irrigation controls, decorative walls, and retaining walls. Figure 8.16 shows an example of common site details. Each is considered a standard detail and could be stored as a WBLOCK that is edited for specific job requirements. The engineer or project coordinator will generally note for the drafter the details that should be inserted into the site plan, and the CAD technician is expected to compile and edit the details to match project requirements.

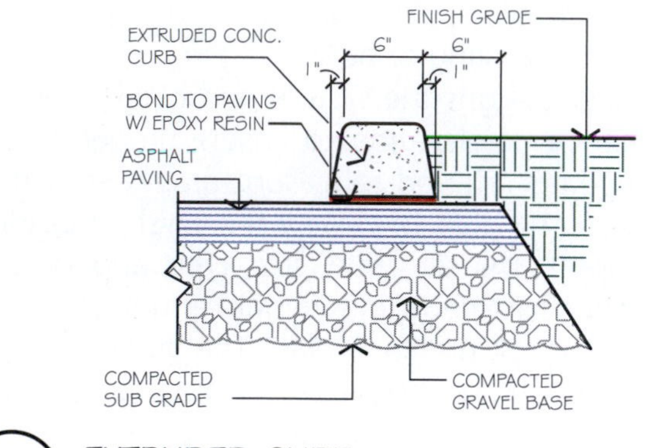

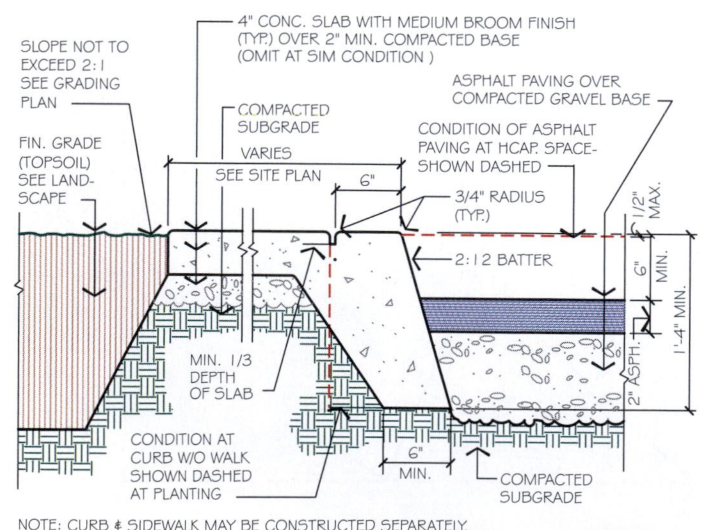

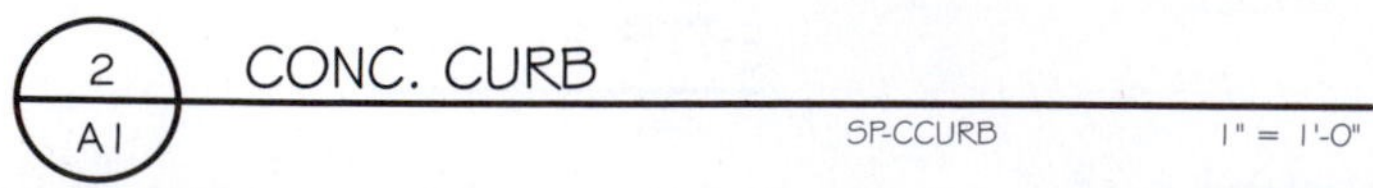

FIGURE 8.16 ■ *Many offices use stock details to explain site-related construction.* Courtesy Architects Barrentine Bates & Lee, AIA.

Annotation

In addition to drawing and locating information, notes must be placed on the site plan to completely specify required construction. General text on a site drawing is placed per the guidelines presented in Chapter 3. Text giving the names of streets and describing the proposed structure should be treated as titles. A notation such as PROPOSED 3000 SQ FT SFR is used to define the proposed construction. Local and general notes are used to define the construction to be completed. On complex drawings, general notes should be broken into categories to make specific information easier to find. Categories such as paving, flatwork, landscaping, and irrigation are common. In addition to the general note, which gives a full specification, a local note is usually placed on the plan, with a partial note to clarify what each symbol represents. Abbreviations exclusive to site-related drawings are also used. Common abbreviations include:

BF	Bottom of footing	PL	Property line
BW	Bottom of wall	ROW	Right-of-way
CB	Catch basin	SD	Storm drain
CO	Clean out	SOV	Shutoff valve
D/W	Driveway	SS	Sanitary sewer line
G	Gas	TC	Top of curb
EG	Edge of gravel	TG	Top of grate
EP	Edge of pavement	TP	Top of paving
FF	Finish floor	TW	Top of wall
FH	Fire hydrant	UP	Utility pole
FP	Flagpole stanchion	W	Water
M	Meter	WCR	Wheel chair ramp
MH	Man hole	WM	Water meter
O/H	Overhead utility line	WV	Water valve

Many plans include common abbreviations used throughout the drawings on a title page. Common abbreviations used on construction drawings are given at the end of this text.

Text to describe the property lines for each site must be provided using distances and bearings. This text is placed parallel to the property lines. Figures 8.7 and 8.9 each show examples of required annotation to describe the property lines. In Figure 8.9, the south property line is described with a distance of 175.00', with no bearing given. If no bearing is listed for the property line, it can be assumed that the property line runs either north-south, or, as in this case east-west. The northwest property line is described with a length of 183.98' and a bearing of N47°12'9''E.

Many sites similar to parcel 4 in Figure 8.7 have property lines that curve. Three notations can be used to describe curved property lines, including R 50.00', L 35.24', and Δ 47°. R is the radius of the curve and L is its length. The symbol Δ represents the delta angle, which is the included angle of the curve. The included angle is the angle formed between the center and the endpoints of the arc. In addition to information used to describe the construction site, general information is also placed on the site plan. General information might include a table of contents, list of consultants, and information about the overall construction project. An alternative to placing this general information on the site plan is to provide a title sheet that includes the vicinity map.

Site Dimensions

Each item represented on the site plan needs to be located by dimensions. Three types of dimensions are often found on a site plan, including:

- Land sizes, which are represented in feet and hundredths of a foot, using notations such as 100.50', or in meters (30.6 m)
- Property line dimensions, which are placed parallel to the property line with no use of dimension or extension lines
- Overall sizes of structures, which are often placed parallel to the side of the structure and specified in feet and inches using notations such as 75'-4'' or 22.8 m for meters

These overall dimensions also are usually placed without the use of dimension lines. Parking boundaries and the structure are located using dimension and extension lines. Objects such as sidewalks or planters can often be described in a note rather than by dimensions. Figure 8.9 shows examples of each type of dimension. Notice the symbol used to designate each property corner. It is often omitted from rectangular lots but is very helpful in locating small changes of angle on irregularly shaped lots.

SITE-RELATED DRAWINGS

In addition to the site demolition plan, several related drawings can accompany a set of working drawings. Related drawings include a topography plan, grading plans, profile drawings, landscaping plan, and sprinkler plan. A ***topography plan*** shows the existing contour of the construction site. This plan is normally prepared by a licensed surveyor developed from notes provided by a field survey. It is based on existing municipal drawings describing the site, on measurements taken by the surveyor, or by aerial photography methods. Once the shape of the site is prepared, the results of the field survey are translated onto the drawing by a drafter working for a civil engineer. A ***grading plan*** is used to show the finished configuration of the building site. A grading plan may be designed and completed by a drafter under the supervision of the architectural team project manager or a civil engineer. A ***landscape plan*** may be provided to show the location, type, size, and quantity of all vegetation required for the project. It will typically show patios, walkways, fountains, pools, sports courts, and other landscaping features. Drafters working for the architectural team or a landscape architect usually complete a landscaping plan. On small projects, drafters working for the architectural team may complete the project under the supervision of the project architect. In arid climates, an ***irrigation plan*** may be required to show how landscaping will be maintained. The same team that provides the landscape plan usually completes the irrigation drawings. Because the landscape and sprinkler drawings usually fall under the supervision of the landscape architect, procedures to create these drawings are not presented in this chapter.

Topography Plans

A topography plan provides a description of land surface using lines to show variation in elevation. Lines used to represent a specific elevation are referred to as ***contour lines.*** A contour line connects points of equal elevation. The elevations may be based on a USGS benchmark or the highest corner on the site. If the elevations are not related to a known benchmark, the highest property corner is usually assigned a height of 100.00'. All other elevations for the site are then expressed relative to this base point. If the nearest property corner were 5.5' lower it would receive an elevation of 94.50'. Figure 8.17 shows the contour lines for the site plan shown in Figure 8.15.

The location of contour lines is determined by a land survey. A survey team will record the elevations of the job site at specified distances in what are referred to as field notes. When the slope is uniform, the survey team will typically record elevations at intervals of 25' (7500 mm). When the grade changes, spot elevations are

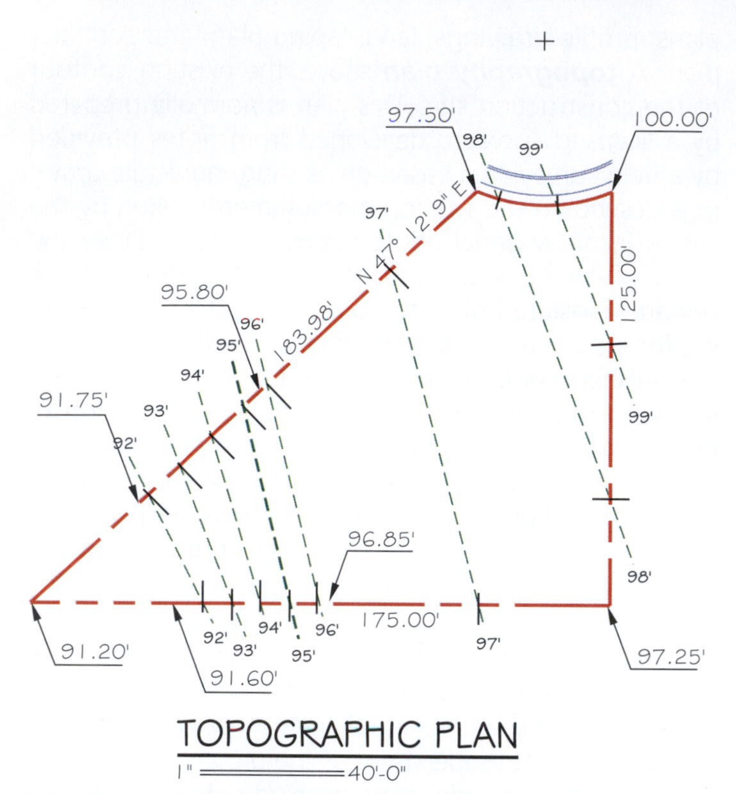

FIGURE 8.17 ■ *Contour lines have been added to the site plan shown in Figure 8.15. Each line represents soil that is at the same elevation. Contour lines help a skilled print reader understand the terrain at the construction site. Contour lines that are close together indicate a steep slope. Lines that are further apart indicate a gentle slope.*

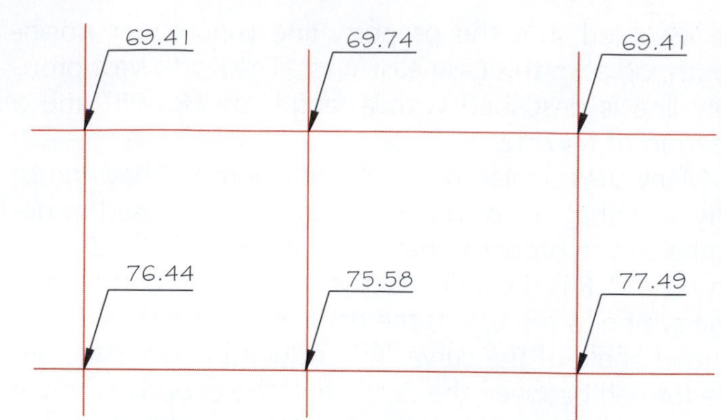

FIGURE 8.18 ■ *Existing topography is determined from a sketch or drawing, provided by the civil engineer, that lists the known elevation at specific points at the site.*

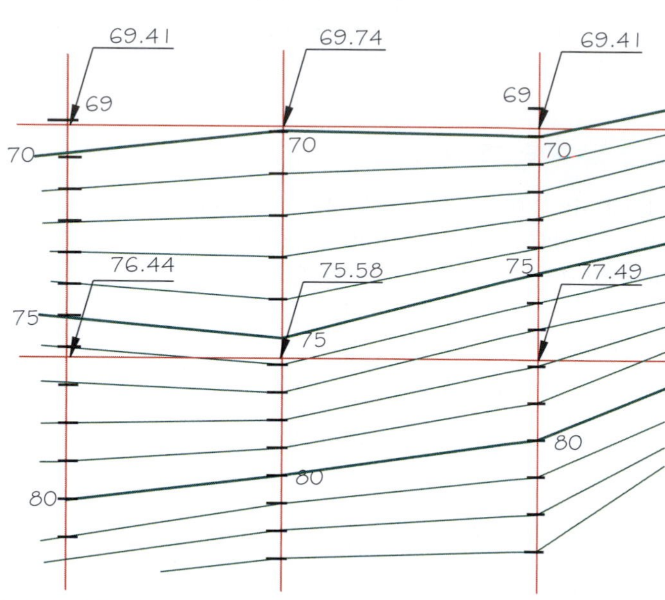

FIGURE 8.19 ■ *Once known elevations are located, lines can be placed to represent specific elevations. Grades can be determined by estimating the rise or fall between two known points. Because lines for grades 70 through 76 occur in the upper left grid, the distance between the two points is divided to represent each elevation.*

recorded at the top and bottom of banks that fall between the base intervals. For small sites, the land will be divided into grids by the survey team, which will take spot elevations at each grid point. For larger sites, the survey may be made only for the area where construction will occur. Once the survey is complete, a drafter working for the civil engineer will convert the field notes to a topography plan. A grid will need to be drawn on the site plan that matches the grid used by the surveyors at the job site.

Figure 8.18 shows an example of a sketch provided to reflect the existing surface elevations. A grid is drawn to represent each known elevation location. Notice the six elevations in the northwest corner of the survey. Between grid 69.41 and 76.44, seven contour lines are required to represent the change of elevation. It can be assumed the contours fall at an even spacing because the surveyor did not change the distance between grids to reflect a rise or depression. The seven lines should be evenly spaced between the two spot grades and will represent the contours for 70 through 76. Because a rise of .6' is required to reach the contour for the next whole foot, the distance between 69.41 and the first line should be slightly more than half the distance between any two of the 1' markers. The distance between the 76' contour and the spot grade for 76.44 should be slightly less than half the distance between any two of the 1' markers.

Six contour lines are required between grids 69.74 and 75.58, and eight contour lines are required to reflect the change in elevation between grids 69.41 and 77.49. Once the distance between grids has been divided into the required divisions, points of equal elevation can be connected, as seen in Figure 8.19. The use of a polyline aids in finishing the drawing. As the distance between contour lines is decreased, the land becomes steeper. As the distances between contour lines increases, the land becomes flatter. The topography plan must be completed before accurate estimates of soil excavation or movement can be planned.

Representing Contours

Once the known elevations have been converted to contour lines, the lines can be curved and altered to provide clarity. Dashed lines are generally used to represent existing contours. Unlike the real contour of the site, in the initial layout stage the lines run from point to point and have distinct directional change at each known point. PEDIT leaves each vertex in its exact location but changes the straight line to a curved line. The WIDTH option of the PEDIT command can also be used to alter the width of the contour lines. The line width used to represent contours will vary based on the drawing scale and the accuracy of the contours to be shown. On many residential sites, contours are shown to represent each 1' of elevation change. Lineweights on such drawings should be selected that provide clear contrast between 1' and 5' (300 and 1500 mm) intervals. Depending on the spacing of the contours, lines representing every 5' or 10' (1500 or 3000 mm) are usually highlighted and labeled as shown in Figure 8.20. Site plans with a large difference in elevation may show the elevation change in 2' or 5' intervals.

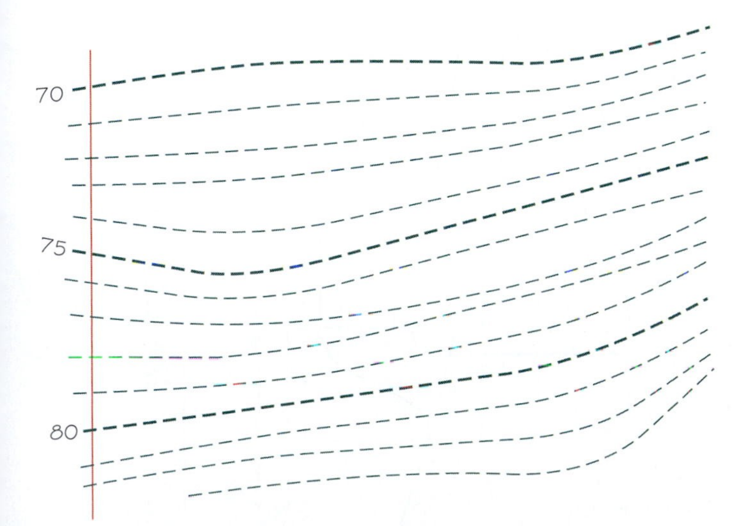

FIGURE 8.20 ■ *A dashed line is generally used to represent existing contours. The angular contour lines from Figure 8.19 were curved using the FIT CURVE option of the PEDIT command.*

Grading Plans

The grading plan shows the proposed structure and its relationship to the contours of the building site. On simple projects, a drafter working for the architectural team may complete the grading plan. A drafter working for the civil engineer translates preliminary designs by the architectural team and the topography drawings into the finished drawings. The grading plan shows the finished contour lines, areas of cut and fill, building footprint, driveways, walkways, patios, steps, catch basins, and drainage provisions. Figure 8.21 shows the grading plan based on the topography shown in Figure 8.17.

A grading plan uses contour lines to represent existing and finished grades. This would include:

- Natural grade: Land in its unaltered state.
- Finish grade: The shape of the ground once all excavation and movement of earth has been completed.
- Cut material: Soil that is removed to lower the original elevation.
- Fill material: Soil that is added to the existing elevation to raise the height.
- Daylight: The point that represents the division between cut and fill.

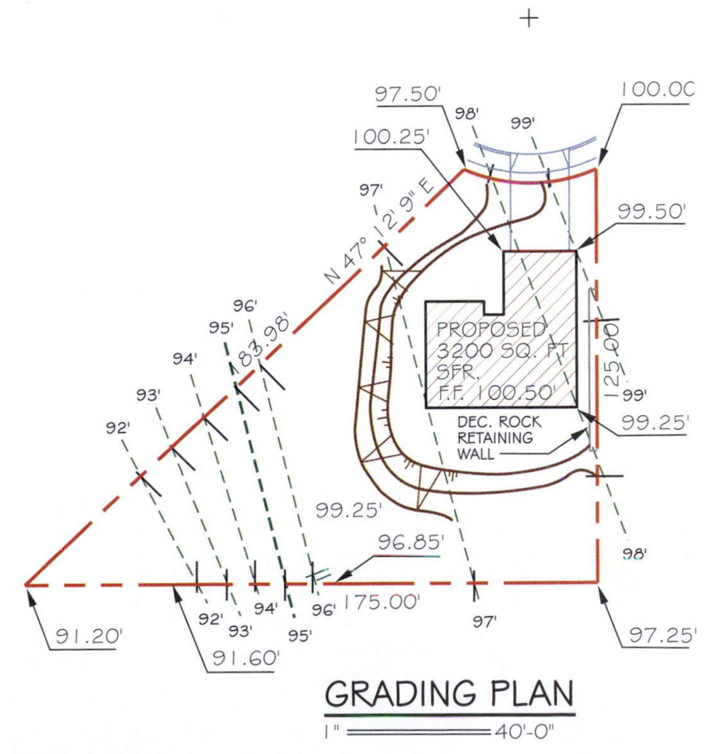

FIGURE 8.21 ■ *A grading plan is used to show changes in topography required by a construction project.*

The project manager determines the finish elevation of floor levels and major area of concern. It is then typically the drafter's job to indicate the extent of cut and fill material based on the desired angle of repose.

Angle of Repose

The angle of the cut or fill bank created is referred to as the angle of repose. The maximum angle of repose is based on the soil type. The municipality that governs the construction project determines the maximum angle. Common limits include:

- For fill banks, a common angle of repose is often set at a 2:1 angle. For every two units of horizontal measurement, one vertical unit of elevation change can occur.
- A common angle for cut banks is 1.5:1. The engineer specifies variations in the angle of repose that will be allowed near footings or retaining walls to minimize loads that must be supported.

Representing Contours

Contours for new and existing elevations are usually combined on one plan. Thin dashed lines are typically used to represent existing 1' (300 mm) contours and thick dashed lines are used to represent 5' (1500 mm) contours. Thin continuous lines are typically used to represent new contours for 1' (300 mm) intervals. New 5' (1500 mm) contours are usually represented by thick continuous lines. Figure 8.22 shows the representation of new cut and fill banks placed on the topography drawing.

The drawing in Figure 8.22 shows the cut-and-fill banks created to place a one level home with a concrete slab on a level pad. To determine what the floor level will be requires that the access to home be considered. Figure 8.23 shows the grading that resulted for a two-level home on the same site used in Figure 8.22. For this example, the following assumptions were made:

A 5% slope for the driveway
A 20' front setback
Cut banks of 1 1/2 / 1
Fill banks of 2 / 1
9' between floor levels.

Placing the home in the desired position and then locating the driveway will allow the floor level to be determined. The home was placed 32' from the front property line to allow for a gentle slope in the driveway. The soil in the front yard was cut to allow the 100' elevation to be pulled close to the property line to help flatten the slope. With a distance of 32' or 420'' (10 500 mm), it would seem that a 21'' (525 mm) fall is allowed. Remember that positive drainage must be provided for

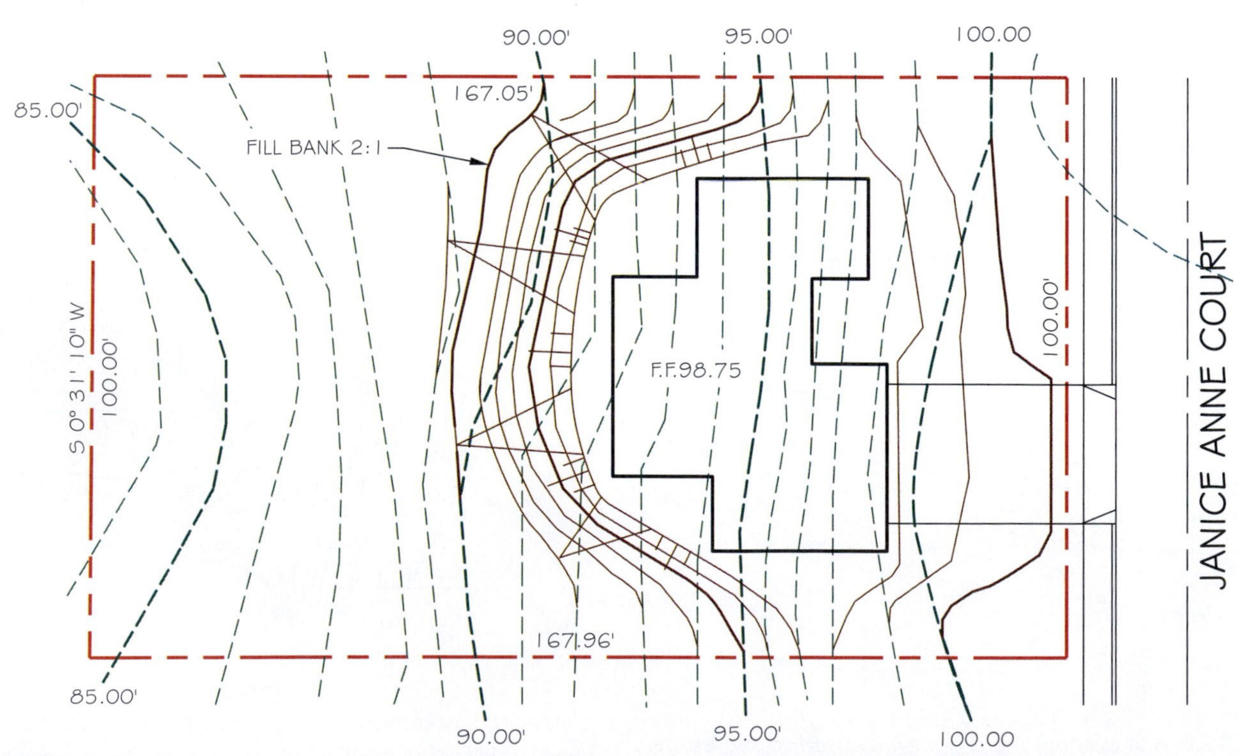

FIGURE 8.22 ▪ *To keep the home as high as possible, minor cutting was required to place the driveway. On the west side of the site, a fill bank was created.*

60'' (1500 mm). This will reduce the effective length to 27' or 324'' (8100 mm) with a maximum slope of 16'' (400 mm). The 16'' must be converted to decimal fractions of a foot to determine the driveway slope (16'' is equal to 1.33'). If the east edge of the driveway is assumed to be 100.5', the west edge of the garage will be set at 99.17' (100.5'-1.33'). The remaining 60" of the driveway must have a slope of 1.25'' (5 × .25'') or .10'. This will set the east edge of the garage slab at 99.27'. This can be rounded to an elevation of 99.25'. A 6'' slope for the garage floor will place the west edge of the garage slab at 99.75'. Adding 8'' (.66') for the step up to the house places the upper floor level at a height of approximately 100.41'. For planning purposes, round up to assume a height of 100.50' for the upper floor and a lower floor elevation of 91.50'. If the lower slab must be a minimum of 6'' above the soil, it will require that the ground be no higher than 91.00'. To provide positive drainage, the edge of the bank was set at 90'. This will provide a gentle slope for a small yard on top of the bank. Figure 8.23 shows the resulting grading plan with the building footprint and the required cut-and-fill banks.

Site Profile Drawings

A ***profile*** is a drawing showing the surface of the ground and underlying earth taken along any desired fixed line. Figure 8.24 shows a profile cut near the center of the site, as in Figure 8.23. The drawing is created by placing a line to represent the cutting plane on the grading plan and then projecting each grade that touches the line into the profile. The horizontal scale of the profile should be plotted at the same scale used for the grading plan. The vertical scale can be the same as the horizontal scale, or it can be exaggerated to give a clearer representation of the contour. The vertical scale in Figure 8.24 is 2X the horizontal scale.

Several profile drawings are often made during the preliminary process to help the owners visualize how the home will relate to the site. Profile drawings are generally not required by most municipalities to obtain a building permit but are used by the architectural team to help define how the home will blend with the site. Figure 8.25 shows the altered profile drawing through the center of the home shown in Figure 8.24.

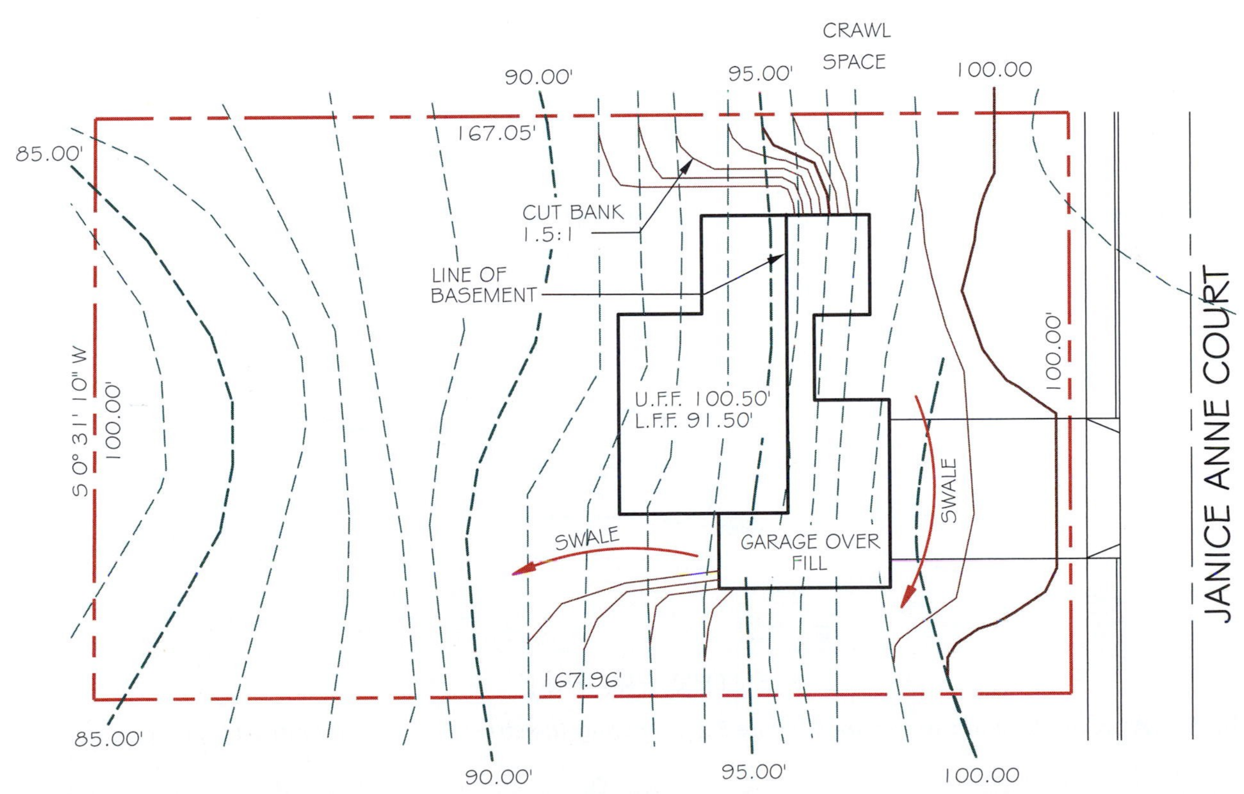

FIGURE 8.23 ■ *The grading required for a two-level home. To allow for the basement, the ground above the 90' contour was cut. Higher grades were run into retaining walls on the north and south ends of the basement. The garage area will require gravel fill to bring it up to the required level. The northeast portion of the home will be over a crawl space, allowing the natural grade to remain.*

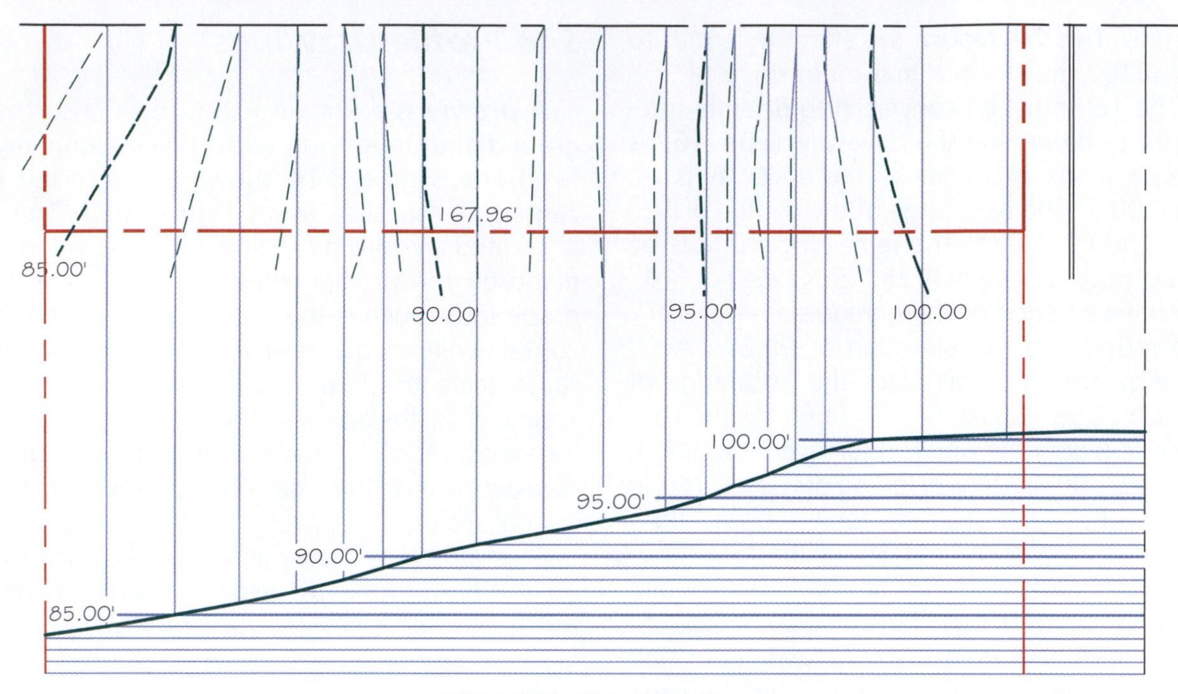

FIGURE 8.24 ■ *A profile for the site shown in Figure 8.23, showing the existing contours.*

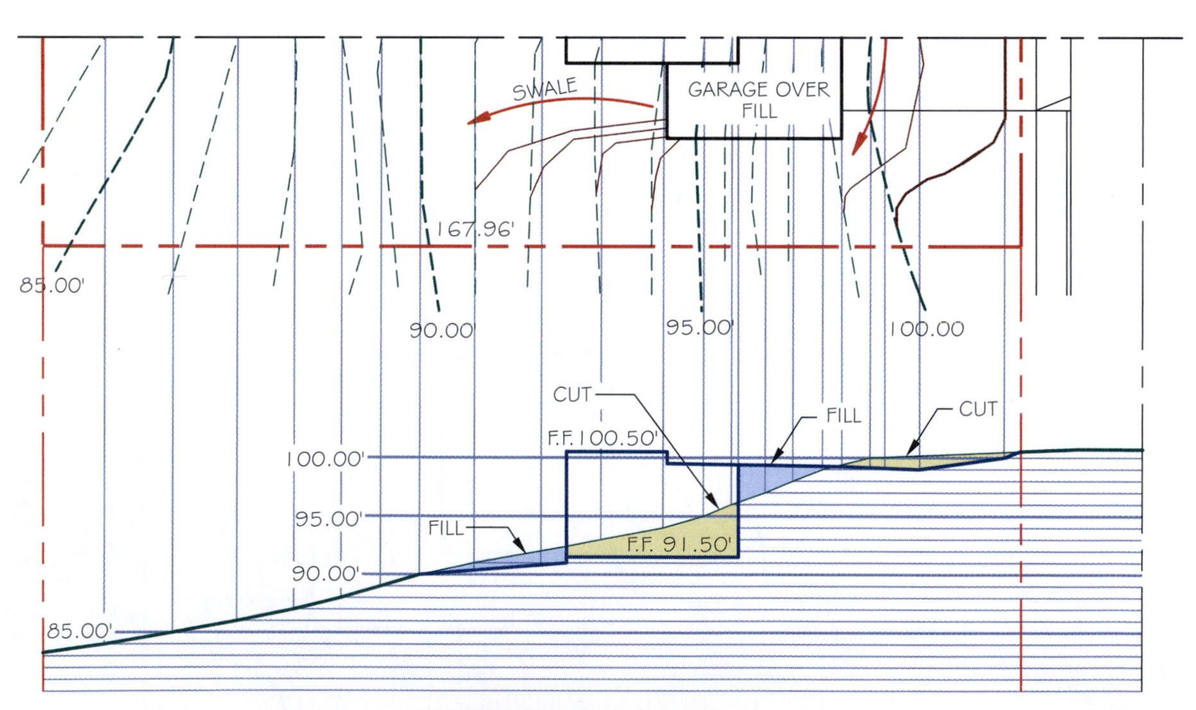

FIGURE 8.25 ■ *A profile for the site shown in Figure 8.23, showing the structure and the finished grading.*

CHAPTER 8

Additional Readings

The following websites can be used as a resource to help you keep current on issues of land use.

ADDRESS	COMPANY OR ORGANIZATION
www.blm.gov	Bureau of Land Management
www.ca.blm.gov/cadastral/meridian.html	California Bureau of Land Management
www.usgs.gov	United States Geological Survey
www.geography.usgs.gov	USGS
http://neic.usgs.gov/index.html	USGS earthquake hazards

CHAPTER 8

Land Descriptions and Drawings Test

DIRECTIONS

Answer the following questions with short complete statements. Type your answers using a word processor or neatly print the answers on lined paper.

1. Type your name, the chapter number, and the date at the top of the sheet.
2. Type the question number and provide the answer in the form of a statement that includes part of the question. You do not need to write out the entire question.

QUESTIONS

8.1. What is a monument as it relates to a site drawing?

8.2. List four units of measurement that might be referred to in a rectangular system legal description.

8.3. What two directions are used to commence a bearing?

8.4. What would a designation of 100.67' S37° 30'45''W represent on a site plan?

8.5. What would a designation of S28AA, T 3S, R1E represent on a site plan?

8.6. List three components of a legal description commonly used for incorporated areas.

8.7. List two common sources of site drawings.

8.8. Explain the difference between a topography plan and a grading plan.

8.9. What is the purpose of a vicinity map?

8.10. Sketch examples of linetypes to represent the following materials: property line, centerline, easement, sewer, and water line.

8.11. List and describe two common hatch patterns that can be used to highlight a building.

8.12. Sketch a bank indicator representing a slope.

8.13. How are existing and new grades typically represented on a plan?

8.14. Describe the process of changing field sketches to a topography map.

8.15. What is an angle of repose?

8.16. Give the common proportions of cut and fill banks.

8.17. Obtain a land map and legal description from a local title company of the property your place of residence is built on or for a site provided by your instructor.

8.18. Interview a local landscape architect, civil engineer, and landscape contractor to access the drafting opportunities in your area and to obtain examples of the types of drawings local professionals expect drafters to draw.

8.19. Take a minimum of 15 photographs representing the installation or completion of work specified on a site, grading, landscape, and sprinkler plan.

8.20. Use the Internet to research the local building department in your area and determine common residential zones and the required front, side, and rear setbacks for each.

CHAPTER 9

Site-Related Drawing Layout

Chapter 8 presented examples of a subdivision, site plan, topography plan, and grading plan. This chapter introduces drawing methods to create site-related drawings. Consideration is given to how files for site drawings are developed; methods to establish common drawing parameters such as scales, paper sizes, and layers; and layout methods for each type of drawing.

PROJECT STORAGE METHODS

The method used to develop the site plan will depend on the complexity of the project and the types of drawings to be created. Because the site plan serves as a basis for so many other drawings, it is important for the drafter to have a clear understanding of what additional site drawings will be required and who will be completing them. Storing all of the site-related drawings in one file, using external referencing, or using a separate drawing file for each drawing are the most common methods used to develop a site plan. See Chapter 3 for a review of drawing and layer prefixes and titles.

One Drawing File

On small-scale projects, one firm often prepares each of the site drawings, allowing them to be stored within one drawing file. Site-related drawings should be stored in one file only if each drawing is stored on a separate layer. You should give layers names describing both the base drawing and the contents of the layer, as outlined in Chapter 3. Titles such as *C BLDG, C PROP,* or *C ANNO* will clearly describe the contents. To plot the topography plan, basic items from the site plan, such as the building and property and all *TOPO* layers, should be set to THAW, with all other layers such as site anno set to FREEZE. The use of separate layouts in one drawing file can also serve to store multiple drawings in one file. Separate layouts can be created for the site plan, topography plan, and landscape plan in one drawing, allowing easy selection of the desired drawing for plotting.

Separate Drawing Files

The use of separate drawing files is the least effective use of disk space, but it can speed up the plotting process. To effectively create separate model and sheet files, place information that is required for each of the related drawing files in a base drawing and store as a wblock. This information can then be reused, saving valuable drafting time on each new drawing. The use of layouts allows this to be done in one drawing file without wasting disk space.

External Reference

Site-related drawings are an excellent example of drawings that can be referenced to other drawings. A drafter working for the architectural team can draw the basic site plan information in a base drawing that will be reflected on all other site drawings. Copies of this drawing file can be given to the other consulting firms that will develop the landscape, sprinkler, and grading drawings. Using external referencing allows each firm to have a current drawing file as a base while progressing with its work independently.

SETTING SITE PLAN PARAMETERS

Site drawings will require a template drawing to be developed that reflects either engineering or metric values instead of architectural values. Parameters such as UNITS, LIMITS, SNAP aspect, GRID sizes, layer parameters, dimension and text variables, and plotting requirements can then be set to meet the specific needs of the site plan. Common layers can also be set up to separate site information from other information that will be stored with the site file. Offices usually have stock template drawings containing common site-related linetypes, dimension and text variables, notes and symbols. As a student, you may need to develop your own template drawing for site plans.

Drawing Scales

Site plans are plotted at a scale factor such as 1'' = 10' (1:120) or 1'' = 20' (1:240). Common architectural and engineering scales for site plans are listed in Table 3.1. Preferred metric scales for civil drawings are listed in Tables 2.3 and 2.4. Factors that influence the scale include building department requirements and the amount of detail required. For single-family projects, many municipalities regulate either the scale of the site plan or the size of the sheet on which the plan is plotted. For multifamily drawings, varied scales may be required for the site plan depending on the stage of the design review or permit process. Verify scale requirements with the governing municipality. If the building department does not mandate a scale, use the largest scale possible for the paper size to be used as well as large enough to show all required information clearly.

Paper Sizes

Site plans are drawn on media ranging in size from A through D. The choice of material is based on building department requirements, lending institution requirements, and the purpose of the plan. Many municipalities require site plans to be drawn on a specific sheet size to allow for ease of filing. Common sizes include 8 1/2 × 11'', 14'', or 17'' (812 × 275, 350, or 425 mm). Building departments and lending institutions often specify the size so that the site plan can be easily filed. When specific paper sizes are not required, use a paper size that will match the size used for the balance of the project. When the site plan is placed on paper larger than A, the sheet is often used as a title sheet. On projects for multifamily developments, the site plan often includes a table of contents, a list of consulting firms involved in the project, and a list of abbreviations used throughout the project.

Layer Guidelines for Civil Drawings

Common layers for site related drawings are listed below. Many of the layers will not be necessary for a single-family residence but will be useful on site related plans for multifamily developments.

C BLDG	Proposed building foot prints
C COMM	Site Communications
C COMM OVHD	Overhead communications lines
C COMM UNDR	Underground communications
C FIRE	Fire protection
C NGAS	Natural gas
C NGAS UNDR	Underground natural gas lines
C PKNG	Exterior parking areas
C PKNG CARS	Graphic illustrations of cars
C PKNG DRAN	Parking lot drainage slope indications
C PKNG ISLD	Parking islands
C PKNG STRP	Parking lot striping, handicapped symbols
C PROP	Property lines
C PROP BRNG	Bearings and distance labels
C PROP CONS	Construction Controls
C PROP ESMT	Easements, rights-of-way and setback lines
C ROAD	Roadways
C ROAD CNTR	Centerlines
C ROAD CURB	Curbs
C SSWR	Sanitary sewer
C SSWR UNDR	Underground sanitary sewer lines
C STRM	Storm drainage catch basins and manholes
C STRM UNDR	Underground storm drainage pipes
C TOPO	Proposed contour lines and elevations
C TOPO RTWL	Retaining walls
C TOPO SPOT	Spot elevations
C WATR	Domestic water, manholes pumping stations, storage tanks
C WATR UNDR	Domestic water underground lines

Gathering Information

Before beginning the layout of the site plan, information must be gathered to explain the site. Information regarding the site size and legal description can be found in legal documents for the property that are provided by the client, a title company, a surveyor's map, the local assessor's office, or the local zoning department.

Additional information that should be determined before starting the site plan includes:

- Legal description
- Zoning information including front, rear, and side yard setbacks; height limitations; and methods of determining building height
- North
- Existing roads and alleys
- All utilities including telephone, gas, water, sewer, or septic disposal
- Drainage and slope requirements
- Size of proposed structures and outbuildings
- Minimum drawing standards based on municipal standards

Figure 8.7 shows a subdivision map that will be used to prepare the site plan. For this example, parcel 1 will be used. The site plan for this example will have a scale of 1'' = 10' (1:50). Other information not found on the subdivision map but required by the building department includes:

- Legal description: Lot 1 of the Schmitke addition in S20, T2S, R5E of the Willamette Meridian, Clark County, Washington.
- Front setback = 20' (6000 mm), rear setback 20' (6000), side yard setback = 5' (1500 mm) one level, 6' (1800 mm) for two levels.
- Gas and water are located in the street. Sewer is available in the street, but the sewer line for this site will connect to a line in the easement.
- 10' sewer easement on west property line.
- 5' utility easement on east property line for phone, with utility box 3' south of the north property line.
- Cut bank= 1.5/1; fill bank = 2/1.

See home in Chapter 11 for footprint size.

Minimum Drawing Standards

The following standards will be assumed to be the minimum requirements for the site based on the "local building department." Although the following standards are common, each time you draw a site plan the requirements of the governing body need to be verified. The site plan must show:

- Site and building sizes.
- Setback dimensions.
- Property corner elevations.
- Contours: if site includes more than a 4' elevation differential, the plan must show 1' contour intervals.
- Location of easements, driveway, footprint of structure and decks.
- Location of all utilities.
- Site area, building coverage area, and building height. The height is to be measured by subtracting the average height of corners of structures above the average height of site corners existing structures.
- Surface drainage.

COMPLETING A SITE PLAN

A site plan can be completed using the following steps:

Step 1. Before starting your drawing at a computer, gather all the information required to complete the drawings.

Step 2. Select or create the proper template to create the site plan. If a new template drawing is required, establish all required parameters for site-related drawings including: UNITS, LIMITS, SNAP aspect, GRID sizes, layer parameters, dimension and text variables, and plotting requirements.

Step 3. Working from a plat map, draw the property lines of the site on the *SITE PROP* layer. For this example the site in Figure 8.7 will be drawn.

Step 4. Establish the center of all access roads, alleys and easements on the *SITE PROP ESMT* layer.

Step 5. Locate all public sidewalks on the *SITE CURB* layer.

Your drawing now shows the existing features of the building site and should resemble Figure 9.1. The next steps will help you proceed in an orderly fashion to show all new features to be included in the project. Steps 6 through 13 can be seen in Figure 9.2.

Step 6. Draw lines to indicate the required setback distances on the *SITE PROP ESMT* layer.

Step 7. Locate the proposed structure on the site plan on the *SITE BLDG* layer. The structure can be located by following the dimensions on the preliminary floor plan or by inserting a block of the home. Create the block by tracing the footprint of the floor plan.

FIGURE 9.1 ■ *Start a site plan by representing the existing features of the building site, including the property lines, easements, access roads, and any public utilities.*

Step 8. Represent all paved areas such as the driveway, curb cuts, ramps, exterior parking spaces, and walkways on the *SITE CURB* layer.
Step 9. Draw all utilities, including, electrical, gas, water, sewer, and telephone lines or easements on the *SITE UTIL* layer.
Step 10. Draw any decks, balconies, or patios on the *SITE BLDG* layer.
Step 11. Draw any required planting areas, fountains, pools, spas, benches, fences, or other landscaping features on the *SITE BLDG* layer.
Step 12. Draw drainage swales and catch basins on the *SITE PROP ESMT* layer.
Step 13. Draw special symbols such as property corner markers, a north arrow, fire hydrants, and cleanouts on the *SITE PROP BRNG* layer.
Step 14. Multifamily projects may require the representation of additional man-made features that are specific to the site, such as trash enclosures, patio furniture, flagpoles, retaining walls, and signs. Draw these features on the *SITE BLDG MISC* layer.

Once all man-made items are drawn, each should be dimensioned. Place dimensions on the *SITE ANNO DIMS*

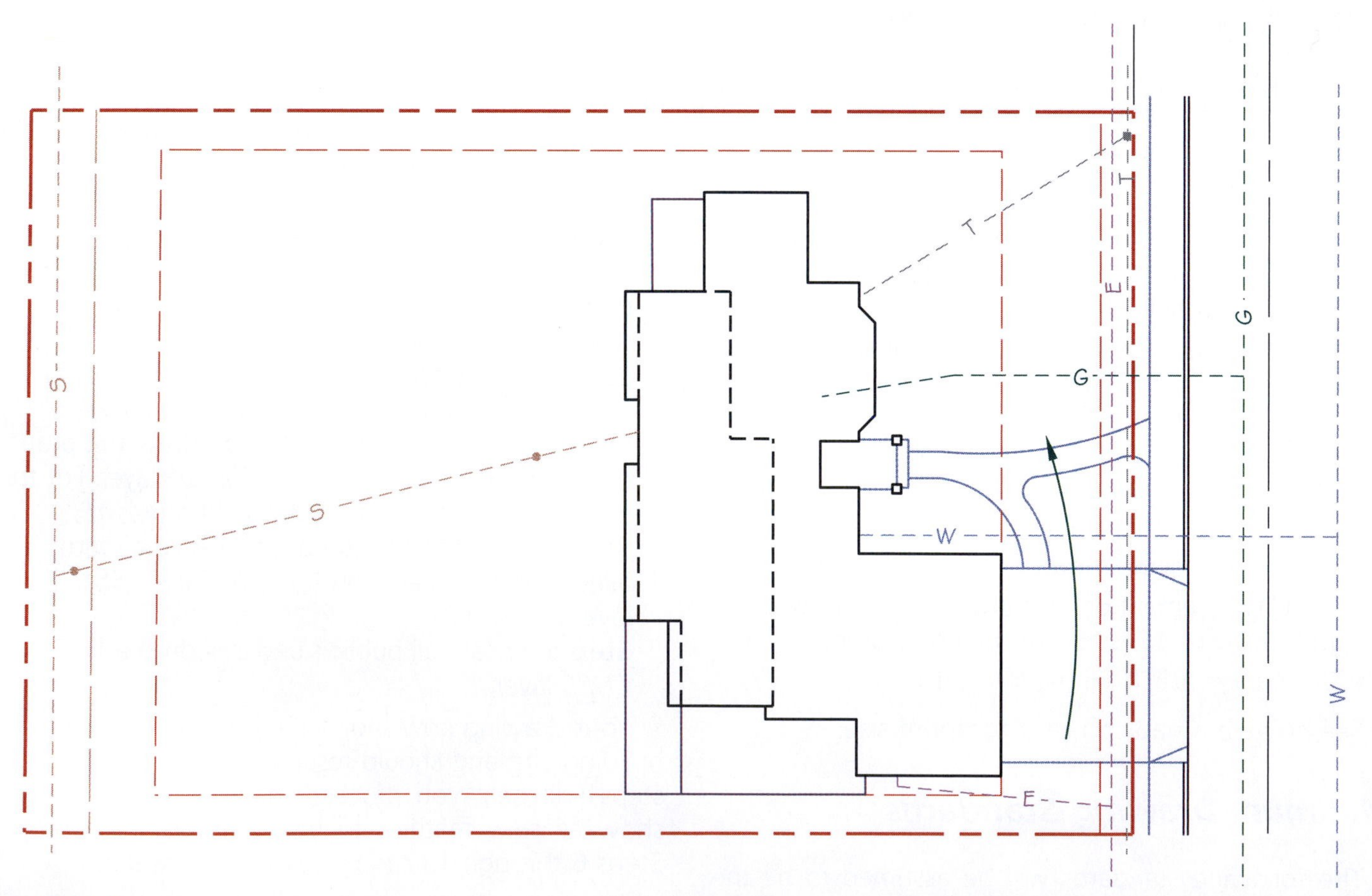

FIGURE 9.2 ■ *Represent all new features to be included in the project, such as the structure, decks, driveway and walkways, public utilities, required planting areas, landscaping features, and drainage patterns.*

layer by locating major items first. Express dimensions as feet and inches unless noted. Use the following order to place dimensions:

Step 15. Dimension property boundaries using decimal feet and list the required bearings.
Step 16. Dimension the location of the structure to the property.
Step 17. Dimension the minimum required yard setbacks.
Step 18. Dimension the overall size of the building footprint.
Step 19. Dimension the locations of streets, public sidewalks, and easements.
Step 20. Dimension all paved areas as well as individual parking spaces and catch basins.
Step 21. Dimension all utility locations.
Step 22. Dimension all other miscellaneous man-made features.

Steps 15 through 22 can be seen in Figure 9.3. With all features drawn and located, add any special symbols necessary to describe the material being used on the *SITE BLDG MISC* layer. Typically this would include hatching concrete walkways with a dot pattern or hatching the footprint of the structure.

The next stage to complete the site plan is to provide annotation to define all man-made features that are added to the site. Care should be taken to distinguish between existing material and material that is to be provided. Occasionally existing material to be removed from the site must also be specified. Place the required notes on the *SITE ANNO NOTE* layer unless noted. Notes that should be included on a site plan include:

Step 23. Proposed building use (SFR can be used to represent a single-family residence), square footage, and building height.
Step 24. Site area and site coverage.
Step 25. Specify elevation markers for each property corner and each building corner. Place this text on the *SITE PROP BRNG* layer.
Step 26. Finished floor elevation.
Step 27. Legal description.

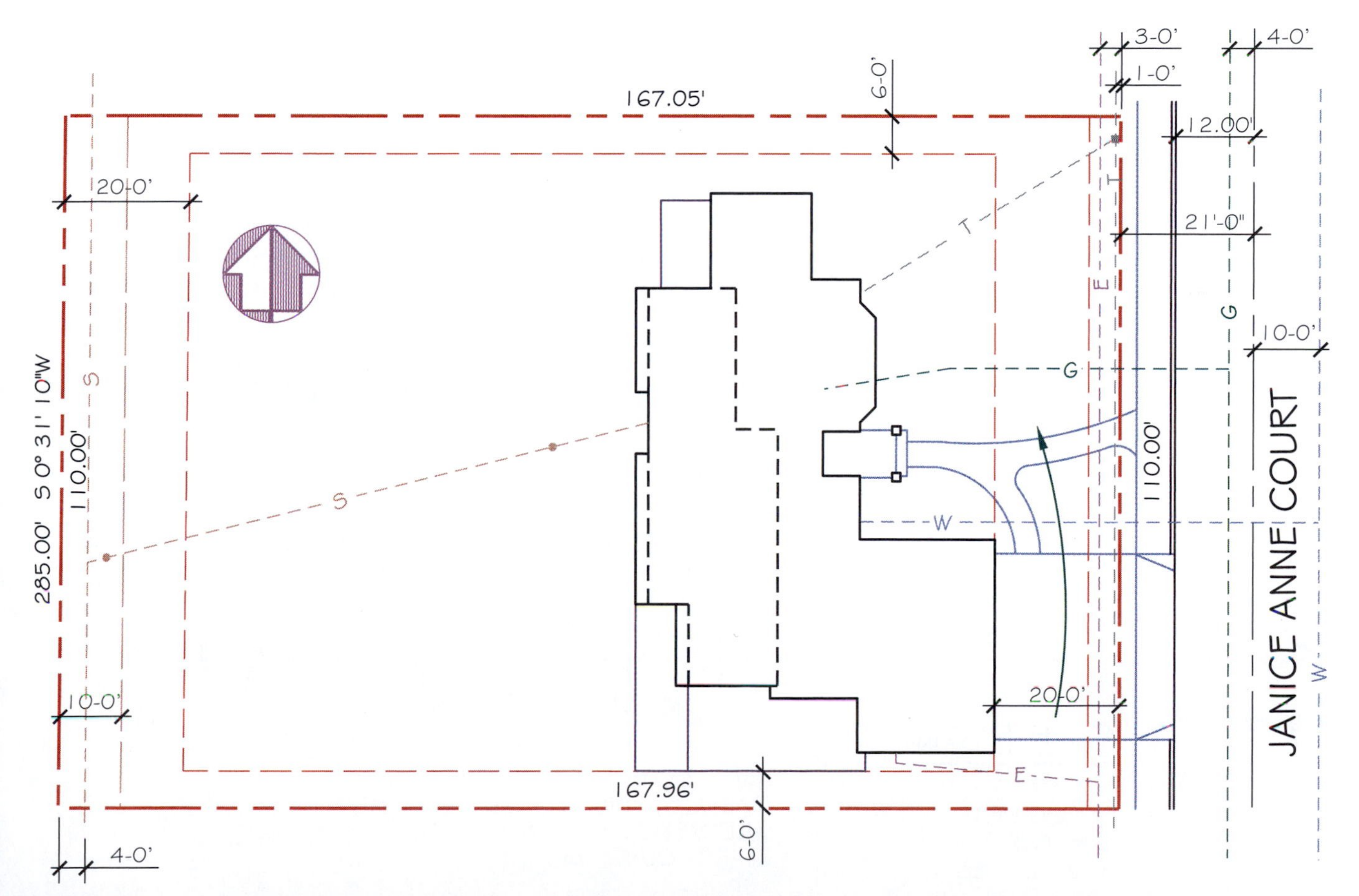

FIGURE 9.3 ■ *All man-made items must be located with dimensions. Express dimensions using feet and inches to locate man-made objects and engineering units to dimension the site.*

Step 28. Specify all streets, public walkways, curbs, and driveways.
Step 29. Describe all utilities.
Step 30. Describe all paved areas and specific parking areas.
Step 31. Describe specific features that are added to the site, including furniture, flagpoles, retaining walls, fencing, and planting areas.

All items on the site plan have now been drawn, dimensioned, and noted. The drawing can be completed using the following steps:

Step 32. Add details to show construction of man-made items.
Step 33. Add general notes to describe typical construction requirements.
Step 34. Specify a drawing title and scale.

The site plan is now complete and should resemble Figure 9.4. Use the drawing checklist from Appendix D to evaluate your drawing before submitting it to your instructor or placing the drawing into production.

COMPLETING A TOPOGRAPHY PLAN

A topography plan is created using field notes from the surveying team. Spot elevations are located on the site plan using the *TOPO OUTL* layer; then points that represent uniform elevations are connected using the *TOPO CONT* layer. Figure 9.5a shows the spot elevations for the site plan that was just created. The topog-

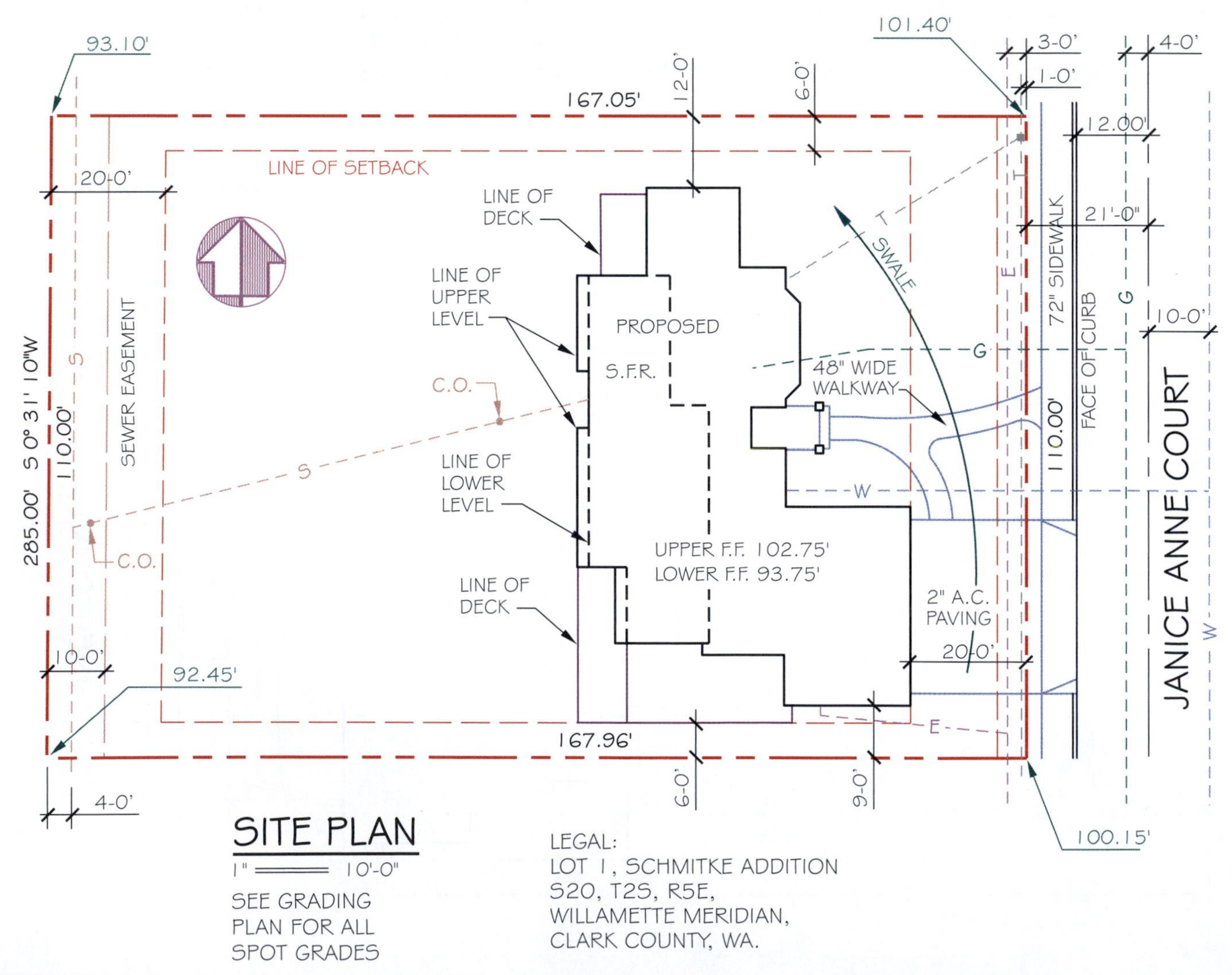

FIGURE 9.4 ■ *Provide annotation to define all man-made features that have been added to the site. Care should be taken to distinguish between new and existing features and material that is to be demolished.*

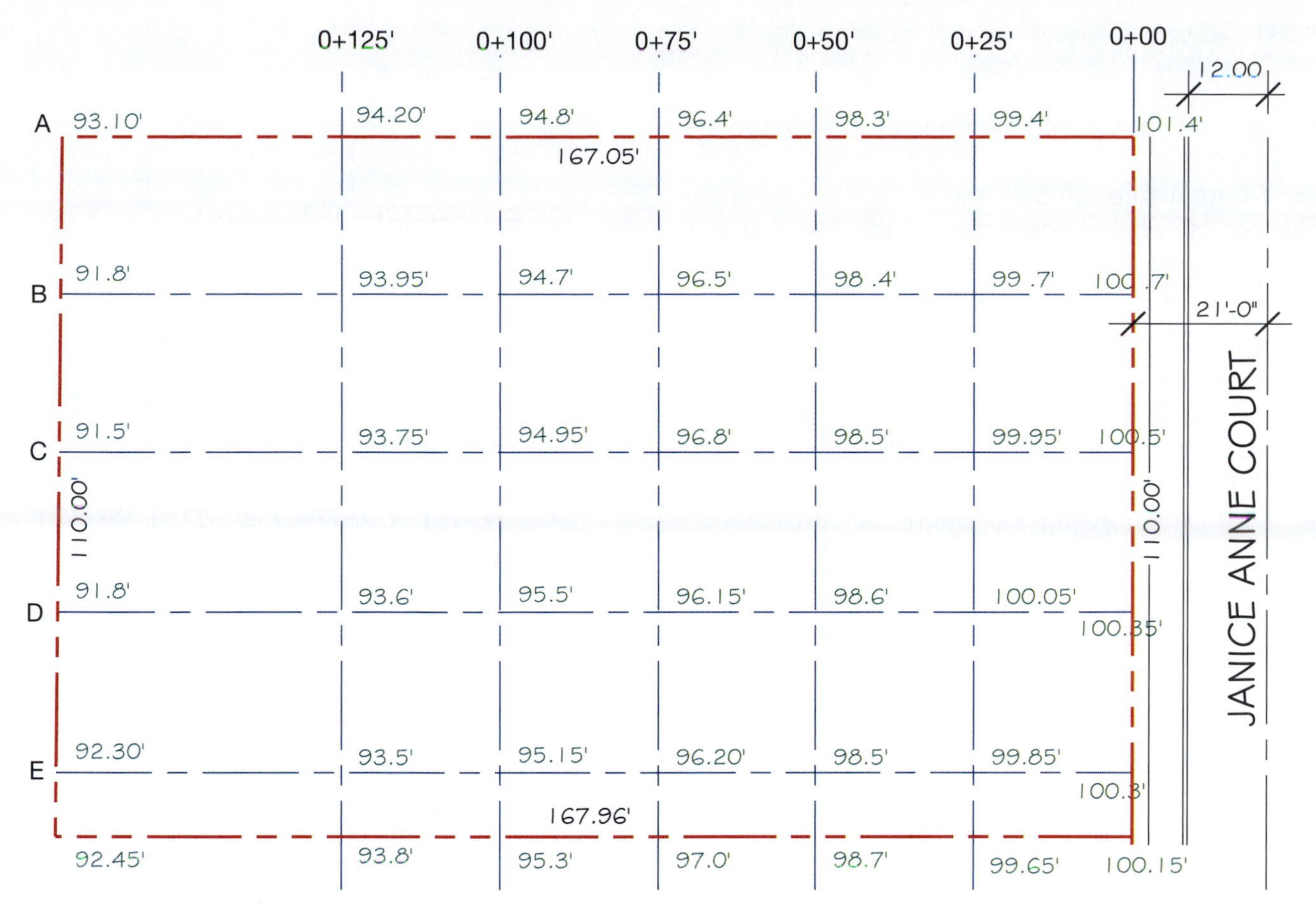

FIGURE 9.5a ■ *A topography plan is created using field notes for the surveying team, locating the spot elevations on the site plan, and then connecting the dots that represent uniform elevations.*

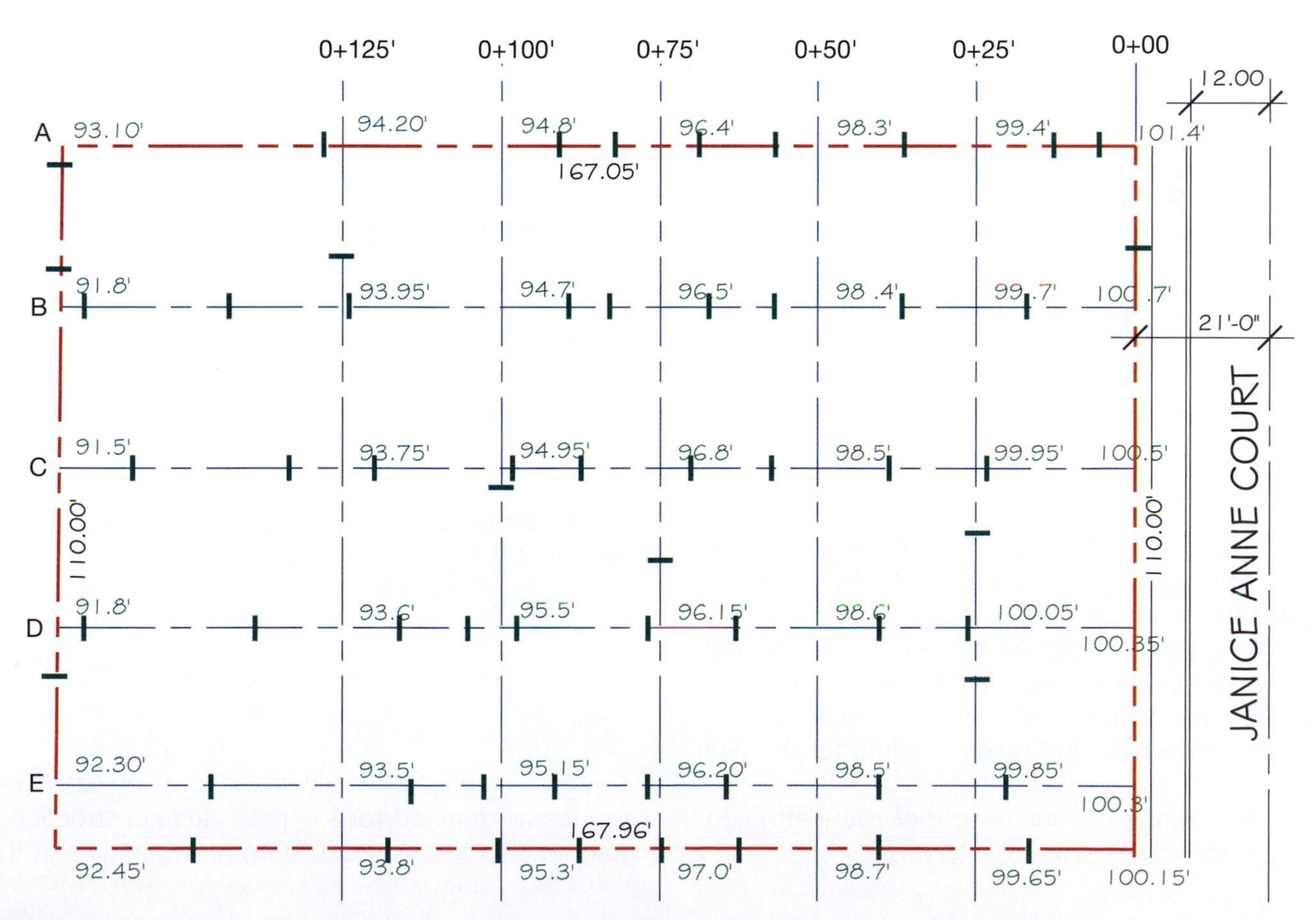

FIGURE 9.5b ■ *Once the field notes are located on the site plan, spot grades can be interpolated.*

raphy plan can be completed using the following steps.

Follow steps 1 through 5 that were used to create the site plan, or use an electronic copy of the site plan as a base for the topography plan. If an electronic copy of the site plan is used, freeze all unnecessary information.

Step 6. Draw grid lines on the site plan that match the grid used on the field drawings. Place the grid on a layer such as *TOPO OUTL* so that it can be frozen and not be part of the finished drawings.
Step 7. Determine the desired contour interval to be used on the drawing. Common intervals include 1, 2, 5, or 10' (300, 600, 1500, or 3000 mm). For this example, contours will be shown for each 1' of elevation change. Accent will be provided to each 5' (1500 mm) contour.
Step 8. Locate points to represent each of the spot grades on the *TOPO OUTL* layer.
Step 9. Using the *TOPO OUTL* layer, divide the space between each spot grade to represent the height between each spot grade.

Note:

Use the PLINE command to place the lines between the grid points to locate each contour. Place these marks on a separate layer so they can be frozen after all contour lines have been established.

The drawing should now resemble Figure 9.5b with each spot grade located and the distance between each spot grade equally divided.

Step 10. Select either the highest or lowest whole-foot elevation and establish a contour line on the *TOPO CONT* layer. Place the lines by picking points on the grid that represent the desired elevation. Use the PLINE command to place the lines between each grid to locate each contour.

Figure 9.6 shows the layout of the 101', 100', and the 99' contours.

Step 11. Draw the balance of the contour lines by connecting points of equal elevation. Work from highest to lowest or lowest to highest.
Step 12. Use the Fit option of the PEDIT command to provide a curve to each contour line.
Step 13. Freeze all layout information such as grids and spot elevations that is not required for the final drawing.
Step 14. Change the line weight of every fifth contour line to highlight the 5' intervals.

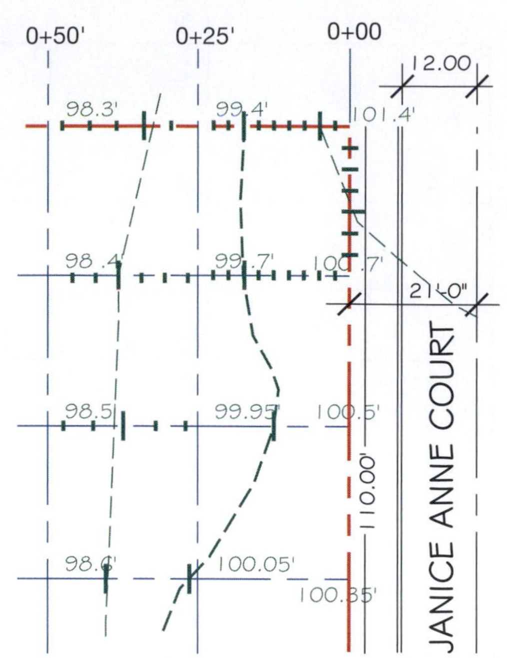

FIGURE 9.6 ■ *A contour line can be represented by picking points on the grid that represent the desired elevation. Knowing that 0 + 50 represents 98.3' and 0 + 25 represents a height of 99.4', it can be assumed that the ground slopes evenly between these two known elevations. If a major change occurred between those two points, the surveyor would have added a spot grade. The first mark to the right of 0 + 50 represents 98.4', with each of the following marks representing 98.6', 98.8', 99.0', and 99.2', respectively. Lines to connect each point of the 101', 100', and the 99' contours are shown.*

Step 15. Provide spot grades for each property corner elevation on the *TOPO ANNO ELEV* layer.
Step 16. Provide labels at the ends of each 5' elevation on the *TOPO ANNO* layer.
Step 17. Display text from the site plan, including property lengths and bearings, labels for the street name, curb, sidewalk, a title and scale, and a north arrow.

The completed topography plan should now resemble Figure 9.7.

COMPLETING A GRADING PLAN

A grading plan is usually provided when extensive excavation is required for the proposed construction. The grading plan shows the existing topography and the el-

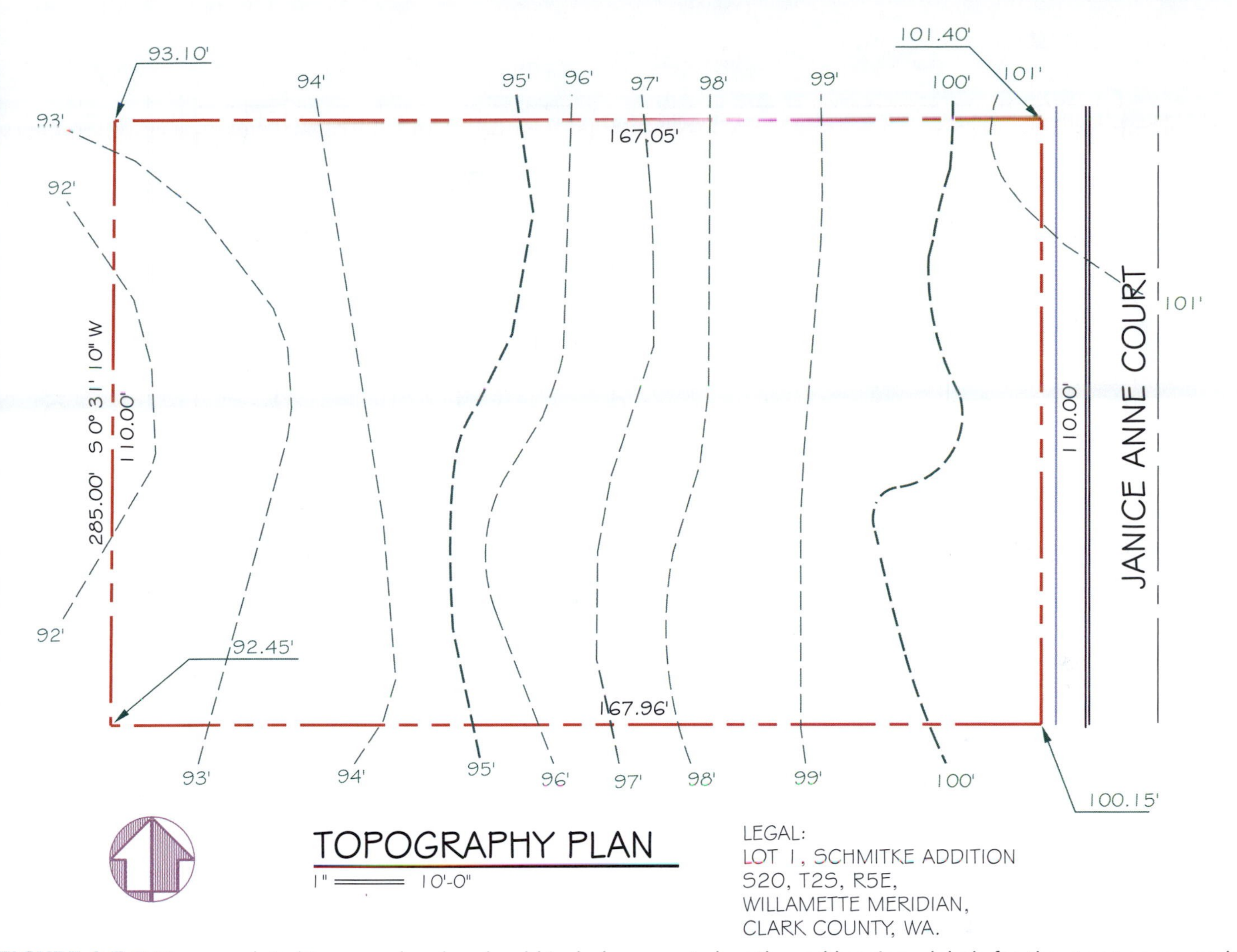

FIGURE 9.7 ■ *The completed topography plan should include property lengths and bearings; labels for the street name, curb, and sidewalk; a title and scale; and a north arrow.*

evations of the site that require excavation. The plan can be completed using the following steps:

Follow steps 1 through 5 that were used to create the site plan or use an electronic copy of the site and topography plans as a base for the grading plan. If an electronic copy is used, freeze all unnecessary information.

Step 6. Determine the angle of repose to be used for cut-and-fill banks based on municipal requirements or standards recommended by the soils engineer who supervised the topography drawings and geology report. The slope of cut surfaces for this site can't exceed 1.5/1 and fill banks shall not exceed 2/1.
Step 7. Insert a block to represent the outline of the upper and lower floor plans. For this project, the floor plan from Chapter 11 will be used.
Step 8. Show the outline of the driveway and proposed walks.

Your grading plan should now resemble Figure 9.8, showing the existing topography and all existing features such as curbs and sidewalks. The information for steps 1 to 8 can be taken from the site and topography plans if an electronic copy of the site plan is not available. If you have access to the electronic drawings, the site and topography plans can be used as a base drawing and unnecessary information on the site or topography plans can be frozen.

Step 9. Make a print of the grading plan in its current state and use it to plan the finished elevations.
Step 10. Use the print to evaluate the finish elevations for the garage, the upper floor plan, and the lower floor. Sketch the proposed changes on your print.
10a. Because the existing grade at the driveway flows down into the garage, a swale will be required.

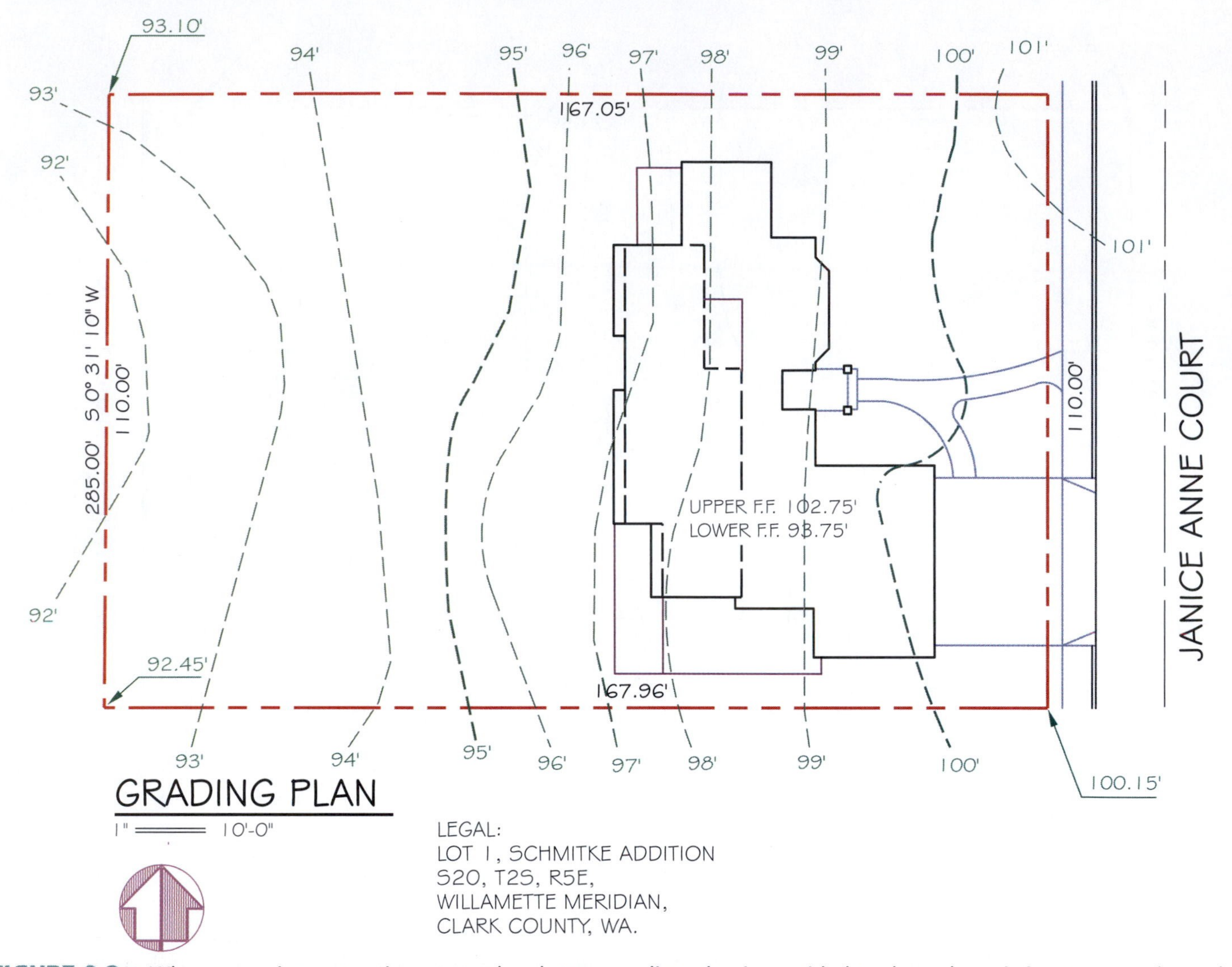

FIGURE 9.8 ▪ *When extensive excavation must take place, a grading plan is provided to show the existing topography and the elevations of the site requiring excavation. Start the plan by using a topography plan as a base, and locate all proposed work, such as the structure, walks, and the driveway.*

10b. If the 100' contour is pulled closer to the property line, the garage can be set at an elevation of 101'. This will allow:

- The driveway and garage to be placed on a small amount of fill creating positive drainage
- Raising the lower floor to minimize cutting
- Front of garage floor at 101.00'
- Rear of garage floor at 101.50'
- Upper finish floor elevation at 102.75'
- Lower finish floor elevation at 93.75'
- A concrete pad at 97.5' in the crawl space for the utilities

10.c Retaining walls will be required on the north, east, and south walls of the basement.

Step 11. Use your free hand sketch as a guide to locate all proposed contour changes on the *TOPO CONT NEW* layer. Assume that no grading will be allowed in the setback area. Use continuous thin lines to represent new contours. Use continuous thick lines to represent 5' contours.

The grading plan should now resemble Figure 9.9. Complete the grading plan using the following steps:

Step 12. Place the needed text to represent the elevation of each contour line on the *TOPO CONT ANNO* layer.

Step 13. Represent swales required to divert water from the residence on the *TOPO DRAN* layer.

Step 14. Provide spot grades for each corner of the site on the *TOPO DRAN ANNP* layer.

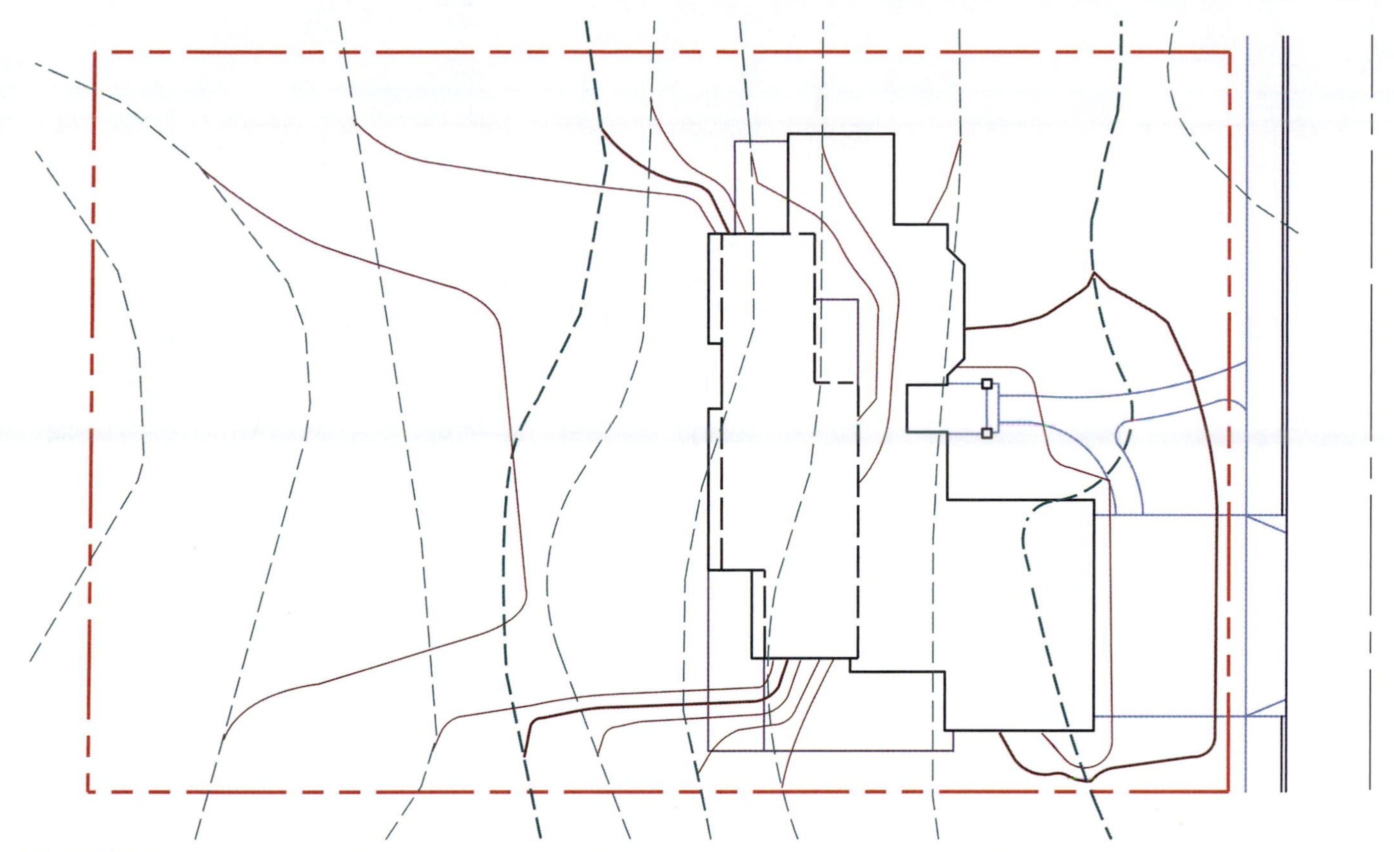

FIGURE 9.9 ■ *Once proposed elevations have been placed on a sketch, new contour locations can be represented using continuous thin lines. Continuous thick lines are used to represent 5' contours.*

Step 15. Provide spot grades for each corner of the residence on the *TOPO BLDG ANNO* layer.
Step 16. Label the elevation of each floor level on the *TOPO BLDG ANNO* layer.
Step 17. Thaw, insert or copy information from the site plan to show information to describe the site such as the property size, legal description, north arrow, street name, sidewalks, driveways, and walks.
Step 18. Place a title and scale below the drawing on the *TOPO ANNO* layer.
Step 19. Establish the proper scale values for the viewport and prepare a check print to evaluate your work.

The completed grading plan can be seen in Figure 9.10.

COMPLETING A PROFILE DRAWING

A profile is a section that shows the contour of the ground along any line extending through the site. Once a line is drawn on the grading plan, the profile is drawn from the contour lines at the section location. The contour map and its related profile are commonly referred to as the plan and profile. The following example was created by placing the viewing line at the north edge of the driveway in Figure 9.10. Information needed to layout the project can be placed on the *SITE OUTL* layer and frozen for plotting. Use the *SITE GRID* layer to display the profile base. A profile can be created using the following steps:

Step 1. Draw a straight line on the topography or grading plan at the location of the desired profile.
Step 2. Draw a horizontal line to represent the base of the drawing.
Step 3. Project a line at 90° from the cutting plane to indicate where each existing contour line touches the cutting plane.
Step 4. Determine the vertical scale to be used. For this drawing the vertical scale is 3X the horizontal scale. Figure 9.11 shows the grid created to establish the profile.
Step 5. Draw a line to connect the corresponding existing contour elevations on the *SITE GRID NEW* layer.

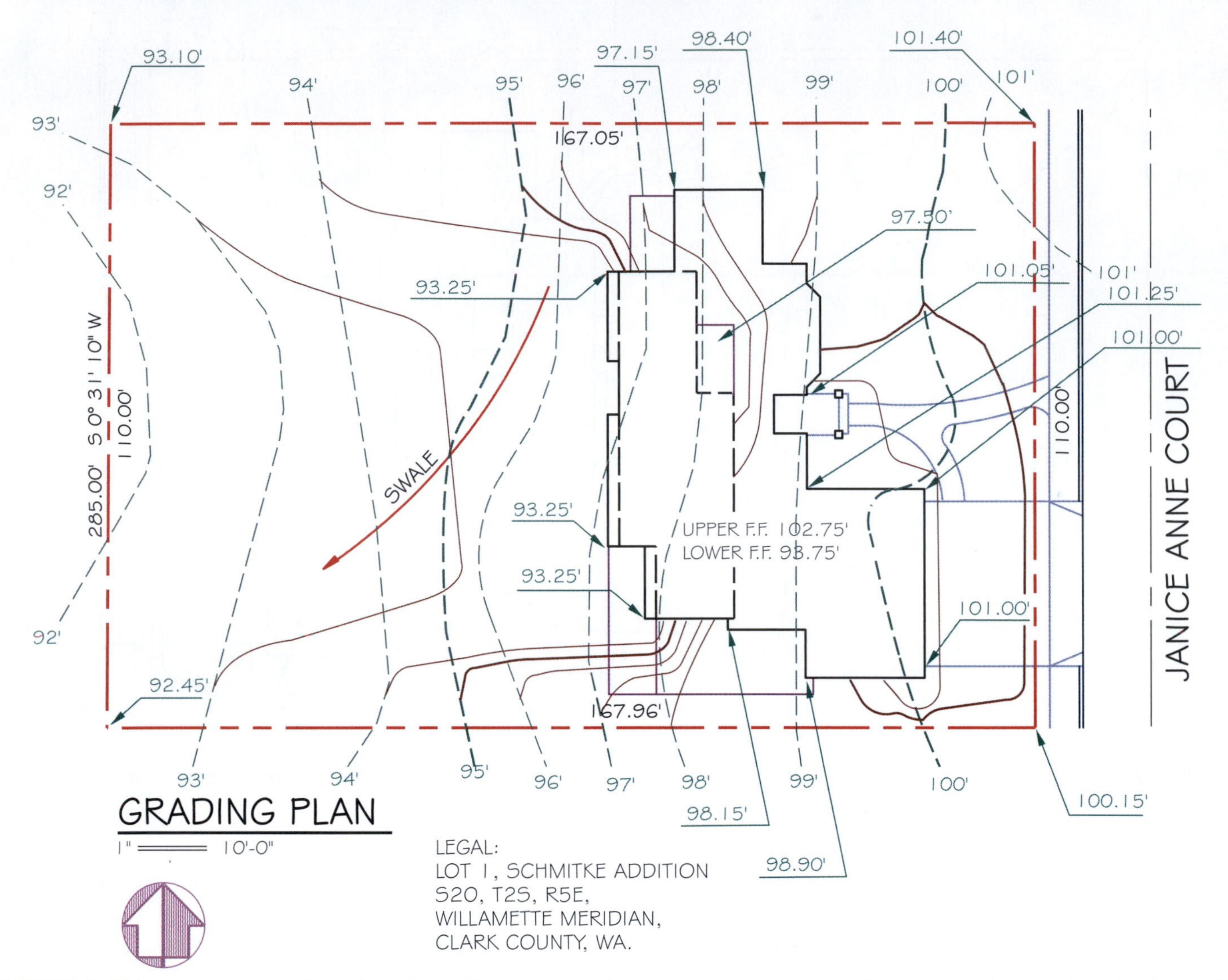

FIGURE 9.10 ■ *The completed grading plan will represent existing and new contours, drainage swales, spot grades for each corner of the site and the residence, and the finish floor elevations. Information should also be provided to describe the site, such as the property size, legal description, north arrow, street name, sidewalks, driveways, and walks.*

Step 6. Freeze the grid used to establish the contours or set the plotting parameters to display it with grayscale.

Step 7. Provide spot grades at the high and low points of the profile and at any significant elevation changes on the *SITE GRID* ANNO layer.

Step 8. Place a title and list the horizontal scale on the *SITE GRID* ANNO layer.

The completed existing profile drawing should resemble Figure 9.12. Use the completed profile to show the proposed grading and the new structure. Complete the profile using the following steps:

Step 9. Thaw the grid to help represent the new grades.

Step 10. Project lines at 90° to the cutting plane to represent the outline of the structure and the new grade contour locations on the *SITE OUTL* layer.

Step 11. Draw a line on the *SITE OUTL* layer to connect the corresponding new contour elevations on the *SITE CONT NEW* layer.

Step 12. Provide spot grades at any significant elevation changes, floor levels, and walkways on the *SITE GRID* ANNO layer.

Step 13. Place a title and list the horizontal scale on the *SITE GRID* ANNO layer.

The completed profile drawing should resemble Figure 9.13.

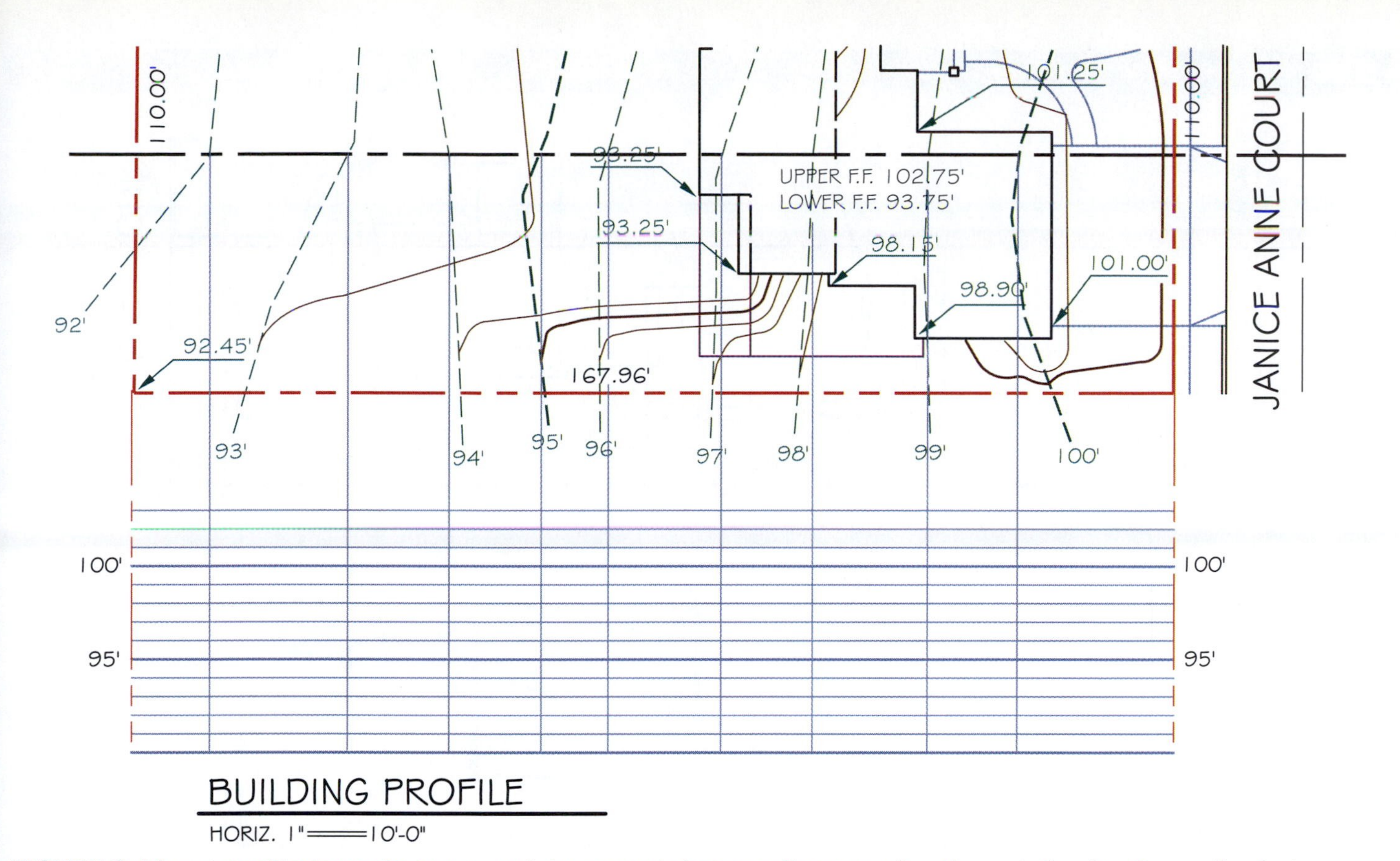

FIGURE 9.11 ■ *A profile shows the contour of the ground along any line extending through the site. Once a line is drawn on the grading plan, the profile is drawn from the contour lines at the section location.*

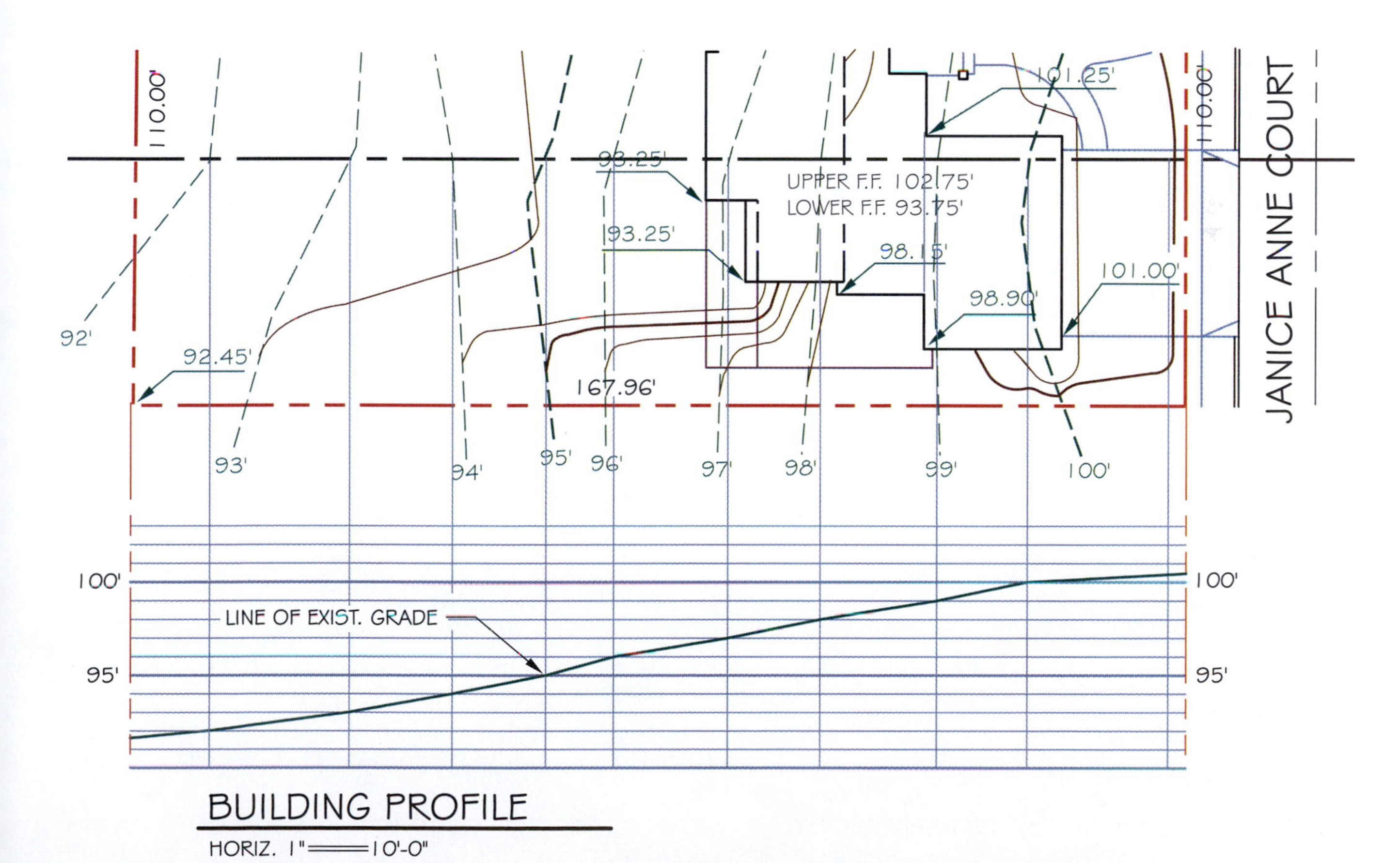

FIGURE 9.12 ■ *A profile drawing is created by drawing a straight line on the topography or grading plan to represent the cutting plane. Once the desired profile location is selected, a line is projected at 90° from the cutting plane where each existing contour line touches the cutting plane.*

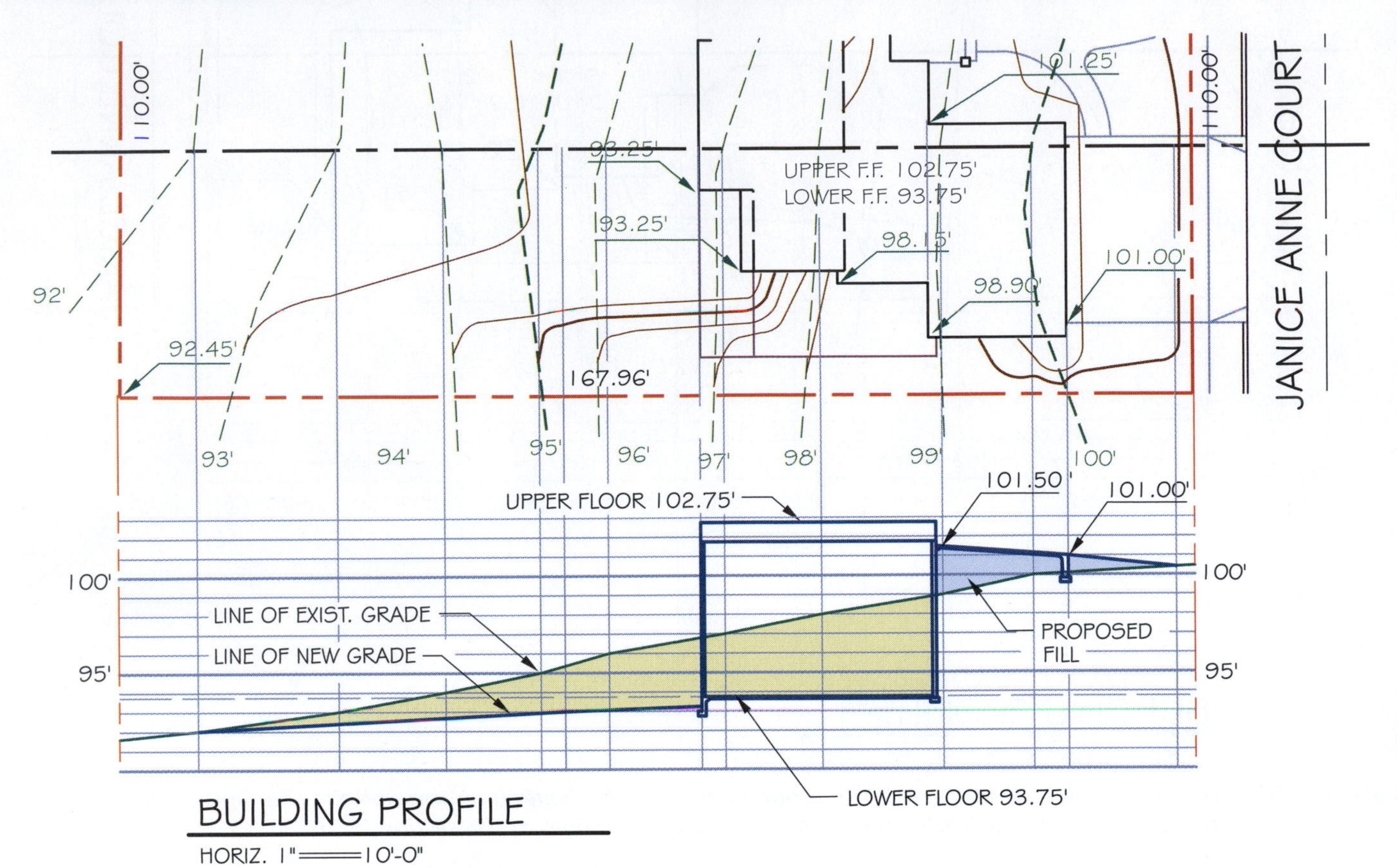

FIGURE 9.13 ■ *The completed profile drawing should show the outline of the structure and the new grade contour locations. Provide spot grades at significant elevation changes, floor levels, and the walkways. Provide a title and list the horizontal scale.*

CHAPTER 9

Site-Related Drawing Layout Test

DIRECTIONS

Answer the following questions with short, complete statements. Type the answers using a word processor or neatly print the answers on lined paper.

1. Type your name, the chapter number, and the date at the top of the sheet.
2. Type the question number and provide the answer in the form of a statement that includes part of the question. You do not need to write out the entire question.
3. Answers should be based on the codes that govern your area unless otherwise noted.

QUESTIONS

9.1. How is the sheet size for a site plan determined?

9.2. What factors influence the drawing scale selection for a site plan, a topography plan, and a grading plan?

9.3. List five pieces of information that should be gathered before starting a site plan.

9.4. List five sources to obtain information for site related drawings.

9.5. Define setbacks and explain how they affect a site and grading plan.

9.6. Explain factors that influence the spacing of contour lines on a grading plan.

9.7. What is a spot grade and how does it relate to a topography?

9.8. What is the distance to station point 0 + 50?

9.9. In addition to the elevation of the site excavation, what does the grading plan show?

9.10. Explain how information on a site plan is transferred to a topography or grading plan.

CHAPTER 9

Drawing Problems

Unless your instructor assigns a specific site, select one of the following site plans using the following guidelines.

Visit the website for the municipality that governs your area and use the setbacks for the given size of the lot you have chosen, or use the following setbacks:

Minimum front setback, 25'-0''
Minimum rear yard setback, 20'-0''
Minimum side yard setback: one level, 5'-0''; two level, 6'-0''

Use the specified paper size for the municipality that governs your area. If your zoning department does not specify a paper size, draw your drawing to fit on an 8.5'' × 14'' sheet of paper.

Select the appropriate scale to draw the required site drawing.

PROBLEMS

Problems 9.1 through 9.7. Use the subdivision map in Figure 8.7 and draw the site plan for one of the parcels of land. Begin the selected site plan problem by representing the given information in preparation to complete a preliminary design study for one of the homes in Chapter 11. Once the floor plan for your preliminary floor plan is approved, complete the required site drawings.

9.8. Draw a vicinity map of your school showing major access routes and important landmarks within a 5-mile radius.

9.9. Use the following legal description and draw the site plan for a tract of land situated in the SW 1/4 of the NE 1/4 of section 17, T2S, R1W of Mount Diablo Meridian, Rancho Santa Barbara:

Beginning at a 5/8'' iron rod marking the true point of beginning of the southwest property corner that lies 32' directly north of the center of Rancho Santa Barbara Place, proceed N 3° 15' W 115.0', thence N8° 30' E 140.0', thence proceed N 90° 00'E a distance of 92.10; thence proceed S38° 30' E 91.5'; thence proceed due south a distance of 181.5'; and thence proceed due West 163.25' back to the true point of beginning.

9.10. Use the following legal description and draw the site plan for the following tract of land:

Parcel 12, tax lot 215682 of the Miller Addition, situated in section 12, T3N, R1E, 6th Principal Meridian, Cloud County, Kansas. Beginning at a 5/8'' iron rod marking the southerly corner and the true point of beginning of the southeast property corner, go a distance of 185.0' N49° 50' W to the far westerly corner. Said property line lies 32.00' north of the center of Miller Drive. Thence N 65° 15' E a distance of 104.50'; thence N49° 00' W a distance of 147.0'; thence N43° 30' E a distance of 93.0' back to the true point of beginning.

9.11. Use the following description to draw and label a site plan that can be plotted at a scale of 1/8'' = 1'–0'':

Beginning at a point that is the NE corner of the G.M. Smith D.L.C., which lies 250 feet north of the centerline of N.E.122 street which is in Section 1AB, Township 6 north, Range 1 west of Los Angles county, California thence south 225.00 feet to a point which is the southwest corner of said property, thence North 85.00 feet 1° 25' 30'' west along the easterly edge which lies along N.E. Sweeney Drive (25' to street centerline) to the northwest property corner, thence 140.25 feet South 89° 15' 00'' East, thence south 85.25 feet 1° 25' 30'' east, thence westerly along the north boundary of N.E. 122 street to the true point of beginning.

9.12. Use the attached drawing and the drawing on the student CD to complete a site plan. After you complete the site plan, determine the square footage of the site and the zone for this property based on local zoning standards. Assign a legal description to the site that would be appropriate for a suburban site in your area.

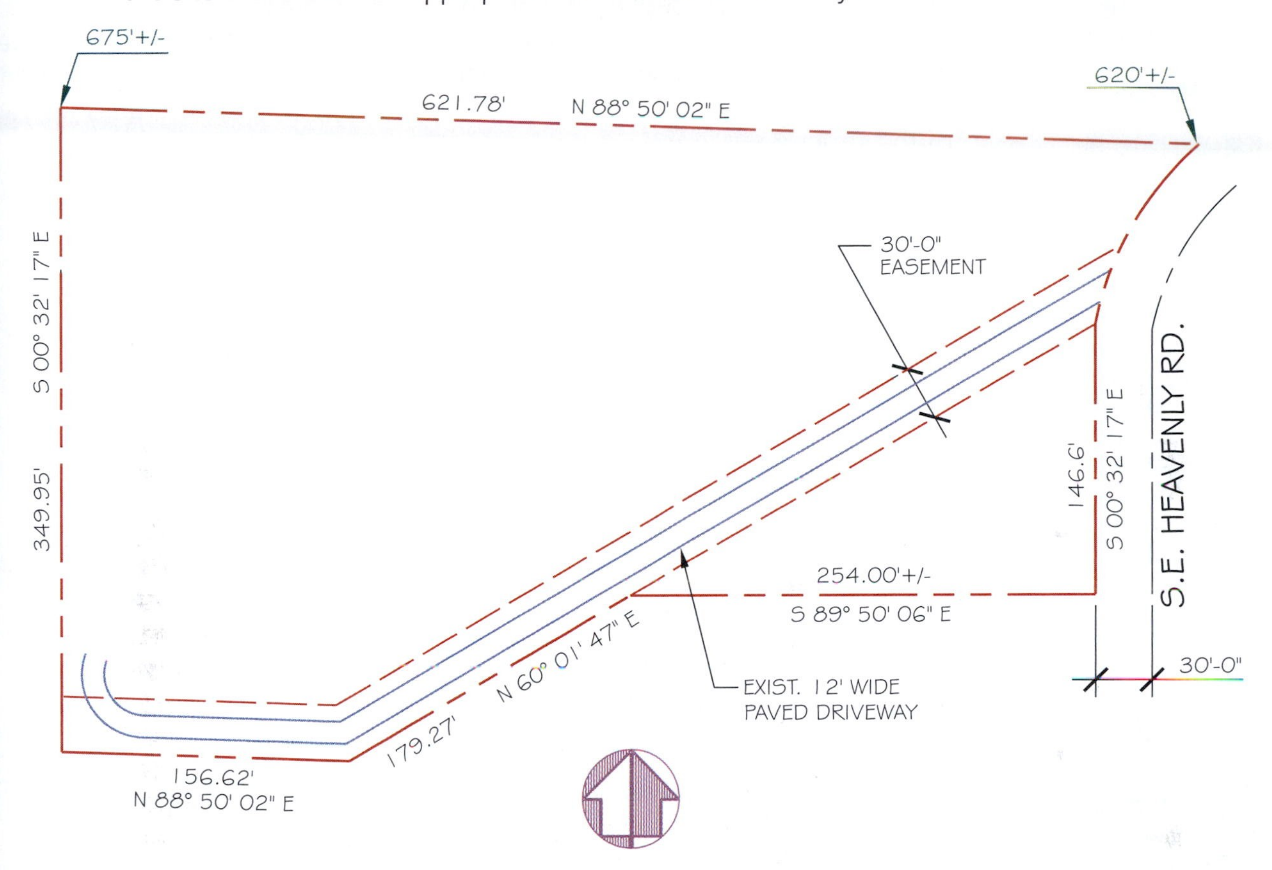

9.13. Use the corresponding drawing on the student CD to lay out the represented grades. Draw the site plan using a scale of 1'' = 10'–0''. Show grades at 1' intervals and highlight contours every 5'.

9.14. Use the site plan created in problem 9.9 and the corresponding field notes on the student CD to create a topography plan for this site. Show grades at 1' intervals and highlight contours every 5'.

9.15. Use the site plan created in problem 9.10 and the field notes on the CD to create a topography plan for this site. Assume point H/0 + 00 to be the south corner of the property and line A-H is the southerly property line.

9.16 through 9.22. Use the field notes on the CD to complete a topography plan for one of the sites created in problems 9.1 through 9.7.

Directions for problems 9.23 through 9.29: Complete one of the following projects once you have selected and done a preliminary layout of one of the homes from Chapter 11. Use the completed topography plan for one of the sites created in problems 9.16 through 9.22 and the footprint of a home from Chapter 11 to create a grading plan. Assume that all fill banks will be at a maximum angle of 2/1 and all cut banks will be at a maximum angle of 1.5/1.

9.30. Use the drawing on the student DC to complete a site plan for four townhouse units. The plan is to be plotted at a scale of 1''=10'–0''. The lot is 120×110' and is Lots 4 and 5 of Arrowhead Estates, Rapid City, South Dakota. The project fronts onto Sioux Parkway with 36' from the property line to centerline of the roadway. Tee water to the property 75' east of the westerly property line from a main line 30' south of the property line. Provide a water shutoff valve in the garage of each unit. Access the city sewer line, which is 25' south of the property, with a new line 25' east of the westerly property line. Provide a clean out in the lower planting strip of each unit. Provide a continuous drainage grate across each driveway. Set each drain 3'' below the finish floor level at each door. Provide a 7.5' wide easement along the northerly property line for a telephone easement. Provide each unit with a fenced yard. Dimension each unit as needed and specify all grades. Interpolate as needed to determine the grades for each corner of the structure.

SECTION 3

Floor Plans and Supplemental Drawings

10 CHAPTER

Floor Plans—Symbols, Annotation, and Dimensions

The floor plan provides a representation of where to locate the major items of a home. The plan shows the location of walls, doors, windows, cabinets, appliances, and plumbing fixtures. Chapter 1 introduced the development of a set of house plans and how the floor plan is developed. The drawing allows the homeowner to evaluate the project to make sure that the home, once constructed, will meet the family's current and future needs. The floor plan also serves as a key tool in the communication process between the design team and the building team. This chapter introduces the symbols, annotation, and dimensions typically associated with residential floor plans.

For the design team, the floor plan will become the skeleton on the basis of which to develop other drawings required for the project, including the electrical plan, fire-protection plan, framing plan, plumbing plan, and HVAC drawings. It will also be used as a base to create the site plan, roof plan, and roof framing plan. Consideration must be given to how the floor plan will be used in order to show the material required for these related drawings. Two common methods are used to place related information on the floor plan. If the entire set of drawings will be drawn by one office, layers are used to control the display of information related to the various plan views. Common layers used to control multiple drawings in one file can be found in Chapter 3 and throughout this chapter. If multiple firms will be consulting on a project, the architectural team will often complete the floor plan and provide an electronic copy of it to each consultant. This allows each firm to bind information to the base drawing using external referencing (XREF).

COMMON FLOOR PLAN SYMBOLS

No matter which method is used, information on the plan must be clearly represented in a way that is consistent with common drawing practices. The floor plan is created by passing an imaginary horizontal cutting plane through the structure approximately 4' (1200 mm) above the floor. The portion of the structure above the cutting plane is then removed, allowing the viewer to look down into the structure. The relationship of the structure to the floor plan is shown in Figure 10.1. To create the plan view, you will be working in model space using full scale. If the home is 70' long, adjust the drawing area so the structure will fit in the viewing area. As the floor plan is developed, the scale used to plot the plan must be considered. Most professionals plot floor plans using a scale of 1/4'' = 1'-0'' (1:50). Other scales, such as 1/8'' = 1'-0'' or 3/16'' = 1'-0'', are commonly used for large projects such as multifamily residential projects. To be consistent with common practices, common symbols must be used to represent common features such as walls, doors, windows, cabinets, appliances, and plumbing features.

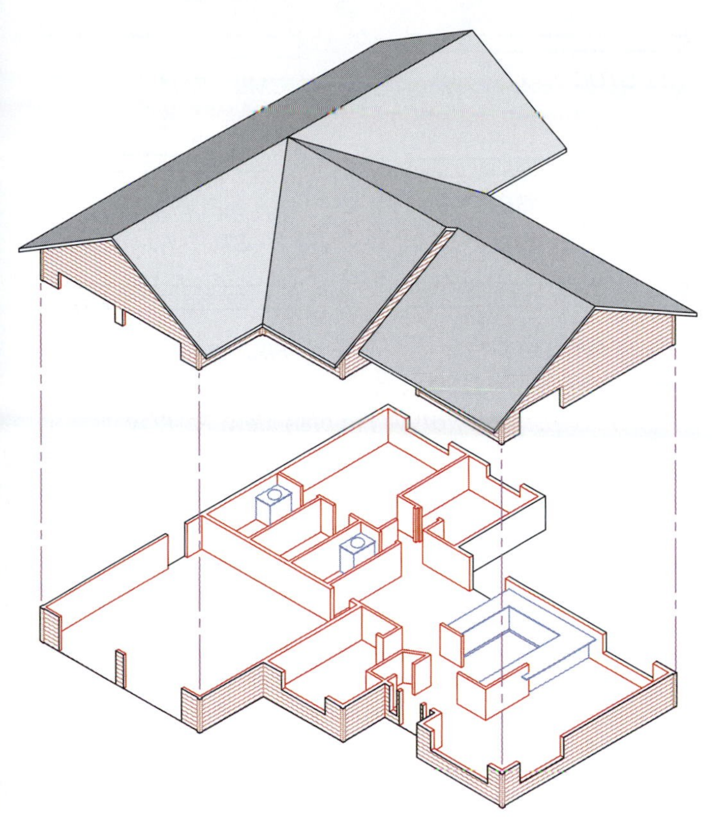

FIGURE 10.1 ■ *The floor plan is created by passing an imaginary horizontal cutting plane through the structure approximately 4' (1200 mm) above the floor. The portion of the structure above the cutting plane is then removed, allowing the viewer to look down into the structure.*

Note:

Most of the symbols presented in this chapter can be found in the Floor Blocks subfolder of the DRAWING BLOCKS *folder of the CD. Any text that is displayed with the block is set for display when plotting at a scale of 1/4'' = 1'-0''. Verify all layers and dimension sizes prior to inserting these blocks or any others into your drawing. Once an object is in your drawing, you assume responsibility that the object is correct.*

Wall Symbols

Use pairs of thick (0.024''/60 mm) parallel lines to represent walls on the floor plan. The distance between the lines will vary based on office practice. Common methods of representing wood walls include:

- Represent the walls based on the nominal thickness of the building material. This will produce 6 or 4'' (150 or 100 mm) wide walls for wood construction.
- Represent the wall thickness based on the thickness of the construction materials to be used.

Wall Thickness for Wood Walls

When the thickness of wood walls is drawn to the exact construction dimensions, the material used establishes these values. Exterior walls can be built using 2 × 6 (50 × 150) or 2 × 4 (50 × 100) studs based on the region where the construction will take place. Studs are the vertical members used to construct walls (see Chapter 20). Their finished size is 1 1/2 × 3 1/2'' (40 × 90 mm) for a 2 × 4 and 1 1/2 × 5 1/2'' (40 × 140 mm) for a 2 × 6 (50 × 150 mm). Exterior and interior construction materials are then applied to the studs. For a 6'' exterior wall, the common thickness would be:

Studs	5.5''	140 mm
Interior sheet rock	.5''	13 mm
Exterior sheathing	.5''	13 mm
Exterior stucco	1.0''	25 mm
Total wall thickness	7.5''	190 mm (191 mathematically)

Since the wall covering may not be known when the preliminary drawings are started, it is common to represent the wall thickness based on the stud width. It's important to remember that although the computer is very accurate, you are creating only a representation of the structure. The contractor will determine the needed sizes based on your dimensions, not on the exact thickness of the walls on the floor plan.

When the wall thickness is based on the size of the studs to be used, four common sizes are usually represented on a floor plan. These sizes include:

6'' (150 mm) wide for exterior heated walls.
4'' (100 mm) wide for exterior unheated walls.
4'' (100 mm) wide for interior walls.
6'' (150 mm) wide for walls to hide plumbing for toilets on the lower level of multilevel structures. (Pipes for toilets in a single level-home and other plumbing fixtures can be placed in a 2 × 4 stud wall, but 2 × 6 studs are often used to aid the plumbers).
2'' (50 mm) wide for furring (furring is material placed over the interior side of concrete walls to provide a smooth finish).
6-8'' (150 to 200 mm) party walls and firewalls for multifamily units.

Although not found in a single-family residence, multifamily projects contain party walls and often contain firewalls. The party wall is the wall between two adjoining units. Two separate walls are usually used to form a party wall in order to minimize sound transmission. A firewall is used to separate units so that the floor area does not exceed a specified size. A firewall is typically required when the floor area exceeds 3000 sq ft, but local codes should be verified. Firewalls are typically constructed from wood studs covered with 5/8'' type "X" gypsum board on each side. Chapter 21 explains party and firewall construction.

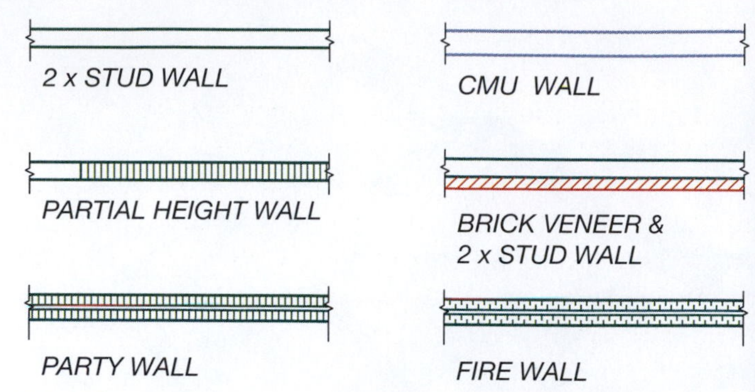

FIGURE 10.2 ■ *Common methods of representing each type of wall in plan view. Party walls and firewalls are found on multifamily projects.*

Wall Thickness for Nonwood Walls

In addition to wood, other materials can be used to frame walls. Walls framed with steel studs are drawn using the same methods used to draw wood walls. Masonry walls are represented in plan view with thick (.024''/60 mm), parallel lines to represent the edges of the masonry. Thin lines for hatching are placed at a 45° angle to the edge of the wall using approximately .125'' (.3 mm) spacing. The AutoCAD ANSI31 hatch pattern is the common method of hatching masonry. The type, size, and reinforcement used to build the wall are not represented on the floor plan but are specified in note form. Common masonry wall thicknesses include:

- 4'' (100 mm) wide for masonry veneer
- 8 to 10'' (200 to 250 mm) wide for structural brick
- 8'' (200 mm) wide for poured concrete walls
- 8'' (200 mm) wide for concrete masonry units (CMUs)

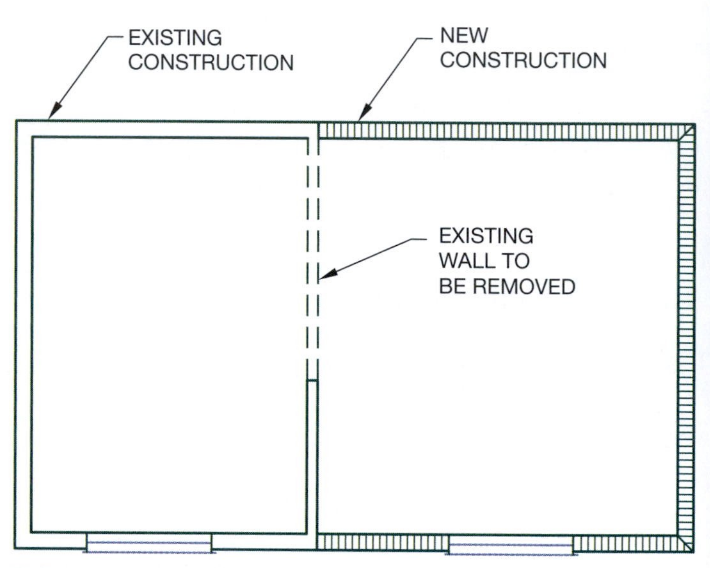

FIGURE 10.3 ■ *Methods of highlighting varied wall construction such as new walls, existing walls, and walls to be demolished are often found on residential remodeling projects.*

Representing Walls

Walls should be represented on a layer that describes their function. A layer name such as *FLOR WALL* clearly represents the contents. A modifier can be added to further describe the contents using names such as *FLOR WALL NEW, FLOOR WALL EXST, FLOR WALL REMOV, FLOR WALL CMU, FLOR WALL PRHT,* or *FLOR WALL* MASN.

Figure 10.2 shows common methods of representing each type of wall in plan view. Although most offices represent walls with no shading, some projects require differences in wall construction to be highlighted. Partial-height walls are often shaded to provide contrast with full-height walls. New construction will need to be contrasted with existing walls and walls that will be removed. Figure 10.3 shows methods of highlighting varied wall construction. In planning wall layouts, common sizes should include:

Hallways: 36'' (900 mm) minimum clearance
Entry hallways: 42'' to 60'' (1060 to 1500 mm) minimum
Bedroom closets: 24'' (600 mm) minimum depth; 48'' (1200 mm) minimum length
Linen closets: 14'' to 24'' (350 to 600 mm) deep (not over 30'', or 760 mm)
Washer/dryer space: 36'' (900 mm) deep, 5'6'' (1675 mm) long minimum
Stairways: 36'' (900 mm) minimum wide

Door Symbols

Standard symbols have been developed by the construction industry to represent common door types. AutoCAD and third-party vendors supply blocks to represent common door symbols, or they can easily be made and saved as part of your architectural template. Common types of doors found on a floor plan include swinging, sliding, folding, and overhead, as shown in Figure 10.4.

The door symbol to be placed on the floor plan will reflect the type of door to be used. The size and material are specified in a schedule. The creation of schedules is discussed later in this chapter. Although a door can be placed where the traffic pattern dictates, doors are usually placed in one of three locations:

- Within 3'' (75 mm) of a corner to allow space for the doorframe and trim
- Approximately 24'' (600 mm) from a corner to allow furniture to be placed behind the open door
- Centered on the wall

Swinging-Door Symbols

Swinging doors are represented on the floor plan by thin lines that show the door and the door swing. Symbols should be placed on a layer with a title such as *FLOR DOOR*. Common types of swinging doors represented on a floor plan may include exterior, interior, double, French, double-acting, and Dutch. Figure 10.5 shows the representation for each type of door. Each of these door types can be used as an exterior or interior door. A *double door* consists of two or more single swinging doors mounted in one frame, similar to Figure 10.6. Double swinging doors are often used for the main entry door of a large formal entry or for access to decks and other outside living areas. If more than two

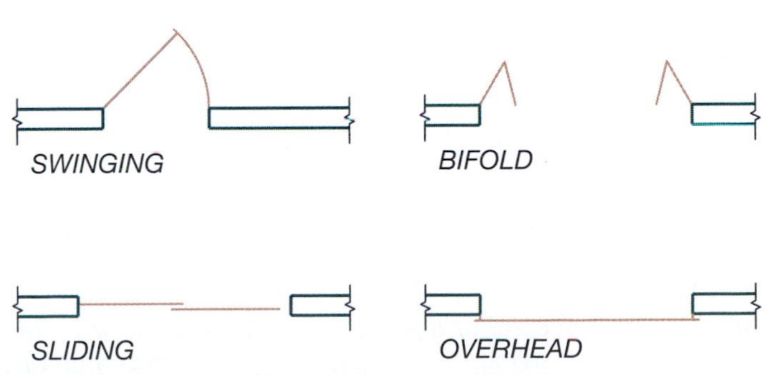

FIGURE 10.4 ■ *Common doors found on a floor plan include swinging, sliding, folding, and overhead types.*

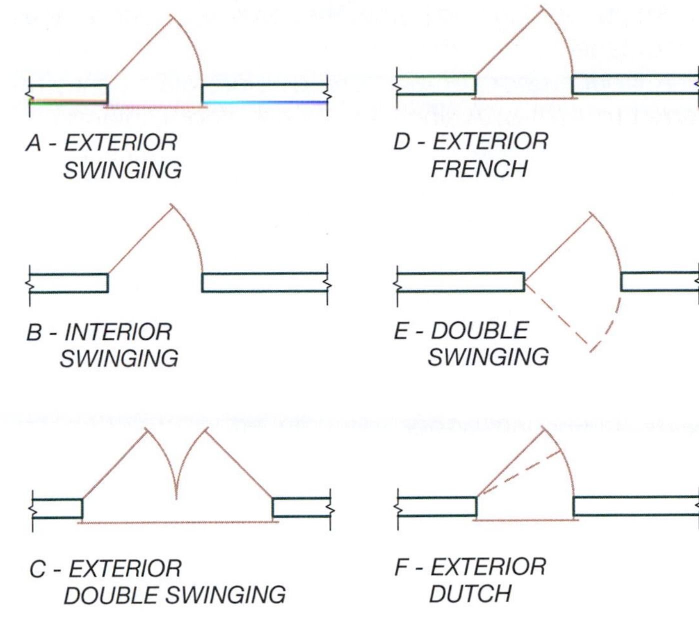

FIGURE 10.5 ■ *Common types of swinging doors include exterior, interior, double, French, double-acting, and Dutch doors.*

FIGURE 10.6 ■ *Double doors consists of two or more single swinging doors mounted in one frame.*

doors are used in one frame, they are usually installed as fixed panels.

A *French door* contains one or more glass panes referred to as lites. A single-lite French door contains one large piece of glass surrounded by a wood frame. A ten-lite door would contain ten glass panels divided by muntins between the glass panes. The most common French doors contain 1, 5, 10, or 15 lites. French doors can have multiple panes of glass or have one panel of glass with surface-mounted ***mullions.*** The surface-mounted ***muntins*** are removable to allow for easy cleaning.

Double-acting doors swing in either direction for easy passage. See Figure 10.5e. The access between kitchen and dining area is a common place for a double-acting door. *Dutch doors* (see Figure 10.5f) are divided in half, allowing one portion to be closed while the other is open. Typically the top portion can be opened and used as a pass-through while the lower portion is closed. Common door widths for single swinging doors range from 2'-0'' through 3'-6'' (600 to 1050 mm) in 2'' (50 mm) increments. Pairs of doors range from 2'-6'' through 12'-0'' (750 to 3600 mm) wide in 6'' (150 mm) increments. Doors are typically 6'-8'' (2000 mm) high, although 8'-0'' (2400 mm) doors are available.

A final type of swinging door that may be found on a floor plan is a *café door.* These doors can be full-height but are generally 36'' to 42'' (900 to 1050 mm) high and range from 28'' to 42'' (700 to 1050 mm) wide. Café doors are often used to block the view into a kitchen. Louvered and raised-panel patterns are normally used for café doors.

Exterior Swinging Doors

Notice, in Figure 10.5, that an exterior door has a thin line across the opening on its outside edge. This line represents the sill, which provides weatherproofing at the bottom of the door, and the step from the finish floor to the landing. (See Chapter 4 for a review of door/step relationships). The sill is commonly drawn projected about 1'' (25 mm) from the exterior side of the wall. Some companies draw the sill line flush with the wall, but this tends to make the symbol harder to find on a complicated plan. Common sizes for swinging doors include:

■ Main entry door	3'-0'' (900 mm) wide
■ Garage to utility room or outside	2'-8'' (800 mm) wide
■ Pairs of swinging doors	4' to 12' (1200 to 3600 mm) wide

Exterior doors should be placed so the door swings in and opens toward the common direction of travel. By having the door swing inward, the hinges are placed on the interior side of the door and avoids swinging the door over a step. Figure 10.7 shows common considerations in door placement. Exterior doors are made of solid wood components, fiberglass, or hollow metal with insulation. Doors may be either smooth (referred to as slab) or have decorative panels.

Interior Swinging Doors

Interior doors use the same symbol as exterior doors, but they have no sill line. The interior door symbol, as shown in Figure 10.5b, is drawn without a sill. Interior doors should swing into the room being entered and against a wall. Common interior door sizes are:

Utility rooms and garages	2'-8'' (800 mm)
Bedrooms, dens, family, and dining rooms	2'-8'' to 2'-6'' (800 to 750 mm)
Bathrooms	2'-6'' to 2'-4'' (750 to 700 mm)
Closets	2'-4'' to 2'-0'' (700 to 600 mm)
ADA access route	2'-8'' (800 mm)

When two sizes are listed, the first, larger size is normally used for custom homes and the smaller is the standard size. Although the added cost for a larger door is minimal, providing wider halls for access can be costly.

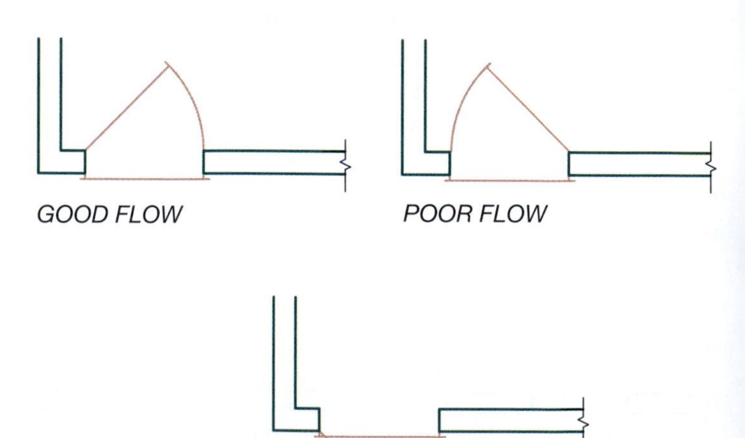

FIGURE 10.7 ■ *Exterior doors should be placed so that they swing in and open toward the common direction of travel. The inward swing places the hinges on the interior side of the door and avoids swinging the door over a step.*

The smaller door sizes are used in homes where space is critical. Keep in mind that the door size plus 6'' (150 mm) for trim [3'' (75 mm) each side] should be provided. Interior doors are usually 6'-8'' (2000 mm) high. Taller doors are available for custom situations. Interior doors are usually made of wood, fiberglass, or vinyl panels that have a hollow core with a smooth surface on each side. This type of door is referred to as a hollow-core flush door. Interior doors can also be purchased with raised panels or glass panels.

ADA Access

The Americans with Disabilities Act (ADA) specifies that all doors must have a minimum opening of 32'' (815 mm) for wheelchair access. This size is the clear, unobstructed dimension, measured to the edge of the door when opened at 90° (see Figure 10.8).

Nonswinging Doors

Common alternatives to swinging doors include pocket, slider, bipass, bifold, accordion, and overhead doors.

Pocket Doors

A *pocket door* is a door that slides into a wall cavity. This type of door is used when space for the door swing is limited. It is also used where a door may occasionally be desired for privacy, but the open position is the normal preference. A pocket door should not be placed in heavy-traffic areas, where the pocket is in an exterior wall, or where it would interfere with plumbing or electrical wiring. Pocket doors may be hollow-core flush, raised-panel, or louvered. They range in width from 2'-0'' through 3'-6'' (600 to 1050 mm) in 2'' (50 mm) increments. Figure 10.9 shows the representation for a pocket door.

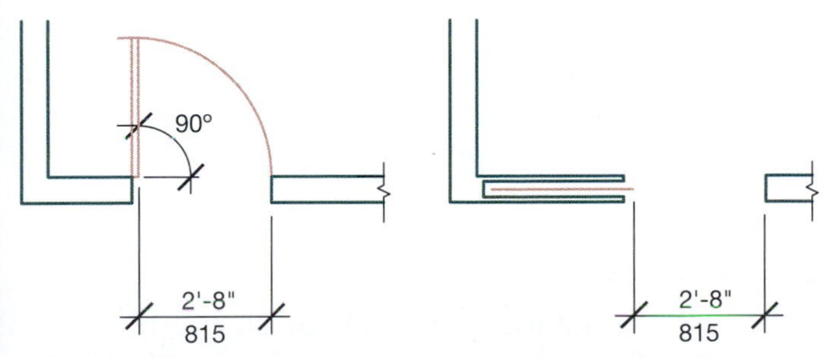

FIGURE 10.8 ■ *All doors for egress must have a minimum opening of 32'' (815 mm) for wheelchair access.*

FIGURE 10.9 ■ *A pocket door that slides into a wall cavity can be used when space for the door swing is limited.*

Sliding Doors

Exterior sliding doors are made with wood, vinyl, or aluminum frames that contain tempered glass panels. Figure 10.10 shows a sliding door and the floor plan symbols for representing exterior sliding doors. These doors are used to provide glass areas to meet code-mandated light and ventilation requirements and are excellent for access to outdoor living areas. Common sizes for exterior sliding doors are 6'-0'' and 8'-0'' (1800 and 2400 mm), but they range in width from 5'-0'' through 12'-0'' (1500 to 3600 mm) in 12'' (300 mm) increments.

Interior sliding doors are referred to as bipass doors. Bipass doors are often used for closets when complete access to the closet is required. Figure 10.11 shows the floor plan symbol for a bipass door. Because the doors are supported on rollers on either the top or bottom, no special framing is required at their edges. They are usually centered between walls, but the door opening can be flush with the closet edge. Bipass doors normally range in width from 4'-0'' (1200 mm) through 12'-0'' (3600 mm) in 1' (300 mm) increments. As doors increase in width, the number of panels also increase. Common panel arrangements include:

Up to 8'	(2400 mm)	two-panel door
8' to 10'	(2400 to 3000 mm)	three-panel door
10' and larger	(3000 mm+)	four-panel door

Folding Doors

Common interior folding doors include bifold and accordion doors. *Bifold doors* are always done in pairs, either with two doors folding to one side or with four doors split in the center of the opening, with two doors folding back to each side. Bifold doors are often used to provide full access to storage areas, but they can also be used to hide appliances such a washer or dryer and in some cases as folding doors between rooms. Bifold doors can be hollow-core slab, raised-panel, louvered, or French doors. Figure 10.12 shows the floor plan sym-

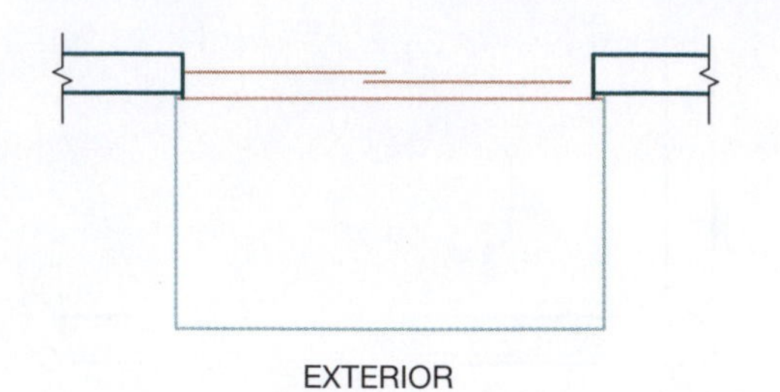

FIGURE 10.10 ■ *An exterior sliding glass door.* Courtesy Marvin Window & Doors.

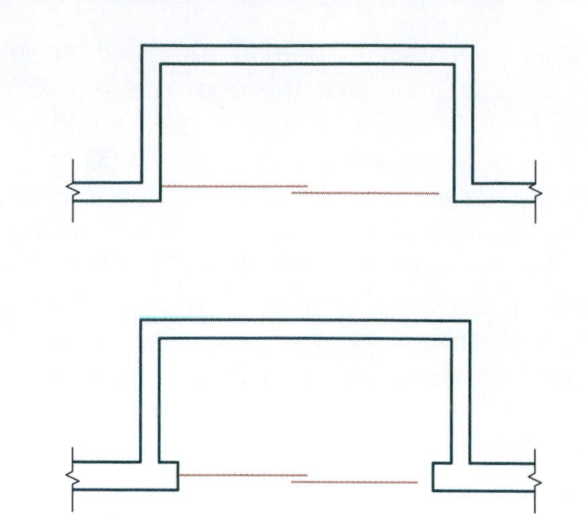

FIGURE 10.11 ■ *A bypass door is often used for a closet.*

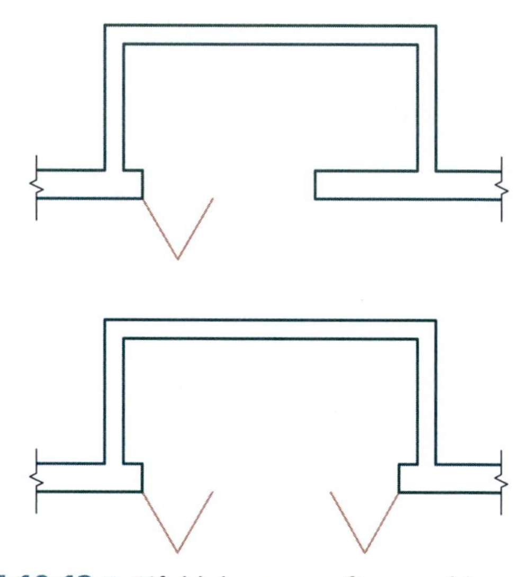

FIGURE 10.12 ■ *Bifold doors are often used to provide full access to storage areas, but they can also be used to hide appliances such as a washer or dryer.*

bols for representing folding doors. Because hinges support the doors, framing support is required at the edges. They are usually centered between walls, but the door opening can be flush with a closet edge if hinge support has been provided. Common widths range from 4'-0'' through 9'-0'' (1200 to 2700 mm) in 6'' (150 mm) increments.

Accordion doors are often used for closets or wardrobes, as a room divider, or as an acoustical barrier between living areas such as basement or family room recreation areas, or galley-style kitchen facilities. Accordion doors can also be used to provide separation between sleeping and living areas within an efficiency-style housing unit. Figure 10.13 shows the floor plan symbol for accordion doors. Accordion doors range in width from 4'-0'' through 15'-0'' (1200 to 4500 mm) in 1' (300 mm) increments, but most suppliers provide custom widths. Door panels are made of vinyl or wood veneer panels and are supported from a ceiling-mounted track.

Garage Doors

Access to a garage is typically by an overhead or a sectional roll-up door. The floor plan symbol for a garage door is shown in Figure 10.14. The dashed lines represent the size and extent of the garage door

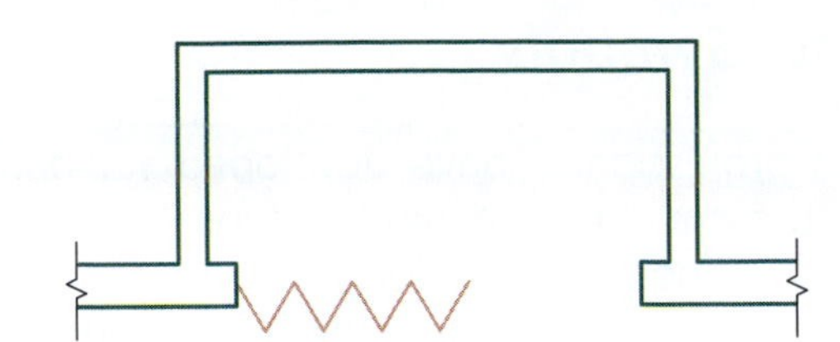

FIGURE 10.13 ■ *Accordion doors are often used at closets or wardrobes, as room dividers, or as acoustical barriers between living areas.*

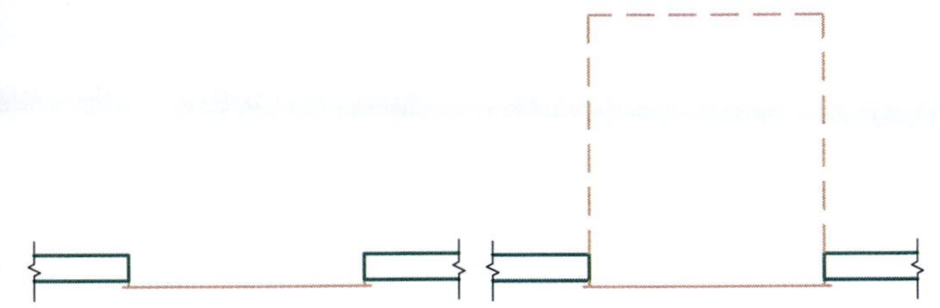

FIGURE 10.14 ■ *An overhead garage door can be simplified as shown with the left symbol, or dashed lines can be used to represent the size and extent of the garage door when the door interferes with something on the ceiling.*

when open. The extent of the garage door should be shown when the door interferes with something on the ceiling. Overhead doors range in width from 8'-0'' through 18'-0'' (2400 to 5400 mm). An 8'-0'' (2400 mm) door is a common width for a single car. A 9'-0'' (2700 mm) width is common for a single door that will accommodate a pickup truck or large van. A door 16'-0'' (4800 mm) wide is common for double-car access. Doors are 7'-0'' (2100 mm) high, although doors 8'-0'' (2400 mm), 10'-0'' (3000 mm), or 12'-0'' (3600 mm) high are common for campers or recreational vehicles.

Creating Door Blocks

Blocks to represent each type of door can be created and stored in a template drawing if third-party blocks are not available. Thin, continuous lines should be used to represent each type of door except pocket doors. Blocks should be created for each size of exterior and interior swinging door. Figure 10.15 shows the process for creating a block for a swinging door. Use the following steps to create an exterior door:

- Create the block on the 0 layer.
- Draw two lines to represent the wall thickness [6'' (150 mm)] for exterior walls. Use the OFFSET command to place the lines at the desired spacing.
- Draw two thick lines perpendicular to the wall lines to represent the door width. Use the TRIM command to place the lines exactly between the walls.
- Use the intersection of the interior wall line and one edge of the door as the center point for a circle (Figure 10.15c). Use the opposite edge of the door to determine the radius of the circle.
- Draw a line at 45° to represent the door swing (Figure 10.15d).
- Use the BREAK command to remove a portion of the circle so that only an arc remains from the wall to just past the door swing (Figure 10.15e).
- Remove the wall lines that cross the door opening.
- Add the line to represent the sill (Figure 10.15f).
- Save the symbol as a block with a name that will describe the size and style (DOOR-SWING-EXT-36).
- In creating the block, pick an insertion point that will be useful for inserting the block into the floor plan. Common insertion points include:
  - The intersection of the door edge and the door swing
  - A point 3'' (75 mm) from the hinge point
  - The midpoint of the door
- Insert the block on a layer with an appropriate title, such as *FLOR-DOOR.*

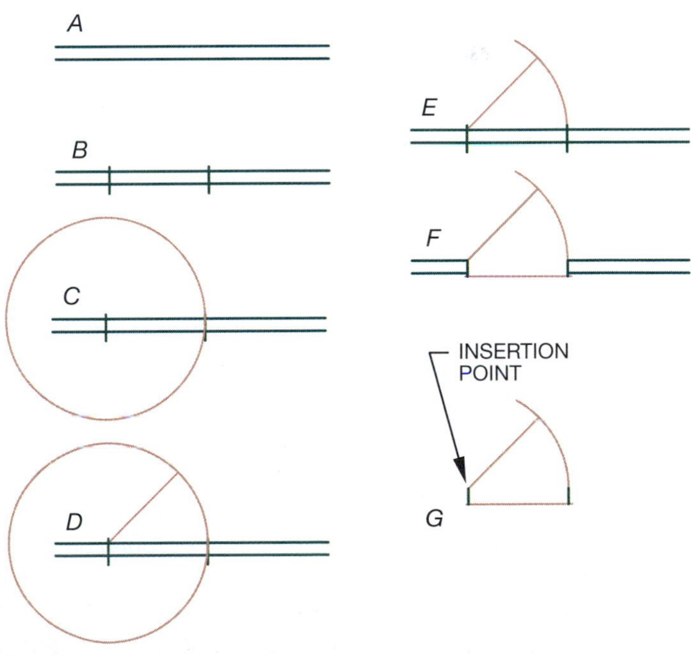

FIGURE 10.15 ■ *Blocks for doors can easily be created and stored for multiple use.*

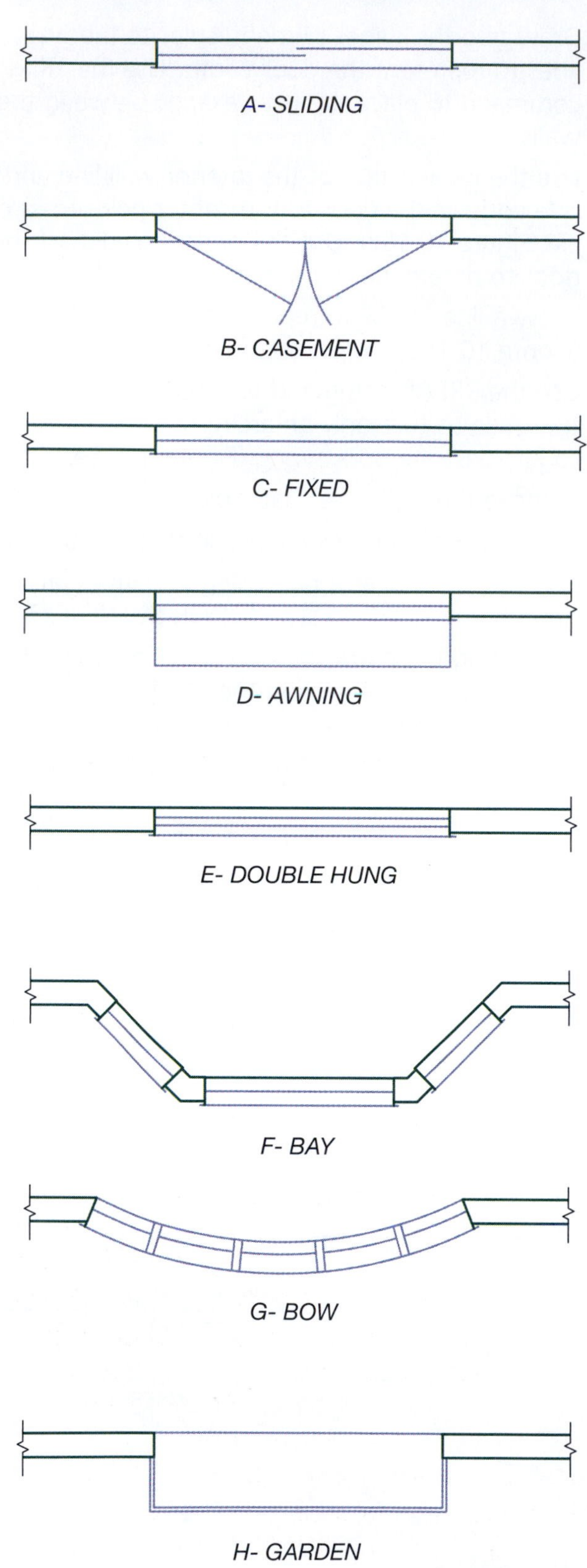

FIGURE 10.16 ■ *Common types of windows found on a floor plan.*

Window Symbols

Design professionals have developed standard symbols to represent common window types. AutoCAD and third-party vendors supply blocks to represent common styles, or they can easily be made and saved as part of your architectural template. Common types of windows found on a floor plan include sliding, picture, casement, single-hung, awning, bay, bow, garden, radius, and specialty windows. Each type of window is shown in Figure 10.16. In planning windows for a project, the designer must consider code requirements for light, ventilation and egress, the type of window to be used, the type of glass, and the size of the window. Chapter 4 discussed code requirements for windows. Only the size of the window is represented on the floor plan by the window symbol. The style, frame, and glass type are specified in a schedule. The creation of schedules is discussed later in this chapter.

Types of Windows

Sliding windows (Figure 10.16a) are popular because of their moderate price and the amount of ventilation they provide. A single sliding window is 50% openable. Windows wider than 6'-0'' (1800 mm) typically include a fixed glass panel with a sliding window on each end. Craftsman, Tudor, Mission, and various twentieth-century styles often have casement windows. *Casement windows* (Figure 10.16b) are 100% openable. Jamb hinges allow the window to open outward for maximum ventilation and egress. Single casement windows usually range in width from 18'' to 42'' (450 to 1050 mm), but most manufacturers will bind multiple units into one frame to create a larger window. A rectangular fixed panel of glass is referred to as a picture or fixed window. *Picture windows* (Figure 10.16c) do not provide ventilation but are used to maximize view exposure. Picture windows can be combined with sliding, casement, or awning windows to meet ventilation requirements. Picture windows usually come in widths ranging from 1'-6'' through 12'-0'' (450 to 3600 mm). Larger windows are available but are difficult to handle. *Awning windows* are hinged at the top and swing outward (Figure 10.16d). They are often placed below picture windows to provide ventilation when no egress is required. A *transom window* is hinged at the top and swings in. These windows are often used on custom homes above a door or another window. A fixed transom window may also be placed over an interior door.

Single- or double-hung windows (Figure 10.16e) replicate the look of the traditional double-hung window having a bottom panel that slides up and an upper panel that slides down. Single-hung windows have a lower panel that slides vertically, with a fixed upper panel. Double hung-tilt windows allow the lower portion of the window to slide vertically and both panels of the window to tilt inward for easy cleaning. Single and double-hung tilt windows typically range in width from 1'-6'' to 4'-0'' (450 to 1200 mm), but multiple units can be bound into one frame to create larger windows. *Bay windows* (Figure 10.16f) project beyond the exterior walls of the structure to increase the illusion of a large interior. Used when a traditional style is desired, they can be purchased as a unit or built by the framer to increase the floor space. When purchased as a unit, a picture window is centered between two fixed single-hung, or casement windows. Usually the side panels are placed at a 45° or 30° angle to the picture window. The projection of the bay from the outer wall is usually between 18'' and 24'' (450 to 600 mm). The total width of a bay is determined by the size of the center window. Bays constructed at the job site can have more than one center panel.

Bow windows (Figure 10.16g) project beyond the exterior wall of the structure to increase the illusion of a larger interior. Bow windows are usually constructed of four to six panels that may be fixed, casement, or single-hung. The windows are arranged to form an arc and typically range in width from 6 to 12' (1800 to 3600 mm). *Garden windows* (Figure 10.16h) project out from the exterior wall to provide an interior shelf at the base of the window. Garden windows are often used above a kitchen sink to provide extra space behind the sink. They usually project between 12'' and 18'' (300 to 450 mm) from the exterior wall and range in width from 24'' to 60'' (600 to 1500 mm). The large glass panel parallel to the wall is fixed with vertical sliding side panels. Depending on the manufacturer, either the side or the top panels open.

Radius windows provide a half-round arc to the top of the window. A radius can be added to a fixed window without creating a mullion to disturb the view, as seen in Figure 10.16b. A radius can also be added to a single-hung, double-hung, or casement window, but a mullion (a horizontal divider between glass panels) is created. Most window manufacturers provide a line of specialty windows. *Specialty windows* include fixed panels that are round, half-round, quarter-round, arched, multiple arched, gable 3-sided, gable 4-sided, gable-doghouse, hexagon, and octagon units in standard and custom sizes. Each can be seen in Figure 10.17. Additional window shapes that are used in trying to copy a specific historical era include:

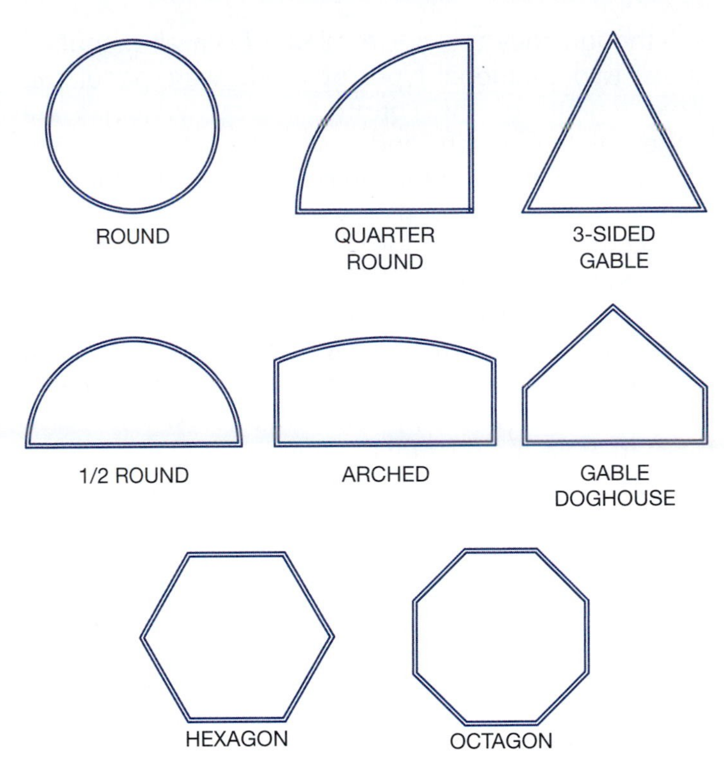

FIGURE 10.17 ■ *Common shapes of specialty windows stocked by most window manufacturers.*

- **Pointed arch:** Rooted in the tradition of medieval cathedrals, narrow windows with pointed arches are common on Victorian Gothic homes. Wider, squat Gothic arches are characteristic of Tudor homes.
- **Rounded arch:** Rounded or Roman arches date from Renaissance Italy. Modeled after ancient Greek and Roman forms, these windows feature a gently curved archway. They are often found on Italian Renaissance and Victorian Italianate homes.
- **Palladian:** A Palladian window is divided into three parts, with rectangular panes on each side of a wide arch. Placed at the center on an upper story, a Palladian window is an elegant focal point in Federal, Queen Anne, and Classical Revival homes.
- **Semicircular and oval:** Like rounded arches, half-circles and ovals are classically inspired. These accent windows are a hallmark of Victorian homes.
- **Triangular and trapezoid:** Angular shapes add drama to contemporary homes. A cathedral window forms a narrow triangle as it stretches across the room, following the line of a slanted roof.

Although they are not available from all manufacturers, two additional types of window are popular in some regions of the country. *Hopper windows* are hinged at the bottom and swing inward. This design allows the window to remain open with a minimum of water penetration. Hopper windows are often used in basements if egress is not required. *Jalousie windows* are made with horizontal vinyl, aluminum, wood, or glass blades that can be opened to allow ventilation. The blades overlap each other to form the panes of a jalousie window. Operated with a crank or turn-screw, the louvers tilt to open, permitting airflow. This is also their greatest disadvantage. They allow ventilation so well that they are almost impossible to seal. When closed, each louver rests against the one below it, rarely if ever making an airtight seal, and the hinges along the sides are almost impossible to seal without covering the entire window. These windows are not energy-efficient and because each pane is easily removed, they are a security risk. Some building codes no longer allow jalousie windows.

Window Locations

The placement of windows can be as important as their shapes. Common window locations based on historical home styles include:

- **Ribbon:** Common in Prairie, Craftsman, and twentieth-century homes, several ribbon windows are placed in a row with their frames abutting.
- **Five-ranked:** Georgian-style homes have five rectangular windows equally spaced across the second story.
- **Sidelights:** Neoclassical and Greek Revival homes often have tall, narrow sidelight windows flanking the entry door.
- **Fanlights:** Many classically styled homes have a semicircular fanlight above the entry door.
- **Projecting windows:** Bay, bow, and oriel windows are the most common types of projecting windows. Bay and oriel windows became popular during the Victorian era. Bay windows jut out from the side of the house. An oriel window projects from an upper story and is supported by decorative brackets.

Common Window Sizes Based on Use

Windows typically come in widths that range from 2' through 12' (600 to 3600 mm) at intervals of about 6'' (150 mm). The exact size will vary based on the manufacturer and the type of frame to be used. Vinyl, aluminum, and fiberglass window frames generally fall within the range of these nominal sizes, but custom sizes are readily available. Wood-frame window sizes are different for each manufacturer and should be confirmed with the manufacturer's specifications. The location of a window in the house and the way the window opens has an effect on the size. Common sizes include:

ROOM	WIDTH	DEPTH
Living, dining, and family rooms	6' to 12' (1800-3600 mm)	4'-5' (1200 to 1500 mm)
Bedrooms	3' to 6'-0'' (900 to 1800 mm)	3'-6'' to 4' (1050 to 1800 mm)
Kitchen	3' to 6'-0'' (900 to 1800 mm)	3' to 3'-6'' (900 to 1050 mm)
Bathrooms	2' to 3' (600 to 900 mm)	18'' to 36'' (450 to 900 mm)

Windows in the main living areas range from 6' to 12' (1800 to 3600 mm) wide and between 4' and 6' (1200 and1800 mm) tall allow the occupants to take advantage of a view while sitting. Windows in bedrooms are often smaller to allow for the placement of furniture. The type of window used in the bedroom is important because of the emergency egress requirements of the IRC. See Chapter 4 for a review of egress codes. The width of the kitchen windows will affect the placement of the upper cabinets. In a small kitchen, the tradeoff between natural light and sufficient cabinet space is critical. Generally placed behind the sink, kitchen windows range between 3' and 5' (900 and 1800 mm) in width and between 3'-0'' and 3'-6'' in height. Wide windows are nice to have in a kitchen for the added light they provide, but some of the upper cabinets will need to be eliminated as the width increases. Because the tops of windows are normally set at 6'-8'' (2000 mm), windows deeper than 42'' (1050 mm) will interfere with the countertop.

The size of bathroom windows will vary depending on the location in the room. When placed over a toilet or in a shower, the window often ranges between 2' and 3' (600 and 900 mm) in width. When the window is placed over a tub or spa, its width will usually closely match the width of the plumbing fixture. Windows installed in showers are usually set higher than the standard header height in order to minimize water buildup on the windowsill. Windows placed by a spa are required by the IRC to be tempered. Tempered glass is also required for all windows located within 18'' (450 mm) of a door or where the window could be used as a back rest at a window seat.

Window Considerations Affecting Green Design

Although the sizes just listed are common, consideration must be given to the relationship of the window to the sun. Increasing the size of south-facing windows will help to achieve direct gain from sun in winter but may cause the home to overheat during the summer. The size of non-south-facing windows and their effect on heat loss also needs to be considered. Balance the size of windows based on the view, orientation, roof overhangs, glass type, and landscaping. With proper planning, windows can provide energy by harnessing the sun in winter, breezes in summer, and natural daylight all year.

Glass Options

Windows are made with many options for glass. Although double glazing is common, single-pane windows are available for special conditions. Other common glazing options include "Low-E" and insulated glass to help control heat loss and obscure and tinted glass to control privacy.

Heating Considerations

Windows are rated for energy efficiency by R values and U values. R values indicate the energy efficiency of the window unit. The U value is the rate of heat flow through the window. Low-E glass is designed to increase the U factor of a window. The glass is coated with a thin, invisible, metallic layer several atoms thick, making it transparent to short-wave solar energy. This coating allows most of the solar spectrum, including visible light, to pass through. The glass is opaque to long-wave infrared energy, reflecting most heat energy. This allows interior heat to be retained in the winter and prevents radiated heat from outside objects from entering. One negative aspect of the Low-E coating is that there is a slight loss of solar contribution, but this loss is offset by its insulated value at night.

In addition to coating the panes of glass, thermal performance can be improved by reducing the conduction of heat in the air space between the glass layers. Manufacturers have introduced the use of argon and krypton gas in the sealed space between windowpanes. Argon is inexpensive, nontoxic, nonreactive, clear, and odorless.

Privacy Considerations

Two common methods to provide privacy are by the use of obscure and tinted glass. Obscure glass has a pattern placed in the glass designed to allow light to pass through but to maintain privacy. The glass can disrupt the view by using a pattern, color, and textures. Tinted glass uses shading to provide protection from direct sunlight and warm conditions.

Glazing Patterns

Although windows are usually made from a single sheet of glass, a variety of glazing patterns or windowpane arrangements are available.

- Windows set in wood frames may be set in six, nine, or twelve panes. Windows with many small square panes suggest a Colonial, Georgian, or Federal influence.
- Diamond-shaped panes are characteristic of Tudor, English Cottage, and some Mission-style homes.
- Leaded glass windows have panes secured with thin strips of lead; pieces of clear, frosted, beveled, or stained glass can be arranged in dazzling patterns. When leaded glass is to be used, it should be installed over a standard fixed glass panel to increase energy efficiency.

Representing Windows on the Floor Plan

A wide range of methods is used to represent windows on a floor plan. Many professionals represent windows using a thin line for the glass and thick lines to represent the surrounding walls, similar to Figure 10.18. Some professionals represent windows with

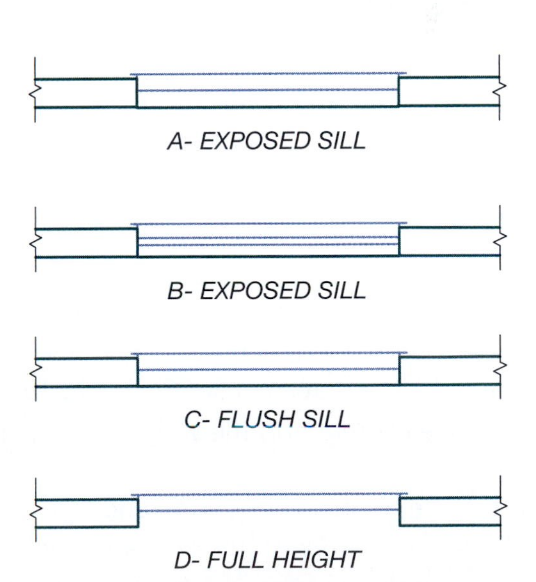

FIGURE 10.18 ■ *Common methods of representing windows on a floor plan. The symbols are used to represent any style of window, with the type specified in a note or schedule.*

two thin lines to represent the panes of glass. These symbols are used to represent any style of window. They are used on the floor plan since the window location must be represented, but the style may not be known when the preliminary and presentation drawings are started. A second consideration in drawing the window symbol is the representation of the sill. Most professionals draw windows, like exterior doors, with a projecting sill. The sill is usually projected about 1'' (25 mm) from the exterior side of the wall, although some companies draw it flush. On the interior side of the window, a thick line is used to represent the wall below the window. If the window extends to the floor, the line is omitted. Figure 10.16 shows alternative methods of representing various types of windows. Although these symbols provide a graphic representation of the window type to be used, the window type can be described clearly using a window schedule. The method used should be consistent throughout the plan and should be determined by the preference of the specific architectural office. Window symbols should be placed on the *FLOR GLAZ* layer.

Creating Window Blocks

Blocks to represent windows can be created and stored in a template drawing if third-party blocks are not available. Thin, continuous lines should be used to represent the glass and thick lines to represent the walls surrounding the window. Only one block needs to be created if the single-line symbol is used. The process for creating a window block is shown in Figure 10.19. Use the following steps to create a window block:

- Create the block on the 0 layer.
- Draw two lines to represent the wall thickness [6'' (150 mm)] for exterior walls. Use the OFFSET command to place the lines at the desired spacing.
- Draw two thick lines 12'' (300 mm) apart to represent the window width. Use the TRIM command to place the lines exactly between the walls. (The true window size will be determined when the block is inserted into the drawing.) See Figure 10.19a.
- Draw a line centered on the wall lines to represent the glass.
- Add the line to represent the sill and trim the exterior wall line between the window edges. See Figure 10.19b.
- Save the symbol as a block with a name that describes the size and style, such as WINDOW FLOOR. See Figure 10.19c.

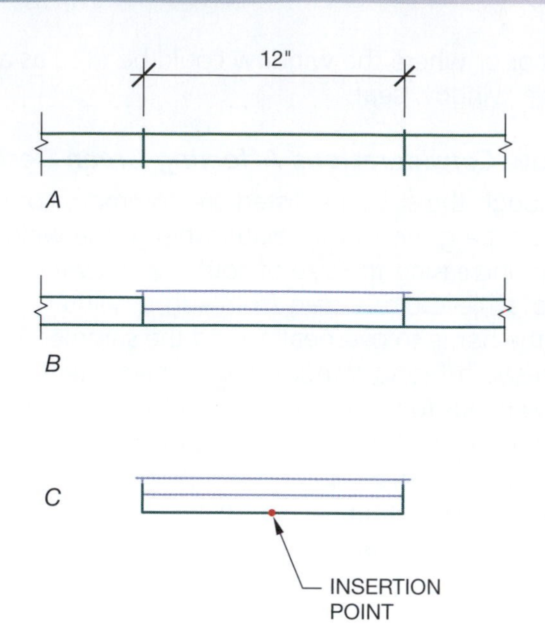

FIGURE 10.19 ■ *Creating a window block.*

- In creating the block, pick an insertion point that will be useful for inserting the block into the floor plan. The center of the wall line of the symbol is a convenient location. This will allow the midpoint of a wall to be used in inserting the block.
- Insert the block on a layer with an appropriate title, such as *FLOR GLAZ*.

Placing Blocks on the Floor Plan

Windows are usually centered in the wall of a room. When one window symbol is being inserted into a room, the center point of the block can be aligned with the center point of the interior wall, as seen in Figure 10.20a. If two or an even number of windows will be provided, the locations of the window blocks should be based on the center of the wall between the windows. See Figure 10.20b. A minimum of 3'' (75 mm) is required to represent the (2) 2× studs (vertical wall framing members) that must be provided to frame a window opening. (See Chapter 20 for a discussion of framing methods.) The center post or wall should be placed relative to the midpoint of the wall. If three or an odd number of windows will be provided, the window blocks should be located based on the middle of the center window relative to the center of the room. See Figure 10.20c. An alternative to having the framer place a post between windows is to have the manufacturer bind multiple windows together. When windows are combined, the vertical support between the windows is referred to as a mullion (Figure 10.20d). The individual panes of a window can be further divided by muntins (Figure 10.20e).

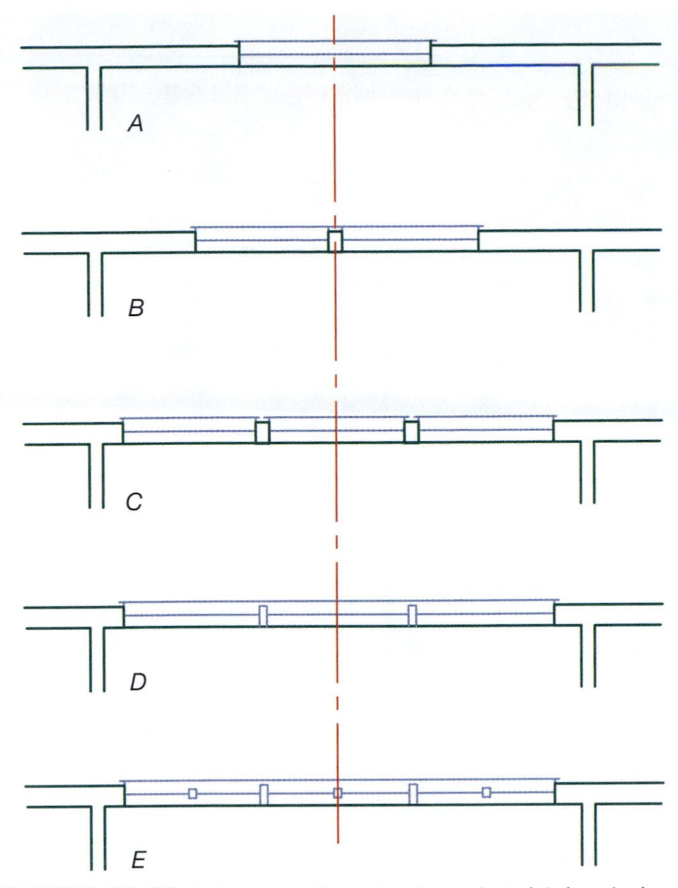

FIGURE 10.20 ■ *Representing single and multiple windows on a floor plan. Examples A, B, and C represent individual openings placed by the framer. Windows D and E represent multiple windows joined together by the manufacturer. Only single openings would be required for windows D and E.*

Roof Openings

Sky windows, skylights, and sun tunnels are common methods of delivering light through the roof and ceiling. Each can be seen in Figure 10.21a. ***Skylights*** are openings in the roof used to bring additional daylight into a room and let natural light enter an interior room. Skylights are available in fixed and openable units. They are made of double-domed plastic or flat tempered glass. Skylights are approximately 24'' (600 mm) wide. The length is usually a minimum of 24'' (600 mm) long. Common sizes include 2' × 2', 2' × 3', and 2' × 4' (600 × 600, 600 × 900, and 600 × 1200 mm). Exact sizes vary based on the manufacturer. A ***Sky window*** combines features of a skylight and a window, using glazing on a wall that extends to meet glazing on a portion of the roof. An alternative to a skylight is a ***sun tunnel,*** which passes light into a room through a reflective tunnel that connects the roof opening to the opening in the ceiling. Unlike the chase connecting the skylight to the ceiling, the tunnel is flexible, allowing sharp bends that will not disrupt the amount of light being delivered. A diffuser is mounted at the ceiling end of the tunnel to disperse the light.

Figure 10.21b shows the representation of a skylight on the floor plan. Place the skylights on a layer with a title such as *FLOR OVHD* (overhead). Skylights will also be represented on the roof plan. If possible, locate the skylight so that it fits between the roof trusses. If the roof is framed with rafters, skylights can be placed based on design and the roof framing can easily be altered to meet the design. Up to four separate skylights can usually be bound together to create one consecutive unit.

REPRESENTING CABINETS, APPLIANCES, AND PLUMBING

Cabinets are found in the living, sleeping, and services areas of a home. Specialized cabinets such as a desk, hutch, bookcase, entertainment unit, window seat, or built-in dresser are found in bedrooms, family rooms, dens, dining rooms, home offices, and garages. In general, the cabinets drawn on the floor plans are built-in units. To complete a floor plan, the cabinets the contractor will provide must be represented. The appliances and plumbing fixtures associated with these cabinets should also be placed on the floor plan when the cabinets are drawn.

Cabinets, appliances, and plumbing features are each represented by thin lines. The linetype used will vary with office practice. Common linetypes include:

- Continuous lines: Lower cabinets, the outline of the upper cabinets, changes in cabinet height, plumbing fixtures, the front of the dishwasher, and closet shelves
- Hidden lines: Appliances that are under the counter, the refrigerator, the furnace, water heater, limits of knee space under counters, washers, dryers, fold-down ironing boards
- Centerlines: Rods in closets

The use of linetypes varies greatly with each office. Figure 10.22 shows common options. The objects drawn with continuous lines are fairly consistent throughout the industry. The upper cabinets may be drawn using continuous lines, but they are sometimes drawn with hidden lines because they are above the cutting plane. Other

FIGURE 10.21a ■ *Skylights, sky windows, and sun tunnels all can be used to provide light and ventilation to habitable rooms.* Courtesy VELUX America Inc.

items, such as a dishwasher or trash compactor, are drawn with hidden lines because they are below the counter. Items such as the refrigerator, washer, and dryer are drawn with hidden lines because the contractor does not provide them. A final group of items, such as the water heater and furnace, may be drawn with hidden lines because of office practice. Your boss or instructor will be the final word on which method to use.

Base cabinets are usually drawn 24'' (600 mm) deep and upper cabinets are 12'' (300 mm) deep. Custom

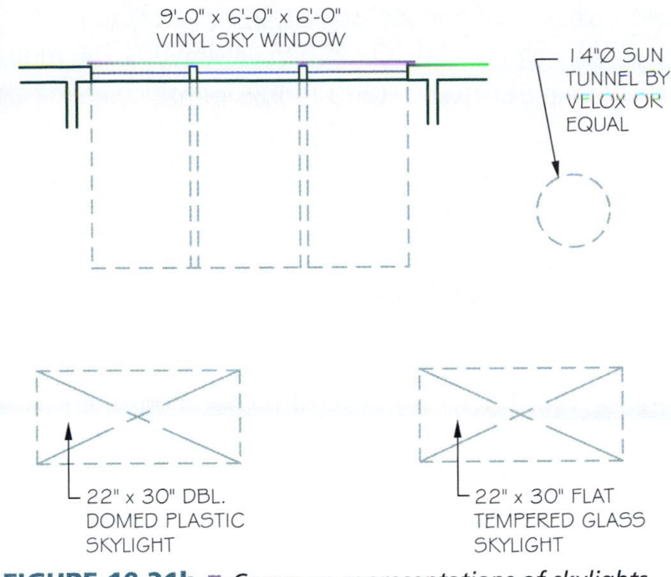

FIGURE 10.21b ■ *Common representations of skylights, sky windows, and sun tunnels on the floor plan. Some professionals use dashed lines to represent the skylights.*

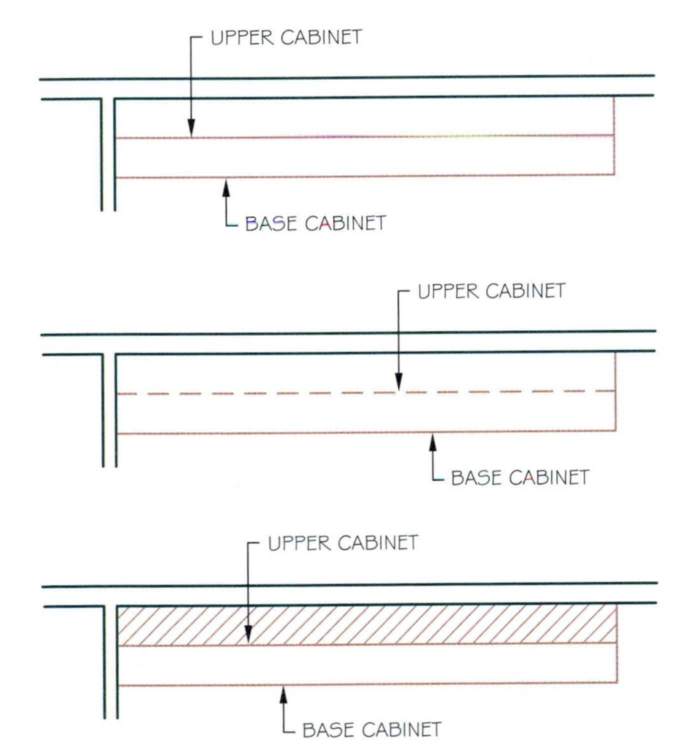

FIGURE 10.22 ■ *Common methods for representing upper and base cabinets.*

base cabinets may be 27'' and 30'' (675 and 750 mm) deep. Base cabinets with varied depths are often used on custom homes to create the illusion of built-in furniture. A width of 12'', 15'', or 18'' (300, 375, or 450 mm) is added to the counter depth if a food bar is to be added. Place the cabinets on a layer with a title such as

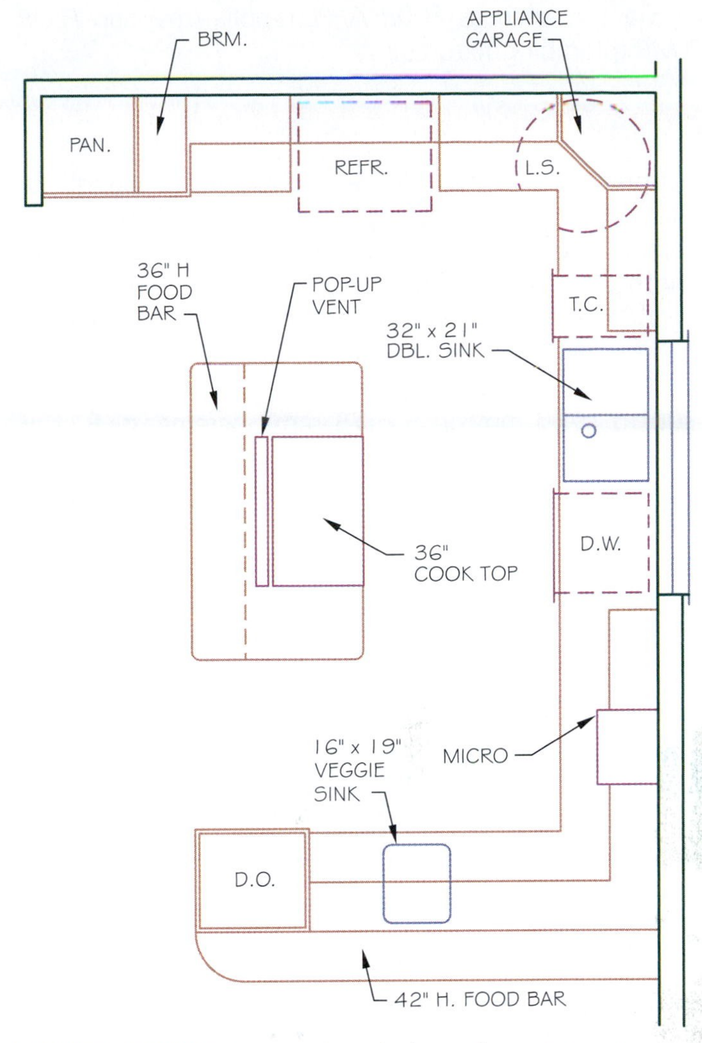

FIGURE 10.23 ■ *Standard symbols used to represent kitchen cabinets, fixtures, and plumbing features.*

FLOR CASE (casework-manufactured cabinets) or *FLOR WDWK* (field-built architectural woodwork).

Kitchens

In addition to representing the cabinets, fixtures such as sinks, butcher-block cutting boards, countertops, appliances, and plumbing fixtures must be located. Common appliances that must be represented include the range, refrigerator, dishwasher, trash compactor, and garbage disposal. Plumbing fixtures such as the main and vegetable sink must be located, as well as water to the refrigerator and possibly a faucet in the wall above the range or cooktop. Figure 10.23 shows common kitchen features. Appliances and plumbing fixtures can be placed on the same layer as the cabinets or on a layer with a title such as *FLOR APPL* (appliances) and *FLOR PLMB* (plumbing fixtures).

Cabinets

If prefabricated cabinets are to be used, the length of each cabinet should be represented in 3'' (75 mm) increments. Custom units can be built to any desired length. Pantries, broom closets, and other cabinets that extend from floor to ceiling are drawn with two parallel lines spaced approximately 1'' (25 mm) apart. Common widths of prefabricated pantries range from 12'' to 48'' (300 to 1200 mm) in 3'' (75 mm) increments. Broom closets generally range from 12'' to 24'' (300 to 600 mm) wide × 24'' (600 mm) deep. A pantry may be designed to be part broom closet and part shelves for storage. If a desk unit is to be represented, the counter should be lowered to a height of approximately 32'' (800 mm). A space between 30'' to 48'' (750 to 1200 mm) wide should be provided. Chapter 19 gives additional information related to cabinet drawings.

Plumbing

Common kitchen plumbing fixtures usually include the main sink, a vegetable sink, and a bar sink. Each can be placed on the *FLOR PLMB* layer. The use and location of each type of sink was introduced in Chapter 5 as work areas in the kitchen were explored. Figure 10.24 shows common kitchen sink symbols. Common kitchen sink options vary by manufacturer but include:

- Single: 19'' × 25'', 19'' × 30'', and 21'' × 24'' (475 × 625, 475 × 750, and 525 × 600 mm)

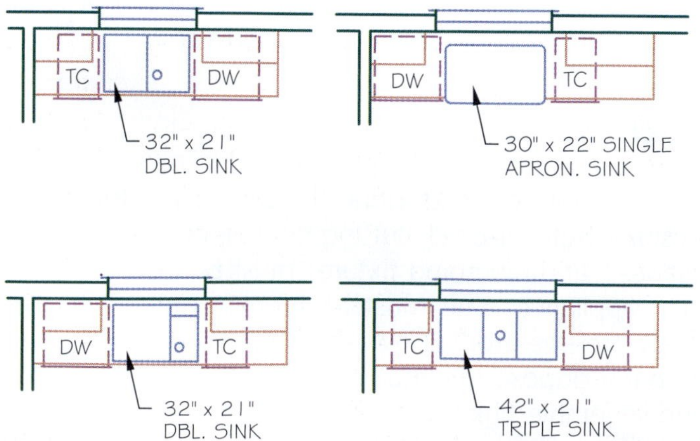

FIGURE 10.24 ■ *Standard kitchen sink symbols. Symbols for a dishwasher and trash compactor have also been added, although the compactor is optional.*

- Single apron front: 22'' × 22'', 22'' × 25'', and 22'' × 30'' (550 × 550, 550 × 625, and 550 × 750 mm)
- Double: 32'' × 21'' (800 × 525 mm)
- Triple: 42'' × 21'' (1050 × 525 mm)
- Vegetable sink and/or bar sink: 16'' × 16'' or 16'' × 21'' (400 × 400 or 400 × 525 mm)

A 1/4'' water line is usually provided to the refrigerator to supply water for an icemaker and for a door-mounted chilled water dispenser. In some custom kitchens, a wall-mounted faucet may be provided in the wall behind the range or cooktop to aid in food preparation. Both features can be represented by a note and usually require no special symbol, as shown in Figure 10.25.

Appliances

Common appliances found in a kitchen include the refrigerator, cooking units, dishwasher, and trash compactor. Each can be represented on the *FLOR APPL* layer. The refrigerator can be either freestanding or built in and may or may not contain the freezer. Figure 10.25 shows common kitchen appliances. Common widths of refrigerators include:

- Freestanding refrigerator: side by side, 36'' (900 mm) typical; 28'' to 42'' (700 to 1050 mm) available; 27'' (675 mm) deep
- Built-in refrigerator: 42'' (1050 mm) common; 36'' to 54'' (900 to1250 mm) available; 27'' (675 mm) deep

In addition to the refrigerator, a custom kitchen may include an icemaker and a small under-counter cooling unit called a wine cellar.

- Freestanding and built-in icemakers are typically 15'' or 18'' (375 or 450 mm) wide.
- Built-in wine cellars are 24'' (600 mm) wide.

Several options are available for cooking units, including a range or cooktop and oven. A range (Figure 10.25) contains the cooking elements over the oven. It can be either gas, electric, or dual-fuel and can be either freestanding or drop-in. The most common size is 30'' × 26'' (750 × 650 mm). Other common range sizes include:

- 36'' × 26'' (900 × 650 mm)
- 48'' × 27'' (1200 × 675 mm)
- 60'' × 28'' (1800 × 700 mm)

A standard cooktop (Figure 10.25) contains gas or electric cooking elements. Larger units can include simmer plates, grills, griddles, and a rotisserie. A standard

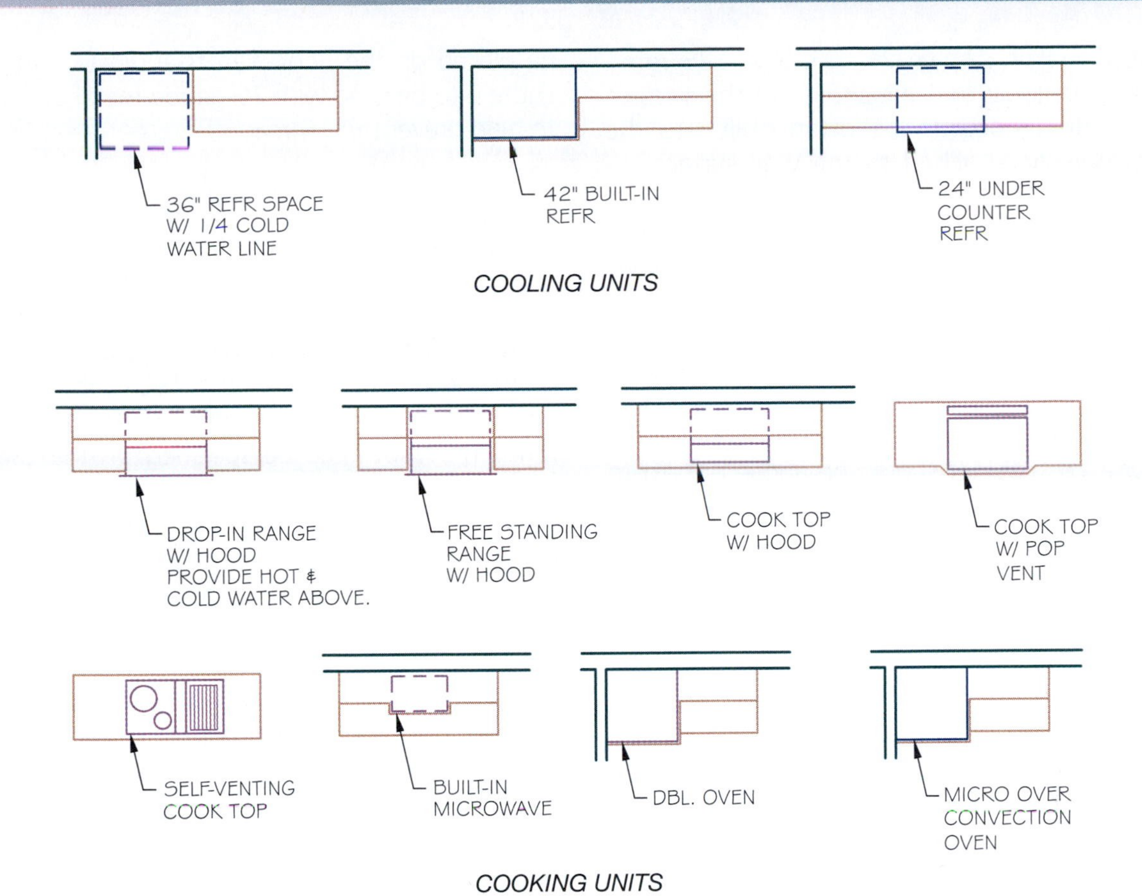

FIGURE 10.25 ■ *Representing standard kitchen cooling and cooking appliances.*

cooktop is 30'' × 21'' (750 × 525 mm), but sizes range from 15'' to 48'' (375 to 1200 mm) wide. A built-in oven (Figure 10.25) may contain one or more baking units. Cooking units can include a standard baking unit, convection oven, microwave, and warming oven. A standard built-in oven is 27'' (675 mm) wide and 24'' deep. Common width options include 24'', 30'', and 36'' (600, 750, and 900 mm).

Some method of ventilation must be provided for a range and a cooktop. Venting typically consists of a hood, a popup vent, or a self-contained vent. No symbol is required for a self-contained vent, but it should be specified on the floor plan. Figure 10.25 shows symbols for a cooktop hood and a popup vent. The width of each type of vent usually matches the width of the cooking unit.

Other common appliances found in a kitchen include the dishwasher, microwave, trash compactor, and range hood. A dishwasher (Figure 10.24) is usually placed beside the sink. A standard dishwasher is 24'' (600 mm) wide and is placed about 3'' (75 mm) from the edge of the sink. It should not be more than 24'' (600 mm) from the sink. Trash compactors can be built-in or freestanding. The symbol for a trash compactor resembles that of a dishwasher. The common width of a trash compactor is 15'' (375 mm). The microwave (Figure 10.25) can be freestanding or built-in. Common sizes for countertop units vary widely but are usually about 22'' × 20'' (550 × 500 mm). Built-in units are generally 27'' or 30'' (675 or 750 mm) wide with a cabinet depth of 15''.

Utility Rooms

The utility room may have upper, lower, and full-height cabinets and can contain several plumbing fixtures and appliances. The upper and lower cabinets should be drawn using the same layer and linetypes used to represent the kitchen. The symbols for the clothes washer, dryer, laundry sink, and fold-down ironing board are shown in Figure 10.26. The symbols for the clothes washer and dryer are drawn with dashed lines, since they are not part of the construction contract. Sizes vary widely, but a size of 28'' (700 mm) square can generally be used to represent the washer and dryer symbols. Stacked units are also available with a typical width of 27'' (675 mm). Vertical dryers are also available that are 36'' (900) wide × 29'' (725 mm) deep × 74'' (1850 mm) high. Laundry sinks are usually drawn using a symbol that is 21'' × 21'' (525 × 525 mm). Larger sizes are available that are similar to the sizes of kitchen sinks. Ironing boards are often built into the laundry room wall or attached to the wall surface, as shown in Figure 10.26. Drawer mounted units are also available. Each type of unit typically ranges in width from 12'' to 15'' (300 to 375 mm). Laundry utilities may be placed in a closet when only minimum space is available, as shown in Figure 10.27.

When a laundry room is below the bedroom area, a laundry chute is often provided from a convenient area near the bedrooms through a ceiling and into a cabinet in the utility room. The cabinet should be above or next to the washing machine. Figure 10.28 shows how a laundry chute can be represented on the floor plan. A note should be placed on the floor plan to indicate that the laundry chute is to be lined with 26-gauge metal or gypsum board to help prevent the spread of fire between floors.

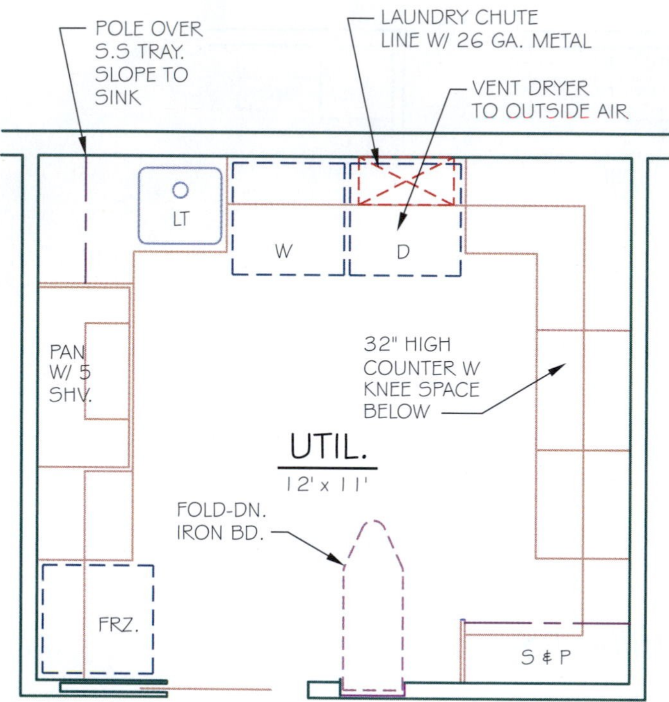

FIGURE 10.26 ■ *Representing appliances and plumbing fixtures of a laundry room.*

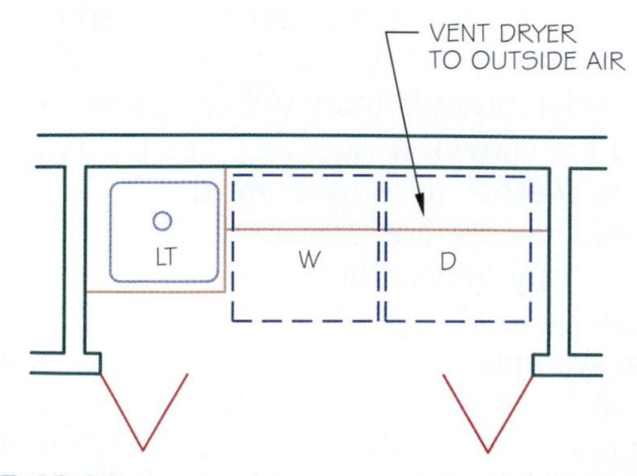

FIGURE 10.27 *Representing laundry facilities in a closet.*

Bathrooms

Common bathroom cabinets and fixtures are shown in several typical floor plan layouts in Figure 10.29. If a bathroom counter or vanity is to be provided, it is represented on the floor plan like any other base cabinet. The counter depth generally ranges from 22'' to 24'' (550 to 600 mm). The width can be specified on the cabinet elevations or in a general note if no interior elevations will be provided. The counter can be eliminated if the lavatory (sink) is wall-mounted or freestanding. If a counter is provided, 30'' (900 mm) is the smallest length that should be used, with a counter length of 36'' to 42'' (900 to 1050 mm) provided for each user. A minimum counter space of 9'' (225 mm) should be provided on the wall side of the lavatory and 12'' (300 mm) between lavatories. If a makeup counter is to be represented, a minimum space 30'' (750 mm) wide and 30'' (750 mm) high is usually specified. See Chapter 4 for additional ADA requirements for vanity sizes. Represent the cabinets and changes in counters with thin lines on the *FLOR CASE* layer.

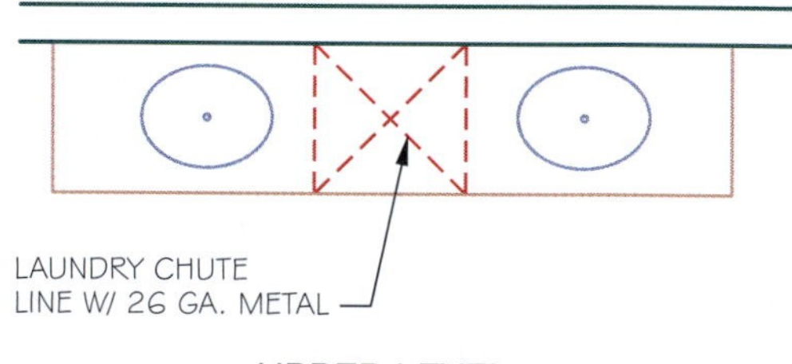

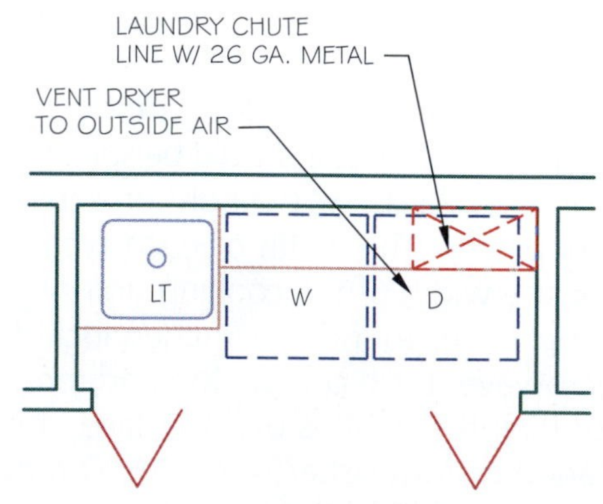

FIGURE 10.28 ■ *Representing a laundry chute on floor plans. Although the upper and lower levels of the chute do not need to match, two surfaces of the chute should align.*

Other standard features in a bathroom include multiple lavatories, tub, shower, a combination tub and shower, spa, water closet, and bidet. Each or these items can be represented with thin lines on the *FLOR PLMB* layer. The bathroom sink or lavatory is usually oval, but rectangular, round, and square fixtures are also available. Common sink sizes include:

ADA wall-hung: 20'' × 27'' (500 × 675 mm)
Oval: 19'' × 16'' and 20'' × 17'' (475 × 400 mm and 500 × 425 mm)
Pedestal: 22'' × 18'' and 26'' × 20'' (550 × 450 mm and 625 × 500 mm)
Round: 19'' (475 mm) diameter

Tubs

Tubs can be freestanding or built in and made of fiberglass, cast iron, or ceramic tile. Within each category of material and style there are a wide variety of options. Sizes vary with manufacturers based on the style and the material used to make the tub. Lengths range from 4' to 7' (1200 to 2100 mm) and widths from 32'' to 48'' (800 to 1200 mm). Common tub sizes include:

Standard fiberglass: 60'' × 32'' (1500 × 800 mm)
Built-in corner: 60'' × 60'' (1500 × 1500 mm)
Freestanding: 72'' × 38'' (1800 × 950 mm)
Oval: 66'' × 36'' (1650 × 900 mm)

Tub sizes also vary if jets are placed in the unit. Common jetted tub sizes include 60'' × 42'' and 72'' × 48'' (1500 × 1050 mm and 1800 × 1200 mm).

Showers

A shower is often added over a tub unit for a bathroom that will be used by children. This type of unit is specified as a tub/shower on the floor plan. The shower in a master bathroom suite is often separate from the tub or bathing fixture. Shower sizes vary based on the manufacturer, style, shape, and material to be used.

- The standard fiberglass tub shower unit is 60'' × 32'' (1500 × 800 mm), which can be purchased as a square, rectangular, or neo-angle type if it is separate from the tub fixture

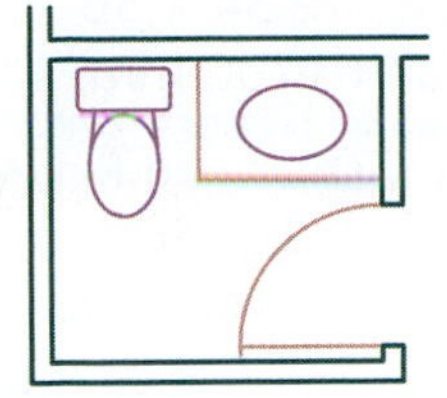
POWDER ROOM

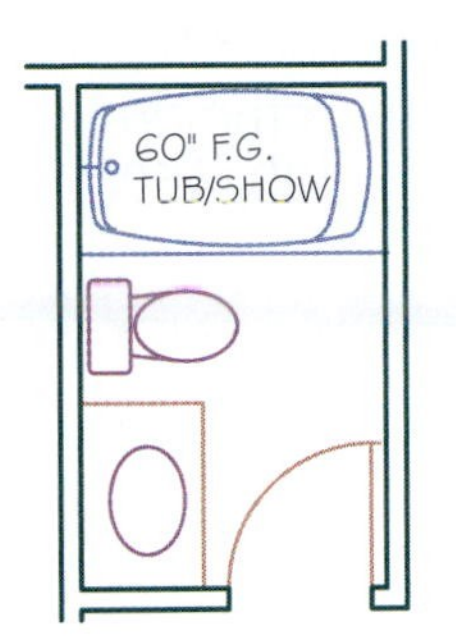

SMALL BATH WITH TUB/SHOWER COMBINATION

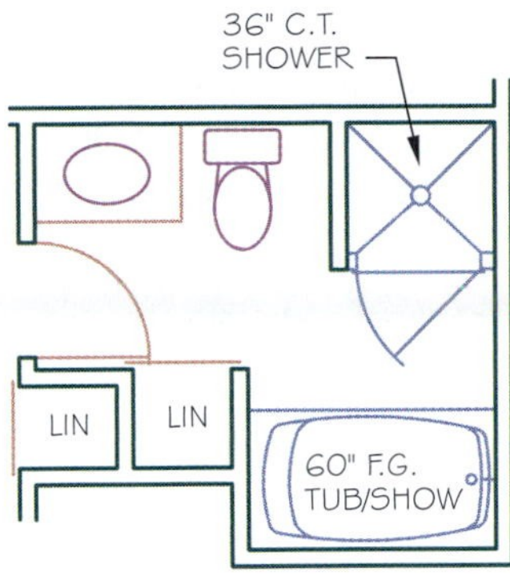

BATH WITH TUB & SHOWER

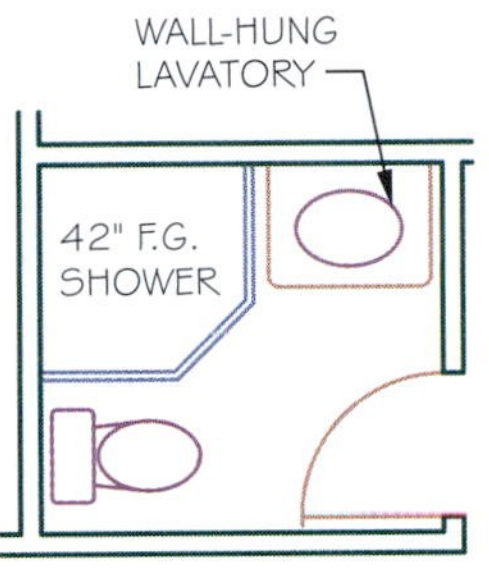

SMALL BATH WITH NEO-ANGLE SHOWER

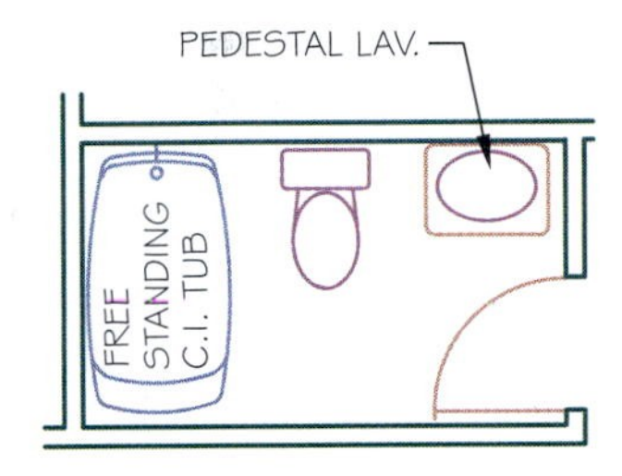

BATH WITH FREE STANDING TUB

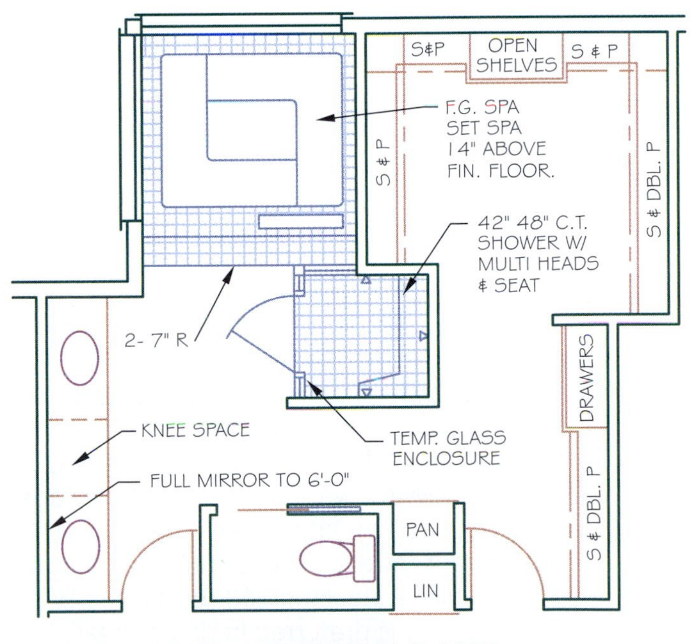

BATH WITH WALK-IN CLOSET

FIGURE 10.29 ■ *Common arrangements of cabinets and plumbing fixtures for small bathrooms. Options for the master suite are endless but typically provide private space for each area of the bathroom.*

- Common sizes for square showers are 34'', 36'', 42'', and 48'' (850, 900, 1050, and 1200 mm)

An example of a neo-angle shower is shown in Figure 10.29. Two surfaces are mounted to the walls in a corner, and the other three sides are freestanding. Neo-angle showers are specified by their overall size and by the size of the opening.

- Common widths for neo-angle units are 34'', 36'', and 42'' (850, 900, and 1050 mm). Common door widths are 16'', 18'', and 23'' (400, 450, and 575 mm).

Custom shower units can also be made of ceramic tile. These showers are made by craftsmen at the job site and may be any shape and size.

Spas

A homeowner often uses the term *spa* when he or she really means a jetted tub. Common sizes for jetted tubs have been introduced earlier. A true spa is typically placed outside the residence but may be placed inside a master bedroom or master bath suite. See Figure 10.29 for common symbols used on a floor plan. The size of a spa is based on the number of seats to be provided. Common seating arrangements and sizes for built-in fiberglass spas include:

- Three-person size: 76'' × 66'' (1900 × 1650 mm)
- Four-person size: 84'' × 76'' (2100 × 1900 mm)
- Five-person size: 84'' × 84'' (2100 × 2100 mm)
- Six-person size: 91'' × 84'' (2275 × 2100 mm)
- Seven-person size: 94'' × 94'' (2350 × 2350 mm)

Spas can also be freestanding and portable. Verify the seating capacity and the location of the pump access prior to drawing a spa on the floor plan. If the spa is to be placed inside the home or on a deck, the size and capacity will also need to be considered when the structural drawings are completed. See Chapters 22 and 23 for structural considerations.

Closets, Wardrobes, and Other Storage

Closets are used in each of the three main areas of the residence to provide various types of storage. Building codes do not regulate closet sizes, but FHA and HUD regulations affect their depth and length. The common depth of a closet is 24'' (600 mm), but a depth of 30'' (750 mm) is preferred in damp areas. The added space allows air to circulate around damp clothes. In the living area, a closet with a shelf and pole should be provided near the entry to store coats for the family and guests. Figure 10.30 shows two common methods of drawing the shelf and pole. Some professionals use two dashed lines to symbolize the shelf and pole. The shelf is typically placed at 72'' (2100 mm), with the rod placed directly below it. Represent the shelves and other storage material with thin lines on a layer with a title such as *FLOR FIXT* or *FLOR WKWD* (woodwork).

Service Area Storage

In the service area, storage should be provided in the kitchen, utility room, and garage. Storage for the kitchen and utility rooms was discussed as cabinets were explored. In addition to the pantry provided with the cabinetry, many custom homes have walk-in pantries for long-term storage (see Chapter 5). This storage may consist of base and upper cabinets that match the kitchen cabinets, exposed shelves, or a com-

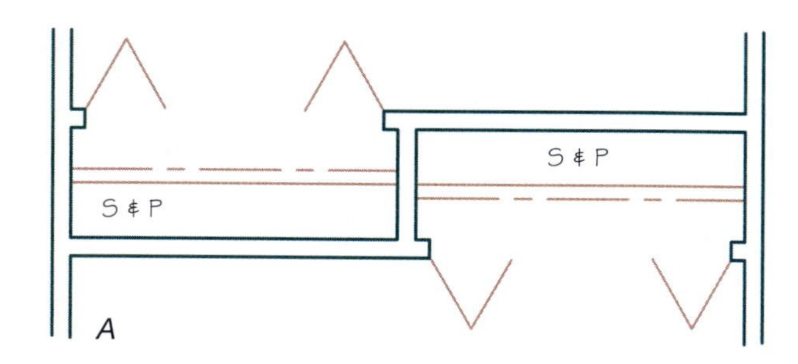

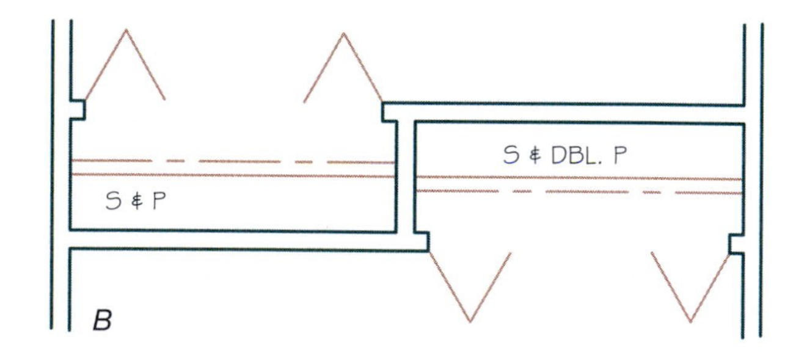

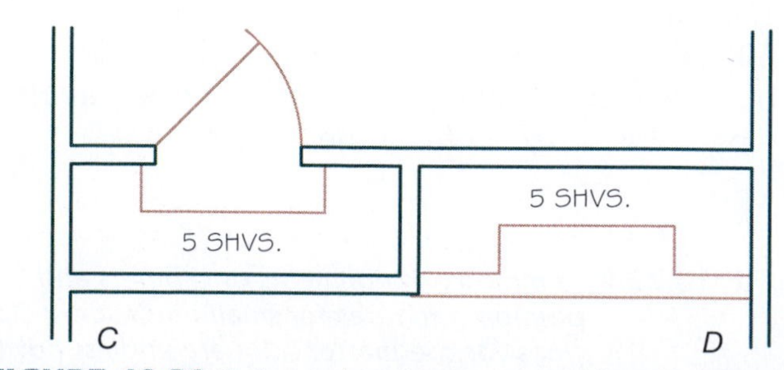

FIGURE 10.30 ■ *Several options are used by professional designers to represent shelves and poles.*

bination of shelves over enclosed cabinets. Shelves ranging in depth from 9'' to 24'' (225 to 600 mm) are common. A space of 36'' to 42'' (900 to 1050 mm) between banks of shelves is desirable, but a distance of 30'' (750 mm) is tolerable if no space for pull-out units is required. Figure 10.30c and d show how to represent shelves on the floor plan.

Sleeping Area Storage

Three types of storage are common in the sleeping area of a home, including a linen closet, a bedroom wardrobe closet, and walk-in closets. A storage closet with either five or six shelves with an approximate depth of 24'' (600 mm) should be provided for bath and bed linens. A linen closet can be represented as shown in Figure 10.30c and d. Each bedroom should also be provided with a wardrobe closet unless it will include an armoire. The FHA recommends a length of 48'' (1200 mm) of space for males and 72'' (1800 mm) for females. The practical minimum for resale should be 6' (1800 mm) if a traditional shelf-and-pole storage system is to be used. Space can be reduced if closet organizers are used. The minimum closet depth is 24'' (600 mm). A 30'' (750 mm) depth is desirable to keep clothing from becoming wrinkled. If a closet must be small, a double-pole system can be used to double the storage space or a premanufactured closet organizer can be used to provide storage (see Figure10.30b).

Master bedrooms often have walk-in closets similar to that in Figure 10.31a. It should be a minimum of 6' × 6' (1800 × 1800 mm) to provide adequate space for clothes storage. A closet of 6' × 8' (1800 × 2400 mm) provides better access to all clothes. Providing multiple rods allows short clothes to be hung above each other. An area with a single pole should be provided to hang dresses and seasonal coats. An area containing shelves, baskets, and drawers is also desirable if space permits. Closet packages are available from companies that customize the wardrobe closet to meet the specific needs of family members. Figure 10.31b shows a well-planned wardrobe area for a master suite.

REPRESENTING STEPS, STAIRWAYS, AND RAILINGS

Steps are required even on a single-level home. Steps or a landing are required at each exterior door to provide a surface for the home's entry. A minimum width of 36'' (900 mm) should be provided for a landing at each exterior door. The step from the landing to walkways or the ground cannot exceed 7 3/4'' (195 mm). These steps can be represented and specified as shown in Figure 10.32. A second area where steps are required is where floor levels change elevation. Figure 10.33 shows

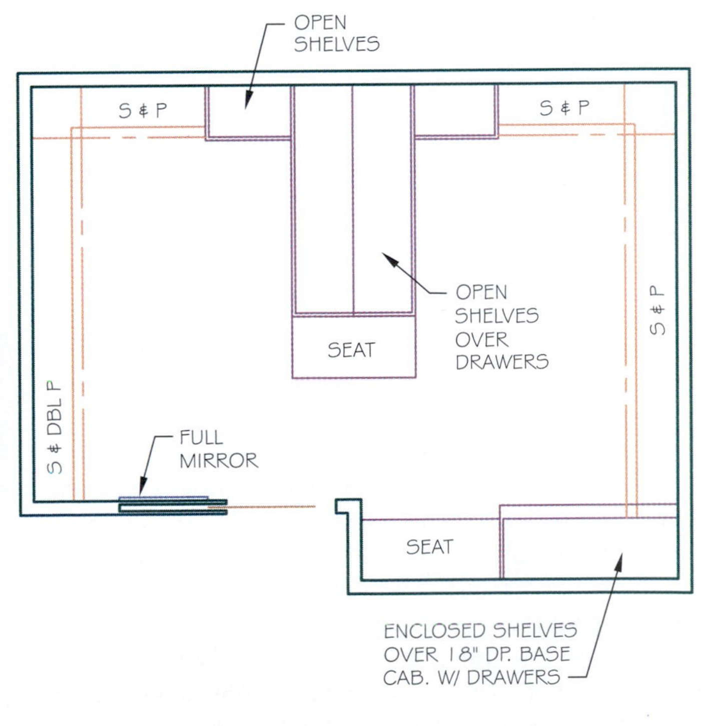

FIGURE 10.31 ■ *The closet of a master bedroom suite will typically comprise multiple types of storage, including the traditional shelf and pole, double poles, shelves, baskets, and drawers. A full-length mirror and seating should also be provided. Each should be clearly labeled on the floor plan.*

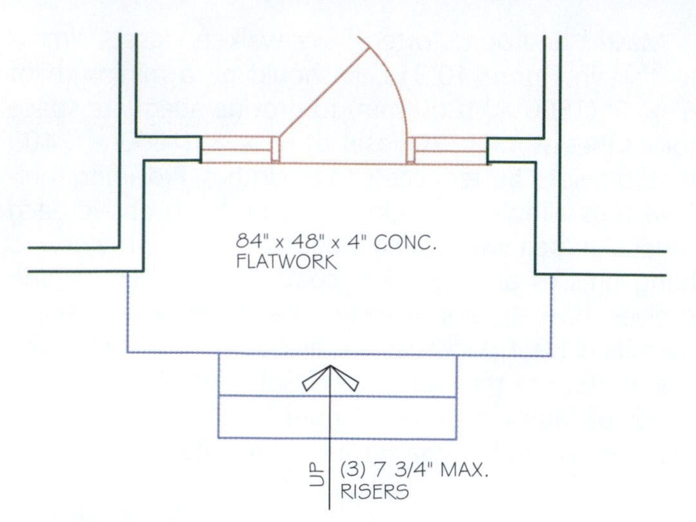

FIGURE 10.32 ■ *Steps from the landing to walkways or the ground cannot exceed 7 3/4''(195 mm).*

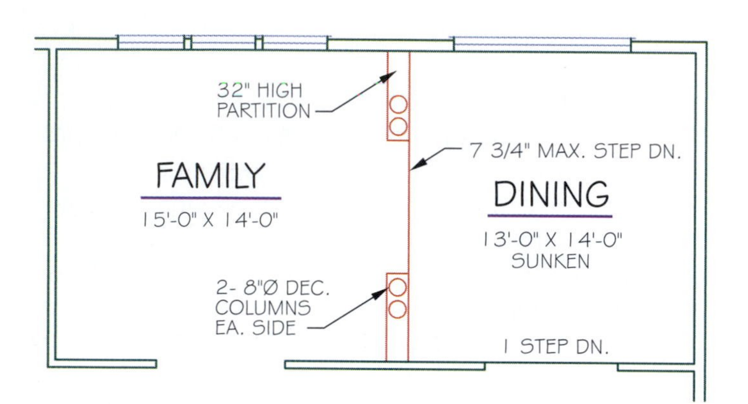

FIGURE 10.33 ■ *Changes in floor elevation can be represented using thin, continuous lines and explained with local notes.*

how changes in floor elevation can be represented and specified on a floor plan. Stairs should be placed on a layer with a name such as *FLOR STRS* (stair). If the home is being designed to meet ADA requirements or for universal living standards, a ramp should be provided instead of a step. The ADA requires the ramp should have a slope of 1:48 or less.

If the difference in floor elevation exceeds 30'' (750 mm), a guardrail must be provided. The IRC requires rails to be 36'' (900 mm) above the floor. Rails can be represented on the floor plan as shown in Figure 10.34, using a layer name such as *FLOR HRAL* (handrail). Many professionals use two thin, continuous lines with a 1'' offset. A partial wall is sometimes used as a guard at the floor edge.

In representing the steps for sunken rooms or split-level homes, each end of the stair run will be shown. The steps between these two types of floor changes can

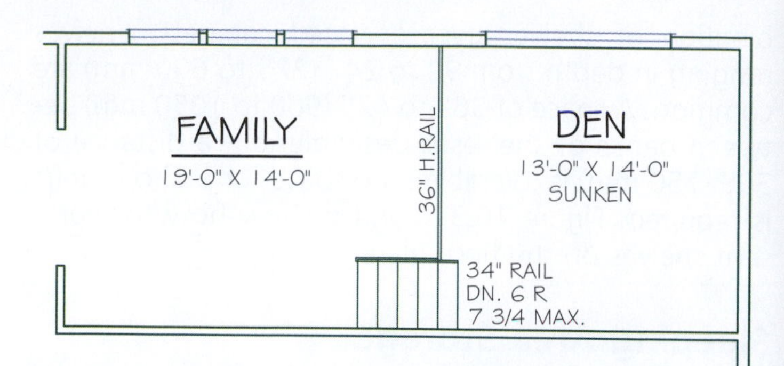

FIGURE 10.34 ■ *In representing the steps for sunken rooms or split-level homes, each end of the stair run must be shown.*

be depicted as shown in Figure 10.34. In addition to representing the steps, the number of required risers and a handrail must be specified. An arrow should also be provided to show the direction of travel. Chapter 30 provides a complete description for planning stairs, but to complete a floor plan a few key sizes will be required. Minimum stair sizes include:

- Stairs must be a minimum of 36'' (900 mm) wide, but a width such as 48'' to 60'' (1200 to 1500 mm) is preferred if space is available.
- The tread depth should be 10'' to 12'' (250 to 300 mm) with a 10'' (250 mm) minimum.
- Individual risers may range from 4'' to 7 3/4'' (100 to 195 mm) in height but must be a consistent size within the stair run.
- Landings at the top and bottom of stairs must be equal in size to the width of the stairs.
- A minimum clear height of 6'-8'' (2000 mm) is required for the length of the stairs.
- The stairs must have a handrail that measures between 34'' and 38'' (850 and 950 mm) above the tread nosing.
- Guardrails at landings above stairs, at balconies, lofts, or any area above another floor must be at least 36'' (900 mm) above the floor.

The design of a multistory home will require more coordination to assure that adequate length and height has been provided while still meeting the aesthetic requirements of the owners. Any of several common methods of arranging the stairs can be used to meet the design needs. Figure 10.35 shows three common arrangements for steps. A key point in drawing a multilevel stair run is that both ends of the stair are never seen on the same floor plan. A second consideration for stair layout is to show the stairs if more than two levels

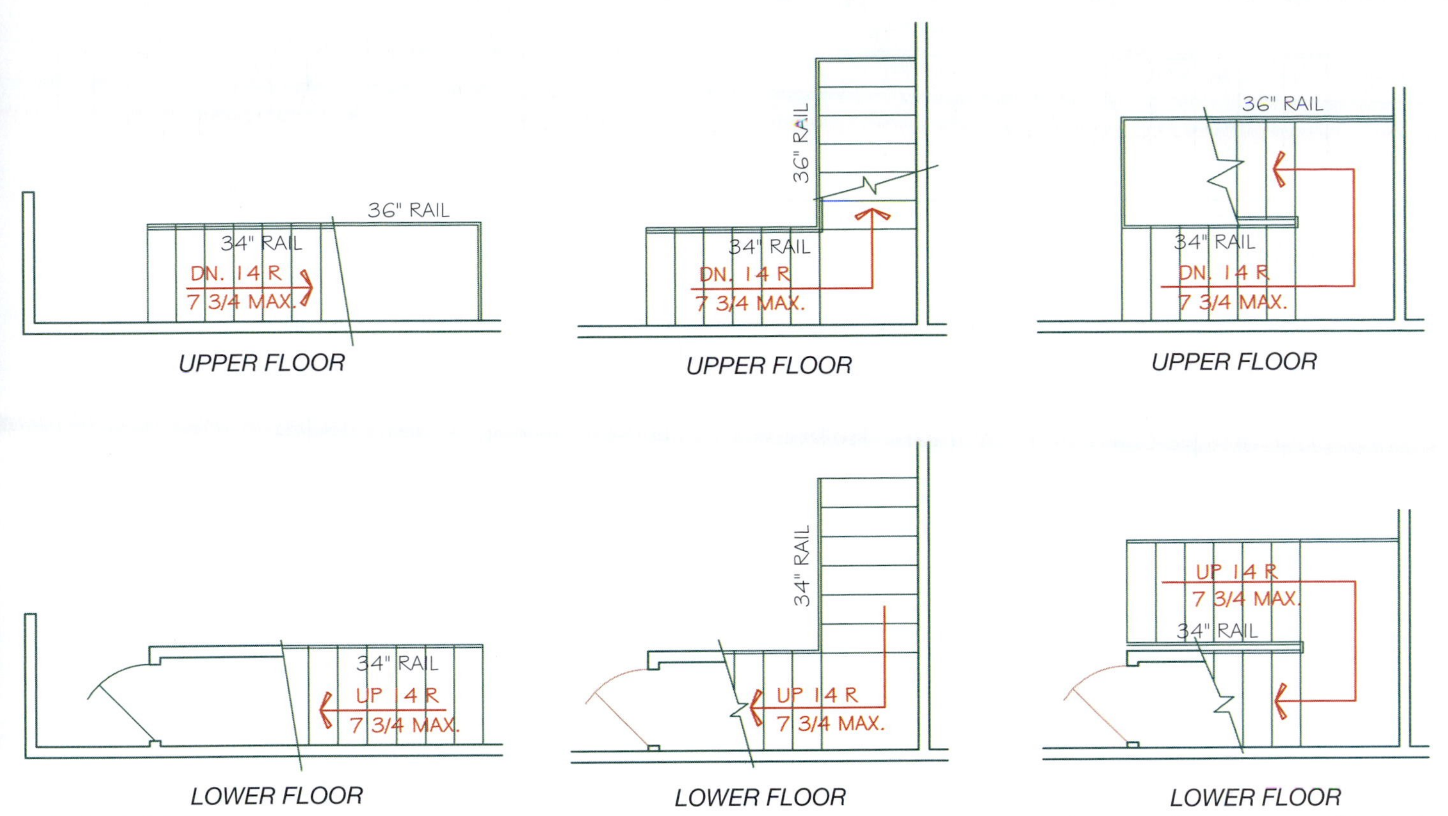

FIGURE 10.35 ■ *Several common methods of arranging the stairs, including straight run, L shape, and U shape, can be used to meet the design needs.*

will be accessed. The middle-level floor plan will show the top end of the lower stair and the bottom portion of the upper stair (see Figure 10.36).

Determining the Number of Steps Required

Keep in mind that the goal here is to represent stairs on the floor plan. See Chapter 30 for a complete discussion of stair design. The height between floor levels must be known to determine how many steps must be represented on the floor plan. This height is a decision made by the designer. For this introduction, the common height of 8' (2400 mm) between floors is used; 12'' (300 mm) for the thickness of the floor framing will be assumed and 7 3/4'' (195 mm) for the rise. Although these heights are not exact, they allow the stairs to be represented on the floor plan. The exact size for the riser (the vertical height of the step) and the tread (the horizontal depth of the step) is determined as the stair section is drawn. (See Chapter 30 for additional information on stairs.)

With a total rise of approximately 9' (2700 mm) and a maximum riser of 7 3/4'' (195 mm), the number of risers can be determined by dividing the total height by the maximum riser size. Dividing 108'' by 7 3/4'' (2700 by 195 mm) gives 13.9 risers. Since all risers must be of equal height, 14 risers (the vertical portion of the step) should be provided in the stair run. There will always be one less tread, the horizontal portion of the step, than there are risers. Once the number of risers and treads are known, the total stair run can be determined. If each tread is 10 1/2'' (263 mm) deep, the total run is found by multiplying 10 1/2'' (263 mm), the width, by 13, the number of treads required. A run of 136 1/2'' (3413 mm) is the minimum length that can be provided for the stairs. With this basic information, the layout of the stairs can be completed.

Use thin lines to represent the members of the stairway and place the stair information on a layer with a title such as *FLOR STRS*. Usually the starting or ending point will be determined by the location of the surrounding walls. In the example in Figure 10.37, the lower end of the stair is based on the location of the first step from the exterior wall. The design requires a hallway 42'' (1050 mm) wide. With the location of the first riser known, the total length of the stairway can be located. Show enough of the treads on the floor plan so a repetitive pattern can be seen. Use a break line to end

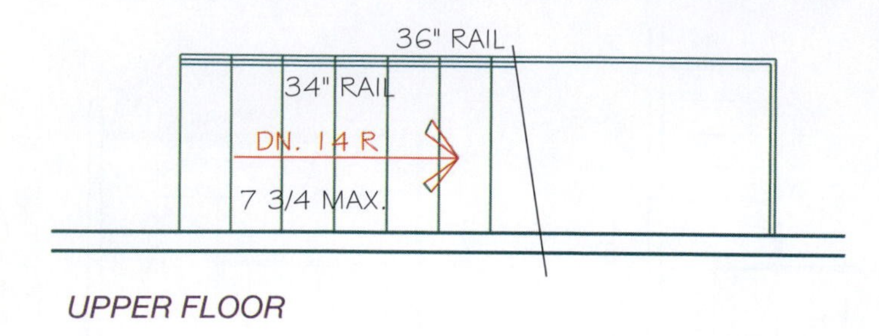

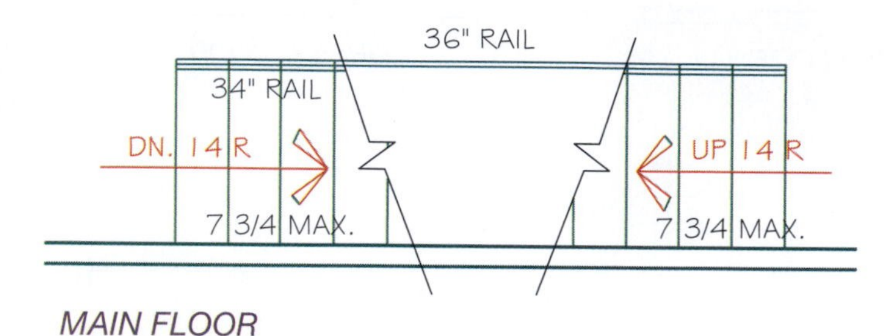

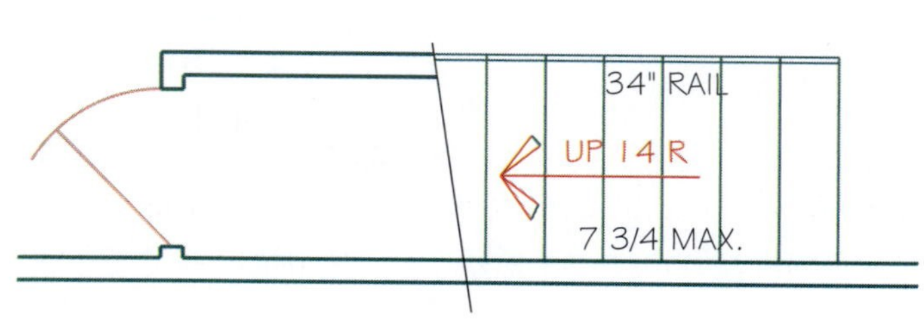

FIGURE 10.36 ■ *In drawing a multilevel stair run, both ends of the stair are never seen on the same floor plan. The middle-level floor plan will show the top end of the lower stair and the bottom portion of the upper stair.*

the stair pattern and show what will be under the upper end of the stair.

On the upper floor plan, the end of the stair run is determined based on the beginning of the run on the lower floor. Show the end of the stair run and enough of the treads to produce a pattern. Provide a break line on the upper run to terminate the stair pattern, but show the guardrail that surrounds the entire opening for the stair. On the upper floor, walls may be placed over the first few steps of the stair run. For planning purposes, the upper floor can extend over the first two steps of the lower floor and still remain level. As seen in Figure 10.38, once the upper floor extends over the third step, the 6'-8'' (2000 mm) minimum headroom no longer exists. Depending on how the floor above the lower steps will be used, an inclined floor may be suitable. If a closet will be placed over the steps, the inclined floor could be used to provide an excellent place to display shoes. An inclined floor will not be suitable if a bathroom is placed above the stairs unless it can be hidden beneath a cabinet.

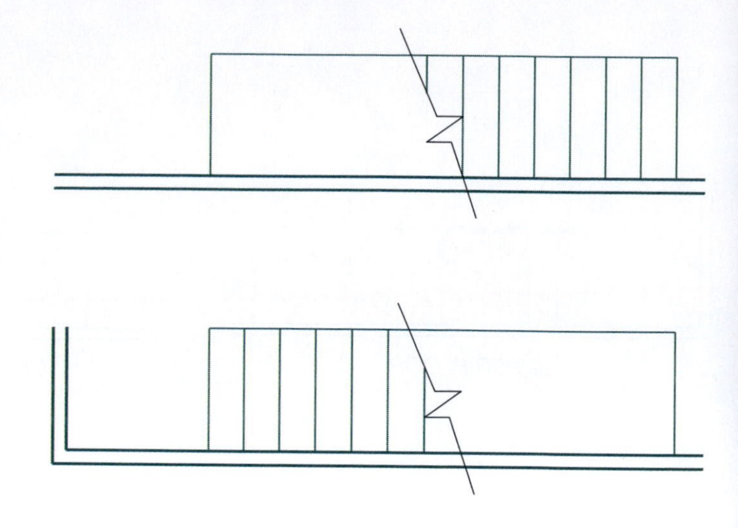

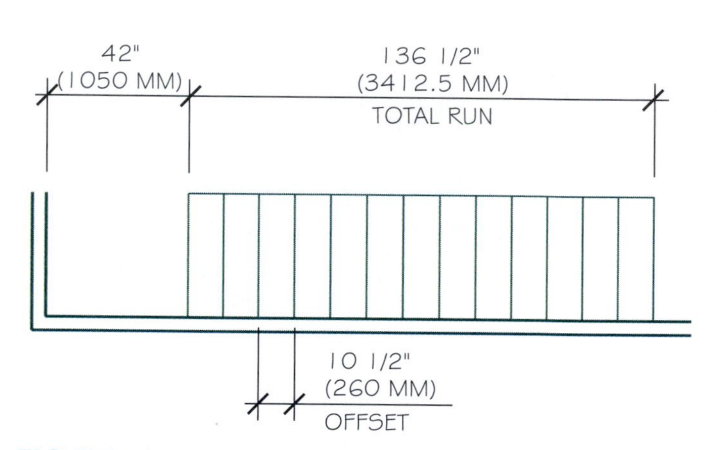

FIGURE 10.37 ■ *The number of risers can be determined by dividing the total height by the maximum rise. Once the number of risers and treads is known, the total stair run can be determined by multiplying the tread width by the number of treads required. With this basic information, the layout of the stairs can be completed.*

REPRESENTING A FIREPLACE AND CHIMNEY

Several types of fireplaces may be represented on a floor plan, including single- and multilevel masonry, prefabricated zero clearance, direct vent, and freestanding wood burning. Chapter 31 provides a detailed description of each type. This chapter provides the information needed to represent each type of fireplace on a floor plan.

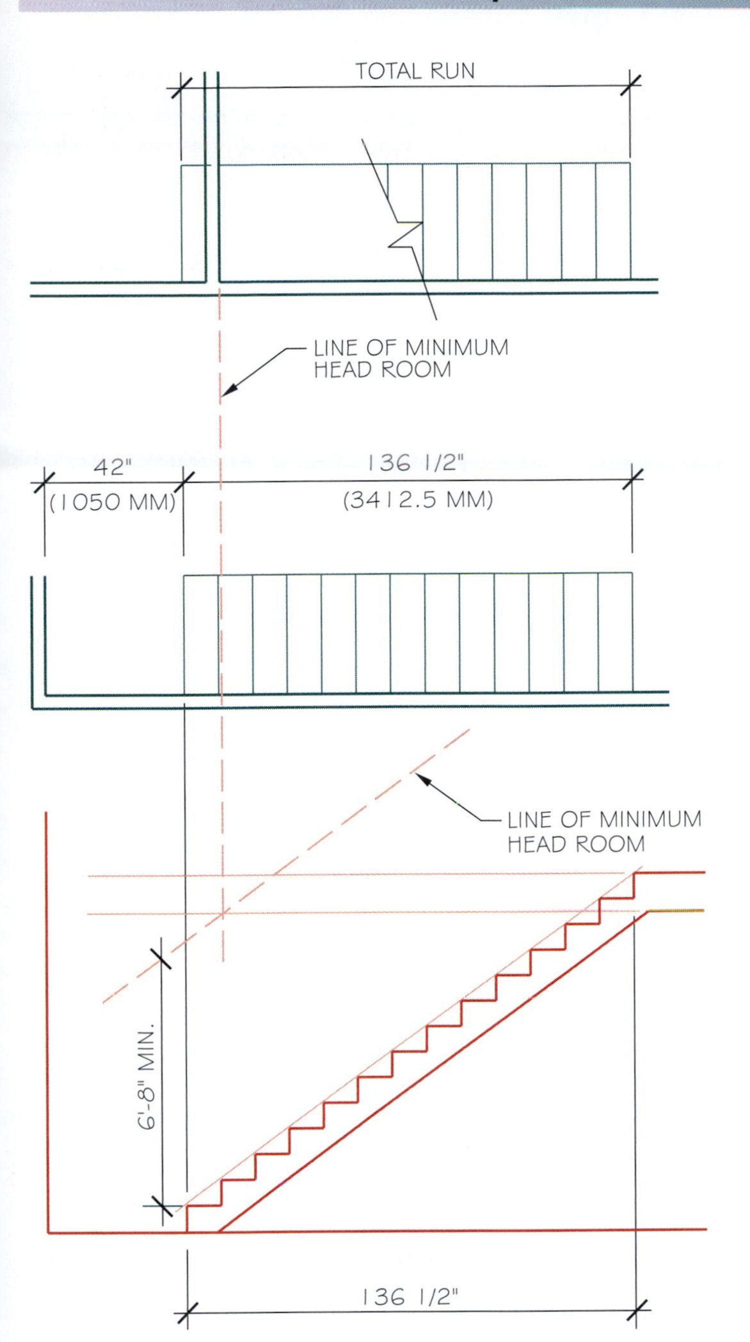

FIGURE 10.38 ■ *Walls on the upper floor can be placed over the first few steps of the stair run. For planning purposes, the upper floor can extend over the first two steps of the lower floor and still remain level. Once the upper floor extends over the third step, the minimum required headroom no longer exists.*

Single-Level Masonry Fireplace

Figure 10.39 shows the symbol and the minimum sizes used to represent a single masonry fireplace. Represent the edge of the fireplace with thick lines that match the wall lineweight. Draw a masonry fireplace on a layer with a name such as *FLOR MASN*. Place the hatch pattern on the *FLOR PATT* layer. Notice that the face of the interior masonry laps over the interior face of the exterior wall to provide a tight seal. The hatch pattern and all other lines are thin lines. The hatch pattern used to represent the masonry can be placed using ANSI31 set at a 0° angle. Adjust the pattern so the hatch lines are approximately 1/16'' (2 mm) apart when plotted. The dimensions that describe the fireplace should not be placed on the floor plan. Common fireplace opening sizes include:

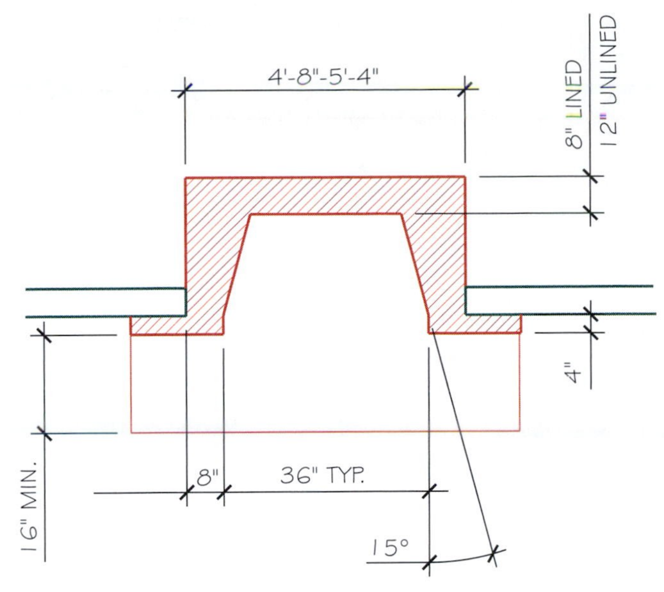

FIGURE 10.39 ■ *Minimum dimensions for a single masonry fireplace. See Chapter 31 for a complete listing of IRC minimum sizes and construction requirements.*

OPENING HEIGHT	OPENING WIDTH	OPENING DEPTH
36'' (900 mm)	24'' (600 mm)	22'' (550 mm)
40'' (1000 mm)	27'' (675 mm)	22'' (550 mm)
48'' (1200 mm)	30'' (750 mm)	25'' (625 mm)
60'' (1500 mm)	33'' (825 mm)	25'' (625 mm)

Natural gas may be provided to the fireplace, either for starting a wood fire or for fuel to provide flames on artificial logs. The gas supply should be noted on the floor plan, as shown in Figure 10.40, with the notation of FG placed by the symbol to represent fuel gas.

Common Shapes

Several common arrangements are available for masonry fireplaces, but shapes are limited by the designer's imagination and the skills of the mason. Figure 10.41

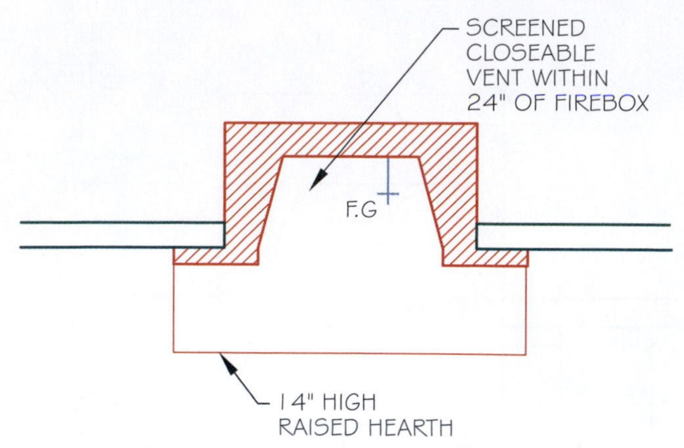

FIGURE 10.40 ■ *Representing gas, venting for combustion air, and the hearth on the floor plan.*

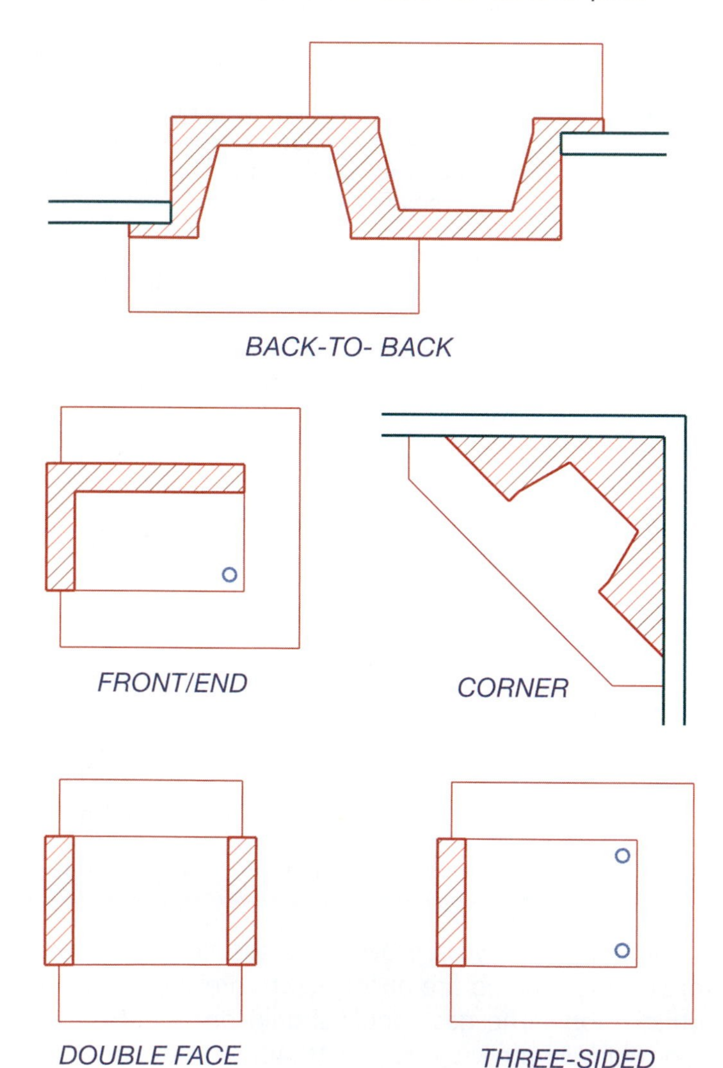

FIGURE 10.41 ■ *Common arrangements of masonry fireplaces.*

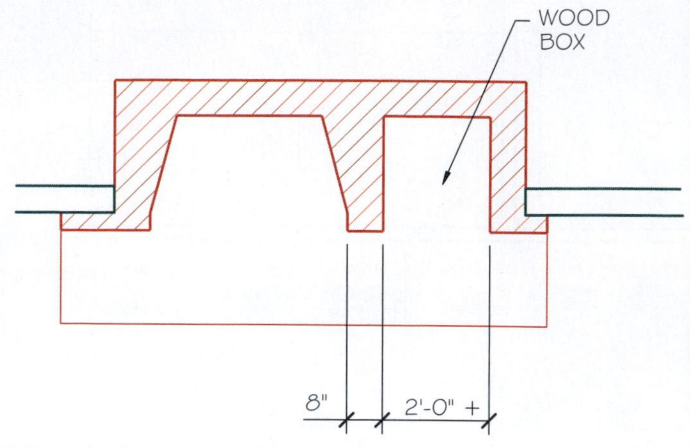

FIGURE 10.42 ■ *Masonry fireplace with wood storage.*

shows several fireplace symbols often found on a floor plan. In addition to considering the shape and location of the firebox, a wood compartment can be provided next to the fireplace opening to store small amounts of wood. The floor plan representation of a wood storage box is shown in Figure 10.42. Support below the fireplace is required to be shown on the foundation plan if a fireplace is represented on the floor plan. See Sections 7 and 8 for a discussion of fireplace support.

Barbecues

When a home has a fireplace in a room next to the dining room, nook, or kitchen, the masonry structure may also incorporate a built-in barbecue. If a barbecue will be incorporated into the exterior side of the fireplace, it can be represented as shown in Figure 10.43. A prefabricated built-in unit can be set into the masonry structure surrounding the fireplace. Gas and electricity to supply the barbecue should be specified. As an alternative, the barbecue unit may be built into the exterior structure of a fireplace for outdoor cooking. The barbecue may also be installed separately from a fireplace.

Multilevel Masonry Fireplaces

A multilevel masonry fireplace is represented using a symbol similar to that for a single-level fireplace. On the lowest level, the firebox must be offset from the firebox on the upper level to allow the lower-chimney to vent. Adequate area must be provided for the lower level to support the additional upper-level fireplaces. On the upper level, an area for the lower flue must be represented separate from the upper firebox. A flue is a heat-

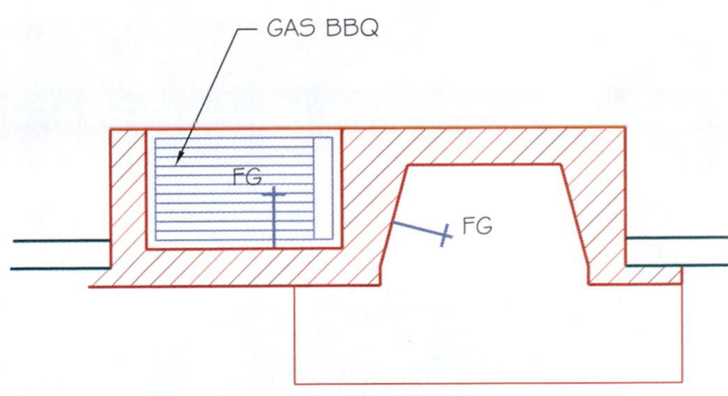

FIGURE 10.43 ■ *A masonry fireplace with an outdoor sink and barbecue. Each item would need to be specified on the floor plan, as well as any utilities that may be required.* Photo courtesy Zachary Jefferis.

resistant, noncombustible passageway in a chimney used to carry combustion gases from the fireplace. There must be a separate flue for each fireplace. Figure 10.44 shows how a multilevel masonry fireplace is represented on the floor plan. Overall sizes will vary based on the opening size of the firebox. The chimney above the upper fireplace can taper to a smaller size, but it needs to be large enough to house the flues and still provide the minimum surrounding masonry material.

Prefabricated Metal Fireplaces

Fireplace fireboxes made of steel are available from various manufacturers. These prefabricated fireplaces are popular because they are efficient and easy to install. Metal wood-burning fireplaces can be installed using a masonry or metal chimney. The metal chimney is usually hidden in a wood-framed chimney called a chase. Figure 10.45 shows a metal fireplace and chimney and how they are represented. Metal fireplaces, referred to as zero-clearance units, can be enclosed by wood walls. Units are available that burn wood, natural gas, or propane, or they may use electricity to produce heat. Sizes vary greatly among the several major suppliers, and there are many models from which to choose. An area of 48'' × 24'' (1200 × 600 mm) will provide ample space on the floor plan during the preliminary design stage. Draw the fireplace on the *FLOR CHIM* layer. Once a specific model has been selected, the exact size is specified based on the manufacturer's specifications. To specify a zero-clearance fireplace, provide the model number and other specific information next to the fireplace symbol on the floor plan or in

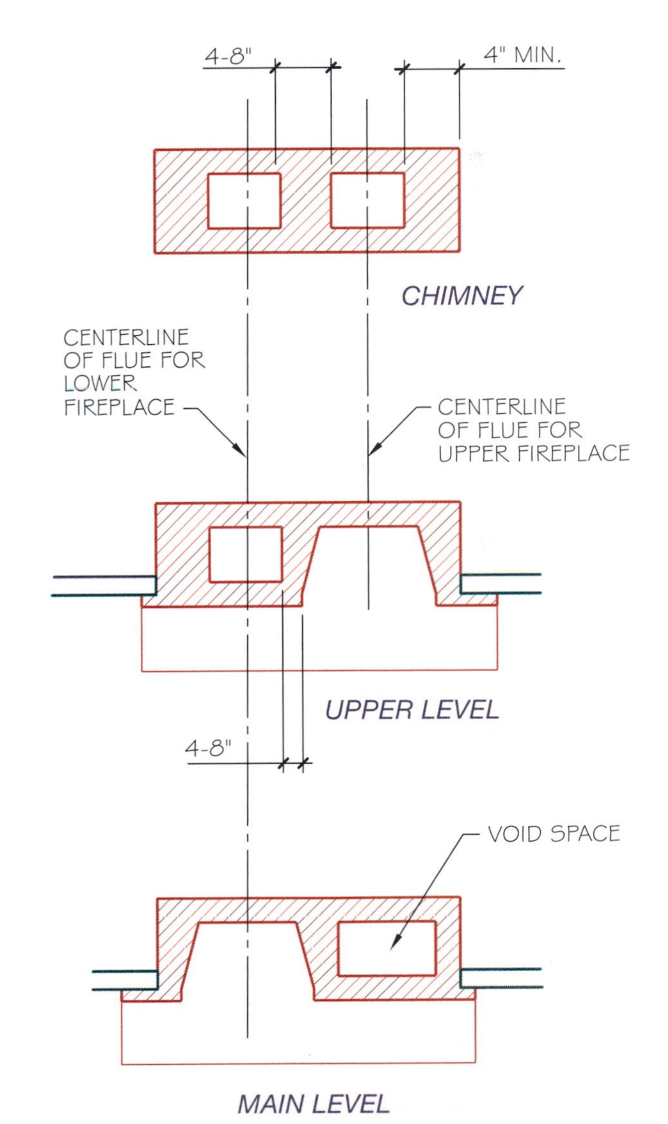

FIGURE 10.44 ■ *A multilevel chimney must allow space for each flue beside the fireplace opening.*

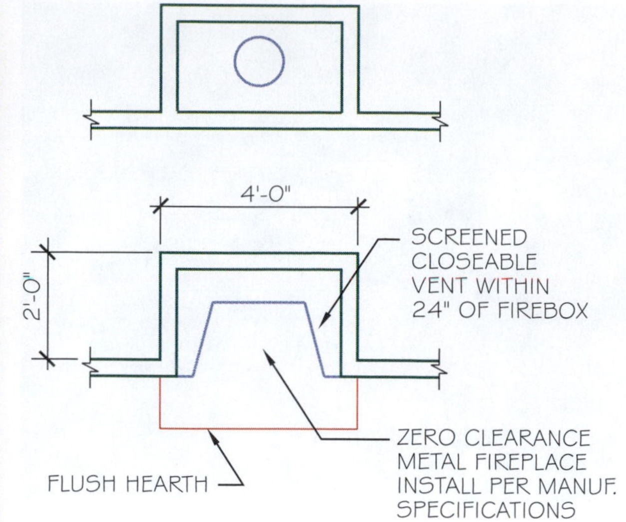

FIGURE 10.45 ■ *A metal fireplace and chimney.* Courtesy Andrea Worcester.

a general note. The gas supply should also be shown with a symbol and a note, similar to the examples in Figure 10.40.

Fireplace Alternatives

Common alternatives to a fireplace include inserts, logs, and stoves. A ***fireplace insert*** is a metal fireplace that is inserted into an existing masonry fireplace and vented using the existing chimney. Inserts are often used for renovations to get better energy efficiency from an old firebox. Inserts can be installed that burn gas, propane, wood, and wood pellets or that use electricity. A chimney liner may need to be provided for the existing chimney and should be specified on the floor plan based on the manufacturer's recommendations. Figure 10.46 shows a metal insert and how it can be represented on a floor plan.

Freestanding stoves are excellent radiant heating units. Models are available that burn wood, pellets, gas, and coal. Sizes vary widely based on the manufacturer and the fuel that is burned. A rectangle 30'' × 27'' (750 × 675 mm) is used to represent a stove on a floor plan. Figure 10.47 shows a freestanding fireplace represented on a floor plan. Gas and propane log units are available that can be installed in existing masonry fireplaces or placed in a freestanding pit to imitate the effect of an open wood fire. The size of the pit varies based on design values. Common widths for log units range from 18'' to 24'' (450 × 600 mm). Draw each type of heating unit on the *FLOR APPL* layer.

Floor and Wall Protection

Floor and wall protection for a fireplace, insert, and stove must also be represented on the floor plan. Combustible floors must be protected from hot embers from an open firebox. A hearth is the noncombustible protection for the floor that is usually constructed from brick, stone, tile, or concrete. The exact size required for the hearth will vary depending on the size and type of the heating unit. A hearth must extend a minimum of 16'' (406 mm) in front and 8'' (203 mm) to the side of the firebox. The hearth may be flush with the floor or raised approximately 14'' (350 mm) to provide a place to sit. Self-contained zero-clearance units are available that do not require special floor protection. Chapter 31 explores hearth options.

Unless a solid masonry fireplace that meets the minimum sizes shown in Figure 10.39 or a zero-clearance

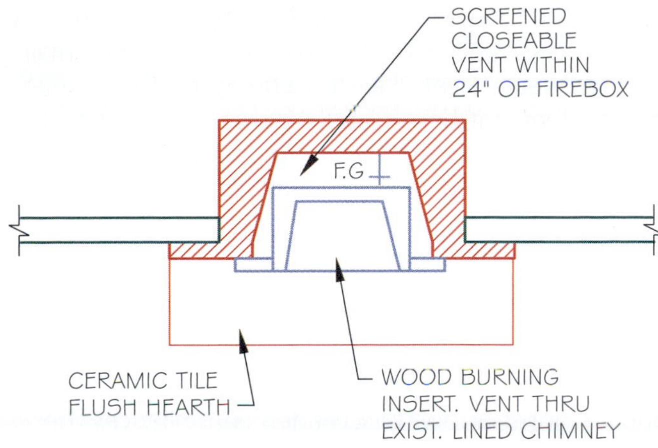

FIGURE 10.46 ■ *Representing an insert on the floor plan.* Courtesy Nancy Hockert.

unit is used, the walls near a heating unit must be protected. Exact distances from the heating unit to an unprotected wall will vary based on the specifications of the manufacturer. A minimum distance of 18'' (450 mm) is required between unprotected walls and the backs of most units. This distance can often be reduced to between 8'' and 12'' (200 and 300 mm) if a noncombustible surface such as masonry, stone, or tile is laid over a cement asbestos board. Draw protective surfaces on a layer with other masonry products and place the hatch pattern with other hatch patterns.

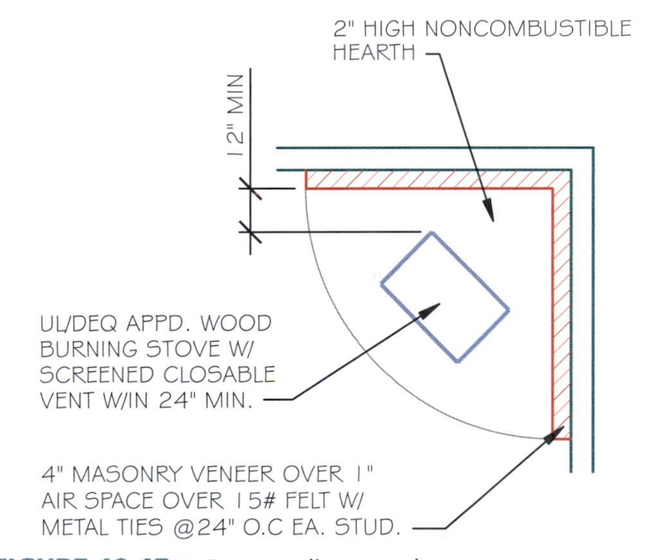

FIGURE 10.47 ■ *Freestanding wood stove.*

Metal Chimneys

The type of chimney must also be considered, based on the manufacturer's recommendations, if a prefabricated metal fireplace or stove is specified. Flues must be designed to provide proper ventilation and draft for the fireplace. Draft is a current of air and gases that pass from the fireplace through the chimney. The size of the flue is determined by the size of the fireplace opening and the chimney height. Common options for venting include a direct vent, rear vent, and self-vent. The direct-vent models have a chimney that is vented through the roof or out the wall and extended above the roof either exposed or in a framed enclosure. A triple-wall metal chimney with an approximate diameter of 14'' (350 mm) is common when the chimney is to be vertical. Draw the metal chimney on the *FLOR CHIM* layer. Metal fireplaces are also available that vent out the back, with a vent similar to a dryer vent. Self-venting units are vented directly out the wall behind the fireplace. A vent-free fireplace is installed without a flue.

Combustion Air

Combustion air is supplied to the firebox to allow in fuel combustion. The IRC requires that a closable vent to outside air be provided within 24'' (600 mm) of the firebox to provide combustion air. Providing outside air ensures that the fireplace does not draw heated air from the interior. The vent allows indoor oxygen levels to be maintained and keeps heated air from going up the chimney. The vent does not have to be shown on the floor plan, but a note should be placed near the fireplace to indicate how the combustion air will be provided.

MISCELLANEOUS FLOOR PLAN SYMBOLS

Common symbols that may be shown on a floor plan include a north arrow, miscellaneous appliances, decks and porches, attic and crawl access openings, miscellaneous plumbing symbols, cross-section symbols, and structural symbols. A legend should be provided to explain the symbols represented on the floor plan.

North Arrow

An arrow must be placed on the floor plan to indicate north. It should be simple and easy to reproduce. It is usually placed near the drawing title using the *FLOR SYMB* layer. Figure 10.48 shows common symbols used for the north arrow.

Miscellaneous Equipment

Common equipment that may be necessary to represent on a floor plan includes a heating unit, air conditioner, water heater, and a built-in vacuum system.

FIGURE 10.48 ■ *A north arrow must clearly define the direction without being a distraction.*

Heating Units

The type of heating unit to be used will reflect what will be placed on the floor plan. Chapter 14 introduces major types of residential heating systems. Generally, only a forced-air furnace needs to be represented on the floor plan. Other types of heaters are represented on the electrical plan. When drawing a very simple residence, the electrical symbols can be placed on the floor plan. For most custom plans, a separate plan is provided to represent all electrical fixtures. Chapter 12 explores how electrical fixtures and symbols are represented on the floor plan. The furnace is often placed in a location central to the house, or it may be in the garage, a basement storage area, in an attic, or in a closet. Units that burn a fuel such as gas cannot be located under a stairway. Figure 10.49 shows a common method of representing a forced-air unit. Sizes will vary depending on the fuel source, but common sizes for representing the heating unit include:

Gas forced-air unit: 18'' (450 mm) square (minimum)
Electric forced-air unit: 24'' × 30'' (600 × 750 mm)

A 6'' (150 mm) space should be provided on each side of the unit for airflow. Space will also be needed to service the unit. Draw the furnace on the *FLOR HVAC* layer.

Air Conditioning

A compressor for the cooling unit should be drawn on the floor plan. The size of the unit will vary greatly depending on the size of the house to be cooled. For planning purposes, a 27'' × 27'' (675 × 675 mm) square can be used to represent the compressor (see Figure 10.49). The compressor should be placed on a concrete pad that is typically about 6'' (150 mm) larger than the unit. Place the unit outside of the house in a location near the heating unit. Draw the air conditioning unit on the *FLOR HVAC* layer. Chapter 14 explores options to cool a residence.

Planning for HVAC Ducts

Ducts for a forced-air system for a single-level structure are usually placed in a crawl space or attic. In a multilevel home, consideration must be given to where the ductwork can be placed. When supply ducts cannot be confined to a crawl space or attic, they must be run inside the occupied areas of the home. When possible, conceal ducts between floor and ceiling members. Figure 10.50 shows a multilevel home, the required HVAC registers, and possible locations for ductwork. As a

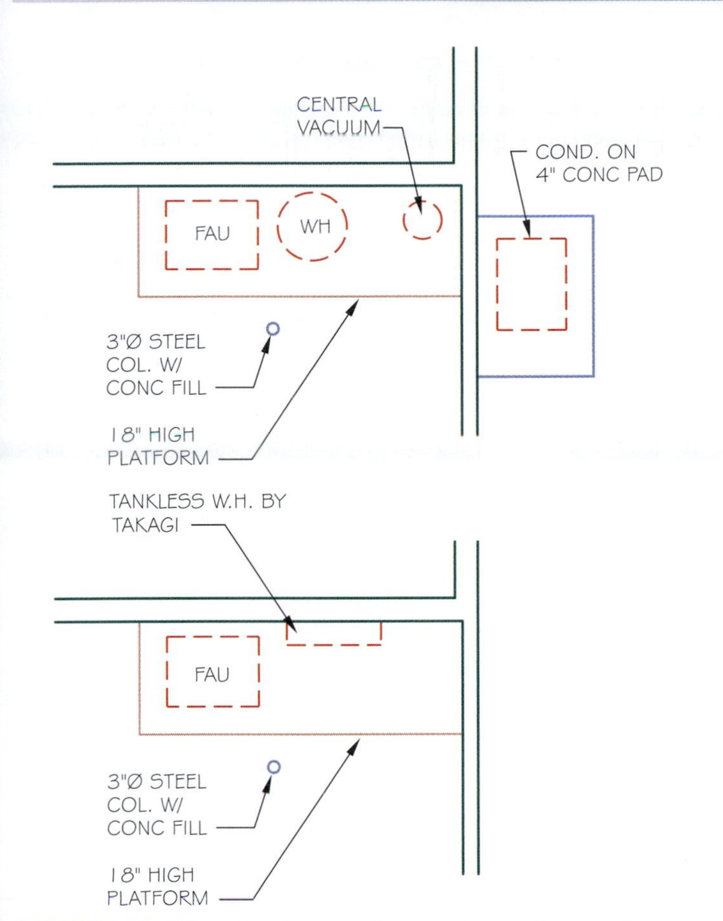

FIGURE 10.49 ■ *Heating, cooling, and cleaning appliances are often located in the garage to conserve living space. The burning element for a gas unit is required by code to be 18'' (450 mm) above the floor level. Most modern appliances meet this requirement, but an 18''-high platform is often specified. Many homeowners are moving away from the traditional round water heater for the more efficient tankless heater with instant hot water.*

drafter, it is usually not your job to plan the ductwork; you indicate only where it can be placed so that treated air can be supplied to each room. In this example, as the main supply ducts leave the furnace, they travel through unfinished rooms and can be run below the ceiling. When the supply ducts run parallel to structural members and the duct size is equal to or smaller than the size of the construction members, they can be placed within the space between the structural members. Framing covered with gypsum or other finish material can be used to enclose ducts that must be placed in habitable areas. When possible, place duct runs in the ceiling of a hallway, kitchen, or bathroom. These rooms can be framed and finished with a 7'-0'' (2100 mm) ceiling. Figure 10.51 shows how a chase can be represented on a floor plan. Show the chase on the *FLOR HVAC* layer. If the chase is to be hatched, place the pattern on the *FLOR PATT* layer.

When ducts must be run between floor levels, they need to be run in an easily concealed location, as in a closet or in the stud space. The stud space can be used for ducts that are 3 1/2'' deep for 2 × 4 studs, 5 1/2'' deep for 2 × 6 studs, or 7 1/2'' deep for 2 × 8 studs. If the duct can be run up through a closet, then it can be placed in a chase. A chase is a continuous recessed area built to conceal ducts, pipes, or other construction products. If the duct cannot be easily concealed, it might need to be framed into the corner of a room. The framing for a chase is shown on the floor plan as a wall surrounding the duct to be concealed; a note is usually placed indicating the size and use. A typical note would read:

CHASE FOR 22 × 24 RETURN DUCT.

In the case of return air ducts, the construction members and enclosing materials can be used as the duct plenum. When this can be done, no extra framing is required to conceal the ducts. When ducts are run in an area such as an unfinished basement, it is possible to leave them exposed.

Water Heaters

The water heater is generally located in a central area of the home. It is often placed near the furnace, not because they have common elements but because they often share leftover space. Figure 10.49 shows a water heater in a garage located by the furnace. Represent the water heater with a circle drawn with dashed lines and a diameter of 18'' to 24'' (450 to 600 mm). The actual size will vary based on the fuel type, the number of bathrooms and bedrooms, and the manufacturer. Chapter 13 provides further guidelines to determine the size of the water heater. Draw the water heater on the *FLOR PFIX* layer. When the water heater is located in a garage, a steel column filled with concrete, called a ballard, is required to protect the unit from damage due to impact by a car. Gas units must be installed so the burning element is 18'' (450 mm) above the floor level. Most gas water heaters are manufactured to meet this minimum height, but many professionals specify that the unit be placed on an 18'' (450 mm) platform to ensure compliance. If the water heater is to be placed in a room with a finished floor, a metal pan with a drain connected to the sewer system should be provided.

A tankless system is a popular alternative to the traditional water heater. Tankless water heaters do not store water but heat it as needed. As the hot water tap

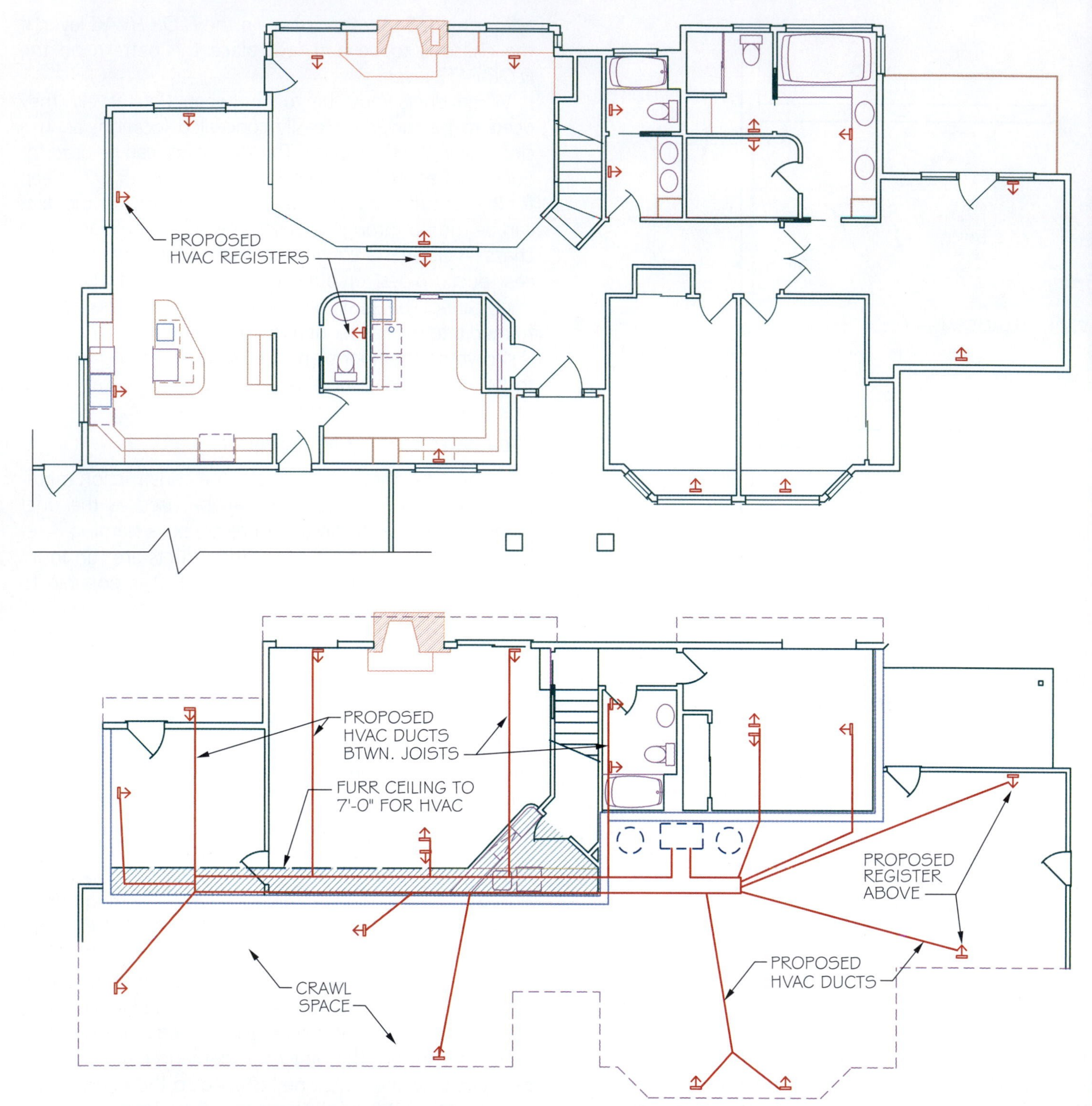

FIGURE 10.50 ■ *The upper floor plan of this home shows where the design team plans to have HVAC registers placed. The lower plan shows how possible ductwork can be placed. Neither the ducts nor the vents are shown on the floor plan, but possible locations must be determined so that the HVAC contractor has room to place needed equipment.*

is opened at a fixture, water flows over a heating element to the fixture. When the fixture is turned off, the heater is turned off. Tankless water heaters can be drawn as shown in Figure 10.49. Sizes vary based on fuel type and manufacturer, but they are approximately 16'' (400 mm) wide and 8'' (200 mm) deep. Draw a tankless water heater on the *FLOR APPL* layer. Use thin continuous lines for wall-mounted units and dashed lines for undercabinet models. Provide a note to specify the manufacturer, model number, and fuel type.

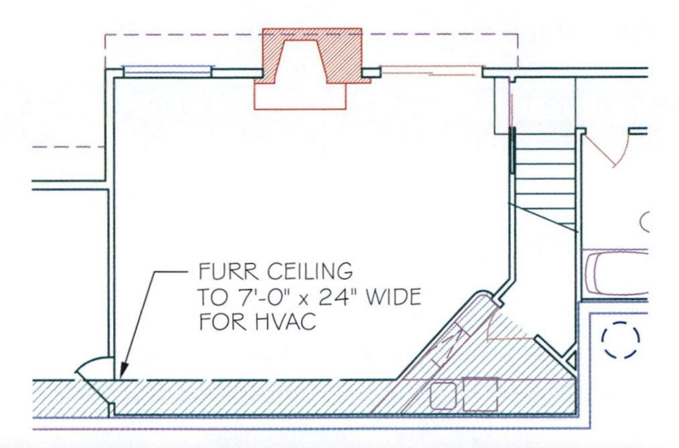

FIGURE 10.51 ■ *Although the main supply duct could be run through the crawl space, the designer has planned to provide a lowered ceiling parallel to the exterior wall. Ductwork can then be run to registers through the joist space. The ceiling was also lowered over the bar to help hide the chase.*

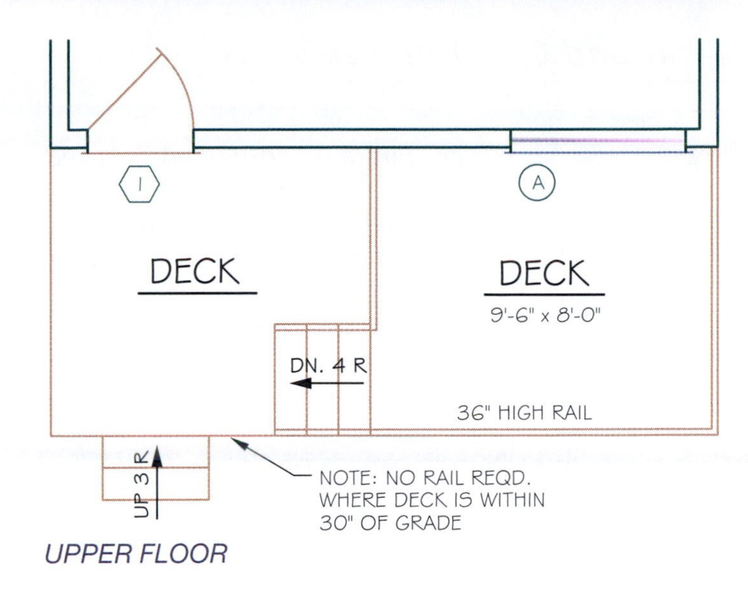

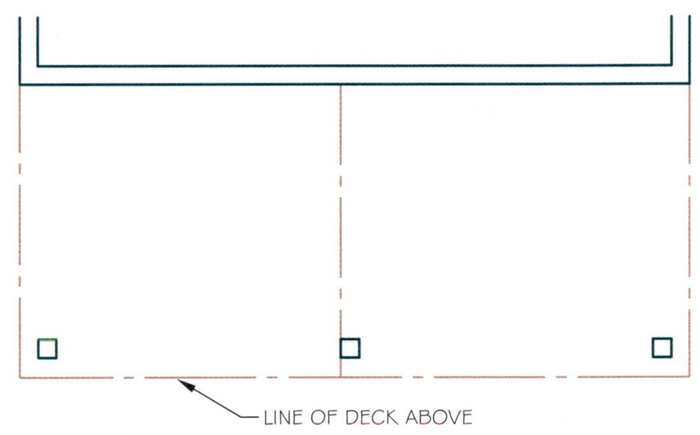

FIGURE 10.52 ■ *On multilevel homes, thin lines should be used to represent the outline of the deck and any required guardrails. A dashed or centerline can be used on the lower plan to represent the outline of the deck above. Continuous bold lines should be used to represent support posts. If exterior stairs are required, they can be represented using the same techniques that were used to represent interior stairs.*

Built-in Vacuum

A central vacuum system should be represented on the floor plan and the electrical plan. The outlets are shown on the electrical plan and are discussed in Chapter 12. The central unit should be shown on the floor and the electrical plans (see Figure 10.49). A central vacuum is about 12'' (300 mm) in diameter and is usually mounted on a wall in a location that provides noise dampening and easy access. This may be in the area where the furnace and water heater are located. Place the central vacuum on the *FLOR APPL* layer.

Decks and Porches

Decks and porches should be shown on a floor plan using thin, continuous lines on a layer such as *FLOR DECK*. When the deck is higher than 30'' (750 mm) above the ground, a guardrail must be placed around the deck using the same methods and layer title used to represent interior rails. On multilevel homes, a dashed, hidden, or centerline should be used on the lower plan to represent the outline of the deck above. Place the outline of the deck on the *FLOR DECK OTLN* layer. Continuous bold lines should be used to represent support posts. The posts should be placed on the same layer as the lower wood walls. If exterior stairs are required, they can be represented using the same techniques used to represent interior stairs. Figure 10.52 shows common methods of representing deck information.

Concrete slabs for patios, walks, or driveways should be noted on the floor plan. Common notes that might be used to describe hardscape include:

4'' THICK CONCRETE WALK

4'' THICK CONCRETE STOOP

4'' THICK CONCRETE FLATWORK

Stoop and *flatwork* are terms used by professionals to describe concrete slabs. Concrete flatwork is often sloped 1/4'' per foot to drain water. Decks, porches, and concrete flatwork can be placed on the *FLOR DECK* layer.

Attic and Crawl Access

Access is necessary to crawl space and attics. The crawl access for homes with a wood-framed floor may be located in an exterior foundation wall or placed on the floor plan in any convenient location. An exterior access provides an excellent method for inserting pipes or other long materials if repairs are required under the structure. An access opening in the interior will help stop wild critters from getting free rent. The master bedroom closet or a utility room is a common place to hide the crawl access. Avoid placing it in a child's closet. The access must be a minimum of 18'' × 24'' (457 × 610 mm) if it is located in the floor. Figure 10.53 shows an example of representing the crawl access. Access located in the foundation wall is explored in Section VII.

Attic access must be provided if an attic exceeds an area of 30 sq ft (2.8 m²) and has a vertical height of 30'' (762 mm). The attic access must be a minimum of 22'' × 30'' (550 × 750 mm), and it must be located in an area of the house that has 30'' (762 mm) of unobstructed headroom above the access opening. The attic access is often placed at the end of a hallway, in the master bedroom walk-in closet, or in a utility room. The attic access cannot be located in a closet that requires shelving to be moved to gain access to the attic. The attic access may include a fold-down ladder if the attic is to be used for storage. A fold-down access door is usually between 48'' and 60'' (1200 and 1500 mm) long. Figure 10.53 shows attic access and crawl access symbols and related notes describing them. The symbols can be drawn with thin lines and may be drawn with continuous or dashed lines depending on the office preference. Place the access opening on the *FLOR OVHD* layer (overhead).

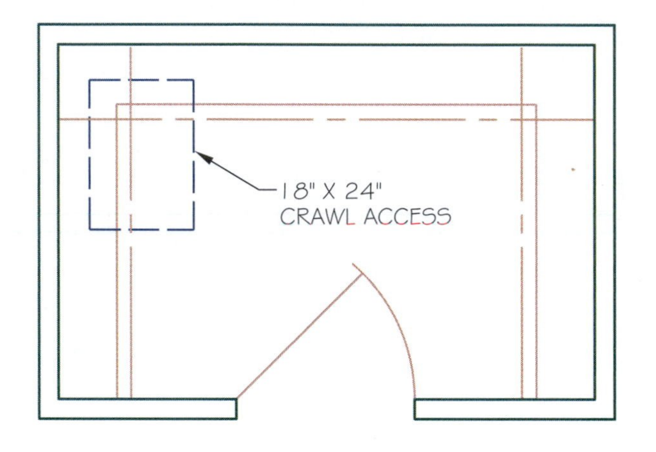

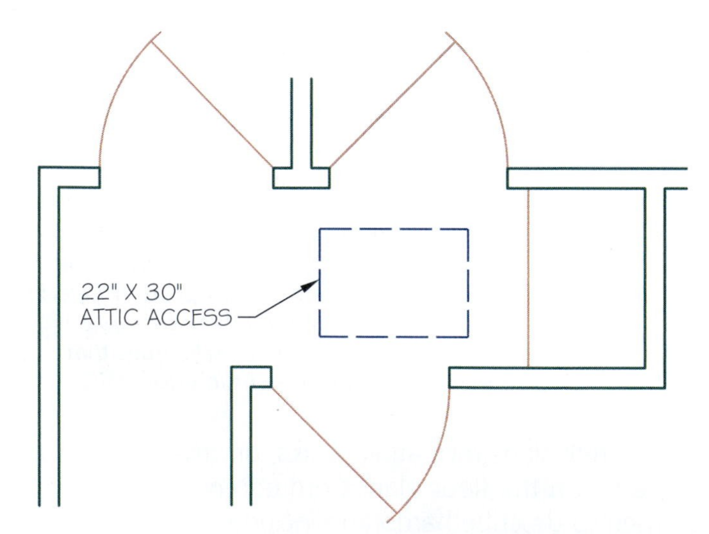

FIGURE 10.53 ■ *The attic access cannot be located in a closet that would require shelving to be moved to gain access to the attic. The attic access may include a fold-down ladder if the attic is to be used for storage. The crawl access for a home with a wood-framed floor may be located in an exterior foundation wall or placed on the floor plan in any convenient location.*

Miscellaneous Plumbing Symbols

Floor drains should be used in any room where water could accumulate and damage the finished floor material. Rooms such as the laundry room, bathroom, and garage are common locations to have a pan placed under a specific appliance that might cause water damage. Figure 10.54 shows a floor drain in a utility room. A

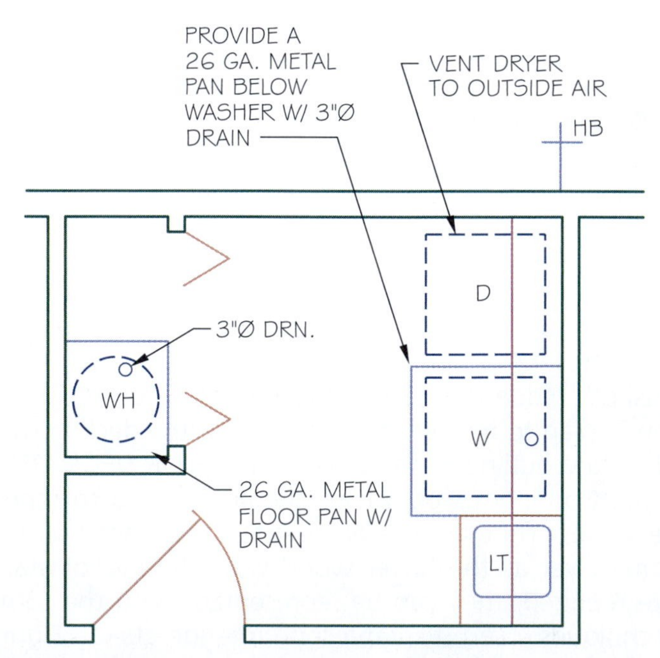

FIGURE 10.54 ■ *Floor drains and hose bibbs may also be specified on a floor plan.*

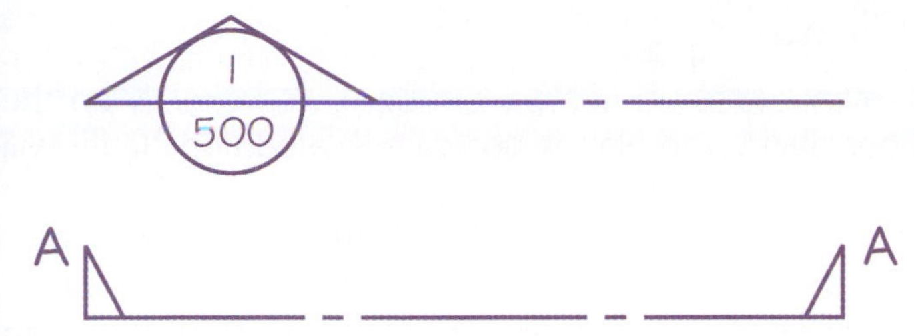

FIGURE 10.55 ■ *Cutting-plane lines are shown on the floor plan, but they may also be placed on a framing plan.*

floor drain should be placed on the *FLOR PFIX* layer and specified with a note such as

30" SQ. METAL PAN BELOW WASHER W/ 3" DIA. DRAIN.

A ***hose bibb*** is an outdoor water faucet used to connect a garden hose to the plumbing system. Hose bibbs should be placed at locations convenient for watering lawns or gardens and washing a car, but they should also be placed near other plumbing runs. The floor plan symbol for a hose bibb is shown in Figure 10.54. Depending on the complexity of the home, each of these fixtures may be shown on a separate plumbing plan. See Chapter 13 for a discussion of plumbing plans.

Cross-Section Symbols

The location on the floor plan where a cross section is taken is identified with symbols known as cutting-plane lines. These symbols are discussed in detail in Chapter 29. Figure 10.55 shows options for the symbol representing the cutting-plane line. The method used depends on your school or office practice and the drawing complexity. Place the symbol on the *FLOR SYMB* (symbol) layer. Cutting planes may be omitted from the floor plan and shown only on the framing plan of complicated plans.

Structural Materials

Structural materials may be identified on simple floor plans with notes and symbols or on the framing plan. Framing information and symbols are discussed in Chapter 24.

DRAWING ANNOTATION

The first portion of this chapter introduced the symbols to represent the features shown on a floor plan. Once drawn, most of the symbols need some form of annotation to explain the size, material, or function of the symbol. Drawing annotation is divided into ***general notes,*** local notes, schedules, legends, annotation symbols, and title block text. The challenge to the architectural drafter is to include all the necessary notes needed for the various phases of construction and yet make the plan easy to read. Annotation can also be divided into the categories of titles and text. Room names and the actual title of the drawings are examples of titles. General notes and ***local notes*** and text in schedules and legends are considered text. Titles are usually placed on the drawing using 1/4"-high text. If you're plotting at a scale of 1/4" = 1'-0" (1:50) and creating the titles in model space, this requires titles to be 12" (300 mm) high. Drawing titles may be as large as 24" (600 mm) high. Secondary headings such as the subtitles in schedules are often placed using letters 8" (200 mm) high. Text is usually 1/8" (3 mm) high. When created in model space and plotted at a scale of 1/4" = 1'-0", text will need to be 6" (150 mm) high. No matter the size, use an architectural font such as StylusBT or a similar third-party font.

General Notes

General notes apply to the overall drawing or to a specific group of items in the drawing. The lower right corner of the sheet is an ideal location for general notes, but they can be placed in any open area of a sheet that surrounds the drawing. Notes may also be placed in large open areas inside the drawings, as in the garage. Notes should not be placed closer than 1/2" (13 mm) from the drawing border or the actual drawing. Common general notes that might be found on a floor plan are shown in Figure 10.56. General notes are typically saved as a block and inserted into the floor plan drawing so they do not have to be typed for each project. Always be sure to edit stock notes to make sure they comply with the current project. Place general notes on the *FLOR ANNO NOTE* layer.

Local Notes

Local or ***specific notes*** relate to specific features within the drawing, such as an appliance or plumbing fixture. A specific note should be connected to a feature with a leader line. Figure 10.57 shows a portion of a floor plan and the local notes required to explain it. Place local notes on the *FLOR ANNO TEXT* layer. When possible, place all notes so lines of text run parallel to the long edge of the drawing paper. Some local notes

GENERAL NOTES:

1. ALL CONSTRUCTION TO BE IN COMPLIANCE W/ THE 2006 IRC.
2. CONTRACTOR & ALL SUBCONTRACTORS TO VERIFY ALL DIMENSIONS BEFORE ORDERING OR INSTALLING MATERIALS.
3. INSTALL ALL MATERIALS PER THE MANUF. SPECIFICATIONS.
4. ALL PENETRATIONS IN THE TOP OR BOTTOM PLATES FOR PLUMBING OR ELECTRICAL RUNS TO BE SEALED. SEE CAULKING NOTES.
5. PROVIDE 1/2" WATER RESISTANT GYPSUM BD. AROUND ALL TUBS, MODULAR SHOWERS, & SPAS.
6. PROVIDE 1/4"Ø COLD WATER LINE TO REFR.
7. VENT DRYER AND ALL FANS TO OUTSIDE AIR THROUGH VENTS W/ DAMPERS.
8. INSULATE THE WATER HEATER TO R-11. GAS W.H. TO BE ON 18" HIGH PLATFORM.
9. PROVIDE 1-HOUR FIREWALL BETWEEN GARAGE AND RESIDENCE BY PROVIDING 5/8" TYPE 'X' GYP. BD. FROM FLOOR TO ROOF SHEATHING. PROVIDE ALTERNATIVE BID FOR 1/2" GYP. BD. ON ALL GARAGE WALLS AND CEILINGS.

FIGURE 10.56 ■ *General notes can be saved as a block and inserted into the floor plan drawing so that they do not have to be typed for each project. Be sure to edit the notes to meet specific needs of each project.*

that are too complex or take up too much space may be placed with the general notes. These notes are then keyed to the floor plan with a short identification symbol or with a phrase such as

SEE NOTE #5.

Common information that may be specified in the form of local notes includes:

- Windows
- Doors
- Room titles, size, and building area
- Appliances
- Plumbing
- Fireplace and chimney
- Stairs
- Storage
- Miscellaneous notes

Windows

Window size and type can be placed directly on the floor plan if the plan is not complicated and clarity will be maintained. For a complex plan, window information should be placed in a schedule. Figure 10.57 shows window symbols that reference the windows to a schedule. The creation of schedules is explored later in this chapter. When placed on the floor plan, specifications should resemble Figure 10.58. Some professionals use this method to save time. There are three common methods of specifying windows on a floor plan:

6036, represents 6'-0'' × 3'-6''
$6^0 \times 3^6$, represents 6'-0'' × 3'-6''
6'-0'' × 3'-6''

Although these methods are easy to use, they should not be used when specific data must be identified. If window text is placed on the floor plan, use the *FLOR ANNO TEXT GLAZ* layer.

Doors

Door size and type can be placed directly on the floor plan or placed in a schedule. When placed on the floor plan, notation should resemble that in Figure 10.58. Place door text on the *FLOR ANNO TEXT DOOR* layer. Some drafters use the same system that is used for windows to place door information. Common methods to represent a door include:

2868 represents 2'-8'' wide × 6'-8'' high
$2^8 \times 6^8$ represents 2'-8'' wide × 6'-8'' high
2'-8'' × 6'-8''

- The height of doors may be omitted from the specification if a note is provided to indicate that all doors are 6'-8'' high unless noted. This simplified method of identification is best suited for housing in which the manufacturer of doors is not specified in the plans. Completing door schedules is explored later in this chapter.

Room Titles, Size, and Building Area

Place the room name in the center of all habitable rooms with the interior size below the name. Use text 1/4'' (6 mm) high for the title and 1/8'' (3 mm) high for the room sizes. Most professionals list the width (left/right) followed by the depth (top/bottom). On an irregularly shaped room similar to the room in Figure 10.59, list the size of the room based on the size that represents most of the room. Place the room title and size on the *FLOR ANNO TEXT IDEN* layer.

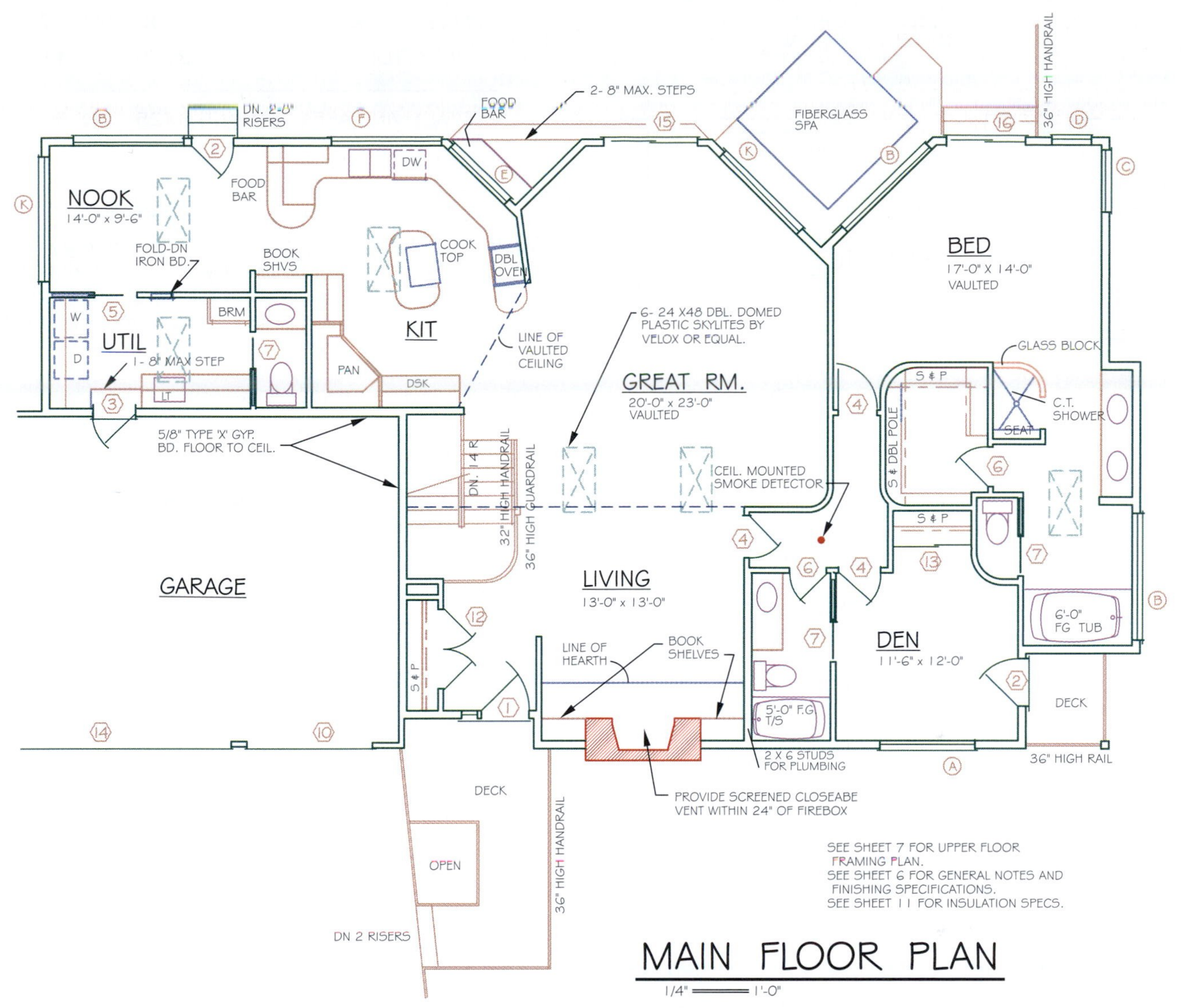

FIGURE 10.57 ■ *Local or specific notes relate to a specific feature within the drawing, such as an appliance or plumbing fixture. A specific note should be connected to a feature with a leader line.*

Determining Room Size

The size can be determined using the DIST (distance) or the LINE command of AutoCAD. The best accuracy is determined by setting OSNAP to ON. If the LINE command is used, select point A in Figure 10.59 as the first point of the line. To determine the distance between A and B, move the cursor to B, but do not select the "to point." As the cursor rests at B, check the coordinate display for the distance between A and B. Once the distance is noted, move to point C and check the coordinate display for that distance. Use this method to move through each room and write each room size on a sheet of paper. Once all sizes are recorded, the data can be placed in the drawing.

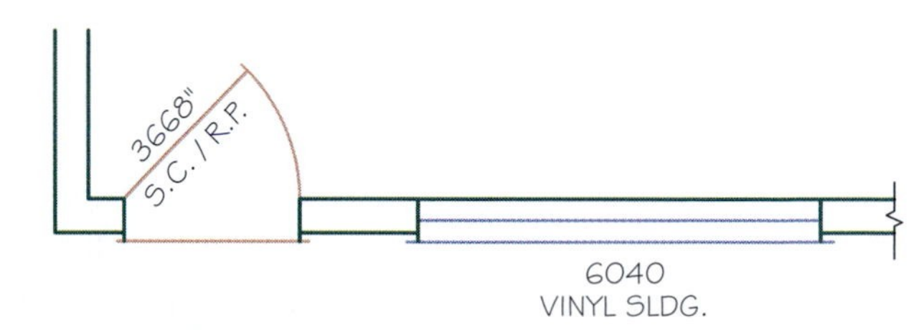

FIGURE 10.58 ■ *On a simple plan, the types and sizes of windows and doors can be placed directly on the floor plan.*

Determining the Building Area

In addition to listing the approximate size of all rooms, the area of each floor should be determined and listed on the drawing. Use the AREA command of AutoCAD to determine the size of the structure. Toggle OSNAP to ON to

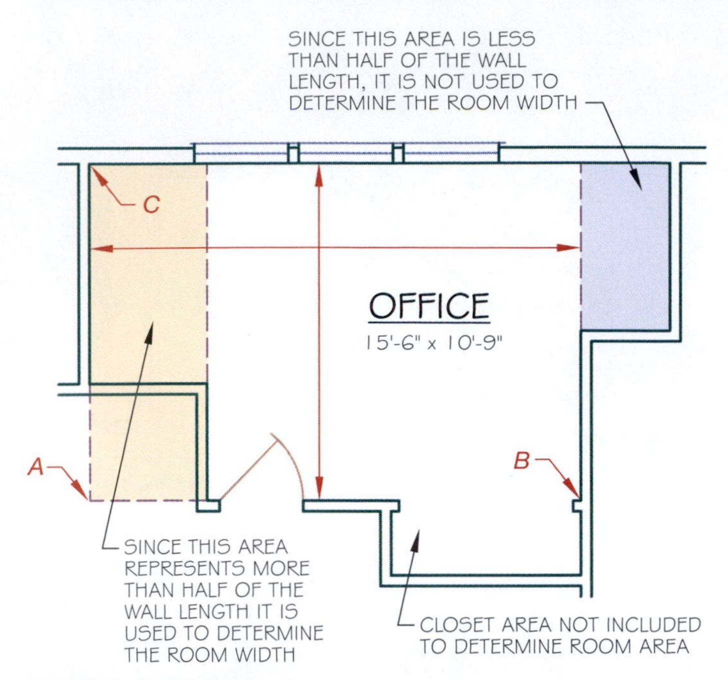
SINCE THIS AREA IS LESS THAN HALF OF THE WALL LENGTH, IT IS NOT USED TO DETERMINE THE ROOM WIDTH
C
OFFICE
15'-6" x 10'-9"
A
B
SINCE THIS AREA REPRESENTS MORE THAN HALF OF THE WALL LENGTH IT IS USED TO DETERMINE THE ROOM WIDTH
CLOSET AREA NOT INCLUDED TO DETERMINE ROOM AREA

FIGURE 10.59 ▪ *The names and sizes of all habitable rooms should be placed in the center of the room. The size can be determined using the LINE command with OSNAP set to ON. Select point A as the first point of the line. To determine the distance between A and B, move the cursor to B, but do not select the "to point." As the cursor rests at B, check the coordinate display for the distance between A and B. Once the distance has been noted, move to point C and check the coordinate display for that distance.*

UPPER FLOOR	1250 SQ. FT.
LOWER FLOOR	2175 SQ. FT.
TOTAL LIVING AREA	3425 SQ. FT.
GARAGE	822 SQ. FT.
TOTAL BLDG. AREA	4247 SQ. FT.

FIGURE 10.60 ▪ *The size of each floor, including all walls, should be listed and a total building area provided. These sizes are often used by the lender and building department to determine minimum safety requirements.*

increase accuracy as each "next point" is selected. Use the exterior corners of the structure to include the walls in the area. Determine the area of each floor and the garage area. Arrange the answers as you would an addition problem similar to the display in Figure 10.60

Appliances and Equipment

Label equipment such as furnace, water heater, dishwasher, compactor, refrigerator, and range. Place this text on the *FLOR ANNO TEXT* layer. Examples of notes to be placed can be seen throughout this chapter. Common notes include:

- **FAU** (forced air unit). List the power/fuel type (gas, electric, propane) manufacturer and size in BTUs if known).
  - **W.H.** (water heater). List the size in gallons if known and the fuel type.
  - **D.W.** (dishwasher).
  - **T.C.** (trash compactor).
  - **REFR.** (refrigerator). List the size, style if known, and specify a water line if an icemaker will be provided. A typical note would read **PROVIDE 38" WIDE SPACE FOR SIDE/SIDE REFR & SUPPLY 1/4" COLD WATER LINE TO REFR.**

For cooking units, list the size, type, venting method, and fuel source. Common options include:

- **30" GAS COOKTOP W/ POP-UP VENT.**
- **30" FREESTANDING RANGE. PROVIDE HOOD W/ LITE AND FAN. VENT TO OUTSIDE AIR.**
- **30" DROP-IN RANGE. PROVIDE HOOD W/ LITE AND FAN. VENT TO OUTSIDE AIR.**
- **27" WIDE DOUBLE OVEN**
- **30" WIDE MICRO OVER CONVECTION OVEN**

Plumbing

Label all tubs, showers, or spas, giving size, type, and material. Items that can be distinguished by shape, such as toilets and sinks, do not need to be identified. Place this text on the *FLOR ANNO TEXT* layer. Common notes include:

- **60" F.G. T/S** (fiberglass tub/shower)
- **60" C.I. FREE-STANDING TUB** (cast iron)
- **42" F.G. SHOWER** (fiberglass)
- **48" C.T. SHOWER W/ MULTIPLE HEADS** (ceramic tile)
- **84" × 96" F.G. SPA W/ TILE SURROUND**

Fireplace

Label each fireplace or solid fuel–burning appliance with a note such as:

- **MASONRY FIREPLACE W/ SCREENED CLOSABLE VENT W/IN 24" OF FIREBOX. PROVIDE 18" WIDE × 14" HIGH C.T. HEARTH.**

- **MAJESTIC 0-CLEARANCE FIREPLACE W/ DIRECT VENT METAL CHIMNEY.**
- **32 × 21" WOOD STOVE BY MAJESTIC W/ TRIPLE LINED METAL CHIMNEY. SET ON FLUSH MASONRY HEARTH. PROVIDE 4" MASONRY WALL PROTECTION OVER 1" AIR SPACE. PROVIDE SCREENED CLOSABLE VENT WITHIN 24" OF FIREBOX.**

Stairs

Label stairs, giving direction of travel, number of risers, and rail height. Common notes include:

- **Up 14 R** (risers)
- **34" HANDRAIL**
- **36" HIGH GUARDRAIL**
- **4" CONC. FLATWORK W/ 7 3/4" MAX. STEP.**
- **48" × 42 × 4" CONCRETE STOOP**

Storage

Label all closets by their use or their storage method. Common notations include: **5 SHELVES, S&P** (shelf and pole), **DBL POLE & SHELF,** closet organizer, linen, broom, or pantry.

Miscellaneous Notes

Specify access openings, masonry veneer, firewalls, changes in ceiling or floor levels, and upper-level projections.

- Specify attic crawl access openings with notes such as:
  - **22 × 30" ATTIC ACCESS**
  - **22 × 48" PULL-DN. ATTIC ACCESS W/ STEPS**
  - **22 × 30" CRAWL ACCESS**
- Specify masonry veneer with a note similar to:
  - **MASONRY VENEER OVER TYVEK AND 1" AIR SPACE W/ METAL TIES @ 24" O.C. EA. STUD**
- Designate a 1-hour firewall between the garage and residence with a note similar to one of the following notes:
  - **1/2 GYP. BD. ON WALLS & CEILING OF GARAGE**
  - **5/8 TYPE X GYP. BD. FROM FLOOR TO CEILING.**
- Specify changes in ceiling and floor levels. Notes can be placed below the room title or placed with a local note similar to the following:
  - **LINE OF VAULTED CEILING**
  - **LINE OF SUNKEN FLOOR**
- On multilevel structures, call out projections with notes such as
  - **LINE OF UPPER FLOOR**
  - **LINE OF BALCONY ABOVE**
  - **LINE OF LOWER FLOOR**

Schedules

Schedules allow drawings to be simplified by removing information from the drawing and placing it in a table. Although window, door, and finish information can be referenced as local notes, it is usually represented in a schedule. Information regarding appliances, fixtures, hardware, and finishes can also be placed in a schedule. To accurately convey information, symbols are placed on the floor plan to represent a material, such as a window; then a schedule is created to explain each specific window. The schedule can be included on the floor plan or placed on a separate sheet that contains all of the schedules and general notes for the entire project. The exact contents of the schedule will vary on the materials being referenced, but common components include:

- Manufacturer's name
- Product name
- Model number
- Type
- Quantity
- Size
- Rough opening size
- Color

During the early stages of plan development, the designer works with the clients to determine their preferences. A drafter may then be assigned the task of research, working with manufacturers' catalogs to find the products to be used. In addition to material supplied directly by vendors, the World Wide Web and Sweets catalogs are used to obtain needed information to complete the schedules. Place schedules on the *FLOR ANNO SCHD* layer and place symbols related to the schedule on the *FLOR ANNO SYMB* layer. The CD contains skeletons to develop a window schedule.

Creating a Window Schedule

Office practice dictates how schedules will be created. One method is to create a schedule in model space as previously discussed. An alternative method is to create the schedule in paper space. This method provides the benefit that the schedule can be inserted directly into a sheet without changing the text and title scale for a drawing of a different scale. The following guidelines can be used to create a window schedule:

- Use text 1/8'' high (6'').
- Place window symbols on a layer separate from the schedule and other text.
- Place all schedules and notes on a separate layer.
- Place all general notes related to windows by the schedule on the *FLOR ANNO SCHD* layer.
- List skylights in the window schedule if they are supplied by the window manufacturer.

Before working at a computer, plan the schedule on a print of the floor plan. Use the following steps to create a window schedule.

1. Start at the front door and work counterclockwise or clockwise to identify window sizes.
2. Represent each different window with a letter starting with A. Do not use the letters I or O.
   - Each window that is similar in every way should have the same letter.
   - If a window is different in any way from other similar windows, it should have a separate symbol from other similar windows.
   - Windows from all floors should be included in one schedule.
3. Keep track of symbols to be used: the size, type, model number, and quantity of the windows. If wood windows are used, you will also need to create a column for rough opening size. Use roman numerals to track the quantity.

Once a rough schedule has been determined, create the schedule using the TABLE command. Figure 10.61 shows an example of a window schedule using vinyl windows. Figure 10.62 shows an example of similar windows using wood frames. Use the following steps to create the schedule.

1. Use text 12'' (300 mm) high for the titles, 8'' (200 mm) high for subtitles, and 6'' (150 mm) high for the text.
2. Use thick lines to represent the schedule box and provide a space equal to half of the height of the lettering between the text and the lines.

WINDOW SCHEDULE

SYM	SIZE	TYPE	QUAN
A	5'-0" X 3'-6"	SLDG. W/ GRIDS	1
B	2'-6" X 4'-0"	CASEMENT W/ GRIDS	2
C	6'-0" X 4'-0"	SLIDING W/ GRIDS	2
D	3'-0" X 5'-0"	CASEMENT - TEMP.	1
E	7'-0" X 5'-0"	FIXED - TEMPERED	1
F	3'-0" X 2'-0"	SLDG.	2
G	1'-0" X 5'-0"	FIXED	2
H	1'-0" X 2'-0"	TRANSOM	2
J	5'-0" X 5'-0"	SLDG.	3
K	5'-0" X 2'-0"	TRANSOM	2
L	6'-0" X 5'-0"	FIXED	1
M	9'-0" X 2'-0"	TRANSOM	1
N	5'-0" X 2'-6"	1/2" ROUND	1
P	6'-0" X 5'-0"	SLDG.	1

WINDOW NOTES:

1. ALL WINDOWS TO BE U-.40 MIN. PROVIDE ALT. BIDD FOR U=.54. IF FLAT CEILINGS ARE INSULATED TO R=49
2. ALL WINDOWS TO BE MILGARD OR EQUAL W/ VINYL FRAMES.
3. ALL BEDROOM WINDOWS TO BE WITH IN 44" OF FIN. FLOOR.
4. ALL WINDOW HEADERS TO BE SET AT 6'-10" UNLESS NOTED.
5. UNLESS NOTED ALL HDRS. OVER EXT. DOORS AND WINDOWS TO BE 4" WIDE WITH 2" RIGID INSUL. BACKER. SEE TYP. SECTION.

SKYLITES NOTES:

1. ALL SKYLITES TO BE VELOX OR EQUAL DOUBLE DOMED PLASTIC OPENABLE SKYLITES. U=.50 MIN.

FIGURE 10.61 ■ *A window schedule is a convenient method of listing required window information and at the same time keeping the floor plan from becoming cluttered. A letter is used to represent each type of window, with all windows from each floor located in one schedule. An additional column may be added to the schedule for comments.*

3. List the width and height of windows in a schedule. Always use foot, inch, and dash symbols, so a listing resembles 6'-0'' × 4'-0''.
4. Place text in a circle 1/4'' (12'') in diameter to represent the window on the floor plan. Place the symbol on the outside of the window and offset from the center. Figure 10.63 shows how window symbols should be represented on the floor plan.

The window schedule should resemble Figure 10.61 when completed. Add notes as necessary to describe windows. Common notes include:

- **ALL WINDOWS TO BE U-0.40 MIN.**
- **ALL WINDOWS TO BE VINYL FRAMED BY MILGARD OR EQUAL.**

WINDOW SCHEDULE	ALL WINDOWS TO BE POZZI OR EQUAL				
SYM	SIZE	TYPE	MODEL *	ROUGH OPENING	QUAN
A	6'-10 1/4" X 4'-0"	CSM/PIC.	CR14/CP24/CR14	6'-10 3/4" X 4'-0 1/2"	2
B	7'-1 3/8" X 4'-0"	CASEMENT	CW14-3	7'-1 7/8" X 4'-0 1/2"	1
C	4'-8 1/2" X 4'-0"	CASEMENT	CW24	4'-9" X 4'-0 1/2"	1
D	4'-8 1/2" X 3'-4 13/16"	CASEMENT	CW235	4'-9" X 3'-5 3/8"	5
E	2'-4 3/8" X 4'-0"	CASEMENT	CW14	2'-4 7/8" X 4'-0 1/2"	1
F	2'-4 3/8" X 1'-4 1/2"	1/2 ROUND	CTCW1	2'-4 7/8" X 1'-5"	2
G	2'-4 3/8" X 2'-11 15/16"	CASEMENT	CW13	2'-4 7/8" X 3'-0 1/2"	
H	5'-11 7/8" X 4'-0"	PICTURE	CP34		
J	4'-0" X 3'-4 3/4"				

FIGURE 10.62 ■ *In addition to the information normally found in a window schedule, a schedule for wood-frame windows must include columns for the model number and the rough opening size. The rough opening size is determined by consulting the manufacture's specifications.*

- **ALL BEDROOM WINDOWS TO BE WITHIN 44" OF THE FIN. FLOOR.**
- **ALL WINDOW HEADERS TO BE SET AT 6'-8" UNLESS NOTED.**
- **ALL WINDOWS WITHIN 18" OF THE FLOOR OR A DOOR ARE TO BE TEMPERED.**
- **UNLESS NOTED, ALL HEADERS OVER EXTERIOR WINDOWS TO BE 4" WIDE W/ 2" RIGID INSULATION BACKER. SEE TYPICAL SECTION.**
- **PROVIDE AN ALTERNATE BID FOR ALL WINDOWS TO BE WOOD-FRAMED BY HILLSDALE POZZI OR EQUAL.**

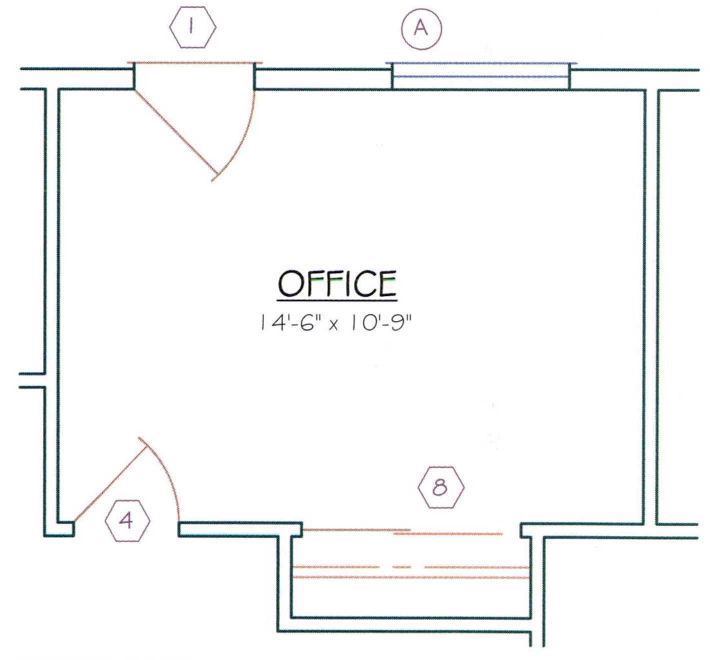

FIGURE 10.63 ■ *Symbols for windows and doors should be placed near the appropriate fixtures. Place symbols on the exterior side of the fixture when it is located in an exterior wall and centered in or near objects in the interior of the house.*

Creating a Door Schedule

A door schedule can be created using similar steps used to create the window schedule:

- Use text 1/8'' high (6'').
- Place door symbols on a layer separate from the schedule and other text.
- Place the door schedule and notes on a layer separate from door symbols and text.
- Place the schedule and all general notes related to the doors by the schedule on the *FLOR ANNO SCHD* layer.

Before working at a computer, plan the door schedule on a print of the floor plan. Use the following steps to create the schedule.

1. Start at the front door and list all exterior swinging doors sequentially from the largest to smallest.

- Each door that is similar in every way should have the same number.
- If a door is different in any way from other similar doors it should have a separate symbol.
- Doors from all floors should be included in one schedule.

2. List all interior swinging doors from largest to smallest.
3. List all doors by group, such as swinging, pocket, folding, bifold, and so on, from largest to smallest. The last door listed should be the overhead garage door.
4. Represent doors on the floor plan with a number 6'' high placed in a hexagon (circumscribed around a 12'' diameter circle). Place the hexagon near the door swing.
5. Common door abbreviations: MI = metal insulated, RP = raised-panel, SC = solid-core, SC/SC = solid-core/self-closing.

DOOR SCHEDULE

SYM.	SIZE	TYPE	QUAN.
1	3'-6" X 8'-0"	S.C. RAISED PANEL	1
2	2'-8" X 6'-8"	METAL INSUL.	3
3	2'-8" X 6'-8"	1-LITE FRENCH	1
4	9'-0" X 6'-8"	SLIDING - TEMPERED	1
5	9'-0" X 6'-8"	1-LITE FRENCH	1
6	6'-0" X 6'-8"	SLIDING - TEMPERED	1
7	2'-8" X 6'-8"	M. I./ SELF CLOSING	1
8	2'-8" X 6'-8"	INTERIOR	7
9	2'-6" X 6'-8"	INTERIOR	4
10	2'-8" X 6'-8"	POCKET	1
11	2'-6" X 6'-8"	POCKET	4
12	PR. 2'-6" X 6'-8"	PR. 1-LITE FRENCH	1
13	6'-0" X 6'-8"	BI-PASS	2
14	4'-0" X 6'-8"	BI-PASS	1
15	9'-0" X 8'-0"	OVERHEAD GARAGE	3

DOOR NOTES:

1. FRONT DOOR TO BE RATED AT U 0.54 OR LESS.
2. EXTERIOR DOORS IN HEATED WALLS TO BE U 0.20 OR LESS.
3. DOORS THAT EXCEED 50% GLASS ARE TO BE U 0.40 OR LESS.
4. ALL INTERIOR DOORS TO BE RAISED 6 PANEL DOORS.

FIGURE 10.64 ■ *A simplified door schedule is often used on homes that are drawn as spec homes. When a home is drawn for a specific owner, the schedule may include the door thickness, the manufacturer's model number, the side of the door to be hinged, finish, threshold type, hardware type, and rough opening size.*

The door schedule should resemble Figure 10.64 when completed. Add notes as necessary to describe doors. Common notes might include:

- **FRONT DOOR TO BE RATED AT .54 OR LESS.**
- **EXTERIOR DOORS IN HEATED WALLS TO BE U.20 OR LESS.**
- **DOORS THAT EXCEED 50% GLASS ARE TO BE U.40 OR LESS.**
- **ALL GLASS IN DOORS TO BE TEMPERED.**

Finish Materials Schedules

Representation of finish materials is usually restricted to the presentation floor plan, and not shown on the floor plan associated with the working drawings. The finish materials are usually specified on the floor plan with notes or key symbols that relate to a finish schedule. Figure 10.65 shows a typical interior finish schedule. Listing each room and then providing a menu of options is a common method of creating a finish schedule. Another option is to provide a column for each wall of each room and a listing of finishing materials. Place the finish schedule on the *FLOR ANNO SCHD* layer and place finish symbols related to the schedule on the *FLOR ANNO LEGN* layer. When floor finishes are identified, the easiest method is to label the material directly under the room designation, as shown in Figure 10.63.

Legends

Legends are used to explain symbols that are referenced to a specific plan. Common legends found on a set of residential plans include those used to explain electrical, HVAC, and plumbing symbols. A legend to explain types of wall construction should be provided on renovations or additions to explain existing, new, and partial walls as well as walls to be removed. If legends are used on the floor plan, place them on the *FLOR ANNO SCHD* layer. Legends may also be placed on a separate sheet with the schedules and general notes.

Title Block Text

The title block consists of two types of text. Some text is based on the office logo and includes the office name, legal disclaimers, and other types of information that appears on every sheet. This text is part of the office template used for all drawings and was discussed in Chapter 3. This text is usually created on layers such as *TITL BLCK ANNO* and should not be altered. Other text

INTERIOR FINISH SCHEDULE															
ROOM	FLOOR					WALLS								CEILING	
	CARPET	CERAMIC TILE	SHEET VINYL	HARDWOOD-OAK	CONCRETE	PAINT N/E	PAINT S/W	PAPER N/E	PAPER S/W	WAINSCOT LOWER 36" N/E	WAINSCOT LOWER 36" S/W	SPRAY N/E	SPRAY S/W	SMOOTH	SPRAY
LIVING				•		ALL				ALL					•
DINING				•		ALL				ALL					•
BED. 1	•					N • E •	S • W •	E •	S •	N •		N •	W •		•
BED.2	•					ALL						ALL			•
BED. 3	•					ALL						ALL			•
FAMILY	•					ALL					S •	ALL			•
DEN				•		ALL						ALL			•
KITCHEN		•						ALL						•	
UTILITY		•												•	
STAIRS															

FIGURE 10.65 ■ *A finish schedule is often provided for custom homes to specify floor, wall, and ceiling treatments. General categories are typically listed in the schedule, but specific manufacturers, styles, and types of materials may also be listed in notes that accompany the schedule.*

must be altered for each project. This includes information such as:

- Sheet number
- Date of completion
- Revision date
- Client name
- Sheet contents

Place text for the title block on the *FLOR ANNO TTLB* layer. By providing a separate layer, text can be frozen, allowing the base floor drawings to be used for the electrical drawings without displaying floor-related text. Because information such as the date of completion and the revision date will be the same for all drawings within the project, this can be placed on the *TITL BLOCK TEXT* layer.

PLACING DIMENSIONS ON A FLOOR PLAN

The use of dimensions was introduced in Chapter 3. The use of dimensions on the floor plan varies with office practice and the complexity of the project. On a simple project similar to the residence in Figure 10.66, dimensions and other information related to the framing of the structure is included with the floor plan. For most residential projects, a separate plan is provided, so dimensions and framing-related information can be separated from the architectural information on the floor plan. Chapter 24 explains the use of dimensions related to the framing plan. Figure 10.67 compares similar portions of a floor plan and a framing plan.

If used, dimensions on the floor plan locate walls, windows, doors, and interior features. Not all windows, doors, and interior furnishing have dimension lines to locate them if they can be located by their position to a known wall. The material being represented also affects the method of placing dimensions.

Dimension Placement

Placing dimensions in plan view are divided into the areas of interior and exterior dimensions. Whenever possible, dimensions should be placed outside of the drawing area.

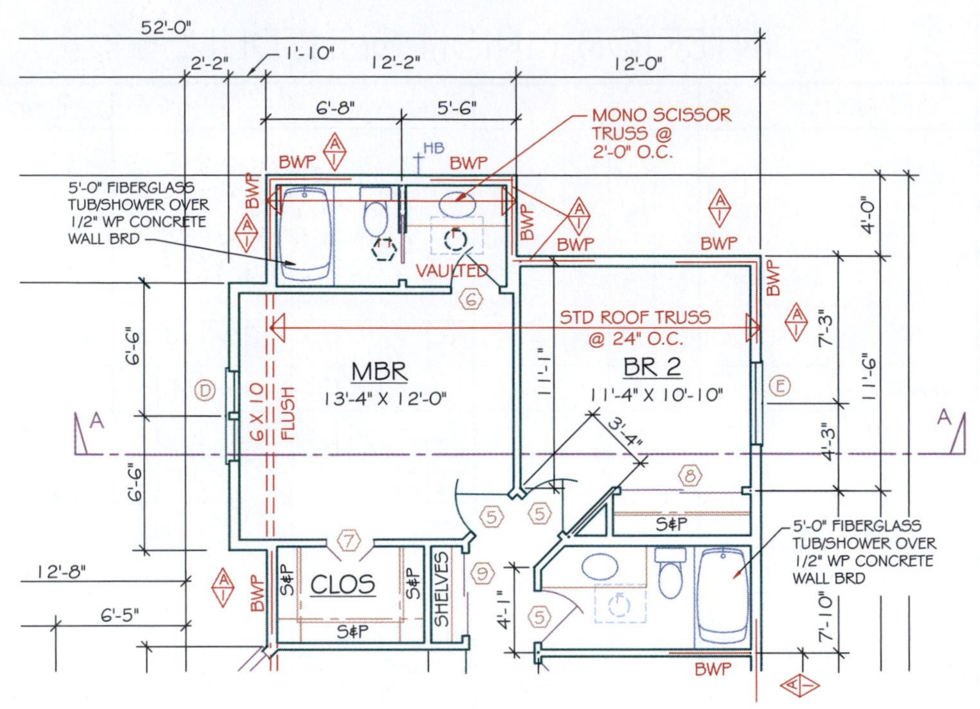

FIGURE 10.66 ■ *Dimensions and other information related to the framing of the structure can be included with the floor plan.*

Exterior Dimensions

Most offices start by placing an overall dimension on each side of the structure that is approximately 2'' (50 mm) from the exterior wall when plotted. Moving inward, approximately 1/2'' (13 mm) is placed between lines that are used to describe major jogs in exterior walls, the distance from wall to wall, and the distance from wall to opening. The exact method to place these dimensions will vary slightly depending on the material being used. Figure 10.68 shows an example of exterior dimensions for a structure.

In placing dimensions, it is important to adjust the use of four dimension lines to the shape of the structure. If one side of the structure has no offsets, the wall-to-wall and wall-to-opening lines of dimensions can be moved out from the structure. If one portion of a structure has no interior walls and several windows, the dimensions for wall-to-windows would be moved out as far as possible from the structure. It is not important that four lines of dimensioning always be maintained. It is critical, however, that each line of dimension adds up to the corresponding dimension on each succeeding dimension line, as seen in Figure 10.68.

Interior Dimensions

The two main considerations in placing interior dimensions are clarity and groupings. Dimension lines and text must be placed so they can be read easily and so neither interferes with other information that must be placed on the drawing. Information should also be grouped together as seen in Figure 10.69, so construction workers can find dimensions easily. Interior dimensions should be placed so they extend between the location of a surface described by the exterior dimensions. Interior walls are generally dimensioned to the center of stud. Some architectural firms dimension interior walls to each face of interior studs. Doors located in the corner or centers of rooms are generally not dimensioned.

Dimensioning Small Spaces

Often small areas that do not allow enough space for the text to be placed between extension lines must be dimensioned. Although options vary with each office, several alternatives to place dimensions in small spaces can be seen in Figure 10.70.

Metric Dimensioning

Structures dimensioned using metric measurement are defined in millimeters. Large distances may be defined in meters. When all units are given in millimeters, the unit of measurement requires no identification. For instance, a distance of 2440 mm would be written as 2440. Dimensions other than millimeters should have the unit description by the dimension, such as 24 m. See Chapter 3 for metric conversions and equivalents.

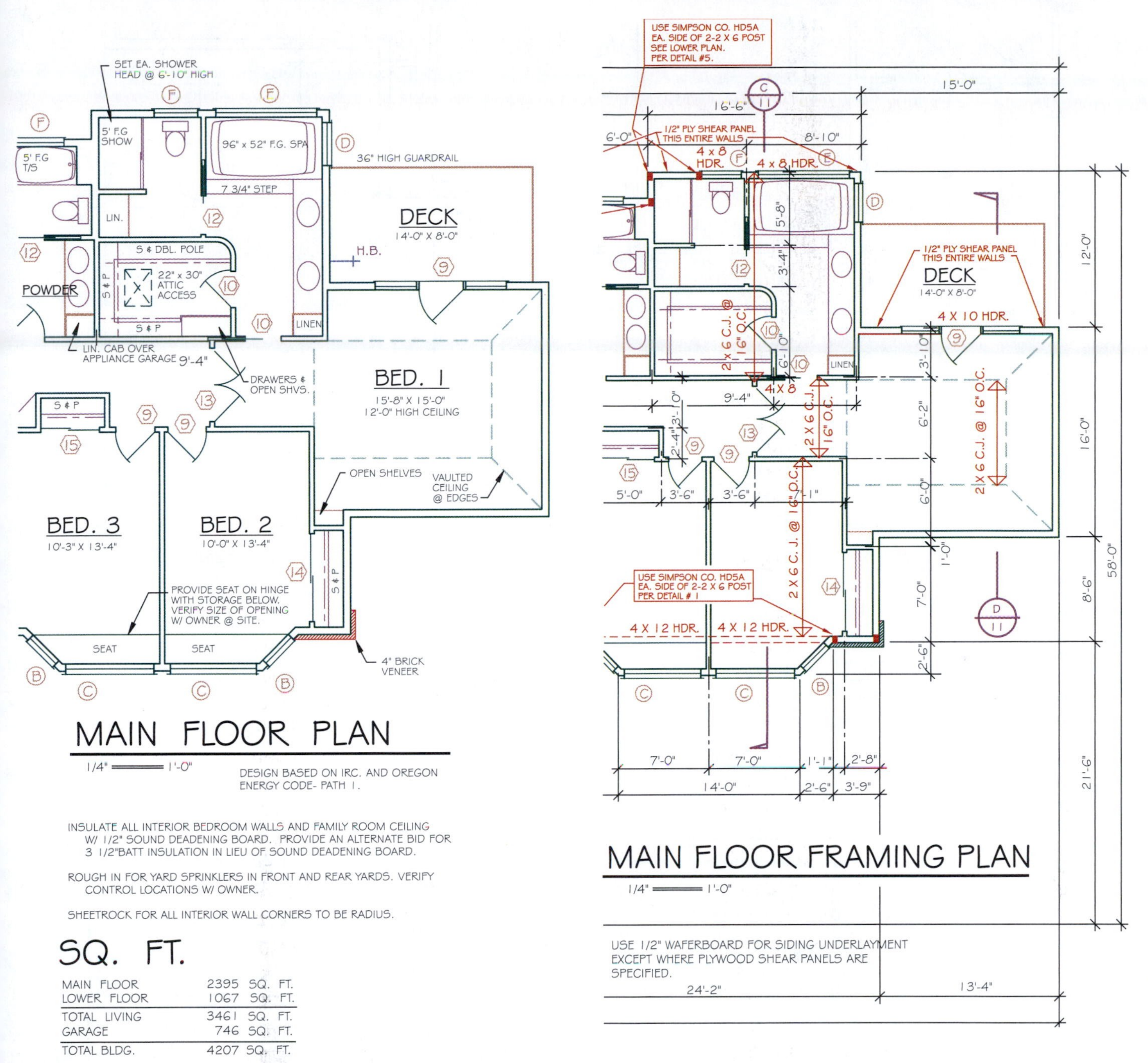

FIGURE 10.67 ■ *(A) A floor plan with architectural information only. (B) A framing plan can be used to display all dimensions and structural information needed by the framing crew.*

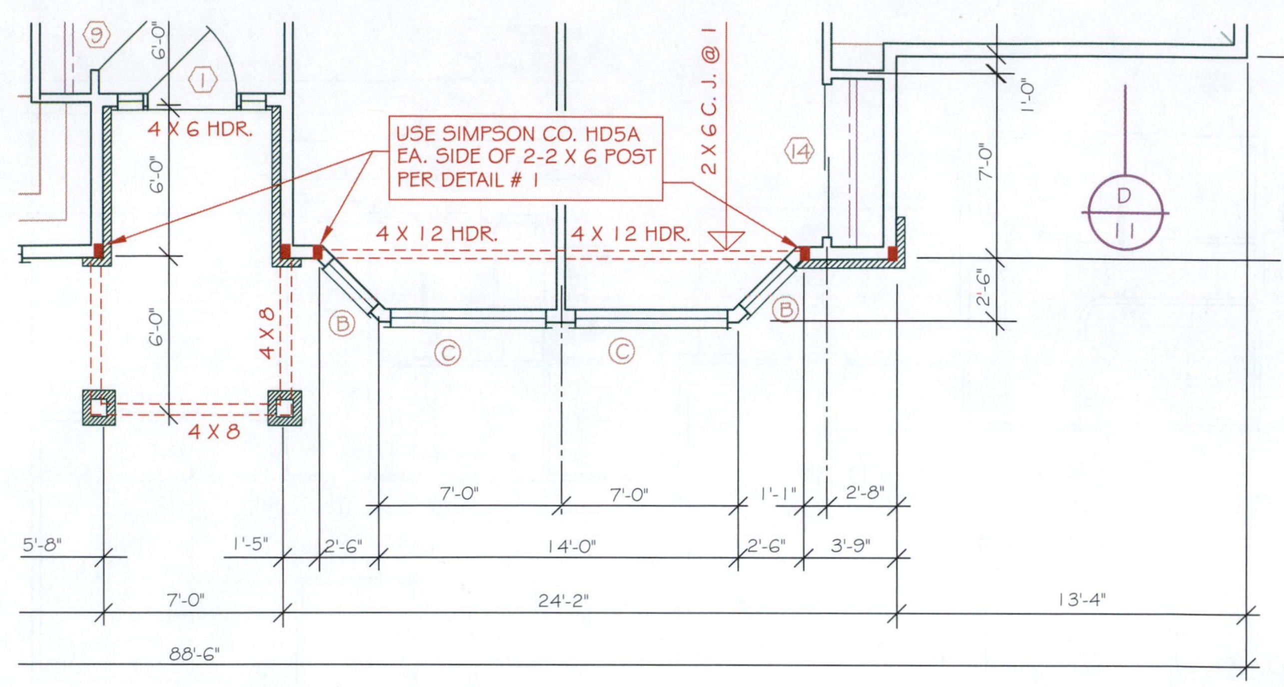

FIGURE 10.68 ■ *Exterior dimensions for light frame construction are placed to reference the exterior sides of the exterior walls and the centers of interior walls.*

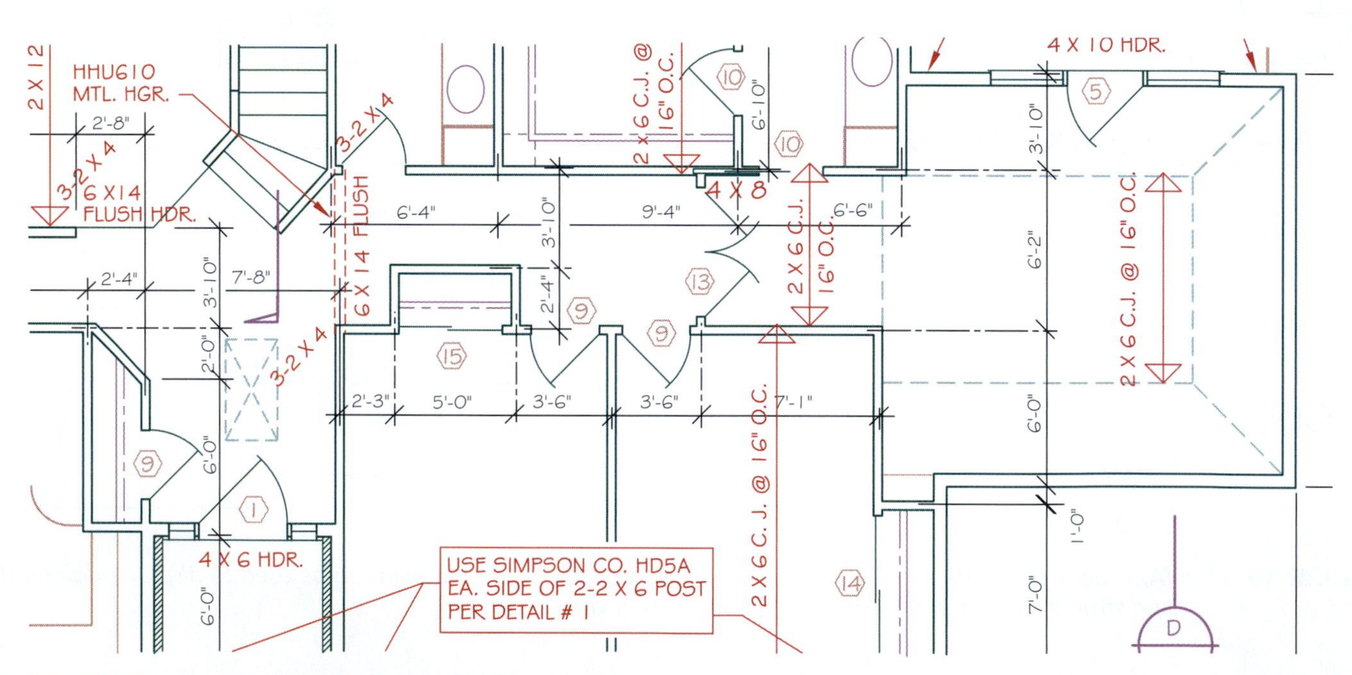

FIGURE 10.69 ■ *Interior dimensions should be placed so that they can be read easily and do not interfere with other information that must be placed on the drawing.*

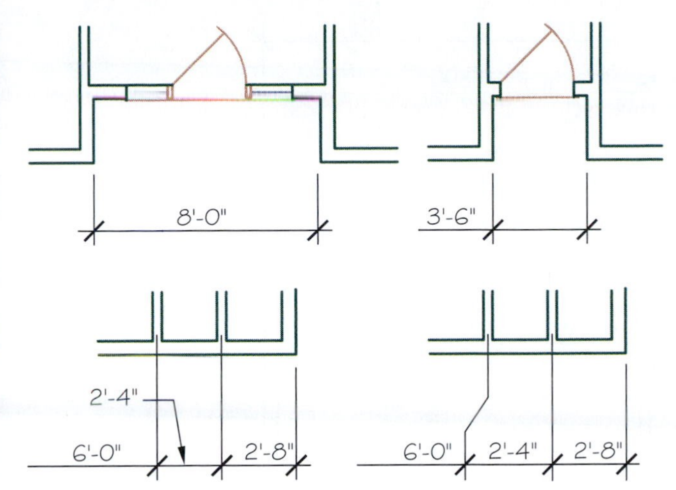

FIGURE 10.70 ■ *Several alternatives are available for placing dimension in small spaces.*

Light Frame Dimensions

Light frame structures are usually dimensioned by using the four dimension groupings just presented. Dimensions are placed on each side of the structure to define:

- The outer line defines the overall limits of the structure.
- The next line of dimensions describes changes in shape or major jogs. The changes are dimensioned from exterior face of the material used to build one wall to the exterior face of the next wall. The extension line is sometimes labeled with text parallel to the extension line that reads F.O.S. or F.S., representing the face of stud.
- The third line of dimensions represents the distance from the edge of exterior walls to the center (edge for some architectural offices) of interior walls.
- The fourth line of dimensions locates openings, and extends from a wall to the center of the openings.

Figures 10.68 and 10.69 show dimensioning methods for a light frame structure.

Masonry Dimensions

Structures made of concrete block, poured concrete, or tilt-up concrete are dimensioned using similar line placement methods as with light frame construction. Differences include the following:

- Masonry walls are always dimensioned to edge and never to center.
- Openings for doors and windows are also dimensioned from edge rather than from center.
- When the structure is made of concrete block, all dimensions should be based on 8'' (200 mm) modules. Distances in odd numbers of feet should always end in a 4'' (100 mm) increment, such as 15'-4''. Even-numbered distances will always end in 0'' or 8'' increments such as 8'-0'' or 10'-8'' to be modular. Interior light frame walls are dimensioned to their centers as previously described. Figure 10.71 shows examples of common dimensioning practices for a masonry structure.

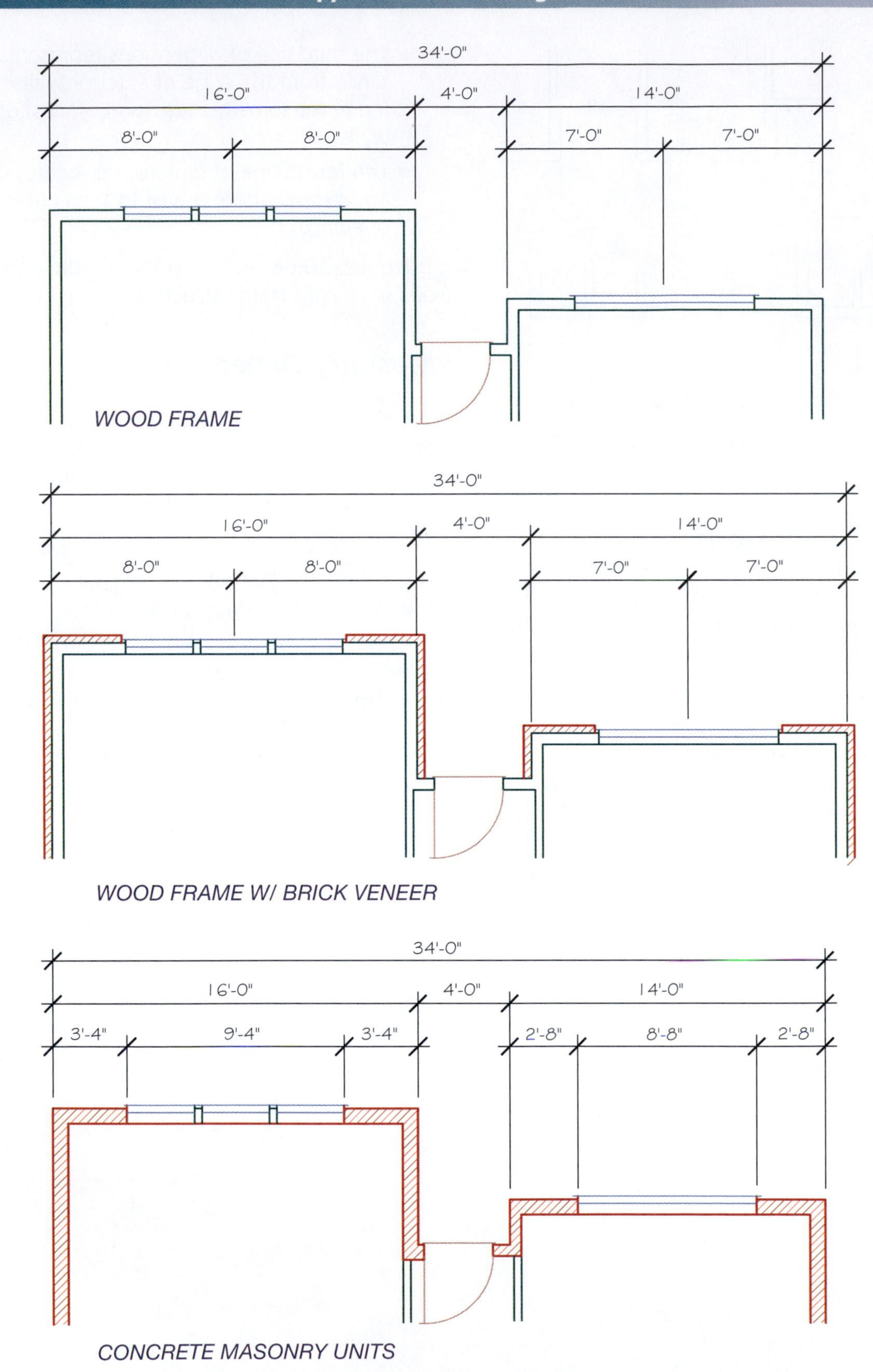

FIGURE 10.71 ■ *Placing dimensions for wood, wood with masonry veneer, and masonry structures.*

CHAPTER 10

Additional Readings

The following websites can be used as a resource to help you keep current with changes in floor plan–related materials.

ADDRESS	COMPANY OR ORGANIZATION
APPLIANCES	
www.geappliances.com	Home appliances
www.kitchenaid.com	Kitchen appliances
www.nutone.com	Nutone
www.subsero.com	Sub-zero refrigerators
www.whirlpool.com	Whirlpool Home Appliances
CABINETS AND FURNITURE	
www.aristokraft.com	Aristokraft Cabinetry
www.homecrestcab.com	HomeCrest (Cabinets)
www.kitchenaid.com	KitchenAid Home Appliances
www.kraftmaid.com	Kitchen and bath cabinets
www.schultestorage.com	Schulte Storage Systems
www.ssina.com	The Stainless Steel Information Center
www.temafurniture.com	Bedroom furniture
www.woodwork-lowes.com	Cabinets
DOORS AND WINDOWS	
www.andersonwindows.com	Anderson Windows
www.crestlineonline.com	Crestline Windows and Doors
www.jeld-wen.com	Jeld-Wen windows and doors
www.milgard.com	Vinyl aluminum, fiberglass and wood windows, doors, and skylights
www.marvin.com	Marvin Windows and Doors
www.pella.com	Pella
FIREPLACES	
www.heatilator.com	Heatilator (Fireplaces)
www.heatnglo.com	Heat-N-Glo (Fireplaces)
www.majesticfireplaces.com	Prefabricated fireplaces

HVAC

www.waterfurnace.com	Water Furnace International, Inc.

MISCELLANEOUS

www.beamvac.com	Beam Central Cleaning Systems
www.builtinvacuum.com	M.D. Manufacturing
www.easyclosets.com	Closet storage units
www.generalshale.com	General Shale Brick
www.homedepot.com	Kitchen, bath, and lighting supplies
www.homecenter.com	Kitchen, bath, and lighting supplies
www.lowes.com	Kitchen, bath, and lighting supplies
www.abracadata.com	Abracadata

PLUMBING

www.americanstandard-us.com	Bath and kitchen plumbing fixtures
www.jacuzzipremium.com	Jacuzzi spas
www.us.kohler.com	Kitchen and bath fixtures
www.peerlesspottery.com	Bath and kitchen fixtures, bath and kitchen hardware

SOFTWARE

www.autodesk.com	Supplier of engineering software great and small
www.builderswebsource.com	Builders Web source for software
www.caddepot.com	Symbol libraries
www.cadprosoftware.com	CADPro drawing program
www.catalog.com	Symbol libraries
www.graphisoft.com	Graphisoft-ArchiCAD software

CHAPTER 10

Floor Plan Symbols and Annotation Test

DIRECTIONS

Answer the following questions with short, complete statements. Type your answers using a word processor or neatly print the answers on lined paper.

QUESTIONS

10.1. How thick are exterior walls for a wood-frame residence usually drawn in your area?
10.2. Interior walls are commonly drawn how thick?
10.3. How does the floor plan symbol for an exterior door differ from the symbol for an interior door?
10.4. What are the recommended spaces for the following?
a. Wardrobe closet depth
b. Fireplace hearth for a 36'' wide opening
c. Stair width
d. Fireplace hearth depth
10.5. Describe an advantage of using schedules for windows instead of placing the information on the floor plan.
10.6. Describe an advantage of placing window and door information on the floor plan.
10.7. Sketch the following floor plan symbols:
a. Pocket door
b. Bifold closet door
c. Casement window
d. Sliding window
e. Skylight
f. Hose bibb
10.8. What is the minimum required height of a guardrail?
10.9. A deck is 32'' above the ground. Is a guardrail required?
10.10. When is a guardrail required for interior floor changes?
10.11. What style of window is 100% openable?
10.12. What does the note 6040 CSM next to a window on a floor plan mean?
10.13. What is the required minimum headroom height for an attic access?
10.14. List possible reasons that an appliance might be shown with dashed lines on the floor plan.
10.15. Explain options for representing the upper cabinets on the floor plan.
10.16. List factors that may influence window sizes.
10.17. What is the purpose of a flue?
10.18. Determine the minimum size for a fireplace based on current code requirements.
10.19. Give the flue size for a standard fireplace opening 36'' W × 28'' H × 20'' D.
10.20. Describe a local note and give an example of a local note that might be found in a bathroom.

CHAPTER 10

Drawing Problems

Use the symbols on the CD\Drawing Blocks\Floor Blocks and convert each symbol to a drawing block. Assign all drawing objects to appropriate layers. Provide appropriate insertion points and names to easily identify each block.

CHAPTER 11

Completing a Floor Plan

A complete floor plan is shown in Figure 11.1. This floor plan is shown without the electrical layout, dimensions, or structural elements. Designing and drawing the electrical plan is discussed in Chapter 12; representing structural members is covered in Chapter 24. The step-by-step method used to draw this floor plan provides you with a method to lay out any floor plan. As you progress through the step-by-step discussion of the floor plan layout, refer back to Chapter 10 to review specific floor plan symbols. Keep in mind that these layout techniques represent a suggested typical method to establish a complete floor plan. You may alter these layout steps to suit your individual preference as your skills develop and your knowledge increases. A floor plan can be completed using the following steps:

1. *Plan the drawing*
2. *Draw the walls*
3. *Insert all window blocks*
4. *Insert all door blocks*
5. *Represent all cabinets*
6. *Insert or draw all symbols*
7. *Place annotation*
8. *Place dimensions*
9. *Place title block text*

This discussion assumes that a drawing template is being used and the drawing will be completed in model space. See Chapter 3 for a review of drawing templates and the use of model and paper space.

PLANNING THE DRAWING

Step 1. Open the drawing template provided by your school or office and verify that the template is appropriate to complete a floor plan. This would include:

- Toggle the template to make MODEL space active.
- Verify the LTSCALE and DIMSCALE settings are appropriate for eventual plotting at a scale of 1/4'' = 1'-0''. The final drawing will be plotted within the selected template.
- Verify that the PSLTSCALE setting is zero.

Step 2. Adjust the VIEWPORT limits to include all space within the borders. The size of the viewport will determine the drawing area.

- The drawing area should include the area for the floor plan plus space to include future dimensions required when completely drawn. Residential floor

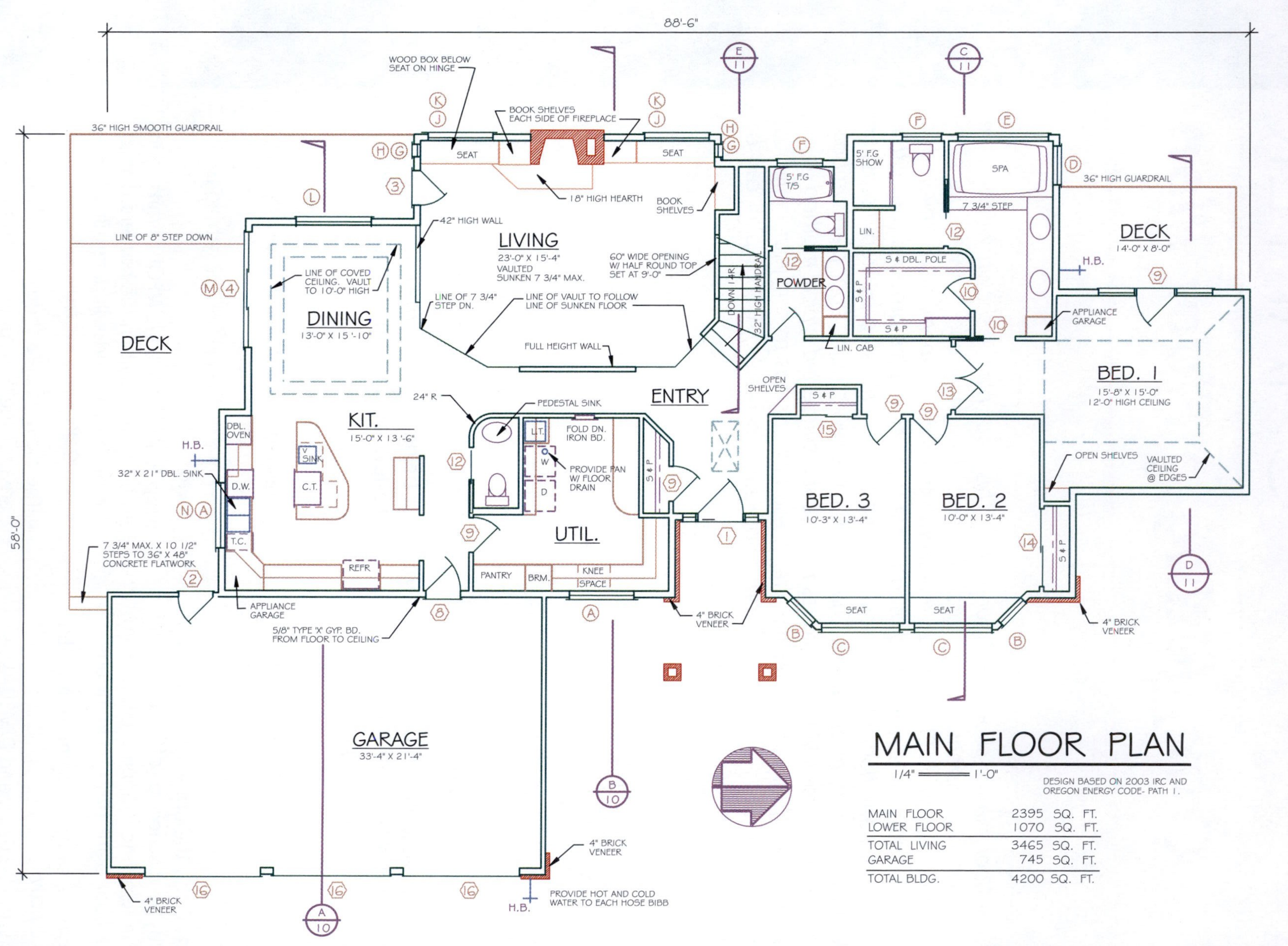

FIGURE 11.1 ■ *A floor plan is used to show all of the architectural information required by the lending institution, building department, and building crew to understand the home.*

plans are usually drawn to a scale of 1/4'' = 1'-0''. The house to be drawn is 89'' long and 58'' wide. This size does not take into consideration any area needed for notes or dimensions. Even though the dimensions will be placed on the framing plan, space should be allowed now. However, in this case there is enough space for dimensions because the house size is much less than the working area. Provide at least 2'' (paper space) on each side of the house for dimensions.

- If the plan does not fit within the working area, increase the sheet size, reduce the drawing scale, or divide the plan to fit on two or more drawing sheets. Confirm your decision with your instructor or job captain to be sure you conform to the governing standards.

Step 3. Consider the orientation of the house or building on the template. There are two standard ways to orient the building to the border.

- Draw the building on the sheet with the entry door parallel to the long border of the template, as in Figure 11.1. This is normally done when the building has not been designed for any specific construction site.
- Place the building on the sheet with the north direction pointing toward the top of the sheet. This is normally done when the building is designed for a specific construction site.

Step 4. Plan your work before drawing with the computer. The houses in problems 1 through 10 at the end of the chapter provide a skeleton but no actual room names. Use the guidelines presented in Chapter 3 and plan:

- Room layout
- Plumbing layouts
- Appliance locations

Create a print of the desired floor plan and mark your intended layout on the print. If you are completing one of the problems from 11 to 20 or a similar project, there is no skeleton to use as a base. Careful planning is imperative.

Step 5. Update your template layers. Your template should include the layers found on the CD\Exercises\Layer names file. If these layers are not part of your template, create each layer, and assign a color of your choice as well as the appropriate lineweight and linetype to each layer. See Appendix F for common layer names.

Step 6. When you've been given only a paper copy of a floor plan, a drawing scale is useful to determine room sizes. Unfortunately a plan from a plan book is proportional but an exact scale is not known. A simple measuring device can be determined using the following steps:

- Use a piece of paper and place it next to a bathtub.
- Place a dot on the edge of the paper to locate each end of the tub. This space is typically 60'' long.
- Divide the space between the two dots into five equal spaces. This now provides you with a 5' long measuring device.
- Check the accuracy using this measuring device to measure the front door. This size is normally 3' wide.

Use the scale to measure unlisted sizes on the floor plan. Use the minimum sizes presented in Chapters 4 and 10 and your measuring scale to estimate unspecified sizes.

Note:

Professional designers and architects have donated the plans in this book knowing that they will be copied and redrawn. It is perfectly legal for you to reproduce these plans for school projects. It is important to realize though, that it is not OK to reproduce these drawings for resale. Laws vary for each state, but as a general rule, unless you've changed over 50% of the plan or significantly altered the appearance of the design, you may be found guilty of copyright infringement. Never take a plan from another professional, make minor changes, and then call it your design. It's lazy, unethical, and illegal.

DRAWING THE WALLS

The method used to complete this step will depend on what you've been given. If you are working with a project similar to Problems 11.1 through 11.10, most of the walls have already been drawn. This drawing will need to be inserted into your drawing template. If you need to start a drawing from a sketch, proceed to the next section.

Inserting a Drawing

Insert a preliminary drawing into a template using the following steps:

Step 1. Open the template and complete the adjustments described in the last section.

Step 2. Use the INSERT command to place the floor plan in the template drawing on the *FLOR WALL* layer.
Step 3. Click the PAPER button on the bottom of the drawing area to enter paper space in the viewport.
Step 4. Use the ALL or EXTENTS options of the ZOOM command to view the drawing.
Step 5. Use the PAN command to center the drawing to center the drawing in the layout.
Step 6. Use the VIEWPORTS toolbar to adjust the scale. For this drawing, select the 1/4'' = 1'-0'' option.

Creating a New Drawing

For the example used in the balance of this chapter, it is assumed that you are starting a new plan based on a drawing the owner has provided. The residence in Figure 11.2 will be used as an example for the balance of this chapter. Although it rarely happens, we'll assume the client likes the plan exactly as it is. Use the following steps to complete a floor plan after the initial planning is completed.

Step 1. Make the FLOR WALL current.
Step 2. Offset lines to represent the overall size of the structure. Convert these lines to 6''-wide walls, moving the new line inward. Whenever possible, keep exterior wall sizes modular. A wall length of 16' can be covered in four sheets of OSB applied vertically. A wall 16'-2'' will require five sheets of OSB, with 3'-10'' left over. Hopefully the remaining sheet will get used, but that is not always the case. If the design is not compromised, delete the 2'' during the layout stage.
Step 3. Draw as many of the 6''- (150 mm) wide exterior walls as possible.

- Start with two perpendicular lines to represent a specific corner of the structure. With these two lines, an entire rectangular structure can be drawn by offsetting lines and then cleaned up using other editing commands, such as TRIM or FILLET. Using the OFFSET command, lines can be drawn in the exact location without considering their required length.
- Start the layout on the side of the house where you can obtain the most information from the sketch.
- The northwest corner and the southeast corner of the home are each fairly boxy. In the example, the northwest corner was selected as the starting point.
- Use the drawing scale developed in the first stage to determine the size of the jogs in the exterior walls.
- Lay out as many of the exterior walls as possible based on the information provided before moving to the interior walls. Your drawing should now resemble Figure 11.3.

Step 4. Lay out all interior walls 4'' (100 mm) wide using the LINE, OFFSET, TRIM, and FILLET commands.

- Draw all interior walls that extend from an exterior wall. This would include walls such as the east wall of the dining room, the west wall of the master bedroom, and the east side of the entry closet.
- Draw interior walls for areas that must be a specific size. This includes walls that form the closet on the east side of bedroom 2, the west side of the entry closet, and the wall at the west end of the spa in the master bath. Your drawing should now resemble Figure 11.4.

To determine the size of the spa, measure the owner's plan to determine the approximate size, and then verify the size by researching spas using the Internet.

- Draw the interior walls that can be located based on a given room size. This includes the wall that divides bedrooms 2 and 3 and the east wall of the kitchen.
- Draw walls based on design standards. This includes:
  - The west wall of the utility room (36'' min./42'' desirable).
  - The wall between the bath and the utility room (30'' min./36'' desirable).
  - The width of the stairs (36'' min./42'' desirable).

When all walls are drawn, your floor plan will resemble Figure 11.5.

INSERTING WINDOW BLOCKS

Window blocks were introduced in Chapter 10 as symbols were discussed. Use the blocks created in Chapter 10, those supplied in DesignCenter or blocks purchased from a third-party vendor, to complete this

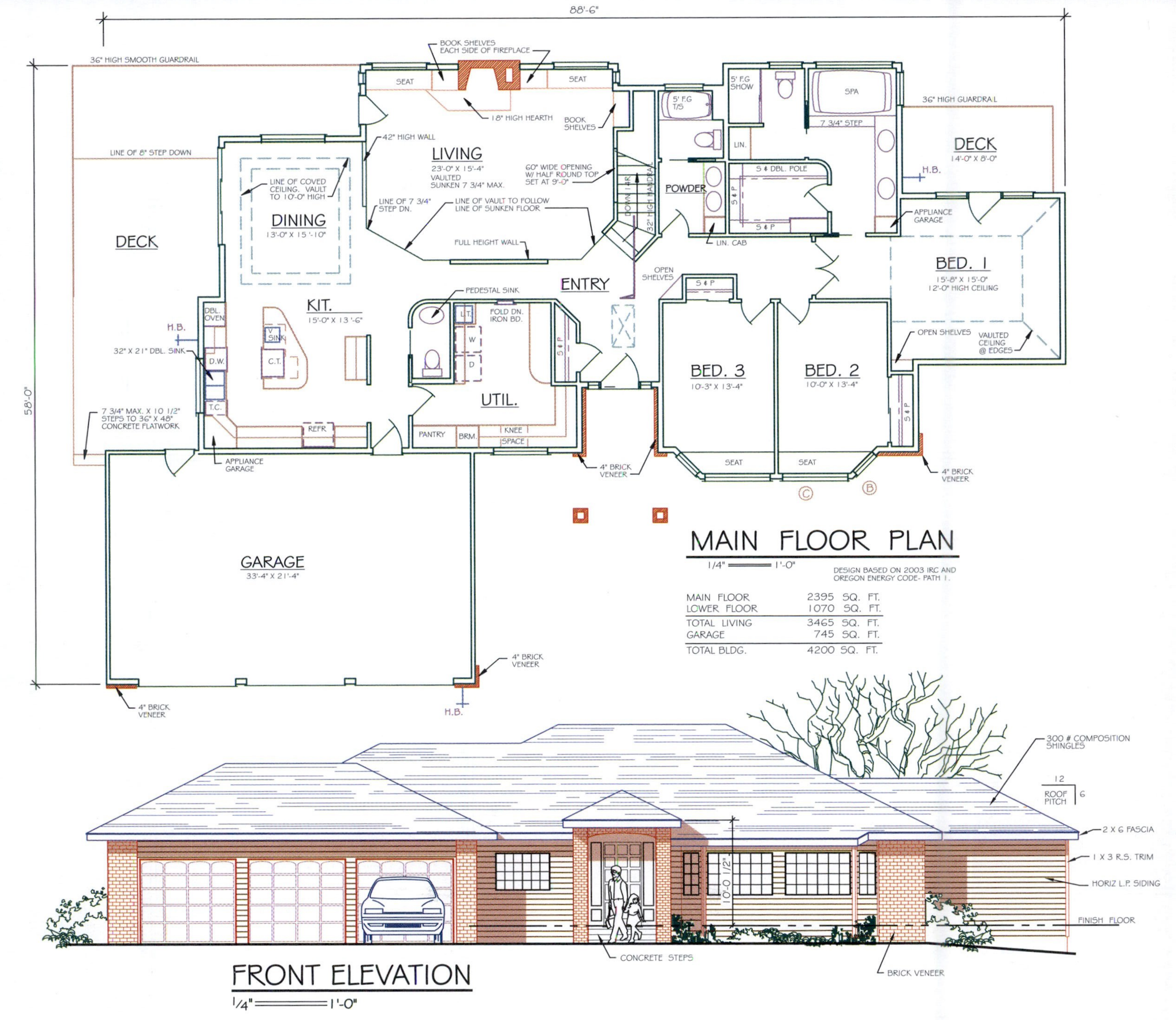

FIGURE 11.2 ■ *A print of a preliminary floor plan, an elevation, and a list of changes is often the starting point for drawing a set of house plans.*

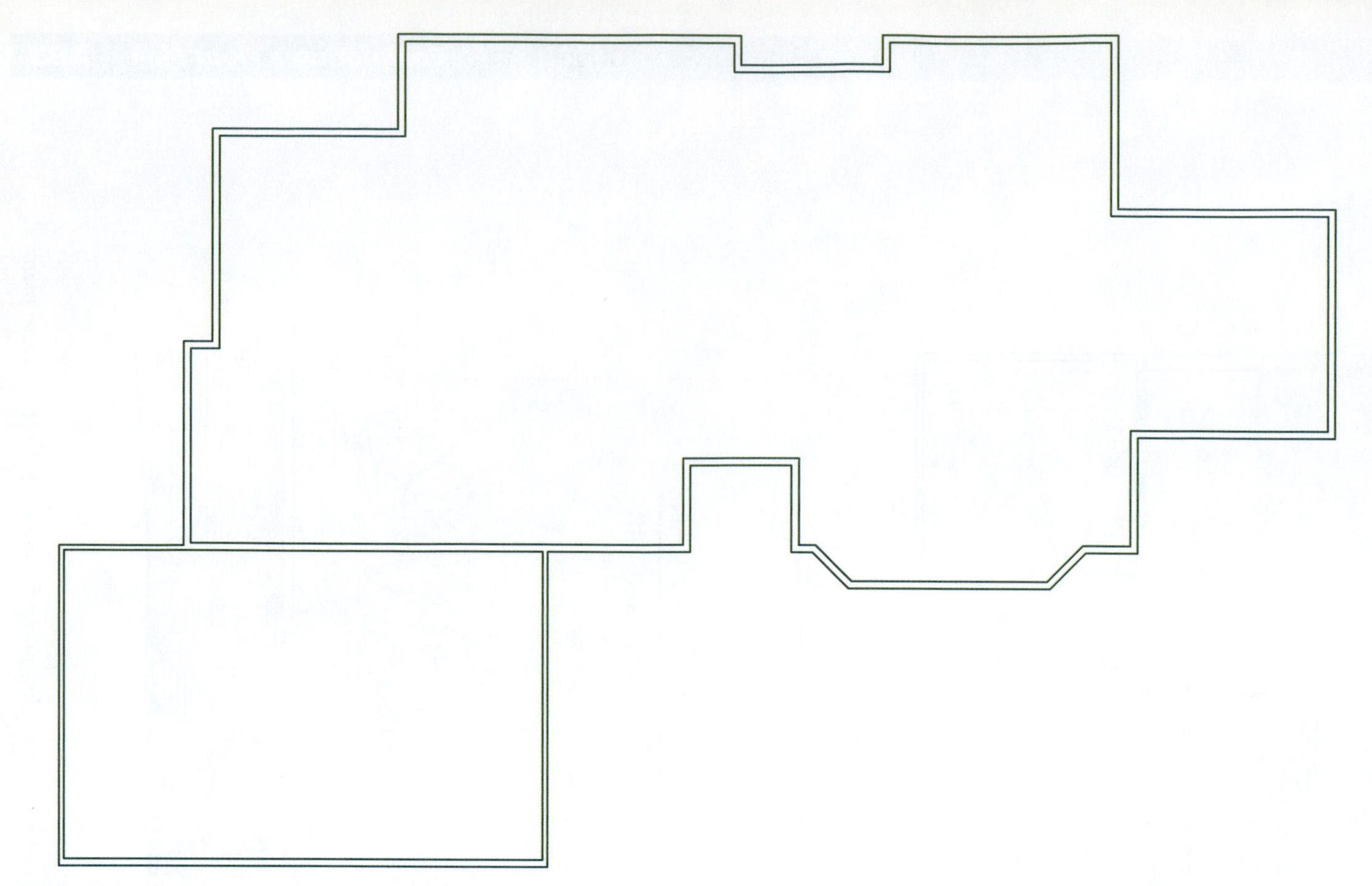

FIGURE 11.3 ■ *The initial stage of the drawing layout is to establish as many of the exterior walls as possible based on site setbacks and basic room dimensions.*

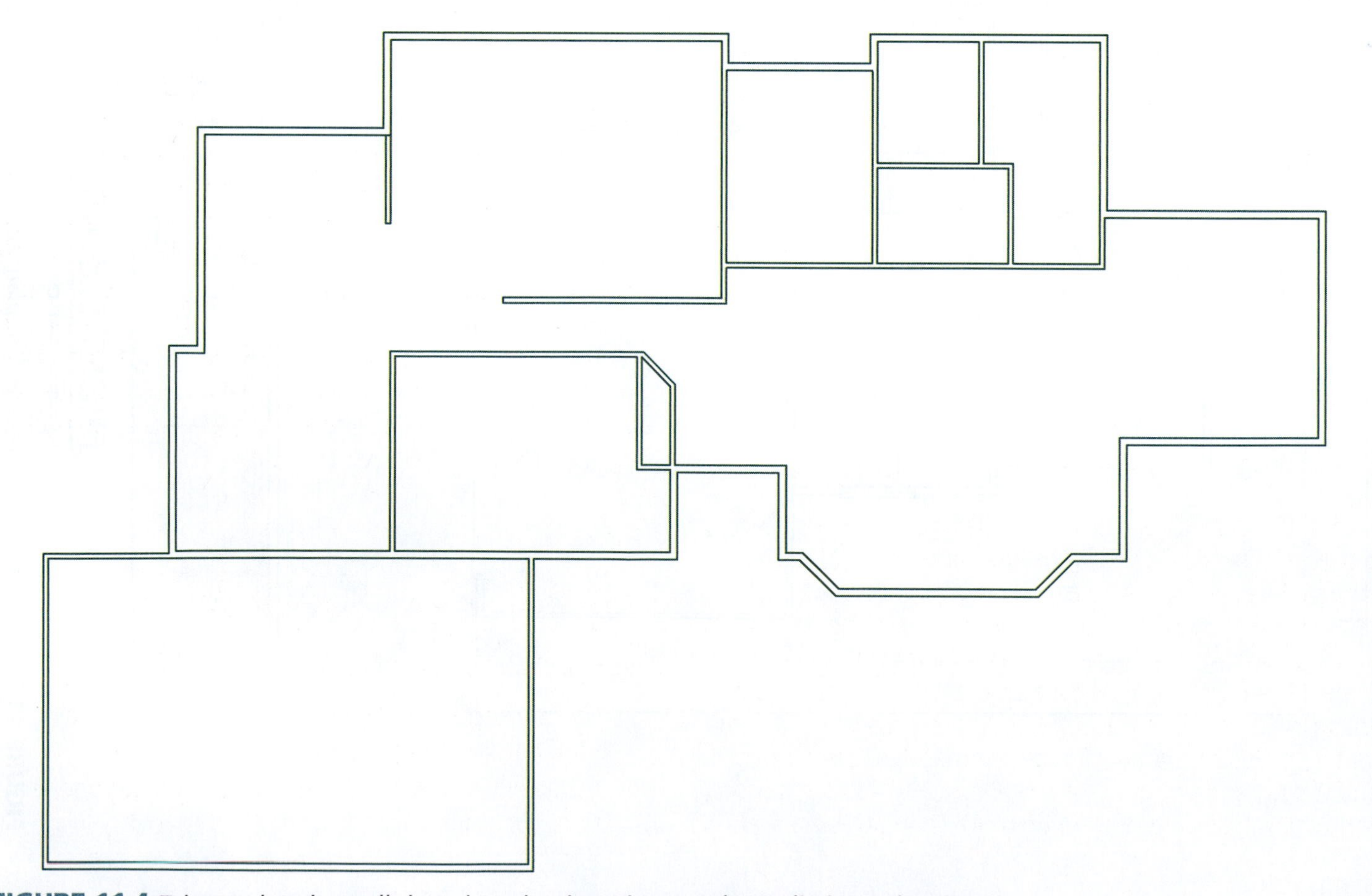

FIGURE 11.4 ■ *Locate interior walls based on the sizes given on the preliminary drawing.*

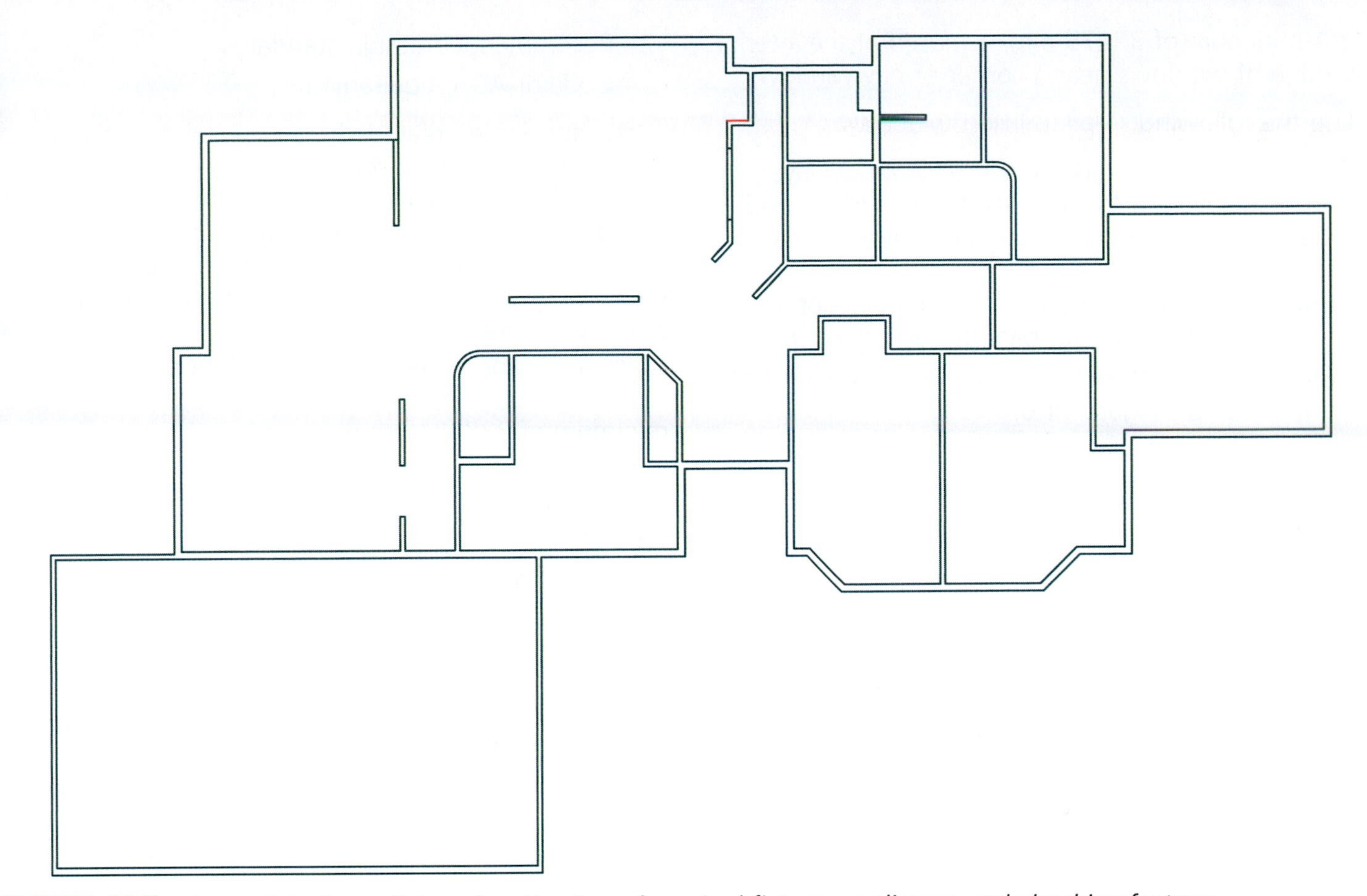

FIGURE 11.5 ■ *Locate interior walls based on the sizes of required fixtures, appliances, and plumbing features.*

portion of a floor plan. In placing windows, assign the window size based on:

- Scaling the sketch provided by the owner
- Recommended design standards (see Chapter 5)
- Legal requirements for light, ventilation, and emergency egress
- The layout required for lateral bracing (see Chapter 24)

Placing Windows

Use the scale created in the planning process to approximate the window size on the sketch provided by the owner. This size can be combined with information provided by the owner or based on common design standards presented in Chapter 10. The windows in bedroom 2 and 3 each scale 6' wide and comply with the guidelines presented in Chapter 10. The rooms are approximately 136 sq ft, requiring approximately 8 sq ft of light and 4 sq ft of ventilation. If a 6'-0'' × 3'-6'' slider is used, ventilation will exceed the minimum required size. If sliding windows are used, the approximately 3' of width easily complies with the egress requirements of the IRC.

The final consideration for window placement will depend on the seismic zone of the construction site. Even if you are drawing plans for a construction site where seismic risk is high, window placement will be altered depending on whether an engineer supervises the design or the design follows the prescriptive path of the IRC. The example in this chapter assumes the construction site is in a low-risk seismic activity area. Chapter 24 reexamines window placement for buildings constructed in high-risk seismic activity areas.

Inserting Blocks

Once the size and location of windows are determined, the symbol can be inserted into the drawing. Common locations include:

- Windows are usually centered in a room. See guidelines in Chapter 10.
- A minimum of 3'' (75 mm) is placed between window openings.

- A minimum of 3'' (75 mm) is placed at the interior edge of windows near a corner of bay windows.

Use the following steps to insert windows:

Step 1. Insert the block on the *FLOR DOOR* layer.
Step 2. Develop a rough draft of a window schedule and add each window to the schedule as each window is inserted.
Step 3. Trim the line representing the exterior side of the wall so it does not pass through the window.

When all the windows have been drawn, your floor plan will resemble Figure 11.6.

INSERTING DOOR BLOCKS

Use the blocks created in Chapter 10, supplied in AutoCAD DesignCenter or purchased from third-part vendors, to complete this portion of a floor plan. In placing doors, assign the sizes based on:

- Scaling the sketch provided by the owner
- Recommended design standards
- Code or ADA requirements

Use the scale created during the planning process and approximate each door size. This size can be combined with information provided by the owner or based on common design standards from Chapter 10. Code requirements are so minimal that they will rarely impact the drawing of the floor plan, and ADA requirements are not required for single-family units. If space allows or the home is being designed for universal appeal, all doors on an access route must have a minimum of 32'' (815 mm) of clearance. Once the size and location of doors are determined, the door symbol can be inserted into the drawing. Common locations include:

- 3'' (75 mm) from a corner for the hinge side of the door so that it swings into the room
- 24'' (600 mm) from a corner
- Centered on a wall

Use the following steps to insert door blocks:

Step 1. Insert the block on the *FLOR DOOR* layer.

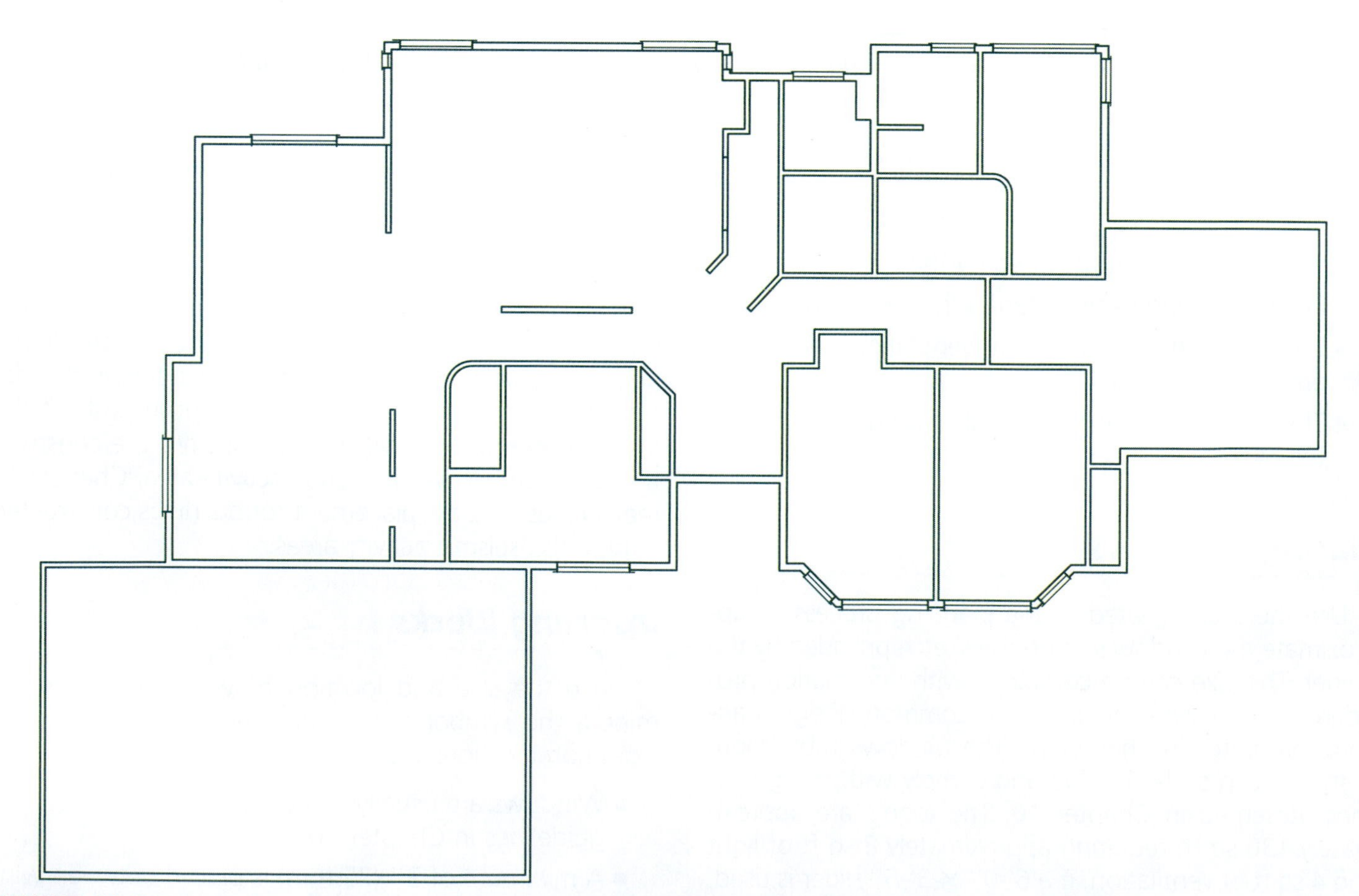

FIGURE 11.6 ■ *Once all walls have been located, window blocks can be inserted.*

Step 2. Develop a rough draft of a schedule and add each door to the schedule as it is inserted.
Step 3. Trim the lines representing the walls at each edge of the door.

When all of the doors have been drawn, your floor plan will resemble Figure 11.7.

REPRESENTING CABINETS

Use the information presented in Chapter 10 to draw all cabinets. Use the following sizes to represent the cabinet outlines.

- Base kitchen cabinets are 24'' (600 mm) deep
- Upper cabinets are 12'' (300 mm) deep
- Bath vanities are 22'' (560 mm) deep

Cabinets can be drawn using the OFFSET and PROPERTIES commands. Offset the wall line the required distance to represent the depth of the cabinet. The new line can then be moved to the proper layer, and assigned the desired linetype and lineweight by using the PROPERTIES command. Use either the *FLOR CASE* or the *FLOR WDWK* layer depending on who will build the cabinets. Represent cabinets that will be constructed off site in 3'' (75 mm) modules, if possible, to reduce expenses.

Use the following steps to complete this stage:

Step 1. Represent all kitchen base cabinets.
Step 2. Represent all kitchen upper cabinets.
Step 3. Represent all storage related cabinets such as built-in entertainment centers, bookcases, window seats, pantries, or broom closets in kitchens or utility rooms, shelving, and miscellaneous closet storage devices such as a shelf and pole in closets or custom storage in wardrobes.
Step 4. Represent all cabinetry in the utility room, garage, or mudrooms.
Step 5. Represent all cabinetry in each bathroom.

When all cabinets are drawn, your floor plan will resemble Figure 11.8.

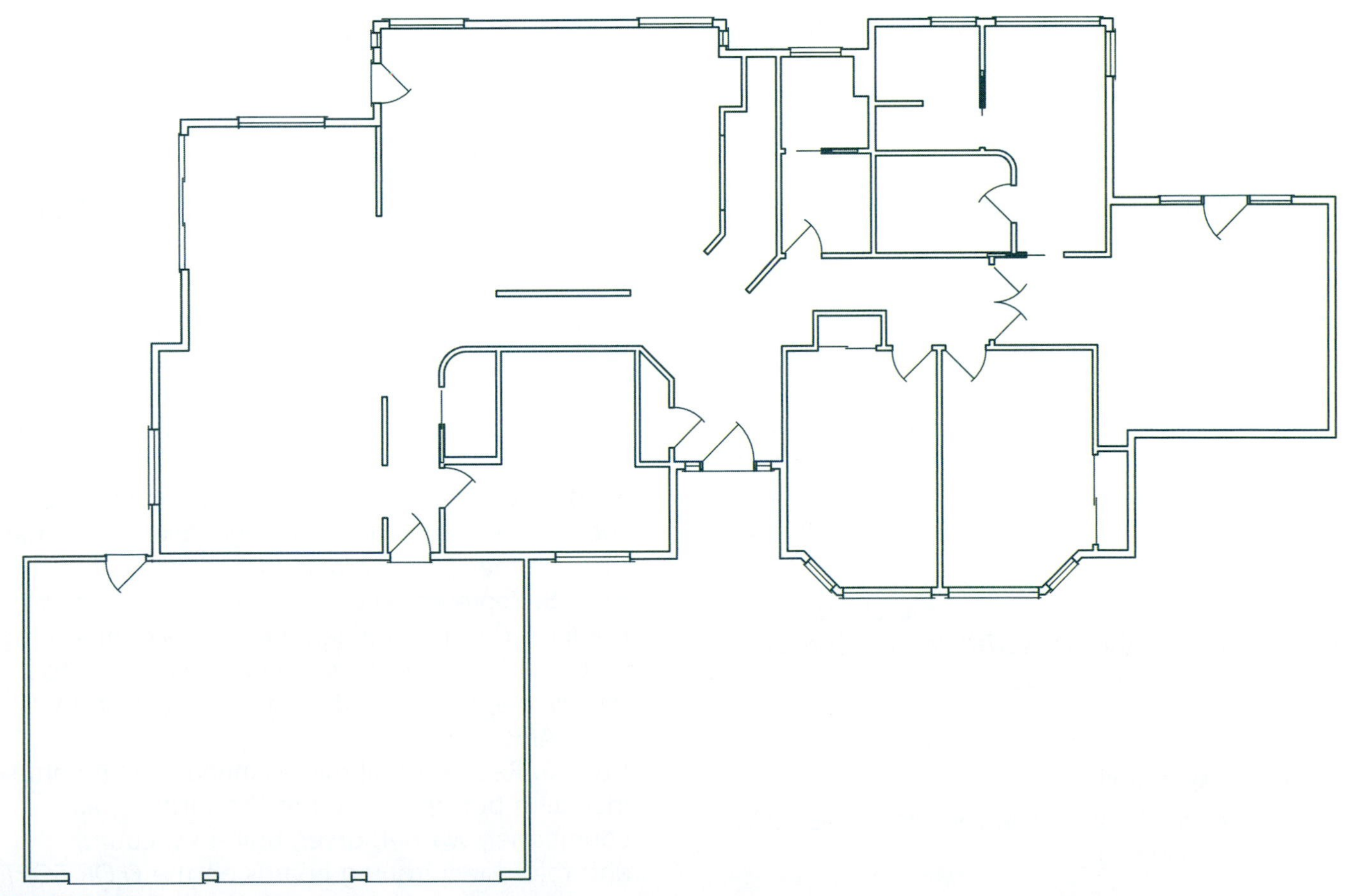

FIGURE 11.7 ■ *Adding door blocks to the floor plan.*

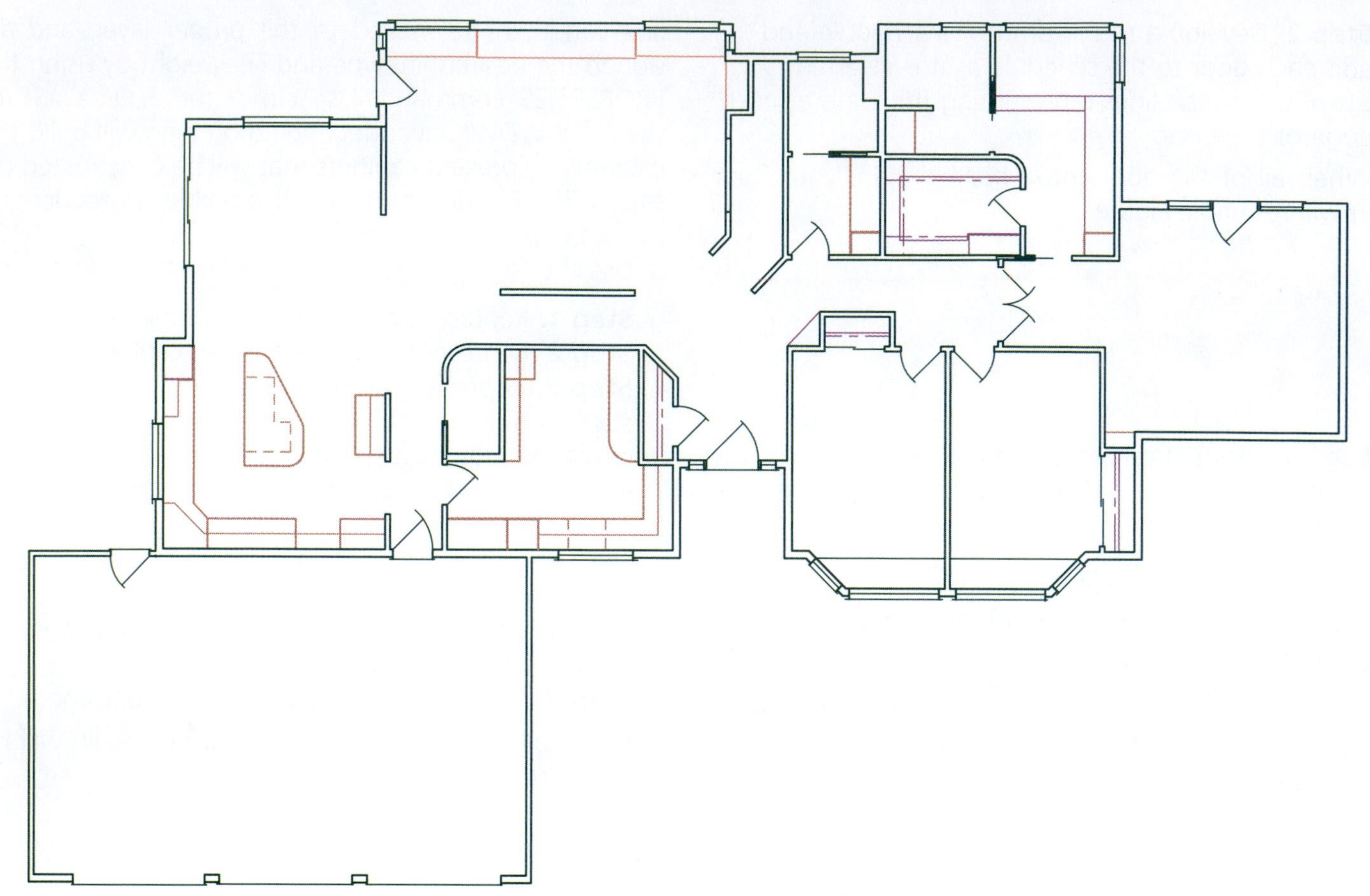

FIGURE 11.8 ■ *Locate all cabinet outlines for the living, sleeping, and service areas of the home.*

PLACING MISCELLANEOUS SYMBOLS

The use of an orderly system to place symbols on the floor plan will ensure that all appliances and fixtures are properly represented. There is nothing magical about the following order, but it provides a guideline to place floor plan symbols. As you gain experience, develop your own system based on how you think and view the problem. Use the following steps to represent symbols:

Step 1. Represent changes in floor levels such as lowered floors on the *FLOR STRS* layer. (Represent stairways between floors later.)

- Represent steps at floor changes.
- Represent guardrails or handrails.

Step 2. Represent changes in ceilings on the *FLOR OVHD* layer including:

- Lines of vaulted ceilings
- Coved, coffered, or tray ceilings
- Soffits created for storage, display, or lighting
- Skylights

Step 3. Represent all decks, patios, flatwork, or stoops by each exterior door on the *FLOR DECK* layer.

Represent appliances in the following steps on the *FLOR APPL* layer. When completed, your drawing should resemble Figure 11.9.

Step 4. Represent the kitchen plumbing fixtures such as the main work sink, vegetable sink, or party sinks on the *FLOR PLMB* layer.

Step 5. Represent kitchen appliances, including but not limited to the refrigerator, under-counter refrigerator, dishwasher, range, double oven, cooktop, microwave, range hood, or pop-up vent on the *FLOR APPL* layer.

Step 6. Represent all miscellaneous equipment, including but not limited to the furnace, air conditioner, washer, dryer, built-in vacuum, and fold-down ironing boards on the *FLOR EQPT* layer.

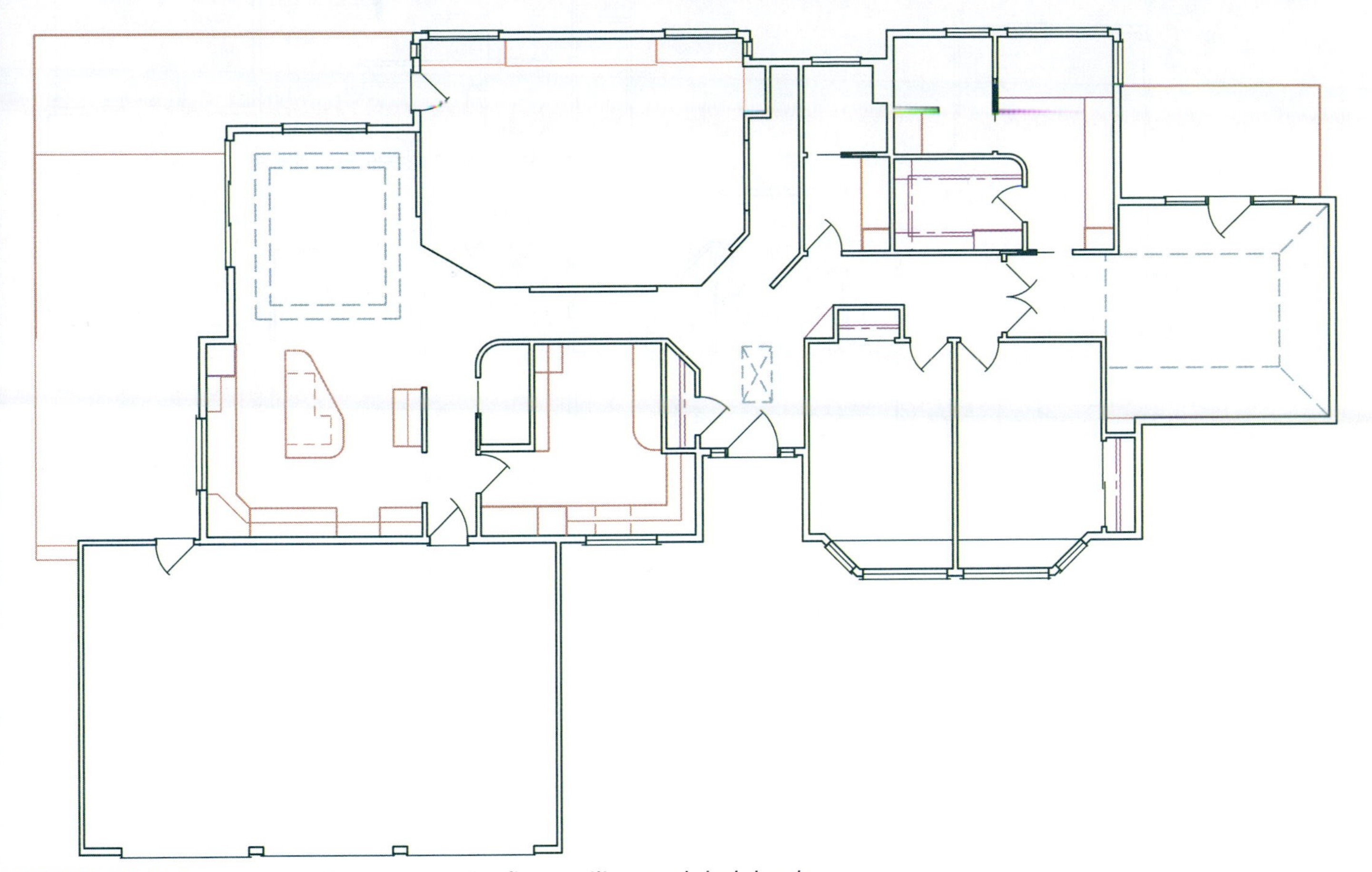

FIGURE 11.9 ■ *Represent all changes in the floor, ceiling, and deck levels.*

Represent the plumbing fixtures in the following steps on the *FLOR PLMB* layer. When completed, your drawing should resemble Figure 11.10.

Step 7. Represent the bathroom plumbing fixtures, such as the vanity sinks, pedestal sinks or freestanding sinks, toilets, bidets, tubs, jetted tubs, or showers.

Step 8. Represent all miscellaneous plumbing fixtures, such as the water heater, water softener, laundry sinks, spas, or hot tubs. If the water heater is located in the garage, show a platform and a steel bollard.

Step 9. Show all stair runs and stair-related information on the *FLOR STRS* layer.

- Determine walls on the upper or lower floor that affect the starting or ending points. On the house in this example, the stairs end 40'' from the inner side of the exterior wall. See Figure 11.13 to see the lower plan.
- Assume 10 1/2'' (265 mm) treads unless your job captain provides other information. Use the ARRAY or OFFSET command to locate the needed treads and risers.
- Show required hand and guardrails.
- Provide a break line so only the steps related to the specific floor are shown.

Step 10. Show all fireplace-related symbols on the *FLOR CHIM* or *FLOR PATT* layer.

- Draw the outline of the chimney and firebox. Plan for a 36'' wide firebox on each floor.
- Plan for and draw a flue to vent the lower firebox.
- Draw a hearth that extends 20'' (500 mm) from the front of the fireplace.
- Place the appropriate hatch pattern to represent a masonry chimney.

With the stair and fireplace information added, your drawing should now resemble Figure 11.11.

Step 11. Show all masonry veneer on the *FLOR MASN* layer. Place the appropriate hatch pattern to represent masonry on the *FLOR PATT* layer.

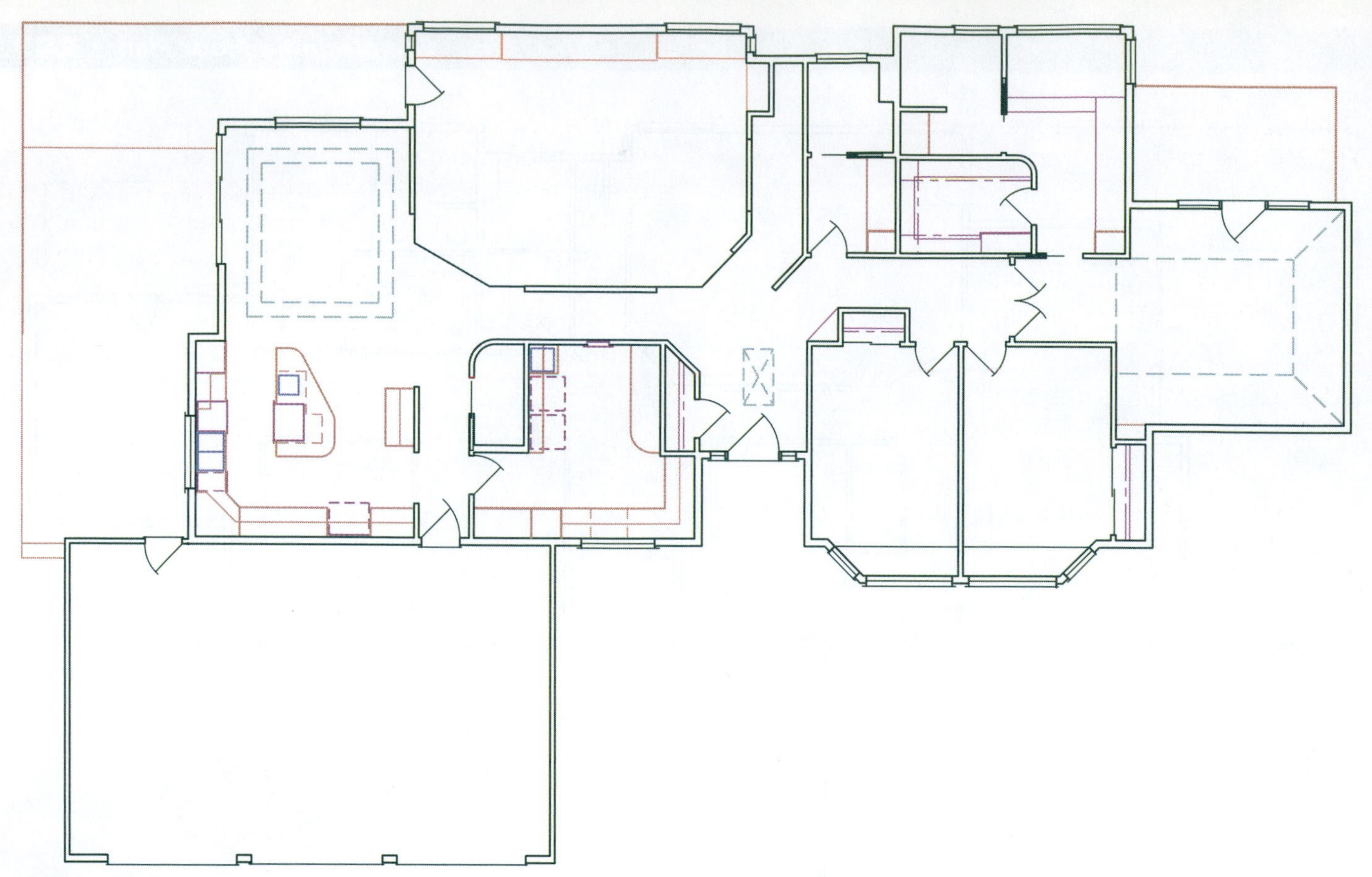

FIGURE 11.10 ■ *Add all appliances and equipment to the kitchen and service areas.*

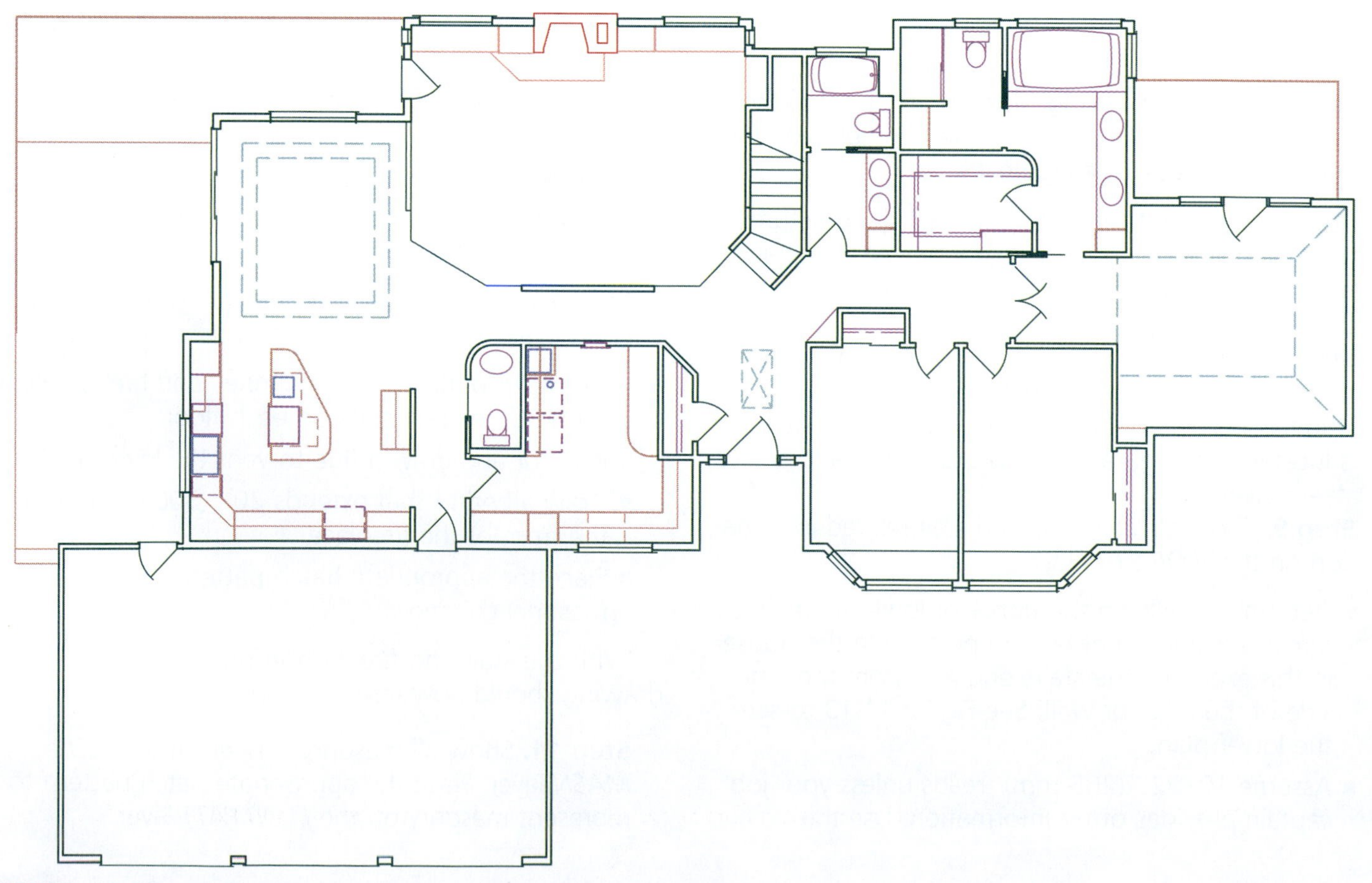

FIGURE 11.11 ■ *Add all plumbing fixtures, steps, and each fireplace.*

Step 12. Show all miscellaneous symbols on the *FLOR SYMB* layer unless noted. Symbols should include but are not limited to:

- North arrow
- Attic access (*FLOR OVHD* layer)
- Optional crawl access if it is not located in the foundation wall
- Miscellaneous plumbing symbols such as hose bibbs and metal pans for interior fixtures such as a water heater or washing machine

The floor plan is now completely drawn, representing the walls, windows, doors, cabinets, appliances, plumbing fixtures, and miscellaneous symbols. The drawing should resemble Figure 11.12. This drawing should now be saved in the appropriate folder, using a name that accurately represent the contents. See Chapter 3 for a review of folder names. This drawing will now serve as a base drawing if separate framing, electrical, plumbing, or HVAC drawings are provided. As a base drawing, the floor plan can be saved and related information can now be attached using the XREF command, or the related information can be displayed on additional layers and controlled using the NAMED LAYER FILTER feature of the LAYER PROPERTIES MANAGER.

Step 13. No matter which method will be used to proceed, this is a good time to pause and check your work. Make a check print of your drawing without the title block or other template-related material. A full-size print is not necessary to verify that you have completely drawn all required symbols. If you're thinking you can skip this step, please note

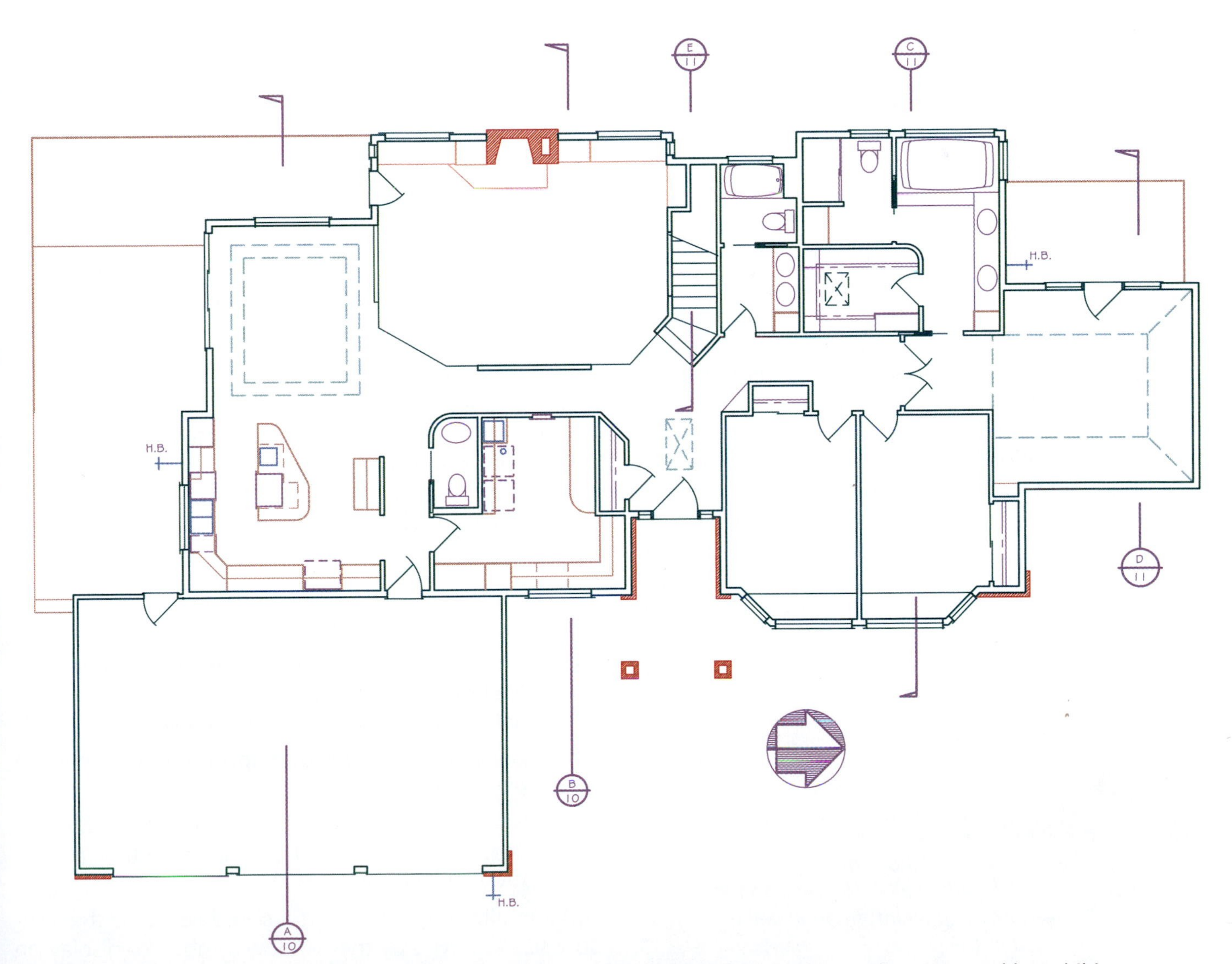

FIGURE 11.12 ■ *Add miscellaneous symbols such as the north arrow, section markers, attic access, and hose bibbs.*

that you must be the second human to walk the face of the earth who is perfect. TAKE TIME TO CHECK YOUR WORK! You'll be more likely to find mistakes if you use the following techniques:

- Take a short break after the print is made. A few minutes away from the computer will do wonders for your mind and your eyesight.
- Rotate your print so the door is now at the top of the drawings. Seeing the drawing in a new orientation will often open your mind to new experiences and help you see things you didn't see as you stared at the computer screen.

Use the checklist from Appendix D to verify the content of the drawing.

DRAWING ADDITIONAL FLOOR LEVELS

In drawing a multilevel home, additional floor levels can be drawn in the same drawing file that was used to create the main floor. Levels must be stacked directly above or below the main floor. Create new layers for the additional floor levels using prefixes to describe the new floor level and to keep each level separate. This might include prefixes such as *UP* or *LOW*. Multilevel drawings might include *UP FLOR WALL, M FLOR WALL,* and *LOW FLOR WALL.* Although any color can be used, using different colors to represent walls on different layers will help track your drawing. To start drawing a second floor, create the needed layers and then freeze the main floor information. The second floor, in this case a daylight basement, can now be drawn directly over the main floor with layer names that identify the lower floor features. When the layout is complete, the main layers used as a reference can be frozen. Figure 11.13 shows the lower floor that was created in the same drawing file as the main floor plan. When a basement plan is very simple, it may be drawn in conjunction with the foundation plan.

Note:

An important key to remember on multilevel homes is to never move one floor level without moving all levels. To keep drawings stacked, thaw all layers prior to moving one layer. This ensures that all information will remain stacked in the proper location.

PLACING ANNOTATION

For this example, annotation will be added to the floor plan and controlled by the use of layers. Annotation can be placed on the drawing in five major stages that include:

- Place general notes
- Place schedules (For this project, that will include a door, windows, and finish schedules.)
- Place local notes
- Place titles
- Place the title block text

Step 1.

- Add general notes.
- Place general notes on the *FLOR ANNO NOTE* layer. Notes are typically standardized within each office. See Figure 10.54 for notes that should be added to your drawing.
- Edit general notes.
- Add any additional notes required to describe the project.

Step 2. Use the rough draft of the door schedule created when doors were inserted into the floor plan to create a finished door schedule. Place the schedule on the *FLOR ANNO SCHD* layer.

- Use bold lines to create the outline for the schedule.
- Create a listing for symbols, size, type, and quantity. Follow the guidelines presented in Chapter 10 to create a door schedule.
- List the appropriate information for each door. On a multilevel home, all doors should be listed in one schedule.
  - Some offices list all doors for both floors using the guidelines presented in the last chapter.
  - An alternative method is to list doors for multiple levels in one schedule, but grouped together according to their floor level.
  - Place general notes that apply to doors near the schedule.
  - Place symbol markers to represent each door near the door. Place the symbols on the *FLOR ANNO SYMB* layer.

As an alternative to using a schedule, write the window size and type by the window symbol for display on the floor and framing plans.

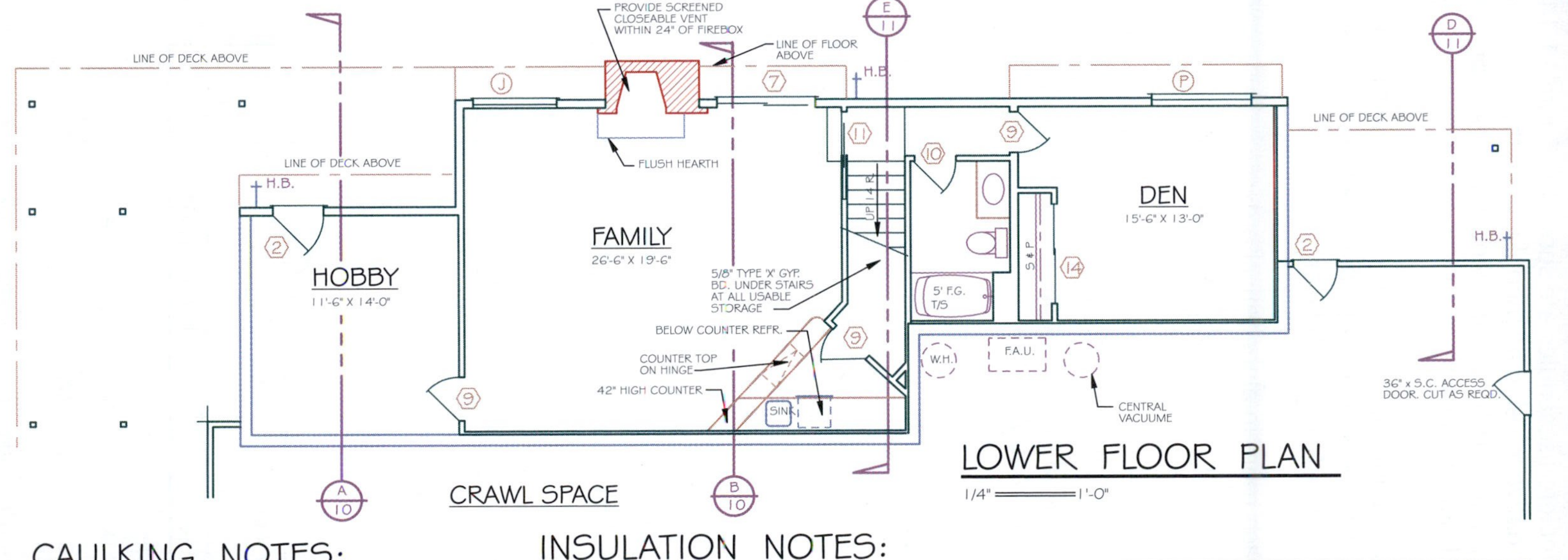

CAULKING NOTES:

CAULKING REQUIREMENTS BASED ON 2006 OREGON RESIDENTIAL ENERGY CODE

1. SEAL THE EXTERIOR SHEATHING @ CORNERS, JOINTS, DOOR & WINDOW AND FOUNDATION SILLS WITH SILICONE CAULKING.

2. CAULK THE FOLLOWING OPENINGS W/ EXPANDED FOAM OR BACKER RODS, POLYURETHANE, ELASTOMERIC COPOLYMER, SILCONIZED ACRYLIC LAYTEX CAULKS MAY ALSO BE USED WHERE APPROPRIATE.
   - ANY SPACE BETWEEN WINDOW AND DOOR FRAMES.
   - BETWEEN ALL EXT. WALL SOLE PLATES & PLY SHEATHING.
   - ON TOP OF RIM JSTS. PRIOR TO PLYWOOD APPLICATION
   - WALL SHEATHING TO TOP PLATE.
   - JOINTS BETWEEN WALL AND FOUNDATION
   - JOINTS BETWEEN WALL AND ROOF
   - JOINTS BETWEEN WALL PANELS
   - AROUND OPENINGS FOR DUCTS, PLUMBING, ELECTRICAL, TELEPHONE & GAS LINES IN CEILINGS, WALLS AND FLOORS. ALL VOIDS AROUND PIPING RUNNING THROUGH FRAMING OR SHEATHING TO BE PACKED W/ GASKETING OR OAKUM TO PROVIDE A DRAFT FREE BARRIER.

INSULATION NOTES:

INSULATION BASED ON PATH # 1 OF 2006 OREGON RESIDENTIAL ENERGY CODE.

1. INSULATE ALL EXTERIOR HEATED FRAME WALLS W/ 5 1/2" HIGH DENSITY FIBERGLASS BATT INSULATION R-21 MIN. W/ PAPER FACE. INSULATE EXTERIOR WALLS PRIOR TO INSULATION OF TUB / SHOWER UNITS.
2. INSULATE ALL FLAT CEILINGS W/ 12" R-38 FIBERGLASS BATT INSULATION (NO PAPER FACE REQUIRED).
3. INSULATE ALL VAULTED CEILINGS TO W/ 10 |" HIGH DENSITY PAPER FACED FIBERGLASS BATTS R-38 MIN. W/ 2" MIN AIR SPACE ABOVE.
4. INSULATE ALL WOOD FLOORS W/ 8" FIBERGLASS BATTS R-25 MIN. W/ PAPER FACE OR 1 PERM FORMULATED VAPOR RETARDER INSTALLED ABOVE FLOOR DECKING PRIOR TO INSTALLING FINISHED FLOORING. INSTALL PLUMBING ON HEATED SIDE OF INSULATION.
5. INSULATE CONCRETE SLAB FLOORS BENEATH HEATED ROOMS W/ 3" EXTRUDED POLYSTYRENE R-15 MIN. x 24" WIDE @ ALL SLAB EDGES IN CONTACT WITH UNHEATED EARTH.
   * PROVIDE ALT. BID FOR R-15 RIGID INSULATION ON EXT. FACE OF STEM WALL FROM TOP OF SLAB TO FND. BTM. PROVIDE FLASHING AND PROTECTION BOARD @ EXPOSED INSULATION ABOVE GRADE.
6. INSULATE BASEMENT WALLS W/ R-21 HIGH DENSITY BATTS IN 2 X 4 FURRING WALL. ** PROVIDE ALTERNATE BID FOR R-21 RIGID INSULATION ON EXTERIOR FACE OF WALL W/ FLASHING AND PROTECTION BOARD FOR EXPOSED INSULATION ABOVE GRADE.
7. COVER THE EXTERIOR FACE OF ALL EXTERIOR HEATED WALLS W/ TYVEK VAPOR BARRIER. LAP ALL JOINTS 6" MIN. AND TAPE ALL JOINTS.
8. WEATHER STRIP THE ATTIC AND CRAWL ACCESS DOORS. INSULATE ATTIC ACCESS DOOR TO R-38.
9. SET ALL MUDSILLS FOR HEATED WALLS ON NON POROUS SILL SEAL.
10. INSULATE ALL HEATING DUCTS IN UNHEATED AREAS TO R-8. INSULATION TO HAVE A FLAME SPREAD RATING OF 50 MAX. W/ AND A SMOKE DEVELOPMENT RATING OF 100 MAX. ALL DUCT SEAMS TO BE SEALED.

FIGURE 11.13 ■ *The lower floor plan can be completed using the same steps used to complete the upper floor.*

Step 3. Use the rough draft of the window schedule created as windows were inserted into the floor plan and create a finished window schedule. Place the schedule on the *FLOR ANNO SCHD* layer.

- Use bold lines to create the outline for the schedule.
- Create a listing for symbols, size, type, and quantity. Add a column to list the rough opening to the table if wood-frame windows are used. Follow the guidelines presented in Chapter 10 to create a window schedule.
- List the appropriate information for each window in the schedule. On a multilevel home, all windows are listed in one schedule using the same methods to list door information.
- Place general notes that apply to windows near the schedule.
- Place symbol markers to represent each window near the center of each window. Place the symbols on the exterior side of the window on the *FLOR ANNO SYMB* layer.

The completed schedules are shown in Figure 11.14. As an alternative to using a schedule, write the door size and type parallel to the door swing for display on the floor and framing plans.

Step 4. If required, create a finish schedule similar to the examples in Chapter 10. Place the schedule on the *FLOR ANNO SCHD* layer.

- List each room in alphabetical order.
- Provide columns for floors, walls, and ceilings materials.
- Place markers using the DONUT command in the appropriate material columns.
- Place general notes that apply to finish material near the schedule.

WINDOW SCHEDULE

SYM	SIZE	TYPE	QUAN
A	5'-0" X 3'-6"	SLDG. W/ GRIDS	1
B	2'-6" X 4'-0"	CASEMENT W/ GRIDS	2
C	6'-0" X 4'-0"	SLIDING W/ GRIDS	2
D	3'-0" X 5'-0"	CASEMENT - TEMP.	1
E	7'-0" X 5'-0"	FIXED - TEMPERED	1
F	3'-0" X 2'-0"	SLDG.	2
G	1'-0" X 5'-0"	FIXED	2
H	1'-0" X 2'-0"	TRANSOM	2
J	5'-0" X 5'-0"	SLDG.	3
K	5'-0" X 2'-0"	TRANSOM	2
L	6'-0" X 5'-0"	FIXED	1
M	9'-0" X 2'-0"	TRANSOM	1
N	5'-0" X 2'-6"	1/2" ROUND	1
P	6'-0" X 5'-0"	SLDG.	1

WINDOW NOTES:

1. ALL WINDOWS TO BE U-.40 MIN. PROVIDE ALT. BIDD FOR U=.54. IF FLAT CEILINGS ARE INSULATED TO R=49
2. ALL WINDOWS TO BE MILGARD OR EQUAL W/ VINYL FRAMES.
3. ALL BEDROOM WINDOWS TO BE WITH IN 44" OF FIN. FLOOR.
4. ALL WINDOW HEADERS TO BE SET AT 6'-10" UNLESS NOTED.
5. UNLESS NOTED ALL HDRS. OVER EXT. DOORS AND WINDOWS TO BE 4" WIDE WITH 2" RIGID INSUL. BACKER. SEE TYP. SECTION.

SKYLITES NOTES:

1. ALL SKYLITES TO BE VELOX OR EQUAL DOUBLE DOMED PLASTIC OPENABLE SKYLITES. U=.50 MIN.

DOOR SCHEDULE

SYM.	SIZE	TYPE	QUAN.
1	3'-6" X 8'-0"	S.C. RAISED PANEL	1
2	2'-8" X 6'-8"	METAL INSUL.	3
3	2'-8" X 6'-8"	1-LITE FRENCH	1
4	9'-0" X 6'-8"	SLIDING - TEMPERED	1
5	9'-0" X 6'-8"	1-LITE FRENCH	1
6			
7	6'-0" X 6'-8"	SLIDING - TEMPERED	1
8	2'-8" X 6'-8"	M. I./ SELF CLOSING	1
9	2'-8" X 6'-8"	INTERIOR	7
10	2'-6" X 6'-8"	INTERIOR	4
11	2'-8" X 6'-8"	POCKET	1
12	2'-6" X 6'-8"	POCKET	4
13	PR. 2'-6" X 6'-8"	PR. 1-LITE FRENCH	1
14	6'-0" X 6'-8"	BI-PASS	2
15	4'-0" X 6'-8"	BI-PASS	1
16	9'-0" X 8'-0"	OVERHEAD GARAGE	3

DOOR NOTES:

1. FRONT DOOR TO BE RATED AT U 0.54 OR LESS
2. EXTERIOR DOORS IN HEATED WALLS TO BE U 0.20 OR LESS.
3. DOORS THAT EXCEED 50% GLASS ARE TO BE U 0.40 OR LESS.
4. ALL INTERIOR DOORS TO BE RAISED 6 PANEL DOORS.

FIGURE 11.14 ■ *The completed door and window schedules for the home in Figures 11.1 and 11.13.*

As an alternative to using a schedule, write the required finish information below the room information on the floor plan. In the example for this chapter, no finish schedule was provided.

Step 5. Add room titles and interior dimensions to all habitable rooms. Place the room name near the center of all habitable rooms with the interior dimensions below the name. Place the information on the *FLOR ANNO IDEN* layer.
Step 6. List the information to describe the building square footage on the *FLOR ANNO IDEN* layer.

Place the following annotation on the *FLOR ANNO TEXT* layer.

Step 7. Label all appliances such as the dishwasher, compactor, cooktop, range, microwave, ovens, refrigerator, and equipment such as the furnace, air conditioner, water heater, and vacuum.
Step 8. Label all plumbing fixtures, such as tubs, showers, or spas: provide the size, type, and material. Items that can be distinguished by shape, such as toilets, do not need to be identified.
Step 9. Label fireplace or solid-fuel-burning appliance. Specify the firebox venting, chimney size, or venting type (self-venting or non-venting), hearth height, U.L.-approved materials for electrical units, barbecue, and wood box if applicable.
Step 10. Label stairs, giving direction of travel, number of risers, guardrail, and handrail height.
Step 11. Label all closets with shelves with **S&P** (shelf and pole) or other methods of storage such as **SLVS, DBL. S & P, DRWS,** or **CLOSET ORGANIZER.**
Step 12. Describe built-in furniture, specialty storage, linen, broom, and pantry storage methods.
Step 13. Specify attic and crawl access openings and sizes.
Step 14. Specify any masonry veneer by type, height, and application.
Step 15. Designate a 1-hour firewall between garage and residence. Specify the material to be applied and its location (floor to roof sheathing or cover walls and ceiling with sheetrock).
Step 16. On multilevel structures, call outline of upper floor, balcony above, line of lower floor, or other projections of one level beyond another.
Step 17. Specify a drawing title (FLOOR PLAN) and scale (1/4'' = 1'-0'').
Step 18. Verify codes and construction methods that must be described by additional local notes.

The floor plan with completed annotation is shown in Figure 11.15. Use the checklist from Appendix D to verify the content of the drawing.

PLACING DIMENSIONS

Depending on the complexity of the project, dimensions may be placed on the floor plan or framing plan. For this example, dimensions will be added to the floor plan and controlled by the use of layers. Exterior dimensions can be placed on the drawing using the following steps:

- Place an overall dimension on each side of the structure.
- Dimension all major jogs.
- Locate all interior walls that intersect with an exterior wall.
- Dimension all openings in the exterior walls.
- Locate all interior walls.

Step 1. Place an overall dimension on each side of the structure.

- Place the overall dimensions on the *FLOR ANNO DIMS BASE* layer. This will allow these dimensions to be displayed on other plans.
- Whenever possible, keep overall dimensions in a whole number of feet.

Step 2. Dimension all major jogs.

- Place the major jog dimensions on the *FLOR ANNO DIMS BASE* layer.
- Locate each jog from the exterior face.
- Verify each line of major jog dimensions equal the corresponding overall dimension.

Figure 11.16 shows an example of the overall and major jog dimensions.

Step 3. Locate all interior walls that intersect with an exterior wall.

- Place the wall-to-wall dimensions on the *FLOR ANNO DIMS EXT* layer.
- Locate each interior wall based on its location from a major jog.
- Verify that each line of wall-to-wall dimensions adds up to the corresponding major jog dimension.

Step 4. Dimension all openings in the exterior walls.

- Place the opening dimensions on the *FLOR ANNO DIMS EXT* layer.
- Locate each opening based on its location from an interior wall.
- Verify that each line of dimensions that describe wall openings adds up to the corresponding wall-to-wall dimension.

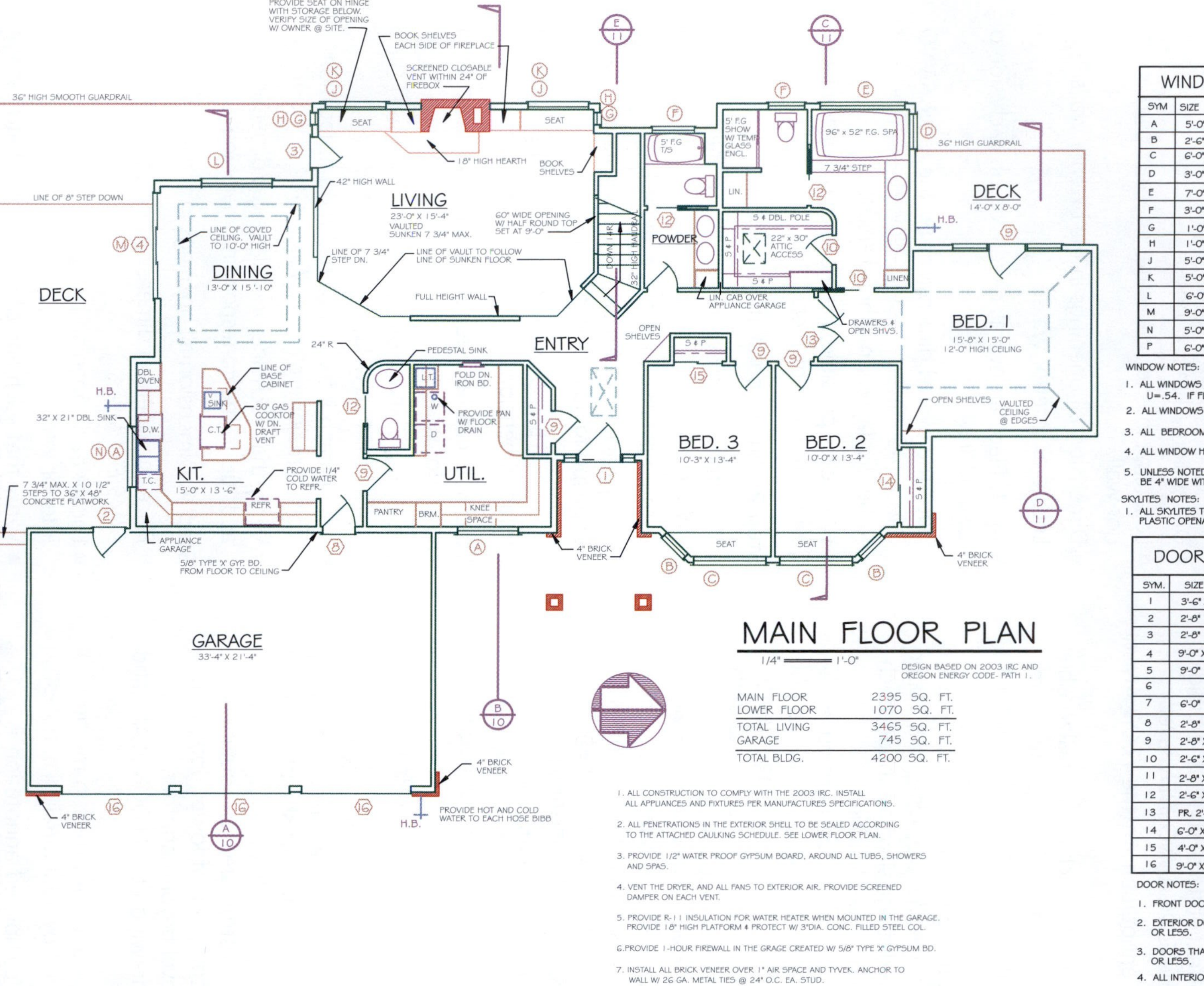

WINDOW SCHEDULE

SYM	SIZE	TYPE	QUAN
A	5'-0" X 3'-6"	SLDG. W/ GRIDS	1
B	2'-6" X 4'-0"	CASEMENT W/ GRIDS	2
C	6'-0" X 4'-0"	SLIDING W/ GRIDS	2
D	3'-0" X 5'-0"	CASEMENT - TEMP.	1
E	7'-0" X 5'-0"	FIXED - TEMPERED	1
F	3'-0" X 2'-0"	SLDG.	2
G	1'-0" X 5'-0"	FIXED	2
H	1'-0" X 2'-0"	TRANSOM	2
J	5'-0" X 5'-0"	SLDG.	3
K	5'-0" X 2'-0"	TRANSOM	2
L	6'-0" X 5'-0"	FIXED	1
M	9'-0" X 2'-0"	TRANSOM	1
N	5'-0" X 2'-6"	1/2" ROUND	1
P	6'-0" X 5'-0"	SLDG.	1

WINDOW NOTES:

1. ALL WINDOWS TO BE U-.40 MIN. PROVIDE ALT. BIDD FOR U=.54. IF FLAT CEILINGS ARE INSULATED TO R=49
2. ALL WINDOWS TO BE MILGARD OR EQUAL W/ VINYL FRAMES.
3. ALL BEDROOM WINDOWS TO BE WITH IN 44" OF FIN. FLOOR.
4. ALL WINDOW HEADERS TO BE SET AT 6'-10" UNLESS NOTED.
5. UNLESS NOTED ALL HDRS. OVER EXT. DOORS AND WINDOWS TO BE 4" WIDE WITH 2" RIGID INSUL. BACKER. SEE TYP. SECTION.

SKYLITES NOTES:

1. ALL SKYLITES TO BE VELOX OR EQUAL DOUBLE DOMED PLASTIC OPENABLE SKYLITES. U=.50 MIN.

DOOR SCHEDULE

SYM.	SIZE	TYPE	QUAN.
1	3'-6" X 8'-0"	S.C. RAISED PANEL	1
2	2'-8" X 6'-8"	METAL INSUL.	3
3	2'-8" X 6'-8"	1-LITE FRENCH	1
4	9'-0" X 6'-8"	SLIDING - TEMPERED	1
5	9'-0" X 6'-8"	1-LITE FRENCH	1
6			
7	6'-0" X 6'-8"	SLIDING - TEMPERED	1
8	2'-8" X 6'-8"	M. 1/ SELF CLOSING	1
9	2'-8" X 6'-8"	INTERIOR	7
10	2'-6" X 6'-8"	INTERIOR	4
11	2'-8" X 6'-8"	POCKET	1
12	2'-6" X 6'-8"	POCKET	4
13	PR. 2'-6" X 6'-8"	PR. 1-LITE FRENCH	1
14	6'-0" X 6'-8"	BI-PASS	2
15	4'-0" X 6'-8"	BI-PASS	1
16	9'-0" X 8'-0"	OVERHEAD GARAGE	3

DOOR NOTES:

1. FRONT DOOR TO BE RATED AT U 0.54 OR LESS
2. EXTERIOR DOORS IN HEATED WALLS TO BE U 0.20 OR LESS.
3. DOORS THAT EXCEED 50% GLASS ARE TO BE U 0.40 OR LESS.
4. ALL INTERIOR DOORS TO BE RAISED 6 PANEL DOORS.

FIGURE 11.15 ■ *The completed floor plan with the door and window schedules added to the drawing for plotting.*

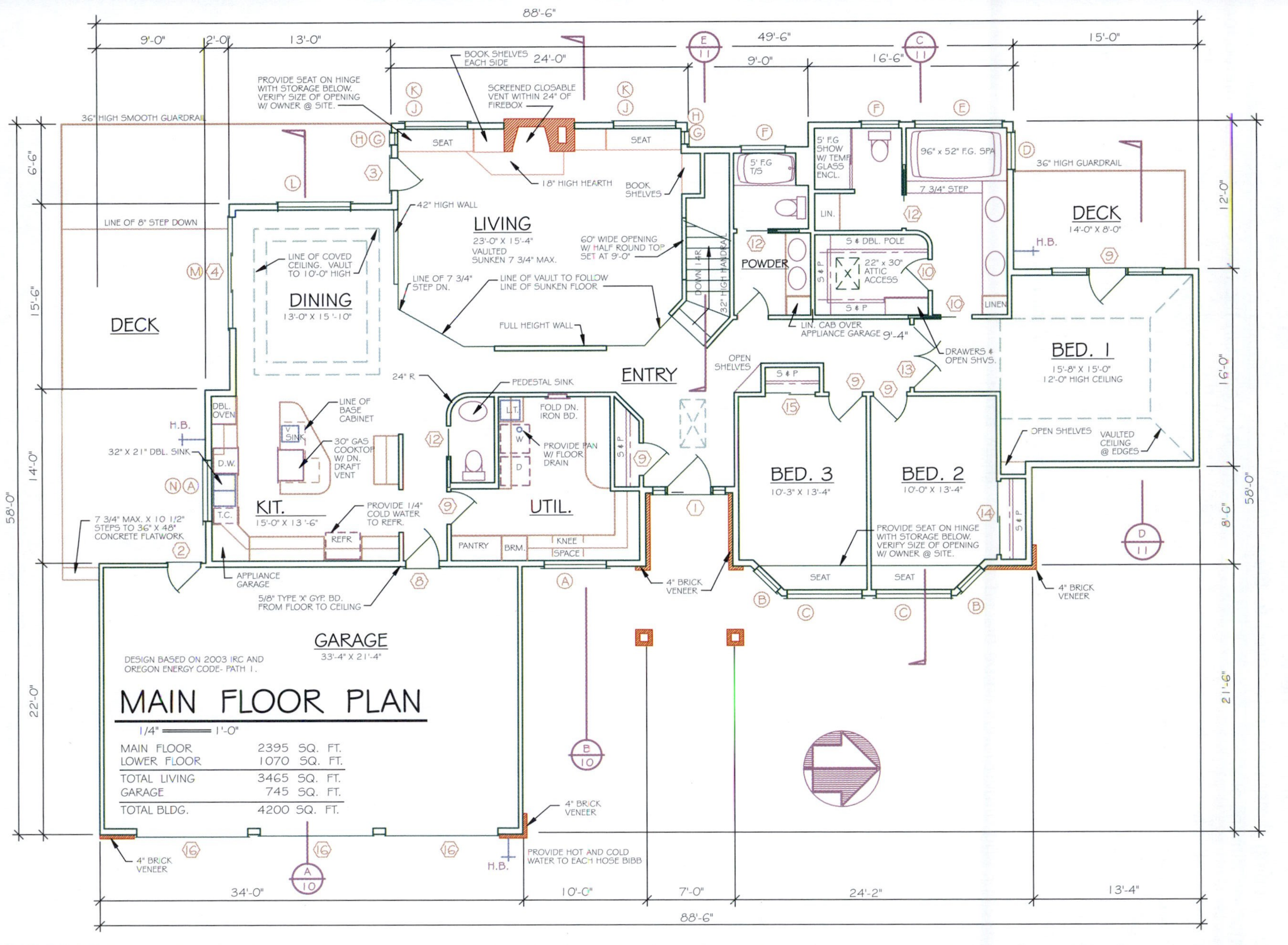

FIGURE 11.16 ■ *Overall and major jog dimensions are added to the floor plan. Keep these dimensions on a separate layer from other dimensions so that they can be displayed on the foundation plan.*

Figure 11.17 shows an example of the wall–to-wall and the wall-to opening dimensions referenced to the overall and major jog dimensions.

Step 5. Dimension all interior walls.

- Place all interior dimensions on the *FLOR ANNO DIMS INT* layer.
- Locate all interior walls that are not referenced to an exterior wall.
  - Use a print of your drawing to identify all unknown walls.
  - Plan where missing dimensions can be placed so they are grouped together.
  - Locate required wall dimensions. Place dimensions so they coordinate with exterior dimensions.
- Locate any interior doors that are not located by walls, cabinets, or other fixed objects.
  - Doors are typically 3'' from a cabinet edge or wall corner. Such doors do not need to be dimensioned.
  - Doors that appear to be centered between two known objects do not need to be dimensioned.
- Locate any interior cabinets or fixtures that must be located based on project requirements.
  - Locate islands, other cabinets, or fixtures if the location of which cannot be determined based on their relationship to other known features.

Figure 11.18 shows an example of the completed dimensions. Use the checklist from the Drawing Checklist\Dimensions Checklist folder on the CD to verify the content of the drawing.

TITLE BLOCK TEXT

Some of the text in the title block must be altered for each sheet of the project. Place text for the floor plan title block on the *FLOR ANNO TTLB TITL* layer. By providing a separate layer, text can be frozen so the base floor drawings can be used for the electrical drawings without displaying floor-related text. This can include information such as:

- Sheet number: Assign the main floor plan a number of A100. The lower floor plan will be listed as sheet A101.
- Sheet contents: List the major contents of the project. For this project the box would display MAIN FLOOR PLAN/SCHEDULES.

Some information such as the client name, date of completion, and revision date will be the same for all drawings within the project. This information should be placed on the *FLOR ANNO TTLB TEXT* layer.

- Date of completion: For school projects, list the date of the drawing completion. Most offices list the date the total project is completed.
- Revision date: Do not assign a revision date until the project has been altered. If placeholder text is placed here, keep it frozen or in a nonplotting state until the project is revised. For a school project, update the date in this box to reflect various submissions. Professionals usually alter this date only once the permit has been obtained.
- Client name: List the information as provided by the client.

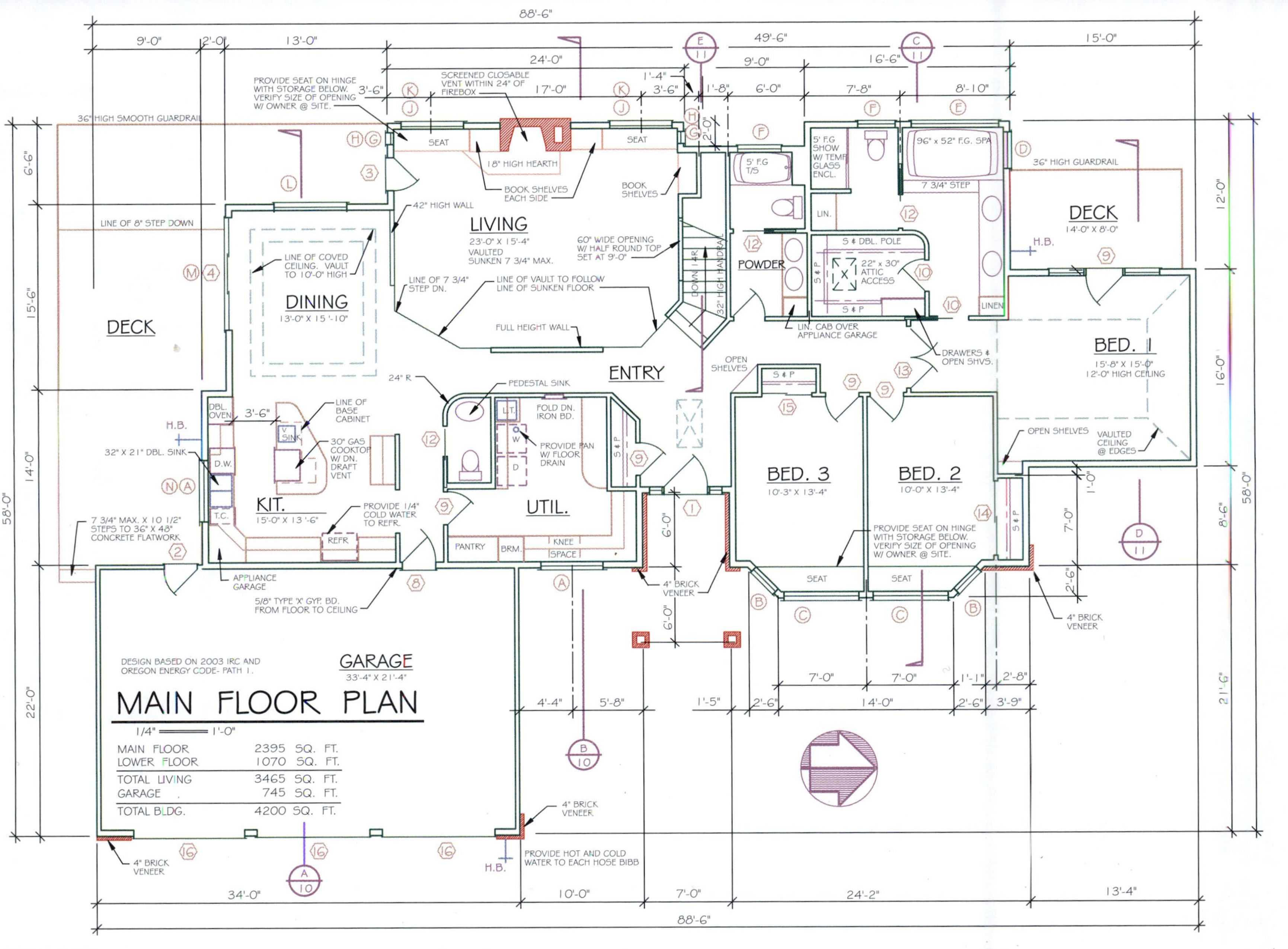

FIGURE 11.17 ■ The wall-to-wall and the wall-to-opening dimensions are referenced to the overall and major jog dimensions.

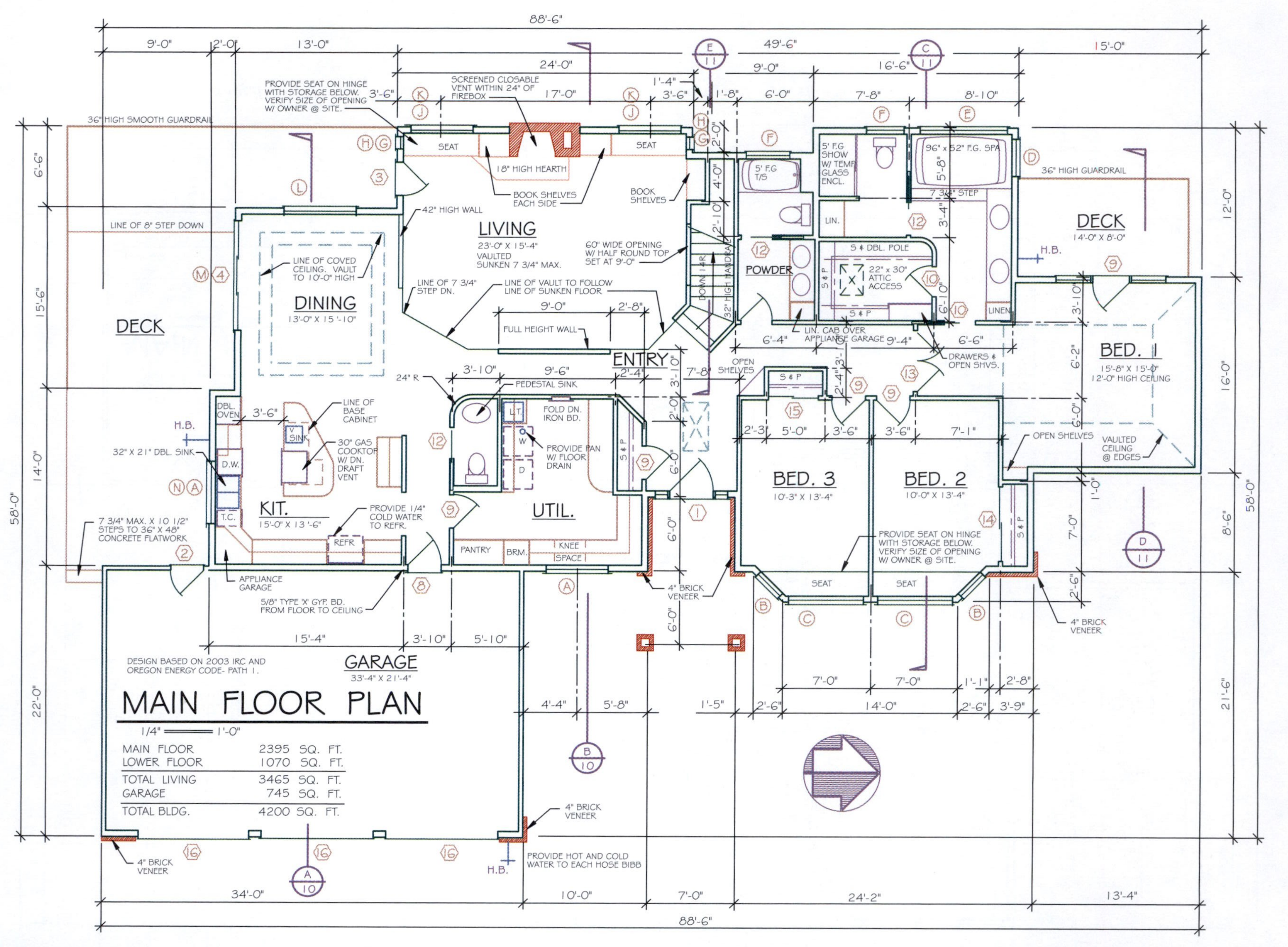

FIGURE 11.18 ■ *The completed floor plan with all text and interior dimensions added. Electrical and structural information will be placed on separate plans.*

CHAPTER
11

Floor Plan Drawing Test

DIRECTIONS

Answer the following questions with short, complete statements. Type your answers using a word processor or neatly print the answers on lined paper.

1. Type your name, the chapter number, and the date at the top of the sheet.
2. Type the question number and provide the answer in the form of a statement that includes part of the question. You do not need to write out the entire question.

QUESTIONS

11.1. Describe the drawing area of a template.

11.2. How tall will text need to be if a floor plan will be plotted at a scale of 1/8''=1'-0''?

11.3. What scale is generally used to draw a residential floor plan?

11.4. Identify two alternatives that could be considered when the drawing space is not adequate to accommodate the drawing.

11.5. Sketch four different patterns that can be used to represent walls.

11.6. Describe the process to block out a floor plan.

11.7. Describe a method to help determine the layout for a second-floor plan quickly.

11.8. Describe the most efficient method to represent a 6'-0''-wide window on the floor plan.

11.9. Describe two ways to orient the floor plan on the sheet.

11.10. Given each of the following drawing types, name a recommended CAD layer name that could be used if you are following the AIA CA Layer Guidelines:

Cabinets built on the job
Common symbols
Doors
Fireplace
Floor plan appliances
Handrails
Manufactured cabinets
Notes
Plumbing fixtures
Stairs
Walls
Walls on first floor
Walls on second floor
Water heater
Windows

CHAPTER 11

Drawing Problems

GENERAL DIRECTIONS

A base floor plan has been provided for projects 1 through 10. These homes are saved in a state similar to what might be given to a junior drafter. House plans 1 and 2 have most of the required blocks inserted and need appropriate annotation to be completed. Homes 3 through 10 need to have many of the basic drawing symbols added to the drawing, which require some planning based on an understanding of basic design principles. These homes are the types of projects a junior drafter might be given as he or she advances. Homes 11 through 35 will require extensive design work to complete the floor plan. These homes are examples of projects an experienced technician can expect to complete.

1. Select, or your instructor will assign, one of the following residential projects. The basic design is given. You need to complete the floor plan drawing(s) by adding the missing and unidentified items.
2. Completely label the floor plan.
3. Use one of the following layout methods for your floor plan problem:
   Include all floor plan symbols, room labels, and notes.
   Do not include the electrical layout.
   Establish layers as identified by this chapter, your class guidelines, or the AIA layer naming system.

PROBLEMS 11.1 THROUGH 11.10: BEGINNING DRAFTING PROJECTS

Use the attached print as a guide to complete the floor plan skeleton that is started and stored on the CD. Select an appropriate template, set all required variables, and complete the drawing to meet the requirements of your local building department. Use a check sheet from the CD to evaluate your work before submitting the assignment to your supervisor. Place the dimensions for projects 1 through 5 on the floor plan. Assume the use of a separate framing plan for problems 6 through 11.

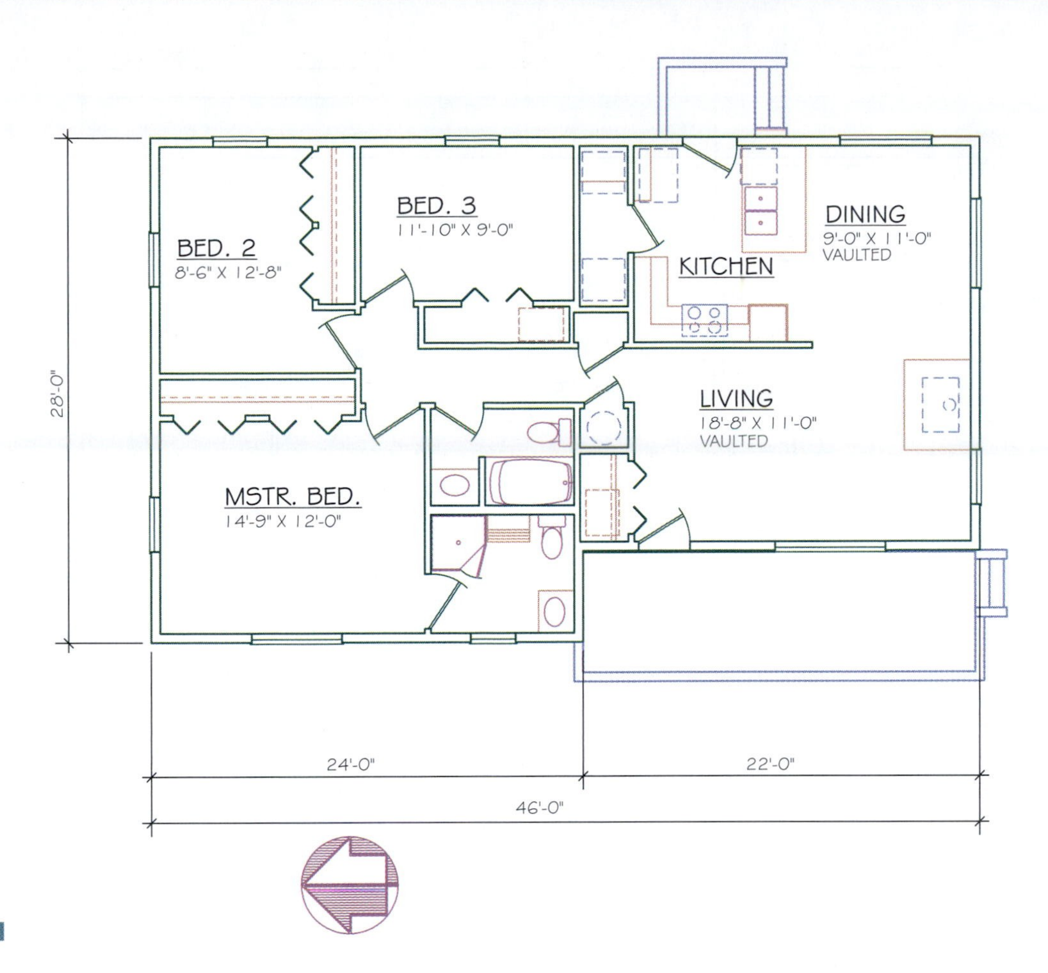

PROBLEM 11.1

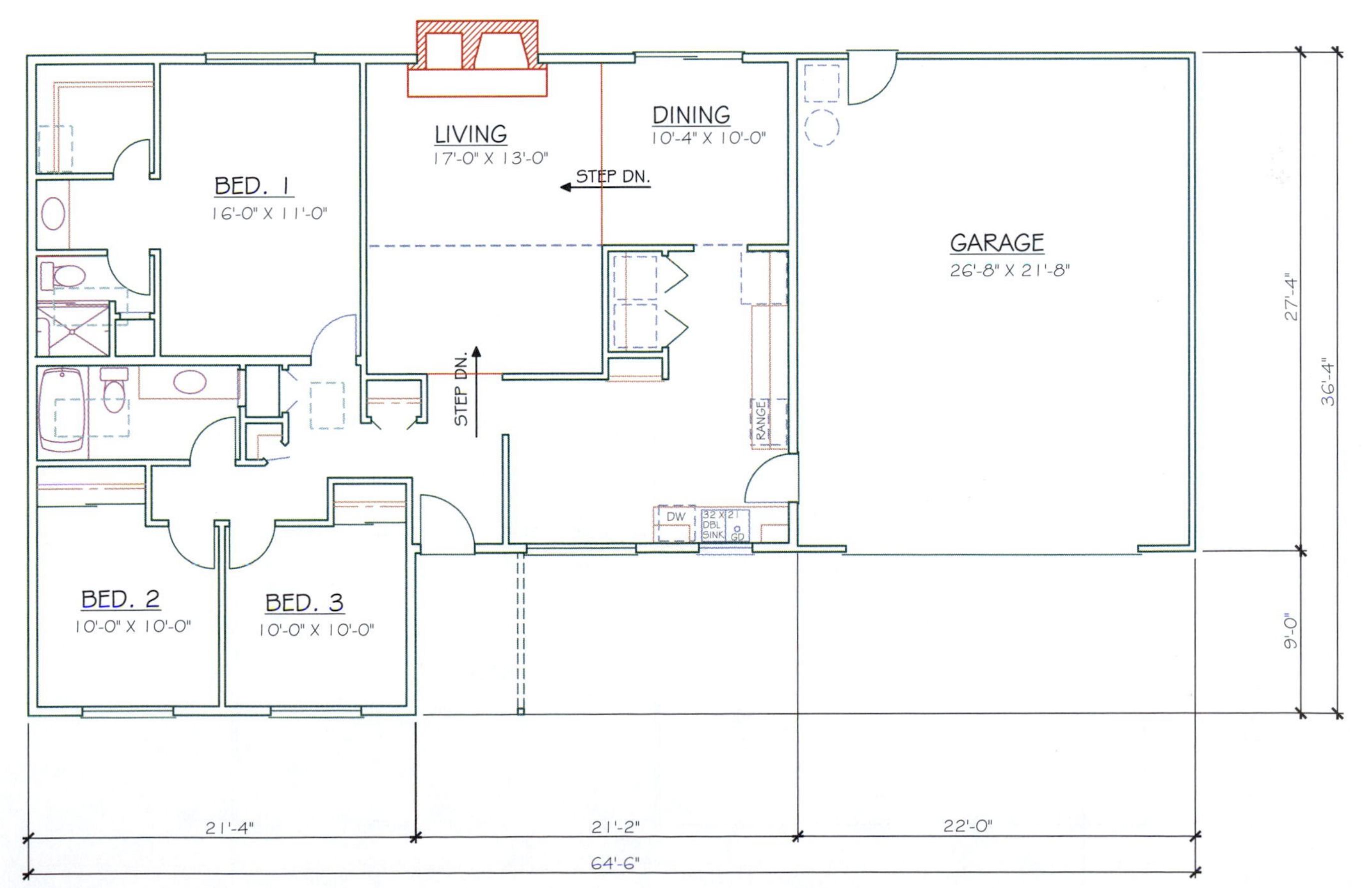

PROBLEM 11.2

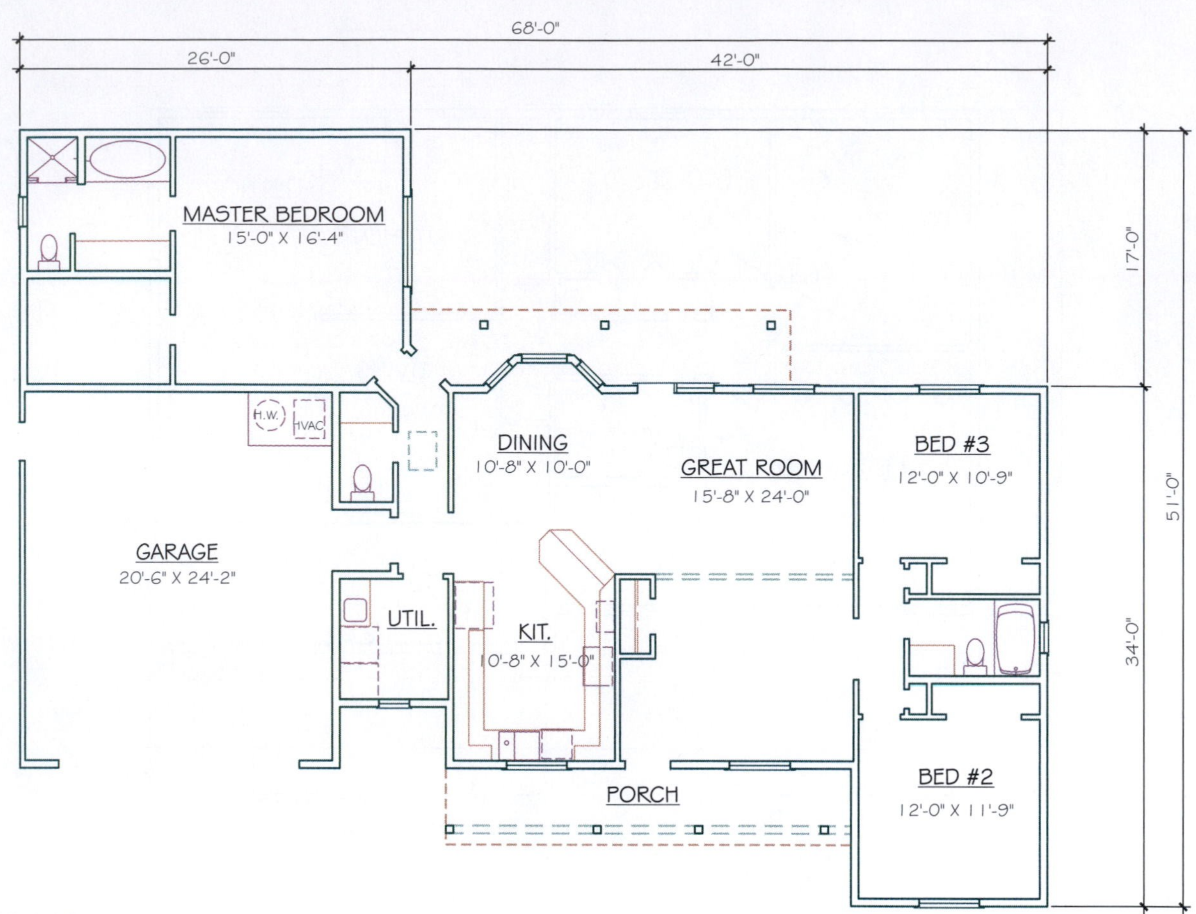

PROBLEM 11.3

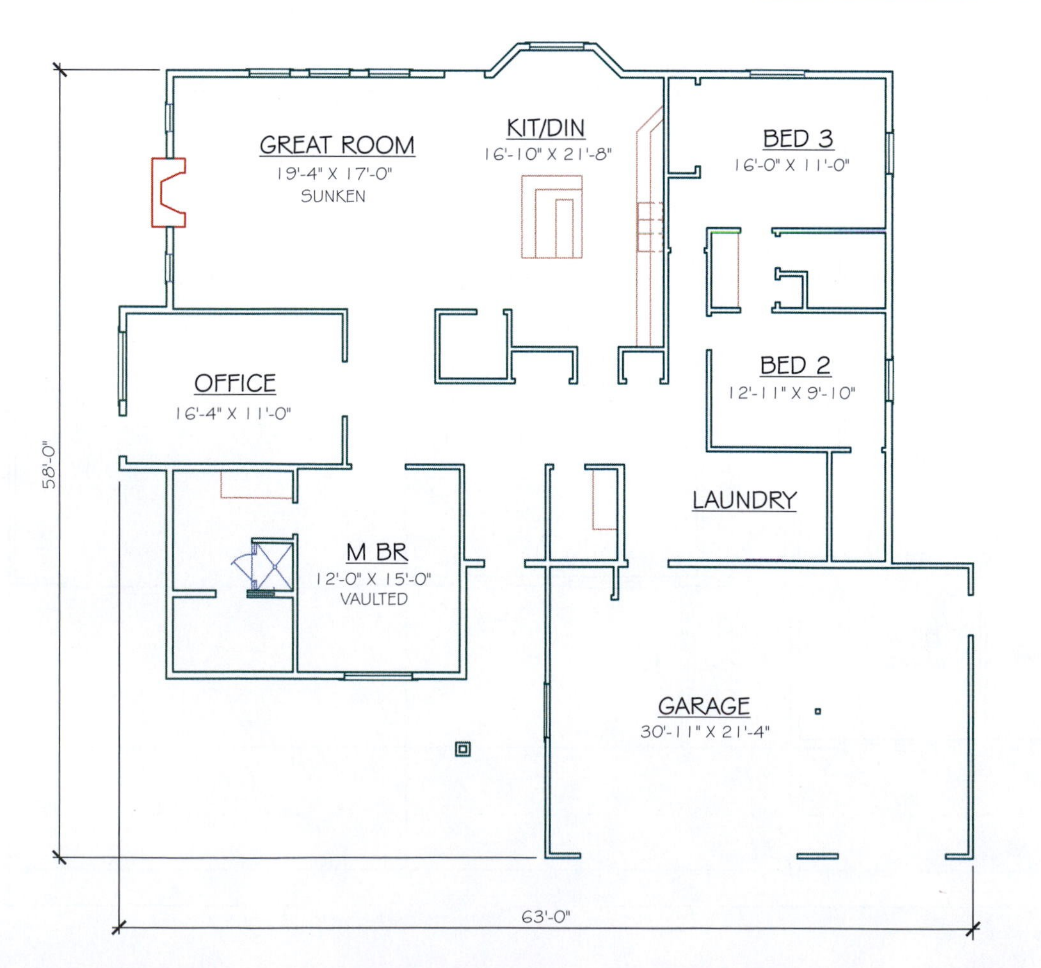

PROBLEM 11.4

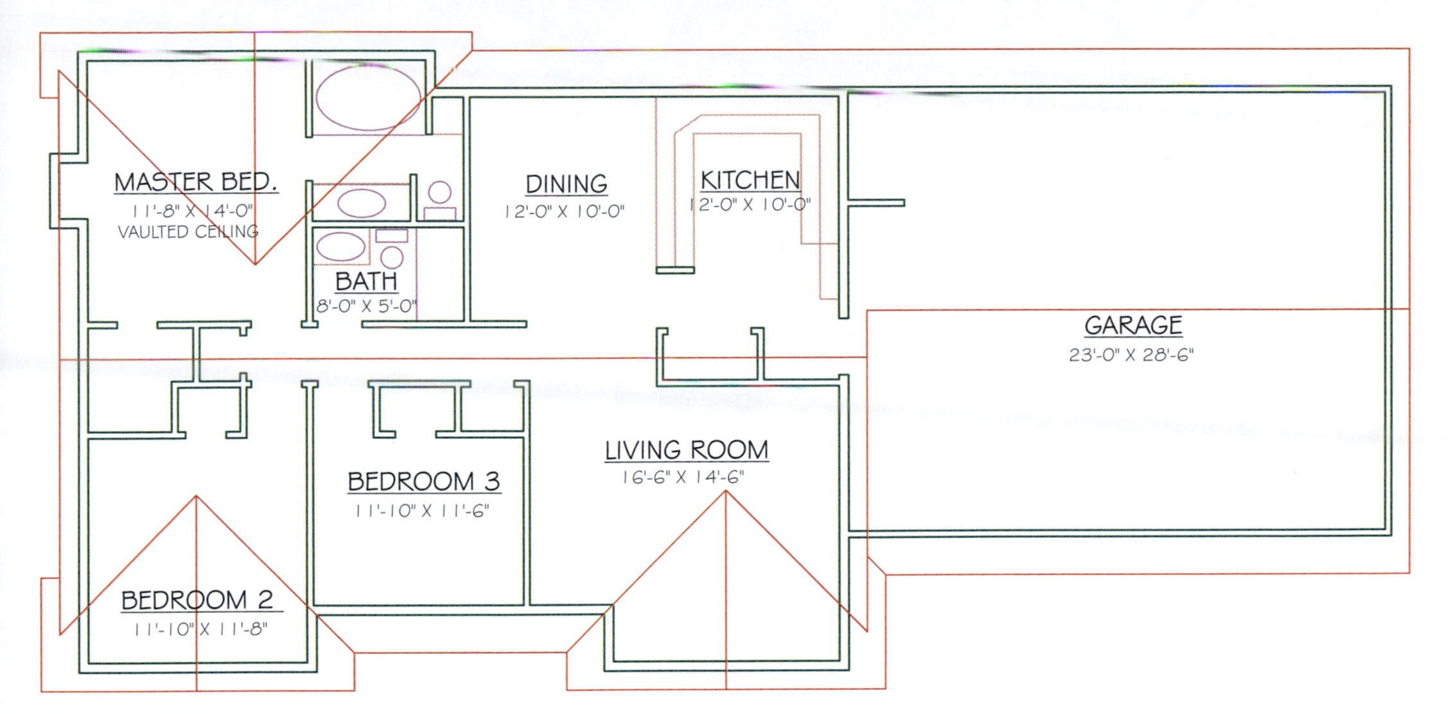

PROBLEM 11.5

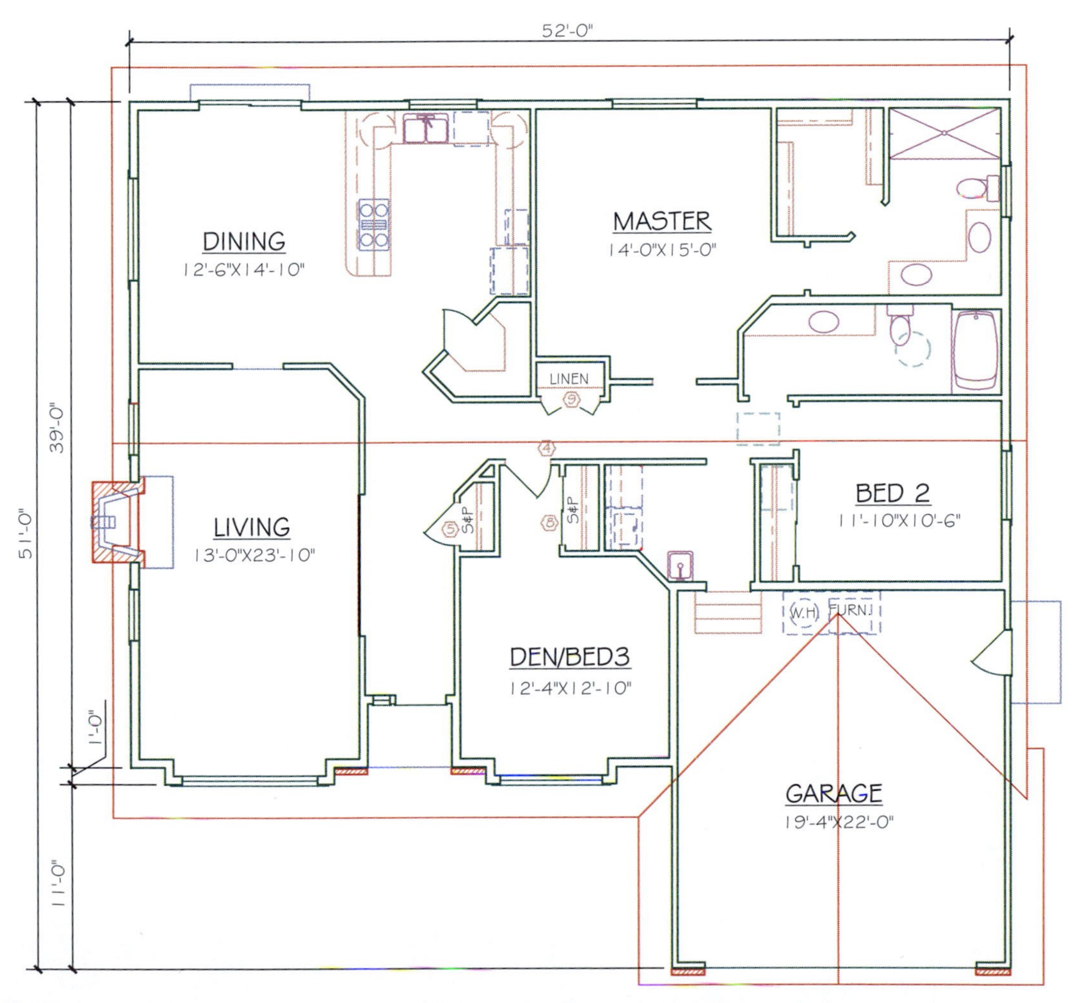

PROBLEM 11.6

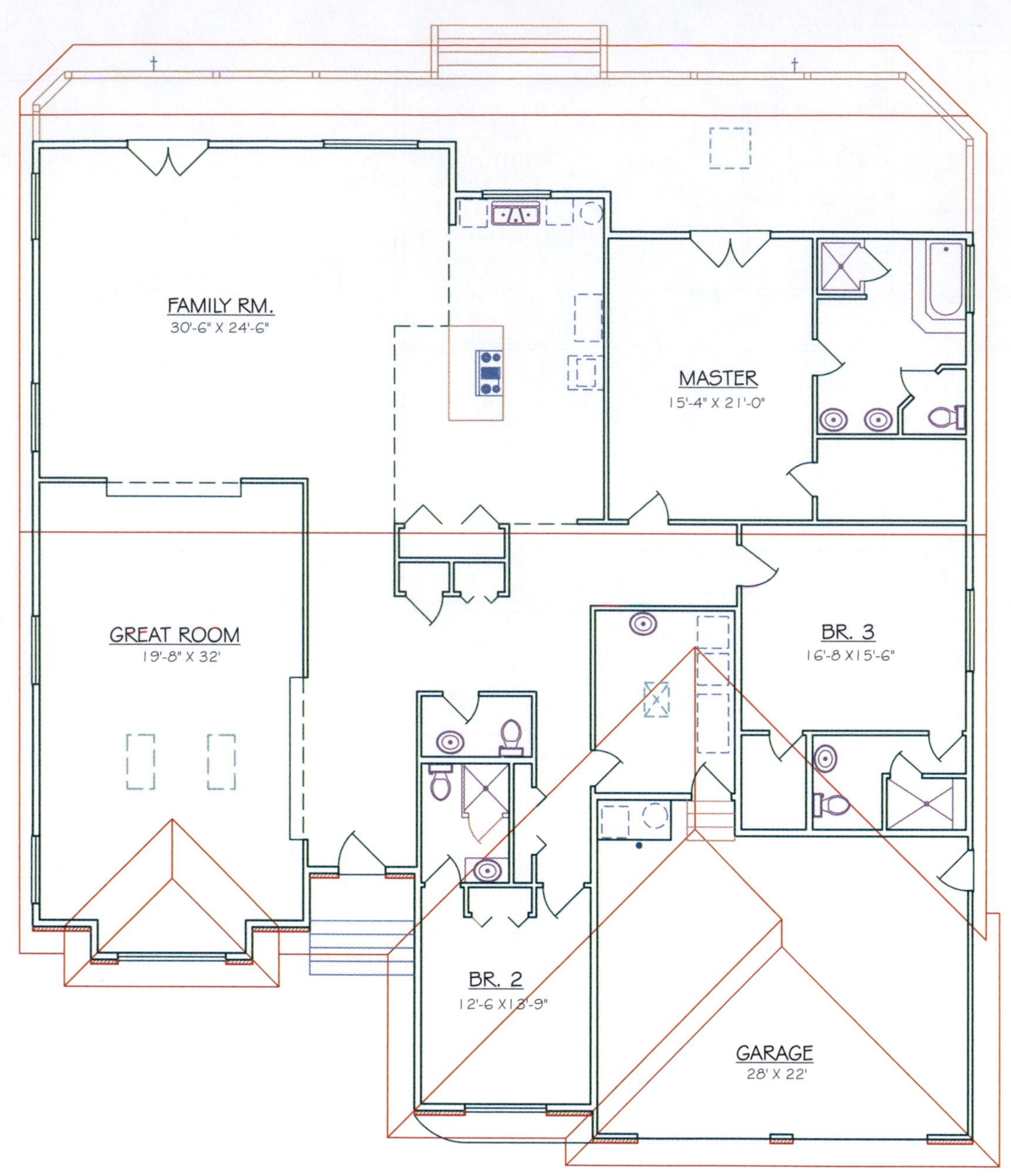
FAMILY RM.
30'-6" X 24'-6"
MASTER
15'-4" X 21'-0"
GREAT ROOM
19'-8" X 32'
BR. 3
16'-8 X 15'-6"
BR. 2
12'-6 X 13'-9"
GARAGE
28' X 22'

PROBLEM 11.7

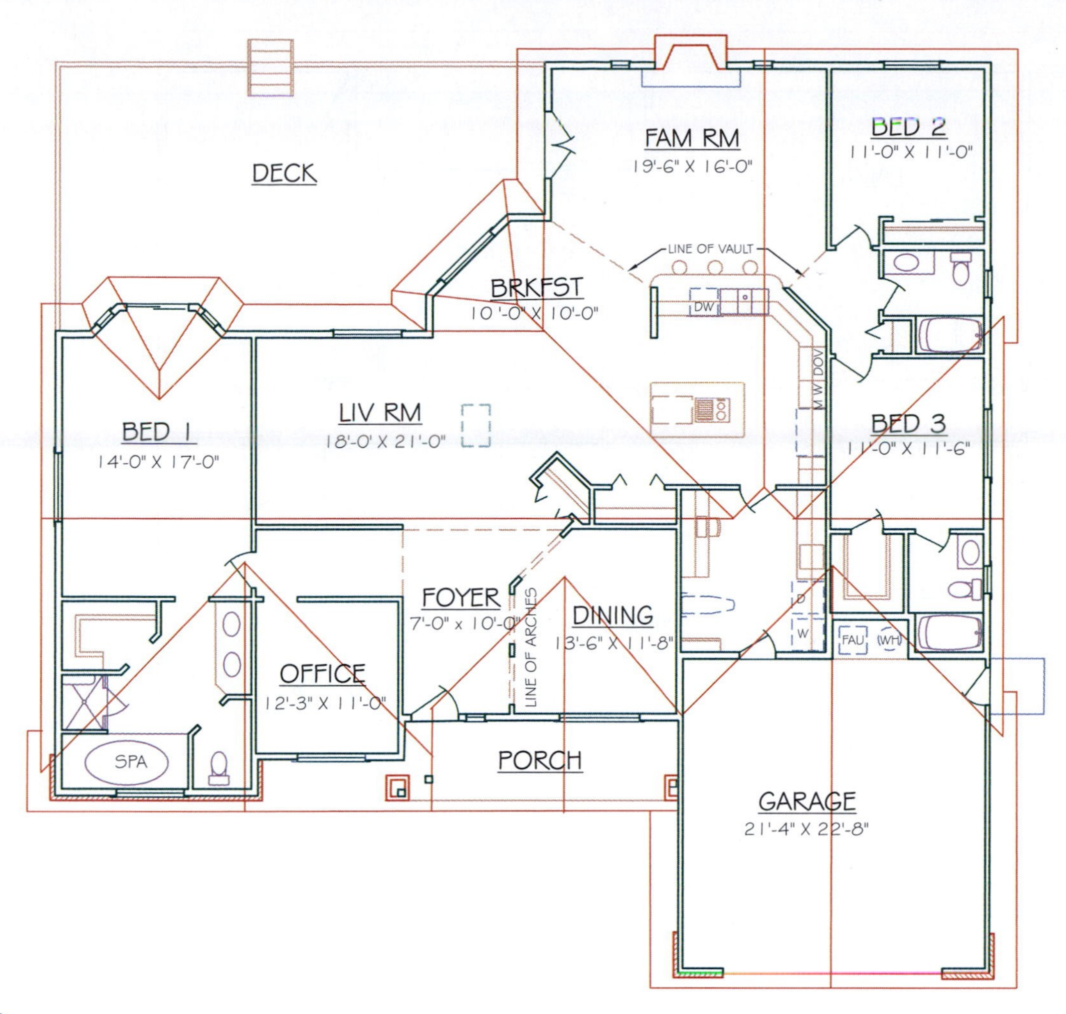
DECK
FAM RM
19'-6" X 16'-0"
BED 2
11'-0" X 11'-0"
LINE OF VAULT
BRKFST
10'-0" X 10'-0"
DW
BED 1
14'-0" X 17'-0"
LIV RM
18'-0' X 21'-0"
BED 3
11'-0" X 11'-6"
FOYER
7'-0" x 10'-0"
DINING
13'-6" X 11'-8"
LINE OF ARCHES
OFFICE
12'-3" X 11'-0"
SPA
FAU
WH
PORCH
GARAGE
21'-4" X 22'-8"

PROBLEM 11.8

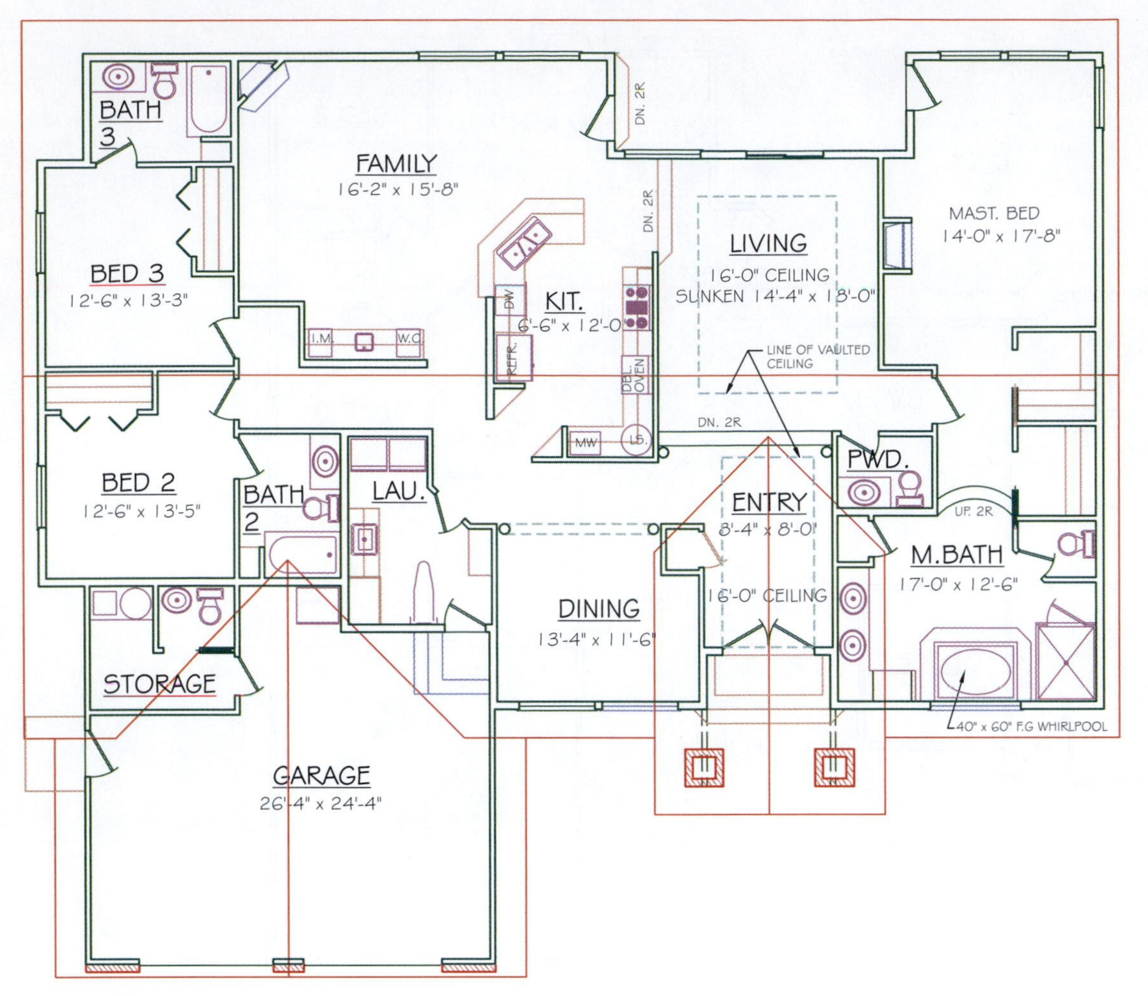
BATH 3
FAMILY
16'-2" x 15'-8"
DN. 2R
MAST. BED
14'-0" x 17'-8"
BED 3
12'-6" x 13'-3"
LIVING
16'-0" CEILING
SUNKEN 14'-4" x 18'-0"
KIT.
6'-6" x 12'-0
DW
I.M.
W.C.
REFR.
DBL. OVEN
LINE OF VAULTED CEILING
DN. 2R
MW
L.S.
BED 2
12'-6" x 13'-5"
BATH 2
LAU.
ENTRY
8'-4" x 8'-0'
16'-0" CEILING
PWD.
UP. 2R
M.BATH
17'-0" x 12'-6"
DINING
13'-4" x 11'-6"
STORAGE
40" x 60" F.G WHIRLPOOL
GARAGE
26'-4" x 24'-4"

PROBLEM 11.9

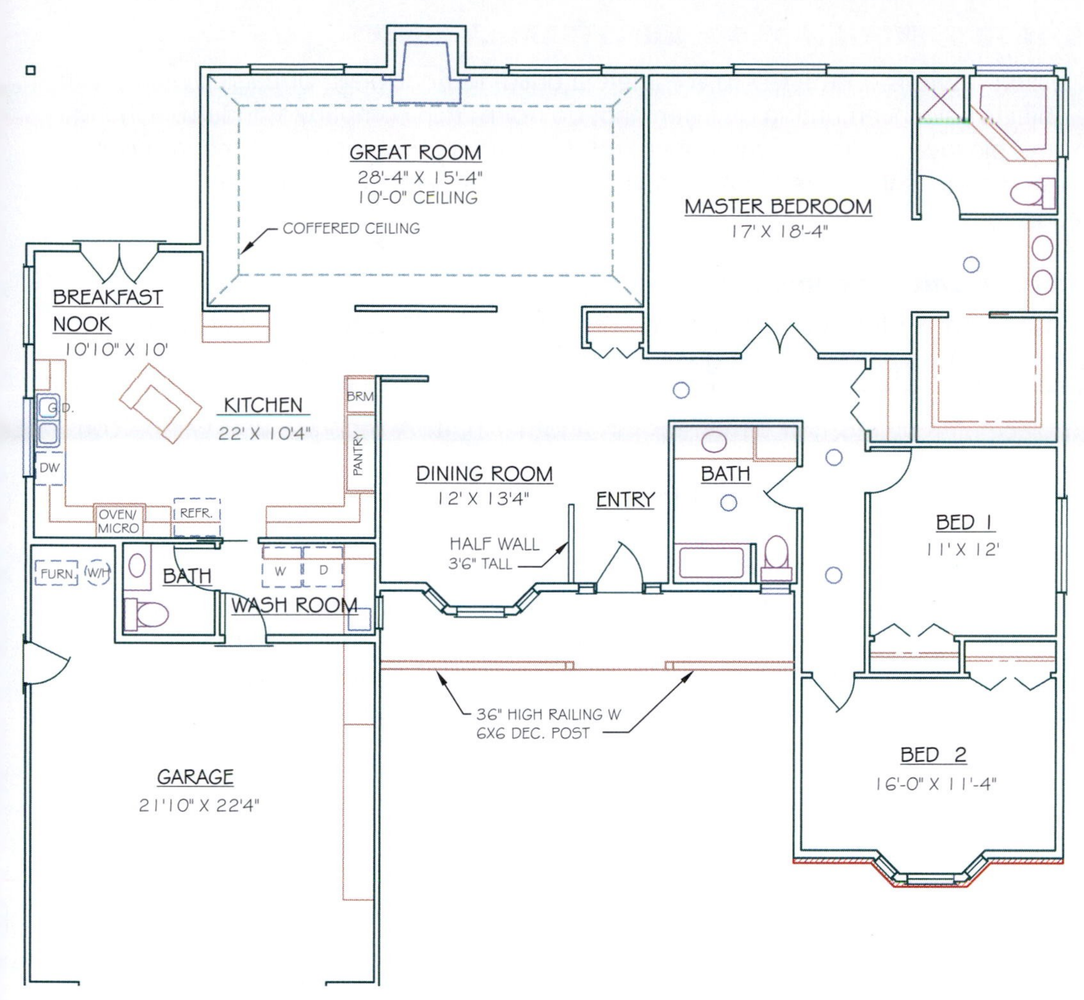

PROBLEM 11.10

PROBLEMS 11.11 THROUGH 11.30: DESIGN PROJECTS

Use the floor plan skeleton that has been started and is stored on the CD. Since there is no owner's sketch to use as a guide, you will be responsible for making all of the required design decisions. Start by planning how each space will be used, and then determine where plumbing fixtures and appliances will be located. Before design changes will be accepted, you must be able to defend your ideas. Provide your job captain with a list of changes and a list of how your changes will be an improvement. Select an appropriate template, set all required variables, and complete the drawing to meet the requirements of your local building department. Some objects have been drawn on layers that do not match company standards. Change these items to suitable layers. Some of the upper and lower floors have been created in separate files. Merge these drawings into one file and stack the levels. Rename layers to keep upper and lower items separate. Use a grade sheet from the CD to evaluate your work before submitting the assignment to your supervisor.

PROBLEMS 11.31 THROUGH 11.37: MULTILEVEL ADVANCED DESIGN

Use one of the electronic files supplied by William E. Poole Designs, Inc., from the CD as a guide to complete the necessary floor plans. Although there are .PDF files for a guide, there are no electronic drawings to use as a base. The proposed owner likes the home as it is but is open to minor changes. Before design changes will be accepted, you must be able to defend your ideas. Select an appropriate template, set all required variables, and complete the drawing to meet the requirements of your local building department. Use a grade sheet from the CD to evaluate your work before submitting the assignment to your supervisor.

PROBLEMS 11.38 THROUGH 11.40: MULTIFAMILY UNITS

Use the skeleton drawings on the CD to complete a building floor plan for each level of each project. Some objects have been drawn on layers that do not match company standards. Change these items to suitable layers. Some of the upper and lower floors have been created in separate files. Merge these drawings into one file and stack the levels. Rename layers to keep upper and lower items separate.

Problem 11.38 Four-Unit Condominium

The building is to contain four units that have aligning front walls. Each end unit is to have a bay window with center units to have flat glass. Place all required text in one unit and specify all door and window sizes using a schedule. Future chapters will add electrical information in another unit and place all required unit dimensions in another unit. A separate unit will be completed at a later time to show all framing material. Use a grade sheet from the CD to evaluate your work before submitting the assignment to your supervisor.

Problem 11.39 Six-Unit Condominium

Use the floor plans on the student CD as a guide to create a floor plan for each level. Place all required text in one unit and specify all door and window sizes using a schedule. Future chapters will add electrical information in another unit, and place all required unit dimensions in another unit. A separate unit will be completed at a later time to show all framing material. Use a grade sheet from the CD to evaluate your work before submitting the assignment to your supervisor.

Problem 11.40 Four-Unit Condominium

Use the site plan from problem 9.30 as a guide to layout the floor plans for each level of the structure. The building is to contain four units that have aligning front walls. Each end unit is to have a bay window with center units to have flat glass. Place all required text in one unit and specify all door and window sizes using a schedule. Future chapters will add electrical information in another unit, and place all required unit dimensions in another unit. A separate unit will be completed at a later time to show all framing material. Use a grade sheet from the CD to evaluate your work before submitting the assignment to your supervisor.

CHAPTER 12

Electrical Plans

The electrical needs of a residence may be the fastest-changing area of residential design. The use of computers to control almost every function of a home, smart appliances, and demands for increasing energy efficiency have fueled the development of electrical and electronic items that must be represented on the construction drawings. Kitchens have moved beyond a microwave oven and electric ovens to include warming drawers, pizza ovens, rotisseries, built-in espresso and cappuccino machines, heated towel drawers, dishwasher drawers, and refrigerators that tell you when the milk is spoiled. The electrical plans for a residence must display all of these electrical systems as well as the specialty systems to be installed in the entire project. This chapter introduces the principles of electrical distribution, key considerations in the design of electrical fixtures, basic code restrictions and design considerations that influence the development of electrical drawings, and skills needed to complete an electrical plan. In addition to learning how to draw an electrical plan, this chapter will help you plan a well-designed home similar to that shown Figure 12.1, which will satisfy a wide variety of needs.

ELECTRICAL DISTRIBUTION

The electricity to power a home can come from the local power company that uses water, fossil fuel, wind, or nuclear power to generate electricity, or it may be generated by private wind, solar, or geothermal methods. With the exception of solar photovoltaic cells, each of these methods is harnessed to produce a rotary mechanical motion that converts the rotary movement of a generator into electricity. Transformers are used at the generation point to increase the electrical power to hundreds of thousands of volts for transmission over long distances. Transformers are then used at local substations to step down the voltage to a few thousand volts for distribution to neighborhoods. Local transformers are also used within the neighborhood to step down the power to 120/240 volts for delivery to a home. Older designations list the service drop as 110/220 volts. Appliances are rated to be 110 or 220 volts. The service is supplied at higher voltages to compensate for increased starting loads. Once in the home, power flows from the service entrance, to the distribution box, through individual circuits, and finally into individual fixtures.

Electrical Measurement

Knowing that electrical power is delivered to a home in 240- and 120-volt supplies is important to understanding residential power usage. Any technician with knowledge of computing can place symbols on a drawing and call it an electrical plan. Understanding how electricity is used in a circuit will help you to advance from a technician placing symbols to a designer who works with a client to plan complex electrical needs. Key terms used to measure the electrical requirements of fixtures include volt, ampere, and watt.

A ***volt*** is a unit of measurement of electrical force or potential. It's the voltage that makes electricity flow through an electrical wire. For a specific load, the higher the voltage, the more electricity will flow. The flow of electricity is referred to as current. The rate of the current

FIGURE 12.1 ■ *A well-designed home must meet a variety of needs and provide a pleasant environment throughout the year in a variety of conditions.* Courtesy William E. Poole, William E. Poole Designs, Inc.

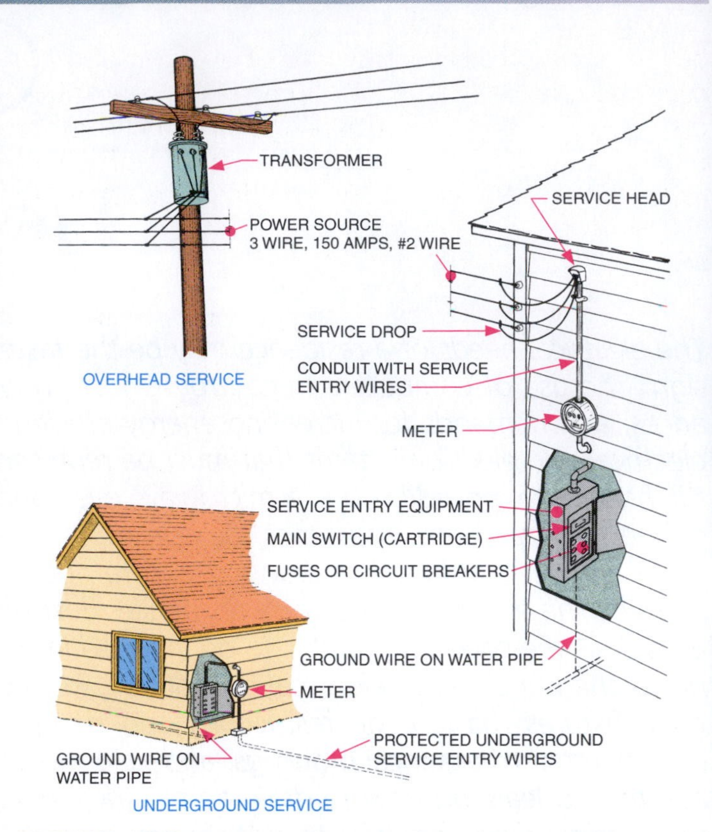

FIGURE 12.2 ■ *Electrical distribution includes 240- and 120-volt service, which is delivered to a service entrance using either an overhead drop or an underground conduit that leads to the electric meter.*

flow is measured in amperes or amps. An amp measures the number of electrons passing a specific location each second. The electrician needs to know the number of amps that will pass through a circuit to determine proper wire sizes and breaker sizes. The actual power required to operate a specific fixture or appliance is measured by watts. A ***watt*** is a unit of power measurement based on the potential and the current. The amount of power to be used per fixture is determined by multiplying the current ***(amps)*** by volts (potential) to determine the watts. In residential construction, the electrician generally determines the electrical loads that need to be considered in the design. Remembering these three terms will allow you to have an informed discussion with the electrician.

Electrical Service Entrance

The 240 and 120 volts that are supplied to a residence enter the home at the service entrance. Three wires are either dropped from a utility pole or run underground to the electric meter of the home to provide power. Overhead wires enter a service head mounted to the residence and pass through a conduit, which leads to the electric meter. A ***conduit*** is a metal or plastic tube used to enclose one or more electrical wires. The contractor is responsible to ensure minimum clearance heights are maintained between the ground and the service head, but the design team may provide details to mount the meter and service head to ensure compliance with local laws. If service to the home is underground, a conduit is used to bring the electrical service up to the meter. Each condition is shown in Figure 12.2. The electric ***meter*** is mounted on the wall where the power enters the structure and is used by the electrical supplier to measure the amount of electricity used by a structure.

All systems must be grounded at the service entrance, and some municipalities require lightning protection and brownout equipment. The contractor and electrician are generally responsible to plan for these needs as well as to determine the exact location for the service drop. On homes designed to appeal to the local market conditions, the contractor will determine the location of the meter. For a custom home, the design team is responsible for coordinating the meter location with the owner and the contractor and representing the location on the electrical drawings. The service location and required trenches are often represented on the site plan, and the meter is represented on the electrical drawings, The design team may also be required to locate a transformer on the site plan for homes in rural settings.

Distribution Panel

From the meter, power flows to the distribution panel, which is also referred to as the service panel or circuit box. The distribution panel contains circuit breakers, which control each of the individual circuits in the home. The minimum size of the panel in amperes is determined by the total load requirements in watts for the entire project. Watts are converted to amperes by dividing the total watts needed by the amount of voltage delivered to the distribution box. In determining the total watts needed, consider future expansion and any possible additions that may be added in the future. The National Electrical Code (NEC) requires a minimum of 60 amps, but most residential projects require a distribution panel with a capacity of between 100 to 200 amps. If heating, cooking, water heating, and similar heavy loads are supplied by gas, a 100-amp distribution panel will be suitable. A 200-amp panel is common for most residential projects and is required by some local codes for each single dwelling.

The distribution box must be located on the electrical plan (see Figure 2.24). The box is generally placed on the interior side of the wall where the meter is mounted. On custom homes, the circuit box may be placed in a more convenient location such as the master bedroom closet, in a closet in the utility room, or by the door leading from the house to the garage. Place the circuit box in a location that will be easy to get to in the dark and away from areas where there is a risk of injury from falling in the dark.

Branch Circuits

The distribution panel contains circuit breakers that control the individual circuits of the home. A ***circuit breaker*** is a safety switch that automatically opens a circuit when excessive amperage occurs. A ***circuit*** is a closed loop that electricity follows from the circuit box to one or more fixtures and then back to the power source. The circuit is controlled by a circuit breaker that is rated in the amount of amps the breaker allows to flow through the circuit. A distribution panel typically contains breakers that control 50-, 40-, 30-, 20-, and 15-amp circuits. Appendix E contains the amps required for common residential appliances. Figure 12.3 lists the amp requirements for common residential circuits. When the breaker is closed, it allows electricity to flow through the circuit. If too many fixtures are used on a circuit at one time, the circuit may overheat and cause a fire. If the circuit requires more amps than the breaker is rated to allow, the breaker will trip, or open. When the breaker is open, the flow of electricity to the circuit is disrupted. If you've lived in an older home, you know the joys of not using the microwave while the dishwasher is on. For most residential projects, the electrician is responsible to determine how many fixtures will be placed on each circuit. For very large custom projects, an electrical consulting company will design the circuits and provide the electrician a plan of how circuits will be established. In considering circuits, it is important to know how the NEC defines branch circuits as lighting circuits, small appliance circuits, and individual circuits.

Lighting Circuits

The lighting circuits control the light fixtures for the home. Outlets that provide power for small appliances such as clocks, radios, and fans can also be placed on lighting circuits. The NEC requires the minimum number of lighting circuits to be based on a lighting load of 3 watts per square foot of floor space. For a 2000-sq-ft home, it would seem that a lighting load of 6000 watts must be provided. The NEC allows this number to be reduced based on what it refers to as the demand factor. The ***demand factor*** assumes that not all lights will be on at the same time, so the lighting load can be reduced. The demand factor allows:

100% based on the first 3000 watts
35% for the next 17,000 watts

Based on the reduced demand factor, the 2000-sq-ft home would require

First 3000 watts = 3000 W
Next 3000 watts × 35% = 1050 W
4050 watts required

If each lighting circuit supplies 2400 watts (120 V × 20 A = 2400 W), a 2000-sq-ft house will require a minimum of two lighting circuits.

Small Appliance Circuits

The small appliance circuits are used to provide power to outlets where small appliances are likely to be located. The NEC considers small appliances to be irons, toasters, skillets, crockpots, and computers. These are

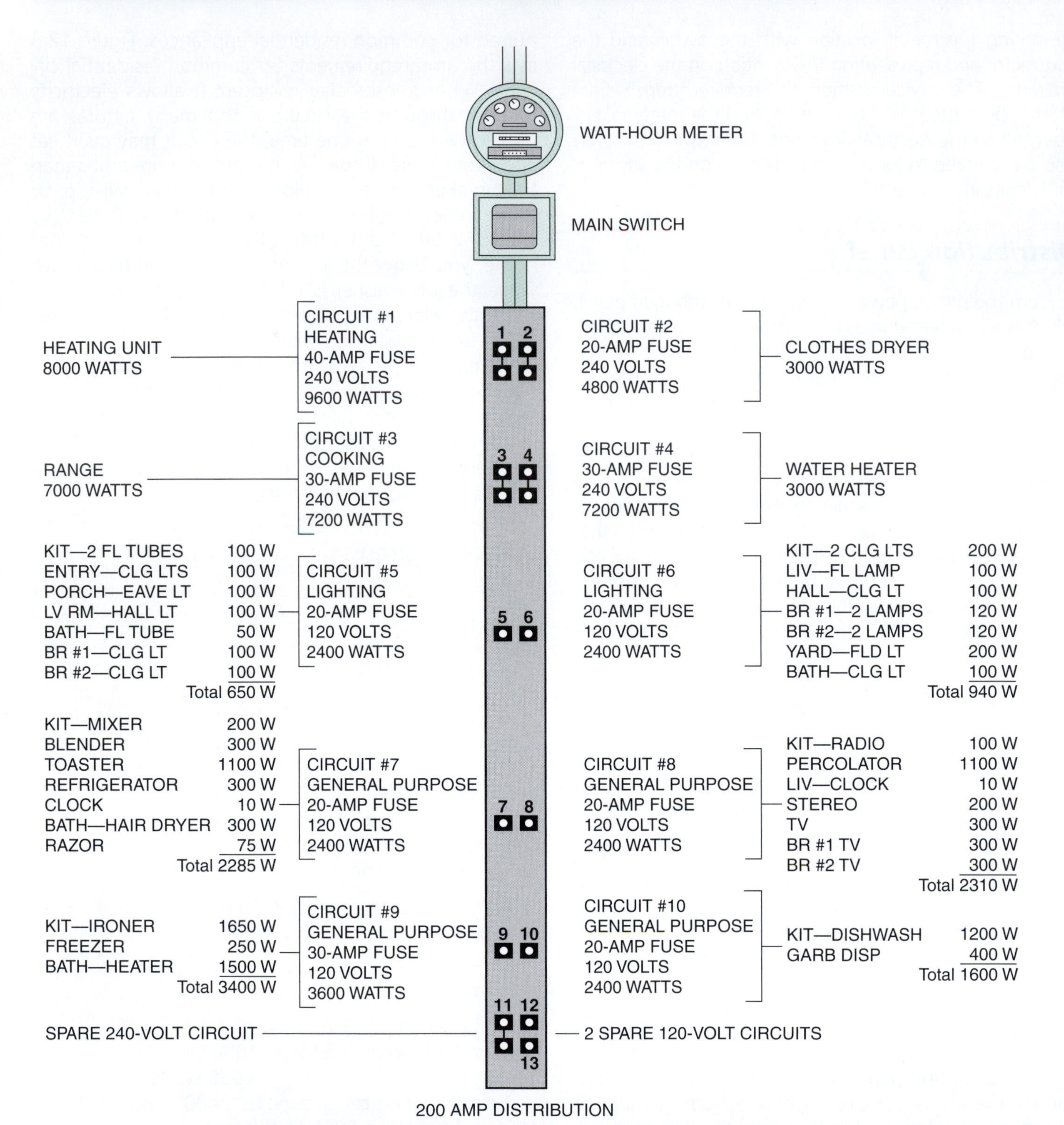

FIGURE 12.3 ■ *The distribution panel, also known as a circuit panel, is used to control the flow of power from the power source into the structure. Although the exact appearance will vary with each structure, each panel will contain 50-, 40-, 30-, and 20-amp circuit breakers to regulate each electrical fixture in the home. See Appendix F for specific requirements for common home appliances.*

appliances that do not produce a power surge when started. Small appliance circuits cannot supply power to lighting fixtures. The NEC requires a minimum of two small appliance circuits per residence. When the locations of home computers are known, it is ideal to provide a separate circuit for each computer station. The actual number of circuits is usually determined using a 3600-watt load (30 A × 120 V = 3600 W).

Dedicated Circuits

A dedicated circuit is designed to serve a single large electrical appliance such as a dryer, range, or heating unit. Any large motor-driven appliance such as a dishwasher or washing machine that produces a surge of power when started (the starting load) should also be provided with its own circuit.

Fixture Boxes

The final destination of each wire in a circuit is a junction box. Such a box, also known as an outlet box or j-box, is used to protect and seal the connection of the wires associated with the fixture and the wiring that runs through the walls. The wires for a ceiling fixture are connected to the circuit wires and hidden in the junction box. Wall-mounted boxes usually contain either a switch or an outlet with a direct connection to the circuit wires. Some appliances, such as a dishwasher and some garbage disposals, have wires without an end plug. These appliances are connected directly to the circuit wires and hidden in the box. When no plug is provided, the appliance is said to be hard-wired. Figure 12.4 shows examples of wall and ceiling boxes.

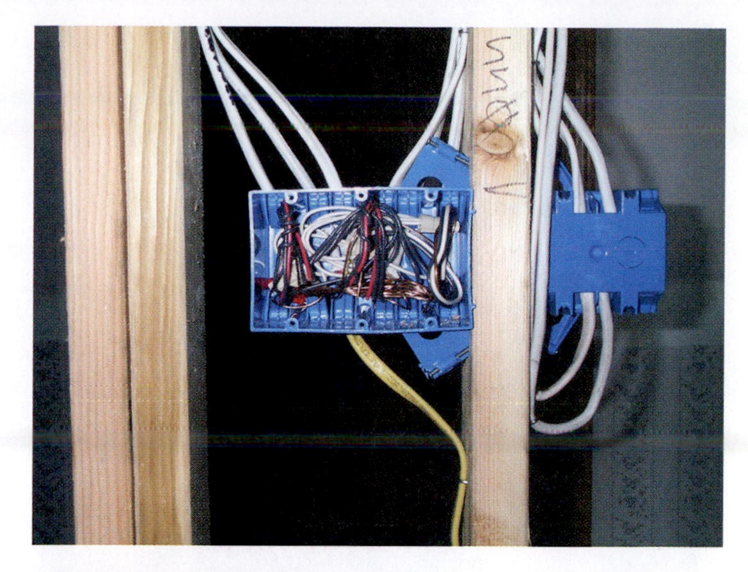

FIGURE 12.4 ■ *Wires from the circuit box to a fixture terminate in a box that is hidden in the wall, ceiling, or floor. The box is used to provide a sealed termination point between the circuit and the fixture. (Top) The wall-mounted box will be used to provide for two switches on the left and a 120-volt convenience outlet on the right. (Bottom) A ceiling-mounted light fixture contains its own termination box.*

PLANNING FOR ELECTRICAL FIXTURES

Your job captain may provide you, as a new employee, with a copy of a floor plan with all the required symbols marked on the plan. As you advance as a CAD technician, you'll be expected to make a print of the floor plan and make your own plan. Many municipalities do not require an electrical plan to obtain a permit for a residence. The contractor or homeowner can apply for an electrical permit and pay fees based on the number of fixtures and outlets to be installed. The general contractor may rely on a skilled electrician when the home is to be built for an unknown buyer. An electrical plan is given careful consideration when a home is designed for a specific owner. Planning for the electrical needs of a home requires considering the type and placement of lighting fixtures, placing outlets based on code and practical use, and placement of switches to control each fixture.

Lighting Design

The type of lights, how light is dispersed, and the method to provide lighting should be considered in planning the lighting requirements of a home. Too much light will cast a glare on objects, ruin a mood, and cause the electric bill to skyrocket. Insufficient lighting can require the use of unwanted table or freestanding lights. Within a room, if lighting is ample in one area and inadequate in another, shadows will result, making the area difficult to use. Good planning allows all areas of a home to be used throughout the day and in each season by carefully balancing the light provided by the sun, outside lights in the neighborhood, and lighting that is part of the home.

The intensity of a light is measured in units of footcandles. A footcandle is the amount of light a candle

SUNLIGHT	
Beaches, open fields	10,000 FC (107 640 LX)
Tree shade	1000 FC (10 764 LX)
Open park	500 FC (5382 LX)
Inside 38 from window	200 FC (2153 LX)
Inside center of room	10 FC (108 LX)
ACCEPTED ARTIFICIAL LIGHT LEVELS	
Casual visual tasks, conversation, watching TV, listening to music	10–20 FC (108–215 LX)
Easy reading, sewing, knitting, house cleaning	20–30 FC (215–323 LX)
Reading newspapers, kitchen & laundry work, keyboarding	30–50 FC (323–538 LX)
Prolonged reading, machine sewing, hobbies, homework	50–70 FC (538–753 LX)
Prolonged detailed tasks such as fine sewing, reading fine print, drafting	70–200 FC (753–2,153 LX)

FIGURE 12.5 ■ *Comparison of natural sunlight and artificial sunlight levels. By providing adequate lighting levels, each area of the home can be efficiently used at any time throughout the day.*

casts on an object 12'' (300 mm) away. Figure 12.5 provides a listing of the minimum amount of footcandles electrical designers have determined should be provided for basic activities. Values are also listed in footcandles with the metric equivalent measured in lux (lx). One lux is equal to .093 fc (footcandles). To convert footcandles to lux, multiply by 10.764. Knowing the light required to perform the expected tasks of a room will aid you to determine the lighting needs of a room.

Types of Lighting

The type of lighting to be provided will affect the amount and placement of lighting fixtures. Lighting needs are considered to be general, specific, or decorative.

General Lighting

General lighting provides a comfortable level of illumination for an entire room. One or more ceiling-mounted lights, a chandelier, a series of recessed ceiling fixtures, or several wall-mounted lights are common methods of providing general light to a room. Each of these methods allows a shade or globe to be used to diffuse the light and avoid direct viewing of the light source. Track lighting and other adjustable lighting fixtures can also be used to light the room. The placement of windows and skylights and their relationship to the movement of the sun must be considered in planning general lighting needs. Figure 12.6 shows a chandelier and recessed lights used to provide general lighting to a dining room.

FIGURE 12.6 ■ *General lighting provides for the overall illumination at a comfortable level for the entire room. The overhead and recessed fixtures provide the general lighting for this dining room.* Courtesy BOWA Builders Inc. Bob Narod, photographer.

Specific Lighting

Specific lighting provides light to do a specific task, such as reading, applying makeup, shaving, or watching television. A ceiling-mounted light directly over a kitchen sink, under-cabinet lighting in a kitchen, and wall-mounted lights over bathroom sinks are examples of specific task lighting. Each type of light provides lighting so shadows do not affect the specific work area. Recessed or wall-mounted fixtures, track lighting, and free-standing lamps are common methods of providing light to a specific work area. Figure 12.7 shows the use of recessed fixtures that provide light to each of the kitchen work areas.

Decorative Lighting

Decorative lighting, also referred to as mood lighting, is used to create a specific atmosphere. Adjustable spot and recessed lights with partial covers are often used to cast light on art, photos, unique wall textures, or other architectural features. Figure 12.8 shows how lighting can be used to highlight a photo over a fireplace.

Light Distribution

No matter which of the three roles the lighting fixture is designed to meet, light is dispersed throughout a room in one of five methods. These lighting methods in-

FIGURE 12.7 ■ *Specific lighting provides lighting to do a task in a specific work area. The lighting in this kitchen is arranged so that a shadow will never be cast on any work area.* Courtesy Alan Mascord AIBD, Alan Mascord Design Associates, Inc. Bob Greenspan, photographer.

clude direct, indirect, semidirect, semi-indirect, and diffused. Each method of casting light is shown in Figure 12.9. Direct lighting allows light to be cast directly from a source such as a ceiling light. Indirect lighting reflects light off of a ceiling or wall surface and then into a room. Semidirect lighting fixtures direct most of the light downward while still allowing some light to go upward. Semi-indirect fixtures reflect most of their light off the ceiling but cast some of it downward as direct light. Diffused light is spread evenly over a room through a translucent shade or globe. Wall- and ceiling-mounted fixtures that provide direct or diffused lighting are often used for general lighting. Under-cabinet fixtures, wall-mounted, and recessed fixtures that provide direct or diffused lighting are also used for specific task lighting. In addition to the fixtures already mentioned, fixtures that provide indirect, semidirect, and semi-indirect lighting are generally used for decorative lighting.

Lighting Fixture Types

As seen in Figure 12.10, the selection and placement of lighting fixtures can dramatically affect the appearance and livability of a home. For a home built to appeal to a wide variety of potential buyers, the fixtures may be selected by the design team, an interior designer, the contractor, or the homeowner prior to the purchase of the home. The budget, market conditions, and timeline for the project will influence who selects the light fixtures. On a custom home, usually the architect, interior designer, or a lighting specialist will work closely with the owner to select each type of fixture. Choices can generally be divided into ceiling, wall, and exterior fixtures.

FIGURE 12.8 ■ *Decorative lighting is used to create a specific atmosphere. Here light from a recessed ceiling light is directed onto the art hung over the mantle.* Courtesy Benigno Molina-Manriquez.

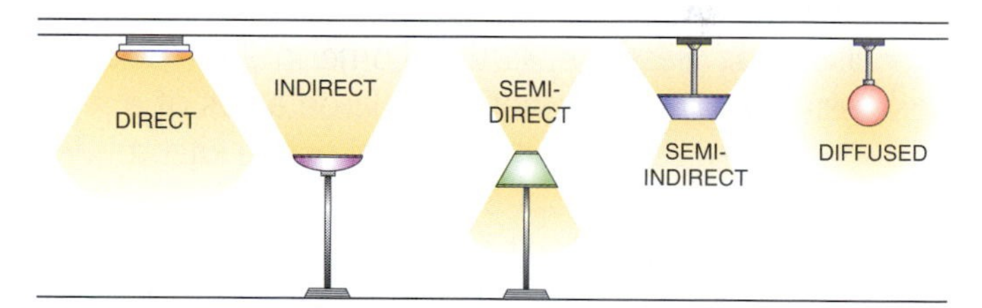

FIGURE 12.9 ■ *Five methods are used to distribute light in a room. Each of these methods can be used to meet general, specific, and decorative lighting needs.*

Ceiling Fixtures

Ceiling fixtures range from a surface-mounted fixture that can be purchased at your local outlet store for $30 to a $15,000 chandelier. Figure 12.11 shows common types of ceiling fixtures, including surface-mounted and recessed fixtures, fluorescent light panels, and tract lighting. Surface-mounted fixtures include chandeliers as well as adjustable, pendant, and reel fixtures. Recessed lights include rectangular fixtures, which are typ-

FIGURE 12.10 ■ *The placement and type of lighting fixtures can greatly affect the appearance and livability of a residence.* Courtesy Frank Serpe, owner. Franco D. Demetrio, architect. Susan Miller, photographer.

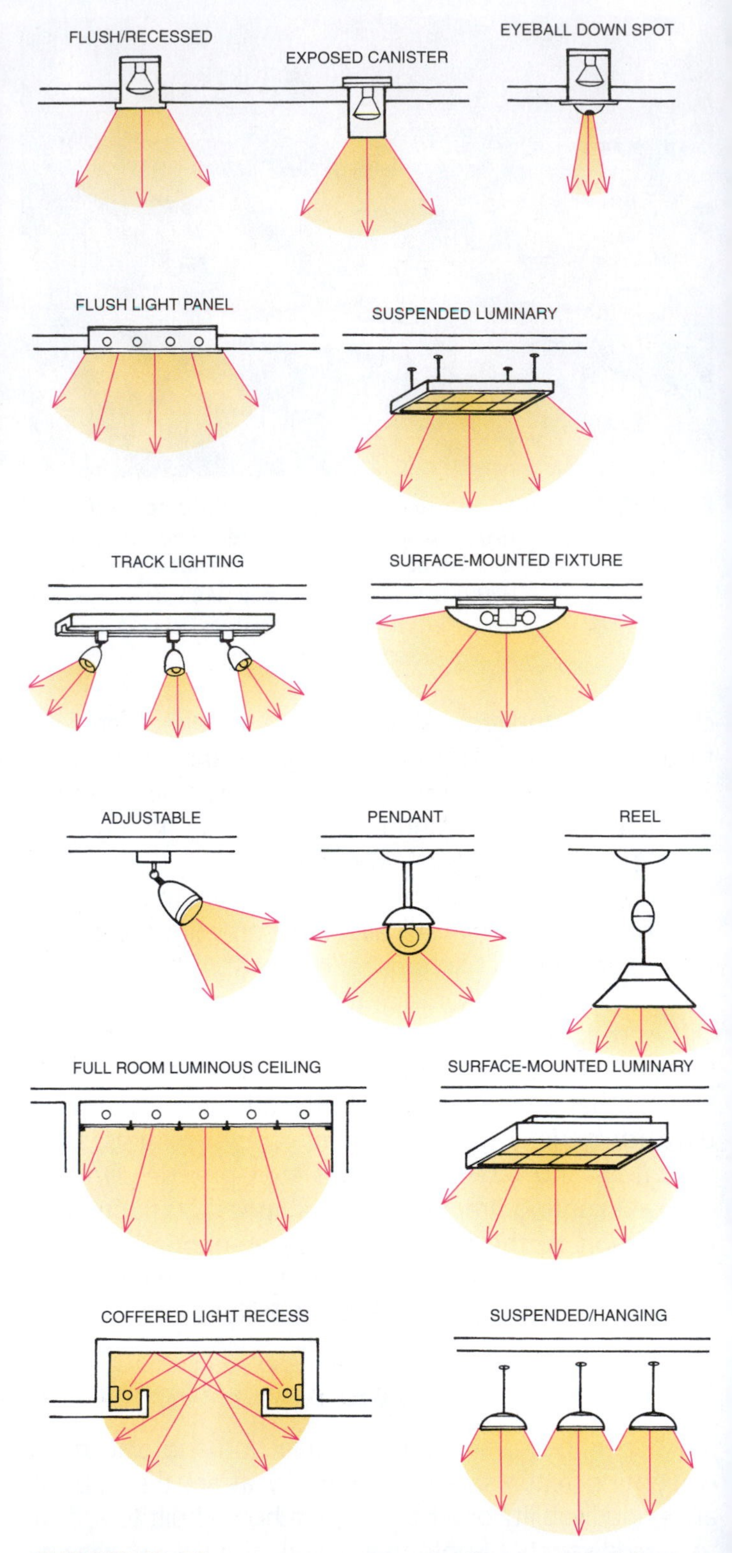

FIGURE 12.11 ■ *Common types of ceiling-mounted fixtures.*

ically used in hallways or closets, and round fixtures, referred to as can lights. Can fixtures can be fully recessed or partially exposed. A movable cover may be placed over a can fixture to focus light in a specific direction. This type of light is referred to as an eyeball fixture and is often used for decorative purposes.

Another common type of ceiling lighting is referred to as soffit lighting. Soffit lighting places the lighting fixture behind a translucent panel and directs the light source downward through the panel. Soffit fixtures can be surface- or flush-mounted and are often used in utility and sewing rooms as well as kitchens. Job-built soffit lighting can be installed between the joists or trusses and covered with plastic or glass panels that come in 24" (600 mm) modules, providing illumination to large areas of a room. An under-cabinet soffit can be added over a kitchen work area by extending the front face of the cabinet to hide the lighting fixture. Figure 12.12 shows an example of lights mounted above and below the upper cabinets to provide both specific and decorative lighting. Cornice lighting places the fixture near the edge of a wall and hides the light behind trim. Light can be directed either up or down depending where the trim is placed. Figure 12.13 shows an example of stepped cornice lighting directed toward the ceiling. Both soffit and cornice lighting can produce direct or indirect lighting, but both hide the fixture so the direction of the light is controlled by the trim.

Wall Fixtures

Wall fixtures similar to those in Figure 12.14 are used to fill the need for general and specific lighting. Depending on the room in which they are used, wall-mounted lights connected to a dimmer switch are also used to provide decorative lighting. Wall-mounted fixtures are usually classed as a globe, sconce, canister, spots, and lanterns, which provide either diffused, indirect, semi-direct, or semi-indirect lighting. Wall-mounted globe lights provide diffused light through some type of shade or translucent cover. A wall sconce

FIGURE 12.12 ■ *Lights hidden behind the front face of the upper cabinet provide direct lighting below the cabinet and indirect decorative lighting above the cabinet.* Courtesy Janet Rowell.

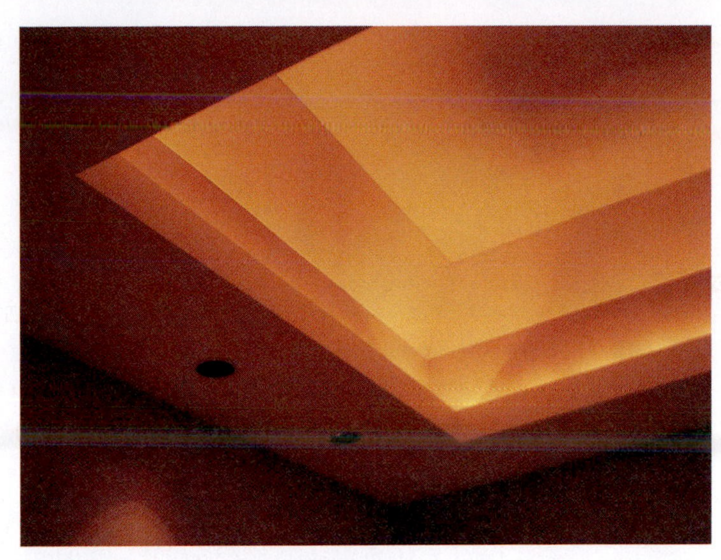

FIGURE 12.13 ■ *A stepped ceiling soffit provides a distinctive finish to this ceiling.* Courtesy Benigno Molina-Manriquez.

FIGURE 12.14 ■ *Wall fixtures are used to provide general and specific lighting needs. If only a ceiling-mounted light were used, work at the sink would be done in shadow. The wall-mounted light will allow any task preformed at the sink or counter to be shadow-free.* Courtesy Bob Frederick.

directs light either up or down, depending on the shape of the sconce, and provides indirect lighting. Figure 12.15 shows wall sconces used to light a home theater. Canister lighting can provide direct, indirect, or diffused lighting depending on the material that is used for the canister. If solid material is used for the "can", light is emitted from its top and bottom. If a translucent canister is used, it will provide diffused, direct, and indirect lighting. Wall-mounted spotlights can be adjusted to provide decorative lighting using direct and indirect methods. A lantern fixture resembles an antique candleholder that has been mounted on a wall. Lantern-type fixtures are often placed by exterior doors to provide direct lighting for safe access to the home.

Fixtures can also be mounted on a wall and hidden behind decorative trim to provide indirect lighting. This type of lighting, referred to as cove lighting, places trim in front of and below the fixture to reflect light upward to the ceiling. Valance lighting places trim in front of the fixture to reflect light upward to the ceiling and allow

FIGURE 12.15 ■ *This home theater uses wall sconces to provide defused light. Using a dimmer allows the level of light to be adjusted for comfortable viewing.* Courtesy Portland Cement Association.

FIGURE 12.16 ■ *Cornice lighting placed behind the crown molding adds to the atmosphere of the room.* Courtesy W. Lee Roland, Builder. Hayman Studios, photographer.

direct lighting below the fixture. Figure 12.16 shows the use of cornice lighting placed behind the crown molding of a bedroom.

Exterior Fixtures

Wall-, ceiling-, and post-mounted as well as ground fixtures similar to those in Figure 12.17 are used to accent the exterior of a structure and enhance the exterior living and landscaping areas. Exterior lights can be installed to meet several needs. Recessed fixtures are often mounted in the eaves to provide accent lighting to specific areas of a home. Wall-mounted fixtures can be used to provide direct or indirect general and decorative lighting for exterior areas. Wall-mounted spots are often connected to a motion detector for security purposes. Recessed floor-mounted lights can be installed in walkways or decks to facilitate nighttime use. Lighting fixtures mounted on posts or columns or ground-mounted fixtures are used to

FIGURE 12.17 ■ *Wall, ceiling, and ground fixtures are used to accent the structure and enhance the exterior living and landscaping areas.* Courtesy W. Lee Roland, builder. Hayman Studios, photographer.

light walkways and driveways or to accent landscaping. Switches mounted on interior walls near doors leading to decks or patios or photovoltaic controls can be used to control exterior lighting. Exterior lighting attached to the structure is always shown on the electrical drawings. Depending on the project, exterior lighting used for landscaping may or may not be shown on the drawings. At a minimum, the switches to control exterior lighting should be represented on the electrical plans. The location of fixtures that are not attached to the structure can be shown on a site plan or on landscaping drawings, depending on who will do the work.

Guidelines for Placing Lighting Fixtures

Use the following guidelines to place light fixtures on an electrical plan:

- Place lights in relation to their use based on general, specific, or decorative needs.
- Every entry should have at least one wall- or ceiling-mounted fixture controlled by three-way switches. Provide one switch near the entry door and one switch as you leave the entry area. Depending on the size of the foyer, additional ceiling- or wall-mounted lighting may be required.
- Use switch-controlled receptacles to provide lighting centered in the living and family rooms.
- Ceiling mounted lights are rarely provided in living rooms for general lighting purposes, but recessed ceiling or wall lights are often provided for accent lighting.

- Use a switch-controlled centrally located overhead light in the dining room, kitchen, office, study, nooks, and baths. Also provide lighting in built-in units or in front of hutches and cabinets.
- Use ceiling- or wall-mounted lights over stairwells. Lights are required by code to illuminate landings at each end of the stair. Show the fixtures on the upper and lower levels and represent three-way switches at each end of the stair.
- Use wall-mounted or recessed ceiling lights in hallways.
- Use a combination of wall-mounted, recessed ceiling, and switched duplex outlets for bedside lamps in a master bedroom.
- Use ceiling-mounted fixtures in children's bedrooms. A bedroom may have a recessed ceiling light in front of a wardrobe closet.
- Provide ceiling-mounted lights centered over the seating areas in bay windows.
- Provide a light over each kitchen sink.
- Provide a ceiling-mounted fixture for general bathroom lighting and a light above each vanity mirror. In a large bathroom, place a recessed light that is rated to be waterproof above a shower or tub. Chord-connected lighting—such as lamps, hanging fixtures, track lighting, pendants, or ceiling-mounted fans—can't have any part of the fixture located in a zone that is 36'' (900 mm) horizontally, or 8' (2400 mm) vertically above the rim of a bathtub or a shower stall threshold.
- Provide surface-mounted or recessed incandescent fixtures with lights that are completely enclosed in closets or any alcove or pantry that requires light. Lights must be a minimum of 12'' (300 mm) above the nearest point of storage. This distance can be reduced to 6'' (1250 mm) if fluorescent lights are used.
- Each enclosed bath or laundry room must have an exhaust fan or an openable window. If no window is provided, a fan with a minimum rating of 50 ft^3 per minute (cfm) (0.024 m^3/s) intermittent or 20 cfm (0.009 m^3/s) continuous must be provided. Kitchens without windows must have a minimum rating of 100 cfm (0.047 m^3/s) intermittent or 25 cfm (0.012 m^3/s) continuous must be provided.
- Place lights and receptacles in garages or shops in relation to their use.
- Place exterior lights to illuminate walks, drives, patios, decks, and other high-use areas.

Outlet Selection and Placement

The placement of electrical receptacles to connect appliances to the house wiring system should be planned based on the type of outlets to be used, NEC requirements, energy efficiency, and commonsense guidelines based on normal usage.

Types of Outlets

Although the terms *outlet* and *receptacle* may be used interchangeably, the NEC defines an outlet as the location in the circuit where electrical devices are connected. The NEC defines a receptacle as the device in the outlet box where the electrical component is actually attached. No matter the term used in your area, the two major types of receptacles are 240- and 120-volt. Older designations of these voltages were 220/110 connections. The official voltage for residential appliances is 125/250V based on the National Electrical Manufactures Association.

240-Volt Receptacles

Large electrical appliances—such as furnace, water heater, spa, clothes dryer, oven, and range—require 240-volt receptacles. Each outlet is placed on an independent circuit for one specific appliance. A key consideration for 240-volt receptacles is the size in amps of the circuit that powers the appliance. A dryer is generally placed on a 30-A circuit, and a range may be placed on either a 30- or 50-A circuit depending on the manufacturer's recommendations. These receptacles may be similar to those in Figure 12.18, although each appliance except the dryer is usually hard-wired directly to the circuit. Each of these appliances is also available as gas-powered. When a gas unit is used, a dedicated 120-volt circuit is provided to the appliance to power blowers, timers, lights, or other components of the unit. Although gas ovens are available, electric ovens tend to be more popular because of their convection and self-cleaning features.

120-Volt Receptacles

A 120-volt receptacle is referred to as a convenience outlet. The standard receptacle consists of two plugs and is referred to as a duplex convenience outlet. Receptacles containing three and four plugs are also available. Common types of 120-volt convenience outlets include:

- A half-hot receptacle, which is a conventional outlet with one receptacle that is always hot. A second receptacle is connected to a wall switch that controls the flow of power through the receptacle. The switch will now control any fixture that is plugged

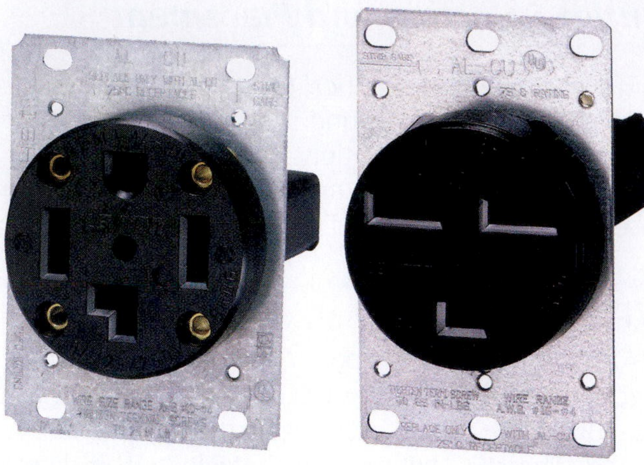

FIGURE 12.18 ■ *Fixtures such as a furnace, water heater, spa, clothes dryer, double oven, and range receive their power from 240-volt receptacles placed on an independent circuit. A key consideration for 240-volt receptacles is the size in amps of the circuit powering the appliance. A dryer is generally placed on a 30-A circuit, and a range may be placed on either a 30- or 50-A circuit depending on the manufacturer's recommendations. The range receptacle on the left is rated as 30A-250V grounding, NEMA 14-30R, with two hot connections as well as a neutral, and a grounding connection. The range receptacle on the right is rated as 30A-125/250V grounding, NEMA 14-30R, with two hot connections and a grounding connection.* Courtesy Pamela Winikoff, Leviton Manufacturing Company.

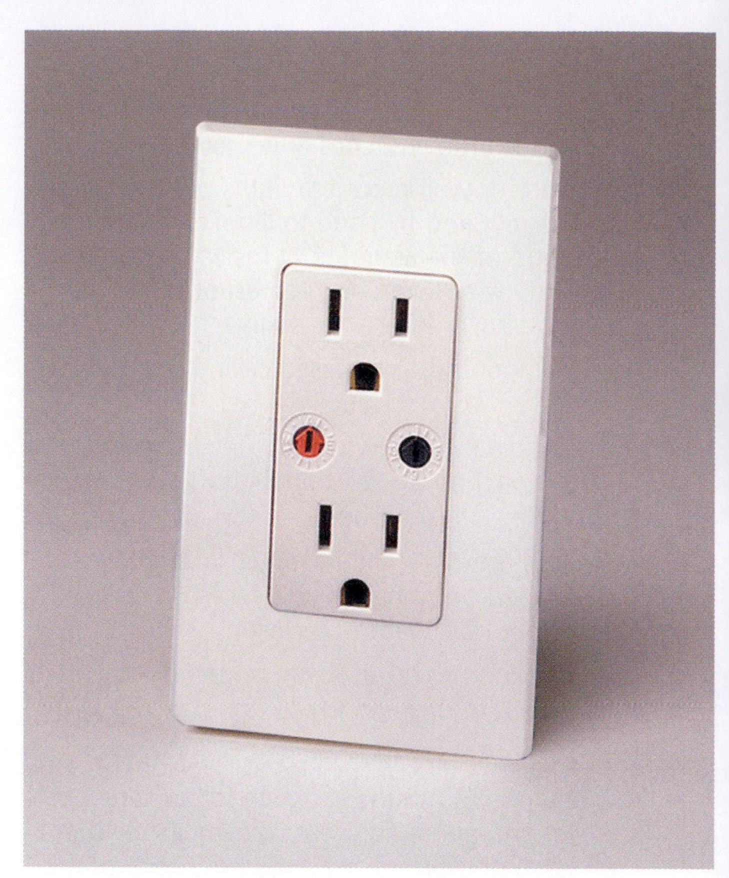

FIGURE 12.19 ■ *Any fixture or appliance that is or can be placed within 60'' (1500 mm) of a water source must be powered by a GFCI circuit breaker or receptacle. The receptacle can break the supply of power in 1/40th of a second.* Courtesy Pamela Winikoff, Leviton Manufacturing Company.

into this receptacle and set in the ON position. Half-hot receptacles are used in rooms such as living rooms or bedrooms in place of a light fixture.

- Ground fault circuit interrupter receptacles (GFCI or GFI) are used when an appliance or fixture is to be used within 60'' (1500 mm) of water. The receptacle, similar to that in Figure 12.19, disrupts the flow of electricity through the circuit with even the smallest change in current flow to protect human life. Because the receptacle trips so easily, it contains a reset button so the circuit can be restored without going all the way to the distribution panel. GFCI receptacles are generally used in bathrooms, and GFCI circuit breakers are used in kitchens, laundry rooms, garages, and for exterior fixtures and outlets. The GFCI circuit breaker is used to ensure that all lights and receptacles near the water source are protected. Exterior receptacles are required to be GFCI and waterproofed. A waterproof receptacle uses metal or plastic cover to protect the receptacle even if it is in use.

Guidelines for Placing Outlets

The placement of 120-volt convenience outlets is based on rules established by the NEC, but energy considerations and common usage patterns will also influence placement.

Code Requirements for Placing Outlets

The placement of receptacles depends on whether a room is habitable, non-habitable, or a kitchen. Habitable rooms are rooms such as living rooms, dining rooms, bedrooms, dens, and family rooms. Non-habitable rooms include the laundry, entry, halls, and rooms intended for a specific nonliving use, such as photo labs. The electrical needs are based on specific uses for the room. Placement of receptacles for habitable rooms is based on the numbers 2, 6, and 12. For a habitable room:

- Any wall longer than 2' (600 mm) must contain an outlet. If a wall is long enough to place a nightstand or small table with a clock or other appliance, provide an outlet to service the appliance.

- An outlet must be located within 6' (1800 mm) of an opening in a wall. An opening is a total disruption of the wall caused by a door or fireplace. This rule is based on the typical length of the chord on an appliance and not on wall corners.
- Outlets must be within 12' (3600 mm) of each other. Receptacles should be placed so that the chord is not stretched to its limits, forcing the use of an extension chord. At the other extreme, each receptacle has costs that far exceed the $3 it takes to purchase the box and the receptacle. Once the cost of the circuit, the wire, and the electrician are considered, care must be taken by the designer to limit the number of receptacles specified on an electrical plan.

Figure 12.20 shows how these minimum standards are applied to a habitable room.

Kitchen receptacles require careful consideration and are not based on the numbers 2, 6, 12 for placement. Kitchen outlets are required:

- Every 4' (1200 mm) of counter space. The receptacle can be placed in the wall or in the face of the cabinet.
- Within 2' (600 mm) of a corner, end of counter, or an appliance.
- At each end of an island or peninsula.

Placement of Outlets Based on Usage

In addition to placing receptacles based on code requirements, consideration should be given to how the receptacle will be used. Use the following suggestions to improve the design when placing receptacles on the electrical plan:

- Consider furniture placement so receptacles do not become inaccessible behind large pieces of furniture.
- Place a receptacle next to or behind a desk.
- Place a receptacle near the front face of a fireplace. Specify a receptacle inside the chase for the fan motor of a zero-clearance fireplace.
- Provide at least one receptacle in the entry (foyer).
- Place a receptacle in each hallway for a vacuum.
- Provide a receptacle for each appliance, including the refrigerator, range, hood light and fan, oven, microwave, dishwasher, and trash compactor.
- Place a receptacle in a pantry for portable appliances.

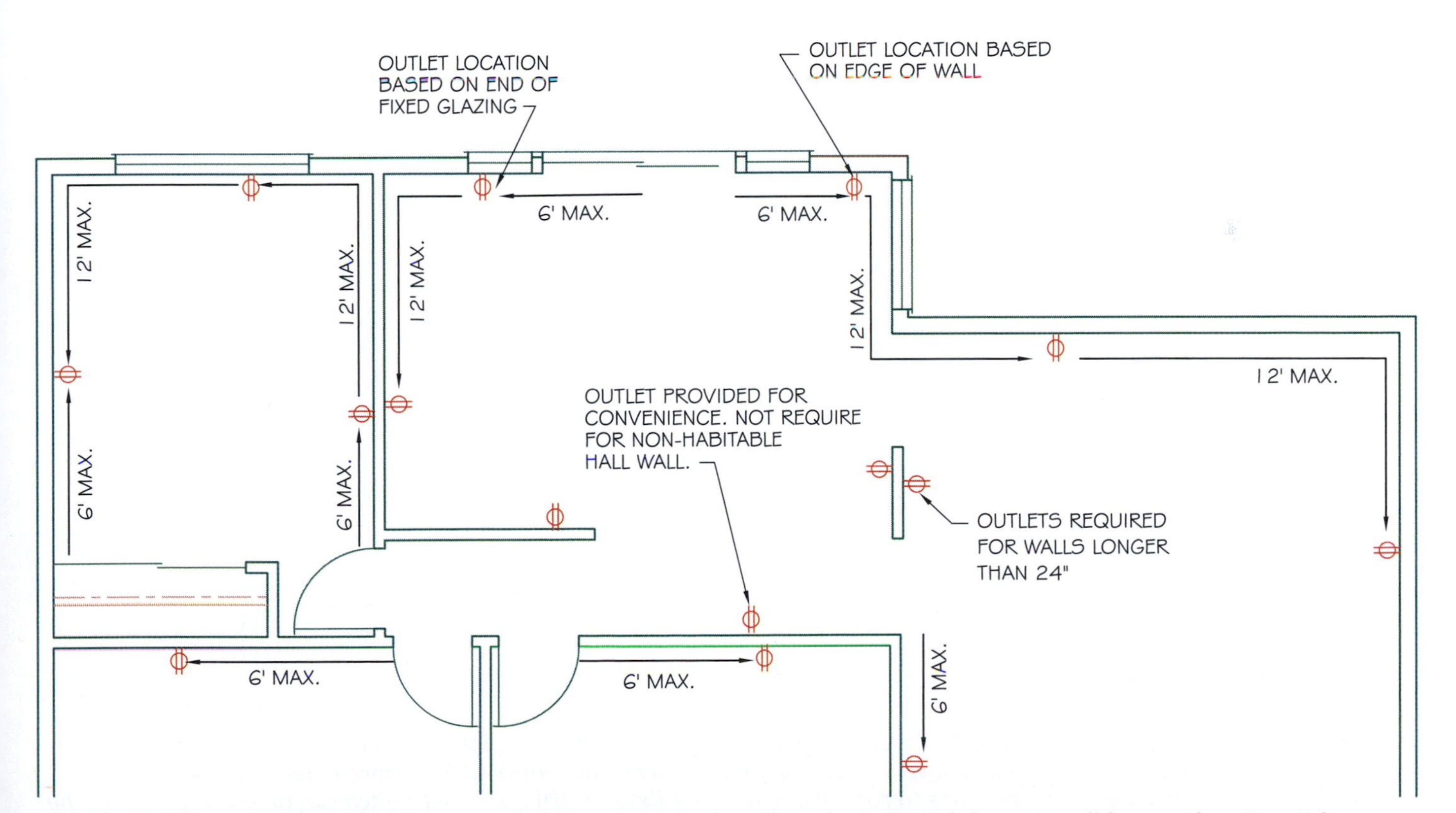

FIGURE 12.20 ■ *Outlet placements for habitable rooms are based on the 2, 6, 12 rule. Any wall longer than 2' needs a receptacle. A receptacle can't be more than 6' from the end of a wall, and no two receptacles should be more than 12' apart.*

- Provide an outlet on a separate circuit at an office desk where a computer will be used.
- Provide a waterproof GFCI outlet near the front door, on the front wall of a garage, and to serve each patio, balcony, and outside living area.
- Provide a waterproof GFCI outlet near the main parking areas.
- Provide a GFCI outlet on a 15- or 20-amp circuit in any crawl space.

Controlling Electrical Circuits with Switches

Switches are used to control the flow of electricity through light fixtures and receptacles. The types of switches to be used and their locations are important considerations in designing the electrical plan.

Types of Switches

The most common types of electrical switches found in a residence include single-pole, three- and four-way, dimmer, timer, photoelectric, master, low-voltage, and motion-detection switches. Examples of each of these types of switches are shown in Figure 12.21. Common uses for these switches include:

- A single-pole switch is used to control one or more fixtures from one location.
- Three-way switches are used in pairs to control one or more fixtures from two locations. Common uses include switches placed at each end of a hallway to control hall lights or switches placed at the top and bottom of a stair to control stair lighting. Three- and four-way switches are named from the number of wires required to make the circuit work, not the number of switches.
- Circuits controlled by four-way switches use three switches to control one or more fixtures from three locations. (Three switches require four wires.) Figure 12.22 shows examples of single pole, three-way, and four-way switches.
- A dimmer switch allows the amount of current flowing through a fixture to be varied using a touch, slider, or rotary control. A dimmer is often used in dining rooms, master bedrooms, home theaters, and other rooms where the need for general lighting varies.

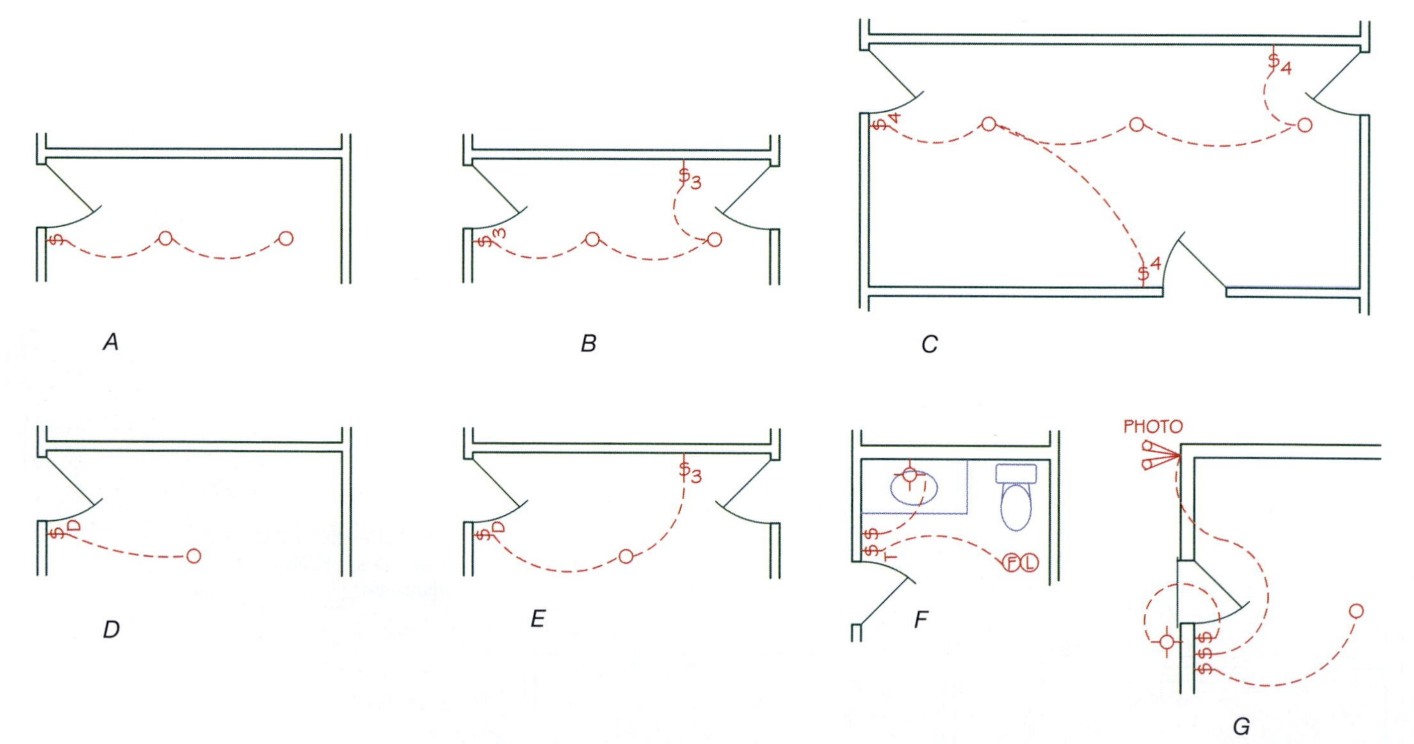

FIGURE 12.21 ■ *Common types of switches found on electrical plans. (A) A single-pole switch can be used to control one or more fixtures. (B) A three-way switch uses two switches to control one or more fixtures. (C) A four-way switch uses three switches to control one or more fixtures. (D) A dimmer switch can be used to control the amount of light produced by a fixture. (E) When a dimmer switch is used on a three-way switch, the symbol is shown by only one switch. (F) A timer switch can be used to control the length of time a circuit will remain closed. (G) Lights controlled by photovoltaic sensors turn on and off based on the amount of available natural light. A manual override is usually provided.*

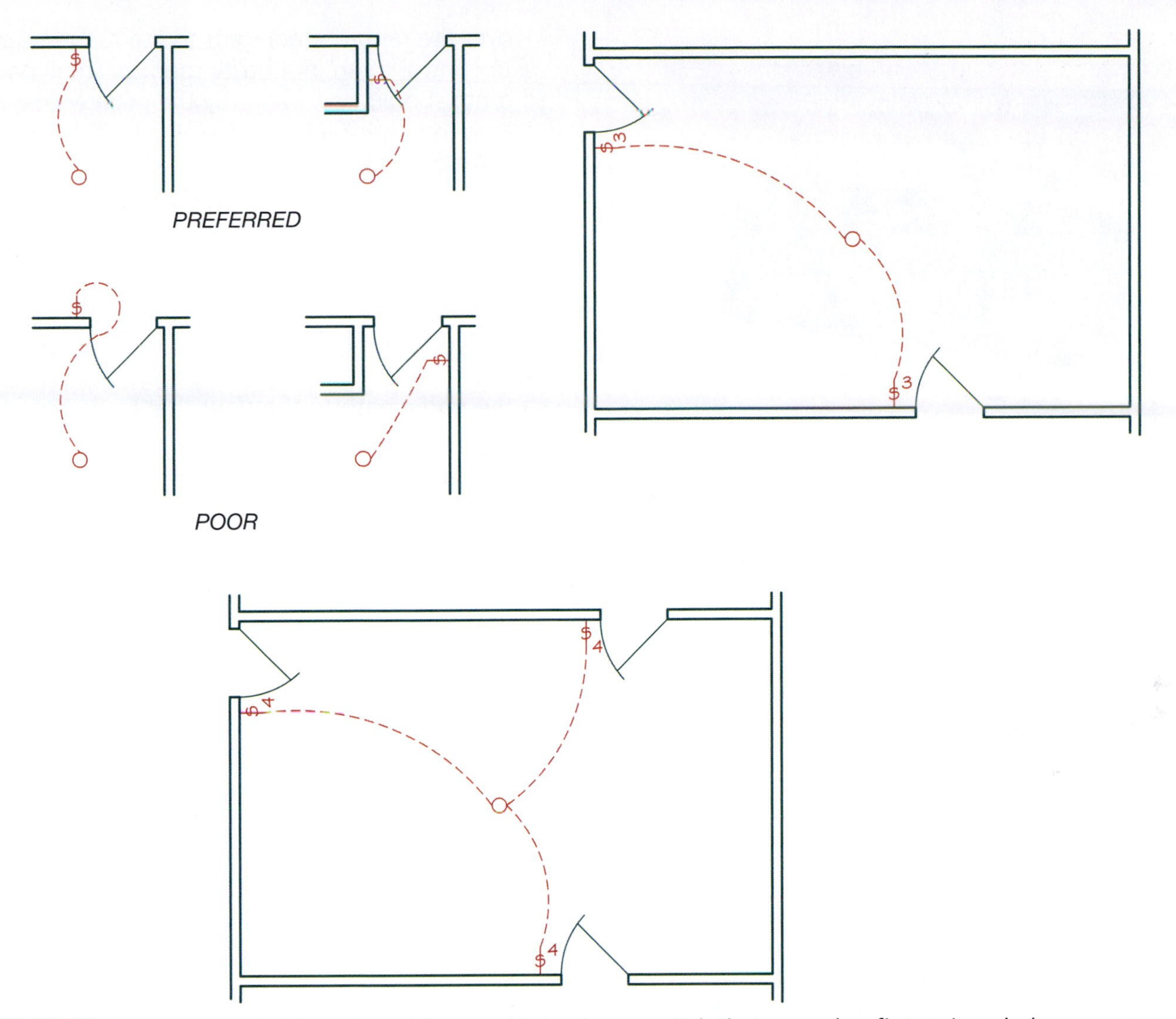

FIGURE 12.22 ■ *Common switch locations. A key goal is to place a switch that controls a fixture in a dark room near a lighted area of that room. Placing a switch in a dark area will lead to accidents.*

- Timers are switches that allow the circuit to be regulated based on a specific amount of time. Bathroom fans, security lighting, and landscaping lighting are often controlled with timer switches.
- Photoelectric switches allow the circuit to be regulated based on a specific amount of natural light. When low light levels are sensed, the switch allows the flow of electricity through the circuit. The switch restricts the flow of current when high levels of light are detected. This type of switch is ideal to turn on outside lights automatically at sunset.
- Master switches similar to those in Figure 12.23 can be used to override all other circuits located in a structure from one location.
- Motion-detection switches emit an electrical beam. When a person or animal passes through the beam, the beam is reflected back to a receiving unit in the switch. As the signal is received, it allows electricity to flow through the switch and light fixtures controlled by the switch to be activated. A motion-detector switch also contains a timer unit that will interrupt the flow of electricity to the fixtures on the circuit after a specific period of time. If no additional motion is detected, the light will remain off. Continued motion will reactivate the lighting fixtures controlled by the switch. Motion-detection switches can be used for interior and exterior circuits.

Switch Locations

The switch location and the number of switches used in a room should be determined based on traffic patterns. The switch for a dark room must be in an area

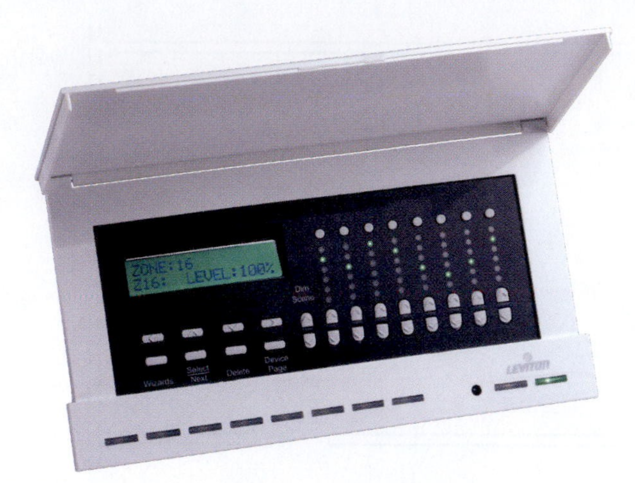

FIGURE 12.23 ■ *A master switch can be used to control multiple circuits. The Leviton Programmer provides one-touch access to up to 64 lighting scenes and remote control to over 256 devices in the home.* Courtesy Pamela Winikoff, Leviton Manufacturing Company.

that is already in the light. Use the following guidelines as additional aids when planning for switch locations.

- Every light fixture should have one or more switches to control the fixture.
- Place a switch on the interior side of the room containing the fixture or outlet to be controlled.
- Place the switch on the latch side of the door, never on the hinge side. Remember that the first 3'' of wall beside a door is filled with support framing and can't be used to place a switch box (see Figure 12.22).
- Provide an adequate turning radius next to the switch for a user who might be in a wheelchair.
- Use three-way switches at the end of each hall or stairway or when two means of egress are provided to a room. A switch by the door and another by the bed is helpful for bedrooms.
- Provide switches that control lights in outside living areas by the door that leads to these areas. If the budget allows, provide a master control in the master bedroom for all outside lights.
- Provide a switch near the kitchen sink to control the garbage disposal. (Such a switch can be placed in the countertop if no walls are near the sink.)
- Provide switches to control under- and over-cabinet lighting and portable appliance compartments or an appliance garage. Place the switch in an area convenient to the work area to be illuminated.
- Provide switches to control exhaust fans. The kitchen exhaust fan can be in a hood with a light over the range or adjacent to the range. A switch to control a fan in a utility room or small bathroom should be located by the main access door. Fans in larger bathrooms may be located near a toilet or shower.
- Locate switches so they will not be placed within 6' (1800 mm) of a tub.
- Use switches with timers in the garage, closets, storage areas, and bathrooms to control unnecessary operation of fixtures.
- Locate switches to control garage door openers in the traffic path to the door that leads from the garage to the house.

In addition to placing switches on the plans, written specifications should be provided to specify the type and location of switches. Consider placing switches 2'-6'' (750 mm) above the floor for easy use by children and people in wheelchairs. Also consider using switches that operate by touch or that are sound- or motion-activated.

Planning for an Efficient System

The final stage of planning before starting the electrical drawings is to consider options that will make the entire system as efficient as possible. Careful planning will contribute to energy savings for the homeowner at little additional cost. Increase the energy efficiency of a home by:

- Placing receptacles in interior walls rather than in exterior walls when code-required distances can be maintained. Receptacles placed in an exterior wall will eliminate or compress the insulation and reduce its insulating value. Specify receptacle gasket covers to be provided to help eliminate air infiltration in cold climates.
- In cold climates, specifying that wires should be run in the bottom 2'' (50 mm) of the wall cavity if electrical wiring must be placed in an exterior wall. When wires are run at the normal height, the wall insulation will be compacted. A notch can be placed in the studs prior to framing the wall.
- Specifying that all holes for electrical wiring in the top and bottom plates are to be sealed and caulked.
- Specifying that caulking be placed around all light and convenience outlets.
- Selecting energy-efficient appliances such as a self-heating dishwasher or a high-insulation water heater.

- Using energy-saving fluorescent lighting fixtures where practical.
- Fully insulating above and around recessed lighting fixtures.
- Specifying that all recessed lights are to be IC (insulation cover) rated to help reduce heat loss.
- Specifying that all fans and other systems exhausting air from the building be provided with back-draft or automatic dampers to limit air leakage.

DRAWING AN ELECTRICAL PLAN

Creating the electrical drawings is a common job given to junior technicians. Because most municipalities do not require electrical drawings, it's a great drawing to allow a technician to complete as a confidence builder and to become familiar with company standards. If a mistake is made and not caught, the skill of the electrician will easily cover the mistake of the technician. On a custom home, the design of the electrical plan is as critical as the design of the framing members and may be completed by an experienced technician. To communicate clearly with the owner, contractor, and electrician, easily recognized symbols must be used. Most offices have a symbols library that contains most of the symbols needed to complete the drawings. Symbols that are not part of the library can be created based on individual project requirements. Consideration must also be given to the layers that will contain the electrical information and how and when these layers will be displayed. The final consideration will be to ensure an orderly process, so that electricity will be provided to all appliances and fixtures.

Common Electrical Symbols

Symbols are used to represent the light fixtures, receptacles, and switches in the home. Common symbols are shown in Figure 12.24. Slight variations may be found with each office. Additional symbols can be found by consulting the National CAD Standard. Most offices will provide drawing blocks that represent standard symbols. Many common symbols are available on the student CD in the PROTO folder. If additional symbols need to be created, use the following guidelines:

- All electrical symbols should be drawn with circles 6'' (150 mm) in diameter. These circles may be slightly larger if text must be placed within them.
- All text for switches should be 6'' (150 mm) high and be created using simple block lettering. Do not use an architectural font that matches the general text font.
- Text used to supplement a symbol such as WP or GFCI may be reduced, depending on space requirements and office practice. Text smaller than 4'' (100 mm) high will be difficult to read when a drawing is plotted at a scale of 1/4'' = 1'-0''.
- Place the switch symbol perpendicular to the wall, using an orientation that allows the symbol to be read when looking from the right side or bottom of the sheet (see Figure 12.25). Use a thin dashed arc or spline to connect the switch and the fixtures the switch controls (see Figure 12.21). Figure 12.25 shows preferred methods of connecting switches to a fixture.
- When a specific location for a fixture or receptacle is required, use a local note or a dimension to locate the fixture. Figure 12.26 shows common locations for bathroom symbols and Figure 12.27 shows common locations of kitchen fixtures.

Preparing the Drawing Base

The electrical information for a simple project may be placed on the floor plan with all the other symbols, information, and dimensions, as shown in Figure 12.28. On a custom home or complicated set of drawings, the electrical plan is drawn as a separate plan. When a separate electrical plan is drawn, the layers containing the base floor plan information—including the *WALLS, DOOR, GLAZ, CASE, STRS, CHIM, PATT, APPL,* and *PLMB*—should be displayed and used as a base for the electrical plan. Before symbols can be placed on the electrical drawings, layers need to be created to separate the electrical information from the information on the floor plan. Layers for the electrical information should start with a prefix of *ELEC.* Layers containing electrical information should be named with a modifier listed in Appendix F. As with any other drawing, create layers as needed to ensure that only layers that will be used are added to the drawing.

Creating a Separate Electrical Plan

Step 1. Freeze all information directly related to the floor plan. Your base drawing should include the walls, doors, windows, stairs, cabinets, and the fireplace. Basic room titles are optional

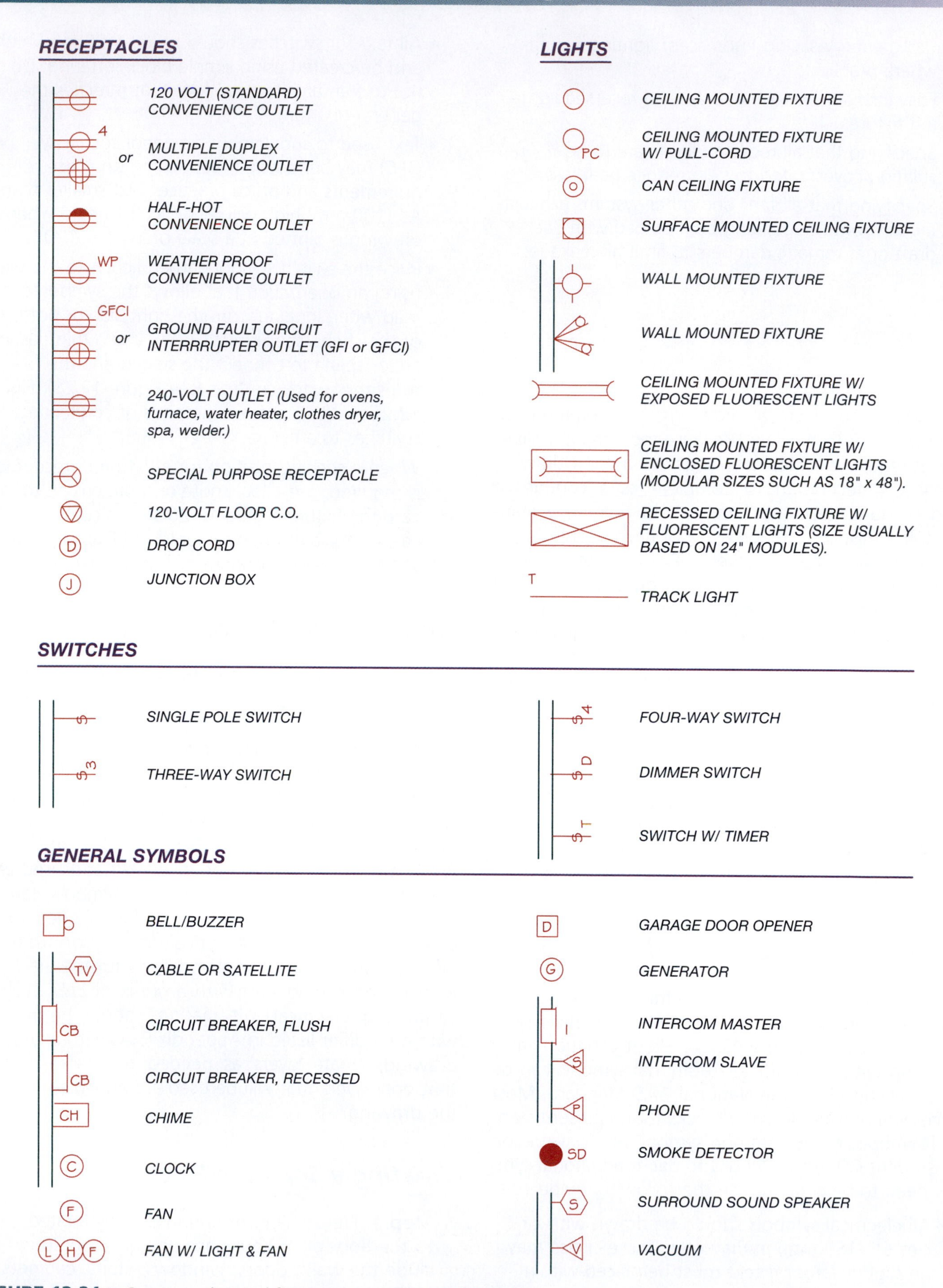

RECEPTACLES
120 VOLT (STANDARD) CONVENIENCE OUTLET
4
or
MULTIPLE DUPLEX CONVENIENCE OUTLET
HALF-HOT CONVENIENCE OUTLET
WP
WEATHER PROOF CONVENIENCE OUTLET
GFCI
or
GROUND FAULT CIRCUIT INTERRRUPTER OUTLET (GFI or GFCI)
240-VOLT OUTLET (Used for ovens, furnace, water heater, clothes dryer, spa, welder.)
SPECIAL PURPOSE RECEPTACLE
120-VOLT FLOOR C.O.
D
DROP CORD
J
JUNCTION BOX
LIGHTS
CEILING MOUNTED FIXTURE
PC
CEILING MOUNTED FIXTURE W/ PULL-CORD
CAN CEILING FIXTURE
SURFACE MOUNTED CEILING FIXTURE
WALL MOUNTED FIXTURE
WALL MOUNTED FIXTURE
CEILING MOUNTED FIXTURE W/ EXPOSED FLUORESCENT LIGHTS
CEILING MOUNTED FIXTURE W/ ENCLOSED FLUORESCENT LIGHTS (MODULAR SIZES SUCH AS 18" x 48").
RECESSED CEILING FIXTURE W/ FLUORESCENT LIGHTS (SIZE USUALLY BASED ON 24" MODULES).
T
TRACK LIGHT
SWITCHES
S
SINGLE POLE SWITCH
S3
THREE-WAY SWITCH
S4
FOUR-WAY SWITCH
SD
DIMMER SWITCH
ST
SWITCH W/ TIMER
GENERAL SYMBOLS
BELL/BUZZER
TV
CABLE OR SATELLITE
CB
CIRCUIT BREAKER, FLUSH
CB
CIRCUIT BREAKER, RECESSED
CH
CHIME
C
CLOCK
F
FAN
L H F
FAN W/ LIGHT & FAN
D
GARAGE DOOR OPENER
G
GENERATOR
I
INTERCOM MASTER
S
INTERCOM SLAVE
P
PHONE
SD
SMOKE DETECTOR
S
SURROUND SOUND SPEAKER
V
VACUUM

FIGURE 12.24 ■ Common electrical fixtures, receptacles, and switches.

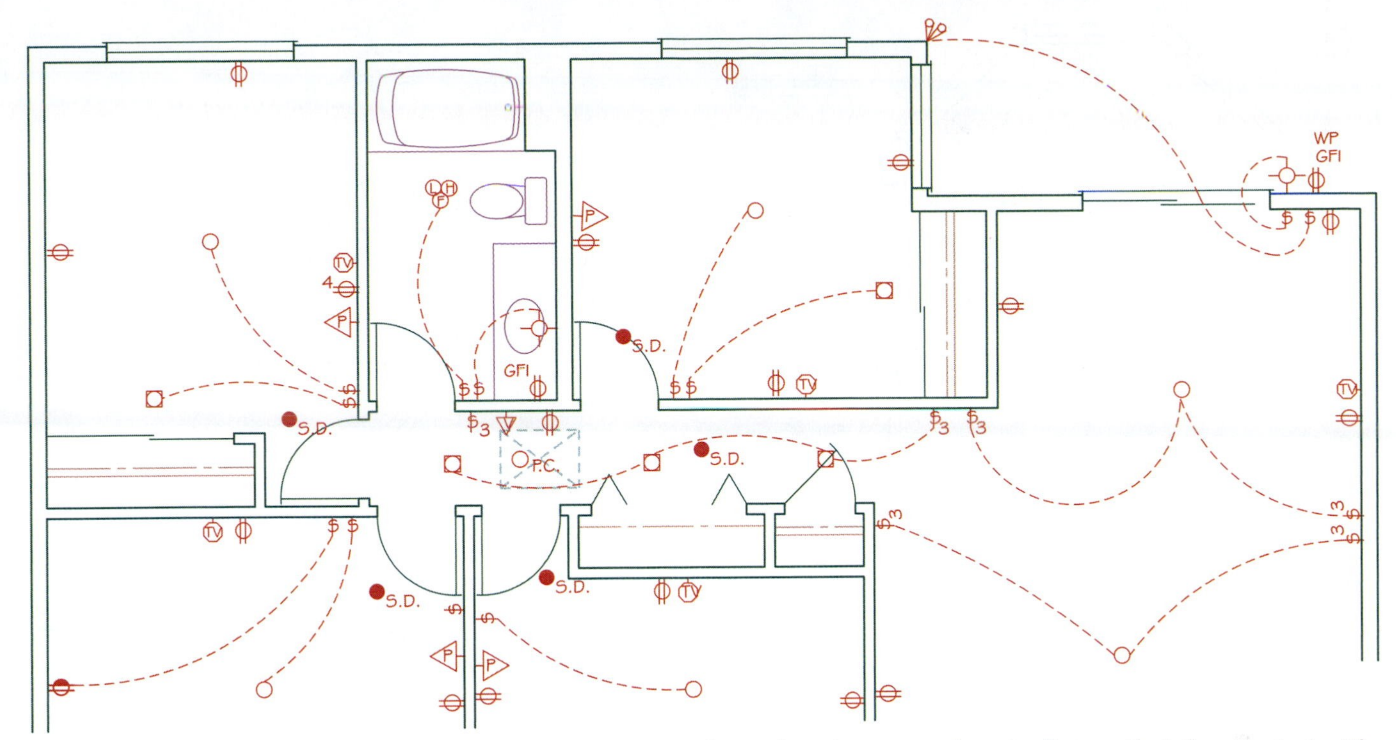

FIGURE 12.25 ■ *Switches should be placed perpendicular to the wall and connected to the fixture that they control with a thin dashed arc. Notice that only one switch, supplied by the manufacturer, controls the light/heat/fan unit.*

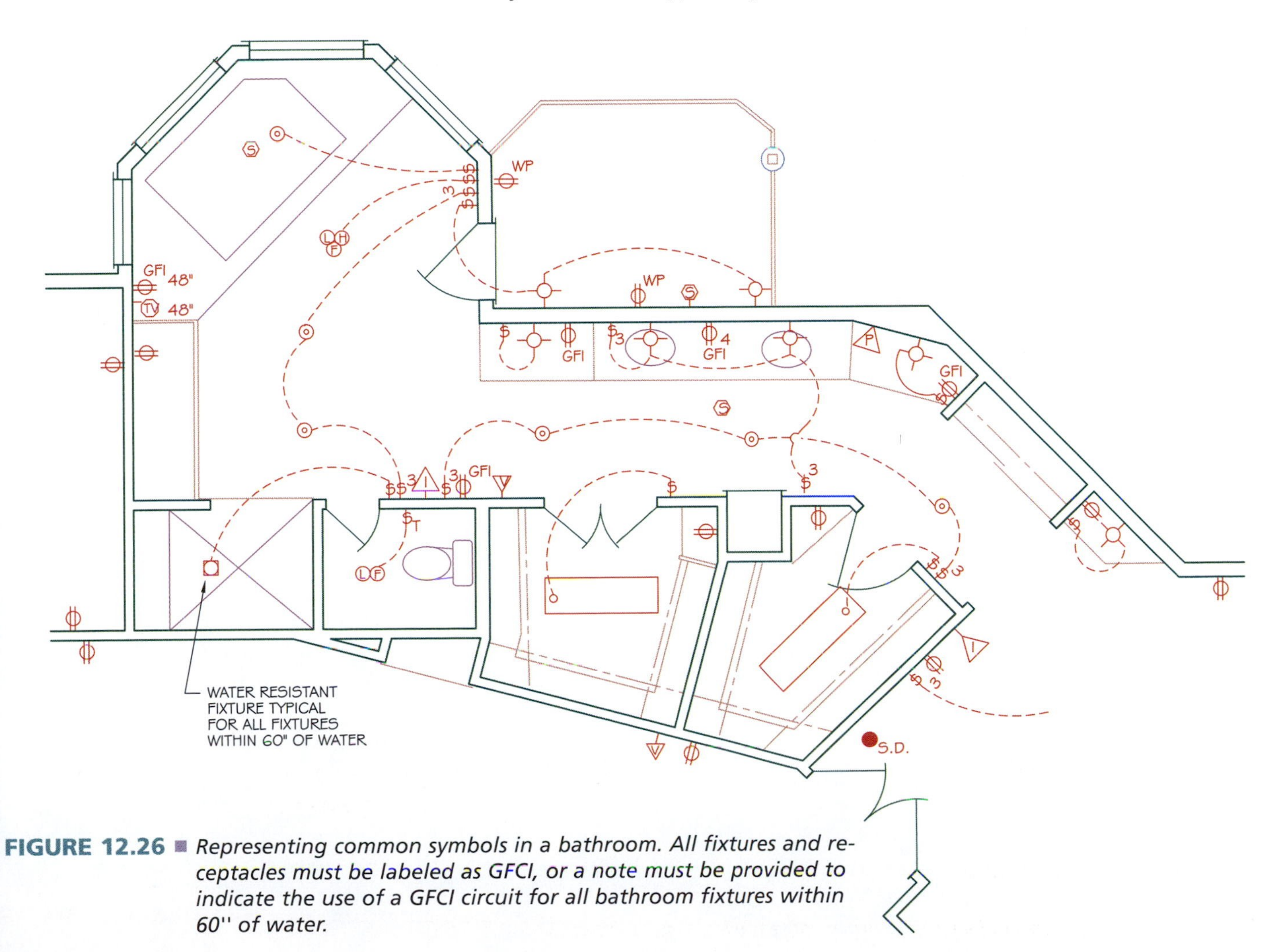

FIGURE 12.26 ■ *Representing common symbols in a bathroom. All fixtures and receptacles must be labeled as GFCI, or a note must be provided to indicate the use of a GFCI circuit for all bathroom fixtures within 60'' of water.*

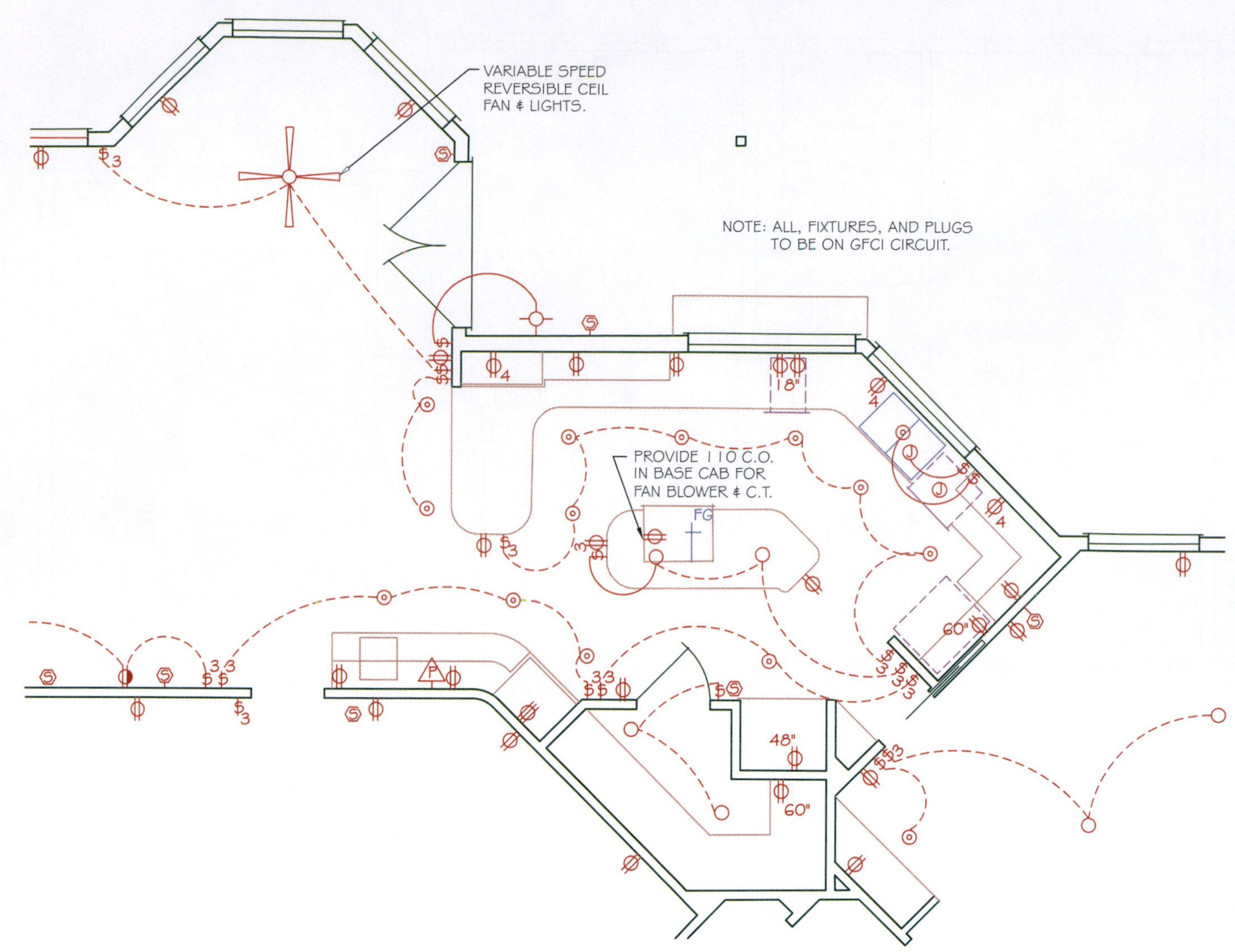

FIGURE 12.27 ■ *Common placement of symbols for the kitchen area. Careful planning is required to light in each work area of the kitchen. Consideration must also be given to traffic flow in the kitchen and from the kitchen to related rooms.*

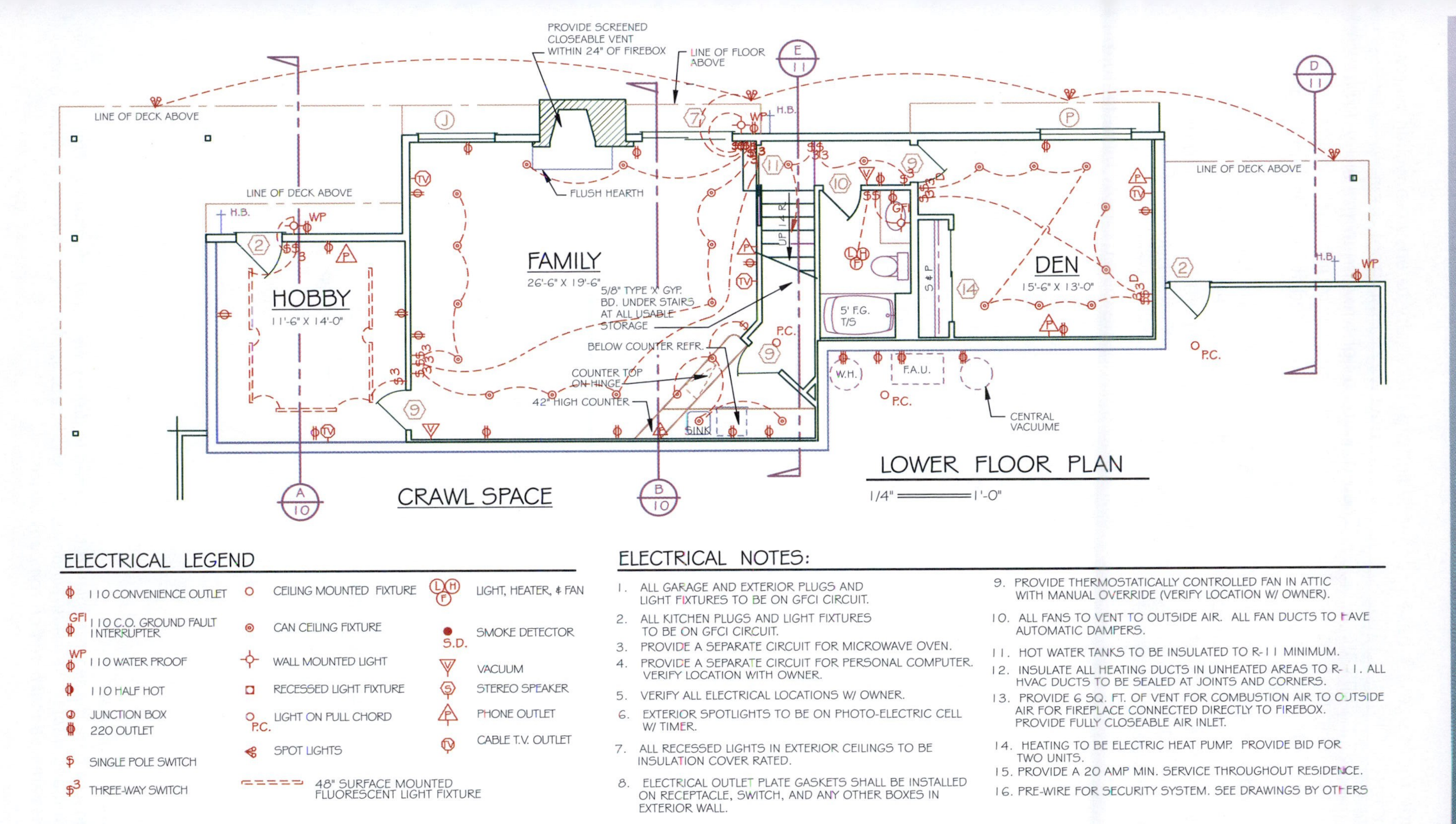

FIGURE 12.28 ■ *Electrical information can be combined with other information and displayed on the floor plan on a simple project. This is the lower level of the home in Chapter 11. Figure 12.34 shows the completed upper floor electrical plan for the same home.*

but if displayed do not include room sizes. A base drawing for the residence that is in Chapter 11 is shown in Figure 12.29. The following steps are given as a guide. The order is not as important as proceeding in an orderly manner. Some technicians prefer to insert all lights and then all receptacles. If you prefer a different order or going room by room, that works too. The goal is to provide for all of the needs of the family. Use the checklist from Appendix C when your drawing is complete to evaluate your success at meeting the goal.

Placing Lighting Fixtures

Place the items described in the following steps on the *ELEC SYMB* layer.

Step 2. Place a light fixture on the interior side of the door.
Step 3. Place a light fixture on the exterior side of each exterior door, including the garage door. Provide lights as necessary to ensure a well-lit walkway from the driveway to the front door.
Step 4. Place a light fixture in the entry. Proceed from the entry through each hallway and place fixtures as needed for general lighting needs.
Step 5. Start at the front door and mentally walk through the residence. Place light fixtures in each room to meet the general, specific, and decorative needs in keeping with the budget of the homeowner. Lights placed in the centers of rooms can be easily placed using the OBJECT TRACKING feature of AutoCAD. Their location can also be determined by eye, since the electrician will have the final say on their location. You may find it easier to keep coming back to the front door and inserting blocks to represent each type of fixture. Using this method, you can place all ceiling-mounted fixtures, all wall fixtures, and all fluorescent fixtures and then all "can" fixtures until all lighting needs have been met.

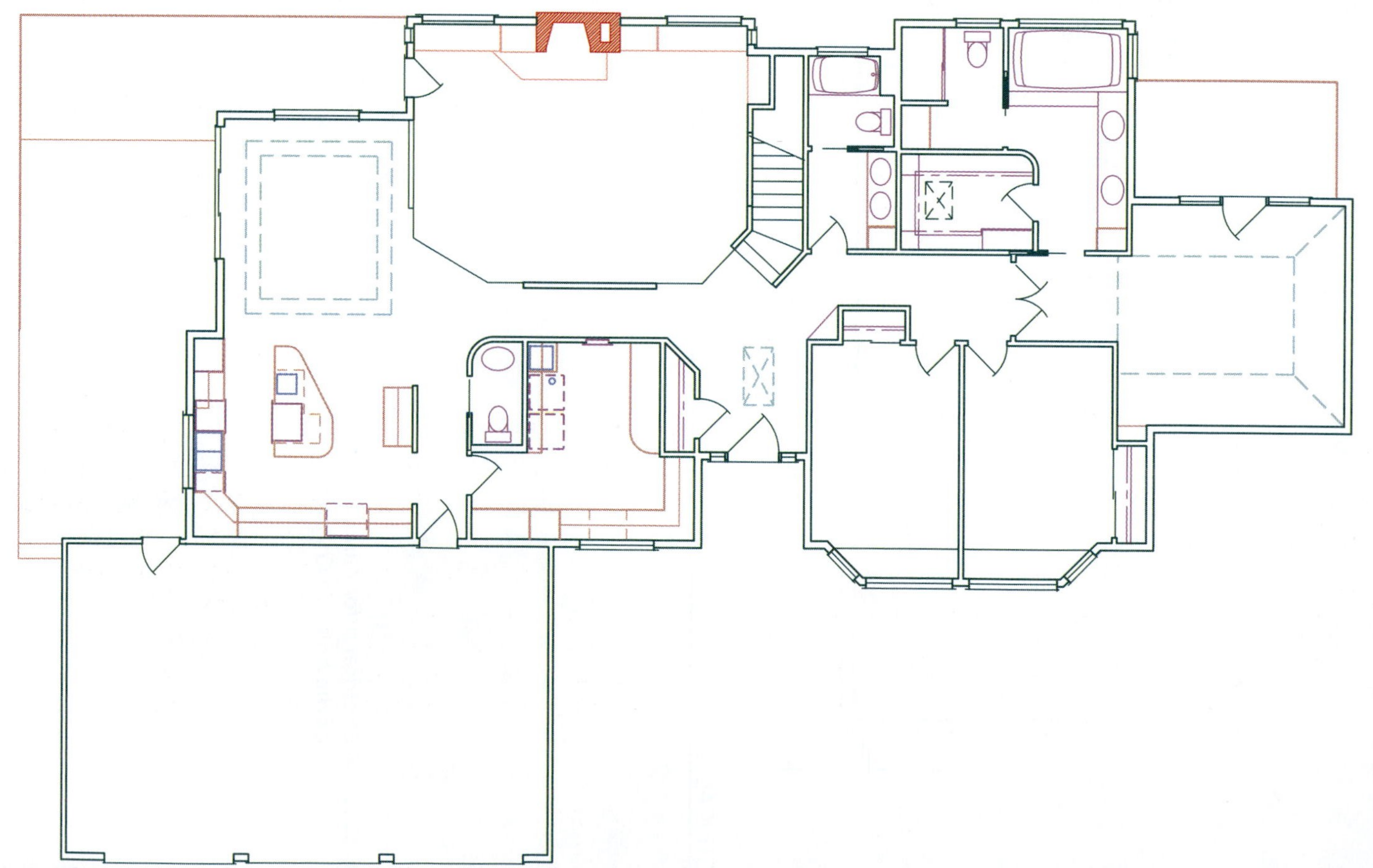

FIGURE 12.29 ■ *Use the floor plan as a base if a separate plan is to be used to display electrical information. Freeze all floor-specific information and display the walls, windows, doors, cabinets, appliances, plumbing, stairs, and fireplace. Room names are optional based on company policy.*

Placing Outlets

Step 6. Start at the front door and place receptacles in the entry.
Step 7. Place receptacles as needed in each hallway.
Step 8. Place receptacles as needed in all habitable rooms to meet the 2, 6, 12 rule.
Step 9. Place receptacles for all non-habitable rooms based on usage and the location of specific appliances.
Step 10. Place half-hot receptacles as needed.
Step 11. Place all receptacles for the kitchen based on the location of appliances and usage.
Step 12. Place GFCI receptacles in each bathroom based on usage and code restrictions.
Step 13. Place GFCI outlets in the garage and waterproof receptacles as needed to provide for all exterior living areas.
Step 14. Place 240-volt receptacles as needed for the furnace, water heater, spa, clothes dryer, oven, range, and specialty equipment in the garage or utility room.
Step 15. Place specialty waterproof GFCI receptacles for landscaping, pools, and spas.

Placing Switches

Step 16. Start at the front door and mentally walk through the home and place all single-pole switches.
Step 17. Place all required 3- and 4-way switches.
Step 18. Place the dashed control lines from each switch to the fixture or receptacle to be controlled on the *ELEC POWR* layer.

Place the items described in the following steps on the *ELEC ANNO* layer.

Step 19. Provide an electrical legend to identify all symbols placed on the drawing. Use the legend on the CD and modify it to meet your project.
Step 20. Place all notes required to explain any special construction for electrical needs. Place a drawing title and scale below the drawings and complete text in the title block to describe the drawing contents.

Figure 12.30 shows an electrical plan with the lights, receptacles, and switches placed.

Placing Electrical Specialty Equipment and Fixtures

Specialty equipment should be placed on the electrical drawings once all lighting fixtures, receptacles, and switches are represented. This includes the placement of smoke detectors, telephones, television jacks for cable or satellite lines, speakers for surround-sound systems and intercoms, inlets for built-in vacuum systems, security systems, and home automation. Symbols to represent these materials can be added to the *ELEC SYMB* layer.

Placing Smoke Alarms

Methods of representing smoke alarms are shown in Figure 12.24. Smoke alarms are generally shown on the electrical plan along with the symbols for other electrical equipment. Key requirements to be met when placing smoke alarms include:

- A smoke alarm must be located at the start of every hall that serves a bedroom.
- A smoke alarm must be placed in each sleeping room near the entry door.
- A smoke alarm is required on every floor, including the basement for multilevel homes.
- A smoke alarm should be located over the stair leading to the upper level.

Review Chapter 4 or the local codes for additional requirements that affect placement. Smoke alarms are generally shown on the electrical plan along with the symbols for other electrical equipment. Figure 12.35 shows the placement of the smoke alarms. The type of smoke alarm can be indicated in the electrical general notes. Occasionally the team that designs the fire suppression system also plans the location of the smoke detectors. Figure 12.31 shows an example of the smoke detector layout for the lower floor.

Representing Telephones

Even in an age of portable and cell phones, the location of telephone jacks should be carefully planned. A phone jack is usually placed in the kitchen, each bedroom, laundry room, family room, and office. Telephone jacks should be rough-wired for future installation in children's bedrooms. Telephone jacks should be represented using the symbol from Figure 12.24 and placed near seating and work areas. Place a jack near the bed location in each bedroom and on each side of where the bed is likely to be placed in the master bedroom.

Television, Stereo, and Intercom Systems

Wiring for television, stereo sound systems, and home intercoms should be designed into the project so electrical needs for these items can be pre-wired

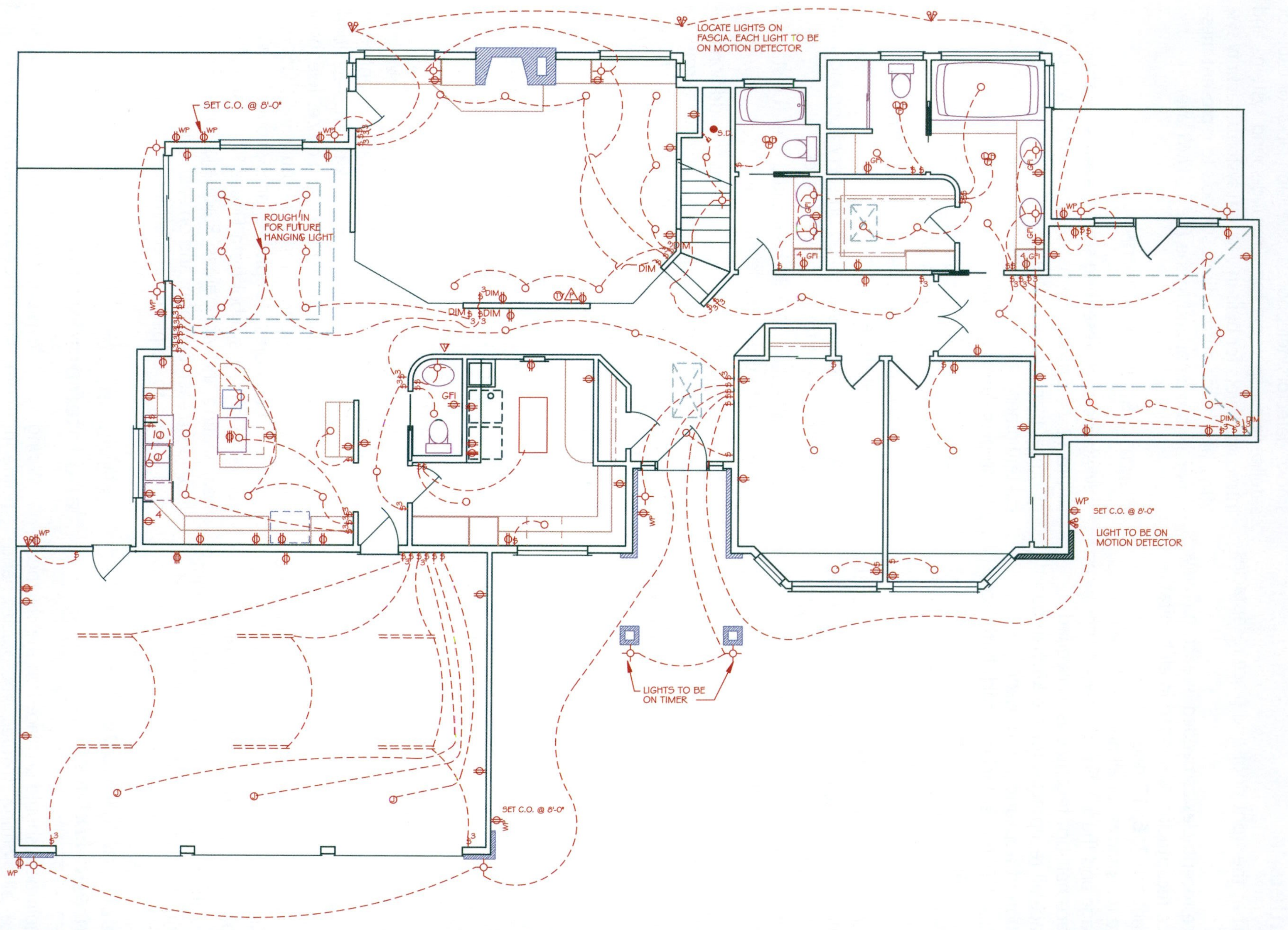

FIGURE 12.30 ■ *Symbols to represent light fixtures, receptacles, and switches are placed using the floor plan as a base. Thin dashed curving lines are used to show fixture control, not circuit layout. Notes to specify specific needs of the project are shown in Figure 12.28.*

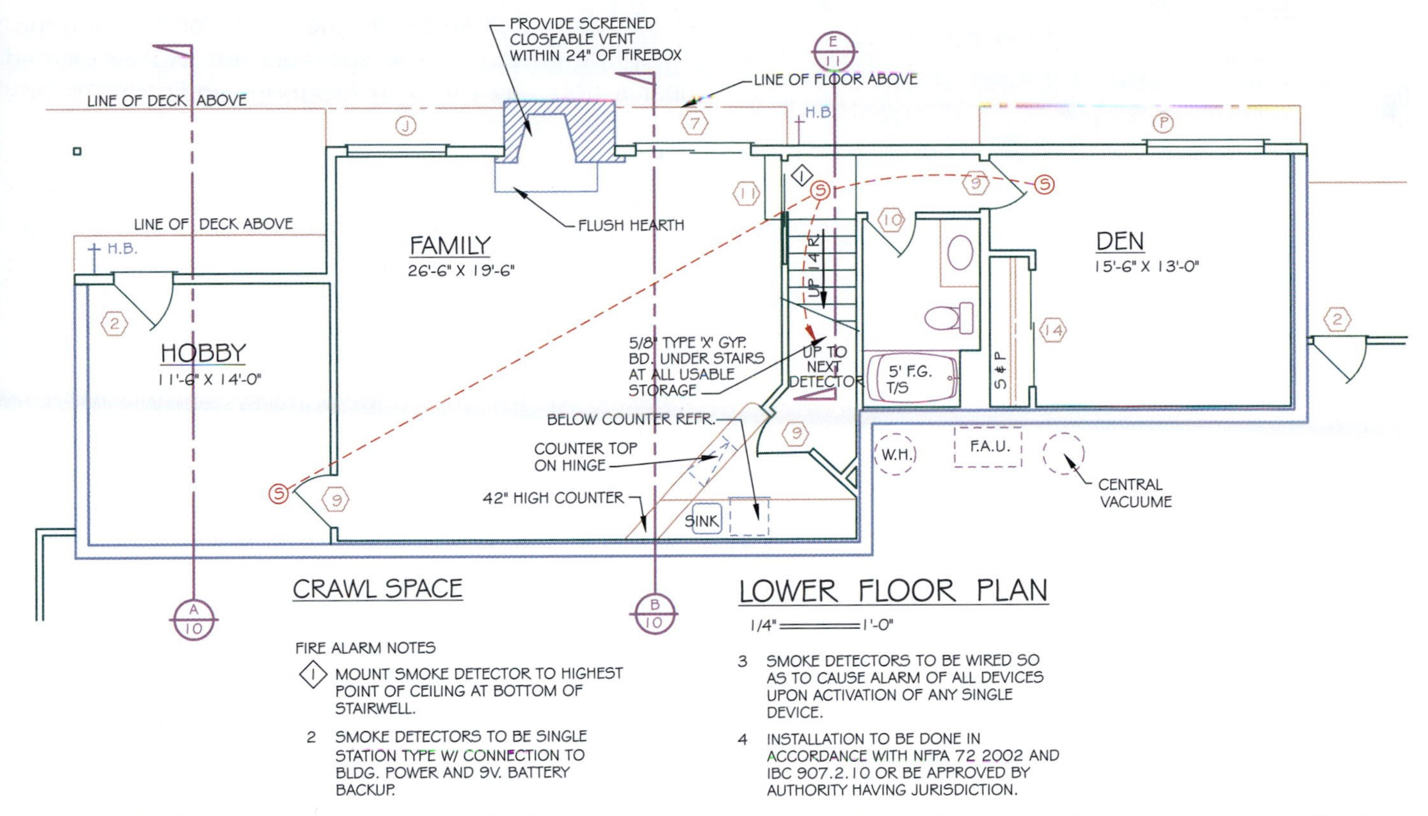

FIGURE 12.31 ■ *The team that designs the fire suppression system may plan the location of the smoke detectors. The layout for the lower floor's smoke detectors was completed by the consultants who designed the fire suppression system shown in Chapter 14.* Courtesy Michael Coffman.

before the walls are covered with sheetrock. Symbols for each system can be found in Figure 12.24. The use of an antenna, cable, or satellite will determine the type of television wiring to be installed. Although the service provider will generally determine the location and run the required cables, planning where service jacks are to be located will aid the owner. TV jacks should be provided in high-usage rooms such as the living, family, and recreation rooms as well as the kitchen, laundry, master bedroom, and master bathroom. Children's rooms are also typically pre-wired for future access.

Stereo installations can be wired separately for sound throughout the home or associated with the cable television. A central intercom system similar to that in Figure 12.32 contains a radio, CD player, and MP3 player. In addition to delivering sound, most systems can also be used for two-way communication and room monitoring or serve as speakers for door chimes or driveway gate monitors. If an intercom is to be installed, a master station and each slave speaker should be represented on the electrical drawings. The intercom master must be located in a central location such

FIGURE 12.32 ■ *A central intercom system can deliver music from a radio, CD, or MP3 player as well as providing controls for two-way communication, room monitoring, door chime amplification, and gate monitors.* Courtesy Tracy Lucht, Broan-NuTone.

as a kitchen desk area or in the family room near a major traffic pattern. Slave speakers are usually located near a doorway for each room, including the garage and major outside living areas.

For a custom home, the television will most likely be wired for the addition of a surround-sound system. If this system is to be represented on the electrical drawings, the designer should work closely with a media consultant to help plan the locations of sound equipment. Figure 12.33 shows four common layouts for home theaters with surround sound. Individual speakers for the surround-sound system will need to be distinguished from intercom speakers. The height above the floor must be given for each wall-mounted speaker, and ceiling mounted speakers may need dimensions to mark their location. In addition to planning the speaker locations, attention must be given to the number and location of receptacles for the equipment to be placed in an entertainment center. If specifications on the floor and electrical plans need to be supplemented in details and interior elevations, a note must be placed on the electrical plan to alert the electrician to additional sources of information. Figure 12.34 shows the electrical plan for a simple entertainment center above a gas fireplace.

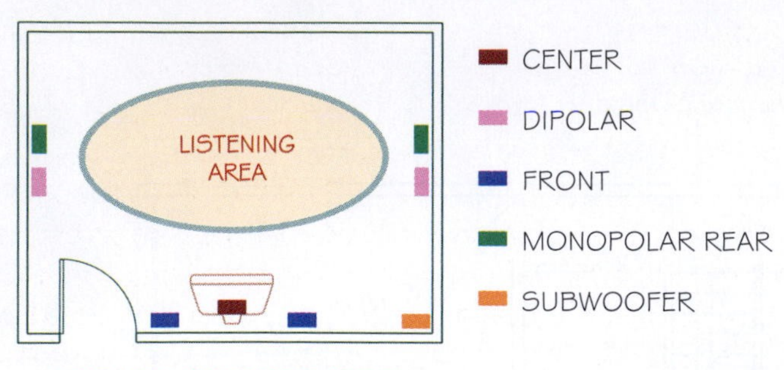

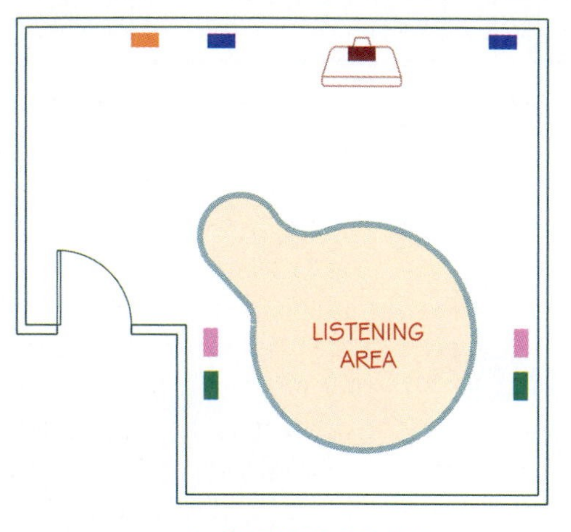

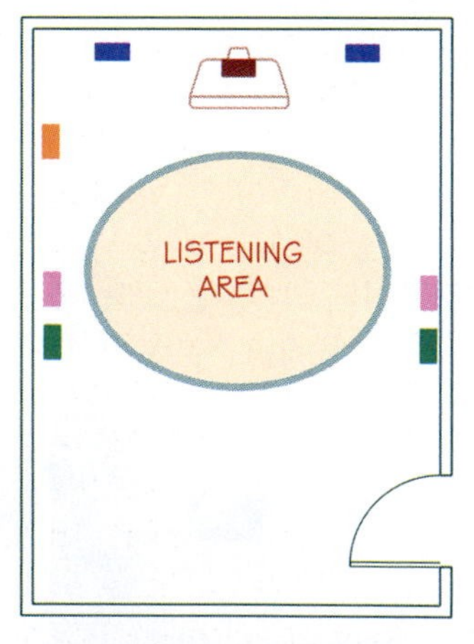

FIGURE 12.33 ■ *Four common layouts for home theater surround-sound systems.*

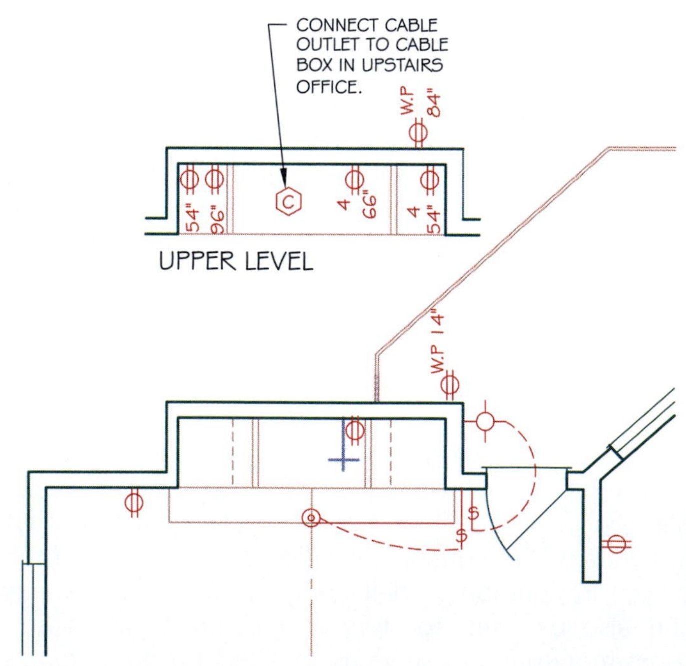

FIGURE 12.34 ■ *The electrical plan for a simple entertainment center that will be above a gas fireplace. The number of receptacles and the their heights above the floor must be specified to receive adequate service for all equipment.*

Built-in Vacuum System

A built-in vacuum system uses a central power unit with individual inlet valves located throughout the home. The central unit is usually located in a garage, basement, or utility room. Wall inlet valves are installed in various locations throughout the home and connected to the power unit through tubing that runs through the wall and floor framing. The system is activated by plugging a portable hose into a wall inlet. The dirt, dust, animal dander, and allergens are then carried out of the room and into the main power unit, where all of the debris is deposited into the canister. The central unit, introduced in Figure 10.49, should be shown on both the floor and electrical plans. Individual inlet valves are shown only on the electrical plan, as seen in Figure 12.25.

As a general guideline, one inlet valve should be installed for every 600 sq ft of a home. Household cleaning needs, the number of stories, and convenience will also impact the number of inlets that need to be provided. Inlet valves must be strategically placed to reach all areas of the home. Unless additional length is required to reach the tops of drapes or ledges, two hose lengths is the maximum distance that should separate two inlet valves. Assume a hose length of 25' (7500 mm). The first inlet valve should be located at a point the farthest distance from the power unit. From this location, additional valve locations can be selected that allow a hose to reach all rooms. Remember, walls and furniture can shorten the distance serviced by a valve in some areas, so be sure to locate inlets with furniture and walls in mind. Figure 12.35 shows the completed electrical drawings for the home drawn in Chapter 11. Figure 12.36 shows the completed plan for the lower level. Use the checklist from Appendix D to evaluate your work.

Security Systems

Home security for a custom home involves more than providing exterior security lighting. Provisions are often made for internal video monitoring, perimeter surveillance, moisture detection, burglary, fire, carbon monoxide, and medical alert monitoring. Each type of system can be an internal system or connected to a monitoring station. Monitoring can be provided by a private security or health care provider or linked directly to a public police or fire department. The home may even include a separate room to provide security from the forces of nature or from intruders. To meet any of these security needs, the system should be designed in cooperation with a security expert to provide the best possible installation needed by the owner. When a secure room is provided, a structural engineer is usually consulted to help plan additional reinforcing and structural features.

Home Wiring and Automation

Some custom homes are built with automation systems to control and operate mechanical devices that regulate the heating, air conditioning, landscape sprinkler systems, lighting, and security systems. These systems can be linked to a personal computer to allow the user to set and monitor a variety of electrical circuits throughout the home. To make efficient use of home computers, structured wiring systems for Internet access must be considered during the planning of the electrical system. Structured wiring systems allow high-speed voice, data lines, and video cables wired to a central service location. These wires and cables optimize the speed and quality of various communication signals coming in and out of the residence by providing each electrical outlet, telephone jack, or computer port with a dedicated line back to the central service location. The central service location allows the wires to be connected as needed for a network configuration or for dedicated wiring from the outside. High-quality structured wiring systems use network connectors and parallel circuits to maintain a strong electrical signal. A parallel circuit is an electrical circuit that contains two or more paths for the electricity or signal to flow from a common source. Structured wiring systems allow the use of a fax, multiline telephone, and computer at the same time. Additional applications include a digital satellite system (DSS), digital broadcast system (DBS), stereo audio, and closed-circuit security systems.

The product supplier or the design team may do the design and drawing of the home automation system. Electrical symbols and specific notes are placed on the floor and electrical plans. The automation center for a home theater is shown in Figure 12.37. Cables from the home computer are routed through the hub and then to each fixture to be controlled.

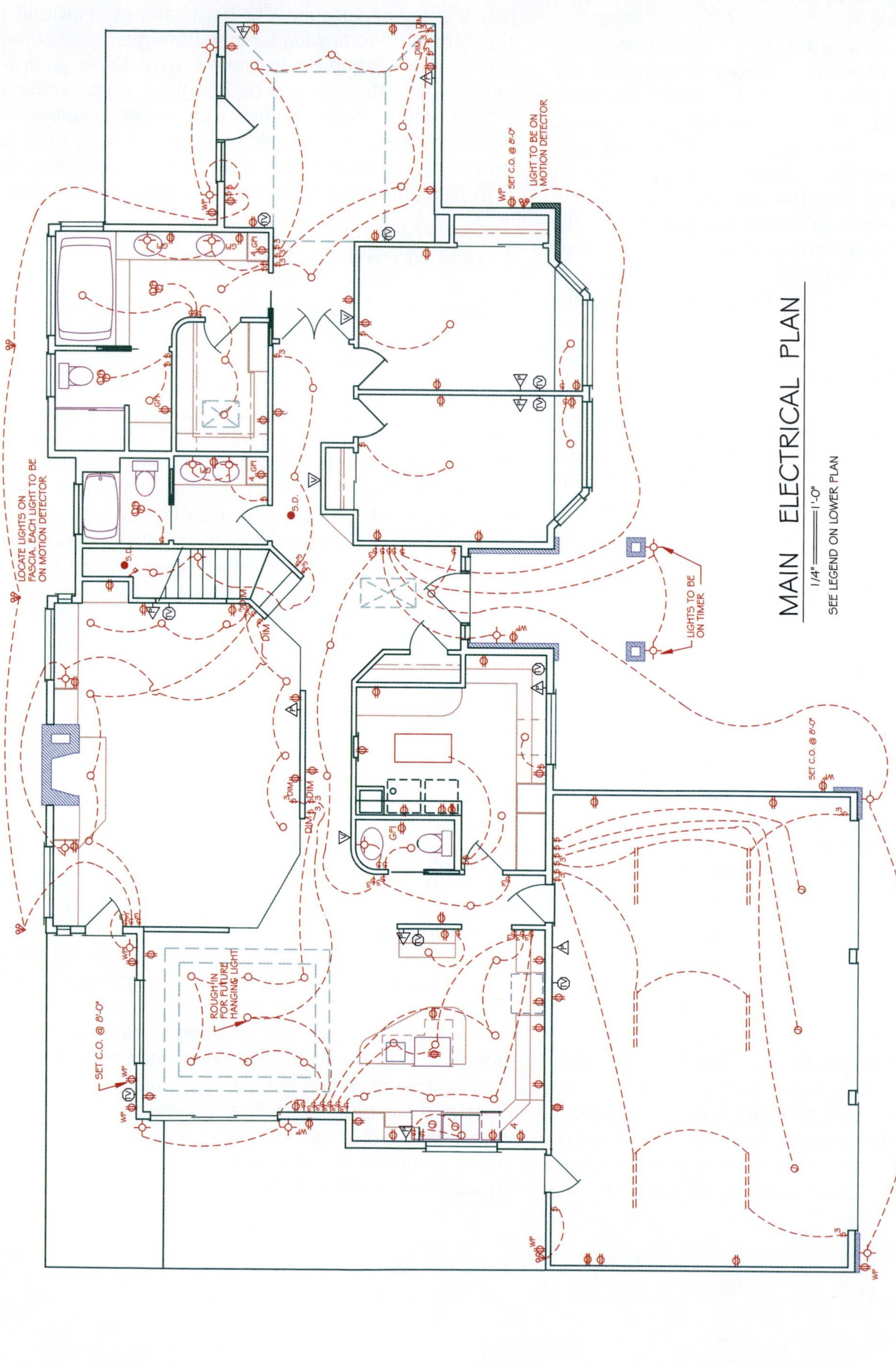

MAIN ELECTRICAL PLAN

1/4" = 1'-0"

SEE LEGEND ON LOWER PLAN

FIGURE 12.35 ■ *The completed electrical plan containing all fixtures, receptacles, switches, and specialty equipment.*

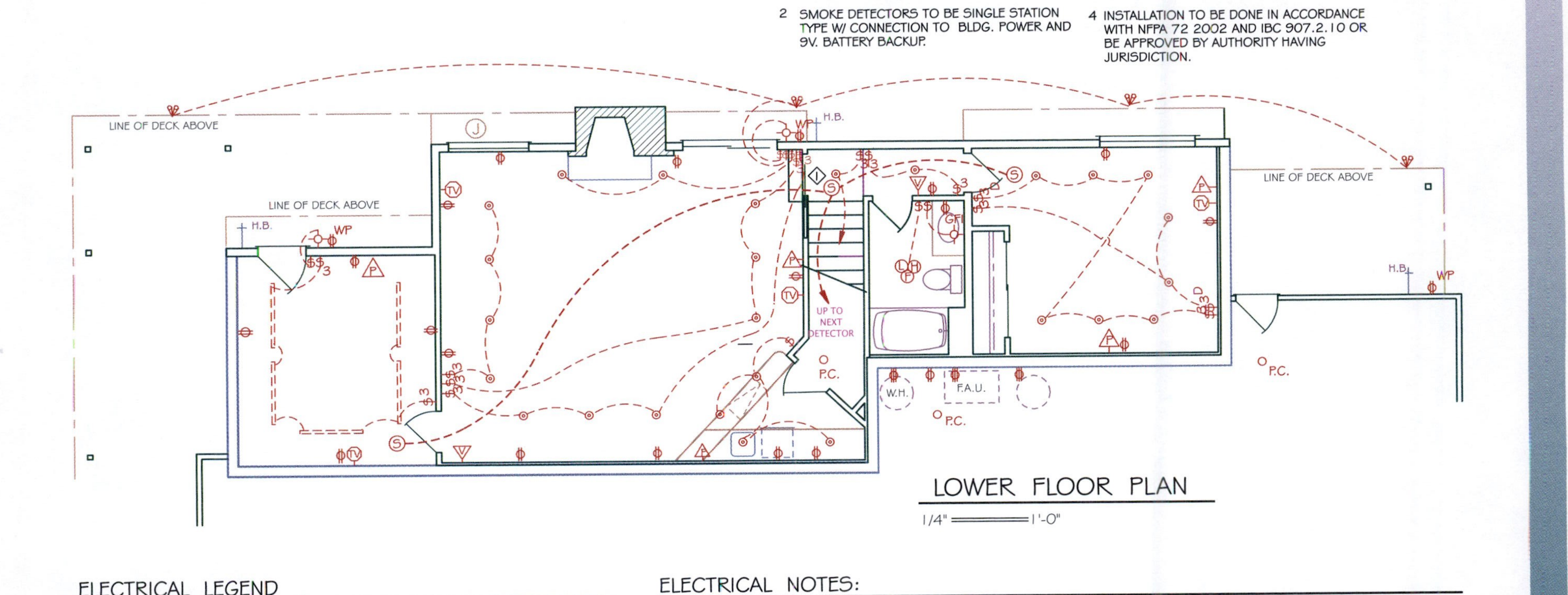
FIRE ALARM NOTES
1 MOUNT SMOKE DETECTOR TO HIGHEST POINT OF STAIRWELL CEILING.
2 SMOKE DETECTORS TO BE SINGLE STATION TYPE W/ CONNECTION TO BLDG. POWER AND 9V. BATTERY BACKUP.
3 SMOKE DETECTORS TO BE WIRED SO AS TO CAUSE ALARM OF ALL DEVICES UPON ACTIVATION OF ANY SINGLE DEVICE.
4 INSTALLATION TO BE DONE IN ACCORDANCE WITH NFPA 72 2002 AND IBC 907.2.10 OR BE APPROVED BY AUTHORITY HAVING JURISDICTION.
LINE OF DECK ABOVE
LINE OF DECK ABOVE
LINE OF DECK ABOVE
H.B.
WP
UP TO NEXT DETECTOR
P.C.
W.H.
F.A.U.
LOWER FLOOR PLAN
1/4" = 1'-0"
ELECTRICAL LEGEND
110 CONVENIENCE OUTLET
GFI 110 C.O. GROUND FAULT INTERRUPTER
WP 110 WATER PROOF
110 HALF HOT
JUNCTION BOX
220 OUTLET
SINGLE POLE SWITCH
THREE-WAY SWITCH
CEILING MOUNTED FIXTURE
CAN CEILING FIXTURE
WALL MOUNTED LIGHT
RECESSED LIGHT FIXTURE
LIGHT ON PULL CHORD
SPOT LIGHTS
48" SURFACE MOUNTED FLUORESCENT LIGHT FIXTURE
LIGHT, HEATER, & FAN
SMOKE DETECTOR
S.D.
VACUUM
STEREO SPEAKER
PHONE OUTLET
CABLE T.V. OUTLET
ELECTRICAL NOTES:
1. ALL GARAGE AND EXTERIOR PLUGS AND LIGHT FIXTURES TO BE ON GFCI CIRCUIT.
2. ALL KITCHEN PLUGS AND LIGHT FIXTURES TO BE ON GFCI CIRCUIT.
3. PROVIDE A SEPARATE CIRCUIT FOR MICROWAVE OVEN.
4. PROVIDE A SEPARATE CIRCUIT FOR PERSONAL COMPUTER. VERIFY LOCATION WITH OWNER.
5. VERIFY ALL ELECTRICAL LOCATIONS W/ OWNER.
6. EXTERIOR SPOTLIGHTS TO BE ON PHOTO-ELECTRIC CELL W/ TIMER.
7. ALL RECESSED LIGHTS IN EXTERIOR CEILINGS TO BE INSULATION COVER RATED.
8. ELECTRICAL OUTLET PLATE GASKETS SHALL BE INSTALLED ON RECEPTACLE, SWITCH, AND ANY OTHER BOXES IN EXTERIOR WALL.
9. PROVIDE THERMOSTATICALLY CONTROLLED FAN IN ATTIC WITH MANUAL OVERRIDE (VERIFY LOCATION W/ OWNER).
10. ALL FANS TO VENT TO OUTSIDE AIR. ALL FAN DUCTS TO HAVE AUTOMATIC DAMPERS.
11. HOT WATER TANKS TO BE INSULATED TO R-11 MINIMUM.
12. INSULATE ALL HEATING DUCTS IN UNHEATED AREAS TO R-11. ALL HVAC DUCTS TO BE SEALED AT JOINTS AND CORNERS.
13. PROVIDE 6 SQ. FT. OF VENT FOR COMBUSTION AIR TO OUTSIDE AIR FOR FIREPLACE CONNECTED DIRECTLY TO FIREBOX. PROVIDE FULLY CLOSEABLE AIR INLET.
14. HEATING TO BE ELECTRIC HEAT PUMP. PROVIDE BID FOR TWO UNITS.
15. PROVIDE A 20 AMP MIN. SERVICE THROUGHOUT RESIDENCE.
16. PRE-WIRE FOR SECURITY SYSTEM. SEE DRAWINGS BY OTHERS

FIGURE 12.36 ■ *The completed electrical plan for the lower floor contains all of the fixtures, receptacles, switches, and specialty equipment.* Courtesy Michael Coffman.

FIGURE 12.37 ■ *The automation center for a home theater. Cables from the home computer are routed through the hub and then to each piece of equipment to be controlled.*
Courtesy Keith Rowell.

CHAPTER 12

Additional Readings

The following websites can be used as a resource to help keep you current with changes in electrical plan–related materials.

ADDRESS	COMPANY OR ORGANIZATION
www.acousticalsolutions.com	Acoustical Solutions
www.avenow.com	Audio Video Environments
www.beamvac.com	Beam Central Cleaning Systems
www.broan.com	Broan (Bath Fans)
www.cedia.net	Custom Electrical Installation Association
www.hometheathermag.com	Home Theater Magazine
www.leviton.com	Leviton (Lighting controls)
www.lucent.com	Lucent Technologies (Wiring systems)
www.nutone.com	Nutone (Central Cleaning Systems, Intercoms)
www.onqtech.com	On Q Home Wiring Systems
www.squared.com	SquareD (Electrical systems)

CHAPTER 12

Electrical Plans Test

Answer the following questions with short, complete statements. Type the answers using a word processor or neatly print the answers on lined paper.

QUESTIONS

12.1. What is the maximum allowable distance between duplex convenience outlets in a living room?
12.2. What is the maximum allowable distance a duplex convenience outlet can be installed from a corner in a kitchen?
12.3. Describe at least four energy-efficient considerations related to electrical design.
12.4. Draw the proper floor plan symbol for:
- a. 120 duplex convenience outlet
- b. 240-volt outlet
- c. Circuit breaker panel
- d. Speakers for surround-sound system
- e. Ceiling-mounted light fixture
- f. Wall-mounted light fixture
- g. Three-way switch
- h. Simplified fluorescent light fixture
- i. Bathroom fan, heat, and light fixture

12.5. Explain where to place a switch in a bedroom as it relates to the door.
12.6. What is a GFCI duplex convenience outlet?
12.7. How many amps are typically provided for a residential distribution box?
12.8. Define a junction box as it relates to electrical wiring.
12.9. What voltages are normally delivered to a residence?
12.10. What level of lighting should be provided to an area intended for reading?
12.11. What size circuit breaker should be provided for an electric range with six burners?
12.12. List four code requirements for the spacing of receptacles in a kitchen.
12.13. What size circuit breaker should be provided for a gas dryer?
12.14. Define structured wiring systems.
12.15. List four locations where smoke detectors are required.

DRAWING PROBLEMS

12.1. Convert each symbol in the CD\Drawing Blocks\FLOOR BLOCKS\ELECTRICAL\ELEC SYMB to a drawing block. Assign all drawing objects to appropriate layers. Provide appropriate insertion points and names to easily identify each block.
12.2. Draw a floor plan representation of the following items assuming plotting at a scale of 1/4'' = 1'-0''.
- a. A room with two means of access with appropriate switches controlling three ceiling-mounted lighting fixtures
- b. A room with three means of access with appropriate switches to control one ceiling-mounted lighting fixture
- c. A room with three single-pole switches that control three different light fixtures
- d. A single-pole switch that controls two half-hot convenience outlets

12.3. Draw a small bathroom layout with a tub/shower, water closet, and vanity with two lavatories. Provide lighting fixtures, receptacles, and switches to control fixtures as needed.

12.4. Draw a U-shaped kitchen with a double sink, dishwasher, gas range, warming drawer, built-in espresso machine, double oven with a separate microwave oven, and a wall-mounted television connected to a satellite receiver. Provide the necessary notes and symbols to make the kitchen legal and functional.

12.5. Use the drawing of the house that you started in Chapter 11 and do one of the following as directed by your instructor: Insert the electrical notes and symbols from the CD\DRAWING BLOCKS\FLOOR BLOCKS\ELECT and edit them to meet the needs of your project. Use a legend to explain each symbol used on the electrical portion of your drawing.

a. Add the needed information to meet local codes and the demands of the general public on the floor plan that was started in Chapter 11.

b. Use the floor plan started in Chapter 11 as a base and draw an electrical plan to meet the local codes and the demands of the general public. Freeze all unnecessary floor information and create an electrical plan that includes:
   - All electrical fixtures, plugs, and switches.
   - Provide a distribution box, doorbell and chime, a maximum of 5 phone jacks, 4 TV jacks, and 3 waterproof receptacles. Provide a wall-mounted light by each exterior door.

c. Use the floor plan started in Chapter 11 as a base and draw an electrical plan to meet local code requirements and the demands of a specific family. Freeze all unnecessary floor information and create an electrical plan that includes:
   - Locate all electrical fixtures, plugs, and switches to meet code and general, specific, and decorative lighting needs.
   - Provide a meter, distribution box, doorbell and chime, a minimum of 5 phone jacks, 4 TV jacks, and 3 waterproof receptacles. Provide a wall-mounted light by each exterior door, an intercom system with slave units in each room, surround sound in the family room, and a built-in vacuum system.

13 CHAPTER

Plumbing Systems

The plumbing system involves the delivery of fresh water and the control and discharge of all liquids, solids, and gases from the residence. Plumbing drawings are not completed by the design team for residential drawings. For most municipalities, even the plumbing contractor may not be required to complete drawings. Instead of drawings, the application for a plumbing permit requires a listing of the number of fixtures being installed. Fees charged for the permit are based on the size and complexity of the plumbing work to be done. Some municipalities base the plumbing fees on the number of fixtures to be installed or the number of feet of water and sewer lines or the number of rain drains to be added. Some municipalities require a one-line diagram showing pipe sizes for fresh- and wastewater lines.

The architectural team will place plumbing-related information on the site plan, floor plan, and occasionally on the foundation plan, but plans to show pipe sizes and locations are completed by the plumbing contractor. The site plan will show the location of the water meter, water and gas supply lines, and sewer lines. If a private sewer disposal system will be used, the septic tank and drain field must also be represented on the site plan. Drains connected to the roof downspouts and foundation drain lines are shown on the foundation plan for most custom homes. Most plumbing information is placed on the floor plan. The floor plan shows symbols that represent fixtures such as sinks, toilets, bidets, the water heater, and tubs. Drawings that show how the system works are not drawn by the architectural team. This fact, however, does not free you from having to understand the drawings created by the plumbing and fire safety contractors. Figure 13.1 shows a simple one-line schematic drawing that a plumber might draw to explain a layout. You also need to understand how each system works. This requires an understanding of delivery methods, fresh- and wastewater systems, fire suppression systems, and other systems used to eliminate waste gas. Most importantly, you need to understand how work done by the plumber will affect other structural portions of the home you are responsible to draw.

DELIVERY AND REMOVAL MATERIALS

The pipe material and size will affect the design of the plumbing system and can affect the framing system. The pipes used in the plumbing system may be made of plastic, copper, galvanized steel, or cast iron, depending on the usage. The usage and material will affect the pipe size to be used.

Common Materials

The common types of pipes referred to by the IRC include the main, branch, and risers. The ***main*** is the water supply line that extends from the water meter into the home to deliver potable (drinkable) water. ***Branch lines*** are feeder lines that branch off the main line to supply fixture groups in the home. A ***riser*** is a water supply pipe that extends vertically one or more stories to carry

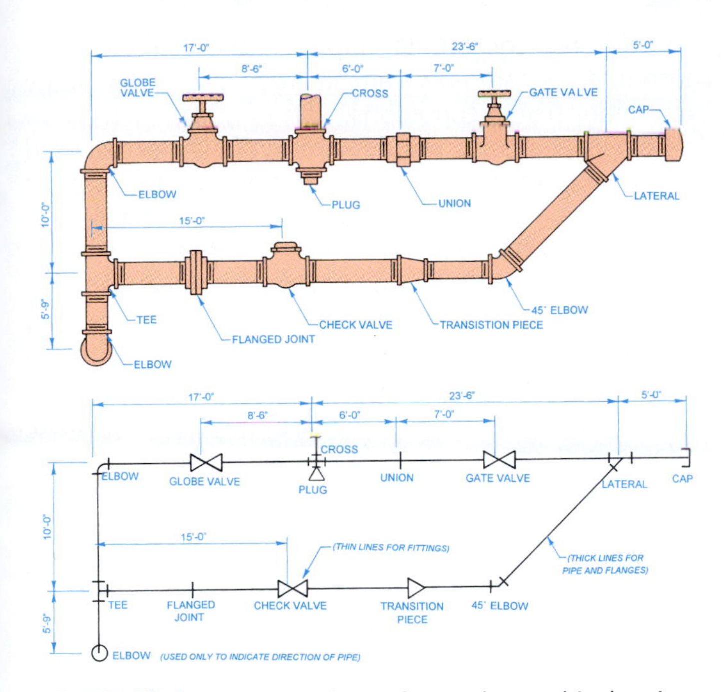

FIGURE 13.1 ■ *A comparison of an orthographic drawing and a one-line schematic drawing. For a residence, someone working for the plumbing contractor will provide any needed one-line diagrams to obtain a building permit.*

water to fixtures. The IRC defines a fixture as a unit used to contain and discharge water. Common fixtures found in a residence include sinks, lavatories, showers, tubs, toilets ***(water closets),*** bidets, spas, and ***hose bibbs.*** The IRC allows copper, plastic, steel, and cast-iron pipes to be used to service fixtures.

Copper Piping

Copper lines similar to those in Figure 13.2 have been a popular choice for delivering hot and cold water in homes for over 50 years. Copper pipes provide a durable material to distribute water and are quickly assembled by the use of soldered joints and ***fittings.*** Because of the expense and the use of other materials, in many areas copper is used for only the main supply line and for a ***manifold*** (see Figure 13.3). Smaller manifolds may also be provided at the ends of branch lines to feed each fixture group. Copper pipes, once a popular choice for branch and riser lines, have been replaced in many areas of the country by plastic pipes inside the structure.

Plastic Pipes

Plastic pipes have glued joints and fittings and are used for fresh water, wastewater, and vent pipes. Plastic piping includes the use of cross-linked polyethylene

FIGURE 13.2 ■ *Copper lines have been used for years in residential plumbing systems to deliver fresh water.* Courtesy Peter Rowell.

FIGURE 13.3 ■ *A manifold is a distribution center between the main, branch, and riser lines. Smaller manifolds are used at the end of feeder lines to distribute water to each fixture.* Courtesy Lisa A. Kotasek, Uponor Wirsbo.

(PEX), polyvinyl chloride (PVC), postchlorinated polyvinyl chloride (CPVC), polybutylene (PB), and acrylonitrile-butadiene-styrene (ABS). Most new homes use plastic polyethylene tubing, similar to the piping shown in Figure 13.4, to deliver fresh water throughout the interior of the home. Plastic tubing, known as PEX or Wirsbo (also a specific brand name), has been used in homes for over 30 years. PEX is a flexible, expandable plastic that can easily be run through framing. Rather than having glued joints, PEX joints are heated and expanded to fit over fittings. Once the tubing cools, a watertight seal is formed, providing a system that withstands high pressure and freezing. If the tube freezes, it will return to its original size when thawed.

FIGURE 13.4 ■ *Plastic polyethylene tubing is used to deliver fresh water throughout the interior of the home that is remodeled in Chapter 32.* Courtesy Megan Jefferis.

PVC pipe is used throughout residential construction for below- and above-ground uses. Common uses for PVC include water mains, fresh- and wastewater lines, drain and waste-vent lines, and irrigation lines. The components of a PVC piping system are manufactured in a variety of colors to help identify the application. A common color scheme (although not universal) is:

- White: Drain, waste and vent and some low-pressure applications
- White, blue, and dark gray: cold water piping
- Green: sewer service
- Dark gray: high-pressure applications

This color scheme has an exception in that much of the white PVC pipe is dual-rated for DWV (drain/waste/vent) and pressure applications.

CPVC is a corrosion-resistant plastic piping. Potable water applications include cold-water services from wells or water mains up to the building as well as hot- and cold-water distribution piping within buildings. Because it is corrosion-resistant, it maintains water purity even under severe conditions.

Polybutylene or PB piping is a form of plastic resin used in the manufacture of water-supply piping. It is popular because of the low material cost and ease of installation. PB piping may fail if oxidants such as chlorine in the water supply react with the piping, causing it to become brittle. As the system becomes weak, it may fail, causing damage to the structure and personal property. Although still approved by the IRC, its popularity has decreased.

ABS pipes and fittings are used throughout the waste and vent systems for in- or above-ground applications. It may be used outdoors if the pipe contains pigments to shield it against ultraviolet radiation, or jurisdictions may require the pipe to be painted with water-base latex paint for outdoor use. ABS is preferable to cast iron, which corrodes, causing pipes to block. ABS pipes will not corrode.

Steel Pipe

Flexible steel pipe protected with a coat of varnish is used to deliver natural gas or propane to a fixture such as a water heater, furnace, or fireplace. Steel pipe is joined by threaded joints and fittings or grooved joints.

Cast-Iron Pipe

Although often replaced by ABS, cast-iron pipe is still used in some areas to carry solid and liquid waste from the structure to the local sewer system. Cast-iron pipe may also be used for the piping in the drain system throughout the structure to help reduce the noise of water flow. It is more expensive than plastic pipe, but quiet piping may be worth the price for many clients.

Pipe Sizes

In addition to pipe material, the size of the pipe is also important. The IRC requires a minimum water main diameter of 3/4'' (20 mm), but diameters of 1'' or 1 1/2'' (25 or 38 mm) are more common. The size of residential branches and risers is determined by the plumber based on:

- IRC requirements for the total load on the pipe
- The psi available to the home
- The amount of water needed at the fixture
- The height of the riser
- The length of the pipe.

FIXTURE	COLD WATER	HOT WATER	SOIL, WASTE	VENT
Sinks (Lav)	1/2"	1/2"	1 1/2"–2"	1 1/4"–1 1/2"
Lavatory	1/2"–3/8"	1/2"–3/8"	1 1/4"–2"	1 1/4"–1 1/2"
Water closet	1/2"–3/8"	—	3"– 4"	2"
Tub (Bath)	1/2"	1/2"	1 1/2"–2"	1 1/4"–1 1/2"
Shower	1/2"	1/2"	2"	1 1/4"
Water heater	3/4"	3/4"	—	4"
Washer	1/2"	1/2"	2"	1 1/2"
Lau Sink	1/2"	1/2"	1 1/2"	1 1/4"

FIGURE 13.5 ■ *Minimum pipe sizes for fresh-water, waste-water, and vent lines.*

- The flow pressure needed at the farthest point from the source

The IRC requires a minimum diameter of 3/8'' (10 mm) for individual feeder lines from branch lines. Other common sizes are seen in Figure 13.5. The manufacturers of spas and other fixtures often dictate the use of larger supply lines. When a larger size is required, the size of the supply line should be specified in the general notes on the floor plan. Supply lines are not allowed to be longer than 60' (18 000 mm). Although the plumber is required to know this for installation of the system, the design team must consider these guidelines in specifying fixtures on the floor plan. Minimum sizes are also listed in the general notes on the floor plan.

FRESH-WATER SYSTEMS

The fresh-water supply system consists of a network of pipes similar to that seen in Figure 13.6, which deliver freshwater using pressure throughout the residence. The IRC uses the term *potable* water to refer to water free of impurities and suitable for drinking. Water is brought from a public water supply main or from a private well through a building main. The term ***main*** is used to describe the primary water delivery line to the residence. If public water is used, the main contains a utility company ***valve,*** a meter, and a building main valve. The main supplies water to the structure and routes it to the water heater or to branch lines. In many locations, the main connects water to a filtering system to filter, soften, and purify water before it is dispersed throughout the house.

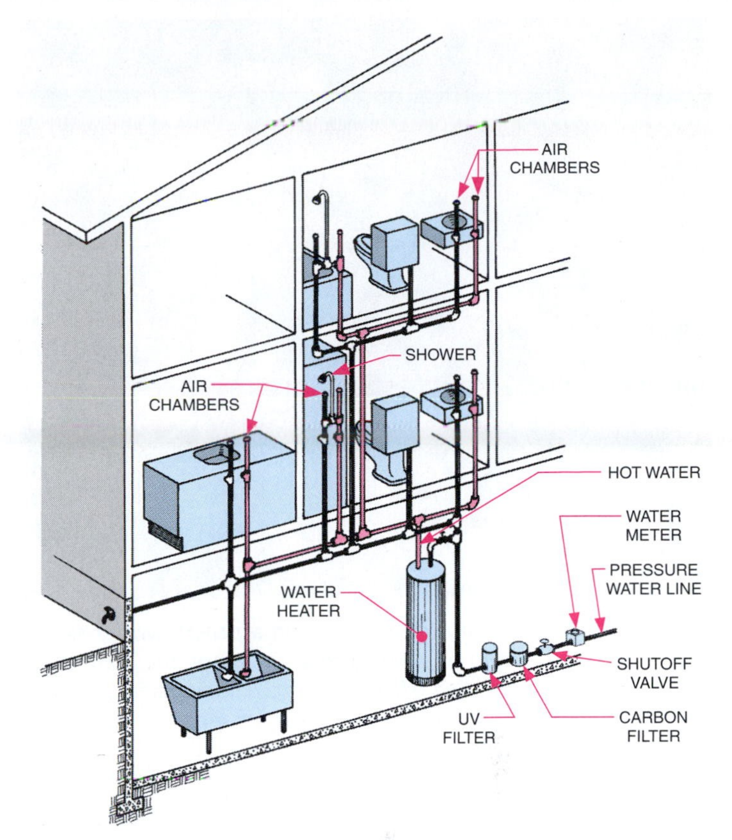

FIGURE 13.6 ■ *Common elements of the fresh-water delivery system include the main, branch, and feeder lines and the air chambers. Air chambers are pipe extensions placed by each fixture that are used to eliminate noises caused by pipe vibration when taps are opened and closed.*

The IRC requires a minimum 40 psi for water entering the home. If water pressure exceeds 80 psi, an approved pressure-reducing valve must be installed on the main or riser at the connection to the water-service pipe. For homes with poor water pressure or those on private wells, a pressurized storage tank is used to store water in sufficient supply for appliances and fixtures. Figure 13.7 shows the use of a storage tank and filters. Hose bibbs, exterior sprinkler systems, a fire-suppression system, and other fixtures that do not require purification are connected to the water main before the water supply line enters the filtering system. After filtering, water is pushed by pressure into the cold-water branch lines and the hot-water main. The cold-water line branches to each fixture. The hot-water line passes through the water heater

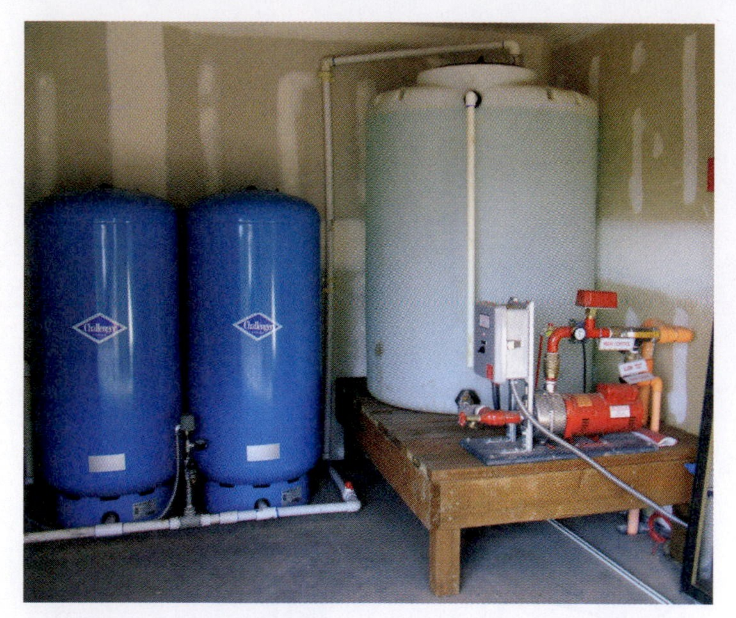

FIGURE 13.7 ■ *When the home is on a private water system or water pressure is low, the main water line often leads to a storage tank. For this home, water is then sent to a filtering system.* Courtesy Matt Goldsberry.

FIGURE 13.8 ■ *Plastic lines are used throughout the home to deliver fresh water. Blue indicates cold water lines and red the hot-water lines.* Courtesy Matthew Jefferis.

and then on to fixtures that require hot water. Because water in the hot and cold lines is under pressure, pipes can run in any direction. Hot-water branch lines are normally located 6'' (150 mm) from and parallel to cold-water lines. Hot-water lines are placed on the left side of cold water lines and may even be color-coded, like the lines in Figure 13.8. As the line reaches the termination point for a fixture, a shutoff valve must be provided.

Hot-Water Systems

Common methods to provide heated water to a system include hot-water storage, tanks, tankless heating systems, and continuous loop systems.

Hot-Water Tanks

Hot-water storage systems have been the traditional method to heat water since people got tired of using buckets to boil water over an open fire. A water heater is a storage tank that heats water and keeps it warm. These systems are typically powered by gas or electricity, with gas units providing a much cheaper fuel source. Water heaters are available in a variety of sizes, including 30-, 40-, 50-, 65-, and 80-gallon (115-, 150-, 190-, 250-, and 300-liter) tanks. Larger-capacity tanks may be ordered for special applications. The size should be selected based on the size of the family and the appliance load to be served (showers, dishwashers, and washing machines). Common guidelines to size a water heater include:

NUMBER OF BATHS	NUMBER OF BEDROOMS	WATER HEATER SIZE (GALLONS)*	
		GAS	ELECTRIC
1 to 1	2	30	30
	3	30	40
2 to 2 1/2	2	30	40
	3	40	50
	4	40	50
	5	50	66
3 to 3 1/2	3	40	50
	4	50	66
	5	50	66
	6	50	80

*Some spas are larger than a typical tub and require a higher-capacity water heater to provide an adequate supply of hot water.

The method of heating the water and the location of the tank will affect what must be specified on the floor plan. Common specifications placed on the floor plan in the form of general notes may include:

- Gas-fired water heaters should be placed on a platform so the flame source is at least 18'' (450 mm) above floor level.
- A water heater in a living area must be placed over a 1 1/2''- (38 mm) deep × 24-gauge overflow tray with a 3/4''- (19 mm) diameter drain.
- Fuel-fired water heaters cannot be installed in a room used as a storage closet. Other regulated areas within the living space include:
  - A bedroom or bathroom containing a water heater that is not a direct-vent model. Such a water heater must be installed in a sealed enclosure so combustion air is not taken from the living area.
  - When a water heater is located in an attic or crawl space and the access is in the closet of a sleeping room or bathroom. The access must have a minimum opening size of 30 × 22'' (750 × 550 mm).
  - Attic passageways serving the water heater or other mechanical equipment cannot exceed 20' (6000 mm) in length and must not be less than 24'' (600 mm) wide.
    - A solid floor must be provided to access the unit as well as a level service platform 30 × 30'' (750 × 750 mm) wide on all sides of the unit that might require service.
- Water heaters should be strapped to the walls of the structure.
- When a water heater is located in a garage, a concrete-filled steel pipe embedded in concrete must be placed in front of the unit to protect it from impact.

FIGURE 13.9 ■ *A tankless water system provides continuous hot water once a tap is turned on.* Courtesy Takagi.

Tankless Hot-Water Systems

Tankless hot-water systems provide continuous hot water when a tap is turned on. Hot water is not stored or kept heated until required. Tankless systems generate hot water on demand, and—depending on the model—can deliver between 200 and 500 gallons (760 and 1900 liters) of hot water per hour on demand. This feature results in a savings of time, money, and space over a traditional water heater. An average tankless water heater is approximately 20 × 14 × 6'' (500 × 350 × 150 mm). See Figure 13.9.

Tankless water heaters start to operate as soon as the hot-water tap at a fixture is opened. The heater detects the flow of water and a computer in the unit automatically ignites the burner. Water flows over a heat exchanger in the unit and, within approximately 5 seconds, is heated to the preset temperature. The unit will continue to provide hot water until the water tap is closed. With the tap closed, the heating unit automatically shuts down until needed again. It is extremely important that the design team size a tankless water-heater system to meet the needs of the family. If multiple fixtures are operated simultaneously, the demand may exceed the ability of the heater to supply hot water at the desired temperature.

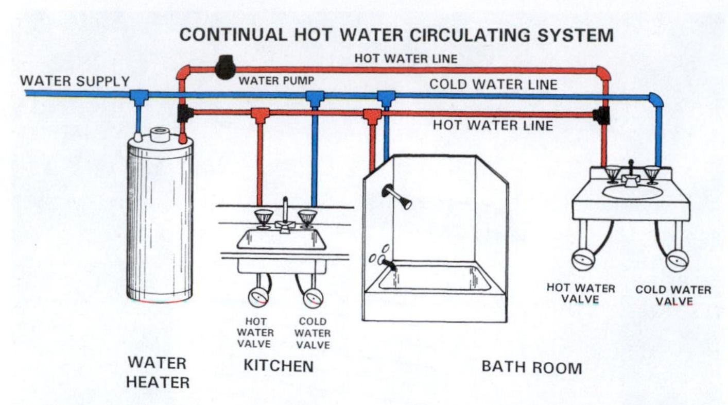

FIGURE 13.10 ■ *A continual hot-water circulating system keeps water moving between the hot-water supply and each fixture on the loop.*

FIGURE 13.11 ■ *Solar heating can be used in most areas of the country to either heat or preheat water for a residence.* Courtesy Wagner & Co.

Circulating Hot-Water Systems

Hot-water recirculation is the process of constantly moving water through a loop in a hot-water pipe. Loops are formed between the water heater and groups of appliances so there will be no wait for the water to warm up. Circulating systems use a small motor located by the water heater to keep hot water moving through a loop formed between the water tank and each group fixture. The system loop conserves water by providing hot water as soon as the hot-water tap is turned on. Figure 13.10 shows the concept of a circulating system.

Solar Hot Water

A solar water-heating system is designed to heat water by capturing energy from the sun. Solar energy is collected by water that passes through a panel system and into a hot-water storage tank. Once collected, heated water is dispersed to the domestic hot-water system or to a swimming pool. Solar collectors may be located on a roof, a wall, or the ground. Figure 13.11 shows a roof-mounted application of solar collectors. Solar systems vary in efficiency but generally are used to preheat water before entering the water heater. The number of south-facing collectors needed to provide hot water to a structure depends on the size of the structure and the volume of water needed. A conventional central heating pump forces water through a coiled pipe in the solar panel, where it is heated by the sun. The heated water then flows down and through a second coil in the hot-water cylinder. Hot water passing through this coil heats the water in the cylinder. A pump then returns the water to the solar panels. If the system is roof-mounted, solar panels would be specified on the roof plan; the storage tank and any required pumps are specified on the floor and electrical plans.

WASTEWATER AND VENTING SYSTEMS

The drainage system moves water and other waste from the plumbing system to the main sewer line. The main sewer line then ***drains*** into a public system or to a private waste system. In addition to the drainage system, the venting system must also be planned for when drawing the floor plan. The vent system prevents vacuum blocks in the drainage system to allow a continuous flow of air through the wastewater system to vent gases and odors out of the house.

Waste Discharge System

The waste discharge system takes water from the fixture drains and moves it to the disposal source. See Figure 13.12. Waste lines ranging in size from 1 1/2 to 4'' (38 to 100 mm) are used to remove gray water and sewage from the residence. ***Soil lines*** are empty until waste is flushed through the system. Because

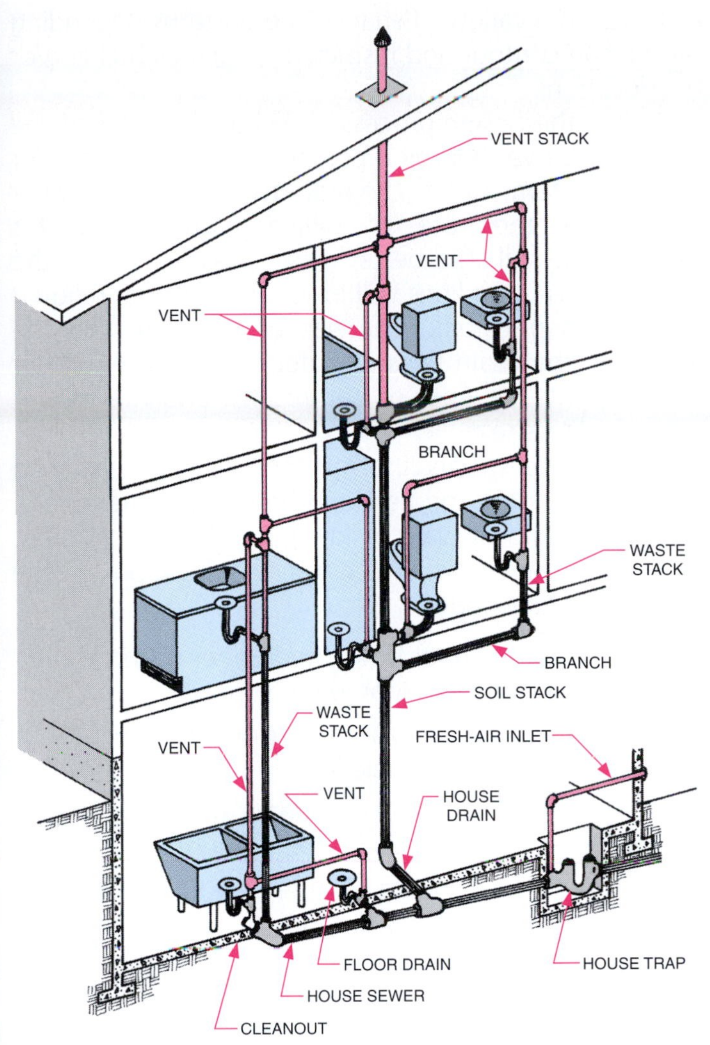

FIGURE 13.12 ■ *The flow of all wastewater begins at a fixture, then runs through a trap into a branch line, and finally into the main sewer line. Once the line extends past the exterior walls of the residence, a cleanout is provided in the main sewer line to allow access for cleaning. The main then extends to the municipal sewer lateral serving the construction site.*

the waste system is a nonpressurized gravity-flow system, waste lines are larger than the water supply lines. Common sizes of waste lines are shown in Figure 13.5. In addition to the size difference, drainage pipes must have a minimum slope of 1/4'' per foot (6/25 mm).

Branch and Stack Lines

Two common types of soil lines include branches and stacks. Branches are the nearly horizontal lines that carry waste from each fixture to the stacks. The ***stack lines*** are the vertical drain lines that carry waste from the home to the sewer main. Stack lines range in size from 3 to 4'' (75 to 100 mm) in diameter, increasing in size as the distance from the fixture increases. The portion of the soil stack above the highest branch intersection is referred to as a ***vent stack.*** Some vent stacks are separate and parallel to the soil stack. Vent stacks are dry pipes that extend through the roof a minimum of 6'' (150 mm) to provide ventilation for the discharge system. Vent stacks permit sewer gases to escape to outside air and equalize the air pressure in the system. Examples of each are shown in Figure 13.12.

The flow of all wastewater begins at a fixture. Each fixture contains a fixture trap to prevent the backflow of sewer gas from the branch lines. Fixture traps—except toilet traps, which are built into the fixture—are exposed for easy maintenance. A total-system house trap is provided in the main sewer line once the line extends past an exterior wall of the residence. In addition to the trap, a cleanout is provided in the main sewer line and for each drain line for sinks.

Gray Water

Some municipalities allow ***gray water*** to be reused. Gray water is captured by running toilet drain lines separate from other waste lines. Drain lines containing gray water are run to a storage tank and do not connect to the main sewer line. Lines from the storage tank can then be connected to yard sprinkler systems or for other uses of non-potable water. Because gray water is not disinfected, it could be contaminated. Although laws governing the use of gray water vary, common guidelines to avoid potential hazards include:

- Gray water is not potable water.
- Gray water must not be used directly on anything that may be eaten.
- Gray water should not be sprayed, allowed to puddle or run off property.
- Gray water from the kitchen sink or water that comes in contact with soiled diapers, meat, or poultry cannot be recirculated.

Sewage Disposal

Once the main sewer line leaves the home it connects either to the public sewer system or to a private disposal system. Each type of system must be represented on the site plan. For a home on a public system, only the main waste line is represented. Homes with private disposal systems require the septic tank, drain field, and the relation of each to a well or other bodies of water to be located. See Section II for information required on a site plan.

Public Sewers

Each municipality provides sewer service to residences. Many cities also extend service to all but the most rural residents. Where provided, ***sanitary sewers*** are located under the street or an easement next to the construction site. Most cities have a separate ***storm sewer*** system to dispose of groundwater, rainwater, surface water, or other nonpolluting waste. Sewer line locations and depths are available at the public works department of the governing body. The municipality generally is responsible for placing the lateral that connects each building parcel to the public system. A ***lateral*** is an underground branch line that extends from the sewer line to the edge of the street or to the property line. The plumbing contractor is responsible for locating the existing lateral and connecting the house to the public system.

Private Sewage Systems

A septic system consists of a storage tank and an absorption field. Solid and liquid waste enters the septic tank, where it is stored and begins to decompose into sludge. Liquid material, or effluent, flows from the tank outlet and is dispersed into a drain field. These lines are sometimes referred to as leach lines.

When the solid waste decomposes, it also dissipates into the soil absorption field. Chemicals must be added to the system periodically to aid in the decomposition of solid waste. Septic tanks also need to have solids pumped from the tank so that the system does not become overloaded. The leach lines are PVC perforated pipes laid in a course gravel bed approximately 12'' (300 mm) below the grade level. See Figure 13.13. Fields are arranged in a variety of shapes and patterns depending on the site contour and restrictions, such as building or tree locations.

During the design process, the characteristics of the soil must be verified by a percolation test suitable for a septic system. To size the drain field, a soils engineer determines the rate at which water percolates through the soil. The better water passes through the soil, the less drain lines will be required to drain liquid from the septic tank. The size of the tank and the length of the drain lines also depend on the number of occupants, number of bathrooms, and the topography of the site.

Local codes specify the minimum distance allowed between the drainage field and bodies of water, wells, roadways, right-of-way, buildings, and property lines. Common restrictions placed on the location of the drainage field include:

- Drains lines cannot be located under paved areas or uphill from a well or water supply.
- Drain fields can be placed no closer than 5' (1500 mm) from a water-table level.
- Regardless of location, soil under the field must be porous enough to absorb the effluent.
- The drain field should be at least 100' (30 m) from a water well, but this distance should be verified with local codes.

FIRE-SUPPRESSION SYSTEMS

Many municipalities require fire-suppression systems for homes that are approximately 4000 sq ft in area and larger. The exact size is based on the location of the home in relation to fire hydrants and the municipal water pressure at the hydrant. The method of fire suppression may also be dependent on the home insurance carrier and fire marshal. Suppression systems may range from the installation of a private fire hydrant

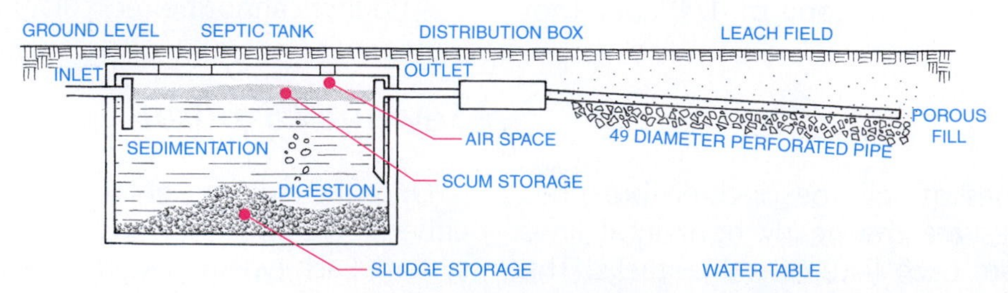

FIGURE 13.13 ■ *The major components of a private septic system include the septic tank, distribution box, and leach lines.*

connected to the public system, a pump connected to a swimming pool, or a fire sprinkler system. If you build in an area governed by the Life and Safety code, this code, as of January 2006, requires all homes to have sprinklers.

Home Fire Sprinklers

Fire sprinklers are most effective during the fire's initial flame growth stage. A properly selected sprinkler detects the fire's heat, initiates an alarm, and begins suppression within seconds after flames appear. Studies by the Fire Suppression System Association show that in most instances, sprinklers control fire advancement within a few minutes of their activation. Reducing the advancement of flames results in significantly less damage than would be the case without sprinklers. If you're thinking less damage from fire, more damage from water, you have bought into a widely held myth about sprinklers. In a home fire sprinkler system, a network of piping filled with water under pressure is installed behind the walls and ceilings, and individual sprinklers are placed along the piping to protect the areas beneath them. The sprinklers work independently. When the temperature from a fire reaches approximately 130 to 150 degrees, the sprinkler closest to the flame automatically opens and sprays water over the area, providing plenty of time for a family to escape unharmed from the fire.

Figure 13.14 shows examples of the rough installation as well as the finished fixture. Water is always in the piping, so the sprinkler system is always available for delivery. If fire breaks out, the temperature of the air above the fire rises and the sprinkler is activated when the temperature gets high enough. The sprinkler sprays water over the flames at a rate of between 10 and 25 gallons (38 and 95 liters) per minute. Only the sprinkler nearest the fire is activated. Smoke will not activate sprinklers. Figure 13.15 shows the sprinkler plan for the lower level of the home drawn in Chapter 11.

FIGURE 13.14 ■ *A network of piping filled with water under pressure is installed behind the walls and ceilings, and individual sprinklers are placed along the piping to protect the areas beneath them. The top photo shows plastic supply and branch lines merging at a sprinkler head. The bottom photo shows a sprinkler head in the finished ceiling.* Courtesy Lisa Kotasek, Uponor Wirsbo.

THE EFFECTS OF THE PLUMBING SYSTEM ON THE FRAMING SYSTEM

You've been introduced to the basic components of the fresh- and wastewater systems and told repeatedly that the architectural team usually doesn't have to worry about the required drawings. You do need to plan for how the plumbing affects the design of the floor plan, where lines can be placed, and how framing members will be affected by the placement of plumbing lines.

Room Design

How plumbing affects the layout of the floor plan is first introduced in Chapters 5 and 10. Design elements that must be considered include:

- Keeping bathroom plumbing away from bedroom walls. If plumbing must be placed in a

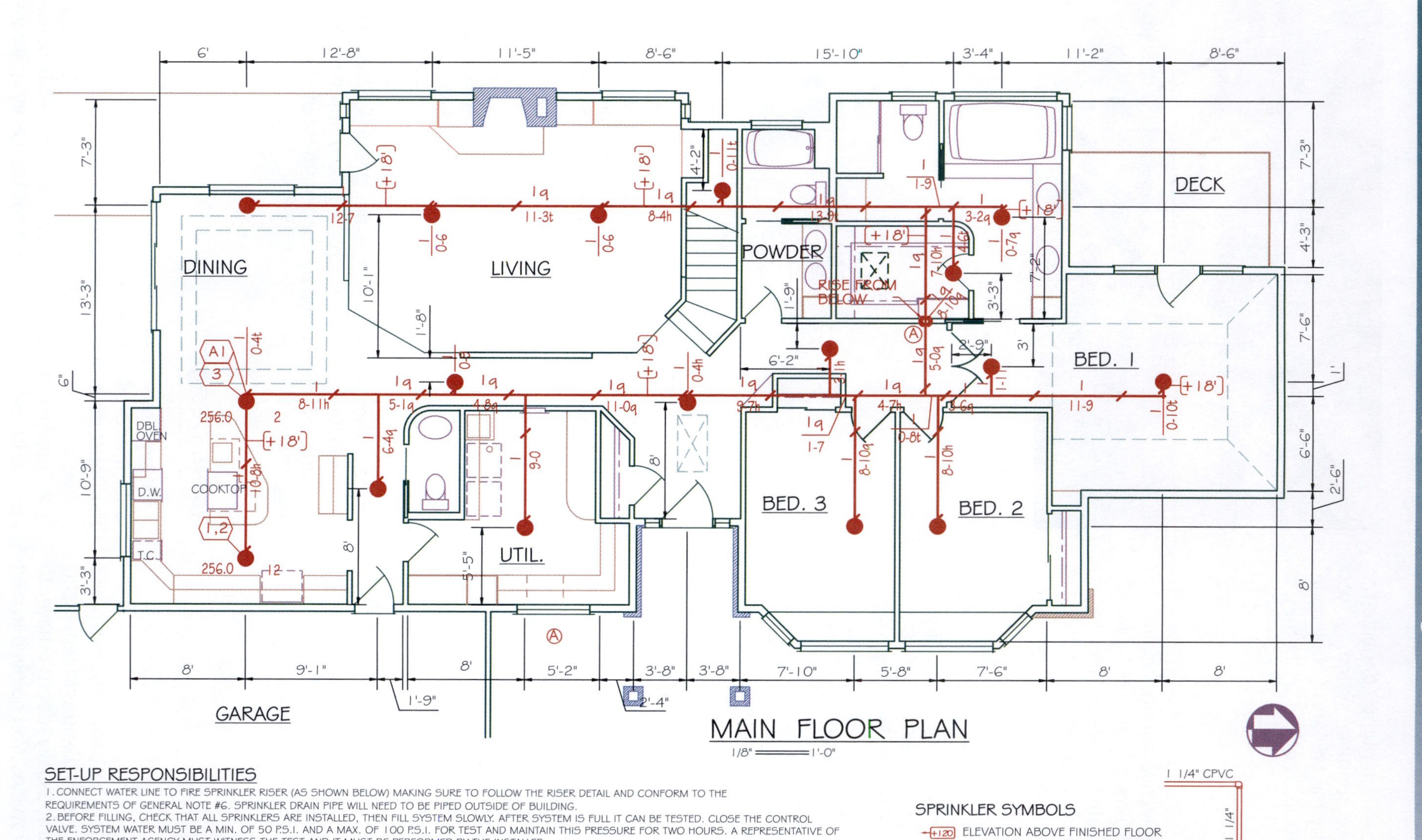

DECK
DINING
LIVING
POWDER
RISE FROM BELOW
BED. 1
BED. 2
BED. 3
UTIL.
GARAGE
DBL OVEN
D.W.
COOKTOP
T.C.
MAIN FLOOR PLAN
1/8" = 1'-0"
SET-UP RESPONSIBILITIES
1. CONNECT WATER LINE TO FIRE SPRINKLER RISER (AS SHOWN BELOW) MAKING SURE TO FOLLOW THE RISER DETAIL AND CONFORM TO THE REQUIREMENTS OF GENERAL NOTE #6. SPRINKLER DRAIN PIPE WILL NEED TO BE PIPED OUTSIDE OF BUILDING.
2. BEFORE FILLING, CHECK THAT ALL SPRINKLERS ARE INSTALLED, THEN FILL SYSTEM SLOWLY. AFTER SYSTEM IS FULL IT CAN BE TESTED. CLOSE THE CONTROL VALVE. SYSTEM WATER MUST BE A MIN. OF 50 P.S.I. AND A MAX. OF 100 P.S.I. FOR TEST AND MAINTAIN THIS PRESSURE FOR TWO HOURS. A REPRESENTATIVE OF THE ENFORCEMENT AGENCY MUST WITNESS THE TEST AND IT MUST BE PERFORMED BY THE INSTALLER.
3. THE SYSTEM FIRE ALARM CAN BE TESTED BY OPENING THE MAIN DRAIN. CLOSING THE DRAIN WILL SILENCE THE ALARM. AFTER ALL TESTS HAVE BEEN COMPLETED, THE SYSTEM CAN BE PUT IN SERVICE BY OPENING THE CONTROL FULLY. VALVE MUST REMAIN IN THE OPEN POSITION FOR THE SYSTEM TO BE OPERATIONAL.
GENERAL NOTES
1. FIRE SPRINKLER SYSTEM DESIGN. MATERIALS AND INSTALLATION PER NFPA 13D 2002
2. SPRINKLER SPACING TO BE A MAXIMUM OF 16'-0" BETWEEN HEADS AND 8'-0" OFF ALL WALLS WITH A MINIMUM DISTANCE OF 8'-0" BETWEEN HEADS
3. ALL PIPING TO BE CPVC U.N.O. PIPE DIMENSIONS ARE CENTER TO CENTER
4. HANGERS TO BE TOLCO CPVC HANGERS (AS SHOWN) AND TO BE SPACED AT A MAXIMUM OF 6'-0" BETWEEN HANGERS PER NFPA 13D
5. WHERE NECESSARY, FREEZE PROTECTION FOR PIPING MUST BE PROVIDED (BY OTHERS)
6. OWNER TO PROVIDE REQUIRED WATER SUPPLY TO OPERATE SYSTEM WITH A MINIMUM OF 36.0 GPM WITH 38.3 PSI AT THE RISER
7. CONNECTION, VALVES, AND PIPING BELOW FLOORLINE TO BE COMPLETED BY SITE CONTRACTOR
8. ALL PIPING TO RUN JUST ABOVE BOTTOM CHORD OF WOOD TRUSS
SPRINKLER SYMBOLS
ELEVATION ABOVE FINISHED FLOOR
ELEVATION BELOW BOTTOM OF DECK
RISER BETWEEN FLOORS
LINE NUMBER
WELD TAG
TOP OF STEEL
HYDRAULIC REFERENCE
SQ.FT./SPKLR REMOTE AREA #
RISER ELEVATION
SCALE:NONE
1 1/4" CPVC
1" DRAIN
WORK BEGINS HERE
1 1/2" DOMESTIC SERVICE

FIGURE 13.15 ■ *The upper level of the fire suppression drawings for the home started in Chapter 11.* Courtesy Michael Easley.

FIGURE 13.16 ■ *In planning a residence, the design team must consider where the rough plumbing will be placed. Space above the fixture for vent lines and below the fixture for drain lines must be carefully planned.* Courtesy Danielle Worcester.

bedroom wall, use insulated water pipes, cast-iron drain lines, or wall insulation to control noise.

- When possible, place plumbing fixtures back to back to save materials and labor costs.
- In designing a two-story structure, it is economical to place plumbing fixtures one above the other. If the functional design of the floor plan does not allow for such economies, arrange plumbing fixtures so that stack and vent lines are shared.
- Place the laundry room next to a bath or other plumbing trees.
- Never use pocket doors where the pocket is behind a plumbing fixture. Always verify that fresh, vent, and drain lines can be placed by inserting a fixture symbol on the floor plan.
- Carefully plan for the placement of vent lines when a window is located in the wall behind a plumbing fixture.

Perhaps the biggest design consideration involves the placement of bathrooms in a multilevel structure. Multilevel designs can present problems for the placement of drain and waste lines. Fresh-water lines are small enough that their placement is not a problem. Because of the drain line's large size and need to slope, the depth of the floor joists will dictate how far it can run before it connects with the riser. Most designers consider it tacky to place a riser in the middle of a family room on the lower floor. Exposed drainpipes could be a great place to hang coats, but they have little other appeal. Carefully plan where lower walls will be placed to hide drain lines on the floor below the fixture. Equally important is to plan where vent lines can be placed on floor levels above the fixture. Figure 13.16 shows the rough plumbing for a vanity with two sinks. You don't have to draw the pipes on the floor plan, but you must plan for them. Figure 13.17 shows the drawings provided by the plumbing contractor for the home started in Chapter 11.

Planning Below-Slab Work

When a concrete slab is used for the lower floor system, great care must be taken to locate plumbing when the foundation plan is drawn. Figure 13.18 shows the rough plumbing for back-to-back bathrooms in a concrete slab. Stacks that intersect with house sewer lines under the slab must be dimensioned accu-

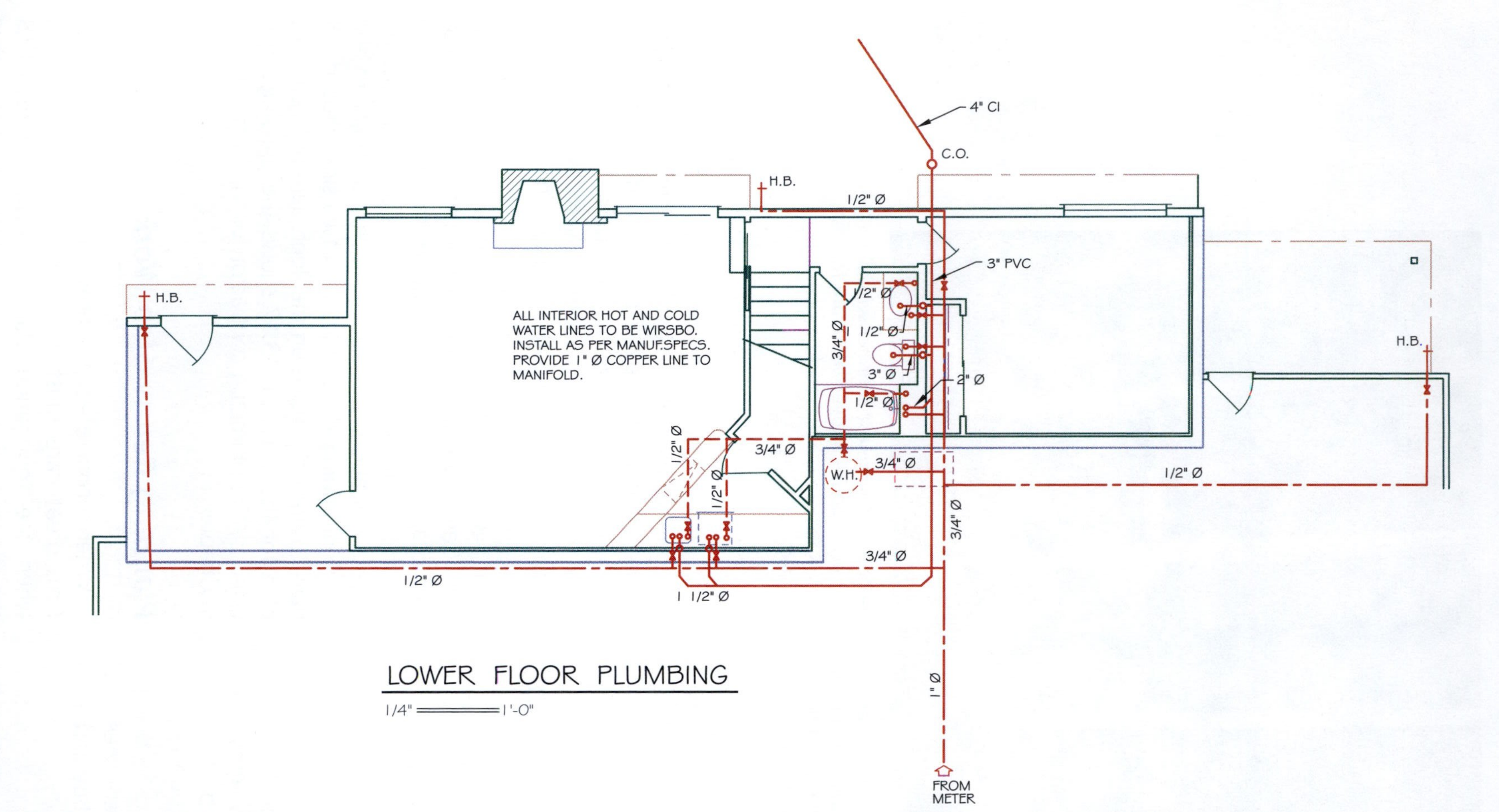

FIGURE 13.17 ■ *The fresh- and wastewater layout for the home started in Chapter 11.* Layout courtesy Tereasa Jefferis.

FIGURE 13.18 ■ *If plumbing will be placed in a concrete slab, the foundation plan will need to accurately dimension the location of all plumbing fixtures and drawings.* Courtesy Margarita Miller.

FIGURE 13.19 ■ *The architectural team must plan how lines can be placed where walls intersect the concrete stem walls.* Courtesy Sherry Benfit.

rately to ensure alignment of the stack pipe with the partition location.

When plumbing is placed in an exterior wall, you must also consider how drain lines are affected at the intersection of the wall, floor, and concrete stem walls. With a joist floor system, the depth of the joist usually provides enough depth to run pipes in the interior side of the stem wall. With a post-and-beam floor, the concrete needs to be notched. Figure 13.19 shows a notch placed in the stem wall to allow placement of the drainpipe. Poor planning by the design team often requires adjustments to be made by the plumber at the job site. Placement of plumbing over a concrete retaining wall is even more critical. If the plumbing is in a nonhabitable room, the pipes can be left exposed. If the pipes enter habitable space, furring needs to be placed over the retaining wall to hide the plumbing. Normally 1x or 2x (25x or 50x) thick fur strips are used to hide the concrete basement walls. The furring thickness is often increased to 4'' (100 mm) if waste pipes need to be hidden.

Placing Plumbing in Wood Framing

A home with a wood floor system allows the plumber ample space to run fresh- and wastewater lines. Lines are often placed in the cavity formed between joists. When lines must run perpendicular to the floor joists, care must be taken so drilled holes do not weaken the joists. The IRC places the following limits on notches placed in sawn joists and beams:

- The notch depth cannot exceed 1/6 of the joist depth
- The notch length cannot be longer than 1/3 the depth of the member being notched.
- The notch cannot be placed in the middle third of a joist span.
- Notches placed at the end of a joist cannot exceed 1/4 of the joist depth.
- The tension side of a member 4'' (100 mm) or wider in thickness shall not be notched except at the ends of the joists.
- Holes bored into joists shall not exceed 1/3 of the joist depth.
- Holes cannot be within 2'' (50 mm) of the top or bottom of the joists.
- Holes must be placed a minimum of 2'' (50 mm) apart from other holes and notches.

Although plumbers are usually aware of these requirements, these notes should be placed on the framing plan to ensure quality construction. Figure 13.20 shows the results of plumbing notches that were placed by an uninformed plumber.

Wood walls will hide most vent and drain lines, but walls hiding the vent lines for toilets need special attention. Plumbing walls that hide toilet vents are usually framed with 2 × 6 (50 × 150) studs to hide plumbing for toilets on the lower level of multilevel structures. Pipes for toilets on a single-level home and other plumbing fixtures can be placed in a 2 × 4 stud wall, but 2 × 6 studs are also used to aid plumbers.

FIGURE 13.20 ■ *The location of notches and holes in joists is highly regulated by the IRC. Notes to specify notches and holes should be placed on the framing plan.* Courtesy Floyd Miller.

CHAPTER 13

Additional Readings

The following websites can be used as a resource to help you keep current with changes in plumbing materials.

ADDRESS	COMPANY OR ORGANIZATION
www.allaroundthehouse.com	All Around the House
www.americanstandard.com	American Standard
www.aquaglass.com	Aqua Glass (Bathing fixtures)
www.homedepot.com	The Home Depot
www.homefiresprinkler.org	Home Fire Sprinkler Coalition
www.kohler.com	Kohler Company
www.lowes.com	Lowe's (home improvement)
www.nfpa.org	National Fire Protection Assoc
www.nibco.com	Nibco (Flow Control)
www.polybutylene.com	General information regarding polybutylene piping
www.wagner-solar.com	Wagner & Co. (European Solar Thermal Industry Federation)

CHAPTER 13

Plumbing Plans Test

DIRECTIONS

Answer the following questions with short, complete statements. Type the answers using a word processor or neatly print the answers on lined paper.

1. Place your name, the chapter number, and the date at the top of the sheet.
2. Type the question number and provide the answer in the form of a statement that includes part of the question. You do not need to write out the entire question.

Note: The answers to some questions may not be contained in this chapter and will require you to do additional research.

QUESTIONS

13.1. What plumbing drawings are required to get a plumbing permit?

13.2. Visit the website of your local building department and determine the requirements for residential fire sprinklers.

13.3. List the maximum length and rise for a residential supply line.

13.4. List three different types of pipes described by the IRC.

13.5. A gas-fired water heater will be placed in a pantry by the kitchen. List any requirements for this location.

13.6. Explain the difference and the relationship of drain and vent pipes.

13.7. Determine the physical size and output of the smallest and largest Takagai tankless water heater.

13.8. List a minimum of four requirements to place a water heater in an attic.

13.9. What is the minimum water main size allowed by the IRC?

13.10. List at least four factors that influence the sizing of water-supply pipes.

CHAPTER 14

Comfort Control Systems

Comfort control requires more than just providing air that is delivered at a comfortable temperature. True comfort includes providing clean, fresh, odorless air at the correct temperature and humidity. The demands of comfort control are met with a heating system, a cooling system, air filters, and humidifiers. The type of system to be used and the requirements of the building department will dictate whether drawings are required to obtain a mechanical permit. If required, the mechanical contractor who will install the system usually completes the drawings that show any required climate control equipment. As with the plumbing drawings, even though the architectural team does not supply the drawings, you must understand their contents to effectively complete the architectural drawings. Climate control plans show the system and the equipment used to maintain temperature, moisture, and the exchange and purification of the air supply.

Air temperature, movement, pollutants, humidity, and odors can all be controlled through the use of mechanical systems. Heating, ventilating, and air conditioning systems, also known as HVAC, include a wide variety of devices and delivery systems shown on the HVAC drawings. The most common types of systems designed to bring comfort control to a building are forced-air, hydronics, radiant, steam, active solar, and passive solar systems.

PRINCIPLES OF HEAT TRANSFER

A key element to understanding the HVAC drawings is to understand the basics of creating and transferring heat from one object to another. Heat inside a building is created not only by natural solar heat gained through roofs, windows, and walls but also by heat-producing equipment such as computers, television sets, and ovens. The occupants of a room also raise its temperature.

Methods of Heat Transfer

Whether heat is inside or outside a structure, it always travels from a warm surface or area to a cooler surface or area. Heat travels by radiation, convection, and conduction. ***Radiant heat*** travels in waves through the atmosphere in the same manner as light. All materials constantly radiate heat because molecules at their surface are constantly moving. Radiant heat travels from a heated surface to the cooler air surrounding the heated surface. An example of radiant heat is a burning coal on a barbecue. Long after the glow of the flame dies, the coal still radiates heat.

With ***convection,*** heat travels from a heated surface to the molecules of liquids or gases surrounding that surface. Convection occurs when heat transfers from a heated surface to a fluid moving over the heated surface or is transferred by molecules in a fluid from one heated molecule to another. Hydronic systems pass heated liquid through tubes buried in the floor. These tubes pass heat from the liquid to the concrete floor and then through radiation to the surrounding air. The heated air rises and cool air moves in to take its place, causing a convective current.

Each material that is struck by the sun's radiation absorbs some solar radiation. As a material absorbs this radiation, heat is conducted between its molecules. The denser the material, the better the conduction rate. Many solar applications depend on heat absorbed into dense materials and radiated back into the room as the air around the surface cools. This principle also works in reverse. Materials with tiny air pockets will not transfer heat well. The use of materials such as insulation, with thousands of tiny air pockets, slows the transfer of heat from heated to cold surfaces.

Heat Measurement

The standard unit of measurement for heat production or loss is the British thermal unit ***(Btu).*** The metric unit of heat measurement is joules (J). Btu are converted to joules by multiplying the Btu value by 1055. A Btu is a measure of heat generated. ***Thermal conductivity*** is the measure of heat flow. Thermal conductivity is the amount of heat that flows from one face of a material to the opposite face. Materials that transfer heat easily are known as conductors. Materials that resist the transfer of heat are known as insulators. The effectiveness of a material to resist heat transfer is indicated as its R value. Thermal resistance is the reciprocal of thermal conductivity.

R-Values and U-Values

The R-value of a material provides a uniform method to rate the resistance of heat flow through it. The higher the R-value, the greater the ability of a material to resist heat transfer to another material. Most major building materials have been tested and assigned an R-value. Common R-values include:

CONCRETE/MASONRY R-RATING	
4'' brick	.44
1'' stone	.08
1'' stucco	.20
1'' poured concrete	.08
8'' concrete block	1.04 (hollow)
8'' concrete block	1.93 (filled)
WOOD	
1'' soft wood	1.25
1'' hardwoods	.91
1'' plywood	1.25
1'' poured concrete	.08
5/8'' particle board	.82
1'' wood fiberboard	1.93
SIDING	
Aluminum siding	.61
Beveled wood	.81
Building paper	.06
Vinyl siding	1.00
Wood shingles	.87
INSULATION	
1''' Glass fiber batt	3.13
1'' Blown cellulose	3.40
1'' Expanded polystyrene	3.85
1'' Expanded polyurethane	6.64
1'' Extruded Polystyrene	4.92 (Styrofoam blue board)
ROOFING	
Built-up	.33
Fiberglass shingles	.44
Slate roofing	.05
Wood shingles	.94

When building materials are combined in layers, the sum of their R-values becomes the total R-value for the component. Figure 14.1 shows the R-value for a 2 × 6 (50 × 100) stud wall.

The U-value is the reciprocal (1/R) of the R-value and indicates the combined thermal conductivity of all materials in a structure, including air spaces. The U-value is the amount of heat conducted in 1 hour through a 1-sq-ft area for each Fahrenheit degree of difference in temperature between inside and outside air. High R-values and low U-values indicate greater efficiency. Because different climates and seasons require different R-value levels to maintain the desired indoor temperature, R-values must be chosen for the average low temperature of a geographic area. Figure 14.2 provides examples of the required R-values to maintain various temperatures.

Windows and doors account for the greatest heat loss in cold climates. The door surface and core mater-

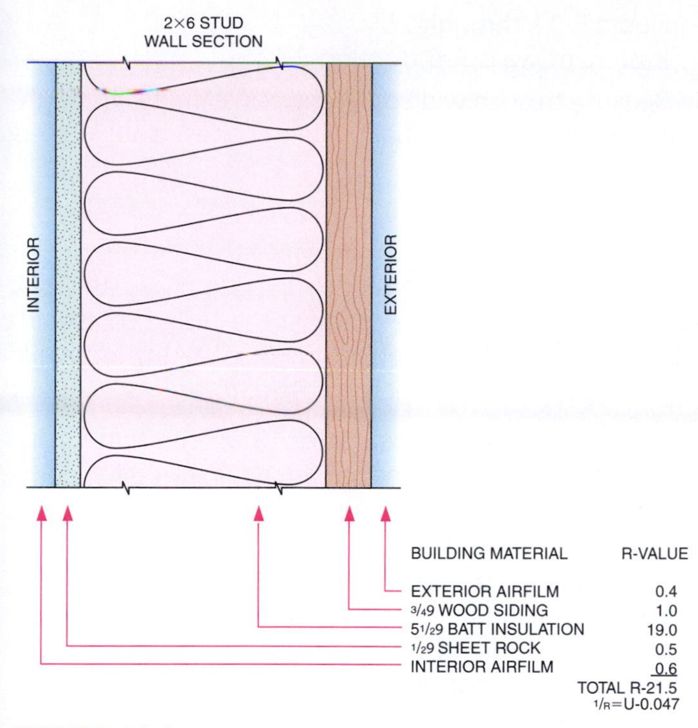

FIGURE 14.1 ■ *When building materials are combined, the sum of their R-values becomes the total R-value for the component. The combined R-value for a 2 × 6 (50 × 100) stud wall is 21.7 with a U-value of .047.*

OUTDOOR TEMP	INDOOR SURFACE TEMPERATURE				
	COOL 60°F	FAIR 64°F	MEDIUM 66°F	WARM 68°F	MIN FOR FLOOR
+30°F	R-2.3	R-3.4	R-5.1	R-10.0	R-1.7
+20°F	R-2.8	R-4.2	R-6.4	R-12.5	R-2.2
+10°F	R-3.4	R-5.1	R-7.8	R-14.5	R-2.6
0°F	R-3.9	R-6.0	R-9.2	R-17.0	R-3.0
−10°F	R-4.4	R-6.8	R-10.1	R-20.0	R-3.4
−20°F	R-5.1	R-7.8	R-11.3	R-23.0	R-3.9
−30°F	R-5.7	R-8.4	R-12.8	R-25.0	R-4.4
−40°F	R-6.4	R-10.2	R-14.5	R-28.0	R-4.8

FIGURE 14.2 ■ *Required R-values needed to maintain various indoor temperatures based on the outdoor temperature.*

MATERIAL	U-FACTOR		R-VALUE	
	COLD CLIMATE (WINTER)	WARM CLIMATE (SUMMER)	COLD CLIMATE (WINTER)	WARM CLIMATE (SUMMER)
SINGLE GLASS	1.13	1.06	0.88	0.94
INSULATED GLASS				
1/4″ Air space	0.65	0.61	1.54	1.64
1/2″ Air space	0.58	0.56	1.72	1.79
STORM WINDOWS 1″–4″ Air space	0.56	0.54	1.79	1.85
LOW EMITTANCE 1/2″ Air space				
$\varepsilon = .20$	0.32	0.38	3.13	2.63
$\varepsilon = .60$	0.43	0.51	2.33	1.96
GLASS BLOCK				
6″ × 6″ × 4″	0.60	0.57	1.67	1.76
12″ × 12″ × 4″	0.52	0.50	1.92	2.00

FIGURE 14.3 ■ *Comparative R- and U-values for various glass products.*

ial greatly affect R- and U-values. Two common materials for exterior doors include:

- Solid core wood R = 2.3 U = 0.43
- Metal with urethane core R = 13.5 U = 0.07

Heat flows through windows in both directions through radiation, convection, and ***conduction.*** Windows with a sealed space between double layers of glazing greatly reduce heat loss. Filling the air space between the glazing with argon or krypton gas will further reduce heat loss. Low-E glass has a transparent coating that acts as a thermal mirror to increase the insulating value. The coating reflects heat energy, which is invisible solar radiation, but allows the transmission of visible light. Most codes now require low-E glazing because of its increased efficiency. These measures increase R-values and decrease U-values. Figure 14.3 compares R- and U-values for various glass products.

Doors and windows also lose heat by air ***infiltration.*** Infiltration is the flow of air through poorly sealed building intersections. Infiltration is greatly reduced and energy efficiency increased by proper use of caulking and ***insulation.*** The DRAWING BLOCKS\FLOOR BLOCKS\NOTES folder of the student CD contains common caulking and insulation notes that can be added to a set of plans. These notes can be inserted on the floor plan, on a sheet containing all general notes, or on any sheet that space allows. If these notes are not located on the floor plan, be sure to list the notes in the title block and on a table of contents on the title sheet. Edit the standard notes to meet code and climate requirements for your area.

Insulation

The IRC refers to the exterior of a home as the building envelope. The insulation, outer walls, ceiling, doors, windows, and floors work together to control airflow in and out of the structure, repel moisture, and prevent heat from being lost or gained. A high-performance envelope maintains a consistent temperature even under extremely hot or cold conditions. Insulation improves the home envelope to make homes more comfortable and energy-efficient. Without insulation an HVAC system must work harder to overcome the loss of treated air through the walls, floors, and ceilings.

Insulation is any material used to slow the transfer of heat. Figure 14.4 shows where insulation and caulking should be applied to a home. Determining the required amounts of insulation to use in a home will depend on the IRC or other applicable codes, the climate, energy cost, and personal desires of the homeowner. Moderate zones of the country require the following minimum values:

Ceilings R-30 through 38
Walls R-13 through 21
Floors R-11 through 25
Basement walls R-5 through 21

Because of such wide differences in climates throughout the country, individual building codes should determine specific needs.

Types of Insulation

Common insulation materials include fiberglass, rock wool, cellulose, urethane foam, and recycled cotton. Insulation is available in the form of blanket, loose fill, rigid board, and expanding spray foam.

Batts and Blanket Insulation: The common form of batts and blanket insulation is roll insulation made of fiberglass fibers produced in widths suitable to place between standard framing members. The thickness of the insulation determines its R-value. Batts and blankets are available with or without vapor-retarded facings. Figure 14.5 shows the application of batt insulation.

Loose-Fill Insulation: Loose-fill insulation consists of fibers or granules primarily made from cellulose, fiberglass, rock wool, or cotton materials. Loose-fill

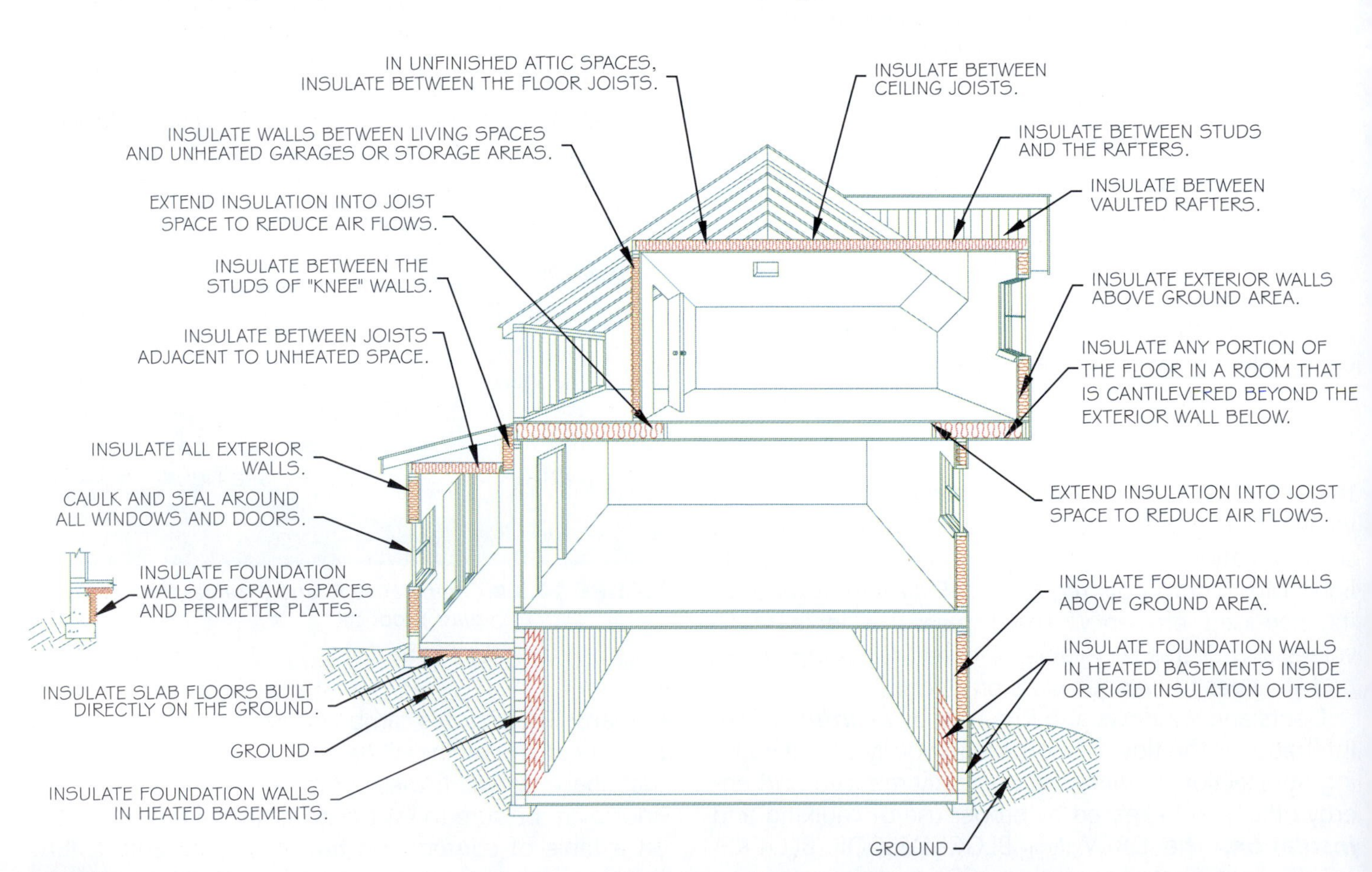

FIGURE 14.4 ■ *Common locations where caulking and insulation should be applied. It is up to the architectural team to specify the location and type of each type of caulking and insulation.*

FIGURE 14.5 ■ *Batt insulation is applied in most new homes because of its ease of installation and low cost.* Courtesy CertainTeed.

insulation is blown into areas and cavities, as shown in Figure 14.6. Used for attic installations, blown-in insulation allows the material to fill the cavity where it is applied. A disadvantage of using loose-fill insulation is that it settles over time, leaving an uninsulated area at the top of the cavities.

Rigid Insulation: Rigid insulation is made from plastic foams and formed into sheets. The thickness varies based on the desired R-value. Rigid foam insulation is primarily used to insulate foundation walls and footings and vaulted ceilings as an overlay to roof decking. When used to insulate foundation walls, a protective layer of concrete board must be

FIGURE 14.6 ■ *Blown-in insulation is a popular choice for open attic spaces, but it tends to compress over time.* Courtesy CertainTeed.

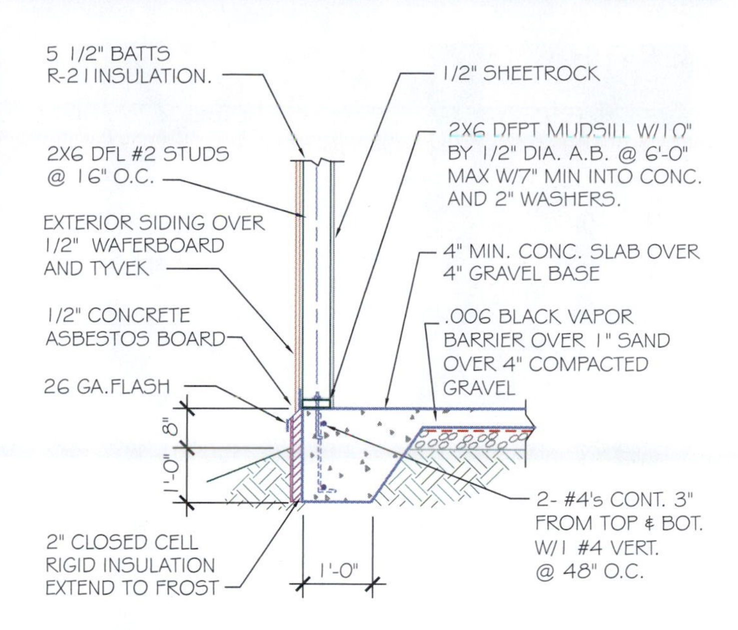

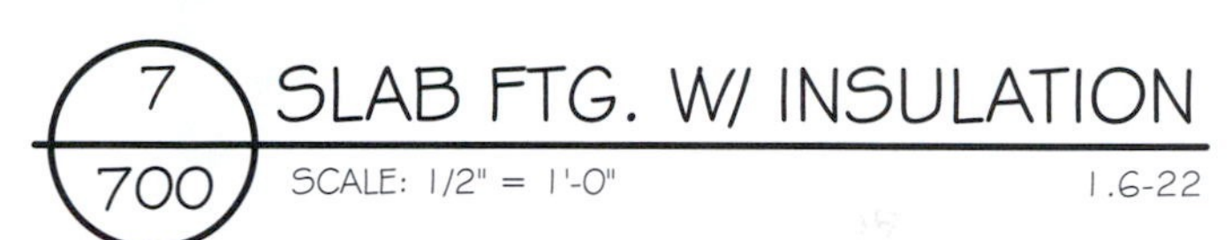

FIGURE 14.7 ■ *Details are usually required to describe where the insulation will be placed and how it will be protected from damage.*

used at exposed areas. Figure 14.7 shows a detail that describes the use of rigid insulation for a foundation application. See Chapter 26 for additional review of foundation insulation.

Spray-in-Place Insulation: Thin layers of polyurethane spray-in foam insulation provide the highest R-value per inch of current insulation materials. With a value of R-7 per inch, the material is sprayed into an open cavity and hard-to-access areas, where it expands to 100 times its original volume, filling the cavity and any existing cracks. Excess foam is trimmed away when it hardens (see Figure 14.8).

HEATING SYSTEMS

Conventional heating methods use a variety of devices to produce heat, including forced-air, hot-water, steam, electrical, heat-pump, and solar systems. Heat pumps are also used for cooling. In addition to the method to produce heat, the drafter must consider the method of distribution. Heat is usually distributed throughout a residence by ***ducts,*** tubes, ***plenums,*** pipes, or wires. Round, square, or rectangular ducts move both heated and cooled air. Typically sheet metal,

FIGURE 14.8 ■ *Spray-in foam insulation provides the highest R-value per inch of all current insulation materials.* Courtesy Icynene Insulation System.

flexible foil-covered fiberglass, and the wood of the framing system are used for ducts. Copper pipes carry hot water or steam to radiators within each room. Flexible plastic tubes are buried in concrete slabs to circulate heated water to warm the concrete. ***Radiant heating*** systems rely on electrical resistance wires embedded in ceilings or floors or connected to floor or wall convection units.

Before any HVAC systems can be considered, you must understand the symbols used by the mechanical contractor to represent HVAC components. Figure 14.9 shows common symbols used on HVAC drawings. An understanding of these symbols will give you a better understanding of the drawings completed by the mechanical contractor and help you coordinate the architectural drawings with the mechanical drawings. These symbols show the location and type of equipment. Arrows show the movement of hot air, cold air, and water. HVAC symbols are available in the DesignCenter of AutoCAD and are also available from third-party vendors to improve productivity. Custom CAD programs are also available to aid in completing HVAC drawings.

Forced-Warm-Air Furnace Units

Warm-air furnace units operate using either gravity or forced air. The furnace for a ***gravity system*** must be located below the lowest level that will be heated. ***Forced-air systems*** heat air in a central furnace and then use a fan to force air through ducts to distribute the heated air to ***diffusers*** placed throughout the residence. Forced-air units are represented and labeled FAU on the floor and electrical plans, but the ducts and diffusers are not represented on either plan. If a mechanical plan is drawn, the unit and the ducting are represented. The complexity of the system dictates whether a drawing will be included with the construction documents. Otherwise the installer must work with drawings or sketches created by the mechanical contractor. Figure 14.10 shows the drawings for the

NAME	ABBREV	SYMBOL	NAME	ABBREV	SYMBOL
DUCT SIZE & FLOW SELECTION	DCT/FD	109 × 159	HEAT REGISTER	R	R
DUCT SIZE CHANGE	DCT/SC		THERMOSTAT	T	T
DUCT LOWERING	DCT/LW	D D	RADIATOR	RAD	RAD
DUCT RISING	DCT/RS	R R	CONVECTOR	CONV	CONV
DUCT RETURN	DCT/RT		ROOM AIR CONDITIONER	RAC	RAC
DUCT SUPPLY	DCT SUP	S	HEATING PLANT FURNACE	HT PLT FUR	FURN
CEILING-DUCT OUTLET	CLG DCT OUT		FUEL-OIL TANK	FOT	OIL
WARM-AIR SUPPLY	WA SUP	WA	HUMIDISTAT	H	H
SECOND-FLOOR SUPPLY	2nd FL SUP		HEAT PUMP	HP	HP
COLD AIR RETURN	CA RET	CA	THERMOMETER	T	T
SECOND-FLOOR RETURN	2 FL RET		PUMP	P	
GAS OUTLET	G OUT	G	GAGE	GA	
HEAT OUTLET	HT OUT		FORCED CONVECTION	FRC CONV	

FIGURE 14.9 ■ *Common climate-control symbols found on HVAC drawings.*

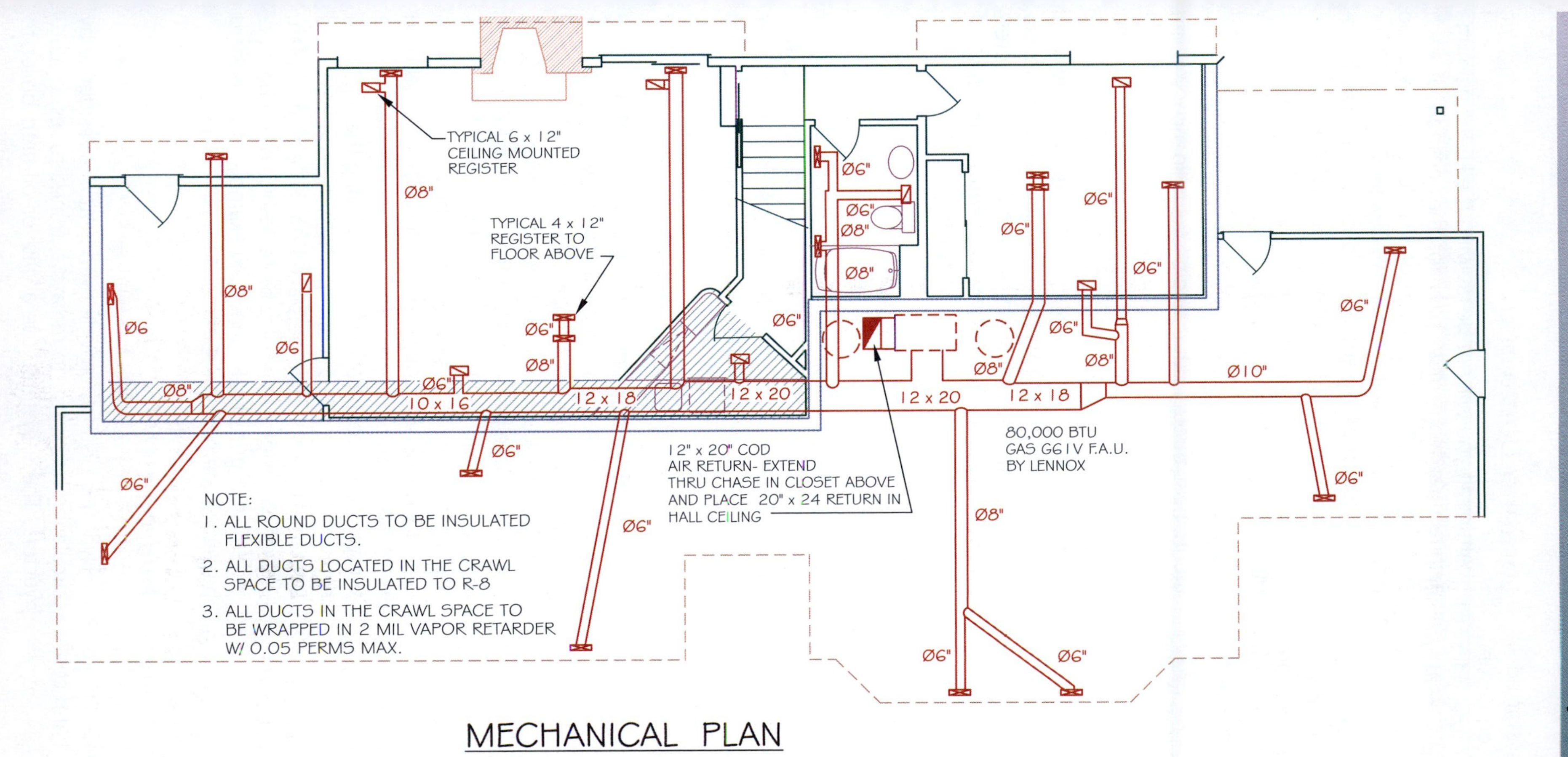

FIGURE 14.10 ■ *The forced-air plan for the home drawn in Chapter 11. CAD technicians working for the mechanical contractor generally draw HVAC drawings.* Courtesy CertainTeed.

forced-air system for the home that was started in Chapter 11. The advantages of using a forced-air unit are that air-cooling systems can use the same ductwork and also that filters and dehumidifiers can be built into the system to be used in the heating and cooling cycles.

Residential furnaces produce heat by burning fuel oil, natural gas, or by using electric heating coils. If heat comes from burning fuel oil or natural gas, the combustion takes place inside a combustion chamber. Air to be heated absorbs heat from the outer surface of the chamber. The combustion gases are vented through a chimney or metal ducting that extends through the roof. When a furnace is powered by electricity, cool air is heated as it passes over the heating coils. Electric furnaces do not require exterior venting.

In the heating cycle, heated air is forced through the system to ducts that lead to floor outlets. Heat outlets are usually placed in front of windows in outside walls. At least one outlet should be provided for each 15' (4500 mm) of exterior wall space. Where heating ducts also serve an air-conditioning system, the locations of the inside air handlers, exhaust hoses, and outside compressor units must be represented. Ceiling locations are preferred for cooling, but separate duct systems are rarely used for heating and cooling.

In addition to the distribution ductwork, forced-air systems use a second set of ducts to bring return air to the furnace to be warmed. A home with an open layout does not require return ducts. Instead, return air moves directly to the return side of a furnace. Forced-air distribution ducts connect to a plenum chamber. This is an enclosed space located between the furnace and distribution ducts. The plenum is larger than any duct and slows the flow of air through the ducts.

The location, size, and Btu capacity of each duct is shown on an HVAC plan. Each duct is represented on the plan by outlining the position of the ducts and providing a notation such as 6/18 (150/450) to represent a duct 6 × 18'' (150 × 450 mm). A single number by a duct represents the diameter. Other notations that might be shown by a duct include:

WA: warm air
RA: return air (cold air)
CFM: cubic feet per minute

The size of all registers is also listed on the drawing. Because vertical ducts pass through the plane of projection, diagonal lines are used to indicate the location of vertical ducts. HVAC plans will also show the location of all control devices, outlets, pipes, and heating and cooling units.

Locating Ductwork

Although you will not be drawing the HVAC plan, the architectural team must plan for the location of each duct. You don't plan the size of the ducts, but you do plan the space available for them. Review Chapter 10 for a discussion of where and how a duct or chase can be located on the drawings created by the architectural team.

Wood Plenum System

Instead of ducts, a plenum system may be used to distribute heat throughout a single-level residence. Plenum systems are based on a simple concept that has been in use since the Romans. The entire crawl space is sealed and used to distribute treated air through floor registers in rooms above the crawl space. This type of system requires the foundation walls to be sealed and insulated. The fan on the forced-air units maintains slight air pressure in the plenum. This ensures a uniform distribution of conditioned air throughout the residence.

Zone-Control Systems

An electric zoned heating system offers much greater flexibility than a central heating system. A zoned heating system requires one heater and one thermostat per room, allowing the heater in an unoccupied room to be turned off. Common types of zone heaters include baseboard and wall-mounted units with fans. Baseboard units have an electric heating element that causes a convection current as the air around the unit is heated. The heated air rises into the room and is replaced by the cooler air that falls to the floor. Baseboard heaters should be placed on exterior walls under or next to windows or other openings. These units project a few inches into the room at floor level. Fan heaters are mounted in a wall recess. A resistance heater is used to generate heat, and a fan circulates the heat into the room. The location of baseboard and wall-mounted heaters and their thermostats can be shown on the electrical plan. Avoid placing the heaters in exterior walls to avoid reducing the insulation. Wall and baseboard units should be placed on the electrical plan. Each thermostat should also be represented on the electrical plan. Figure 14.11 shows an example of baseboard heaters in the home in Chapter 11.

Hydronic Units

Hydronic systems provide even heat in a residence by circulating hot water to radiators, finned tubes, or convectors. These systems use a gas- or oil-fired boiler to

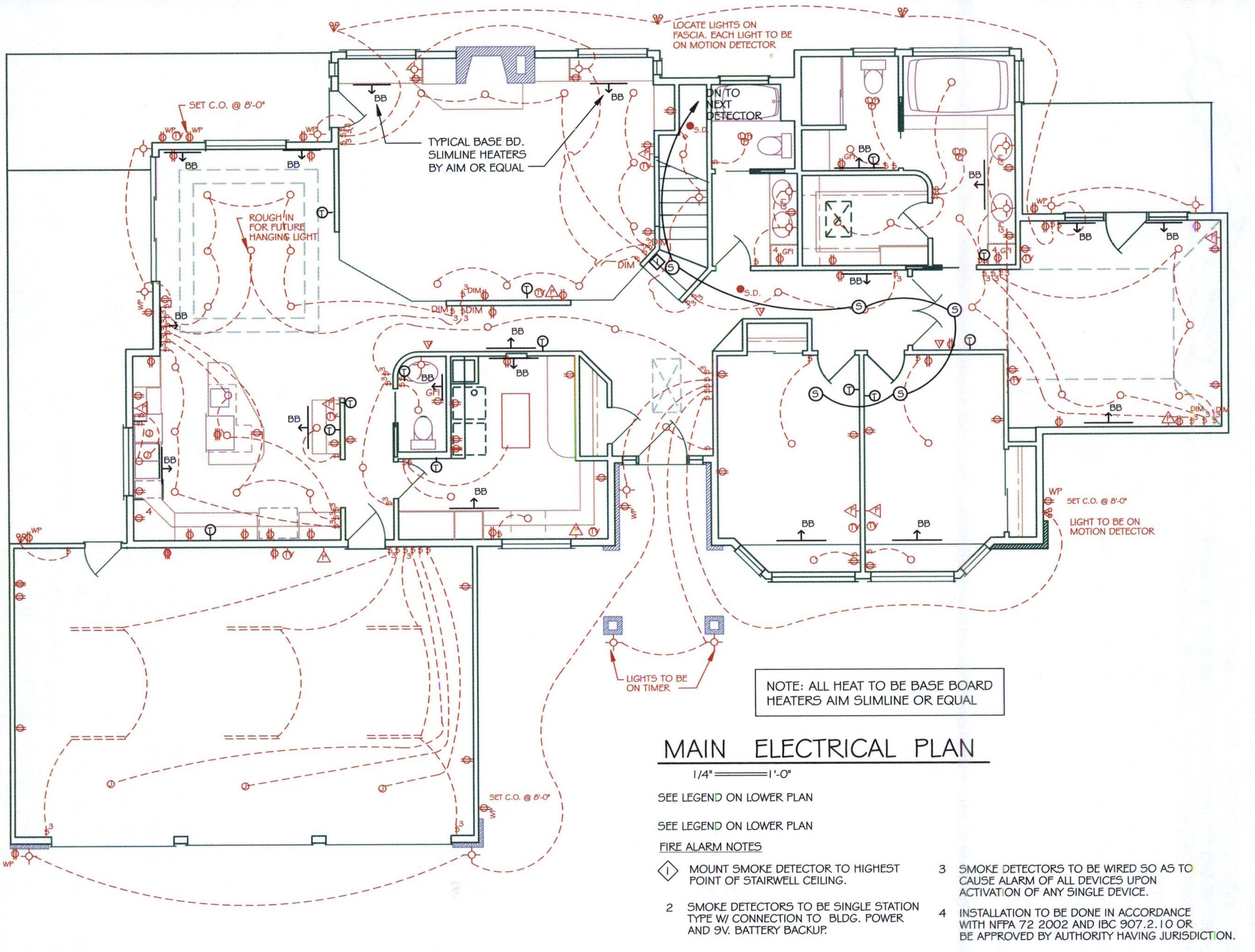
LOCATE LIGHTS ON FASCIA. EACH LIGHT TO BE ON MOTION DETECTOR
SET C.O. @ 8'-0"
TYPICAL BASE BD. SLIMLINE HEATERS BY AIM OR EQUAL
ON TO NEXT DETECTOR
ROUGH IN FOR FUTURE HANGING LIGHT
S.D.
SET C.O. @ 8'-0"
LIGHT TO BE ON MOTION DETECTOR
LIGHTS TO BE ON TIMER
SET C.O. @ 8'-0"
NOTE: ALL HEAT TO BE BASE BOARD HEATERS AIM SLIMLINE OR EQUAL
MAIN ELECTRICAL PLAN
1/4" = 1'-0"
SEE LEGEND ON LOWER PLAN
SEE LEGEND ON LOWER PLAN
FIRE ALARM NOTES
1 MOUNT SMOKE DETECTOR TO HIGHEST POINT OF STAIRWELL CEILING.
2 SMOKE DETECTORS TO BE SINGLE STATION TYPE W/ CONNECTION TO BLDG. POWER AND 9V. BATTERY BACKUP.
3 SMOKE DETECTORS TO BE WIRED SO AS TO CAUSE ALARM OF ALL DEVICES UPON ACTIVATION OF ANY SINGLE DEVICE.
4 INSTALLATION TO BE DONE IN ACCORDANCE WITH NFPA 72 2002 AND IBC 907.2.10 OR BE APPROVED BY AUTHORITY HAVING JURISDICTION.

FIGURE 14.11 ■ *The location of baseboard heaters, wall heaters, and thermostats can be shown on the electrical plan.*

heat fresh water, which is circulated around a combustion chamber, where it absorbs heat. A pump then circulates water heated to temperatures ranging between 150°F and 180°F throughout the system. When heat is needed, a thermostat starts the circulator, which supplies hot water to the room convectors. Although these systems are efficient for heating, they do not provide air filtration or circulation and are not compatible with cooling systems that use air ducts.

Baseboard Heaters

The most common and effective type of hydronic heater is the baseboard outlet. Baseboard units provide most of their heat through convection, with a small amount of heat produced through radiation. Some hydronic system units eliminate the need for a separate water-heating unit by also providing hot water for the fresh hot-water supply and for home heating. Common types of residential hot-water systems include the series-loop system, the one-pipe system, the two-pipe system, and the radiant system.

Common conductors for each system include baseboards and radiators. Baseboard units are generally placed along exterior walls. Radiators are typically made of cast iron or decorative aluminum and provide more mass for heating than any other type of heating device. Heat is predominately provided through convection. Drawings for each system resemble Figure 14.11, showing each heater. The locations of the delivery tubes are not shown.

Series Loop Systems

The series-loop system shown in Figure 14.12 is a continuous loop of pipes containing hot water. Hot water flows continually from the boiler through conductors and then back to the boiler for reheating. The heat in a series-loop system cannot be controlled except at the source of the loop. The entire loop is either conducting heat or is turned off.

One-Pipe Systems

One-pipe heating systems circulate heated water through a continuous pipe loop. Individual radiators connect to the circulating loop using a bypass loop. This system allows individual heating units to be controlled by valves at the radiator. Once the water has been routed through a heating unit, it is routed back to the main loop, through other heating units on the basis of demand, and then back to the boiler to be reheated. Figure 14.13 shows an example of a one-pipe system.

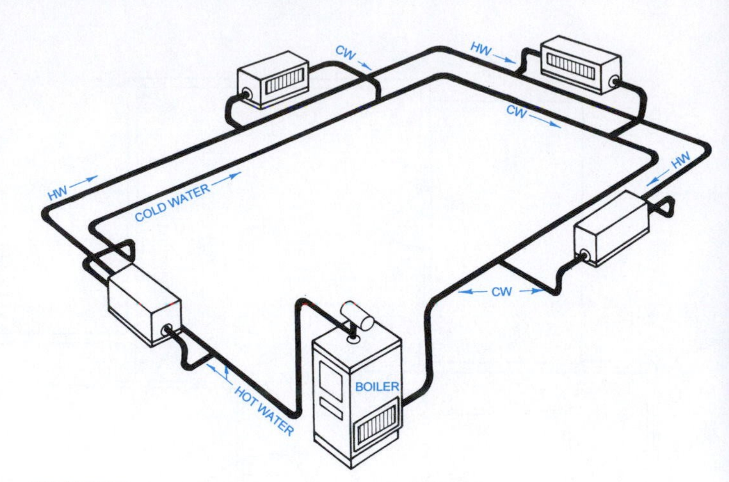

FIGURE 14.12 ■ *The series-loop system is a continuous loop of pipes that circulates hot water to heating units. Hot water flows continually from the boiler through conductors and then back to the boiler for reheating.*

Two-Pipe Systems

A two-pipe system uses two parallel pipes to route water through the heating system. One pipe forms a loop used to supply heated water to each convector, allowing each heating unit to be used as needed. The heating loop circulates water to each heating unit and routs unused water back to the boiler to maintain the water temperature. Once water passes through a conductor rather than being routed back into the heated line, it exits the heater into the second pipe of the system. The parallel pipe routes cooled water back to the boiler rather than to the next conductor. Figure 14.14 shows an example of a two-pipe heating system.

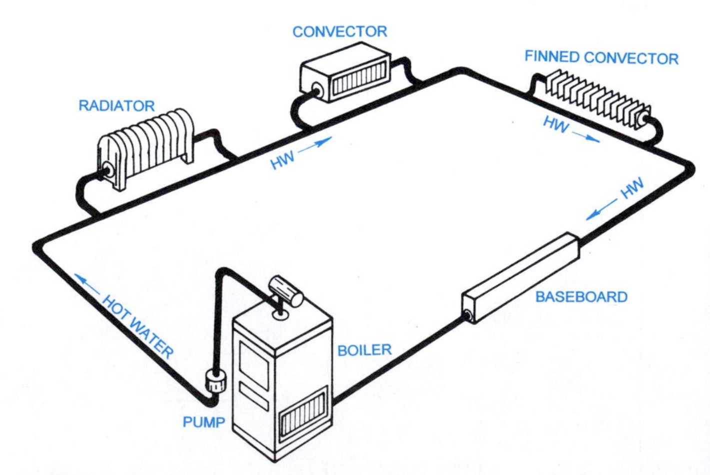

FIGURE 14.13 ■ *One-pipe heating systems circulate heated water through a continuous pipe loop. Individual heaters are connected to the circulating loop using a bypass that allows each heater to be controlled.*

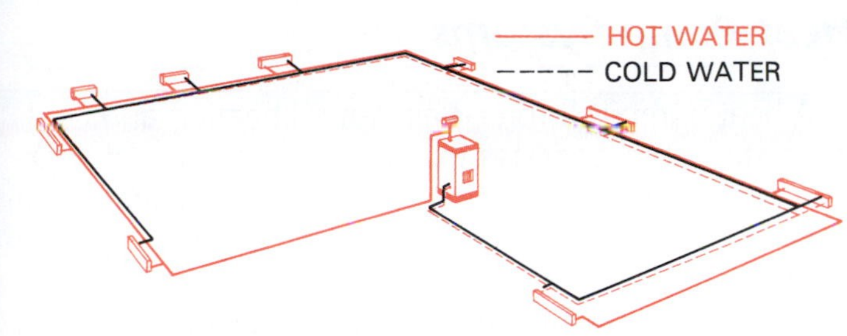

FIGURE 14.14 ■ *Two-pipe hydronic systems use one set of pipes to deliver heat to each heater and another set of pipes to return cooled water to the boiler.*

Radiant Systems

Radiant heating systems were first used by the Romans to provide heat to floors. Radiant heating distributes hot water through a series of continual tubes placed in the floor system, turning the whole floor system into a radiator. Tubes can also be placed in ceilings and walls to provide supplemental heating in colder climates when the floor system cannot match the heat loss of a room. Tubing is generally placed only in the lower 42'' (1050 mm) of a wall to avoid interference with window framing and possible punctures when objects are hung on the wall. Figure 14.15 shows an example of wall-mounted tubing.

A radiant floor system consists of a heating unit that warms the water and pipes laid on a concrete base that is then covered by a concrete slab. Heating can be provided by gas, oil, electricity, geothermal heat, wood, or pellet-burning boilers. The hot pipes conduct heat to the surface, where convection currents take over. The system can be divided into zones so that heat can be provided to rooms as desired. Figure 14.16 shows a radiant hot water system prepared for the pouring of the concrete slab.

Steam Units

A steam-heating unit uses a boiler to make steam that is transported through pipes to radiators and baseboard heaters producing heat through convection. After passing the heating unit, the steam condenses to water and is returned to the boiler to be reheated to steam. Although steam-heating systems function on water vapor rather than hot water, drawings for steam systems are identical to those prepared for hot-water systems. Steam-heated radiators are most likely to be found on larger multifamily projects located in cold climates and in remodeling projects involving older homes.

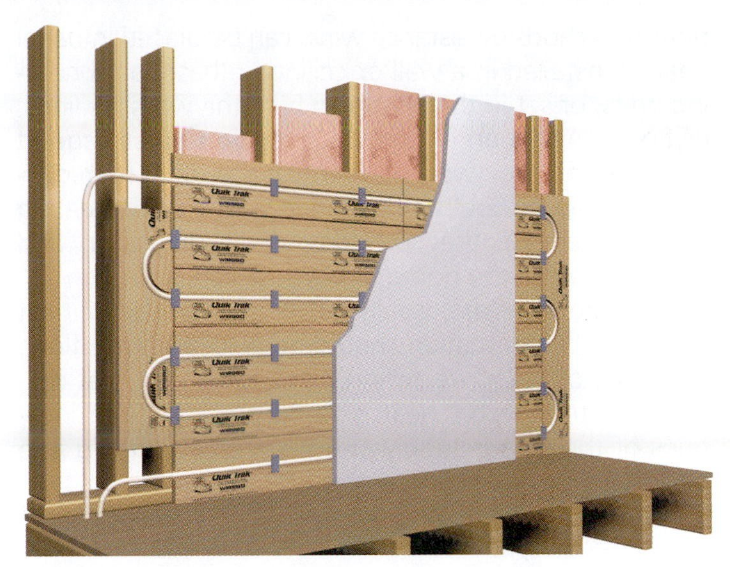

FIGURE 14.15 ■ *Radiant tubes can also be placed in ceilings and walls to provide supplemental heating in colder climates when the floor system cannot match the heat loss of a room.* Courtesy Lisa Kotasek, Uponor Wirsbo.

FIGURE 14.16 ■ *Radiant tubes laid in preparation for the pouring of a concrete slab. The tubes will circulate heated water through the concrete. The concrete floor will then act as a radiator. Notice that the individual lines in the upper right corner lead to different zones, which will allow individual control of each zone.* Courtesy Lisa Kotasek, Uponor Wirsbo.

Electric Resistance Heat

Electric heat produces a very clean heat by passing an electrical current through resistance wires. The heat that is produced is usually radiated using similar methods that were used with radiant floor heat or through fan-blown methods. Resistance wires can be placed in panel heaters installed in a wall or ceiling, in baseboard heating units, or set in the plaster to heat the walls, ceilings, or floors. Although electric heat requires no storage of fuel and no ductwork, a ventilation- and humidity-control system should be provided with electric heating systems because they provide no air circulation and cause the air in the home to become very dry.

Separate plans are usually not drawn for electric heat systems, but annotation should be placed on the floor, electrical, or reflective ceiling plans to explain that ceiling-mounted electric heat is to be provided. The locations of the power supply and thermostats should also be shown on the electrical plan.

COOLING SYSTEMS

Since heat can be transferred only from warm objects to cooler objects, buildings must be cooled by removing heat. To cool a building, warm air is carried away from rooms to an air conditioning unit, where a filter removes dust and other impurities. A cooling coil containing refrigerant absorbs heat from the air passing around it. Then the blower that pulls the warm air from the rooms pushes cooled air back to the rooms, using the same ducts and outlets used with the heating systems.

The size of air-conditioning equipment is rated in Btu. A 2000-sq-ft home can be comfortably cooled with a central air conditioning unit of 24,000 to 36,000 Btu. Larger homes may require 60,000 or more Btu, depending on the components specified. Drawings of combined systems use the same duct patterns and outlets used for the heating unit.

Heat-Pump Systems

A heat pump is a forced-air central heating and cooling system that operates using a compressor and a circulating refrigerant system. Heat is extracted from outside air and pumped inside the residence. In the summer the cycle is reversed and the unit operates as an air conditioner. In this mode, the heat is extracted from the inside air and pumped outside. On the cooling cycle, the heat pump also acts as a dehumidifier. Figure 14.17 shows how a heat pump works. Heat pumps may not be as efficient in some areas of the country as in other areas because of annual low or high temperatures. Verify the product efficiency with local vendors.

Residential heat pumps vary in size from 2 to 5 tons. During minimal demand, the more efficient 3-ton phase is used; the 5-ton phase is operable during peak demand. In planning the size of the unit, it can be assumed that each ton of the rating will remove approximately 12,000 Btu per hour of heat. The total heat-pump system uses an outside compressor, an inside blower to circulate air, a backup heating coil, and a complete duct system. All but the duct system is shown on the floor and electrical plans. The duct system is represented on the HVAC plan that is completed by the mechanical contractor. Show and specify the concrete pad on the foundation plan. Place the unit in a shady location where the noise will not cause a problem. The compressor can be placed on a fiberglass or concrete slab that provides approximately 6'' (150 mm) on all sides of the compressor. To avoid transmitting vibrations from the compressor into the residence, the compressor should be placed on a slab that is not attached to the home.

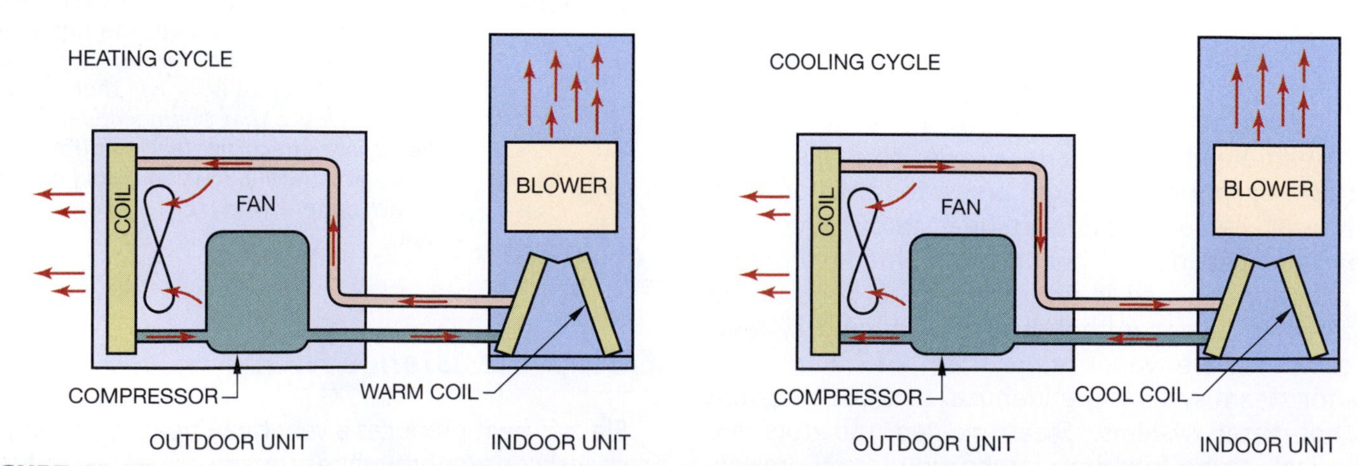

FIGURE 14.17 ■ *A heat pump is a forced-air central heating and cooling system that extracts heat from outside air and pumps the air to the inside of the residence. In the summer the cycle is reversed and the unit operates as an air conditioner.* Courtesy Lennox Industries, Inc.

IMPROVING AIR QUALITY IN THE HOME

In addition to the methods used to heat and cool the residence, some means must be provided to control each system. Systems must also be provided to ventilate, filter, and control the humidity in the residence.

Thermostatic Controls

Each of the systems mentioned in this chapter requires some method of control. A thermostat is an automatic mechanism to control the amount of air treatment provided by a heating or cooling system. Thermostatic controls keep buildings at a constant temperature by turning a climate-control system on or off to maintain a desired temperature range. The type of system used to control the air temperature determines the number of thermostats required. Central heating and cooling systems usually require only one thermostat for each unit servicing the structure. For zoned heating or cooling units, thermostats must be placed to control each zone; alternatively, a central thermostat to control each room may be placed in a convenient location. Systems using electrical heating and some water systems require that room be thermostatically controlled.

The location of thermostats is shown on the electrical plans with a thermostat symbol (see Figure 14.11). Annotation should be placed on the electrical plan to specify that the thermostats are to be placed between 36 and 48'' (900 to 1200 mm) above the floor level. Since thermostats are sensitive to heat and cold, they should be located on interior walls, away from sources of heat or cold such as fireplaces or windows. A location near the center of the residence and close to a return air duct is an ideal location because the temperate air entering the air return will cause only minor temperature variation at the thermostat. A location free from drafts caused by an exterior door, direct sunlight from windows, or heat from a heat register should also be avoided. Avoid locations by stairwells to prevent convective air currents between floors and vibrations caused by traffic on the stairs.

Heat Recovery and Ventilation

In addition to the air temperature, comfort control requires providing well-ventilated clean air. One of the drawbacks of new construction is that many architects and designers spend much of their design time focusing on how to minimize air infiltration, so that they lose track of the need for proper ventilation. The tighter homes become because of super insulation, air wraps, and caulking at every point of entry, the more important good ventilation becomes to keep fresh air circulating and to remove pollutants. Ventilation also plays an important role in keeping the interior air humidity balanced. Air in a residence must be constantly circulated to reduce the risk of allergies, which can develop from exposure to the chemicals used in many building materials. Common problems that can be caused by poor ventilation include:

- High interior humidity levels can cause structural damage as well as health problems. In addition to natural humidity from the atmosphere, steam from showers, cooking and the home's occupants can produce as much as 1 gallon (3.8 liters) of water vapor per day.
- Formaldehyde is found in carpets, furniture, the glue used in plywood, particleboard, and some insulation products. Formaldehyde can cause eye irritation and respiratory problems.
- Carbon monoxide gas—due to incomplete combustion in gas-fired or wood-burning appliances—can be trapped in a residence and prove deadly to the inhabitants.
- The buildup of radon, a naturally occurring radioactive gas, can be a serious health risk. Structures built in areas with high radon levels must be properly ventilated to make sure that residents do not inhale this cancer-causing gas. Chapter 25 introduces specific methods to control the buildup of radon by venting the crawl space.

Blowers built into the heating and cooling units can be used to provide interior air movement. Ceiling fans and exhaust fans remove steam and odors from kitchens and bathrooms. Roof vents equipped with power ventilators reduce the buildup of heat in the attic and circulate air in attics and crawl spaces. The inclusion of crawl space ventilation helps remove excessive moist air from this space before it enters the living space. Section 7 introduces requirements for crawl space ventilation for various types of foundation systems as well as ventilation systems to remove radon.

Exhaust System Requirements

The information required for the exhaust system must be reflected throughout the construction documents. The following are some general IRC requirements that need to be considered during the design process. Unless noted, these specifications can be placed in general notes on the floor plan.

- Range hoods shall be vented to the outside by a single-wall galvanized stainless steel or copper duct. This duct shall have a smooth inner surface, be substantially airtight, and have a backdraft damper.
- Cabinets supporting the range hood cannot be closer than 24'' (600 mm) to the cooking surface (Show the minimum height on the cabinet elevations). Clothes dryer vents must be:
  - Independent of all other systems
  - Directly vented to carry the moisture outside
  - Made of rigid metal 0.016'' (0.406 mm) thick with a smooth inside and joints running in the direction of the airflow and be equipped with a backdraft damper.
  - A clothes dryer duct 4'' (100 mm) in diameter cannot be longer than 25' (7500 mm) from the dryer to the wall or roof vent.
    - The total length shall be reduced by 2'-6'' (750 mm) for each 45° bend.
    - The total length shall be reduced by 5' (1500 mm) for each 90° bend.

Air-to-Air Heat Exchangers

An air-to-air heat exchanger is a heat recovery and ventilation device that pulls polluted air from within the building envelope and transfers the heat by pulling fresh air into the house. A heat exchanger does not produce heat but moves it from one air stream to another. The two streams of air never come in contact with each other, thus preventing the indoor pollutants in the stale air from being added back into the fresh air.

Air Filtration

Moving air throughout a home does not ensure that the home will have a supply of fresh air. To create clean air, airborne pollutants such as dust mites, pollen, bacteria, mold spores, and mildew must constantly be removed from the air inside the residence. For many, these pollutants are just minor irritants. For inhabitants with allergies, they can be the source of ongoing health problems. Each heating or cooling system that relies on ducts to move the treated air can be equipped with air filters. Filters filled with fiberglass or charcoal remove approximately 15% of the household pollutants. Electronic filters remove approximately 80% of the household impurities, but they also dry the air and cause a buildup of static electricity. Electrostatic filters with ionizing wires trap more than 99% of household impurities. To further purify the air, ozonators may be attached to the ductwork. An ozonator adds low levels of electronically charged oxygen to the air supply.

Humidity Control

Moisture in the air is referred to as humidity. The proper amount of moisture in the air is important for good climate control as well as the control of common household allergens. Some heating devices, such as forced air systems, dry the air as they are used. Heating systems that use hot water generally add moisture to the interior of the home. The use of vapor barriers in the walls, ceiling, and foundation can help limit the buildup of excessive moisture within the building envelope. Two electronic devices are also available to control humidity: a humidifier may be added to a home to increase the humidity levels of the interior air and a dehumidifier can be added to remove excessive moisture from the building envelope. Either of these units can be added individually to a home or to any comfort-control system that uses ductwork.

SOLAR SYSTEMS

Solar heating and cooling involves using the sun's energy to provide heat and convective currents within a home. Passive solar systems operate without the use of special mechanical or electronic devices. Passive systems use the structure and orientation of the residence to maximize energy from the sun. Many of the basic design considerations of passive solar design have become common features of good design based on green and Energy Star programs. Active solar systems incorporate mechanical devices to provide power, heat, and hot water for a home. No matter the type of system, solar heating and cooling includes the four basic steps of collecting, storing, distributing, and controlling the solar radiation that is collected. Both methods are still popular means to heat a home, but their use has diminished with the incorporation of other green and Energy Star building advances. Because of the ease of incorporating passive solar concepts into home construction and the expense of active systems, active systems are not examined in this text.

Solar Design Basics

Not all of the sun's energy directed toward the earth strikes the surface of the earth. As much as 35% of the sun's energy is reflected back into space by our atmosphere. As much as 15% of the remaining energy is scattered and reflected by water vapor, dust, and other particles in the air. The distance through the atmosphere through which the remaining radiation must pass and the angle it travels in relation to the earth's surface, determine how effective the remaining energy is. When the sun is directly overhead, radiation travels through the least amount of atmosphere to reach the surface of the earth. During the early hours after sunrise or the later stages of the afternoon, when the sun is closer to the horizon, the radiation passes through more of the atmosphere. Figure 14.18 illustrates this elementary but important concept. The more atmosphere the radiation must pass through, the lower the amount of energy available at the surface of the earth.

The seasonal position of the earth in relation to the sun is a second major consideration affecting the available amount of solar radiation. Because of the earth's tilt and rotation, the amount of atmosphere the solar radiation must pass through varies throughout the year. The earth orbits the sun in an elliptical path and rotates once a day on an axis that extends through the north and south poles. The axis is tilted at 23.47° from vertical of the plane of the earth's orbit around the sun. The constant tilt of the earth's axis produces the seasonal variations in weather. The northern hemisphere is tilted away from the sun during winter, decreasing the hours of available sunlight and increasing the atmosphere through which radiation must pass. For those in the southern hemisphere, the situation is reversed; they experience summer while we experience winter. During the summer, the situation is again reversed. The northern axis of the earth is tilted toward the sun, the length of available daylight increases, and the available radiation from the sun is closer to perpendicular to the earth's surface. As the sun's rays move closer to perpendicular to the surface of the earth, the amount of radiation from the sun increases. This may be third-grade science, but it's important to know if a residence is to be made to work efficiently.

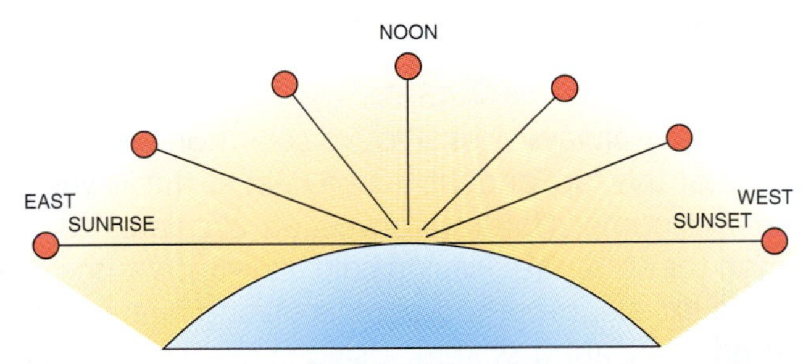

FIGURE 14.18 ■ *The depth of atmosphere that solar radiation must pass through and its angle in relation to the earth's surface determines how concentrated the energy will be. When the sun is directly overhead, radiation travels through the least amount of atmosphere to reach the surface of the earth. During the early hours after sunrise or the later hours of the afternoon, when the sun is closer to the horizon, the radiation must pass through a greater depth of atmosphere.*

Solar Orientation

Chapter 7 introduces the importance of the home's orientation to take advantage of solar heating. A rectangular home with its long surface facing south is ideal to maximize solar gain. Although facing due south is the ideal, the south side of the home may be rotated up to 30° to the east or west and still receive 90% of the available radiation. Equally important to the orientation of the home is the angle of the window surfaces to the sun's rays. Windows perpendicular to the angle of the sun's rays will intercept 100% of the rays. This is why many solar homes shown in magazines and on television have windows placed on an angle rather than in the normal vertical position. As the angle of the glass is decreased from being perpendicular to the rays of the sun, so is the absorption. Glass can be tilted 25° to the angle of the sun and still achieve 90% of the sun's radiation. Tables are available on the Internet and through books to determine the angle of the sun at various latitudes at different times of the year. For instance, at 45° north latitude, the sun's angle is 21° above the horizon on January 21; it is 69° above the horizon during the summer. Armed with this information, a simple section can be drawn of a proposed project to determine the needed roof overhangs to block the sun as well as the depth that the sun will enter a home during the winter (see Figure 14.19). In addition to the atmospheric relationship of the sun to the earth's surface, two other principles of solar physics are used in passive solar planning. One is the greenhouse effect and the other is the natural law of rising warm air.

Greenhouse Effect

The trapping of heat in the atmosphere is known as the greenhouse effect. Glass allows virtually all solar radiation striking its surface to pass through to the cool

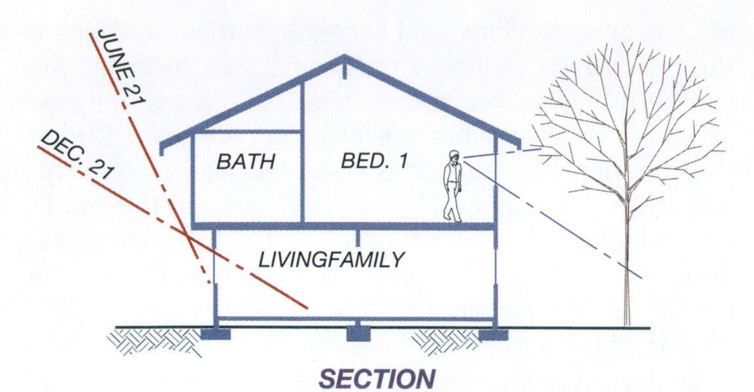

FIGURE 14.19 ■ *A simple section can be used to project the winter and summer sun angles into the residence. The goal is to have the greatest penetration during the winter and the smallest penetration during the summer.*

FIGURE 14.20 ■ *A greenhouse can be used to collect solar heat.* Courtesy Velux America, Inc.

side of the glass. Once the sunlight is transmitted through the glass and absorbed by materials on the interior side, thermal energy radiated by these materials will not pass back out through the glass until the outside air is cooler than the glass. If, during the summer, you've gotten into a car with its windows rolled up, you experienced the greenhouse effect. Although the greenhouse effect can be dangerous to someone left in a sealed automobile, it is very useful in a residence. Heat from the sun that enters through windows can be stored in some type of thermal mass. The heat is released when the home cools to provide heating when the sun's energy is not available.

A thermal mass is any material that absorbs heat from the sun and later radiates the heat back into the air. Walls, floors, and masonry features function as a thermal mass in a building designed for maximum solar effectiveness. The concrete floor of a home, even if covered with carpet or wood, acts as a thermal mass. Masonry or stone veneer on an interior wall or a masonry chimney are all examples of common areas that provide thermal mass in a residence. These items and other materials exposed to the sun's radiation serve both for storage and distribution in a passive solar system. Storing and dissipating the trapped heat to either lower or raise the temperature of a building as needed is a key element of passive solar design. If the budget, the terrain, and the orientation of the house allow, a greenhouse can be attached to the south side of a residence to further increase the solar gain. Don't become trapped by the thought of a green house. A greenhouse or solarium does not have to be a room totally surrounded by glass, as shown in Figure 14.20. Although this is an attractive room, it does not fit in with many historic styles of houses. Figure 14.21 shows a room that blends with the entire design of the residence and also collects solar radiation effectively. A "solarium" used as a family room, nook, or enclosed deck can also double as a solar collector.

Convective Air Loops

Heated air always rises. Passive solar homes take advantage of this fact and introduce heat at the lower levels of the home. Natural convection circulates the air through the structure. Placing cold air returns in the ceiling is an excellent method of recirculating heated air by a forced-air heating system. Using slow-moving ceiling fans is a less expensive method to accomplish a similar effect. Reversible fans can be used to force heated air to the floor level during heating cycles. The fan direction can be reversed to accelerate the rise of heated air during the cooling cycle.

Homes that have large amounts of south-facing glass risk overheating during the summer. In addition to collecting heat, a home must be able to release heat during the cooling cycle. The room in Figure 14.21 has upper windows that can be opened to re-

FIGURE 14.21 ■ *Any room with south-facing glass can be used as a solarium. An efficient collector will also have areas of high thermal mass to store heat as well as openable windows to remove heat during the cooling cycle.* Courtesy BOWA Builders. Greg Hadley, photographer.

FIGURE 14.22 ■ *Openable skylights allow heated air to be vented, providing a convective current in the home.* Courtesy Velux America, Inc.

lease the hot air that accumulates at the ceiling. Placing openable windows on the north face of the home also creates a convective current from south to north through the residence. Openable skylights, such as those in Figure 14.22, allow the heated air from the kitchen to be vented naturally. When the kitchen is in use, it becomes a heat producer. The skylights allow the hot air to be vented without activating the home cooling system. During the heating cycle, the skylights produce a negative effect on heat gain. Although most skylights are double-glazed and contain low-E glass, a protective shade that can be drawn across the opening further reduces heat loss. In placing skylights on a roof plan, remember that skylights on the south, east, and west sides of the roof will have some heat gain. Skylights on the north face of the roof gain heat only during the latest part of the summer, when the sun is at its highest point above the horizon.

Passive Solar Methods

Passive solar heating and cooling methods rely on basic principles of physics to help control the home environment. The direct- and indirect-gain methods take advantage of solar radiation to provide heat and to block the sun's heat when the home is in the cooling cycle.

Direct-Gain Method

Direct gain is the simplest approach to passive solar heat gain. Using the direct-gain method, the inside of a building is directly heated by the sun's rays as they pass through large areas of south-facing glass (see Figure 14.23). Once the sun's rays enter the residence, they are absorbed by each thermal mass. As the room cools through out the evening, the stored heat in the thermal mass radiates back into the room. Water has the highest capacity to retain heat, but steel, aluminum, concrete, masonry, and rock also make excellent thermal storage materials. Wood and similar porous materials are ineffective in retaining heat.

Indirect-Gain Method

The indirect-gain method places a thermal mass between the sun and the interior of the residence. With indirect solar gain, light passes through glass and strikes the thermal mass and then radiates into the liv-

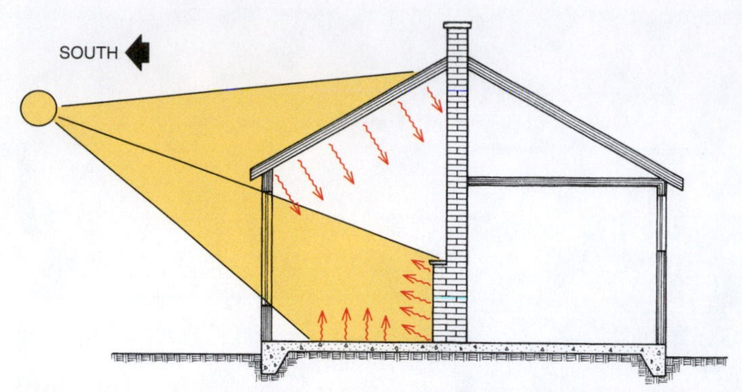

FIGURE 14.23 ■ *Passive solar heating depends on trapping collected heat in a thermal mass. The denser the mass, the longer the heat will be stored before being released back into the room.*

ing space. The most common method of implementing indirect gain is the use of a trombe wall. A trombe wall uses dense materials such as masonry or water containers to store heat (see Figure 14.24). The space between the glass and the wall generates very high temperatures in addition to heating the wall. This air is drawn by convection or the use of mechanical fans into the residence to increase the amount of heat that reaches the living space. An obvious drawback to this system is that it limits visibility on the south side of the home.

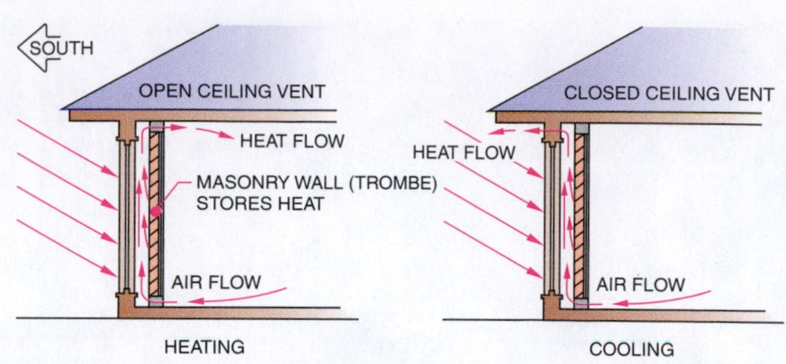

FIGURE 14.24 ■ *A trombe wall uses dense materials such as masonry or water containers to store heat. Heat is stored in the wall and then radiated into the living areas.*

Active Solar Methods

Planning for active solar systems requires knowledge of both mechanical systems and thermal principles. Active solar systems use mechanical devices to drive the components needed for solar heating or cooling. This includes devices and facilities to collect, store, distribute, and control heat. Active solar systems operate more effectively when they are combined with passive solar features. The most frequent use of active systems is to heat hot water for the fresh-water system or for heating pool and spa water. Active systems are most effective when combined with passive solar design features. Active systems generally are designed and drawn by a mechanical consulting firm.

CHAPTER 14

Additional Readings

The following websites can be a resource to help you keep current with changes with HVAC systems.

ADDRESS	COMPANY OR ORGANIZATION
www.ase.org	Alliance to Save Energy
www.ashre.org	American Society of Heating, Refrigerating and Air-Conditioning Engineers, Inc.
www.ases.org	American Solar Energy Society
www.arcat.com	Building Product Information
www.buildingscience.com	Building Science Corporation
www.carrier.com	Carrier
www.certainteed.com	CertainTeed
www.dap.com	Dap
www.earthadvantage.com	Earth Advantage Program
www.energystar.gov	Energy Star
www.eeba.org	Energy Efficient Building Association
www.insulate.com	Insulation Contractors of America
www.lennox.com	Lennox
www.powerfromthesun.net	Solar Concepts
www.seia.org	Solar Energy
www.sprayfoam.org	Spray Polyurethane Foam Alliance
www.comfortfoam.com	Spray Polyurethane Foam Insulation
www.trane.com	Trane
www.doe.gov	U.S. Department of Energy
www.usgbc.org	U.S. Green Building Council

CHAPTER

14 Comfort-Control Systems

Answer the following questions with short, complete statements. Type the answers using a word processor or neatly print the answers on lined paper.

1. Place your name, the chapter number, and the date at the top of the sheet.
2. Type the question number and provide the answer in the form of a statement that includes part of the question. You do not need to write out the entire question.

Note: The answers to some questions may not be contained in this chapter and will require you to do additional research.

QUESTIONS

14.1. Describe and explain the units of heat measurement.

14.2. Describe the four major steps of solar heating and cooling.

14.3. List two advantages and two disadvantages of a zonal heat system as compared with central forced-air heating.

14.4. Describe the difference between a heat pump and a forced-air heating system.

14.5. Describe four factors that influence placement of a thermostat.

14.6. Describe and explain systems that contribute to a healthy interior environment.

14.7. Discuss the function of an air-to-air heat exchanger.

14.8. What is a plenum?

14.9. Define combustion air.

14.10. List passive solar design features that can easily be included in a residence.

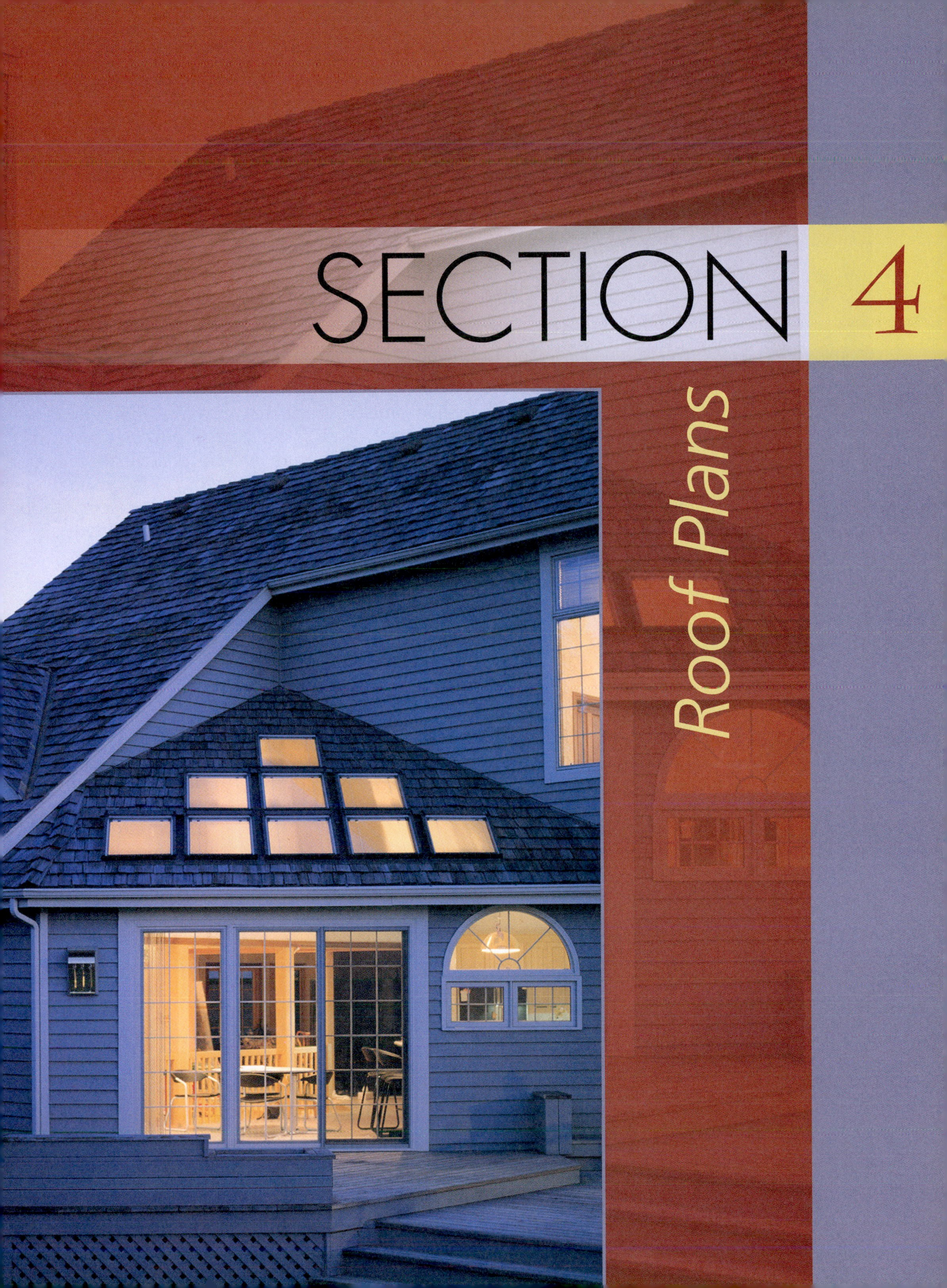

SECTION 4
Roof Plans

15 CHAPTER

Roof Plan Components

The design of the roof must be considered long before the roof plan is drawn. The architect or designer will typically design the basic shape of the roof as the floor plan and elevations are drawn at the preliminary design stage. This does not mean that the designer plans the entire structural system for the roof during the initial stages but only the general shape and type of roofing material to be used will be planned. By examining the structure in Figure 15.1, you can easily see the impact of the roof design on the structure. Often the roof can present a larger visible surface area than the walls. In addition to aesthetic considerations, the roof can also be used to provide rigidity in a structure when wall areas are filled with glass. To ensure that the roof will meet the designer's criteria, a roof plan is usually drawn by the CAD technician to provide construction information. In order to draw the roof plan, knowledge of the types of roof plans, common roof terms, common roof shapes, and common roof materials is required.

TYPES OF ROOF DRAWINGS

The plan that is drawn of the roof area may be either a roof plan or a roof framing plan. For some types of roofs, a roof drainage plan may also be drawn. Roof framing plans are discussed in Chapter 24.

FIGURE 15.1 ■ *The shape of the roof can play an important role in the design of the structure.* Courtesy Monier Lifetile.

Roof Plans

A roof plan is used to show the shape of the roof. Materials such as the roofing material, vents and their location, and the type of underlayment are also typically specified on the roof plan, as seen in Figure 15.2. Roof plans are often drawn at a scale smaller than the scale used for the floor plan. A scale of 1/8'' = 1'-0'' or 1/16'' = 1'-0'' is commonly used for a roof plan. A roof plan is usually displayed on the same sheet as the exterior elevations.

Roof Framing Plans

Roof framing plans are required for complicated residential roof shapes. A roof framing plan shows the size and direction of the construction members required to frame the roof. Figure 15.3 shows an example of a roof framing plan. On very complex projects, every framing member is shown, as seen in Figure 15.4. Framing plans are discussed further in Chapter 24.

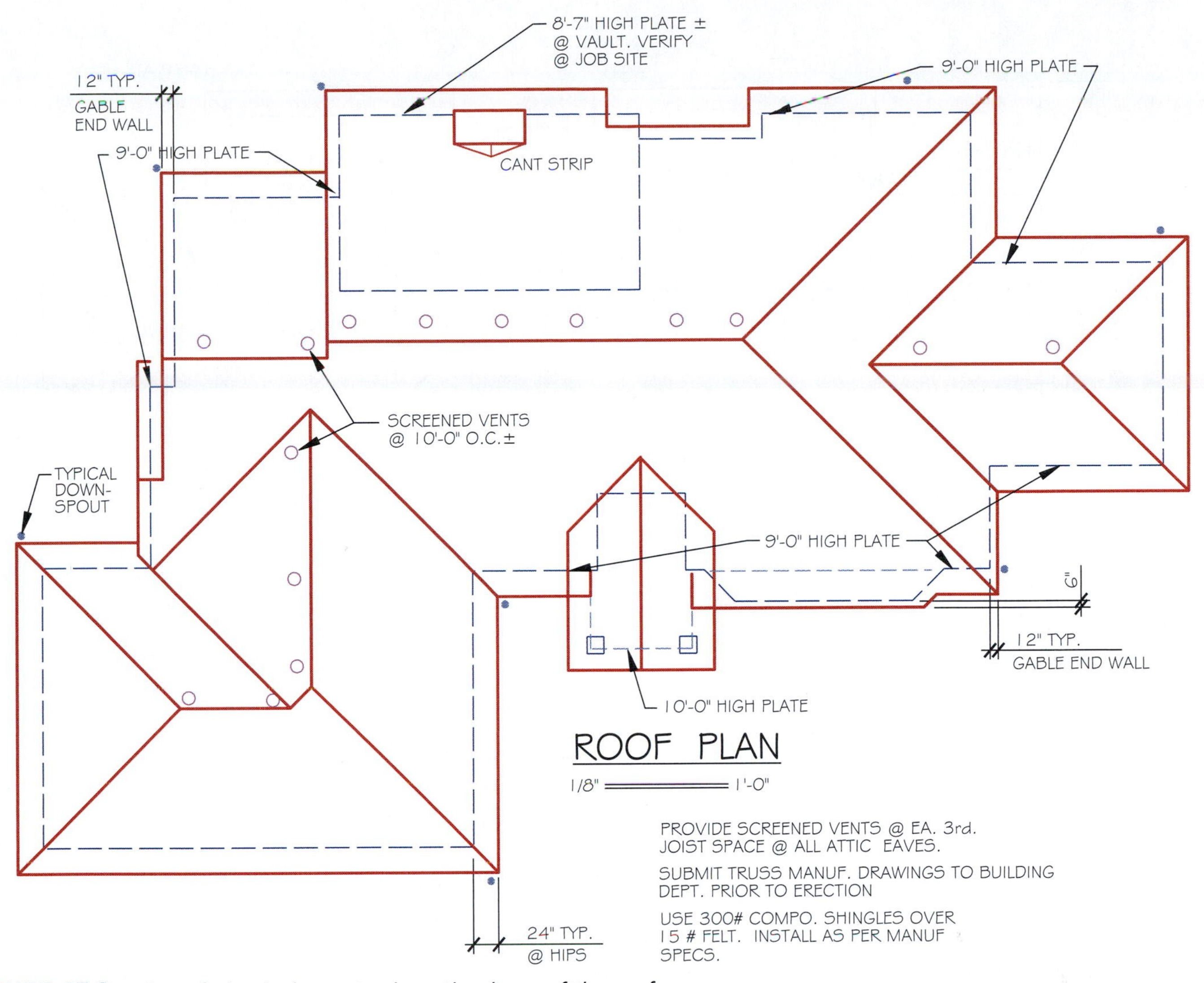

FIGURE 15.2 ■ *A roof plan is drawn to show the shape of the roof.*

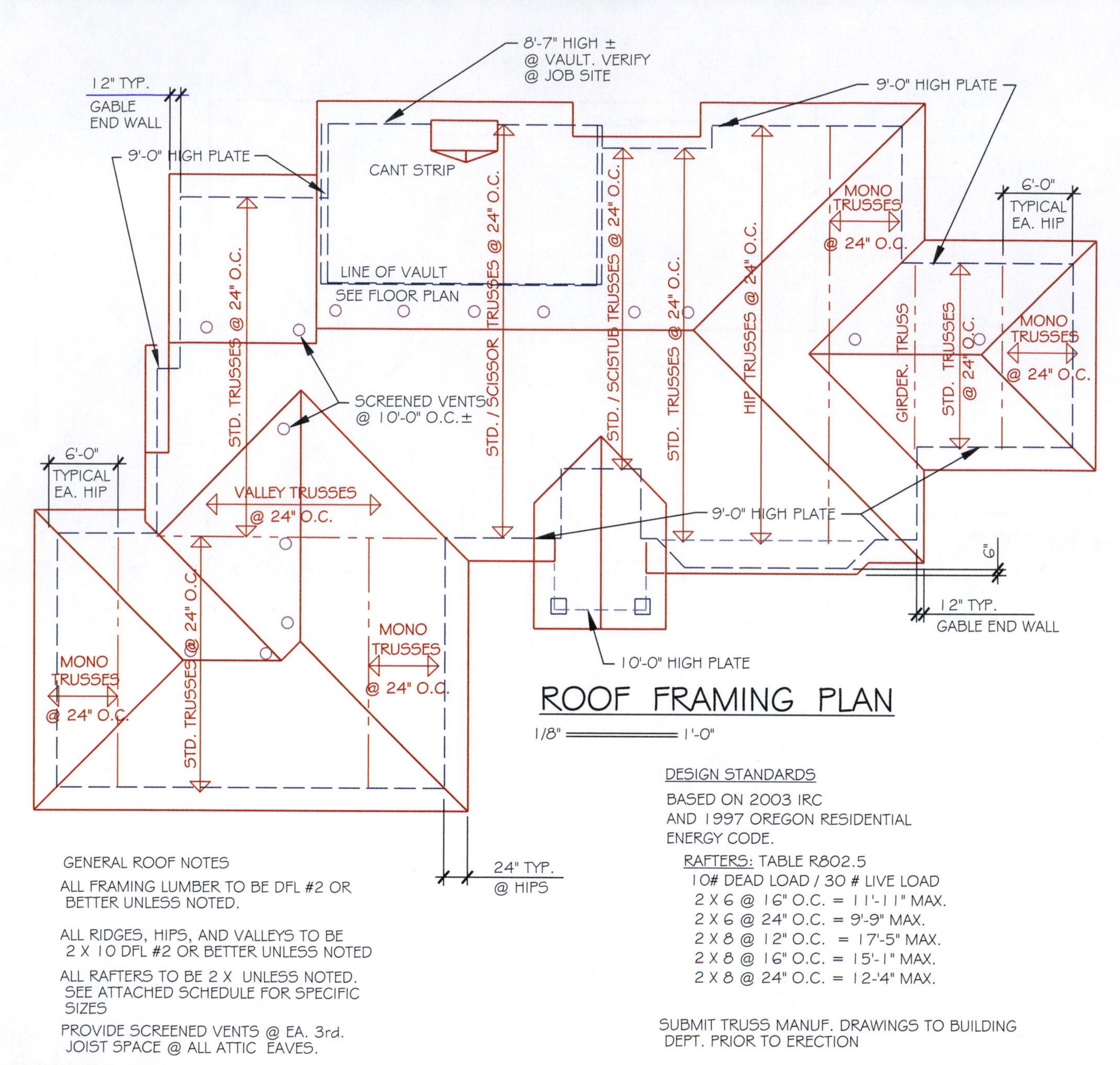
8'-7" HIGH ±
@ VAULT. VERIFY
@ JOB SITE
12" TYP.
GABLE
END WALL
9'-0" HIGH PLATE
9'-0" HIGH PLATE
CANT STRIP
LINE OF VAULT
SEE FLOOR PLAN
STD. TRUSSES @ 24" O.C.
STD. / SCISSOR TRUSSES @ 24" O.C.
STD. / SCISTUB TRUSSES @ 24" O.C.
STD. TRUSSES @ 24" O.C.
HIP TRUSSES @ 24" O.C.
MONO
TRUSSES
@ 24" O.C.
6'-0"
TYPICAL
EA. HIP
GIRDER. TRUSS
STD. TRUSSES @ 24" O.C.
MONO
TRUSSES
@ 24" O.C.
SCREENED VENTS
@ 10'-0" O.C.±
6'-0"
TYPICAL
EA. HIP
VALLEY TRUSSES
@ 24" O.C.
9'-0" HIGH PLATE
6"
12" TYP.
GABLE END WALL
STD. TRUSSES @ 24" O.C.
MONO
TRUSSES
@ 24" O.C.
MONO
TRUSSES
@ 24" O.C.
10'-0" HIGH PLATE
ROOF FRAMING PLAN
1/8" = 1'-0"
24" TYP.
@ HIPS
GENERAL ROOF NOTES
ALL FRAMING LUMBER TO BE DFL #2 OR
BETTER UNLESS NOTED.
ALL RIDGES, HIPS, AND VALLEYS TO BE
2 X 10 DFL #2 OR BETTER UNLESS NOTED
ALL RAFTERS TO BE 2 X UNLESS NOTED.
SEE ATTACHED SCHEDULE FOR SPECIFIC
SIZES
PROVIDE SCREENED VENTS @ EA. 3rd.
JOIST SPACE @ ALL ATTIC EAVES.
DESIGN STANDARDS
BASED ON 2003 IRC
AND 1997 OREGON RESIDENTIAL
ENERGY CODE.
RAFTERS: TABLE R802.5
10# DEAD LOAD / 30 # LIVE LOAD
2 X 6 @ 16" O.C. = 11'-11" MAX.
2 X 6 @ 24" O.C. = 9'-9" MAX.
2 X 8 @ 12" O.C. = 17'-5" MAX.
2 X 8 @ 16" O.C. = 15'-1" MAX.
2 X 8 @ 24" O.C. = 12-'4" MAX.
SUBMIT TRUSS MANUF. DRAWINGS TO BUILDING
DEPT. PRIOR TO ERECTION

FIGURE 15.3 ■ *A roof framing plan is used to show the framing members for the roof.*

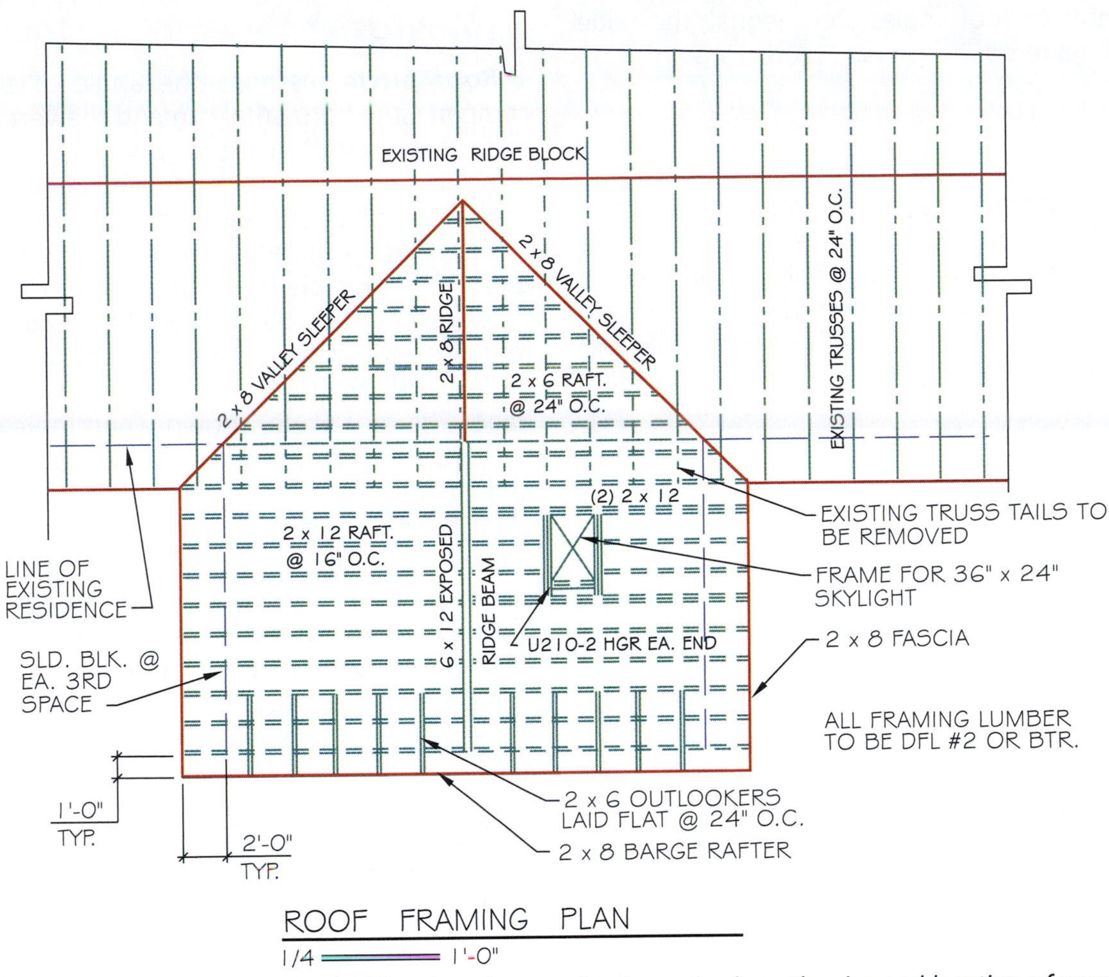

FIGURE 15.4 ■ *For complicated roofs, a roof framing plan may be drawn to show the size and location of every structural member.*

COMMON ROOF TERMS

Several terms that represent common roof components must be understood before a roof plan or roof framing plan can be completed. These terms include *ridge, roof pitch, overhang, eave, soffit, cornice,* and *fascia.* Several of these terms are used interchangeably. The portions of the roof that these terms represent are shown in Figure 15.5.

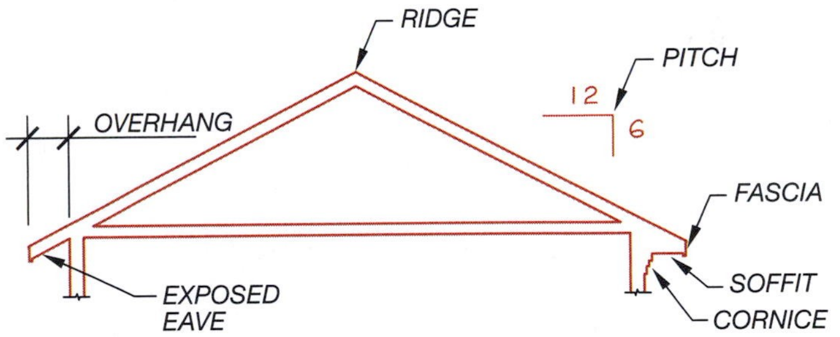

FIGURE 15.5 ■ *Common roof terms associated with roof drawings include* ridge, roof pitch, overhang, eave, soffit, cornice, *and* fascia.

Ridge

The term ***ridge*** is used in several ways that relate to the roof framing methods. Some of these terms include *ridge board, ridge block, ridge beam,* and *ridge brace.* Each of these terms is discussed in Chapters 16 and 20 as roof framing methods are explored. As basic roofing components are explored, the ridge needs to be understood as a specific location. A ridge is a horizontal intersection between two or more roof planes and represents the highest point of a roof. In drawing a roof plan, the ridge is centered between the two walls that support a roof if the following two conditions are met:

- The support walls must be of equal height.
- The angle (pitch) of the roof planes must be equal.

If wall heights or roof angles are unequal, the ridge location will be altered, as seen in Figure 15.6.

Overhang

The term ***overhang*** is used to represent the portion of the roof that extends past the walls. The overhang is used to provide shade for wall openings and provide protection to the walls from weather. The historical style of the home, the amount of shade and protection desired, and the pitch of the roof influence the size of the overhang.

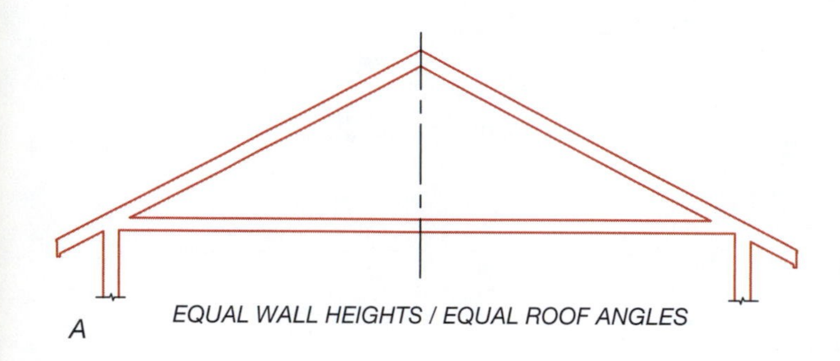

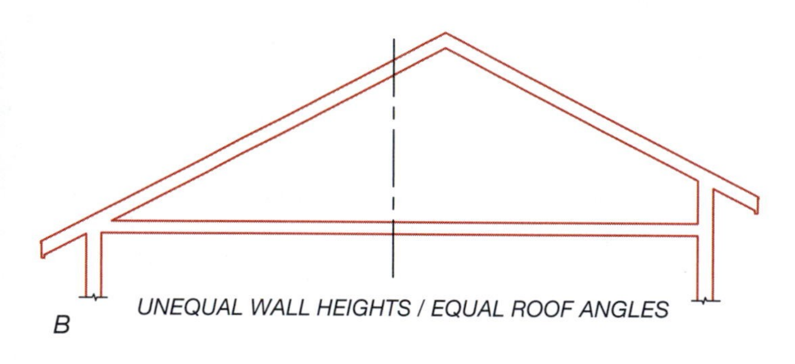

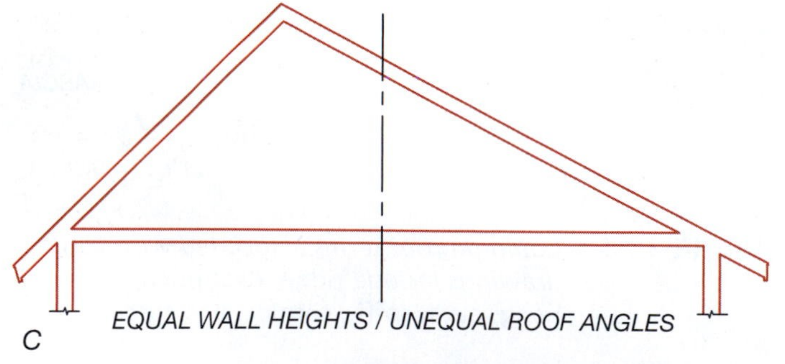

FIGURE 15.6 ■ *Three options for determining the ridge location. (A) If the walls supporting the roof are the same height and the angles of the roof planes are equal, the ridge will be centered between the support walls. (B) If the height of one of the two support walls is altered, the ridge location will be altered. (C) If the angles of the roof planes are not the same, the ridge will no longer be centered between the support walls.*

Roof Pitch

Roof pitch describes the angle of the roof that compares the horizontal run and the vertical rise. Roof pitch is measured in the field with a framer's square, using inches as the unit of measurement. As a CAD technician, you can determine the rise and run using inches, feet, millimeters, meters, or any other equal units. The roof pitch is not represented on the roof drawings, but the intersections that result from various roof pitches must be represented. It is also necessary to understand the effects of pitch on the size of the overhang. The roof slope, shown when the elevations and sections are drawn, is discussed in Chapters 16, 18, and 21.

In order to plot the intersection between two roof surfaces correctly, it is necessary to understand how various roof pitches are drawn. Figure 15.7 shows how the pitch can be visualized. This method of using the rise and run can be used to determine any roof pitch. The roof pitch can also be drawn if the proper angle that represents a certain pitch is known. Knowing that a 4/12 roof equals 18 1/2° allows the drafter to enter the correct angle without having to plot the layout. Figure 15.8 shows angles for common roof pitches.

Roof pitches can be considered as very slight, slight, moderate, or dramatic slopes.

- Homes with a pitch of less than 2/12 are considered to have slight slopes. A very slight sloping roof is not practical for deflecting rain and snow, but in arid parts of the world, slope is less important. The development of more durable roofing materials has allowed flat roofs to become common on modern, International-style homes and many townhouses.

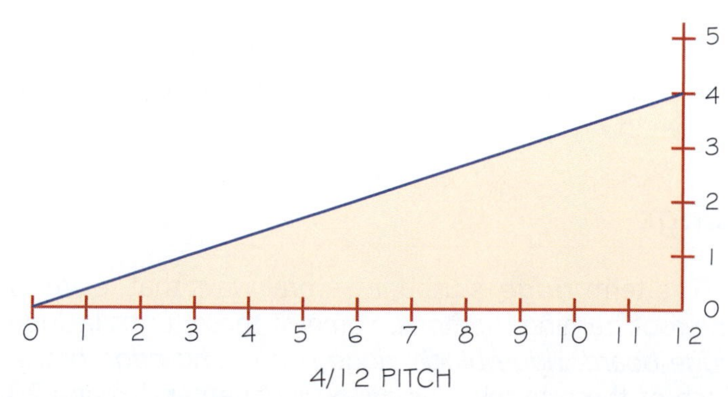

FIGURE 15.7 ■ *In determining the roof slope, the angle is expressed as a comparison of equal units. Units may be inches, feet, meters, etc., as long as the horizontal and vertical units are of equal length.*

COMMON ANGLES FOR DRAWING ROOF PITCHES	
ROOF PITCH	ANGLE
1/12	4°–30′
2/12	9°–30′
3/12	15°–0′
4/12	18°–30′
5/12	22°–30′
6/12	26°–30′
7/12	30°–0′
8/12	33°–45′
9/12	37°–0′
10/12	40°–0′
11/12	42°–30′
12/12	45°–0′

FIGURE 15.8 ■ *Common roof pitches and angles. Angles shown are approximate and are to be used for drawing purposes only.*

- Homes with a pitch between 2/12 to 4/12 are considered to have gradual slopes. Low, gently pitched roofs are characteristic of Mediterranean and Italianate-style homes and many twentieth century styles, such as Craftsman Bungalow-, Prairie-, and Ranch-style homes.
- Moderate roof slopes can be thought of as describing roofs built with a pitch between 4/12 and 7/12. Moderate roof pitches can be found on any style of home.
- Steep roof slopes are roofs that have an 8/12 pitch or greater. Steep gable roofs are common on Gothic-, Greek Revival-, Tudor-, Cape Cod-, French Manor-, Colonial-, and Farmhouse-style homes. A gable roof that is narrow and extremely steep is almost always inspired by Gothic traditions. Imitating the churches of medieval Europe, Gothic Revival and Carpenter Gothic houses create a sense of vaulting height with tall, pointed gables.

Sun Angle and Overhang Size

The size of the overhang varies depending on the pitch of the roof and the amount of shade desired. As seen in Figure 15.9, the steeper the roof pitch, the smaller the overhang that is required to shade an area. If you are drawing a home for a southern location, an overhang large enough to protect glazing from direct

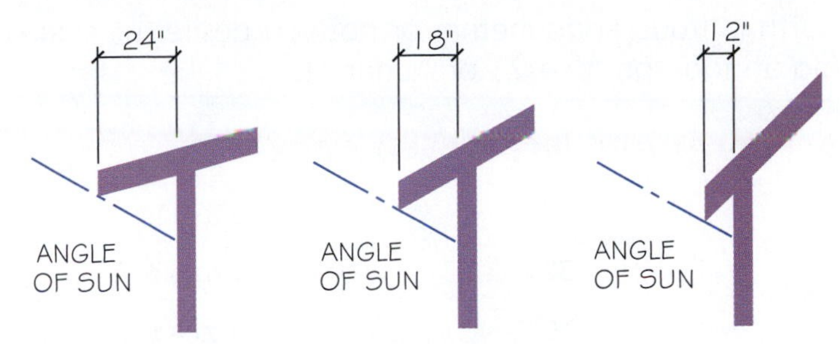

FIGURE 15.9 ■ *The steeper the roof pitch, the more shadow will be cast. Typically the overhang is decreased as the pitch is increased.*

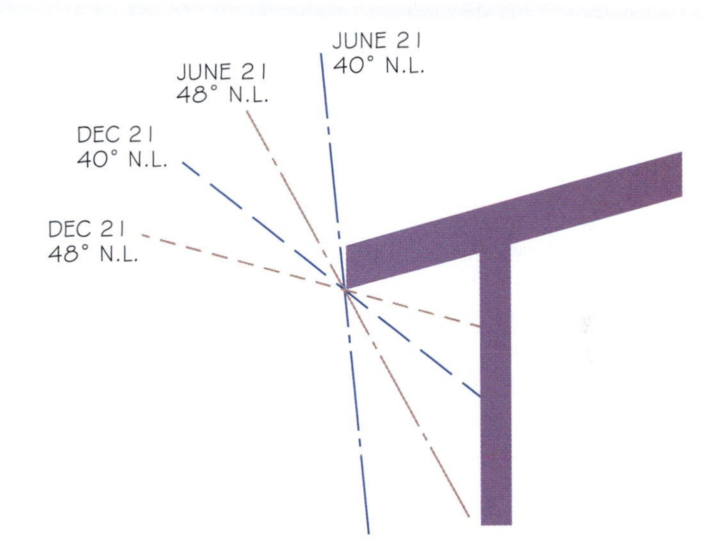

FIGURE 15.10 ■ *The overhang blocks different amounts of sunlight at different times of the year. Notice the difference in the sun's angles at 40° and 48° north latitude.*

sunlight is usually desirable. In northern areas, the overhang is usually restricted to maximize the amount of sunlight received during the winter months. Figure 15.10 compares the effect of an overhang at different times of the year. Another important consideration regarding the size of the overhang is how the view from windows will be affected. As the angle of the roof is increased, the size of the overhang may need to be decreased so that the eave will not extend into the line of sight from a window. The designer will need to base the size of the overhang on the roof pitch so that the view is not hindered. The required length of an overhang can be determined using the following formula:

$$OH = WH / F$$

where OH is the overhang length and WH is the window height. F is a factor that takes into account the north latitude and whether or not you desire full shading at noon on June 21 or August 1.

LATITUDE (IN DEGREES)	F
26	5.6 to 11.1
32	4.0 to 6.3
36	3.0 to 4.5
40	2.5 to 3.4
44	2.0 to 2.7
48	1.7 to 2.2
52	1.5 to 1.8
56	1.3 to 1.5

The F factor with the highest value for your north latitude will provide full shading at noon on June 21. The lower of the two F factors will provide full shading on August 1. These calculations assume that the structure is facing due south. If the structure is situated 15° east or west of true south, the overhang will need to be extended by 10% to ensure proper shading.

Eaves and Rakes

The **eave** is the horizontal edge of a roof that overhangs the exterior wall. The eave area can be open or enclosed. When the eave is left exposed, the rafter or truss tails (the ends of the members) are visible and require painting or other protection from the elements (see the upper portion of Figure 15.11). Plywood on top of the rafter or truss tails that is exposed to the elements at the eave must be rated for weather exposure. When the rafter or truss tails are enclosed, interior grade ply or OSB can be used. Chapter 20 explores eave framing. A **rake** is the sloped end portion of a roof that forms at the gable end wall where the roof intersects the wall. It is the equivalent of an inclined eave.

Boxed and Soffited Eaves

A **boxed eave** is an eave with a covering applied directly to the bottom side of the rafter or truss tails. A **soffit** is the enclosed area below the overhangs that is used to protect the rafter or truss tails from the elements (see the lower portion of Figure 15.11). By enclosing the eave area, future maintenance cost can be reduced. Figure 15.12 shows examples of flat and arched soffits.

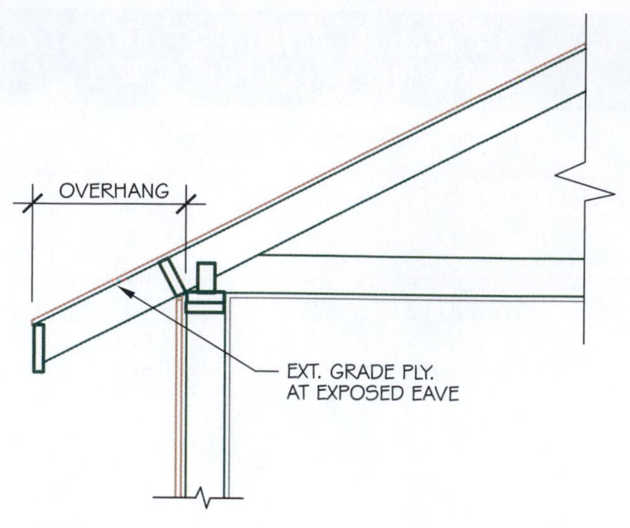

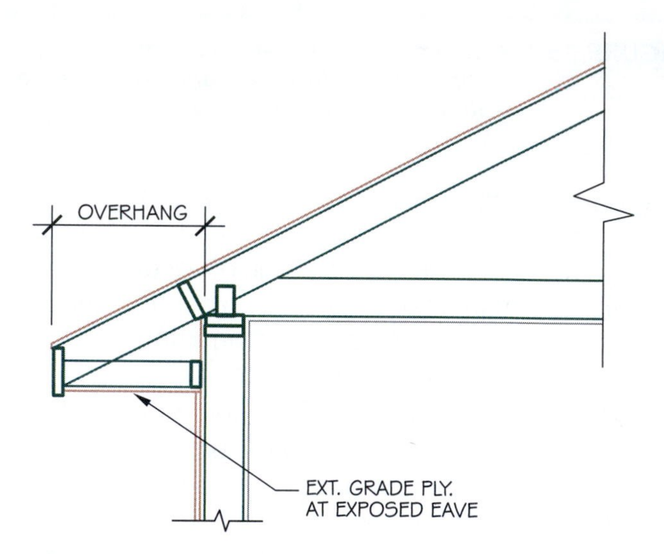

FIGURE 15.11 ■ *The ends of a rafter or truss can be either exposed to the elements or enclosed.*

Cornice

A **cornice** is an ornamental molding or combination of two or more moldings located at the top of an exterior wall just below a roof. Cornice molding is used on many homes in order to match a particular historic style such as Federal or Greek Revival. Figure 15.13 shows the use of cornice molding. Figure 15.14 shows an example of a cornice detail completed by a CAD technician.

FIGURE 15.12 ■ *A soffit is used to enclose the underside of the roof overhang. Flat soffits can be seen on either side of the curved entry soffit.* Courtesy APA–The Engineered Wood Association.

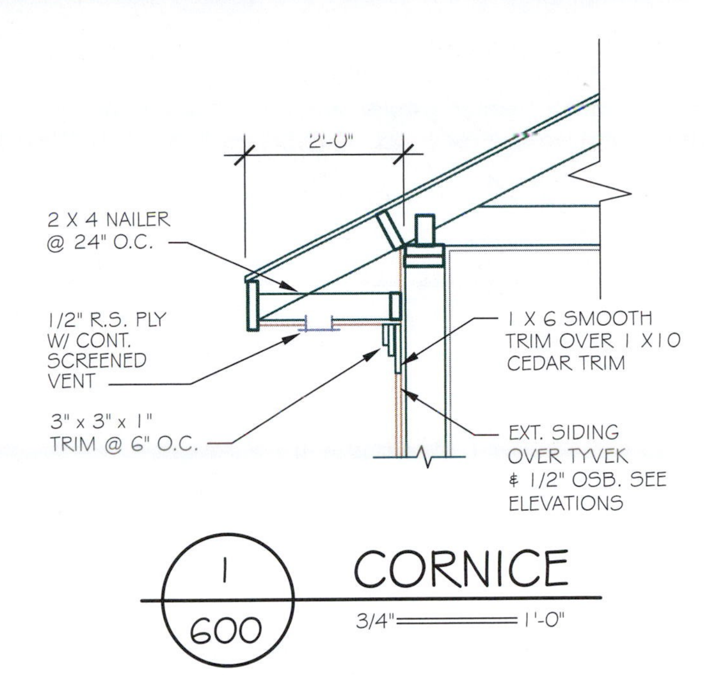

FIGURE 15.14 ■ *A cornice detail completed by a CAD technician.*

FIGURE 15.13 ■ *A cornice is the decorative molding placed at the top of an exterior wall just below a roof.* Courtesy Laine M. Jones–Newbury, Ma.

FIGURE 15.15 ■ *The fascia is the horizontal trim placed at the end of the truss or rafter tails. A barge rafter is the inclined trim that hangs from the projecting edge of a roof.* Courtesy Benigno Molina-Manriquez.

Fascias and Barge Rafters

The ***fascia*** is the trim placed at the end of the truss or rafter tails. The fascia serves to hide the ends of the roof-framing members and provides a mounting service for the gutters. A fascia runs parallel to the ridge and remains parallel to the floor level. The fascia can be seen in Figures 15.11 and 15.14. When the fascia is placed parallel to the rafters or trusses, it is referred to as a ***barge rafter.*** A barge rafter is the inclined trim that hangs from the projecting edge of a roof rake (see Figure 15.15). Depending on your area of the country, a barge may also be referred to as a verge board, gable rafter, or gable board.

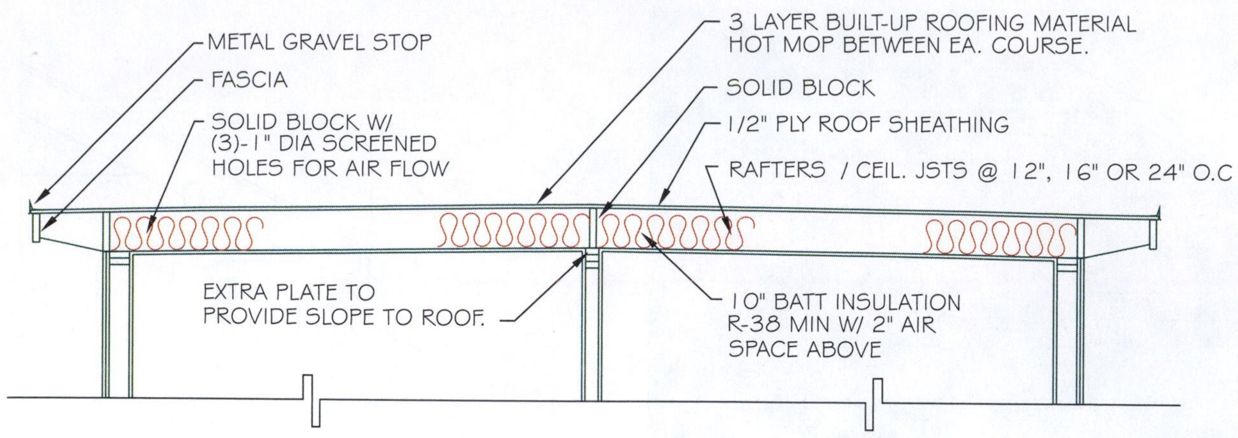

FIGURE 15.16 ■ *Common construction components of a flat roof.*

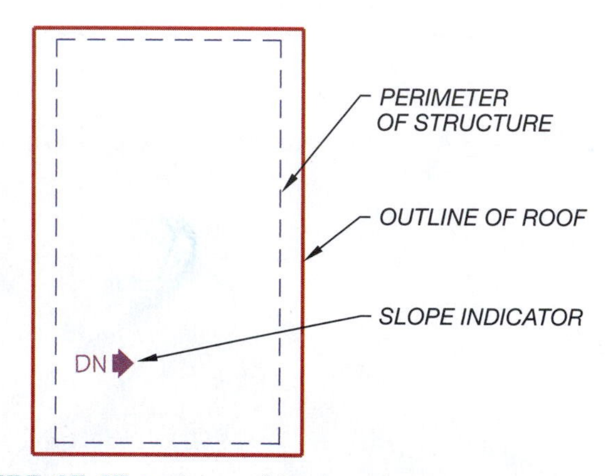

FIGURE 15.17 ■ *Flat roof in plan view.*

FIGURE 15.18 ■ *Many contemporary homes combine shed roofs to create a pleasing design.* Courtesy LeRoy Cook.

COMMON ROOF SHAPES

By changing the roof pitch or adding additional planes, the designer can change the shape of the roof. Common roof shapes include flat, shed, gable, A frame, gambrel, hip, Dutch hip, and mansard. See Chapter 21 for a complete discussion of roof framing terms.

Flat Roofs

The flat roof is a very common style in areas with little rain or snow and is specific to several home styles. Modern, International, Southwestern, Pueblo, and some Spanish home styles usually have low-sloped roofs. The flat roof is economical to construct because ceiling joists are eliminated and rafters are used to support both the roof and ceiling loads. Figure 15.16 shows the materials commonly used to frame a flat roof. Figure 15.17 shows how a flat roof could be represented on the roof plan. Often the flat roof has a slight pitch in the rafters. A pitch of 1/4'' per foot (2% slope) is often used to help prevent water from ponding on the roof. As water flows to the edge, a metal diverter is usually placed at the eave to prevent dripping at walkways. A flat roof will often have a parapet, or false wall, surrounding the perimeter of the roof.

Shed Roofs

The shed roof, as seen in Figure 15.18, offers the same simplicity and economical construction methods as a flat roof but does not have the drainage problems associated with a flat roof. Figure 15.19 shows construction methods for shed roofs. The shed roof may be constructed at any pitch. The roofing material and aesthetic considerations are the only factors limiting the pitch. Drawn in plan view, the shed roof will resemble the flat roof, as seen in Figure 15.20.

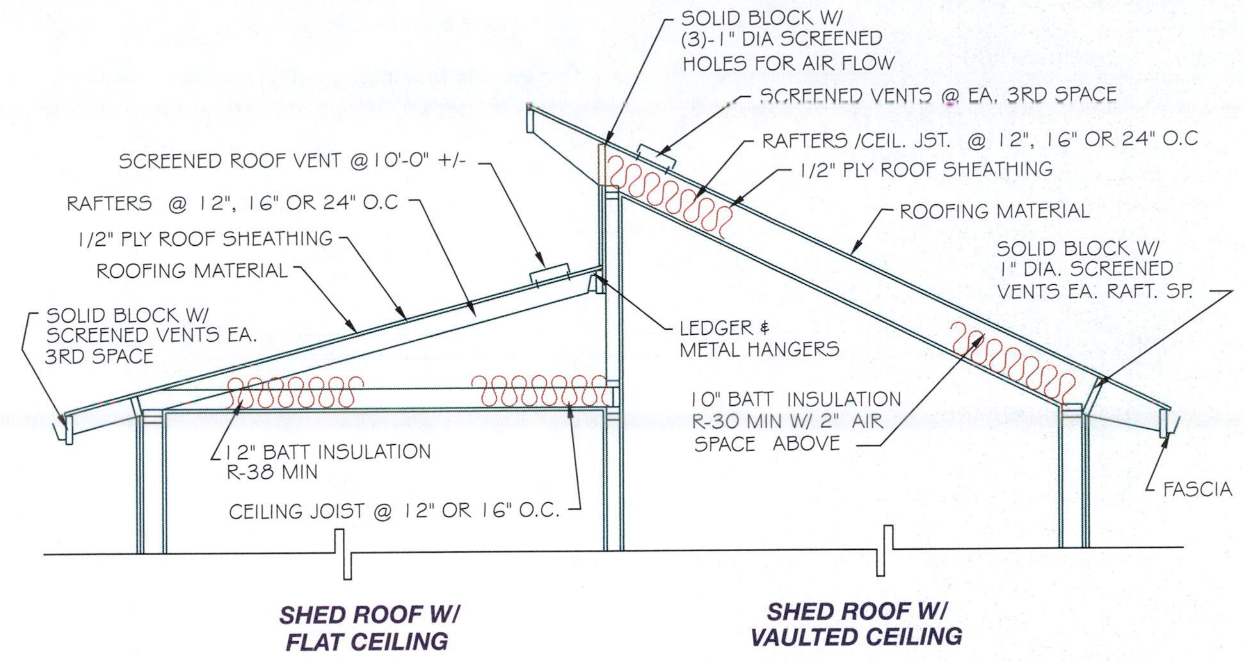

FIGURE 15.19 ■ *Common construction components of shed roofs.*

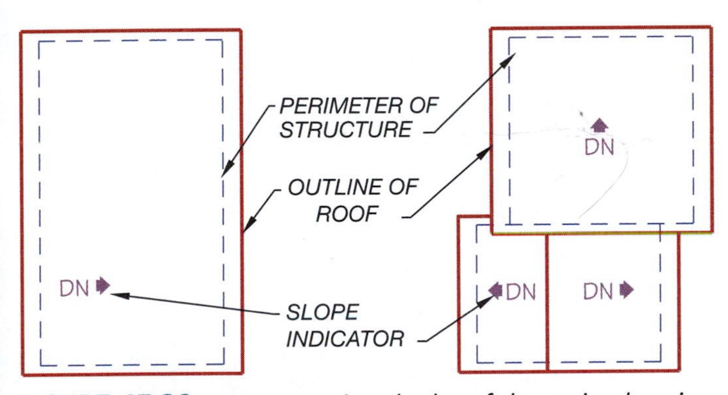

FIGURE 15.20 ■ *Representing shed roof shapes in plan view.*

FIGURE 15.21 ■ *A gable roof is composed of two intersecting planes that form a ridge between them.* Courtesy David Jefferis.

Gable Roofs

A ***gable*** roof is one of the most common roof types in residential construction. As seen in Figure 15.21, it uses two shed roofs that meet to form a ridge between the support walls. Figure 15.22 shows the construction of a gable roof system. The gable can be constructed at any pitch, with the choice of pitch limited only by the roofing material and the effect desired. A gable roof is often used on designs seeking a traditional appearance and formal balance. Figure 15.23 shows how a gable roof is typically represented in plan view. Many plans use two or more gables at 90° angles to each other. The intersections of gable surfaces are called either hips or valleys. A ***hip*** is an exterior corner formed by two intersecting roof planes. A ***valley*** is an interior roof intersection. Each term is explored in Chapters 16 and 21. An additional term associated with gable roofs is a gable end wall. A ***gable end wall*** is the wall perpendicular to the ridge. This type of wall is not specified on the roof plan, but the overhang for a gable end wall is often smaller than the overhang parallel to the ridge. The valley and hip are represented on the roof plan and their size is specified on the roof framing plan.

A-Frame Roofs

A-frame is a method of framing walls as well as a system of framing roofs. An A-frame structure uses rafters to form its supporting walls, as shown in Figure 15.24. The

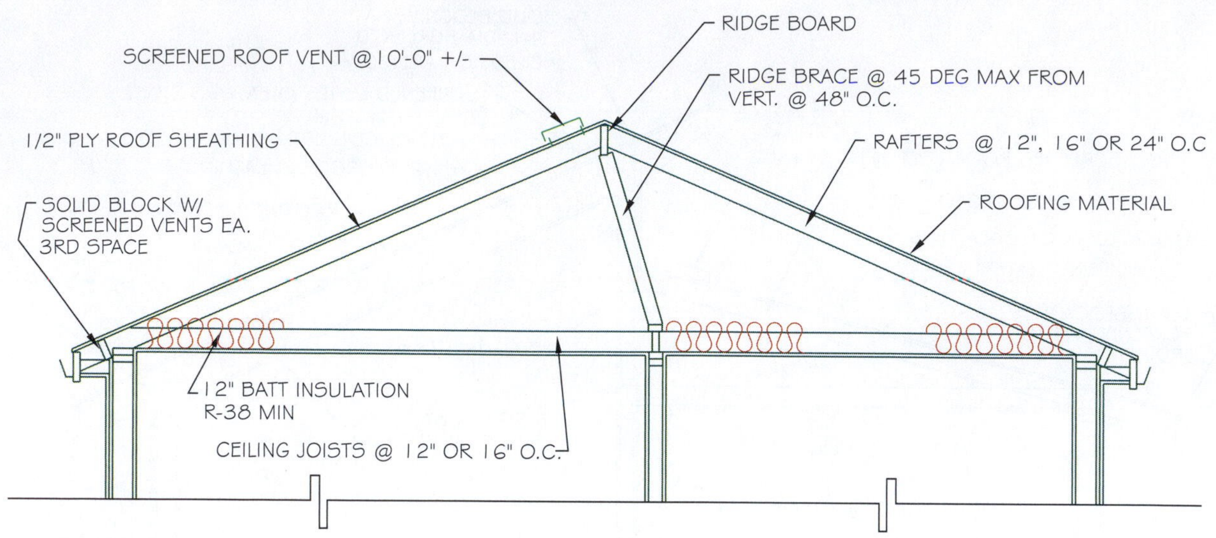

FIGURE 15.22 ■ *Common construction components of a gable roof.*

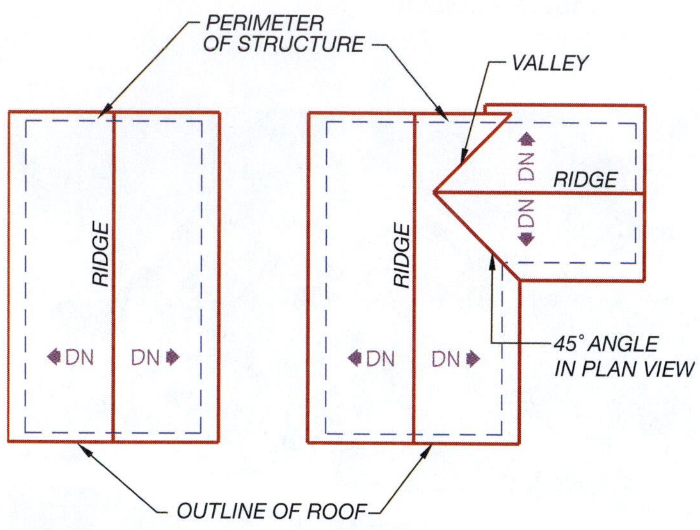

FIGURE 15.23 ■ *A gable roof in plan view. Overhangs, ridges, valleys, and hips must be represented in relation to the exterior walls of the roof plans.*

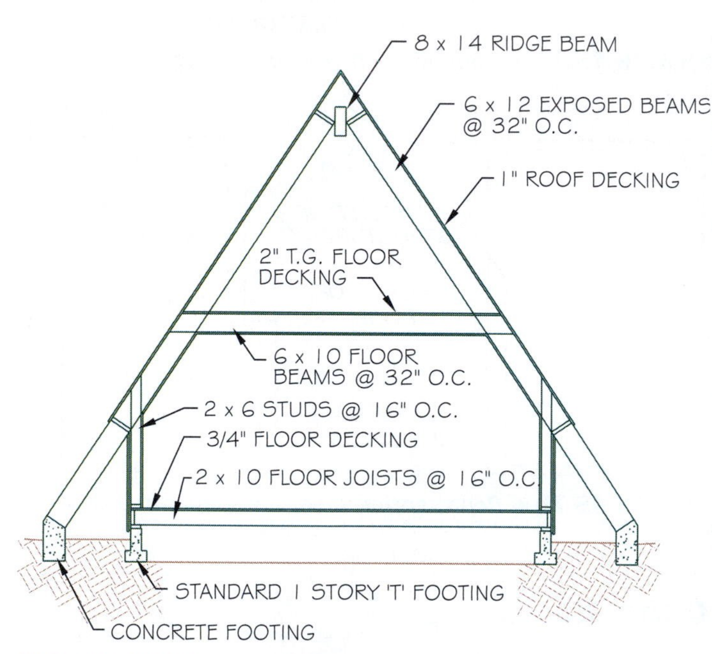

FIGURE 15.24 ■ *An A-frame roof is similar to a very steep gable roof but framed with different framing methods. Common components of A-frame construction typically include large roof beams that extend from the foundation to the ridge.*

structure gets its name from the letter "A," which is formed by the roof and floor systems (see Figure 15.25). The roof plan for an A-frame is very similar to the plan for a gable roof. However, the framing materials are usually quite different. An A-frame is represented on the roof plan using the same methods used to represent a gable roof.

Gambrel Roofs

A gambrel roof is shown in Figure 15.26. The ***gambrel roof*** is a traditional shape that dates back to the colonial period. Figure 15.27 shows construction methods for a gambrel roof. The lower level is covered with a steep roof surface, which connects into the upper roof system with a moderate pitch. By covering the lower level with roofing material rather than siding, the structure is made to appear shorter than it actually is. This roof system can also reduce the cost of siding materials by using less expensive roofing mate-

FIGURE 15.25 ■ *An A-frame uses a steep roof to form the walls of the upper level.*

FIGURE 15.26 ■ *A gambrel roof, which is comprised of four planes, is a common feature of the Dutch Colonial style.* Courtesy Laine M. Jones Design–W. Newbury Ma.

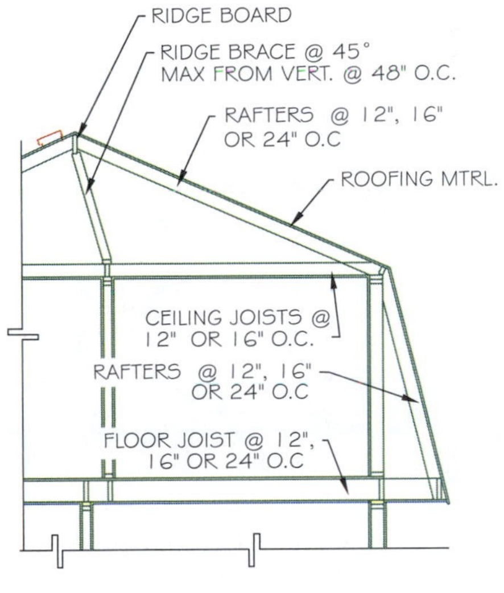

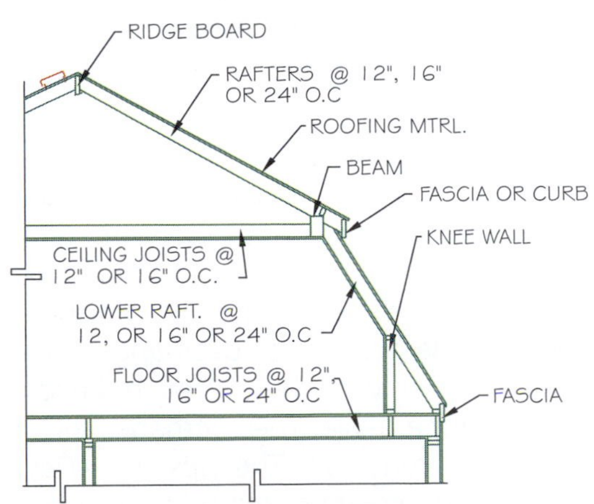

FIGURE 15.27 ■ *A gambrel roof can be constructed with or without a fascia or curb between the upper and lower roofs.*

rials. Figure 15.28 shows a plan view of a gambrel roof.

Hip Roofs

The ***hip*** roof (Figure 15.29) is a traditional shape that has a minimum of four surfaces instead of two. Because a hip roof has no gable end walls, it can be used to help eliminate some of the roof mass and create a structure with a smaller appearance. The inclined intersection between surfaces is called a hip. If built on a square structure, the hips will come together to form a point. If a hip roof is built on a rectangular structure, the hips will form two points with a ridge spanning the distance between

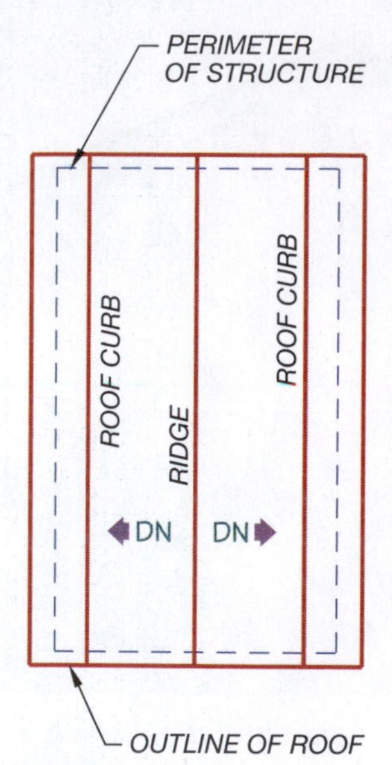

FIGURE 15.28 ■ *Each plane of a gambrel roof, as well as any hips or valleys that result from intersecting roofs, must be represented in plan view.*

FIGURE 15.29 ■ *A hip roof is made of four or more intersecting planes. This home uses a combination of hip roofs with a gable roof over the entry area to create a pleasing roof structure.* Courtesy Benigno Molina-Manriquez.

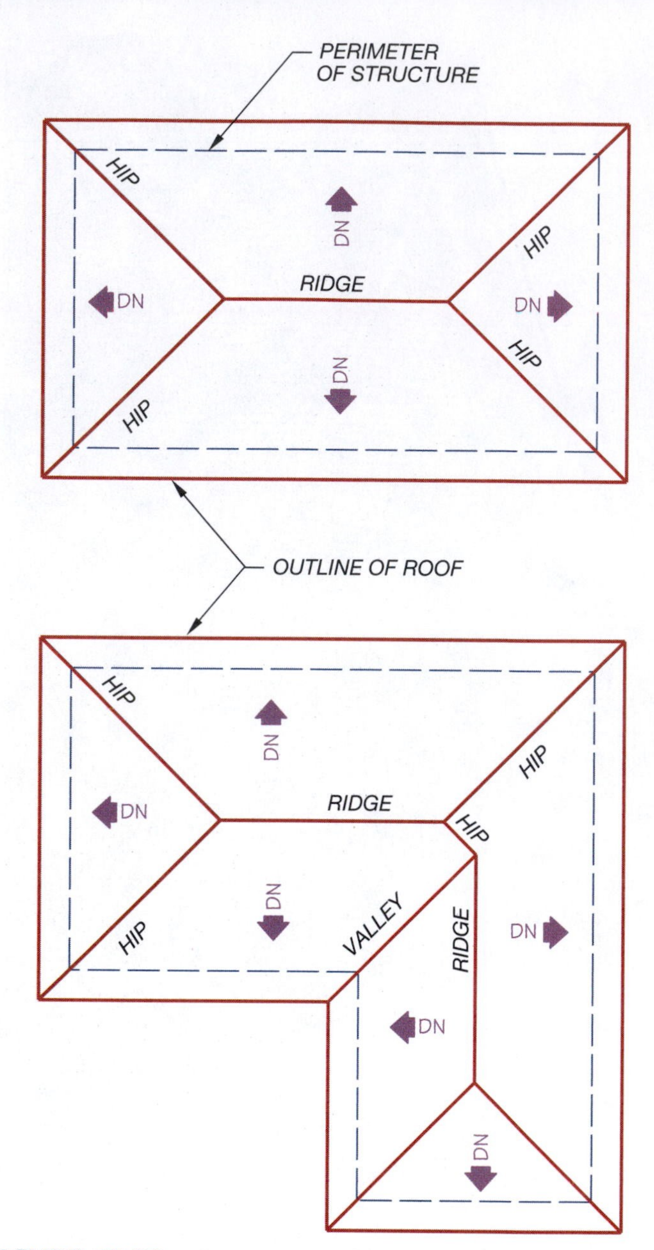

FIGURE 15.30 ■ *Representing hips and valleys in plan view.*

them. When hips are placed over an L- or T-shaped structure, an interior intersection will be formed; this is called a valley. The valley of a hip roof is the same as the valley of a gable roof. Hips and valleys are shown in plan view in Figure 15.30. The elements of a hip roof that must be represented include the ridge, overhangs, valleys, and hips. The ridge for a hip roof is located in the same manner as the ridge for a gable. The overhangs for a hip roof are generally of uniform size. The hips and valleys can be easily drawn if you remember three simple rules:

- Hips and valleys will always pass through the corner of two intersecting walls.
- Hips and valleys will always be represented on the plan view using an angle that is equal to half the angle of the intersecting walls. Generally this will mean that hips and valleys will be drawn at a 45° angle.
- Three lines will always be required to represent the intersections of hips, valleys, ridges, or overhangs on a roof plan.

An alternative method of framing a hip roof is to form the hip at the top of the roof and then terminate the hip at a gable end wall similar to the roofs in Figure 15.31. The size of the hip will depend on

FIGURE 15.31 ■ *A roof with a partial hip places the hip at the top of the roof and then uses a partial-gable end wall to terminate the roof.* Courtesy Katie Rowell.

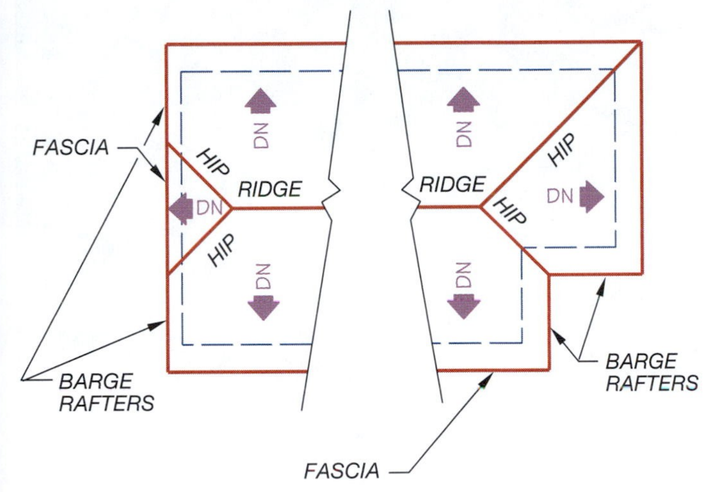

FIGURE 15.32 ■ *Representing a partial hip and a notched hip on roof drawings.*

the spacing of the framing members, but generally the hip will be three or four spaces wide. A second method of altering a hip roof is to remove one or more corners of the hip. This alteration is often done on an irregularly shaped structure to allow more light into what would have been a covered porch. Figure 15.32 shows how each hip shape is represented on a roof plan.

Dutch Hip Roofs

The ***Dutch hip*** roof is a combination of a hip and a gable roof (see Figure 15.33). The center section of the roof is framed using a method similar to that for a gable roof. The ends of the roof are framed with a partial hip that blends into the gable. A small wall (gable end wall)

FIGURE 15.33 ■ *A Dutch hip is a combination of a hip and a gable roof.* Courtesy CertainTeed.

is formed between the hip and the gable roofs, as seen in Figure 15.34. On the roof plan, the shape, distance, and wall location must be shown, as in the plan in Figure 15.35.

Mansard Roofs

The ***mansard*** roof has angled walls on all four sides of the structure to enclose the upper floor. A mansard roof can also be used as a parapet wall to hide mechanical equipment on the roof or to help hide the height of the upper level of a structure. An example is shown in Figure 15.36. Mansard roofs can be constructed in many different ways. Figure 15.37 shows two common methods of constructing a mansard roof. The roof plan for a mansard roof will resemble the plan shown in Figure 15.38.

Dormers

A ***dormer*** is an opening framed in the roof to allow for window placement. Figure 15.39 shows a dormer that has been added to provide light and ventilation to rooms in what would have been attic space. The dormer width can vary in size from the width required to place one window to the entire width of the structure. Figure 15.40 shows alternative dormer shapes. Dormers are most frequently used on traditional roofs such as the gable or hip. Figure 15.41 shows one of the many ways in which dormers can be constructed. Dormers are usually shown on the roof plan, as in Figure 15.42.

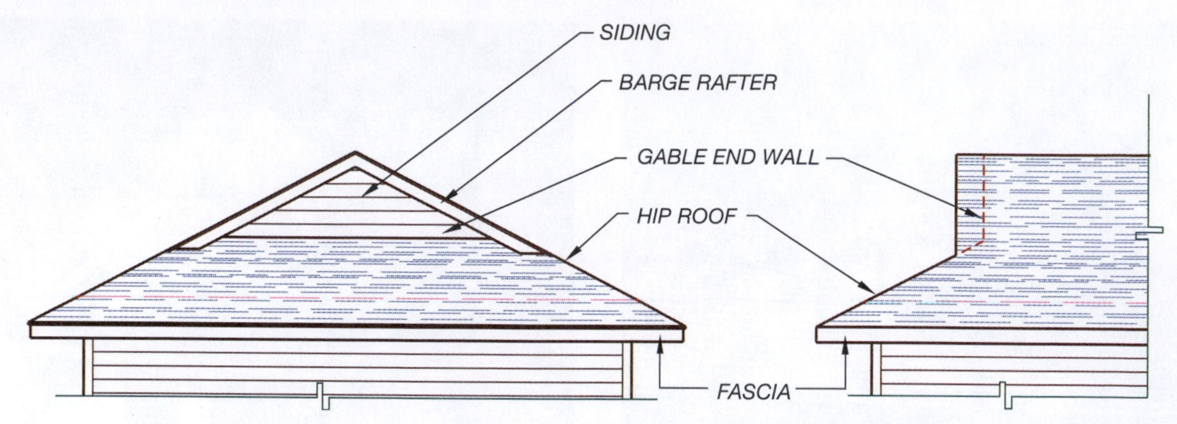

FIGURE 15.34 ■ *A wall is formed between the hip and the gable roof of a Dutch hip roof. The dashed line is used for reference only and can be omitted when drawing the elevations.*

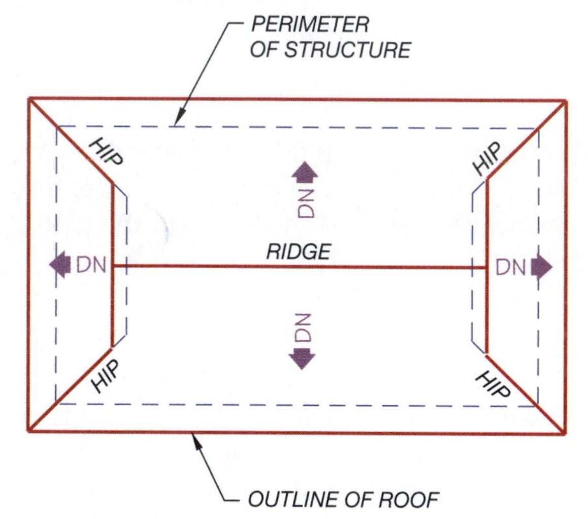

FIGURE 15.35 ■ *Representing a Dutch hip roof in plan view.*

FIGURE 15.36 ■ *Mansard roofs are used to help disguise the height of a structure.* Courtesy Metal Roofing Alliance.

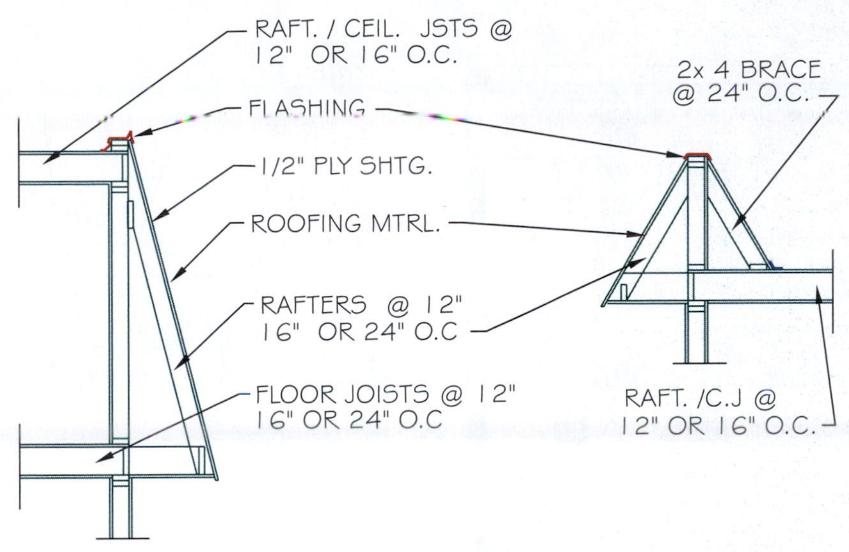

FIGURE 15.37 ■ *Common methods of constructing a mansard roof.*

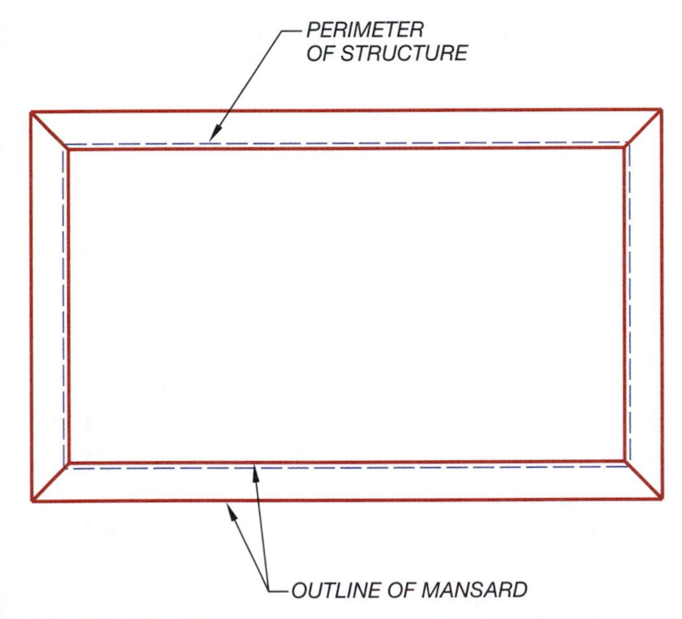

FIGURE 15.38 ■ *Representing a mansard roof in plan view.*

FIGURE 15.39 ■ *Dormers allow windows to be added to attic areas.* Courtesy Brit Frederick.

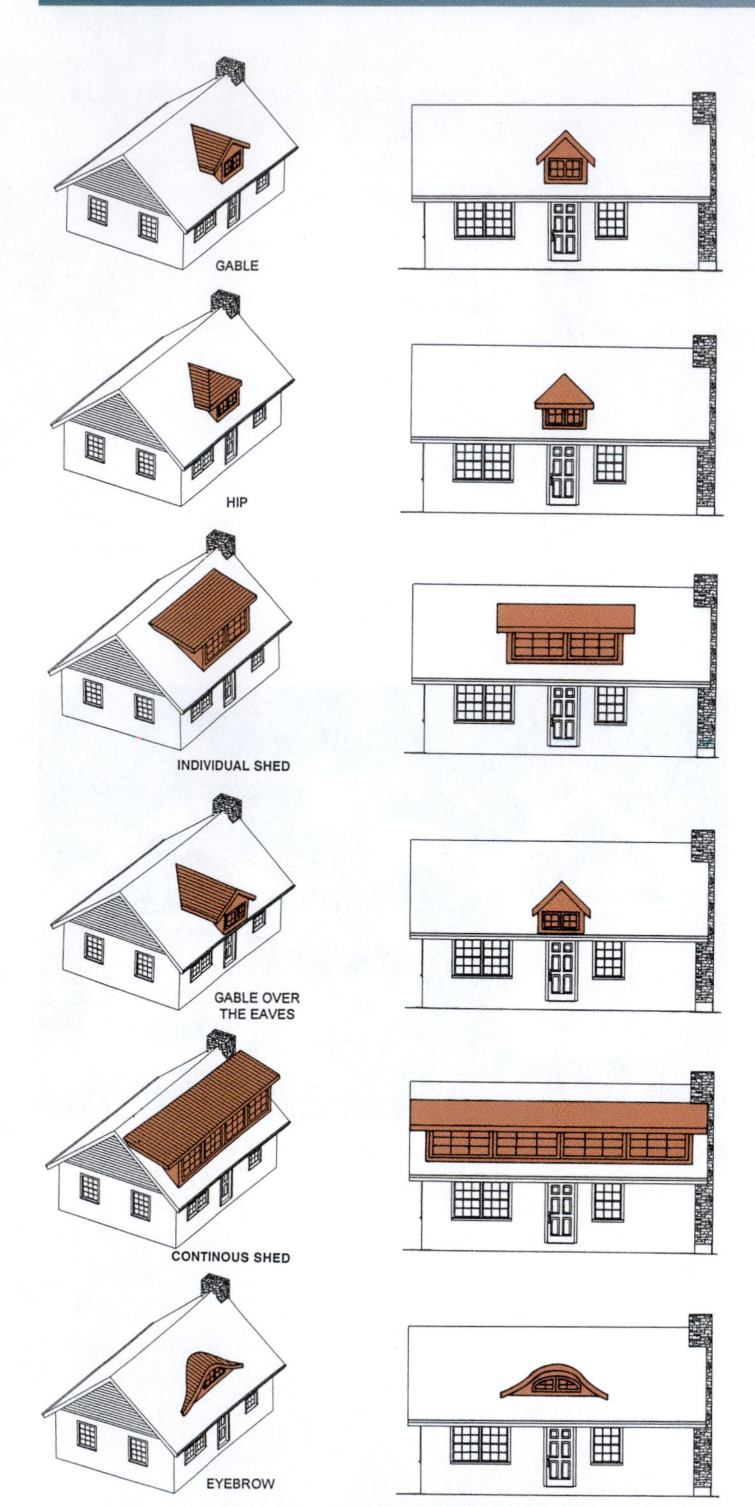

FIGURE 15.40 ■ *The width of a dormer can vary depending on the needs of the design. A typical dormer is based on the width of a window, but the dormer can extend to the full width of the structure.*

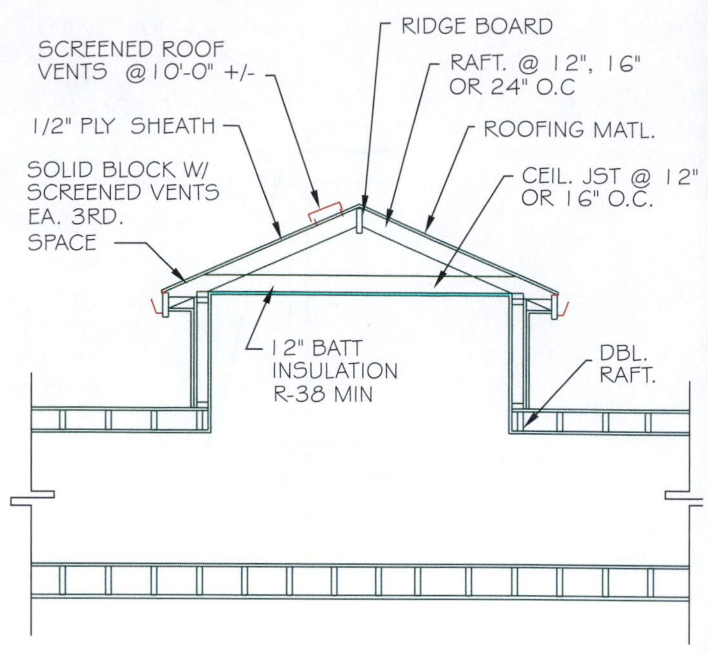

FIGURE 15.41 ■ *Typical components of dormer construction.*

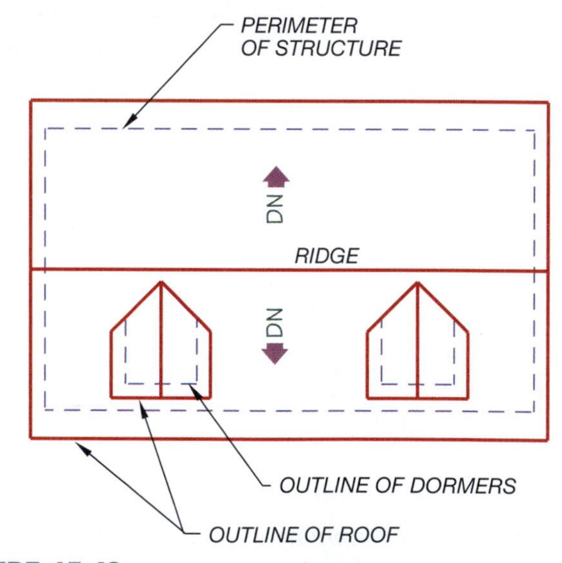

FIGURE 15.42 ■ *Dormers in plan view.*

ROOF MATERIALS

The material to be used on the roof depends on pitch, exterior style, the cost of the structure, and the weather. Common roofing materials include built-up roofing, composition and wood shingles, clay and cement tiles, and metal panels. In ordering or specifying these materials, the term ***square*** is used. This term describes an area of roofing that covers 100 sq ft (9.3 m^2). The drafter will need to be aware of the weight per square and the required pitch as the plan is being drawn. The weight of the roofing material will affect the size of the framing members all the way down to the foundation level. The material will also affect the required pitch and the appearance that results from the selected pitch.

Built-Up Roofing

Built-up roofing of felt and asphalt is used on flat or low-sloped roofs below a 3/12 pitch. When the roof has a low pitch, water will either pond or drain very slowly. To prevent water from leaking into a structure, built-up roofing is used because it has no seams. On a residence, a built-up roof may consist of three alternate layers of felt and hot asphalt placed over solid roof decking, similar to Figure 15.43. The decking is usually plywood. Gravel is often used as a finishing layer to help cover the felt. On roofs with a pitch over 2/12, coarse rocks 2 or 3'' (50 or 75 mm) in diameter are used to protect the roof and for appearance. When built-up roofs are to be specified on the roof plan, the note should include the number of layers, the material to be used, and the size of the finishing material. A typical note would be:

3 LAYER BUILT UP ROOF WITH HOT ASPHALTIC EMULSION BTWN. LAYERS WITH 1/4'' (6 mm) PEA GRAVEL

Other roofing materials suitable for low-sloped (1/4/12 minimum pitch) roofs and typical specifications that might be included on a low-sloped roof include:

- Modified bitumen: **MODIFIED BITUMEN SHEET ROOFING BY JOHNS MANVILLE OR EQUAL OVER LAYERS OF UNDERLAYMENT PER ASTM D226 TYPE I CEMENTED TOGETHER**

FIGURE 15.43 ■ *Built-up roofing of felt and asphalt is used on flat or low-sloped roofs below a 3/12 pitch.* Courtesy CertainTeed.

- Single-ply thermoplastic: **THERMOPLASTIC SINGLE-PLY ROOF SYSTEM BY SARNAFIL OR EQUAL INSTALLED PER ASTM D4434**
- Sprayed polyurethane foam: **SPF ROOFING BY MAINLAND INDUSTRIAL COATINGS, INC., APPLIED PER ASTM 1029**
- Liquid applied coating: **GREENSEAL LIQUID WATERPROOFING MEMBRANE OR EQUAL INSTALLED PER MANUF. SPECS**

Each material can be applied to a roof with minimum pitch of 1/4/12. Mineral-surface roll roofing can be used on roofs with a minimum pitch of 1/12. The extent of the note that is placed on the drawings will vary depending on the use of complete specifications. When specifications are provided, the notes to specify materials will be kept very generic on the drawings.

Shingles

Asphalt, fiberglass, and wood are the most typical types of shingles used as roofing materials. Most building codes and manufacturers require a minimum roof pitch of 4/12 with an underlayment of one layer of 15-lb felt. Asphalt and fiberglass shingles can be laid on roofs as low as 2/12 if two layers of 15-lb felt are laid under the shingles and if the shingles are sealed. Wood shingles must usually be installed on roofs having a pitch of at least 3/12. Asphalt and fiberglass are similar in appearance and application.

Asphalt shingles come in a variety of colors and patterns. Also known as composition shingles, they are typically made of fiberglass backing and covered with asphalt and a filler with a coating of finely crushed particles of stone. The asphalt waterproofs the shingle and the filler provides fire protection. The standard shingle is a three-tab rectangular strip weighing 235 lb per square. The upper portion of the strip is coated with self-sealing adhesive and covered by the next row of shingles. The lower portion of a three-tab shingle is divided into three flaps that are exposed to the weather (see Figure 15.44).

Composition shingles are also available in random widths and thicknesses to give the appearance of cedar shakes. These shingles weigh approximately 300 lb per square based on the shingle style and manufacturer. Both types of shingles styles can be used in a variety of conditions on roofs having a minimum slope of 2/12. The lifetime guarantees for shingles varies from 20 to 40 years (see Figure 15.45).

Shingles are typically specified on drawings in note form, listing the material, the weight, and the underlay-

FIGURE 15.44 ■ *Composition or three-tab shingles are a common roofing material on high-sloped roofs.* Courtesy CertainTeed.

FIGURE 15.45 ■ *Laminated composition shingles with a weight of 300 lb per square or greater are made with tabs of random width and length.* Courtesy CertainTeed.

ment. The color and manufacturer may also be specified. This information is often omitted in residential construction so as to allow the contractor to purchase a suitable brand at the best cost. A typical call-out would be:

- **235LB COMPOSITION SHINGLES OVER 15# FELT**
- **300LB COMPOSITION SHINGLES OVER 15# FELT**

FIGURE 15.46 ■ *Cedar shakes are a rustic but elegant roofing material.* Courtesy Tim Taylor.

- **GEORGIAN BRICK 425# COMPOSITION SHINGLES OVER 15# FELT BY CERTAINTEED LAID W/ 8'' EXPOSURE**
- **ARCHITECT 80 CLASS "A" FIBERGLASS SHINGLES WITH 5 5/8'' EXPOSURE OVER 15# FELT UNDERLAYMENT**

Wood can also be used for shakes and shingles, but their use may be restricted in areas of high fire danger. Wood shakes are thicker than shingles and are also more irregular in their texture (see Figure 15.46). Wood shakes and shingles are generally installed on roofs with a minimum pitch of 3/12 using a base layer of 15-lb felt. An additional layer of 15-lb × 18'' (457 mm) wide felt is also placed between each course or layer of shingles. Wood shakes and shingles can be installed over solid or spaced sheathing. The weather, material availability, and labor practices affect the type of underlayment used.

Depending on the area of the country, shakes and shingles are usually made of cedar, redwood, or cypress. They are also produced in various lengths. When shakes or shingles are specified on the roof plan, the note should usually include the thickness, the material, the exposure, the underlayment, and the type of sheathing. A typical specification for wood shakes would be:

MED. CEDAR SHAKES OVER 15# FELT W/15# × 18'' WIDE FELT BETWEEN EACH COURSE. LAY WITH 10 1/2'' EXPOSURE

Other materials such as Masonite and metal are also used to simulate shakes. Metal is sometimes used for roof shingles on roofs with a 3/12 or greater pitch. Metal provides a durable, fire-resistant roofing material. Metal shingles similar to those in Figure 15.47 are usually installed using the same precautions applied to as-

FIGURE 15.47 ■ *Metal roof shingles are both durable and fire-resistant.* Courtesy Metal Roofing Alliance.

phalt shingles. Metal is typically specified on the roof plan in a note listing the manufacturer, type of shingle, and underlayment.

Clay and Cement Tiles

Clay and cement tiles are often used for homes on the high end of the price scale or where the risk of fire is extreme. Although tile may cost twice as much as the better grades of asphalt shingle, it offers a lifetime guarantee. Tile is available in a variety of colors, materials, and patterns (see Figure 15.48a, b).

Roof tiles are manufactured in both curved and flat shapes. Curved tiles are often called Spanish tiles and come in a variety of curved shapes and colors. Flat or bar tiles are also produced in many different colors and shapes. Tiles are installed on roofs having a pitch of 2 1/2/12 or greater. Tiles can be placed over either spaced or solid sheathing. If solid sheathing is used, wood strips are generally added on top of the sheathing to support the tiles.

When tile is to be used, special precautions must be taken with the design of the structure. Tile roofs weigh between 850 and 1000 lb per square. These weights require rafters, headers, and other supporting members to be larger than normally required for other types of roofing material. Tiles are generally specified on the roof plan in a note, which lists the manufacturer, style, color, weight, fastening method, and underlayment. A typical note on the roof plan might be:

MONIER BURNT TERRA COTTA MISSION'S ROOF TILE OVER 15# FELT AND 1 × 3 SKIP SHEATHING. USE A 3" MINIMUM HEAD LAP AND INSTALL AS PER MANUF. SPECS

FIGURE 15.48a ■ *Tile is an excellent choice of roofing material because of its durability.* Courtesy Travis Frederick.

FIGURE 15.48b ■ *Many tile patterns are made to simulate Spanish clay tiles.* Courtesy Tim Taylor.

Metal Panels

Metal roofing panels often provide savings because of the speed and ease of installation. Metal roof panels provide a water- and fireproof material that comes with a warranty for a protected period that can range from 20 to 50 years. Lapped, nonsoldered metal roofing panels with lap sealant can be laid on a roof with a pitch of 1/2/12 or greater. Nonsoldered seam metal roofing panels without a lap sealant are required to be laid on a pitch of 3/12 or greater. Standing seam metal roof systems can be laid on a roof with a pitch of 1/4/12 or greater. Panels are typically produced in either 22- or 24-gauge metal in widths of either 18 or 24'' (see Figure 15.49). The length of the panel can be specified to meet the needs of the roof in lengths up to 40'. Metal roofing panels typically weigh between 50 and 100 lb

FIGURE 15.49 ■ *Metal is often selected for its durability and pleasing appearance.* Courtesy CertainTeed.

FIGURE 15.51 ■ *Flashing is used to protect valley intersections and also at roof projections such as skylights and chimneys.* Courtesy Michael Jefferis.

FIGURE 15.50 ■ *Special shingles are manufactured to protect the ridge and hip intersections. In addition to the ridge cap shingles, a continuous ridge vent also protects this roof.* Courtesy CertainTeed.

per square. Metal roofs are manufactured in many colors and patterns and can be used to blend with almost any material. Steel, stainless steel, aluminum, copper, and zinc alloys are most typically used for metal roofing. Steel panels are heavier and more durable than other metals but must be covered with a protective coating to provide protection from rust and corrosion. A baked-on acrylic coating typically provides both color and weather protection. Stainless steel does not rust or corrode but is more expensive than steel. Stainless steel weathers to a natural matte-gray finish. Aluminum is extremely lightweight and does not rust. Finish coatings are similar to those used for steel. Copper has been used for centuries as a roofing material. Copper roofs weather to a blue-green color and do not rust. In specifying metal roofing on the roof plan, the note should include the manufacturer, pattern, material, underlayment, and trim and flashing. A typical note would be:

24 GA. × 16" WIDE MEDALLION-LOK STANDING SEAM PANEL SYSTEM BY MCELROY METAL INSTALLED OVER 15# FELT AS PER MANUF. SPECS

Metal Flashing

In addition to using metal as a roof covering, metal is often used as a flashing. Flashing can be found at all levels of a structure. It is material used to protect intersections and joints from moisture intrusion. Roof flashing is an underlayment used to protect joints in the roof. Major joints that must be protected include hips, valleys, and ridges. Special shingles similar to those in Figure 15.50 that are specifically made for these applications usually protect hips and ridges. Valleys may be protected by either extra building paper or by metal flashing in high wind and snow areas. Figure 15.51 shows the use of metal flashing to protect a valley. Metal flashing is also used to protect openings in the roof for chimneys and skylights and to protect the fascia/roofing intersection. Flashing is not shown on the roof plan, but the location and type of flashing is specified in general roof notes.

ROOF VENTILATION AND ACCESS

The size of the attic space must be considered as the roof plan is drawn. The attic is the space formed between the ceiling and the roofing. The attic space must be provided with vents that are covered with 1/8'' (3.2 mm) screen mesh. These vents must have an area equal to 1/150 of the attic area. This area can be reduced to 1/300 of the attic area if a vapor barrier is provided on the heated side of the attic floor or if half of the required vents, but not more than 80% of the vents, are placed in the upper half of the roof area.

The method used to provide the required vents varies throughout the country. Vents may be placed in the gabled end walls near the ridge (see Figure 15.15). This allows the roof surface to remain vent-free. In some areas, a continuous vent is placed in the eaves, or a vent may be placed in each third rafter space. Vents placed near the ridge should be located on planes that are not visible to the line of sight on entering the residence. These vents are normally placed at approximately 10' (3000 mm) intervals, but the exact spacing will vary based on the size of the vent and the size of the attic space. In areas subject to high winds or snow pack, ridge vents can allow wind-driven moisture to enter the attic. Continuous vents can also be used at the ridge to eliminate the need to place holes in the roof (see Figure 15.50). Because of their lower profile, moisture is less likely to be driven into the attic through continuous ridge vents.

Attic Access

Consideration must also be given to how to get into the attic space if the space has an area of 30 sq ft (2.8 m^2) or greater and a height of 30'' (760 mm) or greater. The actual opening into the attic is usually shown on the floor plan, but its location must be considered when the roof plan is being drawn. The minimum size of the access opening is 22 × 30'' (560 × 760 mm) with 30'' (760 mm) minimum headroom. While planning the roof shape, the drafter must find a suitable location for the attic access that meets both code and aesthetic requirements. Code requires that the access be located in a hallway or other accessible location. The access should be placed where it can easily be reached but not where it will dominate a space visually. Avoid placing the access in areas such as the garage; areas with high moisture content, such as bathrooms and utility rooms; or in bedrooms that will be used by young children. A walk-in closet is an excellent location for the access, but it should be placed in an area that does not require the movement of stored material. Hallways provide an area to place an access that is easily accessible but not a focal point of the structure. Avoid placing the access in the garage if the 1-hour rating between the house and the garage will be compromised.

CHAPTER 15

Additional Readings

The following websites can be used as a resource to help you keep current with changes in roof materials.

ADDRESS	COMPANY OR ORGANIZATION
www.asphaltroofing.org	Asphalt Roofing Manufacturers
www.calredwood.org	California Redwood Association
www.cedarbureau.org	Cedar Shake & Shingle Bureau
www.certainteed.org	CertainTeed Corporation Asphalt Shingles
www.elkcorp.com/index.htm	Elk Corporation Asphalt Shingles
www.gaf.com	GAF Corporation Asphalt Shingles
www.jm.com	Johns Manville (roofing)
www.lpcorp.com	Louisiana Pacific (hardboard and wood siding)
www.ludowici.com	Ludowici Roof Tile
www.mca-tile.com	MCA, Inc. (tile roofing)
www.malarkey-rfg.com	Malarkey Corporation High Wind Asphalt Shingles
www.mcelroymetal.com	McElroy Metal
www.metalroofing.com	Metal Roofing Alliance
www.monier.com	Monier Lifetile Concrete Roofing
www.owenscorning.com	Owens-Corning Corporation Asphalt Shingles
www.riei.org	Roofing Industry Educational Institute
www.solatube.com	Solatube (skylights)
www.spri.org	Single Ply Roofing Institute
www.stone-slate.com	Slate/Select Inc.
www.sunoptic.com	Sunoptics (skylights)
www.velux.com	Velux (skylights)
www.zappone.com	Zappone Manufacturing (copper shingles)

CHAPTER 15

Roof Plan Components Test

DIRECTIONS

Answer the following questions with short, complete statements. Type the answers using a word processor or neatly print the answers on lined paper.

1. Place your name, the chapter number, and the date at the top of the sheet.
2. Type the question number and provide the answer in the form of a statement that includes part of the question. You do not need to write out the entire question.

Note: The answers to some questions may not be contained in this chapter and will require you to do additional research.

QUESTIONS

15.1. List and describe two different types of roof plans.
15.2. In describing roof pitch, what do the numbers 4/12 represent?
15.3. What angle represents a 6/12 pitch?
15.4. What is a barge rafter?
15.5. What are two advantages of using a flat roof?
15.6. What is the major disadvantage of using a flat roof?
15.7. List three traditional roof shapes.
15.8. Sketch and define the difference between a hip and a Dutch hip roof.
15.9. What are the two uses for a mansard roof?
15.10. List two common weights for asphalt or fiberglass shingles.
15.11. What are two common shapes of clay roof tiles?
15.12. What advantage do metal roofing panels have over other roofing materials?
15.13. What is the minimum headroom required at the attic access?
15.14. What is the minimum size of an attic access opening?
15.15. What type of roof is both a roof system and a framing system?

16

CHAPTER

Roof Plan Layout

The design of a roof is considered early in the process of designing a structure. The designer will often draw a preliminary roof plan to coordinate key design elements of the floor, roof, and elevations and to project the preliminary elevations. The actual drawing of the roof plan for the working drawings can be completed once the design has been finalized. To complete the roof plan, the technician must be familiar with the lines and symbols used to represent roofing material and the type of roof framing system to be used. This chapter examines the lines and symbols used to represent the roof shape, structural materials, nonstructural materials, and dimensions. This chapter discusses drawings for gable, hip, and Dutch hip roof framing systems.

REPRESENTING ROOF PLAN COMPONENTS

The steps to complete a roof plan include representing the roof shape, representing nonstructural materials, dimensioning major components, and providing text to explain the required equipment. Each area is examined prior to the layout of the roof. Structural materials used to build the roof are explored in Chapters 21 and 24.

Representing Basic Roof Shapes

The first step in determining the roof shape is to trace the outline of the exterior walls and all exterior support posts. Methods and layers for representing walls and other features are explored later in this chapter. Once the walls are drawn, the shape of the roof can be explored. Assuming that a gable roof is to be drawn, the location of all ridges will need to be determined. On a rectangular structure, the ridge can run parallel to the longest wall and be centered between the two longest walls. A second alternative allows the ridge to be located based on the two short walls of the rectangle. Figure 16.1 shows two options for placing a ridge on a rectangular shaped home and the resulting shape that would be seen when looking at the home. Several options are available for an L-shaped structure including designs using one, two, or three ridges. Fig-

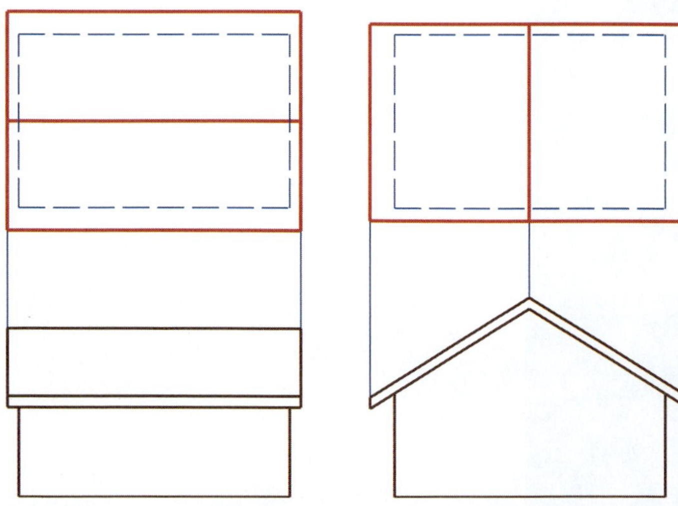

FIGURE 16.1 ■ *The ridge for a rectangular home will usually place the ridge parallel to the two longest walls. Placing the ridge parallel to the two shortest walls will provide a higher ridge line.*

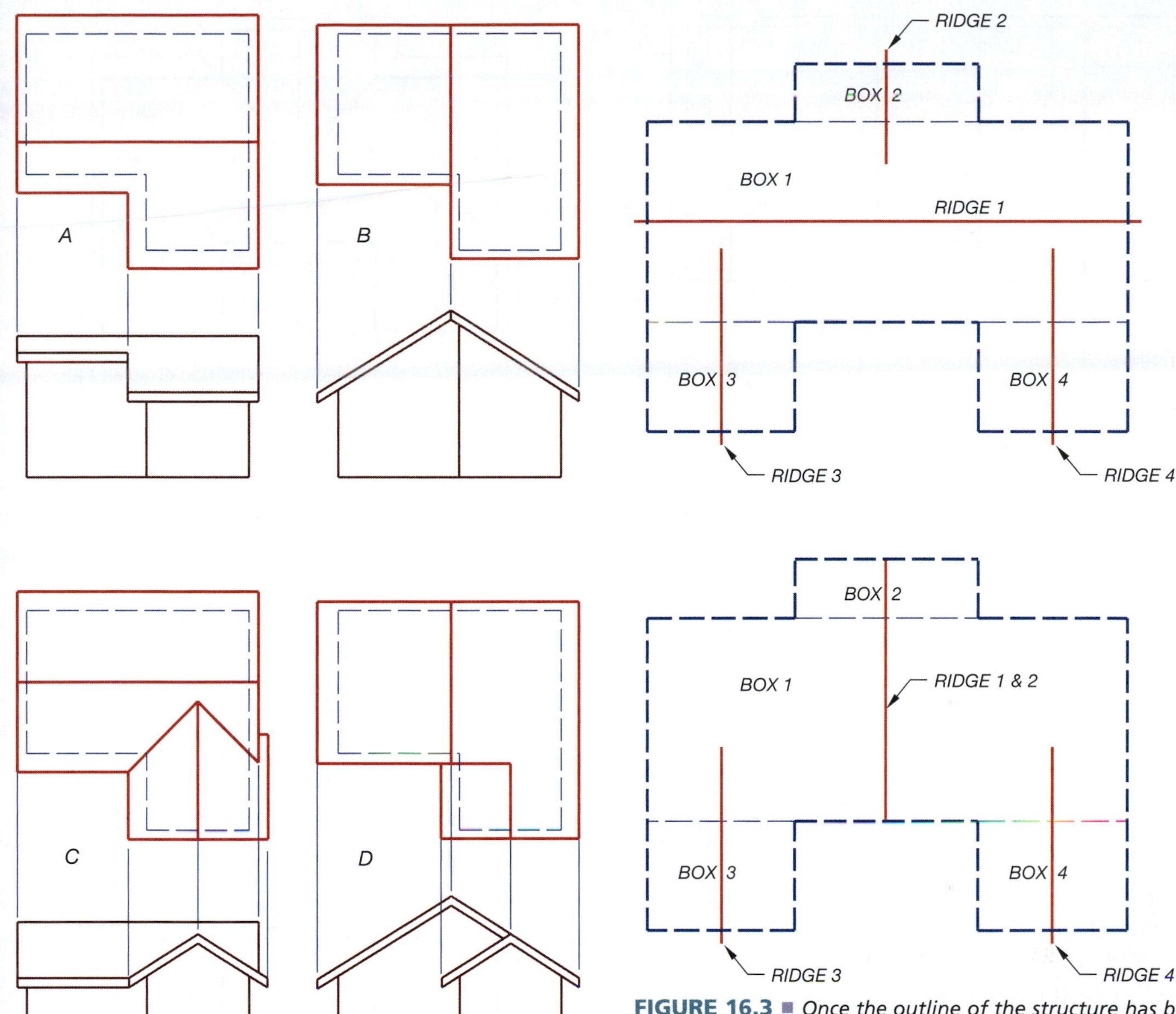

FIGURE 16.2 ■ *Common roof options for an L-shaped structure with designs using single and multiple ridges include (A) one ridge parallel to longest wall. (B) One ridge perpendicular to the longest wall. (C) Two ridges, with the shortest ridge perpendicular to the long ridge. A valley is created at each side of the short ridge where the planes intersect. (D) Two parallel ridges.*

FIGURE 16.3 ■ *Once the outline of the structure has been determined, draw lines to divide the home into the fewest number of boxes. Several options will be possible for each structure. If necessary, draw several options and select the one that presents the most pleasing shape. With the boxes determined, the ridges for each box can be located.*

ure 16.2 shows options for the roof layout and the shapes that will result. The design that is selected will depend on the desires of the owners and the goals to be achieved.

The easiest method to determine the shape of the roof is to trace the outline of the house and then divide the overall shape into the fewest number of boxes. Figure 16.3 shows two alternatives for this preliminary design step for an irregularly shaped home. Once the structure has been broken into boxes, the ridge for each area can be located by drawing a line through the midpoint of the box. When the roof pitches are unequal or the walls are to be framed with different heights, draw a simple box similar to Figure 16.4 and represent the desired roof pitch to determine the location of the ridge.

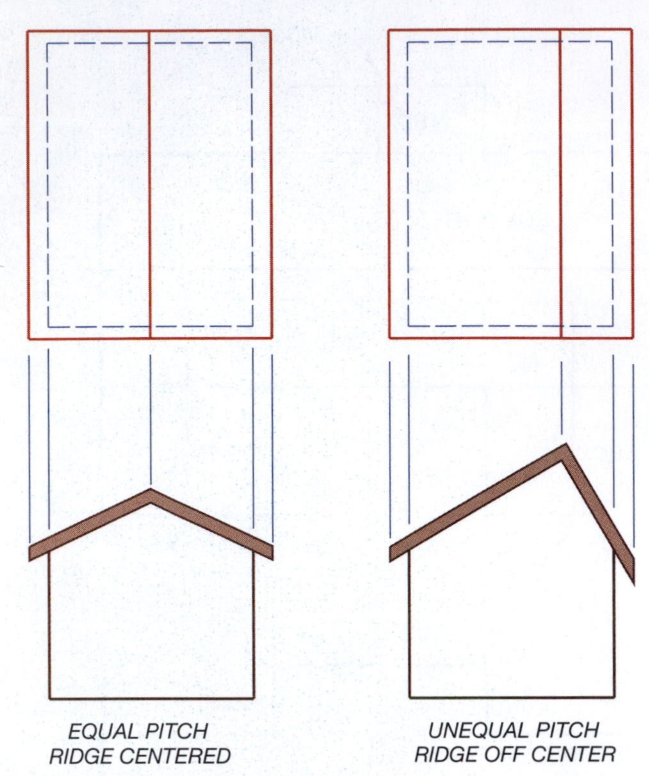

FIGURE 16.4 ■ *Roof pitches affect the location of the ridge. When roof pitches are unequal, draw a simple elevation and project the ridge location to the roof plan.*

The elevation will allow you to project the ridge location to the roof plan.

Once the ridges are located, the overhangs and valleys required for intersecting planes can be represented. The overhang size will be dependent on:

- The style of home to be designed
- The roof pitch
- The area of the country and the resulting natural elements
- The amount of shade desired during heating and cooling cycles

For the examples in this chapter, a 24'' (600 mm) overhang will be assumed for all eaves parallel to the ridge and a 12'' (300 mm) overhang for eaves perpendicular to the ridge.

Remember the three rules from the previous chapter:

- A valley will always pass through the corner of two intersecting walls.
- A valley will always be represented on the plan view using an angle equal to half the angle of the intersecting walls. Generally this will mean that valleys will be drawn at a 45° angle.

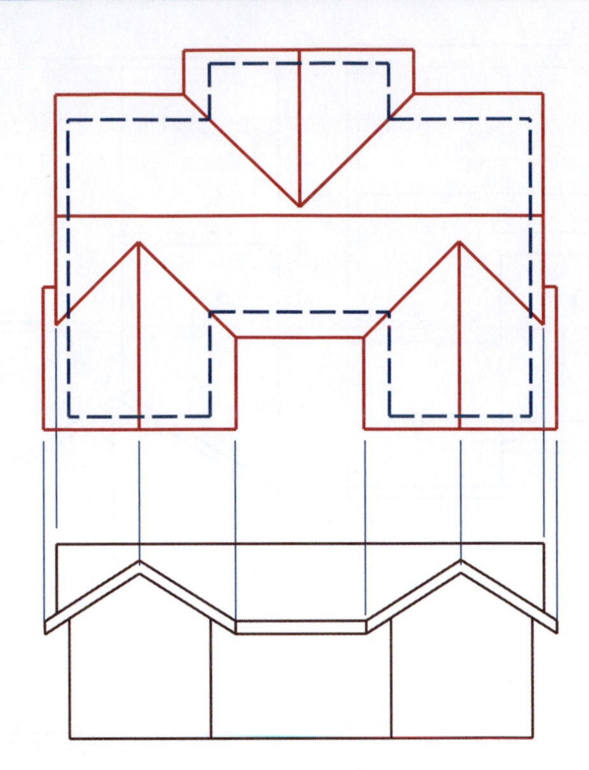

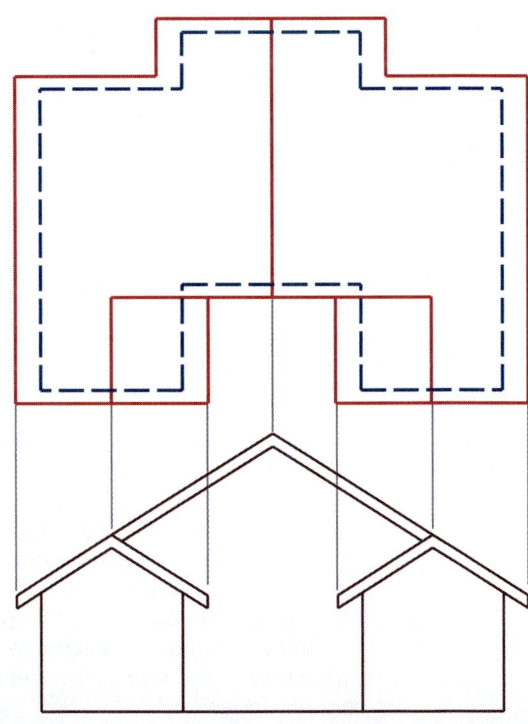

FIGURE 16.5 ■ *Once the ridges have been located, overhangs and valleys can be represented.*

- Three lines will always be required to represent the intersections of valleys, ridges, and overhangs on a roof plan.

The valleys and overhangs that result from applying these guidelines are shown in Figure 16.5.

Representing Varied Ridge Heights

The distance between supporting walls and the pitch must be considered to determine how the roof intersections are represented on the roof plan. Notice in Figure 16.6 that the shape of the roof changes dramatically as the width between the walls is changed. When one roof is taller than another, the change can be made by using a wall similar to the example on the left in Figure 16.7, or by extending the lower plane, as in the example on the right in Figure 16.7. Remember that when the pitches are equal, the wider the distances between the walls, the higher the roof.

When the walls are equally spaced but not perpendicular, the valley and ridge intersection will occur along a line, as shown in Figure 16.8. With unequal distances between the supporting walls, the valleys and ridges can be drawn as shown in Figure 16.9.

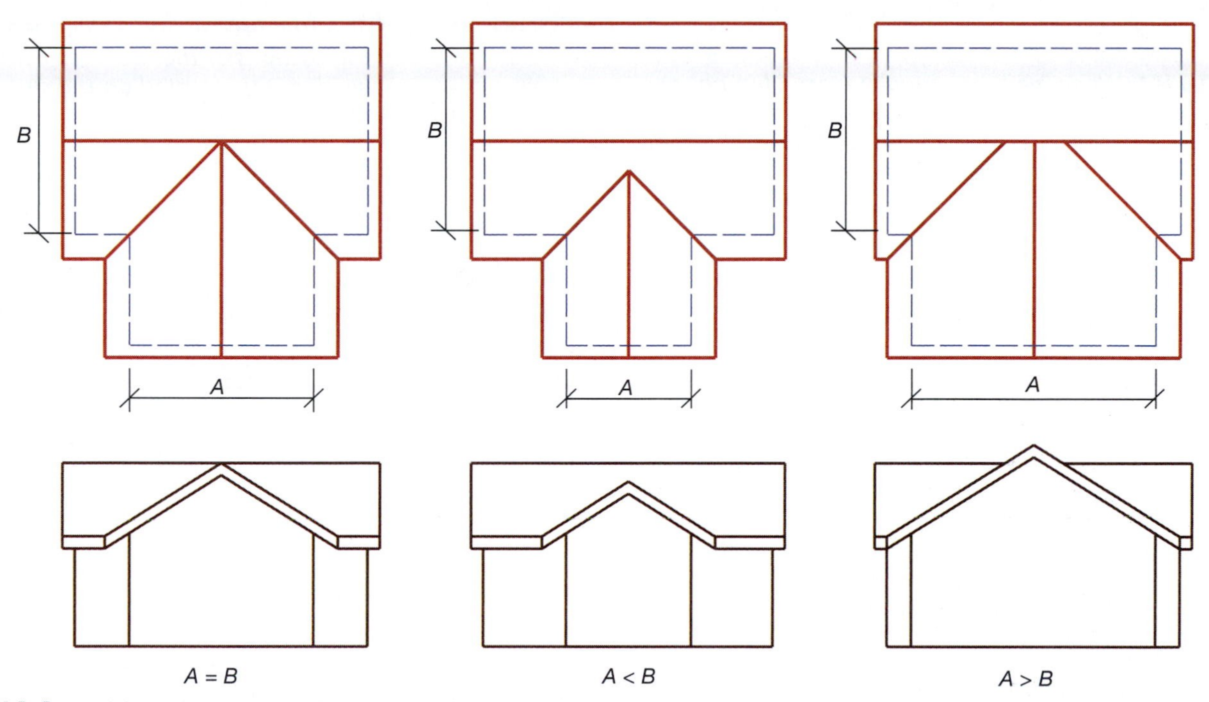

FIGURE 16.6 ■ *Although the basic shape remains the same, the roof plan and elevations change as the distance between walls varies.*

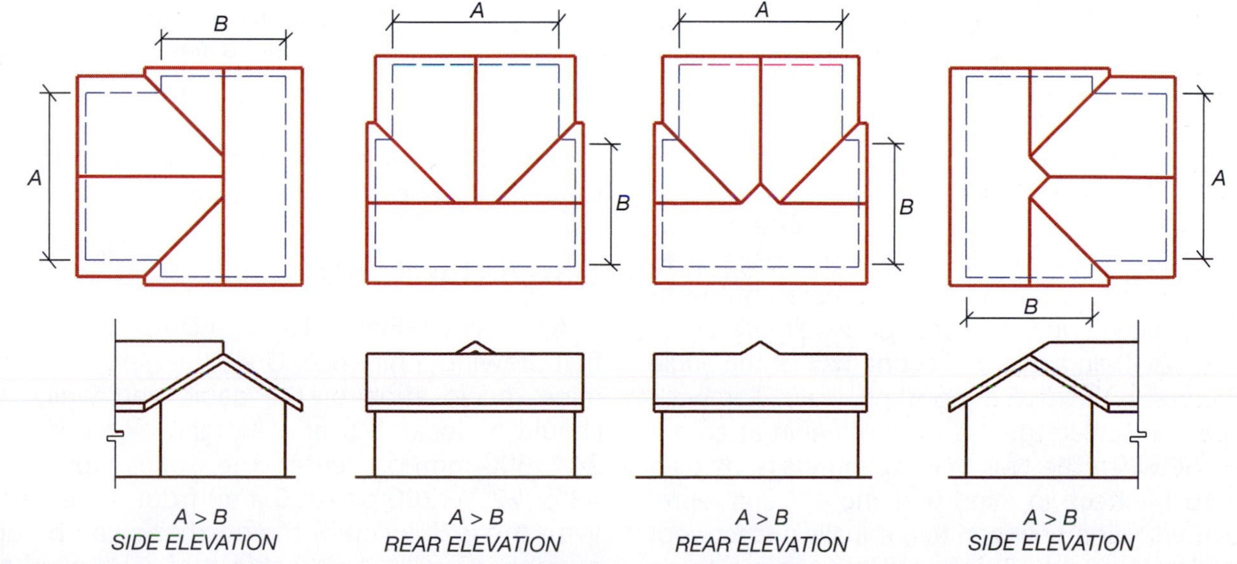

FIGURE 16.7 ■ *When roof heights are unequal, the transition can be made by extending the lower roof pitch until it matches the height of the upper roof (right) or by allowing a gable roof to be formed between the two roofs (left).*

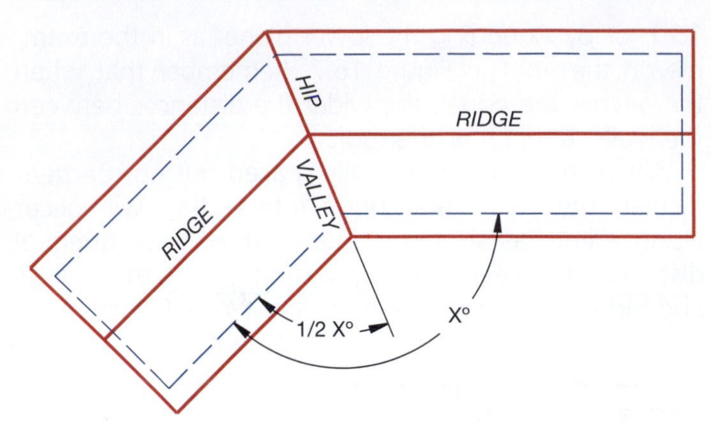

FIGURE 16.8 ■ *When walls are equally spaced but not perpendicular, the roof valleys and hips are formed at an angle that is half of the angle of the intersecting walls.*

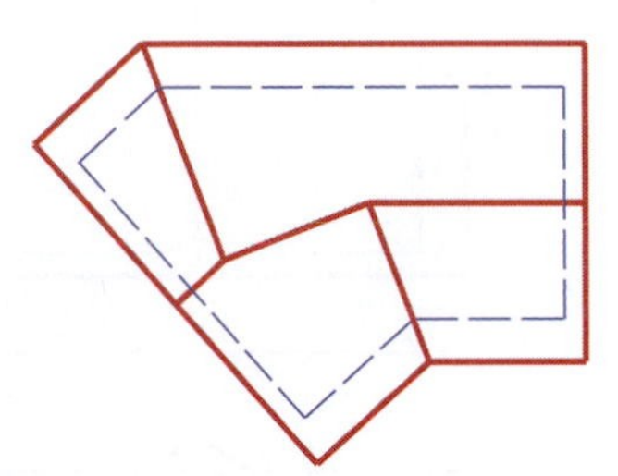

FIGURE 16.9 ■ *When walls are unequally spaced and the ridge remains at the midpoint, each valley is drawn at an angle equal to half of the angle between the supporting walls and a hip is formed between the endpoints of the two ridges.*

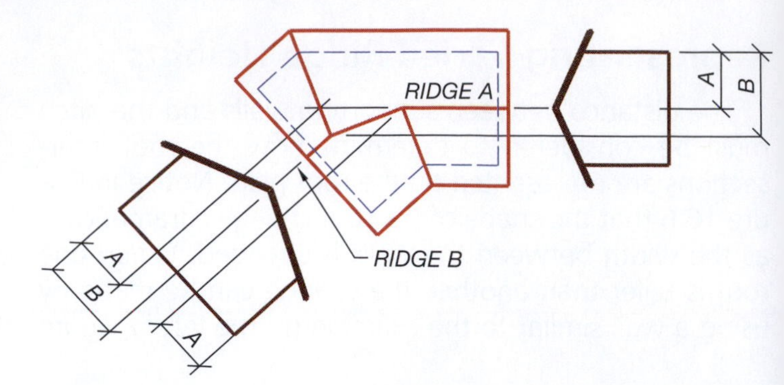

FIGURE 16.10 ■ *The true roof shape can be seen by drawing sections of the roof. Project the intersections represented on the sections onto the roof plan.*

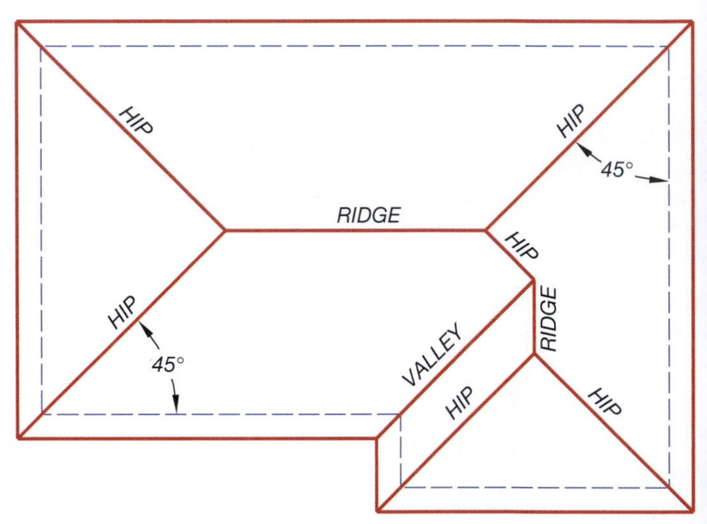

FIGURE 16.11 ■ *A hip (exterior corner) or a valley (interior corner) is formed between two intersecting roof planes. When each plane is framed from walls that are perpendicular to each other and the slope of each roof is identical, the valley is drawn at a 45° angle.*

Making such a drawing is often a difficult procedure for an inexperienced technician. The process is simplified if you draw partial sections by the roof plan, as shown in Figure 16.10. By comparing the heights and distances in the sections, you can often visualize the intersections better.

Hip Roofs

If a hip roof is to be drawn, the hips, or external corners of the roof, can be drawn in a manner similar to that used to locate the valleys. The hips represent the intersection between two roof planes and are represented by a line drawn at an angle that is one-half of the angle formed between the two supporting walls. For walls that are perpendicular, the hip will be drawn at an angle of 45° between the two intersecting walls, as seen in Figure 16.11. Keep in mind that the 45° line represents the intersection between two equally pitched roof planes in plan view only. This angle has nothing to do with the slope of the roof. The angle that represents the roof pitch can only be shown in an elevation of the structure. Drawing elevations will be introduced in Chapters 17 and 18.

Dutch Hip Roofs

As shown in Figure 16.12, a Dutch hip is started by first drawing a hip roof. Once the hip is drawn, determine the location of the gable end wall. The wall should be located over a framing member, spaced at 24'' (600 mm) on center. The wall is typically located 48 or 72'' (1200 or 1800 mm) from the exterior wall. With the wall located, the overhang can be drawn in a manner similar to a gable roof. The overhang line will intersect the hip lines to form the outline of the Dutch hip.

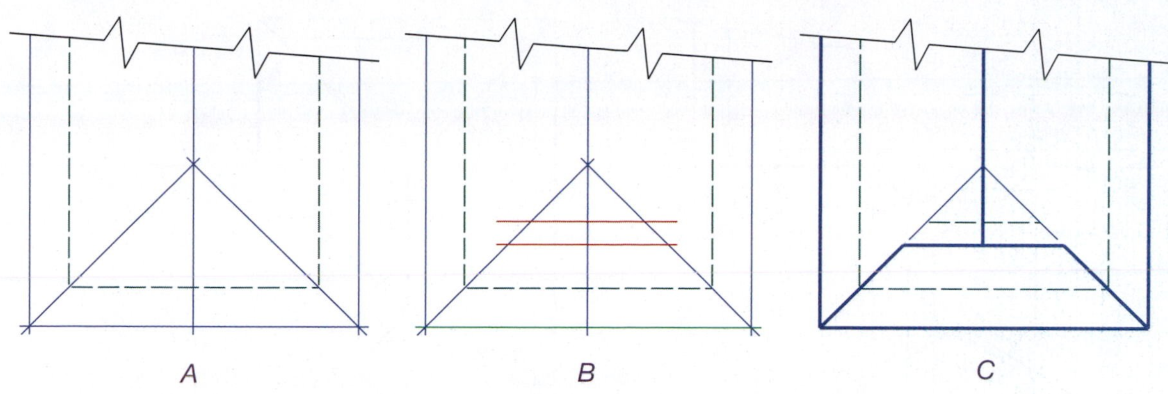

FIGURE 16.12 ■ *The layout of a Dutch hip roof can be completed in three steps: (A) Establish the ridge and hip locations; (B) locate the gable end wall; and (C) converting the lines to the proper layer, lineweight, and linetype to complete the roof outline.*

Bay Projections

If a bay window is to be included on the floor plan, special consideration will be required to draw the roof covering it. Figure 16.13 shows the steps required to lay out a bay roof. After drawing the overhangs, the layout of the roof can be eased by squaring the walls of the bay. Lay out the line of the ridge from the hips and valleys created by the intersecting rectangles. The true hips over the bay are drawn with a line from the intersection of the roof overhangs, which extends up to the end of the ridge. The layout is the same if the window bay is parallel to the gable end wall. Use the same layout procedure from the first two steps. Since the roof terminates at the gable end wall, trim all projection lines that extend past the line that represents the gable end wall. The finished layout can be seen in Figure 16.14.

Roof Intersections with Varied Wall Heights

Another feature common to roof plans is the intersection between two roof sections of different heights. Figure 16.15 shows how an entry portico framed with a 10' (3000 mm) wall would intersect with a residence framed with walls 8' (2400 mm) high. The roof pitch must be known to determine where the intersection between the two roofs will occur. In this example a

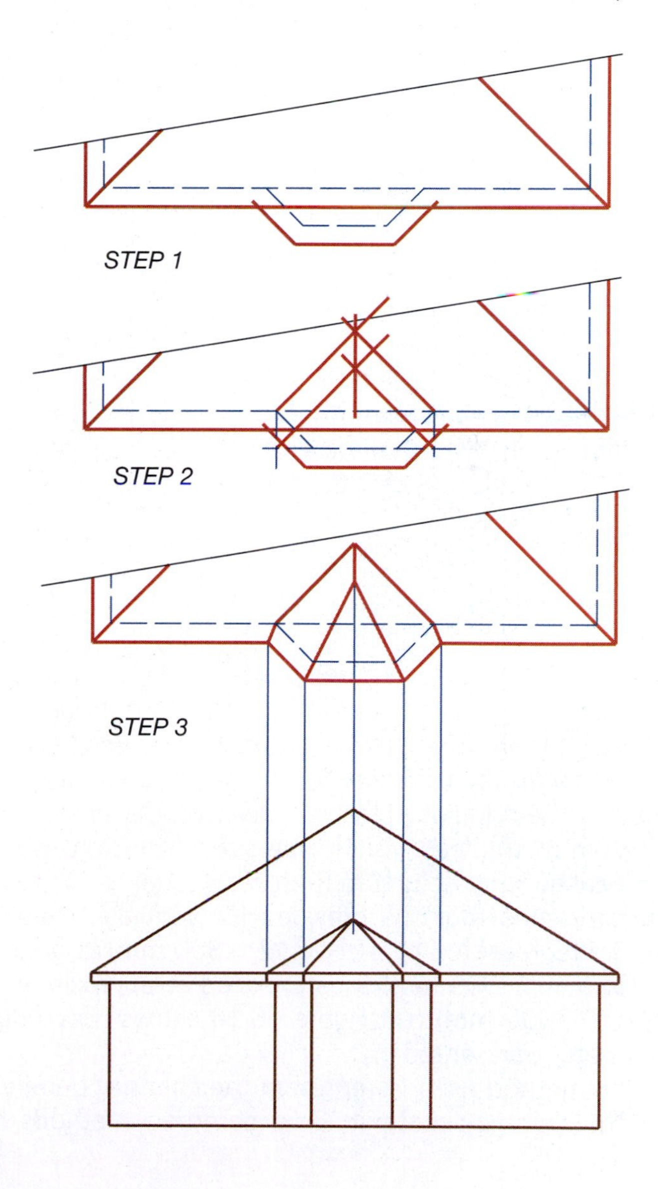

FIGURE 16.13 ■ *A bay roof can be completed in three steps: (1) The roof to cover a bay can be represented by drawing the outline of the bay and the desired overhangs parallel to the bay walls. (2) Draw lines to represent the roof that would be created if the bay were a rectangle. This will establish the ridge and the intersection of the ridge with the main roof. (3) Draw the hips created by a rectangular bay to establish the point where the true hips intersect the ridge. The true hips will start at the intersection of the overhangs, pass through the intersection of the bay walls, and end at the ridge.*

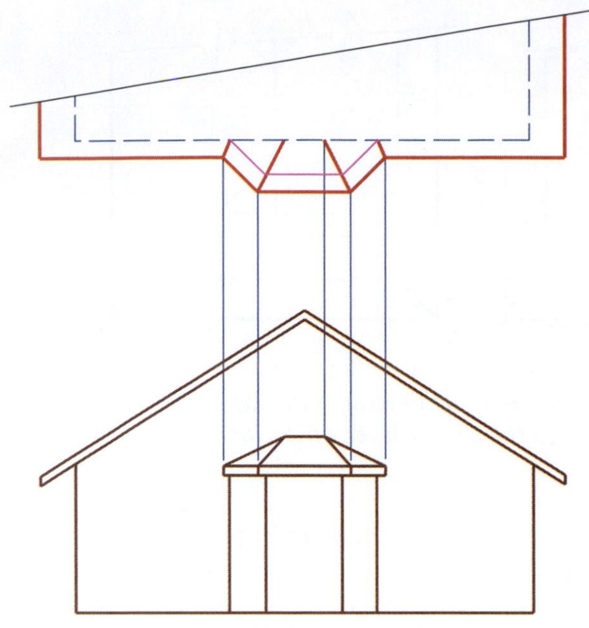

FIGURE 16.14 ■ *A bay parallel to a gable end wall is drawn using the first two steps shown in Figure 16.13. The only difference is that all lines terminate when they reach the gable end wall.*

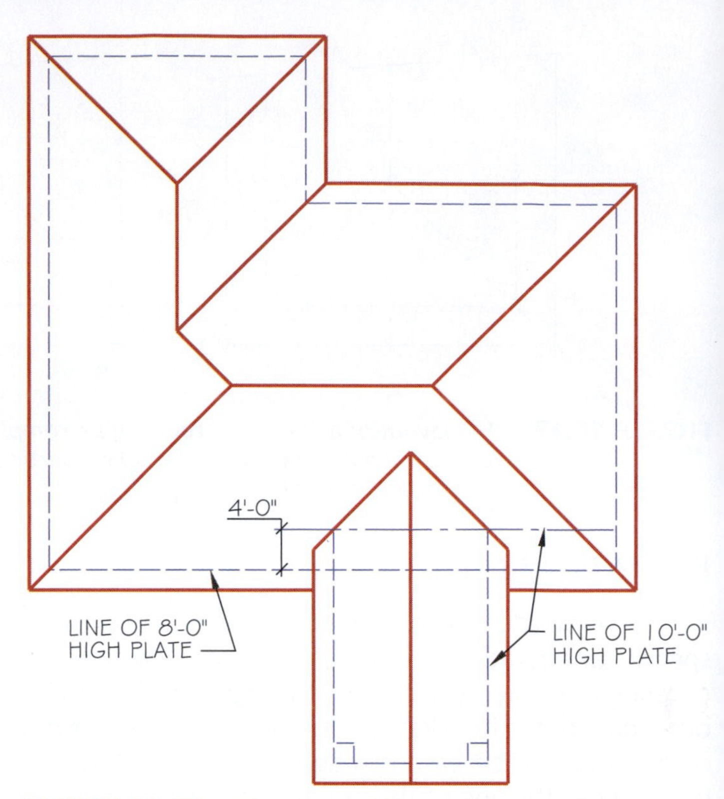

FIGURE 16.15 ■ *The slope of the roof must be known to draw the intersection between roofs built on walls of differing heights. With a 6/12 pitch, a horizontal distance of 48'' (1200 mm) would be required to reach the starting height of the upper roof (6'' per foot × 4' = 24'' rise). By projecting the line of the upper walls to a line 48'' in from the lower wall, the intersection of the two roof planes can be determined. A valley will be formed between the two intersecting roofs starting at the intersection of the wall and the line representing the rise.*

6/12 pitch was used. At this pitch, the lower roof must travel in a horizontal direction 4' (1200 mm) before it will be 10' (3000 mm) high. A line was offset 48'' (1200 mm) from the wall representing the horizontal distance. The line that represents the 10' high plate is extended to the line that was just drawn to represent where the 8' roof reaches 10' high. The valley between the two roofs will occur where the lines representing the higher walls intersect the line representing the horizontal distance.

Representing Nonstructural Material

Vents, chimneys, skylights, solar panels, diverters, cant strips, slope indicators, and downspouts are the most common nonstructural materials that will need to be shown on the roof plan. Vents are typically placed as close to the ridge as possible, usually on the side of the roof that is the least visible. The size of the vent varies with each manufacturer, but an area equal to 1/300 of the attic area must be provided for ventilation. Vents can be represented by a circle 12'' (300 mm) in diameter or a 12'' (300 mm) square placed at approximately 10'-0'' (3000 mm) o.c. Figure 16.16 shows how ridge vents are represented.

The method used to represent the chimney depends on the chimney material. Two common methods of representing a chimney can be seen in Figure 16.16. A metal chimney can be represented using a circle 14'' (300 mm) in diameter. A masonry chimney can range from a minimum of 16'' (400 mm) square, up to matching the size used to represent the fireplace on the floor plan. The location of skylights can usually be determined from their location on the floor plan. As seen in Figure 16.17, the opening in the roof for the skylight does not have to align directly with the opening in the ceiling. The skylight is connected to the ceiling by an enclosed area, called a chase or well. By adjusting the angle of the chase, the size and location of the skylight can be adjusted. If solar panels are to be represented, the size, angle, and manufacturer should be specified.

The amount of rainfall determines the need for gutters and downspouts. In semiarid regions, a metal strip

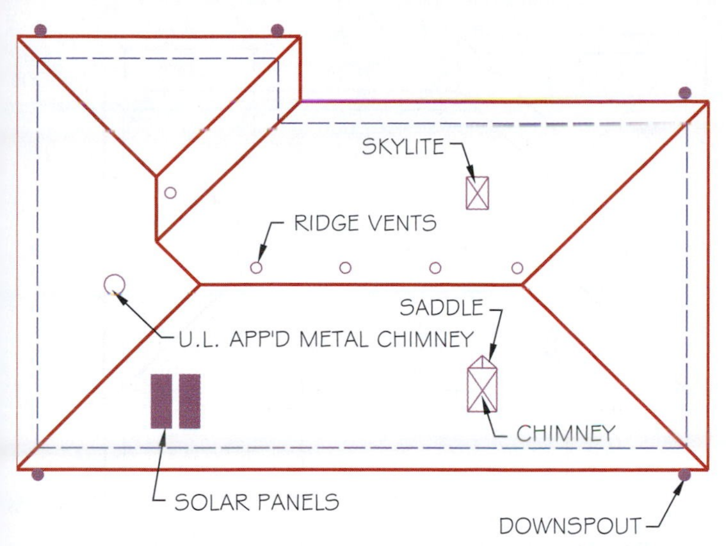

FIGURE 16.16 ■ *Nonstructural materials shown on the roof plan.*

FIGURE 16.17 ■ *A skylight is connected to the ceiling by an enclosed area called a chase. By altering the angle of the chase, the location of the ceiling opening can be altered.* Courtesy Velux.

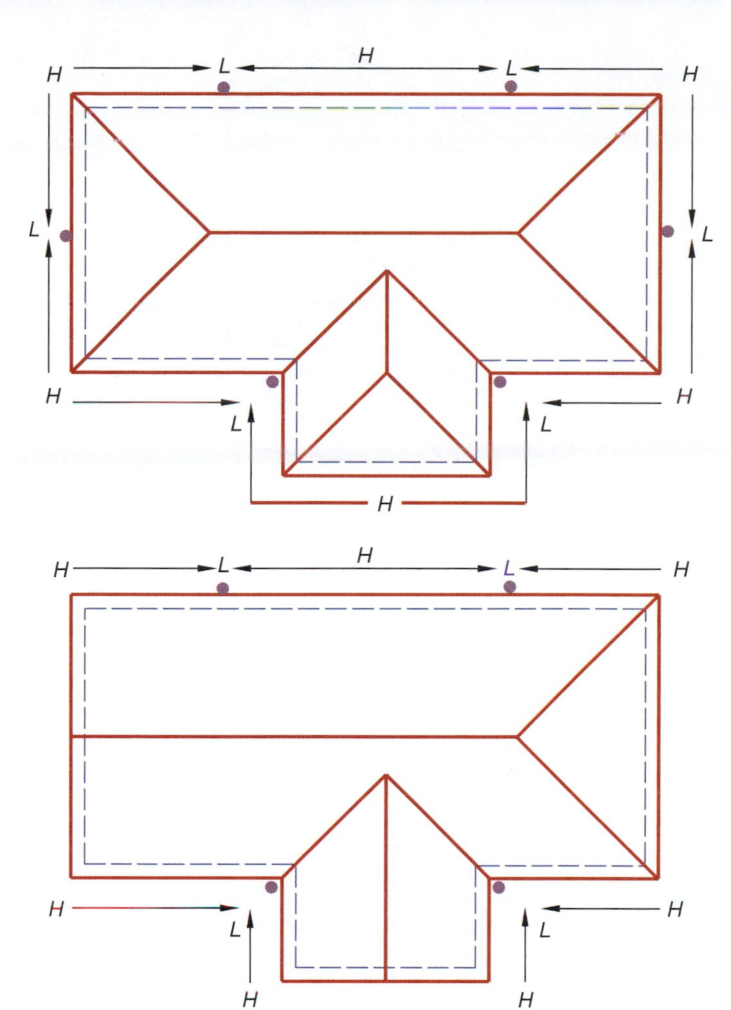

FIGURE 16.18 ■ *The locations of gutters and downspouts will be determined by the type of roof and the amount of rainfall to be drained. Gutters must be sloped to the downspouts, which should be placed so that no view is blocked.*

called a ***diverter*** can be placed on the roof to route runoff on the roof from doorways. The runoff should be diverted to an area of the roof where it can drop to the ground and be adequately diverted from the foundation. In wetter regions, gutters should be provided to collect and divert the water collected by the roof. Local codes will specify if the drains must be connected to storm sewers, private drywell, or a splash block. The downspouts that bring the roof runoff to ground level can be represented by approximately a 3'' (70 mm) circle or square. Care should be taken to keep downspouts out of major lines of sight. Each downspout can drain approximately 20' (6100 mm) of roof on each side of the downspout, allowing the downspouts to be spaced at approximately 40' (12 200 mm) intervals along the eave. This spacing can be seen in Figure

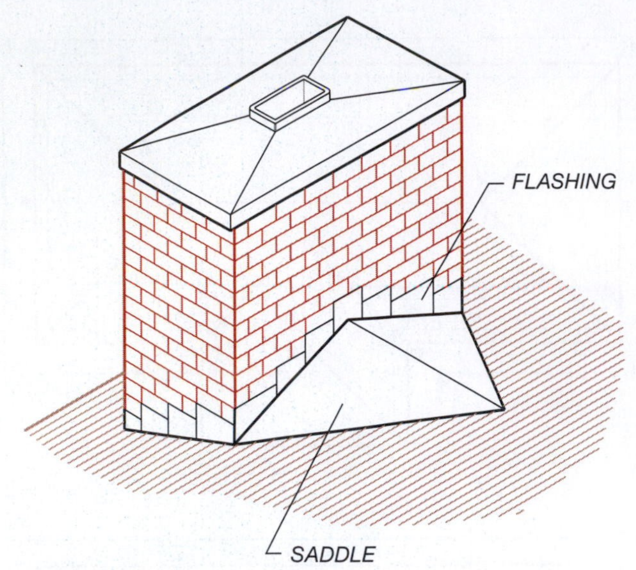

FIGURE 16.19 ■ *A saddle is used to divert water around the chimney.*

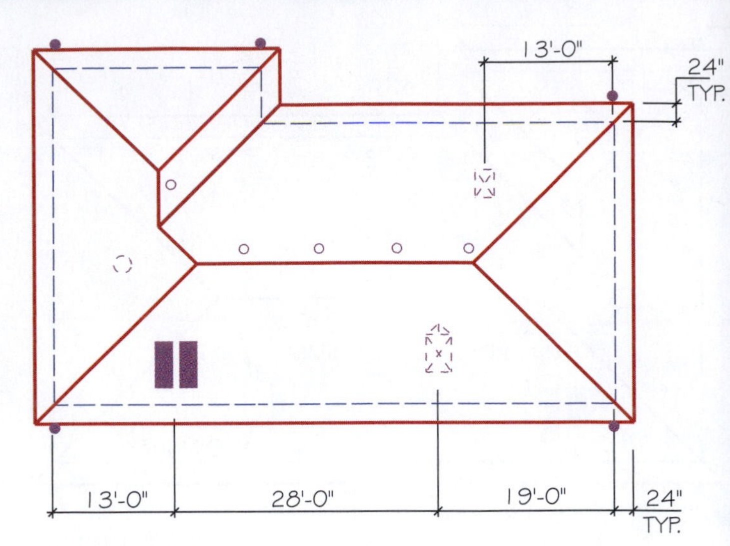

FIGURE 16.20 ■ *Dimensions should be placed by using leader and extension lines or in a note.*

16.18. The distance between downspouts will vary depending on the amount and rate of rainfall and should be verified with local manufacturers. Common methods of showing downspouts can be seen in Figure 16.16. The ***saddle*** is a small gable built behind the chimney to divert water away from the chimney, as seen in Figure 16.19.

Dimensions

The roof plan requires very few dimensions. Typically only the overhangs and openings are dimensioned. These may even be specified in note form rather than with dimensions. If trusses are to be used to frame the roof, be sure and locate items on the roof with modular dimensions. Use the following guidelines to keep dimensions modular:

- Dimensions should be expressed from the outer edge of an exterior wall to the center of the opening.
- Dimensions should be expressed in an odd, whole number of feet so that trusses will not have to be altered.
- Dimensions from opening to opening should be from center to center expressed in an even, whole number of feet.

Figure 16.20 shows how dimensions are placed on roof and framing plans. In drawing framing plans, beam locations are often dimensioned to help in estimating the materials required.

Notes

As with the other drawings, notes on the roof plan can be divided into general and local notes. General notes might include the following:

Vent notes
Sheathing information
Roof covering
Eave sheathing
Pitch

Material that should be specified in local notes includes the following:

Skylight type, size, and material
Chimney caps
Solar panel type and size
Cant strips and saddles

In addition to these notes, a title and scale should be placed below the drawing. An example of a completed roof plan can be seen in Figure 16.21.

CONSIDERATIONS IN DRAWING A ROOF PLAN

The outline to represent the roof may be placed on the floor plan for a very simple home. Related information, such as vents, is then specified on the elevations and building section. A better practice is to place the information on a separate roof plan. The roof plan should be created in the file containing the floor plan and placed directly over the floor plan. The walls of the floor plan can be traced to form

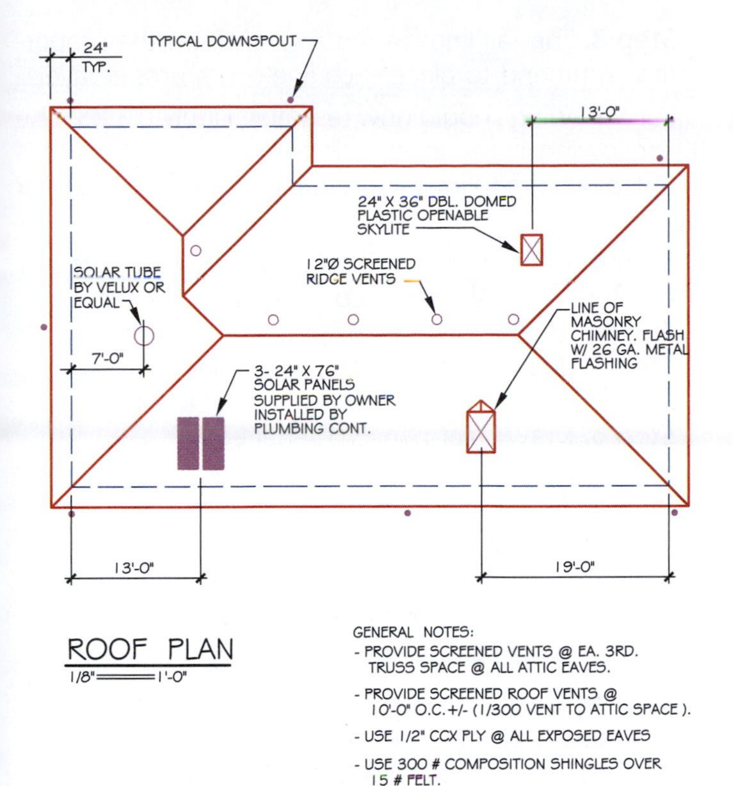

FIGURE 16.21 ■ *A completed roof plan with symbols, text, and dimensions.*

the outline to be used for the roof plan and then frozen.

Before symbols can be placed on the roof drawings, layers need to be created to separate the roof information from the information on the floor plan. Layers for the roof plan should start with the prefix *ROOF* Layers containing roofing information should be named with a modifier listed in Appendix F. As with any other drawing, create layers as needed to ensure that only layers that will be used are added to the drawing. Layers that will be needed to start the layout and their linetypes include:

ROOF LAYT Thin continuous lines; used to represent the roof boxes
ROOF OUTL Thick continuous lines; used to represent the shape
ROOF WALL Thin dashed lines; used to represent floor outlines

Initial Roof Layout Procedures

The initial layout procedure for a roof plan is the same no matter what type of roof is to be drawn. Initial steps include:

Step 1. Open the floor plan and freeze all information directly related to the floor plan except for the exterior walls, fireplace, and skylights.

Step 2. Using OSNAP, trace the outline of the exterior walls and any supports that may be required for covered porches.
Step 3. Using OSNAP, make the *ROOF OUTL* layer current and trace the outlines of the chimney and skylights.
Step 4. Freeze any open floor-related layers.
Step 5. Create a layer titled *ROOF LAYT* and divide the home into the fewest number of boxes so that the ridge locations can be determined.

Your drawing will now resemble Figure 16.22

Step 6. Work on the *ROOF OUTL* layer and find the center point for each box. Draw a line from the midpoint of each box to represent the ridge location for each.
Step 7. Offset the wall line as needed to represent the desired overhangs. For this example, 24'' (600 mm) overhangs will be used parallel to the ridge and 12'' (300 mm) overhangs perpendicular to the ridge.

Your drawing should now resemble Figure 16.23.

Step 8. Use the PROPERTIES command to change the lines that represent the walls to the *ROOF WALL* layer.
Step 9. Use the PROPERTIES command to change the lines that represent the roof overhangs (offset from the lines that represent the wall outlines) to be on the *ROOF OUTL* layer. Use the TRIM and FILLET commands to adjust the lines to match the shape of the residence.

Your drawing will now resemble Figure 16.24.

DRAWING PLANS FOR GABLE ROOFS

Once the preliminary layout procedures have been completed, design characteristics specific to gable roofs can be added. The following instructions are for the

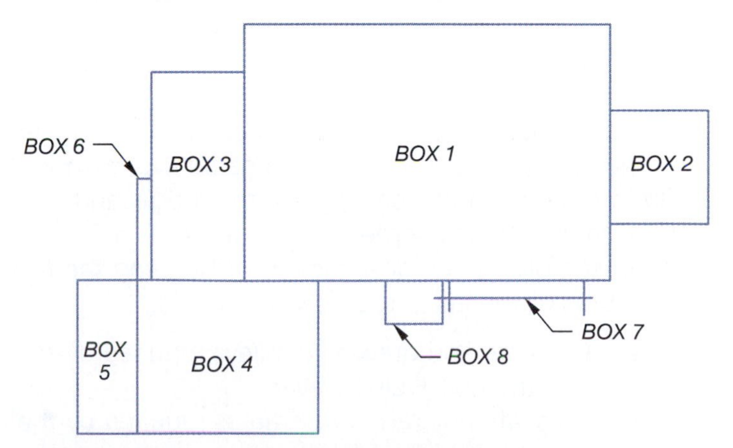

FIGURE 16.22 ■ *Divide the residence into boxes to help determine where the ridges will be formed.*

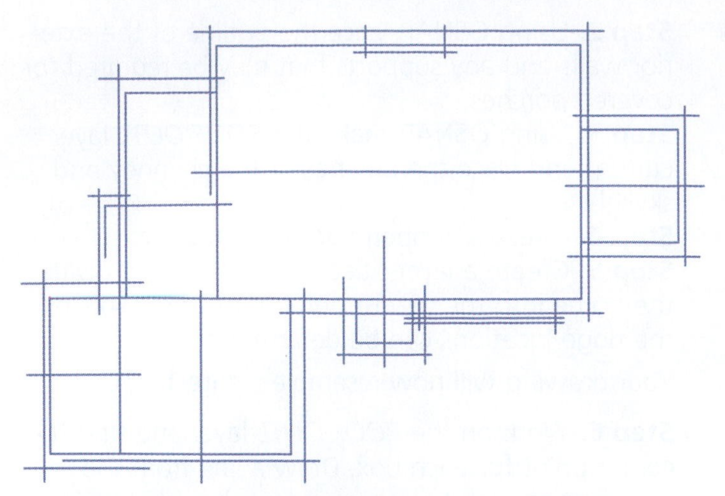

FIGURE 16.23 ■ *Place a line to represent the ridge in the center of each box and place the overhangs based on the home's style, the roof pitch, and regional standards. This home has 12'' overhangs for each gable end wall and 24'' overhangs parallel to each ridge.*

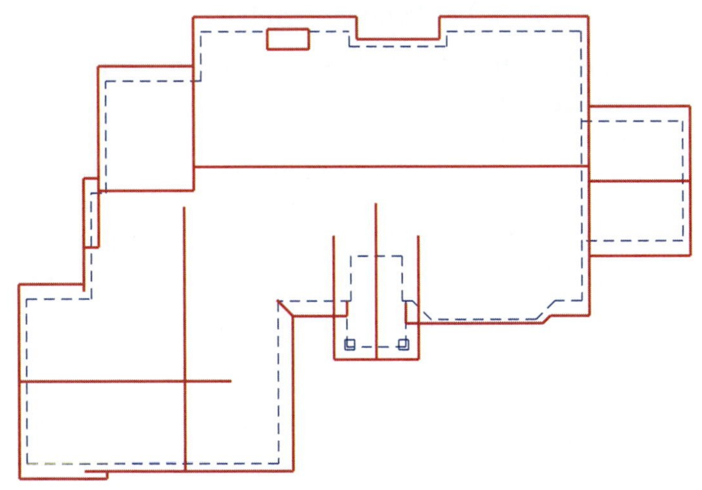

FIGURE 16.24 ■ *Use the PROPERTIES command to place lines on the desired layers. Use the TRIM and FILLET commands to form the desired intersection. Ridges are placed in the proper location, but their true length has not yet been determined.*

roof plan to accompany the residence used in Chapter 11. The plan will be drawn using a gable roof with 24'' (600 mm) overhangs parallel to the ridge and 12'' (300 mm) overhangs perpendicular to the ridge.

Start the plan view for a gable roof by using the first eight steps of the initial roof layout process.

Step 1. Locate and draw any valleys required by the design on the *ROOF OUTL* layer.

Step 2. Locate any raised roof areas required by the design on the *ROOF LEVL* layer.

Step 3. Trim all lines as required and use the properties command to place each line on its proper layer.

Your drawing should now resemble Figure 16.25. Use the following steps to complete the drawings. Unless noted, place the following materials on the *ROOF SYMB* layer.

Step 4. Assume that 12'' (300 mm) round vents will be used. Draw the required vents on a surface of the roof that will make the vents least visible.

Step 5. Draw the saddle by the chimney if a masonry chimney or wood chase is to be used.

Step 6. Draw solar panels if required.

Step 7. Draw the downspouts.

Step 8. Add dimensions for the overhangs, chimney, and skylights on the *ROOF DIMS* layer.

Step 9. Label all materials using local and general notes, a title, and a scale on the *ROOF ANNO* layer.

Your drawing should now resemble Figure 16.26

Step 10. Evaluate your drawing for completeness and make any minor revisions required before giving your drawing to your instructor. Use the checklist from Appendix D to evaluate your work.

DRAWING PLANS FOR HIP ROOFS

A roof plan representing a hip can be drawn by completing the following steps. Although the drawing of hip roof plans may prove frustrating, remember two key pieces of information:

- Lines will always be vertical, horizontal, or at a 45° angle.
- Three lines are always required to represent an intersection of hips, valleys, and ridges.

Start the plan view for a hip roof by using the first five steps of the initial roof layout process. Because the roof is designed for the same house used for the gable layout, your drawing will resemble Figure 16.22. Complete the drawing using the following steps.

Step 1. Offset the wall line as needed to represent the desired overhangs. For this example, 24'' (600 mm) overhangs will be used for all overhangs except for a 6'' (150 mm) overhang at the bay.

Draw the following features on the *ROOF OUTL* layer.

Step 2. Use the PROPERTIES command to change the lines representing the roof overhangs (offset from the lines represent the wall outlines).

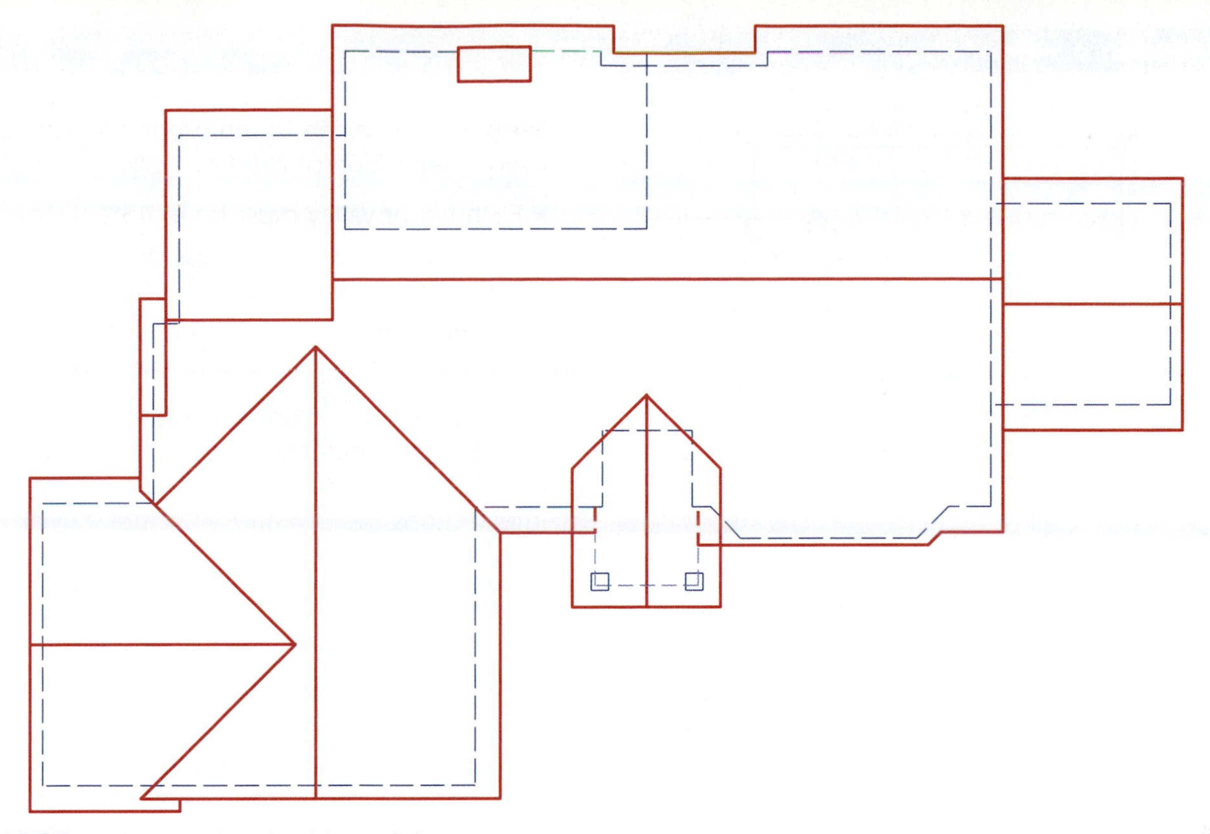

FIGURE 16.25 ■ *The completed gable roof shape.*

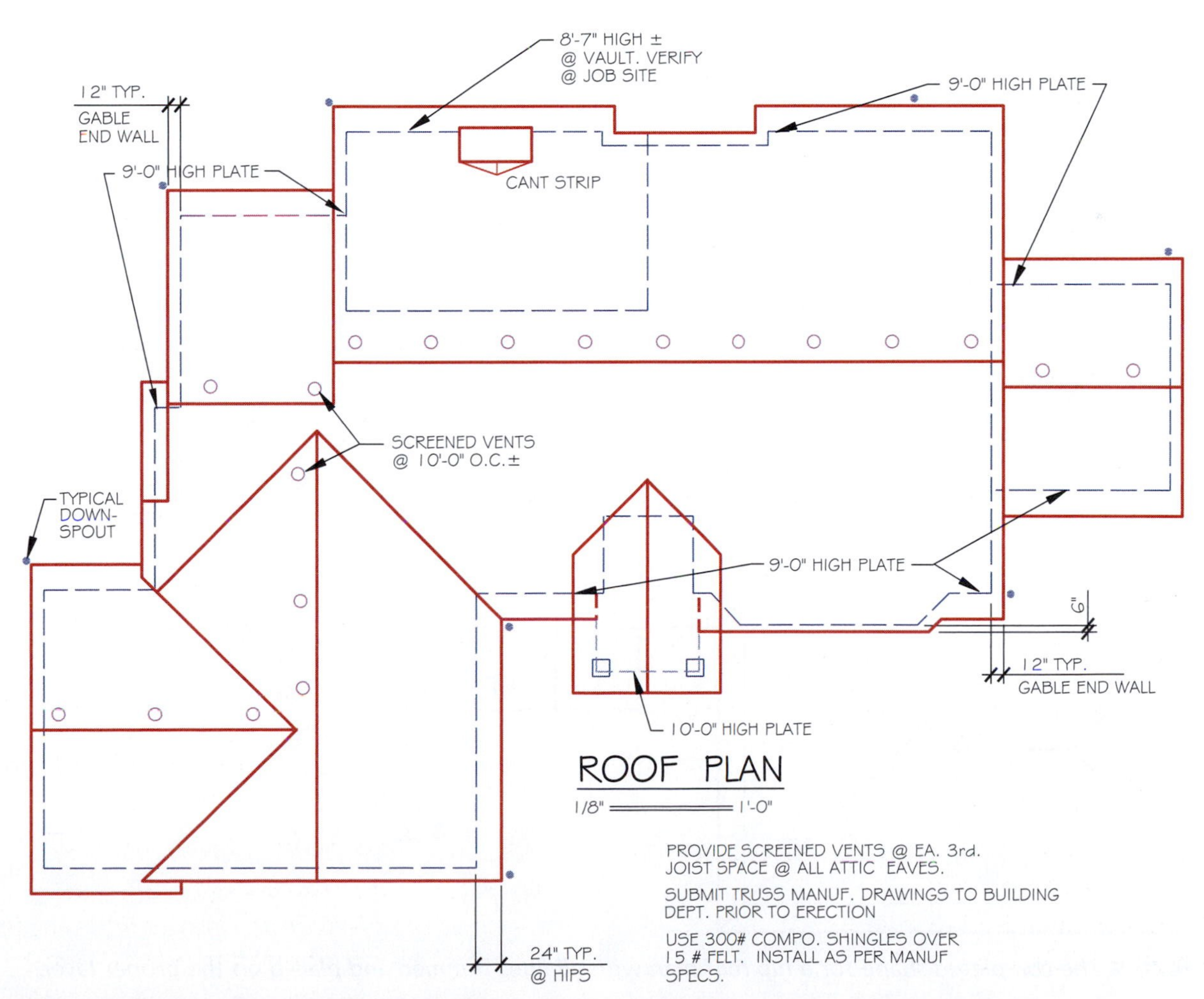

FIGURE 16.26 ■ *The size of the overhangs and the locations of all openings in the roof should be dimensioned. The roof plan is completed by adding the required local and general notes to specify all roofing materials.*

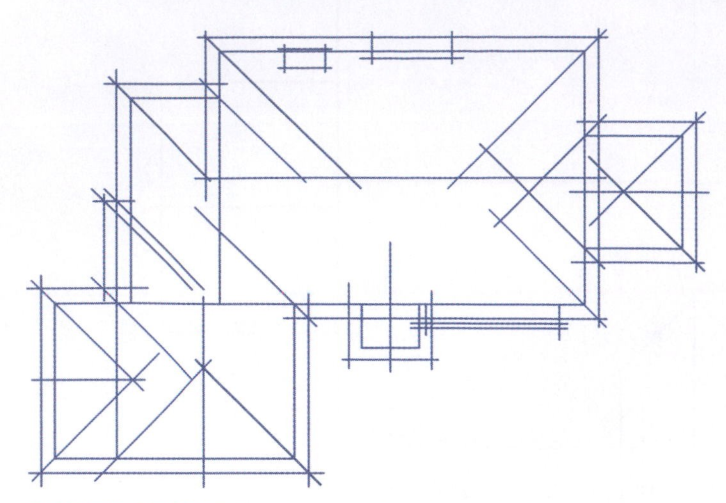

FIGURE 16.27 ■ *The layout of a hip roof plan requires that a hip or valley be placed through each intersection in the exterior walls. Draw the lines without regard to their length, and then use the TRIM or FILLET command to adjust their length. Use the PROPERTIES command to place each line on the proper layer.*

Step 3. Draw a line to represent a hip or valley at every intersection of exterior walls.

- Each hip or valley must be at 45°.
- Each hip or valley must pass directly through a wall intersection and where boxes intersect.

Your drawing will now resemble Figure 16.27. Use the following steps to complete the roof shapes.

Step 4. Locate and draw all ridges. Ridges will extend between hip intersections. To verify your accuracy, three lines should be provided at all intersections.

Step 5. Trim all overhangs, hip, valley, and ridgelines as required.

Step 6. Use the PROPERTIES command to change the lines that represent the walls to be on the *ROOF WALL* layer.

Step 7. Use the PROPERTIES command to change the lines that represent the roof overhangs (offset from the lines represent the wall outlines) to be on the *ROOF OUTL* layer. Use the TRIM and FILLET commands to adjust the lines to match the residence's shape.

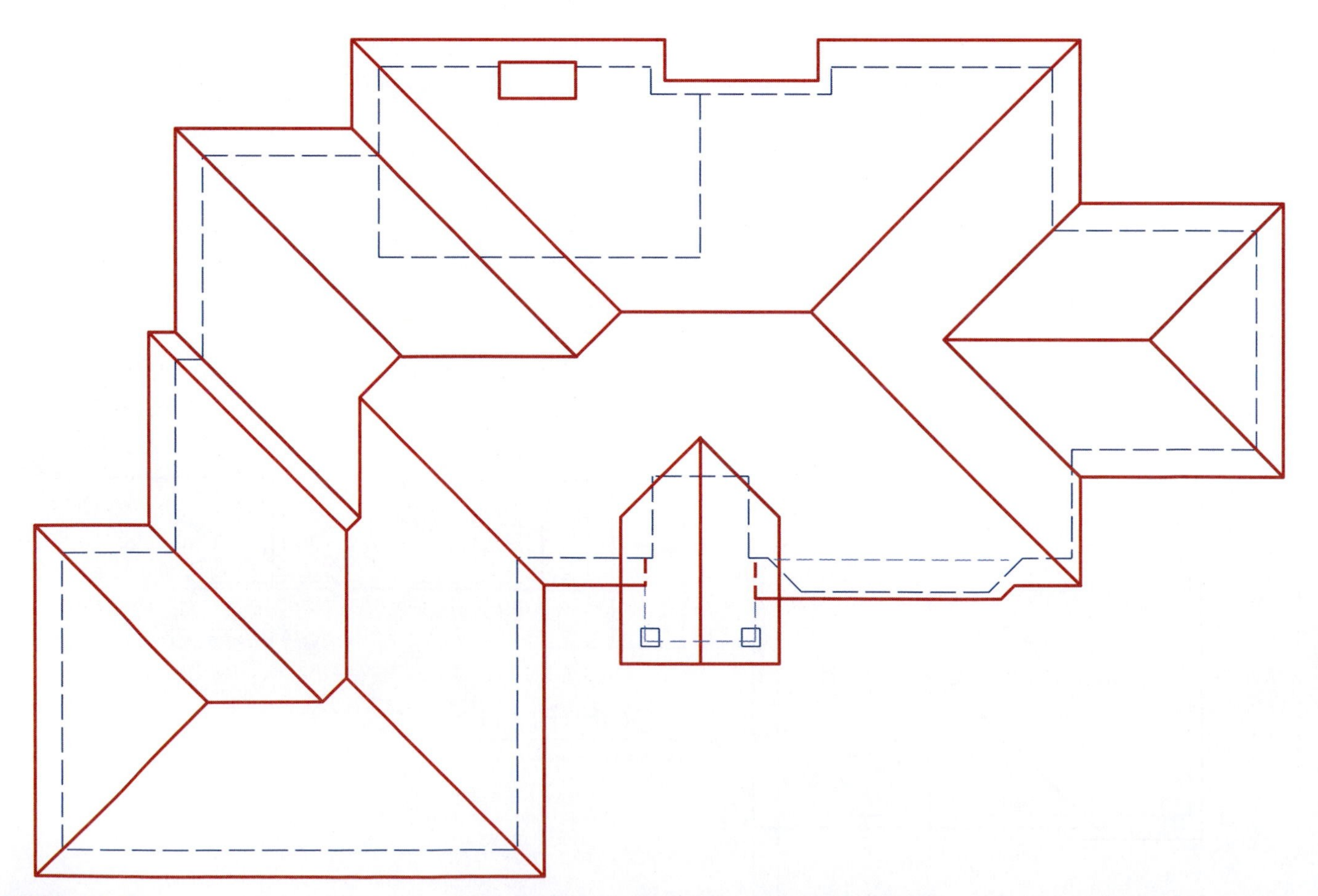

FIGURE 16.28 ■ *The completed shape for a hip roof plan with all lines trimmed and placed on the proper layer.*

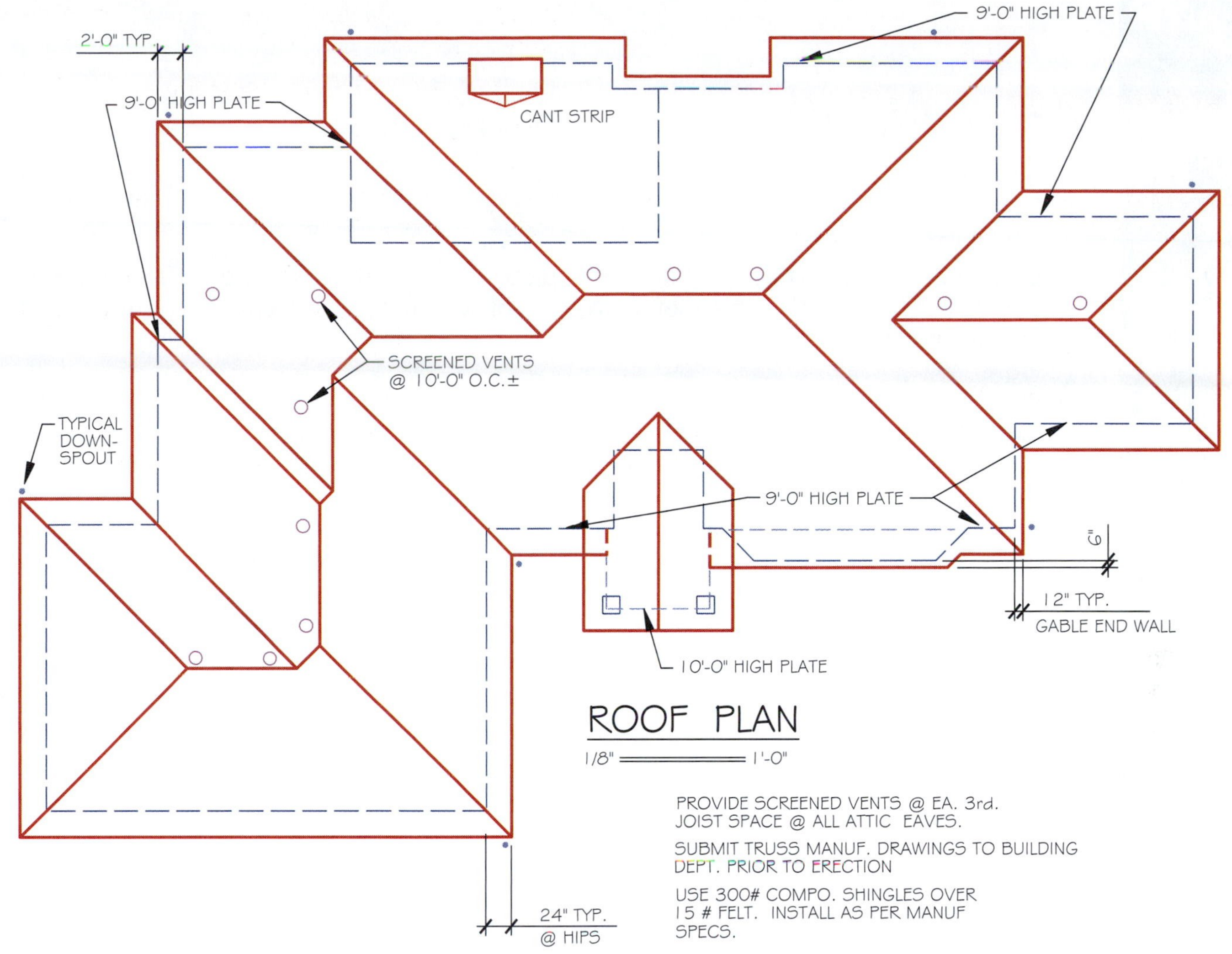

FIGURE 16.29 ■ *All openings in the roof—such as the chimney, skylights, and materials to be mounted on the roof—should be identified. The overhangs and the locations of all openings in the roof should be dimensioned.*

Step 8. Locate any raised roof areas as required by the design.

Your drawing will now resemble Figure 16.28.

Step 9. Assume that 12'' (300 mm) round vents will be used. Draw the required vents on a surface of the roof where the vents will be least visible.

Place the following information on the *ROOF MISC* layer.

Step 10. Draw saddles for any chimneys.
Step 11. Draw solar panels if required.
Step 12. Draw downspouts.

Step 13. Provide dimensions to locate any roof openings and all overhangs on the *ROOF DIMS* layer.
Step 14. Label all materials using local and general notes, a title, and a scale on the *ROOF ANNO* layer. Your drawing should now resemble Figure 16.29.
Step 15. Evaluate your drawing for completeness and make any minor revisions required before giving your drawing to your instructor. Use the checklist from Appendix D to evaluate your work.

CHAPTER 16

Roof Plan Layout Problems

DIRECTIONS

1. Using a print of your floor plan drawn for problems from Chapter 16, draw a roof plan for your house featuring a gable roof. Design a roof system appropriate for your area. Sketch the layout you will use and have it approved by your instructor prior to starting your drawing.
2. Place the plan on the same sheet as the elevations if possible. If a new sheet is required, place the drawing so that other drawings can be put on the same sheet.
3. Set the values for TEXT, DIMVARS, and LTSCALE for plotting at a scale of 1/8'' = 1'-0''.
4. When you've completed your drawing, turn in a copy to your instructor for evaluation.
5. Develop an alternate roof plan for your house using a hip roof.

SECTION 5

Elevations

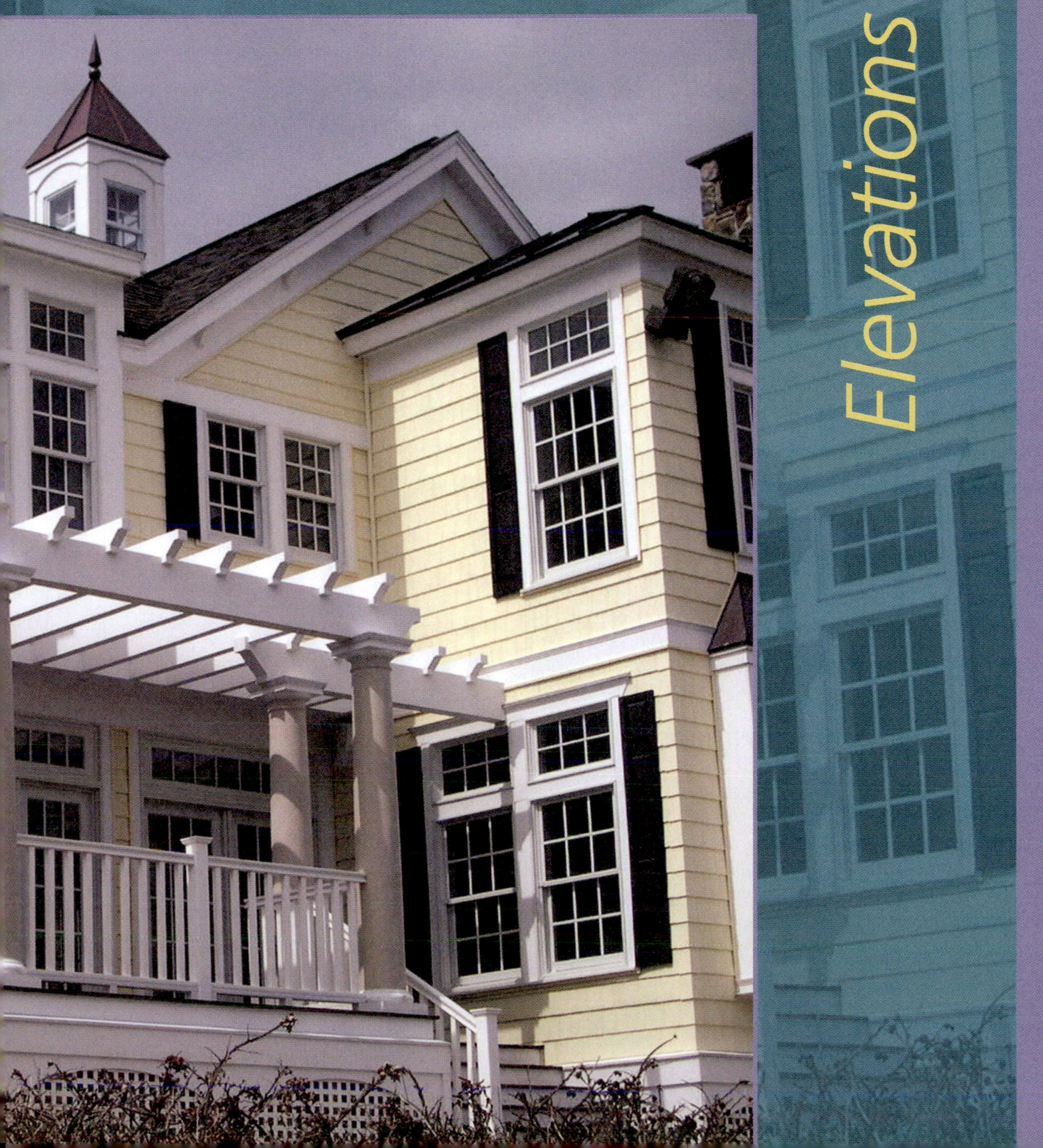

17

CHAPTER

Introduction to Elevations

Elevations are an essential part of the design and drawing process. The elevations are a group of drawings that show the exterior of a building. To communicate clearly, the technician will need to carefully plan the number, type, and scale to be used to complete the drawings. Skill will also be needed to represent materials accurately without spending unnecessary drawing time.

PLANNING ELEVATIONS

An elevation is an orthographic drawing that shows one side of a building. In true orthographic projection, the elevations would be displayed as shown in Figure 17.1a. The true projection is typically modified, as shown in Figure 17.1b, to ease viewing. No matter how elevations are displayed, it is important to realize that between each elevation projection and the plan view is an imaginary 90° fold line. An imaginary 90° fold line also exists between elevations in Figure 17.1b. Elevations are drawn to show exterior shapes and finishes as well as the vertical relationships of the building levels. By using the elevations, sections, and floor plans, the exterior shape of a building can be determined.

Required Elevations

Typically, four elevations will be required to show the features of a building. On a simple building, only three elevations will be needed, as shown in Figure 17.2. When a building with an irregular shape is drawn, parts of it may be hidden. An elevation of each surface should be drawn, as shown in Figure 17.3. If a building has walls that are not at 90° to each other, a true orthographic drawing could be very confusing. In the orthographic projection, part of the elevation will be distorted, as shown in Figure 17.4. Elevations of this type of building are usually expanded so that a separate elevation of each face is drawn. Figure 17.5 shows the layout for a residence with an irregular shape.

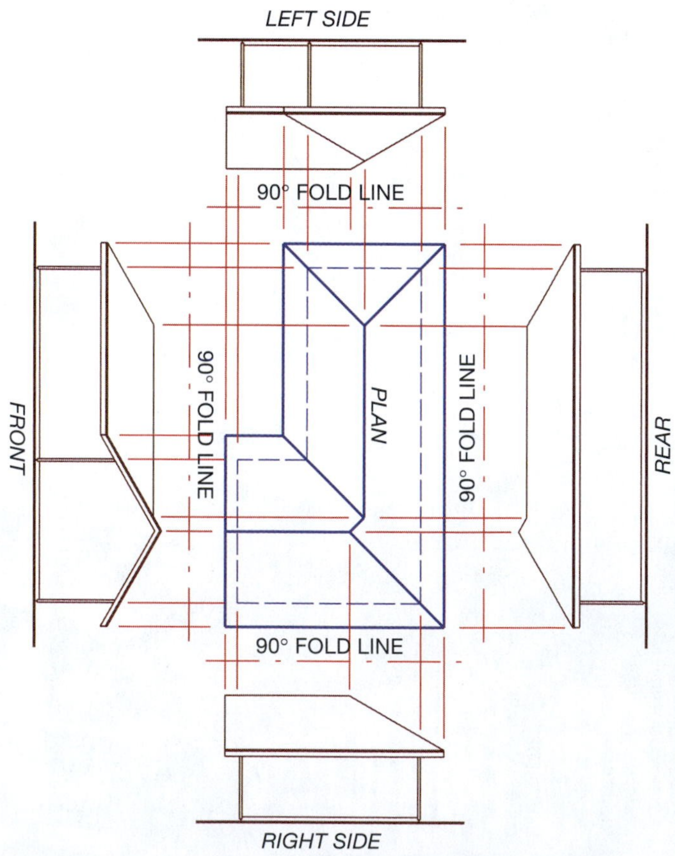

FIGURE 17.1a ■ *Elevations are orthographic projections showing each side of a structure.*

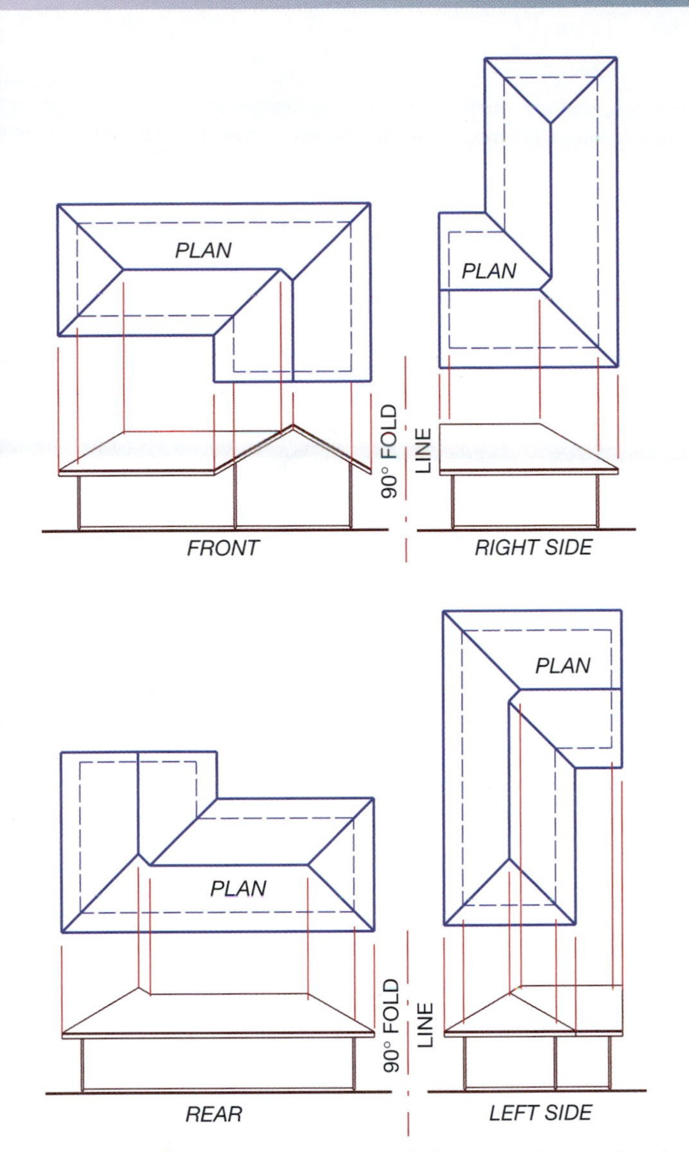

FIGURE 17.1b ■ *The placement of elevations is usually altered to ease viewing. Group elevations so that a 90° rotation exists between views.*

Types of Elevations

Elevations can be drawn as either presentation drawings or working drawings. Presentation drawings were introduced in Chapter 1 and are covered in depth in Chapter 33. These drawings are part of the initial design process and may range from sketches to very detailed drawings intended to help the owner and lending institution understand the basic design concepts (see Figure 17.6). Because the front elevation is drawn as part of the preliminary design process, it is often drawn using rendering methods. Common elements that can be added to a rendered elevation include shade, landscaping, people, and automobiles. Each of these items can be added to a drawing using blocks developed by third-party vendors. Text and dimensions are kept to a minimum. Enough information must be provided to explain the project, but complete annotation is not added until the working drawings are started.

Working elevations are part of the working drawings and provide information for the building team. They include information on roofing, siding, openings, chimneys, land shape, and sometimes even the depth of footings, as shown in Figure 17.7. Although the preliminary elevation serves as a drawing base, the layers containing the artistic material are often frozen to provide better clarity to the building team. Additional information should be added to each note to clearly explain the application of each material. Because the front elevation is drawn to please the owner during the preliminary process, the elevation will generally show all materials. The working elevations are drawn to communicate with the building team and will generally show just enough material to represent the material to be applied. Figure 17.8 shows an example of a working elevation. The floor plans and elevations provide the information the contractor needs to determine surface areas. Once surface areas are known, exact quantities of material can be determined. The heating contractor also uses the elevations if heat loss calculations need to be completed. The elevations are used to determine the surface area of walls and wall openings for the required heat-loss formulas.

Elevation Scales

Elevations should be drawn at full scale in model space. This allows for the elevations to be projected directly from the floor plans. They are generally plotted in paper space

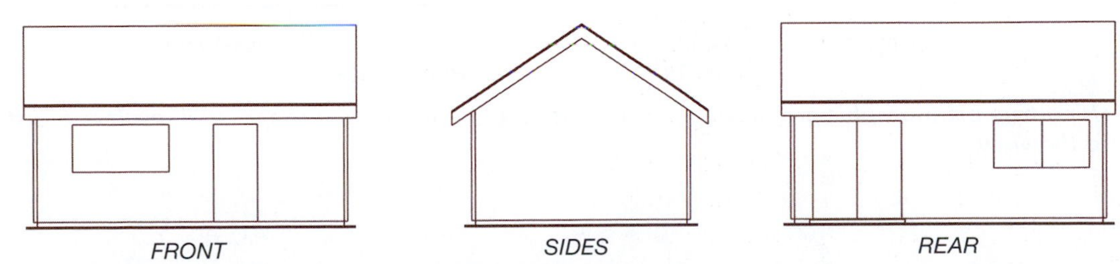

FIGURE 17.2 ■ *Elevations are used to show the exterior shape and material of a building. For a simple structure, only three views are required.*

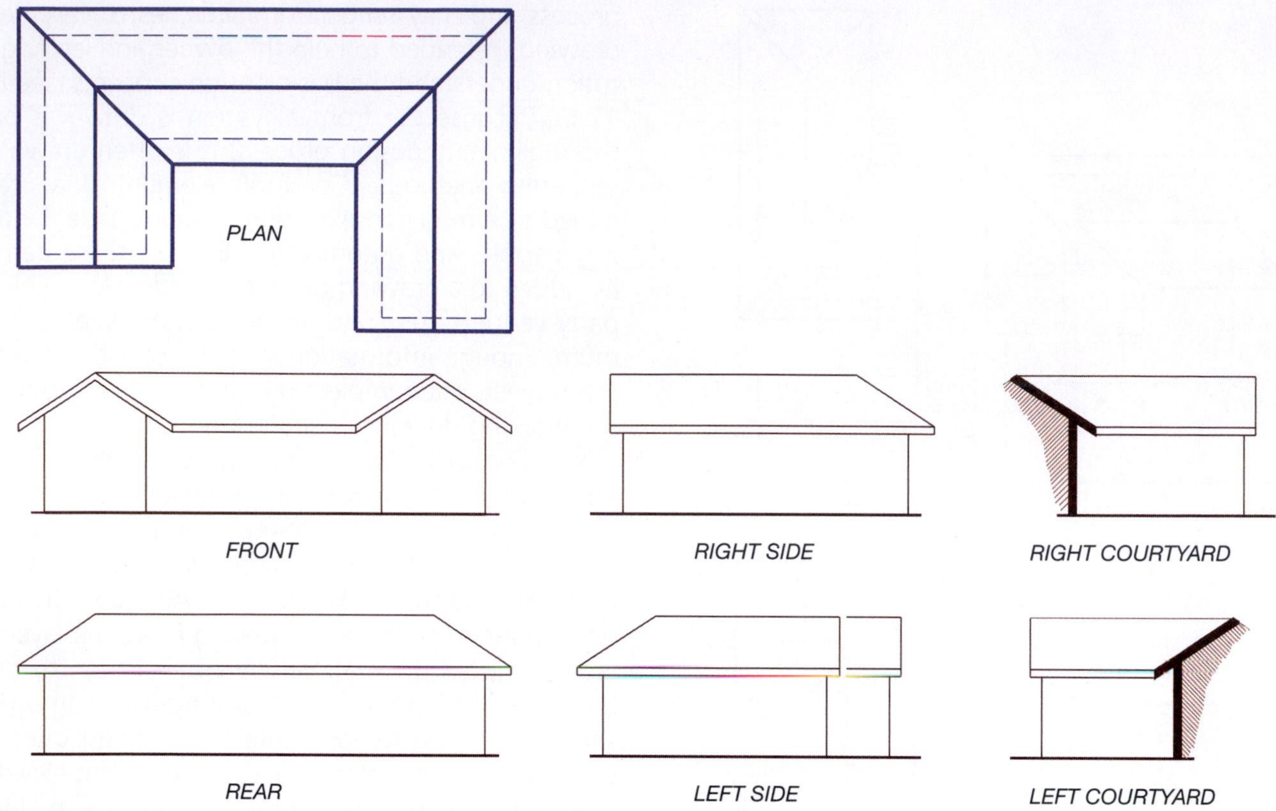

FIGURE 17.3 ■ *Plans of irregular shapes often require an elevation of each surface.*

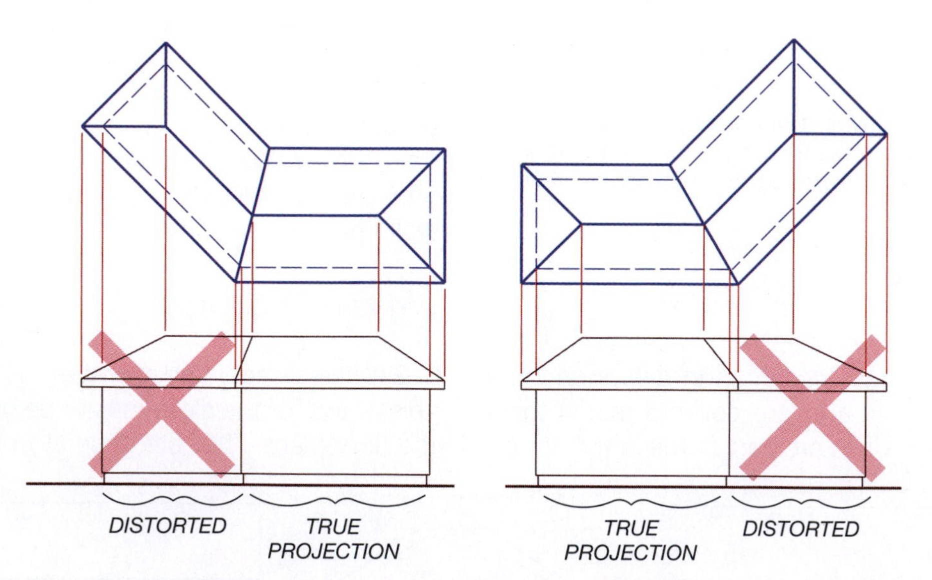

FIGURE 17.4 ■ *Using true orthographic projection methods with a plan of irregular shapes will result in a distortion of part of the view.*

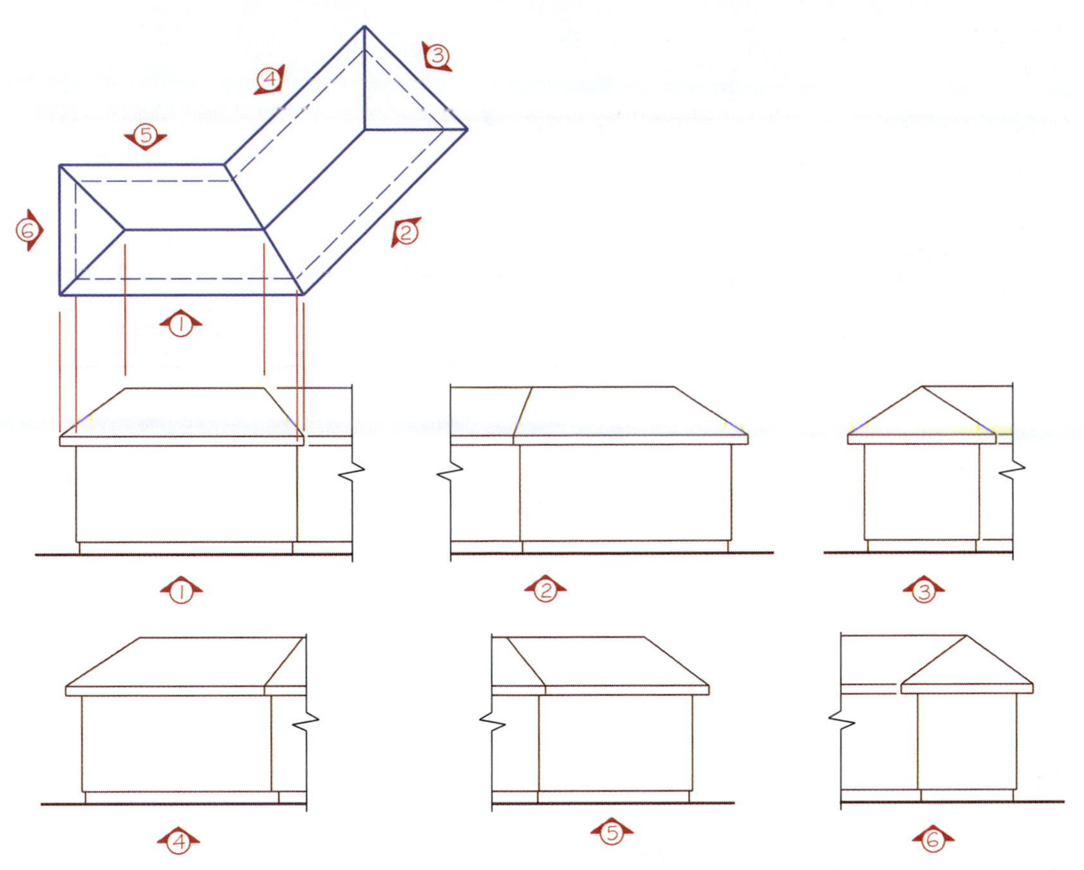

FIGURE 17.5 ■ *A common practice when drawing an irregularly shaped structure is to draw an elevation of each surface. Each elevation is then given a reference number to tie the elevation to the floor plan.*

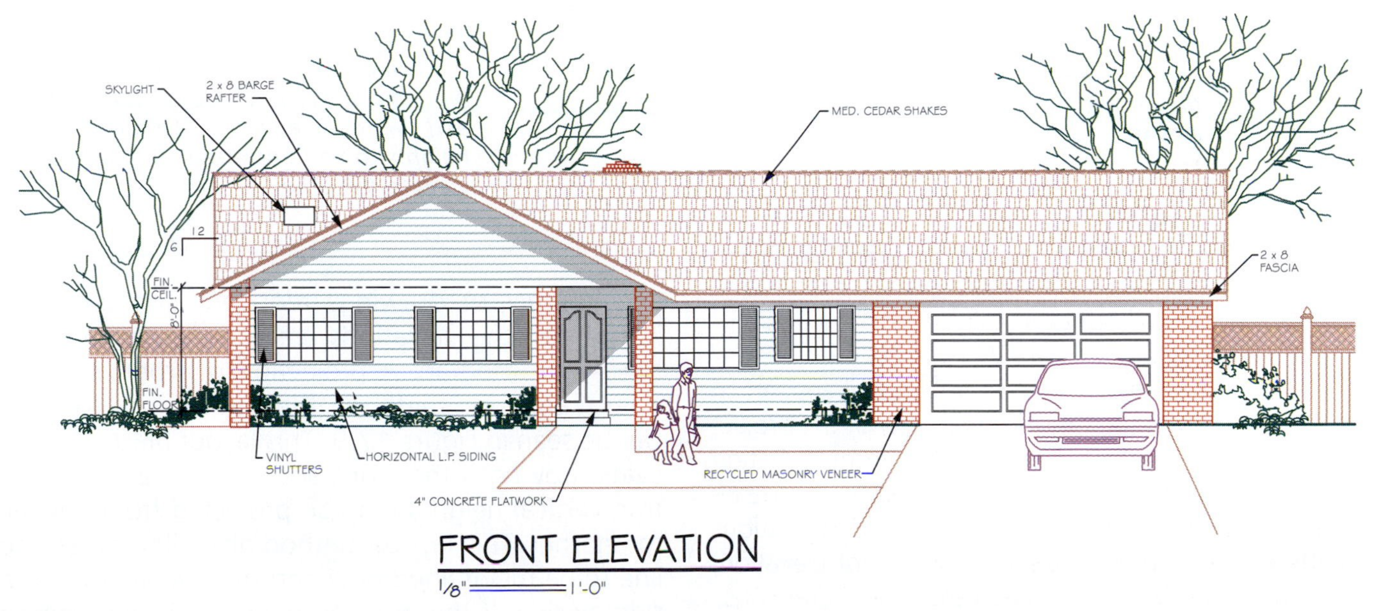

FIGURE 17.6 ■ *A presentation elevation is a highly detailed drawing used to show the exterior shapes and material to be used. Shades, shadows, and landscaping are usually added to enhance the drawing.*

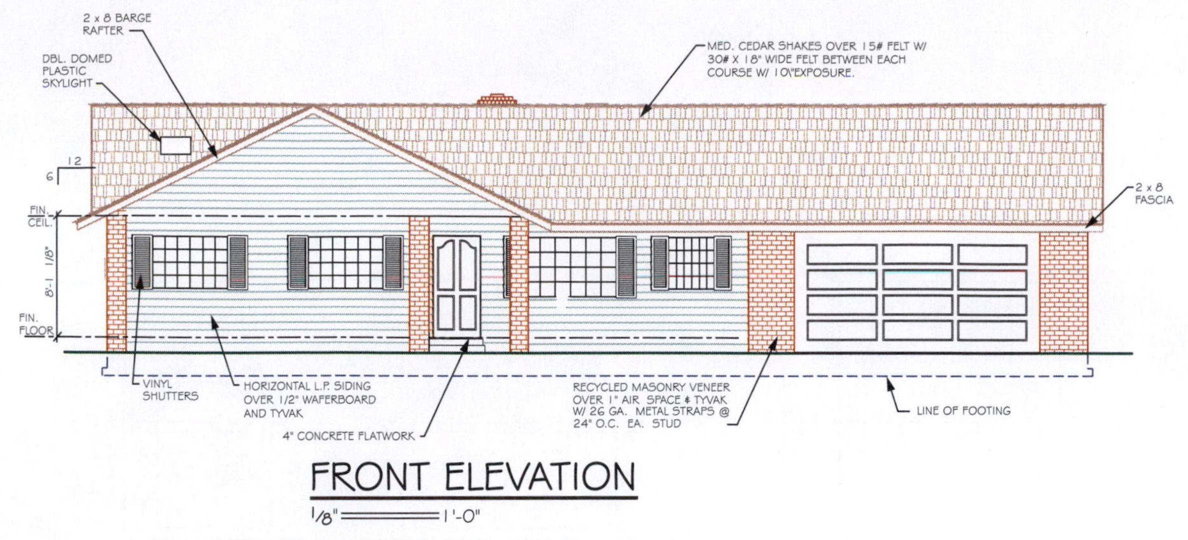

FIGURE 17.7 ■ *Working elevations contain less finish detail but still show the shape of a structure accurately and include specifications for all materials to be provided.*

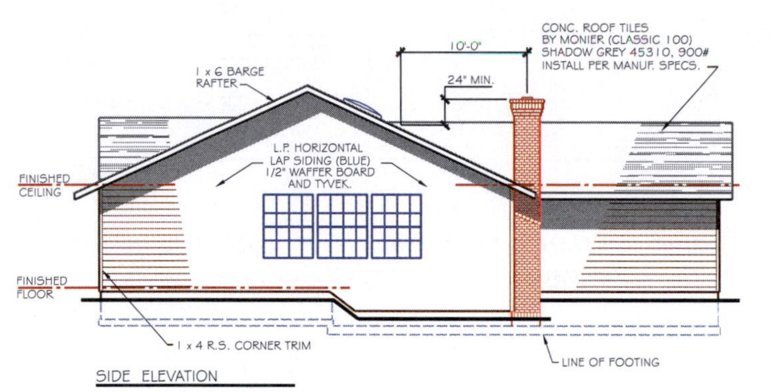

FIGURE 17.8 ■ *The side and rear elevations are typically drawn with less detail than the front elevation.*

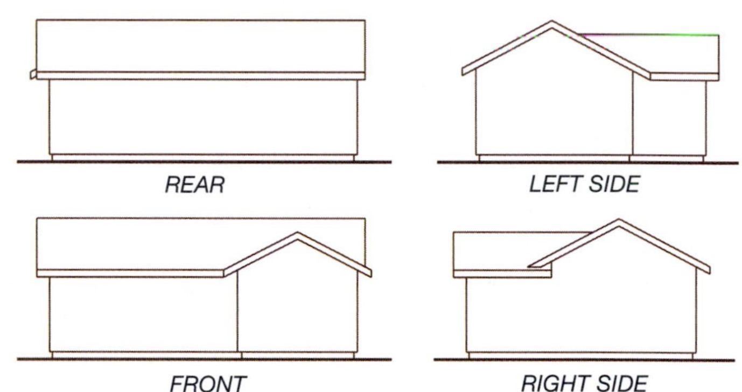

FIGURE 17.9 ■ *A common method of elevation layout is to place a side elevation beside both the front and the rear elevations. This allows for heights to be transferred directly from one view to the other.*

at the same scale as the floor plan. For most plans, this means plotting at a scale of 1/4'' = 1'-0'', with two elevations placed on a sheet. Some floor plans for multifamily projects may be laid out at a scale of 1/16'' = 1'-0'' or even as small as 1/32'' = 1'-0''. When a scale of 1/8'' = 1'-0'' or less is used, generally very little detail appears in the drawings. Depending on the complexity of the project or the amount of space on a page, the front elevation may be drawn at 1/4'' = 1'-0'' and the balance of the elevations at a smaller scale. If the side, rear, side elevations are plotted at a different scale from the front elevation, the scale must be clearly indicated below each drawings.

Elevation Placement

It is usually the drafter's responsibility to plan the layout for drawing the elevations. The layout will depend on the scale to be used, size of drawing sheet, and number of drawings required. Because of size limitations, the elevations are not usually laid out in the true orthographic projection of front, side, rear, side. A common method of layout for four elevations can be seen in Figure 17.9. This layout places a side elevation by both the front and rear elevations so that true vertical heights may be projected from one view to another. This layout method also allows a 90° fold line to be maintained between the two views that are side by side. If the drawing paper is not long enough for this placement, the layout shown in Figure 17.10 may be used. This arrangement is often used when the elevations are placed next to the floor plan to conserve space. Chapter 18 offers aids for the drawing setup no matter how they will be displayed for plotting.

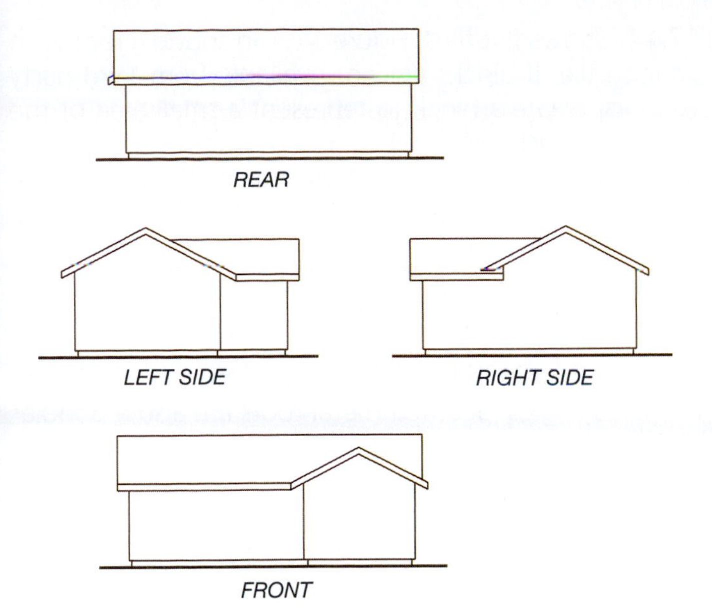

FIGURE 17.10 ■ *An alternative elevation arrangement is to place the two shortest elevations side by side.*

Elevation Names

There are three common methods for naming the elevations. The method that is used will depend on the client, and if the elevations are being developed for a specific site. Common methods include:

- If the drawings are being developed for general use when a specific site is not known, titles such as FRONT, SIDE, REAR, and SIDE are generally used below each elevation.
- If a specific site is known, the elevations are often named for their relationship to the north arrow on the floor plan. If the front wall on the floor plan faces south, instead of being labeled FRONT ELEVATION, this elevation would be labeled SOUTH ELEVATION. The remaining elevations would then be named EAST, NORTH, and WEST.
- Irregularly shaped homes are often numbered rather than labeled. A number symbol is typically placed on the floor plan to identify each elevation plane, and the corresponding number is placed below the elevation. Figure 17.5 shows examples of elevation number symbols for the floor and elevations. A small nonscaled floor plan is often placed on the sheet containing the elevations to aid in referencing the elevation to the structure.

REPRESENTING SURFACE MATERIALS IN ELEVATION

The materials that are used to protect the building from the weather need to be shown on the elevations. This information will be considered in four categories: roofing, wall coverings, doors, and windows. Additional considerations are rails, shutters, eave vents, and chimneys. Many third-party architectural programs have a variety of elevation symbols representing doors, windows, and other common materials.

If the program you are using does not contain architectural blocks, they can easily be created using the examples found throughout this chapter. Each of the materials such as doors, windows, and rails can be drawn, saved, and reused. Once a symbol such as door has been created, the location where it is saved in should be given careful consideration. Creating a folder titled PROTO or BLOCKS will make an excellent storage area for keeping symbols. Be sure to subdivide the folder for each drawing so that you can easily find the desired symbol. Subfolders such as FLOOR, ELECTRICAL and EXT ELEVS will help provide organization to your storage system. If you plan on creating large quantities of symbols use titles such as ROOF, WALL, and LANDSCAPING to further divide your drawing elements. A sample title for a tree to be placed on elevations might read PROTOS/EXT ELEVS/LANDSCAPING/TREE/OAK. Taking the time to get organized as you create symbols will greatly decrease the amount of time you spend looking for a symbol to be inserted into a drawing.

Roofing Materials

Several common materials are used to protect the roof from the elements. Among the most frequently used are asphalt shingles, wood shakes and shingles, clay and concrete tiles, metal sheets, and built-up roofing materials, which were introduced in Chapter 15. It is important to have an idea of what each material looks like so that it can be drawn in a realistic style. It is also important to remember that the elevations are meant for the framing crew. The framer's job will not be made easier by seeing every shingle drawn or other techniques used on presentation drawings and renderings. The elevations need to have materials represented clearly and quickly.

Shingles

Asphalt, fiberglass, and metal shingles come in many colors and patterns. Asphalt shingles are typically drawn using the method seen in Figure 17.11. When these are placed using AutoCAD, use the AR-RROOF hatch pattern to represent shingles. Use a scale that places the lines in the pattern about 1/16 to 1/8'' (1.5 to 3 mm) apart.

Wood Shakes and Shingles

Figure 17.12 shows a roof protected with wood shakes. Other materials such as Masonite are used to simulate wood shakes. Shakes and Masonite create a jagged surface at the ridge and the edge. These types of materials are often represented using the AR-RROOF hatch pattern to represent shingles. Use a thick line placed about 2° less than the rake of the roof to represent the edge of the shingles on the barge rafter. Lines representing shingles will resemble Figure 17.13.

Tile

Concrete, clay, or a lightweight simulated tile material presents a very rugged surface at the ridge and edge as well as many shadows throughout the roof. Figure 17.14a shows flat tiles. Figure 17.14b shows a roof with Spanish tile. If blocks are not available from third-party vendors, create a block to represent a small area of the tiles. The block can then be inserted to fill the desired area. Flat tiles can be drawn as shown in Figure 17.15. Spanish tiles can be represented as shown in Figure 17.16.

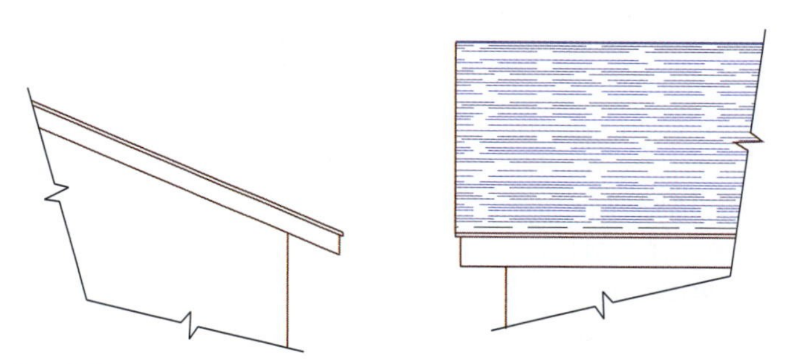

FIGURE 17.11 ■ *The AR-RROOF hatch pattern of AutoCAD can be used to represent composition shingles.*

FIGURE 17.12 ■ *Wood shakes and shingles have much more texture than asphalt shingles.*

Metal

Metal shingles can be drawn in a manner similar to asphalt shingles. Seamed metal roofs are also a popular form of metal used in residential roofing. Figure 17.17 shows an example of metal panel roof sheathing. Metal panels can be drawn with a series of vertical lines to represent the panel seams. Seams can be spaced between

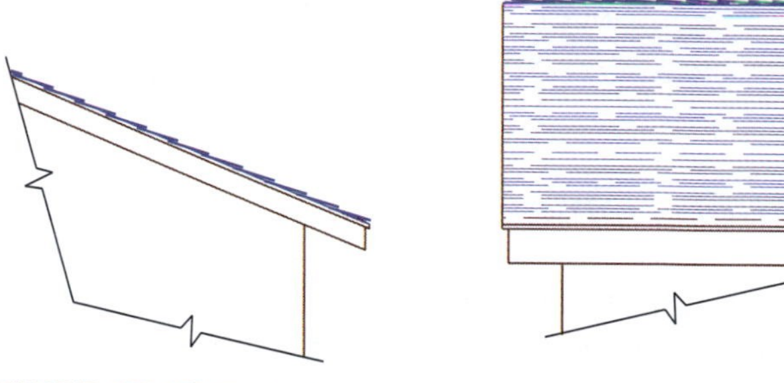

FIGURE 17.13 ■ *Shakes and shingles can be represented using AR-RROOF or AR-RSKE. Notice that an inclined line is added to the eave and to the ridge to represent the contour of the roofing material.*

FIGURE 17.14a ■ *Flat tiles are used in many parts of the country because of their low maintenance, durability, and resistance to the forces of nature.* Courtesy Benigno Molina-Manriquez.

FIGURE 17.14b ■ *Curved or Spanish tiles are a traditional roofing material of Spanish and Mediterranean style homes.*

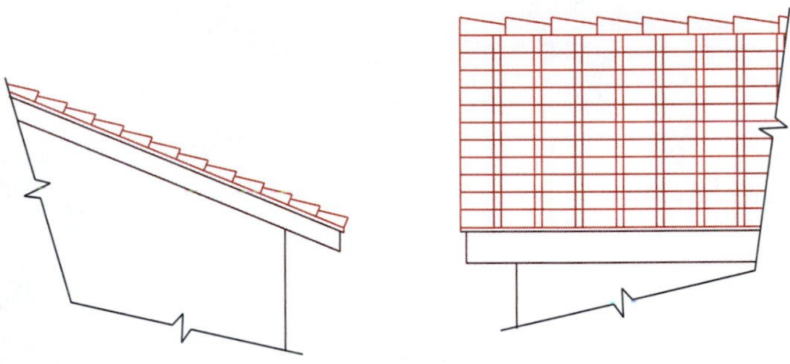

FIGURE 17.15 ■ *Representations of tile may require a hatch pattern to be created if a pattern from third-party software is not available.*

12 and 18'' (300 and 450 mm) on the drawings with the exact spacing specified in a note.

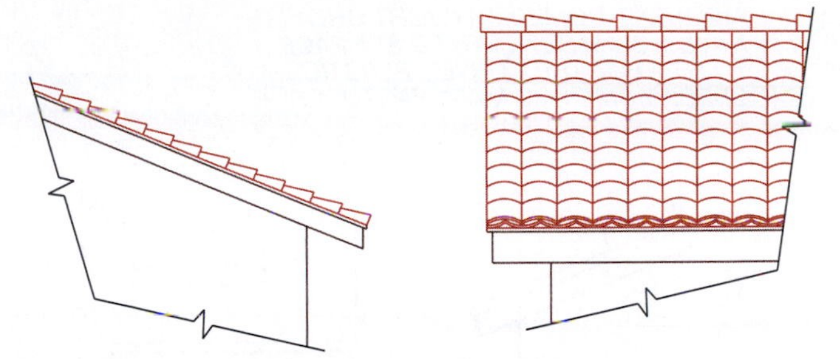

FIGURE 17.16 ■ *Representing Spanish tiles.*

FIGURE 17.17 ■ *Standing metal seamed roofs are often used for homes in rural settings because of their resistance to fires.*

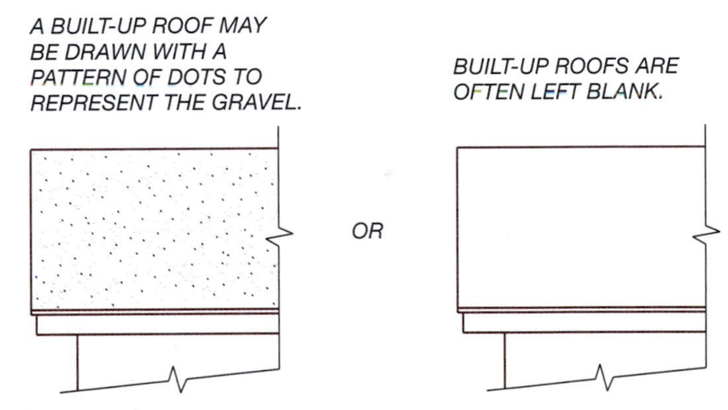

A BUILT-UP ROOF MAY
BE DRAWN WITH A
PATTERN OF DOTS TO
REPRESENT THE GRAVEL.
OR
BUILT-UP ROOFS ARE
OFTEN LEFT BLANK.

FIGURE 17.18 ■ *Drawing built-up roofs.*

Built-up Roofs

Because of the low pitch and the lack of surface texture, built-up roofs are usually outlined and left blank. Occasionally a built-up roof will be covered with rock 2 or 3'' (50 or 75 mm) in diameter. The drawing technique for this roof can be seen in Figure 17.18.

Skylights

Skylights may be made of either flat glass or domed plastic and come in a variety of shapes and styles. Depending on the pitch of the roof, skylights may or may not be drawn. On very low pitched roofs, a skylight may be unrecognizable. On roofs over 3/12 pitch, the shape of the skylight can usually be drawn without creating confusion. Unless the roof is very steep, a rectangular skylight will appear almost square. The flatter the roof, the more distortion there will be in the size of the skylight. Figure 17.19 shows common methods of drawing both flat-glass and domed skylights.

Wall Coverings

Exterior wall coverings are usually made of wood, wood substitutes, masonry, metal, plaster, or stucco. Each has its own distinctive look in elevation.

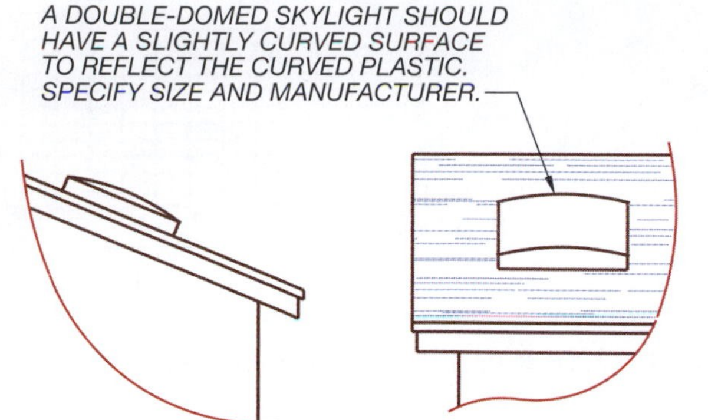
A DOUBLE-DOMED SKYLIGHT SHOULD HAVE A SLIGHTLY CURVED SURFACE TO REFLECT THE CURVED PLASTIC. SPECIFY SIZE AND MANUFACTURER.
SIDE
FRONT
DOUBLE-DOMED SKYLIGHT

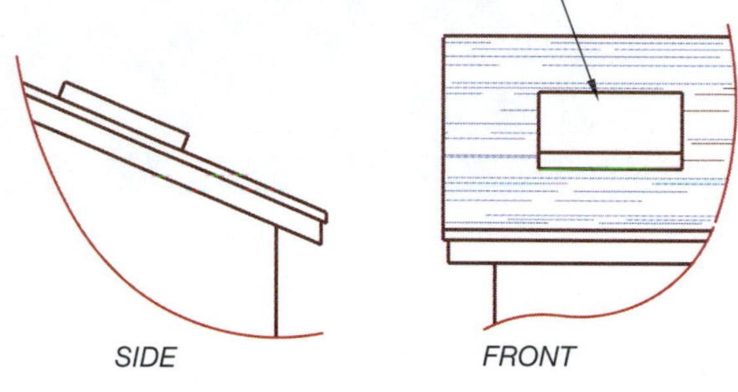
THE FLAT-GLASS SKYLIGHT CAN BE DRAWN AS A RECTANGLE. SPECIFY SIZE AND MANUFACTURER.
SIDE
FRONT
FLAT-GLASS SKYLIGHT

FIGURE 17.19 ■ *Representing domed and flat-glass skylights on elevations.*

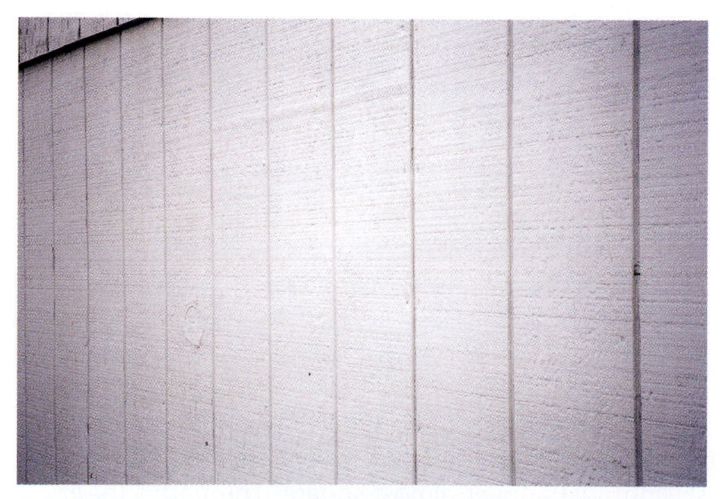

FIGURE 17.20 ■ *T1-11 plywood is a common siding.* Courtesy Ryan McFadden.

Wood

Wood siding can be installed in large sheets or in individual pieces. Plywood sheets are a popular wood siding because of their low cost and ease of installation. Individual pieces of wood provide an attractive finish but usually cost more than plywood. This higher cost results from differences in material and the labor to install each individual piece.

FIGURE 17.21 ■ *Redwood board-on-board siding is used to protect this home from the elements.* Design by James Bischoff of Callister, Gately, Heckmann & Bischoff. Charles Callister, Jr., photographer. Courtesy California Redwood Association.

FIGURE 17.22 ■ *Batt-on-board siding is used to highlight the horizontal siding of this home.*

Plywood siding can have many textures, finishes, and patterns. Textures and finishes are not shown on the elevations but may be specified in a general note. Patterns in the plywood are usually shown. The most common patterns in plywood are T1-11 (Figure 17.20), board on board (Figure 17.21), board and batten (Figure 17.22), and plain or rough-cut plywood (Figure 17.23). Figure 17.23 shows methods for drawing each type of siding.

Lumber siding comes in several types and can be laid in many patterns. Common lumber for siding is cedar, redwood, pine, fir, spruce, and hemlock. Common styles of lumber siding are tongue and groove, bevel, and channel. Various types of wood siding can be seen in Figures

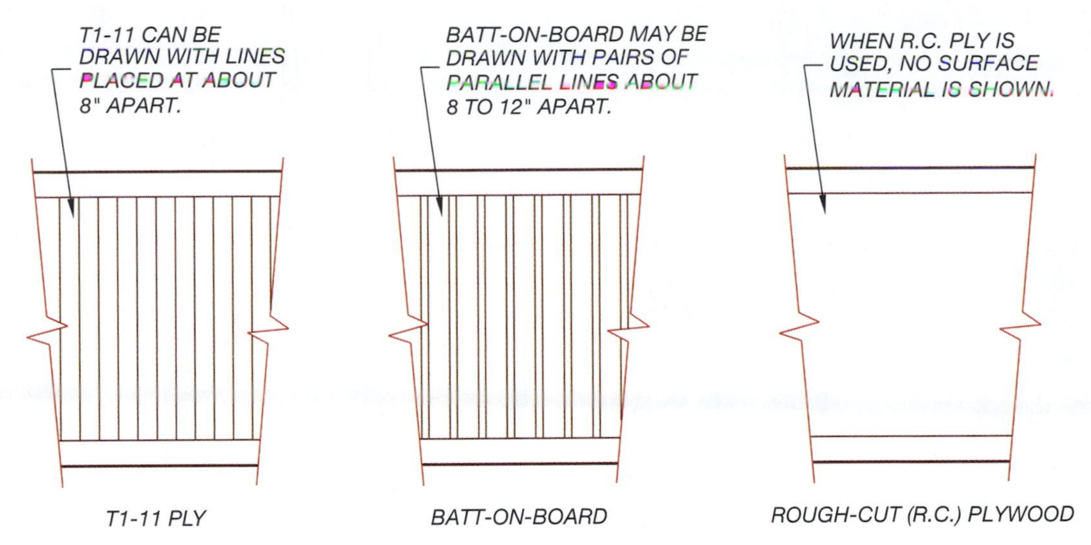
T1-11 CAN BE DRAWN WITH LINES PLACED AT ABOUT 8" APART.
BATT-ON-BOARD MAY BE DRAWN WITH PAIRS OF PARALLEL LINES ABOUT 8 TO 12" APART.
WHEN R.C. PLY IS USED, NO SURFACE MATERIAL IS SHOWN.
T1-11 PLY
BATT-ON-BOARD
ROUGH-CUT (R.C.) PLYWOOD

FIGURE 17.23 ■ *Drawing plywood siding in elevation.*

NORDSTROM & CO.
3050
BROADWAY

FIGURE 17.24 ■ ***Vertical siding is a common natural material for protecting a structure from the elements.*** Designed by Architects Knudson-Williams. T. S. Gordon, photographer. Courtesy California Redwood Association.

17.24 through 17.26. Figure 17.27 shows common shapes of wood siding. Each of these materials can be installed in a vertical, horizontal, diagonal, or laid to match the rake of the roof. The material and type of siding must be specified in a general note on the elevations. The pattern in which the siding is to be installed must be shown on the elevations as in Figure 17.28. The type of siding and the position in which it is laid will affect how the siding appears at a corner. Figure 17.29 shows two common methods of corner treatment.

Wood shingles similar to Figure 17.30 can either be installed individually or in panels. Shingles are often represented as shown in Figure 17.31 using the AR-RSHKE hatch pattern.

FIGURE 17.25 ■ ***Beveled redwood siding is used on this home to create a pleasing blend of the home with the site.*** Designed and built by Cramer-Weir Co. Photo by Saari & Forrai Photography. Courtesy California Redwood Association.

FIGURE 17.26 ■ *Finger-jointed redwood siding is used to create the lower walls of this project.* Designed by William Zimmerman, architect. Courtesy California Redwood Association.

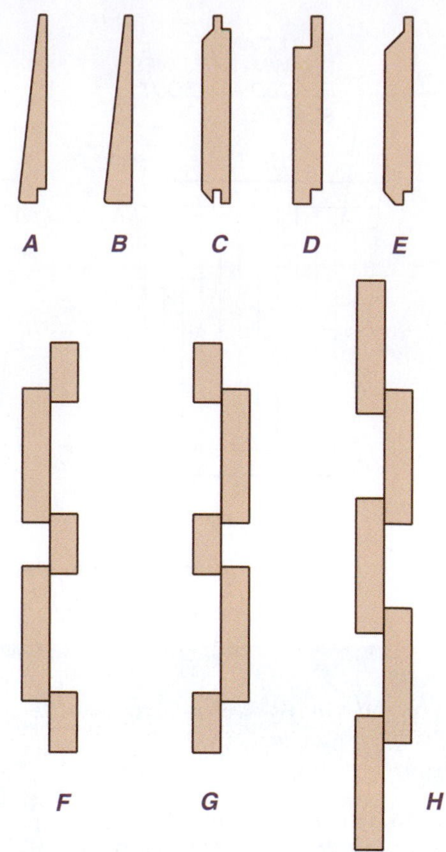

FIGURE 17.27 ■ *Common types of siding: (A) bevel; (B) rabbeted; (C) tongue and groove; (D) channel shiplap; (E) V shiplap; (F, G, H) these types can have a variety of appearances, depending on the width of the boards and battens being used.*

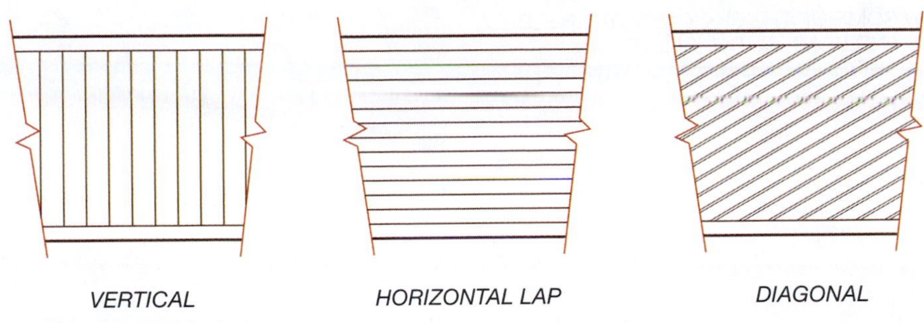

FIGURE 17.28 ■ *Representing siding in elevation.*

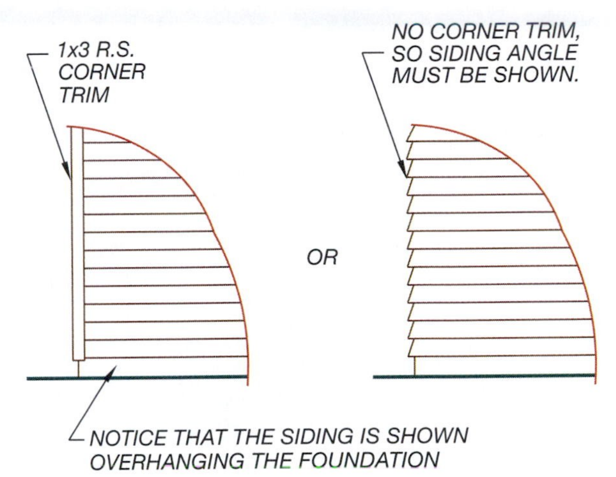

FIGURE 17.29 ■ *Common methods of corner treatment.*

Wood Substitutes

Hardboard, fiber cement, aluminum, and vinyl siding can be produced to resemble lumber siding. Figure 17.32 shows a home finished with hardboard siding. Hardboard siding is generally installed in large sheets similar to plywood but often has more detail than plywood or lumber siding. It is drawn using the same methods used for drawing lumber sidings. Each of the major national wood distributors has also developed siding products made from wood by-products that resemble individual pieces of beveled siding. Strands of wood created during the milling process are saturated with a water-resistant resin binder and compressed under extreme heat and pressure. The exterior surface typically has an embossed finish to resemble the natural surface of cedar. Most engineered lap sidings are primed to provide protection from moisture prior to installation.

Fiber cement siding products are a common wood substitute because of their durability. Engineered to be resistant to moisture, cold, insects, salt air, and fire, fiber cement products such as DuraPress by ABTco and Hardiplank by James Hardie are being used in many areas of the country as an alternative to wood and plaster products. Available in widths of 6 1/2, 7 1/2, and 9 1/2'' (165, 190, and 240 mm) or 4' × 8' (100 × 200 mm) sheets, fiber cement products can reproduce smooth or textured wood patterns as well as cedar plywood panels and stucco. Products are installed in much the same

FIGURE 17.30 ■ *Shingles are often used as a siding material to provide a casual or rustic finish.* Courtesy Laine M. Jones Design–W. Newbury Ma.

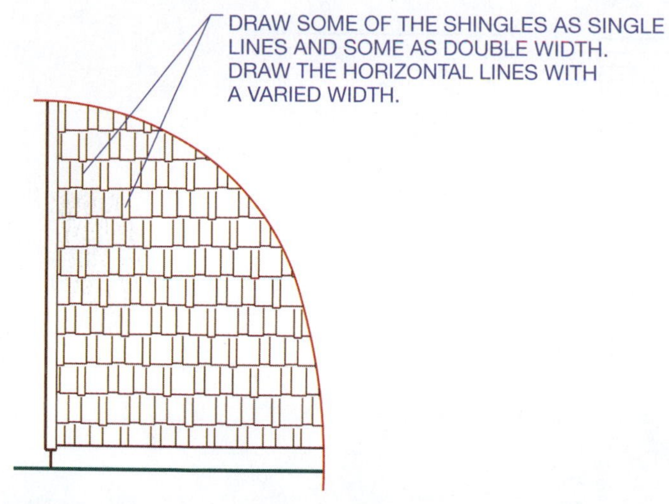
DRAW SOME OF THE SHINGLES AS SINGLE LINES AND SOME AS DOUBLE WIDTH.
DRAW THE HORIZONTAL LINES WITH A VARIED WIDTH.

FIGURE 17.31 ■ *The AR-RSHKE hatch pattern can be used to represent shingle siding.*

FIGURE 17.32 ■ *Hardboard can be used as a siding material if precautions are taken to protect against moisture.*

FIGURE 17.33 ■ *Fiber cement siding provides excellent protection from moisture, insects, and other natural elements.* Courtesy ABTco, Inc.

way as their wood counterparts. Figure 17.33 shows an example of fiber cement bevel siding. Aluminum and vinyl sidings also resemble lumber siding in appearance, as shown in Figure 17.34. Aluminum and vinyl sidings are drawn similarly to their lumber counterpart.

Masonry

Masonry finishes include the materials of brick, concrete block, and stone. Brick is used on many homes, like the one in Figure 17.35, because of its beauty and

FIGURE 17.34 ■ *Vinyl and aluminum sidings are manufactured to resemble their wood counterparts.* Courtesy ABTco, Inc.

FIGURE 17.35 ■ *Brick is used in many traditional designs because of its elegant appearance and resistance to natural forces.* Courtesy William E. Poole, William E. Poole Designs, Inc.

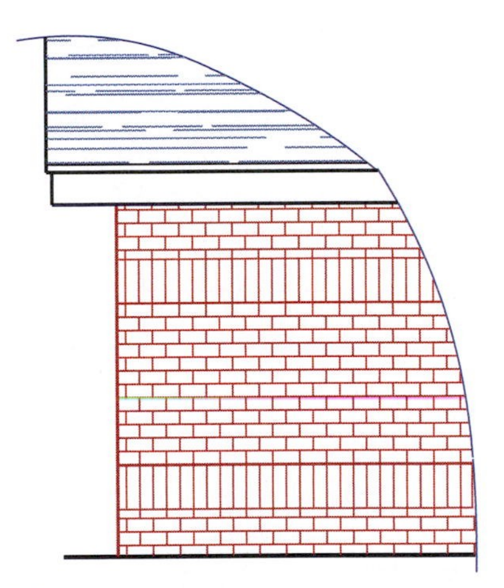

FIGURE 17.36 ■ *Representing brick in elevation.*

durability. Bricks come in a variety of sizes, patterns, and textures. In drawing elevations, represent the pattern of the bricks on the drawing and the material and texture in the written specifications. A common method for drawing bricks is shown in Figure 17.36. The horizontal brick was drawn using the BRICK hatch pattern. The BRSTONE pattern can also be used. Notice, in Figure 17.36, that two rows of decorative brick have been added to add interest to the wall. These vertical bricks were drawn using the LINE and ARRAY commands. Although bricks are not usually drawn exactly to scale, the proportions of the brick must be maintained. Because the hatch pattern is not controlled by the LTSCALE factor, it will help to draw a line that is the approximate length of a brick, and then match the size of that line as the pattern is inserted.

FIGURE 17.37 ■ *Concrete block is used in many areas to provide a long-lasting, energy-efficient building material.*

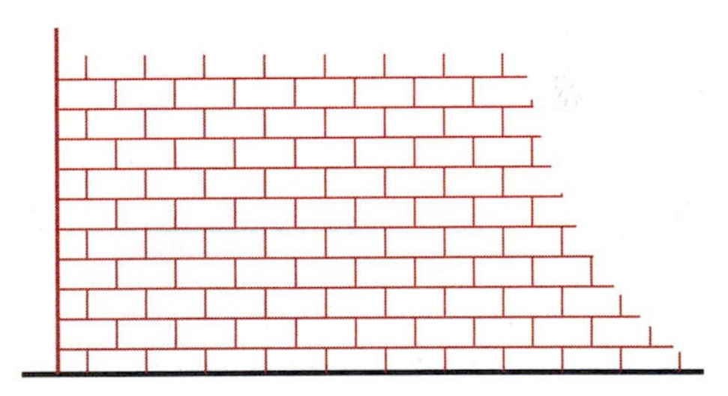

FIGURE 17.38 ■ *Representing 8'' × 8'' × 16'' concrete blocks in elevation.*

Concrete block is often used as a weather-resistant material for above- and below-grade construction. Figure 17.37 shows examples of two types of concrete block used to form above-ground walls. Show the size, pattern, and texture on the elevation when drawing concrete blocks. Figure 17.38 shows an example of simplified concrete blocks placed with the AR-816 hatch pattern. If the AR-816C pattern is used, the grout lines will be represented. Draw a line that is 16'' (400 mm)

FIGURE 17.39 ■ *Stone is used on homes in many traditional styles.* Courtesy LeRoy Cook.

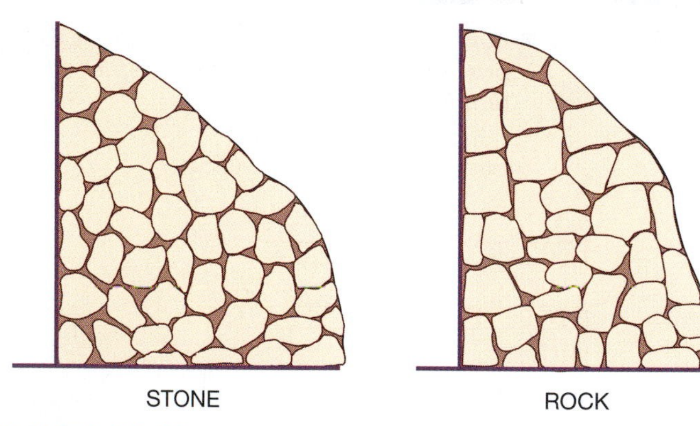

FIGURE 17.40 ■ *Stone is used in a variety of shapes and patterns.*

long and use this line as a guide to set the scale of the hatch pattern during insertion.

Stone is often used to provide a charming, traditional style that is extremely weather-resistant. Figure 17.39 shows an example of stone used to accent a home. Stone or rock finishes also come in a wide variety of sizes and shapes and are laid in a variety of patterns. Stone or rock may be natural or artificial. Both appear the same when drawn in elevation. Rounded stone can be represented using the GRAVEL pattern from AutoCAD. If a pattern is not available through third-party vendors, be careful to represent the irregular shape, as shown in Figure 17.40.

FIGURE 17.41 ■ *Stucco and plaster can be installed in many different patterns and colors to provide a durable finish. EIFS offers the look of stucco while providing added insulation and durability.* Courtesy Pamela Russom, American PolySteel, LLC.

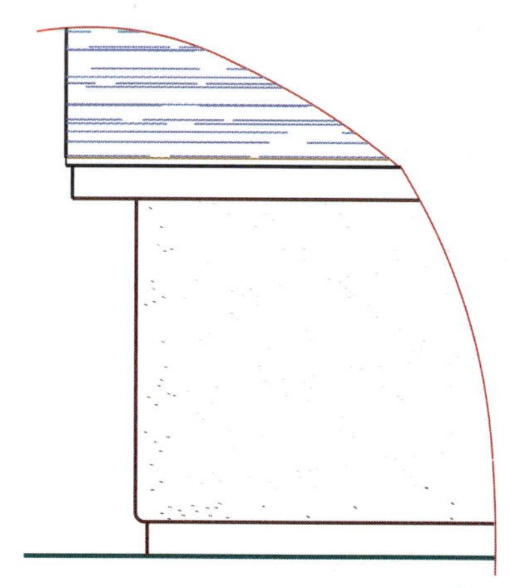

FIGURE 17.42 ■ *Representing stucco, plaster, and EIFS in elevation. Notes are used to explain which product will be applied.*

Metal

Although primarily a roofing material, metal can be used as an attractive wall covering. Drawing metal in elevation uses a method similar to drawing lumber siding.

Plaster or Stucco

Although primarily used in areas with little rainfall, plaster or stucco can be found throughout the country. Figure 17.41 shows an example of a wall covered with stucco. Stucco can be represented as shown in Figure 17.42 using the AR-SAND hatch pattern. Similar in appearance to stucco or plaster are exterior insulation and finishing systems (EIFS). This type of weather protection is installed over a rigid insulation board that is used as a base for a fiberglass-reinforced base coat. A weather-resistant colored finish coat is then applied by trowel to seal the structure. Shapes made out of insulation board can be added wherever three-dimensional details are desired.

Doors

Doors should be drawn to resemble the type of door specified in the door schedule, but be careful not to try to reproduce an exact likeness of a door. This is especially true of entry and garage doors, which typically have a decorative pattern on them. It is important to show this pattern, but do not spend time trying to reproduce the exact pattern. Since the door is manufactured, you'll be wasting your time drawing details that add nothing to the plan. Figure 17.43 shows the layout of a raised-panel door, and Figure 17.44 shows how other common types of doors can be represented. The finished door block from Figure 17.43 and each symbol from Figure 17.44 can be found on the student CD.

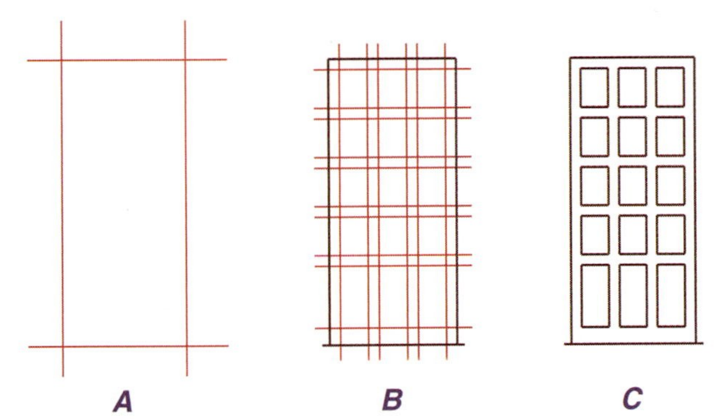

FIGURE 17.43 ■ *Layout steps for drawing a block to represent a raised-panel door.*

Windows

The same precautions about drawing needless details for doors should be taken in drawing windows. Care must be given to the frame material. Wooden frames are wider than metal frames. When the elevations are started in the preliminary stages of design, the drafter may not know what type of frames will be used. In this case, the drafter should draw the windows in the most typical usage for the area. Figure 17.45 shows the layout steps for a vinyl or aluminum sliding window. Figure 17.46 shows how other common types of windows can be represented. The finished window blocks can be found on the student CD. When creating your own blocks, assign a width of 1. With a width of 1, an X factor representing the desired width can be assigned as the block is inserted into the elevation.

Rails

Simple rails can be solid to match the wall material or open. Open rails can be made of wood or wrought iron. Vertical rails are required to be no more than 4'' (100 mm) clear and are often made from 2 × 2 (50 × 50) material. Verticals can be placed using the ARRAY command. Rails are often built using a 2 × 6 (50 × 150) and

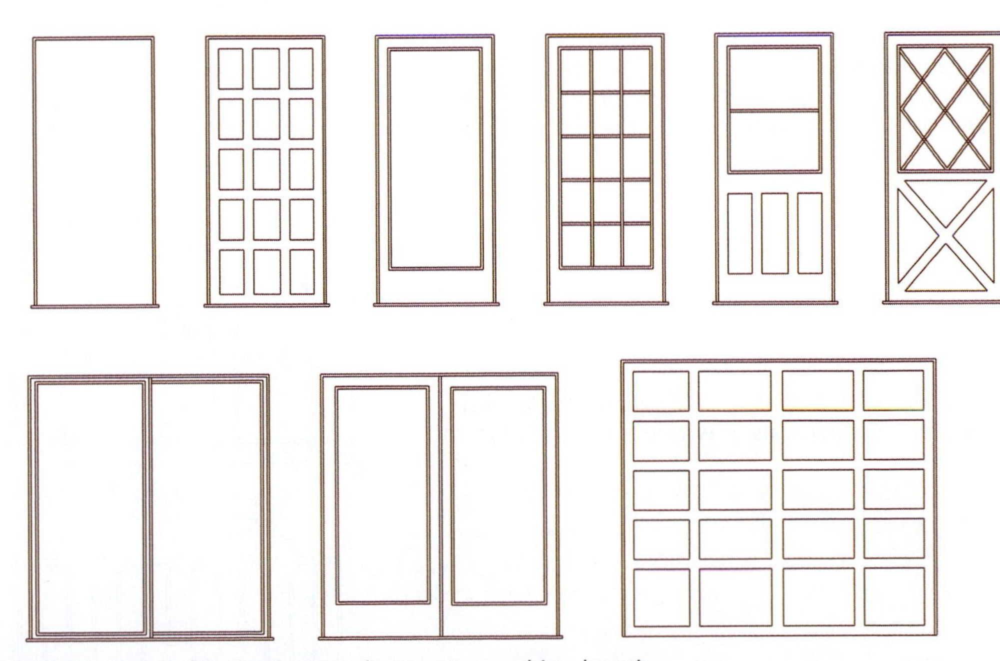

FIGURE 17.44 ■ *Common door blocks that can be represented in elevation.*

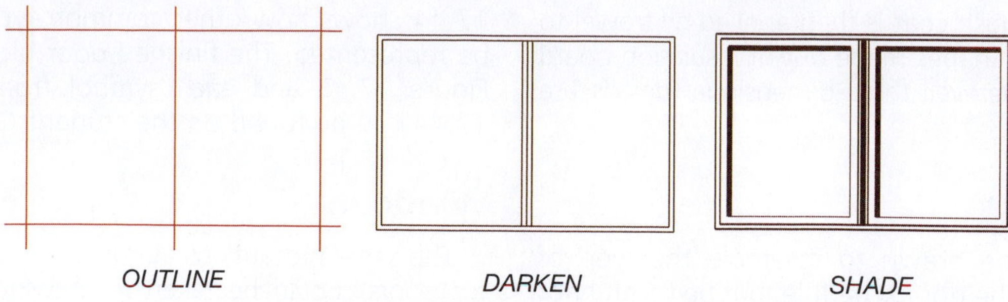

FIGURE 17.45 ■ *Layout steps for creating a block for a sliding window.*

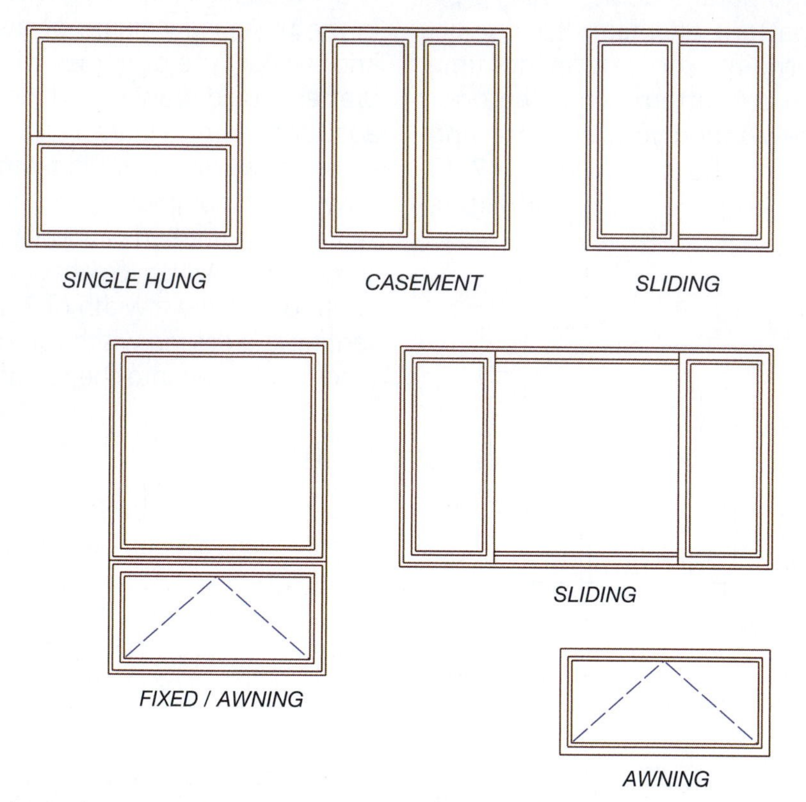

FIGURE 17.46 ■ *Representing common window shapes on elevations. A block with a width of 1 can be created for each style of window. When inserted into the drawing the X factor can be adjusted to meet the required width. The depth can be adjusted by altering the Y scale factor or by using the STRETCH command.*

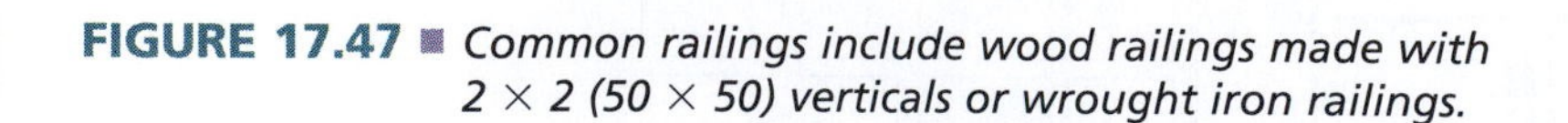

FIGURE 17.47 ■ *Common railings include wood railings made with 2 × 2 (50 × 50) verticals or wrought iron railings.*

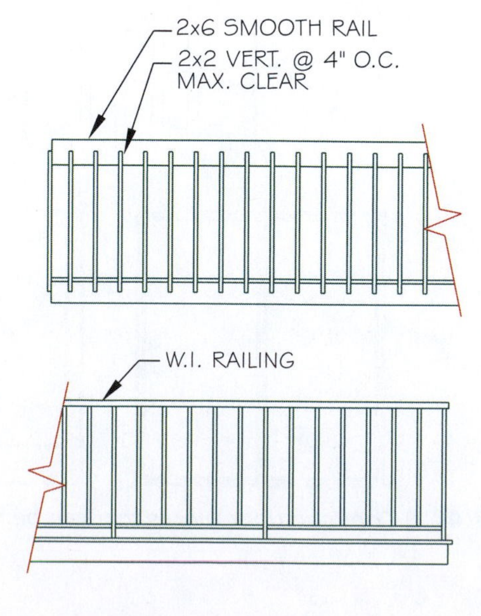

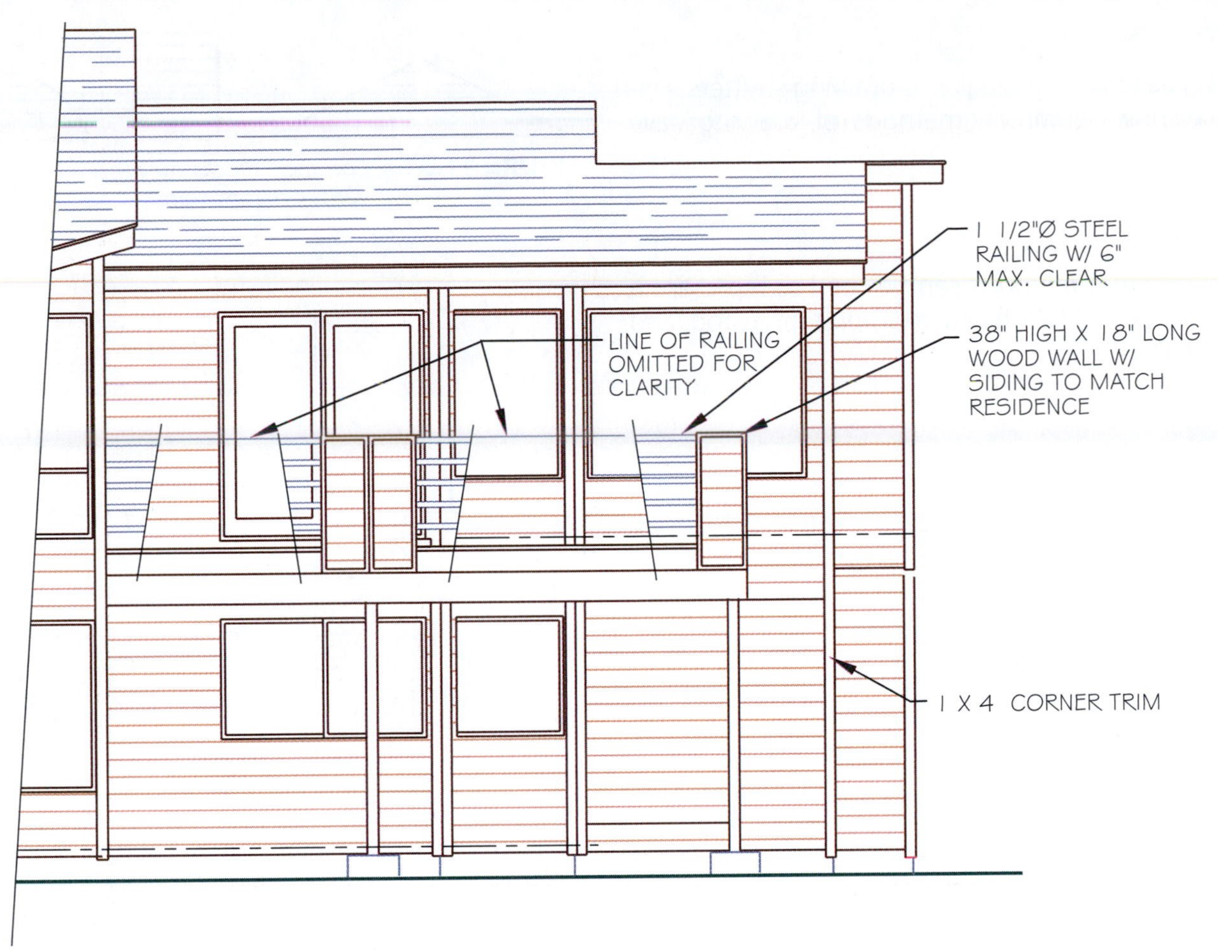

FIGURE 17.48 ■ *Railings can often be omitted to allow objects that lie beyond the rail to be seen. Use a center or phantom line to represent the limits of the rails that have been omitted.*

can be drawn as shown in Figure 17.47. The entire railing should be represented on the front elevation. On the remaining elevations, only a portion of a rail may be drawn as shown in Figure 17.48, with the line used to represent the limits of the rail. More decorative railings similar to Figure 17.49 may be required to match a specific historic style. The MIRROR command can be used once one side of a baluster has been drawn. The balusters should be placed using the ARRAY or COPY command.

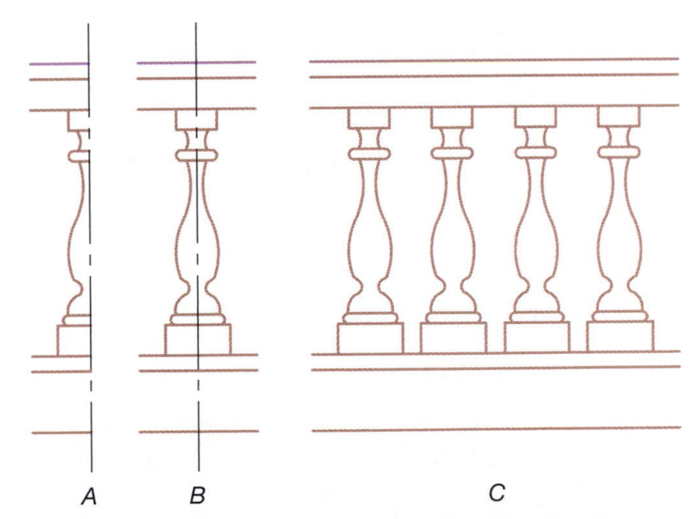

FIGURE 17.49 ■ *The MIRROR command can be a useful tool for creating decorative railings. (A) Half of the baluster is drawn. (B) One full baluster created. (C) Multiple balusters are created using the COPY command.*

Shutters

Shutters are sometimes used as part of the exterior design to match specific historic styles and must be shown on the elevations. Figure 17.50 shows a typical shutter and how it could be drawn. Spacing for the outer frame and the louvers were represented using a 1 1/2'' (38 mm) offset.

Eave Vents

Drawing eave vents is similar to drawing shutters. Figure 17.51 shows common methods of drawing eave vents.

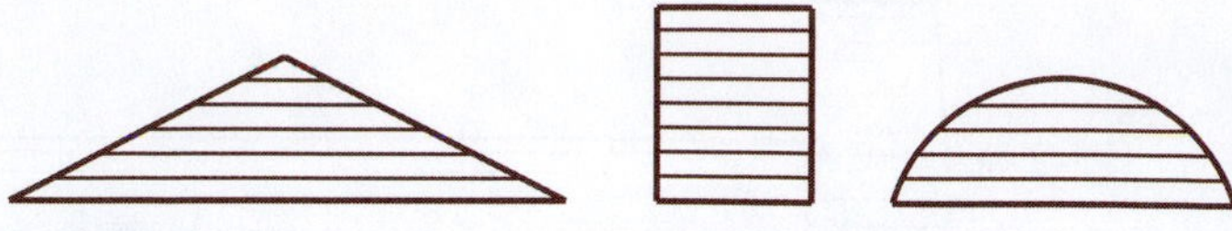

FIGURE 17.51 ■ *Common blocks for representing attic vents on elevations.*

Chimney

Several different methods can be used to represent a chimney. Figure 17.52 shows examples of wood and masonry chimneys.

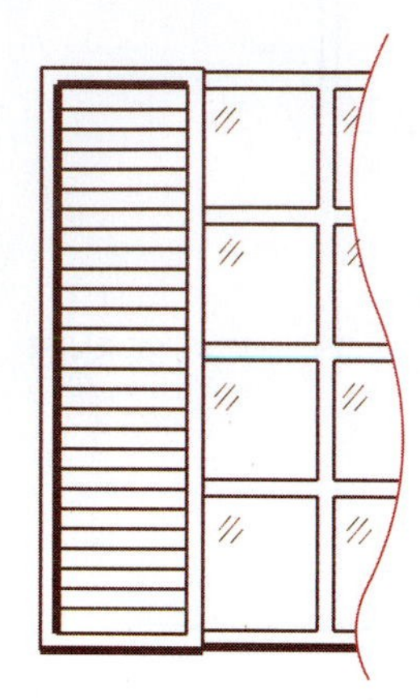

FIGURE 17.50 ■ *Creating a block for a shutter for use in exterior elevation.*

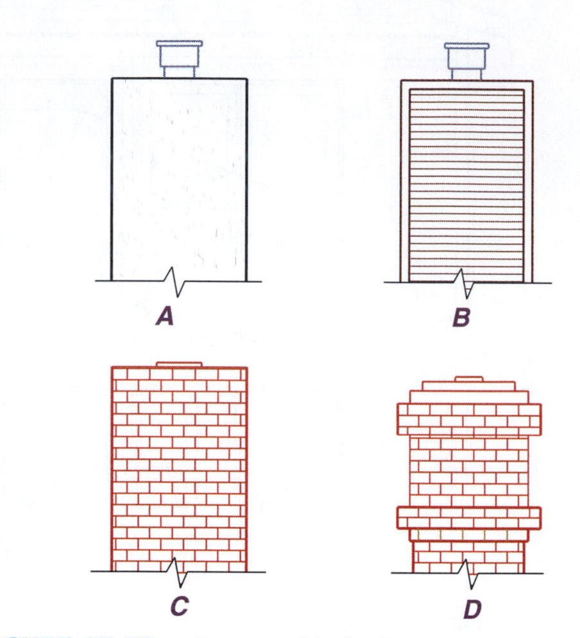

FIGURE 17.52 ■ *Common blocks for representing chimneys include (A) stucco chase with metal cap; (B) horizontal wood siding with metal cap; (C) common brick; (D) decorative masonry.*

CHAPTER 17

Additional Readings

The following websites can be used as a resource to help you keep current with changes in roofing and exterior siding materials.

ADDRESS	COMPANY OR ORGANIZATION
www.abtco.com	Abtco, Inc. (hardboard and vinyl siding products)
www.alcoahomes.com	Alcoa Building Products (aluminum and vinyl siding products)
www.alsco.com	Alsco Building Products (vinyl and aluminum siding)
www.alside.com	Alside (vinyl siding, windows, trim)
www.ambrico.com	American Brick
www.brickinfo.org	The Brick Institute of America
www.canamould.com	Canamould (exterior moldings)
www.cemplank.com	Cemplank (fiber-cement siding)
www.culturedstone.com	Cultured Stone
www.apawood.com	The Engineered Wood Association
www.cwc.ca	Canadian Wood Council
www.caststone.org	Cast Stone Institute
www.cedarbureau.org	Cedar Shake and Shingle Bureau
www.eifsfacts.com	EIFS Industry Members Association
www.fabral.com	Fabral (metal roofing and siding)
www.gp.com	Georgia Pacific (siding)
www.shakertown.com	Hardie Building Products (cedar shingles)
www.jameshardie.com	James Hardie Building Products
www.nailite.com	Nailite International (replica brick, stone, shake products)
www.napcobuildingmaterials.com	NAPCO Building Specialties
www.norandex.com	Norandex (vinyl siding)
www.reynoldsbp.com	Reynolds Building Products
www.senergyeifs.com	Senergy Inc. (EIFS)
www.vinylinfo.org	The Vinyl Institute
www.wrcla.org	Western Red Cedar Lumber Association
www.vinylsiding.com	Wolverine (metal siding)

CHAPTER 17

Introduction to Elevations Test

DIRECTIONS

Answer the following questions with short, complete statements. Type the answers using a word processor or neatly print the answers on lined paper.

1. Place your name, the chapter number, and the date at the top of the sheet.
2. Type the question number and provide the answer in the form of a statement that includes part of the question. You do not need to write out the entire question.

Note: The answers to some questions may not be contained in this chapter and will require you to do additional research.

QUESTIONS

17.1. Under what circumstances could a technician be required to draw only three elevations?

17.2. When are more than four elevations required?

17.3. What are the goals of the exterior elevations?

17.4. What is the most common scale for drawing elevations?

17.5. Would an elevation that has been drawn at a scale of 1/16'' = 1'-0'' require the same methods to draw the finishing materials as an elevation that has been drawn at some larger scale?

17.6. Describe methods for transferring the heights of one elevation to another.

17.7. Describe two different methods of showing concrete tile roofs.

17.8. What are the two major types of wood siding?

17.9. In drawing a home with plywood siding, how should the texture of the wood be expressed?

17.10. Sketch the pattern most typically used for brickwork.

17.11. What problems are likely to be encountered when drawing stone?

17.12. Sketch the way wood shingles appear when one is looking at the gable end of a roof.

17.13. Give the major consideration for drawing doors in elevation.

17.14. What are the most common materials used for rails?

17.15. How should the pattern be expressed in drawing stucco?

CHAPTER 18

Elevation Layout and Drawing Techniques

Many technicians consider the drawing of elevations to be among the most enjoyable projects in architecture. As a student, though, you should keep in mind that a beginning technician rarely gets to draw the main elevation. The architect or project manager usually draws the main elevation in the preliminary stage of design, and then a senior technician gets to finish the elevations. As a beginner, you may be introduced to elevations by making corrections resulting from design changes. Not a lot of creativity is involved in corrections, but it is a start. As you display your ability, you'll most likely be involved in earlier stages of the drawing process. This chapter assumes that you will be completing the elevations. It may not be realistic for a beginning technician, but your day will come. Use the following four steps to complete the elevations:

- *Layout*
- *Drawing completion*
- *Lettering and dimensions*
- *Paper space arrangements*

DRAWING LAYOUT

The layout of the elevations should be completed in model space with no regard to the size of the drawing template. Elevations can be completed at full size in model space using side-by-side arrangements to ease projection of surfaces. The scale to be used when the drawings are plotted can be assigned to the viewport containing the drawings in paper space. Be sure to plan the plotting scale prior to adding text or dimensions to the drawings, because the scale will affect the display size of each item. The size of drawing template will affect the plotting scale and placement of the elevations in paper space. Although the decision about plotting scale can be made after the initial layout of the elevations, the drawing sheet should match the size used for the floor plan. Because D-size paper is so commonly used in architectural settings, these instructions will assume that the drawing is to be made on a D-size template. The following instructions will be given for plotting all of the elevations at a scale of 1/4'' = 1'-0''. A common alternative is to plot the front elevation at the same scale as the floor plan and the rear and both side elevations at a scale of 1/8'' = 1'-0''. Using two scales will require the use of multiple viewports for the plotting display. Plotting all of the elevations at one scale will usually require multiple sheets.

The layout process can be divided into four stages:

- Preparing a floor plan block for projection
- Overall-shape layout
- Roof layout
- Projection of openings

Preparing the Floor Plan

The floor and roof plans contain all of the needed information to draw the elevations for a home to be developed on a flat site. A

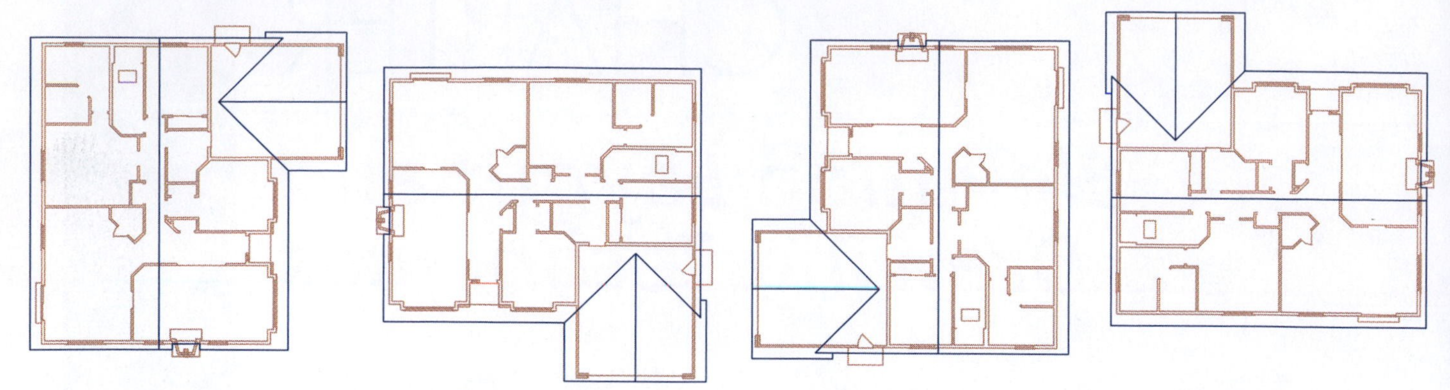

FIGURE 18.1 ■ *Create a block of the floor plan showing the exterior walls, roof outline, skylights, and chimneys. This block can then be inserted into the template where the elevations will be drawn. With four copies of the floor plans placed in the model space layout; horizontal distances can then be projected down to the elevations.*

site or grading plan will be required to aid in the projections of the ground line for a specific site. For this project, a flat site is assumed. Use the following steps to prepare the floor plan for use in projecting the elevations. These steps assume the floor and roof plans are in the same file and stacked over each other. If this is not the case, copy the roof outline into the drawing containing the floor plan. The following procedures demonstrate how to draw a single level home with a post and beam floor system. Examples demonstrate the process for a single level home. Once this procedure is complete, examples are provided to draw a multilevel home.

Step 1. Open the floor plan and freeze all materials. The only materials to be displayed should include the exterior walls, chimneys, skylights, and roof outlines.

Step 2. Using the materials displayed in step 1, create a block. Use a name such as ELEV FLOR when you save the block.

Step 3. Open a new drawing template and enter Model space.

Step 4. Create a new layer titled *ELEV LOUT* and make this layer current. Set the layer to have thin, continuous lines.

Step 5. Set the drawing limits large enough so that four copies of your home can be placed side by side. Use a setting such as 900 × 600' and then use the ZOOM command to display the EXTENTS of the drawing limits.

Step 6. Insert the ELEV FLOR block into model space of the drawing template on the *ELEV LOUT* layer.

Step 7. Make three additional copies of the ELEV FLOR block. Place the copies side by side. Keep the drawings as close as possible without any overlap.

Step 8. Rotate the copies of the ELEV FLOR block so that you now have a block with the left side, front, right side, and rear facing the bottom of your monitor. Your drawing should now resemble Figure 18.1.

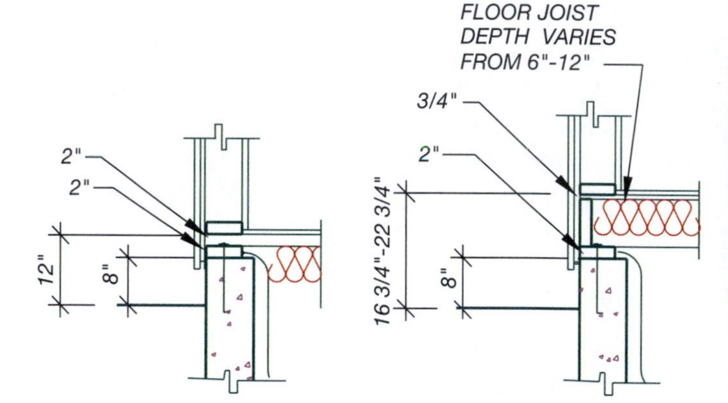

FIGURE 18.2 ■ *The top of concrete slabs must be placed 8'' (200 mm) above the grade. The height for wood floors will be determined by the type of the floor framing system. Exact framing sizes are often rounded up to the nearest inch in elevation drawings.*

Overall Shape Layout

Place the objects from the following steps on the *ELEV LOUT* layer.

Step 9. Draw a line to represent the grade. Place this line approximately 20' below the floor plans. Make the line long enough so that it extends past the edges of the floor blocks.

Step 10. Establish the floor line. The floor line will be a minimum of 8'' (200 mm) above the grade line for a concrete slab. A minimum distance of 12'' (300 mm) is required for a post-and-beam floor system and 20'' (500 mm) will be required for a joist floor system. See Figure 18.2 to determine how these sizes were established. Use the OFFSET command to place the floor line in the proper position relative to the finished grade line.

Step 11. Use the OFFSET command and place a line that is 6'-8'' (2000 mm) above the floor to represent the tops of the windows and doors. Remember the height of the garage door will be based on the location of the finished grade, not the floor level.

Step 12a. Use the OFFSET command to place a line to represent the ceiling level. For most homes, this distance will typically be 8'-0'' (2400 mm) above the floor level.

If the home features multiple floor levels, use the following steps to layout additional levels.

Step 12b. Measure up the depth of the floor joists to establish the floor level of the second floor. If the floor joist size has not been determined, assume that 10'' (250 mm) depth (2 × 10s) will be used.

Step 12c. Measure up 8'-0'' (2400 mm) from the second floor line and draw a line to represent the upper ceiling level.

These same steps can be used if a lower floor is to be placed below the main level. Offset down the required distance to represent the floor joists and then down again to represent the height from the basement floor to the ceiling level.

Step 13. Project lines down from the floor plan to represent each edge of the house. Your drawing should now resemble Figure 18.3.

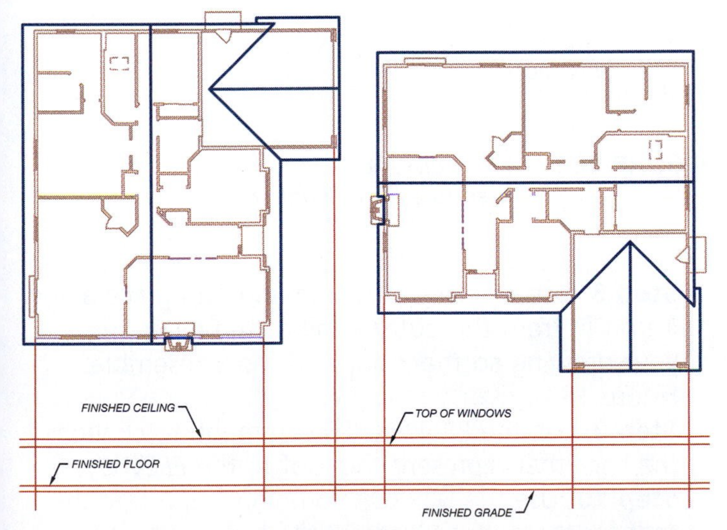

FIGURE 18.3 ■ *The location of major shapes of each elevation and lines to represent the grade, floor, top of windows and the finished ceiling must be represented.*

Roof Layout

Complete the projection for the roof for the front elevation first, followed by the roof layout for the remaining views. For this home, a truss roof and a roof pitch of 6/12 is assumed. See Figure 15.8 to determine the required angle to draw the rake of the roof, or measure over 12 units and up X units to represent the desired roof angle. Continue to use the *ELEV LOUT* layer for the following steps.

Step 1. Working only with the front elevation, project a line from each eave overhang and ridge of the ELEV FLOR block down below the ceiling line. By keeping these lines shorter than the lines projected from the walls, the lines representing the roof and walls can easily be distinguished as the drawing gets more crowded. On a complicated project, layout lines for the roof and walls can be assigned different colors to aid projection.

Step 2. Determine the roof slope. For this home, draw the roof at a 6/12 pitch.

A 6/12 pitch equals 26 1/2°. See Section 4 for a complete explanation of pitches. Start the roof layout where a gable end wall can be seen. For this home, the right side of the home has a gable end wall.

Step 3. Start at the intersection of the wall and the ceiling and draw a line at the desired angle to represent the roof. Start on the left side of the gable end wall. This will allow you to enter positive angles. Use a keyboard entry, such as "@12'<25.5'' to represent the rake of the roof. Enter a random length, such as 12' (3600 mm). If the line is too long, it can be trimmed later. If it is too short, the EXTEND command can be used to adjust the length. This line will represent the bottom of the truss. See Figure 18.4a. If rafters are to be used, follow the guidelines in Figure 18.4b.

Step 4. Use the OFFSET command and place a line 6'' (150 mm) above the first inclined line drawn in Step 3. This will represent the top of the trusses. On the elevations, it is not important that the exact depth of the truss be represented, only that you remain consistent for all elevation views.

Step 5. Use the OFFSET command and place a line 1'' (25 mm) down from the top line.

Step 6. Use the EXTEND command to extend the inclined lines to the overhang.

Step 7. Use the EXTEND or TRIM command to extend or shorten the inclined lines to the line that represents the ridge. Your drawing will now resemble Figure 18.5.

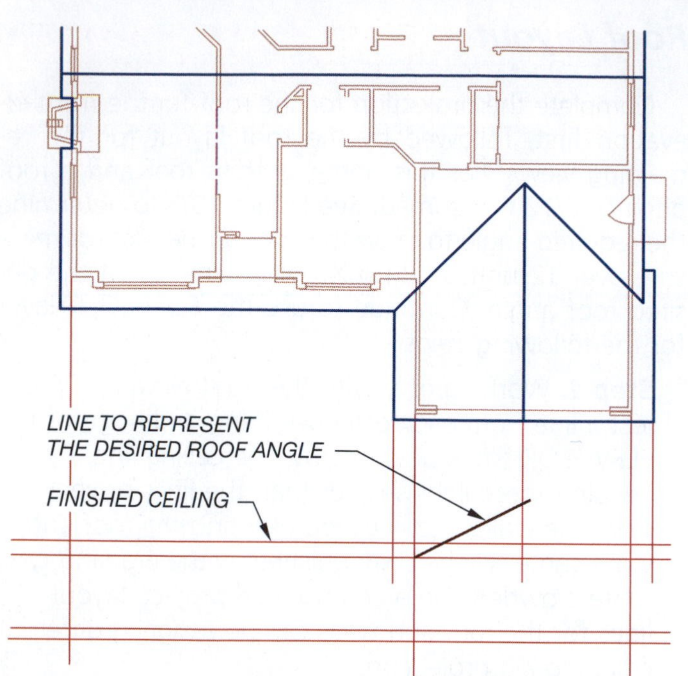

FIGURE 18.4a ■ *The rake of the roof is determined by drawing a line from the intersection of the wall and the ceiling at the desired angle to represent the roof. Start on the left side of the gable end wall. This will allow you to enter positive angles.*

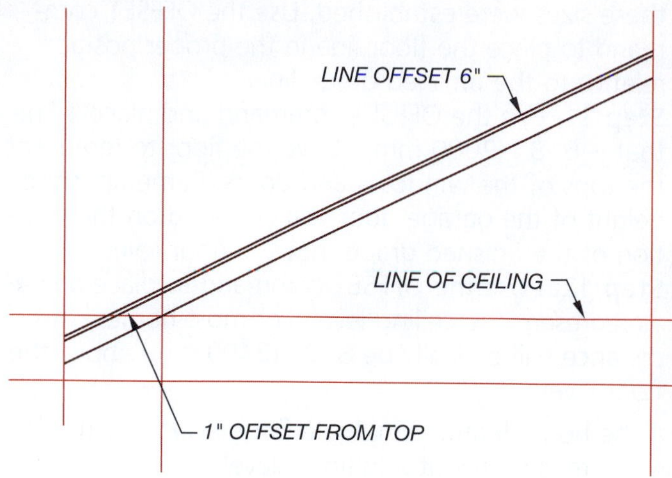

FIGURE 18.5 ■ *Once the angle of the roof is determined, use the OFFSET command and place a line 6'' (150 mm) above the first line to represent the top of the trusses. Offset down 1'' to represent the drip edge of the roofing over the fascia.*

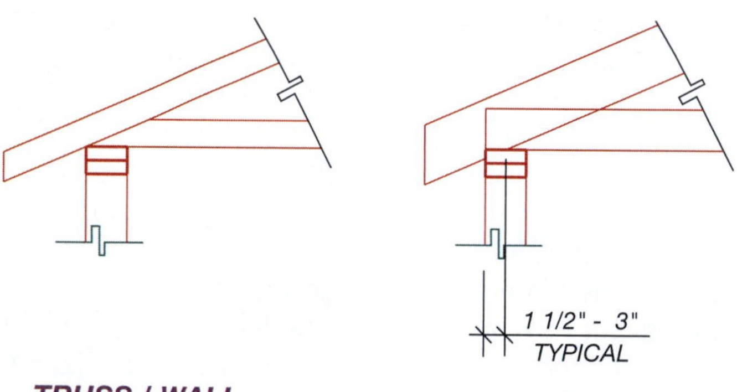

FIGURE 18.4b ■ *The layout of the roof incline will vary slightly, depending on whether trusses or rafters are used to frame the roof. The incline of a truss passes through the intersection of the ceiling and the outer edge of the walls. Because the rafter must have at least 1 1/2'' of bearing surface, the intersection of the rafter and wall must be adjusted slightly. If the notch in the rafter exceeds 3'', the rafter may not be able to support its design loads. Later chapters explore minimum framing requirements.*

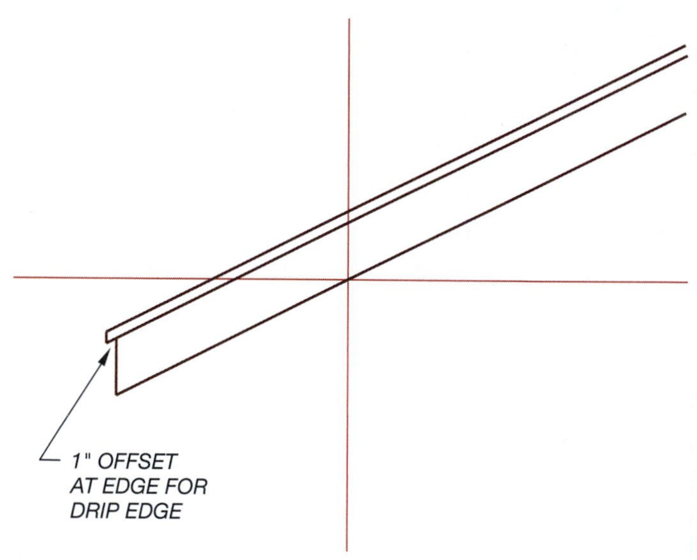

FIGURE 18.6 ■ *The TRIM and FILLET commands can be used to adjust all line intersections.*

Step 8. Use the OFFSET command and place a line 1 unit in from the outer edge of the eave. Trim your drawing so that your eave now resembles Figure 18.6.

Step 9. Use the PROPERTIES command to change the lines that represent the roof to the *ELEV* layer.

Step 10. Use the MIRROR command to create the roof for the opposite side of the home. You now have the gable roof for the right side of the roof drawn. Your drawing should resemble Figure 18.7.

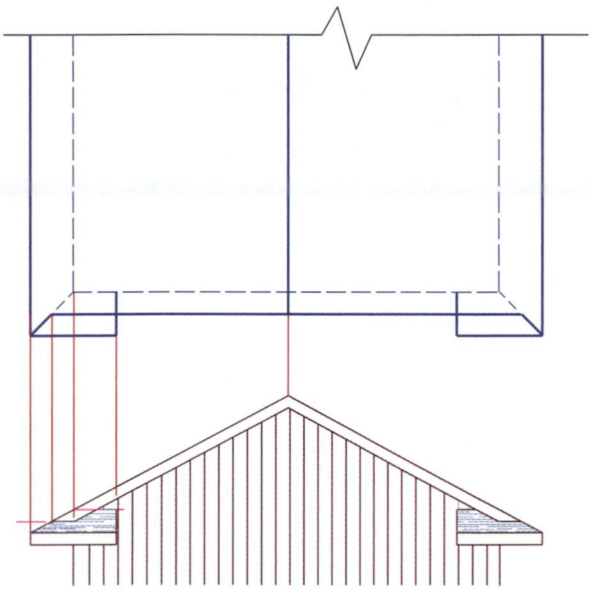

FIGURE 18.30 ■ *Projecting and representing corner eaves.*

Projecting Grades

Projecting ground levels is required on elevations when a structure is being constructed on a hillside. Figure 18.33 shows the layout of a hillside residence. To complete this elevation, a site, floor, and roof plan were used to project locations. Usually the grading plan and the floor plan are drawn at different scales. The easiest method of projecting the ground level onto the elevation is to measure the distance on the site plan from one contour line to the next. This distance can then be marked on the floor plan and projected onto the elevation.

Combining Techniques

Figure 18.34 shows the completed elevations for the home that was started in Chapter 11. These elevations were completed using each of the techniques that were presented throughout this chapter. The front elevation was drawn using the methods described in the initial layout steps. The rear and side elevations required the addition of the grading plan to be added to the ELEV FLOR block so that the existing and finished grades could be projected into the elevations.

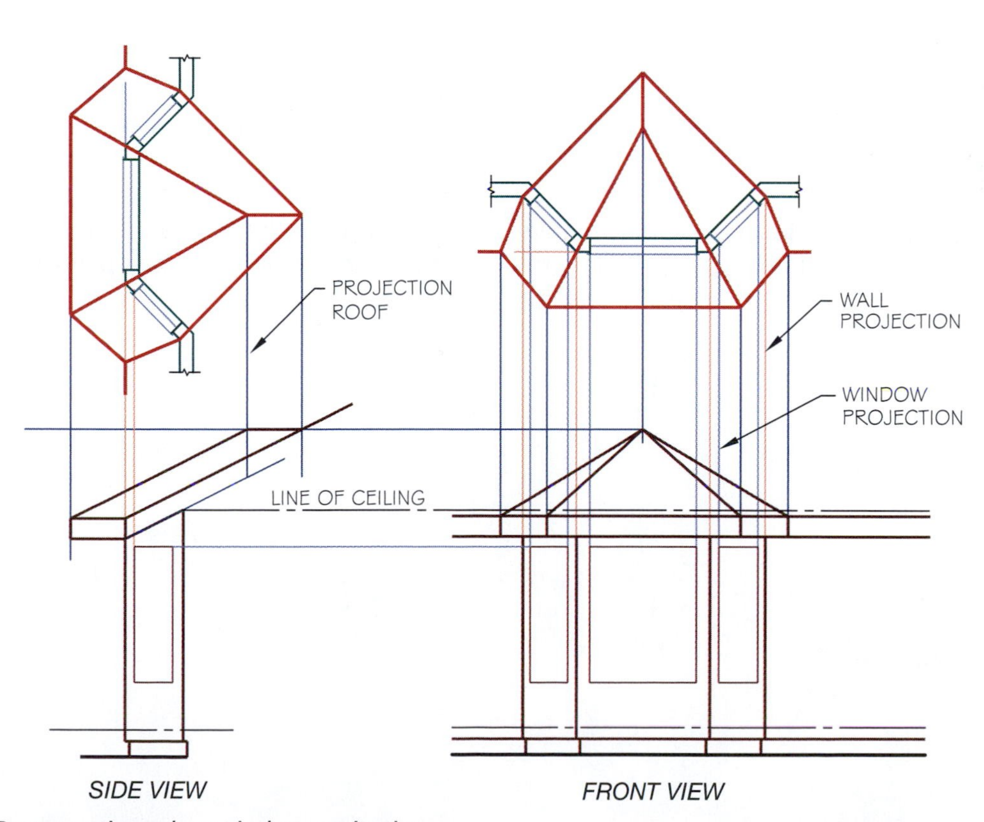

FIGURE 18.31 ■ *Representing a bay window projection.*

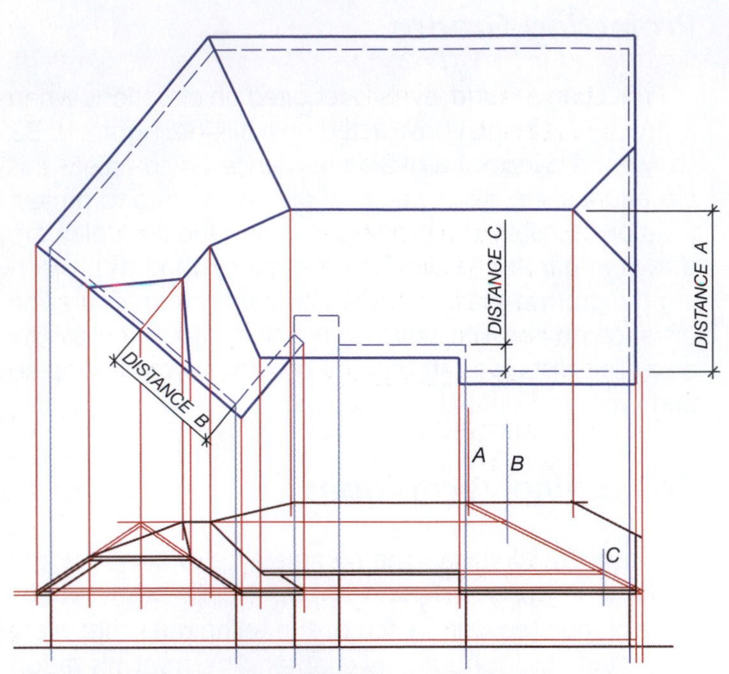

FIGURE 18.32 ■ *In drawing the elevations of a structure with an irregular shape, great care must be taken with the projection methods. This elevation was prepared using the outline of the roof displayed over floor plan and then projecting the needed intersections down to the drawing area. Notice that on the right side of the elevation, a line at the required angle has been drawn to represent the proper roof angle for establishing the true heights.*

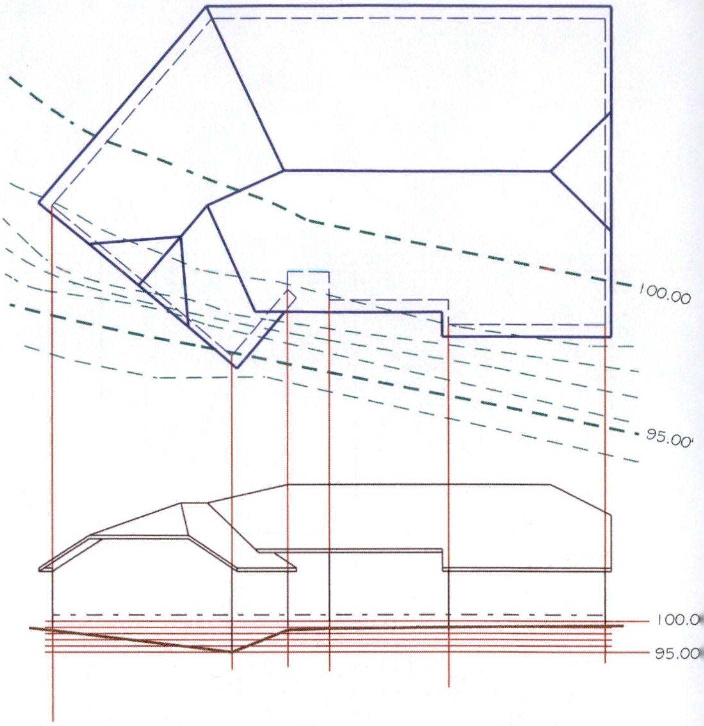

FIGURE 18.33 ■ *Projecting grades is similar to the method used to project roof lines. This elevation was drawn by using the information on the roof, floor, and grading plans. For clarity, only the projection lines for the grades are shown.*

Creating Details

Because of their complexity, elevations often require the use of details to explain complex construction methods. When required, a portion of the elevation can be reproduced using the COPY command. The copy can then be displayed in a separate viewport so that the plotting scale can be enlarged to allow the needed information to be clearly displayed. Figure 18.35a shows a partial elevation for a residence and Figure 18.35b shows the detail used to explain the cornice and eave construction.

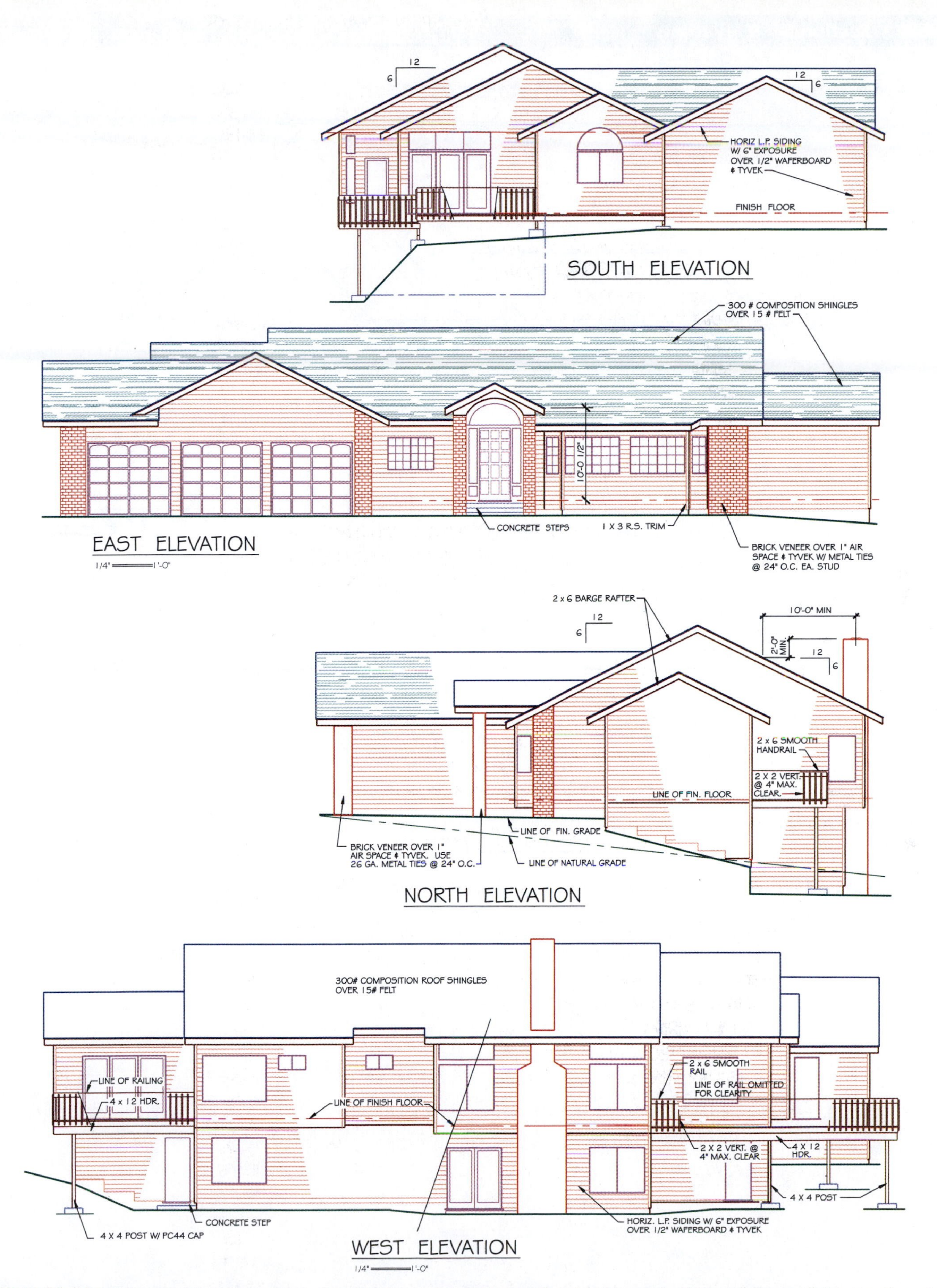

FIGURE 18.34 ■ *The completed elevations for the home completed in Chapter 11 required the addition of the grading plan to the ELEV FLOR block. To complete these drawings, the grades were projected from the site plan and the steps for a multilevel home followed.*

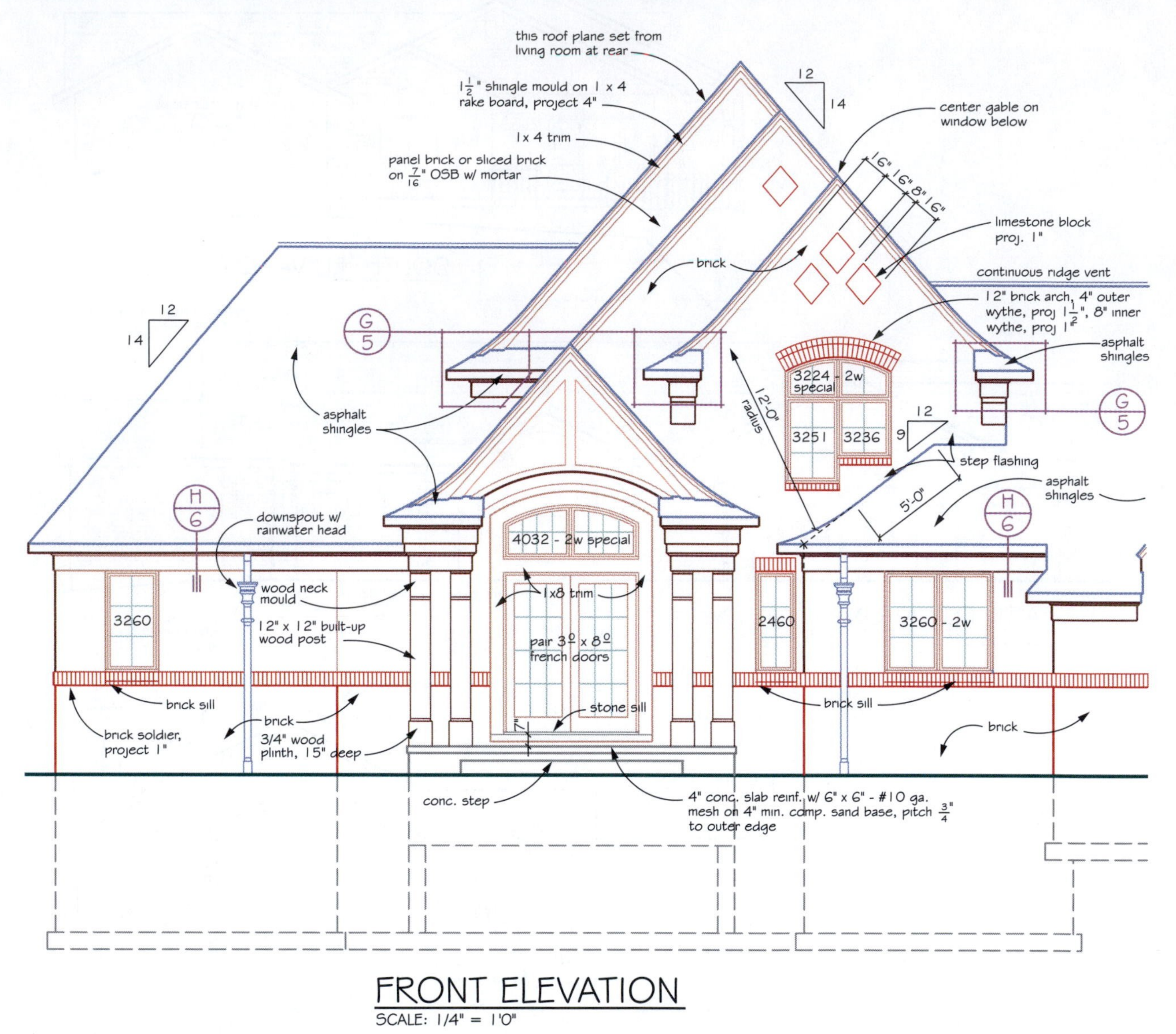

FIGURE 18.35a ■ *A partial elevation with references for two different details.* Dennis Dinser, Principal, Arcadian Designs, LLC.

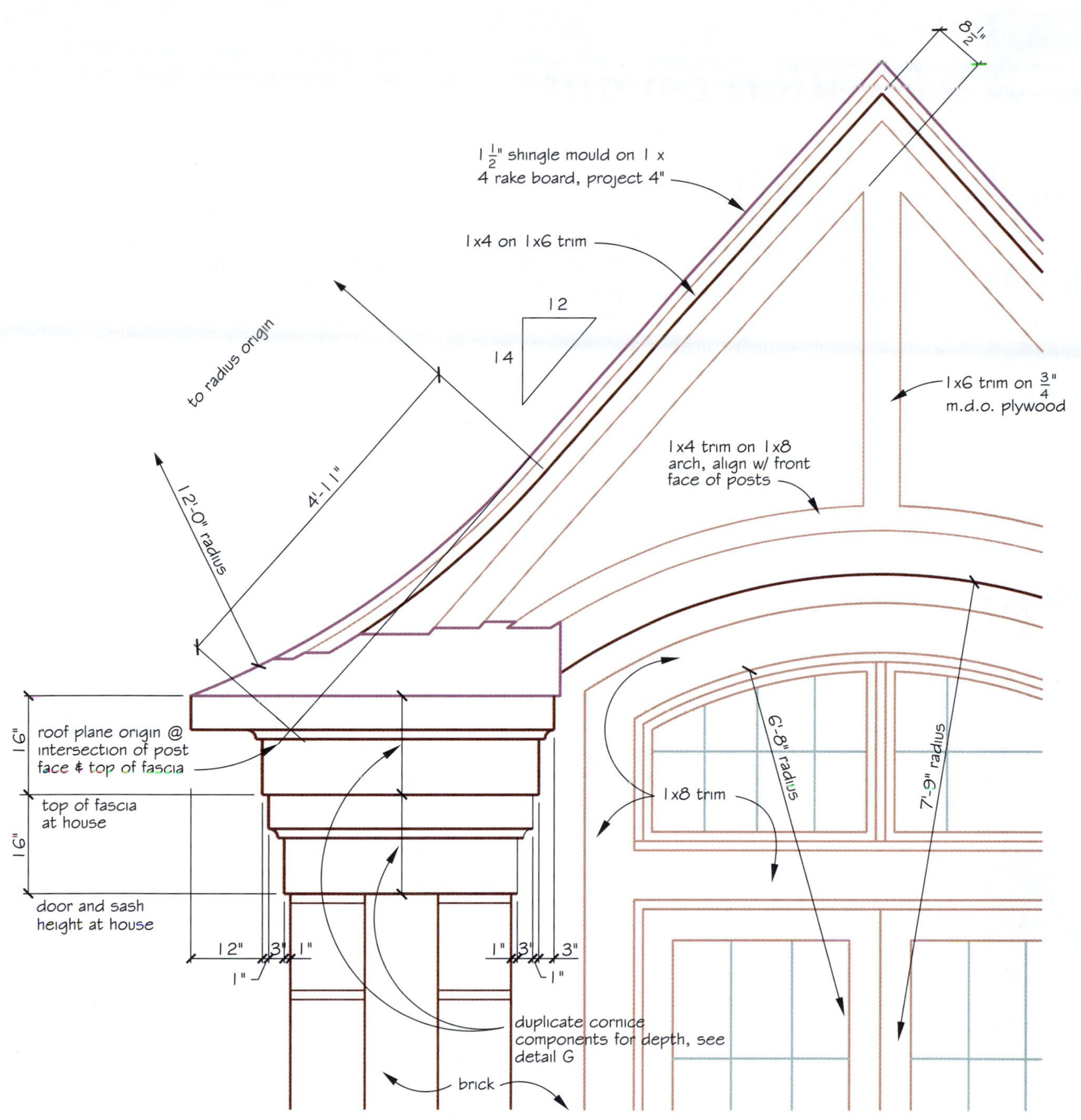

FIGURE 18.35b ■ *An enlarged detail of the entry roof with the information to explain the eave construction.* Dennis Dinser, Principal, Arcadian Designs, LLC.

CHAPTER 18

Elevation Layout and Drawing Techniques Test

DIRECTIONS

Answer the following questions with short, complete statements. Type the answers using a word processor or neatly print the answers on lined paper.

1. Type your name, the chapter number, and the date at the top of the sheet.
2. Type the question number and provide the answer in the form of a statement that includes part of the question. You do not need to write out the entire question.

QUESTIONS

18.1. Who typically draws the main elevations of a structure? Explain.
18.2. List methods of projecting heights from one elevation to another.
18.3. What scales are typically used to draw elevations?
18.4. What is the best method of determining the height of a roof when it is being viewed in a manner that does not show the pitch?
18.5. Explain the advantages of each method of representing the elevations in paper space.
18.6. What drawings are required to project grades onto an elevation? Briefly explain the process.
18.7. What drawings are required to draw a structure with an irregular shape?
18.8. Sketch the method of showing the chimney height and provide the minimum dimensions that must be shown.
18.9. What dimensions are typically shown on elevations?
18.10. What angle is used to represent a 4/12 pitch?

CHAPTER 18

Problems

DIRECTIONS

1. Use the sketch from Chapter 16 and draw the required elevations. The sketches do not contain all of the features represented on the floor plans and elevations, and not all plans have a sketch to use as a guide. Draw the elevations so they coordinate with other drawings to be included in the drawing set. If no sketch is provided, make all decisions regarding the style and materials to be used.
2. Draw the required elevations using the same type and size of drawing material that you used to draw your floor, roof, and site plans.
3. Design the structure using a style and materials suitable for your area.
4. Refer to your floor plan to determine all size and location of all features. If you find material that cannot be located using your floor plan, make additions to your floor plan as required.
5. Refer to the text of this chapter and notes from class lectures to complete the drawings.
6. Use the checklist from Appendix D to evaluate your work.
7. When your drawing is complete, turn in a copy to your instructor for evaluation.

19 CHAPTER

Interior Finishing Materials and Cabinets

The interior trim and cabinetry are often referred to as millwork. Millwork includes any finish trim or finish woodwork installed by the finish carpenters and cabinetmakers. Most drawings for a stock home do not include details of millwork items. The practice of drawing millwork depends on the specific requirements of the project and the price range to which the home is built to appeal. A home designed as an entry-level or starter home will often have very little millwork to keep the price as low as possible. The plans for a starter home may show only a plan view of the cabinetry with no other specifications for millwork. At the other end of the price range, custom homes may show very detailed drawings of the finish woodwork in the form of plan views, elevations, construction details, and written specifications. No matter the amount of millwork, building departments rarely require cabinet drawings, although some lenders require millwork drawings to be part of a set of plans.

FIGURE 19.1 ■ ***Millwork includes any finish trim or woodwork that is installed by the finish carpenters and cabinetmakers.*** Courtesy Alan Mascord AIBD, Alan Mascord Design Associates, Inc. Bob Greenspan, photographer.

MILLWORK

Millwork may be designed simply to hide the intersection of materials or to add warmth and charm to a home. Millwork such as that seen in Figure 19.1, which is designed for appearance, may be ornate trim or created with a group of shaped wooden or plastic forms placed together to capture a specific style of architecture. Molded millwork can be reproduced in a wide variety of shapes, styles, and patterns. Premanufactured millwork moldings are available at lower cost than custom designs.

Functional millwork is trim that is used to hide intersections of varied materials. Trim at the edges of cabinets to wall intersections, door, and floor trim are each examples of functional millwork. These types of millwork may be very plain in appearance and are generally less expensive than standard sculptured forms. Functional wood millwork may be replaced with rubber, vinyl, or plastic products. Common applications of these types of trims include curved ceiling trim or floor trims in a laundry or bathroom. Other common uses include a plastic or rubber base strip used on a wall at floor level to protect the wall.

Wall Treatments

Common wall treatments using millwork include baseboards, wainscots, chair rails, crown molding, and casings. Additional work often done by the cabinet shop includes built-in shelving, mantels, and railings.

Baseboards

Baseboards similar to Figure 19.2 are placed at the intersections of finished walls and floors and are used to hide these intersections. They also serve to protect the walls from damage caused by furniture rubbing against the wall. Figure 19.3 shows some standard molded baseboards. Baseboards can be as ornate or as plain as the specific design or location dictates. In some designs, baseboards are the same shape as other millwork members, such as trim around doors and windows. Baseboards are generally specified in detail. For a large custom home, the room and pattern number of the trim can be specified in a finish schedule.

FIGURE 19.2 ■ *Baseboards are used to hide the intersections of the finished floor and walls.* Courtesy W. Lee Roland, Builder (rolandbuilder.com). Hayman Studios, photographer.

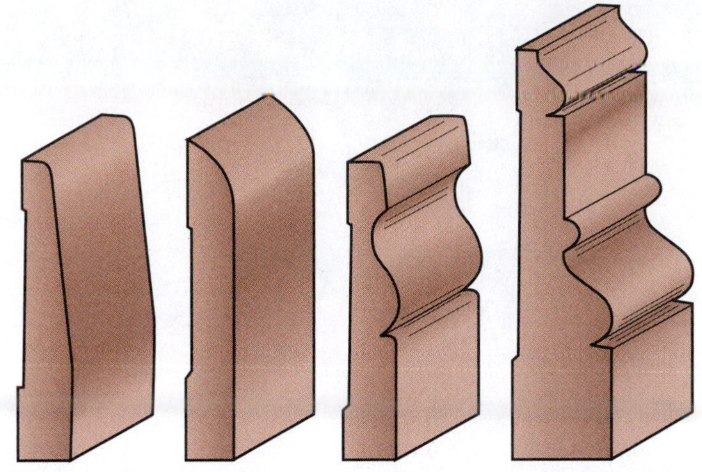

FIGURE 19.3 ■ *Common baseboard styles.* Courtesy Hillsdale Sash and Door Co.

Wainscots

A ***wainscot*** or wainscoting is millwork placed on the bottom portion of the wall. Wainscoting can be any material including wallpaper, sheetrock of varied textures, wood paneling, tile, stone, or masonry used to divide walls into two visual sections. Figure 19.4 shows an example of wood wainscot. Common wood wainscotings include oak and other hardwood veneers.

Chair and Plate Rails

The ***chair rail*** in Figure 19.5 is a horizontal trim placed just below the midheight of a wall. It has traditionally been placed on walls at a height where chair backs would damage the wall. It is often used on custom homes in dining rooms, dens, offices, and other areas where chairs frequently move against a wall. Chair rails may be used with wainscoting or to separate two different materials or wall textures. Chair rails are specified in general notes placed with the floor plan or interior elevations. Notes should specify the molding type and height of the rail.

A plate rail is a horizontal rail set approximately 48 to 60'' (1200 to 1800 mm) above the floor (see Figure 19.6). A piece of trim with a ribbed surface approximately 3 to 4'' (75 to 100 mm) wide provides support to display decorative plates or knickknacks. A detail should be drawn to describe the location, size, and type of trim used to construct the rail.

Crown Molding

Crown moldings have been used since the earliest colonial days to highlight ceilings. In its simplest form, crown molding is a decorative trim placed in the cor-

FIGURE 19.4 ▪ *A wainscot is placed on the bottom portion of the wall.* Courtesy Dean Benfit.

FIGURE 19.5 ▪ *A chair rail is used as a cap for wainscoting or to protect a wall from damage caused by chairs rubbing the wall.* Courtesy Stephen Brown.

ner where the wall meets the ceiling. Oak, mahogany, and walnut are often used to create crown molding with a standard depth of 5'' (125 mm). Crown molding may also be an elaborate component made up of several individual members, similar to Figure 19.7. Cornice boards are often used on custom homes and are especially common in traditional architectural styles, such as English Tudor, Victorian, or Colonial. A detail should be drawn to describe the location, size, and type of trim used to construct the crown molding.

Window and Door Trim

Casings are the trim members used to hide the window- and door-to-wall intersections. To save cost, casings may be omitted, and windows and doors are trimmed with sheetrock. To increase appeal, wood or vinyl casings are used at windows or doorjambs. Casings may be very simple trim to serve the functional purpose of covering the space between the doorjamb or window jamb and the wall. Decorative casings to match other moldings are also used on custom homes. Figure 19.8 shows the use of decorative window trim. For homes at the lower end of the budget scale, the use of casings is specified in a general note on the floor plan. The contractor or the finish framers will supply trim to match the budget. On custom homes, specific trim locations and styles are specified in notes on the floor and interior elevations as well as in details.

FIGURE 19.6 ■ *A plate rail is a common piece of millwork in many traditional dining rooms. It is normally set at or near eye level to display decorative plates or collectables.* Courtesy Sherry Benfit.

FIGURE 19.7 ■ *A multi-piece cornice can be used to add charm and beauty to a room or to match a specific historical style.* Courtesy W. Lee Roland, Builder (rolandbuilder.com). Hayman Studios, photographer.

FIGURE 19.8 ■ *Trim installed by doors and windows is referred to as casings. The decorative trim at the corner is referred to as a rosette.* Courtesy Barbara Conklin.

Specialty Wall Materials

A second category of millwork includes custom mantels, railings, and built-in shelving. The finish carpenters install each item based on drawings provided by the design team.

Mantels

A mantel is an ornamental shelf or structure built above a fireplace opening. Mantel designs vary with individual preference, home style, and the budget of the home. Mantels may be made of masonry as part of the fireplace structure or may be comprised of ornate decorative wood moldings. Figure 19.9 shows a custom mantel application. The material to be used for the mantel affects where information about its construction is placed. Masonry mantles may be left to a mason or specified in details placed near the fireplace section and elevation. Information regarding wood mantels is generally given in interior elevations or in the interior details.

Railings and Dividers

Railings, introduced in Chapter 4, are for use at stairs, landings, decks, and open balconies higher than 30'' (750 mm) above the floor. Figure 19.10 shows rails used

FIGURE 19.9 ■ *The information provided for a mantel will vary depending on the material to be used. Mantels made of wood are often described in details placed with other interior elevations.* Courtesy Lee Gleason.

FIGURE 19.10 ■ *Railings are used to provide protection for hand and guardrails for changes in floor levels over 30'' (750 mm) in height. In addition to the railings, this home features base trim, wainscoting, chair moldings, and other decorative moldings.* Courtesy BOWA Builders, Inc. Bob Norad, photographer.

for handrails and guardrails for the upper balcony. Depending on the budget for the home, they may be detailed on the plans or left to the craftsperson's to create. Even on low-budget homes, the rail height and the baluster spacing must be specified on stair drawings. Railings and partial walls are also used as decorative room dividers or to add accent to a room.

Instead of using a physical railing, columns can be used to effectively divide a room. Figure 19.11 shows

FIGURE 19.11 ■ *Decorative columns can be used to provide a sense of openness, while dividing two rooms.* Courtesy BOWA Builders, Inc. Bob Norad, photographer.

columns dividing a dining room from the entry. If the columns are premanufactured, details are not required to explain their construction. Information must be provided on the floor plan to specify their location and diameter. Rectangular columns that are site-built require both an elevation and a section to explain the materials to be used.

Ceiling Treatments

Based on standard construction practices, ceilings are approximately 8' (2100 mm) high and flat. Most custom homes have taller ceilings to make the home seem more spacious. Ceilings sometimes depart from these norms for structural, spatial, or decorative reasons. Common ceiling options include soffited, vaulted, beamed, coved, and tray.

Soffited Ceilings

A common method to add interest to a ceiling is to drop or soffit the ceiling. A kitchen is a common place to use a soffit. A 12'' (300 mm) deep soffit that follows the shape of the cabinets is often used to reduce the height of the upper cabinets. The soffit usually extends a few inches in front of the upper cabinets, but the size may be increased to install can ceiling lights in the soffit. Figure 19.12 shows a ceiling and a wall soffit in a bedroom. The walls were built-up to provide the soffited effect in order to highlight the bed, and the ceiling was soffited into two tiers to highlight the placement of lights. The width and depth of soffits must be shown on the floor plan and the

FIGURE 19.12 ■ *A soffit is often used to provide a lowered ceiling to accent lighting.* Courtesy W. Lee Roland, Builder. Hayman Studios, photographer.

FIGURE 19.13 ■ *A vaulted ceiling can be added to a room to add a sense of spaciousness.* Courtesy David Jefferis.

interior elevations. Any special moldings or trim must be shown on the interior elevations.

Vaulted Ceilings

A vaulted or cathedral ceiling similar to the ceiling in Figure 19.13 is a common departure from a standard ceiling. Vaulted ceilings angle upward from walls to form a peak. A vaulted ceiling can be formed using:

- Rafter/ceiling joists so the same members frame the roof and the ceiling
- Inclined ceiling joists placed below the rafters to form an air pocket between the rafters and ceiling
- Scissor trusses that have an inclined bottom chord

Chapter 21 explores each option. No matter the framing method, a vaulted ceiling adds drama and a sense of spaciousness to a room. An alternative to a traditional vault is to use a barrel vault, similar to the ceiling in Figure 19.14. A drawback to having cathedral ceilings is the expense of heating the added volume. This drawback can be overcome by placing a ceiling fan at the top of the vault to force hot air down. A second alternative is to place a cold-air return at the top of the vault to circulate the heated air back through the heating system.

FIGURE 19.14 ■ *A traditional method of vaulting a ceiling is to use a barrel vault.* Courtesy Judy Cooper.

Information explaining vaulted ceilings must be placed on the floor plan, interior elevations, framing plan, and sections. The limits of the vaulted area must be represented on the floor plan, as shown in Figure 10.57. The height of the vault must be shown on the interior elevations and the sections. The sections show how the ceiling will be framed. The interior elevations show any effect the ceiling will have on the interior finishes.

Beamed Ceilings

Beams have been used for thousands of years to form ceilings and continue to be a popular material for residential construction. Beamed construction with wood decking is a common construction method for many homes with vaulted ceilings. Decorative beams similar to those in Figure 19.15 are often added to flat ceilings to replicate historic construction methods that relied on the use of beams. Beams can be solid members or made of 1× (25×) material to reduce the cost. Figure 19.16 shows the use of decorative beams that are added to accent the high ceilings. The use of beams on a low, flat ceiling tends to make the room seem smaller. The use of beams on a ceiling with a height of at least 9' (2700 mm) helps bring high ceilings down to a more human scale. The ceiling height and room dimensions also dictate the width and number of beams to be used. A square room with high ceilings may look good with many beams. Fewer beams are advisable in a rectangular room with a lower ceiling.

Structural beams exposed to the interior must be located on the framing plan. The framing crew will place these beams, and will need all information about their size and location along with other framing information. When decorative beams are a part of the interior finish, their location should be specified on the floor plan, interior elevations, and sections. The finish carpenters will place decorative beams. Their location must be noted on the floor plan but can be omitted if notes clearly describe their location. Exposed structural beams and decorative beams must be shown on the interior elevations if their location affects the placement of other interior materials.

Coffered Ceilings

A ***coffered*** ceiling uses beams to create a pattern of recessed panels or grid-like compartments in a ceiling. Coffered ceilings are a common element of a traditional Tudor style home and are now found in many upscale dens as well as family and living rooms. Figure 19.17 shows the use of a coffered ceiling. This type of ceiling treatment is very expensive; the investment pays off in a permanent architectural feature that adds height, space, and drama to a room. The cost is affected by the type of wood selected and the amount of detailing

FIGURE 19.15 ■ ***Decorative beams are often added to flat ceilings to replicate historic construction methods. Beams can be solid members or made of 1× (25×) material to reduce the cost.*** Courtesy BOWA Builders, Inc. by Jim Tetro, photographer.

FIGURE 19.16 ■ ***This bathroom combines vaulted ceilings and decorative beams to create an elegant bathroom.*** Courtesy Alan Mascord AIBD, Alan Mascord Design Associates, Inc. Bob Greenspan, photographer.

FIGURE 19.17 ■ *A coffered ceiling uses beams to create a pattern of recessed panels in a ceiling reminiscent of traditional Tudor style homes.* Courtesy BOWA Builders, Inc. Greg Hadley, photographer.

FIGURE 19.18 ■ *A tray ceiling gets its name from its resemblance to an inverted tray. This type of ceiling is often used hide recessed lighting in living, dining and bedrooms.* Courtesy Benigno Molina-Manriquez.

added beside the beams. The use of narrower beams that extend only about 3'' from the ceiling will reduce cost and create the illusion of height in rooms with low ceilings. The size and location of the pattern will need to be shown on the floor plan. Details must also be drawn to show the size, type, and location of the trim.

Trayed Ceilings

Tray ceilings are often used to hide recessed lighting in living rooms, dining rooms, and bedrooms. A tray ceiling gets its name from its resemblance to an inverted tray. These ceilings are constructed with the sides angling at approximately 45% or curving to a higher flat ceiling. Another option to form the inclined edges is to follow the roofline at the wall intersection, then angle from the wall in one or more steps. Remember that if this method is to be used, the rafters must be able to contain the needed insulation. Adding a tray to a first-floor ceiling in a multilevel home is usually not done unless the ceilings are taller than 8' (2400 mm). Trays can easily be added to a one-level home or to the upper level of a multilevel home. Figure 19.18 shows the use of a tray ceiling.

FIGURE 19.19 ■ *An open ceiling allows viewing of an area from an upper walkway and lets sunlight to enter further into a home.* Courtesy William E. Poole, William E. Poole Designs, Inc.

Open Ceilings

Open ceilings are often provided in multilevel homes. A typical form of open ceiling is found in rooms that have a full two-story ceiling height. This allows rooms on the upper level, similar to the home in Figure 19.19, to look directly into the lower level. Rooms that are two stories high should have a note on the lower plan indicating that there is no floor level above. Figure 19.20a and b show how open ceilings are noted on a plan. An alternative to having the full ceiling open is to provide a

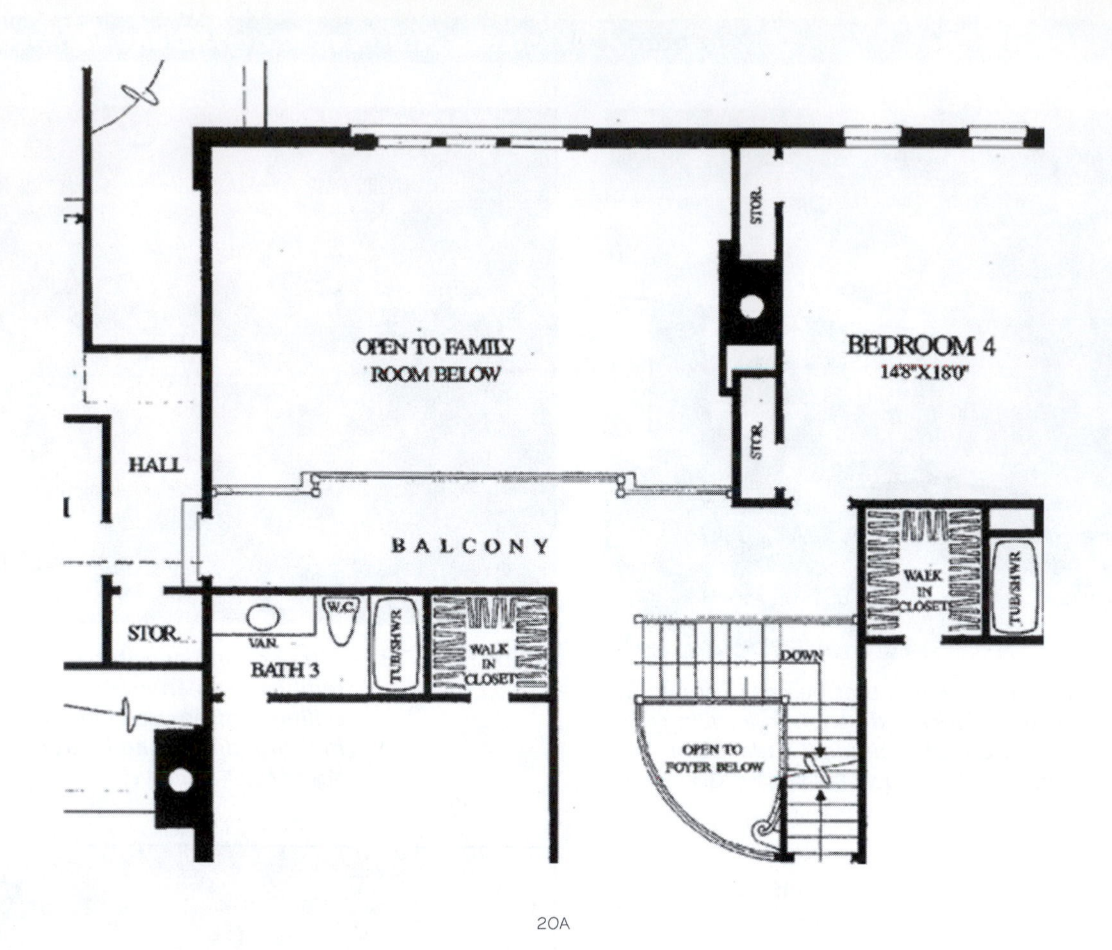

20A

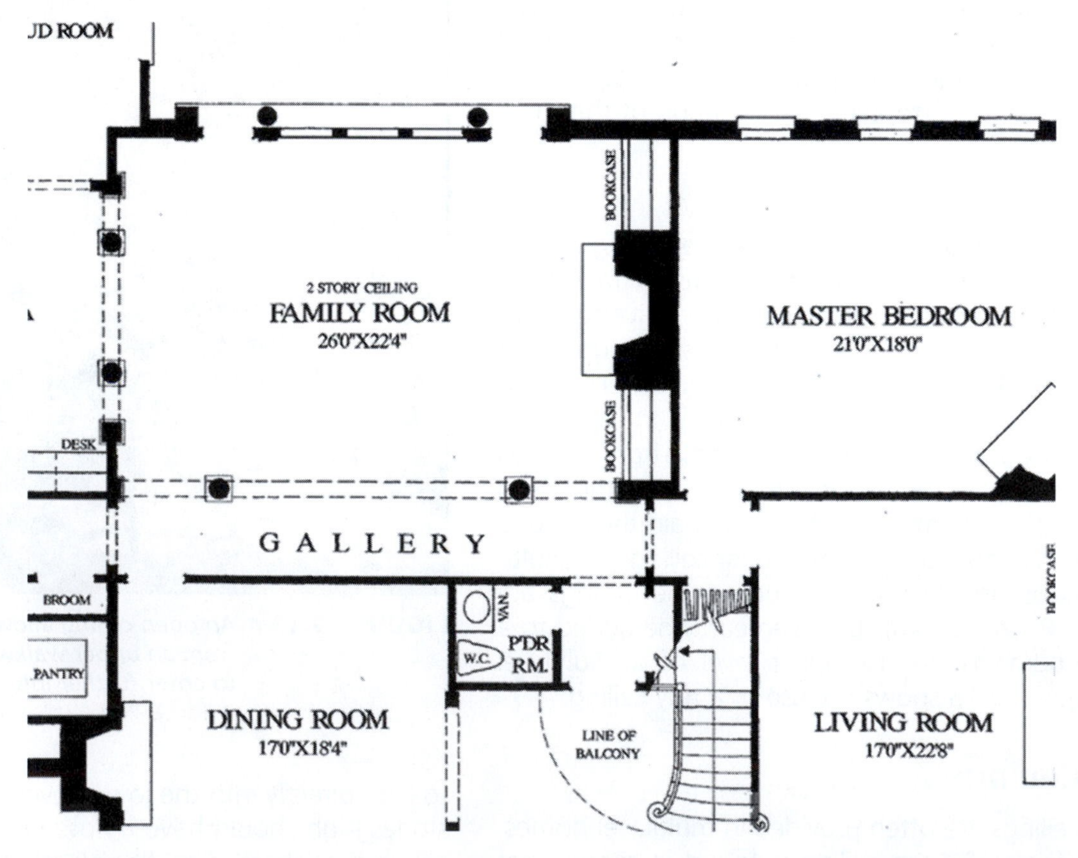

20B

FIGURE 19.20 ■ *If a room on the lower floor is open to the upper level, a note should be provided on each to indicate the opened area.* Courtesy William E. Poole, William E. Poole Designs, Inc.

FIGURE 19.21 ■ ***An opening in the floor adds a very dramatic feature to this entry.*** Courtesy Alan Mascord AIBD, Alan Mascord Design Associates, Inc. Photographer Bob Greenspan.

limited opening, such as that shown in Figure 19.21. The opening needs to be noted on the floor plan, and the structural members used to support the opening must be shown on the framing plan. Trim required to complete the opening must be shown in detail.

CABINETS

More than any other area of millwork, cabinetry has a major impact on the sale of a home. Chapters 5 and 10 introduced important considerations for the design and layout of features that will be included in the cabinets. This chapter explores common types of cabinets and features that should be represented on the construction drawings. The most common locations for cabinets are the kitchen, utility rooms, and bathrooms, but custom homes also have cabinets in offices, dens, and bedrooms.

Cabinet Types and Sources

Cabinets are differentiated as base, upper, and full-height cabinets. Base cabinets rest on the floor or on short decorative legs to resemble furniture. Base cabinets are generally 36'' (900 mm) high and 24'' (600 mm) deep in the kitchen and utility room and 32'' (800 mm) high and 22'' (550 mm) deep in bathrooms. Upper or wall cabinets are generally installed 18 to 24'' (450 to 600 mm) above the base cabinets, but some occasionally extend to the counter to add interest to the design. Upper cabinets are typically 12'' (300 mm) deep, but depths of 15 and 18'' (375 and 450 mm) are also used to add interest. Full-height cabinets are generally 24'' (600 mm) deep and extend from the floor to the height of the wall-hung cabinets. Full-height cabinets are used to house oven units or for storage, such as a pantry or broom closet. Cabinets are represented on the floor plan so that the print reader understands their shape and general layout. Two common sources are available for cabinets, including modular and custom cabinets.

Modular Cabinets

Visit any large home improvement store such as Lowe's or Home Depot and you'll find large displays of modular or stock cabinets. The term ***modular*** refers to prefabricated cabinets that are constructed in specific sizes and designed to meet a wide variety of needs. Figure 19.22 shows a small sample of the options available with modular cabinets. Modular cabinets are used on most spec homes and are best suited when a group of boxes can be placed side by side in a given space. If a series of cabinet boxes do not fill the entire space, pieces of wood called fillers are spliced between the modules. Modular cabinets should not be thought of as cheap. The cost of upper-end modular cabinets can exceed the cost of similar custom cabinets. Modular cabinet suppliers offer different wood species, door styles, hardware options, finish colors, and a wide variety of drawer and tray options. Modular cabinets are usually based on a module 3'' (75 mm) wide and are sized to fit standard appliance applications. Many manufacturers of modular cabinets also make components by special order that satisfy unique design situations. Figure 19.23 shows one face of a kitchen cabinet drawn to specify modular cabinets.

Custom Cabinets

Cabinets made in a cabinet shop for a specific client are referred to as custom cabinets. Custom cabinets are fabricated to meet the specifications of the design team and allow cabinets to meet very specific applications. Figure 19.24 shows custom cabinets designed for a home office. These cabinets include features that are not available in premanufactured packages and also include custom millwork that will be installed by the finished carpenters. One of the advantages of custom cabinets is that their design is limited only by the imag-

WALL CABINETS Wall Cabinets are 12" deep (excluding doors). Most wall cabinets available in 3" width increments from 9" to 48".

Single Door

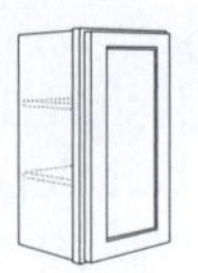

Available in 24", 30", 36", 42" heights.

Double Door

Available in 24", 30", 36", 42" heights.

Wall End

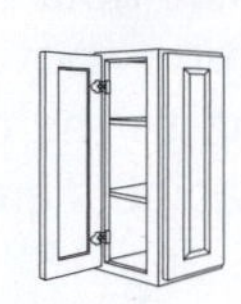

Available in 30", 42" heights.

45° Corner Glass Mullion Door

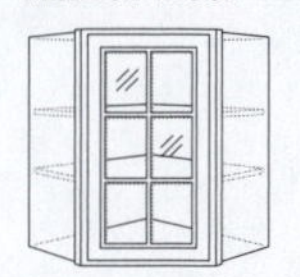

Available in 30" height.

18" High Double Door

18" High Double Door

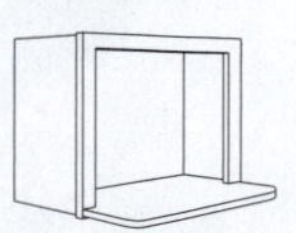

Available in 30" width.

BASE CABINETS Base Cabinets are 24" deep (excluding doors) and 34½" high except where noted.

Base Tray

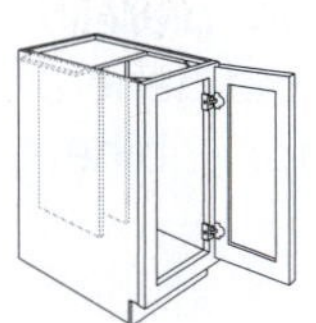

Available in 9" width Left or right hinging.

Single Door

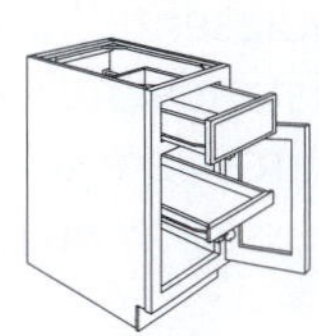

Available in 12", 15", 18", 21", 24" widths. Left or right hinging.

Double Door

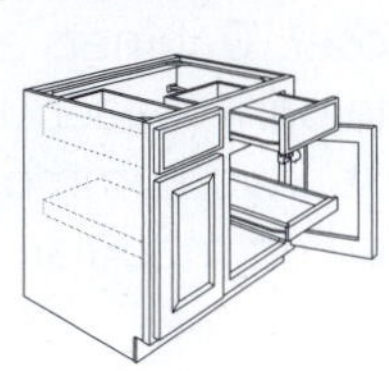

Available in 27", 30", 33", 36", 39", 42", 45", 48" widths.

Single Drawer

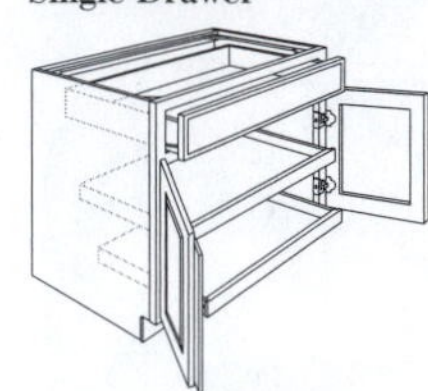

Available in 30", 36" widths.

Base Blind Corner

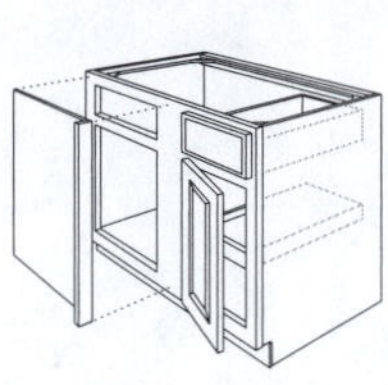

Available in 36", 39", 42", 45", 48" widths.

Sink Base Double Door

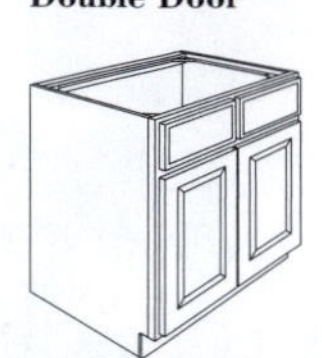

Available in 24", 27", 30", 33", 36", 39", 42", 48" widths.

TALL CABINETS Tall Cabinets are 24" deep (excluding doors) except where noted.

Single Oven

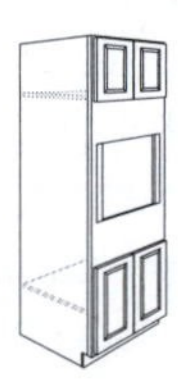

Available in 27", 30", 33" widths. Available in 84", 90", 96" heights.

96" High Utility Cabinet

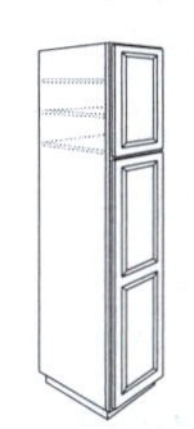

Available in 18", 24" widths. Available in 12" or 24" depths.

90" High Pantry Cabinet

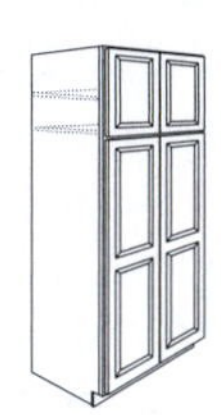

Available in 36" width.

VANITY CABINETS Vanity Base Cabinets are 31½" high and 21" deep except where noted.

Vanity Bowl

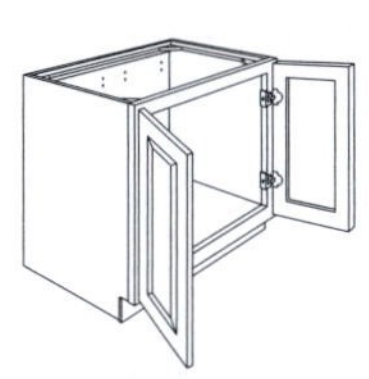

Available in two door 24" to 42" widths in 3" increments. Three door available in 48" width. Four door available in 60" width.

Vanity Bowl-Two Drawer

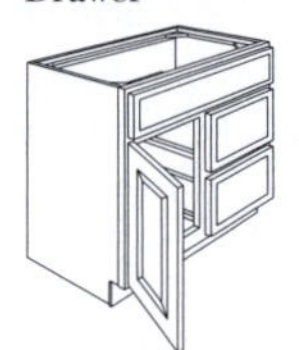

Available in 24", 30", 36" widths. Available in 18" (space saver) depth.

84" Vanity Linen

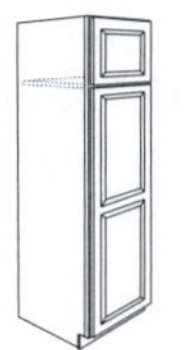

Available 18" wide. Left or right hinging.

FIGURE 19.22 ■ *Modular cabinets are prefabricated in a variety of sizes.* Courtesy Merillat Industries, Inc.

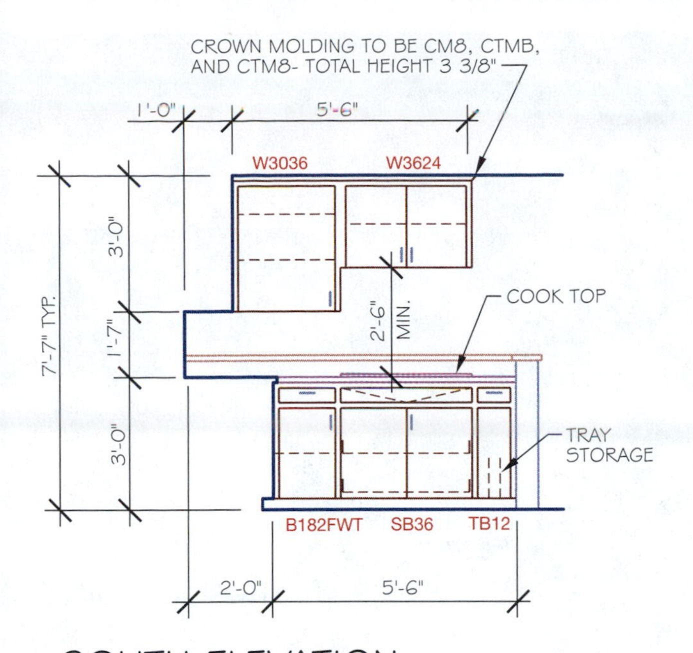

FIGURE 19.23 ■ *Cabinet elevations for modular cabinets are similar to the elevations for custom cabinets. If the supplier is unknown, dimensions are provided for each box. If the supplier is known, dimensions can be replaced by the product's model number.*

FIGURE 19.24 ■ *Custom cabinets allow the use of features that are not available in premanufactured packages. The cabinets also include custom millwork that will be installed by the finish carpenters.* Courtesy W. Lee Roland, Builder (rolandbuilder.com). Photographer Hayman Studios.

ination of the design team and the cabinet shop. Custom cabinets can be built to any height, space, and geometric shape, with any type of material or hardware, or to meet any other design criteria. The many options available with custom cabinets include:

- Self-closing hinges
- A variety of drawer slides, rollers, and hardware
- Clear, leaded, and stained glass cabinet fronts
- Varied styles and materials for range hood covers
- Specially designed storage options including pantries, appliance hutches, and a lazy Susan
- Specialty bath, linen storage, den, study, and family room cabinets

After consulting with the owner, the design team draws an elevation of each cabinet face similar to Figure 19.25. This drawing shows the basic sizes of each

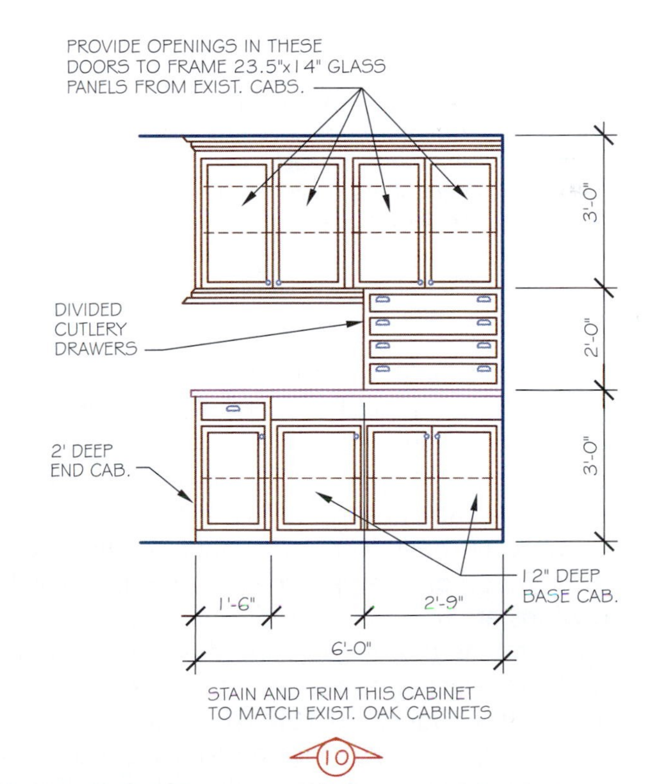

FIGURE 19.25 ■ *Cabinet elevations prepared by the design team convey the owner's ideas to the cabinetmaker.*

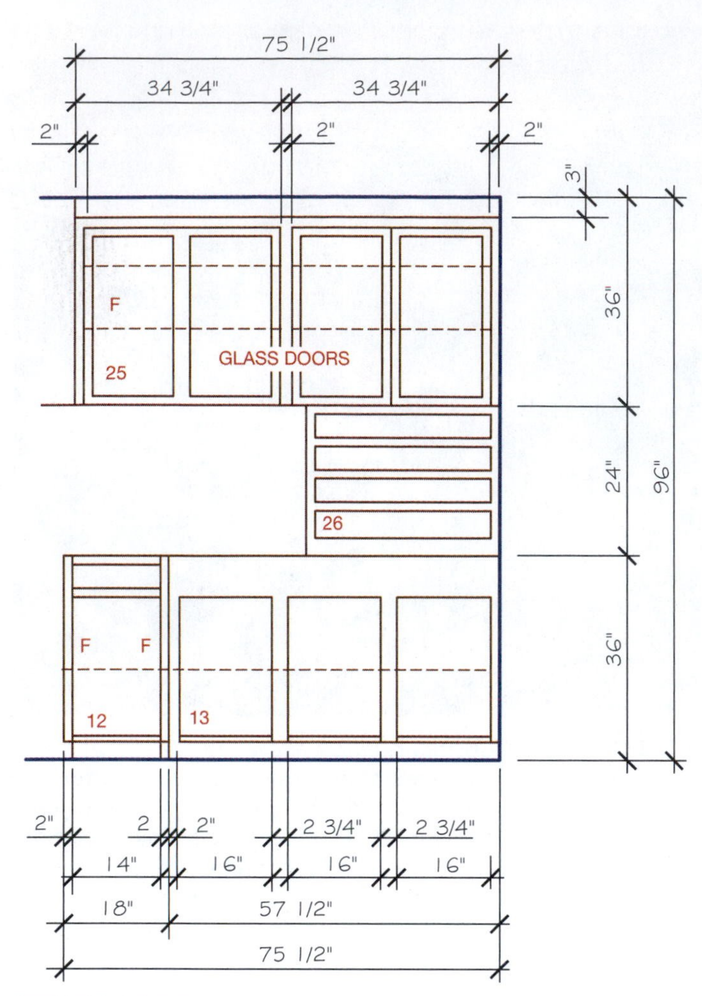

FIGURE 19.26 ■ *The cabinetmaker redraws the elevations to plan all required components. Many cabinet shops have CNC machines that will plan and cut each component to optimize each sheet of material.* Courtesy John Earp, Mountain Home Custom Cabinets.

element of the cabinet and each type of storage to be provided. Once approved by the owner, the drawings are sent to the cabinet shop. Although the architectural team produces highly accurate drawings, a representative of the cabinet shop will visit the job site after the sheetrock has been installed to take the final measurements. Based on the final measurements, the cabinet drawings are redrawn by the cabinetmaker. These drawings combine the features specified by the design team and the exact sizes required to achieve the desired results (see Figure 19.26). These final drawings are used to guide a CNC machine to lay out and cut each piece required to construct the cabinets. Figure 19.27 shows the cabinet built on the basis of the cabinet drawing.

FIGURE 19.27 ■ *The cabinets that resulted from the drawings in Figures 19.25 and 19.26.* Courtesy Leonard Conklin.

Representing Millwork and Cabinets

Various degrees of details may be needed to fully describe millwork, depending on its complexity and the design specifications. Because cabinets, railings, door, and window trims are standard items, millwork is almost an automated part of many of today's CAD systems. A designer only needs to decide which style is the default and let the program use it where it normally goes. In special cases, these defaults can be overridden as required. When something needs to be hand-crafted, it is possible to create new moldings, panels, and other special effects. Figure 19.28 shows an example of how millwork drawings can define very clear and precise construction techniques.

The representation of cabinets on the floor plan was discussed in Chapter 10. Cabinet elevations are developed directly from the floor plan using methods similar to those that were used to develop the exterior elevations. The elevations show how the cabinet face will look when completed and give general dimensions, notes, and spec-

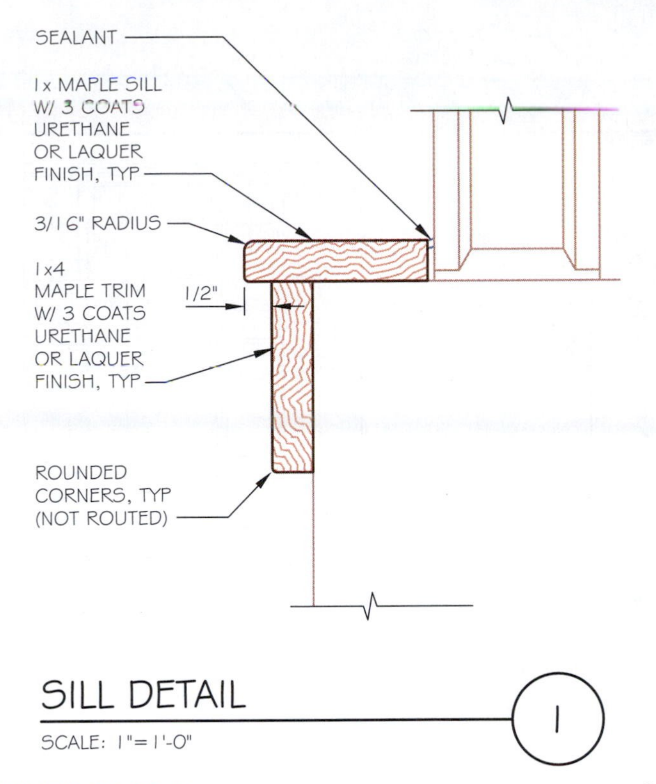

FIGURE 19.28 ■ *Drawings describing window trim are often provided for custom homes.* Courtesy Eric Hess, O.H. Architecture.

ifications. The detail placed on cabinet drawings will vary greatly based on the detail the architectural team considers necessary to get the desired results. Figure 19.29 shows cabinet elevations that are very clear and well done without providing unneeded detail or artwork. Simplified cabinets can be drawn when the quality of the cabinetmaker is known and the project will not go out to bid. In Figure 19.30, because the drawings will go out to bid, the cabinet elevations are drawn with more detail to ensure the desired end result. Many cabinetmakers use specialized computer software to redraw elevations. The software develops material lists and cutting plans to limit material waste.

Keying Cabinet Elevations to Floor Plans

Several methods are used to reference the cabinet elevations to the floor plan. In Figure 19.29, the cabinet elevations are referenced to the floor plans with room titles such as BATH. In Figure 19.30, a symbol with a letter inside is used to correlate the elevations to the floor plan. Figure 19.31 shows a floor plan with identification numbers. The symbols are pointing to various areas that correlate with the same symbols below the elevations. A

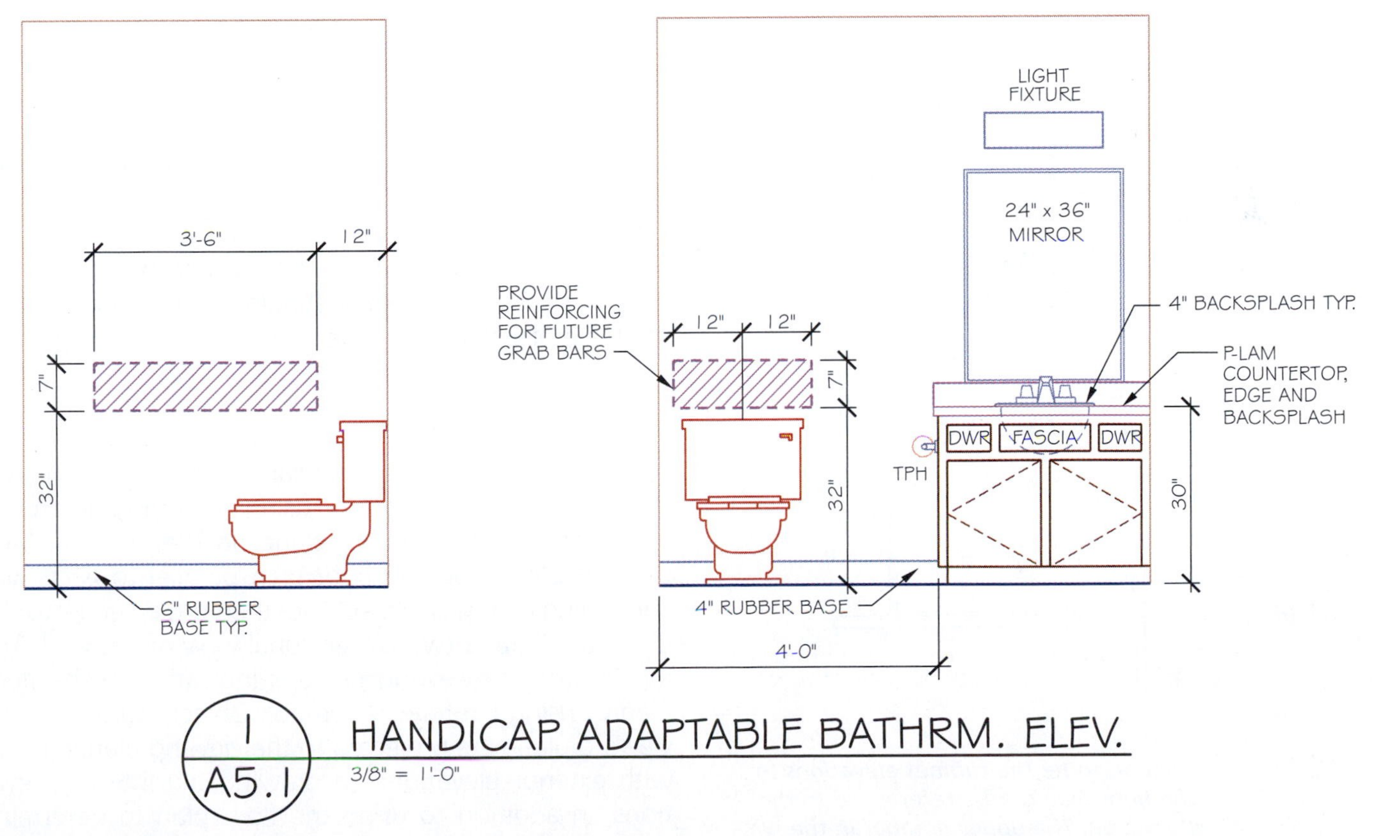

FIGURE 19.29 ■ *Cabinet drawings can be simplified if they are not going out for bid and the skill of the craftsperson is known.* Courtesy Scott Beck, Scott R. Beck Architect.

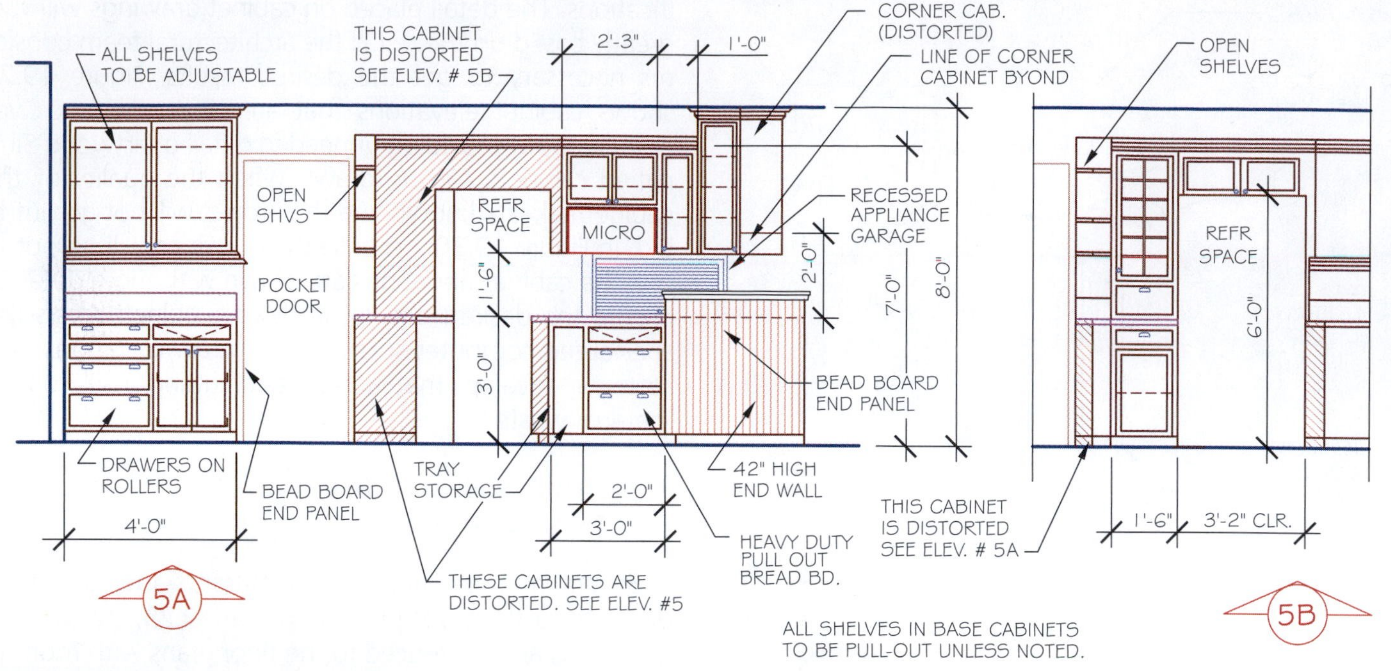

FIGURE 19.30 ■ *The elevations must contain more detail if the cabinetmaker is unknown.*

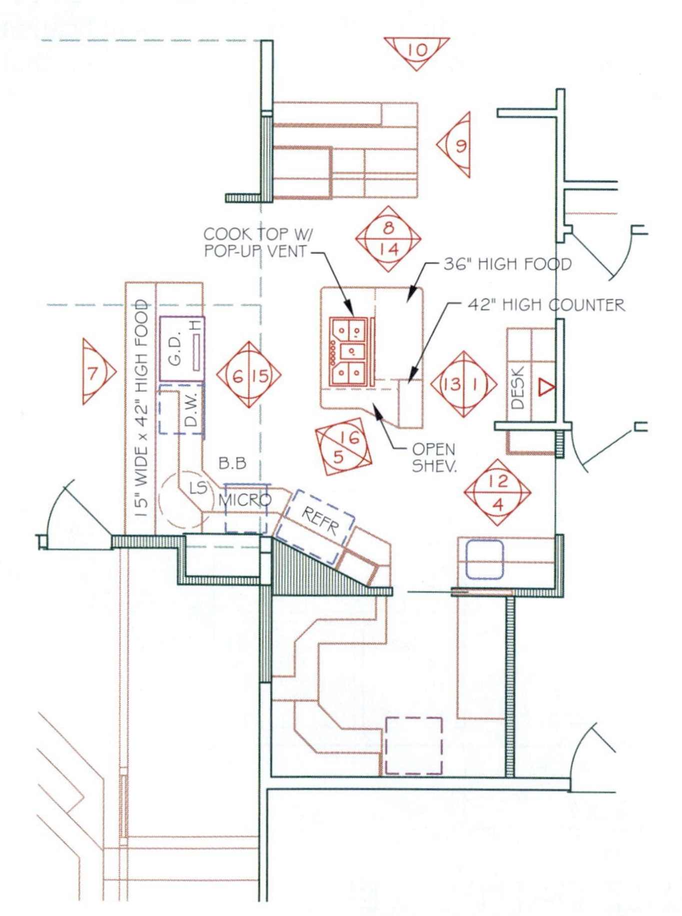

FIGURE 19.31 ■ *Symbols to tie the cabinet elevations to the floor plan can be referenced to the floor plan. The upper number in the symbol represents a specific detail, and the bottom number represents the page number of the sheet containing the detail.*

similar method may be used to relate cabinet construction details to the elevations.

Although interior elevations are referenced to the floor plan, their location in the architectural drawings will vary greatly. When a separate sheet is required to show interior elevations, they are often placed immediately behind the floor plan in the drawing set and assigned an A-200 page number. They can also be placed on a sheet containing any other architectural drawing as long as space allows. Every effort should be made to keep all interior elevations on the same sheet. Regardless where they are placed within the drawings, the location of interior elevations, like other drawings, should be clearly indicated in the table of contents on the title page.

Drawing Projection

Interior elevations are similar to the exterior elevations in that they are drawn using orthographic projections based on a viewing plane. In theory, a viewing plane is placed on the floor plan parallel to each wall containing cabinets. Objects are then projected onto the plane to create a two-dimensional view of the wall. The location of the viewing plane is indicated on the floor plan by using a reference symbol. Objects parallel to the viewing plane are projected to the viewing plane, just as with exterior elevations. In drawing the interior elevations, in addition to using the floor plan to determine the shape, the sections can be used to determine interior heights. Base cabinets are usually 36'' (900 mm) high with upper cabinets typically placed 18, 24, or 30'' (450, 600, or 750 mm) above the base cabinets. Because cabinet placements and sizes are so dependent on the equipment that will be installed in the cabinets, rely on vender catalogs and vender specifications found on the Internet.

Choosing a Drawing Scale

Like other drawings, the interior elevations are drawn full scale in model space. The complexity of the materials to be drawn determines the scale used to plot the cabinet elevations and any required details. A scale of 1/4'' = 1'-0'' (1:50) can be used for very simple elevations, but larger scales such as 3/8''= 1'-0'' or 1/2''= 1'-0'' (1:32 or 1:24) are often found on professional examples. Select a drawing scale that shows sufficient detail to meet the demands of the drawings. If the elevations represent premanufactured materials, a small scale showing little detail can be used. If the drawings are used to construct the item, a scale sufficient to show all detail must be used.

Material Representation

The symbols used to represent materials in the exterior elevations are also used to represent material drawn in cabinet details. Continuous lines represent materials represented in elevations with varied thickness. Thick lines usually represent the floor, walls, and ceiling, which define the view. On L-shaped cabinets, the line that represents the drawing limits will follow the contour of the cabinets when a cabinet is shown in end view. Figure 19.32a shows the outline of a room with L-shaped cabinets. Common features represented in the elevations—such as the range, dishwasher, and refrigerator—are usually inserted as blocks.

Cabinetry

The outlines of cabinets and the lines representing doors, drawers, and exposed shelves are thin, continuous lines. A single thin dashed line is used to represent shelves hidden behind doors. On larger-scale drawings, pairs of continuous lines are used to represent shelves. A short, thick line or a circle can be used to represent the approximate style of hardware. Hinges are usually not shown, but door swings are represented by continuous or dashed lines in the shape of a V drawn through the door. The point of the V represents the hinged side of the door. Figure 19.29 shows an elevation of cabinets with door swings represented. Although sizes vary greatly, common sizes that must be represented on cabinet drawings include:

- Base cabinets 36'' high × 24'' deep (900 × 600 mm).
  - Desk counters are generally set at 30'' (750 mm).
  - Bath cabinets are generally set at 32'' (800 mm) high.
- Counter thickness between 1 and 1 1/2'' (25 and 38 mm). Show a 1'' (25 mm) overhang over the cabinets.
- Backsplash height: 4'' (100 mm) or full height (extend to the bottom of the upper cabinets).
- A space of 18'' (450 mm) between base and upper cabinets.
  - A space of 24'' (600 mm) between base and upper cabinets at a peninsula or island.
  - A clear space of 30'' (750 mm) from the top of the cooktop to the bottom of cabinets above.
- Refrigerator hole size: 36'' wide × 72'' high (900 × 1800 mm).
- Double sink box: 36 or 42 (900 or 1050 mm) wide.
- Dishwasher space: 24'' (600 mm) wide.
- Range openings: 30, 36, and 42'' wide (750, 900, and 1050 mm) wide.
- Cooktop box: 30, 36, and 42'' wide (750, 900, and 1050 mm) wide.
- Wall oven box: 27 and 30'' wide (675 and 750 mm) wide.
- Microwave opening: 18 wide × 14'' high (450 × 350 mm).
- Toe space: 3 × 3" (75 × 75 mm).
- Common drawer depths: 3, 4, 9, and 12'' (75, 100, 225, and 300 mm) deep.
- Common door widths: 9, 12, 15, 18, and 24'' (225, 300, 375, 450, and 600 mm) wide.
  - Doors wider than 24'' (600 mm) should be divided into pairs.

Plumbing Fixtures

Because of the repetitive nature of elevations, plumbing symbols should be created as a block for quick installation. Cabinet-mounted lavatories and sinks are represented by hidden or continuous lines to define the outline of the unit. Fixtures built into a counter are drawn using the same methods as a cabinet-mounted unit. Wall-mounted and pedestal lavatories are drawn using thin continuous lines to represent the outline and some detailing of the unit.

Thin lines are used to represent water closets, showing the basic components of the fixture. Blocks should be developed to show both a front and side view of a toilet. Figure 19.29 shows common methods to represent water closets. A thin continuous line should be used to represent the outlines of tubs and showers, with the interior shape represented by dashed lines. When the unit is perpendicular to the viewing plane, both the outline and the interior shape are shown with continuous lines.

Appliances

Kitchen appliances are represented by a box that outlines the shape or by a detailed representation. The detailed representation should not be drawn to represent an exact model but rather a generic representation. Appliances should be drawn so that they are easily recognized but are not of a specific unit. In creating blocks, show doors or drawers that open and the general location of control features, but don't waste time adding features that don't aid in the installation of the appliance.

Drawing Notations

The materials represented in elevations require both written and dimensional clarification. Add notations to each elevation to clearly define objects that are drawn. The larger the drawing scale, the more detailed the specifications should be. Information to be specified on the elevation is provided by the architect's preliminary sketches and check prints and by consulting vendor catalogs.

Vertical dimensions must be provided to locate the limits of all counter heights, cabinets, special equipment, and soffits. Use horizontal dimensions to locate special equipment from walls and divisions in cabinets, as shown in Figure 19.30.

Layer Guidelines for Interior Elevations

Cabinet elevations should be placed on layers with a prefix of *I ELEV.* Modifiers such as *APPL, CABS, CNTR, EQIP, PLMB,* or *HDWR* can be used to describe layers such as appliances, cabinets, counters, equipment, plumbing fixtures, and hardware. See Appendix D for additional layer names.

Cabinet Layout

Figure 19.32a shows a portion of the floor plan for the home drawn in Chapter 11. This plan will be used to complete the cabinets for the kitchen. By making a copy of the floor plan, the cabinet elevations can be projected directly from the copy, assuring that the floor plan from the working drawings will not be accidentally altered. For this drawing, assume that:

- The cabinets will be custom-made.
- The kitchen ceiling height is 9'-0'' (2800 mm).
- The base cabinets are 36'' (900 mm) high with a 3'' (75 mm) toe kick.
- The bottom of the upper cabinets is set at 4'-6'' (1350 mm).
- Set the top of the upper cabinets at 8'-0'' (2400 mm) above the floor.
- There will be a soffit 12'' (300 mm) deep × 21'' (525 mm) wide.

Step 1. Make a block of the floor plan to project the outline of all walls that enclose the elevations.
Step 2. Make copies of the floor plan for each elevation that will be drawn.

- Trim unneeded materials from each view.
- Rotate each view into its proper orthographic position (see Figure 19.32b).

Use the *I ELEV OUTL THCK* layer to draw the following steps.

Step 3. Use thick lines to represent the finish floor.
Step 4. Use thick lines to represent the ceiling.

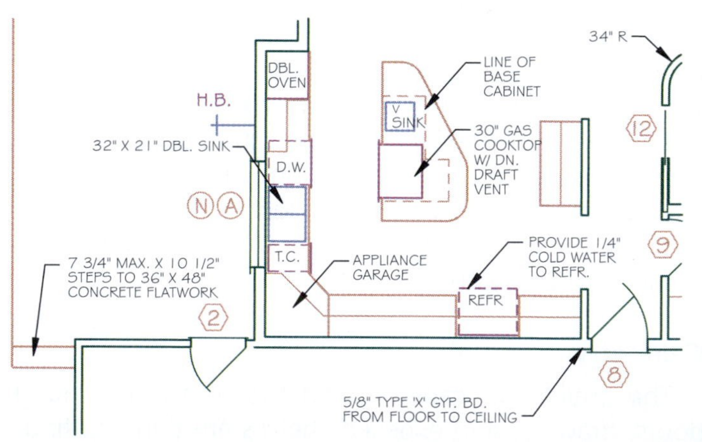

FIGURE 19.32a ■ *The floor plan serves as a base for projecting the cabinets. Make a copy of the kitchen and insert the block into the template to be used for the cabinet drawings.*

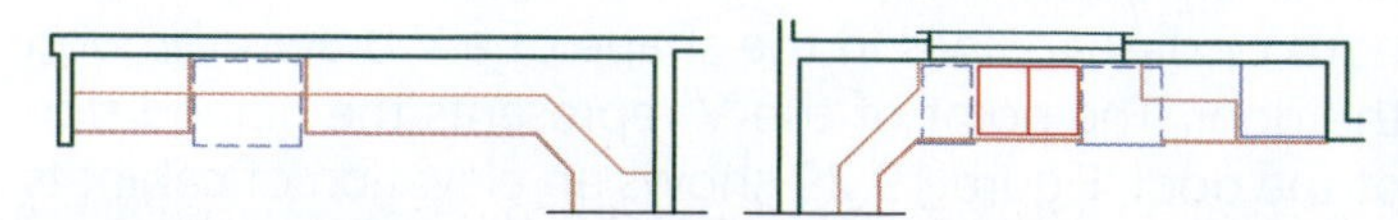

FIGURE 19.32b ■ *Make copies of the floor block for each elevation that needs to be drawn.*

Step 5. Use thick lines to represent any required soffits.

Step 6. Use thick lines to represent any walls surrounding the cabinets that are cut by the viewing plane.

Use the *I ELEV OUTL THIN* layer to draw the following steps.

Step 7. Use thin lines to represent any walls surrounding the cabinets.

Step 8. Use the OFFSET command to represent the top of the toe space, base cabinet height, bottom of the upper cabinets, and top of the upper cabinets that do not extend to the soffit.

Step 9. Use thin lines to represent each cabinet shape on the *I ELEV CASE* layer.

The initial layout should resemble Figure 19.33. Use the TRIM and FILLET commands to remove any unwanted lines. Use the PROPERTIES command to assign lines to their proper layer.

Step 10. Draw lines to represent the outlines of any windows or doors surrounding the cabinets.

Step 11. Draw lines to represent the sizes of all appliances. Use the window and door schedules to determine the required size. Assume a header height of 6′-8″ (2000 mm).

Step 12. Insert blocks to represent each appliance or draw a box with an X on the *I ELEV APPL* layer to represent the appliance.

Step 13. Use the OFFSET command to represent the counter thickness.

The drawing should resemble Figure 19.34. Place the following information on the *I ELEV CASE* layer.

Step 14. Draw the back splash required for each base cabinet.

Step 15. Draw all required drawers.

Step 16. Draw doors for the upper and base cabinets.

Step 17. Represent door swings and represent all shelves and pull-outs in the base and upper cabinets.

Step 18. Draw all equipment such as sinks, breadboards, and fixtures.

Step 19. Draw all required interior finishes such as tile, wainscoting, and masonry.

The cabinet elevations should now resemble Figure 19.35. Place the following information on the *I ELEV DIMS* layers.

Step 20. Provide horizontal dimensions to locate each cabinet length. (A unit number may be substituted for the horizontal dimensions of premanufactured cabinets.)

Step 21. Place all vertical dimensions to locate the heights of all cabinets and equipment. (A unit number may be substituted for the cabinet heights for premanufactured cabinets, but the height of all equipment should still be located.)

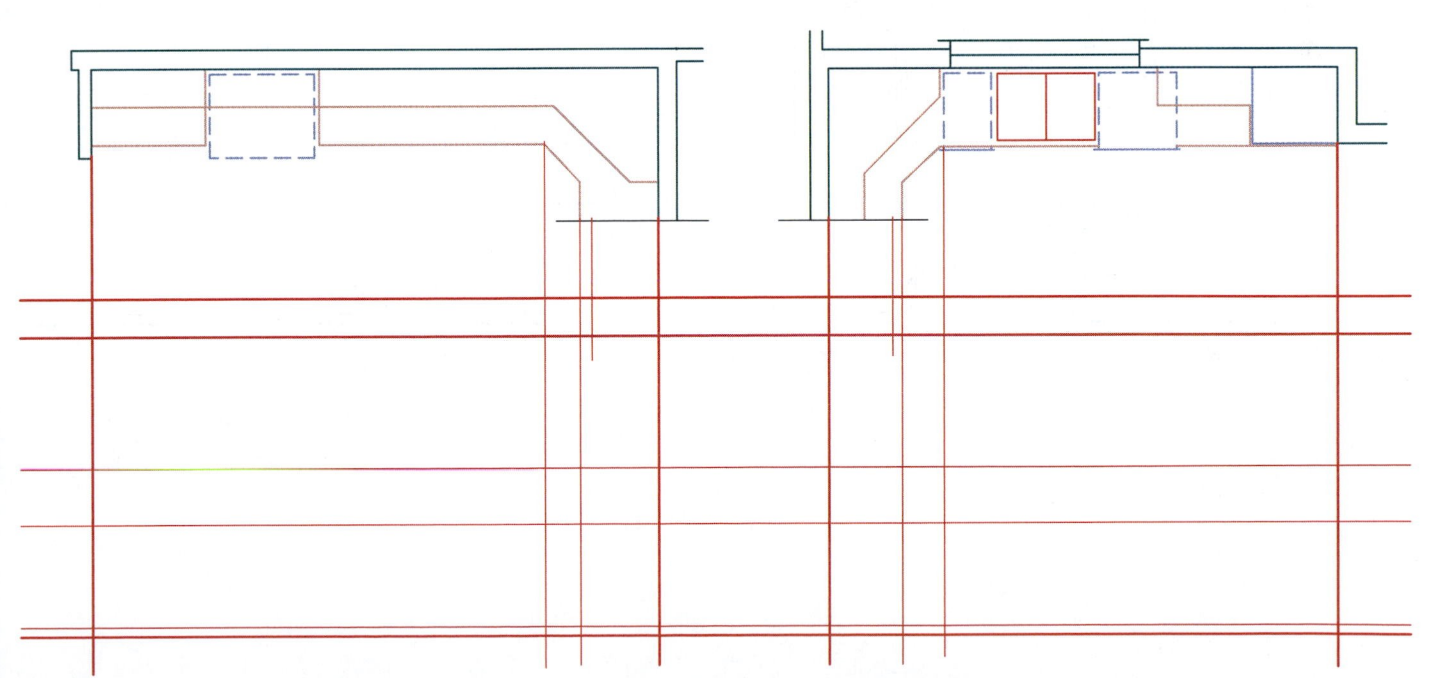

FIGURE 19.33 ■ *Use the floor plan to project the outlines of each cabinet face. After drawing a line to represent the floor, use the OFFSET command to represent the toe space, countertop, bottom of the upper cabinets, any soffits, and the ceiling.*

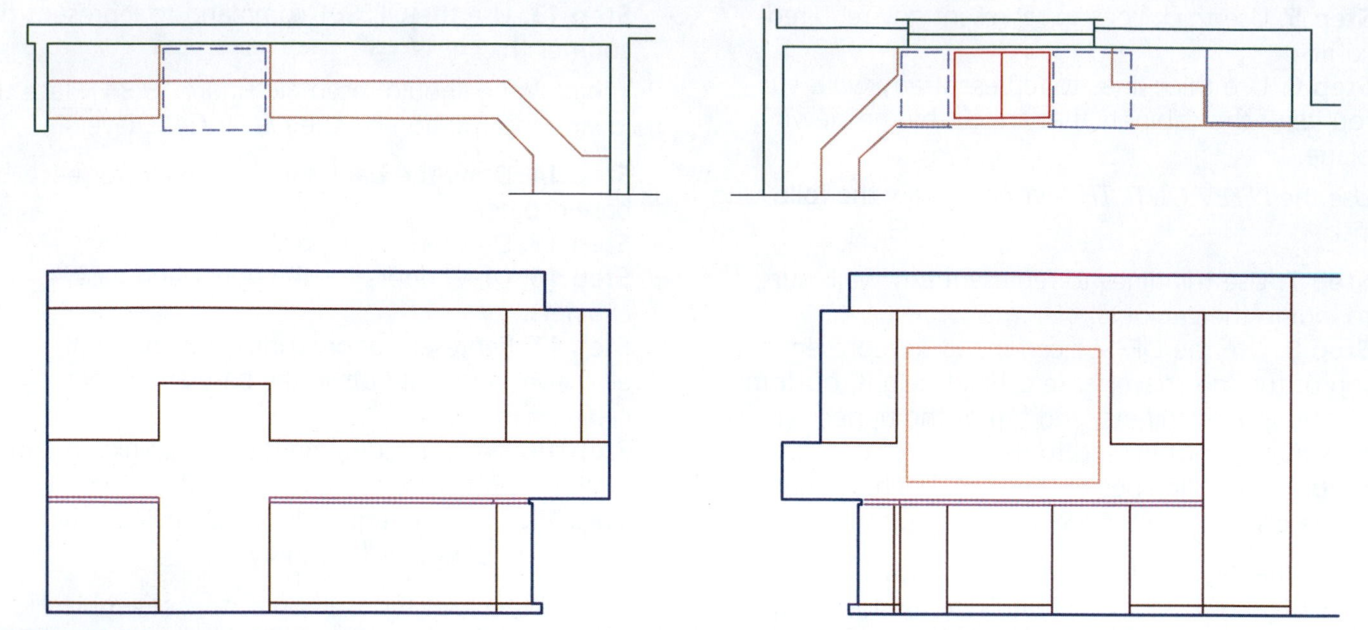

FIGURE 19.34 ■ *Draw lines to represent the outlines of any windows or doors surrounding the cabinets, all appliances, and the outline of each cabinet.*

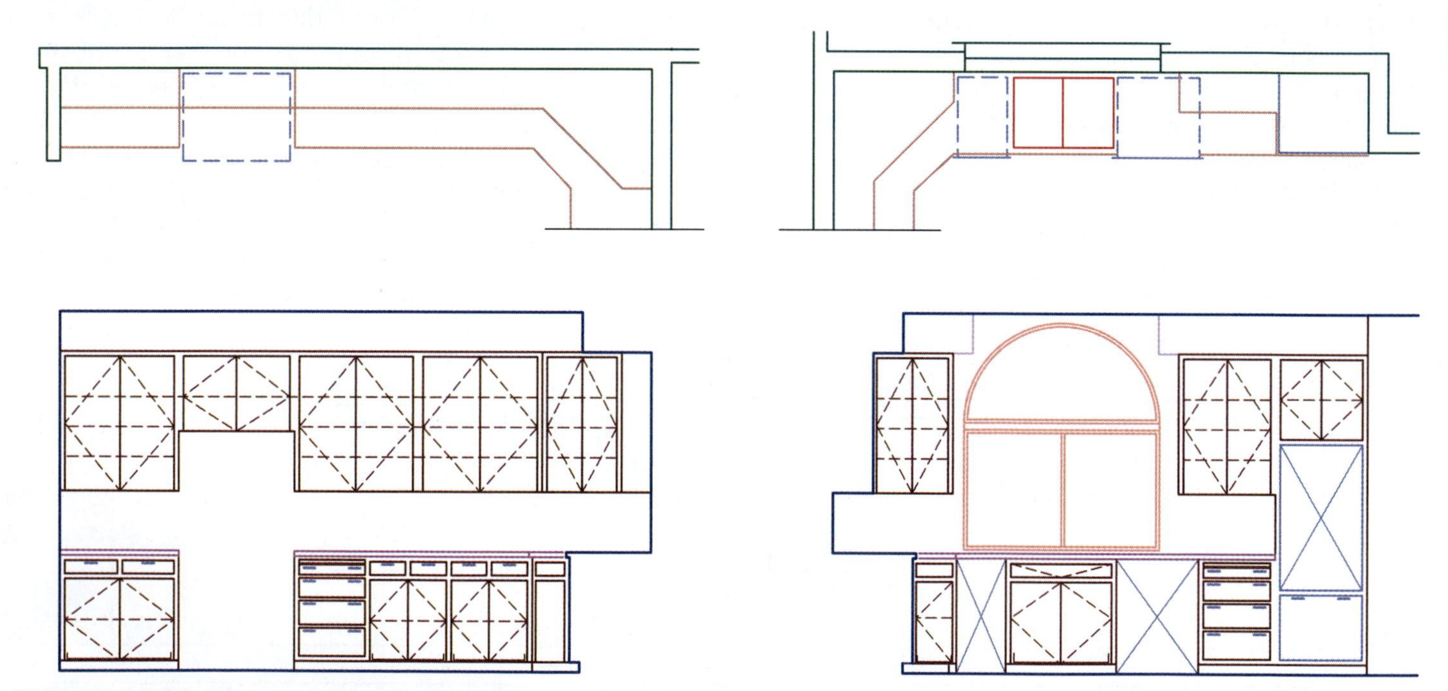

FIGURE 19.35 ■ *Representing all doors, drawers, and shelving.*

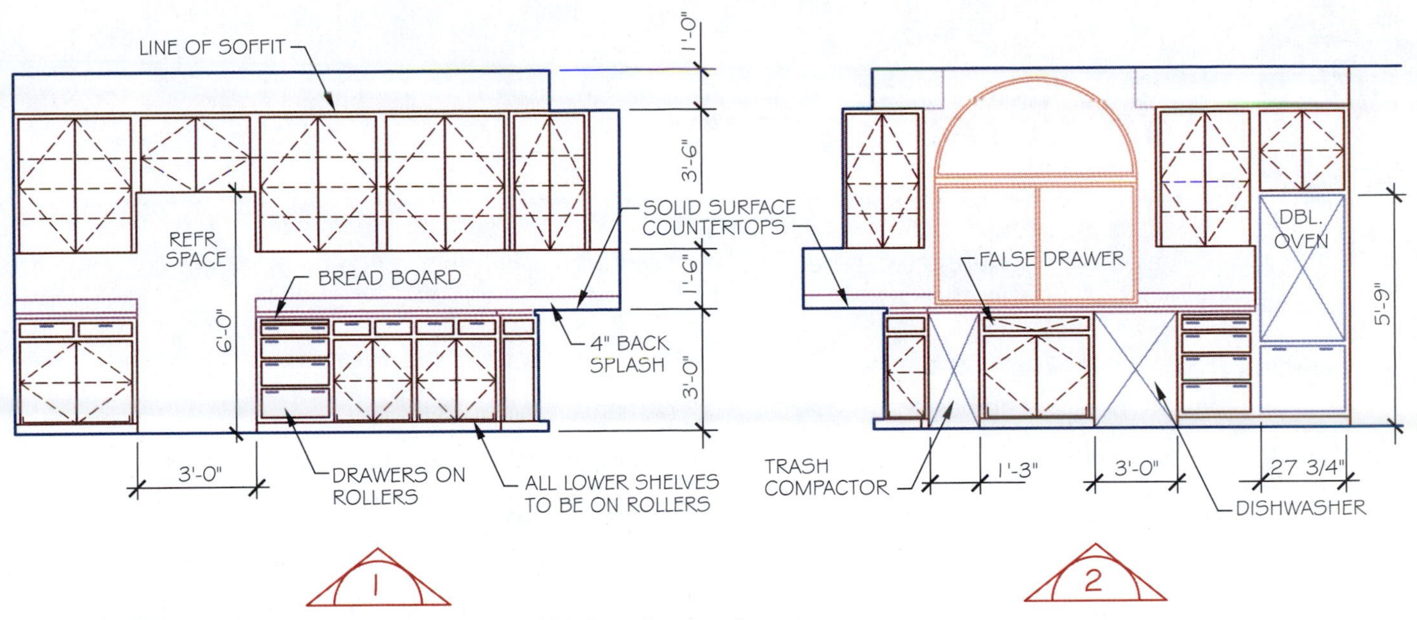

FIGURE 19.36 ■ *Dimensions and annotation added to the drawings.*

Place the following annotation on the *I ELEV ANNO* layer.

Step 22. Place notes to specify all materials and equipment.
Step 23. Reference any required details.
Step 24. Provide titles and drawing scales.

Your drawing should now resemble Figure 19.36.

Bathroom Cabinet Layout

The bathroom cabinets are drawn using the same procedures used to draw kitchen elevations. Although sizes may vary, common elements of bathrooms include:

- Counters are typically set 32'' (800 mm) high.
- Mirrors are generally set so that the top of the mirror is set at a height of 6'-0'' (1800 mm).
- A counter for a makeup area is typically set at 30'' (750 mm) above the floor.

Figure 19.37 shows the elevations for the master bathroom for the home drawn in Chapter 11.

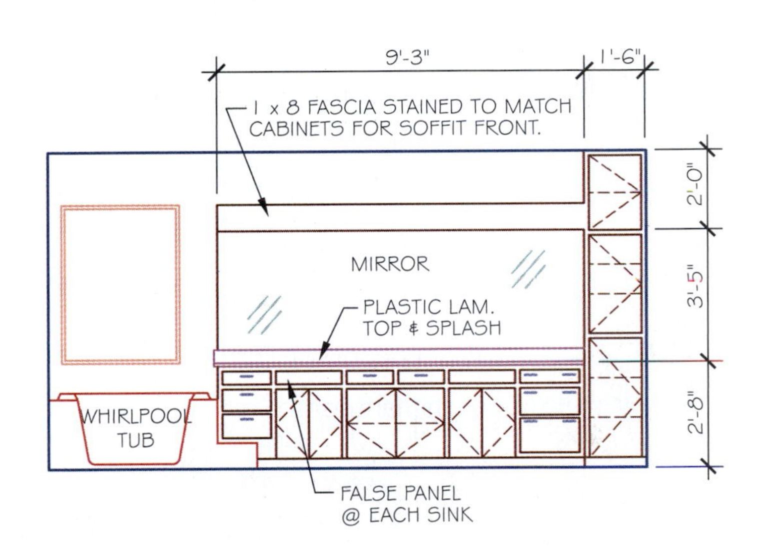

FIGURE 19.37 ■ *The master bath elevation for the home drawn in Chapter 11.*

CHAPTER 19

Additional Readings

The following websites can be used as a resource to help you keep current with changes in millwork and cabinetry.

ADDRESS	COMPANY OR ORGANIZATION
www.woodmark-lowes.com	American Woodmark
www.enkeboll.com	Architectural Components
www.brosco.com	Brosco (millwork)
www.bmcwest.com	BMC West (millwork)
www.cnamac.com	CN Architectural Millwork and Construction, LTD
www.cumberlandwoodcraft.com/	Cumberland Woodcraft (trim & millwork)
www.diamond2.com	Diamond Cabinets
www.diamondcutmillwork.com	Diamond Cut Millwork & Cabinetry
www.gossencorp.com	Gossen Corporation (molding)
www.kraftmaid.com	Kraftmaid Cabinetry
www.lowes.com	Lowe's Home Improvement
www.homedepot.com	The Home Depot
www.garysengraving.com	Decorative moldings and mantles
www.shenandeah.com	Shenandoah

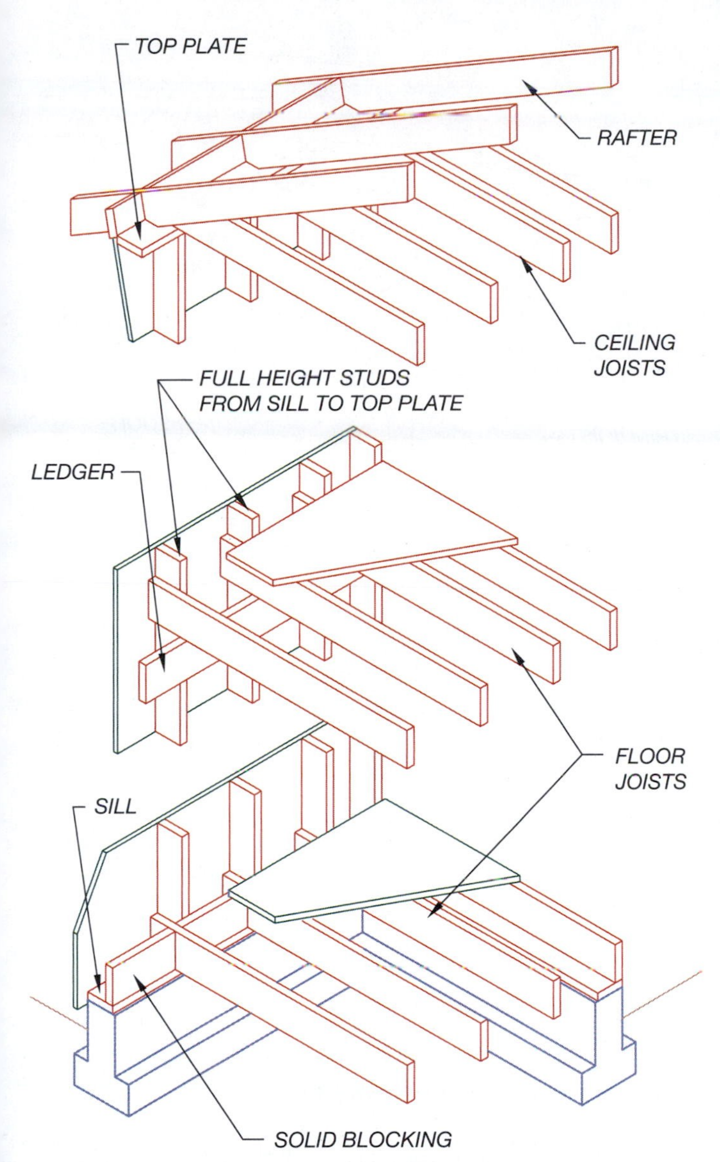

FIGURE 20.1 ■ *Balloon framing wall members extend from the foundation to the roof level in one continuous piece.*

FIGURE 20.2 ■ *Western platform framing allows the framers to use the floor to construct the walls that will support the next level. Once completed, walls can be tilted into position and the next level can be started.* Courtesy Aaron Jefferis.

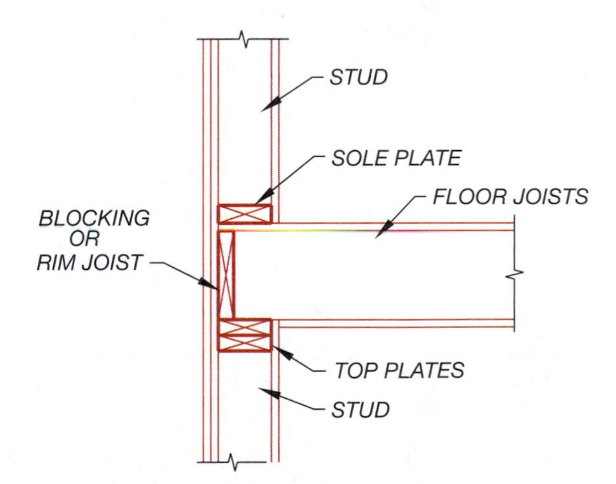

FIGURE 20.3 ■ *The fireblocks of the balloon system gave way to continuous supports (top plates) at each floor and ceiling level. Blocking is placed between each floor joist to stop the joists from rolling over.*

of the need for fireblocks in the balloon framing system. The fireblocks that had been put in individually between the studs in balloon framing became continuous members placed over the studs to form a solid bearing surface for the floor or roof system (see Figure 20.3).

Building with the Platform System

Once the foundation is in place, the framing crew sets the girders and the floor members. Figure 20.4 shows a foundation with the floor joists set in place. The major components of platform construction are shown in Figure 20.5. Sawn floor joists ranging in size from 2 × 6 through 2 × 14 (50 × 150 to 50 × 350) can be used, depending on the distance they are required to span. If engineered joists are used, 9 1/2 and 11 7/8'' (240 and 300 mm) are common depths. Chapter 21 explores the use of engineered joists. Once the joists are in place, plywood ranging in thickness from 1/2, 5/8, or 3/4'' (12.5, 15.5, or 18.5 mm) is installed over the floor joists. The plywood, referred to as subfloor, usually has a ***tongue-and-groove (T&G)*** pattern in the edge to help minimize floor squeaking. Gluing the plywood to the supporting members in addition to the normal nailing also helps eliminate squeaks. Occasionally T&G lum-

FIGURE 20.4 ■ *A western platform floor system is constructed with floor joists resting on a pressure treated sill. Floor joists are typically placed at 16'' (400 mm) o.c.* Courtesy Benigno Molina-Manriquez.

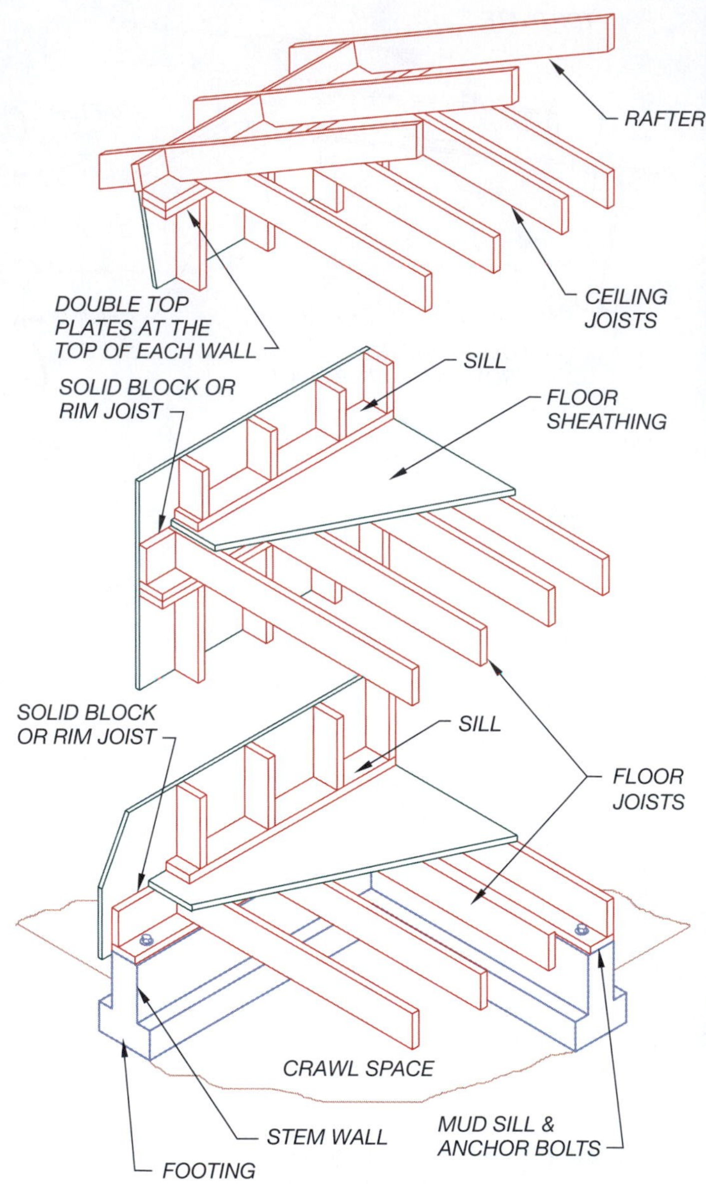

FIGURE 20.5 ■ *Structural members of the western platform framing system.*

ber 1 × 4 or 1 × 6 (25 × 100 or 25 × 150 mm) will be laid diagonal to the floor joists to form the floor system, but plywood is primarily used because of the speed with which it can be installed. Figure 20.6 shows the plywood being installed over the floor joists.

With the floor in place, the walls are constructed using the floor as a clean, flat layout surface. Walls are typically built flat on the floor using a bottom plate, studs, and two top plates. With the walls squared, sheathing can be nailed to the exterior face of the wall, and then the wall is tilted up into place. When all of the bearing walls are in place, the next level can be started. As in the balloon system, studs are typically placed at 12, 16, or 24'' (300, 400, or 600 mm), with 16'' (400 mm) o.c. being the most common spacing. Although in theory, with the platform framing method, the height of a structure is limitless, the IRC does not allow studs for each level to be taller than 10' (3000 mm) (or 12' [6000 mm] under special conditions presented in Chapter 24) and does not allow a wood-framed residence to be more than three stories above grade. The height restriction is due to the combustible nature of wood and the resulting risk of fire, and the risk of lateral failure from high winds and earthquakes.

Energy-Efficient Platform Framing

Typically, exterior walls have been framed with 2 × 4 (50 × 100 mm) studs at 16'' (400 mm) spacing. The IRC allows framers to substitute 2 × 6 (50 × 150 mm) studs at 24'' (600 mm) spacing to allow for added insulation in the wall cavity. Many long-held framing practices have been altered to allow for greater energy savings. The framing practices are referred to as advanced framing techniques (AFT).

AFT systems eliminate nonstructural wood from the building shell and replace it with insulation. Wood has an average resistive value for heat loss of R-1 per inch of wood compared with R-3.5 through R-8.3 per inch of insulation. Reducing the amount of wood in the shell increases the energy efficiency of the structure. Advanced framing methods include 24'' (600 mm) stud spacing, insulated corners, insulation in exterior walls behind partition intersections, and insulated headers. Many

FIGURE 20.6 ■ *Platform floors are constructed using 1/2, 5/8, or 3/4'' (12.5, 15.5, or 18.5 mm) thick floor sheathing nailed over the floor joists.* Courtesy APA–The Engineered Wood Association.

municipalities limit advanced framing methods to one-level construction. For a multilevel residence, advanced methods can be used on the upper level, and the spacing can be altered to 16'' (400 mm) for the lower floor.

The insulation that is added to corners, wall intersections, and headers must be equal to the insulation value of the surrounding wall. Figure 20.7 shows examples of framing intersections using advanced framing techniques. Figure 20.8 shows examples of how insulated headers can be framed. Advanced framing methods also affect how the roof will intersect the walls. Figure 20.9 compares standard roof and wall intersections with those of advanced methods.

Engineered Lumber Framing

In addition to using advanced framing techniques, engineered materials affect both the framing method and the impact the framing system will have on the environment. Engineered lumber products are made by turning small pieces of wood into framing members. Structural engineered members are made from fast-growing tree species grown on tree farms specifically for the purpose of being used to make structural materials. Depending on the product to be made, sawdust, wood scraps, small pieces of lumber, or whole pieces of sawn lumber can be joined by adhesives applied under heat and pressure to produce engineered building products.

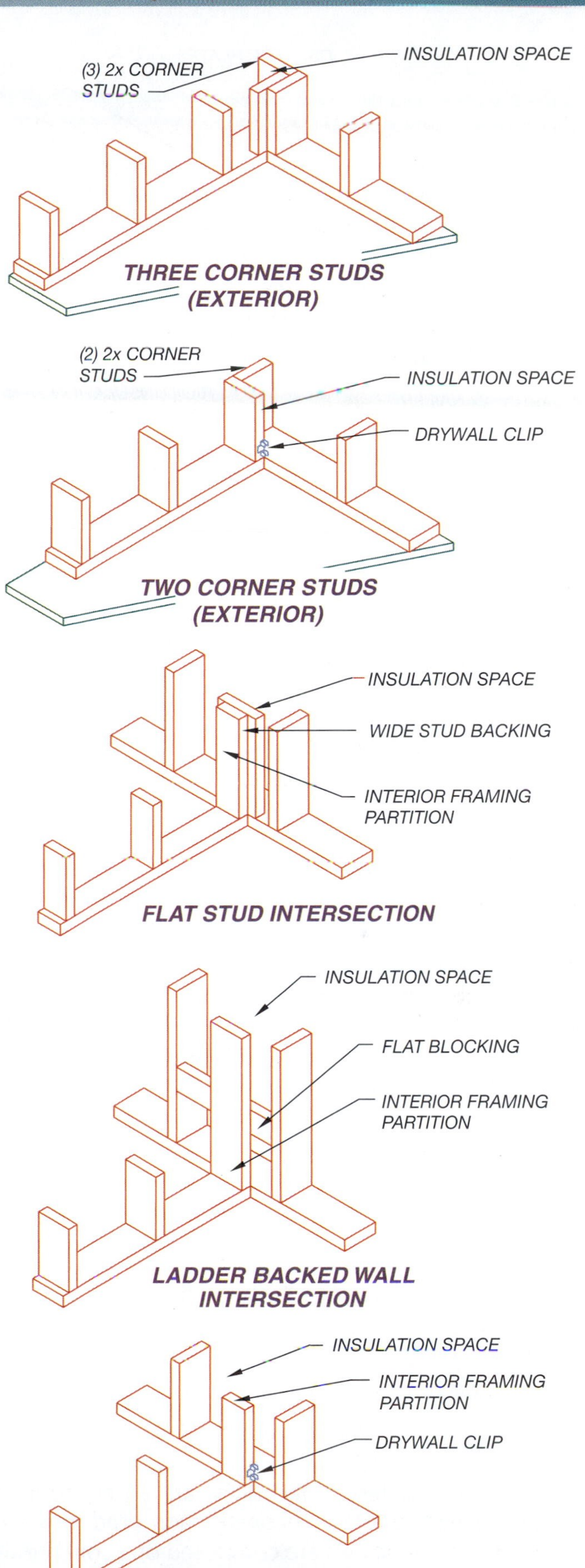

FIGURE 20.7 ■ *Reducing the amount of lumber used at wall intersections reduces the framing cost and allows added insulation to be used.*

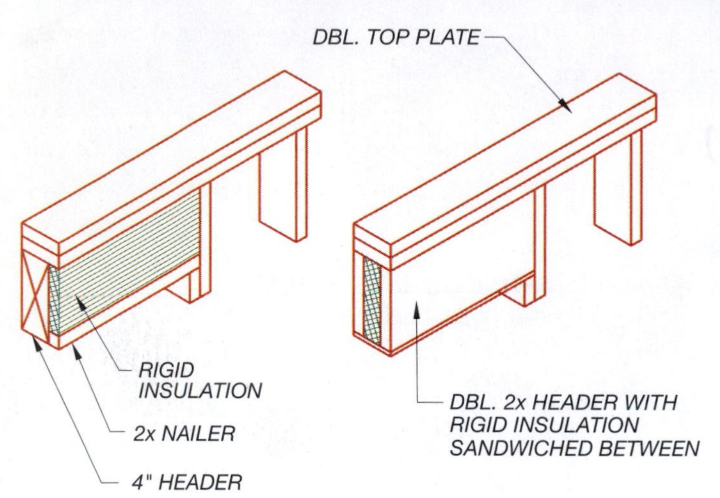

FIGURE 20.8 ■ *Rigid insulation placed behind or between headers increases the insulation value of a header from R-4 to R-11.*

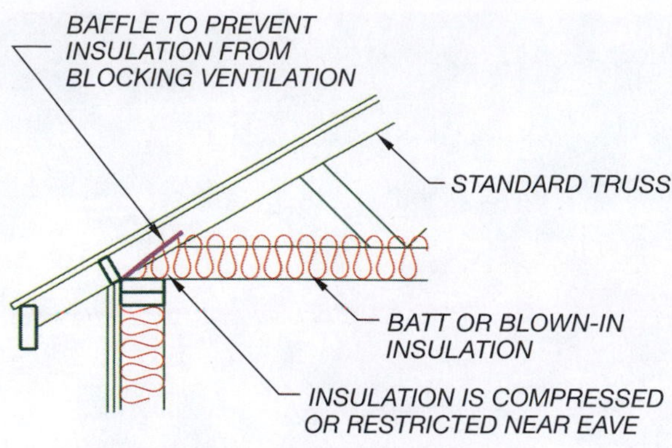

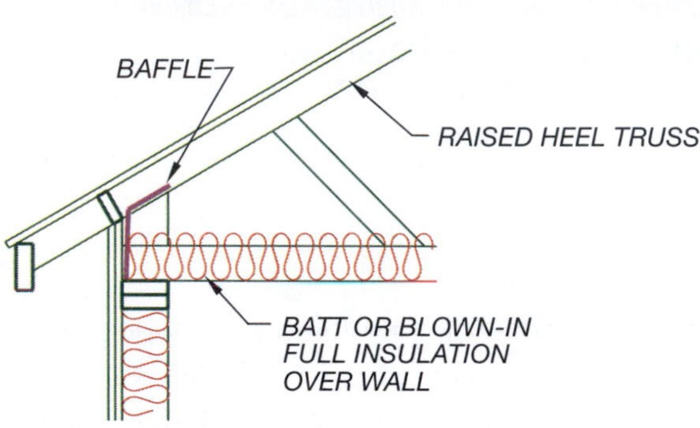

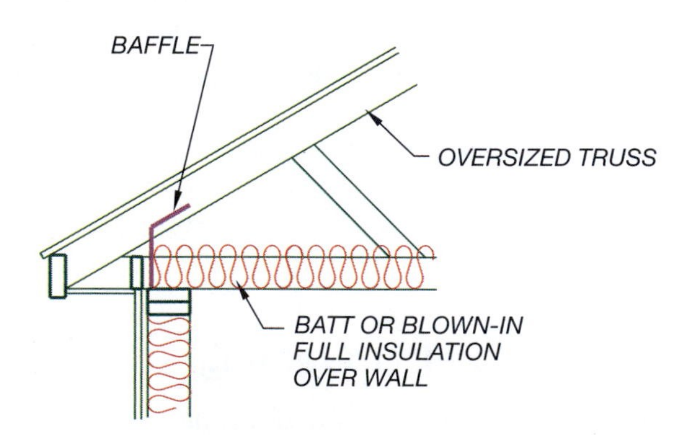

FIGURE 20.9 ■ *By altering the bearing point of a truss or rafter, the ceiling insulation can be carried to the outer edge of the exterior wall, thus reducing heat loss.*

Common engineered products found throughout a residence include laminated veneered lumber, oriented strand board, engineered studs, I-joists, and laminated beams. Production of engineered framing products provides efficient use of each log that enters the mill and predictable, superior structural quality, and reduced construction waste at the job site. Figure 20.10 shows a platform floor system constructed of engineered lumber.

Engineered lumber for framing consists of laminated veneer lumber (LVL) and laminated strand lumber (LSL). The lumber is made by combining small pieces of second-growth trees to create wood that is free of knots and splits. LSL is created by assembling small sections of wood approximately 12'' (300 mm) long into larger members. LVL is created by stacking thin veneers of wood peeled from a log and cutting them into lumber-sized members. Framing members can also be created by using small pieces of lumber that would have been scraps and joining them together with finger joints and adhesive. Once joined with finger joints, this lumber will be as strong as a standard piece of lumber of the same grade. Engineered materials and how they are used in the construction industry are explored in Chapter 21.

STRUCTURAL ENGINEERED PANELS

Structural insulated panels (SIPs) are an economical method of energy-efficient construction used in many parts of the country. SIPs are composed of a continuous core of rigid foam insulation laminated between two layers of structural board with an adhesive to form a single solid pane. Typically, expanded polystyrene (EPS) foam is used as the insulation material, and oriented strand board (OSB) is used for the outer shell of the panel. By using OSB made from fast-growing trees, SIPs are a very environmentally friendly product suitable for wall, floor, and roof construction. ESP panels are also available with a steel structural framework with no OSB

FIGURE 20.10 ■ *Engineered lumber used for the floor joists, rim joist, and mudsill.* Courtesy W. Lee Roland, Builder (rolandbuilder.com). Brent W. Roland, photographer.

FIGURE 20.11 ■ *Structural insulated panels are made with a continuous core of rigid foam insulation that is laminated between two layers of OSB. Panels are joined to each other with tongue-and-groove seams.* Courtesy Insulspan.

facing, or panels can be shipped with 1/2'' (13 mm) gypsum board preinstalled over the OSB panel. Panels are also available that use structural gypsum board as the interior face. Figure 20.11 shows a SIP being prepared for installation.

The EPS used to fill the SIP offers increased R-values over a wood wall of similar size with batt insulation. Although values differ somewhat for each manufacturer, an R-value of 4.35 per inch of EPS is common. In addition to the increased R-value, because the panels do not contain studs, the SIPs do not create a thermal bridge from the exterior face to the interior face. At the edges of panels, joint techniques allow the insulation to be virtually continuous. In addition to the insulation value, SIPs are environmentally friendly. Although the foam core is made from a nonrenewable resource, it does offer a very efficient use of the resource. One quart of an oil-based product is expanded to create approximately 40 quarts of foam panel filler. SIPs also offer increased construction savings by combining framing, sheathing, and insulation procedures into one step during construction. Panels can be made in sizes ranging from 4 through 24' (1200 through 7300 mm) long and 8' (2400 mm) high, depending on the manufacturer and design requirements. SIPs can be precut and custom-fabricated for use on the most elaborately shaped structures. Figure 20.12 shows an example of SIPs used to form the walls of a home. Door and window openings can be precut during assembly or cut at the job site. A chase is typically installed in the panels to allow for electrical wiring.

FIGURE 20.12 ■ *SIPs can be used for walls and roofs. Openings can be formed at the factory or cut into the panel at the job site.* Courtesy Insulspan.

POST-AND-BEAM FRAMING

Post-and-beam framing places framing members at greater distances apart than platform methods do. In residential construction, posts and beams are usually spaced at 48'' (1200 mm) o.c. Although this system uses less lumber than other methods, it requires larger members. Sizes vary depending on the span, but beams 4 or 6'' (100 or 150 mm) wide are typically used. The subflooring and roofing over the beams are usually 2 × 6, 2 × 8, or 1 1/8 (50 × 150, 50 × 200, or 28 mm) T&G plywood. Figure 20.13 shows typical construction

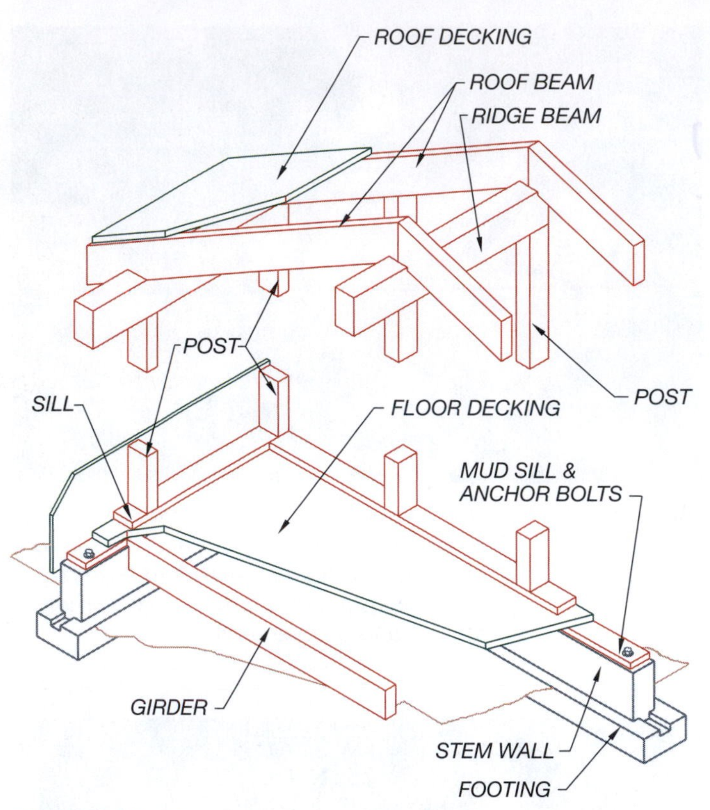

FIGURE 20.13 ■ *Structural members of a post-and-beam framing system. In residential construction, structural members are usually placed at 48'' (1200 mm) o.c.*

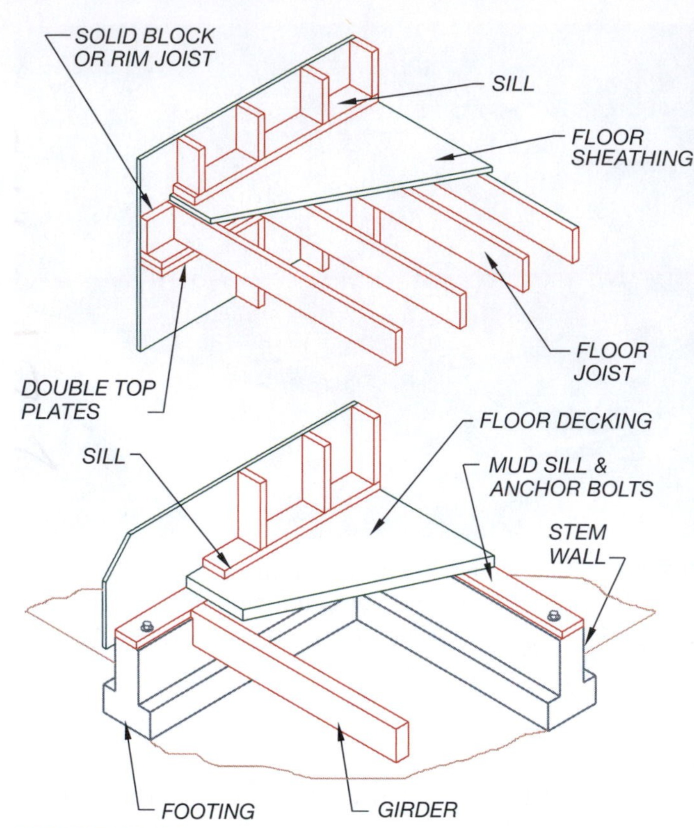

FIGURE 20.14 ■ *Post-and-beam construction (lower floor) is often mixed with platform construction for upper floors and roof framing.*

members of post-and-beam construction. Post-and-beam construction can offer great savings in both lumber and nonstructural materials. Saving results from careful planning of the locations of the posts and the doors and windows that will be located between them. Savings also result from having the building conform to the modular dimensions of the material being used. Although an entire home can be framed with post-and-beam or timber methods, many contractors use the post-and-beam system for supporting the lower floor (when no basement is required) and then use conventional framing methods for the walls and upper levels, as seen in Figure 20.14. Figure 20.15 shows the beams of a post-and-beam floor system.

TIMBER CONSTRUCTION

Although timber framing has been used for over 2000 years, the system has not been widely used for the last 100 years. The development of balloon framing methods with its smaller materials greatly reduced the desire for timber methods. Many homeowners are now returning to timber framing methods for the warmth and coziness timber homes tend to create. Figure 20.16 shows the roof of a home with timber framing set in place.

The length and availability of lumber affect the size of the frame. Although custom sizes are available for added cost, many mills no longer stock timbers longer than 16' (4900 mm). If laminated timbers are used, spans will not be a consideration. The method for lifting the timbers into place may also affect the size of the frame. Beams are often lifted into place by brute force, by winch, by forklift, or by crane. The method of joining the beams at joints affects the frame size. Figure 20.17 shows common timber components.

STEEL FRAMING

Steel framing has become increasingly competitive with wood framing techniques in residential construction. Lower energy costs, greater strength, and insurance

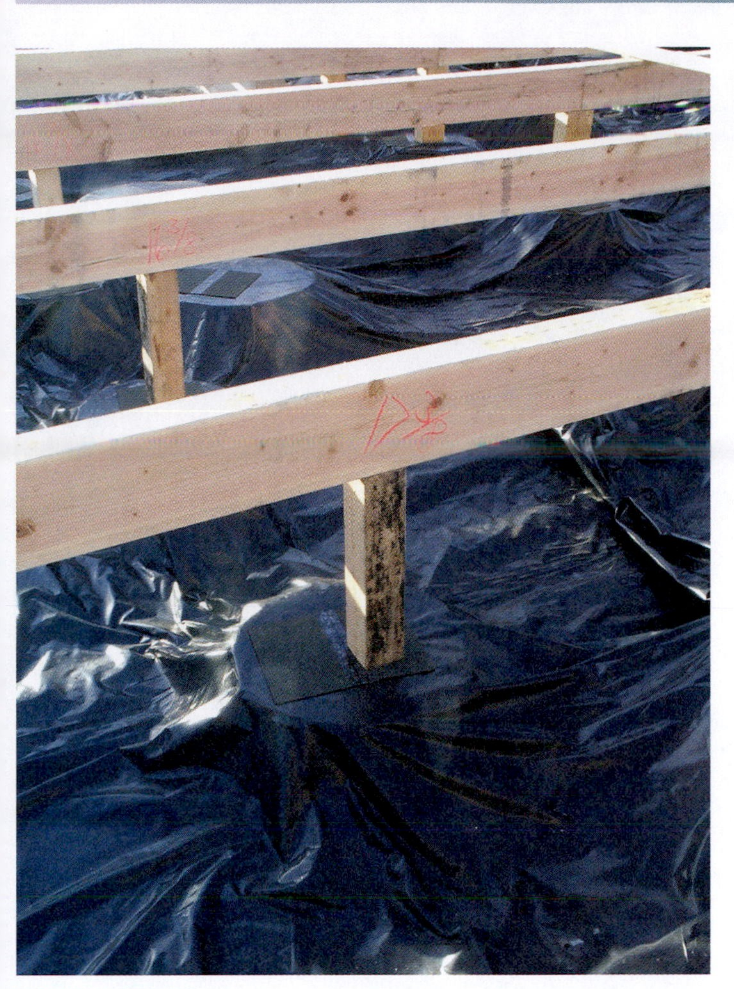

FIGURE 20.15 ■ *Beam placement for a post-and-beam floor system.* Courtesy Cary Gleason.

considerations are helping steel framing companies make inroads into residential markets. Because exterior walls are wider, insulation of R-30 can be used for exterior walls. Steel framing also has excellent properties for resisting stress from snow, wind, and seismic forces as well as termite and fire damage.

Steel offers several earth-friendly benefits as a construction material. Standard steel used for construction contains approximately 50% recycled metal. Steel can be reused if the building is altered or remodeled, or it can be recycled after the end of the structure's life. A disadvantage to steel framing materials is the amount of energy required in the initial manufacturing process. Because steel is very conductive, steel framing increases the risk of thermal bridging through the exterior walls. This conductivity can be overcome by using insulated exterior sheathing. Many residential steel structures incorporate techniques similar to western platform construction methods (see Figure 20.18), using steel studs, steel joists, and open-web steel trusses.

FIGURE 20.16 ■ *The use of exposed timbers throughout the residence creates a warm, cozy feeling.* Courtesy Sara Jefferis.

FIGURE 20.17 ■ *Typical components of a post-and-beam home.* Courtesy Ross Chappel, Timberpeg Post and Beam Homes.

Steel Studs

Steel studs, as in Figure 20.19, are used to frame walls. Walls are normally framed in a horizontal position on the floor and then tilted into place and bolted together. Steel studs offer lightweight, noncombustible, corrosion-resistant framing for interior walls and for load-bearing exterior walls up to four stories high. Steel members are available for use as studs or joists. Members are designed for rapid assembly and are predrilled for electrical and plumbing conduits. The standard

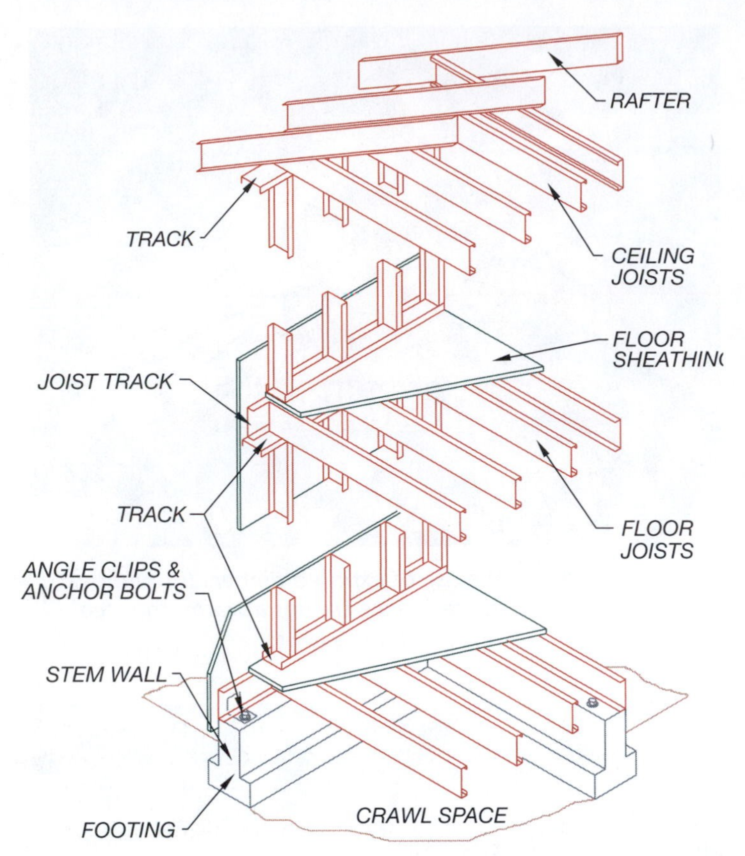

FIGURE 20.18 ■ *Structural members of a western platform framing system composed of steel members.*

FIGURE 20.19 ■ ***A home framed with steel framing members.*** Courtesy North American Steel Framing Alliance.

24'' (600 mm) spacing reduces the number of studs required by about one-third, compared with common studs spacing of 16'' (400 mm) o.c. Widths of studs range from 3 5/8 to 10'' (93 to 250 mm), but they can be manufactured in any width. The material used to produce studs ranges from 12 to 20 gauge steel, depending on the loads to be supported. Figure 25.18 shows steel studs as they are typically used. Steel studs are mounted in a channel track at the top and bottom of the wall. This channel is similar to the top and bottom plates of a standard stud wall.

Stud Specifications

Section properties and steel specifications vary among manufacturers, so the material to be specified on the plans will vary. Catalogs of specific vendors must be consulted to determine the structural properties of the desired stud. The project manager will determine the size of stud to be used, but vendors' catalogs will need to be consulted to completely specify the studs. As a minimum, the gauge and usage are specified on the framing plan, details, and sections. If a manufacturer is to be specified, the callout will typically include the size, style, gauge, and manufacturer of the stud. An example of a steel stud specification would read:

362SJ20 STEEL STUDS BY UNIMAST

Steel Wall Panels

Steel wall panels are often used in place of wood framed walls to meet code requirements for lateral bracing. The wall panels are manufactured by companies such as Simpson Strong-Tie to provide resistance to the lateral stress caused by wind and earthquakes. Wall panels typically range from 12 through 24'' (300 through 600 mm) wide and are available to be used with 4 and 6'' (100 and 150 mm) deep wood wall members. See chapter 24 for additional information.

Steel Joists

Steel joists provide the same advantages over wood as steel studs. The gauge and yield point for joists are similar to stud values. Many joists are manufactured so that a joist may be placed around another joist or nested. ***Nested joists*** allow the strength of a joist to be greatly increased without increasing the size of the framing members. A nested joist is often used to support a bearing wall placed above the floor. Joists are available in lengths up to approximately 40' (12 000 mm), depending on the manufacturer.

Open-Web Steel Joists

Open-web steel joists are manufactured in the same configurations as open-web wood trusses. Open-web steel joists offer the advantage of being able to support greater loads over longer spans than their wood counterparts, with less dead load than solid members. Open-web joists are referred to by series number, which defines their use. The Steel Joist Institute (SJI) designations include K, LH, and DLH. The K-series joists are parallel web members that range in depth from 8 through 30'' (200 through 750 mm) in 2'' (50 mm) increments for spans up to approximately 60' (18 300 mm).

Steel joists should be specified on the framing plans, structural details, and sections by providing the approximate depth, the series designation, chord size, and spacing. An open-web steel joist specification would resemble:

14K4 OPEN WEB STEEL JOIST @ 32'' O.C.

Just as with a wood truss, span tables are available to determine safe working loads of steel joists. The engineer will determine the truss to be used, but the drafter will often be required to find information to complete the truss specification.

CONCRETE MASONRY CONSTRUCTION

Concrete masonry units (CMUs) are a durable, economical building material that provides excellent structural and insulation values. Concrete blocks are used primarily in warmer climates, from Florida to southern California, as an above-ground building material. CMUs can be waterproofed with cement-based paints and used as the exterior finish, or they can be covered with stucco or other masonry products. Waterproof wood furring strips are normally attached to the interior side of the block to support sheetrock. In areas with cooler climates, concrete blocks are often used only for below-grade construction.

Four classifications used to define concrete blocks for construction include:

- Hollow load-bearing (ASTM C 90)
- Solid load-bearing (ASTM C 145)
- Non-load-bearing blocks (ASTM C 129), which are either solid or hollow blocks.

Blocks are also classed by their weight as normal, medium, and light. The type of aggregate used to form the unit affects the weight of the block. Normal aggregates such as crushed rock and gravel produce a block weight of between 40 and 50 lb for a 8 × 8 × 16'' (200 × 200 × 400 mm) block. Lightweight aggregates produce approximately a 50% saving in weight.

Modular Block Sizes

CMUs come in a wide variety of patterns and shapes including:

- The most common size is 8'' × 8 × 16 (200 × 200 × 400 mm).
- Nominal block widths are 4, 6, 8, 10, and 12'' (100, 150, 200, 250, and 300 mm).
- Lengths include 6, 8, 12, 16, and 24'' (150, 200, 300, 400, and 600 mm).

Each dimension of a concrete block is actually 3/8'' (10 mm) smaller to allow for a mortar joint. Although the project manager is responsible for the design and size of the structure, it is important that each member of the team be aware of the modular principles of concrete block construction. Lengths of walls, locations of openings in a wall, and heights of walls and openings must be based on the modular size of the block being used. Failure to maintain the modular layout can result in a tremendous increase in the cost of labor to cut and lay the blocks. To minimize cutting and labor cost, a concrete block structure should be kept to its modular size. For the standard 8 × 8 × 16'' (200 × 200 × 400 mm) CMU:

- Structures that are an even number of feet long should have dimensions that end in 0'' or 8''. For instance, structures that are 24'-0'' or 32'-8'' are modular.
- Dimensions for walls that are an odd number of feet should end with 4''. Walls that are 3'-4'', 9'–4'', and 15'-4'' are modular, but a wall that is 15'-8'' will require blocks to be cut.

Concrete blocks come in a variety of shapes, as shown in Figure 20.20. These shapes allow for the placement of steel reinforcement bars, but steel reinforcing mesh can also be used. Chapter 25 provides further information on concrete reinforcement. Reinforcing is typically required at approximately 48'' (1200 mm) o.c. vertically. Horizontal reinforcement is approximately 16'' (400 mm) o.c., but the exact spacing depends on the seismic or wind stresses to be resisted. Reinforcement is typically specified on the framing plan and also shown in a cross section and detail. Figure 20.21 shows components of CMU construction.

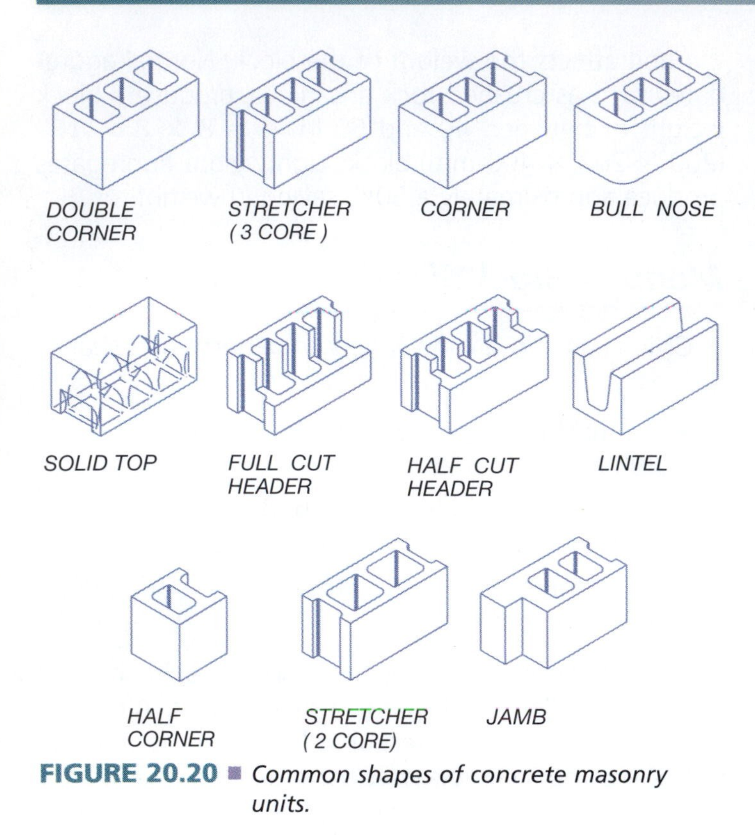

FIGURE 20.20 ■ *Common shapes of concrete masonry units.*

Masonry Representation

Concrete blocks are represented in plan view as shown in Figure 20.22. Type, size, and reinforcement are specified on the plan views, with bold lines to represent the edges of the masonry. Thin lines for hatching are placed at a 45° angle to the edge of the wall. The size and location of concrete block walls must be dimensioned on the floor plan. Concrete block is dimensioned from edge to edge of walls. Openings in the wall are also located by dimensioning to the edge. Masonry is also shown in sections and details, but the scale of the drawing will affect the drawing method. At scales under 1/2'' = 1'-0'', the wall is typically drawn just as in plan view. At larger scales, details typically reflect the cells of the block. Figure 20.23 shows methods of representing concrete blocks in cross section.

Steel Reinforcement

Masonry units are excellent at resisting forces from compression but very weak at resisting forces in tension. Steel is excellent at resisting forces of tension but tends to buckle under forces of compression. The combination of these two materials forms an excellent unit for resisting great loads from lateral and vertical forces. Reinforced masonry structures are stable because the

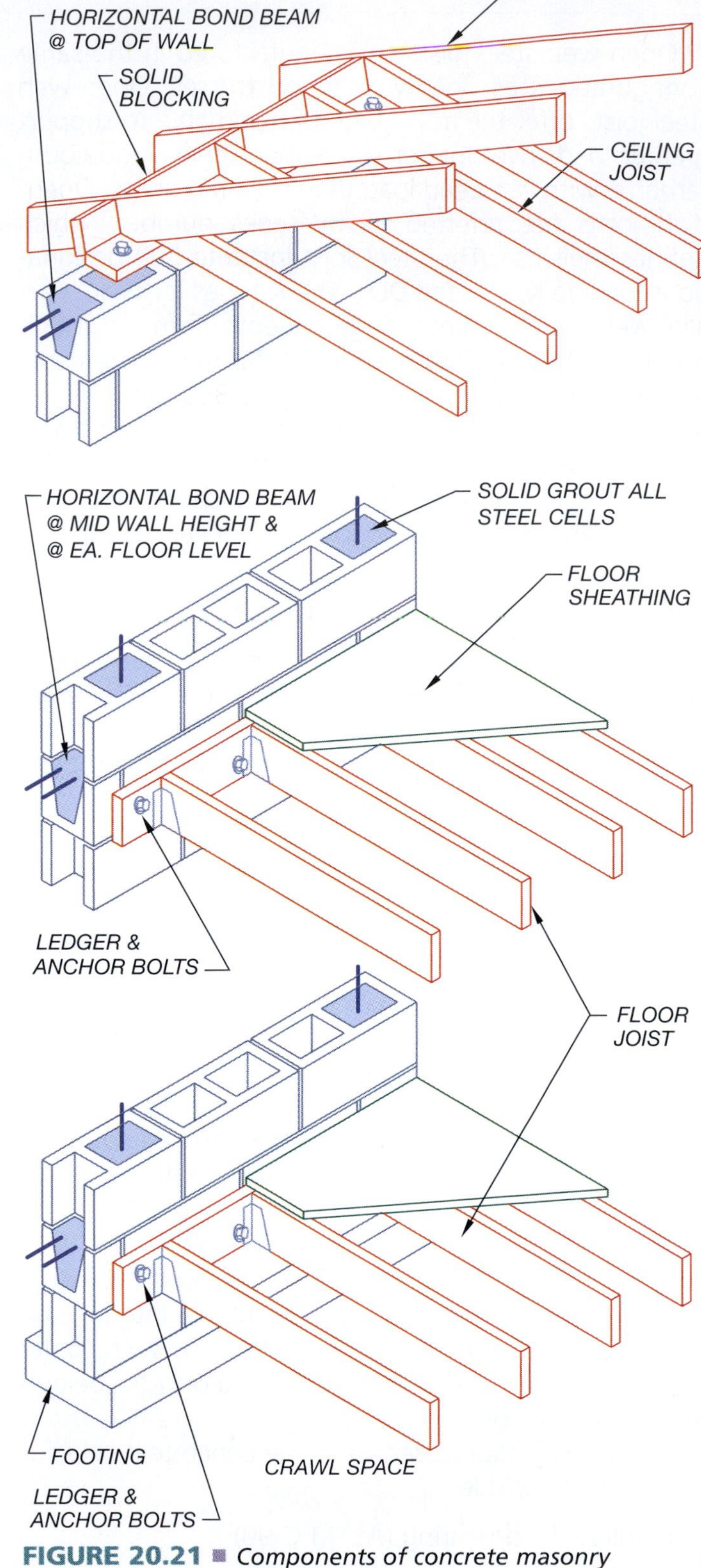

FIGURE 20.21 ■ *Components of concrete masonry construction.*

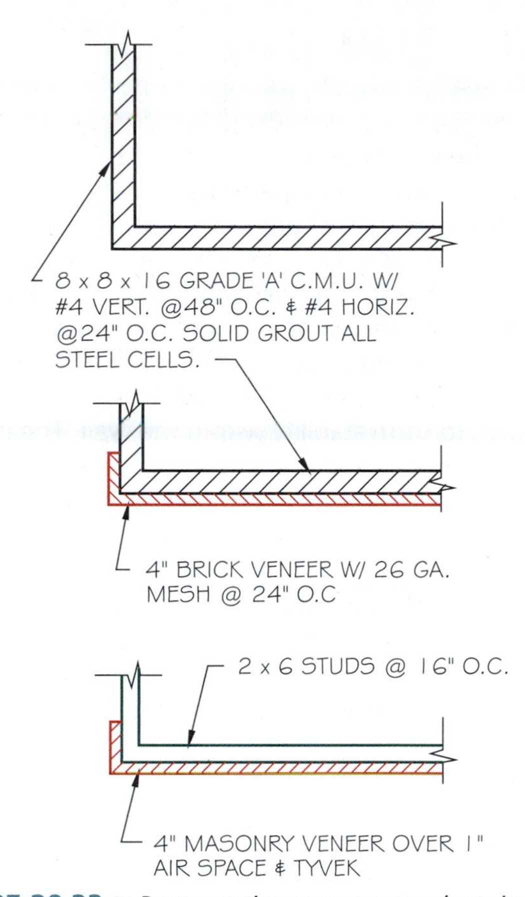

FIGURE 20.22 ■ *Representing masonry products in plan view.*

masonry, steel, grout, and mortar bond together effectively. Loads that create tension on the masonry are effectively transferred to the steel. If the steel is carefully placed, the forces will result in tension on the steel and will be safely resisted.

Reinforcing Bars

Steel reinforcing bars, called rebar, are used to reinforce masonry walls. The IRC sets specific guidelines for the size and spacing of rebar based on the seismic zone of the construction site. Reinforcement will be specified by the engineer throughout the calculations and sketches and must be represented accurately by the CAD technician. Bars used for reinforcing masonry walls are usually deformed so that the concrete will bond more effectively with the bar and not allow slippage as the wall flexes under pressure. Deformations consist of small ribs that are placed around the surface of the bar. Deformed bars range in size from 3/8 through 2 1/4'' diameter. Inch sizes do not readily convert to metric sizes. Bars are referenced on plans by a number rather than a size. A number represents the size of the steel in approximately 1/8'' increments. Common sizes of steel include:

BAR SIZE	METRIC BAR SIZE	DIAMETER IN INCHES, NOMINAL
# 3	#10	0.375 – 3/8''
# 4	#13	0.500 – 1/2''
# 5	#16	0.625 – 5/8''
# 6	#19	0.750 – 3/4''
# 7	#22	0.875 – 7/8''
# 8	#25	1.000 – 1''
# 9	#29	1.128 – 1 1/8''
# 10	#32	1.270 – 1 1/4''
# 11	#36	1.410 – 1 3/8''
# 14	#43	1.693 – 1 3/4''
# 18	#57	2.257 – 2 1/4''

Steel Placement

Wall reinforcing is held in place by filling each masonry cell containing steel with grout. The placement of steel varies with each application, but steel reinforcement is typically placed on the side of the wall that is in tension. Because a wall will receive pressure from each

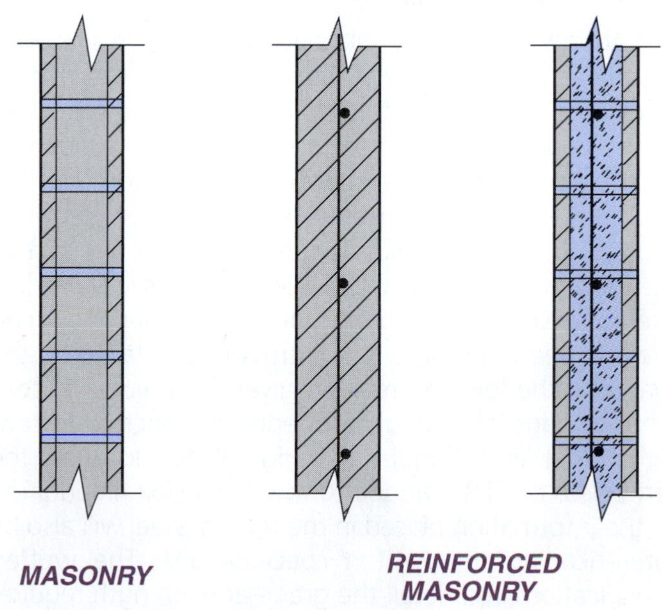

FIGURE 20.23 ■ *Representing concrete blocks in section view.*

side, masonry wall construction usually has the steel centered in the wall cavity. For retaining walls, steel is normally placed near the side that is opposite the soil. The location of steel relative to the edge of the wall must be specified in the wall details.

Rebar Representation

The quantity of bars, the bar size by number, the direction that the steel runs, and the steel grade will need to be specified on the drawings. Common grades associated with residential construction are 40, 50, 60, and 75. On drawings such as the framing plan, walls will be drawn, but the steel is not drawn. Steel specifications are generally included in the wall reference, as in Figure 20.22. When shown in section or detail, steel is represented by a bold line which can be solid or dashed, depending on office practice. Steel is represented by a solid circle when shown in end view. In detail, steel is often drawn at an exaggerated size so that it can be clearly seen. It should not be so small that it blends with the hatch pattern used for the grout, or so large that it is the first thing seen in the detail. Figure 20.23 shows how steel can be represented in detail.

Locating Steel

Dimensions will be required to show the location of the steel from the edge of the concrete. The engineer may provide a note in the calculations such as:

(7)- #5 HORIZ. @ 3'' UP/DN

Although this note could be placed on the drawings exactly as is, a better method would be to use dimensions to specify the distance from the top and bottom of the concrete rather than using a note. The quantity and size of the steel are specified in the detail, but the locations are placed using separate dimensions. Although this requires slightly more work on the part of the drafter, the more visual specifications will be less likely to be overlooked or misunderstood. Depending on the engineer and the type of stress to be placed on the masonry, the location may be given from edge of concrete to edge of steel or from edge of concrete to center of steel. To distinguish the edge of steel location, the term *clear* or CLR is added to the dimension. In addition to the information placed in the details, steel will also be referenced in the written specifications. The written specifications will detail the grade and strength requirements for general areas of the structure such as walls, floors, columns, and retaining walls.

SOLID MASONRY CONSTRUCTION

One of the most popular features of brick is the wide variety of positions and patterns in which it can be placed. These patterns are achieved by placing the bricks in various positions to each other. The position in which the brick is placed will alter what it is called. Figure 20.24 shows the names of common brick positions. Bricks can be placed in various positions to form a variety of bonds and patterns. A ***bond*** is the connecting of two wythes to form stability within the wall. The pattern is the arrangement of the bricks within one ***wythe.*** The Flemish and English bonds in Figure 20.25 are the most common methods of bonding two wythes or vertical section of a wall that is one brick thick. The Flemish bond consists of alternating headers and stretchers in every course. A course of brick is one row in height. An English bond consists of alternating courses of headers and stretchers. The headers span between wythes to keep the wall from separating.

Masonry walls must be reinforced using methods similar to those used with concrete blocks. The loads to be supported and the stress from wind and seismic

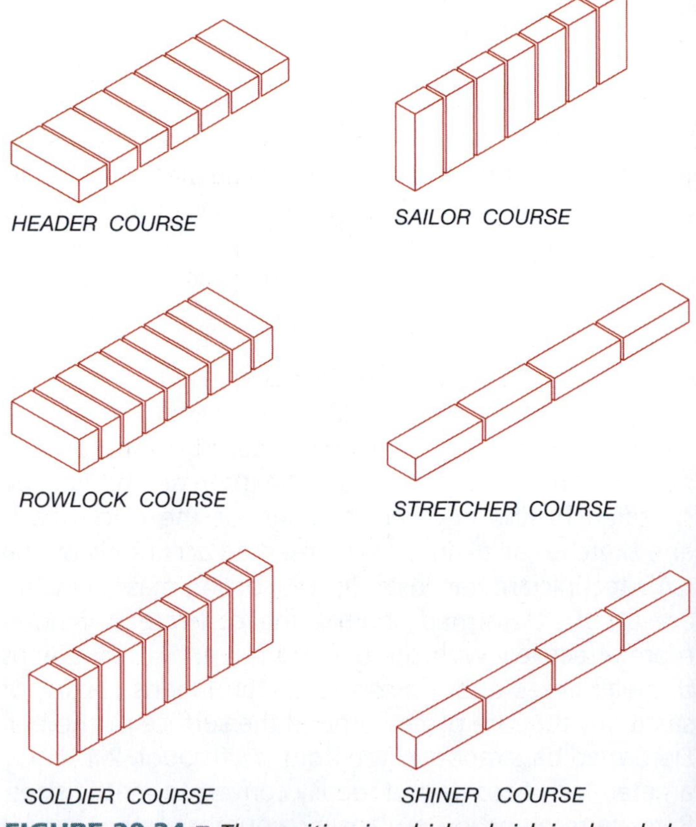

FIGURE 20.24 ■ *The position in which a brick is placed alters the name of the unit.*

loads determine the size and spacing of the rebar to be used and are established by the architect or engineer. If a masonry wall is to be used to support a floor, a space one wythe wide will be left to support the joist. Joists are usually required to be strapped to the wall so that the wall and floor will move together under lateral stress. The end of the joist must be cut on an angle, called a fire cut. If the floor joist is damaged by fire, the fire cut will allow the floor joist to fall out of the wall without destroying the wall.

Because brick is very porous and absorbs moisture easily, some method must be provided to protect the end of the joist from absorbing moisture from the masonry. Typically the end of the joist is wrapped with 55# felt and set in a 1/2'' (13 mm) air space. The interior of a masonry wall also must be protected from moisture. A layer of hot asphaltic emulsion can be applied to the inner side of the interior wythe and a furring strip attached to the wall. In addition to supporting sheetrock or plaster, the space between the furring strips can be used to hold batt or rigid insulation. When a roof framing system is to be supported on masonry, a pressure-treated plate is usually bolted to the brick, similar to how a plate is attached to a concrete foundation.

Another method of using brick is to form a cavity between each wythe of brick. The cavity is typically 2'' (50 mm) wide and creates a wall approximately 10'' (250 mm) wide with masonry exposed on the exterior and interior surfaces. The air space between wythes provides an effective barrier to moisture penetration to the interior wall. Weep holes in the lower course of the exterior wythe will allow moisture that collects to escape. Rigid insulation can be applied to the interior wythe to increase the insulation value of the air space in cold climates. Care must be taken to keep the insulation from touching the exterior wythe so that moisture is not transferred to the interior. Metal ties are typically embedded in mortar joints at approximately 16'' (400 mm) to tie each wythe together.

Masonry Veneer

A common method of using brick and stone in residential construction is as a veneer, a nonstructural covering material similar to the home in Figure 20.26. Using brick as a veneer offers the charm and warmth of brick with a lower construction cost than structural brick. If brick is used as a veneer over a wood bearing wall, for example, the amount of brick required is half that of structural brick. Care must be taken to protect the wood frame from moisture in the masonry. Brick

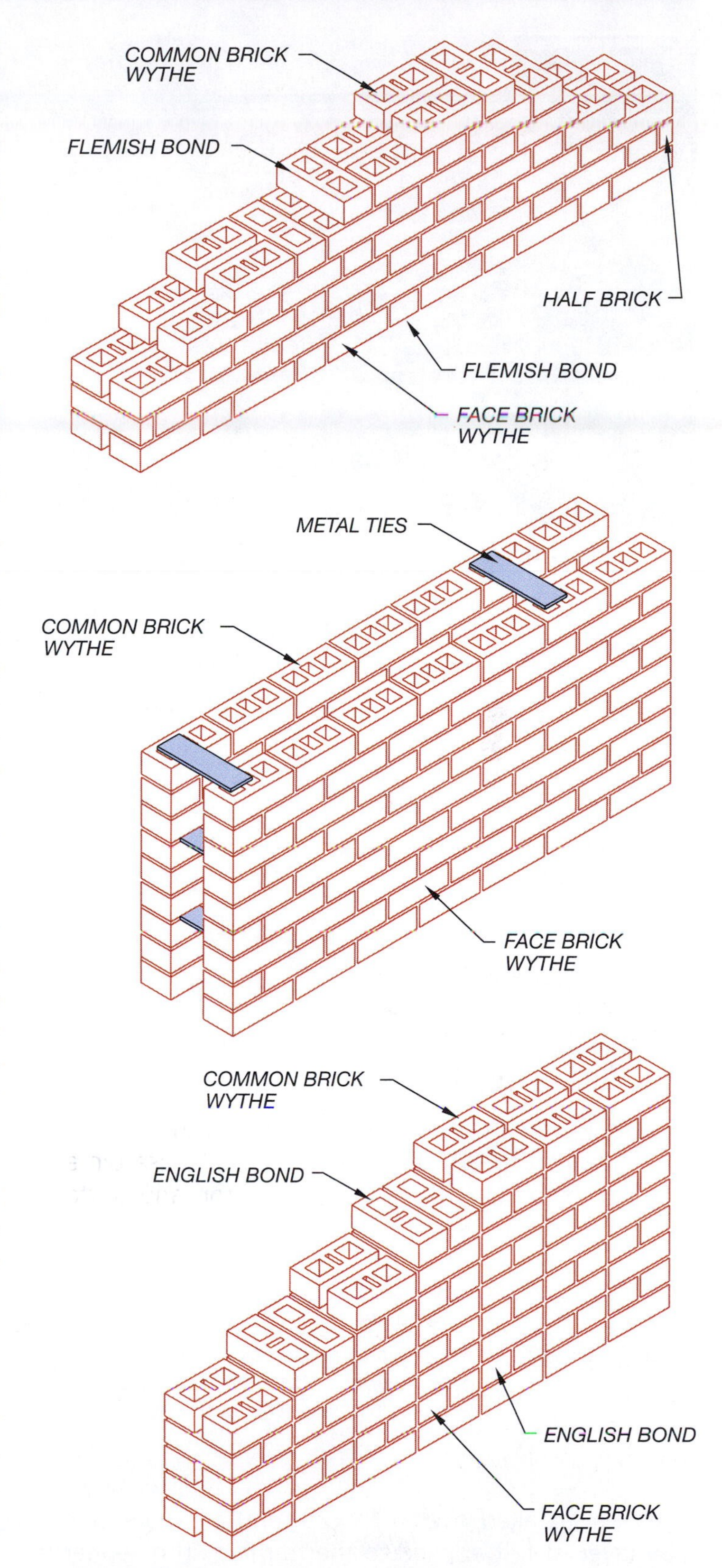

FIGURE 20.25 ■ *Brick walls can be strengthened by metal ties or bricks connecting the wythes. The most common types of brick bonds are the Flemish and English.*

FIGURE 20.26 ■ *Brick veneer is often added to a home.* Courtesy Benigno Molina-Manriquez.

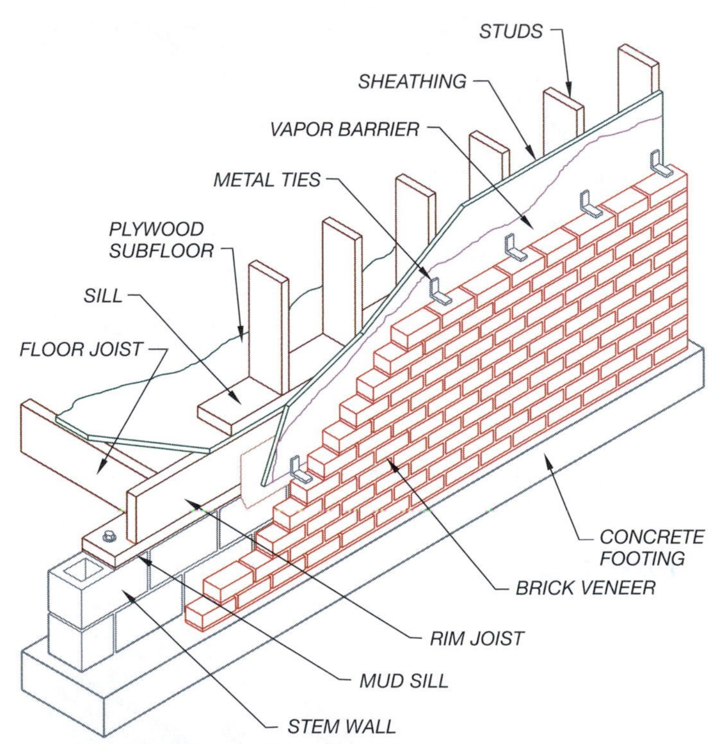

FIGURE 20.27 ■ *Brick is typically used as a non-load-bearing veneer attached to wood or concrete construction with metal ties.*

must be installed over a 1'' (25 mm) air space and a 15# layer of felt applied to the framing. The veneer is attached to the framing with 26-gauge metal ties at 24'' (600 mm) o.c. along each stud. Figure 20.27 shows an example of how masonry veneer is attached to a wood-frame wall.

FIGURE 20.28 ■ *Insulated concrete forms can be easily placed. Reinforcing and concrete is then added to complete the wall.* Courtesy Carol Ventura.

INSULATED CONCRETE FORM CONSTRUCTION

Originally used to form foundation walls, insulated concrete forms (ICFs) are now being used to provide an energy-efficient wall framing system for an entire structure. Poured concrete is placed in expanded polystyrene (EPS) forms that are left in place to create a superinsulated concrete wall system. Forms are placed in a pattern similar to concrete blocks or bricks and then filed with steel reinforcing and concrete (see Figure 20.28). Forms 6 and 8'' (150 and 200 mm) wide × 16'' (400 mm) high × 48'' (1200 mm) long are available from most manufacturers. Figure 20.29 shows a typical detail of an insulated concrete form supplied by the manufacturer. This detail can be inserted into the drawings provided by the design team to explain the construction process. The concrete in the forms creates a pattern similar to post-

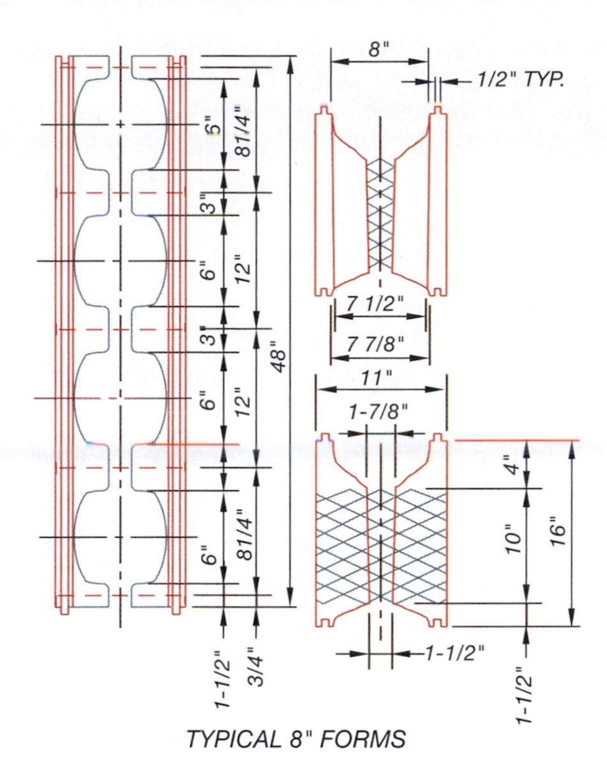

FIGURE 20.29 ■ *The manufacturer of the ICFs provides details that can be added to the construction documents.* Courtesy Pamela Russom, American PolySteel, LLC.

and-beam construction. Vertical posts are created at 12'' (300 mm) o.c., and horizontal beams are created at 16'' (400 mm) o.c. Solid concrete webs are created between the posts and beams at the center of the forms.

Expanded polystyrene forms (EPFs) provide a stable base for attaching interior and exterior finishing materials and help to create the high energy efficiency of the system. The thermal mass of the concrete and the R-20 value of the EPF produce the energy efficiency. Depending on the width of the wall and finishing materials, the total R-value can range from R-30 to R-50. The system also excels by reducing air leakage and air infiltration into the structure. Structures constructed from ICF average 0.1 air exchange per hour (ACH), compared with 0.4 ACH for wood-framed walls. Figure 20.30 shows wall construction using ICF methods.

MODULAR FRAMING METHODS

Many people confuse the term *modular home* with *mobile* or *manufactured home.* Manufactured homes are built on a steel chassis and are built to meet codes

FIGURE 20.30 ■ *Finishing work for a wall made with insulated concrete forms is similar to poured concrete construction.* Courtesy Pamela Russom, American PolySteel, LLC.

FIGURE 20.31 ■ *Modular construction allows as much as 90% of the construction work to be completed inside an environmentally controlled factory.* Courtesy Kelly L. Derin, Cardinal Homes, Inc.

other than the IRC. Modular homes are governed by the same codes that govern site-built homes but are built off site.

Modular construction is a highly engineered method of constructing a building or building components in an efficient and cost-effective manner. Modular homes begin as components designed, engineered, and assembled in the controlled environment of a factory using assembly-line techniques; the home in Figure 20.31 is an example. Work is never slowed by weather, and materials are not subject to warping due to moisture ab-

FIGURE 20.32 ■ *Components ranging in size from a wall panel to a 12'-wide module of a home can be shipped to the site and then assembled.* Courtesy Portland Cement Association.

sorption. A home travels to different workstations, where members of each building trade perform the required assembly. Rooms that require plumbing can be assembled in an area separate from other rooms and then assembled into a module. Modules can be shipped to the construction site, where they are assembled on a conventional foundation. Modules can be shipped complete or may require additional site construction, depending on the complexity of the module. Figure 20.32 shows a wall component of a modular home being delivered to the construction site.

Structures can also use preassembled components. Panelized walls—which include windows, doors, and interior and exterior finish—can be assembled in a factory but installed at the job site. Floor and roof framing members can be precut in a factory, numbered for their exact location, and then shipped for assembly at the job site. This system of construction can offer economical construction in remote areas where materials are not readily available.

GREEN CONSTRUCTION

The final point to be considered in this chapter is not a framing method but a building mindset. Whether it's called earth-friendly, green, ecological, or sustainable construction, the concept is to build in a manner that will produce a structure using energy efficiently, which is built of materials that have a low impact on the environment, and that will be healthy to live in. Environmentally friendly framing is not just a matter of selecting green materials but that these materials, once selected, are used in a manner that will reduce the environmental impact of the home. Products that are not considered green can be used in a home, but they may still benefit the homeowner. Creating an environmentally friendly home requires the matching of materials to a specific design and site that minimizes the effect on that site. Five questions should be considered that will effect the selection of framing and finish materials to make a sustainable structure:

- Can products be selected that are made from environmentally friendly materials?
- Can products be selected because of what they do not contain?
- Will the products to be used reduce the environmental impact during construction?
- Will the products to be used reduce the environmental impact of operating the building?
- Will the products to be used contribute to a safe, healthy indoor environment?

Environmentally Friendly Materials

One of the major considerations that affect framing materials is the selection of products and construction materials that are made from environmentally friendly components. The materials used to produce a building product and where those materials come from are key determinants in labeling a product a green building material. Points to consider in selecting building materials include products that can be salvaged or recycled, certified green products, quick-growth products, agricultural waste materials, and products that require minimal processing. Chapter 21 contains additional information about sustainable building materials.

Salvaged Products

A major goal of sustainable construction is to reuse a product whenever possible instead of producing a new one. Common salvaged materials used in buildings include bricks, millwork, framing lumber, plumbing fixtures, and period hardware. All of these materials are sold on a local or regional basis by salvage yards.

Products with Recycled Content

Recycled content is an important feature of many green products because materials are more likely to be diverted from landfills. Industrial by-products such as iron-ore slag can be used to make mineral wool insulation, fly ash can be used to make concrete, and PVC pipe scraps can be used to make shingles.

Certified Wood Products

Third-party forest certification, based on standards developed by the Forest Stewardship Council (FSC), will help to ensure that wood products come from well-managed forests. Wood products must go through a certification process to carry an FSC stamp. Manufactured wood products can meet the FSC certification requirements with less than 100% certified wood content.

Quick-Growth and Waste By-Products

The use of quick-growth products for framing materials allows old and second-growth trees to remain in the forest. Rapidly renewable materials are made from wood from tree farms with a harvest rotation of approximately 10 years. Wood products such as LVL, OSB, and laminated beams have already been considered. Interior finish products are also produced from agricultural crops or their waste products. Examples of green products made from agricultural crops include linoleum, form-release agents made from plant oils, natural paints, textile fabrics made from coir and jute, cork, organic cotton, wool, and sisal. Building products can also be produced from agricultural waste products made from straw (the stems left after harvesting cereal grain), rice hulls, and citrus oil. These products can be used to improve the interior environment and create an ecologically friendly structure.

Minimally Processed Products

Products that are minimally processed can be green because of low energy use and low risk of chemical release during manufacture. These can include wood products, agricultural or nonagricultural plant products, and mineral products such as natural stone and slate shingles.

Removing Materials to Become Earth-Friendly

Some building products are considered green not because of their content but because they allow material savings elsewhere or they are better alternatives to conventional products containing harmful chemicals. Chemicals that deplete the ozone, CCA wood preservative, polyvinyl chloride (PVC), and polycarbonate are products that should be avoided in a structure, but products with these chemicals may be considered earth-friendly because the product has significant environmental benefits. Some examples include drywall clips that allow the elimination of corner studs; engineered lumber that reduces lumber waste; the piers for a joist floor system that minimize the use of concrete compared to the post-and-beam system; and concrete pigments that eliminate the need for conventional finish flooring by letting the concrete slabs serve as the finished floor.

Reducing the Impact of Construction

Some building products produce their environmental benefits by avoiding pollution or other environmental impacts during construction. Products that reduce the impacts of new construction include various erosion-control products, foundation products that eliminate the need for excavation, and exterior stains that result in lower VOC (volatile organic compound) emissions into the atmosphere. The greatest impact from construction can come from careful design that creates a home that suits the site and a foundation system that requires minimal excavation. Review Chapters 10 and 11 for methods of reducing site impact during construction.

Reducing the Impact After Construction

The ongoing environmental impact that results from operating a structure will far outweigh the impact associated with the construction phase. It is important during the design phase to select components that will reduce heating and cooling loads and conserve energy, select fixtures and equipment that conserve water, choose materials with exceptional durability or low-maintenance requirements, select products that prevent pollution or reduce waste, and specify products that reduce or eliminate pesticide treatments.

Reducing Energy Demands

SIPs, ICFs, autoclaved aerated concrete (AAC) blocks, and high-performance windows are examples of materials that can be used during construction to reduce HVAC loads over the life of the structure. Other energy-consuming equipment such as water heaters, furnaces, dishwashers, refrigerators, washing machines, and dryers should be carefully selected for their ability to conserve energy after construction. Most home appliances are now rated by Energy Star™ standards that have been adopted nationally. Energy Star is a government-backed program designed to help consumers obtain energy efficiency. Using compact fluorescent lamps and occupancy or day lighting-control equipment can save additional energy. Installing efficient equipment as the home is constructed will reduce energy needs over the life of the structure.

Renewable Energy

Equipment that uses renewable energy instead of fossil fuels and conventional electricity are highly beneficial from an environmental standpoint. Examples include solar water heaters and photovoltaic systems (see Chapter 14). Natural gas fuel cells or cells that use other fossil fuels such as a hydrogen source are considered green because emissions are lower than combustion-based equipment they replace.

Conserving Water

All toilets and showerheads are required to meet the federal water efficiency standards. Other products, such as rainwater storage systems will also contribute to making a residence earth friendly (see Chapter 13).

Reducing Maintenance

Products that reduce maintenance make a residence environmentally attractive because they need to be replaced less frequently, or their maintenance has very low impact. Sometimes durability is a contributing factor to the green designation but not enough to distinguish the product as green on its own. Included in this category are such products as fiber-cement siding, fiberglass windows, slate shingles, and vitrified clay waste pipes.

Preventing Pollution

Methods of controlling substances from entering the environment contribute to making a home earth-friendly. Alternative wastewater disposal systems reduce groundwater pollution by decomposing organic wastes more effectively. Porous paving products and vegetated roofing systems result in less storm water runoff and thereby reduce surface water pollution. Providing convenient recycling centers within the home will allow homeowners to safely store recyclables for collection. Planning for a compost system will further allow homeowners to reduce the generation of solid waste.

Eliminating Pesticide Treatments

Although they may be needed to increase livability, periodic pesticide treatment around buildings can be a significant health and environmental hazard. The use of products such as termite barriers, borate-treated building products, and bait systems that eliminate the need for pesticide application all contribute to a sustainable structure.

Contributing to the Environment

Product selection has a significant effect on the quality of the interior environment. Green building products that help ensure a healthy interior living space can be separated into several categories including products that don't release pollutants, products that block the spread of indoor contaminants, and products that warn occupants of health hazards.

Nonpolluting Products

One of the dangers of modern construction is the ability to make a structure practically airtight. This in itself is not a problem, but the adhesives in most products can be harmful when constant air changes are not provided. Products that don't release significant pollutants into a structure contribute to an earth-friendly home. Interior products that contribute to improving the interior environment include zero- and low-VOC paints, caulks, and adhesives as well as products with very low emissions, such as nonformaldehyde manufactured wood products.

Blocking, Removing, and Warning of Contaminates

Certain materials and products are green because they prevent the contaminants from entering the interior environment. Linoleum is available that helps control microbial growth. Coated duct board is available that helps control mold growth, and products are available for blocking the entry of mold-laden air into a duct system. Other products can help remove pollutants from the shoes of people entering the residence. Each of these types of products can help provide an environmentally friendly home.

Once contaminants have entered the residence, several products are available to warn the homeowner. This would include the use of carbon monoxide (CO) detectors and lead paint test kits. Once a homeowner becomes aware of certain environmental dangers, several products are available to remove them. This would include the use of ventilation products, filters, radon mitigation equipment, and other equipment that can remove pollutants or introduce fresh air.

CHAPTER 20

Additional Readings

The following websites can be used as a resource to help you keep current with changes in framing materials.

ADDRESS	COMPANY OR ORGANIZATION
www.aci-int.com	American Concrete Institute
www.steel.org	American Iron and Steel Institute
www.alhloghomes.com	Appalachian Log Homes
www.apawood.org	APA—The Engineered Wood Association
www.betterbricks.com	Better Bricks
www.Bia.org	Brick Industry Association
www.buildinggreen.com	Building Green Inc.
www.cardinalhomes.com	Cardinal Homes, Inc (modular homes)
www.crbt.org	Center for Resourceful Building Technology
www.cmpc.org	Concrete Masonry Promotions Council
www.eere.energy.gov	DOE Integrated Building for Energy Efficiency
www.enercept.com	Enercept, Inc (SIPs)
www.energydesignresources.com	Energy Design Resources
www.energystar.gov	Energy Star (appliance energy standards)
www.fscus.com	Forest Stewardship Council
www.generalshale.com	General Shale Brick
www.greenbuilder.com	Greenbuilder
www.imiwebi.org	International Masonry Institute
www.iza.com	International Zinc Association
www.lpcorp.com	Louisiana-Pacific Corporation
www.masonrysociety.org	Masonry Society
www.nahbrc.org	National Association of Home Builders Research Center
www.newbuildings.org	New Buildings Institute
www.steelframingalliance.com	North American Steel Framing Alliance
www.pge.com	Pacific Energy Center
www.polysteel.com	Polysteel Forms
www.concretehomes.com	Portland Cement Association (Concrete homes site)

www.portcement.org	Portland Cement Association
www.reddi-wall.com	Reddi-wall
www.southernpine.com	Southern Pine Council
www.spacejoist.com	SpaceJoist (open web trusses)
www.ssina.com	Specialty Steel Industry of North America
www.steeljoist.org	Steel Joist Institute
www.housingzone.com	Sustainable Buildings Industry Council
www.timberpeg.com	Timberpeg (post and beam homes)
www.trimjoist.com	TrimJoist Engineered Wood Products
www.trusjoist.com	TrusJoist MacMillan
www.usgbc.com	U.S. Green Building Council
www.wwpa.org	Western Wood Products Association
www.wbdg.org	Whole Building Design Guide

CHAPTER 20

Framing Methods Test

DIRECTIONS

Answer the following questions with short, complete statements. Type your answers using a word processor or neatly print your answers on lined paper.

1. Place your name, the chapter number, and the date at the top of the sheet.
2. Type the question number and provide the answer in the form of a statement that includes part of the question. You do not need to write out the entire question.

QUESTIONS

20.1. Explain two advantages of platform framing.
20.2. Sketch a section view showing platform construction methods for a one-level house.
20.3. Why is balloon framing not commonly used today?
20.4. List five methods that can be used to frame a residence.
20.5. Sketch and label a section showing a post-and-beam foundation system.
20.6. How is brick typically used in residential construction?
20.7. What factor dictates the reinforcing in masonry walls?
20.8. Why would a post-and-beam construction roof be used with platform construction walls?
20.9. What would be the R value for a wood wall 4'' wide?
20.10. What is the typical spacing of posts and beams in residential construction?
20.11. List three qualities that make steel framing a popular residential method.
20.12. List four classifications of CMUs.
20.13. Sketch two methods of framing a header with advance framing technology. Label each major component.
20.14. What is a fire cut, and how is it used?
20.15. How is moisture removed from a masonry cavity wall?
20.16. What are two materials typically used to reinforce concrete block construction?
20.17. What is the maximum height of studs allowed by the IRC without special framing conditions?
20.18. What do the letters AFT mean in relationship to framing?
20.19. What makes steel framing such an earth-friendly material?
20.20. Explain the difference between traditional corner framing and a corner built using AFT.
20.21. Why are engineered lumber products considered earth-friendly?
20.22. List the various parts of a tree that can be used for engineered lumber products.
20.23. Explain the term *SIPS* as it relates to framing.
20.24. Explain the term *VOC* and explain how it relates to construction.
20.25. Use the Internet to research major environmental concerns in the region where you live.

21 CHAPTER

Structural Components

As with every other phase of drafting, construction drawings have their own terminology. The terms used in this chapter are basic for structural components of residential construction. These terms refer to floor, wall, and roof components.

FLOOR CONSTRUCTION

Two common methods of framing the floor system are with conventional joist and post-and-beam. In some areas of the country it is common to use post-and-beam framing for the lower floor when a basement is not used and conventional framing for the upper floor. The project designer chooses the floor system or systems to be used. The size of the framing crew and the shape of the ground at the job site are the two main factors in the choice of framing methods. ***Floor joists*** are often used when the slope of the site or other design considerations dictate minimal pier placement for floor support. A post-and-beam system requires more concrete but eliminates the need for floor joist.

Conventional Floor Framing

Conventional or stick framing involves the use of 2'' (50 mm) wide framing members placed one at a time in a repetitive manner. Basic terms in this system are ***mudsill, floor joist, girder,*** and ***rim joist.*** Each can be seen in Figure 21.1 and throughout this chapter.

Floor joists can range in size from 2 × 6 (50 × 150) through 2 × 14 (50 × 350) and may be spaced at 12, 16, or 24'' (300, 400,

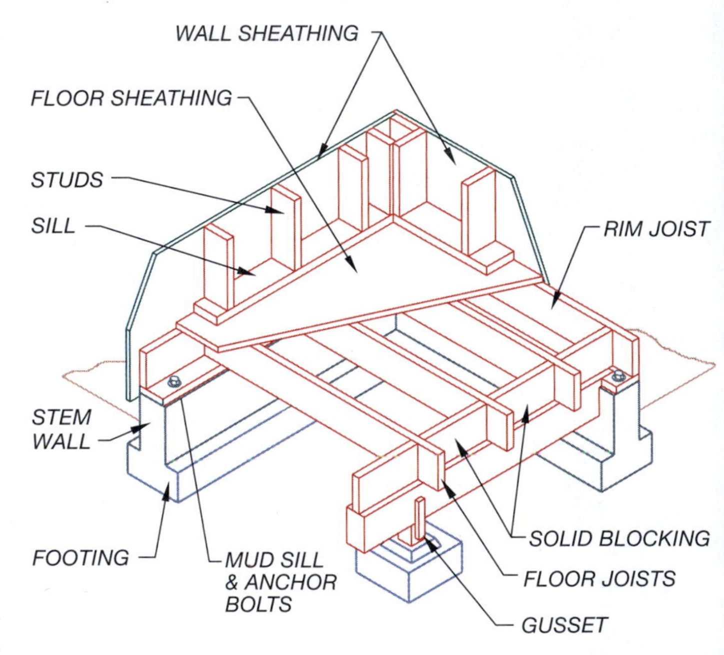

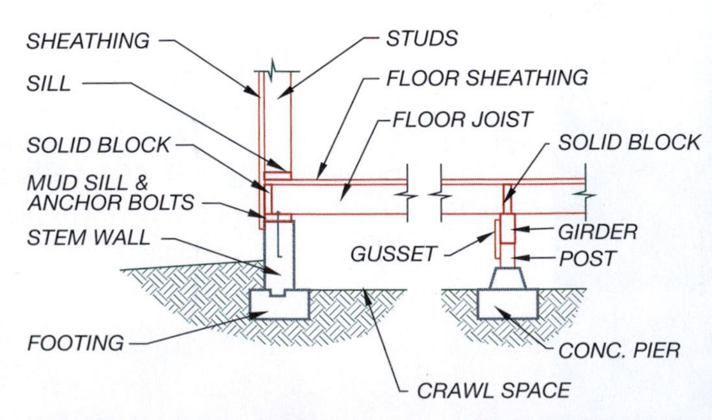

FIGURE 21.1 ■ *Conventional, or stick floor framing at the concrete perimeter wall.*

or 600 mm) o.c., depending on the load, span, and size of joists to be used. A spacing of 16'' (400 mm) is most common. See Chapter 22 for an explanation of sizing joists. Because of the decreasing supply and escalating price of sawn lumber, several alternatives to floor joists have been developed. Four common substitutes are open web trusses, I-joists, laminated veneer lumber, and steel joists.

Mudsill

The mudsill, or base plate, is the first of the structural lumber used in framing the home. The mudsill is the plate that rests on the masonry foundation and provides a base for all framing. Because of the moisture content of concrete and soil, the mudsill is required to be pressure-treated or made of foundation-grade redwood or cedar. A 2 × 6 (50 × 150) mudsill is usually set along the entire perimeter of the foundation and attached to the foundation with anchor bolts. Anchor bolts are required by the IRC to be a minimum of 1/2 × 10'' (13 × 250 mm). Bolts 5/8 and 3/4'' (16 and 19 mm) in diameter are also common. Bolt spacing of 6'-0'' (1800 mm) is the maximum allowed by code for 1/2'' (13 mm) bolts and is common for most single-level residential homes. Areas of a structure that are subject to lateral loads or uplifting, multilevel homes, and retaining walls will require bolts at a much closer spacing based on the engineer's load calculations. Section 7 explores the use of anchor bolts further.

Girders

With the mudsills in place, the girders can be set to support the floor joists. A girder is used to support the floor joists as they span across the foundation. Girders are usually sawn lumber, 4 or 6'' (100 or 150 mm) wide, with the depth determined by the load to be supported and the span to be covered. Sawn members 2'' (50 mm) wide can be joined together to form a girder. If only two members are required to support the load, they can be nailed, according to the nailing schedule provided in the IRC (see Figure 1.21). If more than three members must be joined to support a load, the IRC requires them to be bolted together. Another form of built-up beam is a ***flitch beam.*** Chapter 23 introduces formulas that designers use to determine beam sizes.

Laminated Girders

In areas such as basements, where a large open space is desirable, laminated (glu-lam) beams are often used. They are made of sawn lumber that is glued together under pressure to form a beam stronger than its sawn lumber counterpart. Beam widths of 3 1/8, 5 1/8, and 6 3/4'' (80, 130, and 170 mm) are typical in residential construction. Although much larger sizes are available, depths typically range from 9 through 16 1/2'' (225 through 412 mm) at 1 1/2'' (38 mm) intervals (see Figure 21.2). As the span of the beam increases, a camber is built in to help resist the tendency of the beam to sag when a load is applied. Glu-lams are often used where the beam will be left exposed because they do not drip sap and do not twist or crack, as a sawn beam does as it dries.

Engineered Wood Girders

Engineered wood girders and beams are common throughout residential construction. Unlike glu-lam beams, which are laminated from sawn lumber, parallel-strand lumber (PSL) is laminated from veneer strips of fir and southern pine, which are coated with resin and then compressed and heated. Typical widths are 3 1/2, 5 1/4, and 7'' (90, 130, and 180 mm). Depths range from 7 through 18'' (180 through 460 mm). PSL beams have no crown or camber.

Laminated veneer lumber (LVL) is made from ultrasonically graded Douglas fir veneers, which are lami-

FIGURE 21.2 ■ *Glu-lam beams span longer distances than sawn lumber and eliminate problems with twisting, shrinking, and splitting.* Courtesy Georgia Pacific Corporation.

FIGURE 21.3 ■ *Laminated veneer lumber beams (LVL, on left) made from ultrasonically graded Douglas fir veneers are a popular alternative to glu-lam beams (right) supporting engineered joists.* Courtesy Katja Poschwatta.

nated with all grains parallel to each other with exterior-grade adhesives under heat and pressure (see Figure 21.3). LVL beams come in widths of 1 3/4'' (45 mm) and 3 1/2'' (90 mm) and depths ranging from 5 1/2 through 18'' (140 through 460 mm). LVL beams provide performance and durability superior to that of other engineered products and often offer the smallest and lightest wood girder solution.

FIGURE 21.4 ■ *Steel girders allow greater spans with fewer supporting columns compared to a wood framed floor system.* Courtesy Lisa Echols,

Steel Girders

Steel girders such as those in Figure 21.4 are often used where foundation supports must be kept to a minimum; they offer a tremendous advantage over wood for total load that can be supported on long spans. Steel beams are often the only type of beam that can support a specified load and still be hidden within the depth of the floor framing. They also offer the advantage of no expansion or shrinkage due to moisture content. Steel beams are named for their shape, depth, and weight. A steel beam with the designation W16 × 19 would represent a wide-flange steel beam with the shape of an I. The 16 represents the approximate depth of the beam in inches, and the 19 represents the approximate weight of the beam in pounds per linear foot. Floor joists are usually set on top of the girder, as shown in Figure 21.4, but they also may be hung from the girder with joist hangers.

Girder Supports

Posts are used to support the girders. As a general rule of thumb, a 4 × 4 (100 × 100) post is used below a girder 4'' (100 mm) wide, and a post 6'' (150 mm) wide is used below a girder 6'' (150 mm) wide. Sizes can vary, depending on the load and the height of the post. LVL posts ranging in size from 3 1/2 through 7'' (90 through 180 mm) are also used for their ability to support loads. A 1 1/2'' (38 mm) minimum bearing surface is required to support a girder resting on a wood or steel support; a 3'' (75 mm) bearing surface is required if the girder is resting on concrete or masonry. Steel columns may be used in place of a wooden post, depending on the load to be transferred to the foundation. Because a wooden post will draw moisture out of the concrete foundation, it must rest on 55-lb felt or an asphalt roofing shingle. If the post is subject to ***uplift*** or lateral forces, a metal post base or strap may be specified by the engineer to attach the post firmly to the concrete. Figure 21.5 shows the steel connectors used to keep a post from lifting off the foundation. Forces causing uplift are discussed further in Chapter 22.

FIGURE 21.5 ■ *Steel connectors are often used to resist stress from lateral and uplift by providing a connection between the residential frame and the foundation.* Courtesy Benigno Molina-Manriquez.

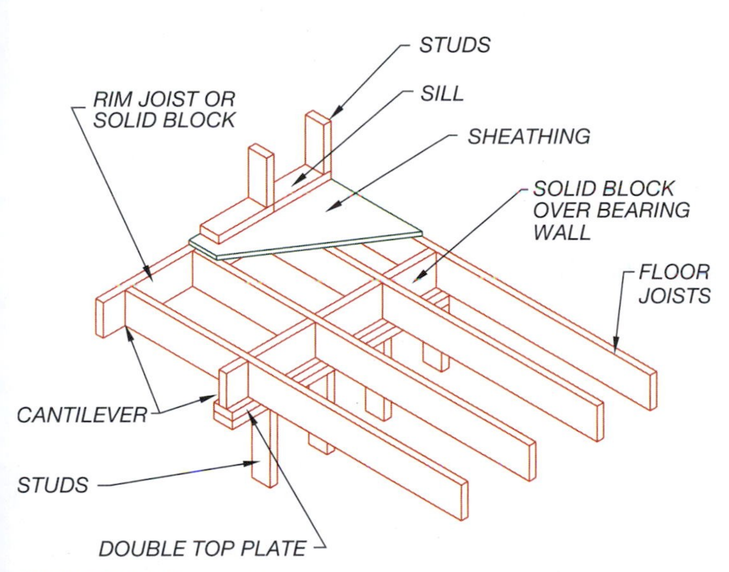

FIGURE 21.6 ■ *A floor joist or beam that extends past its supporting member is cantilevered.*

FIGURE 21.7a ■ *I-joists with LVL flanges and OSB webs.* Courtesy W. Lee Roland, Builder (rolandbuilder.com). Brent W. Roland, photographer.

FIGURE 21.7b ■ *Engineered lumber is a common material for residential floor joists. The web (the center of the joist) can be made from oriented strand board (OSB), laminated veneer lumber (LVL), or plywood. This floor is framed using OSB I-joists supported by metal hangers and a PSL girder.* Courtesy Lisa Echols.

Floor Joists

Once the framing crew has set the support system, the floor joists can be set in place. Floor joists are the structural members used to support the subfloor, or rough floor. Floor joists usually span between the foundation and a girder, but, as shown in Figure 21.6, a joist may ***cantilever*** past its support.

Floor joists that resemble an I (see Figure 21.7a) are a high-strength, lightweight, cost-efficient alternative to sawn lumber. I-joists form a uniform size, have no crown, and do not shrink. They come in depths of 9 1/2, 11 7/8, 14, and 18'' (240, 300, 360, and 460 mm). I-joists are able to span greater distances than comparable-sized sawn joists and are suitable for spans up to 40' (12 000 mm) for residential uses. Figure 21.7b shows an example of a solid web truss. Webs can be made from plywood, oriented strand board (OSB), or laminated veneer lumber (LVL). OSB joists like those in Figure 21.7 are made from wood fibers arranged in a precisely controlled pattern. The fibers are coated with resin and then compressed and cured under heat. Holes can be placed in the web to allow for HVAC ducts and electrical requirements based on the manufacturer's specifications. Joists made of laminated veneer lumber (LVL) are made using the same methods used to make LVL beams. LVL joists are 1 3/4'' (45 mm) wide and range in depths of 5 1/2 through 18'' (140 through 460 mm).

Figure 21.8 shows typical floor construction using engineered floor joists. Depending on the loads to be supported, blocks may need to be placed beside each joist to provide additional support. Figure 21.9 shows another common method of supporting engineered floor joists. The requirements of the manufacturer will need to be verified prior to completing any joist details. Most joist manufacturers supply standard construction details similar to Figure 21.10 that can be modified and inserted into the construction drawings.

Open-Web Floor Trusses

Open-web floor trusses are a common alternative to using 2 × sawn lumber for floor joists. Open-web trusses are typically spaced at 24'' (600 mm) o.c. for residential floors. Open-web trusses are typically available for spans up to 38' (11 590 mm) for residential floors. Figure 21.11 shows open-web floor trusses used to frame a floor system. The horizontal members of the truss are called top and bottom chords respectively and are typically made from 1.5 × 3'' (40 × 75 mm) lumber laid flat. The diagonal members, called webs, are made of wood or tubular steel approximately 1'' (25 mm) in diameter. When drawn in section or details, the webs can be represented with a bold centerline drawn at a 45° angle. The exact angle is determined by the manufacturer and is unimportant to the detail. If sections are drawn showing floor trusses, it is important to work with the manufacturer's details to find ex-

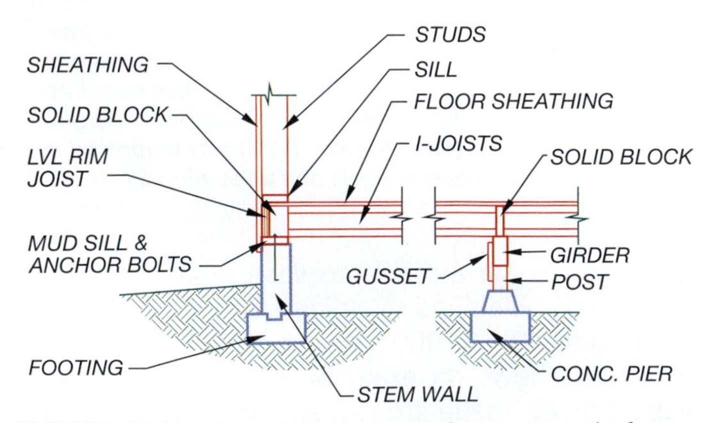

FIGURE 21.8 ■ *Floor framing using the western platform system and engineered lumber. Depending on the loads to be supported, blocks may need to be placed beside each joist to provide additional support.*

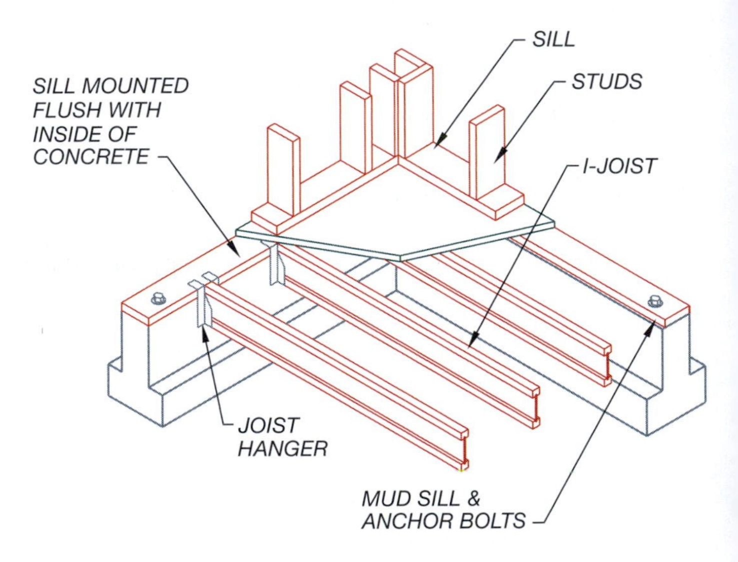

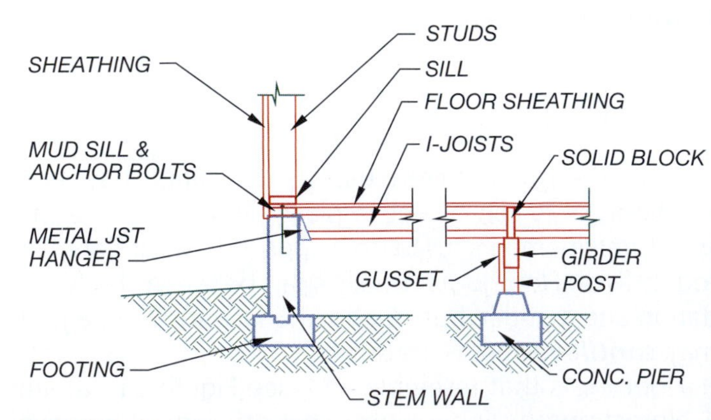

FIGURE 21.9 ■ *Engineered and sawn floor joists can be supported using metal hangers, allowing the floor level to be closer to the finished grade.*

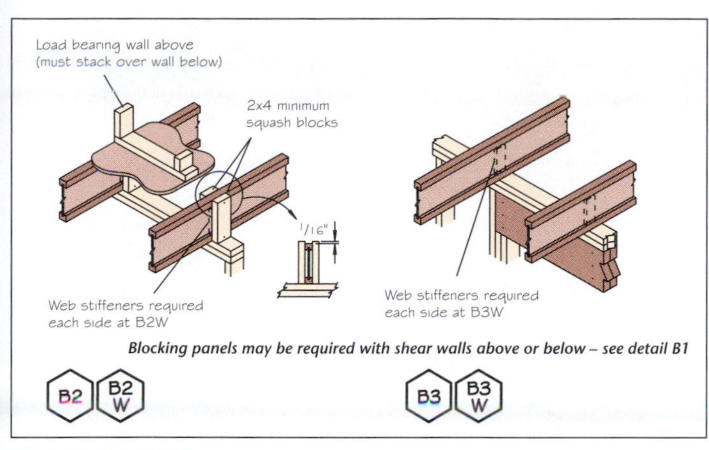

FIGURE 21.10 ■ *Most manufacturers provide details that explain joist connections which can be added to the construction documents.* Courtesy Trus Joist, A Weyerhaeuser Business.

FIGURE 21.11 ■ *Open-web floor joists can be used to span large distances for residential designs.* Courtesy Ray Wallace Southern Pine Council.

act sizes and truss depth. Most truss manufacturers supply copies of common truss connections. Figure 21.12 shows an example of a floor truss detail supplied by the manufacturer.

Steel Floor Joists

When steel joists are used to support the floor, a 6'' × 6'' × 54-mil L clip angle is bolted to the foundation to support the track which will support the floor joist, as shown in Figure 21.13. The IRC requires a minimum of 1/2'' bolts and clip angle at 6'-0'' o.c. to be used to attach the angle to the concrete. Eight #8 screws are required to attach the angle to the floor joist track. Floor joists are then bolted to the track using two #8 screws at each joist. Additional fasteners should be used if winds exceed 90 mph. Figure 21.14 shows the use of steel joists to frame a residential floor.

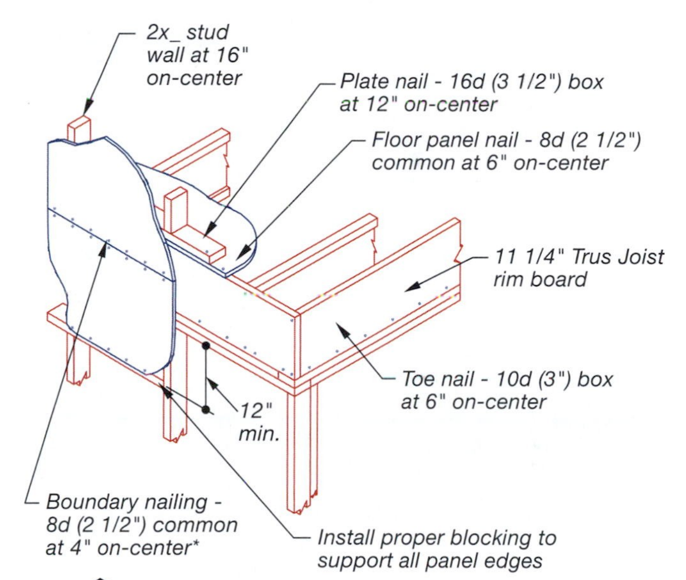

FIGURE 21.12 ■ *The intersection of the solid-web floor truss to the support members must be included in the construction documents. This manufacturer provided this detail for insertion. The text font can be altered to match other details in the project.* Courtesy Trus Joist, A Weyerhaeuser Business.

Floor Blocking

Because of the height-to-depth proportions of a joist, it will tend to roll over onto its side as loads are applied. To resist this tendency, a rim joist or blocking is used to support the joist. A rim joist is aligned with the outer face of the foundation and mudsill. Some framing crews set a rim joist around the entire perimeter and then end-nail floor joists that are perpendicular to the rim joist. An alternative to the rim joist is to use solid blocking at the sill between floor joists. Figure 21.15 shows the difference between a rim joist and solid blocking at an exterior wall.

Blocking is used in the floor system at the end of the floor joists, and at their bearing points. Blocking is also used at the center of the joist span, as shown in Figure 21.16, if joists spans are longer than 10' (3000 mm). The block at the center span helps to transfer ***lateral*** forces from one joist to another and then to the foundation system. These blocks help keep the entire floor system rigid and are used in place of the rim joist. Blocking can be used to provide added support to the floor

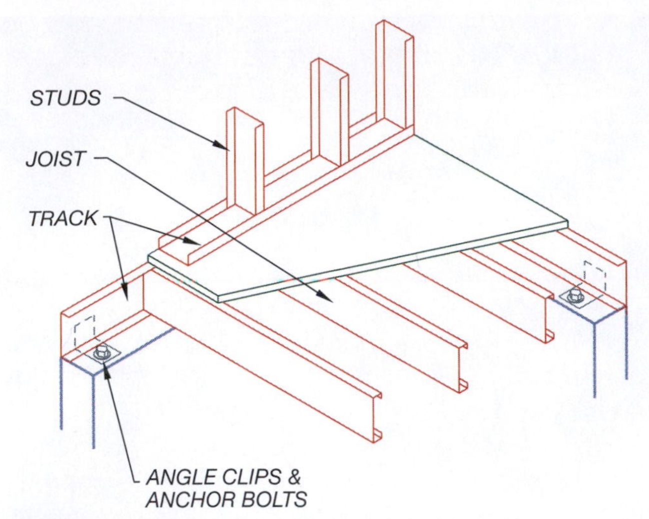

FIGURE 21.13 ■ *Floor construction using steel floor joists.*

FIGURE 21.14 ■ *Steel joists used to support the plywood decking for a lightweight concrete floor slab.* Courtesy Carol Ventura.

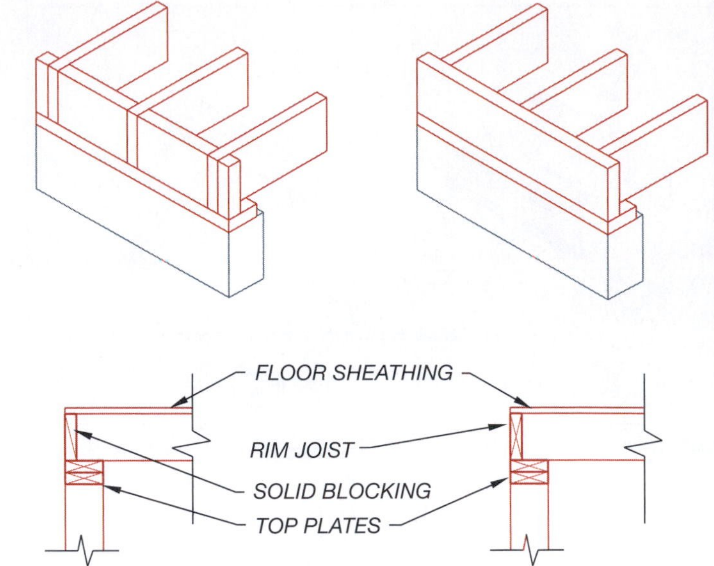

FIGURE 21.15 ■ *Floor joist blocking is placed at the end of each span and at the center. At the mid span of each joist, solid or cross-blocking may be used. At the ends of the floor joists, solid blocking or a continuous rim joist may be used to provide stability.*

FIGURE 21.16 ■ *When the joists rest on a girder, solid blocking is placed between floor joists to help resist lateral loads and to transfer loads from the floor into the girder.* Courtesy Jordan Jefferis.

sheathing when lateral loads are transferred from the walls, through the floor, and into the foundation. Blocking is also used to reduce the spread of fire and smoke through the floor system.

Floor Sheathing

Floor sheathing is installed over the floor joists to form the subfloor or rough floor. The sheathing provides a surface for the base plate of the wall to rest on (see Figure 21.17). Plywood and OSB are the most common materials used for floor sheathing. Depending on the spacing of the joists, plywood with a thickness of 15/32,

19/32, or 23/32'' (12, 15, or 18 mm) with an APA grade of STURD-I-FLOOR EXP 1 or 2, EXT, STRUCT I-EXP-1, or STRUCT 1-EXT is used for sheathing. EXT represents exterior grade, STRUCT represents structural, and EXP represents exposure. Plywood also is printed with a number to represent the span rating, which represents the maximum spacing from center to center of supports. The span rating is listed as two numbers, such as 32/16, separated by a slash.

- The first number represents the maximum recommended spacing of supporters if the panel is used for roof sheathing and the long dimension of the sheathing is placed across three or more supports.
- The second number represents the maximum recommended spacing of supports if the panel is used for floor sheathing and the long dimension of the sheathing is placed across three or more supports.

Plywood is usually laid so that the face grain on each surface is perpendicular to the floor joists. This provides a rigid floor system without having to block the edges of the plywood. Floor sheathing that will support ceramic tile is typically 32/16'' APA span rated 15/32'' (12 mm) or span rated 40/20'' and 19/32'' (15 mm) thick.

Oriented strand board (OSB) is also a popular alternative to traditional plywood subfloors. OSB is made from layers of small strips of wood that have been saturated with a binder and then compressed under heat and pressure. The exterior layers of strips are parallel to the length of the panel, and the core layer is laid in a random pattern.

FIGURE 21.17 ■ ***T&G floor sheathing is screwed and glued to the supporting floor joists to provide a quiet floor system.*** Courtesy APA–The Engineered Wood Association.

Underlayment

Once the subfloor has been installed, ***underlayment*** for the finished flooring is installed. The underlayment is not installed until the walls, windows, and roof are in place, making the house weather-resistant. The underlayment provides a smooth impact-resistant surface on which to install the finished flooring. Underlayment is usually 3/8 or 1/2'' (9 or 13 mm) APA underlayment GROUP 1, EXPOSURE 1 plywood, hardboard, or wafer board. Hardboard is typically referred to as medium- or high-density fiberboard (MDF or HDF) and is made from wood particles of various sizes that are bonded together with a synthetic resin under heat and pressure. The underlayment may be omitted if the holes in the plywood are filled. APA STURD-I-FLOOR rated plywood 19/32'' through 1 3/32'' (15 through 28 mm) thick can also be used to eliminate the underlayment.

Post-and-Beam Construction

Terms to be familiar with in working with post-and-beam floor systems include ***girder, post, decking,*** and ***finished floor.*** Each can be seen in Figure 21.18. Notice that there are no floor joists with this system.

A mudsill is installed with post-and-beam construction, just as with platform construction. Once set, the girders are also placed so that the top of the girder is flush with the top of the mudsill. With post-and-beam construction, the girder rather than floor joists supports the floor decking. Girders are usually 4 × 6 (100 × 150) beams spaced at 48'' (1200 mm) o.c., but the size and distance will vary depending on the loads to be supported. As with conventional methods, posts are used to support the girders. Typically 4 × 4 (100 × 100) is the minimum size used for a post, with a 4 × 6 (100 × 150) post used to support joints in the girders. With the support system in place, the floor system can be installed. Figures 21.19 shows the components of a post-and-beam floor system.

Decking is the material laid over the girders to form the subfloor. Typically decking is 2 × 6 or 2 × 8 (50 × 150 or 50 × 200) tongue-and-groove (T&G) boards similar to those shown in Figure 21.20 or 1 3/32'' (28 mm) APA

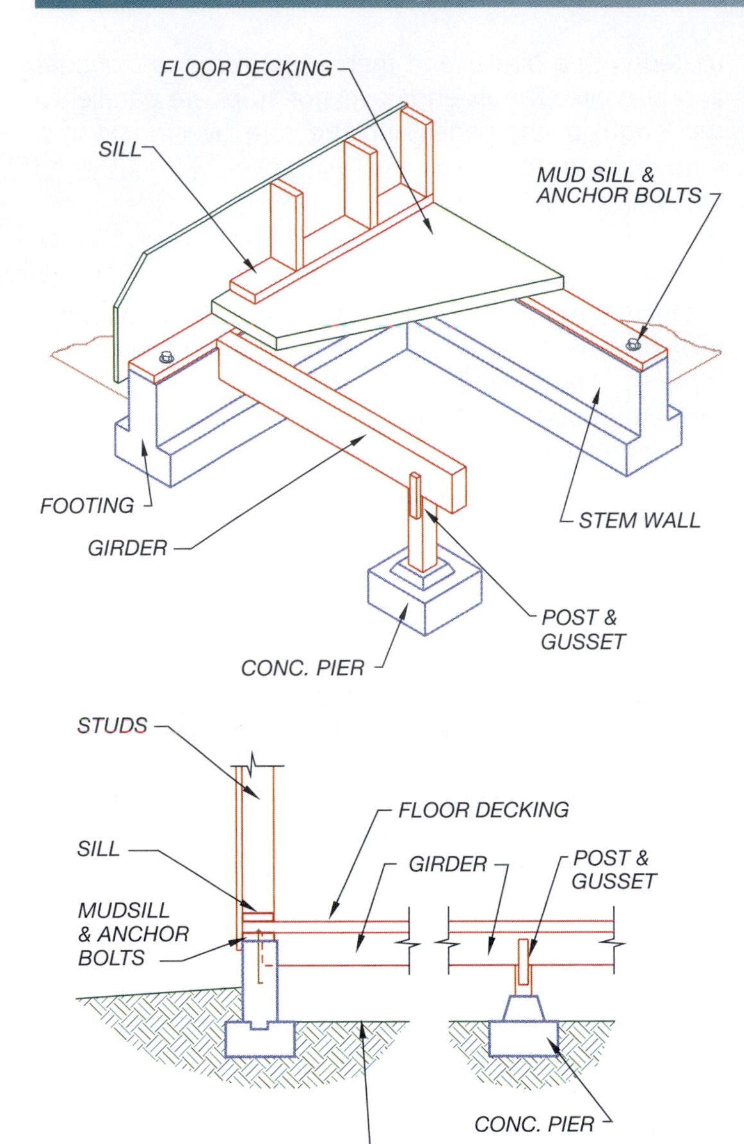

FIGURE 21.18 ■ *Components of post-and-beam construction.*

FIGURE 21.19 ■ *Floors built with a post-and-beam system require no floor joists to support the floor sheathing. Girders are placed at 4'-0'' o.c. with support posts placed at 8'-0'' o.c. The posts are attached to the girders with an OSB gusset. In some areas the posts must be attached to the concrete with steel rebar.* Courtesy Benigno Molina-Manriquez.

rated 2-4-1 STURD-I-FLOOR EXP-1 plywood T&G floor sheathing. The decking is usually finished similarly to conventional decking with a hardboard overlay.

FIGURE 21.20 ■ *Decking for a post-and-beam floor system is typically 2 × 6 tongue-and-grove material supported by the girders and mudsill. The rough plumbing has been completed prior to placing the decking, with holes provided in the decking for fresh- and wastewater lines.* Courtesy Zachary Jefferis.

FRAMED WALL CONSTRUCTION

Two types of walls, bearing and nonbearing, must be understood as you explore construction methods. A ***bearing wall*** supports not only itself but also the weight of the roof or other floors constructed above it. A bearing wall requires some type of support under it at the foundation or lower floor level in the form of a girder or another bearing wall. Nonbearing walls are sometimes called partitions. A ***nonbearing wall*** serves no structural purpose. It is a partition used to divide rooms and could be removed without causing damage to the building. Bearing and nonbearing walls are shown in Figure 21.21. If the roof system is framed with trusses, any interior walls placed between exterior walls are nonbearing walls. Major components of wall construction include the framing members and sheathing.

Framing Members

Bearing and nonbearing walls made of wood or engineered lumber are both constructed using a ***sole plate, studs,*** and a ***top plate.*** Each can be seen in Figure 21.22. The sole, or bottom plate, is used to help disperse the loads from the wall studs to the floor system. The sole plate also holds the studs in position as the wall is being tilted into place. Studs are the vertical framing members used to transfer loads from the top of the wall to the floor system. Typically 2 × 4' (50 × 100) studs are spaced at 16'' (400 mm) o.c. and provide a nailing surface for the wall sheathing on the exterior side and the sheetrock on the interior side. Studs 2 × 6 (50 × 150) are often substituted for 2 × 4 (50 × 100) studs to provide added resistance for lateral loads and a wider area to install insulation. Western platform construction allows wall components to be assembled on the platform provided by the floor. Once the wall members are assembled, exterior sheathing can be applied. For small walls, windows can also be installed before the wall is lifted into place. Figure 21.23 shows the construction of an exterior wall.

Engineered studs are available as an alternative to sawn lumber. Engineered studs are made from short sections of stud-grade lumber that have had the knots and splits have been removed. Sections of wood are joined together with 5/8'' (16 mm) finger joints. The top plate is located on top of the studs and is used to hold the wall together. The top plate also provides a bearing surface for the floor joists from an upper level

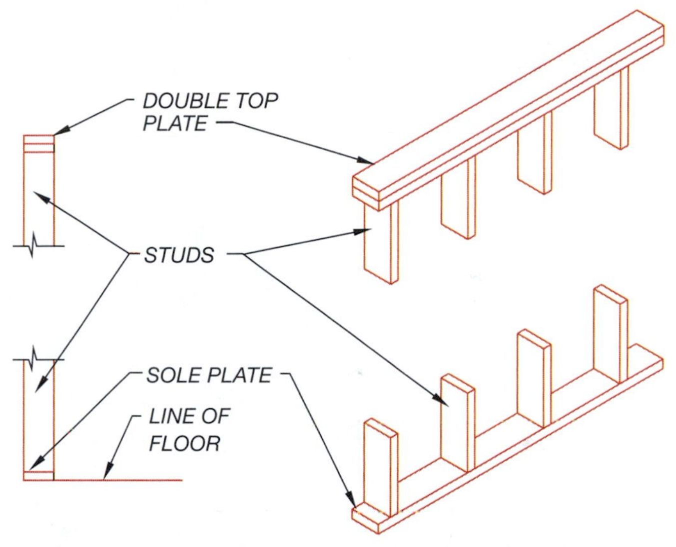

FIGURE 21.22 ■ *Standard wall construction uses a double top plate on the top of the wall, a sole plate at the bottom of the wall, and studs.*

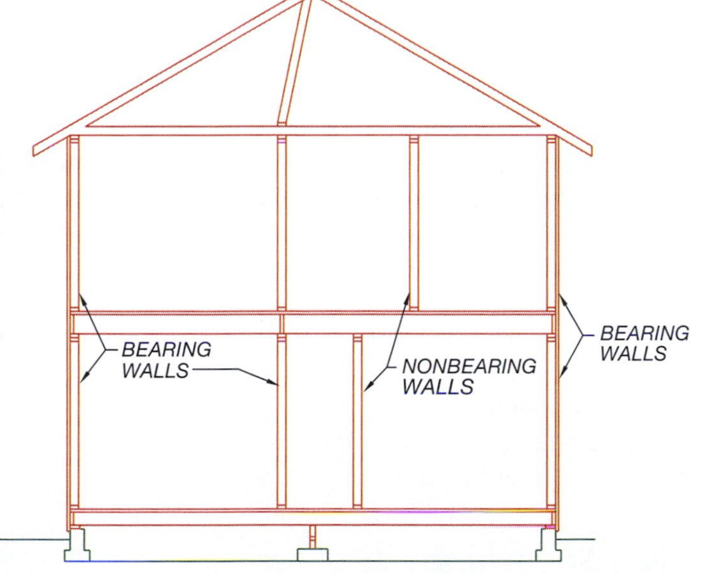

FIGURE 21.21 ■ *Bearing and nonbearing walls. A bearing wall supports its own weight and the weight of floor and roof members. A nonbearing wall supports only its own weight. The IRC allows ceiling weight to be supported on a wall and still be considered a nonbearing wall.*

FIGURE 21.23a ■ *Western platform construction allows wall components to be assembled on the platform provided by the floor.* Courtesy Megan Jefferis.

FIGURE 21.23b ■ *Once assembled, walls can be lifted into place.* Courtesy Megan Jefferis.

FIGURE 21.24 ■ *Steel studs and joists provide the support for the upper floor.* Courtesy Lisa Echols.

or for the roof members. Two top plates are required on bearing walls, and each must lap the other a minimum of 48'' (1200 mm). This lap distance provides a continuous member on top of the wall to keep the studs from separating. An alternative to the double top plate is to use one plate with a steel strap at each joint in the plate.

When steel studs are used, a track is placed above and below the studs and fastened with a minimum of two #8 screws to each stud. The size of the screws will vary depending on the gauge of the track and the thickness of the steel studs. Figure 21.24 shows an example of a steel-framed wall.

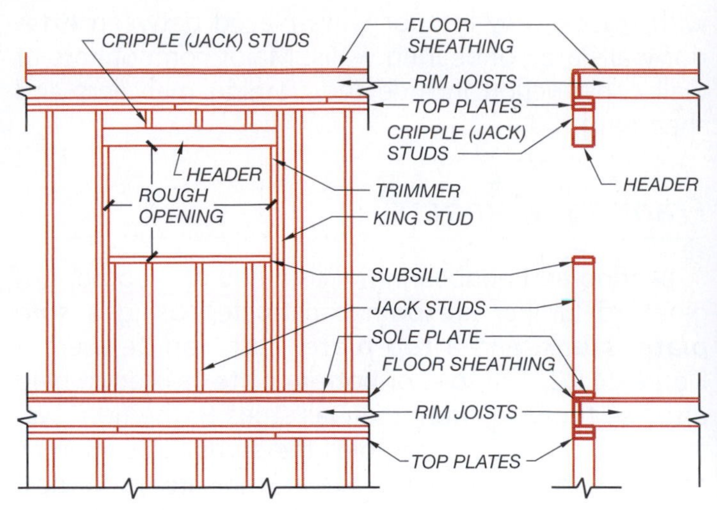

FIGURE 21.25 ■ *Common construction components at a wall opening.*

Door and Window Framing

In addition to the major wall components, several other terms are used to describe the members used to frame an opening in a wall. These members include headers, subsill, trimmers, king studs, and jack studs. Each can be seen in Figure 21.25.

A ***header*** is used over an opening such as a door or window when one or more studs must be omitted. A header is used to support the weight that the missing studs would have carried. A ***trimmer*** supports each end of the header. Depending on the weight the header is supporting, double trimmers may be required. The trimmers also provide a nailing surface for the window and the interior and exterior finishing materials. A ***king stud*** is placed beside each trimmer and extends from the sill to the top plates. It provides support for the trimmers so that the weight imposed from the header can go only downward, not sideways. Between the trimmers is a ***subsill*** located on the bottom side of a window opening. It provides a nailing surface for the window and the interior and exterior finishing materials. ***Jack studs,*** or cripples, are studs that are not full height. They are placed between the subsill and the sole plate and between a header and the top plates. Figure 21.26 shows each component of a window opening.

Wall Sheathing

OSB and plywood sheathing are primarily used as insulators against the weather, to provide resistance to lateral stress, and also as backing for the exterior siding. Sheathing may be considered optional, depending on your area of the country. When sheathing is used on ex-

FIGURE 21.26 ■ *Framing members for a wall opening include the header, king stud, trimmer, sill, and jack studs.* Courtesy Floyd Miller.

FIGURE 21.27 ■ *The siding material can be placed directly over the vapor barrier and studs in what is referred to as single-wall construction. Wire mesh is being installed over the building paper in preparation for the base coat of exterior stucco.* Courtesy Matthew Jefferis.

terior walls, it provides what is called ***double-wall construction.*** In ***single-wall construction,*** wall sheathing is not used, and the siding is attached over a vapor barrier placed over the studs, as in Figure 21.27. The cost of the home and its location will have a great influence on whether wall sheathing is to be used. In areas where double-wall construction is required for weather or structural reasons, 1/2'' (13 mm) OSB is used for wall sheathing below the finished siding. For exterior walls that will be used to resist seismic or wind loads, plywood underlayment may be required on the exterior walls.

Structural Sheathing

The design of the home may require the use of plywood sheathing for its ability to resist the tendency of a wall to twist or rack. Racking, which can be caused by wind or seismic forces, will try to turn a rectangular wall into a parallelogram (see Figure 21.28). The IRC refers to plywood used to resist these forces as a braced wall line. Panels designed by engineers to resist lateral loads are referred to as ***shear panels.*** Chapter 24 explores the use of braced wall lines and shear panels. Plywood sheathing should be APA-rated EXP 1 or 2, EXT, STRUCT 1, EXT 1, or STRUCT 1 EXT. For single-wall construction, an alternative to plywood shear panels is to use let-in braces. A notch is cut into the studs, and a 1 × 4' (25 × 100 mm) is laid flat in this notch at a 45° angle to the studs. The let-in brace forms a triangle between the brace, the studs, and floor system. If plywood siding rated APA Sturd-I-Wall is used, no underlayment or let-in braces are required.

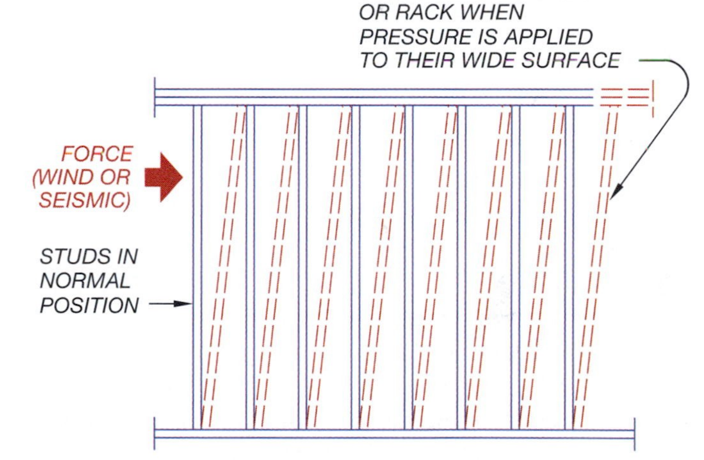

FIGURE 21.28 ■ *Wall racking occurs when wind or seismic forces push the studs out of their normal position. Plywood panels can be used to resist these forces and keep the studs perpendicular to the floor. These plywood panels are called shear panels when designed by an engineer or braced wall panels when constructed to meet the prescriptive path of the IRC (see Chapter 24).*

Wall Blocking

Blocking is common in walls framed with balloon construction methods. Blocking is generally not required when walls are framed using western platform methods. Blocking for structural or fire reasons is required if a wall exceeds 10' (3000 mm) in height. Blocking is often installed for a nailing surface for mounting cabinets and plumbing fixtures. Blocking is sometimes used to provide extra strength in some seismic zones.

Exterior Wall Protection

Prior to installing the siding material, the home must be provided with a weather-resistant exterior wall envelope. This envelope can consist of a water vapor or air barrier and the use of caulking and weather stripping.

Water Barriers

A key goal of the building envelope is to keep water out of the structure. Leaking water rots wood, grows mold, corrodes steel, and lowers insulating R-values. The IRC requires that a water barrier be provided to help keep water from entering the living environment. Although many building programs focus on keeping water vapor out, most structural problems are caused by water intrusion. Water can enter the structure at any break in a wall or roof surface, such as window or skylight, butt joints in sidings, knots, and siding overlaps. Water is driven through these leakage points by wind, gravity, and capillary forces. During a storm, air pressure causes water to move up, down, and sideways, moving from areas of high pressure to areas of low pressure. The area directly behind a windblown wall surface is at a lower pressure than its exterior face. This pressure difference creates siphon points pulling water into the building. Once water is wicked into the wood sheathing or framing material, it can cause structural problems. The IRC requires the use of a minimum of 15# per sq ft (0.683 kg/m^2) of asphalt-saturated felt to resist water intrusion. Also referred to as 15# building paper, the felt must be installed in horizontal rows, with the upper layer lapped 2'' (50 mm) over the lower row. In addition to carefully covering the wall, care must be taken to ensure that water can't enter the envelope at the tops of doors or windows. Small strips of felt are generally provided along each edge of openings to provide overlapping layers of protection (see Figure 21.29). Providing metal ***flashing*** to direct water outward around the tops of openings provides added protection. Grade D Kraft paper instead of 15# felt can be used to protect the structural members behind stucco, brick, stone, and other porous veneers. Panel siding with shiplap joints, battens, and paper-backed stucco lath are not required to have building paper.

Moisture and Air Barriers

The IRC allows other approved water-resistant materials to be used to seal the exterior envelope. In place of the building paper, many designers use plastic house wraps that comply with ASTM standards. Plastic house wraps are engineered materials designed to keep out liquid water and prevent air ***infiltration*** while allowing water vapor to escape from inside of the home. Common wraps include Tyvek, Pinkwrap, and Typar, but many other brands are available. These materials have microscopic pores small enough to resist water and air molecules but large enough to let smaller moisture vapor molecules pass through. These house wraps are available in rolls up to 12' (3600 mm) wide, allowing for fewer seams than with building felt. Any seams required in the house wrap are taped, providing a seamless envelope around the home. Figure 21.30 shows the application of wood siding over a plastic house wrap. Equally important to specifying the underlayment to be used is to specify that the underside of the siding material is to be primed. Because water will get behind the siding, the interior side of the siding must be protected from water intrusion.

FIGURE 21.29 ■ *Small strips of felt are generally provided along each edge of openings to provide overlapping layers of moisture protection.* Courtesy Benigno Molina-Manriquez.

FIGURE 21.30 ■ *Wood siding can be installed over building paper or a house wrap. Because house wrap comes in larger sizes, fewer seams are required, allowing less opportunity for moisture, vapor, or air to infiltrate the structure.* Courtesy Dick Schmitke.

If masonry siding is to be installed, a 1'' (25 mm) air space must be specified between the masonry and the vapor barrier to allow moisture to drain to the bottom of the wall and out through weep holes placed in the masonry (see Figure 21.31). A ***weep hole*** is an opening left in the grout to allow water to escape from the wall cavity. The IRC requires weep holes to be at least 3/16'' (5 mm) in diameter at a maximum spacing of 33'' (825 mm) o.c. Holes must be placed immediately above any required flashing. In addition to specifying moisture protection methods for masonry, ties for attaching masonry veneer must also be specified. The IRC requires that No. 22 gauge by 7/8'' (22 mm) corrugated corrosion-resistant metal ties at 24'' (600 mm) o.c. vertically at each stud be provided. A maximum area of 2.67 sq ft (0.248 m^2) is to be supported by each tie. This area is reduced to 2 sq ft (0.186 m^2) in high wind and seismic risk areas. No. 9 strand wire with hoods embedded in the mortar joints can be used in place of wire straps. If cultured stone is used as a veneer, an alternative to providing the air space is to place the veneer in mortar placed over lath. Wire mesh installed over the vapor barrier is used to help bond the mortar to the framing system (see Figure 21.32).

FIGURE 21.31 ■ *Masonry veneer installed over the moisture barrier must have a 1'' (25 mm) minimum air space to allow water to drain away from the structure. No. 9 strand wire with hooks embedded in the mortar joints can be used in place of wire straps.* Courtesy Sam Griggs.

FIGURE 21.32 ■ *Stone veneer can be installed over grout with no air space. Wire mesh installed over the vapor barrier is used to bond the mortar to the framing system.* Courtesy T.J. Southard.

Exterior Caulking

In addition to the siding and building felt or wrap, caulking provides a third line of defense against water and air intrusion. The type and location of caulking and weather stripping is generally specified in the construction documents. The IRC requires that the following areas be filled with caulking:

- All joints, seams, and penetrations in the exterior shell

- Doors, window, and skylight assemblies and their respective frames
- Any other source of air infiltration

Because this final category for the building code is very general, most offices provide specific locations that must be caulked or weather-stripped. Common areas include:

- Joints between chimney and siding
- Joints between eaves and gable molding
- Joints between window sill and siding
- Joints between window drip cap and siding
- Joints between window sash and siding
- Joints between windows and masonry
- Joints between masonry or concrete parts (steps, porches, etc.) and main part of house
- Inside corners formed by siding

A list of common caulking notes can be found on the CD in the BLOCKS folder.

Interior Moisture Protection

A ***vapor barrier*** is a membrane that is placed on the warm side of the walls and ceilings between the drywall and the insulation. This barrier is used to prevent water vapor from inside the home from entering the insulated wall cavity, where it can condense and cause structural damage. The barrier can be provided by the insulation facing, paint, or primer applied to the drywall or a sheet of 4-mil polyethylene film applied to the wall (see Figure 21.33). If a film is used, the seams between sheets should be taped. Any penetrations for electrical fixtures should also be sealed.

In addition to the vapor barrier, interior caulking should be specified to reduce air infiltration. Common locations for interior caulking include:

- Where pipes enter the structure. Especially critical are any holes placed in the sole plate.
- Outlets and switch plates located in exterior walls. Gaskets for outlets and switch plates can help prevent air leakage through the walls as well as through caulking holes where wiring passes through the sill or top plates.
- Where vents exit the structure. Provide caulking around the vent, but also provide self-sealing ducts that restrict airflow when exhaust fans are not in use.
- Between the fireplace and exterior walls
- Between sheets of paneling

FIGURE 21.33 ■ *A 4-mil vapor barrier is placed over insulation to keep moisture from the interior from entering the wall cavity. Once the barrier is in place, 1/2" Sheet rock is attached to the structural frame using screws.*

- The interior side of door frames
- The interior side of window frames
- Baseboards (even if you have wall-to-wall carpeting)

Interior Finish

The final element of wall construction to be considered is the interior finish. For most homes the interior finish consists of ***gypsum board.*** Gypsum board is the generic name for a family of sheet products such as sheetrock, drywall, or wallboard that consist of a noncombustible core primarily of gypsum with paper surfacing. Gypsum wallboard is a type of gypsum board used for walls, ceilings, or partitions; it affords a surface suitable to receive decoration. Each name is used to describe a manufactured panel that is typically 1/2" (13 mm) thick made of gypsum plaster and encased in a thin cardboard. Panels that are 5/8" (16 mm) thick are recommended for ceiling use with trusses spaced at 24" (600 mm) o.c. The panels are nailed or screwed onto the framing and the joints are taped and covered with a joint compound. Figure 21.34a shows the application of joint compound and taping of joints. Figure 21.34b shows the joints and nail holes prepared for the final texture to be applied. The other types of sheetrock often found in a residence include green board, type X, and sound-deadening board.

- Green board is a water-resistant material used in areas with high moisture content, such as those near a shower, tub, or spa.

FIGURE 21.34a ■ *Once the sheetrock has been installed, joint compound is placed over screw heads and in all joints. Tape is applied in the "mud" to resist cracking.*

FIGURE 21.34b ■ *With all joints and screw heads filled and sanded, the finished texture can be applied.* Courtesy Michael Jefferis.

- Type X gypsum board 5/8'' (16 mm) thick is often used on the wall that separates the garage from the living areas. Municipalities vary on the need to use type X gypsum board. The IRC allows standard sheetrock to be used if it extends the full height of the separating wall or if the separating wall and ceiling are covered with sheetrock. Although the use of standard gypsum board is allowed to separate the garage from the home, type X offers superior protection. Type X gypsum board is also used at the bottom of stairs to protect any usable space below the stairs (see Chapter 30).

- Sound-deadening board is used in many custom homes to muffle mechanical and plumbing sounds near living and sleeping areas. Specially designed drywall composed of gypsum, elastic polymers, and sound-isolation layers (used to absorb sound) are often used in home theaters.

ROOF CONSTRUCTION

Roof framing includes both conventional and truss framing methods. Each has its own special terminology, but many terms that apply to both systems. These common terms are described first, followed by the terms for conventional and truss framing methods.

Basic Roof Terms

Roof terms common to conventional and trussed roofs are eave, cornice, eave blocking, fascia, ridge, sheathing, finishing roofing, flashing, and roof pitch dimensions.

The ***eave*** is the portion of the roof that extends beyond the walls. The ***cornice*** is the ornamental molding located at the top of an exterior wall just below a roof. Each term was introduced in Chapter 15. Common methods for constructing the eave are shown in Figure 21.35. Eave or ***bird blocking*** is a spacer block placed between the ***rafters*** or truss tails at the eave. This block keeps the spacing of the rafters or trusses uniform and keeps small animals from entering the attic. It also provides a cap to the exterior siding, as seen in Figure 21.36.

A ***fascia*** is a trim board placed at the end of the rafter or truss tails that are parallel to the building wall. It hides the truss or rafter tails and also provides a surface where the gutters may be mounted. The fascia can be made from either 1 × or 2 × (25 × or 50 ×) material, depending on the need to resist warping (see Figure 21.36). The fascia is typically 2'' (50 mm) deeper than the rafter or truss tails. On roofs with a gable end wall, the fascia is referred to as a ***barge rafter.*** A barge rafter is an inclined trim board that extends from the fascia to the ***ridge.*** Boards called outlookers or lookouts support the barge rafter. Figure 21.37 shows the use of outlookers to support the barge rafter. On some historic styles, decorative brackets are used to support the barge rafter. At the end of the barge rafter is the ridge. The ridge is the highest point of a roof and is formed by the intersection of the rafters or the ***top chords*** of a truss.

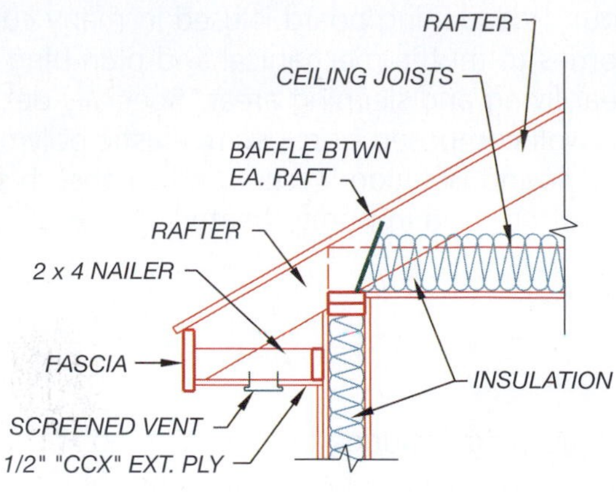

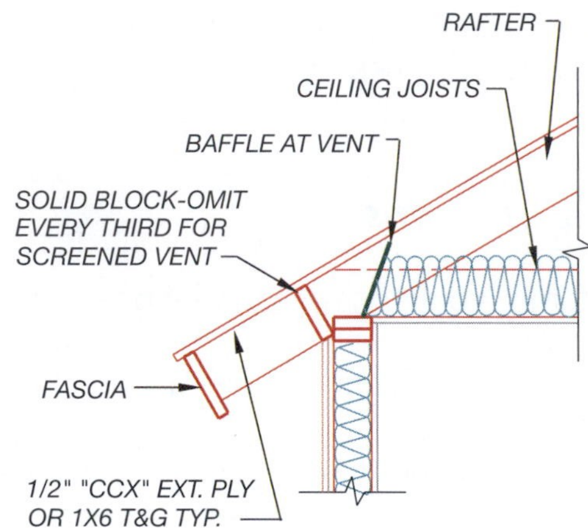

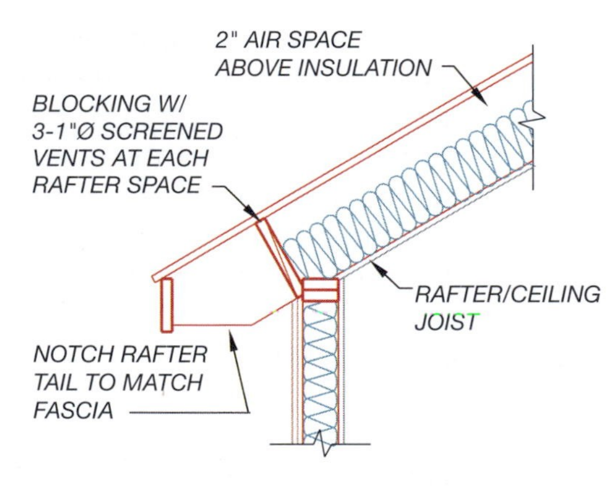

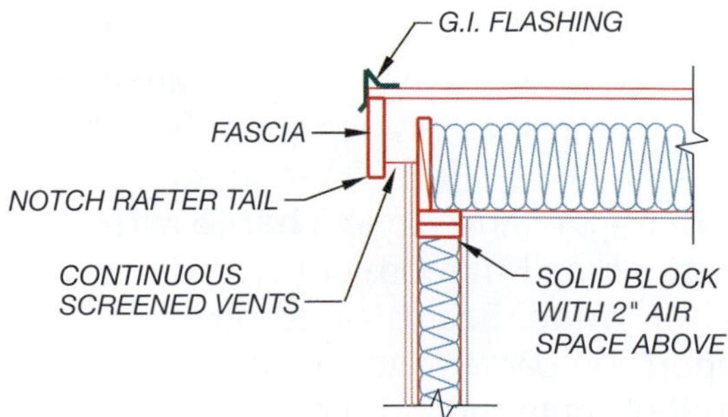

FIGURE 21.35 ■ *Typical methods for constructing eaves. Methods shown in each example are interchangeable. The rafter/ceiling joists could be enclosed similar to the top example, or the tails could be cut similar to the second example.*

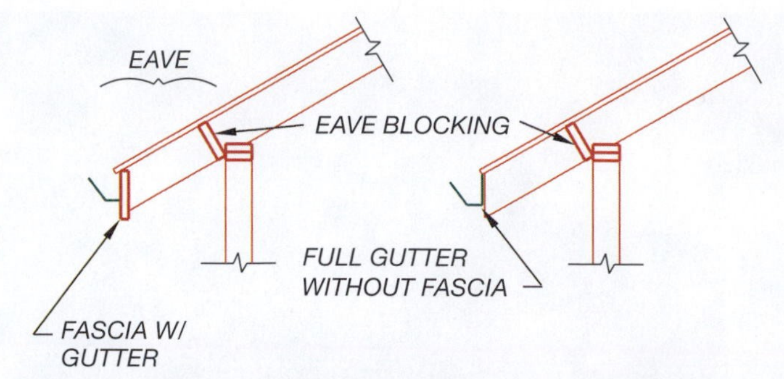

FIGURE 21.36 ■ *Key eave components include the bird blocking, fascia, gutter, and roof sheathing.*

FIGURE 21.37 ■ *A barge rafter is the outer most rafter, which extends from the fascia to the ridge. It can be supported by braces referred to as outlookers or a lookout. Some home styles use decorative brackets to support the barge rafter.* Courtesy David Jefferis.

Roof sheathing is similar to wall and floor sheathing in that it is used to cover the structural members. Roof sheathing may be either solid or skip. The area of the country and the finished roofing to be used will determine which type of sheathing is used. For solid sheathing, 1/2'' (13 mm) OSB or CDX plywood is generally used. CDX is the specification given by the Engineered Wood Association (APA) to designate standard grade plywood. It provides an economical, strong covering for the framing as well as an even base for installing the finished roofing. Common span ratings used for residential roofs are 24/16, 32/16, 40/20, and 48/24. Figure 21.38 shows plywood sheathing being applied to a conven-

FIGURE 21.38 ■ *Plywood with a span rating of 24/16 is typically placed over rafters to support the finish roofing material.* Courtesy Louisiana-Pacific Corporation.

FIGURE 21.39 ■ *Skip or spaced sheathing is used under shakes and tile roofs. In colder climates, skip sheathing is placed over OSB sheathing.* Courtesy Tom Santillanes.

tionally framed roof. Skip sheathing is used with either tile or shakes. In colder climates, the skip sheathing is installed over OSB or plywood sheathing. Typically 1 × 4s (25 × 100s) are laid perpendicular to the rafters with a 4'' (100 mm) space between pieces of sheathing (see Figure 21.39). A water-resistant sheathing must be used when the eaves are exposed to weather. This usually consists of plywood rated CCX or 1'' (25 mm) T&G decking. CCX is the specification for exterior-grade plywood. It is designated for exterior use because of the glue and the types of veneers used to make the panel.

The finished roofing is the weather protection system. Roofing might include built-up roofing, asphalt shingles, fiberglass shingles, cedar, tile, or metal panels (see Chapter 15). Flashing is generally 20 to 21 gauge metal used at wall and roof intersections to keep water out. Pitch, span, and overhang are dimensions needed to define the angle, or steepness, of the roof. Each is shown in Figure 21.40. Pitch, used to describe the slope of the roof, is the ratio between the horizontal ***run*** and the vertical ***rise*** of the roof. The run is the horizontal measurement from the outside edge of the wall to the centerline of the ridge. The rise is the vertical distance from the top of the wall to the highest point of the rafter being measured. Review Chapter 15 for a complete discussion of roof pitch. The ***span*** is the horizontal measurement between the inside edges of the supporting walls. Chapter 23 discusses spans further. The ***overhang*** is the horizontal measurement between the exterior face of the wall and the end of the rafter tail.

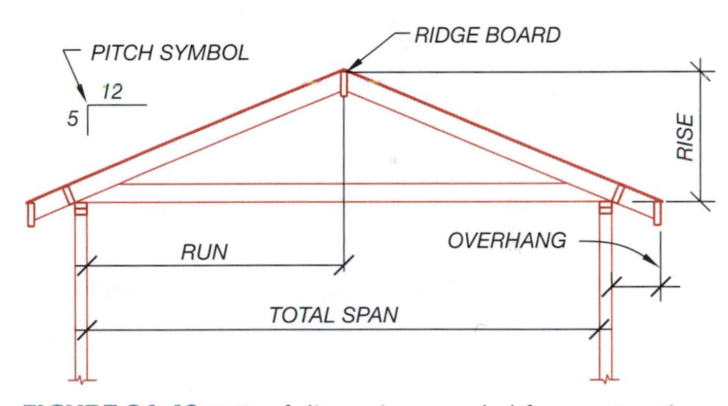

FIGURE 21.40 ■ *Roof dimensions needed for construction.*

Conventionally Framed Roofs

Conventional, or stick, framing methods involve the use of wood members placed in repetitive fashion. Stick framing involves the use of members such as a ridge board, rafter, and ceiling joists. The ***ridge board*** is the horizontal member at the ridge that runs perpendicular to the rafters. It is centered between the exterior walls when the pitch on each side is equal. The ridge board resists the downward thrust resulting from the force of gravity, which tends to push the rafters into a "V" shape between the walls. The ridge board

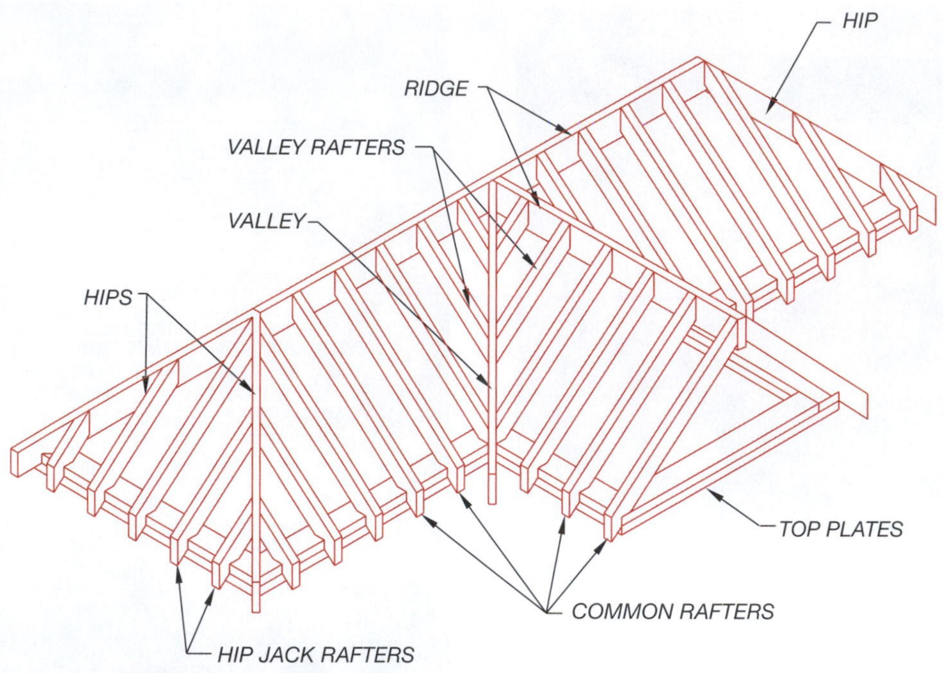

FIGURE 21.41 ■ *Common roof members in stick (conventional) construction.*

FIGURE 21.42 ■ *Rafters are used to span and support the roof loads from the ridge to the top plate.* Courtesy Ray Wallace, Southern Pine Council.

FIGURE 21.43 ■ *A bird's mouth is a notch cut into the rafter to increase the bearing surface.* Courtesy Pavel Adi Sandu.

does not support the rafters but is used to align the rafters so that their forces are pushing against each other.

Rafters are the sloping members used to support the roof sheathing and finished roofing. Rafters are typically spaced at 24'' (600 mm) o.c., but rafters spaced at 12 and 16'' (300 and 400 mm) centers are also used. Engineered rafters similar to those used for floor joists are also used for residential roof construction. No matter what the material, there are various kinds of rafters: common, hip, valley, and jack. Each is shown in Figure 21.41.

Rafters, Hips, and Valleys

Common rafters similar to those in Figure 21.42 are used to span and support the roof loads from the ridge to the top plate. Common rafters run perpendicular to both the ridge and the wall supporting them. The upper end rests squarely against the ridge board and the lower end receives a ***bird's mouth notch*** and rests on the top plate of the wall. This notch increases the contact area of the rafter by placing more rafter surface against the top of the wall, as shown in Figure

FIGURE 21.44 ■ *A hip rafter is used when adjacent roof planes meet to form an inclined edge. Each rafter is longer than the previous rafter. The hip rafter extends diagonally across the common rafters and provides support to the upper end of the rafters. A valley rafter is located where adjacent roof slopes meet to form a valley.* Courtesy Jason Mills.

21.43. ***Hip rafters*** are used when adjacent slopes of the roof meet to form an inclined edge. The hip rafter extends diagonally across the common rafters and provides support to the upper end of the rafters (see Figure 21.44). The hip is inclined at the same pitch as the rafters. A ***valley rafter*** is similar to a hip rafter. It is inclined at the same pitch as the common rafters that it supports. Valley rafters get their name because they are located where adjacent roof slopes meet to form a valley. ***Jack rafters*** span from a wall to a hip or valley rafter. They are similar to common rafters but span a shorter distance. Typically, a section will show only common rafters, with hip, valley, and jack rafters reserved for a very complex section.

Roof Bracing

Rafters settle because of the weight of the roof. As the rafters settle, they push supporting walls outward. These two actions, downward and outward, require special members to resist them. These members are ***ceiling joists,*** ridge bracing, ***collar ties,*** purlins, and purlin blocks and braces. Each is shown in Figure 21.45. When the ceiling joists are laid perpendicular to the rafters, in addition to using a collar tie, a metal strap may be required over the ridge to keep the rafters from separating, as shown in Figure 21.46. Usually double joists are used to support the purlin brace.

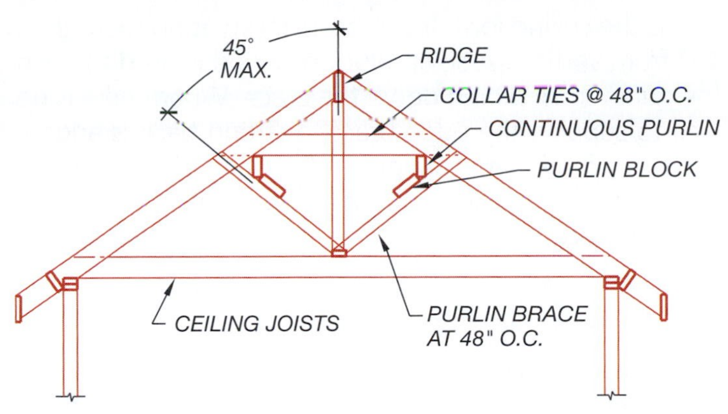

FIGURE 21.45 ■ *Common roof supports include collar ties, purlins, and purlin braces.*

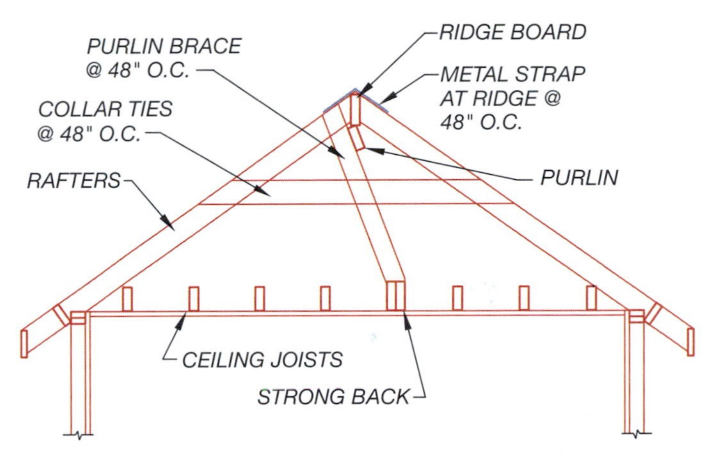

FIGURE 21.46 ■ *When ceiling joists are perpendicular to the rafters, the purlin brace may need to be supported on a strong back. A metal strap may also be required to resist the outward force of the rafters.*

Ceiling joists span between the top plates of bearing walls to resist the outward force placed on the walls from the rafters. The ceiling joists also support the finished ceiling. Collar ties are also used to help resist the outward thrust of the rafters. They are usually the same cross-section size as the rafter and placed in the upper third of the roof. Ridge braces are used to support the downward action of the ridge board. The brace is usually a 2 × 4 (50 × 100) spaced at 48'' (1200 mm) o.c. maximum. The brace must be set at 45° maximum from vertical. A ***purlin*** is a brace used to provide support for the rafters as they span between the ridge and the wall. The purlin is usually the same cross-section size as the rafter and is placed below the rafter to reduce the span. As the rafter span is reduced, the size of the rafter can be reduced. See Chapter 22 for a further explanation of rafter sizes. ***Purlin braces*** are typically 2 × 4s (50 × 100s) spaced at 48'' (1200 mm) o.c. along the purlin and transfer weight from the purlin to a supporting wall. The brace is supported by an interior wall, or a 2 × 4 (50 × 100) lying across the ceiling joist. It can be installed at no more than 45° from vertical. A scrap block of wood is used to keep the purlin from sliding down the brace. When there is no wall to support the ridge brace, a strong back is added. A ***strong back*** is a beam placed between the ceiling joist to support the ceiling and roof loads. Figure 21.47 shows a strong back.

Vaulted Ceilings

If a vaulted ceiling is to be represented, two additional terms understood: rafter/ceiling joist and ridge beam. Both are shown in Figure 21.48. A ***rafter/ceiling joist*** is a combination of rafter and ceiling joist. It is used to support both the roof loads and the finished ceiling. Typically a 2 × 12 (50 × 300) rafter/ceiling joists are used to allow room for 10'' (250 mm) of insulation and 2'' (50 mm) of air space above the insulation. The size of the rafters/ceiling joists must be determined by the load and span. A ***ridge beam*** is used to support the upper end of the rafter/ceiling joist. Because there are no horizontal ceiling joists, metal joist hangers must be used to keep the rafters from separating from the ridge beam.

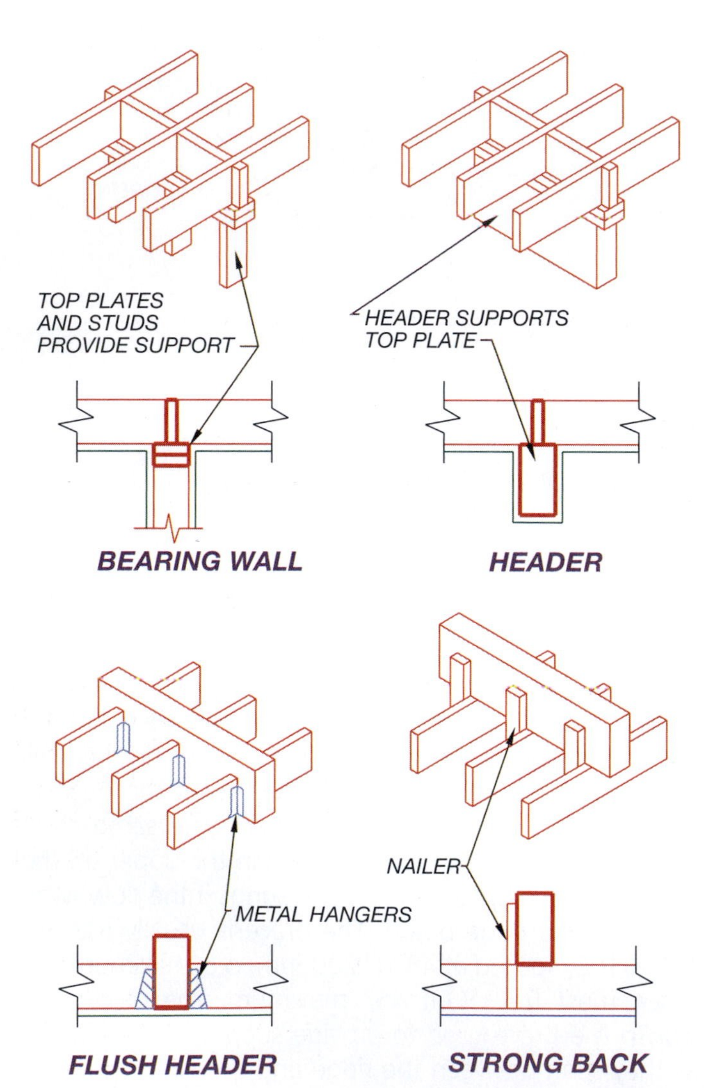

FIGURE 21.47 ■ *Common methods of supporting loads include bearing walls, headers, flush headers, and the strong back. When a header is used, the loads rest on the header. With a flush header, the ceiling joists are supported by metal hangers. A piece of scrap wood can be used to hang the ceiling joist to the strong back.*

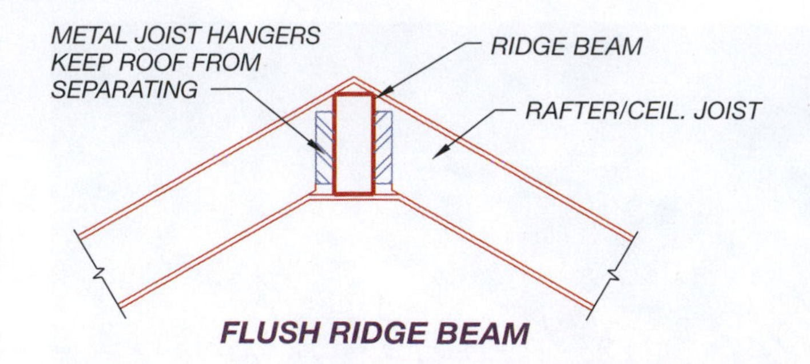

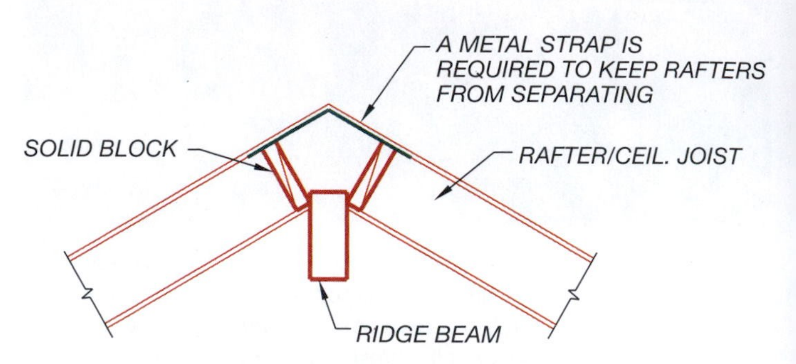

FIGURE 21.48 ■ *Common connections between exposed and hidden ridge beams and rafters. The ridge may be exposed or hidden.*

The final terms that you will need to be familiar with to draw a stick roof are header and trimmer. Both terms are used in wall construction, and they have a similar function when used as roof members (see Figure 21.49). A ***header*** at the roof level consists of two members nailed together and laid perpendicular to the rafters. It is used to support rafters around an opening such as a skylight or chimney. ***Trimmers*** are two rafters nailed together to support the roofing on the inclined edge of an opening (parallel to other rafters).

Truss Roof Construction

A ***truss*** is a component used to span large distances without intermediate supports. Residential trusses can be as short as 15' (4500 mm) or as long as 50'

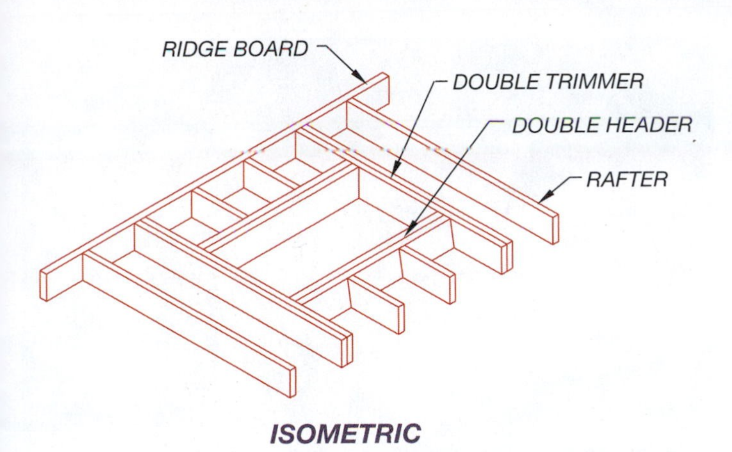

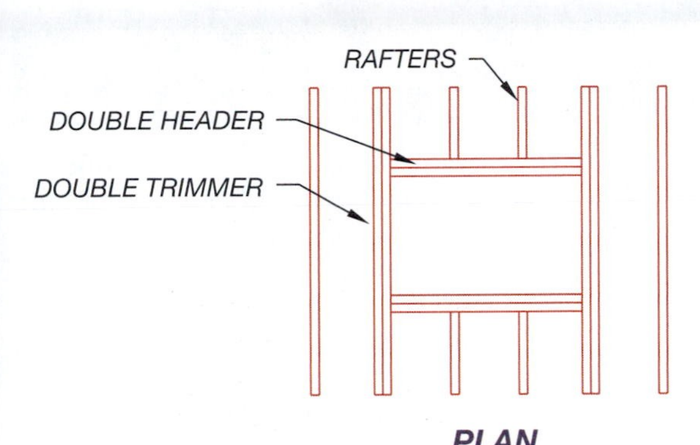

FIGURE 21.49 ■ *Typical construction at a roof opening includes the use of headers and trimmers.*

FIGURE 21.50 ■ *Assembled at the truss company and shipped to the job site, trusses can quickly be set in place.* Courtesy Ray Wallace, Southern Forest Products Association.

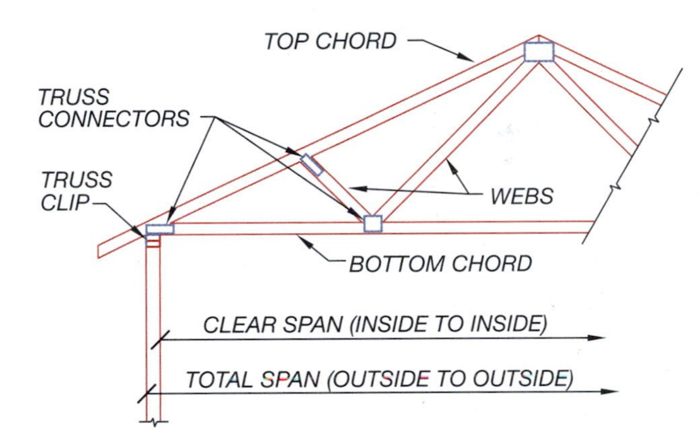

FIGURE 21.51 ■ *Common truss members include the top and bottom chords, webs, and truss clips.*

(15 000 mm). Trusses can be either prefabricated or job-built. Prefabricated trusses are commonly used in residential construction. Assembled at the truss company and shipped to the job site, the truss roof can quickly be set in place, as seen in Figure 21.50. A roof that might take 2 or 3 days to frame using conventional framing can be set in place in 2 or 3 hours using trusses, which are set in place by crane. The size of material used to frame trusses is smaller than with conventional framing. Typically, truss members need be only 2 × 4s (50 × 100s) set at 24'' (600 mm) o.c. An engineer working for the truss manufacturer will determine the exact size of the truss members. When drawing a structure framed with trusses, your only responsibility is to represent the span of the trusses on the framing plan and show the general truss shape and bearing points in the section drawings. Chapter 24 introduces framing plans and Chapter 29 introduces sections. Knowledge of truss terms is helpful in making these drawings. Terms that must be understood are top chord, bottom chord, webs, ridge block, and truss clips. Each is shown in Figure 21.51.

The ***top chord*** serves a function similar to a rafter. It is the upper member of the truss and supports the roof sheathing. The ***bottom chord*** serves a purpose similar to a ceiling joist. It resists the outward thrust of the top chord and supports the finished ceiling material. ***Webs*** are the interior members of the truss that span between the top and bottom chords. The manufacturer attaches the webs to the chords using metal plate connectors. ***Ridge blocks*** are blocks of wood used at the peak of the roof to provide a nailing surface for the roof sheathing and as a spacer in setting the trusses into position. ***Truss clips,*** also known as hurricane ties, are used to strengthen the connection between the truss and top plate or header, which is used to support the truss. The truss clips transfer wind forces applied to the roof, which cause uplift down through the wall framing into the foundation. A block is also used where the trusses intersect a support. Common truss intersections with ***blocking*** and hurricane ties are shown in Figure 21.52.

A truss gains its strength from triangles formed throughout it. The shape of each triangle cannot be

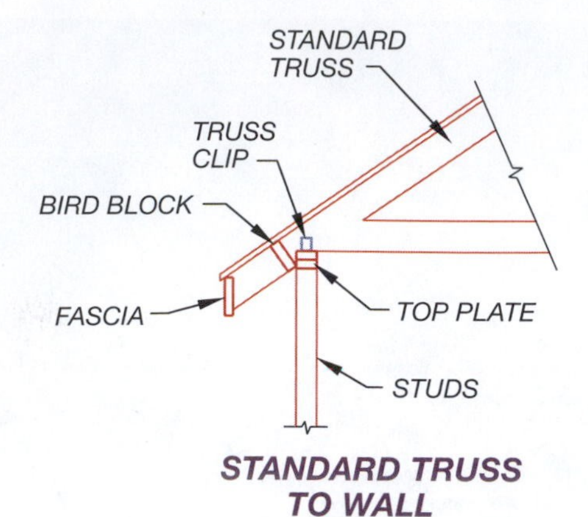

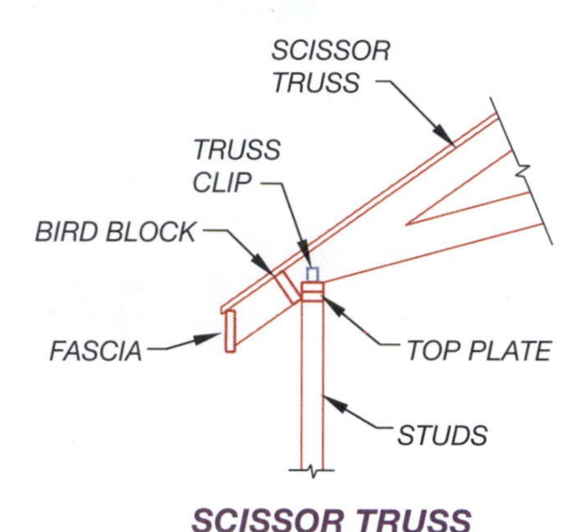

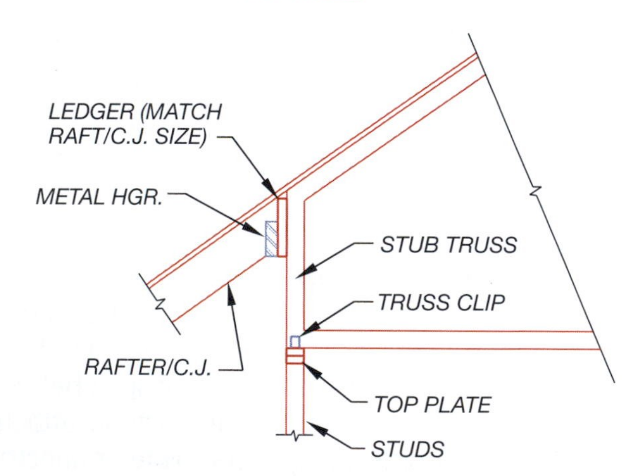

FIGURE 21.52a ■ *Common truss connections normally shown in the sections. Notice that the top chord aligns with the outer face of the top plate. When detailing a scissor truss, the bottom chord is typically drawn a minimum of two pitches less than the top chord.*

FIGURE 21.52b ■ *The rectangular plate is used to bond truss components together and does not need to be specified in details. The truss tie or hurricane tie is used to connect each truss to the top plate. The connector must be specified on the construction documents.* Courtesy Ray Wallace, Southern Forest Products Association.

changed unless the length of one of the three sides is altered. The entire truss will tend to bend under the roof loads as it spans between its bearing points. Most residential trusses can be supported by a bearing point at or near each end of the truss. As spans exceed 40' (12 000 mm) or as the shape of the bottom chord is altered, it may be more economical to provide a third bearing point at or near the center. For a two-point bearing truss, the top chords are in compression from the roof loads and tend to push out the heels and down at the center of the truss. The bottom chord is attached to the top chords and is in tension as it resists the outward thrust of the top chords. The webs closest to the center are usually stressed by tension, and the outer webs are usually stressed by compression. Figure 21.53 shows how the tendency of the truss to bend under the roof loads is resisted and the loads are transferred to the bearing points.

Computers and sophisticated design software have enabled trusses to be manufactured easily in nearly any shape. Several of the common shapes and types of trusses available for residential roof construction are shown in Figure 21.54. Common trusses include:

Standard truss: The truss most often used in residential construction is a standard truss, which is used to frame gable roofs.

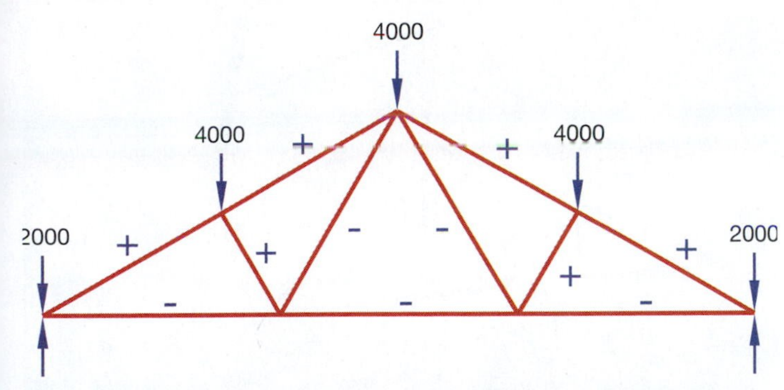

FIGURE 21.53 ■ *Trusses are designed so that the weight to be supported is spread to the outer walls. This is done by placing some members in tension and some in compression. A member in compression is indicated by a plus sign (+) and one in tension is represented by a minus sign (−).*

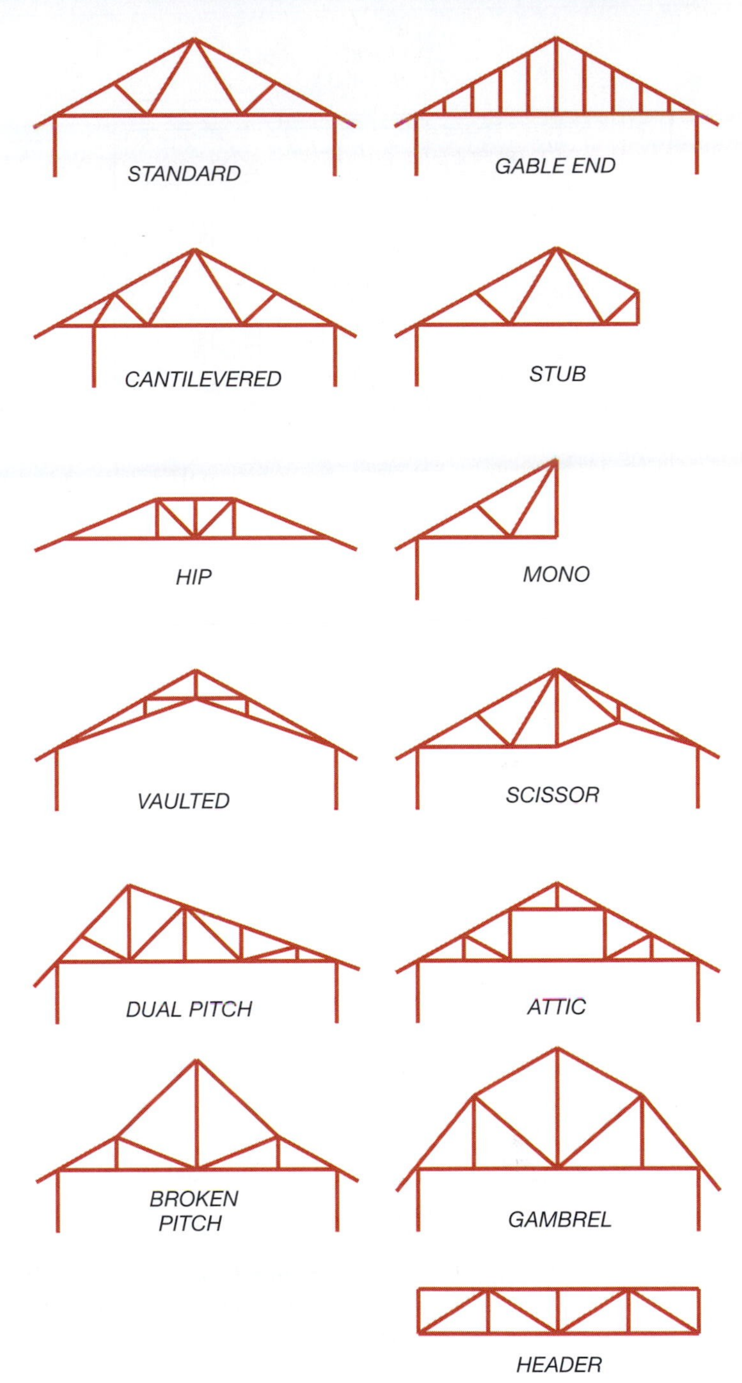

FIGURE 21.54 ■ *Common types of roof trusses used for residential construction.*

Valley truss: A valley truss is used where two roofs are perpendicular to each other. Valley trusses are standard trusses that decrease in height as they get closer to the ridge of the roof they are intersecting.

Gable end-wall truss: A gable end-wall truss is used to form the exterior ends of the roof system and is aligned with the exterior side of the end walls of a structure. This truss is more like a wall than a truss, with the vertical supports typically spaced at 24'' (600 mm) o.c. to support exterior finishing material.

Girder trusses: A girder truss is used on houses with an L- or U-shaped roof where the roofs intersect. To form a girder truss, the manufacturer typically bolts two or three standard trusses together. The manufacturer determines the size and method of constructing the girder truss. Figure 21.54 shows how a girder truss would appear on a roof framing plan.

Cantilevered trusses: The cantilevered truss is used where a truss must extend past its support to align with other roof members. Cantilevered trusses are typically used where walls jog to provide an interior courtyard or patio.

Stub trusses: A stub truss can be used where an opening will be provided in the roof or the roof must be interrupted. Rooms with glass skywalls, as in Figure 21.55, can often be framed using stub trusses. The shortened end of the truss can be supported by either a bearing wall, a beam, or a header truss.

Header trusses: A header truss has a flat top used to support stub trusses. The header truss has a depth to match that of the stub truss and is similar in function to a girder truss. The header truss spans between and is hung from two girder trusses. Stub trusses are hung from the header truss.

Hip trusses: Hip trusses are used to form hip roofs. Each truss has a horizontal top chord, and each succeeding truss increases in height. As the height of the truss increases, the length of the top chord decreases until the full height of the roof is achieved and standard trusses can be used. The height of each hip truss decreases as they get closer to the ex-

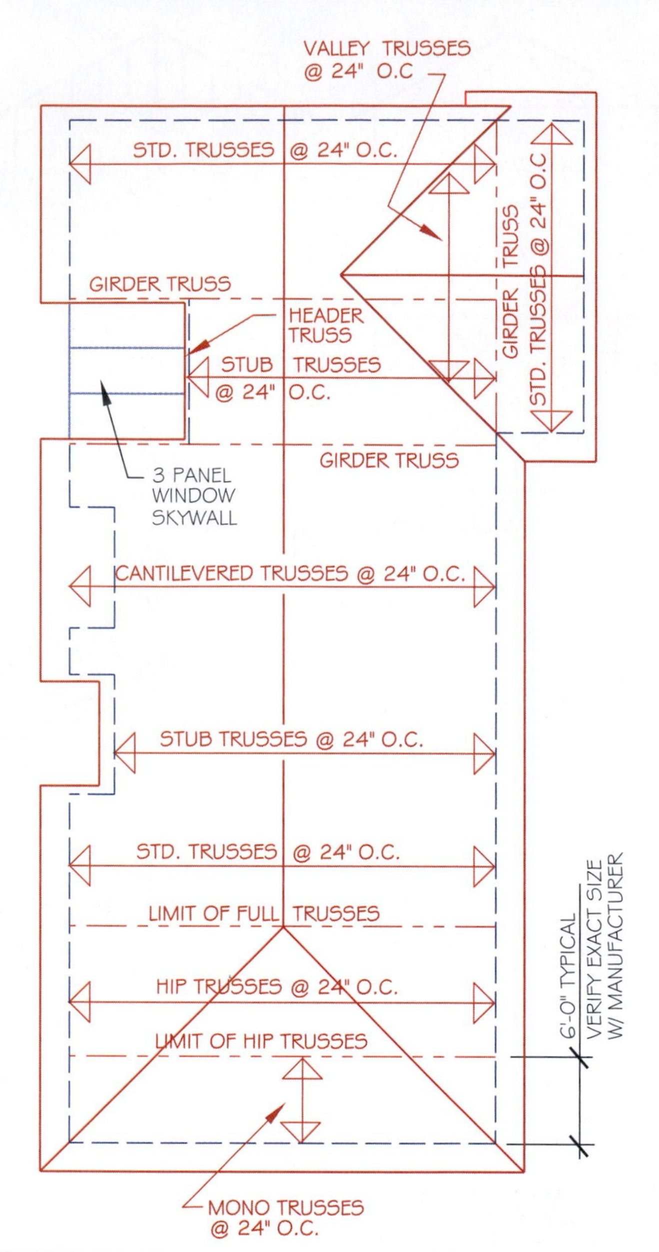

FIGURE 21.55 ■ *Standard, valley, girder, header, stub, cantilever, hip, and mono trusses can each be used to form different roof shapes.*

FIGURE 21.56 ■ *Hip trusses have a horizontal top chord and are used to form hip roofs. Each succeeding truss increases in height until the full height of the roof is achieved and standard trusses can be used.* Courtesy Ray Wallace, Southern Forest Products Association.

terior wall, which is perpendicular to the ridge. Typically hip trusses must be 6' (1500 mm) from the exterior wall to achieve enough height to support the roof loads. Figure 21.56 shows where hip trusses can be used to frame a roof. Figure 21.55 shows a hip truss being lifted into position. The exact distance will be determined by the truss manufacturer and is further explained in Chapter 24.

Mono trusses: A mono truss is a single pitched truss, which can often be used in conjunction with hip trusses to form the external 6' (1500 mm) of the hip. A mono truss is also useful in blending a one-level structure with a two-level structure.

If a vaulted roof is desired, vaulted or scissor trusses can be used. Vaulted and scissor trusses have inclined bottom chords. Typically there must be at least a two-pitch difference between the top and bottom chord. If the top chord is set at a 6/12 pitch, the bottom chord usually cannot exceed a 4/12 pitch. A section will need to be drawn to give the exact location for the vaulted portion of the truss. If a portion of the truss needs to have a flat ceiling, it may be more economical to frame the lowered portion with conventional framing materials, a process often referred to as "scabbing on." Although this process requires extra labor at the job site, it can eliminate a third bearing point near the center of the truss. An alternative to a scissor is to have a barrel vault. A barrel vault provides a curved ceiling (see Figure 21.57).

COMMON CONNECTION METHODS

The stress to be resisted and the materials to be connected will determine which connection method will be used. Common connection methods include nails, staples, ***power-driven studs,*** screws, metal connectors, bolts, and welds.

Nails

Nails are the common connectors for wood-to-wood members with a thickness of less than 1 1/2'' (40 mm). A nail is required by code to penetrate into the sup-

FIGURE 21.57 ■ *A barrel truss can be used to provided a curved ceiling.* Courtesy Ray Wallace, Southern Forest Products Association.

porting member by half the depth of the supporting member. If two 2x (50x) members are being attached, a nail would be required to penetrate 3/4'' (20 mm) into the lower member. Thicker wood assemblies are normally bolted. The types and sizes of nails to be used as well as the placement and nailing pattern will all affect the joint. The IRC provides a schedule for specifying the method, quantity, and size of nailing to be used. A nailing schedule is shown in Figure 21.58.

Types of Nails

The most common nails specified by engineers in the calculations for framing include common, deformed, box, and spike. Most nails are made of stainless steel, but copper and aluminum are also used. Figure 21.59 shows common types of nails used for construction. A common nail is typically used for most rough framing applications. Box nails are slightly thinner and have less holding power than common nails. Box nails are used because they generate less resistance in penetration and are less likely to split the lumber. Box nails come in sizes up to 16d. Spikes are nails larger than 20d. Nailing specifications are found throughout the written specifications, on framing plan and details.

Nail Sizes

Nail sizes are described by the term *penny,* which is represented by the symbol d. Standard nails range from 2d through 60d. Penny is a weight classification; it compares pounds per 1000 nails. For instance, one thousand 8d nails weigh 8 lb. Figure 21.60 shows common sizes of nails. The size, spacing, and quantity of nails to be used are typically specified in structural details.

Nailing Specifications

Nails smaller than 20d are typically specified by penny size and by spacing for continuous joints, such as attaching a plate to a floor system. An example of a nailing note found on a framing plan or detail is:

2 × 6 DFL SILL W/ 20d's @ 4'' O.C.

Nails are specified by a quantity and penny size for repetitive joints such as a joist to a sill. The specification on a section or detail would simply read "3-8d's" and would point to the area in the detail where the nails will be placed. Depending on the scale of the drawing, the nails may or may not be shown. Specifications for spikes are given in a manner similar to those for nails, but the penny size is replaced by spike diameter.

Nailing Placement

Nailing placements are described by the manner in which they are driven into the members being connected. Common methods of driving nails include face, end, toe and blind nailing (see Figure 21.61). The type of nailing to be used is determined by how accessible the head of the nail is during construction and by the type of stress to be resisted. The designer will specify if nailing other than that recommended by the nailing schedule of the prevailing code is to be used. Nails that are required to resist shear are strongest when perpendicular to the grain. Nails that are placed parallel to the end grain, such as end nailing, are weakest and tend to pull out as stress is applied. Common nailing placements include:

- **Face nailing**—driving a nail through the face or surface of one board into the face of another. Face nailing is used to connect sheathing to rafters or studs, nail a plate to the floor sheathing, or to nail a let-in brace to a stud.
- **End nailing**—driving a nail through the face of one member into the end of another member. A plate is end-nailed into the studs as a wall is assembled.
- **Toe nailing**—driving a nail through the face of one board into the face of another. With face and end nailing, nails are driven in approximately 90° to the face. With toe nailing, the nail is driven in at approximately a 30° angle. Connections of rafters to top plates or a header to a trimmer are toe-nailed joints.
- **Blind nailing**—used where it is not desirable to see the nail head. Attaching wood flooring to the subfloor is done with blind nailing. Nails are driven at approximately a 45° angle through the tongue

NAILING SCHEDULE

TAKEN FROM 1997 UBC TABLE 23-II-B-1 CONNECTION	NAILING (1)
1 JOIST TO SILL OR GIRDER, TOE NAIL	3- 8d
2 BRIDGING TO JOIST, TOENAIL EACH END	2- 8d
3 1 X 6 SUBFLOOR OR LESS TO EACH JOIST, FACE NAIL	2-8d
4 WIDER THAN 1 X 6 SUBFLOOR TO EA. JOIST, FACE NAIL	3-8d
5 2" SUBFLOOR TO JOIST OR GIRDER, BLIND AND FACE NAIL	2-16d
6 SOLE PLATE TO JOIST OR BLOCKING TYPICAL FACE NAIL SOLE PLATE TO JOIST OR BLOCKING, @ BRACED WALL PANELS	16d @ 16" O.C. 3- 16d PER 16"
7 TOP PLATE TO STUD, END NAIL	2-16d
8 STUD TO SOLE PLATE	4- 8d, TOENAIL OR 2- 16d, END NAIL
9 DOUBLE STUDS, FACE NAIL	16d @ 24" O.C.
10 DOUBLED TOP PLATES, TYPICAL FACE NAIL DOUBLE TOP PLATES, LAP SPLICE	16d @ 16" O.C. 8- 16d
11 BLOCKING BETWEEN JOIST OR RAFTERS TO TOP PLATE, TOENAIL	3- 8d
12 RIM JOIST TO TOP PLATE, TOENAIL	8d @ 6" O.C.
13 TOP PLATES, LAPS & INTERSECTIONS, FACE NAIL	2-16d
14 CONTINUOUS HEADER, TWO PIECES	16d @ 16" O.C. ALONG EACH EDGE
15 CEILING JOIST TO PLATE, TOE NAIL	3-8d
16 CONTINUOUS HEADER TO STUD, TOE NAIL	4-8d
17 CEILING JOIST, LAPS OVER PARTITIONS, FACE NAIL	3-16d
18 CEILING JOIST TO PARALLEL RAFTERS, FACE NAIL	3-16d
19 RAFTERS TO PLATE, TOE NAIL	3-8d
20 1" BRACE TO EA. STUD & PLATE FACE NAIL	2-8d
21 1" x 8" SHEATHING OR LESS TO EA. BEARING, FACE NAIL	2-8d
22 WIDER THAN 1" x 8" SHEATHING TO EACH BEARING, FACE NAIL	3-8d
23 BUILT-UP CORNER STUDS	16d @ 24" O.C.
24 BUILT-UP GIRDER AND BEAMS	20d @ 32" O.C. @ TOP & BTM. AND STAGGERED 2-20d @ ENDS & @ EACH SPLICE
25	2-16d 2" PLANKS @ EACH BEARING
26 WOOD STRUCTURAL PANELS AND PARTICLEBOARD (2) SUBFLOOR, ROOF AND WALL SHEATHING TO FRAMING:	
19/32" - 3/4"	8d COMMON OR 6d DEFORMED SHANK
1/2" AND LESS	6d COMMON OR DEFORMED SHANK
7/8 "- 1"	8d COMMON OR DEFORMED SHANK
1 1/8"- 1 1/4"	10d COMMON OR 8d DEFORMED SHANK
COMBINATION SUBFLOOR-UNDERLAYMENT TO FRAMING:	
3/4" AND LESS	6d DEFORMED SHANK
7/8" - 1"	8d DEFORMED SHANK
1 1/8" - 1 1/4"	10d COMMON OR 8d DEFORMED SHANK
27 PANEL SIDING TO FRAMING (2)	
1/2" OR LESS CASING NAILS.	6d CORROSION-RESISTANT SIDING OR
5/8"	8d CORROSION-RESISTANT SIDING OR
28 FIBERBOARD SHEATHING (3)	
1/2"	No. 11 GA. (4) 6d COMMON NAILS No. 16 GA. (5)
25/32"	No. 11 GA. (4) 8d COMMON NAILS No. 16 GA. (5)
29. INTERIOR PANELING:	
1/4"	4d (6)
25/32"	6d (7)

1. COMMON OR BOX NAILS MAY BE USED WHERE OTHERWISE STATED.
2. NAILS SPACED @ 6" O.C. @ EDGES, 12" O.C. @ INTERMEDIATE SUPPORTS EXCEPT 6" @ ALL SUPPORTS WHERE SPANS ARE 48" OR MORE. FOR NAILING OF WOOD STRUCTURAL PANEL AND PARTICLEBOARD DIAPHRAGMS AND SHEAR WALLS SEE SPECIFIC NOTES ON DRAWINGS. NAILS FOR WALL SHEATHING MAY BE COMMON, BOX OR CASING.
3. FASTENERS SPACED 3" O.C. @ EXTERIOR EDGES & 6" O.C. @ INTERMEDIATE SUPPORTS.
4. CORROSION -RESISTANT ROOFING NAILS W/ 7/16" DIA. HEAD & 1 1/2" LENGTH FOR 1/2" SHEATHING & 1 3/4" LENGTH FOR 25/32" SHEATHING..
5. CORROSION-RESISTANT STAPLES W/ NOMINAL 7/16" CROWN & 1 1/8" LENGTH FOR 1/2" SHEATHING & 1 1/2" LENGTH FOR 25/32" SHEATHING..
6. PANEL SUPPORTS @ 16" (20" IF STRENGTH AXIS IN THE LONG DIRECTION OF THE PANEL, UNLESS OTHERWISE MARKED). CASING OR FINISH NAILS SPACED 6" ON PANEL EDGES, 12" @ INTERMEDIATE SUPPORTS.
7. PANEL SUPPORTS @ 24". CASING OR FINISH HAILS SPACED 6" ON PANEL EDGES, 12" AT INTERMEDIATE SUPPORTS.

WOOD STRUCTURAL PANEL ROOF SHEATHING NAILING SCHEDULE (1)

WIND REGION	NAILS	PANEL LOCATION	ROOF FASTENER ZONE (2) 1	2	3
			FASTENING SCHEDULE (INCHES ON CENTER) X 25.4 FOR MM		
GREATER THAN 90 MPH (145 km/h)	8d COMMON	PANEL EDGES (3)	6	6	4(4)
		PANEL FIELD	6	6	6(4)
GREATER THAN 80 MPH (129 km/h) TO 90 MPH (145 km/h)	8d COMMON	PANEL EDGES (3)	6	6	4
		PANEL FIELD	12	6	6
80 MPH (129 km/h) OR LESS	8d COMMON	PANEL EDGES (3)	6	6	6
		PANEL FIELD	12	12	12

1. APPLIES ONLY TO MEAN ROOF HEIGHTS UP TO 35 FT (10 700 MM). FOR MEAN ROOF HEIGHTS 35 FEET (10 700 MM), THE NAILING SHALL BE DESIGNED.
2. ROOF FASTENING ZONES ARE SHOWN BELOW.
3. EDGE SPACING ALSO APPLIES OVER ROOF FRAMING AT GABLE-END WALLS.
4. USE 8d RING-SHANK NAILS IN THIS ZONE IF MEAN ROOF HEIGHT IS GREATER THAN 25' (7600mm).

FIGURE 21.58 ■ *The IRC provides specifications that regulate each connection specified on the construction documents. The provision of a nailing schedule with the working drawings will eliminate having to specify nailing for each detail.* Courtesy Residential Designs.

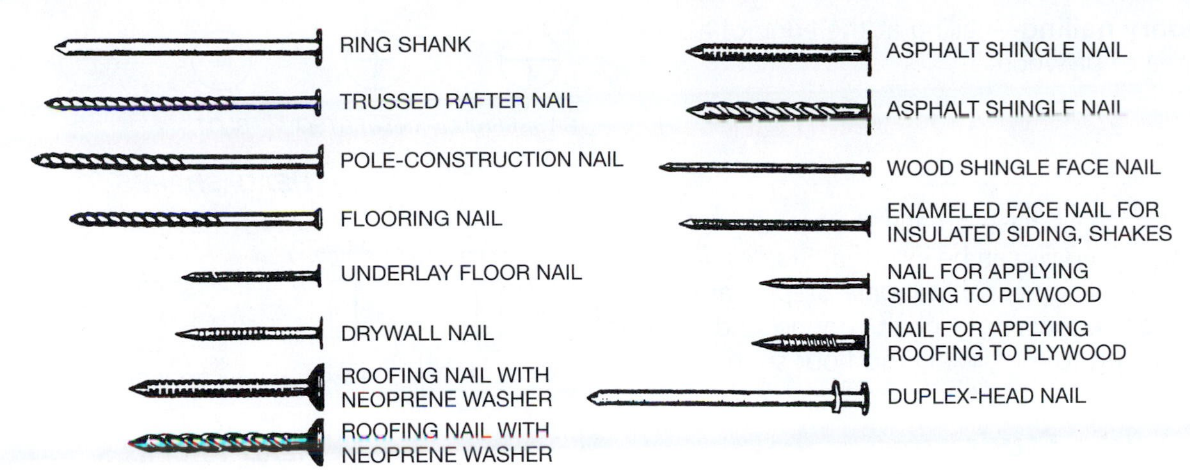

FIGURE 21.59 ■ *Common types of nails used in construction.*

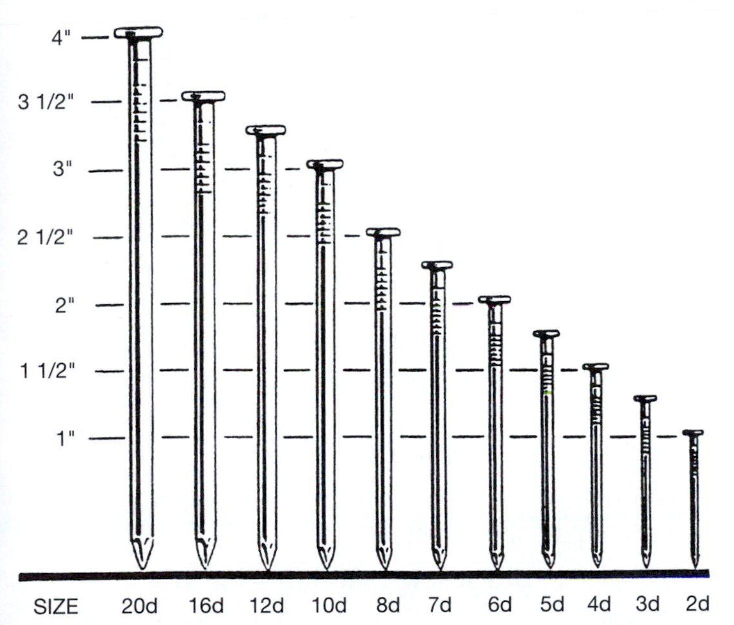

FIGURE 21.60 ■ *Standard sizes of common and box nails. The term penny (d) represents the weight per 1000, not the length of the nail.*

of tongue-and-groove flooring and hidden by the next piece of flooring.

Nailing Patterns

Nail specifications for sheathing and other large areas of nailing often refer to nail placement along an edge, boundary, or field. Common placements include:

- **Edge nailing**—nails placed at the edge of a sheet of plywood
- **Field nailing**—nails placed in the supports for a sheet of plywood excluding the edges

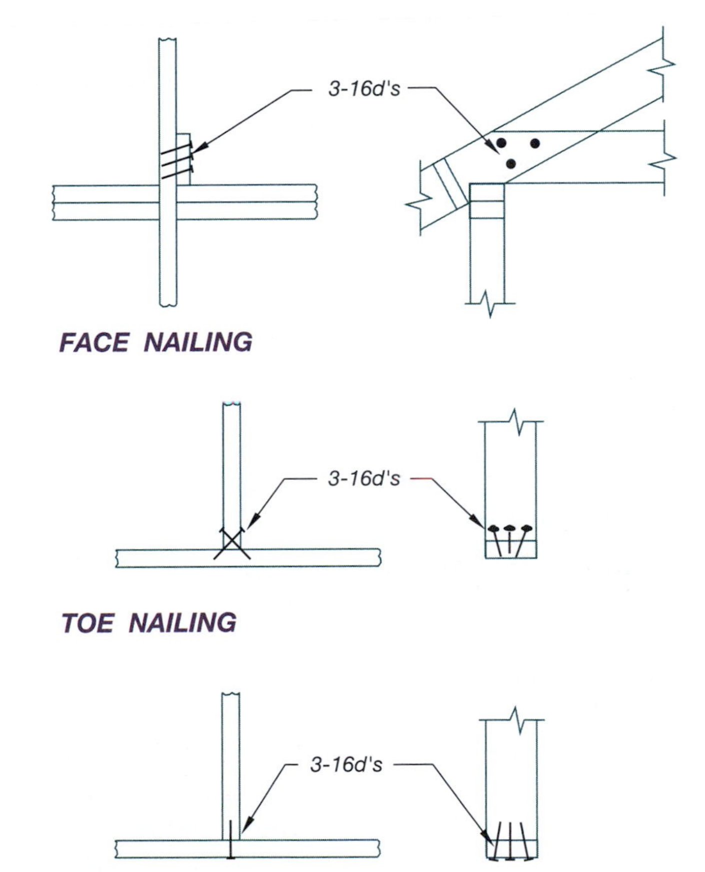

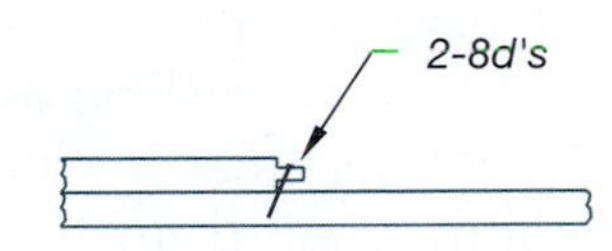

FIGURE 21.61 ■ *Methods of placing nails affect the holding power of the nail.*

- **Boundary nailing**—nailing at the edge of a specific area of plywood

Staples

Power-driven nails have greatly increased the speed and ease in which nails can be inserted. Staples have replaced nailing for some applications. Staples are most often used for connecting asphalt roofing and for attaching sheathing to roof, wall, and floor supports.

Power-Driven Studs

Metal studs can be used to ***anchor*** wood or metal to masonry. These studs range in diameter from 1/4 through 1/2'' (6 to 13 mm) and in length from 3/4 to 6'' (20 to 150 mm). Power-driven studs are made from heat-treated steel and inserted by a powder charge from a gun-like device. They are typically used where it would be difficult to insert the anchor bolts at the time the concrete is poured. Studs can also be used to join wood to steel construction.

Screws

Screws are specified for use in wood connections that must be resistant to withdrawal. Three common screws are used throughout the construction industry. Each is identified by its head shape (see Figure 21.62). Common screws include:

- **Flathead (countersunk) screws**—specified in the architectural drawings for finish work where a nail head is not desirable.
- **Roundhead screws**—used at lumber connections where a head is tolerable. Roundhead screws are also used to connect lightweight metal to wood.
- **Lag screws**—have hexagonal or square heads designed to be tightened by a wrench rather than a screwdriver. Lag screws are used for lumber connections 1 1/2'' (40 mm) and thicker. Lag bolts are available with diameters ranging from 1/4'' to 1 1/4'' (6 to 30 mm). A washer is typically used with a lag ***bolt*** to guard against crushing wood fibers near the bolt. A pilot hole that is approximately three-quarters of the shaft diameter is often specified to reduce wood damage and increase resistance to withdrawal.

Flathead and roundhead screws are designated by the gauge, which specifies a diameter, by length in inches, and by head shape. A typical specification might be:

#10 × 3'' F.H.W.S. (flathead wood screw).

Lag screws are designated by their length and diameter. A typical specification might be:

5/8'' Ø × 6'' LAG SCREW THROUGH 1 1/2'' Ø WASHER.

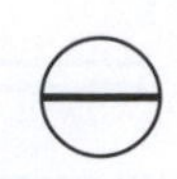

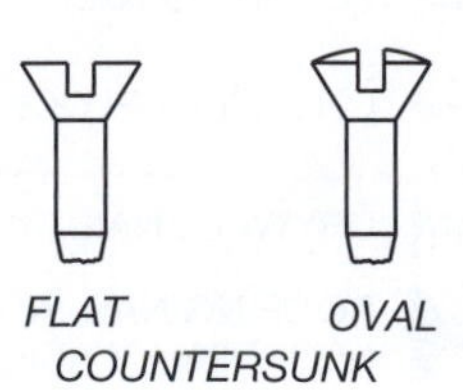

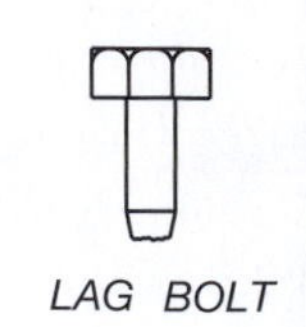

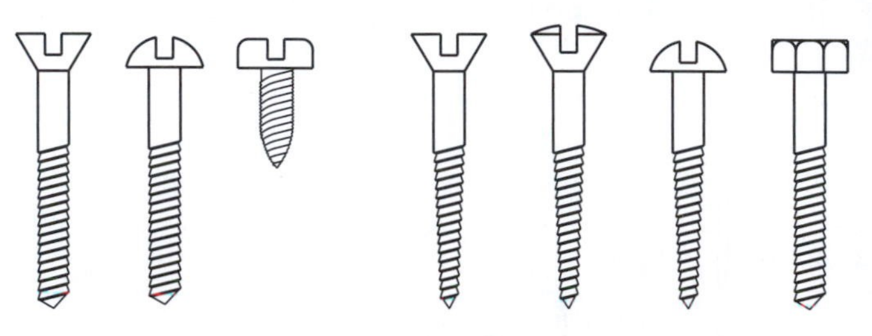

FIGURE 21.62 ■ *Common types of screws used in construction.*

Metal Framing Connectors

Premanufactured metal connectors by companies such as Simpson Strong-Tie are used at many wood connections to strengthen nailed connections. Joist hangers, post caps, post bases, and straps are some of the most common lightweight metal hangers used with wood construction (see Figures 21.05, 21.07, 21.10b and 21.52b). Each connector comes in a variety of gauges of metal with sizes to fit a wide variety of lumber. Metal connectors are typically specified on the framing plans, sections, and details by listing the model number and type of connector. A metal connector specification for connecting 2-2 × 12 joists to a beam would resemble:

SIMPSON CO. HHUS212-2TF JST. HGR.

The supplier is typically specified in general notes and within the written specifications. Depending on the connection, the nails or bolts used with the metal fastener may or may not be specified. If no specification is given, it is assumed that all nail holes in the connector will be filled. If bolts are to be used, they will normally be specified with the connector based on the manufacturer's recommendations. Nails associated with metal connectors are typically labeled with the letter n instead of d. These nails are equal to their d counterpart, but the length has been modified by the manufacturer to fit the metal hanger.

Bolts and Washers

Bolts used in the construction industry include anchor bolts, carriage bolts, and machine bolts. Each is shown in Figure 21.63. A washer is used under the head and nut for most bolting applications. Washers keep the bolt head and nut from pulling through the lumber and reduce damage to the lumber by spreading the stress from the bolt across more wood fibers. Typically a circular washer, specified by its diameter, is used. Common bolts used with construction include anchor, carriage, machine, and miscellaneous bolts.

Anchor Bolts

An anchor bolt is an L-shaped bolt used to attach lumber to the concrete (Figure 21.63a). The short leg of the L is inserted into the concrete to resist withdrawal. The upper end of the long leg is threaded to receive a nut. A 2'' (50 mm) washer is typically used with anchor bolts. Anchor bolts are represented in structural drawings with the letters A.B., along with a specification including the size of the member to be connected, the bolt diameter, length, embedment into concrete, spacing and washer size. A typical note might resemble:

2 × 6 DFPT SILL W/ 5/8'' Ø × 12'' A.B. @ 48'' O.C. W/2'' Ø WASHERS. PROVIDE 9'' MIN. EMBEDMENT.

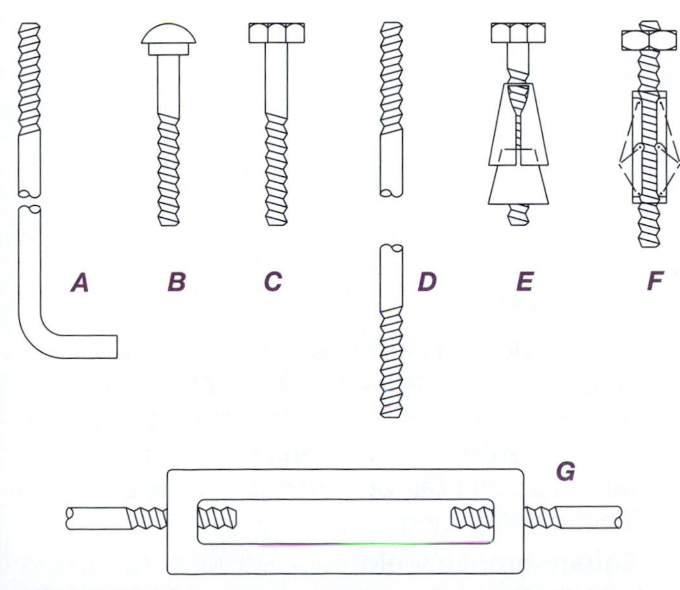

FIGURE 21.63 ■ *Common bolts used in residential construction.*

Carriage Bolts

A ***carriage bolt*** is used for connecting steel and other metal members as well as timber connections (Figure 21.63B). Carriage bolts have a rounded head with the lower portion of the shaft threaded. Directly below the head, at the upper end of the shaft, is a square shank. As the shank is pulled into the lumber, it will keep the bolt from spinning as the nut is tightened. Diameters range from 1/4 to 1'' (6 to 25 mm), with lengths typically available to 12'' (300 mm). The specification for a carriage bolt will be similar to that for an anchor bolt except for the designation of the bolt type.

Machine Bolts

A ***machine bolt*** is a bolt with a hexagonal head and a threaded shaft (Figure 21.63C). Machine bolts are used for attaching steel to steel, steel to wood, or wood to wood. Machine bolts are often used in steel joints to provide a temporary connection while field welds are completed. Bolts are referred to with a note such as:

USE (3)-3/4'' Ø × 8'' M.B. @ 3'' O.C. W/ 1 1/2'' Ø WASHERS.

The project manager or engineer will determine the bolt locations based on the stress to be resisted. Common spacings are:

- 1 1/2'' (40 mm) in from an edge parallel to the grain
- 3'' (75 mm) minimum from the edge when perpendicular to the grain
- 1 1/2'' (40 mm) from the edge of steel members
- 2'' (50 mm) from the edge of concrete

Miscellaneous Bolts

Several other types of bolts are used in special circumstances. These include studs, drift bolts, expansion bolts, and toggle bolts.

- **Stud**—a bolt that has no head. A stud is welded to a steel beam so that a wood plate can be bolted to the beam.

- **Drift bolt** (Figure 21.63D)—a steel rod that has been threaded. Threaded rods can be driven into one wood member with another member bolted to the threaded protrusion. Threaded rods may also be used to span between metal connectors on two separate beams.
- **Expansion bolts**—bolts with a special expanding sleeve (Figure 21.63E). These bolts are designed so that the sleeve, once inserted into a hole, will expand to increase holding power. Expansion bolts are typically used for connecting lumber to masonry.
- **Toggle bolt**—a bolt that has a nut designed to expand when it is inserted through a hole, so that it cannot be removed. Toggle bolts (Figure 21.63F) are used where one end of the bolt may not be accessible because of construction parameters.

Welds

Structural steel is increasingly being used in residential construction to meet the stress imposed by lateral loads. The use of steel members generally requires welded connections and an understanding of basic welding methods. Welding is the method of providing a rigid connection between two or more pieces of steel. In welding, metal is heated to a temperature high enough to cause melting. The parts that are welded become one, with the welded joint actually stronger than the original material. Welding offers better strength, better weight distribution of supported loads, and a greater resistance to shear or rotational forces than a bolted connection. The most common welds in residential construction are shielded metal-arc welding, gas tungsten-arc welding, and gas metal-arc welding. In each case, the components to be welded are placed in contact with each other and the edges are melted to form a bond. Additional metal is also used to form a sufficient bond.

Welds are specified in detail, as shown in Figure 21.64. A horizontal reference line is connected to the parts to be welded by an inclined line with an arrow. The arrow touches the area to be welded. It is not uncommon to see a welding line bend several times to point into difficult-to-reach places or to see more than one leader line extending from the reference line. Information about the type of weld, the location of the weld, the welding process, and the size and length of the weld is all specified on or near the reference line. Figure 21.65 shows a welding symbol and the proper location of information.

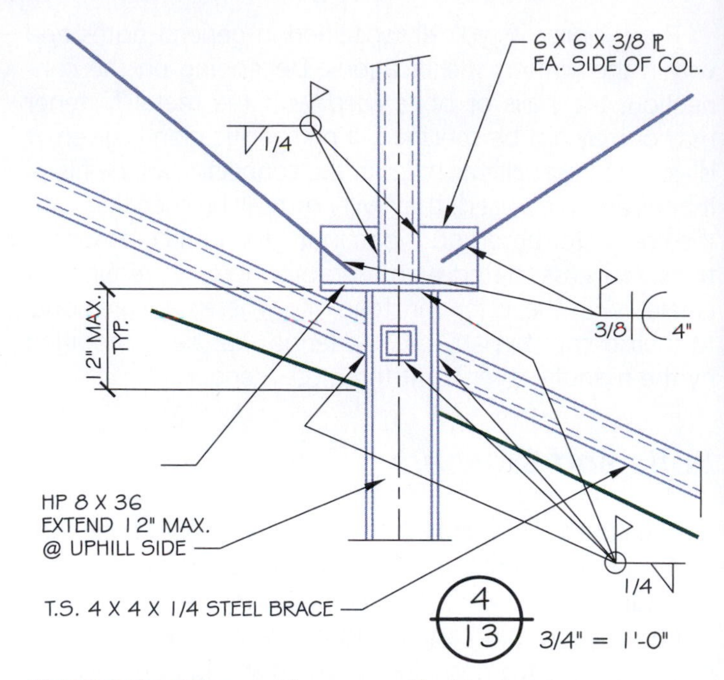

FIGURE 21.64 ■ *Welds are specified in details by the use of specialized symbols that are referenced to the area to receive the weld. Several fillet and two U-groove welds have been specified.* Courtesy Residential Designs.

Welded Joints

The way that steel components intersect greatly influences the method used to weld the materials together and is often included in the specification. Common methods of arranging components to be welded are butt, lap, tee, outside corner, and edge joint (see Figure 21.66).

Types of Welds

The type of weld is associated with the weld shape or the type of groove in the metal components that will receive the weld or both. Welds typically specified on the construction drawings include fillet, groove, and plug welds. The method of joining steel and the symbol for each method are shown in Figure 21.67.

- **Fillet weld**—The most common weld used in construction, a fillet weld (Figure 21.67A) is formed at the internal corner of two intersecting pieces of steel. The fillet can be applied to one or both sides and can be continuous or of a specified length and spacing.
- **Square-groove weld**—applied when two pieces with perpendicular edges are joined end to end. The spacing between the two pieces of metal is

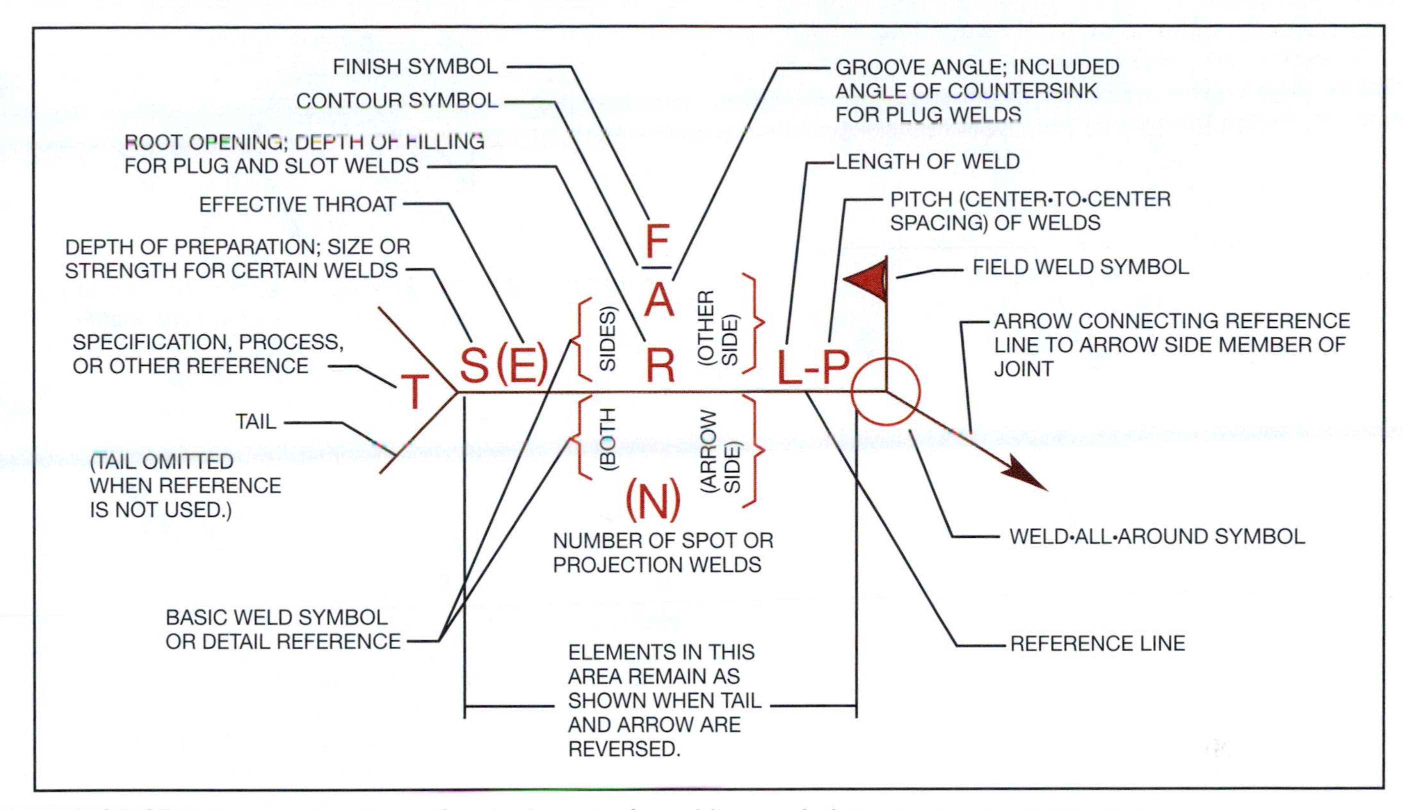

FIGURE 21.65 ■ *Common locations of each element of a welding symbol.* Courtesy American Welding Society.

called the root opening. The root opening is shown to the left of the symbol in Figure 21.67B.

- **V-groove weld**—applied when each piece of steel to be joined has an inclined edge that forms a V. The included angle is often specified, as well as the root opening. See Figure 21.67C.
- **Beveled weld**—created when only one piece of steel has a beveled edge. An angle for the bevel and the root opening is typically given.
- **U-groove weld**—created when the groove between the two mating parts forms a U. See Figure 21.67D.
- **J-groove weld**—results when one piece has a perpendicular edge and the other has a curved grooved edge. See Figure 21.67E. The included angle, the root opening, and the weld size are typically given for U- and J-groove welds.

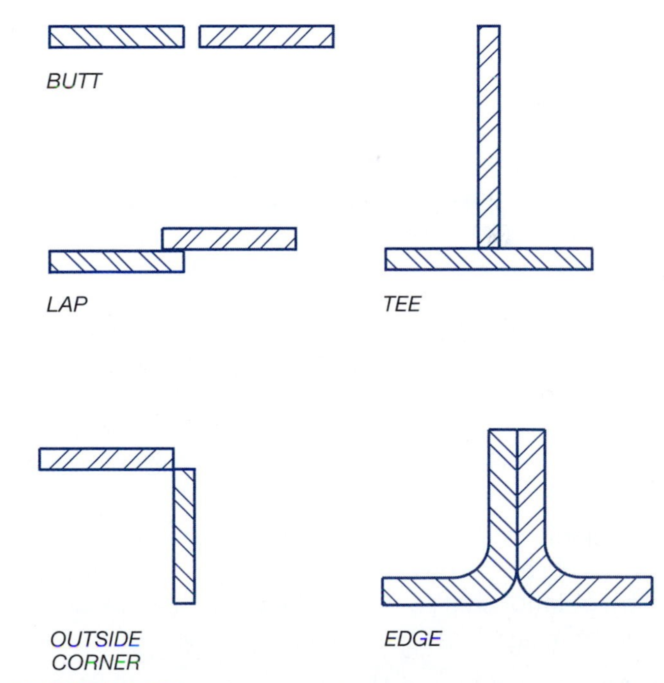

FIGURE 21.66 ■ *The type of welded joint to be used depends on how the members to be welded intersect.*

Welding Locations and Placements

Welds specified on structural drawings may be done away from the job site and then shipped ready to be installed. Large components that must be assembled at the job site are called ***field-welded.*** Two symbols used to refer to a field weld are shown in Figure 21.68.

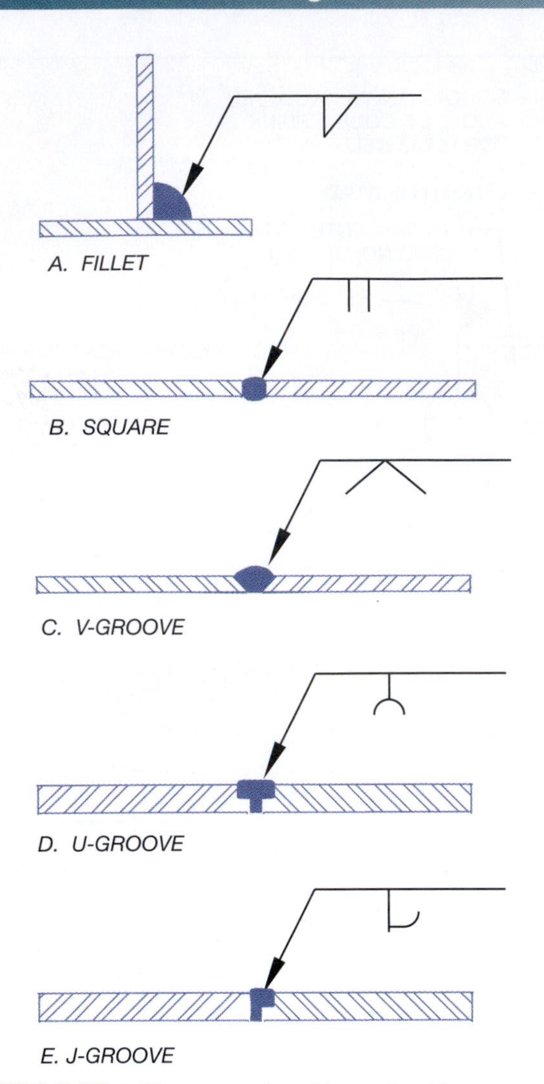

FIGURE 21.67 ■ *The type of weld required is represented by the shape of the material to be welded.*

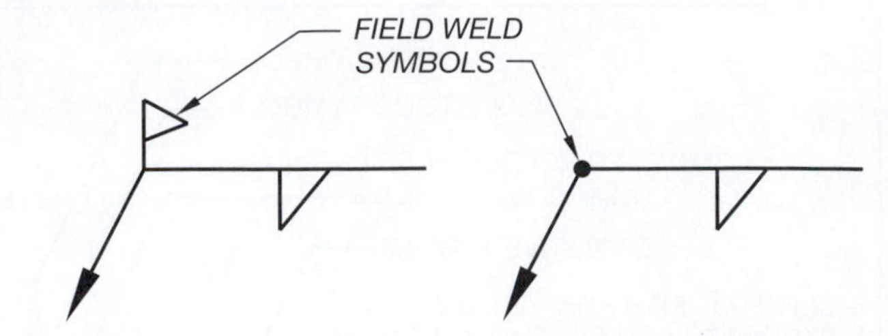

FIGURE 21.68 ■ *The welding symbol can be used to represent on-site (field) or off-site welding.*

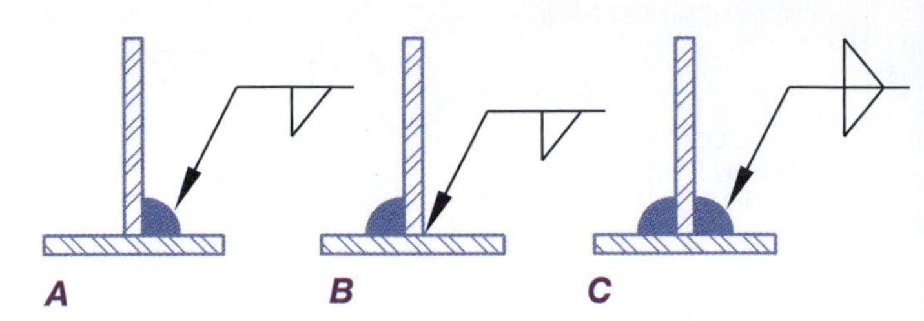

FIGURE 21.69 ■ *The welding symbol can be used to describe the placement of the weld. (A) Symbol below the reference line places the weld on this side. (B) Symbol above the reference line places the weld on the other side. (C) Symbols on each side of the reference line places the welds on each side of the material.*

The placement of the welding symbol in relationship to the reference line is critical in explaining where the actual weld will take place. Figure 21.69 shows three examples and the effects of placing a fillet weld symbol on the reference. The distinction of symbol placement can be quite helpful if adequate space for a symbol is not available on the proper side of the detail. The welding symbol can be placed on either side of the drawing, and the relationship to the reference line can be used to clarify the exact location. Options include:

- **All-around**—used for circular and rectangular parts to indicate that a weld is to be placed around the entire intersection. A circle placed at the intersection of the leader and reference line indicates that a feature is to be welded all around.
- **Weld length and increment**—If a weld does not surround the entire part, the length and spacing of the weld should be placed beside the weld symbol. The number preceding the weld symbols indicates the size of the weld. The number following the weld symbol indicates the length each weld is to be. The final size indicates the spacing of the weld along a continuous intersection of two mating parts.

ENVIRONMENTALLY FRIENDLY CONSTRUCTION

Throughout this chapter, many building components have been discussed. Most of us live and work in structures that are constructed of these components and never even think about the building materials. The Environmental Protection Agency (EPA) list over 48,000 chemicals, with no information on the toxic effects of 79% of them. Of the chemicals that have been tested, many are found in the residential construction industry and can cause severe medical problems to individuals who are chemically sensitive. Allergies and diseases such as chronic fatigue syndrome are being linked to construction materials. The indoor air of new homes often contains as much as six times the acceptable outdoor levels of pollutants. The greatest indoor health risks come from airborne pollutants from products containing formaldehyde-based resins and solvents used in the construction process containing volatile organic compounds.

Formaldehyde-Based Resins

Formaldehyde is a colorless gas compound composed of carbon, hydrogen, and oxygen and is found in most resin-based construction products. Resin is used in products such as plywood, HDF, MDF, PSL, OSB, LVL, linoleum, lacquer, gypsum board, paneling, wallpaper, caulking compounds, insulation, adhesives, upholstery, and carpet. The toxins contained in the products can be present in levels harmful to healthy adults for up to 20 years after insulation. Many household cleaning products also contain small amounts of formaldehyde. Exposure to low-level concentrations can cause irritation of the nose, throat, and eyes as well as headaches, coughing, and fatigue. Higher levels of exposure can cause severe allergic reactions, skin rash, and possibly some types of cancer.

Alternatives to formaldehyde-based products are often expensive and difficult to find. The AIA is becoming a leader in providing education in environmentally friendly construction methods. Concern is now being expressed by many groups for the need to examine the effects of manufacturing, using, and disposing of specific products in relation to the environment.

Alternative Construction Products

Pressure-treated lumber can be removed from a residence by substituting products with a natural ability to repel moisture, such as cedar or redwood. Low-toxic products such as AM Penetrating Seal, Protek, and Boracare Timbor Impel can be used to protect lumber from moisture.

Sheets of grass-based boards such as Meadowboard or Medite can be used in place of plywood sheathing. These boards resemble OSB in appearance but contain no dangerous resins. If plywood must be used, exterior grades contain lower levels of formaldehyde than most interior grades. The sides, edges, and interior side of plywoods can be sealed with sealers such as AM Safeseal or Crystalaire to prevent fumes from leaking into the home. Fiberglass insulation materials are also harmful to chemically sensitive people. Products such as AirKrete can be sprayed into the stud space to provide a toxin-free insulation. Foil-based insulation materials such as Kshield or Dennyfoil, with taped joints, can also be used to shield the interior of a structure from toxins contained in wall cavity material.

Adhesives, joint compound, strippers, paints, and sealants all contain volatile organic chemicals, which can be harmful for years. Such products as AM, Murco, Crystalaire, and Auro are nontoxic or have low toxicity.

Interior Product Alternatives

Carpeting is one of the worst causes of indoor pollution. Typically, carpeting contains over 50 chemicals, which are used as bonding and protective agents. In addition to the chemical content, carpet harbors pollutants such as dust, mold, and pet dander, which affect many people with allergies. All-natural carpets of untreated wool, cotton, or other natural blends such as coir or sea grass are the best choices for allergy-sensitive people. Nylon is the most inert of the synthetic materials used for carpets. Stain treatments, foam pads, and the adhesives used for attaching the pad must be carefully considered because of their chemical content. AFM Carpet Guard seals noxious fumes into most carpets.

For a chemically sensitive person, hard-surface floors are a better alternative than carpet. Tile, hardwood, linoleum, and slate are safe choices. Hardwood floors made of natural wood are safe, but the protective coating should be carefully considered. Wood parquet flooring should not be used because of the resins in the products and the adhesives used to bond them to the subfloor. Vinyl flooring is typically chemically saturated. Natural linoleum made from linseed oil and wood floor with jute backing provide nontoxic, durable floorings.

Equipment and Appliance Alternatives

In addition to the material used for construction, heating equipment, appliances, and furnishings must be considered if the toxins used in construction are to be reduced. Although gas and oil furnaces are cost-effective, the fumes created by the burning of fossil fuels cause allergies for many people. Electric forced air and radiant heating are nontoxic.

Many appliances within a residence should also be carefully considered. Gas appliances should be avoided; they create the same problems as do gas heaters. Ovens should not be self-cleaning and should initially be heated outside the residence to burn off oils, paints, and plastic fumes. Dishwashers and clothes dryers also typically contain materials that affect chemically sensitive users.

CHAPTER 21

Additional Readings

The following websites can be used as a resource to help you keep current with changes in building materials.

ADDRESS	COMPANY OR ORGANIZATION
www.hardboard.org	American Hardboard Association
www.apawood.org	APA the Engineered Wood Association
www.alsc.org	American Lumber Standards Commission, Inc.
www.bc.com	Boise Cascade
www.csa.ca	Canadian Standards Association
www.cwc.ca	Canadian Wood Council
www.celotex.co.uk	Celotex Corporation (Insulating sheathing)
www.gp.com	Georgia-Pacific Corporation
www.gypsum.org	Gypsum Association
www.hpva.org	Hardwood Plywood & Veneer Association
www.iilp.org	International Institute for Lath, Plaster, & Drywall
www.internationalpaper.com	International Paper (high-performance building products)
www.simplex-products.com	Ludlow Coated Products (Thermo-ply sheathing and house wraps)
www.masonrysociety.org	The Masonry Society
www.naad.org	National Association of Aluminum Distributors
www.nahb.com	National Commercial Builders Council of the National Association of Home Builders
www.owenscorning.com	Owens Corning (house wrap and insulation)
www.typarhousewrap.com	Reemay (Typar house wrap)
www.strongtie.com	Simpson Strong-Tie Company, Inc.
www.sfpa.org	Southern Forest Products Association
www.sips.org	Structural Insulated Panel Association
www.trimjoist.com	TrimJoist Engineered Wood Products
www.trusjoist.com	Trus-Joist A Weyerhaeuser Business
www.weyerhaeuser.com	Weyerhaeuser
www.woodtruss.com	Structural Building Components Portal

CHAPTER 21

Structural Components Test

DIRECTIONS

Answer the following questions with short, complete statements. Type your answers using a word processor or neatly print your answers on lined paper.

1. Place your name, the chapter number, and the date at the top of the sheet.
2. Type the question number and provide the answer in the form of a statement that includes part of the question. You do not need to write out the entire question.

Note: The answers to some questions may not be contained in this chapter and will require you to do additional research.

QUESTIONS

21.1. List two different floor framing methods and explain the differences. Provide sketches to illustrate your answer.

21.2. How do a girder, a header, and a beam differ?

21.3. List the differences between a rim joist and solid blocking at the sill.

21.4. How wide are girders typically for a residence framed with a conventional floor system?

21.5. What are let-in braces and wall sheathing used for?

21.6. A let-in brace must be at what maximum angle?

21.7. Define the following abbreviations:

PSL	MDF	APA
OSB	EXP	EXT
LVL	HDF	STRUCT

21.8. How is a foundation post protected from the moisture in the concrete support?

21.9. Blocking is required in walls over how many feet high?

21.10. List common materials suitable for residential beams and girders.

21.11. What purpose does the bird's mouth of a rafter serve?

21.12. What do the numbers 4/12 mean when placed on a roof pitch symbol?

21.13. How do ridge, ridge blocking, and ridge board differ?

21.14. List three types of truss web materials.

21.15. Sketch, list, and define four types of rafters.

21.16. List two functions of a ceiling joist.

21.17. Define the four typical parts of a truss.

21.18. What two elements are typically applied to wood to make an engineered wood product?

21.19. Define mudsill.

21.20. What function does a purlin serve?

21.21. Explain the difference between a bearing and a nonbearing wall.

21.22. What is the minimum lap for top plates in a bearing wall?

21.23. Sketch, label, and define the members supporting the loads around a window.

21.24. What is the most common spacing for rafters?

21.25. What is the function of the top plate of a wall?

21.26. What is the advantage of providing blocking at the edge of a floor diaphragm?

21.27. List the common sizes of engineered wood studs.

21.28. What are the common span ratings for plywood suitable for roof sheathing?

21.29. What grades of plywood are typically used for floor sheathing?
21.30. List eight different qualities typically given in an APA wood rating.
21.31. Sketch the proper symbol for a 3/16'' fillet weld, opposite side, field weld.
21.32. List five pieces of information that might be included in a bolt specification.
21.33. What type of nail is most typically used for wood framing?
21.34. List the names of the following bolts: A.B. C.B. H.S. M.B.
21.35. How long is an 8-penny nail?
21.36. Use the nailing schedule in this chapter to determine (1) how a rafter will be connected to a top plate and (2) how to secure 3/4'' plywood to the floor joist.
21.37. List the three possible locations for placing a weld and give the symbol that represents each location.
21.38. Why is a washer used in placing a bolt?
21.39. List five types of joints where welds can be applied.
21.40. Sketch the symbol used for a 3/16'' × 3'' long V groove weld applied to this side.

CHAPTER 22

Design Criteria for Structural Loading

To advance in the field of architecture requires a thorough understanding of how the weight of the materials used for construction will be supported, including knowledge of loads and determining how they are dispersed throughout a structure. Several types of forces or loads affect structures. The most common are ***gravity, lateral loads,*** *uplift, temporary loads, and moving loads. Gravity is a uniform force that affects all structures. Gravity causes a downward motion on each building component. Lateral loads produce a sideways motion on a structure. Lateral loads are caused by wind and seismic activity and vary in intensity according to the location of the structure and seismic activity. In addition to producing lateral force, wind can also produce uplift. Temporary loads are loads that must be supported for only a limited time. Storing 30 sheets of OSB roof sheathing on a few trusses is an example of a temporary load. Moving loads are loads that are not stationary. The moving loads most common in residential construction are produced by automobiles or construction equipment. This chapter explores gravity loads, how they are transferred throughout a structure, and how to achieve equilibrium with the forces acting on the structure.*

TYPES OF LOADS

As you start to evaluate the loads on a structure, you will need to be concerned with several types of loads acting on a building: dead, live, and dynamic loads. Because of the complexity of determining these loads exactly, building codes have tables of conventional safe loads, which can be used to help determine the amount of weight or stress acting on any given member.

Dead Loads

Dead loads consist of the weight of the structure: walls, floors, and roofs, plus any permanently fixed loads such as fixed service equipment. Building codes typically require design values to be based on a minimum dead load of 10 lb per square foot (psf) (0.48 kN/m^2) for floors and ceilings. A design value of between 7 and 15 psf (0.48 and 0.72 kN/m^2) for rafters is used, depending on the weight of the finished roofing. The design load for dead loads can be verified in the design criteria section of each joist and rafter table (see Chapter 23). The symbol DL is often used to represent the values for dead loads.

Live Loads

Live loads are those superimposed on the building through its use. These loads include such things as people and furniture and weather-related items such as wind, ice, snow, and water (rain). The most commonly encountered live loads are moving, roof, and snow loads. Live loads are represented by the symbol LL.

Floor Live Loads

Buildings are designed for a specific use or occupancy. Depending on the occupancy, the floor live load will vary. The IRC requires floor members to be designed to support a minimum live load of 30 psf (1.44 kN/m^2) for sleeping rooms and 40 psf for all other rooms. Exterior balconies must be able to support a minimum live load of 60 psf (2.88 kN/m^2), and decks must be designed to support a minimum live load of 40 psf (1.92 kN/m^2).

Moving Live Loads

In residential construction, moving loads typically occur only in garage areas and are due to the weight of a car or truck. When the weight is being supported by a slab over soil, these loads do not usually cause concern. When moving weights are being supported by wood, the designer needs to take special care in the design. The IRC requires a minimum load of 50 psf (2.40 kN/m^2) to be used in the design of residential garages, but you should consult with the local building department to determine what design weight should be used for moving loads.

Roof Live Loads

Roof live loads vary from 20 to 40 psf (0.96 to 1.92 kN/m^2) depending on the pitch and the use of the roof. Some roofs also are used for sundecks and are designed in a manner similar to floors. Other roofs are so steep that they may be designed like a wall. Many building departments use 30 psf (1.44 kN/m^2) as a safe live load for roofs. Consult the building department in your area for roof live load values.

Snow Loads

Snow loads may or may not be a problem in the area for which you are designing. You may be designing in an area where snow is something you dream about, not design for. If you are designing in an area where snow is something you shovel, it is also something you must allow for in your design. Because snow loads vary so greatly, the designer should consult the local building department to determine what amount of snow load should be considered. In addition to climatic variables, the elevation, wind frequency, duration of snowfall, and the exposure of the roof all influence the amount of live load design.

Dynamic Loads

Dynamic loads are those imposed on a structure from a sudden gust of wind or from an earthquake.

Wind Loads

Although wind design should be done only by a competent architect or engineer, understanding the areas of a structure that are subject to failure will help you advance in the office. The IRC and many municipalities include maps detailing the basic wind speeds that should be used in the design of a structure. Wind pressure creates wind loads on a structure. These loads vary greatly, depending on the location and the height of the structure above the ground. IRC defines four basic wind exposure categories based on ground surface irregularities of the job site. These include:

- Exposure A—A construction site in large city centers where 50% of the surrounding structures have a height of 70' (21,300 mm) or greater for a distance of 0.5 mi (0.8 km) or 10 times the building height.
- Exposure B—This wind exposure is the assumed design standard for the IRC. It includes urban, suburban, wooded areas, or other terrain with multiple closely spaced structures that are the size of a single-family residence or larger.
- Exposure C—Open terrain with scattered obstructions with a height of less than 30' (9100 mm) that extend more than 1500' (457,200 mm) in any direction from the building site.
- Exposure D—Structures built on flat unobstructed sites within 1500' (457,200 mm) of shorelines exposed to wind flowing over open bodies of water 1 mi (1.6 km) wide or larger. Sites built near shorelines in hurricane-prone regions are excluded from this category. Because such wide variations in wind speed can be encountered, the designer must rely on the local building department to provide information.

Figure 22.1 shows a simplified explanation of how winds can affect a house. Wind affects a wall just as it would the sail of a boat. With a boat, the desired effect is to move the boat. With a structure, this tendency to move must be resisted. The walls resisting the wind will tend to bow under the force of the wind pressure. The tendency can be resisted by roof and foundation members and perpendicular support walls. These supporting walls will tend to become parallelograms and collapse. The designer of the structure determines the anticipated wind speed and design walls, typically referred to as ***braced wall lines*** or ***shear walls,*** to resist this pressure.

In planning for loads from wind, prevailing wind direction cannot be assumed. Winds are assumed to act in any horizontal direction and will create a positive pressure on

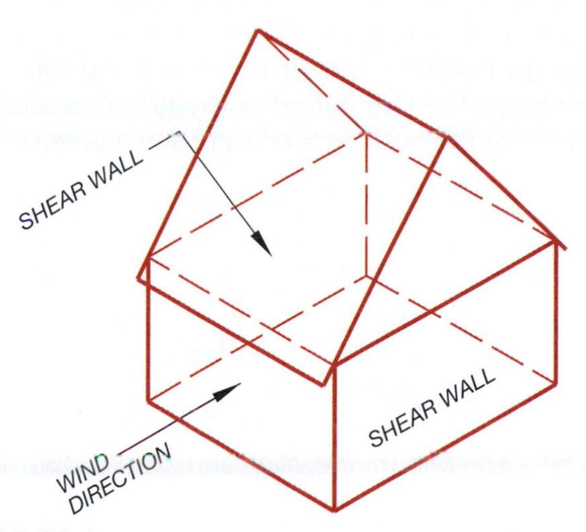

FIGURE 22.1 ■ *Wind pressure on a wall will be resisted by the bolts connecting the wall sill to the foundation and by the roof structure. The wind is also resisted by shear walls, which are perpendicular to the wall under pressure.*

the windward side of a structure. A negative pressure on the leeward (downwind) side of the structure creates a partial vacuum. Design pressures for a structure are based on the total pressure a structure might be expected to encounter, which is equal to the sum of the positive and negative pressure. A design value of 30 psf (1.44 kN/m²) is common for residential projects, but this value should be verified with local building departments. Thirty psf (1.44 kN/m²) equals a wind speed of 108 miles per hour (mph). Other common values include:

PSF	N/M²	MPH	KM/H	PSF	KN/M²	MPH	KM/H
15	0.72	76	122	50	2.40	140	225
20	0.96	88	142	55	2.64	147	237
25	1.20	99	159	60	2.88	153	246
30	1.44	108	174	70	3.36	165	266
35	1.68	117	188	80	3.84	177	285
40	1.92	125	201	90	4.32	188	303
45	2.16	133	214	100	4.80	198	319

Areas subject to high winds from hurricanes or tornados should be able to withstand winds of 125 mph (201 km/h), with 3-second gusts as high as 160 mph (257 km/h). Using these values, the designer can determine existing wall areas and the resulting wind pressure that must be resisted. This information is then used to determine the size and spacing of anchor bolts and any necessary metal straps or ties needed to reinforce the structure.

Wind pressure is also critical to the design and placement of doors and windows. The design wind pressure will affect the size of the glazing area and the method of framing the rough openings for doors and windows. Openings in shear walls will reduce the effectiveness of the wall. Framing around openings must be connected to the frame and foundation to resist forces of uplift from wind pressure. Uplift, the tendency of members to move upward in response to wind or seismic pressure, is typically resisted by the use of steel straps or connectors that join the trimmer and king studs beside the window to the foundation or to framing members in the floor lever below. Wall areas with large areas of openings must also have design studies to determine the amount of wall area that will be required to resist the lateral pressure created by wind pressure. As a general rule of thumb, a wall 4'-0'' (1200 mm) long is required for each 25 lineal feet (7500 mm) of wall. The amount of wall to be reinforced is based on the wall height, the seismic zone, the size of openings by the shear panel, and the method of construction used to reinforce the wall. Chapter 24 explores methods of resisting the forces of shear using the prescriptive methods of the building codes.

Another problem created by wind pressure is the tendency of a structure to overturn. Structures built on pilings or other posted foundations allow wind pressure under the structure to exceed the pressure on the leeward side of the structure. Codes require that the resistance of the dead loads of a building to overturning be 1 1/2 times the overturning effect of the wind. Metal ties are used to connect the walls to the floor, the floor joist to support beams, the support beams to supporting columns, and columns to the foundation. These ties are determined by an architect or engineer for each structure and are based on the area exposed to the wind and the wind pressure.

Seismic Loads

Seismic loads result from earthquakes. The IRC contains seismic maps of the United States and the risk in each area of damage from earthquakes. The IRC has specific requirements for seismic design based on the location and soil type of the building site and the building shape. Chapter 24 explores building shape as it relates to construction materials. Chapter 25 discusses soil types and their effect on structures. Six seismic zones are defined by IRC, ranging from A through E. Zone A is the least prone to seismic damage, and structures in zone E are most likely to be damaged by earthquakes. Zones A, B, and C do not require special construction methods. Special provisions are provided in the IRC for structures built in zones D1 and D2. Structures built in zone E are governed by the IBC (International Building Code).

Stress that results from an earthquake is termed a ***seismic load*** and is usually treated as a lateral load that involves the entire structure. Lateral loads created by wind affect only certain parts of a structure. Lateral forces created by seismic forces affect the entire structure. As the ground moves in varying directions at varying speeds, the entire structure is set in motion. Even after the ground comes to rest, the structure tends to wobble like Jell-O. Typically, structures fall to seismic forces in much the same way as they do to wind pressure. Intersections of roofs-to-wall, wall-to-floor, and floor-to-foundation are critical to the ability of a structure to resist seismic forces. An architect or engineer should design these connections. Typically, a structure must be fluid enough to move with the shock wave but so connected that individual components move as a unit and all units of a structure move as one. Figure 22.2 shows a detail of an engineer's design for a connection of a garage-door king stud designed to resist seismic forces. The straps that are attached to the wall cause the wall and foundation to move as one unit.

LOAD DESIGN

Once the floor plan and elevations have been designed, the designer can start the process of determining how the structure will resist the loads that will be

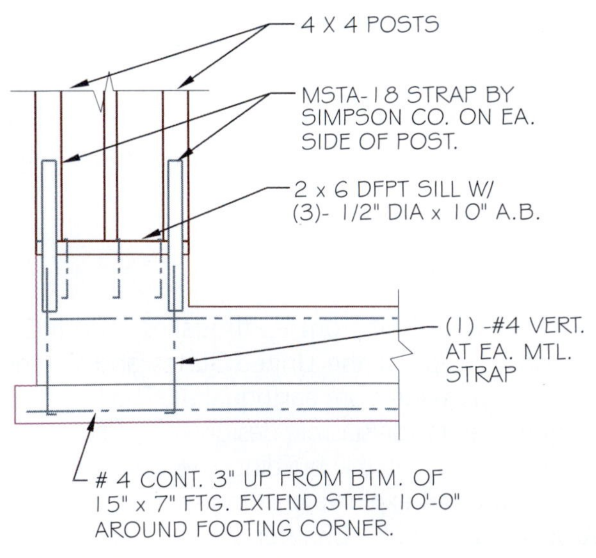

FIGURE 22.2 ■ *Hold-down anchors are used to resist seismic forces and to help form a stable intersection between the floor and foundation.*

imposed on it. The goal of load design is to achieve equilibrium between the structure and the forces that will act upon the structure. For the gravity loads pushing down on the structure, an opposite and equal force or reaction must resist the gravity loads. The structural members of the home form a load path to transfer gravity loads into the foundation and then into the soil. The load path is the route that is used to transfer the roof loads into the walls, then into the floor, and then to the foundation. The structural members must be of sufficient size to resist the loads that are above the member.

To determine the sizes of material required, it is always best to start at the roof and work down to the foundation. When you calculate from the top down, the loads will be accumulating; and when you work down to the foundation, you will have the total loads needed to size the footings. As a new employee, you are not expected to design the structural components. In most offices, the structural design of even the simplest buildings is done by a designer or an engineer. The information in this chapter is a brief introduction into the size of structural members.

LOAD DISTRIBUTION

A beam may be simple or complex. A ***simple beam*** has a uniform load evenly distributed over the entire length of the beam and is supported at each end. With a simple beam, the load on it is equally dispersed to each support. An individual floor joist, rafter, truss or a wall may be thought of as a simple beam. For instance, the wall resisting the wind load in Figure 22.1 can be thought of as a simple beam because it spans between two supporting walls and has a uniform load. If a beam is supporting a uniformly distributed load of 10,000 lb (4536 kg), each supporting post would be resisting 5000 lb (2268 kg). A ***complex beam*** has a nonuniform load at any point of the beam, or supports that are not located at the end of the beam. Chapter 23 introduces methods of determining beam sizes when the loads are not evenly distributed.

Figure 22.3 shows a summary of typical building weights based on minimum design values from the IRC. Figure 22.4 shows the bearing walls for a one-story structure framed with a truss roof system and a post-and-beam floor system. Remember that with a typical truss system, all interior walls are nonbearing. Half of the roof weight will be supported by the left wall, and half of the roof weight will be supported by the right wall. For a structure 32' wide with 2' overhangs, the entire roof will weigh 1440 lb (36' × 40#). Each linear foot of wall will hold 720 lb (18 × 40# psf), which is half the total roof weight.

If the walls are 8' tall, each wall will weigh 80 lb (8' × 10#) per linear feet of wall. The foundation will hold 100 lb of floor load (2' × 50#) per linear foot. Only 2' of floor will be supported, because beams are typically placed at 48'' o.c. Half of the floor weight (2' × 50#) is placed on the stem wall, and half of the weight (2' × 50#) will be supported by the girder parallel to the stem wall. Each interior girder will support 200 lb (4' × 50#) per linear foot of floor weight. The total weight on the stem wall per foot will be the sum of the roof, wall, and floor loads, which equals 900 lb.

Figure 22.5 shows the bearing walls of a two-level structure framed using western platform construction methods. This building has a bearing wall located approximately halfway between the exterior walls. For this

MINIMUM DESIGN LOADS			
MEMBER	DL	LL	TOTAL
Floor (nonsleeping)	10	40	50
Floors (sleeping rooms)	10	30	40
Exterior balconies	10	60	70
Decks	10	40	50
Walls	10	—	—
Attics (without storage)	5	10	15
Attics (with storage)	10	20	30
Roofs (light coverings)	10	30	40
Roofs (tile)	25	30	55

FIGURE 22.3 ■ *Typical live, dead, and total loads for residential construction.*

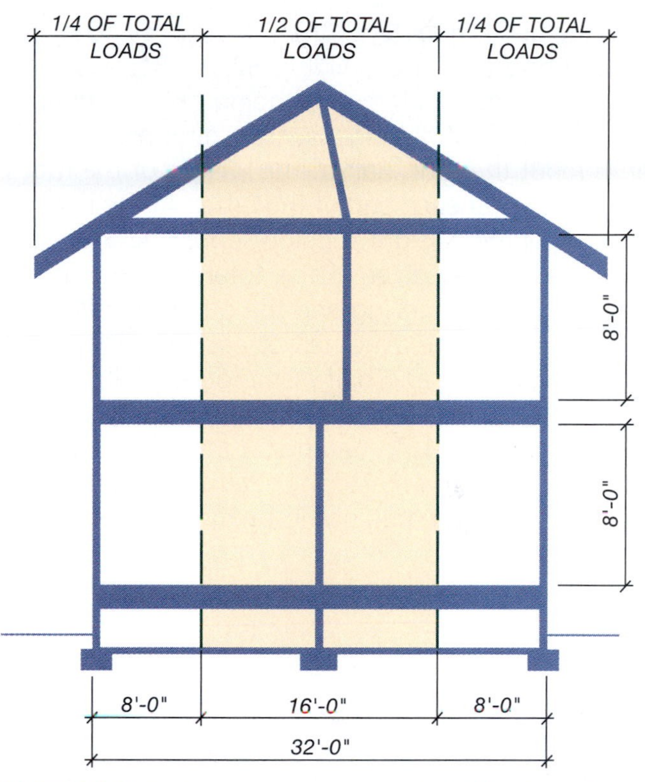

FIGURE 22.5 ■ *Bearing walls and load distribution of a two level home framed using Western platform construction methods.*

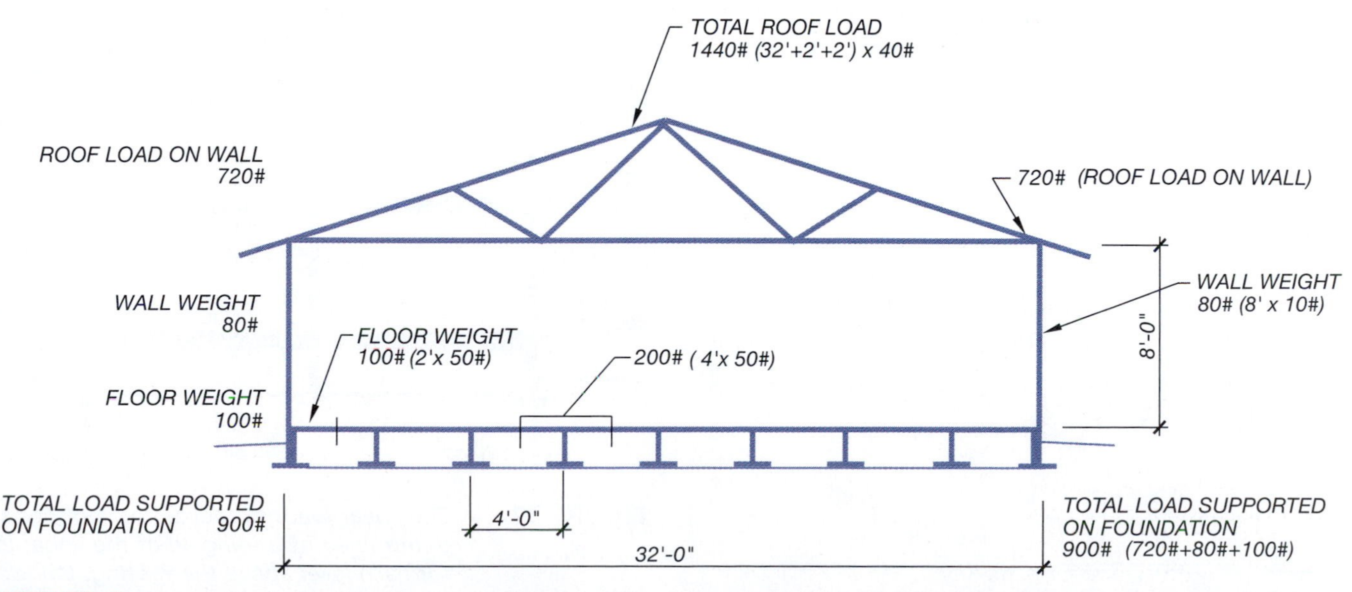

FIGURE 22.4 ■ *Loads for a one-level structure framed with a truss roof with 2' overhangs and a post-and-beam foundation system.*

type of building, one-half of the total building loads will be on the central bearing wall, and one-quarter of the total building loads will be on each exterior wall. Examine the building one floor at a time and see why.

Upper Floor

At the upper floor level, the roof and ceiling loads are being supported. In a building 32 × 15', as shown in Figure 22.6, each rafter is spanning 16' (the horizontal measurement). If the loads are uniformly distributed throughout the roof, half of the weight of the roof will be supported at each end of the rafter. At the ridge, half of the total roof weight is being supported. At each exterior wall, one-quarter of the total roof load is being supported. At the ceiling and the other floor levels, the loading is the same. One-half of each joist is supported at the center wall and half at the outer wall. Using the loads from Figure 22.3, the weights being supported can be determined. To determine the total weight that a wall supports is a matter of determining the area being supported and multiplying by the weight.

Roof

The area being supported at an exterior wall is 15' long by 10' wide (8' of rafter plus 2' overhang), which is 150 sq ft. The roof load equals the sum of the live and dead loads. Assume a live load of 30 psf. For the dead load assume the roof is built with asphalt shingles, 1/2'' ply, and 2 × 8 rafters, for a dead load of approximately 8 psf. For simplicity, 8 can be rounded up to 10 psf for the dead load. By adding the dead and live loads, you can determine that the total roof load is 40 psf. The weight for one linear foot of roof is 400 lb plf. (10 × 40). Multiply the linear load by 15' and it can be determined that the total roof weight on the wall is 6000 lb (15 × 400 lb = 6000 lb).

Because the center area in the example is twice as big, the weight will be twice as big. But just to be sure, check the total weight on the center wall. The linear load is 640 lb (16' × 40 lb) The wall holds 8' of rafters on each side for a total of 16'. Using a LL of 40 lb, the wall is supporting 640 plf. To determine the total load, you should be using 15' × 640 lb for your calculations. Because the wall is 15' long, the total load on the wall is 9600 lb.

Ceiling

The procedure for calculating a ceiling is the same as for a roof, but the loads are different. A typical loading pattern for a ceiling with storage is 30 lb. At the outer walls, the formula to determine the linear load would be 8 × 30 lb, or 240 lb. The total load would be 15' × 240 or 3600 lb. The center wall would be holding 16 × 30 lb, or 480 plf and the total load would be 15 × 480 lb, or 7200 lb.

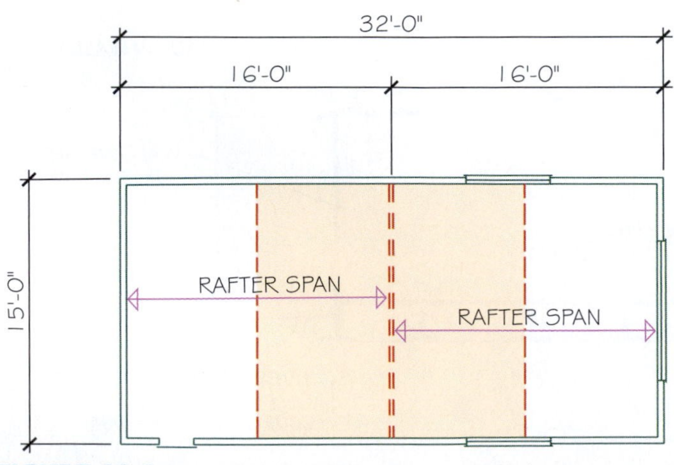

FIGURE 22.6 ■ *Load distribution on a simple beam.*

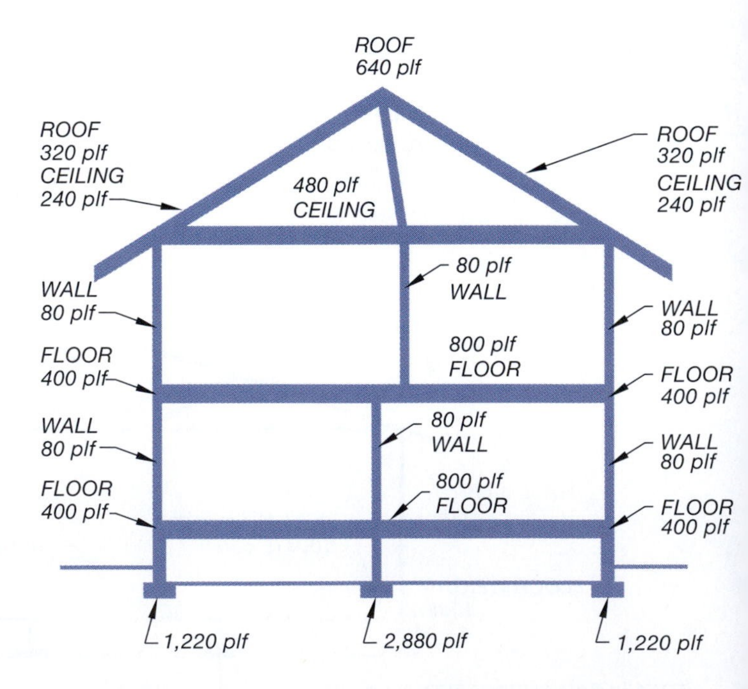

FIGURE 22.7 ■ *The linear loads supported by footings are determined by adding all of the linear loads for each level above the footing. Values are expressed as pounds per linear foot (psf).*

Lower Floor

Finding the weight of a floor is the same as finding the weight of a ceiling, but the loads are much greater. The LL for residential floors is 40 lb and the DL is 10 lb, for a total of 50 lb/sq ft. One linear foot of wall would be support 8' × 50 lb or 400 plf. The floor load at the center wall is 16' × 50 lb or 800 plf.

Walls

The only other weight left to be determined is the weight of the walls. Generally, walls are 8'-0'' high and average about 10 lb psf. Determine the height of the wall and multiply by the length of the wall to find its area. Multiply the area by the weight per square foot, and you will have the total wall weight for one linear foot of wall. Figure 22.7 shows the total weight that will be supported by the footings.

CHAPTER 22

Design Criteria for Structural Loading Test

DIRECTIONS

Answer the following questions with short, complete statements. Type your answers using a word processor or neatly print your answers on lined paper.

1. Type your name, the chapter number, and the date at the top of the sheet.
2. Type the question number and provide the answer in the form of a statement that includes part of the question. You do not need to write out the entire question.

Note: The answers to some questions may not be contained in this chapter and will require you to do additional research.

QUESTIONS

22.1. What are the two major categories of loads that affect buildings?

22.2. What is the safe design live load for a residential floor?

22.3. What is the safe design total load for a residential floor?

22.4. What factors cause snow loads to vary so widely within the same area?

22.5. On a uniformly loaded joist, how will the weight be distributed?

22.6. What load will a floor 15' wide × 25' long generate per linear foot on the stem wall if floor joists are parallel to the long wall? What will be the load if the joists are parallel to the short wall?

22.7. If the floor in question 22.6 had a girder to support the floor joists 7'-6'' in from the edge of the long wall (centered), how much weight would each foot of the girder be supporting?

22.8. A building is 20' wide with 24'' overhangs and a tile roof supported by trusses. The roof shape is a gable, and the building is 30' long. How much roof weight are the walls supporting? How much weight is a footing at the bottom of the 8'-0'' high wall supporting? Express all answers in pounds per linear foot.

22.9. A one-story home with a post-and-beam floor needs a foundation designed. Girders will be placed at 4' o.c. with support every 8'. What weight will the girders be supporting in nonsleeping areas? What weight will the stem wall at the end of the girder be supporting in sleeping areas?

PROBLEMS

Show all work and provide a sketch of the plan view and a simple section showing how loads will be supported.

22.1. If a joist with a total uniform load of 500 lb is supported at each end, how much weight will each end be supporting?

22.2. If a joist with a total uniform load of 700 lb is supported at the midpoint and at each end, how much weight will be supported at each support?

22.3. A 16' rafter ceiling joist at a 6/12 pitch will be supporting a cedar shake roof. What will be the total weight supported if the rafters are at 12'' spacings? 16''? 24''?

22.4. A stub truss with 24'' overhangs will span 28' over a residence and support 300 lb. composition shingles. It will be supported by a wall at one end and a girder truss with a metal hanger at the other end. How much weight will the hanger support?

22.5. Steel columns will be used at each end to support a girder 16' long with a weight of 1260 plf. How much weight will the columns support?

22.6. A residence will be built with a truss roof spanning 24', with built-up roofing and 30'' overhangs. How much weight would a header over a window 8' wide in a bearing wall need to support? How much weight would be supported on an 8' header if it were in the wall perpendicular to the ridge. Provide the answers in plf and for the total load.

Use the following information to complete problems 22.7 through 22.11. Answer the questions by providing the weight per linear foot. A one-level residence will be built using trusses with 36'' overhangs. The residence will be 24' wide. The roof is 235 lb composition shingles. Floor joists will be placed at 16'' o.c. with girders spaced at 12'-0'' o.c. All walls will be 8'-0'' high.

22.7. How much weight will be supported by a header over a window 6' wide in an exterior bearing wall?

22.8. A wall is to be built 11'-9'' from the right bearing wall. If a 2'-6'' pocket door is to be installed in the wall, how long will the header need to be and how much total weight will it be supporting?

22.9. Determine the load to be supported and specify the width of footing required if a wall is placed 14' from the left bearing wall.

22.10. What will be the total dead load per linear foot that the bearing wall on the right side will be supporting?

22.11. What will be the total roof load that the bearing wall on the left side will be supporting?

23 CHAPTER

Sizing Joists, Rafters, and Beams

This chapter introduces loads that must be considered, defines structural lumber, and explores how the sizes of wood framing members are determined. The complexity of the structure and the experience of the CAD technician will determine who will size the framing members. Even if an engineer will determine each framing member, knowledge of the methods used to determine the sizes of framing members will aid you in advancing. There are several skills that you will need to develop in order to determine the sizes of structural members, including the ability to:

- *Understand span tables found in the IRC and in vendor catalogues*
- *Distinguish loading patterns on beams*
- *Recognize standard engineering symbols used in beam formulas*
- *Recognize common causes of beam failure*
- *Understand how to select beams*

WOOD CHARACTERISTICS OF STRUCTURAL LUMBER

No matter how the loads are placed or what the size of the load is, all beams will bend to some extent. How and to what extent a beam will bend depends on its natural properties. These properties vary depending on the type of wood to be used, the size of the member, the grade of lumber used, specific properties of wood, the moisture content of the wood, and defects in the wood.

Types of Framing Lumber

Before the sizes of framing members can be determined, you must be familiar with the species and grade of wood used for framing in your area. The most common types of framing lumber are Douglas fir-larch, southern pine, spruce-pine-fir, and hemlock-fir.

Douglas Fir

Douglas fir is found throughout the western United States and Canada. The species is an all-purpose wood of great dependability and appearance. Douglas fir is used for structural members because it has a high strength-to-weight ratio and excellent dimensional stability. Douglas fir is also used for many woodworking applications such as windows, doors, roller blinds, moldings, ceilings, furniture, and interior trim. The wood is easy to dry and has little tendency to check, warp, cup, twist, or split. Douglas fir is one of the few species where large-sized timber products are available.

Southern Pine

Southern pine is a general name of several pine species that grow across the southern United States from east Texas to Virginia. Its lumber has very high strength, stiffness, and nail-holding capability; it has a unique cellular structure that helps it accept preservative treatments readily. The strength and stiffness of Southern pine make it ideal for framing lumber. Timbers in 4 × 4 and 4 × 6 sizes are used extensively as structural components in framing applications. It is also highly resistant to wear, making it suitable for decks, patios, marinas, boardwalks, and other high-traffic applications, especially after treatment with a preservative.

Spruce-Pine-Fir

SPF (spruce-pine-fir) is the name given to a combination of spruces, pines, and firs that share similar characteristics and are found throughout the forests of Canada. SPF's high fiber strength, light weight, and easy workability make it an excellent choice for framing lumber. SPF dimension lumber is kiln-dried to make it straight and dimensionally stable. Its clean, white appearance and small knots give it an attractive look for interior projects.

Hem-Fir

Hem-fir is the name given to a combination of hemlocks or firs that share similar characteristics. These include western hemlock and the true firs such as noble, California red, grand, Pacific silver, and white fir. These species are primarily found commercially in the northwestern states and British Columbia. Hem-fir design values are nearly as strong as those of Douglas fir, resulting in an economical and excellent structural product. High ratios for strength and stiffness make it a good choice for framing, and its wood is among the lightest in color of the western softwoods. Hem-fir is a species with excellent preservative treatment characteristics, which makes it an economical alternative for naturally durable species like western cedars and redwoods. Hem-fir products are available in structural, appearance, and remanufacturing grades.

Lumber Sizes

Sawn lumber is described using a nominal size and a net size. ***Nominal size*** describes the width and depth of a piece of wood in whole inches, such as a 6 × 10. The 6 represents the width and the 10 represents the depth of the beam. A 6 × 12 will shrink in size as it dries, and it will be reduced in size as it is smoothed with a planer. The final size of wood after planing is referred to as the ***net size.*** The net size of a 6 × 10 is 5 1/2 × 9 1/2'' (140 × 235 mm). The actual or net size is used in designing structural members. Common net sizes include widths of 1 1/2, 3 1/2, 5 1/2 and 7 1/4'' (38, 89, 140, and 184 mm). The net depth will vary depending on the width. Common sizes of net lumber are listed in Figure 23.1.

Lumber Grading

Most structural lumber is visually graded when it is sawn at a mill. With visual inspection methods, wood is evaluated by ASTM standards in cooperation with the U.S. Forest Products laboratory. Some structural lumber is tested nondestructively by machine and graded as mechanically evaluated lumber (MEL). The current standard for rating lumber is the NDS standard (National Design Specifications for Wood Construction), based on destructive testing of visually graded dimension lumber. The lumber guidelines can be found on the American Forest and Paper Association Web site. With NDS guidelines, wood is given a designation such as #1 to define its quality and a base value for the species. The base value is then adjusted, depending on the size of the lumber. The grading process takes into account what portion of the tree the member was sawn from and natural defects in the wood such as knots and checks. In addition to NDS lumber guidelines, LRFD (load reduc-

2'' MATERIAL		4'' MATERIAL		6'' MATERIAL	
1½'' × 3½''	38 × 89	3½'' × 3½''	89 × 89	5½'' × 3½''	140 × 89
1½'' × 5½''	38 × 140	3½'' × 5½''	89 × 140	5½'' × 5½''	140 × 140
1½'' × 7¼''	38 × 184	3½'' × 7¼''	89 × 184	5½'' × 7½''	140 × 190
1½'' × 9¼''	38 × 235	3½'' × 9¼''	89 × 235	5½'' × 9½''	140 × 241
1½'' × 11¼''	38 × 286	3½'' × 11¼''	89 × 286	5½'' × 11½''	140 × 292
1½'' × 13¼''	38 × 337	3½'' × 13½''	89 × 343	5½'' × 13½''	140 × 343

FIGURE 23.1 ▪ *Common net sizes of rectangular-sawn lumber.*

tion factor design standards) guidelines have been adopted by some municipalities. Always verify the current code and lumber standard that will govern the project. No matter how the wood is evaluated, common terms used to define it include *dimension lumber, timbers, post and timbers,* and *beams and stringers.*

Dimension Lumber

Dimension lumber ranges in thickness from 2 to 4'' (50 to 100 mm) and has a moisture content of less than 19%. Common grades of dimension lumber include:

- Stud: Width of 2 to 6'' (50 to 150 mm) and a maximum length of 10' (3000 mm)
- Select structural, No. 1 and better: No. 1, 2, and 3, construction, and utility

Timbers

Timber is both a general classification for large sizes of structural lumber and the name of a specific grade and size. Structural lumber that is at least 5'' (125 mm) thick is considered to be timber. The two basic categories of timber include post and timbers and beams and stringers.

- **Post and timbers:** This classification refers to structural lumber that is 5 × 5 (125 × 125) and larger with a thickness and width the same as or within 2'' (50 mm) of each other, such as a 6 × 6 or 6 × 8 (150 × 150 or 150 × 200) beams.
- **Beams and stringers:** This classification refers to structural lumber that is 5 × 5 (125 × 125) and larger with a thickness 2'' (50 mm) or greater than the width, such as a 6 × 10 or 6 × 12 (150 × 250 or 150 × 300).

Properties of Lumber

In addition to the terms used to describe lumber, several key terms must be understood to determine the size of wood framing members. These terms include *breadth, depth, area, extreme fiber stresses, neutral axis, moment of inertia, section modulus,* and *grain* (see Figure 23.2).

b: Breadth of a rectangular beam in inches
d: Depth of a rectangular beam in inches
A: Area of the beam, determined using b × d
F_b: Extreme fiber stresses. As a beam deflects from the load it supports, the surface of the beam supporting the load will be placed in compression. The bottom of the beam will be placed in tension.

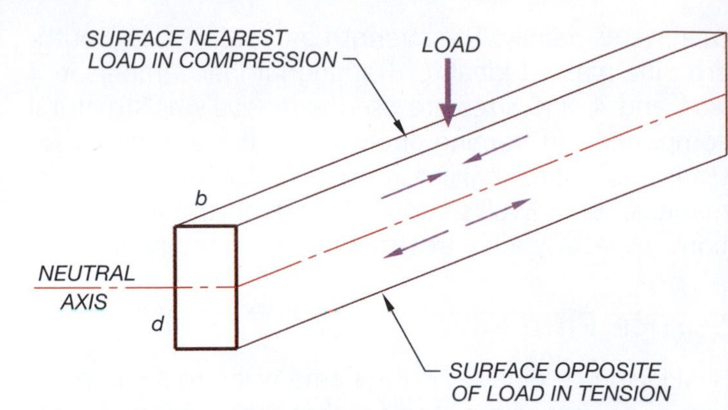

FIGURE 23.2 ■ *Common properties of rectangular lumber. When a load is placed on the top of a beam, the surface nearest the load is placed in compression and the surface away from the beam is in tension.*

Neutral axis: The axis formed where the forces of compression and tension in a beam reach equilibrium. It is formed at the midpoint between the forces of compression and tension. Because the upper and lower surfaces of the beam resist the most stress, holes can be drilled near the neutral axis without greatly affecting the strength of the beam. This chapter will explore the placement of loads on a beam and help determine the loads that must be resisted so that equilibrium can be achieved.
I: Moment of inertia is the sum of the products of each of the elementary areas of a beam multiplied by the square of their distance from the neutral axis of the cross section. This sounds technical, but a value for I is listed in tables that are presented later in this chapter for use in deflection calculations.
S: Section modulus is the moment of inertia divided by the distance from the neutral axis to the extreme fiber of the cross section. Once again, this sounds technical, but a value for S is listed in tables presented later in this chapter for use in determining bending strength calculations.
Grain: Wood grain is composed of fibers that can be visualized as grains of rice that are aligned in the same direction. The wood fibers are strongest in their long direction. Common relationships of loads to grain include loads placed parallel and perpendicular to the grain. How a beam reacts will depend on the relationship of the load to the grain of the wood.

- When a load is applied in the same direction as the fibers of a structural member, the force is said to be parallel to the grain. Loads that are parallel to the grain are represented by the symbol //.

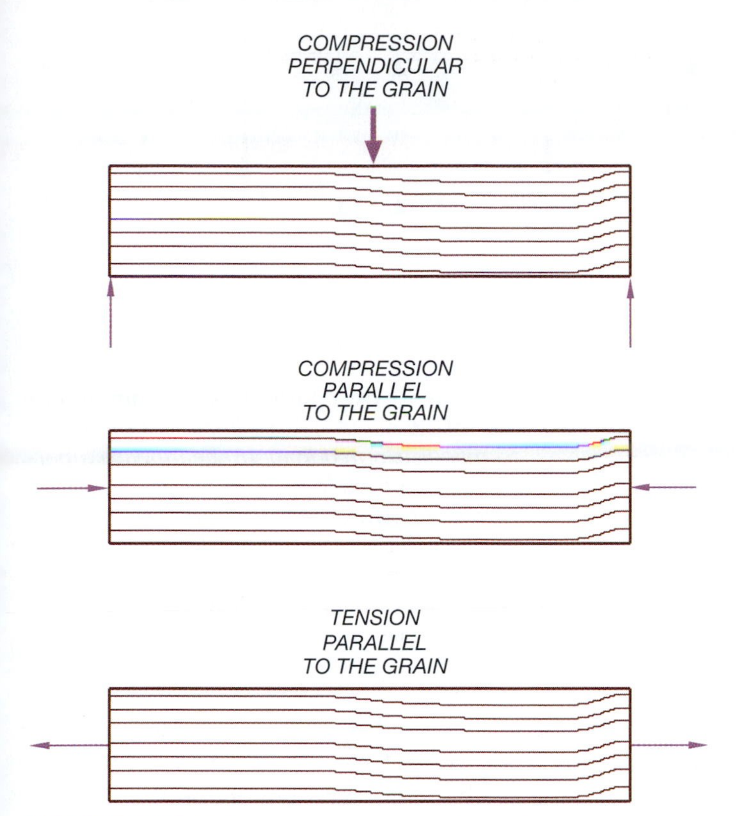

FIGURE 23.3 ■ *Forces acting on horizontal members.*

- When a load is applied across the direction of the fibers of a structural member, the force is said to be perpendicular to the grain and is represented by the symbol ⊥. Figure 23.3 shows a force applied to a beam in each direction.

Loads typically affect structural members in five different ways: bending strength, deflection, horizontal shear, vertical shear, and bearing area. These forces will create stress that affects the fibers inside the beam. The forces of tension and compression from outside of a beam are also considered. Each force is shown in Figure 23.4.

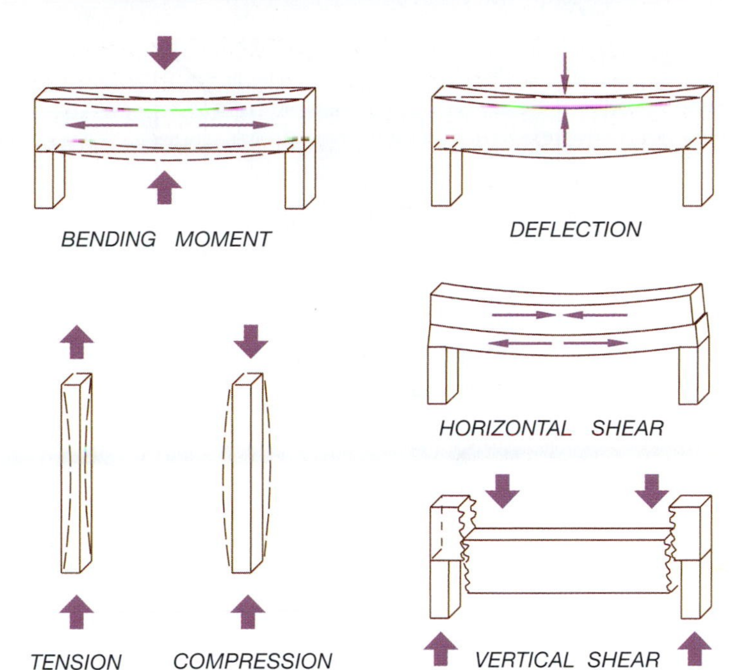

FIGURE 23.4 ■ *Forces acting on structural members.*

LOADING AND SUPPORT PATTERNS OF STRUCTURAL MEMBERS

There are two common ways to load and support a joist or beam. Loads can be uniformly distributed over the entire span of a structural member or can be concentrated in one small area of a beam. In Chapter 22, uniformly distributed loads were discussed. A concentrated load is one that is placed on only one area of a beam. A support post from an upper floor resting on a beam on a lower floor or the weight of a car being transferred through a wheel into the floor system are examples of concentrated loads. Occasionally you will need to deal with increasing loads that result from triangular loading patterns. A beam used to frame a hip or valley supports a triangular load. The triangular load that results from a hip is shown in Figure 23.5. The load starts at zero at the low end of the hip. The maximum load is at the high end of the hip, where it intersects the ridge.

Usually a beam is supported at each end. Common alternatives are to support a beam at the center of the span in addition to the ends or to provide a support at one end and near the other end. The type of beam that extends past the support is called a cantilevered beam. Examples of each type of load and the support system are shown in Figure 23.6.

Loading Reactions of Wood Members

For every action there is an equal and opposite reaction. This law of physics affects every structure. There are several actions or stresses that must be understood before beam reactions can be considered. These stresses include fiber bending stress, deflection, horizontal shear, vertical shear, and compression. You do not need to know how these stresses are generated, only that they exist, to determine the size of a framing member using standard loading tables. Design loads have been discussed in earlier chapters. Review Figure 22.3 for a partial list of design live (LL) and dead (DL) loads. These design loads will be useful in determining the load on a

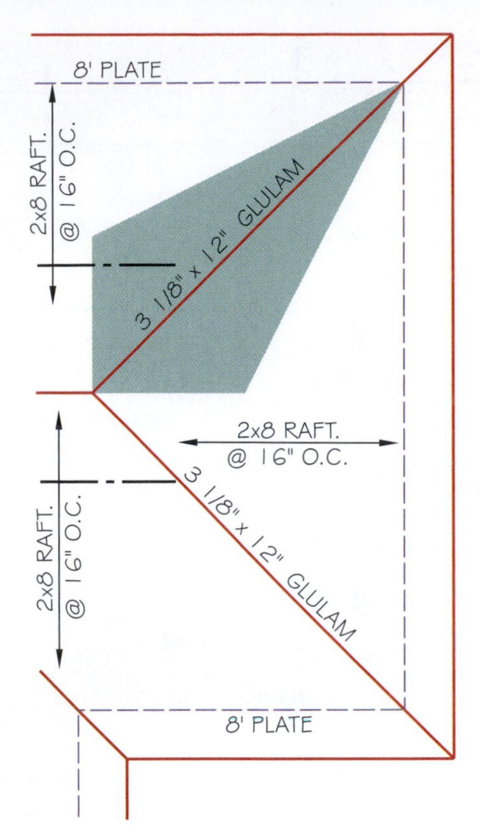

FIGURE 23.5 ■ *The beam used to frame a hip or valley supports a triangular load. The triangle is formed because half of the weight of each rafter is supported on the beam.*

beam. Figure 23.7 shows a partial listing of base design values needed for the design of beams. This table includes a column of values for fiber bending stress (**F_b**), modulus of elasticity (**E**), and horizontal shear (**F_v**).

Fiber Bending Stress

The bending strength of a wood beam is measured in units of fiber bending stress. Earlier you read about extreme fiber stress (compression) that occurs on the beam surface supporting the load. Extreme fiber stress (tension) also occurs on the surface opposite the surface in compression. The relationship of allowable fiber bending stress to the maximum bending stress is used to determine the required strength of a framing member. Fiber bending stress is represented by the symbol **F_b**. The values listed in beam tables indicate the safe, allowable fiber bending.

Bending Strength

Bending strength is the determination of the beam strength required to resist the force applied to a beam measured in moments. The ***bending moment*** at any

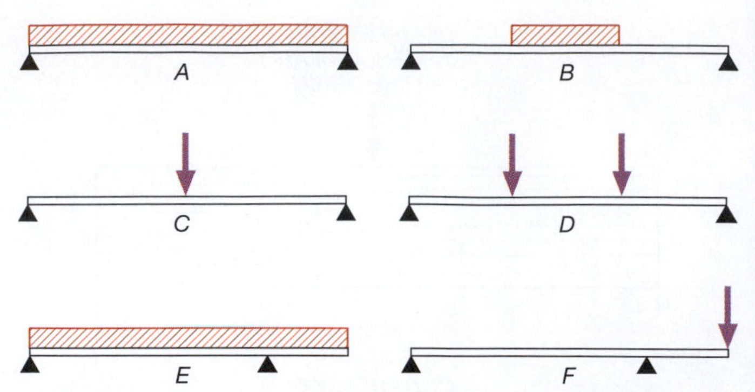

FIGURE 23.6 ■ *Common loading patterns on a beam include the following: (A) simple beam with a uniformly distributed load; (B) beam with a partially distributed load at the center; (C) beam with a concentrated load at the center; (D) beam with two equal concentrated loads placed symmetrically; (E) cantilevered beam with a uniform load; (F) cantilevered beam with a concentrated load at the free end.*

point of the beam is the measure of the tendency of the beam to bend due to the force acting on the beam. The magnitude of the bending moment varies throughout the length of the beam. The maximum bending stress occurs at the midpoint of a simple beam with a uniform load. The location for maximum bending moment in complex beams is discussed later in this chapter. The letter **M** represents bending moment in engineering formulas. The relationship of the allowable extreme fiber bending stress (**F_b**) to the maximum bending moment (**M**) equals the required section modulus (**S**) of a beam. The design equation is $S = M/F_b$. The **F_b** base design value listed in psi for various lumber species can be determined using Figure 23.7. The base **F_b** value for 4× members varies depending on the depth of the beam. The base design value for 6× material remains fixed up to and including 6 × 12. Determining the required section modulus of a beam will provide one of the pieces of information needed to determine the size of beam needed to support a specific load over a given span.

Deflection

Deflection deals with the stiffness of a beam. Deflection measures the tendency of a structural member to bend under a load and as a result of gravity. As a load is placed on a beam, the beam will sag between its supports. As the span increases, the tendency to deflect increases. Deflection rarely causes a beam to break, but greatly affects the materials that the beam supports. When floor joists sag too much, sheetrock will crack or doors and windows will stick. Two formulas will be used

DESIGN VALUES FOR BEAMS					
SIZE, SPECIES, AND COMMERCIAL GRADE	EXTREME FIBER BENDING F_b			HORIZONTAL SHEAR F_v	MODULUS OF ELASTICITY (E)
DFL #2	Base Value	Modifier	Increased Value		
4 × 8	875	(1.3)	1138	85	1.6
4 × 10	875	(1.2)	1050	85	1.6
4 × 12	875	(1.1)	963	85	1.6
DFL #1					
6 × 8	1350			85	1.6
6 × 10	1350			85	1.6
6 × 12	1350			85	1.6
Reduce the base value of all members larger than 6 × 12 by $(12/d)1/9=C_f$.					
Hem-Fir					
6 × _	1050			70	1.3

FIGURE 23.7 ■ *Partial listing of safe design values of common types of lumber used for beams. All values are based on National Design Specifications for Wood Construction (NDS), published by the American Forest and Paper Association.*

to determine deflection. The first is the legal limit of deflection that determines how much the building code will allow a specific beam to bend. Limits are set by the IRC and are expressed as a ratio of the length of the beam in inches over a deflection value. Allowable deflection for a floor beam is represented by the ratio L/360. For a 10' long floor beam, it is allowed to deflect 0.33'' (120''/360). The allowable deflection value defines the maximum amount of deflection (can sag). The IRC allowable deflections include:

L/360 (floors)
L/240 (roofs supporting ceilings)
L/180 (roofs with a slope of 3/12 or greater with no ceiling loads)

Modulus of Elasticity

Modulus of elasticity also deals with the stiffness of a structural member. ***Modulus of elasticity*** is a ratio of the amount a member will deflect in proportion to the applied load. The letter **E** is used in beam tables and design formulas to represent the value for the modulus of elasticity. The **E** value represents how much the member will deflect (the will sag value). For a beam to be safe, the **E** value (will sag) cannot exceed the deflection value (can sag). Different formulas for varied loading conditions are introduced throughout the balance of this chapter. These formulas consider the load and span of the beam, the modulus of elasticity, and the moment of inertia of the beam. Deflection is represented by the symbol Δ. The value for the modulus of elasticity is expressed as **E** for the species and grade of the beam to be used, and the moment of inertia is expressed as **I.**

Horizontal Shear

Horizontal shear is the tendency for the wood fibers to slide past each other at the neutral axis and fail along the length of the beam. Horizontal shear is a result of forces that affect the beam fibers parallel to the wood grain. Under severe bending pressure, adjacent wood fibers are pushed and pulled in opposite directions. The top portion of the beam nearest the load is in compression. The edge of the beam away from the load is under stress from tension. The line where the compression and tension forces meet is the point at which the beam will fail. Horizontal shearing stresses are greatest at the neutral surface. The maximum horizontal shear stress for a rectangular beam is one and one half times the average unit shear stress. The design formula is v = (1.5)(V)/A, where:

v = maximum unit horizontal shear stress in psi
V = total ***vertical shear*** in pounds
A = area of the structural member in square inches

Figure 23.7 lists the maximum allowable horizontal shear values in psi for major species of framing lumber. The value for **V** must be less than the maximum safe F_v value.

Horizontal shear has the greatest effect on beams with a relatively heavy load spread over a short span. Visualize a yardstick with supports at each end supporting a 2-lb load at the center. The yardstick will bow but will probably not break. Move the supports toward the center of the beam so they are 1' apart. The beam will not bend but will be more prone to failure from breaking from the stress of horizontal shear.

Vertical Shear

Vertical shear is the tendency of a beam to fail perpendicular to the fibers of the beam from two opposing forces. Vertical shear will cause a beam to break and fall between its supporting posts. Vertical shear is rarely a design concern in residential construction, since a beam will first fail by horizontal shear.

Tension

In addition to the forces of tension that act within a beam, tension stresses attempt to lengthen a structural member. Tension is rarely a problem in residential beam design. Tension is more likely to affect the intersection of beams than the beam itself. Beam details typically reflect a metal strap to join beams laid end to end, or a metal seat may be used to join beams to a column.

Compression

Compression is the tendency to compress a structural member. The fibers of a beam resting on a column tend to compress, but residential loads are usually not sufficient to cause a structural problem. Posts are another area where compression can be seen. Wood is strongest along the grain. The loads in residential design typically are not large enough to cause problems in the design of residential columns. However, the length of the post is important in relation to the load to be supported. As the length of the post increases, the post tends to bend rather than compress. Building departments often require structures with posts longer than 10' to be approved by a structural engineer or architect. The forces of tension and compression are not analyzed in this book.

USING SPAN TABLES FOR DIMENSIONAL LUMBER

Standard framing practice is to place structural members such as joists and rafters at 12, 16, or 24'' (300, 400, or 600 mm) o.c. Because of this practice, standard tables have been developed for the sizing of repetitive members. Joists and rafters are considered simple beams with uniform loads, allowing sizes to be determined from span tables contained in the IRC. Span tables for dimensional lumber are also available from major lumber manufacturers such as the Western Wood Products Association or the Southern Forest Products Association. To use these tables you must understand a few basic facts, including typical loading reactions of framing members and the structural capabilities of various species and grades of wood.

The most common types of framing lumber are Douglas fir-larch (DFL #2), southern pine (SP #2), spruce-pine-fir (SPF #2), and hemlock-fir (Hem-Fir #2). Notice that each species is followed by #2. This number refers to the grade value of the species. Usually only #1 or #2 grade lumber is used as structural lumber. Once the type of framing lumber is known, the safe span can be determined.

Determining Size and Span

Once the species of the framing lumber is known, the size and span can be determined. The IRC provides allowable span tables for floor joists, ceiling joists, and rafters. Determine how the wood is to be used and proceed to the proper table.

A few simple headings and sub-headings must be located at the top of the table before any spans can be identified. See Figure 23.8 to become familiar with the basics of the span table. Some of the key information to be gleaned from this part of the table follows:

- **Title:** Most codes include several different span charts, and it is extremely easy to use the wrong table. Double check the title to see that you have the right table.
- **Loads:** Within some categories of tables, the values for the table are determined by the loads assumed to be supported. In Figure 23.8, the table is based on an assumed LL value of 40 lb per sq ft and dead loads of 10 or 20 lb. The values in this table will work well for a residence.
- **Deflection:** Deflection describes the stiffness of a beam by its tendency to bend under a load. Allowable deflection is represented by a symbol such as $L/\Delta = 360$, where L equals the length of the joist in feet and Δ represents the deflection value. Other allowable options include $L/\Delta = 240$ and $L/\Delta = 180$. The IRC lists the allowable deflection in the design criteria of each span table. The allowable deflection listed in the design criteria defines how much a member is allowed to sag (can sag).

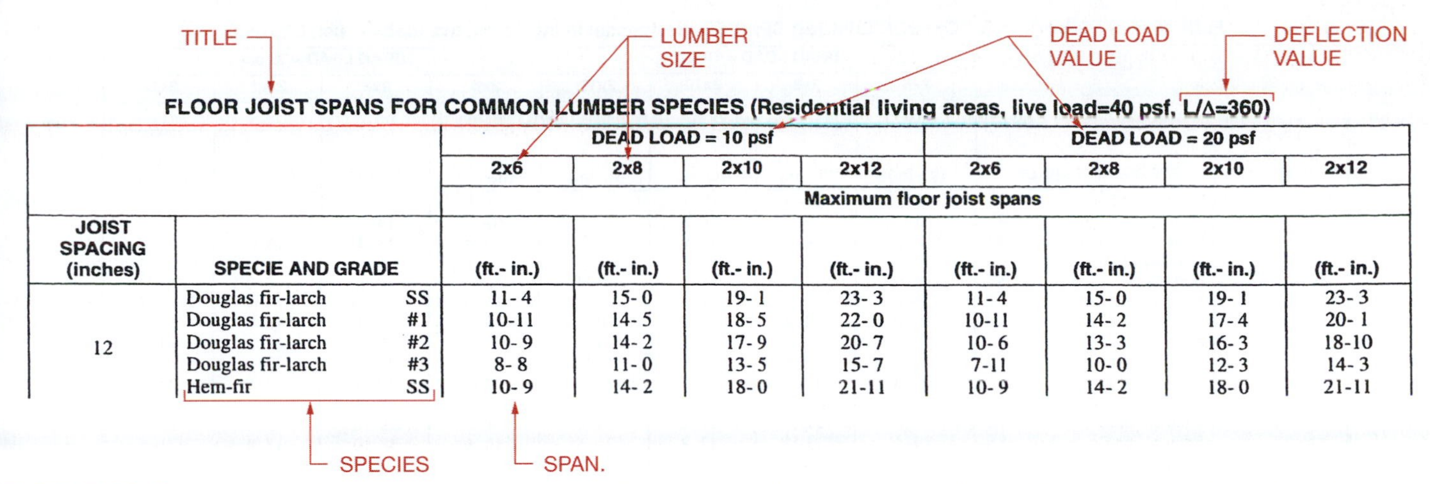

FLOOR JOIST SPANS FOR COMMON LUMBER SPECIES (Residential living areas, live load=40 psf, L/Δ=360)

JOIST SPACING (inches)	SPECIE AND GRADE		DEAD LOAD = 10 psf				DEAD LOAD = 20 psf			
			2x6	2x8	2x10	2x12	2x6	2x8	2x10	2x12
			Maximum floor joist spans							
			(ft.- in.)	(ft.- in.)	(ft.- in.)	(ft.- in.)	(ft.- in.)	(ft.- in.)	(ft.- in.)	(ft.- in.)
12	Douglas fir-larch	SS	11-4	15-0	19-1	23-3	11-4	15-0	19-1	23-3
	Douglas fir-larch	#1	10-11	14-5	18-5	22-0	10-11	14-2	17-4	20-1
	Douglas fir-larch	#2	10-9	14-2	17-9	20-7	10-6	13-3	16-3	18-10
	Douglas fir-larch	#3	8-8	11-0	13-5	15-7	7-11	10-0	12-3	14-3
	Hem-fir	SS	10-9	14-2	18-0	21-11	10-9	14-2	18-0	21-11

FIGURE 23.8 ■ *Because of a wide variety of span tables with a similar appearance, it is critical that a few basic facts be studied before using each table. Before determining a span, verify the table title, the live and dead loads, the deflection limits, the lumber size, and spacing. Reproduced from the International Residential Code / 2006. Copyright © 2006.* Courtesy International Code Council, Inc.

- **Size and spacing of lumber:** The left side of the span tables is the size and spacing of the framing lumber. In Figure 23.9 you will notice that each size of structural member has a value for the spacing of 12, 16, 19.2, or 24'' o.c.

Sizing Floor Joists

To use Figure 23.9, find the column that represents proper species and grade for the lumber used in your area. For this example, Douglas fir #2 is used. The first listing in the column is 11-4. The 11-4 represents a maximum allowable span of 11'-4''. Because the most common spacing for floor joists is 16'' o.c., drop down to the 16'' spacing box. Using the Douglas fir-larch #2 option of the 16'' row and the 10-lb column, you will find spans listed for 2 × 6 = 9-9, 2 × 8 = 12-7, 2 × 10 = 15-5, and 2 × 12 = 17-10. If you were looking for floor joists to support a living room floor that is 14'-6'' wide, the 2 × 10 floor joists would be suitable. Using 2 × 12 floor joists placed at 24'' o.c. would also be suitable, with a maximum span of 14'-7''.

Using Figure 23.9, determine the distance that 2 × 8 hem-fir #2 floor joists will span at 16'' o.c. Work to the right until you come to 2 × 8. Now work down to the 16'' spacing row. The span for a 2 × 8 at 16'' o.c. with a 10# DL is 12'-0''. Use the same procedure for determining the span for a 2 × 10 SP #2 floor joist at 16'' o.c. The listed span is 16'-1''.

Typically, the span will be known, and the size of member will need to be determined. For instance, determine the size of floor joists needed for a living area 15' wide. Assume southern pine with a dead load of 10 lb will be used. Using joists spaced at 24'' or 19.2'' o.c., 2 × 12 could be used. With a spacing of 16'' o.c., either 2 × 10 or 2 × 12 joists could be used. Even if select structural lumber is used, 2 × 8 joists cannot be used to span 15'-0''. If a girder is used at the midpoint to reduce the span, 2 × 6 SP #2 joists placed at 24'' o.c. could be used safely. The designer will now need to determine if it is more cost effective to use 2 × 6 joists at 24'' o.c. and a girder, or 2 × 10 joists at 16'' o.c. with no girder. For most markets, the cost of labor will be the determining factor rather than the cost of the lumber. To reduce labor cost, the 2 × 10 SP #2 at 16'' o.c. will be more economical.

The IRC allows the floor live load of sleeping areas to be reduced from 40 to 30 lb. The table in Figure 23.10 can be used to determine floor joists with a live load of 30 lb and a dead load of either 10 or 20 psf. Values from this table are determined using the same methods that were used with Figure 23.9. Notice that to determine the DFL #2 floor joists needed to support a bedroom with a 14'-6'' span, that 2 × 8 at 12'' o.c. or 2 × 10 at 16'' o.c. can be used.

Sizing Ceiling Joists

To determine the size of a ceiling joist, use the proper table and the same procedure that was used to size a floor joist. Because they are both horizontal members, they will have similar loading patterns. A table for sizing ceiling joists is shown in Figure 23.11. For ceiling joists, the live load is only 10 lb. Span tables are available for ceiling joists with and without storage. Using Figure 23.11, determine the smallest size DFL #2 ceiling joists that will span 16' at 16'' o.c. You will find that 2 × 6 joists at 16'' o.c. will span 17'-8''. Figure 23.12 shows a

FLOOR JOIST SPANS FOR COMMON LUMBER SPECIES (Residential living areas, live load=40 psf, L/Δ=360)

			DEAD LOAD = 10 psf				DEAD LOAD = 20 psf			
			2x6	2x8	2x10	2x12	2x6	2x8	2x10	2x12
			Maximum floor joist spans							
JOIST SPACING (inches)	SPECIE AND GRADE		(ft.- in.)	(ft.- in.)	(ft.- in.)	(ft.- in.)	(ft.- in.)	(ft.- in.)	(ft.- in.)	(ft.- in.)
12	Douglas fir-larch	SS	11-4	15-0	19-1	23-3	11-4	15-0	19-1	23-3
	Douglas fir-larch	#1	10-11	14-5	18-5	22-0	10-11	14-2	17-4	20-1
	Douglas fir-larch	#2	10-9	14-2	17-9	20-7	10-6	13-3	16-3	18-10
	Douglas fir-larch	#3	8-8	11-0	13-5	15-7	7-11	10-0	12-3	14-3
	Hem-fir	SS	10-9	14-2	18-0	21-11	10-9	14-2	18-0	21-11
	Hem-fir	#1	10-6	13-10	17-8	21-6	10-6	13-10	16-11	19-7
	Hem-fir	#2	10-0	13-2	16-10	20-4	10-0	13-1	16-0	18-6
	Hem-fir	#3	8-8	11-0	13-5	15-7	7-11	10-0	12-3	14-3
	Southern pine	SS	11-2	14-8	18-9	22-10	11-2	14-8	18-9	22-10
	Southern pine	#1	10-11	14-5	18-5	22-5	10-11	14-5	18-5	22-5
	Southern pine	#2	10-9	14-2	18-0	21-9	10-9	14-2	16-11	19-10
	Southern pine	#3	9-4	11-11	14-0	16-8	8-6	10-10	12-10	15-3
	Spruce-pine-fir	SS	10-6	13-10	17-8	21-6	10-6	13-10	17-8	21-6
	Spruce-pine-fir	#1	10-3	13-6	17-3	20-7	10-3	13-3	16-3	18-10
	Spruce-pine-fir	#2	10-3	13-6	17-3	20-7	10-3	13-3	16-3	18-10
	Spruce-pine-fir	#3	8-8	11-0	13-5	15-7	7-11	10-0	12-3	14-3
16	Douglas fir-larch	SS	10-4	13-7	17-4	21-1	10-4	13-7	17-4	21-0
	Douglas fir-larch	#1	9-11	13-1	16-5	19-1	9-8	12-4	15-0	17-5
	Douglas fir-larch	#2	9-9	12-7	15-5	17-10	9-1	11-6	14-1	16-3
	Douglas fir-larch	#3	7-6	9-6	11-8	13-6	6-10	8-8	10-7	12-4
	Hem-fir	SS	9-9	12-10	16-5	19-11	9-9	12-10	16-5	19-11
	Hem-fir	#1	9-6	12-7	16-0	18-7	9-6	12-0	14-8	17-0
	Hem-fir	#2	9-1	12-0	15-2	17-7	8-11	11-4	13-10	16-1
	Hem-fir	#3	7-6	9-6	11-8	13-6	6-10	8-8	10-7	12-4
	Southern pine	SS	10-2	13-4	17-0	20-9	10-2	13-4	17-0	20-9
	Southern pine	#1	9-11	13-1	16-9	20-4	9-11	13-1	16-4	19-6
	Southern pine	#2	9-9	12-10	16-1	18-10	9-6	12-4	14-8	17-2
	Southern pine	#3	8-1	10-3	12-2	14-6	7-4	9-5	11-1	13-2
	Spruce-pine-fir	SS	9-6	12-7	16-0	19-6	9-6	12-7	16-0	19-6
	Spruce-pine-fir	#1	9-4	12-3	15-5	17-10	9-1	11-6	14-1	16-3
	Spruce-pine-fir	#2	9-4	12-3	15-5	17-10	9-1	11-6	14-1	16-3
	Spruce-pine-fir	#3	7-6	9-6	11-8	13-6	6-10	8-8	10-7	12-4
19.2	Douglas fir-larch	SS	9-8	12-10	16-4	19-10	9-8	12-10	16-4	19-2
	Douglas fir-larch	#1	9-4	12-4	15-0	17-5	8-10	11-3	13-8	15-11
	Douglas fir-larch	#2	9-1	11-6	14-1	16-3	8-3	10-6	12-10	14-10
	Douglas fir-larch	#3	6-10	8-8	10-7	12-4	6-3	7-11	9-8	11-3
	Hem-fir	SS	9-2	12-1	15-5	18-9	9-2	12-1	15-5	18-9
	Hem-fir	#1	9-0	11-10	14-8	17-0	8-8	10-11	13-4	15-6
	Hem-fir	#2	8-7	11-3	13-10	16-1	8-2	10-4	12-8	14-8
	Hem-fir	#3	6-10	8-8	10-7	12-4	6-3	7-11	9-8	11-3
	Southern pine	SS	9-6	12-7	16-0	19-6	9-6	12-7	16-0	19-6
	Southern pine	#1	9-4	12-4	15-9	19-2	9-4	12-4	14-11	17-9
	Southern pine	#2	9-2	12-1	14-8	17-2	8-8	11-3	13-5	15-8
	Southern pine	#3	7-4	9-5	11-1	13-2	6-9	8-7	10-1	12-1
	Spruce-pine-fir	SS	9-0	11-10	15-1	18-4	9-0	11-10	15-1	17-9
	Spruce-pine-fir	#1	8-9	11-6	14-1	16-3	8-3	10-6	12-10	14-10
	Spruce-pine-fir	#2	8-9	11-6	14-1	16-3	8-3	10-6	12-10	14-10
	Spruce-pine-fir	#3	6-10	8-8	10-7	12-4	6-3	7-11	9-8	11-3
24	Douglas fir-larch	SS	9-0	11-11	15-2	18-5	9-0	11-11	14-9	17-1
	Douglas fir-larch	#1	8-8	11-0	13-5	15-7	7-11	10-0	12-3	14-3
	Douglas fir-larch	#2	8-1	10-3	12-7	14-7	7-5	9-5	11-6	13-4
	Douglas fir-larch	#3	6-2	7-9	9-6	11-0	5-7	7-1	8-8	10-1
	Hem-fir	SS	8-6	11-3	14-4	17-5	8-6	11-3	14-4	16-10[a]
	Hem-fir	#1	8-4	10-9	13-1	15-2	7-9	9-9	11-11	13-10
	Hem-fir	#2	7-11	10-2	12-5	14-4	7-4	9-3	11-4	13-1
	Hem-fir	#3	6-2	7-9	9-6	11-0	5-7	7-1	8-8	10-1
	Southern pine	SS	8-10	11-8	14-11	18-1	8-10	11-8	14-11	18-1
	Southern pine	#1	8-8	11-5	14-7	17-5	8-8	11-3	13-4	15-11
	Southern pine	#2	8-6	11-0	13-1	15-5	7-9	10-0	12-0	14-0
	Southern pine	#3	6-7	8-5	9-11	11-10	6-0	7-8	9-1	10-9
	Spruce-pine-fir	SS	8-4	11-0	14-0	17-0	8-4	11-0	13-8	15-11
	Spruce-pine-fir	#1	8-1	10-3	12-7	14-7	7-5	9-5	11-6	13-4
	Spruce-pine-fir	#2	8-1	10-3	12-7	14-7	7-5	9-5	11-6	13-4
	Spruce-pine-fir	#3	6-2	7-9	9-6	11-0	5-7	7-1	8-8	10-1

For SI: 1 inch = 25.4 mm, 1 foot = 308.4 mm.

NOTES:

a. Check sources for availability of lumber in lengths greater than 20 feet.

b. End bearing length shall be increased to 2 inches.

FIGURE 23.9 ■ *This span table is suitable for determining floor joists for living areas with a live load of 40 psf and dead load of either 10 or 20 psf. Reproduced from the International Residential Code / 2006. Copyright © 2006.* Courtesy International Code Council, Inc.

FLOOR JOIST SPANS FOR COMMON LUMBER SPECIES
(Residential sleeping areas, live load=30 psf, L/Δ=360)

			DEAD LOAD = 10 psf				DEAD LOAD = 20 psf			
			2x6	2x8	2x10	2x12	2x6	2x8	2x10	2x12
			Maximum floor joist spans							
JOIST SPACING (inches)	SPECIE AND GRADE		(ft.- in.)	(ft.- in.)	(ft.- in.)	(ft.- in.)	(ft.- in.)	(ft.- in.)	(ft.- in.)	(ft.- in.)
12	Douglas fir-larch	SS	12-6	16-6	21-0	25-7	12-6	16-6	21-0	25-7
	Douglas fir-larch	#1	12-0	15-10	20-3	24-8	12-0	15-7	19-0	22-0
	Douglas fir-larch	#2	11-10	15-7	19-10	23-0	11-6	14-7	17-9	20-7
	Douglas fir-larch	#3	9-8	12-4	15-0	17-5	8-8	11-0	13-5	15-7
	Hem-fir	SS	11-10	15-7	19-10	24-2	11-10	15-7	19-10	24-2
	Hem-fir	#1	11-7	15-3	19-5	23-7	11-7	15-2	18-6	21-6
	Hem-fir	#2	11-0	14-6	18-6	22-6	11-0	14-4	17-6	20-4
	Hem-fir	#3	9-8	12-4	15-0	17-5	8-8	11-0	13-5	15-7
	Southern pine	SS	12-3	16-2	20-8	25-1	12-3	16-2	20-8	25-1
	Southern pine	#1	12-0	15-10	20-3	24-8	12-0	15-10	20-3	24-8
	Southern pine	#2	11-10	15-7	19-10	18-8	11-10	15-7	18-7	21-9
	Southern pine	#3	10-5	13-3	15-8	18-8	9-4	11-11	14-0	16-8
	Spruce-pine-fir	SS	11-7	15-3	19-5	23-7	11-7	15-3	19-5	23-7
	Spruce-pine-fir	#1	11-3	14-11	19-0	23-0	11-3	14-7	17-9	20-7
	Spruce-pine-fir	#2	11-3	14-11	19-0	23-0	11-3	14-7	17-9	20-7
	Spruce-pine-fir	#3	9-8	12-4	15-0	17-5	8-8	11-0	13-5	15-7
16	Douglas fir-larch	SS	11-4	15-0	19-1	23-3	11-4	15-0	19-1	23-0
	Douglas fir-larch	#1	10-11	14-5	18-5	21-4	10-8	13-6	16-5	19-1
	Douglas fir-larch	#2	10-9	14-1	17-2	19-11	9-11	12-7	15-5	17-10
	Douglas fir-larch	#3	8-5	10-8	13-0	15-1	7-6	9-6	11-8	13-6
	Hem-fir	SS	10-9	14-2	18-0	21-11	10-9	14-2	18-0	21-11
	Hem-fir	#1	10-6	13-10	17-8	20-9	10-4	13-1	16-0	18-7
	Hem-fir	#2	10-0	13-2	16-10	19-8	9-10	12-5	15-2	17-7
	Hem-fir	#3	8-5	10-8	13-0	15-1	7-6	9-6	11-8	13-6
	Southern pine	SS	11-2	14-8	18-9	22-10	11-2	14-8	18-9	22-10
	Southern pine	#1	10-11	14-5	18-5	22-5	10-11	14-5	17-11	21-4
	Southern pine	#2	10-9	14-2	18-0	21-1	10-5	13-6	16-1	18-10
	Southern pine	#3	9-0	11-6	13-7	16-2	8-1	10-3	12-2	14-6
	Spruce-pine-fir	SS	10-6	13-10	17-8	21-6	10-6	13-10	17-8	21-4
	Spruce-pine-fir	#1	10-3	13-6	17-2	19-11	9-11	12-7	15-5	17-10
	Spruce-pine-fir	#2	10-3	13-6	17-2	19-11	9-11	12-7	15-5	17-10
	Spruce-pine-fir	#3	8-5	10-8	13-0	15-1	7-6	9-6	11-8	13-6
19.2	Douglas fir-larch	SS	10-8	14-1	18-0	21-10	10-8	14-1	18-0	21-0
	Douglas fir-larch	#1	10-4	13-7	16-9	19-6	9-8	12-4	15-0	17-5
	Douglas fir-larch	#2	10-1	12-10	15-8	18-3	9-1	11-6	14-1	16-3
	Douglas fir-larch	#3	7-8	9-9	11-10	13-9	6-10	8-8	10-7	12-4
	Hem-fir	SS	10-1	13-4	17-0	20-8	10-1	13-4	17-0	20-7
	Hem-fir	#1	9-10	13-0	16-4	19-0	9-6	12-0	14-8	17-0
	Hem-fir	#2	9-5	12-5	15-6	17-1	8-11	11-4	13-10	16-1
	Hem-fir	#3	7-8	9-9	11-10	13-9	6-10	8-8	10-7	12-4
	Southern pine	SS	10-6	13-10	17-8	21-6	10-6	13-10	17-8	21-6
	Southern pine	#1	10-4	13-7	17-4	21-1	10-4	13-7	16-4	19-6
	Southern pine	#2	10-1	13-4	16-5	19-3	9-6	12-4	14-8	17-2
	Southern pine	#3	8-3	10-6	12-5	14-9	7-4	9-5	11-1	13-2
	Spruce-pine-fir	SS	9-10	13-0	16-7	20-2	9-10	13-0	16-7	19-6
	Spruce-pine-fir	#1	9-8	12-9	15-8	18-3	9-1	11-6	14-1	16-3
	Spruce-pine-fir	#2	9-8	12-9	15-8	18-3	9-1	11-6	14-1	16-3
	Spruce-pine-fir	#3	7-8	9-9	11-10	13-9	6-10	8-8	10-7	12-4
24	Douglas fir-larch	SS	9-11	13-1	16-8	20-3	9-11	13-1	16-2	18-9
	Douglas fir-larch	#1	9-7	12-4	15-0	17-5	8-8	11-0	13-5	15-7
	Douglas fir-larch	#2	9-1	11-6	14-1	16-3	8-1	10-3	12-7	14-7
	Douglas fir-larch	#3	6-10	8-8	10-7	12-4	6-2	7-9	9-6	11-0
	Hem-fir	SS	9-4	12-4	15-9	19-2	9-4	12-4	15-9	18-5
	Hem-fir	#1	9-2	12-0	14-8	17-0	8-6	10-9	13-1	15-2
	Hem-fir	#2	8-9	11-4	13-10	16-1	8-0	10-2	12-5	14-4
	Hem-fir	#3	6-10	8-8	10-7	12-4	6-2	7-9	9-6	11-0
	Southern pine	SS	9-9	12-10	16-5	19-11	9-9	12-10	16-5	19-11
	Southern pine	#1	9-7	12-7	16-1	19-6	9-7	12-4	14-7	17-5
	Southern pine	#2	9-4	12-4	14-8	17-2	8-6	11-0	13-1	15-5
	Southern pine	#3	7-4	9-5	11-1	13-2	6-7	8-5	9-11	11-10
	Spruce-pine-fir	SS	9-2	12-1	15-5	18-9	9-2	12-1	15-0	17-5
	Spruce-pine-fir	#1	8-11	11-6	14-1	16-3	8-1	10-3	12-7	14-7
	Spruce-pine-fir	#2	8-11	11-6	14-1	16-3	8-1	10-3	12-7	14-7
	Spruce-pine-fir	#3	6-10	8-8	10-7	12-4	6-2	7-9	9-6	11-0

For SI: 1 inch = 25.4 mm, 1 foot = 304.8 mm.

NOTE: Check sources for availability of lumber in lengths greater than 20 feet.

FIGURE 23.10 ■ *This span table is suitable for determining floor joists for sleeping areas with a live load of 30 psf and a dead load of either 10 psf or 20 psf. Reproduced from the International Residential Code / 2006. Copyright © 2006.* Courtesy International Code Council, Inc.

CEILING JOIST SPANS FOR COMMON LUMBER SPECIES
(Uninhabitable attics without storage, live load = 10 psf, L/Δ = 240)

CEILING JOIST SPACING (inches)	SPECIE AND GRADE	DEAD LOAD = 5 psf			
		2x4	2x6	2x8	2x10
		Maximum ceiling joist spans			
		(ft. - in.)	(ft. - in.)	(ft. - in.)	(ft. - in.)
12	Douglas fir-larch SS	13-2	20-8	(a)	(a)
	Douglas fir-larch #1	12-8	19-11	(a)	(a)
	Douglas fir-larch #2	12-5	19-6	25-8	(a)
	Douglas fir-larch #3	10-10	15-10	20-1	24-6
	Hem-fir SS	12-5	19-6	25-8	(a)
	Hem-fir #1	12-2	19-1	25-2	(a)
	Hem-fir #2	11-7	18-2	24-0	(a)
	Hem-fir #3	10-10	15-10	20-1	24-6
	Southern pine SS	12-11	20-3	(a)	(a)
	Southern pine #1	12-8	19-11	(a)	(a)
	Southern pine #2	12-5	19-6	25-8	(a)
	Southern pine #3	11-6	17-0	21-8	25-7
	Spruce-pine-fir SS	12-2	19-1	25-2	(a)
	Spruce-pine-fir #1	11-10	18-8	24-7	(a)
	Spruce-pine-fir #2	11-10	18-8	24-7	(a)
	Spruce-pine-fir #3	10-10	15-10	20-1	24-6
16	Douglas fir-larch SS	11-11	18-9	24-8	(a)
	Douglas fir-larch #1	11-6	18-1	23-10	(a)
	Douglas fir-larch #2	11-3	17-8	23-0	(a)
	Douglas fir-larch #3	9-5	13-9	17-5	21-3
	Hem-fir SS	11-3	17-8	23-4	(a)
	Hem-fir #1	11-0	17-4	22-10	(a)
	Hem-fir #2	10-6	16-6	21-9	(a)
	Hem-fir #3	9-5	13-9	17-5	21-3
	Southern pine SS	11-9	18-5	24-3	(a)
	Southern pine #1	11-6	18-1	23-1	(a)
	Southern pine #2	11-3	17-8	23-4	(a)
	Southern pine #3	10-0	14-9	18-9	22- 2
	Spruce-pine-fir SS	11-0	17-4	22-10	(a)
	Spruce-pine-fir #1	10-9	16-11	22- 4	(a)
	Spruce-pine-fir #2	10-9	16-11	22- 4	(a)
	Spruce-pine-fir #3	9-5	13-9	17-5	21-3
19.2	Douglas fir-larch SS	11-3	17-8	23-3	(a)
	Douglas fir-larch #1	10-10	17-0	22-5	(a)
	Douglas fir-larch #2	10-7	16-7	21-0	25-8
	Douglas fir-larch #3	8-7	12-6	15-10	19-5
	Hem-fir SS	10-7	16-8	21-11	(a)
	Hem-fir #1	10-4	16-4	21-6	(a)
	Hem-fir #2	9-11	15-7	20-6	25-3
	Hem-fir #3	8-7	12-6	15-10	19-5
	Southern -pine SS	11-0	17-4	22-10	(a)
	Southern pine #1	10-10	17-0	22-5	(a)
	Southern pine #2	10-7	16-8	21-11	(a)
	Southern pine #3	9-1	13-6	17-2	20-3
	Spruce-pine-fir SS	10-4	16-4	21-6	(a)
	Spruce-pine-fir #1	10-2	15-11	21-0	25-8
	Spruce-pine-fir #2	10-2	15-11	21-0	25-8
	Spruce-pine-fir #3	8-7	12-6	15-10	19-5
24	Douglas fir-larch SS	10-5	16-4	21-7	(a)
	Douglas fir-larch #1	10-0	15-9	20-1	24-6
	Douglas fir-larch #2	9-10	14-10	18-9	22-11
	Douglas fir-larch #3	7-8	11-2	14-2	17-4
	Hem-fir SS	9-10	15-6	20-5	(a)
	Hem-fir #1	9-8	15-2	19-7	23-11
	Hem-fir #2	9-2	14-5	18-6	22-7
	Hem-fir #3	7-8	11-2	14-2	17-4
	Southern pine SS	10-3	16-1	21-2	(a)
	Southern pine #1	10-0	15-9	20-10	(a)
	Southern pine #2	9-10	15-6	20-1	23-11
	Southern pine #3	8-2	12-0	15-4	18-1
	Spruce-pine-fir SS	9-8	15-2	19-11	25-5
	Spruce-pine-fir #1	9-5	14-9	18-9	22-11
	Spruce-pine-fir #2	9-5	14-9	18-9	22-11
	Spruce-pine-fir #3	7-8	11-2	14-2	17-4

For SI: 1 inch = 25.4 mm, 1 foot = 304.8 m, 1 psf = 0.0479 kN/m^2.

a. Span exceeds 26 feet in length. Check sources for availability of lumber in lengths greater than 20 feet.

FIGURE 23.11 ■ *This span table is suitable for determining ceiling joists without storage areas with a live load of 10 psf and a dead load of 5 psf. Reproduced from the International Residential Code / 2006. Copyright © 2006.* Courtesy International Code Council, Inc.

CEILING JOIST SPANS FOR COMMON LUMBER SPECIES
(Uninhabitable attics with limited storage, live load = 10 psf, L/Δ = 240)

CEILING JOIST SPACING (inches)	SPECIE AND GRADE		DEAD LOAD = 10 psf			
			2x4	2x6	2x8	2x10
			Maximum ceiling joist spans			
			(ft. - in.)	(ft. - in.)	(ft. - in.)	(ft. - in.)
12	Douglas fir-larch	SS	10-5	16-4	21-7	(a)
	Douglas fir-larch	#1	10-0	15-9	20-1	24-6
	Douglas fir-larch	#2	9-10	14-10	18-9	22-11
	Douglas fir-larch	#3	7-8	11-2	14-2	17-4
	Hem-fir	SS	9-10	15-6	20-5	(a)
	Hem-fir	#1	9-8	15-2	19-7	23-11
	Hem-fir	#2	9-2	14-5	18-6	22-7
	Hem-fir	#3	7-8	11-2	14-2	17-4
	Southern pine	SS	10-3	16-1	21-2	(a)
	Southern pine	#1	10-0	15-9	20-10	(a)
	Southern pine	#2	9-10	15-6	20-1	23-11
	Southern pine	#3	8-2	12-0	15-4	18-1
	Spruce-pine-fir	SS	9-8	15-2	19-11	25-5
	Spruce-pine-fir	#1	9-5	14-9	18-9	22-11
	Spruce-pine-fir	#2	9-5	14-9	18-9	22-11
	Spruce-pine-fir	#3	7-8	11-2	14-2	17-4
16	Douglas fir-larch	SS	9-6	14-11	19-7	25-0
	Douglas fir-larch	#1	9-1	13-9	17-5	21-3
	Douglas fir-larch	#2	8-9	12-10	16-3	19-10
	Douglas fir-larch	#3	6-8	9-8	12-4	15-0
	Hem-fir	SS	8-11	14-1	18-6	23-8
	Hem-fir	#1	8-9	13-5	16-10	20-8
	Hem-fir	#2	8-4	12-8	16-0	19-7
	Hem-fir	#3	6-8	9-8	12-4	15-0
	Southern pine	SS	9-4	14-7	19-3	24-7
	Southern pine	#1	9-1	14-4	18-11	23-1
	Southern pine	#2	8-11	13-6	17-5	20-9
	Southern pine	#3	7-1	10-5	13-3	15-8
	Spruce-pine-fir	SS	8-9	13-9	18-1	23-1
	Spruce-pine-fir	#1	8-7	12-10	16-3	19-10
	Spruce-pine-fir	#2	8-7	12-10	16-3	19-10
	Spruce-pine-fir	#3	6-8	9-8	12-4	15-0
19.2	Douglas fir-larch	SS	8-11	14-0	18-5	23-4
	Douglas fir-larch	#1	8-7	12-6	15-10	19-5
	Douglas fir-larch	#2	8-0	11-9	14-10	18-2
	Douglas fir-larch	#3	6-1	8-10	11-3	13-8
	Hem-fir	SS	8-5	13-3	17-5	22-3
	Hem-fir	#1	8-3	12-3	15-6	18-11
	Hem-fir	#2	7-10	11-7	14-8	17-10
	Hem-fir	#3	6-1	8-10	11-3	13-8
	Southern pine	SS	8-9	13-9	18-1	23-1
	Southern pine	#1	8-7	13-6	17-9	21-1
	Southern pine	#2	8-5	12-3	15-10	18-11
	Southern pine	#3	6-5	9-6	12-1	14-4
	Spruce-pine-fir	SS	8-3	12-11	17-1	21-8
	Spruce-pine-fir	#1	8-0	11-9	14-10	18-2
	Spruce-pine-fir	#2	8-0	11-9	14-10	18-2
	Spruce-pine-fir	#3	6-1	8-10	11-3	13-8
24	Douglas fir-larch	SS	8-3	13-0	17-1	20-11
	Douglas fir-larch	#1	7-8	11-2	14-2	17-4
	Douglas fir-larch	#2	7-2	10-6	13-3	16-3
	Douglas fir-larch	#3	5-5	7-11	10-0	12-3
	Hem-fir	SS	7-10	12-3	16-2	20-6
	Hem-fir	#1	7-6	10-11	13-10	16-11
	Hem-fir	#2	7-1	10-4	13-1	16-0
	Hem-fir	#3	5-5	7-11	10-0	12-3
	Southern pine	SS	8-1	12-9	16-10	21-6
	Southern pine	#1	8-0	12-6	15-10	18-10
	Southern pine	#2	7-8	11-0	14-2	16-11
	Southern pine	#3	5-9	8-6	10-10	12-10
	Spruce-pine-fir	SS	7-8	12-0	15-10	19-5
	Spruce-pine-fir	#1	7-2	10-6	13-3	16-3
	Spruce-pine-fir	#2	7-2	10-6	13-3	16-3
	Spruce-pine-fir	#3	5-5	7-11	10-0	12-3

For SI: 1 inch = 25.4 mm, 1 foot = 304.8 mm, 1 psf = 0.0479 kN/m^2.

a. Span exceeds 26 feet in length. Check sources for availability of lumber in lengths greater than 20 feet.

FIGURE 23.12 ■ *This span table is suitable for determining ceiling joists with limited storage and a live load of 10 psf and a dead load of 10 psf. Reproduced from the International Residential Code / 2006. Copyright © 2006.* Courtesy International Code Council, Inc.

table for ceiling joists with limited attic storage. This table will be more realistic for most residential uses. Notice that the dead load has been increased from 5 to 10 psf. Using DFL #2 joists at 16'' o.c. to span 16' will require 2 × 8 joists to be used. These joists are suitable for spanning distances of up to 16'-3''.

Sizing Rafters

Although the selection of rafters is similar to the selection of joists, two major differences will be encountered. Tables are available for rafters with and without the ceiling attached to the rafters and based on the ground snow load. Load tables with ground snow loads of 30, 50, and 70 psf are available in the IRC. Remember that the required load to be supported should be verified with the building department that will govern the job site. Your local building department will determine which live load value is to be used. The dead load is determined by the construction materials to be used. The table in Figure 23.13 with a dead load of 10 psf would be suitable for rafters supporting most materials except tile, with no interior finish. The column in Figure 23.14 with a load of 20 psf is suitable for supporting most tiles. Vendor catalogs should always be consulted to determine the required dead loads.

Once the live and dead loads have been determined, choose the appropriate table. Determine what size hem-fir rafter should be used to support a roof over a room 18'' wide framed with a gable roof. Assume the roof is supporting 235-lb composition shingles with a live load of 20 psf, a dead load of 10 psf, and 24'' spacing. Because a gable roof is being framed, the actual horizontal span is only 9'-0''. Using the table in Figure 22.13, it can be seen that 2 × 6 HF #2 rafters spaced at 24'' o.c. are suitable for spans of up to 11'-7''. If the same room were to be framed with DFL #2 lumber, 2 × 6 members would be suitable for spans of up to 11'-9''. If the spacing is reduced to 16'' o.c., 2 × 4 DFL #2 rafters can be used for spans of up to 9'-10''. Even though they are legal, many framers will not use 2 × 4 members because they tend to split as they are attached to the wall top plates.

TABLE R802.5.1(1)[a]
RAFTER SPANS FOR COMMON LUMBER SPECIES
(Roof live load=20 psf, ceiling not attached to rafters, L/Δ=180)

RAFTER SPACING (inches)	SPECIE AND GRADE		DEAD LOAD = 10 psf					DEAD LOAD = 20 psf				
			2x4	2x6	2x8	2x10	2x12	2x4	2x6	2x8	2x10	2x12
			Maximum rafter spans[b]									
			(ft. - in.)	(ft. - in.)	(ft. - in.)	(ft. - in.)	(ft. - in.)	(ft. - in.)	(ft. - in.)	(ft. - in.)	(ft. - in.)	(ft. - in.)
12	Douglas fir-larch	SS	11-6	18-0	23-9	(b)	(b)	11-6	18-0	23-5	(b)	(b)
	Douglas fir-larch	#1	11-1	17-4	22-5	(b)	(b)	10-6	15-4	19-5	23-9	(b)
	Douglas fir-larch	#2	10-10	16-7	21-0	25-8	(b)	9-10	14-4	18-2	22-3	25-9
	Douglas fir-larch	#3	8-7	12-6	15-10	19-5	22-6	7-5	10-10	13-9	16-9	19-6
	Hem-fir	SS	10-10	17-0	22-5	(b)	(b)	10-10	17-0	22-5	(b)	(b)
	Hem-fir	#1	10-7	16-8	21-10	(b)	(b)	10-3	14-11	18-11	23-2	(b)
	Hem-fir	#2	10-1	15-11	20-8	25-3	(b)	9-8	14-2	17-11	21-11	25-5
	Hem-fir	#3	8-7	12-6	15-10	19-5	22-6	7-5	10-10	13-9	16-9	19-6
	Southern pine	SS	11-3	17-8	23-4	(b)	(b)	11-3	17-8	23-4	(b)	(b)
	Southern pine	#1	11-1	17-4	22-11	(b)	(b)	11-1	17-3	21-9	25-10	(b)
	Southern pine	#2	10-10	17-0	22-5	(b)	(b)	10-6	15-1	19-5	23-2	(b)
	Southern pine	#3	9-1	13-6	17-2	20-3	24-1	7-11	11-8	14-10	17-6	20-11
	Spruce-pine-fir	SS	10-7	16-8	21-11	(b)	(b)	10-7	16-8	21-9	(b)	(b)
	Spruce-pine-fir	#1	10-4	16-3	21-0	25-8	(b)	9-10	14-4	18-2	22-3	25-9
	Spruce-pine-fir	#2	10-4	16-3	21-0	25-8	(b)	9-10	14-4	18-2	22-3	25-9
	Spruce-pine-fir	#3	8-7	12-6	15-10	19-5	22-6	7-5	10-10	13-9	16-9	19-6
16	Douglas fir-larch	SS	10-5	16-4	21-7	(b)	(b)	10-5	16-0	20-3	24-9	(b)
	Douglas fir-larch	#1	10-0	15-4	19-5	23-9	(b)	9-1	13-3	16-10	20-7	23-10
	Douglas fir-larch	#2	9-10	14-4	18-2	22-3	25-9	8-6	12-5	15-9	19-3	22-4
	Douglas fir-larch	#3	7-5	10-10	13-9	16-9	19-6	6-5	9-5	11-11	14-6	16-10
	Hem-fir	SS	9-10	15-6	20-5	(b)	(b)	9-10	15-6	19-11	24-4	(b)
	Hem-fir	#1	9-8	14-11	18-11	23-2	(b)	8-10	12-11	16-5	20-0	23-3
	Hem-fir	#2	9-2	14-2	17-11	21-11	25-5	8-5	12-3	15-6	18-11	22-0
	Hem-fir	#3	7-5	10-10	13-9	16-9	19-6	6-5	9-5	11-11	14-6	16-10
	Southern pine	SS	10-3	16-1	21-2	(b)	(b)	10-3	16-1	21-2	(b)	(b)
	Southern pine	#1	10-0	15-9	20-10	25-10	(b)	10-0	15-0	18-10	22-4	(b)
	Southern pine	#2	9-10	15-1	19-5	23-2	(b)	9-1	13-0	16-10	20-1	23-7
	Southern pine	#3	7-11	11-8	14-10	17-6	20-11	6-10	10-1	12-10	15-2	18-1
	Spruce-pine-fir	SS	9-8	15-2	19-11	25-5	(b)	9-8	14-10	18-10	23-0	(b)
	Spruce-pine-fir	#1	9-5	14-4	18-2	22-3	25-9	8-6	12-5	15-9	19-3	22-4
	Spruce-pine-fir	#2	9-5	14-4	18-2	22-3	25-9	8-6	12-5	15-9	19-3	22-4
	Spruce-pine-fir	#3	7-5	10-10	13-9	16-9	19-6	6-5	9-5	11-11	14-6	16-10
19.2	Douglas fir-larch	SS	9-10	15-5	20-4	25-11	(b)	9-10	14-7	18-6	22-7	(b)
	Douglas fir-larch	#1	9-5	14-0	17-9	21-8	25-2	8-4	12-2	15-4	18-9	21-9
	Douglas fir-larch	#2	8-11	13-1	16-7	20-3	23-6	7-9	11-4	14-4	17-7	20-4
	Douglas fir-larch	#3	6-9	9-11	12-7	15-4	17-9	5-10	8-7	10-10	13-3	15-5
	Hem-fir	SS	9-3	14-7	19-2	24-6	(b)	9-3	14-4	18-2	22-3	25-9
	Hem-fir	#1	9-1	13-8	17-4	21-1	24-6	8-1	11-10	15-0	18-4	21-3
	Hem-fir	#2	8-8	12-11	16-4	20-0	23-2	7-8	11-2	14-2	17-4	20-1
	Hem-fir	#3	6-9	9-11	12-7	15-4	17-9	5-10	8-7	10-10	13-3	15-5
	Southern pine	SS	9-8	15-2	19-11	25-5	(b)	9-8	15-2	19-11	25-5	(b)
	Southern pine	#1	9-5	14-10	19-7	23-7	(b)	9-3	13-8	17-2	20-5	24-4
	Southern pine	#2	9-3	13-9	17-9	21-2	24-10	8-4	11-11	15-4	18-4	21-6
	Southern pine	#3	7-3	10-8	13-7	16-0	19-1	6-3	9-3	11-9	13-10	16-6
	Spruce-pine-fir	SS	9-1	14-3	18-9	23-11	(b)	9-1	13-7	17-2	21-0	24-4
	Spruce-pine-fir	#1	8-10	13-1	16-7	20-3	23-6	7-9	11-4	14-4	17-7	20-4
	Spruce-pine-fir	#2	8-10	13-1	16-7	20-3	23-6	7-9	11-4	14-4	17-7	20-4
	Spruce-pine-fir	#3	6-9	9-11	12-7	15-4	17-9	5-10	8-7	10-10	13-3	15-5
24	Douglas fir-larch	SS	9-1	14-4	18-10	23-4	(b)	8-11	13-1	16-7	20-3	23-5
	Douglas fir-larch	#1	8-7	12-6	15-10	19-5	22-6	7-5	10-10	13-9	16-9	19-6
	Douglas fir-larch	#2	8-0	11-9	14-10	18-2	21-0	6-11	10-2	12-10	15-8	18-3
	Douglas fir-larch	#3	6-1	8-10	11-3	13-8	15-11	5-3	7-8	9-9	11-10	13-9
	Hem-fir	SS	8-7	13-6	17-10	22-9	(b)	8-7	12-10	16-3	19-10	23-0
	Hem-fir	#1	8-4	12-3	15-6	18-11	21-11	7-3	10-7	13-5	16-4	19-0
	Hem-fir	#2	7-11	11-7	14-8	17-10	20-9	6-10	10-0	12-8	15-6	17-11
	Hem-fir	#3	6-1	8-10	11-3	13-8	15-11	5-3	7-8	9-9	11-10	13-9
	Southern pine	SS	8-11	14-1	18-6	23-8	(b)	8-11	14-1	18-6	22-11	(b)
	Southern pine	#1	8-9	13-9	17-9	21-1	25-2	8-3	12-3	15-4	18-3	21-9
	Southern pine	#2	8-7	12-3	15-10	18-11	22-2	7-5	10-8	13-9	16-5	19-3
	Southern pine	#3	6-5	9-6	12-1	14-4	17-1	5-7	8-3	10-6	12-5	14-9
	Spruce-pine-fir	SS	8-5	13-3	17-5	21-8	25-2	8-4	12-2	15-4	18-9	21-9
	Spruce-pine-fir	#1	8-0	11-9	14-10	18-2	21-0	6-11	10-2	12-10	15-8	18-3
	Spruce-pine-fir	#2	8-0	11-9	14-10	18-2	21-0	6-11	10-2	12-10	15-8	18-3
	Spruce-pine-fir	#3	6-1	8-10	11-3	13-8	15-11	5-3	7-8	9-9	11-10	13-9

For SI: 1 inch = 25.4 mm, 1 foot = 304.8 mm, 1 psf = 0.0479 kN/m^2.

a. The tabulated rafter spans assume that ceiling joists are located at the bottom of the attic space or that some other method of resisting the outward push of the rafters on the bearing walls, such as rafter ties, is provided at that location. When ceiling joists or rafter ties are located higher in the attic space, the rafter spans shall be multiplied by the factors given below:

FIGURE 23.13 ■ *This span table is suitable for determining rafters that do not support ceiling loads. Design loads include a live load of 20 psf and a dead load of either 10 or 20 psf. Reproduced from the International Residential Code / 2006. Copyright © 2006.* Courtesy International Code Council, Inc.

RAFTER SPANS FOR COMMON LUMBER SPECIES
(Ground snow load=30 psf, ceiling not attached to rafters, L/Δ=180)

RAFTER SPACING (inches)	SPECIES AND GRADE		DEAD LOAD = 10 psf					DEAD LOAD = 20 psf				
			2x4	2x6	2x8	2x10	2x12	2x4	2x6	2x8	2x10	2x12
			Maximum rafter spans[a]									
			(feet - inches)	(feet - inches)	(feet - inches)	(feet - inches)	(feet - inches)	(feet - inches)	(feet - inches)	(feet - inches)	(feet - inches)	(feet - inches)
12	Douglas fir-larch	SS	10-0	15-9	20-9	Note b	Note b	10-0	15-9	20-1	24-6	Note b
	Douglas fir-larch	#1	9-8	14-9	18-8	22-9	Note b	9-0	13-2	16-8	20-4	23-7
	Douglas fir-larch	#2	9-5	13-9	17-5	21-4	24-8	8-5	12-4	15-7	19-1	22-1
	Douglas fir-larch	#3	7-1	10-5	13-2	16-1	18-8	6-4	9-4	11-9	14-5	16-8
	Hem-fir	SS	9-6	14-10	19-7	25-0	Note b	9-6	14-10	19-7	24-1	Note b
	Hem-fir	#1	9-3	14-4	18-2	22-2	25-9	8-9	12-10	16-3	19-10	23-0
	Hem-fir	#2	8-10	13-7	17-2	21-0	24-4	8-4	12-2	15-4	18-9	21-9
	Hem-fir	#3	7-1	10-5	13-2	16-1	18-8	6-4	9-4	11-9	14-5	16-8
	Southern pine	SS	9-10	15-6	20-5	Note b	Note b	9-10	15-6	20-5	Note b	Note b
	Southern pine	#1	9-8	15-2	20-0	24-9	Note b	9-8	14-10	18-8	22-2	Note b
	Southern pine	#2	9-6	14-5	18-8	22-3	Note b	9-0	12-11	16-8	19-11	23-4
	Southern pine	#3	7-7	11-2	14-3	16-10	20-0	6-9	10-0	12-9	15-1	17-11
	Spruce-pine-fir	SS	9-3	14-7	19-2	24-6	Note b	9-3	14-7	18-8	22-9	Note b
	Spruce-pine-fir	#1	9-1	13-9	17-5	21-4	24-8	8-5	12-4	15-7	19-1	22-1
	Spruce-pine-fir	#2	9-1	13-9	17-5	21-4	24-8	8-5	12-4	15-7	19-1	22-1
	Spruce-pine-fir	#3	7-1	10-5	13-2	16-1	18-8	6-4	9-4	11-9	14-5	16-8
16	Douglas fir-larch	SS	9-1	14-4	18-10	23-9	Note b	9-1	13-9	17-5	21-3	24-8
	Douglas fir-larch	#1	8-9	12-9	16-2	19-9	22-10	7-10	11-5	14-5	17-8	20-5
	Douglas fir-larch	#2	8-2	11-11	15-1	18-5	21-5	7-3	10-8	13-6	16-6	19-2
	Douglas fir-larch	#3	6-2	9-0	11-5	13-11	16-2	5-6	8-1	10-3	12-6	14-6
	Hem-fir	SS	8-7	13-6	17-10	22-9	Note b	8-7	13-6	17-1	20-10	24-2
	Hem-fir	#1	8-5	12-5	15-9	19-3	22-3	7-7	11-1	14-1	17-2	19-11
	Hem-fir	#2	8-0	11-9	14-11	18-2	21-1	7-2	10-6	13-4	16-3	18-10
	Hem-fir	#3	6-2	9-0	11-5	13-11	16-2	5-6	8-1	10-3	12-6	14-6
	Southern pine	SS	8-11	14-1	18-6	23-8	Note b	8-11	14-1	18-6	23-8	Note b
	Southern pine	#1	8-9	13-9	18-1	21-5	25-7	8-8	12-10	16-2	19-2	22-10
	Southern pine	#2	8-7	12-6	16-2	19-3	22-7	7-10	11-2	14-5	17-3	20-2
	Southern pine	#3	6-7	9-8	12-4	14-7	17-4	5-10	8-8	11-0	13-0	15-6
	Spruce-pine-fir	SS	8-5	13-3	17-5	22-1	25-7	8-5	12-9	16-2	19-9	22-10
	Spruce-pine-fir	#1	8-2	11-11	15-1	18-5	21-5	7-3	10-8	13-6	16-6	19-2
	Spruce-pine-fir	#2	8-2	11-11	15-1	18-5	21-5	7-3	10-8	13-6	16-6	19-2
	Spruce-pine-fir	#3	6-2	9-0	11-5	13-11	16-2	5-6	8-1	10-3	12-6	14-6
19.2	Douglas fir-larch	SS	8-7	13-6	17-9	21-8	25-2	8-7	12-6	15-10	19-5	22-6
	Douglas fir-larch	#1	7-11	11-8	14-9	18-0	20-11	7-1	10-5	13-2	16-1	18-8
	Douglas fir-larch	#2	7-5	10-11	13-9	16-10	19-6	6-8	9-9	12-4	15-1	17-6
	Douglas fir-larch	#3	5-7	8-3	10-5	12-9	14-9	5-0	7-4	9-4	11-5	13-2
	Hem-fir	SS	8-1	12-9	16-9	21-4	24-8	8-1	12-4	15-7	19-1	22-1
	Hem-fir	#1	7-9	11-4	14-4	17-7	20-4	6-11	10-2	12-10	15-8	18-2
	Hem-fir	#2	7-4	10-9	13-7	16-7	19-3	6-7	9-7	12-2	14-10	17-3
	Hem-fir	#3	5-7	8-3	10-5	12-9	14-9	5-0	7-4	9-4	11-5	13-2
	Southern pine	SS	8-5	13-3	17-5	22-3	Note b	8-5	13-3	17-5	22-0	25-9
	Southern pine	#1	8-3	13-0	16-6	19-7	23-4	7-11	11-9	14-9	17-6	20-11
	Southern pine	#2	7-11	11-5	14-9	17-7	20-7	7-1	10-2	13-2	15-9	18-5
	Southern pine	#3	6-0	8-10	11-3	13-4	15-10	5-4	7-11	10-1	11-11	14-2
	Spruce-pine-fir	SS	7-11	12-5	16-5	20-2	23-4	7-11	11-8	14-9	18-0	20-11
	Spruce-pine-fir	#1	7-5	10-11	13-9	16-10	19-6	6-8	9-9	12-4	15-1	17-6
	Spruce-pine-fir	#2	7-5	10-11	13-9	16-10	19-6	6-8	9-9	12-4	15-1	17-6
	Spruce-pine-fir	#3	5-7	8-3	10-5	12-9	14-9	5-0	7-4	9-4	11-5	13-2
24	Douglas fir-larch	SS	7-11	12-6	15-10	19-5	22-6	7-8	11-3	14-2	17-4	20-1
	Douglas fir-larch	#1	7-1	10-5	13-2	16-1	18-8	6-4	9-4	11-9	14-5	16-8
	Douglas fir-larch	#2	6-8	9-9	12-4	15-1	17-6	5-11	8-8	11-0	13-6	15-7
	Douglas fir-larch	#3	5-0	7-4	9-4	11-5	13-2	4-6	6-7	8-4	10-2	11-10
	Hem-fir	SS	7-6	11-10	15-7	19-1	22-1	7-6	11-0	13-11	17-0	19-9
	Hem-fir	#1	6-11	10-2	12-10	15-8	18-2	6-2	9-1	11-6	14-0	16-3
	Hem-fir	#2	6-7	9-7	12-2	14-10	17-3	5-10	8-7	10-10	13-3	15-5
	Hem-fir	#3	5-0	7-4	9-4	11-5	13-2	4-6	6-7	8-4	10-2	11-10
	Southern pine	SS	7-10	12-3	16-2	20-8	25-1	7-10	12-3	16-2	19-8	23-0
	Southern pine	#1	7-8	11-9	14-9	17-6	20-11	7-1	10-6	13-2	15-8	18-8
	Southern pine	#2	7-1	10-2	13-2	15-9	18-5	6-4	9-2	11-9	14-1	16-6
	Southern pine	#3	5-4	7-11	10-1	11-11	14-2	4-9	7-1	9-0	10-8	12-8
	Spruce-pine-fir	SS	7-4	11-7	14-9	18-0	20-11	7-1	10-5	13-2	16-1	18-8
	Spruce-pine-fir	#1	6-8	9-9	12-4	15-1	17-6	5-11	8-8	11-0	13-6	15-7
	Spruce-pine-fir	#2	6-8	9-9	12-4	15-1	17-6	5-11	8-8	11-0	13-6	15-7
	Spruce-pine-fir	#3	5-0	7-4	9-4	11-5	13-2	4-6	6-7	8-4	10-2	11-10

FIGURE 23.14 ■ *This span table is suitable for determining rafters that do not support ceiling loads with a ground snow load of 30 psf. Reproduced from the International Residential Code / 2006. Copyright © 2006.* Courtesy International Code Council, Inc.

WORKING WITH ENGINEERED LUMBER

Determining the span of engineered lumber is similar to sizing sawn lumber, although span tables will vary slightly for each manufacturer. Most suppliers of engineered joists provide materials for determining floor joists and rafters. The balance of this section explores material supplied by Trus Joist, A Weyerhaeuser Business.

Sizing Engineered Floor Joists

Figure 23.15 shows a floor span table for engineered floor joists. Notice that this table actually consists of two tables. The lower portion of the table describes spans based on code-allowed deflections. The upper portion of the table describes spans based on the manufacturer's suggested deflection limits of 1/480. The increased deflection limits will produce an excellent floor system with little or no vibration or squeaking. To determine the joist size required to span 16'-0'' would require the following steps:

- Determine the deflection limit to be used. For this example, L/360 will be used.
- Identify the loading condition to be used. For this example a 40 psf LL and a 10 psf DL will be used.
- Select the spacing to be used. For this example 24'' o.c. will be used.
- Scan down the spacing column until a distance that exceeds the span is located. For this example, the first joist that meets the design criteria is the 11 7/8'' TJI/210 joist with a span of 16'-10''.

Compare the same joist with the top half of the table. Using the stricter deflection limits, the 11 7/8'' TJI/210 joist will span 16'-5''. If the spacing is changed to 12'' o.c., a 9 1/2'' TJI/110 could be used with a maximum span of 16'-5''.

Selecting Engineered Rafters

Figure 23.16 shows an example of a table from Trus Joist that can be used to determine the sizes of engineered rafters. Before using the table, notice the deflections and the divisions in the table. Deflection is limited to L/180, and the table is divided into nonsnow and snow load areas and low- and high-sloped roofs. The low listing should be used for roofs with a slope of less than 6/12, and the high listing should be used for slopes greater than 6/12. The following steps can be used to determine the required rafter size to span 16'-0'' on a 5/12 pitch supporting composition shingles:

- Determine the roof loading to be used. For this example, a LL of 20 lb. and a DL of 15 lb. with no snow will be used.

FLOOR SPAN TABLES

Not all products are available in all markets. Contact your iLevel representative for information.

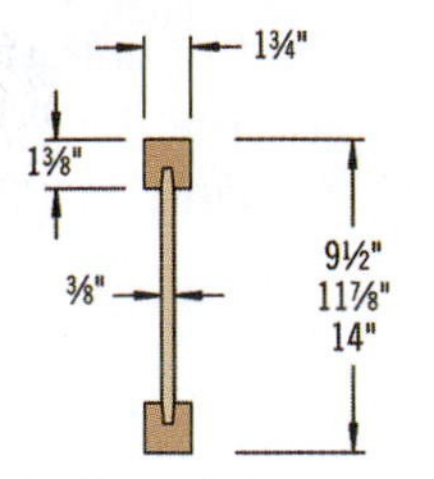

TJI® 110 Joists

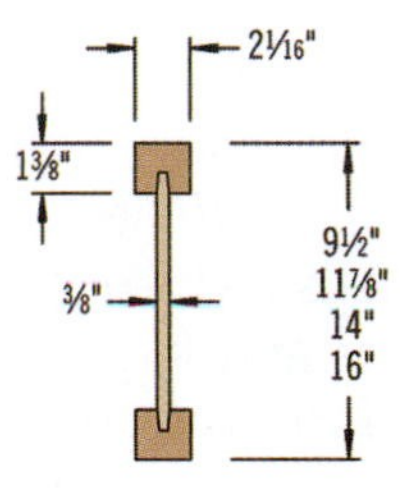

TJI® 210 Joists

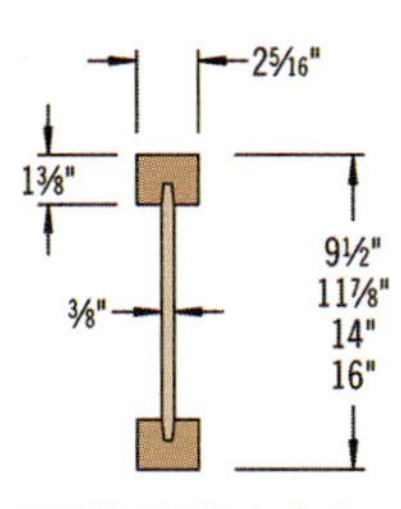

TJI® 230 Joists

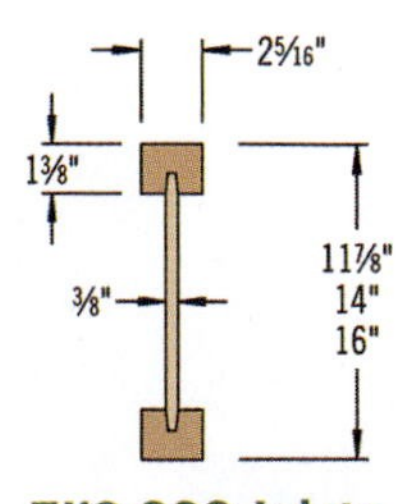

TJI® 360 Joists

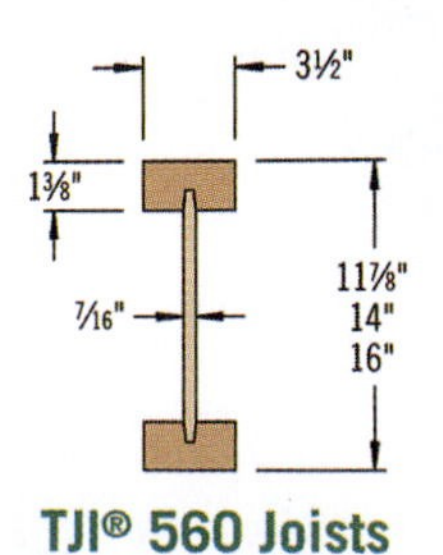

TJI® 560 Joists

L/480 Live Load Deflection

Depth	TJI®	40 PSF Live Load / 10 PSF Dead Load				40 PSF Live Load / 20 PSF Dead Load			
		12" o.c.	16" o.c.	19.2" o.c.	24" o.c.	12" o.c.	16" o.c.	19.2" o.c.	24" o.c.
9½"	110	16'-5"	15'-0"	14'-2"	13'-2"	16'-5"	15'-0"	13'-11"	12'-5"
	210	17'-3"	15'-9"	14'-10"	13'-10"	17'-3"	15'-9"	14'-10"	13'-8"
	230	17'-8"	16'-2"	15'-3"	14'-2"	17'-8"	16'-2"	15'-3"	14'-2"
11⅞"	110	19'-6"	17'-10"	16'-10"	15'-5"[1]	19'-6"	17'-3"	15'-8"	14'-0"[1]
	210	20'-6"	18'-8"	17'-8"	16'-5"	20'-6"	18'-8"	17'-3"	15'-5"[1]
	230	21'-0"	19'-2"	18'-1"	16'-10"	21'-0"	19'-2"	18'-1"	16'-3"[1]
	360	22'-11"	20'-11"	19'-8"	18'-4"	22'-11"	20'-11"	19'-8"	17'-10"[1]
	560	26'-1"	23'-8"	22'-4"	20'-9"	26'-1"	23'-8"	22'-4"	20'-9"[1]
14"	110	22'-2"	20'-3"	18'-9"	16'-9"[1]	21'-8"	18'-9"	17'-1"[1]	14'-7"[1]
	210	23'-3"	21'-3"	20'-0"	18'-4"[1]	23'-3"	20'-7"	18'-9"[1]	16'-2"[1]
	230	23'-10"	21'-9"	20'-6"	19'-1"	23'-10"	21'-8"	19'-9"	17'-1"[1]
	360	26'-0"	23'-8"	22'-4"	20'-9"[1]	26'-0"	23'-8"	22'-4"[1]	17'-10"[1]
	560	29'-6"	26'-10"	25'-4"	23'-6"	***29'-6"***	***26'-10"***	25'-4"[1]	20'-11"[1]
16"	210	25'-9"	23'-6"	22'-0"[1]	19'-5"[1]	25'-5"	22'-0"[1]	20'-1"[1]	16'-2"[1]
	230	26'-5"	24'-1"	22'-9"	20'-7"[1]	***26'-5"***	23'-2"	21'-2"[1]	17'-1"[1]
	360	28'-9"	26'-3"	24'-8"[1]	21'-5"[1]	***28'-9"***	26'-3"[1]	22'-4"[1]	17'-10"[1]
	560	32'-8"	29'-8"	28'-0"	25'-2"[1]	***32'-8"***	***29'-8"***	26'-3"[1]	20'-11"[1]

L/360 Live Load Deflection (Minimum Criteria per Code)

Depth	TJI®	40 PSF Live Load / 10 PSF Dead Load				40 PSF Live Load / 20 PSF Dead Load			
		12" o.c.	16" o.c.	19.2" o.c.	24" o.c.	12" o.c.	16" o.c.	19.2" o.c.	24" o.c.
9½"	110	18'-2"	16'-7"	15'-3"	13'-8"	17'-8"	15'-3"	13'-11"	12'-5"
	210	19'-1"	17'-5"	16'-6"	15'-0"	19'-1"	16'-9"	15'-4"	13'-8"
	230	19'-7"	17'-11"	16'-11"	15'-9"	19'-7"	17'-8"	16'-1"	14'-5"
11⅞"	110	21'-7"	18'-11"	17'-3"	15'-5"[1]	19'-11"	17'-3"	15'-8"	14'-0"[1]
	210	22'-8"	20'-8"	18'-11"	16'-10"	21'-10"	18'-11"	17'-3"	15'-5"[1]
	230	23'-3"	21'-3"	19'-11"	17'-9"	***23'-0"***	19'-11"	18'-2"	16'-3"[1]
	360	25'-4"	23'-2"	21'-10"	20'-4"[1]	***25'-4"***	***23'-2"***	***21'-10"[1]***	17'-10"[1]
	560	28'-10"	26'-3"	24'-9"	23'-0"	***28'-10"***	***26'-3"***	***24'-9"***	20'-11"[1]
14"	110	23'-9"	20'-6"	18'-9"	16'-9"[1]	21'-8"	18'-9"	17'-1"[1]	14'-7"[1]
	210	25'-8"	22'-6"	20'-7"	18'-4"[1]	23'-9"	20'-7"	18'-9"[1]	16'-2"[1]
	230	26'-4"	23'-9"	21'-8"	19'-4"[1]	***25'-0"***	21'-8"	19'-9"	17'-1"[1]
	360	28'-9"	26'-3"	24'-9"[1]	21'-5"[1]	***28'-9"***	***26'-3"[1]***	22'-4"[1]	17'-10"[1]
	560	32'-8"	29'-9"	28'-0"	25'-2"[1]	***32'-8"***	***29'-9"***	***26'-3"[1]***	20'-11"[1]
16"	210	27'-10"	24'-1"	22'-0"[1]	19'-5"[1]	25'-5"	22'-0"[1]	20'-1"[1]	16'-2"[1]
	230	29'-2"	25'-5"	23'-2"	20'-7"[1]	***26'-9"***	23'-2"	21'-2"[1]	17'-1"[1]
	360	31'-10"	29'-0"	26'-10"[1]	21'-5"[1]	***31'-10"***	***26'-10"[1]***	22'-4"[1]	17'-10"[1]
	560	36'-1"	32'-11"	31'-0"[1]	25'-2"[1]	***36'-1"***	***31'-6"[1]***	26'-3"[1]	20'-11"[1]

Long term deflection under dead load, which includes the effect of creep, has not been considered. ***Bold italic*** spans reflect initial dead load deflection exceeding 0.33".

(1) Web stiffeners are required at intermediate supports of continuous-span joists when the intermediate bearing length is *less* than 5¼" and the span on either side of the intermediate bearing is greater than the following spans:

TJI®	40 PSF Live Load / 10 PSF Dead Load				40 PSF Live Load / 20 PSF Dead Load			
	12" o.c.	16" o.c.	19.2" o.c.	24" o.c.	12" o.c.	16" o.c.	19.2" o.c.	24" o.c.
110	N.A.	N.A.	N.A.	15'-4"	N.A.	N.A.	16'-0"	12'-9"
210	N.A.	N.A.	21'-4"	17'-0"	N.A.	21'-4"	17'-9"	14'-2"
230	N.A.	N.A.	N.A.	19'-2"	N.A.	N.A.	19'-11"	15'-11"
360	N.A.	N.A.	24'-5"	19'-6"	N.A.	24'-5"	20'-4"	16'-3"
560	N.A.	N.A.	29'-10"	23'-10"	N.A.	29'-10"	24'-10"	19'-10"

FIGURE 23.15 ■ *Engineered floor joist span tables.* Courtesy Trus Joist, A Weyerhaeuser Business.

ROOF SPAN TABLE

Maximum Horizontal Clear Spans—Roof

O.C. Spacing	Depth	TJI®	Design Live Load (LL) and Dead Load (DL) in PSF											
			Non-Snow (125%)				Snow Load Area (115%)							
			20LL + 15DL		20LL + 20DL		25LL + 15DL		30LL + 15DL		40LL + 15DL		50LL + 15DL	
			Low	High	Low	High	Low	High	Low	High	Low	High	Low	High
16"	9½"	110	19'-3"	17'-2"	18'-4"	16'-3"	18'-5"	16'-6"	17'-9"	15'-11"	16'-7"	15'- 0"	15'-6"	14'-3"
		210	20'-5"	18'-2"	19'-5"	17'-3"	19'-6"	17'-6"	18'-9"	16'-11"	17'-7"	15'-11"	16'-7"	15'-1"
		230	21'-0"	18'-9"	20'-0"	17'-9"	20'-2"	18'-0"	19'-4"	17'-5"	18'-1"	16'-4"	17'-1"	15'-6"
	11⅞"	110	23'-0"	20'-6"	21'-11"	19'-5"	22'-0"	19'-9"	20'-11"	19'-1"	19'- 0"	17'-11"	17'-6"	16'-11"
		210	24'-4"	21'-9"	23'-3"	20'-7"	23'-4"	20'-11"	22'-5"	20'-2"	20'-10"	19'- 0"	19'-2"	18'-0"
		230	25'-1"	22'-5"	23'-11"	21'-3"	24'-1"	21'-7"	23'-1"	20'-10"	21'-7"	19'-7"	20'-3"	18'-7"
		360	27'-9"	24'-9"	26'-5"	23'-5"	26'-7"	23'-10"	25'-6"	23'- 0"	23'-11"	21'-7"	22'-7"	20'-6"
		560	31'-11"	28'-6"	30'-5"	27'- 0"	30'-7"	27'-5"	29'-5"	26'-5"	27'-6"	24'-10"	26'- 0"	23'-7"
	14"	110	26'-3"	23'-5"	25'- 0"	22'-2"	24'-1"	22'-6"	22'-9"	21'-9"	20'-8"	19'-11"	19'-1"	18'-5"
		210	27'-9"	24'-9"	26'-5"	23'-5"	26'-5"	23'-9"	25'- 0"	22'-11"	22'-8"	21'-7"	20'-11"	20'-3"
		230	28'-7"	25'-6"	27'-2"	24'-2"	27'-4"	24'-6"	26'-4"	23'-8"	23'-11"	22'-3"	22'-0"	21'-1"
		360	31'-6"	28'-2"	30'-0"	26'-8"	30'-2"	27'-1"	29'-0"	26'-1"	27'-2"	24'-7"	25'-8"	23'-4"
		560	36'-3"	32'-4"	34'-6"	30'-7"	34'-8"	31'-1"	33'-4"	30'-0"	31'-2"	28'-3"	29'-6"	26'-9"
	16"	210	30'-9"	27'-5"	29'-4"	26'- 0"	28'-3"	26'-5"	26'-9"	25'-6"	24'-3"	23'-4"	22'-4"	21'-8"
		230	31'-8"	28'-3"	30'-2"	26'-9"	29'-10"	27'-2"	28'-2"	26'-3"	25'-7"	24'-7"	23'-7"	22'-10"
		360	34'-11"	31'-2"	33'-3"	29'-6"	33'-5"	30'- 0"	32'-2"	28'-11"	30'-1"	27'-2"	26'- 0"	25'-10"
		560	40'-1"	35'-9"	38'-2"	33'-11"	38'-4"	34'-5"	36'-11"	33'-2"	34'-6"	31'-3"	31'-8"	29'-8"
19.2"	9½"	110	18'-1"	16'-1"	17'-3"	15'-3"	17'-4"	15'-6"	16'-8"	15'- 0"	15'-5"	14'-1"	14'-2"	13'-4"
		210	19'-2"	17'-1"	18'-3"	16'-2"	18'-4"	16'-5"	17'-8"	15'-10"	16'-6"	14'-11"	15'-7"	14'-2"
		230	19'-9"	17'-7"	18'-10"	16'-8"	18'-11"	16'-11"	18'-2"	16'-4"	17'- 0"	15'-4"	16'-1"	14'-7"
	11⅞"	110	21'-7"	19'-3"	20'-7"	18'-3"	20'-3"	18'-6"	19'-1"	17'-11"	17'-4"	16'-8"	16'- 0"	15'-5"
		210	22'-11"	20'-5"	21'-10"	19'-4"	21'-11"	19'-8"	20'-11"	18'-11"	19'-0"	17'-10"	17'-6"	16'-11"
		230	23'-7"	21'-1"	22'-6"	19'-11"	22'-7"	20'-3"	21'-8"	19'-6"	20'-0"	18'-4"	18'-5"	17'-5"
		360	26'-1"	23'-3"	24'-10"	22'-0"	24'-11"	22'-4"	24'- 0"	21'-7"	22'-5"	20'-3"	21'-2"	19'-3"
		560	30'- 0"	26'-9"	28'-7"	25'-4"	28'-8"	25'-9"	27'-7"	24'-10"	25'-9"	23'-4"	24'-4"	22'-2"
	14"	110	24'-6"	22'- 0"	22'-9"	20'-10"	22'-0"	20'-11"	20'-9"	19'-10"	18'-10"	18'-2"	17'- 0"	16'-10"
		210	26'-0"	23'-3"	24'-10"	22'-0"	24'-2"	22'-4"	22'-10"	21'-7"	20'-8"	19'-11"	18'-10"	18'-5"
		230	26'-10"	23'-11"	25'-7"	22'-8"	25'-5"	23'-0"	24'-0"	22'-3"	21'-10"	20'-11"	20'-1"	19'-5"
		360	29'-7"	26'-5"	28'-2"	25'-0"	28'-4"	25'-5"	27'-3"	24'-6"	25'-6"	23'-1"	21'-7"	21'-8"
		560	34'-0"	30'-4"	32'-5"	28'-9"	32'-7"	29'-2"	31'-4"	28'-2"	29'-3"	26'-6"	26'-5"	25'-2"
	16"	210	28'-8"	25'-9"	26'-9"	24'-5"	25'-10"	24'-6"	24'-5"	23'-4"	22'-1"	21'-4"	18'-10"	19'-8"
		230	29'-9"	26'-7"	28'-2"	25'-2"	27'-3"	25'-6"	25'-9"	24'-7"	23'-4"	22'-6"	21'-2"	20'-9"
		360	32'-10"	29'-3"	31'-3"	27'-9"	31'-5"	28'-2"	30'-2"	27'-2"	25'-7"	25'-3"	21'-7"	21'-8"
		560	37'-8"	33'-7"	35'-10"	31'-10"	36'-0"	32'-4"	34'-8"	31'-2"	31'-3"	29'-4"	26'-5"	25'-5"
24"	9½"	110	16'-9"	14'-11"	15'-11"	14'-2"	16'-0"	14'-4"	15'-2"	13'-10"	13'-9"	13'-0"	12'-8"	12'-3"
		210	17'-9"	15'-10"	16'-11"	15'- 0"	17'- 0"	15'-3"	16'-4"	14'-8"	15'-1"	13'-10"	13'-11"	13'-1"
		230	18'-3"	16'-4"	17'-5"	15'-5"	17'-6"	15'-8"	16'-10"	15'-2"	15'-8"	14'-3"	14'-8"	13'-6"
	11⅞"	110	20'-0"	17'-10"	18'-9"	16'-11"	18'-1"	17'-2"	17'-1"	16'-4"	15'-6"	14'-11"	13'-7"	13'-10"
		210	21'-2"	18'-11"	20'-2"	17'-11"	19'-10"	18'-2"	18'-9"	17'-7"	17'- 0"	16'-4"	15'-0"	15'-2"
		230	21'-10"	19'-6"	20'-10"	18'-5"	20'-11"	18'-9"	19'-9"	18'-1"	17'-11"	17'- 0"	16'-6"	16'- 0"
		360	24'-1"	21'-6"	23'- 0"	20'-5"	23'-1"	20'-8"	22'-2"	20'- 0"	20'-5"	18'-9"	17'-3"	17'-4"
		560	27'-9"	24'-9"	26'-5"	23'-6"	26'-7"	23'-10"	25'-6"	23'- 0"	23'-10"	21'-7"	21'-1"	20'-3"
	14"	110	21'-10"	20'-4"	20'-4"	19'-1"	19'-8"	18'-8"	18'-7"	17'-9"	16'-0"	16'-3"	13'-7"	14'-2"
		210	24'- 0"	21'-6"	22'-4"	20'-5"	21'-7"	20'-6"	20'-4"	19'-6"	17'-10"	17'-9"	15'-0"	15'-8"
		230	24'-10"	22'-2"	23'-7"	21'- 0"	22'-9"	21'-4"	21'-6"	20'-6"	19'-6"	18'-9"	16'-11"	16'-7"
		360	27'-5"	24'-6"	26'-1"	23'-2"	26'-3"	23'-6"	25'-0"	22'-8"	20'-5"	20'-2"	17'-3"	17'-4"
		560	31'-6"	28'-1"	30'- 0"	26'-8"	30'-2"	27'-0"	29'- 0"	26'-1"	24'-11"	23'-7"	21'-1"	20'-3"
	16"	210	25'-8"	23'-11"	23'-11"	22'-4"	23'-1"	21'-11"	21'-9"	20'-10"	17'-10"	18'-3"	15'-0"	15'-8"
		230	27'-1"	24'-7"	25'-2"	23'-3"	24'-4"	23'-1"	23'- 0"	22'- 0"	20'-0"	19'-4"	16'-11"	16'-7"
		360	30'-4"	27'-1"	28'-11"	25'-8"	28'-2"	26'-1"	25'-0"	24'-1"	20'-5"	20'-2"	17'-3"	17'-4"
		560	34'-10"	31'-2"	33'-2"	29'-6"	33'-4"	29'-11"	30'-6"	28'-3"	24'-11"	23'-7"	21'-1"	20'-3"

FIGURE 23.16 ■ *Engineered rafter span tables.* Courtesy Trus Joist, A Weyerhaeuser Business.

- Determine the appropriate slope column to be used. For this example, the low column will be used.
- Move down the 20LL + 15DL low column into a row that reflects the desired spacing until a value is found that equals or exceeds the required span. All of the values for 16'' and 19.2 rafters exceed the required 16'-0'' span. A 9 1/2'' TJI110@ 24'' o.c. can be used for the required span.

As with other types of engineered materials, consult all of the manufacturer's instructions.

DETERMINING SIMPLE BEAM SIZES

Throughout this chapter you've explored methods of determining the size of structural members using tables. The balance of this chapter explores methods of solving beams sizes without the use of tables. It examines methods used to solve beams, common notations used in beam formulas, and the process for solving wood beams.

Methods of Beam Design

Four methods of beam design are used by professionals: a computer program, a span computer, wood design books, and of course, the old-fashioned way of pencil, paper, and a few formulas.

Computer Programs

Computer programs that will size beams are available for many personal and business computers. Each national wood distributor supplies a program that can determine spans and needed materials and then print out all appropriate stresses. Software that will size beams made of wood, engineered lumber, and steel is also available through private developers. Two popular programs include StruCalc by Cascade Consulting Associates and BeamChek by AC Software, Inc. Figure 23.17 shows the prompts for solving a hip using BeamChek. Websites for each firm are listed at the end of the chapter. These programs typically ask for loading information and then proceed to determine the span size within seconds. Using a computer program to solve beam spans is extremely easy, but you do have to be able to answer questions that require

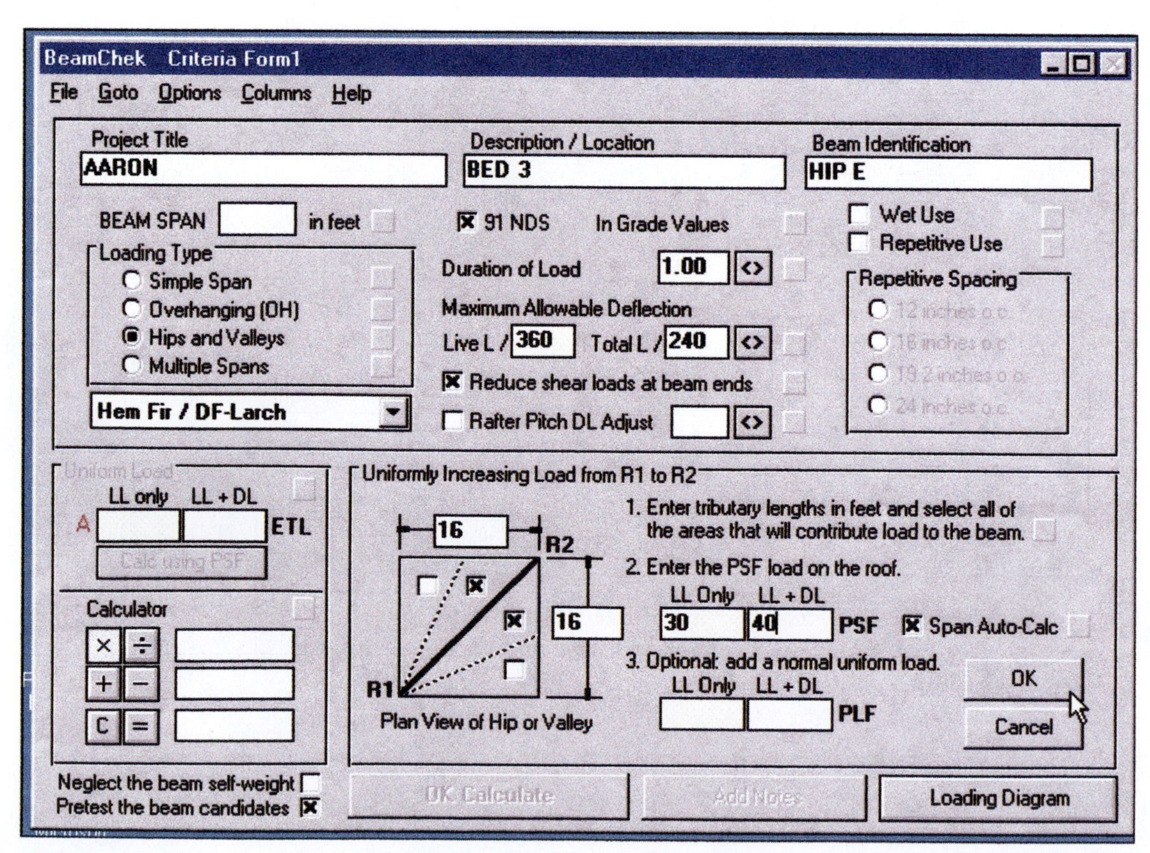

FIGURE 23.17 ■ *Display for determining a hip beam using the program BeamChek developed by AC Software, Inc.*

an understanding of the basics of beam design and loading.

Span Computers

A span computer is very similar to a slide rule. Western Wood Products Association has a span computer that can be used to determine beam sizes (see Figure 23.18). Beam sizes are determined by lining up values of the lumber to be used with a distance to be spanned. Western Wood Products also has a span calculator, named SpanMaster, available for sizing joists, rafters, and beams. Ordering information for each product can be found at the Western Wood Products website.

Span Tables for Beams

Another practical method of sizing beams is the use of books published by various wood associations. Two of the most common span books used in architectural offices are Wood Structural Design Data from the National Forest Products Association and Western Woods Use Book from the Western Wood Products Association. These books contain design information of wood members, standard formulas, and design tables that provide beam loads for a specific span. Figure 23.19 shows a partial listing for the 8' span table. By following the instructions provided with the table, information on size, span, and loading patterns can be determined.

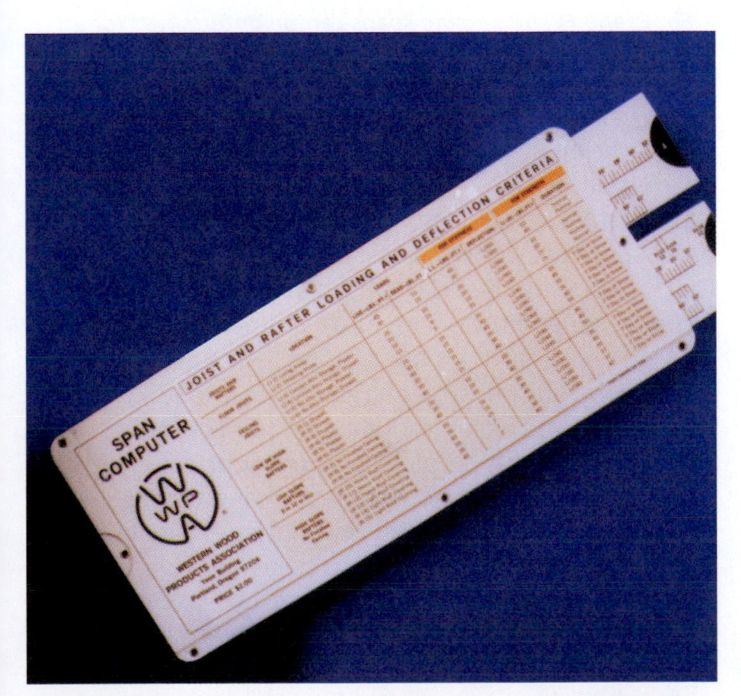

FIGURE 23.18 ■ *A span computer can be used to determine joist and beam sizes.* Courtesy Western Woods Products Association.

Beam Formulas

The final method of beam design is to use standard formulas to determine how the beam will be stressed. Figure 23.20 shows the formulas and the loading, shear, and moment diagrams for a simple beam. These are diagrams that can be drawn by the designer to determine where the maximum stress will occur. With a simple beam and a uniformly distributed load, the diagrams typically are not drawn because the results remain constant.

Notations for Formulas

Standard symbols have been adopted by engineering and architectural communities to simplify the design of beams. Many symbols have been introduced throughout this chapter. Figure 23.21 gives a list of the notations that will be needed to design residential beams. Notice that some are written in uppercase letters and some in lowercase letters. Uppercase letters such as **W** and **L** represent measurements expressed in feet. **W** represents the total load in pounds per square foot; **L** represents the beam span in feet. Lowercase letters represent measurements expressed in inches. The letters **b** and **d** represent beam size in inches, and ℓ represents beam span in inches. As a beginning technician, you need not understand why the formulas work, but it is important that you know the notations to be used in the formulas.

Sizing Wood Beams

Wood beams can be determined by following these steps:

1. Determine the area to be supported by the beam.
2. Determine the weight supported by 1 linear foot of beam.
3. Determine the reactions.
4. Determine the pier sizes.
5. Determine the horizontal shear.
6. Determine the bending moment.
7. Determine the deflection.

Determining the Area to Be Supported

To determine the size of a beam, the weight the beam is to support must be known. To find the weight, find the area the beam is to support and multiply by the total loads. Figure 23.22 shows a sample floor plan with a beam of undetermined size. The beam is supporting an area 10'-0'' long (the span). Floor joists are being supported on each side of the beam. Chapter 22 introduced methods of load dispersal. In this example, each joist on

WOOD BEAMS—SAFE LOAD TABLES

Symbols used in the tables are as follows:

F_b = Allowable unit stress in extreme fiber in bending, psi.
W = Total uniformly distributed load, pounds
w = Load per linear foot of beam, pounds
F_v = Horizontal shear stress, psi, induced by load W
E = Modulus of elasticity, 100 psi, induced by load W for l/360 limit

Beam sizes are expressed as nominal sizes, inches, but calculations are based on net dimensions of S4S sizes.

SIZE OF BEAM		F_b									
		900	1000	1100	1200	1300	1400	1500	1600	1800	2000
8'–0" SPAN											
2 × 14	W	3291	3657	4023	4389	4754	5120	5486	5852	6583	7315
	w	411	457	502	548	594	640	685	731	822	914
	F_v	124	138	151	165	179	193	207	220	248	276
	E	489	543	597	652	706	760	815	869	978	1086
6 × 8	W	3867	4296	4726	5156	5585	6015	6445	6875	734	8593
	w	483	537	590	644	698	751	805	859	966	1074
	F_v	70	78	85	93	101	109	117	125	140	156
	E	864	960	1055	1152	1247	1343	1439	1535	1727	1919
4 × 10	W	3743	4159	4575	4991	5407	5823	6238	6654	7486	8318
	w	467	519	571	623	675	727	779	831	935	1039
	F_v	86	96	105	115	125	134	144	154	173	192
	E	700	778	856	934	1011	1089	1167	1245	1401	1556
3 × 12	W	3955	4394	4833	5273	5712	6152	6591	7031	7910	8789
	w	494	549	604	659	714	769	823	878	988	1098
	F_v	105	117	128	140	152	164	175	187	210	234
	E	576	640	704	768	832	896	959	1024	1151	1279

FIGURE 23.19 ■ *Partial listing of a typical span table. Once W, w, Fb, E, and Fv are known, spans can be determined for a simple beam. See Appendix G on the CD. From Wood Structural Design Data.* Courtesy National Forest Products Association.

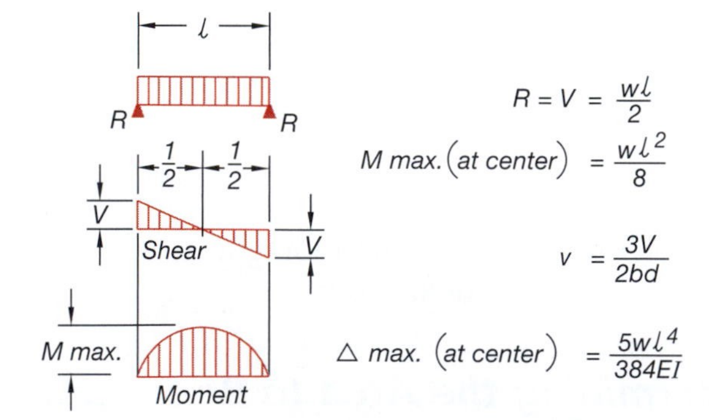

FIGURE 23.20 ■ *Loading, shear, and moment diagrams for simple beam with uniform loads, showing where stress will affect a beam and the formulas for computing these stresses. From Wood Structural Design Data.* Courtesy National Forest Products Association.

each side of the beam can be thought of as a simple beam. Half of the weight of each joist on side *A* will be supported by a wall, and half of the weight of joist *A* will be supported by the beam. Because the total length of the joist on side *A* is 10', the beam will be supporting 5' (half the length of each joist) of floor on side *A*. This 5'-wide area that the beam supports is referred to as a tributary area. In this case, it is tributary load *A*. On the right side of the beam, tributary load *B* would be 5' wide (half of the total span of 10'). A ***tributary width*** is defined as the accumulation of loads that are directed to a structural member. The tributary width will always be half the distance between the beam to be designed and the next bearing point. A second point to remember about tributary loads is that the total tributary width will always be half the total distance between the bearing points on each side of the beam, no matter where the beam is located. Figure 23.23 shows why this is true.

BEAM FORMULA NOTATIONS		
b	=	breadth of beam in inches
d	=	depth of beam in inches
D	=	deflection due to load
E	=	modulus of elasticity
F_b	=	allowable unit stress in extreme fiber bending
F_v	=	unit stress in horizontal shear
I	=	moment of inertia of the section
ℓ	=	span of beam in inches
L	=	span of beam in feet
M	=	bending or resisting moment
P	=	total concentrated load in pounds
S	=	section modulus
V=R	=	end reaction of beam
W	=	total uniformly distributed load in pounds
w	=	load per linear foot of beam in pounds

FIGURE 23.21 ■ *Common notations used in beam formulas.*

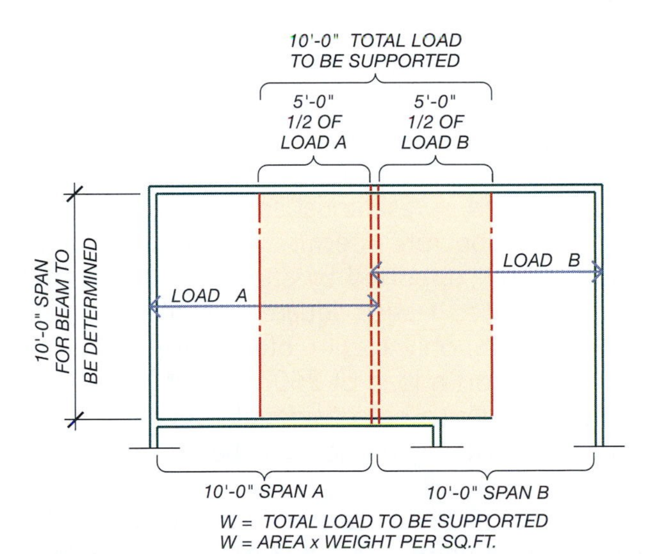

FIGURE 23.22 ■ *A floor plan with a beam to be determined. This beam is supporting an area of 100 sq ft, with an assumed weight of 40 psf LL and 10 psf DL, and 50 psf, total load. When solving problems manually, the total load of* **W** *= 5000 lb can be used. Computer programs often require loads to be divided by the dead and live load for each side of the beam. The tributary loads A and B are each 2,000 lb LL and 500 lb DL.*

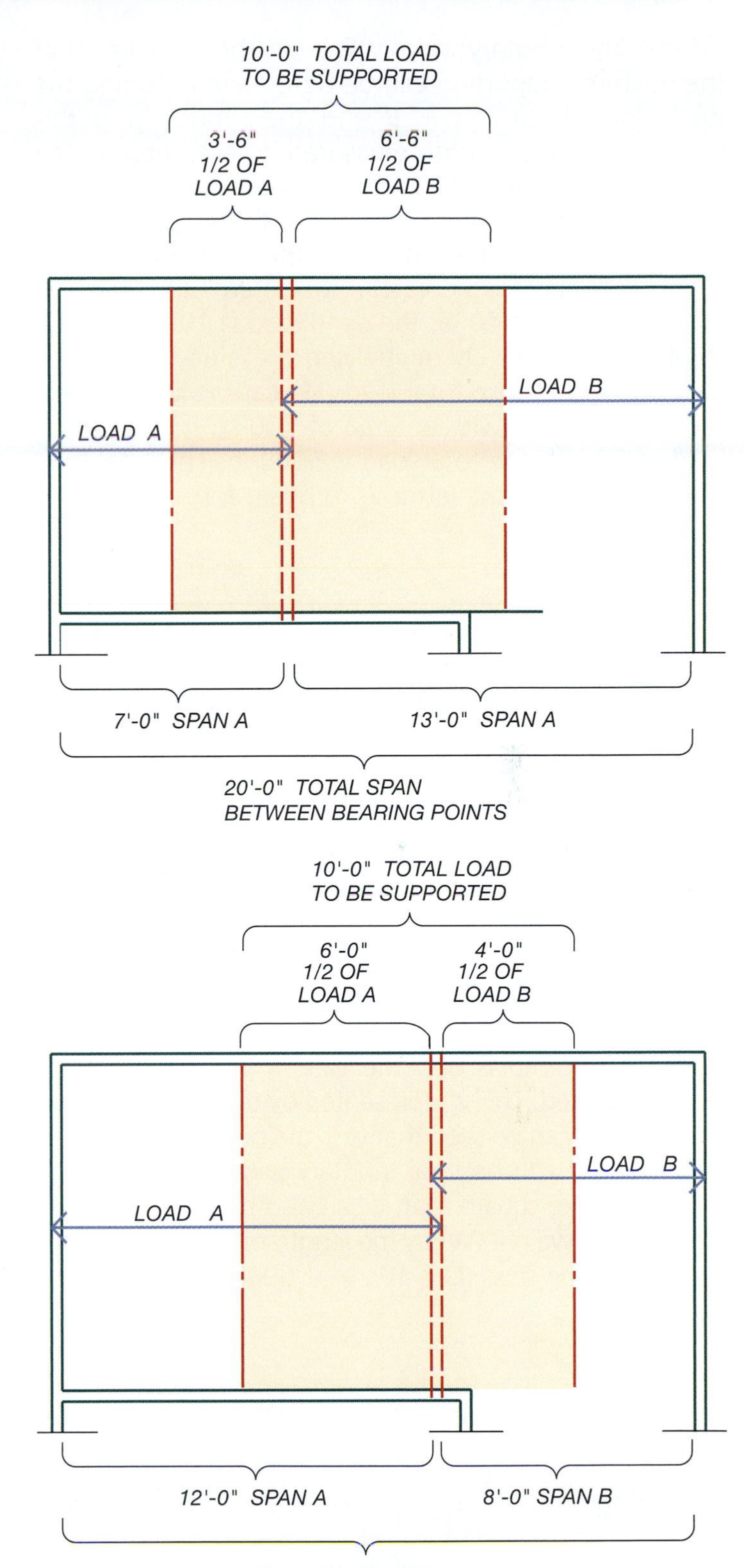

FIGURE 23.23 ■ *The sum of the tributary widths is equal to half the total distance between the bearing point on each side of the beam being determined. With a total width of 20', the tributary width supported by the beam will be 10' no matter where the beam is placed.*

Once the tributary width is known, the total area that the beam is supporting can be determined. Adding the width of tributary *A* + *B* and then multiplying this sum by the beam length measured in feet, represented by *L,* will determine the area that the beam is supporting. For the beam in Figure 23.22, the area would be (5' + 5')(10') = 100 sq ft. Using the loads from Figure 22.3, the loads for floors can be determined. The live load for a floor is 40 lb, the dead load is 10 lb, and the total load is 50 lb. By multiplying the area by the load per square foot, the total load (***W***) can be determined. For this example, with an area of 100 sq ft and a total load of 50 psf, the total load is 5000 lb. The total load is represented by the letter W in beam formulas.

Note:

If you're using a computer program to solve beam sizes, some programs ask for the total load and some ask for the total live load and the total dead load. For this example, the total dead load is 10 lb × 100 sq ft, or 1000 lb. The total live load is 40 lb × 100 sq ft, or 4000 lb. Although this may seem obvious, the total live load and the total dead load should add up to the total load supported. Taking the time to check the obvious can help to eliminate human errors.

Determining Linear Weight

In some formulas only the weight per linear foot of beam is desired. This is represented by the letter w. In Figure 23.24, it can be seen that w is the product of the area to be supported (the total tributary width), multiplied by the weight per square foot. It can also be found by dividing the total weight (**W**) by the length of the beam. In our example, if the span (**L**) is 10', w = 5000/10 or 500 lb.

Note:

Last reminder! If you're using a computer program to solve beam sizes, some programs ask for the combined loads and some ask for the live and dead loads separately. Take time to verify what the prompt is asking for.

Determining Reactions

Even before the size of a beam is known, the supports for the beam can be determined. These supports are represented by the letter **R,** for reactions. The letter **V** is also sometimes used in place of the letter **R.** On a simple beam, half of the weight it is supporting (**W**) will be dispersed to each end. In the example, **W** = 5000 lb and **R** = 2500 lb. At this point **W, w,** and **R** have been determined (**W** = 5000 lb, **w** = 500 lb, and **R** = 2500 lb).

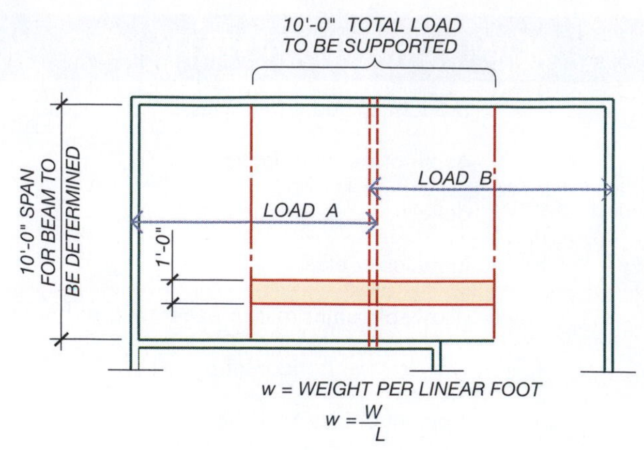

FIGURE 23.24 ■ *Determining linear weights. For an area of 10 sq ft, with an assumed weight of 50 psf,* ***w*** *= 500 lb.*

Determining Pier Sizes

The reaction is the load from the beam that must be transferred to the soil. The reaction from the beam will be transferred by a post to the floor and foundation system. A concrete pier is used to support the loads at the foundation level. To determine the size of the pier, the working stress of the concrete and the bearing value of the soil must be considered.

Concrete with a working stress of between 2000 to 3500 psi is used in residential construction. The working stress of concrete specifies how much weight in pounds can be supported by each square inch (psi) of concrete surface. If each square inch of concrete can support 2500 lb, only 1 sq in. of concrete would be required to support a load of 2500 lb. The bearing value of the soil must also be considered. The pier size is determined by dividing the load to be supported (**R**) by the soil pressure.

See Figure 23.25 for safe soil-loading values. Many building departments use either 1500 or 2000 psf for the assumed safe working value of soil. Don't skim over these numbers. The concrete is listed in pounds per square inch. Soil is listed in pounds per square foot. A pier supporting 2500 lb must be divided by the assumed soil-bearing value to find the area of the pier needed. Using a bearing value of 2000 lb will result in an area of 1.25 sq ft of concrete needed to support the load. See Figure 23.26 to determine the size of pier needed to obtain the proper soil support area.

PRESUMPTIVE LOAD-BEARING VALUES OF FOUNDATION MATERIALS[a]	
CLASS OF MATERIAL	LOAD-BEARING PRESSURE (pounds per square foot)
Crystalline bedrock	12,000
Sedimentary and foliated rock	4,000
Sandy gravel and/or gravel (GW and GP)	3,000
Sand, silty sand, clayey sand, silty gravel and clayey gravel (SW, SP, SM, SC, GM and GC)	2,000
Clay, sandy clay, silty clay, clayey silt, silt and sandy silt (CI, ML, MH and CH)	1,500[b]

For SI: 1 psf = 0.0479 kN/m^2.

a. When soil tests are required by Section R401.4, the allowable bearing capacities of the soil shall be part of the recommendations.

b. Where the building official determines that in-place soils with an allowable bearing capacity of less than 1,500 psf are likely to be present at the site, the allowable bearing capacity shall be determined by a soils investigation.

FIGURE 23.25 ■ *Safe soil-bearing values. Reproduced from the International Residential Building Code/2006. Copyright © 2006.* Courtesy International Code Council.

PIER AREAS AND SIZES	
ROUND PIERS (interior)	SQUARE PIERS (exterior)
15″ DIA. = 1.23 SQ FT	15″ SQ = 1.56 SQ FT
18″ DIA. = 1.77 SQ FT	18″ SQ = 2.25 SQ FT
21″ DIA. = 2.40 SQ FT	21″ SQ = 3.06 SQ FT
24″ DIA. = 3.14 SQ FT	24″ SQ = 4.00 SQ FT
27″ DIA. = 3.97 SQ FT	27″ SQ = 5.06 SQ FT
30″ DIA. = 4.90 SQ FT	30″ SQ = 6.25 SQ FT
36″ DIA. = 7.07 SQ FT	36″ SQ = 9.00 SQ FT
42″ DIA. = 9.60 SQ FT	42″ SQ = 12.25 SQ FT

FIGURE 23.26 ■ *Common pier areas and sizes. The area of the concrete pier can be determined by dividing the load to be supported by the soil-bearing pressure. Areas are shown for common pier sizes. Round piers are typically used for interiors, and square piers are generally used when piers are added to support the exterior foundation.*

Note:

Piers located under the stem wall are typically square because they are dug with a shovel or a backhoe. Piers located in what will be the crawl space or below a slab are typically round because they are usually formed with pre-manufactured forms.

Determining Horizontal Shear

To compute the size of the beam required to resist the forces of horizontal shear, use the following formula.

$$F_v = \frac{(3)(v)}{(2)(b)(d) = \# < 85}$$ (the safe design value for DFL lumber)

This formula will determine the minimum **2bd** value needed to resist the load of 5000 lb. For a simple beam **V = R,** and **R** has been determined to be 2500 lb. To determine the **b** and **d** values, use the actual size of the beam. Since you don't know what size beam to use, the **b** and **d** values remain unknown. Dividing the **3V** value, by the safe limits for the wood (85) can solve this problem, and determine the needed **2bd** value to resist the 5000-lb load. Use the following formula.

$$F_v = \frac{(3)(v)}{85} = \# < 2bd$$ (the required 2 bd value to resist the load).

$$F_v = \frac{(3)(2500)}{85} = 88.23$$ (the minimum 2bd value needed. The 2 bd value of the selected beam must exceed this value).

Use the table in Figure 23.27 to find a **2bd** value that exceeds 88.23. A 4 × 14 or 6 × 10 could be used to meet the requirements of horizontal shear. Now you can solve the required bending moment.

Determining the Bending Moment

A ***moment*** is the tendency of a force to cause rotation about a certain point. In figuring a simple beam, **W** is the force and **R** is the point around which the force rotates. Think of a simple beam supported on two posts. If the post at the right end of beam is removed, the right end of the beam would rotate downward,

PROPERTIES OF STRUCTURAL LUMBER						
NOMINAL SIZE	(b)(d)	(2)(b)(d)	S	A	I	(384)(E)(I) *
2 × 6	1.5 × 5.5	16.5	7.6	8.25	20.8	12,780
2 × 8	1.5 × 7.25	21.75	13.1	10.875	47.6	29,245
2 × 10	1.5 × 9.25	27.75	21.4	13.875	98.9	60,764
2 × 12	1.5 × 11.25	33.75	31.6	16.875	177.9	109,302
2 × 14	1.5 × 13.25	39.75	43.9	19.875	290.8	178,668
4 × 6	3.5 × 5.5	38.5	17.6	19.25	48.5	29,798
4 × 8	3.5 × 7.25	50.75	30.7	25.375	111.0	68,198
4 × 10	3.5 × 9.25	64.75	49.9	32.375	230.8	141,804
4 × 12	3.5 × 11.25	78.75	73.8	39.375	415.3	255,160
4 × 14	3.5 × 13.5	94.5	106.3	47.250	717.6	440,893
6 × 8	5.5 × 7.5	82.5	51.6	41.25	193.4	118,825
6 × 10	5.5 × 9.5	104.5	82.7	32.25	393.0	241,459
6 × 12	5.5 × 11.5	126.5	121.2	63.25	697.1	428,298
6 × 14	5.5 × 13.5	148.5	167.1	74.25	1127.7	692,859

*384 × E × I values are listed in units per million, with E value assumed to be 1.6 for DFL.

FIGURE 23.27 ■ *Structural properties of wood beams. Columns (2)(b)(d) and (384)(E)(I) can be used in the horizontal shear and deflection formulas. These two columns are set up for Douglas fir no. 2. For different types of wood, use NDS values for these columns.*

while the left end of the beam and the left post remained fixed. The rotation, or moment of the right end is determined based on the load, and distance from the left end. Determining the bending moment will calculate the size of the beam needed to resist the tendency for **W** to rotate around **R.**

To determine the size of the beam required to resist the force (**W**), use the following formula:

$$M = \frac{(w)(\ell^2)}{8}$$

Once the moment is known, it can then be divided by the $\mathbf{F_b}$ of the wood to determine the size of beam required to resist the load ($S = M/F_b$). This whole process can be simplified by using the formula:

$$S = \frac{(3)(w)(L^2)}{(2)(F_b)}$$

Because $\mathbf{F_b}$ values for 4 × members are size-dependent, it is important to solve for horizontal shear before solving the bending moment. If you solve for **S** first, you'll need to guess a beam size for a starting point. To determine if a 4 × 10 would be suitable, use the values for Douglas fir from Figure 23.7, and apply this formula to the beam span from Figure 23.22.

$$S = \frac{(3)(500)(100)}{(2)(1050)} = \frac{150{,}000}{2100} = 71.4 = S$$

This formula will determine the section modulus required to support a load of 5000 lb. Look at the table in Figure 23.27 to determine if a 4 × 10 is adequate. A beam must be selected that has an **S** value larger than the **S** value found. By examining the **S** column of the table, you will find that a 4 × 10 has an **S** value of only 49.9, so the beam is not adequate. Because we already know that a 6 × 10 is safe for horizontal shear, check to see if will provide the needed section modulus. The $\mathbf{F_b}$ value for a 6 × 10 of 1350 is determined by using the Table in Figure 23.7. Using the formula 150,000/2700 will require a minimum **S** value of 55.6. It can be seen in Figure 23.27 that a 6 × 8 has an **S** value of only 51.6, so it is not suitable. A 6 × 10 has an **S** value of 82.7, which is suitable to resist the forces of bending moment.

Determining Deflection

Deflection is the amount of sag in a beam. Deflection limits are determined by building codes and are expressed as a fraction of an inch in relation to the span of the beam in inches. Limits set by the IRC are:

$\ell/360$ Floors and ceilings
$\ell/240$ Roofs under 3/12 pitch, tile roofs, and vaulted ceilings
$\ell/180$ Roofs over 3/12 pitch

To determine deflection limits, the maximum allowable limit must first be known. Use the formula:

$$D_{max} = \frac{L \times 12}{360} = \text{maximum allowable deflection}$$

This formula requires the span in feet **L** to be multiplied by 12 (12'' per ft) and then divided by 360 (the safe limit for floors and ceiling).

In the example used in this chapter, L equals 10'. The maximum safe limit would be:

$$D_{max} = \frac{L \times 12}{360} = \frac{10 \times 12}{360} = D = \frac{120}{360} = 0.33''$$

$$= \text{maximum allowable deflection}$$

The maximum amount the beam is allowed to sag is 0.33''. Now you need to determine how much the beam will actually sag under the load it is supporting.

The values for **E** and **I** need to be known before the deflection can be determined. These values are shown in Figure 23.27. You will also need to know $\boldsymbol{\ell^3}$ and **W,** but these values are not available in tables. **W** has been determined to be 5000 lb. For our example:

$$\ell = 10 \times 12 = 120$$

$$\ell^3 = (\ell)(\ell)(\ell) = (120)(120)(120) = 1{,}728{,}000 \text{ or } 1.728$$

Because the **E** value was reduced from 1,600,000 to 1.6, change the $\boldsymbol{\ell^3}$ value from 1,728,000 to 1.728 or 1.73. Each of these numbers is much easier to use for calculating. To determine how much a beam will sag, use the following formula:

$$D = \frac{5(W)(\ell^3)}{(384)(E)(I)} \text{ or } D = 22.5\,\frac{(W)\,(\ell^3)}{(E)(I)}$$

The values determined thus far are W = 5000, w = 500, R = V = 2500, L = 10, ℓ = 120, ℓ^3 = 1.73 and I = 393 (see Figure 23.27, column I). Find the value of (384)(E)(I) for a 6 × 10 beam from the table in Figure 23.27 (241,459). Now insert the values into the formula.

$$D = \frac{5(W)(\ell^3)}{(384)(E)(I)} = \frac{5 \times 5000 \times 1.73}{241{,}459} = \frac{43{,}200}{241{,}459} = 0.178''$$

Because 0.178'' is less than the maximum allowable deflection of 0.33'', a 6 × 10 beam can be used to safely support the load.

Review

In what may have seemed like an endless string of formulas, tables, and values, you have determined the loads and stresses on a 10' beam. The following seven basic steps were required:

1. Determine the values for W, w, L, ℓ, ℓ^3 and R.
   W = area to be supported × weight (LL + DL)
   w = W/L
   L = span in feet
   ℓ = span in inches
   ℓ^3 = (ℓ) (ℓ) (ℓ) (express this value in parts per million)
   R = V = W/2
2. Determine F_v:

$$F_v = \frac{(3)(V)}{(2)(b)(d)} = \# < 85 \text{ or } \frac{(3)(V)}{85} = \# > (2)(b)(d)$$

3. Determine S:

$$S = \frac{(3)(w)(L^2)}{(2)(F_b)}$$

4. Determine the value for D_{max} based on what the beam will support:

$$D = \ell/360 \text{ or } \ell/240 \text{ or } \ell/180$$

5. Determine D:

$$D = \frac{(5)(W)(\ell^3)}{(384)(E)(I)}$$

6. Determine post supports: R = W/2
7. Determine piers size: R/assumed soil bearing pressure

Figure 23.28 shows a floor plan with a ridge beam that needs to be determined. The beam will be DFL with no snow loads. Using the seven steps, determine the size of the beam.

Step 1. Determine the values.
W = 10 × 12 × 40 = 4800 lb
w = W/L = 4800/12 = 400 lb
R = V = 2400
L = 12'
L2 = 144
ℓ = 12' × 12'' = 144
ℓ^3 = 2.986
F_b = (see Figure 23.7 or 23.29)
F_v = 85 max (see Figure 23.7 or 23.29)
E = 1.6 max (see Figure 23.7 or 23.29)

Step 2. Determine horizontal shear. The value from Table 23.7 or 23.29 is 85.

$$\frac{(3)(V)}{85} = \frac{3 \times 2400}{85} = \frac{7200}{85} = 84.70$$

By looking in the table in Figure 23.27, it can be seen that a 4 ×14 and a 6 × 10 each have a 2bd value larger than the required value of 84.70. Each of these beams,

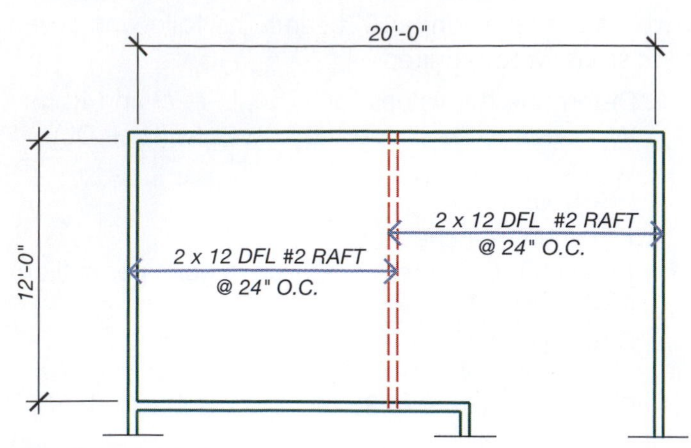

FIGURE 23.28 ■ *A sample floor plan with a beam of undetermined size.*

and any larger beam will meet the stress from horizontal shear.

Step 3. Determine the section modulus.

$$S = \frac{(3)(w)(L^2)}{(2)(F_b)} = \frac{3 \times 400 \times 144}{2 \times 1350} = \frac{172{,}800}{2700} =$$

required section modulus = 64

Because a 6 × 10 beam has a section modulus of 82.7 units of stress, it will be safe.

Step 4. Determine D_{max}.

$$D_{max} = \frac{\ell}{180} = \frac{144}{180} = 0.8''$$

Step 5. Determine D:

$$D = \frac{(5)(W)(\ell^3)}{(384)(E)(I)} = \frac{5 \times 4800 \times 2.986}{384 \times 1.6 \times 393}$$

$$D = \frac{71{,}664}{241{,}459} = D = 0.296$$

Step 6. Determine reactions:

$$R = W/2 = 2400$$

Step 7. Determine piers. R/soil value (assume 2000 lb)

$$\frac{2400}{2000} = 1.2 \text{ sq ft}$$

According to the table (Figure 23.26) use either a 15'' diameter or 15'' square pier.

Wood Adjustment Factors

To this point you've been introduced to the basics of solving simple beams using the NDF base design values for the particular species based on the size of the member. These values will provide an accurate but conservative description of how a specific member can be used. Building codes and the NDS guidelines allow the beam values to be adjusted based on conditions that will reflect the true use of the structural member. These reductions include repetitive use, load duration, moisture and temperature content, size, and shear stress.

Repetitive Use Factor

When a load is spread over several members, a ***reduction*** in the $\mathbf{F_b}$ value is allowed. The fiber bending value of lumber 2'' and 4'' thick placed repetitively can be applied if all of the following conditions are met:

- The structural members receiving the reductions must be placed at 24'' maximum spacing.
- A minimum of three members are used in the pattern
- Each of the structural members are joined by sheathing or decking

When all three conditions are met, an increase in the $\mathbf{F_b}$ value can be used. The increase is $F_b \times 1.15$. If you're using a computer program to solve beams, this adjustment will usually be made automatically if repetitive use is specified.

Load Duration Factor

Tests have shown that a piece of lumber can carry greater maximum loads over a short period of time than over a long period of time. When a load is supported by a wood member for a short duration, the member will return to its original shape when the load is reduced. When the load is supported for a long period of time, the member will become permanently deformed. Design standards generally recognize several common duration periods that affect the maximum live load to be supported. These load factors include permanent, normal, 2 months, 7 days, wind or earthquake, and impact. When two or more loads of different duration will be supported, the factors are not cumulative. Use the load duration factor that best represents the lifetime of the load. Because wind, earthquake, and impact loads can vary so widely, these should be based on local codes and considered with the supervision of a licensed professional. If you're using a computer program to determine beam sizes, this adjustment will usually be made automatically.

Laminated Beams

Two beam problems have now been solved using the tables in this chapter. Both beams were chosen from standard sawn lumber. Often, glu-lams are used because of their superior strength. Using a glu-lam can greatly reduce the depth of a beam as compared with conventional lumber. Glu-lam beams are determined in the same manner as standard lumber, but different values are used based on values provided by the American Institute of Timber Construction (see Figure 23.29). Figure 23.30 gives values for glu-lams made of Douglas fir. You will also need to consult the safe values for glu-lam beams. For beams constructed of Douglas fir, the values are

$$F_b = 2400,\ F_v = 165, \text{ and } E = 1.7.$$

Using Engineered Beams

Most manufacturers of engineered joists and rafters also supply beams made of LVL (laminated veneer lumber), PSL (parallel-strand lumber), and LSL (laminated-strand lumber). Figure 23.31 shows an example of a table for sizing Parallam beams made of PSL. To use this table, follow these steps:

- Determine the roof loading (snow duration and live and dead loads) and find the appropriate section of the table that represents the design loads.
- Find the house width from the appropriate loading section that meets or exceeds the span of the trusses.
- Locate the opening size in the "Rough Opening" column that meets or exceeds the required window or door rough opening.
- The beam listed at the intersection of the rough opening and the house width/roof load is the required header size.

Because sizes vary with each manufacturer, specific standards and tables should be obtained from the beam supplier.

DETERMINING BEAMS WITH COMPLEX LOADS

Up to this point, you've been introduced to sizing structural members by using span tables and mathematical formulas. The procedures worked because the loading patterns and support methods remained constant. The final portion of this chapter will introduce for-

BASE DESIGN VALUES FOR BEAMS & STRINGERS

SPECIES AND COMMERCIAL GRADE	EXTREME FIBER BENDING F_b	HORIZONTAL SHEAR F_v	MODULUS OF ELASTICITY E
DFL #1	1350	85	1,600,000
24 F_b V4 DF/DF	2400	165	1,700,000
Hem Fir #1	1050	70	1,300,000
24 F_b E-2 HF/HF	2400	155	1,700,000
SPF #1	900	65	1,200,000
22F_b E-2 SP/SP	2200	200	1,700,000

BASE DESIGN VALUES FOR POSTS & TIMBERS

SPECIES AND COMMERCIAL GRADE	EXTREME FIBER BENDING F_b	HORIZONTAL SHEAR F_v	MODULUS OF ELASTICITY E
DLF #1	1200	85	1,600,000
Hem Fir #1	950	70	1,300,000
SPF #1	800	65	1,200,000

FIGURE 23.29 ■ *Comparative values of common framing lumber with laminated beams of equal materials. Values based on the Western Wood Products Association.*

PROPERTIES OF GLU-LAM BEAMS

SIZE (b)(d)	S	A	(2)(b)(d)	I	(384)(E)(I) *
3⅛ × 9.0	42.2	28.1	56.3	189.8	123,901
3⅛ × 10.5	57.4	32.8	65.6	301.5	196,819
3⅛ × 12.0	75.0	37.5	75.0	450.0	293,760
3⅛ × 13.5	94.9	42.2	84.4	640.7	418,249
3⅛ × 15.0	117.2	46.9	93.8	878.9	573,746
5⅛ × 9.0	69.2	46.1	92.0	311.3	203,217
5⅛ × 10.5	94.2	53.8	107.6	494.4	322,744
5⅛ × 12.0	123.0	61.5	123.0	738.0	481,766
5⅛ × 13.5	155.7	69.2	138.4	1,050.8	685,962
5⅛ × 15.0	192.2	76.9	153.8	1,441.4	940,946
5⅛ × 16.5	232.5	84.6	169.0	1,918.5	1,252,397
6¾ × 10.5	124.0	70.9	141.8	651.2	425,103
6¾ × 12.0	162.0	81.0	162.0	972.0	634.522
6¾ × 13.5	205.0	91.1	182.6	1,384.0	903,475
6¾ × 15.0	253.1	101.3	202.0	1,898.4	1,239,276
6¾ × 16.5	306.3	111.4	222.8	2,526.8	1,649,495

*All values for 384 × E × I are written in units per million. All E values are figured for Doug fir @ E = 1.7. Verify local conditions.

FIGURE 23.30 ■ *Structural properties of glu-lam beams. Columns (2)(b)(d) and (384)(E)(I) are set up for Douglas fir. If a different type of wood is to be used, you must use different values for these columns. Values based on the Western Wood Products Association.*

mulas that are needed when the loading patterns are altered. Don't skip over the word *introduced.* This chapter is not intended to make you an engineer but to acquaint you with some of the basic formulas needed for determining light framing members sizes with irregular loading patterns. Consult information provided by Western Wood Products Association for the formulas for additional loading patterns.

Sizing a Simple Beam with a Concentrated Load at the Center

Often in light construction, load-bearing walls of an upper floor may not line up with the bearing walls of the floor below, as shown in Figure 23.32. This type of loading occurs when the function of the lower room dictates that no post be placed in the center. Since no post can be used on the lower floor, a beam will be required to span between the bearing walls and be centered below the upper-level post. This beam will have no loads to support other than the weight from the post.

In order to size the beam needed to support the upper area, use the formulas in Figure 23.33. Notice that a new symbol, **P,** has been added to represent a point, or concentrated load in the formula used to determine the maximum bending moment (**M**). The point load is the sum of the reactions from each end of the beam.

Figure 23.34 shows a sample worksheet for this beam. Notice that the procedure used to solve for this beam is similar to the one used to determine a simple

Note:

A new formula is provided in Figure 23.33 that was not used to determine deflection for a simple beam in 23.20. Notice that a formula for Δ_x (deflection at any point) has been provided. Formulas for other loading patterns that follow throughout the balance of this chapter will also introduce formulas for solving additional moment values, and deflection values at multiple points. Don't get distracted by all of the options. Remember this is an introduction to what an engineer does, not the CAD technician. You'll make most employers happy if you can identify structural members that carry or transfer loads, and if you can follow the load path from the roof to the ground. It sounds simple, but as your projects get more complicated, so does the load path. That's why many architects hire engineers to provide structural calculations.

Headers Supporting Roof

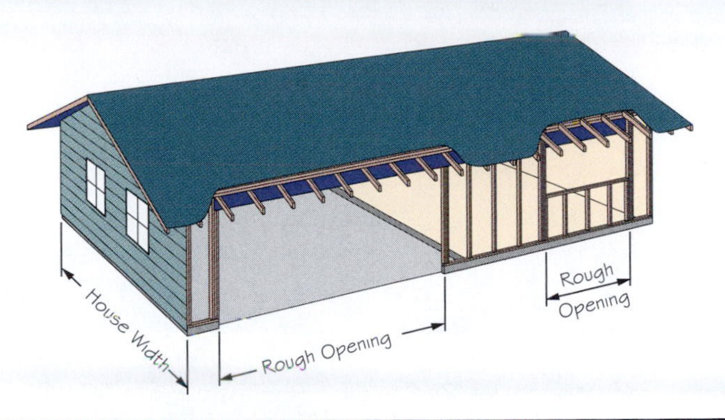

	ROOF LOAD (psf)	HOUSE WIDTH	ROUGH OPENING 8'-0"	9'-3"	10'-0"	12'-0"	14'-0"	16'-3"	18'-3"
NON-SNOW AREA 125%	20LL + 15DL	24'-0"	3 1/2" x 9 1/4"	3 1/2" x 9 1/4"	3 1/2" x 9 1/4"	3 1/2" x 9 1/4"	3 1/2" x 11 1/4" 5 1/4" x 9 1/4"	3 1/2" x 11 1/4" 7" x 9 1/4"	3 1/2" x 14" 5 1/4" x 11 1/4"
		30'-0"	3 1/2" x 9 1/4"	3 1/2" x 9 1/4"	3 1/2" x 9 1/4"	3 1/2" x 9 1/4"	3 1/2" x 11 1/4" 5 1/4" x 9 1/4"	3 1/2" x 11 7/8" 5 1/4" x 11 1/4"	3 1/2" x 14" 5 1/4" x 11 7/8"
		36'-0"	3 1/2" x 9 1/4"	3 1/2" x 9 1/4"	3 1/2" x 9 1/4"	3 1/2" x 9 1/4"	3 1/2" x 11 1/4" 5 1/4" x 9 1/2"	3 1/2" x 14" 5 1/4" x 11 1/4"	3 1/2" x 14" 7" x 11 1/4"
	20LL + 20DL	24'-0"	3 1/2" x 9 1/4"	3 1/2" x 9 1/4"	3 1/2" x 9 1/4"	3 1/2" x 9 1/4"	3 1/2" x 11 1/4" 5 1/4" x 9 1/4"	3 1/2" x 11 7/8" 5 1/4" x 11 1/4"	3 1/2" x 14" 5 1/4" x 11 1/4"
		30'-0"	3 1/2" x 9 1/4"	3 1/2" x 9 1/4"	3 1/2" x 9 1/4"	3 1/2" x 9 1/4"	3 1/2" x 11 1/4" 5 1/4" x 9 1/4"	3 1/2" x 14" 5 1/4" x 11 1/4"	3 1/2" x 14" 7" x 11 1/4"
		36'-0"	3 1/2" x 9 1/4"	3 1/2" x 9 1/4"	3 1/2" x 9 1/4"	3 1/2" x 11 1/4" 5 1/4" x 9 1/4"	3 1/2" x 11 1/4" 7" x 9 1/4"	3 1/2" x 14" 5 1/4" x 11 1/4"	3 1/2" x 16" 5 1/4" x 14"
SNOW AREA 115%	25LL + 15DL	24'-0"	3 1/2" x 9 1/4"	3 1/2" x 9 1/4"	3 1/2" x 9 1/4"	3 1/2" x 9 1/4"	3 1/2" x 11 1/4" 5 1/4" x 9 1/4"	3 1/2" x 11 7/8" 5 1/4" x 11 1/4"	3 1/2" x 14" 5 1/4" x 11 7/8"
		30'-0"	3 1/2" x 9 1/4"	3 1/2" x 9 1/4"	3 1/2" x 9 1/4"	3 1/2" x 9 1/4"	3 1/2" x 11 1/4" 5 1/4" x 9 1/2"	3 1/2" x 14" 5 1/4" x 11 1/4"	3 1/2" x 14" 7" x 11 1/4"
		36'-0"	3 1/2" x 9 1/4"	3 1/2" x 9 1/4"	3 1/2" x 9 1/4"	3 1/2" x 11 1/4" 5 1/4" x 9 1/4"	3 1/2" x 11 7/8" 5 1/4" x 11 1/4"	3 1/2" x 14" 5 1/4" x 11 7/8"	3 1/2" x 16" 5 1/4" x 14"
	30LL + 15DL	24'-0"	3 1/2" x 9 1/4"	3 1/2" x 9 1/4"	3 1/2" x 9 1/4"	3 1/2" x 9 1/4"	3 1/2" x 11 1/4" 5 1/4" x 9 1/4"	3 1/2" x 14" 5 1/4" x 11 1/4"	3 1/2" x 14" 5 1/4" x 11 7/8"
		30'-0"	3 1/2" x 9 1/4"	3 1/2" x 9 1/4"	3 1/2" x 9 1/4"	3 1/2" x 11 1/4" 5 1/4" x 9 1/4"	3 1/2" x 11 1/4" 7" x 9 1/4"	3 1/2" x 14" 5 1/4" x 11 7/8"	3 1/2" x 16" 5 1/4" x 14"
		36'-0"	3 1/2" x 9 1/4"	3 1/2" x 9 1/4"	3 1/2" x 9 1/4"	3 1/2" x 11 1/4" 5 1/4" x 9 1/4"	3 1/2" x 14" 5 1/4" x 11 1/4"	3 1/2" x 16" 5 1/4" x 14"	3 1/2" x 16" 5 1/4" x 14"
	40LL + 15DL	24'-0"	3 1/2" x 9 1/4"	3 1/2" x 9 1/4"	3 1/2" x 9 1/4"	3 1/2" x 11 1/4" 5 1/2" x 9 1/4"	3 1/2" x 11 1/4" 7" x 9 1/4"	3 1/2" x 14" 5 1/4" x 11 7/8"	3 1/2" x 16" 5 1/4" x 14"
		30'-0"	3 1/2" x 9 1/4"	3 1/2" x 9 1/4"	3 1/2" x 9 1/4"	3 1/2" x 11 1/4" 5 1/4" x 9 1/4"	3 1/2" x 14" 5 1/4" x 11 1/4"	3 1/2" x 16" 5 1/4" x 14"	3 1/2" x 18" 5 1/4" x 14"
		36'-0"	3 1/2" x 9 1/4"	3 1/2" x 9 1/4"	3 1/2" x 9 1/2" 5 1/4" x 9 1/4"	3 1/2" x 11 7/8" 5 1/4" x 9 1/2"	3 1/2" x 14" 5 1/4" x 11 1/4"	3 1/2" x 16" 5 1/4" x 14"	5 1/4" x 16" 7" x 14"

General Notes

Table is based on:

- Uniform loads
- Worst case of simple or continuous span. When sizing a continuous span application, use the longest span. Where ratio of short span to long span is less than 0.4, use the TJ-Beam™ software program or contact your Trus Joist MacMillan representative.
- Roof truss framing with 24" soffits
- Deflection criteria of L/240 live load and L/180 total load. All members 7 1/4" and less in depth are restricted to a maximum deflection of 5/16".

BEARING REQUIREMENTS

Minimum header support to be double trimmers (3" bearing).

In shaded areas, support headers with triple trimmers (4 1/2" bearing).

FIGURE 23.31 ■ *Design tables for solving exterior headers for a single-level residence using non-treated Parallam PSL beams.* Courtesy Trus Joist, A Weyerhaeuser Business.

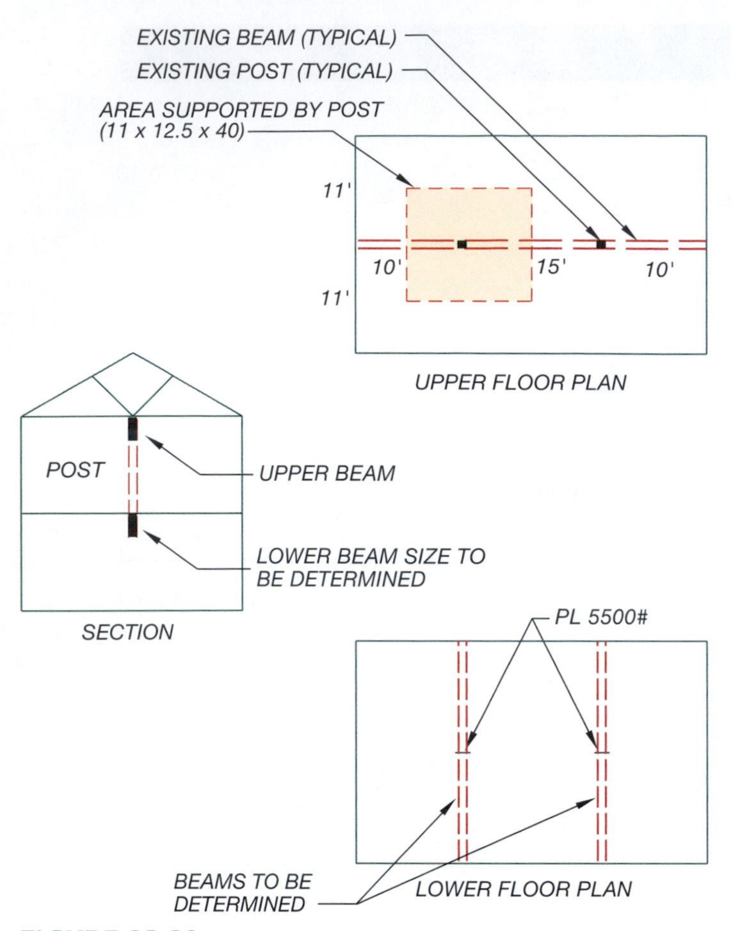

FIGURE 23.32 ■ *The sketch created to help determine the loads on a beam with a concentrated load at the center. Sketching a simple floor plan helps to determine the area to be supported. Sketching a simple section helps determine the load path so that all loads can be safely transferred to the soil.*

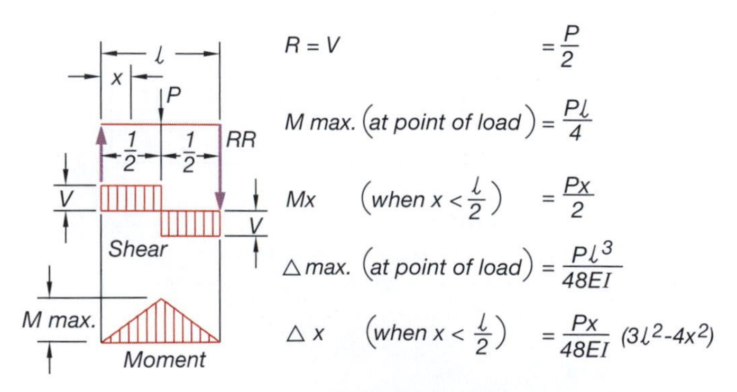

FIGURE 23.33 ■ *Shear and moment diagrams for a simple beam with a load concentrated at the center. Data reproduced.* Courtesy of Western Wood Products Association.

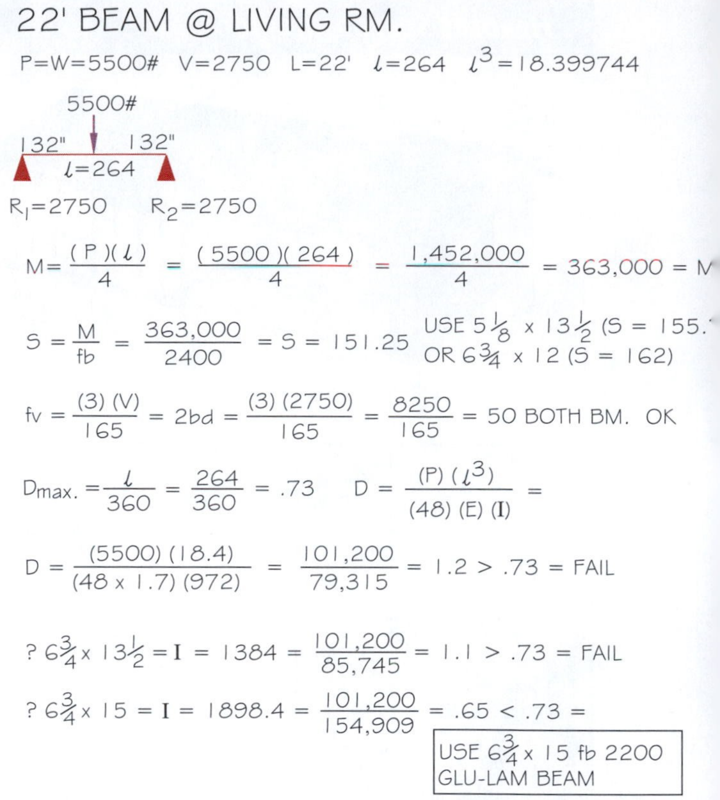

22' BEAM @ LIVING RM.

P=W=5500# V=2750 L=22' ℓ=264 ℓ^3=18.399744

5500#

132" 132"

ℓ=264

R_1=2750 R_2=2750

$$M = \frac{(P)(\ell)}{4} = \frac{(5500)(264)}{4} = \frac{1,452,000}{4} = 363,000 = M$$

$$S = \frac{M}{fb} = \frac{363,000}{2400} = S = 151.25$$

USE $5\frac{1}{8} \times 13\frac{1}{2}$ (S = 155.7) OR $6\frac{3}{4} \times 12$ (S = 162)

$$fv = \frac{(3)(V)}{165} = 2bd = \frac{(3)(2750)}{165} = \frac{8250}{165} = 50 \text{ BOTH BM. OK}$$

$$D_{max.} = \frac{\ell}{360} = \frac{264}{360} = .73 \quad D = \frac{(P)(\ell^3)}{(48)(E)(I)} =$$

$$D = \frac{(5500)(18.4)}{(48 \times 1.7)(972)} = \frac{101,200}{79,315} = 1.2 > .73 = \text{FAIL}$$

$$?\ 6\tfrac{3}{4} \times 13\tfrac{1}{2} = I = 1384 = \frac{101,200}{85,745} = 1.1 > .73 = \text{FAIL}$$

$$?\ 6\tfrac{3}{4} \times 15 = I = 1898.4 = \frac{101,200}{154,909} = .65 < .73 =$$

USE $6\frac{3}{4} \times 15$ fb 2200 GLU-LAM BEAM

FIGURE 23.34 ■ *The worksheet for a beam with a load concentrated at the center. Always include the location of the beam, a sketch of the loading, needed formulas, and the selected beam as part of your calculations.*

beam with uniform loads. With this procedure, you must determine the amount of point loads for the upper beams. The sketch shows the upper floor plan with each point load.

Once the point loads are known, determine **M.** Once the value for **M** has been determined, it can be used to find the required section modulus (**S**). Use the formula $M/F_b = S$. This will provide the **S** value for the beam. Try to use a beam that has a depth equal to the depth of the floor joists.

Note:

Notice in Figure 23.33 that there is no formula for determining the value for F_v. This will also be true of the formulas for solving other complex beams. F_v is always determined using the formula $3V/(2)(b)(d)$ = required F_v value or $3V/max\ F_v$ value = the required minimum 2bd value.

In Figure 23.34, two different beams have been selected that have the minimum required section modulus. Once the section modulus is known, the beam can be checked for horizontal shear and deflection.

Sizing a Simple Beam with a Concentrated Load at Any Point

Figure 23.35 shows a sample floor plan that will result in concentrated off-center loads at the lower floor level. Figure 23.36 shows the diagrams and formulas that would be used to determine the beams of the lower floor plan. Figure 23.37 shows the worksheet for this beam. Just as with the previous example, once the point loads are known, determine **M.** Once the value for **M** has been determined, it can be used to find the required section modulus (**S**) from the tables in Figures 23.27 or 23.30. Use the formula $M/F_b = S$. This will provide the **S** value for the beam.

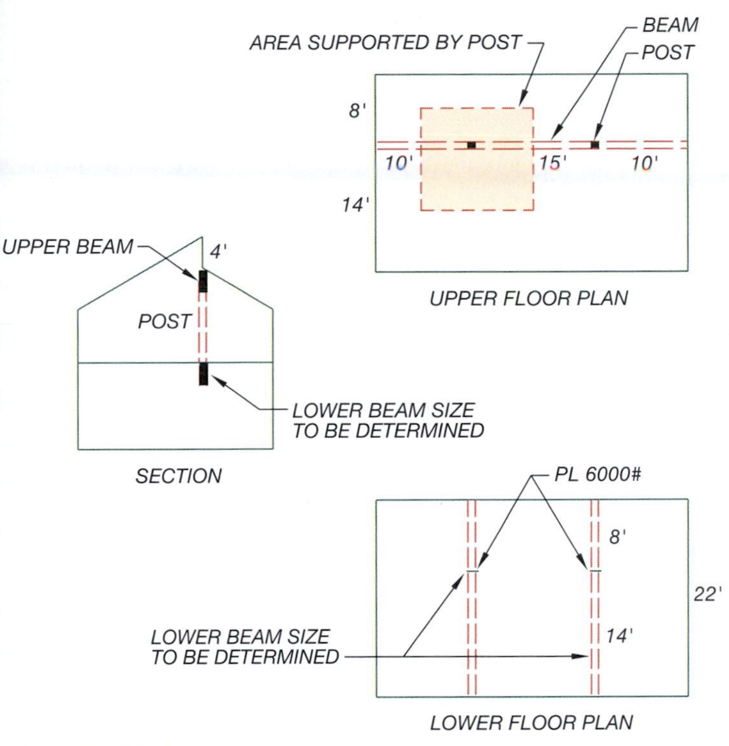

FIGURE 23.35 ■ *A sketch for a beam with a concentrated load placed at any point on the beam.*

Cantilevered Beam with a Uniform Load

This loading pattern typically occurs where floor joists of a deck cantilever past the supporting wall. One major difference of this type of loading is that the live loads of exterior balconies are almost double the normal floor live loads. A second consideration is that cantilevered joists are required by the IRC to have a backspan ratio into the structure of 3:1. The size of cantilevered floor joists can be determined from tables in the IRC. Figure 23.38 shows the diagrams and formulas that can be used to determine the size for a cantilevered beam or joist. Figure 23.39 shows the worksheet to determine the required size of floor beams placed at 24'' o.c. with a 3' balcony cantilever. Once the value for **M** has been determined, use the formula $M/F_b = S$ to provide the **S** value for the beam from the tables in Figure 23.27 or 23.30. Engineered joists that cantilever often require addition support at the cantilever. Be sure to follow the manufacturer's in-

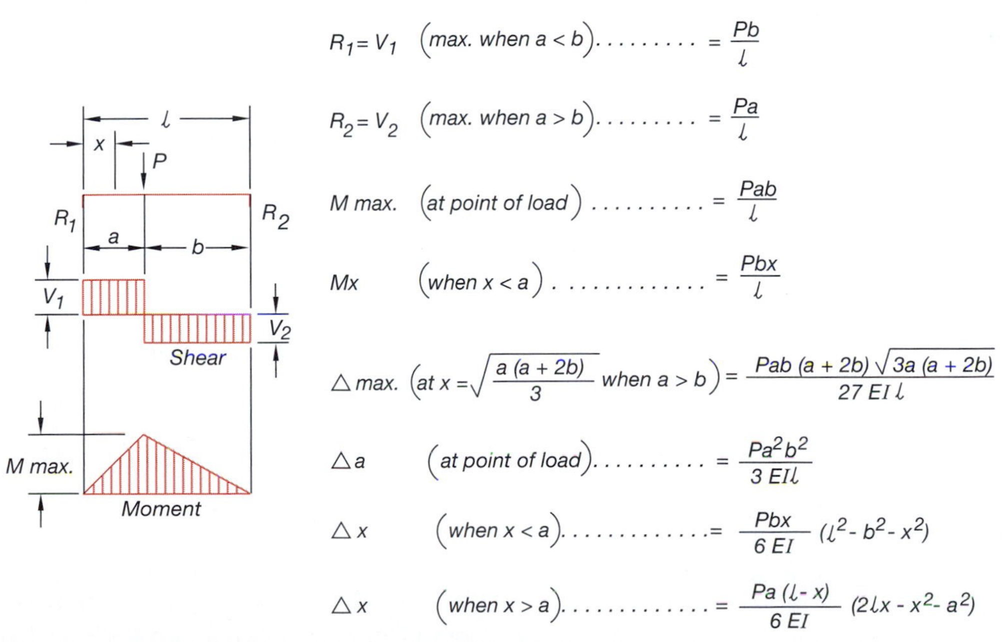

FIGURE 23.36 ■ *The shear and moment diagrams for a simple beam with a load concentrated at any point on the beam. Data reproduced.* Courtesy of Western Wood Products Association.

22' SPAN @ LOWER FLOOR

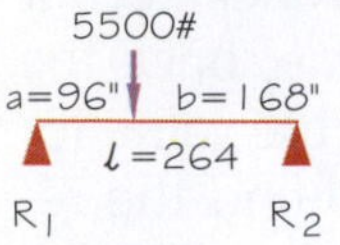

5500#

a=96" b=168"

ℓ=264

R_1 R_2

W = (4 + 7) (5 + 7.5) (40#) = W = 5500#

ℓ = 264"

$$R_1 = \frac{(P)(b)}{\ell} = \frac{(5500)(168)}{264} = \frac{924{,}000}{264} = 3500\# = R_1$$

$$R_2 = P - R_1 = 5500 - 3500 = 2000 = R_2$$

$$M = \frac{(P)(a)(b)}{\ell} = \frac{(5500)(96)(168)}{264} = \frac{88{,}704{,}000}{264} = 336{,}000 = M$$

$$S = \frac{M}{fb} = \frac{336{,}000}{2200} = 152.7$$

USE $5\frac{1}{8} \times 13\frac{1}{2}$
OR $6\frac{3}{4} \times 12$

$$fv = \frac{(3)(V)}{165} = \frac{(3)(3500)}{165} = \frac{10{,}500}{165} = 63.6 \text{ BOTH BM. OK}$$

$$D = \frac{\ell}{360} = \frac{264}{360} = .73 \text{ max.} \quad E = \frac{(P)(a)(b)(a+2b)\sqrt{3a(a+2b)}}{27(E)(\ell)(I)} =$$

$$D = \frac{(5500)(96)(168)(432)\sqrt{(288)(432)}}{(27)(1.7)(264)(I) = (12118)(I)} = \frac{13{,}516{,}525}{12{,}117.6\,(I)} =$$

$$?\ 5\tfrac{1}{8} \times 13\tfrac{1}{2} = \frac{13{,}516{,}525}{12{,}733{,}174} = 106 > .73 = \text{FAIL}$$

$$?\ 5\tfrac{1}{8} \times 15 = \frac{13{,}516{,}525}{17{,}466{,}308} = .77 > .73 = \text{FAIL}$$

$$?\ 6\tfrac{3}{4} \times 15 = \frac{13{,}516{,}525}{23{,}004{,}051} = .58 < .73 = \text{ok}$$

USE $6\frac{3}{4}$ x 15 fb 2200
GLU-LAM BEAM

FIGURE 23.37 ■ *A sample worksheet for a beam with an off-center concentrated load.*

FLOOR JOIST, 36" CANTILEVER

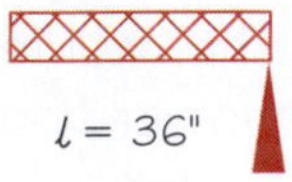

W = 210#

ℓ = 36"

W = 210# w = 70# V = 210

L = 3' ℓ = 36"

ℓ^2 = 1296 ℓ^4 = 1.678,616 (1.68)

$$M = \frac{(w)(\ell^2)}{2} = \frac{(70)(1296)}{2} = \frac{90{,}720}{2} = 45{,}360$$

$$S = \frac{M}{fb} = \frac{45{,}360}{1495} = 30.34$$

USE 4 x 12 OR 6 x 8 @ 24" O.C.

$$2bd = \frac{(3)(V)}{fv} = \frac{(3)(210)}{85} = \frac{630}{85} = 7.4 \quad 2 \times 12 \text{ OK}$$

$$D_{max} = \frac{\ell}{360} = \frac{36}{360} = .1 \text{ max.}$$

$$D = \frac{(w)(\ell^4)}{(8)(E)(I)} = \frac{(70)(1.68)}{(8)(1.6)(178)} = \frac{118}{2278} = .05 < .1 \text{ OK}$$

BOTH BEAMS SAFE @ 24" SPACING

FIGURE 23.39 ■ *A sample worksheet for a cantilevered joist with a uniform load.*

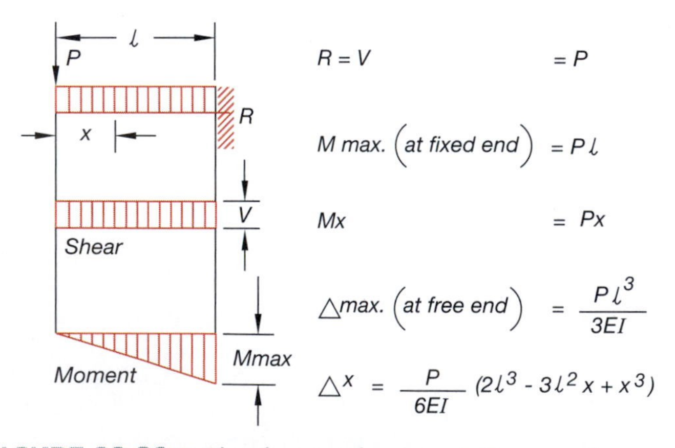

FIGURE 23.38 ■ *The shear and moment diagrams for a cantilevered beam with a uniform load.* Courtesy of Western Wood Products Association.

structions and span tables when using engineered lumber in cantilevered situations.

Cantilevered Beam with a Point Load at the Free End

This type of beam results when a beam or joist is cantilevered and is supporting a point load. Floor joists that are cantilevered and support a wall and roof can be sized using tables in the IRC or the formulas in Figure 23.40. Figure 23.41 shows the worksheet for determining the size and spacing of floor joists to cantilever 2' and support a bearing wall.

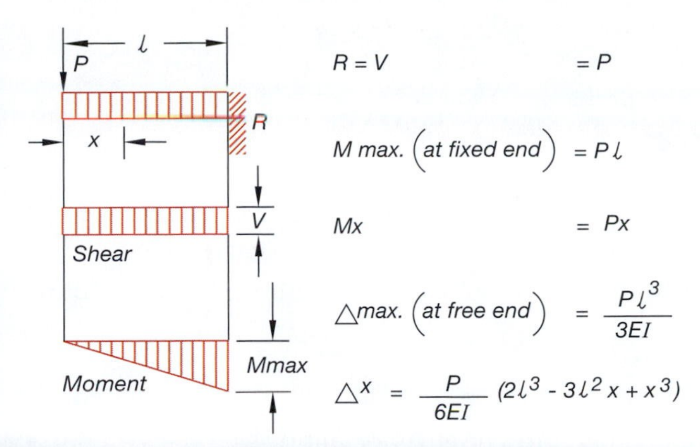

FIGURE 23.40 ■ *The shear and moment diagrams for a cantilevered beam with a concentrated load at the free end.* Courtesy of Western Wood Products Association.

24" CANTILEVER @ WALL

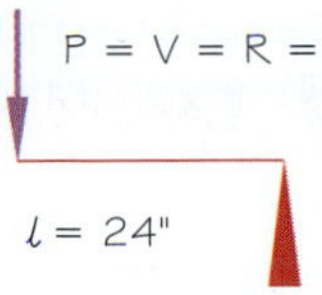

P = V = R = 780#

l = 24"

$$M = (P)(l) = (780)(24) = 18{,}720$$

$$S = \frac{M}{fb} = \frac{18{,}720}{1140} = 16.4$$

USE 2 x 10 F.J. @ 12" O.C.

$$fv = \frac{(3)(V)}{85} = \frac{(3)(780)}{85} = \frac{2380}{85} = 28 = 2bd$$

USE 2 x 10 FAIL
USE 2 x 12 F.J. OK

$$D_{max.} = \frac{l}{360} = \frac{24}{360} = .06 \text{ max.}$$

$$D = \frac{(P)(l^3)}{(3)(E)(I)} = \frac{(780)(.013824)}{(3)(1.6)(177.9)} = \frac{10.78}{853.92} = .013 < .06$$

USE 2 x 10 F.J. @ 12" O.C.

FIGURE 23.41 ■ *A sample worksheet for a cantilevered beam with a concentrated load at the free end.*

CHAPTER 23

Additional Readings

The following websites can be used as a resource to help you keep current on lumber manufacturers.

ADDRESS	COMPANY OR ORGANIZATION
www.afandpa.org	American Forest and Paper Association
www.aitc.glulam.org	American Institute of Timber Construction
www.awc.org	American Wood Council (NDS standards)
www.beamchek.com	AC Software, Inc.
www.bc.com	Boise
www.bcewp.com	Boise (engineered wood products)
www.cwc.ca	Canadian Wood Council
www.forestdirectory.com	Directory of forest products
www.gp.com	Georgia Pacific
www.inpa.org	International Wood Products Association
www.lpcorp.com	Louisiana-Pacific
www.strucalc.com	Cascade Consulting Engineers
www.sfpa.org	Southern Forest Products Association
www.trusjoist.com	Trus Joist, A Weyerhaeuser Business
www.weyerhaeuser.com	Weyerhaeuser building products
www.woodcom.com	Wood Industries Information Center
www.woodtruss.com	Wood Trus Council of America
www.wwpa.org	Western Wood Products Association

CHAPTER 23

Sizing Joists, Rafters, and Beams Test

DIRECTIONS

Answer the following questions with short, complete statements. Type your answers using a word processor or neatly print your answers on lined paper.

1. Place your name, the chapter number, and the date at the top of the sheet.
2. Type the question number and provide the answer in the form of a statement that includes part of the question. You do not need to write out the entire question.
3. Unless something else is noted, answer all questions using the type of lumber common to your area, and assume the use of #2 material.
4. Provide the table number that is used for all appropriate questions.
5. Provide sketches for all questions that require the use of formulas to provide an answer. Write each formula necessary to find your solution and show all work.

Note: The answers to some questions are not contained in this chapter and will require you to visit sites that provide engineered lumber in your area.

23.1. List four common types of lumber used for framing throughout the country.

23.2. How is modulus of elasticity represented in engineering formulas?

23.3. What is deflection, and how do building codes express its limits?

23.4. What are two features that define a simple beam?

23.5. Explain the difference between a uniform and a concentrated load.

23.6. Determine the size of floor joist that will be required to span 15'-0'' if spaced at 16'' o.c.

23.7. You are considering the use of 2 × 8 floor joists to span 14'-0''. Will they work?

23.8. Using 2 × 6 at 16'' o.c. for ceiling joists, determine their maximum safe span for a ceiling with no storage and for a ceiling with limited storage.

23.9. Determine the size of lumber required to span 16'-6'' with 16'' and 24'' spacings used for rafter/ceiling joists.

23.10. Determine the floor joist size needed to span 13'-0'' to support a kitchen floor.

23.11. Determine the smallest size of floor joist needed to span 13'-0'' to support a living room floor.

23.12. What is the smallest size floor joist that can be used to span 15'-9'' beneath a dining area?

23.13. What size floor joist would be needed to support a den 12'-2'' wide? A bedroom of the same size?

23.14. Determine the smallest floor joist that could be used to span a living area that is 12'-8'' wide.

23.15. A bedroom is 17'-9'' wide. What is the smallest ceiling joist that could be used to provide limited storage?

23.16. A contractor bought a truckload of 2 × 6s that are 18' long. Can they be used for ceiling joists with no attic storage if they are spaced at 16'' o.c.?

23.17. What size rafter is needed for a home 28' wide with a gable roof, 4/12 pitch with 235 lb composition shingles if a 10 lb dead load and a 30 lb snow load is assumed?

23.18. What is the maximum span allowed using 2 × 10 rafters supporting built-up roofing if a 10 lb dead load and a 20# snow load is assumed?

23.19. Determine the rafter size to be used to span 14' and support a 30 lb live load and a 10 lb dead load at a 6/12 pitch.

23.20. Determine the required sizes for engineered floor joists to span 17'-0'' using L/360 and L/480 using a spacing of 16'' o.c. and a 10 lb dead load and a 40 lb live load is assumed.

23.21. What is the advantage of using a deflection value of L/480 if the building codes allow a greater deflection?

23.22. List the smallest TJI size and spacing that can be used to span 17-0'' if L/480 is to be used and a 10 lb dead load and a 40 lb live load is assumed?

23.23. A 9 1/2'' TJI-110 at 19.2'' o.c. will be used to cantilever 24''. The joist will support a floor load of 45 psf and a wall that supports a truss 26' wide with 1' overhangs. How should the cantilevered joist be reinforced?

23.24. If the joists in Question 23.23 were placed at 16'' o.c., how should they be reinforced?

23.25. Engineered rafters are required to span 19' using 20LL+ 15DL for a high-slope roof in an area where no snow is expected. What size TJI should be used if 24'' spacing is desired?

23.26. Engineered rafters are required to span 17'-6'' and support a DL of 20 lb on a 5/12 pitch in an area where no snow is expected. What is the smallest size TJI that can be used? What is the smallest size rafter that can be used if 24'' spacing is desired?

23.27. Use the Internet, Sweets catalogs, or local vendor resources to locate five or more different manufacturers of engineered floor joists.

23.28. List three common categories of stress that must be known before a beam size can be determined.

23.29. Define the following notations: W, w, R, L, ℓ, E, I, S, F_v, and F_b.

23.30. List two methods of determining beams other than using formulas.

23.31. List the values for W, w, and R for a floor beam that supports living areas and has an L of 10' and is centered under a room that is 20' wide. Provide a sketch to show the loading.

23.32. A 4 × 6 DFL beam with an L of 6', and a W of 2800 lb will be used to span an opening for a window. Will the beam fail in F_v?

23.33. A 4 × 14 is being used as a ridge beam. L = 12', W = 4650 lb. Will this beam have a safe deflection value? Assume $D_{max} = \ell/360$.

23.34. If the soil bearing pressure is 1500 psf and the concrete has a strength of 2500 psi, what size pier is needed to support a load of 4600 lb?

23.35. What size #2 girder will be required if w = 600 and L = 6'? Determine all necessary values to find the needed S, F_v and deflection stresses.

23.36. What size glu-lam beam will be required for a ridge beam with L = 14.5' and W = 5500 lb. $\ell = 240$ if a F_b value of 2400 is used?

23.37. A 16' long laminated ridge beam will support 800 lb per linear foot. Determine what depth 5 1/8'' wide laminated beam should be used assuming an F_b value of 2400.

23.38. The ridge beam in Question 23.37 will be supported at one end by a post and at the other end by a beam 4' long over a door. The point load from the ridge beam will be 12'' from one end of the 4' header. What size door header should be used?

23.39. Determine the minimum size floor beam needed to span 14' with a concentrated load of 3200 lb at the center. What size piers will be needed to resist the reactions if the assumed safe soil loads are 2000 psf?

23.40. A two-level residence is 28' wide with a gable roof and 24'' wide overhangs. The roof is cedar shakes at a 5/12 pitch, and local codes require a 30-lb live load. Wall heights are 9' at the upper floor. There are several windows in the upper exterior walls that are 60'' wide. A wall will be placed 13' from the left exterior bearing wall on the upper floor. The floor joists supporting the upper floor will be TJI and cantilever 18'' past the lower wall on the right side of the residence so that the lower floor is only 26'-6'' wide. A bearing wall on the main floor will be 16' from the left exterior wall. Part of the wall will be an opening. The walls of the lower floor will be 8' high. This home also has windows on the left side of the house on the lower floor that are directly below the windows on the upper floor. The lower floor joists will be sawn lumber supported by a girder directly below the bearing wall, with supports at 5'-0'' maximum spacing. The owners of the house have two kids and a dog named Spot.

Determine the size of all framing members, assuming that the roof is framed with trusses. Provide the size of the following materials:

Upper headers =
Upper floor joists (left) =
Lower window header =
Upper floor joists (right) =
Lower floor joist =
Girder =

24 CHAPTER

Drawing Framing Plans

A framing plan is a drawing used to show the dimensions, framing members, and methods of resisting seismic and wind loads for a specific level of a structure. Office practice and the complexity of the structure determine the exact contents of the plan. This chapter explores common methods of representing structural information and provides step-by-step instruction for preparing a framing plan for each required level of a home.

FRAMING DRAWINGS

When a simple one-level house using truss roof construction is drawn, architectural and structural information can often be combined on the floor plan, as in Figure 24.1. For a custom multilevel structure, a separate plan is usually developed to explain the roof framing as well as the architectural and structural requirements of each level. The floor plans are used to explain architectural information (see Figure 24.2). The framing plan is used to show the location of all walls and openings, header sizes, and structural connectors. Section markers to indicate where sections and details have been drawn are also placed on a framing plan. Figure 24.3 shows the framing plan that corresponds to the floor plan in Figure 24.2. Key elements that must be represented on a framing plan include representing framing members, lateral bracing, dimensions, annotation explaining all structural materials, and section and detail reference markers.

Framing Plan Orientation

The framing plan can be created from a the base drawing that was used to create the floor plan, showing the walls, doors, windows, cabinets, appliances and plumbing symbols. The symbols used to represent door and window sizes should also be represented. When the floor plan is completed, it is with the thought that a cutting plane passed through the structure and removed the upper portion of the structure. The floor plan is drawn as if the viewer is looking down at the floor. When the framing plan is drawn, the information is placed on the floor plan, but with the premise that the viewer is lying on the floor looking up to see the framing that supports the level above. When the main framing plan is drawn for a two-level structure, the plan shows the walls for the main floor, and the beams and floor joists needed to frame the floor/ceiling above the main floor. If western platform construction methods are used, the framing plan for the upper level shows the walls on the upper floor and the members used to form the ceiling.

Representing Roof Members

Rafters and other roof members are shown on the roof framing plan. If a room has a vaulted ceiling, the rafter/ceiling joist is shown on the framing plan. If the roof system is framed using trusses, these are often shown on the upper framing plan. If a complex roof shape is used, the trusses are shown on the roof framing plan. The floor system used to support the main floor is shown on the foundation plan. Section 8 introduces the foundation and floor framing systems.

Generally, as the framing plans are drawn, the structure is completed beginning with the

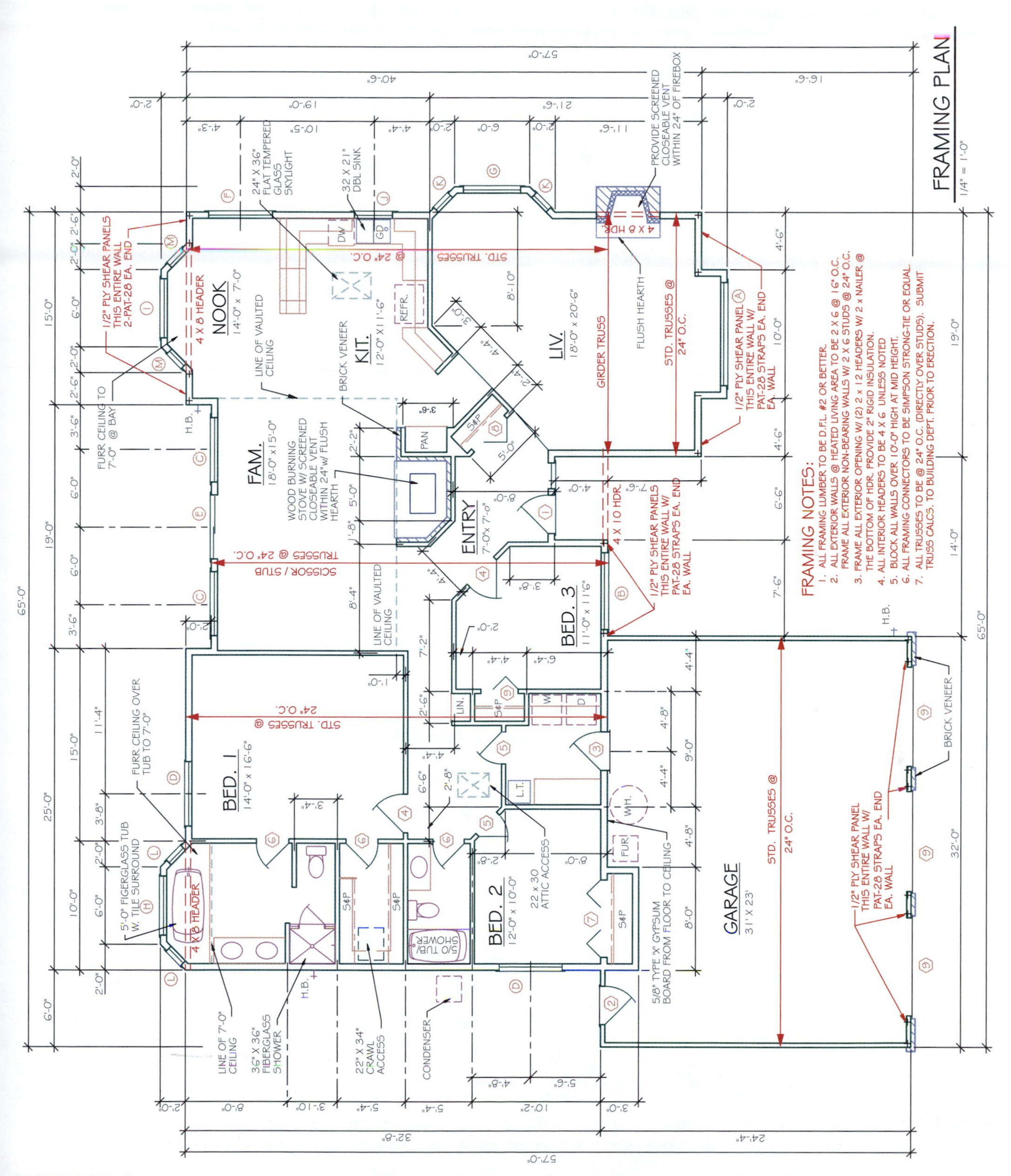
FRAMING PLAN
1/4" = 1'-0"
FRAMING NOTES:
1. ALL FRAMING LUMBER TO BE D.FL. #2 OR BETTER.
2. ALL EXTERIOR WALLS @ HEATED LIVING AREA TO BE 2 X 6 @ 16" O.C. FRAME ALL EXTERIOR NON-BEARING WALLS W/ 2 X 6 STUDS @ 24" O.C.
3. FRAME ALL EXTERIOR OPENING W/ (2) 2 X 12 HEADERS W/ 2 x NAILER @ THE BOTTOM OF HDR. PROVIDE 2" RIGID INSULATION.
4. ALL INTERIOR HEADERS TO BE 4 X 6 UNLESS NOTED
5. BLOCK ALL WALLS OVER 10'-0" HIGH AT MID HEIGHT.
6. ALL FRAMING CONNECTORS TO BE SIMPSON STRONG-TIE OR EQUAL.
7. ALL TRUSSES TO BE @ 24" O.C. (DIRECTLY OVER STUDS). SUBMIT TRUSS CALCS. TO BUILDING DEPT. PRIOR TO ERECTION.
NOOK 14'-0" x 7'-0"
KIT. 12'-0" X 11'-6"
LIV. 18'-0" x 20'-6"
FAM. 18'-0" x 15'-0"
ENTRY 7'-0" x 7'-0"
BED. 3 11'-0" x 11'6"
BED. 1 14'-0" x 16'-6"
BED. 2 12'-0" x 10'-0"
GARAGE 31' X 23'
STD. TRUSSES @ 24" O.C.
SCISSOR / STUB TRUSSES @ 24" O.C.
GIRDER TRUSS
4 X 8 HEADER
4 X 8 HDR.
4 X 10 HDR.
1/2" PLY SHEAR PANELS THIS ENTIRE WALL 2-PAT-28 EA. END
1/2" PLY SHEAR PANEL THIS ENTIRE WALL W/ PAT-28 STRAPS EA. END EA. WALL
PROVIDE SCREENED CLOSEABLE VENT WITHIN 24" OF FIREBOX
FLUSH HEARTH
24" X 36" FLAT TEMPERED GLASS SKYLIGHT
32 X 21" DBL SINK
LINE OF VAULTED CEILING
BRICK VENEER
WOOD BURNING STOVE W/ SCREENED CLOSEABLE VENT WITHIN 24" w/ FLUSH HEARTH
FURR CEILING TO 7'-0" @ BAY
FURR CEILING OVER TUB TO 7'-0"
5'-0" FIBERGLASS TUB W. TILE SURROUND
LINE OF 7'-0" CEILING
36" X 36" FIBERGLASS SHOWER
22" X 34" CRAWL ACCESS
CONDENSER
22 x 30 ATTIC ACCESS
5/8" TYPE 'X' GYPSUM BOARD FROM FLOOR TO CEILING
57'-0"
65'-0"

FIGURE 24.1 ■ *The structural information for a simple residence can be shown on the floor plan.*

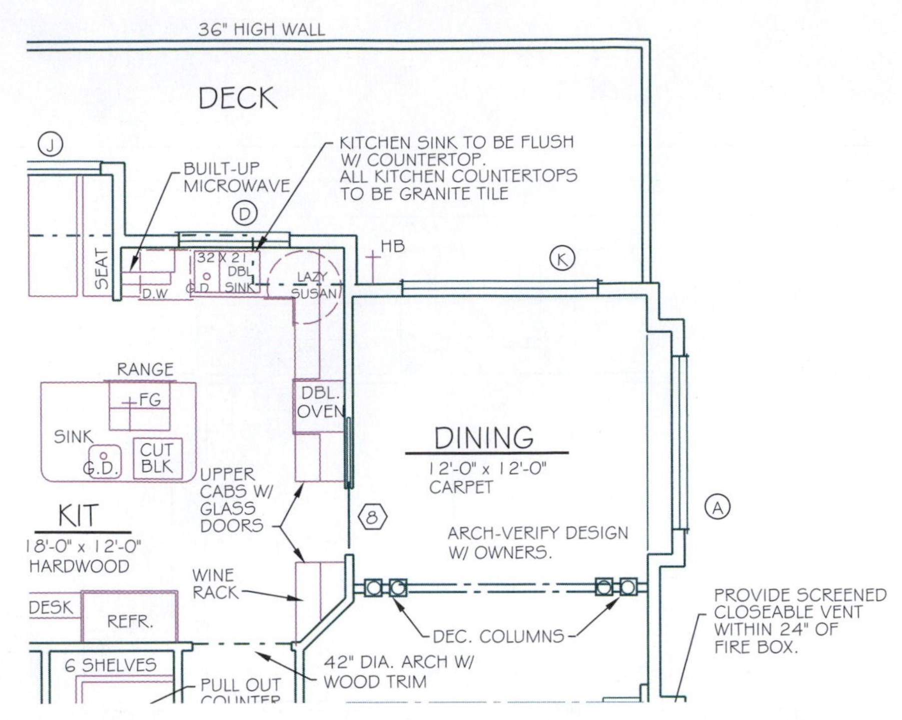

FIGURE 24.2 ■ *For a detailed set of plans, typically only architectural information is placed on the floor plan. All structural information including dimensions is placed on a separate plan.* Courtesy Residential Designs.

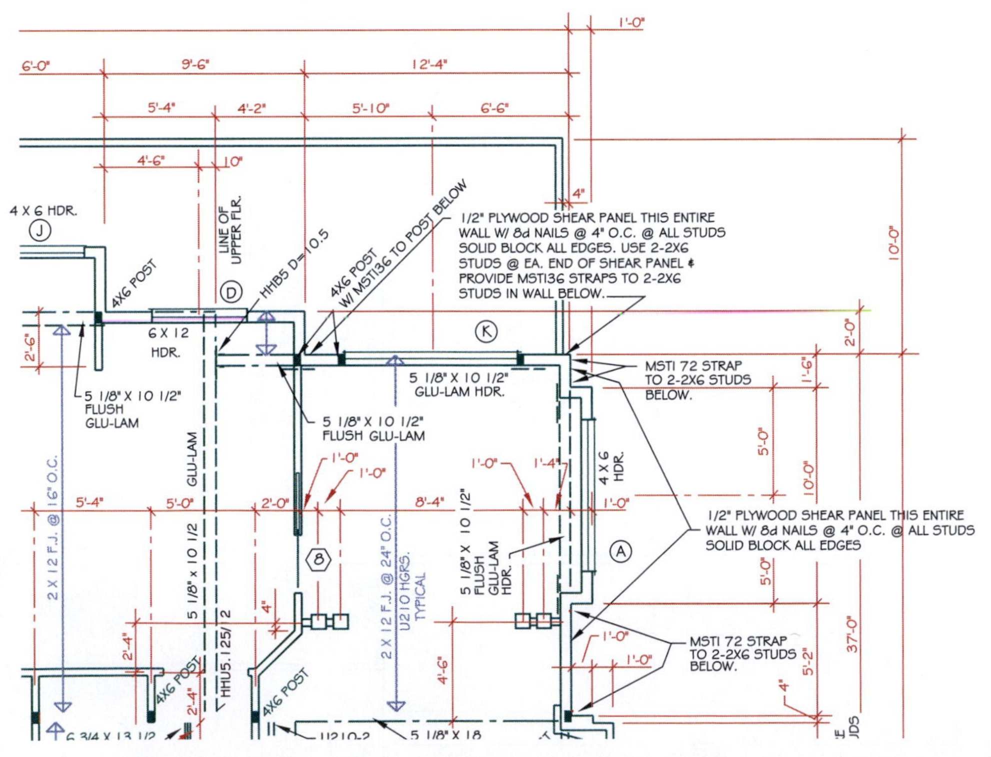

FIGURE 24.3 ■ *The framing plan for the structure shown in Figure 24.2 contains only the structural information. All nonstructural information from the floor plan is frozen on the framing plans.* Courtesy Residential Designs.

highest level and then moving down to the lowest. This method allows the loads to be accumulated, so that lower-level support will be accurately sized. The residence that was drawn in Chapter 11 is used later in this chapter to demonstrate this procedure.

Representing Framing Members

A key element of the framing plan is to show and specify the location of headers, beams, posts, joists, and trusses used to frame the skeleton by the use of notes and symbols.

Headers are located over every opening in a wall. Headers over a door or window are typically not drawn and are referenced by a note. A beam is placed to control the span of a joist or rafter or under a concentrated load and is represented by parallel dashed lines. The distance between the lines is based on the thickness of the beam. If a built-up beam is to be provided, thin dashed lines representing the width of the individual members are drawn. Often the specification for the size and strength is placed parallel to the member.

Each beam and header will be supported by multiple studs, a post, or a steel column, depending on the loads to be supported. When a post or column is hidden in a wall, it may or may not be drawn. Posts or columns inside a room must be drawn using line quality to match that used to represent walls. Specifications for the post size or for connecting hardware are often specified on a 45° angle. Figure 24.4 shows common methods of specifying beams and posts.

Joists, rafters, and ***trusses*** are represented by a thin line with an arrow at each end. Many companies draw the line representing these members so that it extends from bearing point to bearing point. An alternative method is to show the joist symbol centered between the bearing points. The specification for the member is written parallel to the line and should include the size, type, and spacing of the member. If metal hangers are to be used to connect joists to a beam, a note listing the connector is typically placed on a 45° angle and should include the size or model number and the manufacturer of the hanger. If a vaulted ceiling is to be provided, the line dividing the vault from the flat ceiling must be drawn, located, and specified. Figure 24.5 shows common methods of representing joists, trusses, and vaulted ceilings.

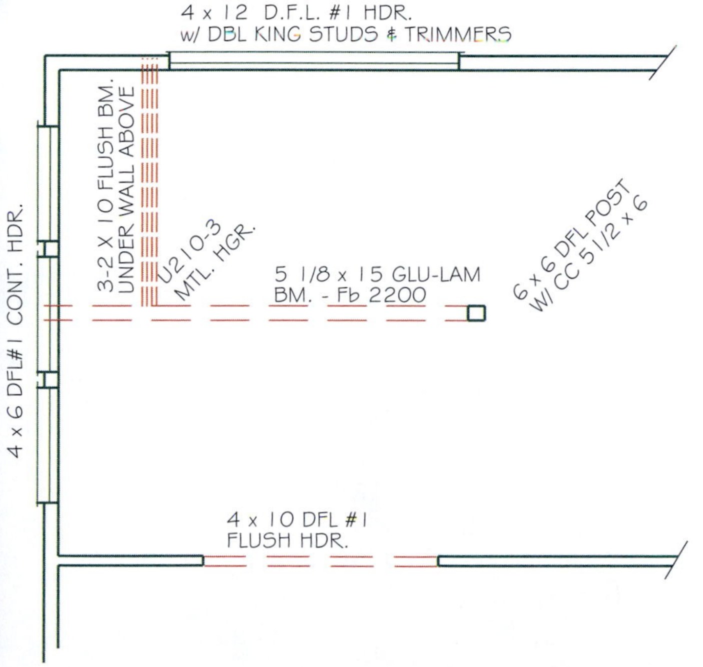

FIGURE 24.4 ■ *Headers over a door or window can be specified by a note, but not drawn. Framers realize that a door or window must have a header and will look for a note. Headers over wall opening or those located between bearing walls must be drawn and specified.*

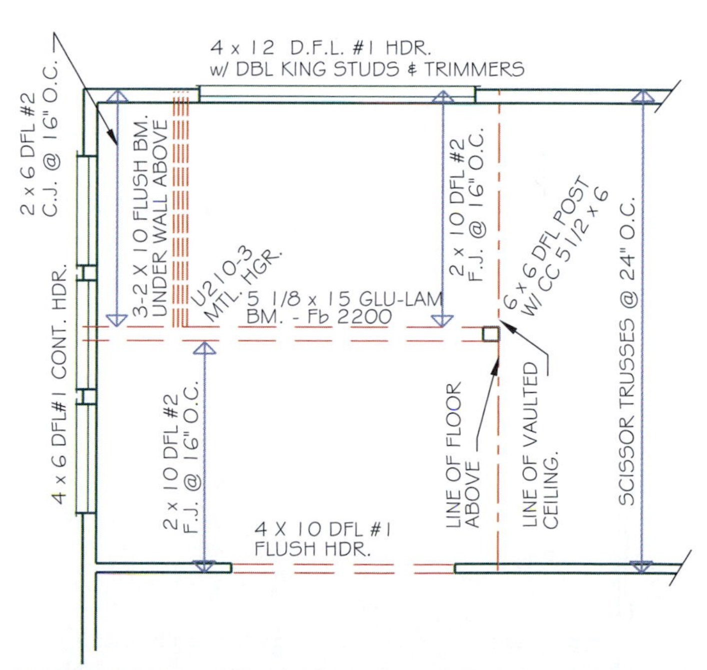

FIGURE 24.5 ■ *Joists, rafters, and trusses are each represented by an arrow that shows the direction of the span, with a note to specify the type, size, and spacing of the framing member. Post sizes are often specified on an angle so that the notation is obvious.*

RESISTING SEISMIC AND WIND LOADS

The severity of lateral loads will vary greatly depending on the area of the country where you work. Plywood shear panels, blocking, metal angles, and metal connectors to tie posts of one level to members on another level are common methods of resisting the forces caused by earthquakes or high winds. These members are determined by an engineer or project manager and placed on the framing plan by the technician.

Plywood shear panels can be specified by a local note pointing to the area of the wall to be reinforced. If several walls are to receive shear panels, a note to explain the construction of the panel can be placed as a general note, with a shortened local note used to locate their occurrence. Horizontal metal straps may be required to strengthen connections of the top plate or to tie the header into wall framing. These straps can be represented by a thick line placed on the side of the wall with a note to specify the manufacturer, size, and required blocking. Where walls can be reinforced with a let-in brace, a bold diagonal line is usually placed on the wall.

Metal straps that are used to tie trimmers, king studs, or posts of one level to another can be represented on each layer of the framing plan by a bold line where they occur. This vertical strap should be shown and specified on the framing plan for each level being connected. The specification should include the manufacturer, size, and required nailing. Figure 24.6 shows how to represent shear panels, let-in braces, and metal straps. Straps are also shown in sections and details, as in Figure 24.7, which are referenced to the framing plan, and they can also be shown on the exterior elevations as shown in Figure 24.8. When the lowest floor level is to be attached to the foundation level, the metal connector is often represented by a cross. Figure 24.9 shows common methods of representing these ties on the foundation plan.

Floors and roof members are often tied into the wall system to help resist lateral loads. In addition to the common nailing used to connect each member, metal angles are used to secure the truss or rafter to the top plate. Where lateral loads are severe, blocking is added between the trusses or rafters to keep these members from bending under lateral stress. These areas where joists or rafters are stiffened to resist bending from lateral loads are often referred to as a ***diaphragm.*** The size, location, and framing method are specified by an engineer. The angles and blocking for the diaphragm must be clearly specified on the framing plan. The angles are typically represented in a section or detail but are not drawn on the framing plan. They should be specified by a note

1/2" PLYWOOD SHEAR PANEL THIS ENTIRE WALL W/ 8d NAILS @ 4" O.C. @ EDGES & 8d NAILS @ 8" O.C. FIELD. USE COMMON WIRE NAILS.
PROVIDE SIMPSON CO. MSTI36 STRAPS TO 2-2X6 STUDS IN WALL BELOW (3 PLACES). W/ 36 10d NAILS
4 x 12 D.F.L. #1 HDR. w/ DBL KING STUDS & TRIMMERS
2 x 6 DFL #2 C.J. @ 16" O.C.
3-2 X 10 FLUSH BM. UNDER WALL ABOVE
MSTI48 STRAPS FROM EA. HDR. TO TOP PL
U210-3 MTL. HGR.
5 1/8 x 15 GLU-LAM BM. - Fb 2200
2 x 10 DFL #2 F.J. @ 16" O.C.
6 x 6 DFL POST W/ CC 5 1/2 x 6
4 x 6 DFL#1 CONT. HDR.
2 x 10 DFL #2 F.J. @ 16" O.C.
4 X 10 DFL #1 FLUSH HDR.
LINE OF FLOOR ABOVE
LINE OF VAULTED CEILING.
SCISSOR TRUSSES @ 24" O.C.
1 X 4 DIAG. LET-IN BRACE ACROSS 3 STUDS MIN.

FIGURE 24.6 ■ *Material for resisting lateral loads is specified on the framing plan by note. Bold lines can be used to represent metal straps.*

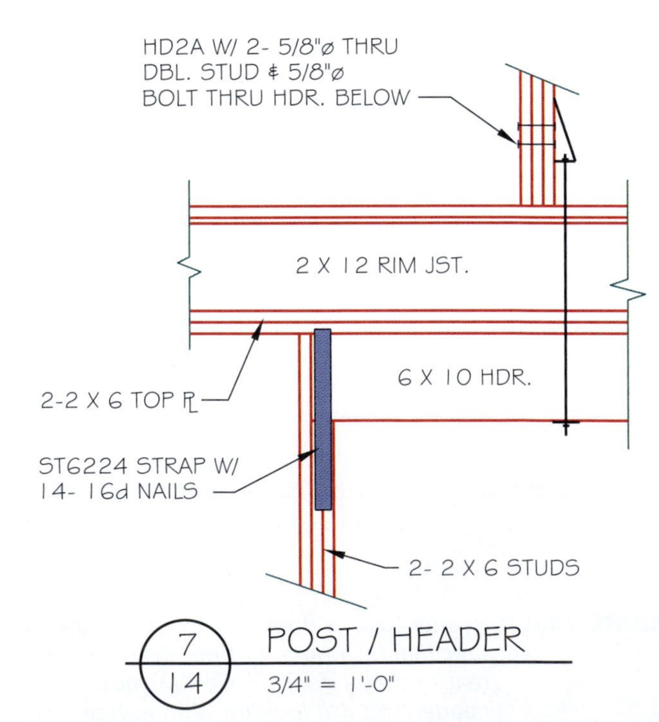

FIGURE 24.7 ■ *Metal straps and hold-down anchors, which are specified on the framing plan, are typically detailed to show exact placement.*

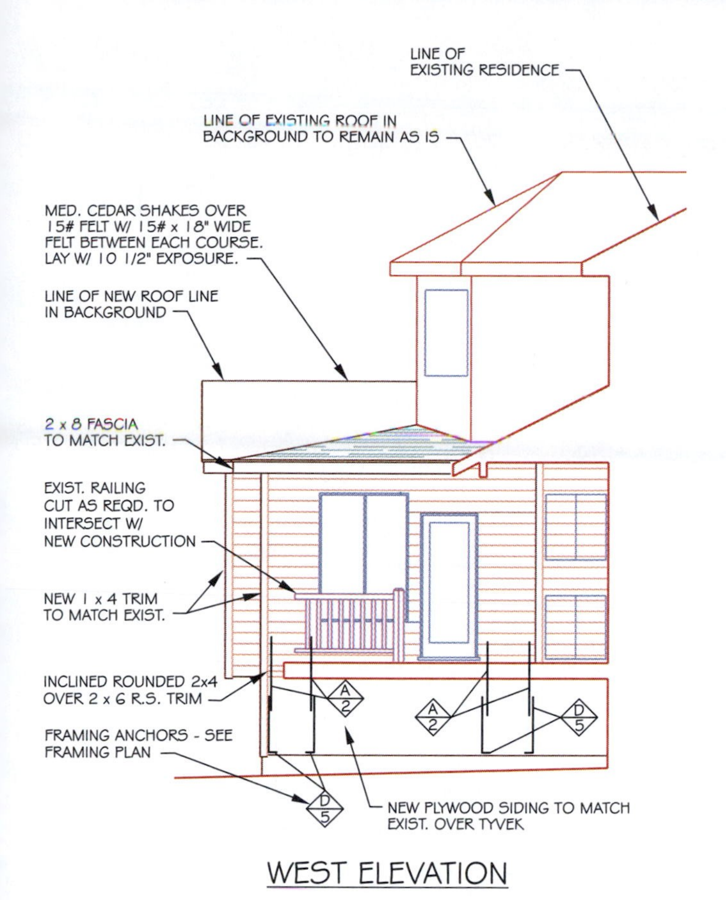

FIGURE 24.8 ■ *Materials used to resist lateral loads can be represented on the elevation to aid the framing crew make accurate placement of materials.* Courtesy Residential Designs.

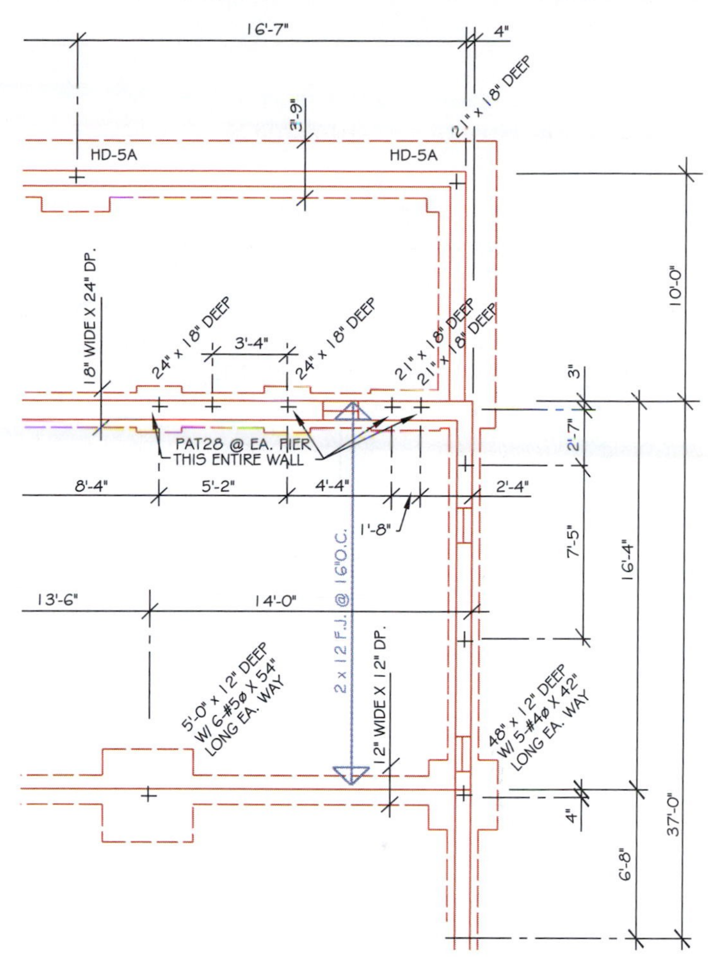

FIGURE 24.9 ■ *Metal anchors must be located and specified on the foundation plan so that lateral and gravity loads are transferred into the concrete.* Courtesy Residential Designs.

that indicates the manufacturer, model number, spacing, and which members are to be connected. Areas that are to receive special blocking should be indicated on the framing plan. Figure 24.10 shows how a diaphragm can be represented on a framing plan.

Resisting Lateral Loads Using Prescriptive Paths

Throughout this chapter it is assumed that someone will be calculating the exact loads on each surface of a structure. An alternative to having an engineer design the method of resisting lateral loads is to use the prescriptive path provided by the IRC. Prescriptive design methods provided by the IRC are specific requirements for construction to ensure that a structure will be able to resist lateral loads. Prescriptive design methods are intended for fairly simple structures and tend to limit design alternatives. The prescriptive path limits ceiling heights, restricts where an opening can be placed in walls, and also limits wall placement. The prescriptive path of the IRC limits the design criteria for basic wind speeds below 130 mph. The IBC provides information for structures that must resist greater wind loads, but these codes are not the subject of this chapter. Homes that must resist greater winds should be designed under the supervision of an engineer. When prescriptive design methods are used, wall heights are limited to a maximum of 10' (3000 mm). Window and wall placement are discussed throughout the balance of this chapter. Before applying the prescriptive path method, several key terms that are used by the codes must be understood.

Note:

Prescriptive methods and specific design methods can't be used on the same house. If a structure can't meet all of the requirements for using the prescriptive path, a specific design to meet the lateral loads must be provided.

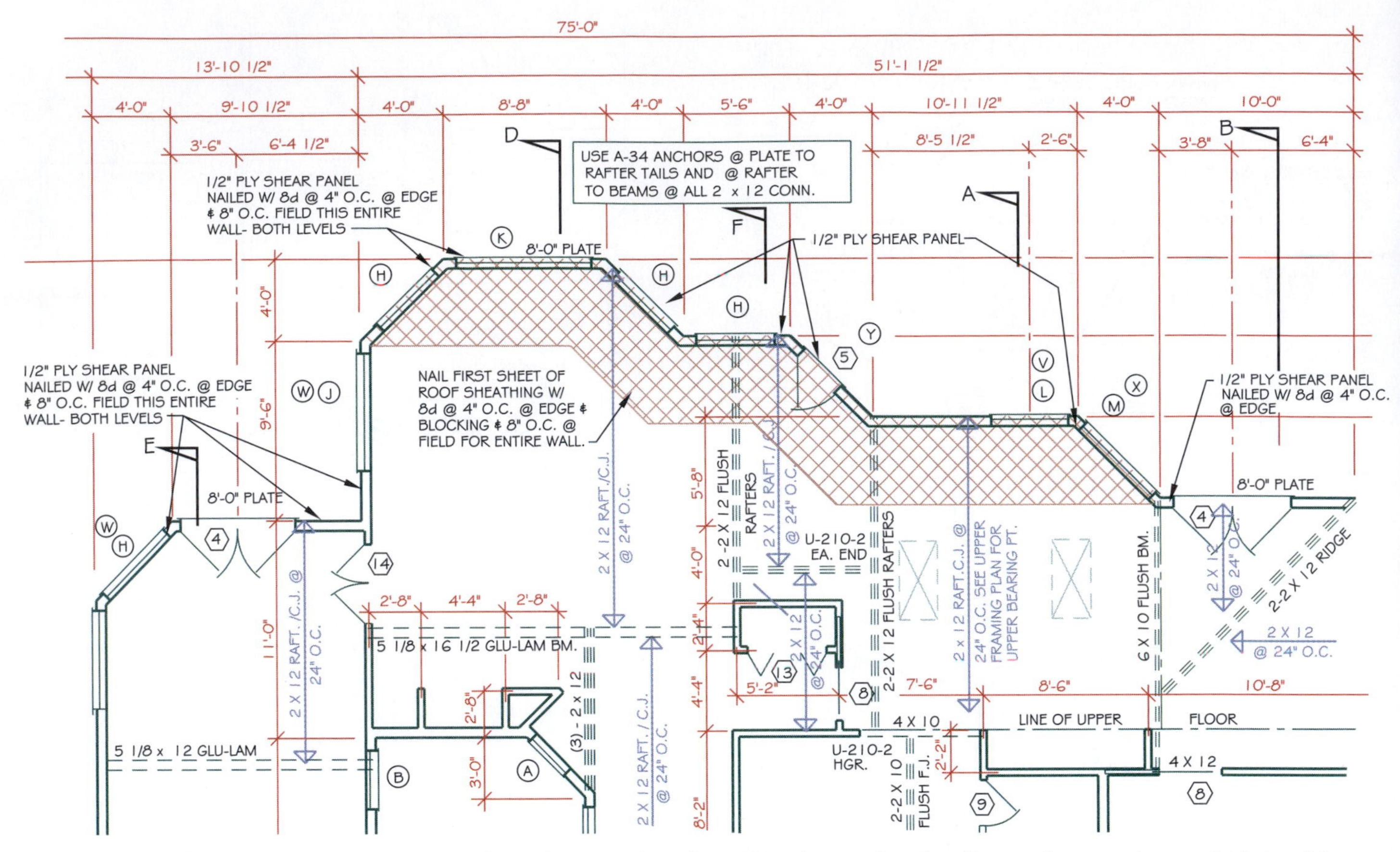

FIGURE 24.10 ■ *Special nailing and reinforced connections between the roof and wall are often used to resist lateral forces. This information must be shown on the framing plan. Because the rear face of this home has very few walls, the engineer is using the roof to keep the walls square. The walls support the weight of the roof, and the roof resists the lateral loads.* Courtesy Residential Designs.

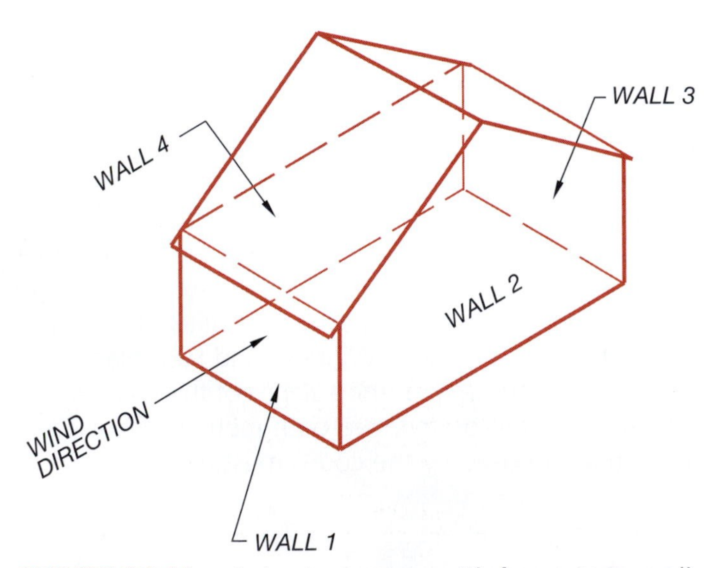

FIGURE 24.11 ■ *A simple structure with four exterior walls (braced wall lines). As wind is applied to the structure, the natural tendency is for the wall to change from a rectangle into a parallelogram. If wind force is applied to surface 1, wall lines 2 and 4 will be used to resist the force.*

Basic Terms of Prescriptive Design

The terms braced wall lines and braced wall panels are used to describe methods of resisting wall loads. Each exterior surface of a residence is considered a ***braced wall line.*** Figure 24.11 shows a simple structure with four exterior walls (braced wall lines). As wind is applied to the structure, the natural tendency is for the wall to change from a rectangle to a parallelogram. If wind force is applied to surface 1, wall lines 2 and 4 will be used to resist the force. Because the wind force can come from any direction, each surface must be reinforced to resist the force that may be applied to the adjoining surface.

Because very few homes are simple rectangles, the IRC allows some modifications to braced wall lines. These alterations include the following:

- A wall may be offset up to 48'' (1200 mm) and still be considered one plane.
- Multiple offsets are allowed in one plane provided that the total distance from each wall line is not greater than 8' (2400 mm).

Three methods are approved by most building departments for reinforcing a braced wall line. These methods include braced wall panels, alternative braced wall panels, and portal frames.

Braced Wall Panels

A **braced wall panel** (BWP) is a method of braced wall line reinforcement that uses panels with a minimum length of 48'' (1200 mm) to resist lateral loads. The minimum length of braced panels can be reduced if the entire wall is sheathed. The size of the panel reduction is based on the wall height and the height of the wall opening (see Figure 24.12).

Alternative Braced Wall Panels

An ***alternative braced wall panel (ABWP)*** is a method of braced wall line reinforcement that uses panels with a minimum length of 2'-8'' (800 mm) to resist lateral loads.

Portal Frames

A ***portal frame (PF)*** is a method of braced wall line reinforcement that uses pairs of panels joined by a header to resist lateral loads. Each panel must be 22 1/2'' (570 mm) wide. Portal frames are not mentioned in the IRC, but are accepted by most building departments when based on APA specifications. Many lumber manufactures also have specifications for building portal frames.

Simpson Strong-Tie Company provides premanufactured, steel skinned, portal frame panels that are set in a wood frame. The steel skin resembles the side of a containerized cargo box. Panels are shipped to the jobsite ready to be bolted to the foundation, eliminating the placement of the hundreds of nails required to make a wood portal frame. Although actual construction methods are considered later in the chapter as placement is considered, it is important to understand that portal frames must be placed in pairs.

Components of Braced Wall Construction

Several methods are allowed by building codes for constructing the panel that will reinforce a braced wall line. IRC-approved construction methods for a braced wall panel include:

1. Nominal 1 × 4 (25 × 100) continuous diagonal let-in brace that is let into the top and bottom plates and a minimum of three studs. The brace cannot be placed at an angle greater than 60° or less than 45° from horizontal.
2. Diagonally placed wood boards with a net thickness of 5/8'' (16 mm) that are applied to studs spaced at a maximum of 24'' (600 mm) o.c.
3. Structural wood sheathing with a minimum thickness of 5/16'' (8 mm) over studs placed at 16'' (400 mm) o.c. or 3/8'' (9.5 mm) structural sheathing over studs spaced at 24'' (600 mm).
4. 1/2'' (12.7 mm) or 25/32'' (19.8 mm) structural fiberboard sheathing panels 4 × 8' (1200 × 2400 mm) applied vertically over studs placed at 16'' (400 mm) o.c.
5. Gypsum board 1/2'' (13 mm) thick × 4' (1200 mm) wide over studs spaced at 24'' (600 mm) o.c. and nailed at 7'' (175 mm) o.c. maximum.
6. Particleboard wall sheathing panels.
7. Portland cement plaster on studs placed at 16'' (400 mm) o.c.
8. Hardboard panel siding.

For these conditions to be allowable, each material must be applied using the connection methods listed in the fastener schedule shown in Figure 24.13. Other limitations include:

- 1 × 4 (25 × 100) continuous diagonal let-in braces (method 1) are not allowed in seismic zones D1, D2, and E.

TABLE R602.10.5
LENGTH REQUIREMENTS FOR BRACED WALL PANELS IN A CONTINUOUSLY SHEATHED WALL[a,b]

LENGTH OF BRACED WALL PANEL (inches)			MAXIMUM OPENING HEIGHT NEXT TO THE BRACED WALL PANEL (% of wall height)
8-foot wall	9-foot wall	10-foot wall	
48	54	60	100%
32	36	40	85%
24	27	30	65%

For SI: 1 inch = 25.4 mm, 1 foot = 305 mm, 1 psf = 0.0479 kN/m².

a. Linear interpolation shall be permitted.

b. Full-height sheathed wall segments to either side of garage openings that support light frame roofs with roof covering dead loads of 3 psf or less shall be permitted to have a 4:1 aspect ratio.

FIGURE 24.12a ■ *Length requirements for braced walls designed using prescriptive methods. Reproduced from the International Residential Code / 2006. Copyright © 2006.* Courtesy International Code Council, Inc.

WALL BRACING

SEISMIC DESIGN CATEGORY OR WIND SPEED	CONDITION	TYPE OF BRACE[b, c]	AMOUNT OF BRACING[a, d, e]
Category A and B ($S_s \leq 0.35g$ and $S_{ds} \leq 0.33g$) or 100 mph or less	One story Top of two or three story	Methods 1, 2, 3, 4, 5, 6, 7 or 8	Located in accordance with Section R602.10 and at least every 25 feet on center but not less than 16% of braced wall line for Methods 2 through 8.
	First story of two story Second story of three story	Methods 1, 2, 3, 4, 5, 6, 7 or 8	Located in accordance with Section R602.10 and at least every 25 feet on center but not less than 16% of braced wall line for Method 3 or 25% of braced wall line for Methods 2, 4, 5, 6, 7 or 8.
	First story of three story	Methods 2, 3, 4, 5, 6, 7 or 8	Located in accordance with Section R602.10 and at least every 25 feet on center but not less than 25% of braced wall line for Method 3 or 35% of braced wall line for Methods 2, 4, 5, 6, 7 or 8.
Category C ($S_s \leq 0.6g$ and $S_{ds} \leq 0.50g$) or less than 110 mph	One story Top of two or three story	Methods 1, 2, 3, 4, 5, 6, 7 or 8	Located in accordance with Section R602.10 and at least every 25 feet on center but not less than 30% of braced wall line for Method 3 or 45% of braced wall line for Methods 2, 4, 5, 6, 7 or 8.
	First story of two story Second story of three story	Methods 2, 3, 4, 5, 6, 7 or 8	Located in accordance with Section R602.10 and at least every 25 feet on center but not less than 16% of braced wall line for Method 3 or 25% of braced wall line for Methods 2, 4, 5, 6, 7 or 8.
	First story of three story	Methods 2, 3, 4, 5, 6, 7 or 8	Located in accordance with Section R602.10 and at least every 25 feet on center but not less than 45% of braced wall line for Method 3 or 60% of braced wall line for Methods 2, 4, 5, 6, 7 or 8.
Categories D_0 and D_1 ($S_s \leq 1.25g$ and $S_{ds} \leq 0.83g$) or less than 110 mph	One story Top of two or three story	Methods 2, 3, 4, 5, 6, 7 or 8	Located in accordance with Section R602.10 and at least every 25 feet on center but not less than 20% of braced wall line for Method 3 or 30% of braced wall line for Methods 2, 4, 5, 6, 7 or 8.
	First story of two story Second story of three story	Methods 2, 3, 4, 5, 6, 7 or 8	Located in accordance with Section R602.10 and at least every 25 feet on center but not less than 45% of braced wall line for Method 3 or 60% of braced wall line for Methods 2, 4, 5, 6, 7 or 8.
	First story of three story	Methods 2, 3, 4, 5, 6, 7 or 8	Located in accordance with Section R602.10 and at least every 25 feet on center but not less than 60% of braced wall line for Method 3 or 85% of braced wall line for Methods 2, 4, 5, 6, 7 or 8.
Category D_2 or less than 110 mph	One story Top of two story	Methods 2, 3, 4, 5, 6, 7 or 8	Located in accordance with Section R602.10 and at least every 25 feet on center but not less than 25% of braced wall line for Method 3 or 40% of braced wall line for Methods 2, 4, 5, 6, 7 or 8.
	First story of two story	Methods 2, 3, 4, 5, 6, 7 or 8	Located in accordance with Section R602.10 and at least every 25 feet on center but not less than 55% of braced wall line for Method 3 or 75% of braced wall line for Methods 2, 4, 5, 6, 7 or 8.
	Cripple walls	Method 3	Located in accordance with Section R602.10 and at least every 25 feet on center but not less than 75% of braced wall line.

For SI: 1 inch = 25.4 mm, 1 foot = 304.8 mm, 1 pound per square foot = 0.0479kPa, 1 mile per hour = 0.477 m/s.

a. Wall bracing amounts are based on a soil site class "D." Interpolation of bracing amounts between the S_{ds} values associated with the seismic design categories shall be permitted when a site specific S_{ds} value is determined in accordance with Section 1613.5 of the *International Building Code*.

b. Foundation cripple wall panels shall be braced in accordance with Section R602.10.2.

c. Methods of bracing shall be as described in Section R602.10.3. The alternate braced wall panels described in Section R602.10.6.1 or R602.10.6.2 shall also be permitted.

d. The bracing amounts for Seismic Design Categories are based on a 15 psf wall dead load. For walls with a dead load of 8 psf or less, the bracing amounts shall be permitted to be multiplied by 0.85 provided that the adjusted bracing amount is not less than that required for the site's wind speed. The minimum length of braced panel shall not be less than required by Section R602.10.3.

e. When the dead load of the roof/ceiling exceeds 15 psf, the bracing amounts shall be increased in accordance with Section R301.2.2.2.1. Bracing required for a site's wind speed shall not be adjusted.

FIGURE 24.12b ■ *Wall bracing methods for prescriptive design. Reproduced from the International Residential Code / 2006. Copyright © 2006.* Courtesy International Code Council, Inc.

NAILING SCHEDULE

TAKEN FROM 1997 UBC TABLE 23-II-B-1 CONNECTION	NAILING (1)
1 JOIST TO SILL OR GIRDER, TOE NAIL	3- 8d
2 BRIDGING TO JOIST, TOENAIL EACH END	2- 8d
3 1 X 6 SUBFLOOR OR LESS TO EACH JOIST, FACE NAIL	2-8d
4 WIDER THAN 1 X 6 SUBFLOOR TO EA. JOIST, FACE NAIL	3-8d
5 2" SUBFLOOR TO JOIST OR GIRDER, BLIND AND FACE NAIL	2-16d
6 SOLE PLATE TO JOIST OR BLOCKING TYPICAL FACE NAIL SOLE PLATE TO JOIST OR BLOCKING, @ BRACED WALL PANELS	16d @ 16" O.C. 3- 16d PER 16"
7 TOP PLATE TO STUD, END NAIL	2-16d
8 STUD TO SOLE PLATE	4- 8d , TOENAIL OR 2- 16d , END NAIL
9 DOUBLE STUDS, FACE NAIL	16d @ 24" O.C.
10 DOUBLED TOP PLATES, TYPICAL FACE NAIL DOUBLE TOP PLATES, LAP SPLICE	16d @ 16" O.C. 8- 16d
11 BLOCKING BETWEEN JOIST OR RAFTERS TO TOP PLATE, TOENAIL	3- 8d
12 RIM JOIST TO TOP PLATE, TOENAIL	8d @ 6" O.C.
13 TOP PLATES, LAPS & INTERSECTIONS, FACE NAIL	2-16d
14 CONTINUOUS HEADER, TWO PIECES	16d @ 16" O.C. ALONG EACH EDGE
15 CEILING JOIST TO PLATE, TOE NAIL	3-8d
16 CONTINUOUS HEADER TO STUD, TOE NAIL	4-8d
17 CEILING JOIST, LAPS OVER PARTITIONS, FACE NAIL	3-16d
18 CEILING JOIST TO PARALLEL RAFTERS, FACE NAIL	3-16d
19 RAFTERS TO PLATE, TOE NAIL	3-8d
20 1" BRACE TO EA. STUD & PLATE FACE NAIL	2-8d
21 1" x 8" SHEATHING OR LESS TO EA. BEARING, FACE NAIL	2-8d
22 WIDER THAN 1" x 8" SHEATHING TO EACH BEARING, FACE NAIL	3-8d
23 BUILT-UP CORNER STUDS	16d @ 24" O.C.
24 BUILT-UP GIRDER AND BEAMS	20d @ 32" O.C. @ TOP & BTM. AND STAGGERED 2-20d @ ENDS & @ EACH SPLICE
25	2-16d 2" PLANKS @ EACH BEARING
26 WOOD STRUCTURAL PANELS AND PARTICLEBOARD (2) SUBFLOOR, ROOF AND WALL SHEATHING TO FRAMING:	
19/32" - 3/4"	8d COMMON OR 6d DEFORMED SHANK
1/2" AND LESS	6d COMMON OR DEFORMED SHANK
7/8 "- 1"	8d COMMON OR DEFORMED SHANK
1 1/8"- 1 1/4"	10d COMMON OR 8d DEFORMED SHANK
COMBINATION SUBFLOOR-UNDERLAYMENT TO FRAMING:	
3/4" AND LESS	6d DEFORMED SHANK
7/8" - 1"	8d DEFORMED SHANK
1 1/8" - 1 1/4"	10d COMMON OR 8d DEFORMED SHANK
27 PANEL SIDING TO FRAMING (2)	
1/2" OR LESS CASING NAILS.	6d CORROSION-RESISTANT SIDING OR
5/8"	8d CORROSION-RESISTANT SIDING OR
28 FIBERBOARD SHEATHING (3)	
1/2"	No. 11 GA. (4) 6d COMMON NAILS No. 16 GA. (5)
25/32"	No. 11 GA. (4) 8d COMMON NAILS No. 16 GA. (5)
29. INTERIOR PANELING:	
1/4"	4d (6)
25/32"	6d (7)

1. COMMON OR BOX NAILS MAY BE USED WHERE OTHERWISE STATED.
2. NAILS SPACED @ 6" O.C. @ EDGES, 12" O.C. @ INTERMEDIATE SUPPORTS EXCEPT 6" @ ALL SUPPORTS WHERE SPANS ARE 48" OR MORE. FOR NAILING OF WOOD STRUCTURAL PANEL AND PARTICLEBOARD DIAPHRAGMS AND SHEAR WALLS SEE SPECIFIC NOTES ON DRAWINGS. NAILS FOR WALL SHEATHING MAY BE COMMON, BOX OR CASING.
3. FASTENERS SPACED 3" O.C. @ EXTERIOR EDGES & 6" O.C. @ INTERMEDIATE SUPPORTS.
4. CORROSION -RESISTANT ROOFING NAILS W/ 7/16" DIA. HEAD & 1 1/2" LENGTH FOR 1/2" SHEATHING & 1 3/4" LENGTH FOR 25/32" SHEATHING..
5. CORROSION-RESISTANT STAPLES W/ NOMINAL 7/16" CROWN & 1 1/8" LENGTH FOR 1/2" SHEATHING & 1 1/2" LENGTH FOR 25/32" SHEATHING..
6. PANEL SUPPORTS @ 16" (20" IF STRENGTH AXIS IN THE LONG DIRECTION OF THE PANEL, UNLESS OTHERWISE MARKED). CASING OR FINISH NAILS SPACED 6" ON PANEL EDGES, 12" @ INTERMEDIATE SUPPORTS.
7. PANEL SUPPORTS @ 24". CASING OR FINISH HAILS SPACED 6" ON PANEL EDGES, 12" AT INTERMEDIATE SUPPORTS.

WOOD STRUCTURAL PANEL ROOF SHEATHING NAILING SCHEDULE (1)

WIND REGION	NAILS	PANEL LOCATION	ROOF FASTENER ZONE (2) 1	2	3
			FASTENING SCHEDULE (INCHES ON CENTER) X 25.4 FOR MM		
GREATER THAN 90 MPH (145 km/h)	8d COMMON	PANEL EDGES (3)	6	6	4(4)
		PANEL FIELD	6	6	6(4)
GREATER THAN 80 MPH (129 km/h) TO 90 MPH (145 km/h)	8d COMMON	PANEL EDGES (3)	6	6	4
		PANEL FIELD	12	6	6
80 MPH (129 km/h) OR LESS	8d COMMON	PANEL EDGES (3)	6	6	6
		PANEL FIELD	12	12	12

1. APPLIES ONLY TO MEAN ROOF HEIGHTS UP TO 35 FT (10 700 MM). FOR MEAN ROOF HEIGHTS 35 FEET (10 700 MM),
THE NAILING SHALL BE DESIGNED.
2. ROOF FASTENING ZONES ARE SHOWN BELOW.
3. EDGE SPACING ALSO APPLIES OVER ROOF FRAMING AT GABLE-END WALLS.
4. USE 8d RING-SHANK NAILS IN THIS ZONE IF MEAN ROOF HEIGHT IS GREATER THAN 25' (7600mm).

FIGURE 24.13 ■ *Building codes provide specifications that regulate each connection specified on the framing plans. A nailing schedule, based on the code provisions, inserted into the working drawings will eliminate having to specify nailing for each drawing.* Courtesy Residential Designs.

- Each panel must be a minimum of 48'' (1200 mm) long covering a minimum of three studs placed at 16'' (400 mm) o.c. or two studs placed at 24'' (600 mm) o.c. for methods 2, 3, 4, 6, 7, and 8.
- Each wall panel must be 96'' (2400 mm) long if 1/2'' (12.7 mm) gypsum board (method 5) is used on one wall face. The length can be reduced to 48'' (1200 mm) if the gypsum board is placed on each face of the stud wall.

If the walls are made rigid to keep their rectangular shape, a second tendency will occur. The edge of the wall panel nearest the load (the wind) will start to lift and rotate around the edge of the panel that is farthest from the load. Figure 24.14 shows this tendency. In addition to specifying the material to cover the wall, building codes require several other types of information to be specified about the reinforced wall panel. Required information includes:

- Framing members to be used to reinforce the wall panel (show on the floor and foundation plan)
- Methods of attaching the wall panel to the sill plate (show in a schedule that is placed with the framing plan and foundation plans)
- Methods of attaching the sill to the stem wall (show on the foundation plan)
- Methods of reinforcing the foundation that will support the wall panel (show on the foundation plan)

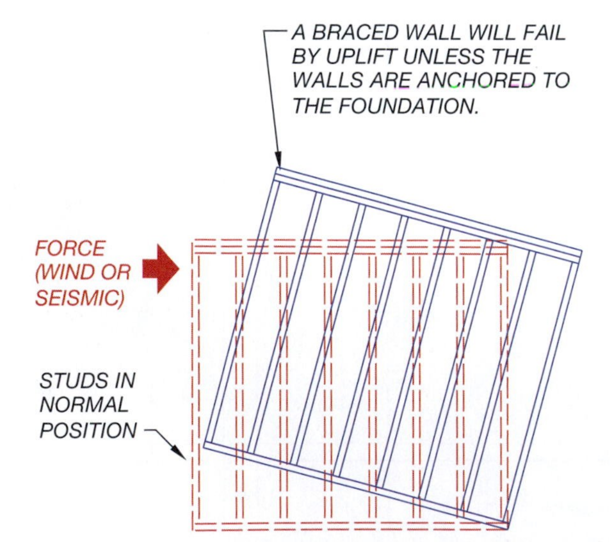

FIGURE 24.14 ■ *If walls are made rigid to keep their rectangular shape, a second tendency will occur. The edge of the wall panel nearest the load (the wind) will lift and rotate around the edge of the panel that is farthest from the load.*

Because this is to be an introduction to reinforcing walls using prescriptive methods, each method is not covered in depth. Remember that walls must be reinforced according to the governing building department and that research will be needed.

Representing Wall Reinforcement on a Framing Plan

One of the requirements for designing using prescriptive methods is that the location of wall panels must be clearly marked on the framing drawings. Prescriptive construction methods can be summarized and listed in a table to represent the materials that will be required to reinforce the walls and foundation. Figure 24.15 shows an example of a table for specifying braced walls, alternative braced walls, and portal frames for a one-level home. The table allows the required framing materials to be listed once and then specified in each required location by the use of a symbol. Figure 24.16 shows how lateral materials can be referenced on a framing plan.

Guidelines for Placement of Wall Reinforcement

The final stage of placing lateral bracing using prescriptive methods is to consider the actual placement of the reinforced wall panels. Although placement may vary depending on local conditions, the IRC requires the following minimum standards for placement of reinforced wall panels:

Placement of BWP/ABWP within a Braced Wall Line

- The reinforcement (the braced wall, alternative braced wall, or portal frame) must start within 12.5' (3750 mm) from each end of a braced wall line. The distance may vary depending on the design wind speed or the seismic design category. Figure 24.12b lists other requirements for placement. This distance is reduced to 8' (2400 mm) by many municipalities. Verify the required distances with your building department.
- The maximum space between reinforcement panels is 25' (7500 mm) Figure 24.12b lists additional requirements for panel spacing along the braced wall line based on the seismic and wind speed category and the method of panel construction. Figure 24.17 shows an example of the reinforcement

FRAMING PLAN	BRACED WALL SCHEDULE					
	MARK	WALL COVERING	EDGE NAILING	FIELD NAILING	PANEL SUPPORT	STRUCTURAL CONNECTORS
BRACED WALL PANEL	A	48" MIN. x 1/2" CD EXT. PLY 1 SIDE UNBLOCKED	8d @ 6" O.C	8d @ 12" O.C.	2 - 2 x 6 STUDS @ EA. END OF PANEL	PAHD42 TO CONC.
ALTERNATIVE BRACED WALL PANEL	B	2'-8" MIN. x 1/2" CD EXT. PLY 1 SIDE UNBLOCKED	8d @ 6" O.C	8d @ 12" O.C.	2 - 2 x 6 STUDS @ EA. END OF PANEL	HPAHD22 TO CONC.
PORTAL FRAME	C	CD EXT. PLY 1 SIDE -BLOCKED EDGE	2 ROWS OF 8d @ 3" O.C	8d @ 12" O.C.	4 x 4 POST @ EA. END OF PANEL	1-MSTA18 STRAP EA. SIDE EA. END OF PANEL TO 4 x 12 MIN HDR. (4 STRAPS MIN EA. END OF FRAME).
INTERIOR BWP	D	1/2" x 48" GYPSUM BOARD EACH SIDE OF WALL W/ 5d COOLER NAILS @ 7" O.C. MAXIMUM W/ BLOCKED EDGES (1/2" x 96" LONG IF 1 SIDE ONLY).				

FOUNDATION PLAN	HOLD - DOWN SCHEDULE			
	MARK	HOLD - DOWN	CONNECTIONS	FOUNDATION REINFORCING
	0	NONE REQUIRED	————	DBL JOIST OR BEAM BELOW. SEE FOUNDATION PLAN
BRACED WALL PANEL	1	PAHD42 STRAP	(16) 16d OR (3) 1/2"Ø M.B.	PROVIDE MIN. OF (2) 1/2"Ø A. B.
ALTERNATIVE BRACED WALL PANEL	2	HPAHD22 STRAP	(19) 16d OR (3) 1/2"Ø M.B.	PROVIDE (2) 1/2"Ø A.B. @ 1/4 POINTS OF SILL W/ 1- #4 3" DN FROM TOP OF WALL & 1 - #4 3" UP FROM BOTTOM OF FOOTING. PROVIDE 1-#4 VERT @24" O.C. IN STEM WALL.
PORTAL FRAME	3	HTT22 STRAP & SSTB24 BOLT TO CONCRETE	(32) 16d SINKERS W/ 2" MAX OFFSET FROM POST FACE TO BOLT CENTERLINE	PROVIDE (3) 1/2"Ø A.B. @ MUDSILL W/ 2-2x6 SILL W/ 2 - #4 - 3" UP FROM BOTTOM OF 15" x 7" FTG. EXTEND STEEL 10'-0" AROUND FOOTING CORNER. PROVIDE 1 #4 VERT. @EA. SSTB24 CONNECTOR

FIGURE 24.15 ■ *Tables can be used to specify the materials to be used to resist lateral shear. This table shows the material needed to resist lateral loads for a one-level residence based on the IRC prescriptive path. If a table is to be used, it should be near the framing plan. If space is not available, the lower portion of the table can be placed with the foundation plan.*

required for a braced wall panel per option A-1 of the table shown in Figure 24.15.

Placement of PF within a Braced Wall Line

Figure 24.18 shows the use of portal frames on a floor plan. Key requirements include:

- Portal frames must be placed in pairs.
- A minimum distance of 8' (2400 mm) between panels is required.
- The maximum distance between interior edges of portal frame panels is 25' (7500 mm).

Much of the strength associated with a portal frame is from how the foundation is constructed and how the wall is tied to the foundation. In addition to referencing portal frames on the framing plan, a detail should be added to the sections to describe construction methods. An example of a portal frame detail is shown in Figure 24.19. Figure 24.20 shows the steel that will be added to the foundation as a result of planning for lateral loads.

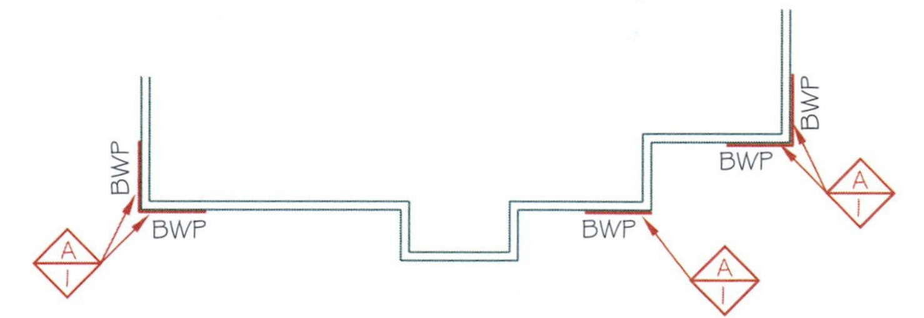

FIGURE 24.16 ■ *Material for resisting lateral loads based on the IRC prescriptive path can be specified on the framing plan. Symbols can be used to reference the braced wall line reinforcement to a table similar to the schedule shown in Figure 24.15.*

Guidelines for Placement of Braced Wall Lines

Don't confuse braced wall panels with braced wall lines. These guidelines apply to the placement of braced wall lines (the exterior walls). As shown in Fig-

FIGURE 24.17 ■ *Metal straps are used to bind wood materials to concrete to resist uplift. These are the PAHD42 straps that are required by #1 of the table in Figure 24.15.* Courtesy Connie Willmon.

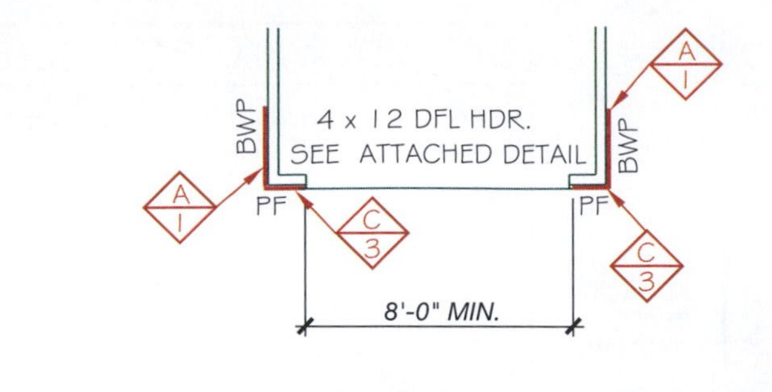

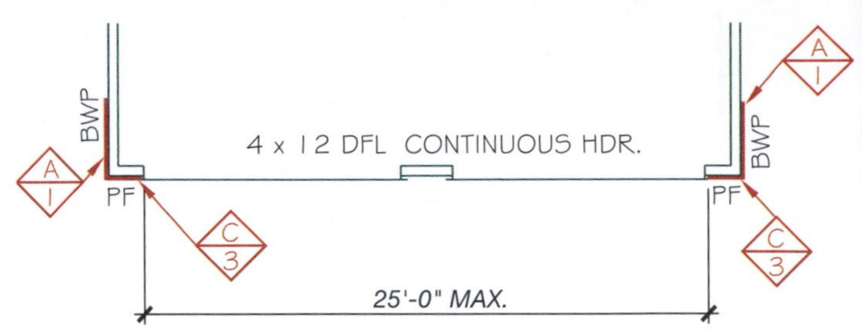

FIGURE 24.18 ■ *Placement of braced walls and portal frame wall panels to meet prescriptive code requirements.*

ure 24.21, a maximum space of 35' (10 500 mm) o.c. between braced wall lines is allowed in both the longitudinal and transverse directions of the structure. The distance between braced wall lines can be increased to 50' (10 000) if the bracing provided equals or exceeds the bracing specified in Figure 30.15 multiplied by a factor equal to the braced wall line spacing divided by 35'. For 50' spacing, this factor would be 50'/35' = 1.43. A second requirement to exceed the 35' (10 500 mm) spacing is that the length-to-width ratio for the floor/wall diaphragm can't exceed 3:1.

Support of Braced Wall Lines

Support must be provided below each braced wall line to transfer loads into the foundation. Guidelines for support of braced wall lines include:

- An exterior braced wall line must be supported over a continuous footing.
- When floor joists are perpendicular to a braced wall line above, solid blocking or a continuous beam must be provided below the braced wall line.

Additional support that may be required based on seismic or wind loads includes:

- Floors with cantilevers or setbacks not exceeding four times the nominal depth of the floor joist may support braced wall panels if the following conditions are met:
  - Minimum joist size of 2 × 10 (50 × 250) with a maximum spacing of 16'' (400 mm) o.c.
  - The ratio of back span to cantilever is a minimum of 2:1.
  - Floor joists at each end of a braced wall panel must be doubled.
  - A continuous rim joist is connected to ends of all cantilevered joists.
  - Gravity loads carried at the end of the cantilevered joist are limited to uniform wall and roof load and the reactions from headers having a span of 8' (2400 mm) or less.
  - An interior braced wall line must be supported over a continuous footing at intervals of 50' (15 000 mm).
  - In a building more than one story in height, all interior braced wall panels must be supported on a continuous footing unless:
    - Cripple wall height does not exceed 48'' (1200 mm).
    - First-floor braced wall panels are supported on double floor joists, continuous blocking, or floor beams.

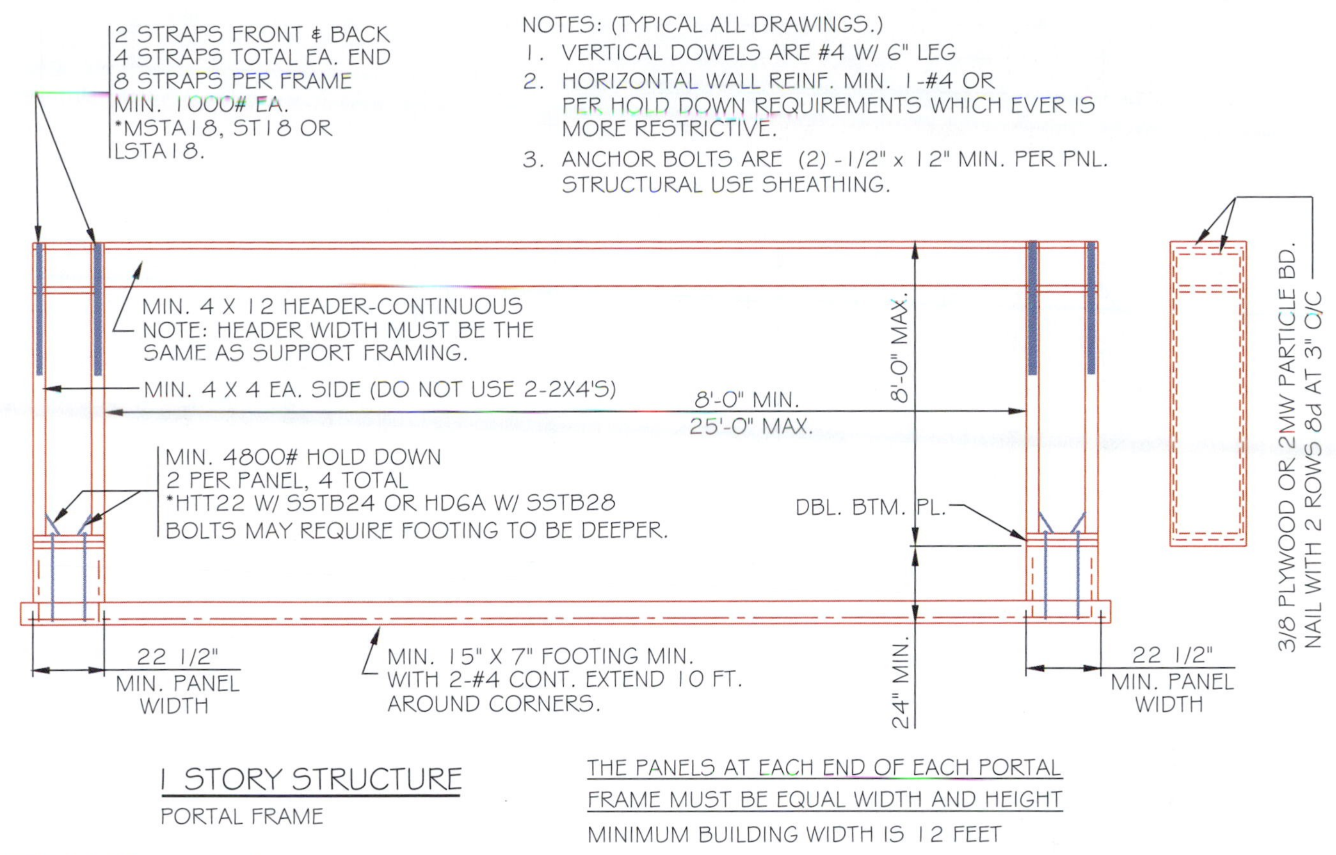

FIGURE 24.19 ■ *A detail used to explain the construction of a portal frame based on APA recommendations.* Courtesy Residential Designs.

FIGURE 24.20 ■ *As the uplift stress increases, connections able to resist greater loads are required, such as these HD5A anchors specified by #4 of Figure 24.41.* Courtesy Sara Santillanes.

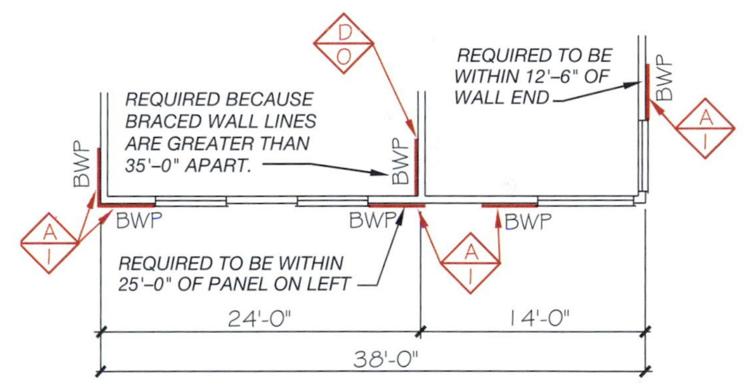

FIGURE 24.21 ■ *Placement of braced wall lines using prescriptive design methods. When braced wall lines (exterior walls) are more than 35'-0'' apart, an interior wall that is perpendicular to the exterior wall must be reinforced.*

- The distance between braced lines does not exceed twice the building width parallel to the braced wall line.

Figure 24.22 shows a one-story residence with braced wall panels. Compare the drawing with the home in Figure 24.1. The residence in Figure 24.1 shows lateral support designed by calculating the actual wind loads that will affect the structure. Comparing that home with the results of using the prescriptive path in Figure 24.22 will show that minor changes were required in the placement of some windows, and the addition of the storage area in the upper left corner of the home. These changes were required to provide adequate space for braced panels. As a general rule, using the prescriptive code may save the initial cost of engineering but require more material to resist the loads.

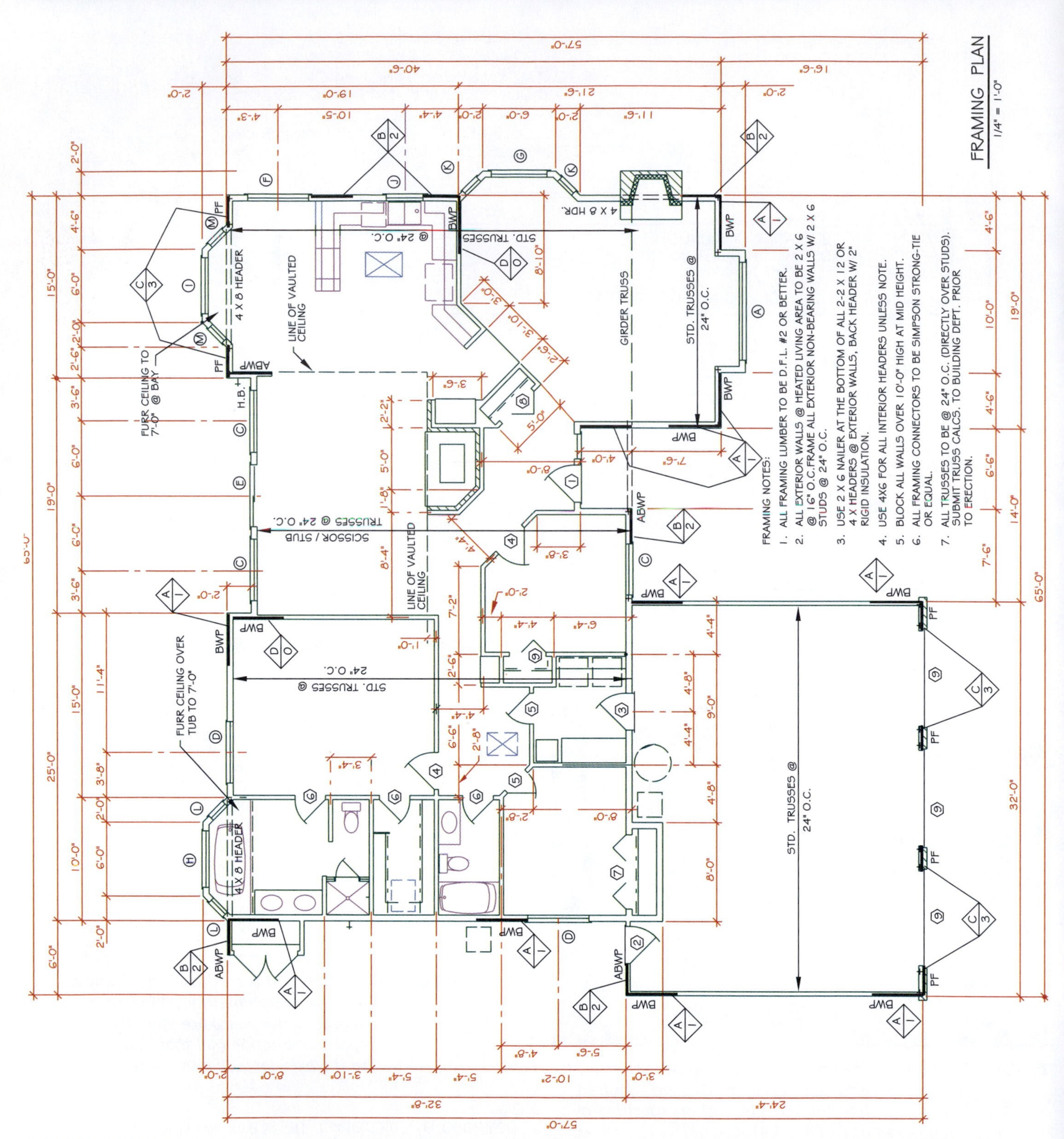
FRAMING PLAN
1/4" = 1'-0"
FRAMING NOTES:
1. ALL FRAMING LUMBER TO BE D.F.L. #2 OR BETTER.
2. ALL EXTERIOR WALLS @ HEATED LIVING AREA TO BE 2 X 6 @ 16" O.C.FRAME ALL EXTERIOR NON-BEARING WALLS W/ 2 X 6 STUDS @ 24" O.C.
3. USE 2 X 6 NAILER AT THE BOTTOM OF ALL 2-2 X 12 OR 4 X HEADERS @ EXTERIOR WALLS, BACK HEADER W/ 2" RIGID INSULATION.
4. USE 4X6 FOR ALL INTERIOR HEADERS UNLESS NOTE.
5. BLOCK ALL WALLS OVER 10'-0" HIGH AT MID HEIGHT.
6. ALL FRAMING CONNECTORS TO BE SIMPSON STRONG-TIE OR EQUAL.
7. ALL TRUSSES TO BE @ 24" O.C. (DIRECTLY OVER STUDS). SUBMIT TRUSS CALCS. TO BUILDING DEPT. PRIOR TO ERECTION.
STD. TRUSSES @ 24" O.C.
GIRDER TRUSS
SCISSOR / STUB TRUSSES @ 24" O.C.
LINE OF VAULTED CEILING
FURR CEILING TO 7'-0" @ BAY
FURR CEILING OVER TUB TO 7'-0"
4 X 8 HEADER
4 X 8 HDR.
57'-0"
65'-0"

FIGURE 24.22 ■ *Representing the materials used to resist lateral loads using prescriptive design methods of the IRC. The same home is shown in Figure 24.1, where a CAD technician specified materials after the actual loads to be resisted were calculated by an engineer.*

DIMENSIONS, ANNOTATION, AND DETAILS

Once the roof plan is complete, information must be placed on the framing plan to locate, specify, and explain the attachment of major structural members.

Dimensions

Chapter 11 introduced the process of placing dimensions on the floor plan. The process for placing dimensions on the framing plan is exactly the same. If a separate framing plan is to be drawn, the floor plan is usually not dimensioned; instead, all dimensions are placed on the framing plans. In addition to walls, doors, and windows, the location of framing members often needs to be dimensioned. Beams that span between an opening in a wall are located by the dimension that locates the wall. Beams in the middle of a room must be located by dimensions.

Where joists extend past a wall, the length of the cantilever must be dimensioned. Joist size or type often varies when an upper level only partially covers another level. The limits of the placement of joists should also be dimensioned on a floor plan. Figure 24.23 shows how the location of structural members can be clarified for the framer.

Annotation

The use of local notes to specify materials on the framing plan has been mentioned throughout this chapter. Many professionals also place general notes on the framing plan to ensure compliance with the code

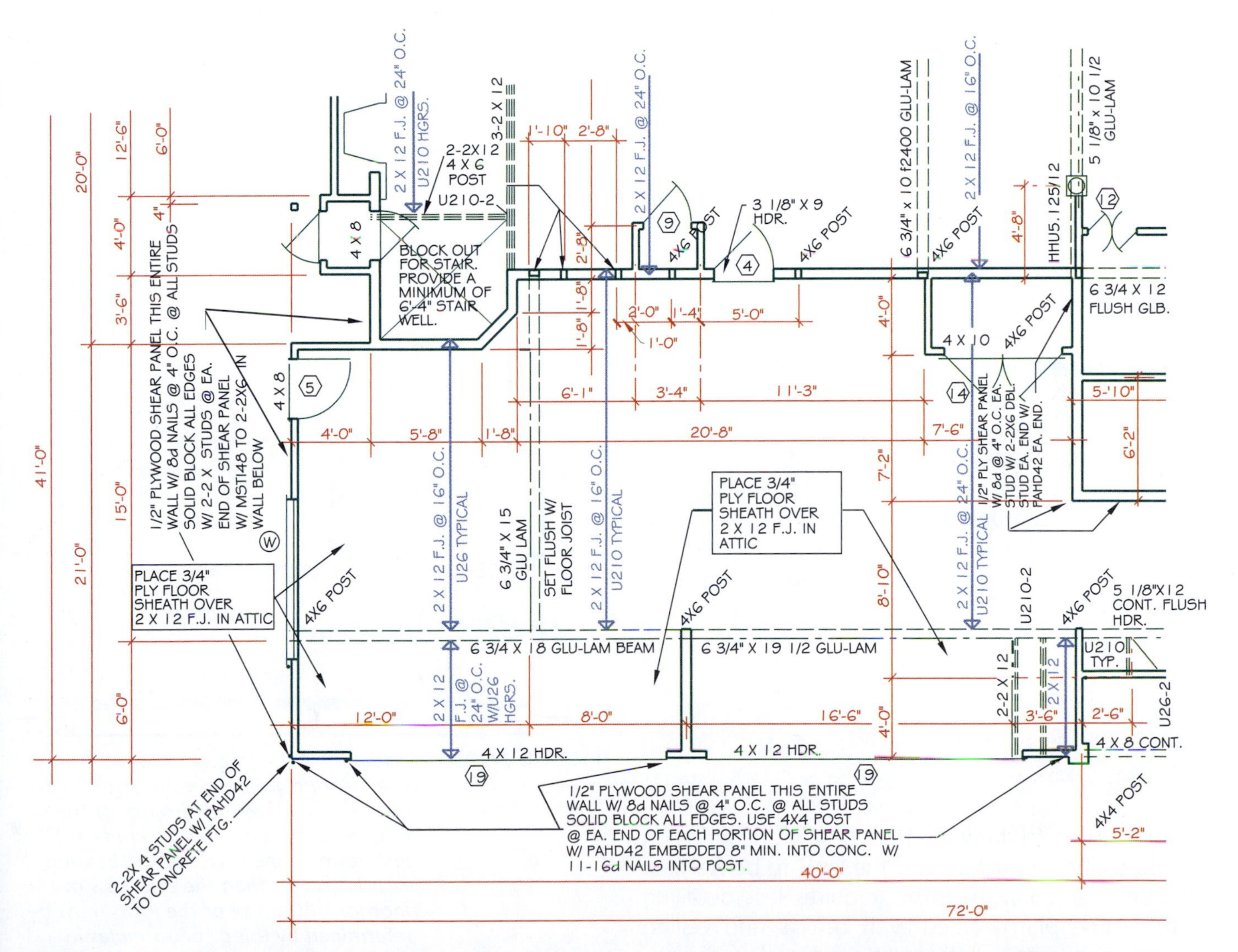

FIGURE 24.23 ■ *The location of beams and posts must be dimensioned on the framing plan. Local notes must also be provided to explain all structural materials.* Courtesy Residential Designs.

FRAMING NOTES: NOTES SHALL APPLY TO ALL LEVELS. FRAMING STANDARDS ARE BASED ON 2003 I.R.C.

1. ALL FRAMING LUMBER TO BE DFL #2 MIN.
   ALL GLU-LAM BEAMS TO BE fb2400, V-4, DF/DF
2. FRAME ALL EXTERIOR WALLS W/ 2 X 6 STUDS @ 16" O.C.
   FRAME ALL EXTERIOR NON-BEARING WALLS W/ 2 X 6 STUDS @ 24" O.C.
3. FRAME ALL EXTERIOR HEADERS W/ (2) -2X12 OR 4 X HEADERS W/ 2 X 6 NAILER AT THE BOTTOM OF ALL HDRS. BACK ALL HEADERS W/ EXTERIOR WALLS. BACK HEADER W/ 2" RIGID INSULATION.
4. FRAME ALL INTERIOR HEADERS W/ 4 x 6 DFL # 2 HEADERS.
5. BLOCK ALL WALLS OVER 10'-0" HIGH AT MID HEIGHT.
6. ALL SHEAR PANELS TO BE 1/2" PLY NAILED W/ 8 d'S @ 4" O.C. AT EDGE AND BLOCKING AND 6 d'S @ 8" O.C. @ FIELD.
7. LET-IN BRACES TO BE 1 X 4 DIAG. BRACES @ 45 DEG. FOR ALL INTERIOR LOAD-BEARING WALLS.
8. PLYWOOD ROOF SHEATHING TO BE 1/2" STD. GRADE 32/16 PLY. LAID PERP. TO RAFT. NAIL W/ 8 d'S @ 6" O.C. @ EDGES AND 12" O.C. AT FIELD PERP. TO RAFTERS. NAIL W/ 8 d'S @ 6" O.C. @ EDGES AND 12" O.C. AT FIELD

FIGURE 24.24 ■ *General notes can be used to ensure that minimum standards established by building codes will be maintained. These notes can be found on the CD for insertion into your drawing. Always be sure to edit standard notes to meet the needs of specific projects.*

that governs construction. These framing notes can be placed on the framing plan, with the sections and details, or on a separate specifications page that is included with the working drawings. Figure 24.24 shows common notes that might be included with the framing plan. These notes can also be found on the CD.

Section References

The framing plans are used as reference maps to show where cross sections have been cut. Detail reference symbols are also placed on the framing plan to help the print reader understand material that is being displayed in the sections. Chapter 28 will further explain section tags and their relationship to the sections and framing plans. Figure 24.10 shows examples of section markers.

COMPLETING A FRAMING PLAN

The order used to draw the framing plan depends on the method of construction and the level to be framed. A structure framed with trusses requires less detailing than the same plan framed with rafters and ceiling joists. Interior beams will be greatly reduced, if not entirely eliminated, when trusses are used. On multilevel structures, because a lower level is supporting more weight, beams tend to be shorter, requiring more and larger posts. Despite differences, framing plans also have similarities that can help in drafting the plan, no matter what level is to be drawn.

Once the walls have been located, dimensions can be placed to locate all walls and openings in them. Then individual framing members can be determined. Rafter direction is based on the shape of the roof, and its length is determined by the span for a specific size of the particular member. Rafter spans can be determined using the tables in Chapter 23. The process is similar if trusses are to be drawn, but the sizes of the trusses are determined by the manufacturer.

After the size and direction of the roof framing members are selected, load bearing walls can be identified and headers selected for openings in the load-bearing walls. With the structural members specified, any dimensions required to locate structural members can be completed. Materials that can be specified with local notes can be added and the drawing completed by adding general notes.

Once the framing for the roof has been determined, framing for lower levels can be drawn. Bearing walls on the upper level need to have some method of supporting the loads as they are transferred downward. Support can be provided by stacking bearing walls or by transferring loads through floor joists to other walls. Figure 24.25 shows a simplified drawing of how loads can be trans-

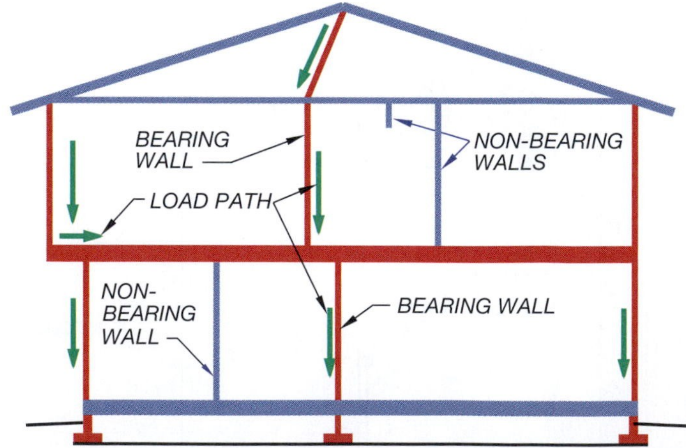

FIGURE 24.25 ■ *Following the load path through a structure. Bearing walls are considered to be aligned as long as the distance of the offset does not exceed the thickness of the floor joists. If the offset of the bearing walls is greater than the thickness of the floor joists, the size of the joists must be determined by using the formulas in Chapter 23.*

ferred. Load-bearing walls can be offset by the thickness of the floor joists and still be considered aligned. For example, if 2 × 10 (50 × 250) floor joists are used, an upper wall could be offset 10'' (250 mm) from a lower wall and still be considered aligned. When floor joists are cantilevered or used to transfer vertical loads to lower walls that are not aligned, the joist tables in Chapter 23 cannot be used to determine joist sizes. These members need to be determined using the formulas. Once the bearing walls for the lowest floor level have been determined, the foundation can be completed.

Roof Framing Plans

The roof plan was introduced in Chapter 16. The roof framing plan appears similar, but in addition to showing the shape of the structure and the outline of the roof, it also shows the size and direction of the framing members used to frame the roof, as in Figure 24.26. Another alternative for creating a roof framing plan is to use the floor plan as a base for the roof drawings. This method can be especially helpful on a complicated stick framed roof. If the floor plan is used as a base, show all of the walls printed in matted lines using gray scale, and show all roofing materials using black lines (see Figure 1.22).

The complexity of the roof framing plan will vary depending on whether trusses or western platform construction methods are used. Figure 24.27 shows the framing plan for a stick-built roof. Each method will be considered for gable and hip roofs for the residence drawn in Chapter 11, using separate roof plans and framing plans. Each roof can be drawn by using the same steps to form the base drawing. If a roof plan has been drawn, it can be copied and altered to form the base of the roof framing plan. Information that pertains to making the roof weather tight must be frozen. This would include removing materials and notes about finish materials, vents, downspouts, and roofing material.

Editing the Roof Plan

Once the layout method is selected, layers can be added to contain materials for the roof plan. For this example, the drawing will be created using the roof plan as a base. All nonstructural material are controlled using layers within the base drawing. Each new layer was given the prefix of *ROOF FRAM* to make it easy to identify required layers for plotting. Sub names for *ANNO, DIMN, JOST,* and *BEAM* were also added. Unless noted, thin continuous lines can be used for the following steps. Complete the drawing by working from the top down, adding major framing members

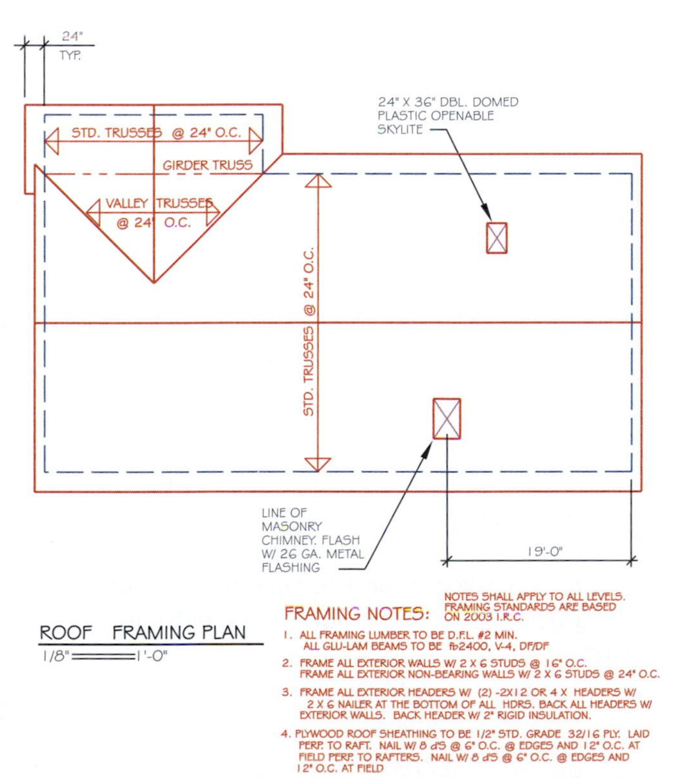

FIGURE 24.26 ■ *A roof framing plan for a simple structure framed with trusses. The roof plan was used as a base for this drawing, and all nonstructural information was frozen.*

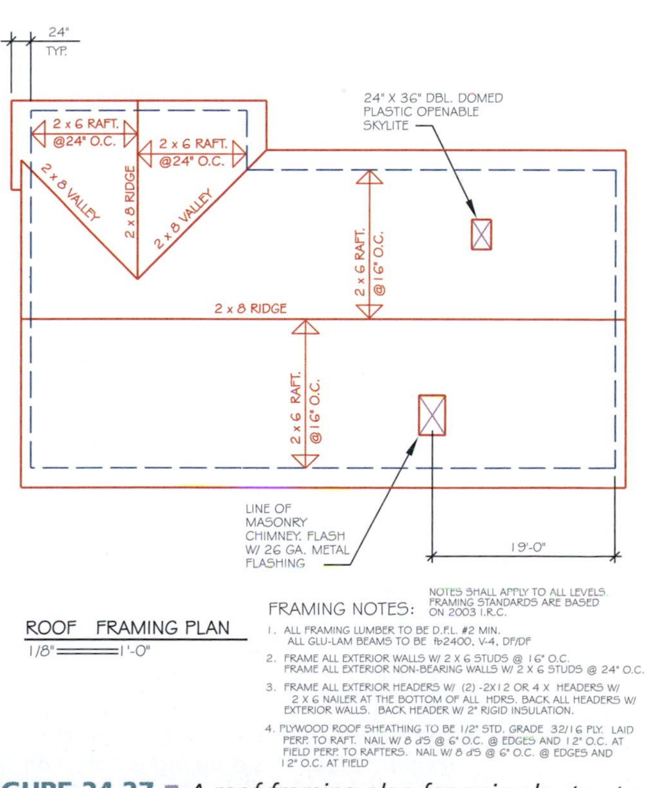

FIGURE 24.27 ■ *A roof framing plan for a simple structure framed with rafters.*

first and then working to smaller members. Girder trusses were drawn, and then the standard trusses were represented. With all materials represented, dimensions should be placed to locate material limitations, beam locations, and overhang sizes. With all materials drawn and located, notation should be provided to specify each material and framing member size that has been represented.

Drawing Roof Framing Plan for a Gable Roof

Step 1. Make a copy of the roof plan.
Step 2. Insert the copy of the roof plan into a new drawing template.
Step 3. Establish a viewport to display the completed drawing at a scale of 1/8'' = 1'-0''.
Step 4. Freeze all material not related to the roof framing. Figure 24.28 shows the drawing base.

Gable Roof Framing Plan with Truss Framing

Draw the following materials on the *ROOF FRAM TRUS* layer.

Step 5. Use thin phantom lines to draw the boundaries of any areas to be vaulted.
Step 6. Use thick phantom lines to draw all girder trusses.
Step 7. Draw or insert truss markers for each type of truss to be used.
Step 8. Label the truss markers to indicate the type and spacing of framing members.

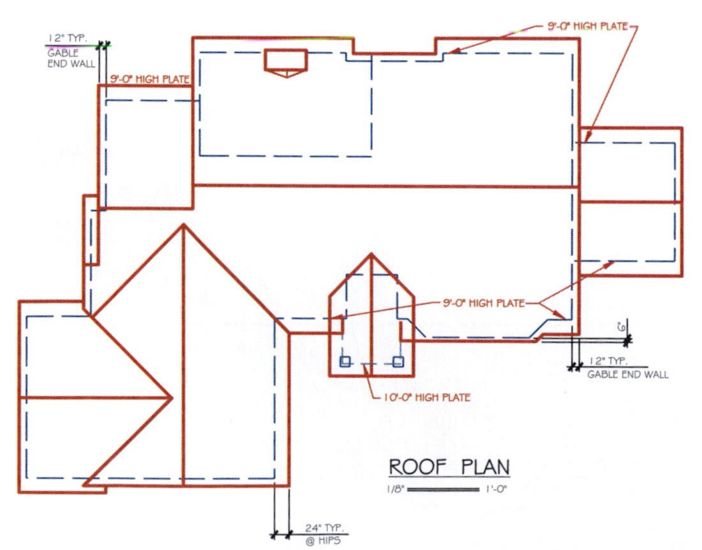

FIGURE 24.28 ■ *The roof plan serves as a base for the framing plan. This drawing is based on the roof plan completed in Figure 16.26, with all nonstructural material frozen.*

Use the *ROOF FRAM DIMS* layer for the following steps:

Step 9. Dimension all roof penetrations such as skylights and chimneys.

Use the *ROOF FRAM ANNO* layer for the following steps:

Step 10. Represent and label any differences in plate heights.
Step 11. Insert and edit the roof framing notes.
Step 12. Provide local notes to specify any necessary straps or metal connectors.
Step 13. Edit the title to ROOF FRAMING PLAN.

Your drawing should now resemble Figure 24.29.

Gable Roof Framing Plan with Rafter Framing

Follow steps 1 through 4 from the previous section to create the base for this plan. Complete the plan for a roof framed using rafters, using the following steps:

Step 5. Determine the location of interior walls, which are parallel to the ridge. Although the walls are not drawn on the framing plan, they can be used to support purlin braces. Some offices display all interior walls to aid in showing purlin supports. If walls are to be displayed, place them on a layer such as *ROOF FRAM WALL* and set the layer for plotting using gray scale.

Place the following structural members on a layer with a title such as *ROOF FRAM RAFT.*

Step 6. Use thin phantom lines to draw the boundaries of any areas to be vaulted. Provide dimensions to specify the limits.

Your drawings should now resemble Figure 24.30. Use the following steps to complete the drawings.

Step 7. Draw and label the arrows to indicate the framing members. Use the span tables from Chapter 23 to determine the required rafter sizes. For this residence, DFL #2 rafters will be used. Use 2 × 6 rafters at 24'' o.c. where possible.
Step 8. Use thick phantom lines to draw purlins and strong backs at approximately the midpoint of rafter spans.

Use the *ROOF FRAM ANNO* layer for the following steps:

Step 9. Determine the required sizes, and label all hips, valleys, and ridges.
Step 10. Represent and label any differences in plate heights.

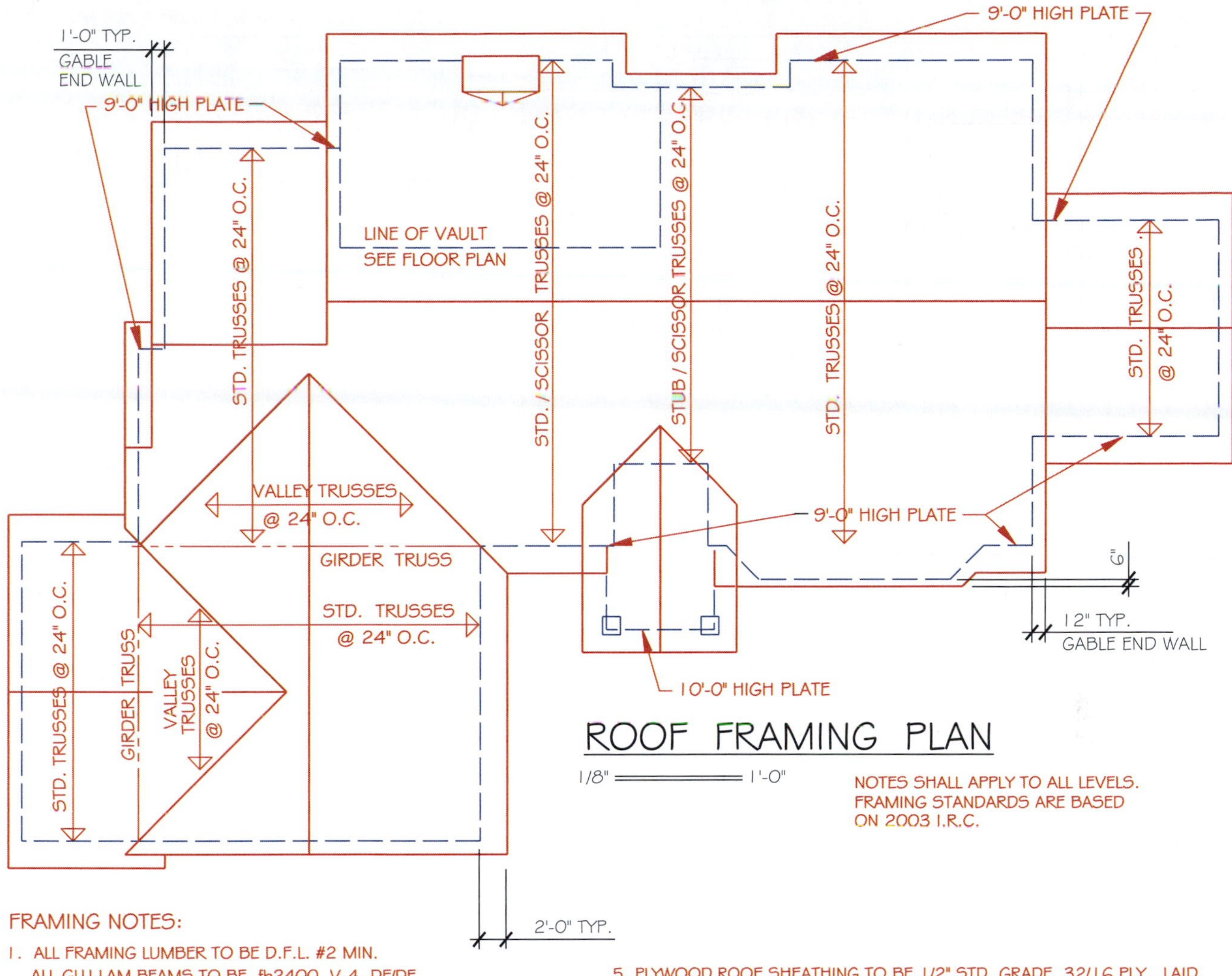

1'-0" TYP.
GABLE END WALL
9'-0" HIGH PLATE
9'-0" HIGH PLATE
LINE OF VAULT
SEE FLOOR PLAN
STD. TRUSSES @ 24" O.C.
STD. / SCISSOR TRUSSES @ 24" O.C.
STUB / SCISSOR TRUSSES @ 24" O.C.
STD. TRUSSES @ 24" O.C.
STD. TRUSSES @ 24" O.C.
VALLEY TRUSSES
@ 24" O.C.
GIRDER TRUSS
STD. TRUSSES
@ 24" O.C.
STD. TRUSSES @ 24" O.C.
GIRDER TRUSS
VALLEY TRUSSES @ 24" O.C.
9'-0" HIGH PLATE
6"
12" TYP.
GABLE END WALL
10'-0" HIGH PLATE
ROOF FRAMING PLAN
1/8" = 1'-0"
NOTES SHALL APPLY TO ALL LEVELS.
FRAMING STANDARDS ARE BASED ON 2003 I.R.C.
2'-0" TYP.
FRAMING NOTES:
1. ALL FRAMING LUMBER TO BE D.F.L. #2 MIN.
ALL GLU-LAM BEAMS TO BE fb2400, V-4, DF/DF
2. FRAME ALL EXTERIOR WALLS W/ 2 X 6 STUDS @ 16" O.C.
FRAME ALL EXTERIOR NON-BEARING WALLS W/ 2 X 6 STUDS @ 24" O.C.
3. FRAME ALL EXTERIOR HEADERS W/ (2) -2X12 OR 4 X HEADERS W/
2 X 6 NAILER AT THE BOTTOM OF ALL HDRS. BACK ALL HEADERS W/
EXTERIOR WALLS. BACK HEADER W/ 2" RIGID INSULATION.
4. FRAME ALL INTERIOR HEADERS W/ 4 x 6 DFL # 2 HEADERS.
5. PLYWOOD ROOF SHEATHING TO BE 1/2" STD. GRADE 32/16 PLY. LAID
PERP. TO RAFT. NAIL W/ 8 d'S @ 6" O.C. @ EDGES AND 12" O.C. AT
FIELD PERP. TO RAFTERS. NAIL W/ 8 d'S @ 6" O.C. @ EDGES AND
12" O.C. AT FIELD
6. SUBMIT TRUSS MANUF. DRAWINGS TO BUILDING
DEPT. PRIOR TO ERECTION
7. PROVIDE SCREENED VENTS @ EA. 3rd.
TRUSS SPACE @ ALL ATTIC EAVES.

FIGURE 24.29 ■ *The completed roof framing plan for a gable roof framed with trusses.*

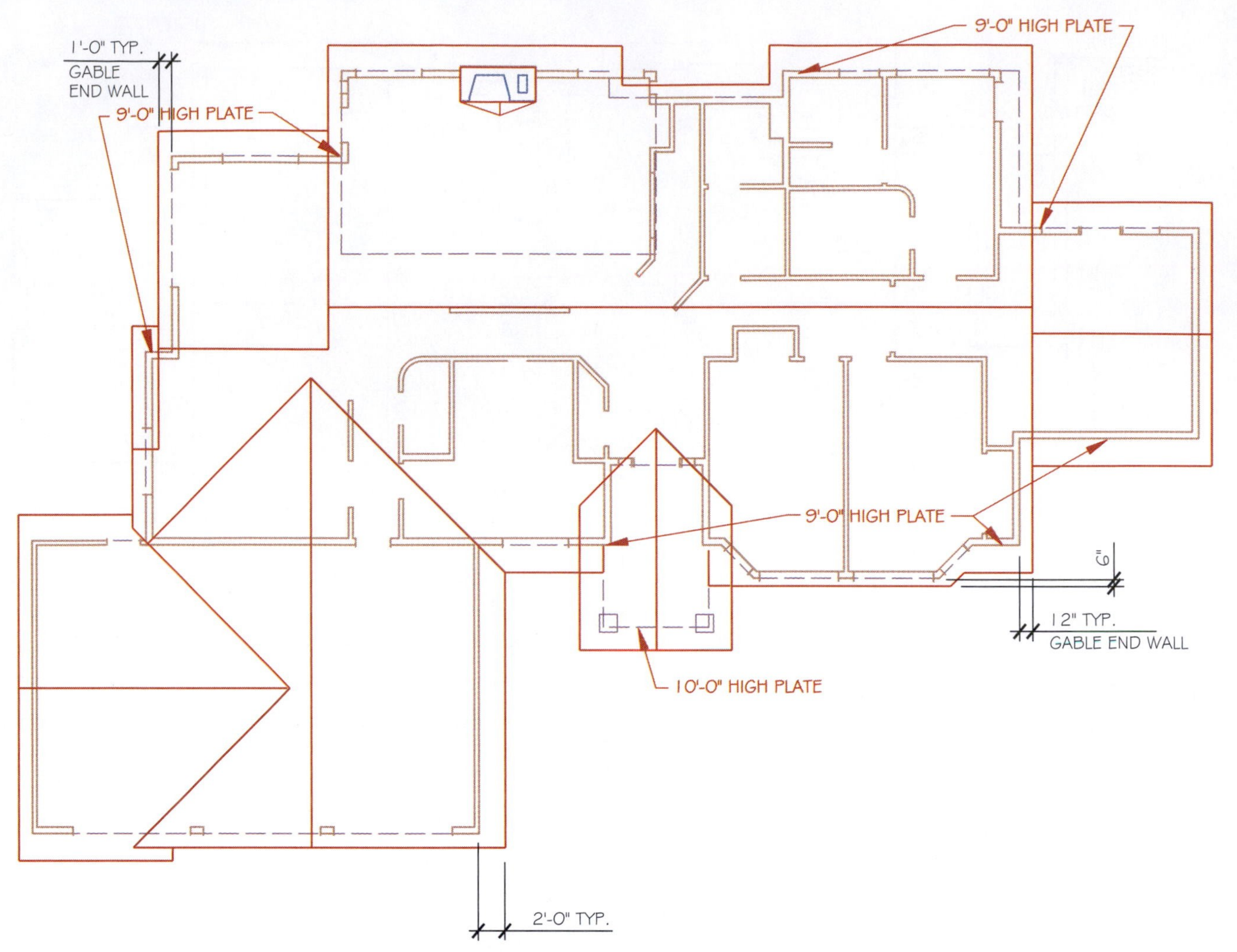

FIGURE 24.30 ■ *When a roof is to be framed with rafters, the walls may be shown on a roof framing plan to aid in displaying the roof supports. If walls are to be displayed, place them on a layer that can be plotted using gray scale.*

Step 11. Provide local notes to specify any necessary straps or metal connectors.
Step 12. Insert and edit the roof framing notes.
Step 13. Dimension all roof penetrations, such as skylights and chimneys.
Step 14. Edit the title to ROOF FRAMING PLAN.

Your drawing should now resemble Figure 24.31.

Hip Roof Framing Plans

Step 1. Make a copy of the roof plan.
Step 2. Insert the copy of the roof plan into a new drawing template.
Step 3. Establish a viewport to display the completed drawing at a scale of 1/8'' = 1'-0''.
Step 4. Freeze all material not related to the roof framing. Your drawing should now resemble Figure 24.32.

Hip Roof Framing Plan with Truss Framing

Use the following steps to draw the hip roof plan framed using trusses. Use thin, continuous lines unless noted to complete the following steps. Place the following information on the *ROOF FRAM TRUS* layer.

Step 5. Determine the limits of the standard trusses based on the intersection of the hips with the ridge.
Step 6. Determine the limits of any required girder trusses. Use thick phantom lines to represent each girder truss.
Step 7. Use thin phantom lines to draw the boundaries and dimension the limits of any areas to be vaulted. On this framing plan, the vaulted ceiling in the family room was deleted to allow a wider use of trusses.
Step 8. Determine the limits of the hip trusses. This distance is typically determined by the truss manu-

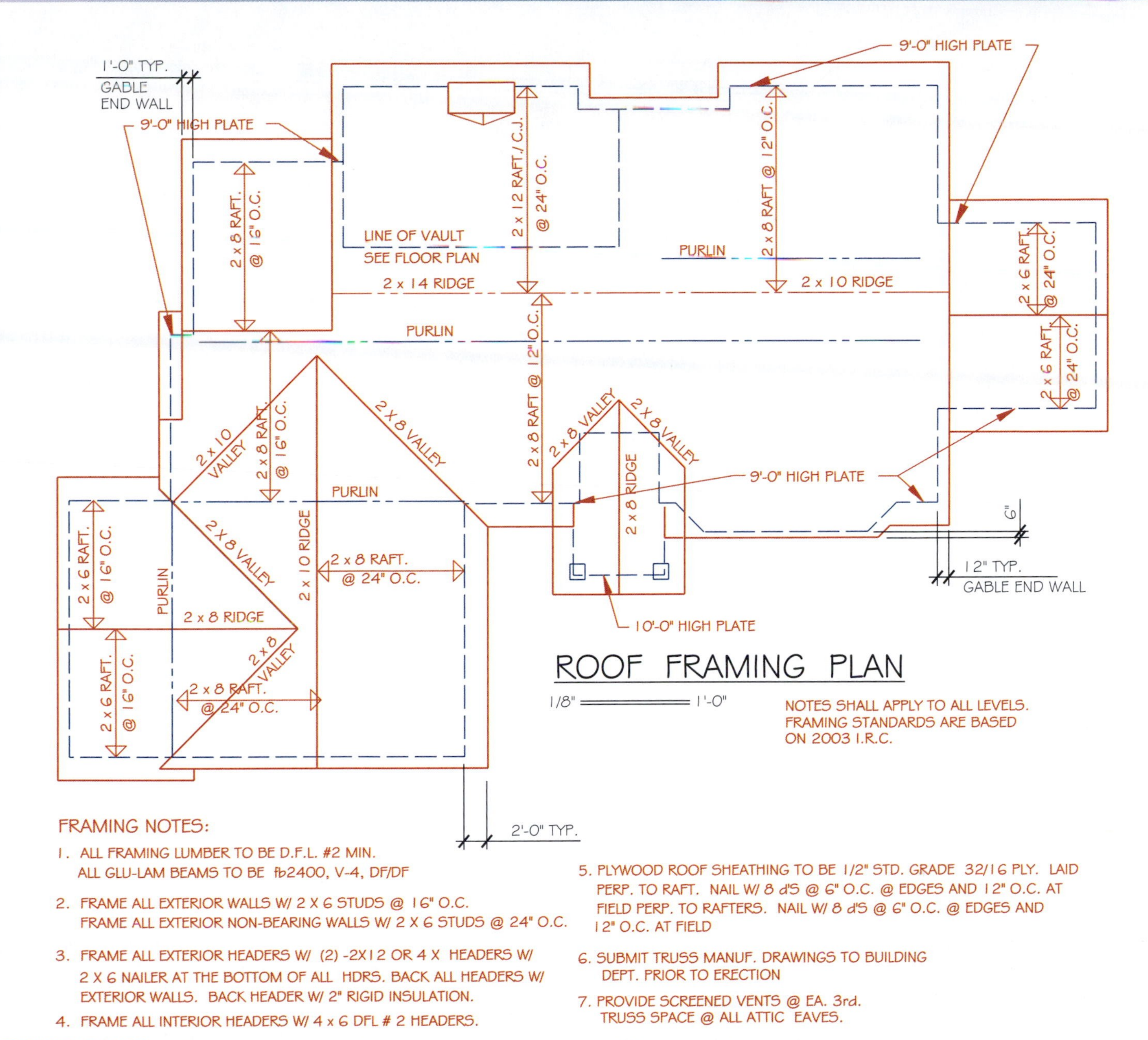

FIGURE 24.31 ■ *The completed roof framing plan for a gable roof framed with rafters.*

facturer and the pitch of the roof. For this plan assume 6'-0'' is required before hip trusses can be used. Represent the limits of trusses with thin dashed lines.

Step 9. Draw and label the arrows to indicate any required stick-framing members. Highlight any areas that require stick framing over the trusses to obtain the desired shape.

Use the *ROOF FRAM ANNO* layer for the following steps:

Step 10. Insert and edit roof framing notes.

Step 11. Determine the required sizes and label all hips, valleys, and ridges. (This step is not always required with trusses, but will be required if some areas of the roof are stick framed.)

Step 12. Represent and label any differences in plate heights.

Step 13. Provide local notes to specify any necessary straps or metal connectors.

Step 14. Insert and edit the roof framing notes.

Step 15. Dimension all roof penetrations, such as skylights and chimneys.

Step 16. Edit the title to ROOF FRAMING PLAN.

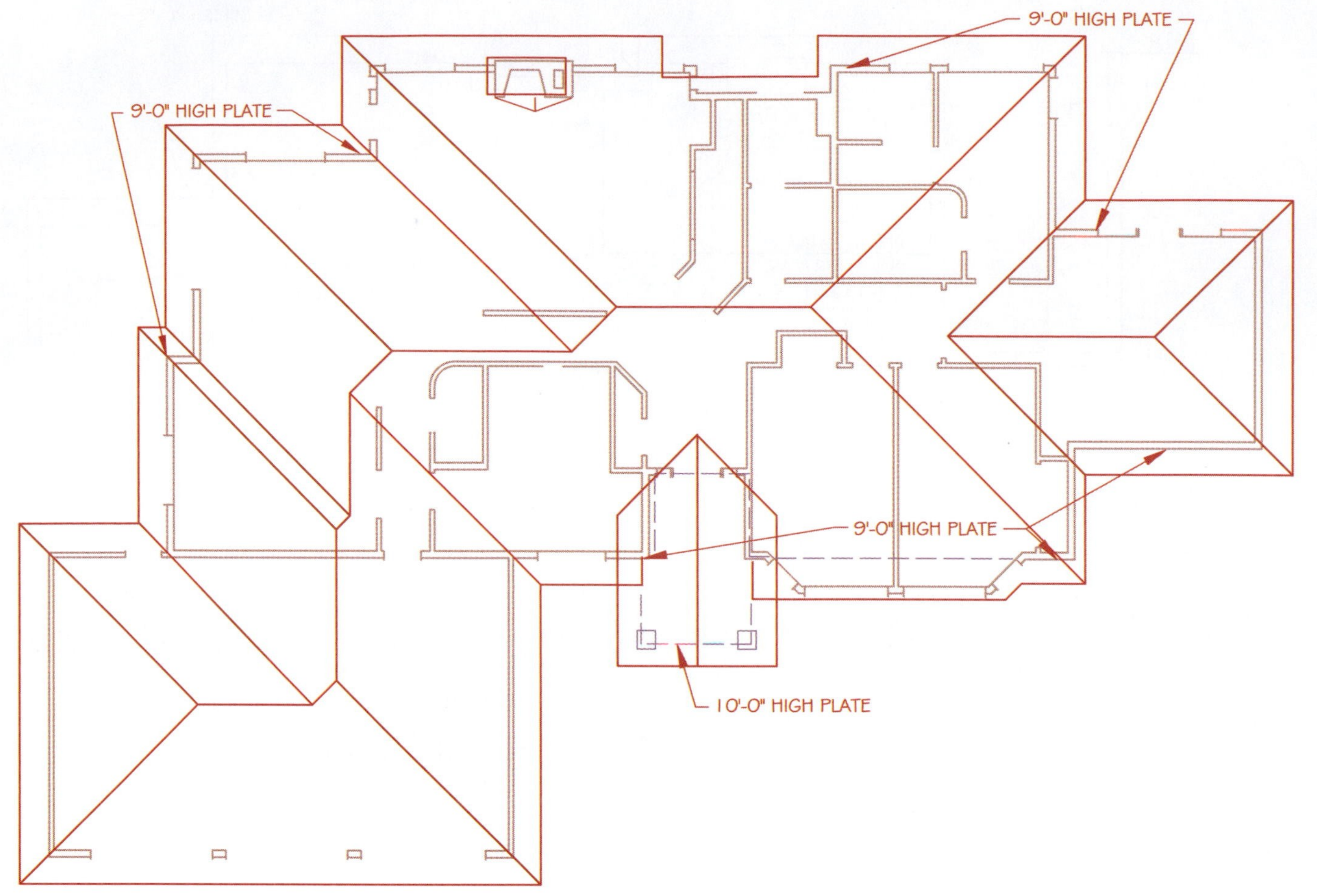

FIGURE 24.32 ■ *The roof plan for a hip roof serves as a base for the roof framing plan. Walls from the floor plan displayed in grayscale will aid in representing supports for each hip, valley, and ridge.*

Your drawing should now resemble Figure 24.33.

Hip Roof Framing Plan with Rafter Framing

Follow steps 1 through 4 to create the base for this plan.

Step 1. Make a copy of the roof plan.
Step 2. Insert the copy of the roof plan into a new drawing template.
Step 3. Establish a viewport to display the completed drawing at a scale of 1/8'' = 1'-0''.
Step 4. Freeze all material not related to the roof framing. Your drawing should now resemble Figure 24.32. Complete the plan for a roof framed using rafters, using the following steps:
Step 5. Determine the location of interior walls, which are parallel to the ridge. Display all interior walls to aid in showing purlin supports. Place the walls on a layer such as *FRAM WALL* and set the layer for plotting using gray scale.

Place the following structural member on a layer with a title such as *ROOF FRAM RAFT.*

Step 6. Use thin phantom lines to draw the boundaries of any areas to be vaulted. Provide dimensions to specify the limits.

Follow Steps 1 through 7 to draw the base for this plan. The drawing should resemble Figure 24.27. The layout of the roof will be similar to the layout of a stick-framed gable roof.

Step 7. Use thick phantom lines to draw purlins and strong backs at approximately the midpoint of rafter spans.
Step 8. Draw and label the arrows to indicate the framing members. Use the span tables from Chapter 23 to determine the required size. For this residence, DFL #2 will be used. Use 2 × 6 rafters at 24'' o.c. where possible. Develop a schedule to reference all rafter sizes.

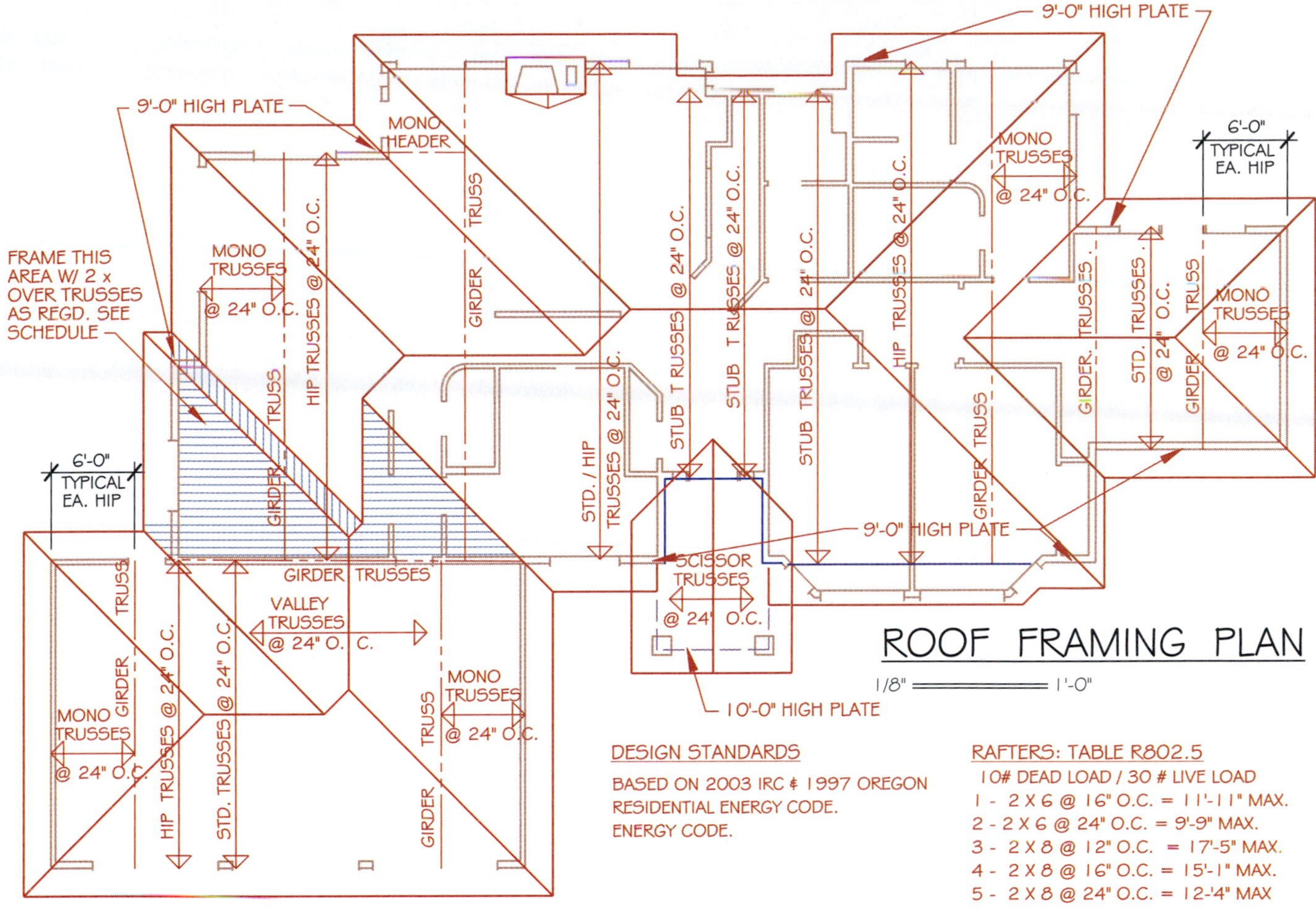

FRAMING NOTES:

1. ALL FRAMING LUMBER TO BE D.F.L. #2 MIN.
ALL GLU-LAM BEAMS TO BE fb2400, V-4, DF/DF
2. FRAME ALL EXTERIOR WALLS W/ 2 X 6 STUDS @ 16" O.C.
FRAME ALL EXTERIOR NON-BEARING WALLS W/ 2 X 6 STUDS @ 24" O.C.
3. FRAME ALL EXTERIOR HEADERS W/ (2) -2X12 OR 4 X HEADERS W/ 2 X 6 NAILER AT THE BOTTOM OF ALL HDRS. BACK ALL HEADERS W/ EXTERIOR WALLS. BACK HEADER W/ 2" RIGID INSULATION.
4. FRAME ALL INTERIOR HEADERS W/ 4 x 6 DFL # 2 HEADERS.
5. PLYWOOD ROOF SHEATHING TO BE 1/2" STD. GRADE 32/16 PLY. LAID PERP. TO RAFT. NAIL W/ 8 d'S @ 6" O.C. @ EDGES AND 12" O.C. AT FIELD PERP. TO RAFTERS. NAIL W/ 8 d'S @ 6" O.C. @ EDGES AND 12" O.C. AT FIELD
6. PROVIDE SCREENED VENTS @ EA. 3rd. TRUSS SPACE @ ALL ATTIC EAVES.
7. ALL RAFTERS TO BE 2 X UNLESS NOTED. SEE ATTACHED SCHEDULE FOR SPECIFIC SIZES
8. ALL RIDGES, HIPS, AND VALLEYS TO BE 2 X 10 DFL #2 OR BETTER UNLESS NOTED

FIGURE 24.33 ■ *The completed roof framing plan for a hip roof framed with trusses. Notice that a portion of the roof on the left side of the structure requires the use of stick framing over the main truss roof.*

Use the *ROOF FRAM ANNO* layer for the following steps:

Step 9. Insert and edit roof framing notes.
Step 10. Determine the required sizes and label all hips, valleys, and ridges.
Step 11. Represent and label any differences in plate heights.
Step 12. Provide local notes to specify any necessary straps or metal connectors.
Step 13. Insert and edit the roof framing notes.
Step 14. Dimension all roof penetrations, such as skylights and chimneys.
Step 15. Edit the title to read ROOF FRAMING PLAN.

Your drawing should now resemble Figure 24.34.

Floor Framing Plans

The framing plan can be completed by using the base drawings for the floor plans. The base drawings should include the walls, doors, windows, cabinets, appliances, plumbing, and door and window schedule symbols. All materials unrelated to the framing plan must be frozen. With the proper material displayed, the framing plan can be completed using similar methods used to complete the

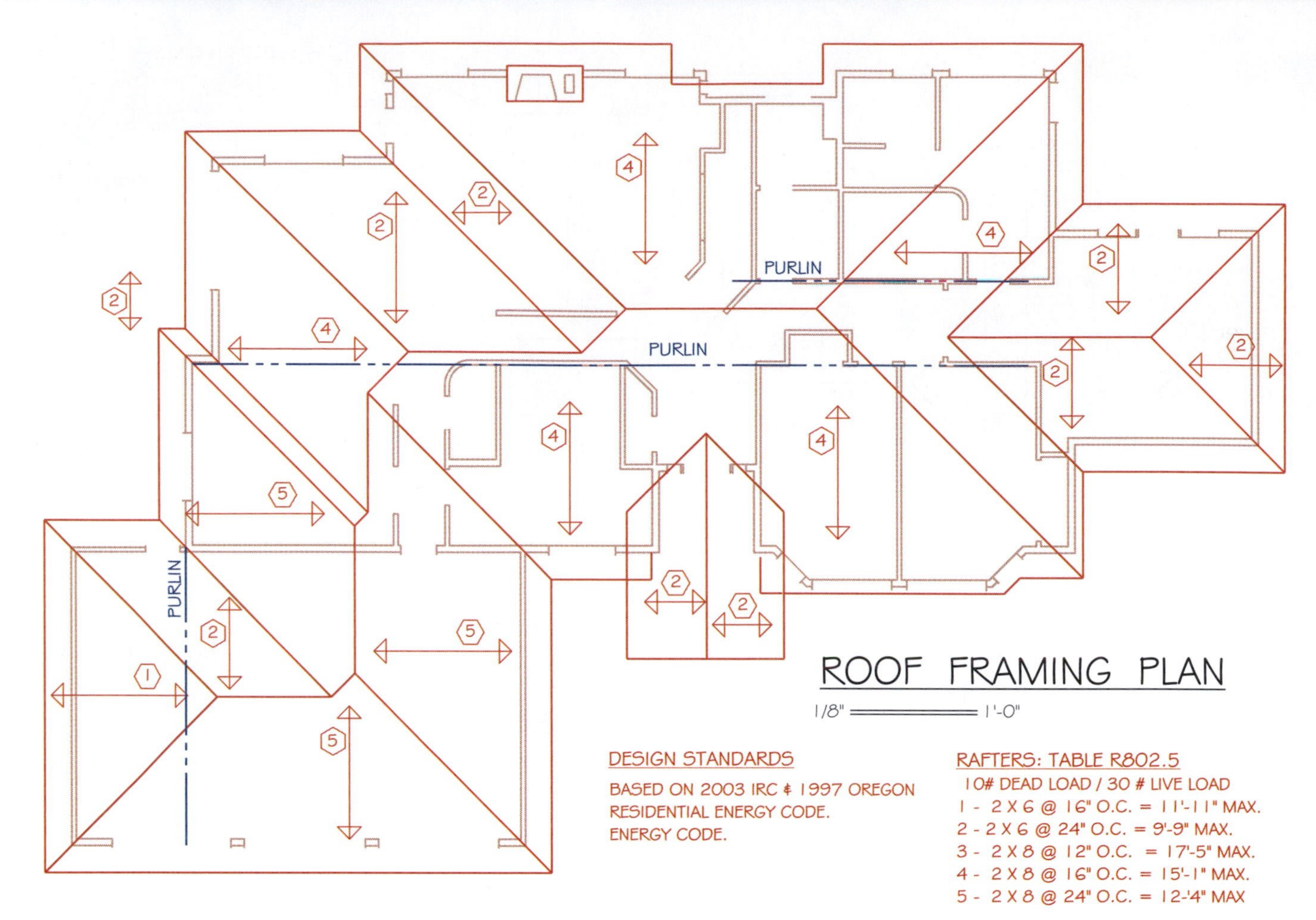

FRAMING NOTES:

1. ALL FRAMING LUMBER TO BE D.F.L. #2 MIN. ALL GLU-LAM BEAMS TO BE fb2400, V-4, DF/DF
2. FRAME ALL EXTERIOR WALLS W/ 2 X 6 STUDS @ 16" O.C. FRAME ALL EXTERIOR NON-BEARING WALLS W/ 2 X 6 STUDS @ 24" O.C.
3. FRAME ALL EXTERIOR HEADERS W/ (2) -2X12 OR 4 X HEADERS W/ 2 X 6 NAILER AT THE BOTTOM OF ALL HDRS. BACK ALL HEADERS W/ EXTERIOR WALLS. BACK HEADER W/ 2" RIGID INSULATION.
4. FRAME ALL INTERIOR HEADERS W/ 4 x 6 DFL # 2 HEADERS.
5. PLYWOOD ROOF SHEATHING TO BE 1/2" STD. GRADE 32/16 PLY. LAID PERP. TO RAFT. NAIL W/ 8 d'S @ 6" O.C. @ EDGES AND 12" O.C. AT FIELD PERP. TO RAFTERS. NAIL W/ 8 d'S @ 6" O.C. @ EDGES AND 12" O.C. AT FIELD
6. PROVIDE SCREENED VENTS @ EA. 3rd. TRUSS SPACE @ ALL ATTIC EAVES.
7. ALL RAFTERS TO BE 2 X UNLESS NOTED. SEE ATTACHED SCHEDULE FOR SPECIFIC SIZES
8. ALL RIDGES, HIPS, AND VALLEYS TO BE 2 X 10 DFL #2 OR BETTER UNLESS NOTED

FIGURE 24.34 ■ *A completed roof framing plan for a hip roof framed with rafters. Sizes can be specified in a table using a rafter symbol on the plan to make the drawing less cluttered.*

roof framing plan. Layers can be added to contain materials for the framing plan. If each level will be stored in the same model file, layers will need to be given a prefix to define each level. Titles such as *UPPER FRAME, MAIN FRAME,* or *LOWER FRAME* can be used to identify required layers for plotting. Sub names for *TEXT, DIMEN, TRUSSES, LATERAL,* and *BEAMS* layers must also be added.

Representing Framing Materials

Complete the drawing by adding major framing members first and then working to smaller members. Place beams that will be required to support joists or trusses, and then place a marker to represent each joist type. The joist markers can be stored as a block and then inserted and stretched to the needed size. Lateral bracing information can be added to the drawing using blocks. Schedules for one and multilevel construction bracing can be inserted into the drawing and edited. Symbols to represent each type of bracing can also be created as blocks and inserted into the required position throughout the drawing. This will require the creation of layers such as *UPPER FRAME LAT SYMB, MAIN FRAME LAT SYMB,* or *LOWER FRAME LAT SYMB.* Place the symbol information on a different layer than the lateral schedule.

Planning Dimension Placement

With the framing and lateral information represented, dimensions can be placed to represent all walls, openings, and framing materials. Dimensions layers should be divided by level. Some dimensions such as the overall and major jogs will be the same on all levels. This information can be placed on a layer titled *FRAM BASE DIMS.* Wall-to-wall and wall-to-opening dimensions should be placed on a layer that is specific to each level of the structure, such as *MAIN FRAM EXT DIMS.* Annotation can now be placed to locate all framing materials. Annotation should also be placed on layers that are level-specific. Local notes should also be divided by layers to represent title block text and text that is level-specific.

Note:

Remember that the information shown on the framing plan is the material that would be seen if you were lying on the floor and looking up. On the upper floor level, the ceiling joists should be represented. On the lower framing plan, the joists that frame the main floor level are represented.

If your floor plan was dimensioned, skip to step 6. If not, place an overall dimension on each side of the structure. Establish lines for overall and major jogs from the exterior face of the wall. Place the following dimensions on the FRAM DIMS BASE layer. This will allow these dimensions to be displayed on other plans.

Placing Dimensions

Step 1. Place an overall dimension on each side of the structure.

Step 2. Dimension all major jogs.

- Locate each jog from the exterior face.
- Verify that each line of major jog dimensions add up to the corresponding overall dimension.

Place the following dimensions on the *FRAM DIMS EXT* layer:

Step 3. Locate all interior walls that intersect with an exterior wall.

- Locate each interior wall based on its location from a major jog.
- Verify that each line of wall-to-wall dimensions adds up to the corresponding major jog dimension.

Step 4. Dimension all openings in the exterior walls.

- Locate each opening based on its location from an interior wall.
- Verify that each line of opening dimensions add up to the corresponding wall-to-wall dimension.

Figure 24.35 shows an example of the wall-to-wall and the wall-to-opening dimensions referenced to the overall and major jog dimensions. Place the following dimensions on the *FRAM DIMS INT* layer.

Step 5. Dimension all interior walls.

- Identify all interiors walls that are not referenced to an exterior wall.
  - Use a print of your drawing to identify all unknown walls.
  - Plan where missing dimensions can be placed so that they are grouped together.
  - Locate required wall dimensions. Place dimensions so that they coordinate with exterior dimensions.
- Locate any interior doors that are not located by walls, cabinets, or other fixed objects.
  - Doors are typically 3'' from a cabinet edge or wall corner. Such doors do not need to be dimensioned.
  - Doors that appear to be centered between two known objects do not need to be dimensioned.

Figure 24.36 shows an example of the completed dimensions. Use the checklist from the Drawing Checklist/Dimensions Checklist folder on the CD to verify the content of the drawing.

Representing Framing Members

Step 6. Use a layer such as *MAIN FRAM JSTS* to draw and label the arrows to indicate the framing members. Because trusses will be used, they have been represented on the roof framing plan shown in Figure 24.29.

Note:

If the roof is to be stick-framed, only the members used to show the support for the ceiling are shown on the framing plan. Figure 24.38 shows the ceiling framing for a stick-framed roof.

Step 7. Use a layer such as *MAIN FRAM BEAM* to draw and specify all beams, headers, and posts.

Step 8. Provide local notes to specify any materials needed for resisting lateral loads caused by wind or seismic forces. Place the notes on the *MAIN FRAM ANNO SHEAR* layer.

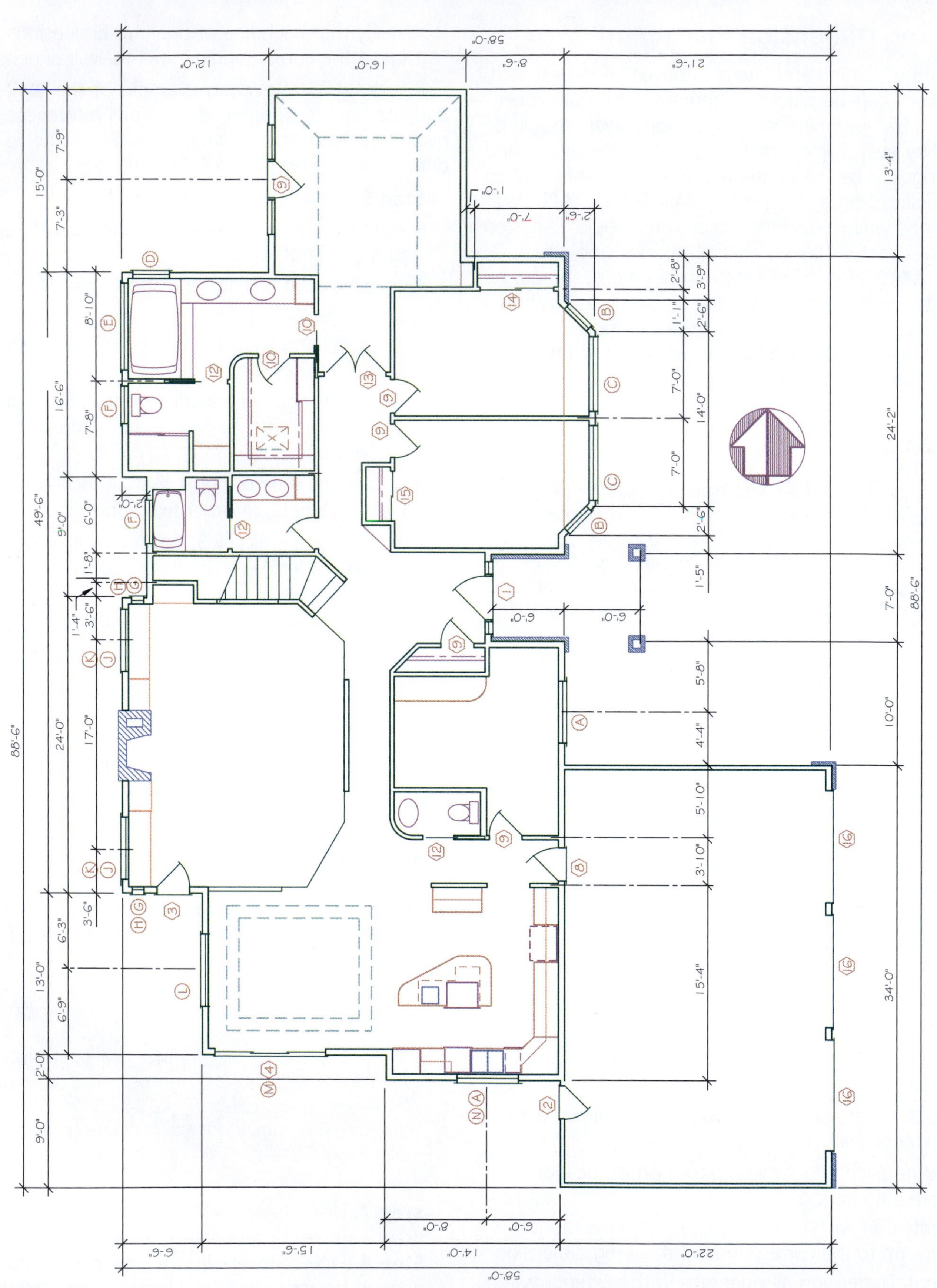

FIGURE 24.35 ■ *The main floor plan for the residence started in Chapter 11 is used as the base for the main floor framing plan. By separating the structural information from the architectural information, greater clarity is achieved. Floor related information is frozen and new layers are created for the framing material. Dimensions for the overall size and major jogs have been placed referenced to the outside edge of exterior walls.*

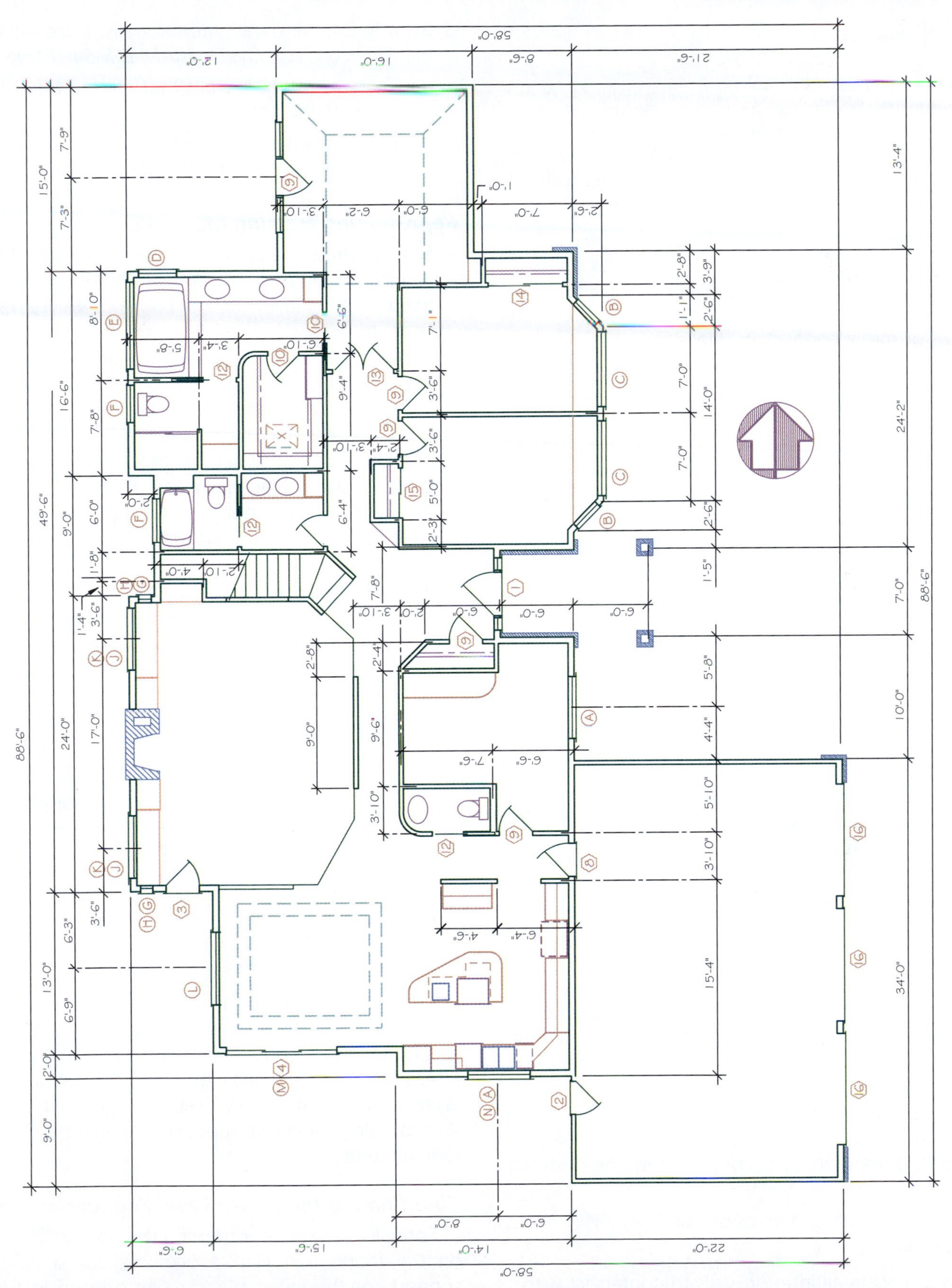

FIGURE 24.36 ■ *Dimensions to locate interior walls that touch the exterior walls are placed. Dimensions are then added to locate openings in the exterior walls that reference the interior wall dimensions. Door openings in the garage are omitted on the framing plan and referenced on the foundation plan.*

Note:

Because requirements vary widely for each area of the country, verify with your instructor materials that are typical for your area. Figure 24.37 shows shear panels designed by an engineer. See Figure 24.38 for an example of this home designed per the IRC prescriptive path.

Placing Annotation

Step 9. Provide local notes to specify any necessary straps or metal connectors. Place the notes on the *MAIN FRAM ANNO* layer.

Step 10. Provide general notes to indicate any necessary framing material. Place required notes on the *MAIN FRAM ANNO* layer. Because of space limitations for this home, all general framing notes will be placed on the lower floor plan. Your completed drawing should now resemble Figure 24.37. If the upper level is framed using western platform construction techniques and the prescriptive path of the IRC, the plan would resemble Figure 24.38. Figure 24.31 shows the roof framing plan that would be needed to accompany this framing plan.

Creating Plans for Additional Levels

The framing plan for the lower floor can be completed following the same procedure that was used to draw the main floor. Because the main floor of the residence drawn in Chapter 11 is larger than the lower floor, the floor framing that is not directly over the lower floor framing will be shown on the foundation plan.

Placing Dimensions

Step 1. Provide overall dimensions for the lower floor. Because the lower floor is so much smaller than the main floor, not all of the overall dimensions will be required. Use the PROPERTIES command to move the needed base dimensions to a new layer titled *LOWER FRAM BASE DIMS* to display the overall dimensions.

Step 2. Dimension all major jogs from their exterior face.

Place the following dimensions on the *LOWER FRAM DIMS EXT* layer.

Step 3. Locate all interior walls that intersect with an exterior wall.

Step 4. Dimension all openings in the exterior walls.

Step 5. Dimension all interior walls. Place the dimensions on the *LOWER FRAM DIMS INT* layer.

- Identify all interiors walls that are not referenced to an exterior wall.
- Locate any interior doors that are not located by walls, cabinets, or other fixed objects.

Representing Framing Materials

Step 6. Draw the boundaries of any areas to be vaulted or any areas to be furred down.

Step 7. Use a layer such as *LOWER FRAM JSTS*. Draw to draw and label the arrows to indicate the framing members. Use the tables in Chapter 23 to determine the sizes of members for species of framing lumber common to your area.

Step 8. Use a layer such as *LOWER FRAM BEAM* to draw and specify all beams, headers, and posts. Because of lateral loads from the upper floor, this home has double floor joists to support the posts on each side of the windows located in the west wall.

Placing Annotation

Step 9. Provide local notes to specify any materials needed for resisting lateral loads caused by wind or seismic forces. Place the notes on the *LOWER FRAM ANNO SHEAR* layer.

Note:

Because requirements vary widely for each area of the country, verify with your instructor materials that are typical for your area.

Step 10. Provide general notes to indicate any necessary framing material. Place required notes on the *LOWER FRAM ANNO* layer. Your completed drawing should now resemble Figure 24.39.

Homes designed per the IRC prescriptive path will resemble the framing plan shown in Figure 24.40. Place the notes to describe lateral bracing on the *LOWER FRAM ANNO* layer. Place the schedule on a separate layer such as *FRAM ANNO SCHD*. Figure 24.41 shows an example of a lateral support schedule based on IRC requirements.

Coordinating the Lower Level/Foundation Plan

Some information for framing the floor will be placed on the foundation plan. Procedures for showing the support for the lower plan are introduced in Chapter 26. Because this plan has a partial basement, the drafter will have the choice of where the floor framing over the

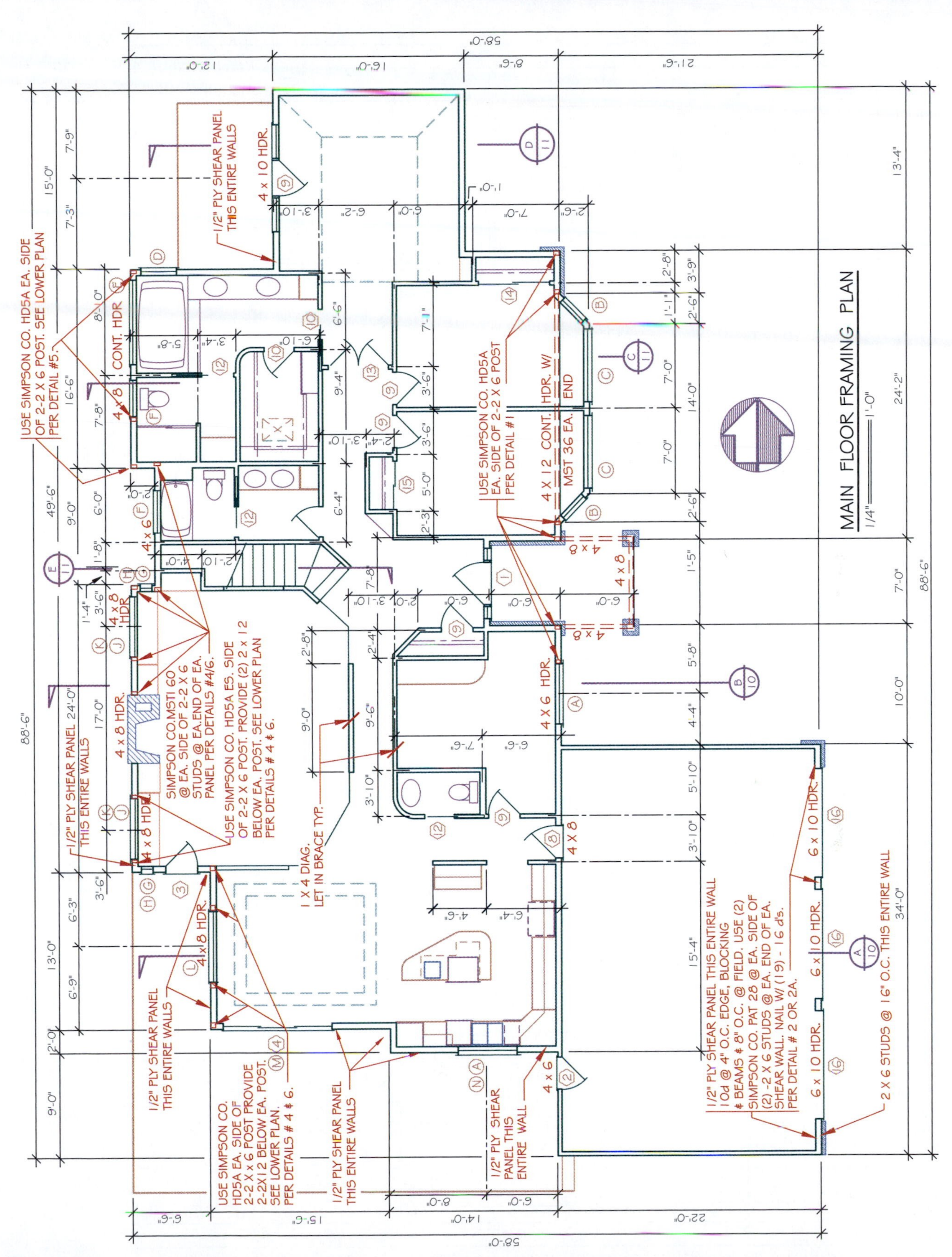

FIGURE 24.37 ■ *The completed main floor framing plan with all interior walls located. Framing members for truss construction are located on the roof framing plan shown in Figure 24.29. Lateral bracing is based on the design-specific method.*

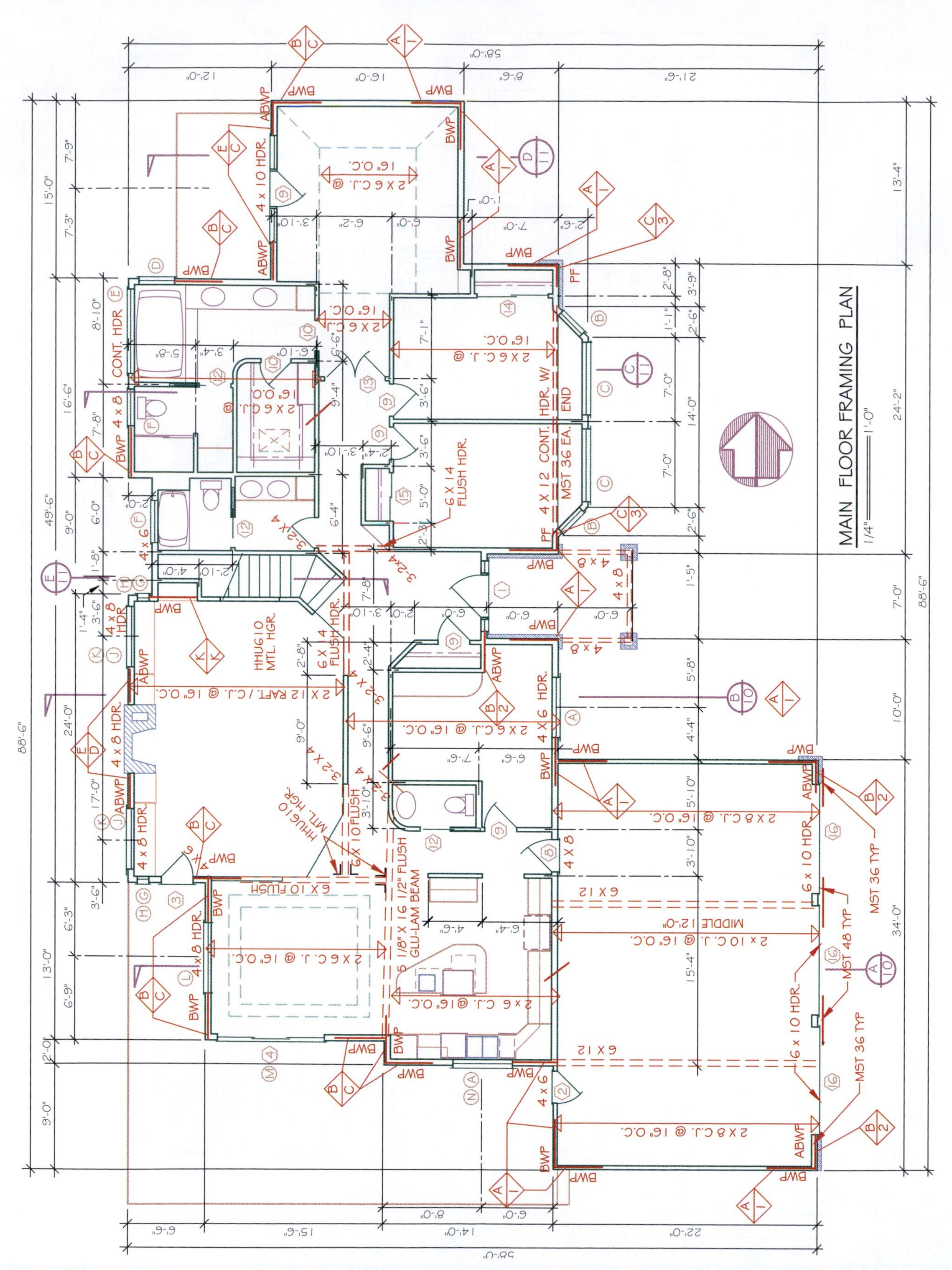

FIGURE 24.38 ■ *The completed main floor plan using truss construction for the structure started in Chapter 16. Lateral bracing is based on the prescriptive design method. Lateral supports are based on the prescriptive methods of the IRC using the table shown in Figure 24.41.*

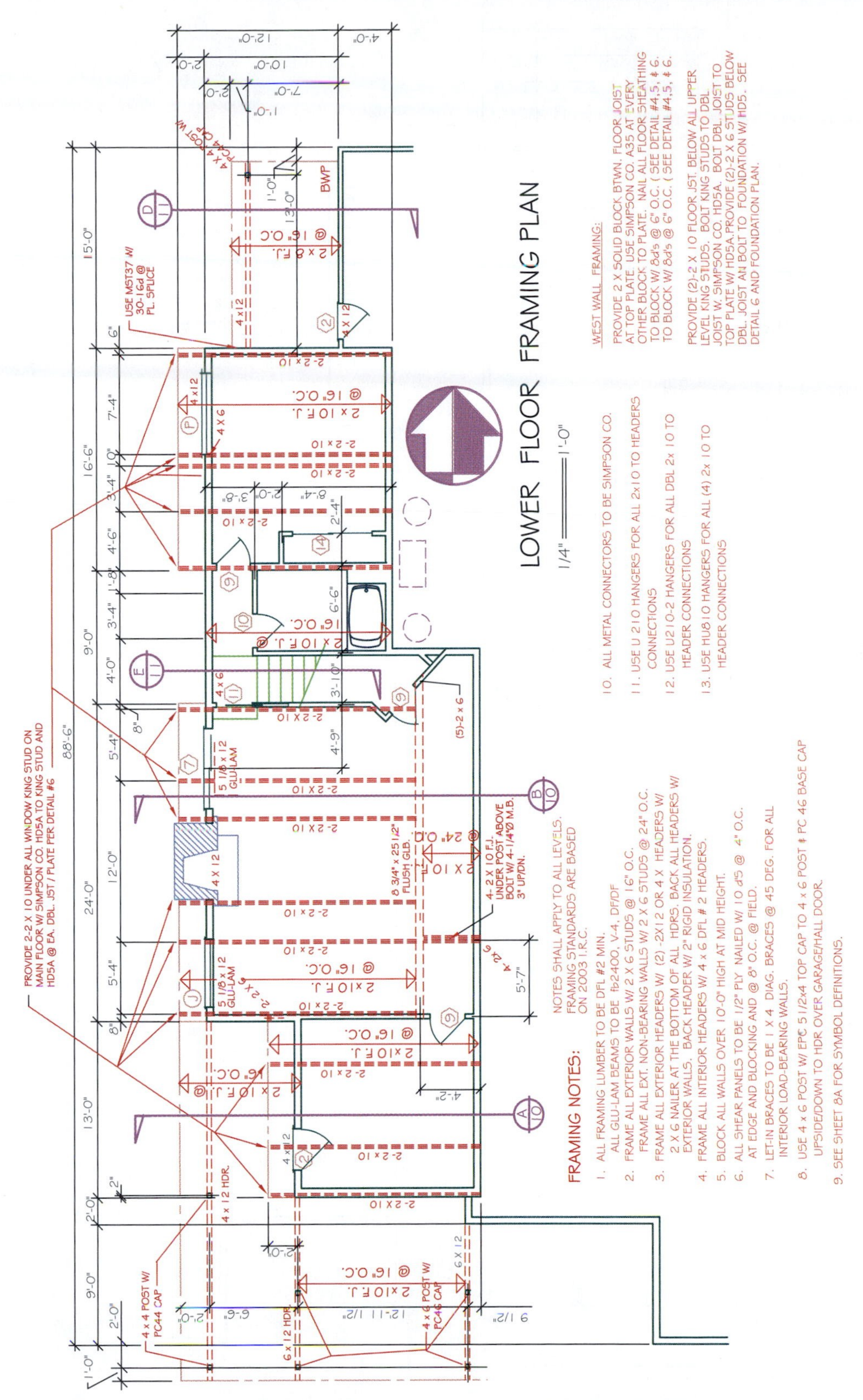

LOWER FLOOR FRAMING PLAN
1/4" = 1'-0"
FRAMING NOTES:
1. ALL FRAMING LUMBER TO BE DFL #2 MIN.
ALL GLU-LAM BEAMS TO BE fb2400, V-4, DF/DF
2. FRAME ALL EXTERIOR WALLS W/ 2 X 6 STUDS @ 16" O.C.
FRAME ALL EXT. NON-BEARING WALLS W/ 2 X 6 STUDS @ 24" O.C.
3. FRAME ALL EXTERIOR HEADERS W/ (2) -2X12 OR 4 X HEADERS W/ 2 X 6 NAILER AT THE BOTTOM OF ALL HDRS. BACK ALL HEADERS W/ EXTERIOR WALLS. BACK HEADER W/ 2" RIGID INSULATION.
4. FRAME ALL INTERIOR HEADERS W/ 4 x 6 DFL # 2 HEADERS.
5. BLOCK ALL WALLS OVER 10'-0" HIGH AT MID HEIGHT.
6. ALL SHEAR PANELS TO BE 1/2" PLY NAILED W/ 10 d'S @ 4" O.C. AT EDGE AND BLOCKING AND @ 8" O.C. @ FIELD.
7. LET-IN BRACES TO BE 1 X 4 DIAG. BRACES @ 45 DEG. FOR ALL INTERIOR LOAD-BEARING WALLS.
8. USE 4 x 6 POST W/ EPC 5 1/2x4 TOP CAP TO 4 x 6 POST & PC 46 BASE CAP UPSIDE/DOWN TO HDR OVER GARAGE/HALL DOOR.
9. SEE SHEET 8A FOR SYMBOL DEFINITIONS.
10. ALL METAL CONNECTORS TO BE SIMPSON CO.
11. USE U 210 HANGERS FOR ALL 2x10 TO HEADERS CONNECTIONS
12. USE U210-2 HANGERS FOR ALL DBL 2x 10 TO HEADER CONNECTIONS
13. USE HU810 HANGERS FOR ALL (4) 2x 10 TO HEADER CONNECTIONS
NOTES SHALL APPLY TO ALL LEVELS. FRAMING STANDARDS ARE BASED ON 2003 I.R.C.
WEST WALL FRAMING:
PROVIDE 2 X SOLID BLOCK BTWN. FLOOR JOIST AT TOP PLATE. USE SIMPSON CO. A35 AT EVERY OTHER BLOCK TO PLATE. NAIL ALL FLOOR SHEATHING TO BLOCK W/ 8d's @ 6" O.C. (SEE DETAIL #4,5, & 6. TO BLOCK W/ 8d's @ 6" O.C. (SEE DETAIL #4,5, & 6.
PROVIDE (2)-2 X 10 FLOOR JST. BELOW ALL UPPER LEVEL KING STUDS. BOLT KING STUDS TO DBL JOIST W. SIMPSON CO. HD5A. BOLT DBL. JOIST TO TOP PLATE W/ HD5A. PROVIDE (2)-2 X 6 STUDS BELOW DBL. JOIST AN BOLT TO FOUNDATION W/ HD5. SEE DETAIL 6 AND FOUNDATION PLAN.
PROVIDE 2-2 X 10 UNDER ALL WINDOW KING STUD ON MAIN FLOOR W/ SIMPSON CO. HD5A TO KING STUD AND HD5A @ EA. DBL. JST / PLATE PER DETAIL #6
USE MST37 W/ 30-16d @ PL. SPLICE
4-2 X 10 F.J. UNDER POST ABOVE BOLT W/ 4-1/4"Ø M.B. 3" UP/DN.
8 3/4" x 25 1/2" FLUSH GLB.
5 1/8 x 12 GLU-LAM
4 x 12 HDR.
6 x 12 HDR.
4 x 4 POST W/ PC44 CAP
4 x 6 POST W/ PC46 CAP
2 x 10 F.J. @ 16" O.C.
2-2 x 10
BWP
88'-6"

FIGURE 24.39 ■ *The completed lower framing plan for the home shown in Figure 24.37. Lateral bracing is based on the design-specific method.*

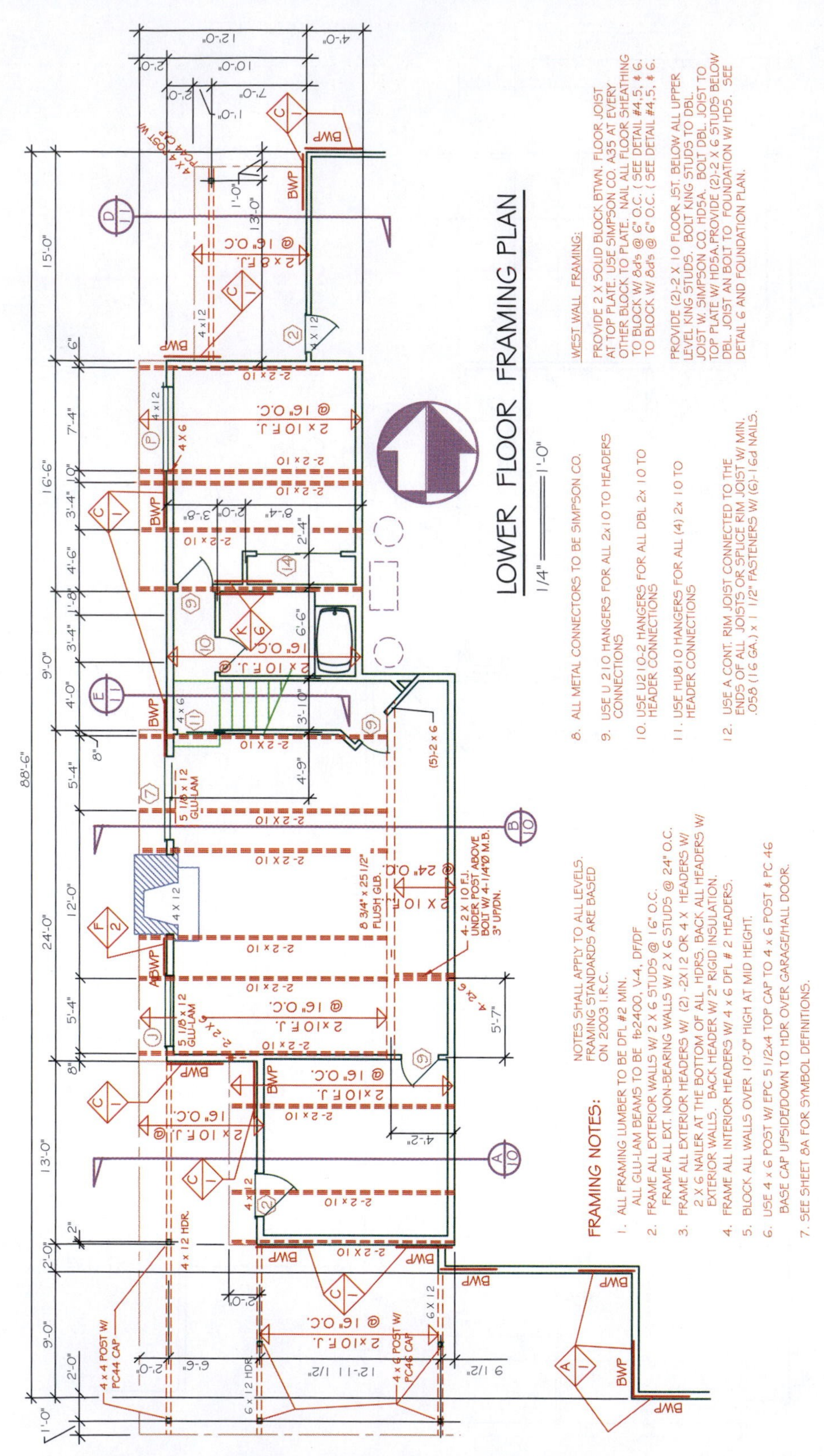
LOWER FLOOR FRAMING PLAN
1/4" = 1'-0"
FRAMING NOTES:
NOTES SHALL APPLY TO ALL LEVELS. FRAMING STANDARDS ARE BASED ON 2003 I.R.C.
1. ALL FRAMING LUMBER TO BE DFL #2 MIN. ALL GLU-LAM BEAMS TO BE fb2400, V-4, DF/DF
2. FRAME ALL EXTERIOR WALLS W/ 2 X 6 STUDS @ 16" O.C. FRAME ALL EXT. NON-BEARING WALLS W/ 2 X 6 STUDS @ 24" O.C.
3. FRAME ALL EXTERIOR HEADERS W/ (2) -2X12 OR 4 X HEADERS W/ 2 X 6 NAILER AT THE BOTTOM OF ALL HDRS, BACK ALL HEADERS W/ EXTERIOR WALLS. BACK HEADER W/ 2" RIGID INSULATION.
4. FRAME ALL INTERIOR HEADERS W/ 4 x 6 DFL # 2 HEADERS.
5. BLOCK ALL WALLS OVER 10'-0" HIGH AT MID HEIGHT.
6. USE 4 x 6 POST W/ EPC 5 1/2x4 TOP CAP TO 4 x 6 POST & PC 46 BASE CAP UPSIDE/DOWN TO HDR OVER GARAGE/HALL DOOR.
7. SEE SHEET 8A FOR SYMBOL DEFINITIONS.
8. ALL METAL CONNECTORS TO BE SIMPSON CO.
9. USE U 210 HANGERS FOR ALL 2x10 TO HEADERS CONNECTIONS
10. USE U210-2 HANGERS FOR ALL DBL 2x 10 TO HEADER CONNECTIONS
11. USE HU810 HANGERS FOR ALL (4) 2x 10 TO HEADER CONNECTIONS
12. USE A CONT. RIM JOIST CONNECTED TO THE ENDS OF ALL JOISTS OR SPLICE RIM JOIST W/ MIN. .058 (16 GA.) x 1 1/2" FASTENERS W/ (6)-16d NAILS.
WEST WALL FRAMING:
PROVIDE 2 X SOLID BLOCK BTWN. FLOOR JOIST AT TOP PLATE. USE SIMPSON CO. A35 AT EVERY OTHER BLOCK TO PLATE. NAIL ALL FLOOR SHEATHING TO BLOCK W/ 8d's @ 6" O.C. (SEE DETAIL #4.5, & 6. TO BLOCK W/ 8d's @ 6" O.C. (SEE DETAIL #4.5, & 6.
PROVIDE (2)-2 X 10 FLOOR JST. BELOW ALL UPPER LEVEL KING STUDS. BOLT KING STUDS TO DBL. JOIST W. SIMPSON CO. HD5A. BOLT DBL. JOIST TO TOP PLATE W/ HD5A.PROVIDE (2)-2 X 6 STUDS BELOW DBL. JOIST AN BOLT TO FOUNDATION W/ HD5. SEE DETAIL 6 AND FOUNDATION PLAN.

FIGURE 24.40 ■ *The completed lower framing plan for the upper level shown in Figure 24.38. Lateral bracing is based on the prescriptive design-specific method.*

BRACED WALL SCHEDULE

	MARK	WALL COVERING	EDGE NAILING	FIELD NAILING	PANEL SUPPORT	STRUCTURAL CONNECTORS
BWP 1 LEVEL	A	48" MIN. x 1/2" CD EXT. PLY 1 SIDE UNBLOCKED	8d @ 6" O.C	8d @ 12" O.C.	2 - 2 x 6 STUDS @ EA. END OF PANEL	PAHD42 TO CONC.
BWP UPPER	B	48" MIN. x 1/2" CD EXT. PLY 1 SIDE UNBLOCKED	8d @ 6" O.C	8d @ 12" O.C.	2 - 2 x 6 STUDS @ EA. END OF PANEL	MST37 UPPER TO LOWER POST
BWP LOWER	C	48" MIN. x 1/2" CD EXT. PLY 2 SIDES UNBLOCKED	8d @ 3" O.C	8d @ 12" O.C.	4 x 6 POST @ EA. END OF PANEL	HD5A TO CONC. THRU DBL SILL
ABWP 1 LEVEL	D	2'-8" MIN. x 1/2" CD EXT. PLY 1 SIDE UNBLOCKED	8d @ 6" O.C	8d @ 12" O.C.	2 - 2 x 6 STUDS @ EA. END OF PANEL	HPAHD22 TO CONC.
ABWP UPPER	E	2'-8" MIN. x 1/2" CD EXT. PLY 1 SIDE UNBLOCKED	8d @ 6" O.C	8d @ 12" O.C.	4 x 6 POST @ EA. END OF PANEL	MST37 UPPER TO LOWER POST
ABWP LOWER	F	2'-8" MIN. x 1/2" CD EXT. PLY 2 SIDES UNBLOCKED	8d @ 3" O.C	8d @ 12" O.C.	2 - 2 x 6 STUDS @ EA. END OF PANEL	HD5A TO CONC. W/ (2) 3/4" M. B. TO POST & STTB20 TO CONCRETE
PF 1 LEVEL	G	22 1/2" MIN. x 1/2" CD EXT. PLY 1 SIDE -BLOCKED EDGE	2 ROWS OF 8d @ 3" O.C	8d @ 12" O.C.	4 x 4 POST @ EA. END OF PANEL	1-MSTA18 STRAP EA. SIDE EA. END OF PANEL TO 4 x 12 MIN HDR. (4 STRAPS MIN EA. END OF FRAME).
P.F. UPPER	H	22 1/2" MIN. x 1/2" CD EXT. PLY 1 SIDE -BLOCKED EDGE	2 ROWS OF 8d @ 3" O.C	8d @ 12" O.C.	4 x 4 POST @ EA. END OF PANEL	HTT22 UPPER TO LOWER POST W/ (4) 1/2"M. B. EA POST @ 5/8" STUD BOLT THROUGH FLOOR.
P.F. LOWER	J	22 1/2" MIN. x 1/2" CD EXT. PLY 2 SIDES -BLOCKED EDGE	2 ROWS OF 8d @ 3" O.C	8d @ 12" O.C.	4 x 4 POST @ EA. END OF PANEL	HD8A TO CONC. W/ (3) 7/8" M. B. TO POST & STTB28 TO CONCRETE
INTERIOR	K	1 x 4 DIAG CONT LET IN BRACE LET-INTO TOP AND BOTTOM PLATE AND INTERVENING STUDS AT NOT MORE THAN 60 ° OR LESS THAN 45° FROM HORIZONTAL. ATTACH TO EACH MEMBER W/ (2) 8d'S.				

HOLD - DOWN SCHEDULE

	MARK	HOLD - DOWN	CONNECTIONS	FOUNDATION REINFORCING
ALL BWP	1	PAHD42 STRAP	(16) 16d OR (3) 1/2"Ø M.B.	PROVIDE MIN. OF (2) 1/2"Ø A. B.
1 STORY ABWP	2	HPAHD22 STRAP	(19) 16d OR (3) 1/2"Ø M.B.	PROVIDE (3) 1/2"Ø A.B. @ 1/5 POINTS OF SILL W/ 1 - #4 3" DN FROM TOP OF WALL & 1 - #4 3" UP FROM BOTTOM OF FOOTING. PROVIDE 1-#4 VERT @24" O.C. IN STEM WALL.
1 STORY P.F.	3	HTT22 STRAP & SSTB24 BOLT TO CONCRETE	(32) 16d SINKERS W/ 2" MAX OFFSET FROM POST FACE TO BOLT CENTERLINE	PROVIDE (3) 1/2"Ø A.B. @ MUDSILL W/ 2-2x6 SILL W/ 2 - #4 - 3" UP FROM BOTTOM OF 15" x 7" FTG. EXTEND STEEL 10'-0" AROUND FOOTING CORNER. PROVIDE 1 #4 VERT. @EA. SSTB24 CONNECTOR
2 STORY ABWP	4	HD5A	(2) 3/4" M. B. THRU POST	STTB20 W/ MIN. OF (3) A. B. THRU 2 - 2 x 6 SILL
2 STORY P.F.	5	HD8A	(4) 7/8" M. B. THRU POST	STTB28 W/ MIN. OF (3) A. B. THRU 2 - 2 x 6 SILL
	0	NONE REQUIRED	————	DBL JOIST OR BEAM BELOW. SEE FOUNDATION PLAN

FIGURE 24.41 ■ *The braced wall schedule required for the framing plans in Figures 24.38 and 24.40. This table can be found on the CD for insertion into your drawing. Edit the table to meet variations caused by local code requirements.*

basement can be shown. One option is to show a separate framing plan for just the basement, with the floor for the basement and the balance of the structure shown on the foundation plan. This plan would be drawn using the steps for the main framing plan. Because the plan is relatively simple, many professional offices would show the material for framing the main floor over the basement on the foundation plan. See Chapter 27 for the layout of the required foundation plan.

CHAPTER 24

Additional Readings

The following websites can be used as a resource to help you keep current with changes in framing materials.

ADDRESS	COMPANY OR ORGANIZATION
www.apawood.org	APA–The Engineered Wood Association
www.intlcode.org	International Code Council
www.strongtie.com	Simpson Strong-tie Company, Inc.

CHAPTER 24

Framing Plan Test

DIRECTIONS

Answer the following questions with short, complete statements. Type your answers using a word processor or neatly print your answers on lined paper.

1. Place your name, the chapter number, and the date at the top of the sheet.
2. Type the question number and provide the answer in the form of a statement that includes part of the question. You do not need to write out the entire question.

Note: Unless noted, base all answers on the prescriptive path of the IRC. The answers to some questions may not be contained in this chapter and will require you to do additional research.

QUESTIONS

24.1. What are the two major methods of determining lateral loads?

24.2. Describe options for where rafters should be specified on the working drawing.

24.3. How are the alternatives for representing trusses different from the alternatives for representing rafters?

24.4. What are the options for locating ceiling joist on the construction drawings?

24.5. You're looking at the upper-level framing plan of a two-story residence. What framing members would you expect to see specified?

24.6. What is the maximum basic wind speed allowed by your building code for prescriptive wall design?

24.7. An engineer has specified that a metal strap be used to tie a header to a top plate. How should the strap be represented, and what material should be specified on the framing plan?

24.8. What is the maximum wall height allowed if prescriptive method of wall reinforcement is used?

24.9. List the dimensional requirements for a portal frame.
Minimum width:
Minimum distance between panels:
Maximum distance between panels:

24.10. How is a single-level braced wall panel attached to the foundation?

24.11. What foundation reinforcement is required for a single-level portal frame?

24.12. What is the most common method of constructing a braced wall panel in your area?

24.13. What are possible reactions walls might have as wind is applied perpendicular to the wall?

24.14. What is the maximum distance between braced wall lines for a wind not exceeding 85 mph?

24.15. What nailing is required for interior wall braces using 1/2'' gypsum board?

24.16. What size, method, and spacing are required if two studs will be nailed together to form a post?

24.17. What size nails are required to attach the edge of 1/2'' plywood to trusses?

24.18. What is the maximum distance two bearing walls can be offset and still be considered to be aligned?

24.19. What is the maximum distance between two braced wall lines before a continuous footing is required to support the bracing?

24.20. Describe common methods of supporting interior braced wall panels that are perpendicular to the floor supports.

FRAMING PLAN LAYOUT PROBLEM

Use information from the floor plan, roof plan, and elevations to draw the framing plan for the residence that was started in Chapter 11. Use trusses and engineered floor joists if possible. If sawn joists or rafters must be used, consult the span tables in Chapter 23 to determine all sizes. Use the formulas from Chapter 23 to determine all beam sizes. Insert and edit appropriate framing notes from the CD to supplement your framing plan. Verify with your instructor the risk of wind and seismic damage that should be considered in the design. If required, insert and edit the appropriate lateral table for your framing plan, and place symbols to represent the appropriate wall bracing.

SECTION 7

Foundation Plans

25 CHAPTER

Foundation Systems

All structures are required to have a foundation to provide a base to distribute the weight of the structure onto the soil. The weight, or load, must be evenly distributed over enough soil to prevent the soil from becoming compressed. In addition to resisting gravity loads, the foundation must resist floods, winds, and earthquakes. Where flooding is a problem, the foundation system must be designed for the possibility that much of the supporting soil may be washed away. The foundation must also be designed to resist any debris that may be carried by floodwaters.

The forces of wind on a structure can cause severe problems for a foundation. The walls of a structure act as a large sail. If the structure is not properly anchored to the foundation, the walls can be ripped away by the wind. Wind tries to push a structure not only sideways but upward as well. Because the structure is securely fastened at each intersection, wind pressure is transferred into the foundation. Proper foundation design will resist this upward movement. Figure 25.1 shows an example of anchor bolts and straps used to resist uplift.

Depending on the risk of seismic damage, special design may be required for a foundation. Although earthquakes cause both vertical and horizontal movement, it is the horizontal movement that causes the most damage to structures. The foundation system must be designed so that it can move with the ground yet keep its basic shape. Steel reinforcing and welded wire mesh are often required to help resist or minimize damage due to the movement of the earth.

FIGURE 25.1 ■ *In addition to resisting the forces of gravity, a foundation must be able to resist the forces of uplift created by wind pressure acting on the structure. Anchor bolts and metal framing straps embedded in the concrete are two common methods of resisting the forces of nature.*

SOIL CONSIDERATIONS

In addition to the forces of nature, the nature of the soil supporting the foundation must also be considered. The texture of the soil and the tendency of the soil to freeze will influence the design of the foundation system.

Soil Texture

The texture of the soil will affect its ability to resist the load of the foundation. Before a foundation can be designed for a structure, the bearing capacity of the soil must be known. This is a design value specifying the amount of weight a square foot of soil can support. The bearing capacity of soil depends on its composition and the moisture content. The IRC provides five basic classifications for soil:

Crystalline bedrock	12,000 psf (574.8 kPa)
Sedimentary and foliated rock	4000 psf (191.6 kPa)
Sandy gravel and/ or gravel	3000 psf (143.7 kPa) (SG and GP)
Sand, silty sand, clayey sand	2000 psf (95.8 kPa) silty gravel, and clayey gravel (SW, SP, SM, SC, GM and GC)
Clay, sandy clay, silty clay	1500 psf (47.9 kPa) clayey silt, silt, sandy silt (CL, ML, MH and CH)

These listings are assumed bearing values. The soil bearing capacity for a specific site can often be determined from the building department. The allowable bearing capacity will need to be determined by a soils engineer when the building site is in an unincorporated area, or if the building department believes that soil with an allowable bearing capacity of less than 1500 psf is likely to be present. A soil bearing pressure of 2000 psf is the design value used for most stock home designs when the site conditions are not known.

Designing Foundations on Poor Soil

Structures built on soils with low bearing capacity require footings that extend into stable soil or are spread over a wide area. Both options typically require the design to be approved by an engineer. The type of soil can usually be determined from the local building department. For large custom homes a soils engineer is often required to study the various types of soil at the job site and make recommendations for foundation design. The soil bearing values must be determined before a suitable material for the foundation can be selected. In addition to the texture, the tendency of freezing must also be considered.

Cut and Fill Materials

Construction sites often include soil that has been brought to the site or moved on the site. Soil that is placed over the natural grade is called fill material (review Chapter 7). ***Fill material*** is often moved to a lot when an access road is placed in a sloping site, as in Figure 25.2. After a few years, vegetation covers the soil and gives it the appearance of natural grade. Footings resting on fill material will eventually settle under the weight of a structure. All discussion of foundation depths in this text refers to the footing depth into the natural grade.

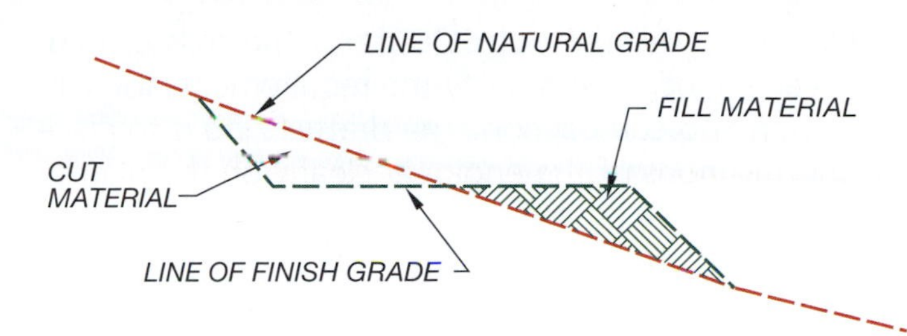

FIGURE 25.2 ■ *As access roads are created, soil is often cut away and pushed to the side of the roadway, creating areas of fill. Unless the foundation extends into the natural grade, the structure will settle.*

Compaction

Fill material can be compacted to increase its bearing capacity. Vibrating, tamping, rolling, or adding a temporary weight are the common methods for compacting soil. There are three major ways to compact soil.

1. Static force: A heavy roller presses soil particles together.
2. Impact forces: A ramming shoe strikes the ground repeatedly at high speed.
3. Vibration: High-frequency vibration is applied to soil through a steel plate.

The type of compaction to be preformed is often specified by the soils engineer, and must be placed on the foundation drawings by the CAD technician. Large job sites are usually compacted by mobile equipment. Small areas in the construction site are generally compacted by handheld mobile equipment. Granular soils are compacted by vibration, and soils containing large amounts of clay are best compacted by force. Each of these methods will reduce air voids between grains of soil. Proper compaction lessens the effect of settling and increases the stability of the soil, which increases the load-bearing capacity. The effects of frost damage are minimized in compacted soil because penetration of water into voids in the soil is minimized.

Moisture content is the most important factor in efficient soil compaction because moisture acts as a lubricant to help soil particles move closer together. Compaction should be completed under the supervision of a geotechnical engineer or another qualified expert who understands the measurements of soil moisture. Compaction is typically accomplished in lifts of from 6 through 12'' (150 through 300 mm). The soils or geotechnical engineer will specify the requirements for soils excavation and compaction. The drafter's job is to place the specifications of the engineer clearly on the plans.

Freezing

Don't confuse ground freezing with blizzards. Even in the warmer southern states, ground freezing can be a problem. Frost penetration depths can range from 12 to 80'' (300 to 2000 mm) for the United States. Exact design parameters must be determined with the local building department as the plans are being drawn. A foundation must be built to a depth where the ground is not subject to freezing. Water in the soil expands as it freezes and then contracts as it thaws. Expansion and shrinking of the soil will cause heaving in the foundation. As the soil expands, the foundation can crack. As the soil thaws, water that cannot be absorbed by the soil can cause the soil to lose much of its bearing capacity, causing further cracking of the foundation. In addition to geographic location, the type of soil also affects freezing. Fine-grained soil is more susceptible to freezing because it tends to hold moisture.

A foundation must rest on stable soil so that it does not crack. The designer will have to verify the required depth of foundations with the local building department. Once the soil's bearing capacity and the depth of freezing are known, the type of foundation system to be used can be determined.

Water Content

The amount of water the foundation will be exposed to, as well as the permeability of the soil, must also be considered in the design of the foundation. The soil expands as it absorbs water, causing the foundation to heave. In areas with extended periods of rainfall, there is little variation in the soil's moisture content; this minimizes the risk of heaving caused by soil expansion. Greater danger results in areas of the country that receive only minimal rainfall followed by extended periods without moisture. Soil shrinkage from these conditions can cause severe foundation problems because of the moisture differential. To aid in the design of the foundation, the IRC has included the Thornthwaite Moisture Index, which lists the amount of water that would be returned to the atmosphere by evaporation from the ground surface and transpiration if there were an unlimited supply of water to the plants and soil.

On-grade concrete slabs are used primarily in dry areas from southern California to Florida, where the contrast between the dry soil under the slab and the damp soil beside the foundation creates a risk of heaving at the edge of the slab. If the soil beneath the slab experiences a change of moisture content after the slab is poured, the center of the slab can heave. Heaving can be resisted by proper drainage and reinforcing placed in the foundation and throughout the floor slab. The effects of soil moisture are investigated in Chapter 27, which considers concrete slab construction.

Surface water and groundwater must be properly diverted from the foundation so that the soil can support the building load. Proper drainage also minimizes water leaks into the crawl space or basement, thus reducing mildew and rotting. The IRC requires the finish grade to slope away from the foundation at a minimum slope of 6'' (150 mm) within the first 10' (3000 mm). A 3% minimum slope is preferable for planted or grassy areas; a 1% slope is acceptable for paved areas.

Gravel or coarse-grained soils can be placed beside the foundation to increase percolation. In damp climates, a drain is often required beside the foundation at the base of a gravel bed to facilitate drainage (see Figure 25.3). As the amount of water in the soil surrounding the foundation is reduced, the lateral loads imposed on the foundation are also reduced. The pressure of the soil becomes increasingly important as the height of the foundation wall is increased. Foundation walls enclosing basements should be waterproofed, as in Figure 25.4. Asphaltic emulsion is often used to prevent water penetration into the basement. Floor slabs below grade are required to be placed over a vapor barrier.

Radon

Structures built in areas of the country with high radon levels need to provide protection from this cancer-causing gas. The IRC and EPA (Environmental Protection Agency) have mapped the continental United States by county and identified areas where there is a high risk of exposure to radon. The buildup of radon can be reduced by making minor modifications to the gravel placed below basement slabs. A 4'' (100 mm) PVC vent can be placed in a minimum layer of 4'' (100 mm) of gravel covered with 6-mil polyethylene. The plastic barrier should have a minimum lap of 12'' (300 mm) at intersections. Any joints, cracks, or penetrations in the floor slab must be caulked. The vent must run under the slab until it can be routed up through the framing system to an exhaust point, which is a minimum of 10' (3000 mm) from other openings in the structure and 12'' (300 mm) above the roof. The system should include rough-in electrical wiring for future in-

FIGURE 25.3 ■ *A gravel bed and a drain divert water away from the foundation.* Courtesy Boccia, Inc.

FIGURE 25.4 ■ *A waterproof emulsion will reduce the risk of water penetrating the concrete wall.* Courtesy Boccia, Inc.

stallation of a fan located in the vent stack and a system-failure warning device. This can usually be accomplished by placing an electrical junction box in an attic space for future installation of a fan.

TYPES OF FOUNDATIONS

The foundation is usually constructed of pilings, continuous footings, or grade beams.

Pilings

A piling foundation system uses beams placed between vertical supports, called pilings, to support structural loads. The columns may extend into the natural grade or may be supported on other material that extends into stable soil. Piling foundations are typically used:

- On steep hillside sites where it may not be feasible to use traditional excavating equipment.
- Where the load imposed by the structure exceeds the bearing capacity of the soil.
- On sites subject to flooding or other natural forces that might cause large amounts of soil to be removed.

Coastal property and sites near inland bodies of water subject to flooding often use a piling foundation to keep the habitable space of the structure above the flood plain level. Typically, a support beam is placed under or near each bearing wall. Beams are supported on a grid of vertical supports, which extend down to a more stable stratum of rock or dense soil. Beams can be steel, sawn or laminated wood, or prestressed concrete. Vertical supports may be concrete columns, steel tubes or beams, wood columns, or a combination of these materials.

Piling Construction

On shallow pilings a hole can be bored, and poured concrete with steel reinforcing can be used. Figure 25.5 shows a detail of a shallow poured-concrete piling. If the vertical support is required to extend deeper than 10' (3000 mm), a pressure-treated wood timber or steel column can be driven into the soil. Figure 25.6 shows a steel girder resting on a steel piling that will be used to

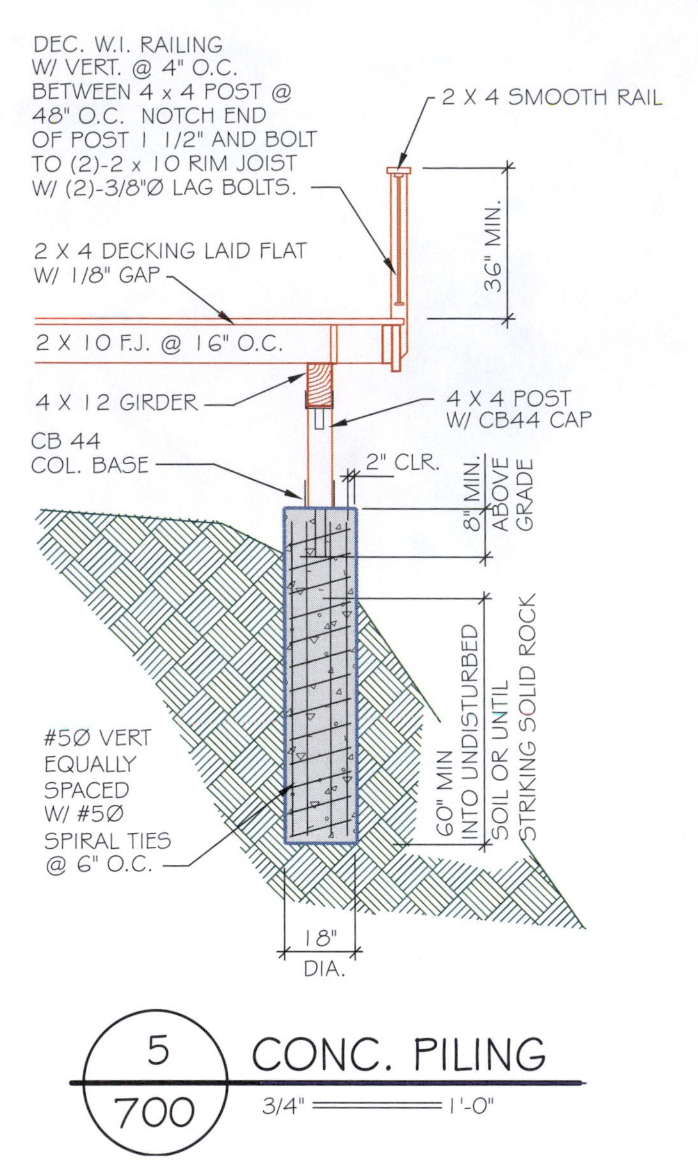

FIGURE 25.5 ■ *A shallow piling made of concrete is used to transfer loads into stable soil. A detail will be required to explain building and steel reinforcement methods.* Courtesy Residential Designs.

FIGURE 25.6 ■ *If a site is too steep for shallow pilings to be dug or drilled, a steel piling can be driven until it reaches stable ground. For this hillside home, pilings averaged a depth of 30' (9000 mm) before striking solid rock.*

support the floor system of a hillside home. Figure 25.7 shows the components of a piling plan for the residence. In addition to the vertical columns supported by the pilings, diagonal steel cables or rods are placed between the columns to resist lateral and rotational forces. An engineer must design the piling foundation and the connection of the pilings to the superstructure. In addition to resisting gravity loads, a piling foundation must be able to resist forces from uplift, lateral force, and rotation. Figure 25.8 shows a detail of a concrete piling used to support steel columns, which in turn support the floor system of the residence; it also shows a detail of a steel piling used to support steel columns above grade. Notice that the engineer has specified a system that uses braces approximately parallel to the ground (see Figure 25.6) to stabilize the tops of the pilings against lateral loads. In addition to showing these braces on the foundation plan, they are also shown and specified on the elevations, sections, and details.

Continuous or Spread Foundations

The most typical type of foundation used in residential construction is a continuous or spread foundation. The two major components of a spread footing are the

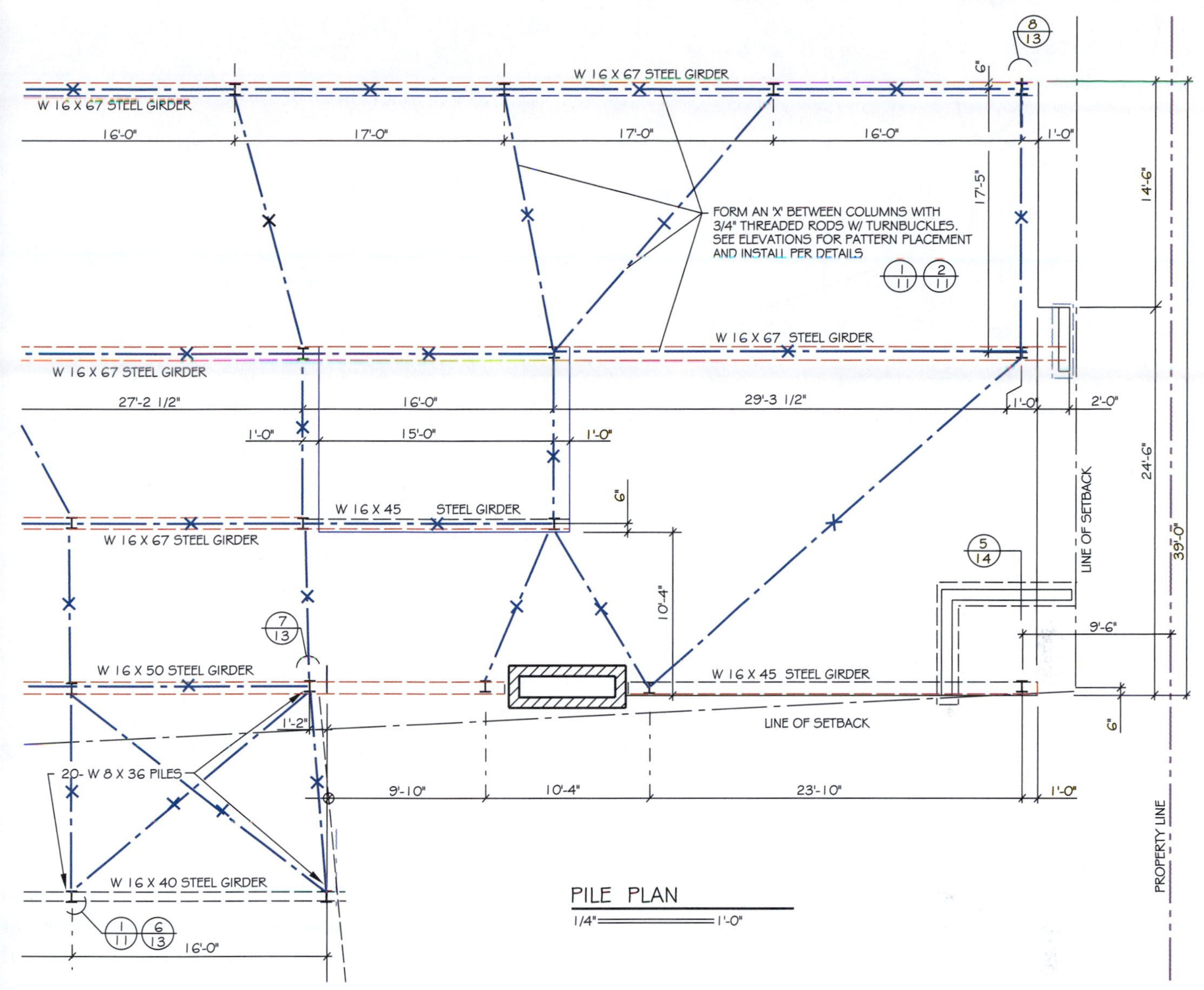

FIGURE 25.7 ■ *The foundation plan for a hillside home supported on pilings shows the location of all pilings, the beams that span between the pilings, and any diagonal bracing required to resist lateral loads.* Courtesy Residential Designs.

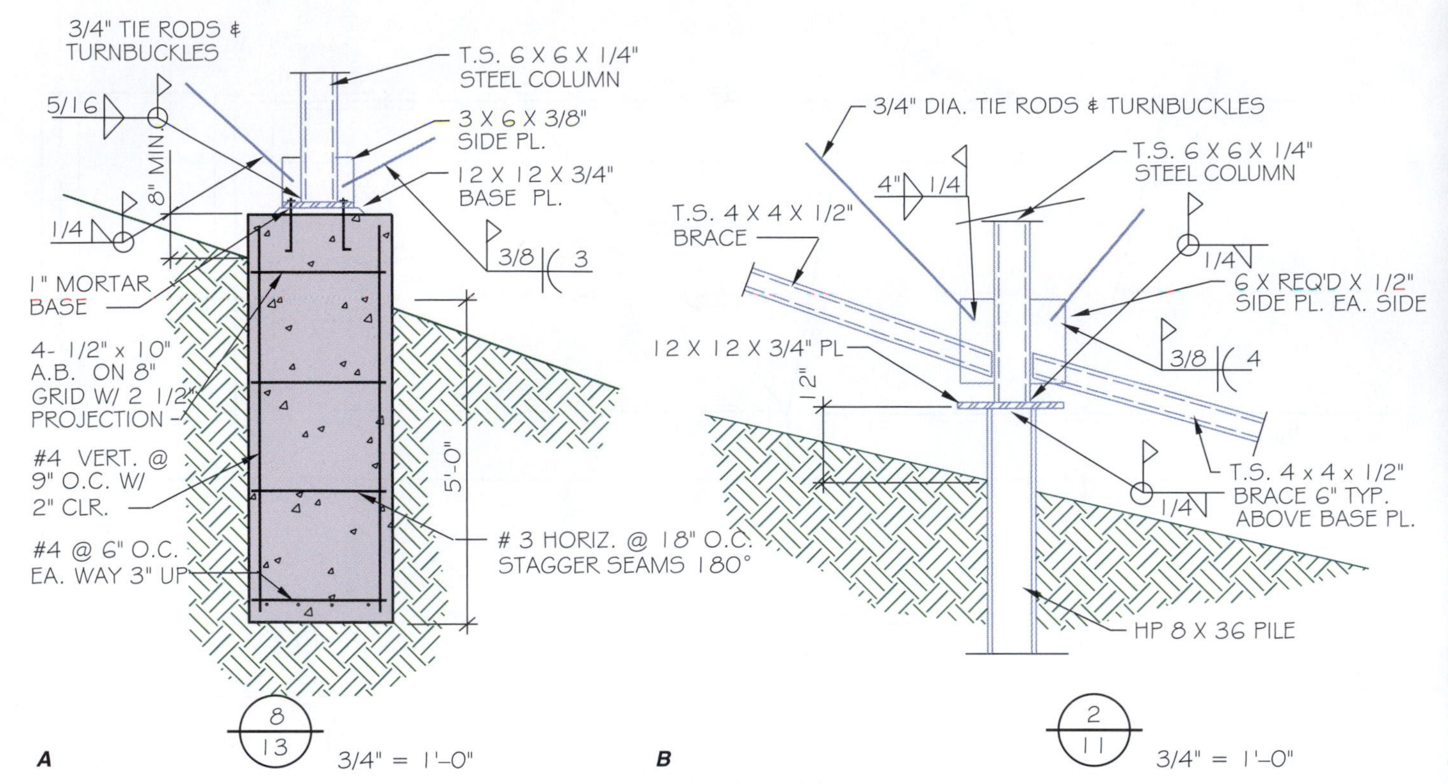

FIGURE 25.8 ■ *Details of the pilings are required to supplement the foundation plan to ensure that all of the engineer's specifications can be clearly understood. (A) The concrete piling on the left supports a steel column and (B) the piling on the right supports a steel column that extends up to the superstructure of the residence. Because of the loads to be supported, the depth of the pilings, and a severe risk of seismic damage, the engineer has specified horizontal steel-tube bracing to the top of the pilings. See Figure 25.6.* Courtesy Residential Designs.

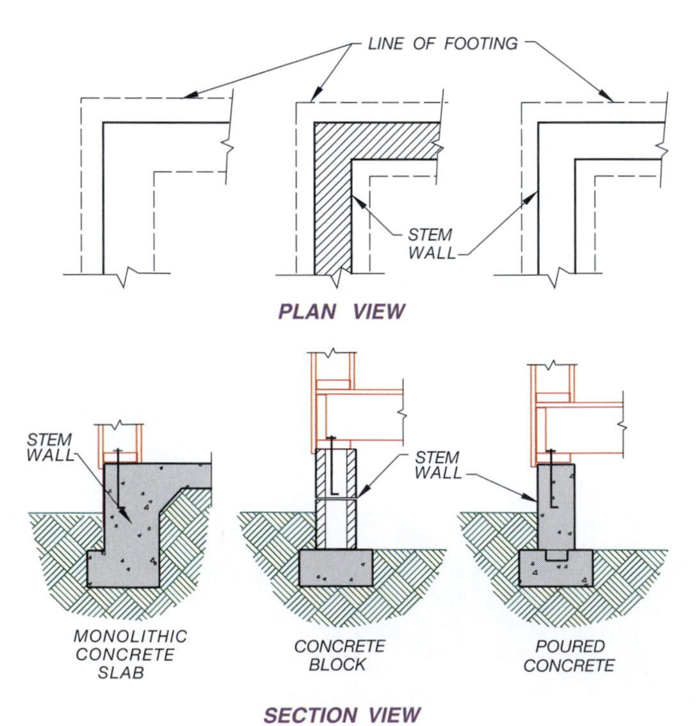

FIGURE 25.9 ■ *A footing, used to spread building loads evenly into the soil, is represented on the foundation plan by hidden lines.*

footing and stem wall. Other components of the foundation include anchor bolts, termite protection, beam supports, crawl space ventilation and access, foundation insulation, interior supports, reinforcement, and metal connectors.

Footings

The footing is the base of the foundation system and is used to displace the building loads over the soil. Figure 25.9 shows typical footings and how they are usually drawn on foundation plans. Footings are made of poured concrete and placed so that they extend below the freezing level. Fully grouted masonry and wood foundations are also allowed by the IRC. The size of the footing is based on the soil's bearing value and the load to be supported. Figure 25.10a and b shows common footing sizes and depths required by the IRC. The strength of the concrete must also be specified; this is based on the location of the concrete in the structure and its chance of freezing. Figure 25.11 lists IRC minimum values.

MINIMUM WIDTH OF CONCRETE OR MASONRY FOOTINGS (INCHES) (A)				
LOAD-BEARING VALUE OF SOIL (PSF)				
	1,500	2,000	3,000	≥4,000
Conventional wood-frame construction				
1-story	12	12	12	12
2-story	15	12	12	12
3-story	23	17	12	12
4-inch brick veneer over wood frame or 8-inch hollow concrete masonry				
1-story	12	12	12	12
2-story	21	16	12	12
3-story	32	24	16	12
8-inch solid or fully grouted masonry				
1-story	16	12	12	12
2-story	29	21	14	12
3-story	42	32	21	16

For SI: 1 inch = 25.4 mm, 1 pound per square foot = 47.88 Pa.

MINIMUM WIDTH OF STEM WALL (B)	
Plain Concrete	Minimum Width
Walls less than 4′–6″ (1372 mm)	6″ (152 mm)
Walls greater than 4′–6″ (1372 mm)	7.5″ (191 mm)
Plain Masonry	
Solid grout or solid units	6″ (152 mm)
Non grouted units	8″ (203 mm)

MINIMUM FOOTING DEPTH (C)*	
Minimum 1-story	6″ (152 mm)
Minimum 2-story	7″ (178 mm)
Minimum 3-story	8″ (203 mm)

*Values vary based on the type of soil. Verify local requirements with your building department.

MINIMUM FOOTING DEPTH INTO NATURAL GRADE (D)*	
Minimum code value	12″ (305 mm)
Recommended 2-story	18″ (457 mm)
Recommended 3-story	24″ (610 mm)

*Values vary based on the frost depth and the type of soil. Verify local requirements with your building department.

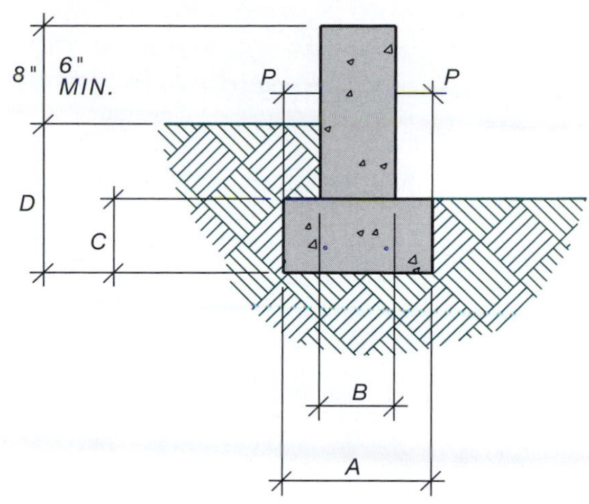

FIGURE 25.10 ■ *The IRC minimum footing size requirements shown in the table on the left vary based on the strength of the soil. Footing projections (P) must be a minimum of 2'' (50 mm) but cannot exceed the depth of the footing. Although the IRC requires the stem wall to extend 6'' (150 mm) above the finished grade, many municipalities require an 8'' (200 mm) minimum projection. Table reproduced from the International Residential Code/2006. Copyright © 2006.* Courtesy International Code Council, Inc.

Veneer Footings

If masonry or stone veneer is used, the footing must be wide enough to provide adequate support for the veneer. The footing is usually 4'' (100 mm) wider than a standard footing, but the exact size will depend on the type of veneer to be supported. Figure 25.12 shows common methods of providing footing support for veneer. The footing size does not need to be altered if cultured stone is to be applied to the exterior wall.

Fireplace Footings

Additional support must be provided if the footing is to support a masonry fireplace. The IRC requires the footing to be a minimum of 12'' (300 mm) deep and to extend 6'' (150 mm) past the face of the fireplace on each side. The footing is not required to extend under the hearth, it is used only to provide support for the chimney. Figure 25.13 shows how fireplace footings are shown on the foundation plan.

TYPE OR LOCATION OF CONCRETE	MINIMUM COMPREHENSIVE STRENGTH WEATHER POTENTIAL		
	NEGLIGIBLE	MODERATE	SEVERE
Basement walls and foundations not exposed to weather	2500	2500	2500
Basement slabs and interior slabs on grade, except garage floor slabs	2500	2500	2500
Basement walls, foundation walls, exterior walls, and other vertical concrete work exposed to weather	2500	3000	3000
Porches, concrete slabs and steps exposed to the weather, and garage floor slabs	2500	3000	3500

FIGURE 25.11 ■ *Compressive strength of concrete. Reproduced from the International Residential Code / 2006. Copyright © 2006.* Courtesy International Code Council, Inc.

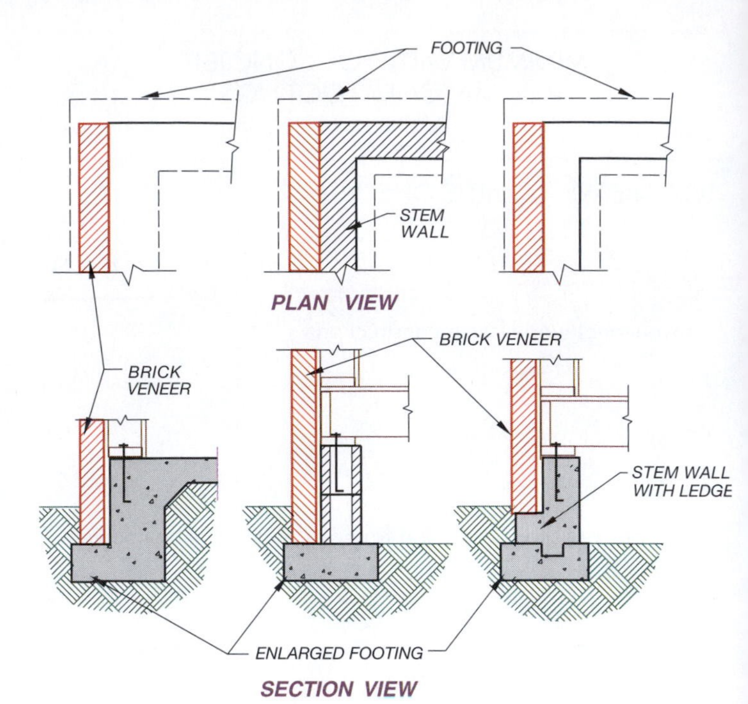

FIGURE 25.12 ■ *When a masonry veneer is added to a wall, additional footing width is needed to provide support.*

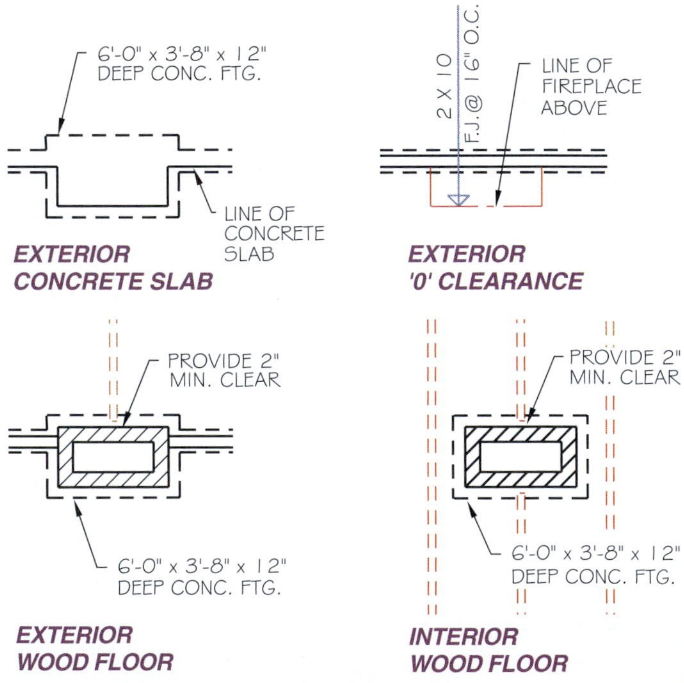

FIGURE 25.13 ■ *A masonry fireplace is required to have a 12'' (300 mm)-deep footing that extends 6'' (150 mm) past the face of the chimney. The footing does not need to extend under the hearth projection. A wood stove, a zero-clearance fireplace, or a gas fireplace is not required to have a footing, but the outline of the unit should be represented on the foundation plan if it extends beyond the foundation.*

Foundation Wall

The foundation wall is the vertical wall extending from the top of the footing up to the first-floor level of the structure, as shown in Figures 25.9, 25.10b, and 25.12. The foundation wall is usually centered on the footing to help spread the loads being supported. The height of the wall must extend 6'' (150 mm) above the ground to provide separation between wood supported by the concrete and the soil. Many municipalities require an 8'' (200 mm) minimum distance between wood and the finished grade. The height can be reduced to 4'' (100 mm) if masonry veneer is used. A stem wall width of 6'' (150 mm) is standard for plain concrete and 8'' (200 mm) for plain masonry walls. The required width of the wall varies depending on the wall height and the type of soil. Figure 25.10 shows common wall dimensions. Common alternatives to the

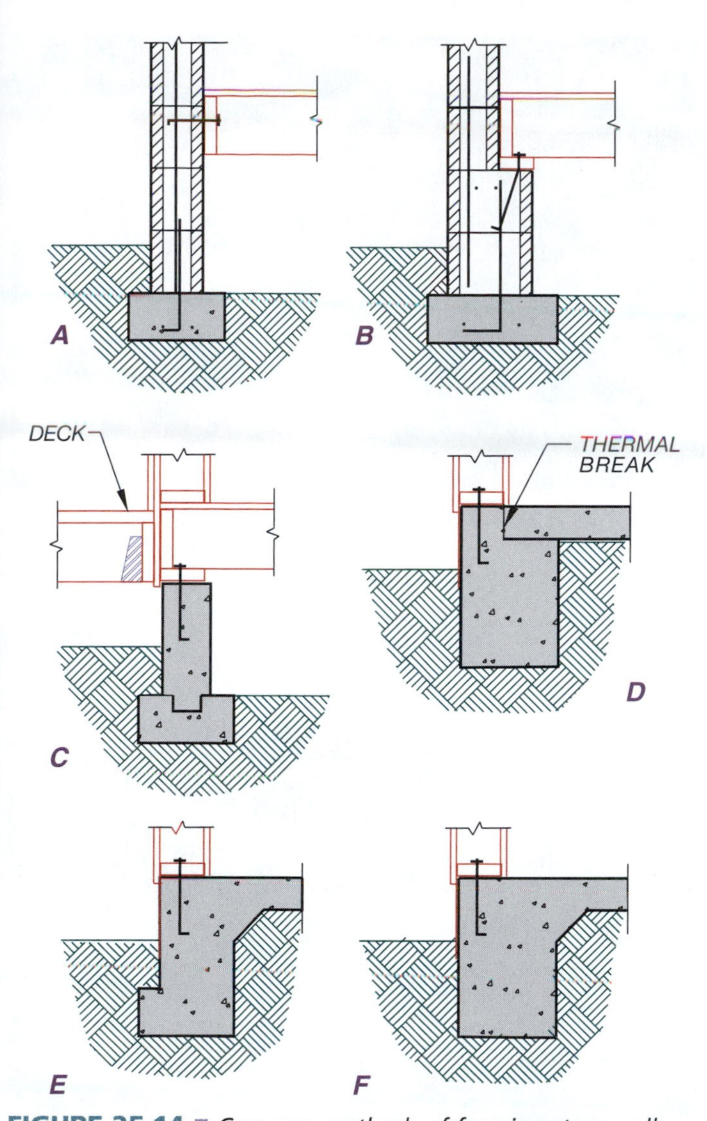

FIGURE 25.14 ■ *Common methods of forming stem walls. (A) Concrete masonry units with a pressure-treated ledger to support wood floor joists. (B) Floor joists supported on a pressure-treated sill with concrete masonry units. (C) Floor joists supported by pressure-treated sill with joists supporting a wood deck hung from a ledger. (D) Concrete floor slab supported on foundation with an isolation joint between the stem wall and slab to help the slab maintain heat. (E) Stem wall with projected footing. (F) Stem wall and foundation of equal width. Although more concrete is used, it can be formed quickly, saving time and material.*

minimum code requirements based on local standards include:

- 8'' (200 mm) wide stem walls supporting two-story construction.
- 10'' (250 mm) wide stem walls supporting three-story construction.

FIGURE 25.15 ■ *Once the concrete is poured, anchor bolts are placed in preparation for placement of the wood sill.* Courtesy Lisa Echols.

FIGURE 25.16 ■ *Once the concrete has cured, the forms can be removed. A strip of insulating foam has been placed in preparation for placing the mudsill.* Courtesy Lisa Mills.

Verify the minimum required stem wall sizes for your area prior to drawing a foundation plan.

Common methods of forming the footing and stem wall are shown in Figure 25.14. Figure 25.15 shows the forms for a poured concrete stem wall. Figure 25.16 shows the completed stem wall. Figure 25.17a, b, and c shows the process of forming a wall with concrete blocks.

In addition to concrete block and poured concrete foundation walls, blocks made of expanded polystyrene foam (EPS) or other lightweight materials can

FIGURE 25.17a ▪ *Forms are set and leveled in preparation for pouring the footings.* Courtesy Dr. Carol Ventura.

FIGURE 25.17c ▪ *The walls of a residence made of concrete blocks. The upper floor can either be supported directly on top of the wall or hung from the side of the wall using a ledger and metal hangers, as in Figure 25.14a.*

FIGURE 25.17b ▪ *The concrete footing with reinforcing steel set to extend into the stem wall. Steel for the wall will be attached to the exposed steel to securely tie the wall to the footing.* Courtesy Dr. Carol Ventura.

FIGURE 25.18a ▪ *Blocks made of expanded polystyrene foam provide both a permanent form for pouring the stem wall and insulation to prevent heat loss.* Courtesy Reward Wall Systems.

be stacked into the desired position and fitted together with interlocking teeth. EPS block forms can be assembled in a much shorter time than traditional form work and remain in place to become part of the finished wall. Reinforcing steel can be set inside the block forms in patterns similar to traditional block walls. Once the forms are assembled, concrete can be pumped into the forms using any of the common methods of pouring. The finished wall has an R-value between R-22 and R-35, depending on the manufacturer. Figure 25.18 shows a foundation being built using EPS forms.

Foundation Wall Construction

When the building site is not level, the foundation wall is often stepped. This helps reduce the material needed to build the foundation wall. As the ground slopes downward, the height of the wall is increased, as

FIGURE 25.18b ■ *EPF block walls are reinforced using the same methods used with CMU block.* Courtesy American Polysteel Forms.

FIGURE 25.19a ■ *The footing and foundation wall that will support the floor system for this residential addition have been stepped to save material.* Courtesy Tom Worcester.

FIGURE 25.19b ■ *As the stem wall steps to match the grade, short studs called jack studs are used to span between the stem wall and the floor system. Jack studs must be a minimum of 14'' (350 mm) in length or the wall must be formed of solid wood.* Courtesy Tom Worcester.

shown in Figure 25.19. Guidelines for stepped wall construction include:

- The foundation walls may not step more than 24'' (600 mm) in one step.
- Each step must be a minimum of 32'' (800 mm) long.
- Wood framing between the wall and any floor being supported may not be less than 14'' (350 mm) in height.

Lines representing steps in the wall should be drawn for custom homes if a topography plan is available. Steps in the foundation wall are not represented for stock plans because grade contours are generally not known. Wall steps will be placed by the concrete crew as the foundation forms are set. The foundation wall will also change heights for a sunken floor. Steps required in the foundation wall based on floor level changes must be represented on the foundation. Figure 25.20 shows how a sunken floor can be represented on a foundation plan.

Anchor Bolts

Steel anchor bolts are placed in the top of the stem wall to secure the wood mudsill to the concrete. The mudsill is required to be a 2 × 4 (50 × 100) and be made from pressure-treated or some other water-resistant wood so that it will not absorb moisture from

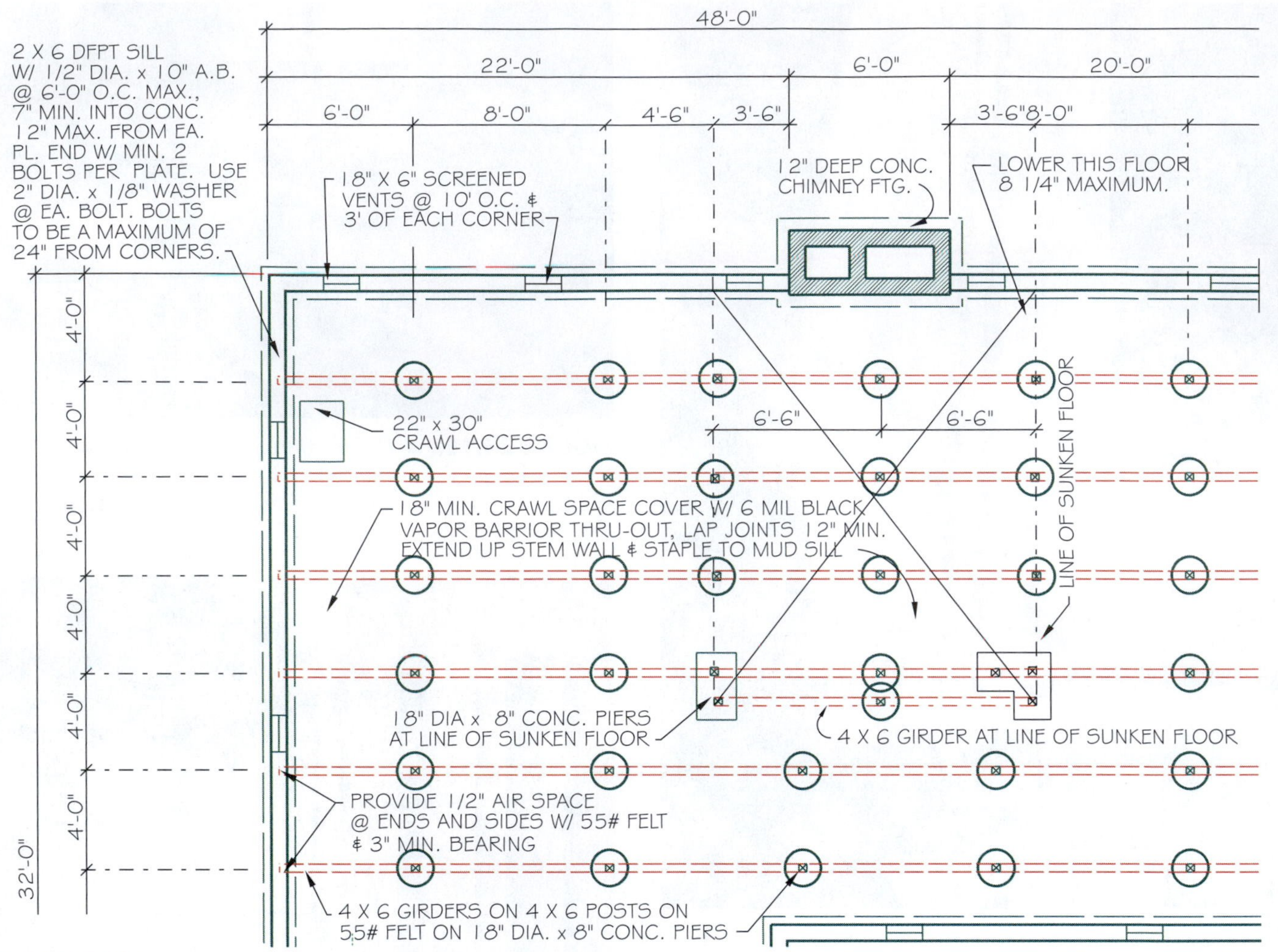

FIGURE 25.20 ■ *Common components of a foundation plan are vents, crawl access, fireplace footings, sunken floors, beam pockets, stem walls, footings, and piers.*

the concrete. Anchor bolts from the concrete extend through the mudsill. If concrete blocks are used, the cell containing the bolt must be filled with grout. An anchor bolt is required to be within 12'' (300 mm) but not less than seven bolt diameters from each end of each plate section. A 2'' (50 mm) round washer is placed over the bolt before the nut is installed to increase the holding power of the bolts. Other requirements for anchor bolts include:

PLACEMENT OF ANCHOR BOLTS	
Min. diameter	1/2'' Ø (13 mm)
Min. depth into concrete or masonry	7'' (175 mm)
Max. spacing (1 story)	6'-0'' (1800 mm)
Max. spacing (2 stories) Zone D1 and D2	4'-0'' (1200 mm)

The spacing of anchor bolts is reduced for retaining walls to approximately 24'' (600 mm). The exact spacing will vary based on the height of the wall, and the soil bearing capacity. The mudsill and anchor bolts are not drawn on the foundation plan but are specified with a note, as in Figure 25.20. Figure 25.21 shows the use of anchor bolts and the pressure-treated sill.

Termite Protection

In many parts of the country the mudsill must be protected from termites. Among the most common methods of protection are metal caps between the mudsill and the wall, chemical treatment of the soil around the foundation, and chemically treated wood near the ground. The metal shield is not drawn on the foundation plan but is specified in a note near the foundation plan or in a foundation detail.

FIGURE 25.21 ■ *Anchor bolts are used to attach the mudsill to the top of the stem wall. The tall bolt on the left end of the sill will be used to attach a metal connector to the floor system to provide additional resistance to uplift.* Courtesy Benigno Molina-Manriquez.

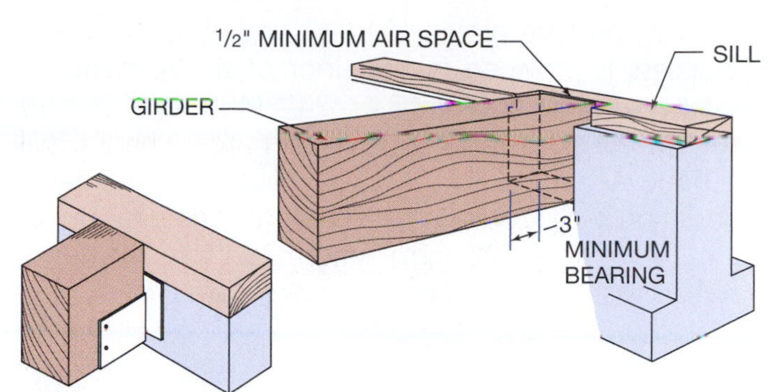

FIGURE 25.22 ■ *A beam seat, or pocket, can be recessed into the foundation wall, or the beam may be supported by metal connectors.* Courtesy Simpson Strong-Tie Company, Inc.

FIGURE 25.23 ■ *The beam seat must provide a minimum of 3'' (75 mm) of bearing surface for the girder and must allow a minimum space of 1/2'' (13 mm) for air flow. The girder must be wrapped with 55# felt to resist rotting. Notice that the girder is set flush with the top of the mudsill, and the use of anchor bolts and the foundation vent that has been placed in the foundation wall to provide for ventilation.*

Wood Floor Support

If the house is to have a wood floor system, some method of securing the girders to the foundation must be provided. Typically a cavity or beam pocket is built into the foundation wall. The cavity provides a method of supporting the beam and helps tie the floor and foundation system together. A 3'' (75 mm) minimum bearing surface must be provided for the beam where it rests on concrete. A 1/2'' (13 mm) airspace must be provided around the beam in the pocket for air circulation. A second common method of beam support is to use a metal beam hanger. The top of the beam hanger extends over the mudsill and supports a U-shaped bracket that supports the girder. Figure 25.22 shows common methods of beam support at the foundation wall.

Crawl Space Ventilation and Access

If the foundation will support a wood floor system, air must be provided to circulate under the floor system. The IRC requires that vents be provided to supply 1 sq ft of ventilation for each 150 sq ft (0.67 m^2/100 m^2) of crawl space. To supply ventilation under the floor, vents must be set into the foundation wall, as shown in Figure 25.23. In addition to minimum size requirements, one vent must be provided within 3'-0'' (900 mm) of each corner to provide air current throughout the crawl space.

Note:

The requirement for placing corner vents often conflict with the requirements for placing lateral bracing. Place the corner vents as close as possible, but so that they do not interfere with lateral bracing panels.

Access must be provided to all underfloor spaces. If the access is provided in the floor of the residence, it must be a minimum of 18 × 24'' (450 × 600 mm). Place the access in a closet or in an area where it will not be in a traffic path. If it is provided in an exterior foundation wall, the access opening is required to be 16 × 24'' (400 × 600 mm). Figure 25.20 shows how a crawl access, vents, and girder pockets are typically represented on a foundation plan.

Foundation Insulation

When required by the IRC or if the foundation is for an energy-efficient structure, insulation is added to the wall. Two-inch (50 mm) rigid insulation is used to protect the wall from cold weather, as with the wall shown in Figure 25.24. This can be placed on either side of the wall. If the insulation is placed on the exterior side of the wall, the wall will retain heat from the building, but the insulation must be protected from damage. Figure 25.24 shows exterior insulation placement.

Interior Supports

Foundation walls and footings support the exterior walls of the structure. Interior loads are supported on spot footings or piers, as seen in Figure 25.25. Piers are formed by using either preformed or framed forms or by excavation. Pier depth is generally required to match that of the footings. Although Figure 25.25 shows the piers above the finished grade, many states require the mass of the footing to be placed below grade to prevent problems with freezing and shifting from seismic activity. The placement of piers will be determined by the type of floor system to be used and is discussed in Chapter 26. The size of the pier will depend on the load being supported and the soil's bearing pressure. Piers are usually drawn on the foundation plan with dashed lines, as shown in Figure 25.26.

Support for Lateral Loads

In addition to the exterior footings and interior piers, footings may also be required under braced walls. Braced walls and braced wall lines were introduced in Chapter 24. On the framing plan, when the distance between braced wall lines exceed 35'-0'' (10 500 mm) an interior braced wall line must be provided. A beam or double joist is required to support the interior braced wall line. At the foundation level, when the distance between the stem walls supporting braced wall lines exceeds 50'-0'' (15 000 mm), a continuous footing must be provided below the braced wall. A continuous foot-

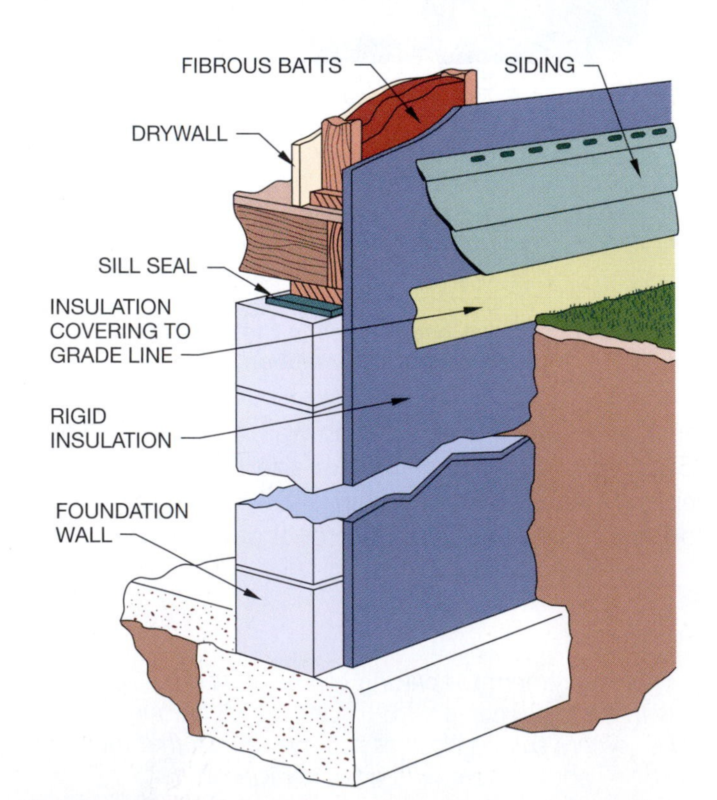

FIGURE 25.24 ■ *Insulation is often placed on the foundation wall to help cut heat loss.*

FIGURE 25.25 ■ *Foundation walls and footings support the exterior walls of the structure. Interior loads are supported on piers formed by using either preformed or framed forms or by excavation. Pier depth is generally required to match that of the footings. Notice the continuous footing extending across the crawl space that will be used to support loads from multilevel floor systems.* Courtesy Benigno Molina-Manriquez.

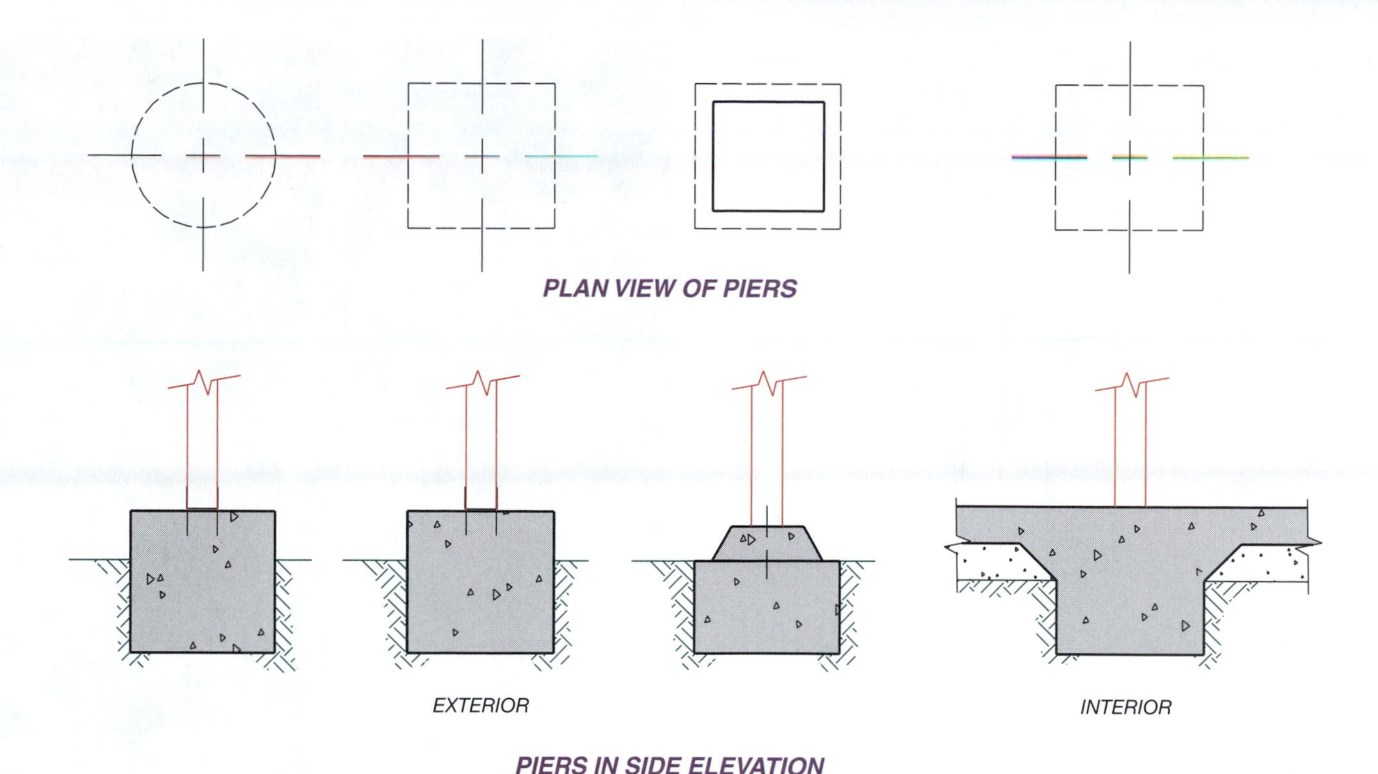

FIGURE 25.26 ■ *Concrete piers are used to support interior loads. Codes require wood to be at least 6'' (150 mm) above grade, which means that piers must also extend 6'' (150 mm) above grade.*

ing must be provided below all multilevel braced walls in seismic zones D1 and D2.

Resisting Lateral Loads at a Garage Opening

A final consideration of foundation walls and footings is the placement of each by a garage door. For small openings such as a 36'' (900 mm) door, the footing is continued under the door opening. The stem wall is omitted at the door to allow access to the garage without having to step over the wall each time the garage is entered. For large openings, such as the main garage door, in areas of the country with low seismic risk, the footing is not continuous across the door opening. In areas at risk of seismic activity, the footing must extend across the entire width of the door. The continuous footing helps the walls on each side of the door to act as one wall unit. The stem wall is cut to allow the concrete floor to cover the stem wall, providing a smooth entry into the garage.

Grade Beams

To provide added support for a foundation in unstable soil, a grade beam may be used in place of the foundation, as in Figure 25.27. The grade beam is similar to a wood beam that supports loads over a window. The grade beam is placed under the soil below the stem wall and spans between stable supports. The support may be stable soil or pilings. The depth and reinforcing required for a grade beam are determined by the load to be supported and are sized by an architect or engineer. A grade beam resembles a footing when drawn on a foundation plan. Steel reinforcing may be specified by notes and referenced to details rather than on the foundation plan.

Foundation Reinforcement

Steel ***rebar*** was first introduced in Chapter 20 as masonry wall reinforcement was explored. For areas of soft soil or fill material, reinforcement steel is placed in the footing. Concrete is extremely durable when it supports a load and is compressed, but it is very weak under tension. If the footing is resting on fill material, the bottom of the concrete footing will bend. As the footing bends, the concrete will be in tension. When required, steel is generally placed 3'' (75 mm) up from the bottom of the footing, to resist the forces of tension in concrete. Figure 25.28 shows the forms set for a reinforced footing. This reinforcing steel, or rebar, is not shown on the foundation plan but is specified in a note giving the size, grade, and spacing of the steel. Common grades associated with residential construction are grades 40, 50, 60, and 75. The material used to construct the foundation wall and the area in which the building is to be located will affect how the wall and footing are tied

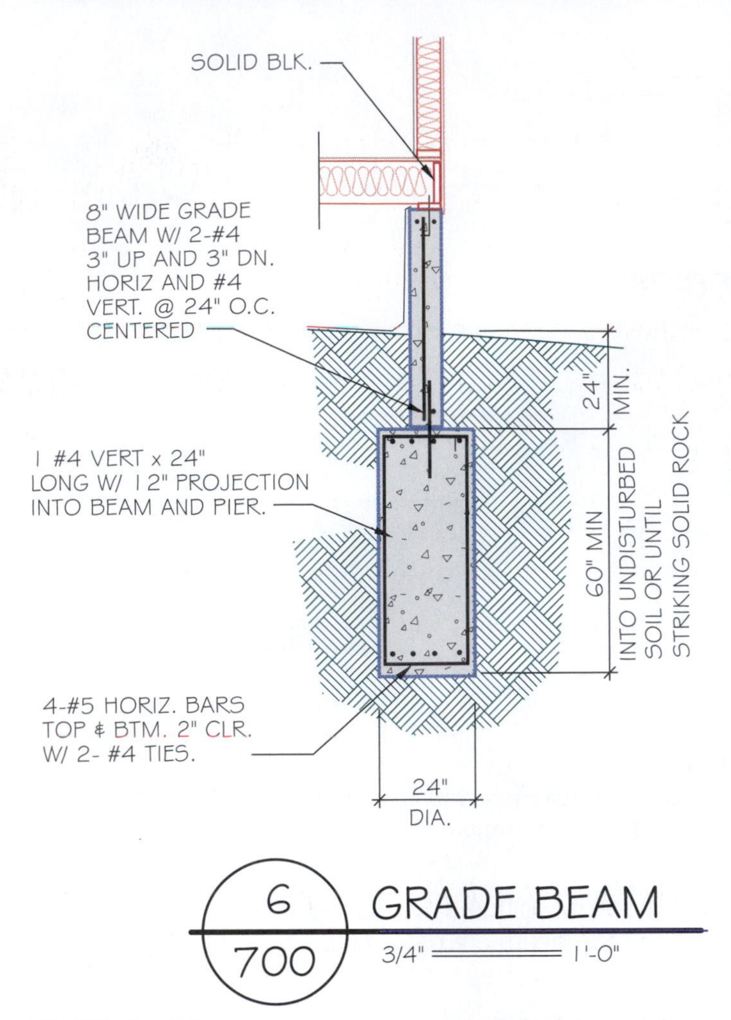

FIGURE 25.27 ■ *A grade beam may be used in place of the foundation to provide added support for the stem wall in unstable soil. The grade beam is placed below the stem wall and spans between stable supports such as pilings.*

FIGURE 25.28 ■ *When concrete will be subjected to stress from tension, steel reinforcing is added to it. This footing has three continuous horizontal bars that will be held in place with circular ties placed at 24'' (600 mm) along the footing.* Courtesy Aaron Jefferis.

together. If the wall and footing are made at different times, a #4 bar must be placed at the top of the stem wall and another near the bottom of the footing. A concrete slab resting on a footing must have a #4 bar at the top and bottom of the footing. The joint between the foundation wall and footing is strengthened by placing a keyway in the footing. The keyway is formed by placing a 2 × 4 (50 × 100) in the top of the concrete footing while the concrete is still wet. Once the concrete has set, the 2 × 4 (50 × 100) is removed, leaving a keyway in the concrete. When the concrete for the wall is poured, it will form a key by filling in the keyway. If a stronger bond is desired, steel is often used to tie the footing to the foundation wall. Both methods of attaching the foundation wall to the footing are shown in Figure 25.29. Footing steel is not drawn on the foundation plan but is specified in general notes and shown in a footing detail similar to Figure 25.30.

Note:

The IRC requires steel to be placed in all footings in seismic zones D1 and D2. Requirements include:

- *One #4 horizontal bar, 3'' (75 mm) down from the top of the stem wall.*
- *One #4 horizontal bar 3'' (75 mm) up from the bottom of the foundation.*
- *One #4 vertical bar placed @ 48'' (1200 mm) o.c., that has a standard hook at its low end. The bar must extend to within 3'' (75 mm) of the bottom of the footing.*
- *Slabs on grade with turned-down footings are required to have one #4 horizontal bar located 3'' (75 mm) up from the bottom of the footing, and one #4 bar located 3'' (75 mm) down from the top of the slab. On-grade slabs that are poured monolithically with the footing can have one #5 or two #4 bars located in the middle third of the footing.*

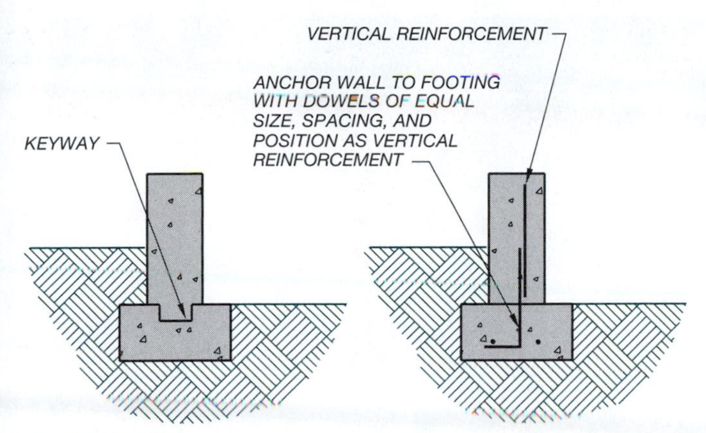

FIGURE 25.29 ■ *The footing can be bonded to the foundation wall with a key or steel reinforcing.*

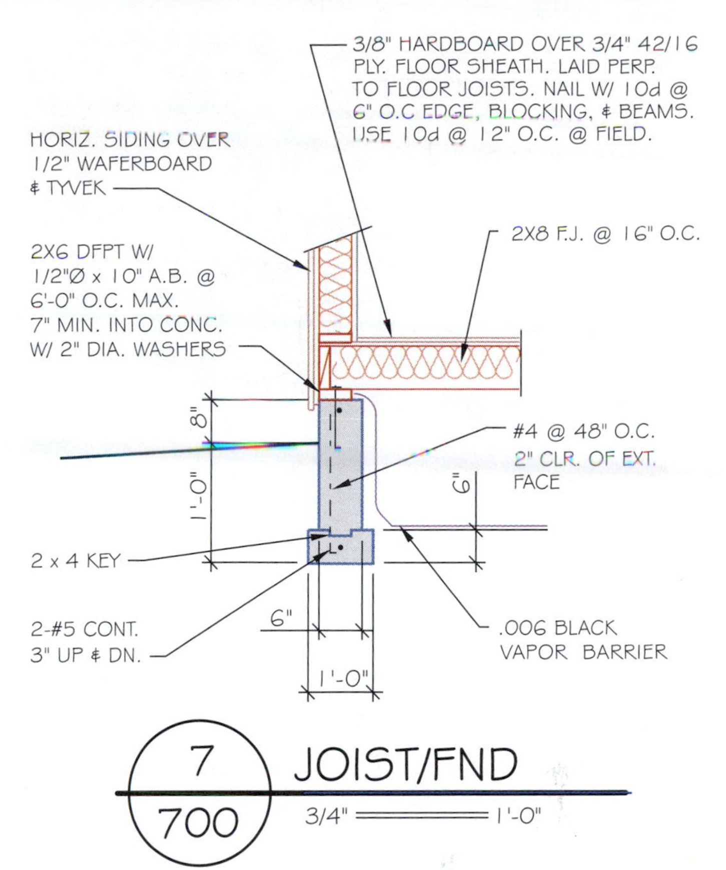

FIGURE 25.30 ■ *When steel is required in the foundation, it is usually specified in a general note on the foundation plan and in a detail.*

Metal Connectors

Metal connectors are often used at the foundation level to resist stress from wind and seismic forces. Three of the most commonly used metal connectors are shown in Figure 25.31. The senior drafter or designer will determine the proper connector and specify it on the foundation plan. How the connector is used will determine how it is specified. Figure 25.32 shows how these connectors might be specified on the foundation plan.

RETAINING WALLS

Common types of retaining walls include full-height masonry walls, full-height wood walls, and partial-height walls.

Full-Height Masonry and Concrete Walls

Retaining or basement walls are primarily made of concrete blocks or poured concrete, although wood walls are used in some areas. The material used will depend on labor trends in your area and will affect the height of the wall. If concrete blocks are used, the wall is approximately 12 blocks high from the top of the footing. If poured concrete is used, the wall will normally be 8' (2400 mm) high from the top of the footing. This will allow 4 × 8' (1200 × 2400 mm) sheets of plywood to be used as forms for the sides of the wall.

Wall Reinforcement

Regardless of the material used, basement walls serve the same function as the shorter foundation walls. Because of the added height of these walls, the lateral forces acting on their sides are magnified. As seen in Figure 25.33, lateral soil pressure bends the wall inward, thus placing the soil side of the wall in compression and the interior face of the wall in tension. To resist this tensile stress, steel reinforcing may be required by the building department. The seismic zone will affect the size and placement of the steel. Figures 25.34 and 25.35 show common patterns of steel placement. Figure 25.36 shows a typical foundation detail used to represent the steel placement of a retaining wall. Figure 25.37 shows an example of steel placement for a poured concrete retaining wall.

Wall-to-Floor Connections

The footing for a retaining wall is usually 16'' (400 mm) wide and either 8 or 12'' (200 or 300 mm) deep. The depth depends on the weight to be supported and the soil bearing capacity. Steel is extended from the footing into the wall. At the top of the wall, anchor bolts are placed in the wall, using the same method as for a foundation wall. Anchor bolts are

A

B

C

FIGURE 25.31 ■ *Three common types of metal connectors used on the foundation systems to resist (A) uplift, (B) gravity loads, and (C) lateral loads. Vendor catalogs or the Internet are common sources of information that must be consulted to complete the foundation plan or details.* Courtesy Simpson Strong-Tie Company, Inc.

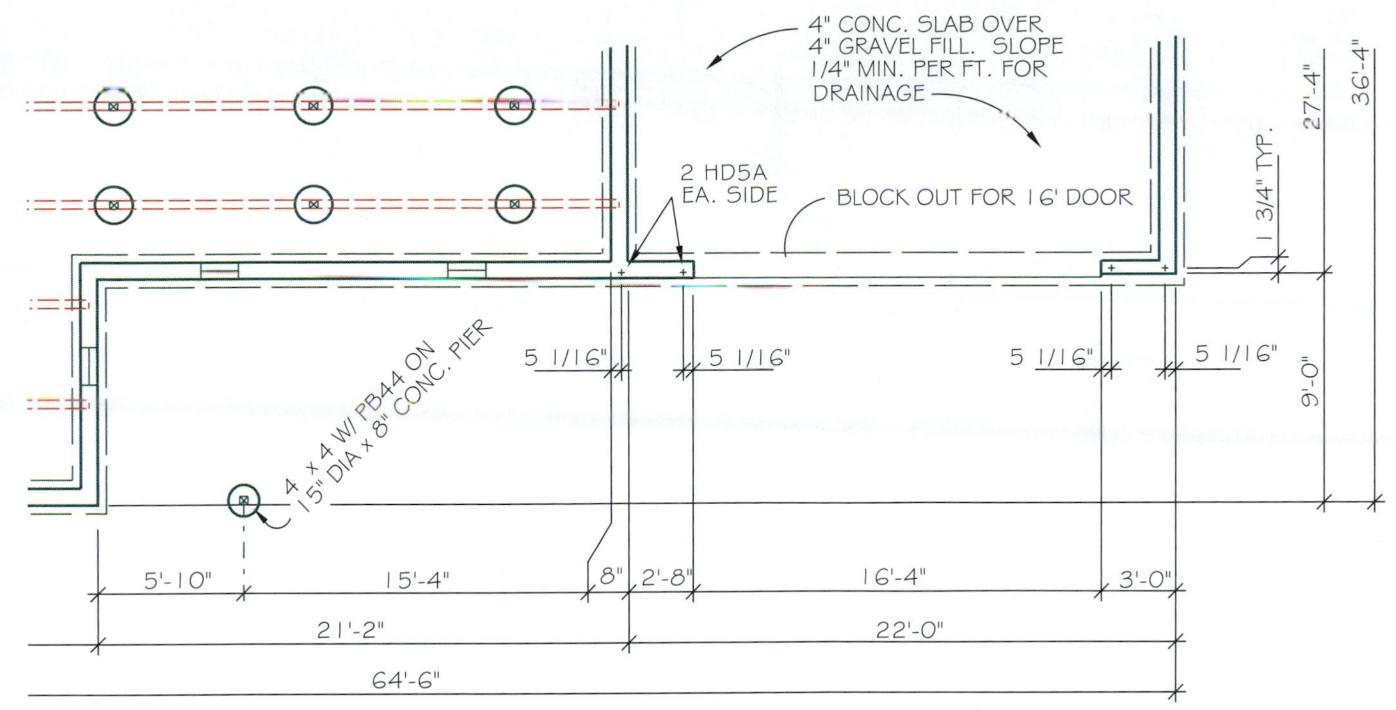

FIGURE 25.32 ■ *Metal connectors used at the foundation level must be clearly specified on the foundation plan.*

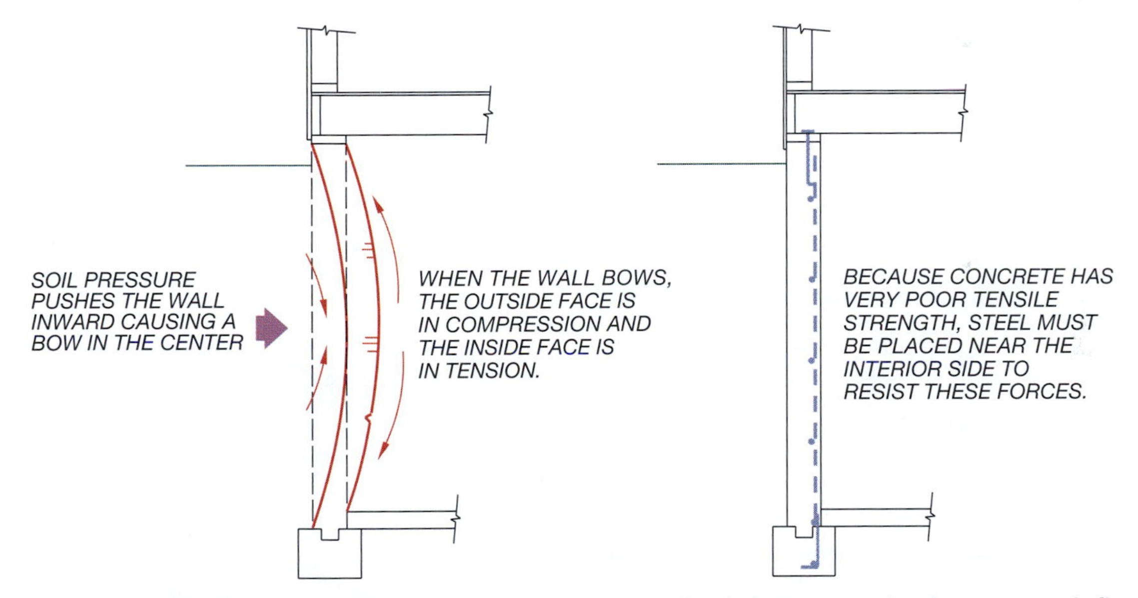

FIGURE 25.33 ■ *Stresses acting on a retaining wall cause it to act as a simple beam spanning between each floor. The soil is the supported load causing the wall to bow. The side of the wall closest to the soil will be in compression, and the opposite side of the wall will be in tension. Steel is added to the tension side of the wall to resist the tendency to stretch.*

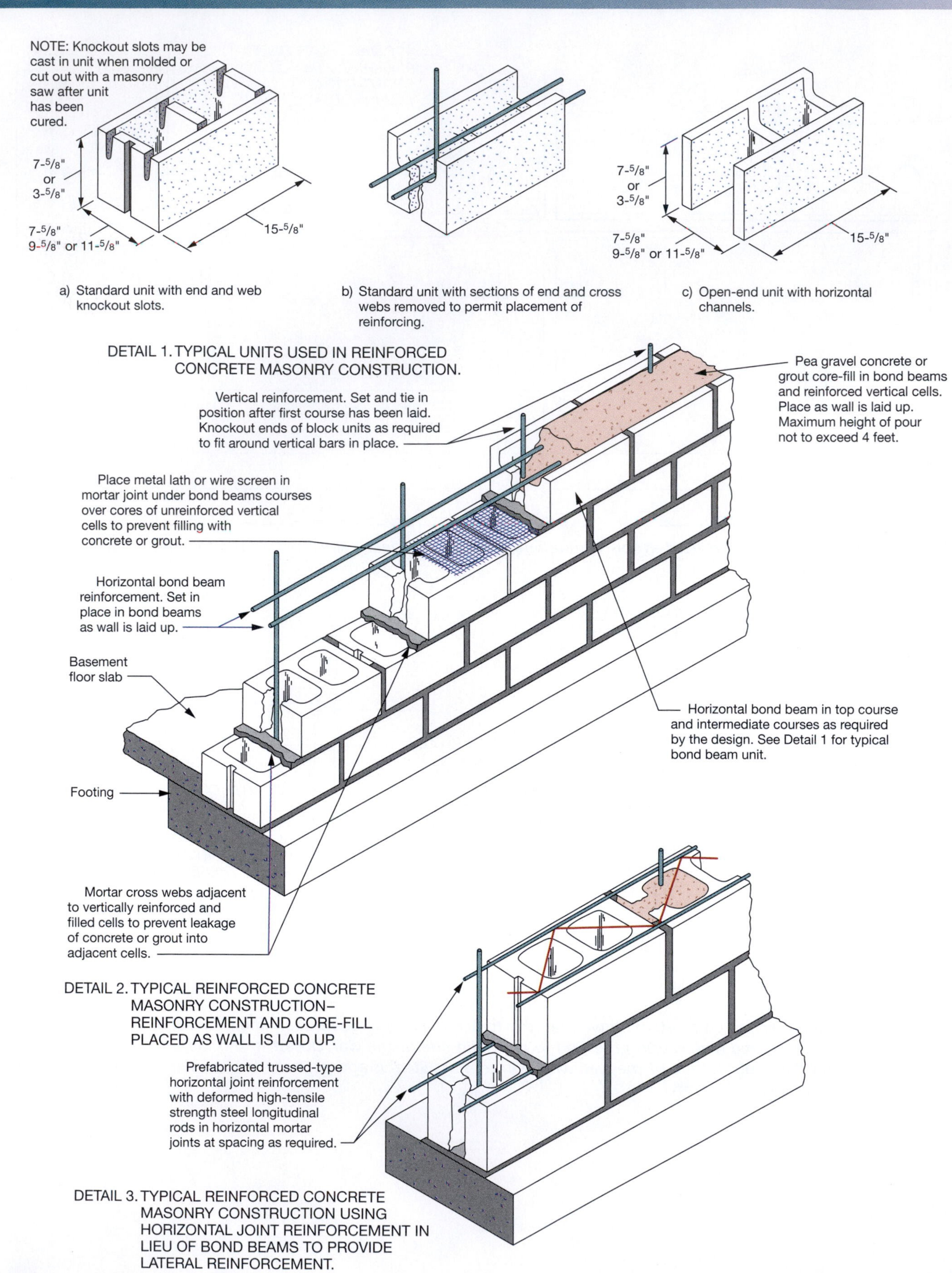

NOTE: Knockout slots may be cast in unit when molded or cut out with a masonry saw after unit has been cured.
7-5/8" or 3-5/8"
7-5/8" 9-5/8" or 11-5/8"
15-5/8"
7-5/8" or 3-5/8"
7-5/8" 9-5/8" or 11-5/8"
15-5/8"
a) Standard unit with end and web knockout slots.
b) Standard unit with sections of end and cross webs removed to permit placement of reinforcing.
c) Open-end unit with horizontal channels.
DETAIL 1. TYPICAL UNITS USED IN REINFORCED CONCRETE MASONRY CONSTRUCTION.
Vertical reinforcement. Set and tie in position after first course has been laid. Knockout ends of block units as required to fit around vertical bars in place.
Pea gravel concrete or grout core-fill in bond beams and reinforced vertical cells. Place as wall is laid up. Maximum height of pour not to exceed 4 feet.
Place metal lath or wire screen in mortar joint under bond beams courses over cores of unreinforced vertical cells to prevent filling with concrete or grout.
Horizontal bond beam reinforcement. Set in place in bond beams as wall is laid up.
Basement floor slab
Horizontal bond beam in top course and intermediate courses as required by the design. See Detail 1 for typical bond beam unit.
Footing
Mortar cross webs adjacent to vertically reinforced and filled cells to prevent leakage of concrete or grout into adjacent cells.
DETAIL 2. TYPICAL REINFORCED CONCRETE MASONRY CONSTRUCTION– REINFORCEMENT AND CORE-FILL PLACED AS WALL IS LAID UP.
Prefabricated trussed-type horizontal joint reinforcement with deformed high-tensile strength steel longitudinal rods in horizontal mortar joints at spacing as required.
DETAIL 3. TYPICAL REINFORCED CONCRETE MASONRY CONSTRUCTION USING HORIZONTAL JOINT REINFORCEMENT IN LIEU OF BOND BEAMS TO PROVIDE LATERAL REINFORCEMENT.

FIGURE 25.34 ■ *Suggested construction details for reinforced concrete masonry foundation walls.* Courtesy National Concrete Masonry Association.

PLAIN MASONRY FOUNDATION WALLS

MAXIMUM WALL HEIGHT (feet)	MAXIMUM UNBALANCED BACKFILL HEIGHT[c] (feet)	PLAIN MASONRY[a] MINIMUM NOMINAL WALL THICKNESS (inches)		
		Soil classes[b]		
		GW, GP, SW and SP	GM, GC, SM, SM-SC and ML	SC, MH, ML-CL and inorganic CL
5	4	6 solid[d] or 8	6 solid[d] or 8	6 solid[d] or 8
	5	6 solid[d] or 8	8	10
6	4	6 solid[d] or 8	6 solid[d] or 8	6 solid[d] or 8
	5	6 solid[d] or 8	8	10
	6	8	10	12
7	4	6 solid[d] or 8	8	8
	5	6 solid[d] or 8	10	10
	6	10	12	10 solid[d]
	7	12	10 solid[d]	12 solid[d]
8	4	6 solid[d] or 8	6 solid[d] or 8	8
	5	6 solid[d] or 8	10	12
	6	10	12	12 solid[d]
	7	12	12 solid[d]	Footnote e
	8	10 solid[d]	12 solid[d]	Footnote e
9	4	6 solid[d] or 8	6 solid[d] or 8	8
	5	8	10	12
	6	10	12	12 solid[d]
	7	12	12 solid[d]	Footnote e
	8	12 solid[d]	Footnote e	Footnote e
	9	Footnote e	Footnote e	Footnote e

For SI: 1 inch = 25.4 mm, 1 foot = 304.8 mm, 1 pound per square inch = 6.895 Pa.

a. Mortar shall be Type M or S and masonry shall be laid in running bond. Ungrouted hollow masonry units are permitted except where otherwise indicated.

b. Soil classes are in accordance with the Unified Soil Classification System. Refer to Table R405.1.

c. Unbalanced backfill height is the difference in height between the exterior finish ground level and the lower of the top of the concrete footing that supports the foundation wall or the interior finish ground level. Where an interior concrete slab-on-grade is provided and is in contact with the interior surface of the foundation wall, measurement of the unbalanced backfill height from the exterior finish ground level to the top of the interior concrete slab is permitted.

d. Solid grouted hollow units or solid masonry units.

e. Wall construction shall be in accordance with Table R404.1.1(2) or a design shall be provided.

FIGURE 25.35a ■ *Required sizes of plain concrete and masonry foundation walls. Reproduced from the International Residential Code / 2006. Copyright © 2006.* Courtesy International Code Council, Inc.

8-INCH MASONRY FOUNDATION WALLS WITH REINFORCING WHERE d > 5 INCHES[a]

WALL HEIGHT	HEIGHT OF UNBALANCED BACKFILL[e]	MINIMUM VERTICAL REINFORCEMENT[b,c]		
		Soil classes and lateral soil load[d] (psf per foot below grade)		
		GW, GP, SW and SP soils 30	GM, GC, SM, SM-SC and ML soils 45	SC, ML-CL and inorganic CL soils 60
6 feet 8 inches	4 feet (or less)	#4 at 48″ o.c.	#4 at 48″ o.c.	#4 at 48″ o.c.
	5 feet	#4 at 48″ o.c.	#4 at 48″ o.c.	#4 at 48″ o.c.
	6 feet 8 inches	#4 at 48″ o.c.	#5 at 48″ o.c.	#6 at 48″ o.c.
7 feet 4 inches	4 feet (or less)	#4 at 48″ o.c.	#4 at 48″ o.c.	#4 at 48″ o.c.
	5 feet	#4 at 48″ o.c.	#4 at 48″ o.c.	#4 at 48″ o.c.
	6 feet	#4 at 48″ o.c.	#5 at 48″ o.c.	#5 at 48″ o.c.
	7 feet 4 inches	#5 at 48″ o.c.	#6 at 48″ o.c.	#6 at 40″ o.c.
8 feet	4 feet (or less)	#4 at 48″ o.c.	#4 at 48″ o.c.	#4 at 48″ o.c.
	5 feet	#4 at 48″ o.c.	#4 at 48″ o.c.	#4 at 48″ o.c.
	6 feet	#4 at 48″ o.c.	#5 at 48″ o.c.	#5 at 48″ o.c.
	7 feet	#5 at 48″ o.c.	#6 at 48″ o.c.	#6 at 40″ o.c.
	8 feet	#5 at 48″ o.c.	#6 at 48″ o.c.	#6 at 32″ o.c.
8 feet 8 inches	4 feet (or less)	#4 at 48″ o.c.	#4 at 48″ o.c.	#4 at 48″ o.c.
	5 feet	#4 at 48″ o.c.	#4 at 48″ o.c.	#5 at 48″ o.c.
	6 feet	#4 at 48″ o.c.	#5 at 48″ o.c.	#6 at 48″ o.c.
	7 feet	#5 at 48″ o.c.	#6 at 48″ o.c.	#6 at 40″ o.c.
	8 feet 8 inches	#6 at 48″ o.c.	#6 at 32″ o.c.	#6 at 24″ o.c.
9 feet 4 inches	4 feet (or less)	#4 at 48″ o.c.	#4 at 48″ o.c.	#4 at 48″ o.c.
	5 feet	#4 at 48″ o.c.	#4 at 48″ o.c.	#5 at 48″ o.c.
	6 feet	#4 at 48″ o.c.	#5 at 48″ o.c.	#6 at 48″ o.c.
	7 feet	#5 at 48″ o.c.	#6 at 48″ o.c.	#6 at 40″ o.c.
	8 feet	#6 at 48″ o.c.	#6 at 40″ o.c.	#6 at 24″ o.c.
	9 feet 4 inches	#6 at 40″ o.c.	#6 at 24″ o.c.	#6 at 16″ o.c.
10 feet	4 feet (or less)	#4 at 48″ o.c.	#4 at 48″ o.c.	#4 at 48″ o.c.
	5 feet	#4 at 48″ o.c.	#4 at 48″ o.c.	#5 at 48″ o.c.
	6 feet	#4 at 48″ o.c.	#5 at 48″ o.c.	#6 at 48″ o.c.
	7 feet	#5 at 48″ o.c.	#6 at 48″ o.c.	#6 at 32″ o.c.
	8 feet	#6 at 48″ o.c.	#6 at 32″ o.c.	#6 at 24″ o.c.
	9 feet	#6 at 40″ o.c.	#6 at 24″ o.c.	#6 at 16″ o.c.
	10 feet	#6 at 32″ o.c.	#6 at 16″ o.c.	#6 at 16″ o.c.

For SI: 1 inch = 25.4 mm, 1 foot = 304.8 mm, 1 pound per square foot per foot = 0.157 kPa/mm.

a. Mortar shall be Type M or S and masonry shall be laid in running bond.

b. Alternative reinforcing bar sizes and spacings having an equivalent cross-sectional area of reinforcement per lineal foot of wall shall be permitted provided the spacing of the reinforcement does not exceed 72 inches.

c. Vertical reinforcement shall be Grade 60 minimum. The distance from the face of the soil side of the wall to the center of vertical reinforcement shall be at least 5 inches.

d. Soil classes are in accordance with the Unified Soil Classification System and design lateral soil loads are for moist conditions without hydrostatic pressure. Refer to Table R405.1.

e. Unbalanced backfill height is the difference in height between the exterior finish ground level and the lower of the top of the concrete footing that supports the foundation wall or the interior finish ground level. Where an interior concrete slab-on-grade is provided and is in contact with the interior surface of the foundation wall, measurement of the unbalanced backfill height from the exterior finish ground level to the top of the interior concrete slab is permitted.

FIGURE 25.35b ■ *Required sizes of reinforced concrete and masonry foundation walls. Reproduced from the International Residential Code / 2006. Copyright © 2006.* Courtesy International Code Council, Inc.

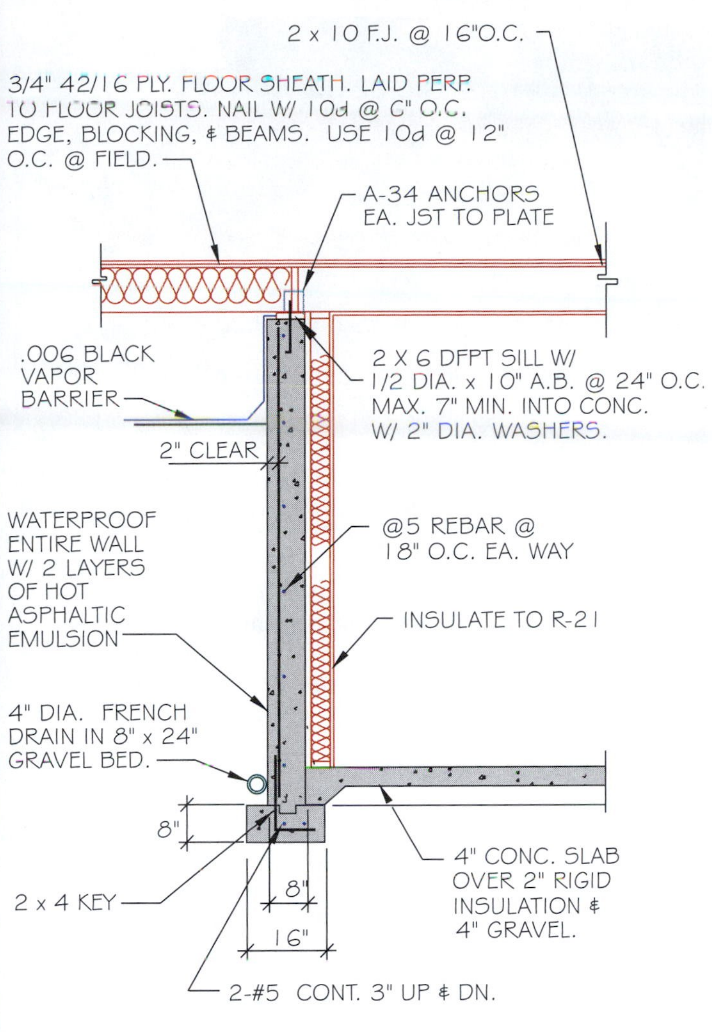

FIGURE 25.36 ■ *The components of a concrete retaining wall are specified in a section or detail. Key components are the building material of the wall, reinforcing material, floor attachment, waterproofing, and drainage method.* Courtesy Residential Designs.

FIGURE 25.37 ■ *Reinforcing steel extends from the footing and ties to the wall steel to provide strength for the retaining wall. Once all of the wall steel has been placed, the balance of the wood forms can be placed and then the concrete can be poured.*

placed much closer together for retaining walls than for shorter walls. Common bolt spacing for retaining walls include:

- 24'' (600 mm) o.c. when floor joists are perpendicular to the wall
- 32'' (800 mm) o.c. where joists are parallel to the retaining wall

It is very important that the wall and the floor system be securely tied together. The floor system is used to help strengthen the wall and resist soil pressure. Where seismic risk is great, metal angles are added to the anchor bolts to make the tie between the wall and the floor joist extremely rigid. These connections are shown on the foundation plan, as seen in Figure 25.38a.

Soil Drainage

To reduce soil pressure next to the footing, a drain is installed. The drain is set at the base of the footing to collect and divert water from the face of the wall. Figure 25.36 shows how the drain is placed. The area above the drain is filled with gravel, so that subsurface water will percolate to the drain and away from the wall. Reducing the water content of the soil reduces

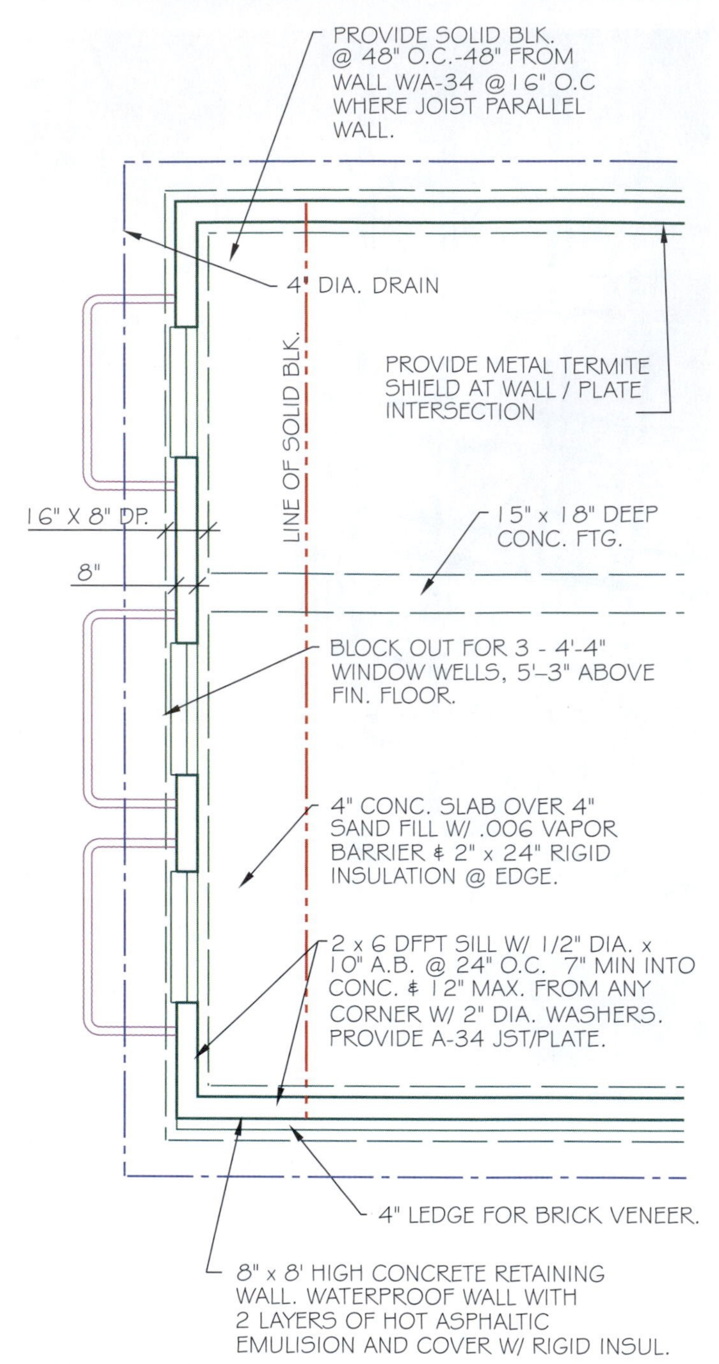

FIGURE 25.38a ■ *Common elements shown on a foundation with a basement. If a window is to be placed in a full-height basement wall, a window well to restrain the soil around the window is required. On the interior side of the wall, blocking is placed between floor joists that are parallel to the retaining wall. The blocking provides rigidity to the floor system, allowing lateral pressure from the wall to be transferred into the floor system.*

FIGURE 25.38b ■ *Windows framed into an EPS foundation wall. Once the wall has been reinforced and the concrete has cured, window wells can be placed around the windows and backfill can be placed around the foundation.* Courtesy Reward Wall Systems.

the lateral pressure on the wall. The drain is not drawn on the foundation plan but must be specified in a note on the foundation plan, sections, and foundation details.

Water and Moisture Protection

No matter what the soil's condition, the basement wall must be protected to minimize moisture passing through the wall into the living area. The IRC specifies damp proofing and waterproofing as the two levels of basement wall protection. Each method must be applied to the exterior face of a foundation wall surrounding a habitable space. The protection must be installed from the top of the footing to the grade. A masonry or concrete wall can be damp proofed by adding 3/8'' (9.5 mm) of Portland cement parging to the exterior side of the wall. The parging must then be protected by a bituminous coating, acrylic modified cement, or a coat of surface-bonding mortar. Materials used to waterproof a wall can also be applied. In areas with a high water table or other severe soil/water conditions, basement walls surrounding habitable space must be waterproofed. Concrete and mortar walls are waterproofed by adding 2-ply hot-mopped felts, 6-mil polyethylene or 40-mil polymer-modified asphalt, 6-mil (0.15 mm) polyvinyl chloride, or 55-lb (25 kg) roll roofing. Additional provisions are provided by the IRC for wood foundation walls.

Adding Windows

Adding windows to the basement can help cut down the moisture content of the basement. Figure 25.38b shows the framing for a window in a EPS foundation wall. This will sometimes require adding a window well to prevent the ground from being pushed in front of the window. Figure 25.38 shows how a window well—or areaway, as it is sometimes called—can be drawn on the foundation plan.

Like a foundation wall, the basement wall needs to be protected from termites. The metal shield will not be drawn on the foundation plan but should be called out in a note.

Treated-Wood Basement Walls

Pressure-treated lumber can be used to frame both crawl space and basement walls. Treated wood basement walls allow for easy installation of electrical wiring, insulation, and finishing materials. Instead of a concrete foundation, a gravel bed is used to support the wall loads. Gravel is required to extend 4'' (100 mm) on each side of the wall and be approximately 8'' (200 mm) deep. A 2 × (50 ×) pressure-treated plate is laid on the gravel and the wall is built of pressure-treated wood. Pressure-treated 1/2'' (13 mm) C-D grade plywood with exterior glue is laid perpendicular to the studs, covered with 6-mil polyethylene, and sealed with an adhesive. Figure 25.39 shows a detail for a wood retaining wall.

Partial-Height Retaining Walls

As seen in Figure 25.40, when a structure is built on a sloping site, the masonry wall may not need to be full height. Although less soil is retained than with a full-height wall, more problems are encountered. Figure 25.41 shows the tendency of this type of wall to bend. Because the wall is not supported at the top by the floor, the soil pressure must be resisted through the footing. This requires a larger footing than for a full-height wall. If the wall and footing connection is rigid enough to keep the wall in position, the soil pressure will try to overturn the whole foundation. The extra footing width is required to resist the tendency to overturn. Figure 25.42 shows a detail required to explain the construction of the wall. Figure 25.43 shows how a partial-height retaining wall will be represented on a foundation plan. Depending on the slope of the ground being supported, a key may

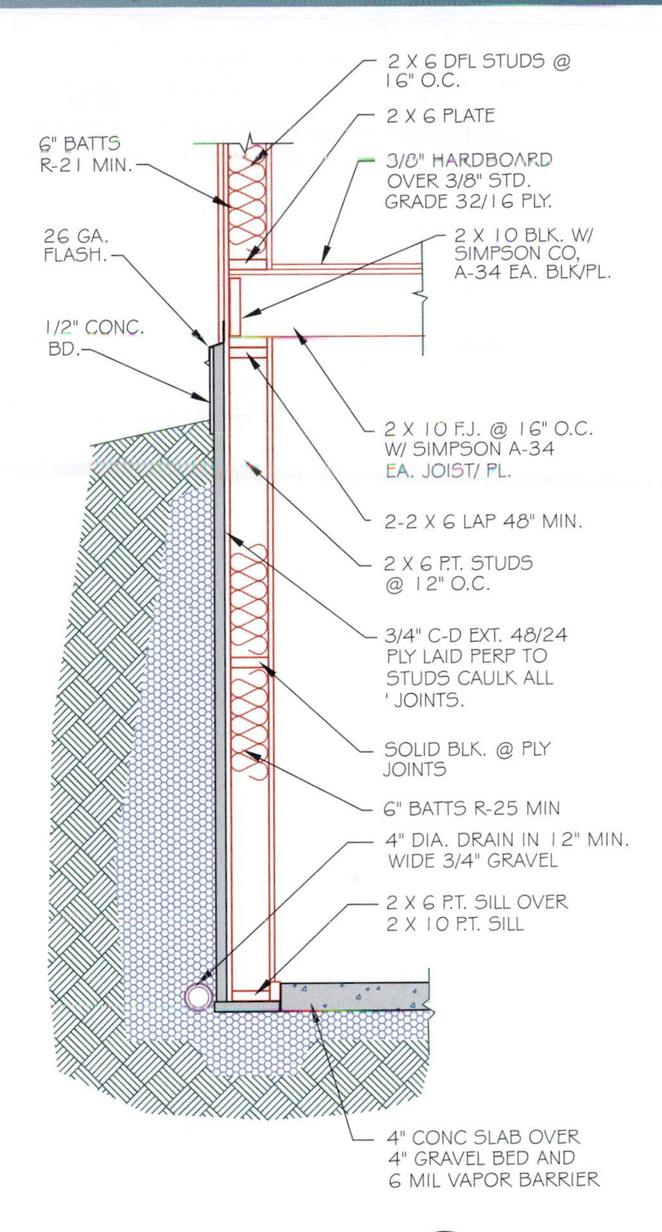

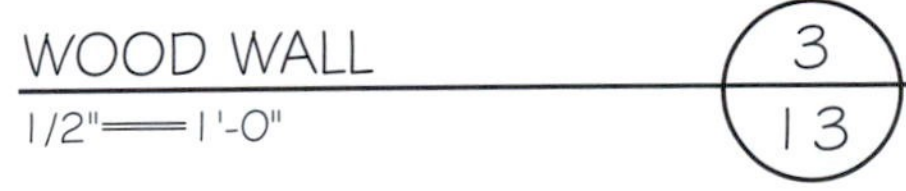

ASSUMED SOIL PRESSURE 30# PER CUBIC FOOT.

GENERAL NOTES:

WATER PROOF EXTERIOR SIDE OF WALL WITH 2 LAYER HOT ASPHALTIC EMULSION COVERED WITH 6 MIL. VAPOR BARRIER.

PROVIDE ALTERNATE BID FOR 2" RIGID INSULATION ON EXTERIOR FACE FOR FULL HEIGHT.

ALL MATERIAL BELOW UPPER TOP PLATE OF BASEMENT WALL TO BE PRESSURE PRESERVATIVELY TREATED IN ACCORDANCE WITH AWPA-C22 AND SO MARKED.

FIGURE 25.39 ■ *Stem walls and basement retaining walls can be framed using pressure-treated wood with no foundation. The size and spacing of members are determined by the loads being supported and the height of the wall.*

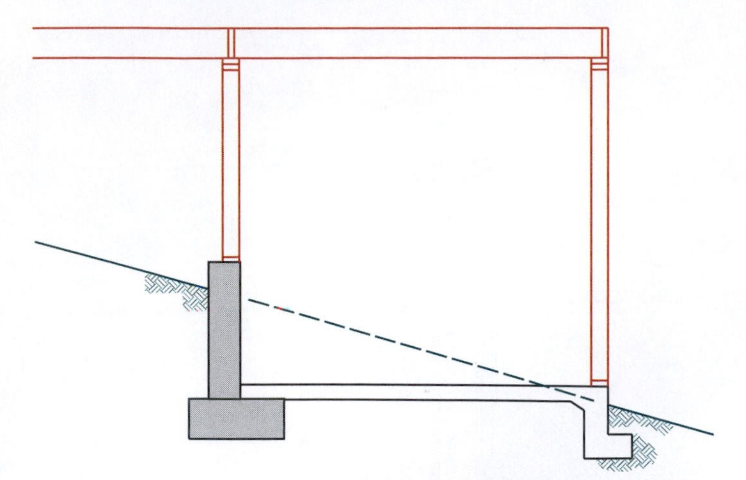

FIGURE 25.40 ■ *An 8' (2400 mm) high retaining wall may not be required on sloping lots. A partial restraining wall with wood-framed walls above it is often used.*

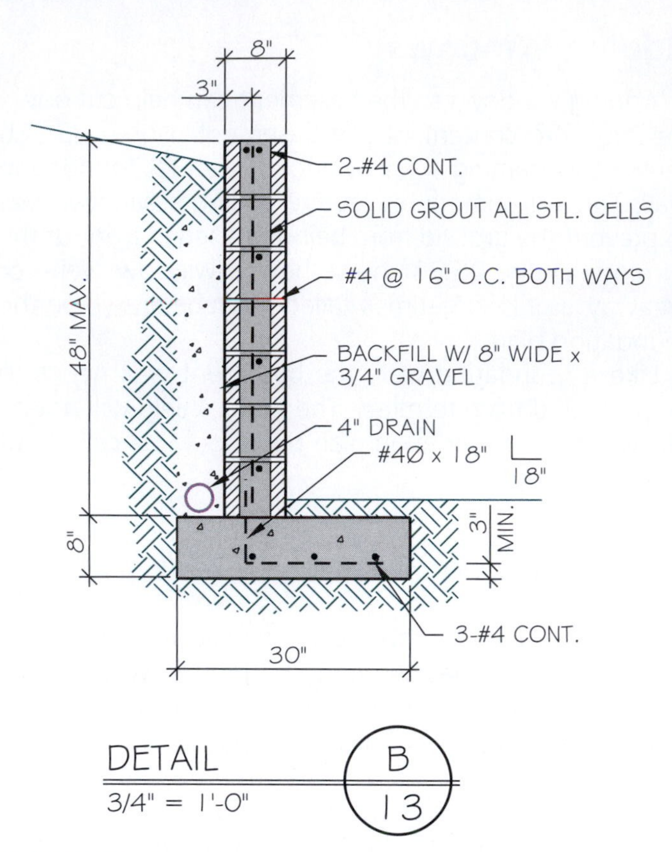

FIGURE 25.42 ■ *A detail must be drawn to explain the construction of a partial-height retaining wall.*

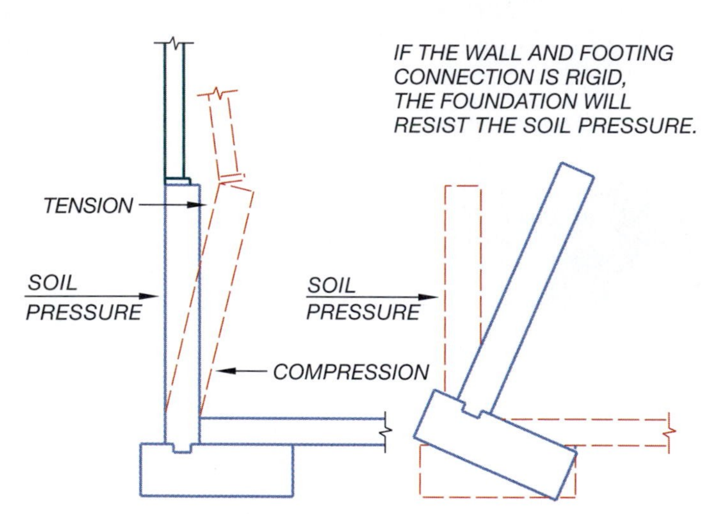

FIGURE 25.41 ■ *When a wall is not held in place at the top, soil pressure will attempt to move the wall inward. The intersection of the wood and concrete walls is called a hinge point because of the tendency to move. The width of the footing must be increased and the hinge point must be rigid to resist overturning.*

be required, as in Figure 25.44. A keyway on the top of the footing has been discussed. This key is added to the bottom of the footing to help keep it from sliding as a result of soil pressure against the wall. The key is not shown on the foundation plan but is shown in a detail of the wall.

DIMENSIONING FOUNDATION COMPONENTS

This chapter has presented components of a foundation system and how they are drawn on the foundation plan. The manner in which they are dimensioned is equally important to the construction crew. The line quality for the dimension and leader lines is the same as was used on the floor plan. Jogs in the foundation wall are dimensioned using the same methods used on the floor or framing plan. Most of the dimensions for major shapes will be exactly the same as the corresponding dimensions on the framing plan. The foundation plan is typically placed in the same drawing file as the floor and framing plans. This will allow the overall dimensions for the framing plan to be displayed on the foundation plan if different layers are used to place the dimensions. Layers such as *DIM BASE, DIM FRAM,* and *DIM FND* will help differentiate between dimensions for the floor and foundation.

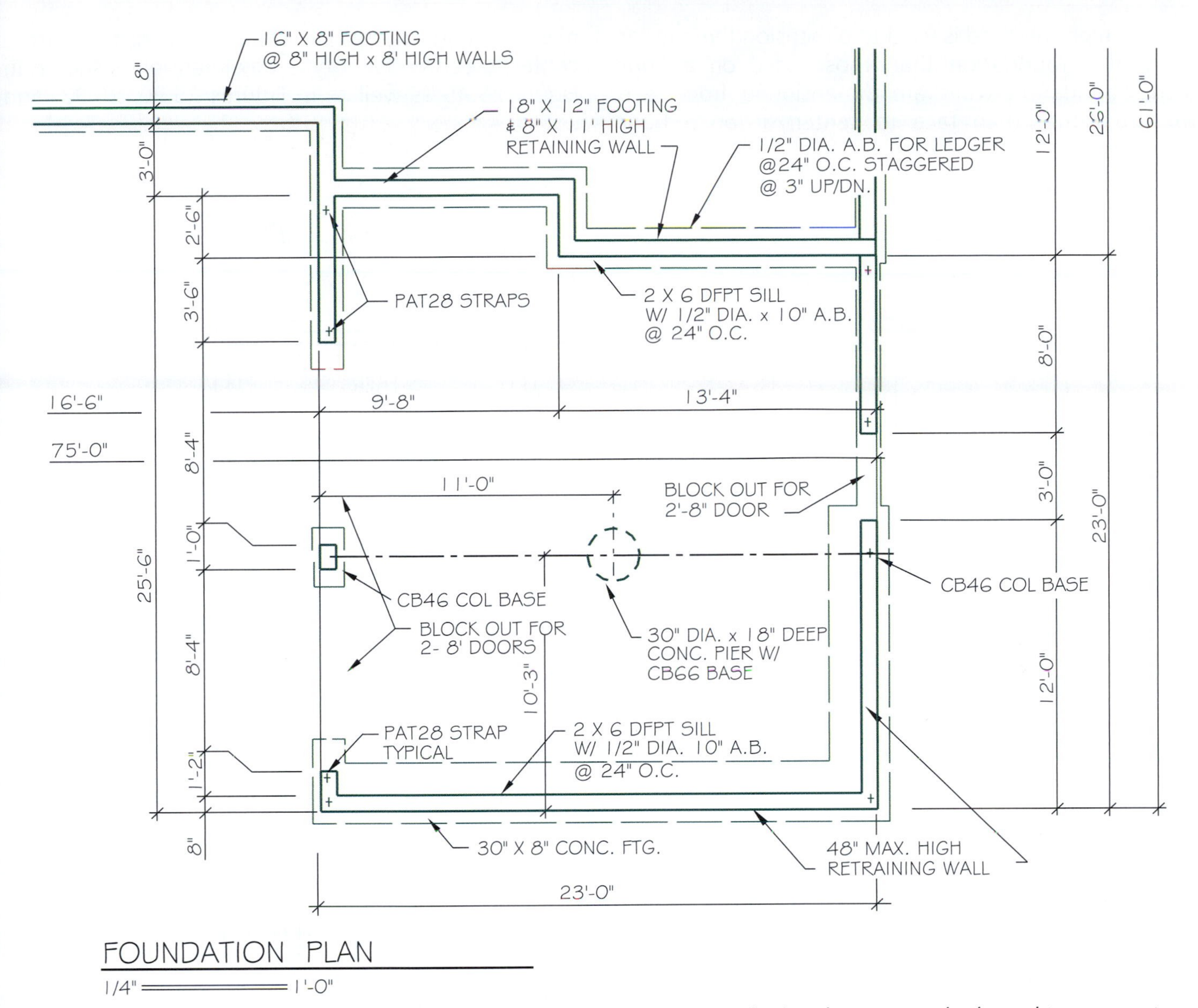

FIGURE 25.43 ■ *Full-height and partial-height retaining walls are represented using the same methods used to represent footings and stem walls. Because of the added height, the width is typically specified on the foundation plan and the building components are specified in details or sections.*

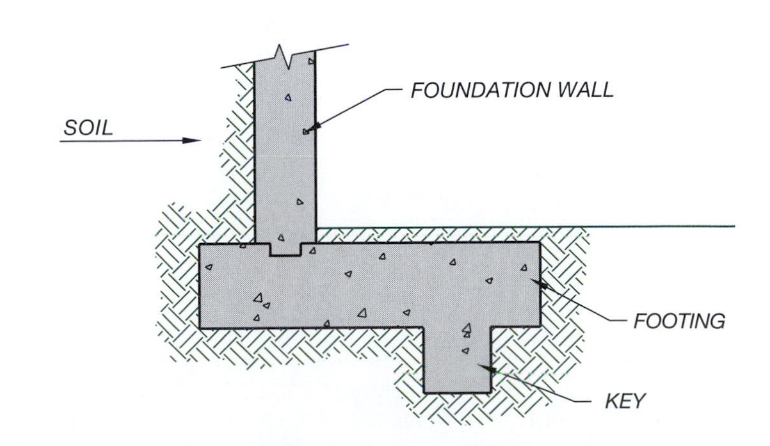

FIGURE 25.44 ■ *Soil pressure will attempt to make the wall and footing slide across the soil. A key may be provided on the bottom of the footing to provide added surface area to resist sliding.*

A different method is used to dimension the interior walls of a foundation than those used on a floor plan. Foundation walls are dimensioned from face to face rather than face to center, as on a floor plan. Footing widths are usually dimensioned from center to center. Each type of dimension is shown in Figure 25.45 as well as in Figures 25.20, 25.32, and 25.44.

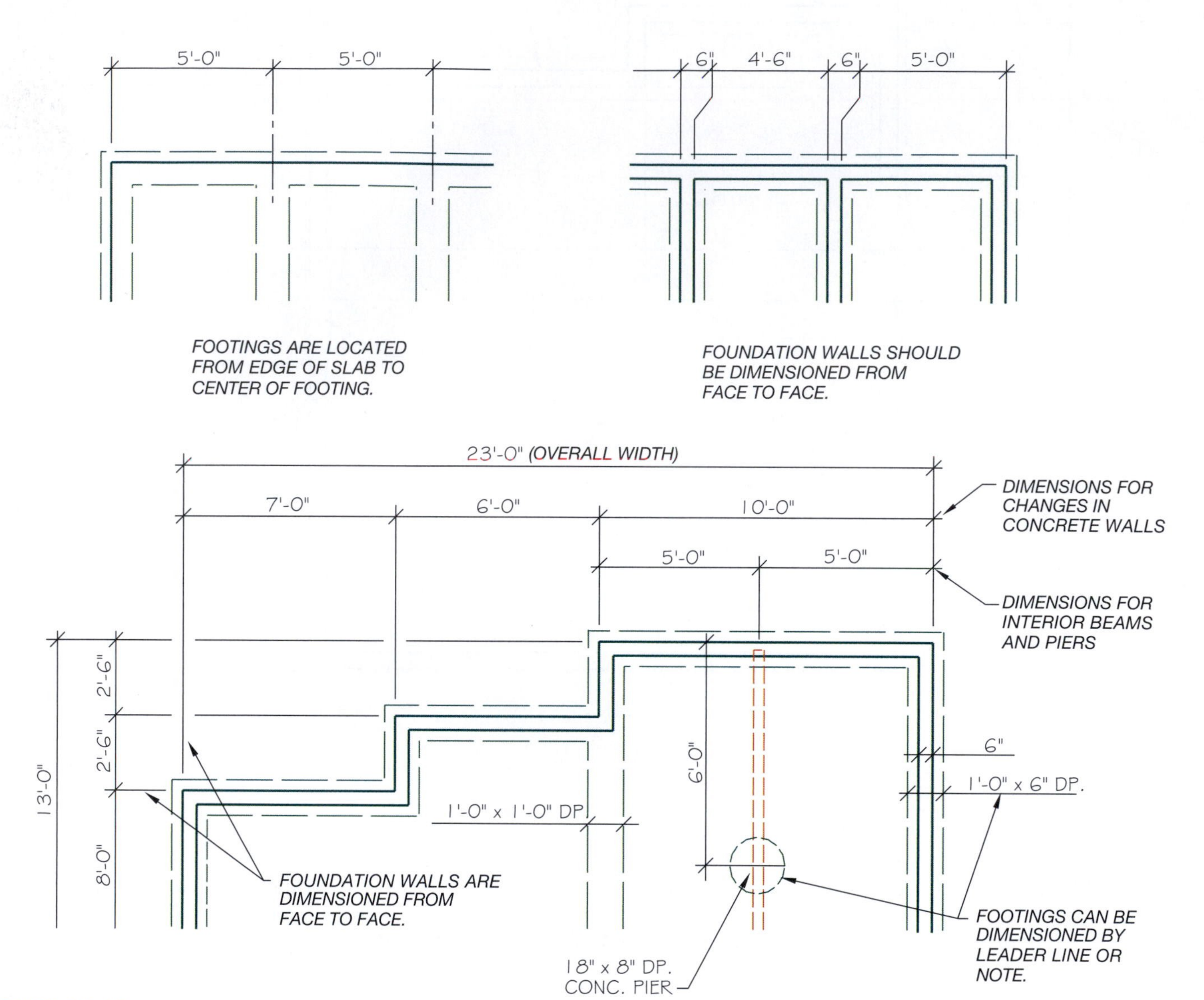

FIGURE 25.45 ■ *Dimensioning techniques for foundation plans. Always dimension to the outside edge of concrete walls and to the center of concrete piers.*

Additional Readings

The following websites can be used as a resource to help you keep current with changes in foundation materials:

ADDRESS	COMPANY OR ORGANIZATION
www.bluemaxxaab.com	AAB Building Systems Inc. (ICFs)
www.afmcorp-epsfoam.com	AFM Corporation
www.bocciabros.com	Boccia Inc.
www.aci-int.org	American Concrete Institute International
www.polysteel.com	American Poly Steel
www.arxxbuild.com	Arxx Walls and Foundations
www.eco-block.com	Eco-Block
www.haenerblock.com	Haener Block Inc. (mortar interlocking blocks)
www.portcement.org	Portland Cement Association
www.pwf.com	Permanente Wood Foundations
www.woodbasement.com	Permanente Wood Foundations
www.woodfoundation.com	Permanente Wood Foundation Systems
www.post-tensioning.org	Post-tensioning Institute
www.rewardwalls.com	Reward Wall Systems
www.strongtie.com	Simpson Strong-Tie Connections
www.soils.org	Soil Science Society of America
www.vobb.com	VOBB—Verot Oaks Building Blocks

CHAPTER 25

Foundation Systems Test

DIRECTIONS

Answer the following questions with short, complete statements. Type your answers using a word processor or neatly print your answers on lined paper.

1. Place your name, the chapter number, and the date at the top of the sheet.
2. Type the question number and provide the answer in the form of a statement that includes part of the question. You do not need to write out the entire question.

Note: The answers to some questions may not be contained in this chapter and will require you to do additional research.

QUESTIONS

25.1. What are the major parts of a continuous foundation system?
25.2. What factors influence the size of footings?
25.3. List five forces that a foundation must withstand.
25.4. How can the soil texture influence a foundation?
25.5. List the major types of material used to build foundation walls.
25.6. Why is steel placed in footings?
25.7. Describe when a stepped footing might be used.
25.8. What size footing should be used with a basement wall?
25.9. What influences the size of piers?
25.10. Describe the difference between a full-height retaining wall and a partial-height retaining wall.

Note: See Chapter 28 for foundation-related details.

CHAPTER 26

Floor Systems and Foundation Support

The foundation plan shows not only the concrete footings and walls but also the members used to form the floor. Two common types of floor systems are used in residential construction: those built at grade level and those with a crawl space or basement below the floor system. Each has its own components and information that must be represented on a foundation plan.

ON-GRADE FLOOR SYSTEMS

A concrete slab is often used for the floor system of residences in warm, dry climates. A concrete slab provides a firm floor system with little or no maintenance and generally requires less material and labor than a conventional wood floor system. The floor slab is usually poured as an extension of the foundation wall and footing, in what is referred to as monolithic construction (see Figure 26.1). Other common methods of pouring the foundation and floor system are shown in Figure 26.2. A 3 1/2'' (90 mm) concrete slab is the minimum thickness allowed by the IRC for residential floor slabs. The slab is used only as a floor surface, not to support the weight of the walls or roof. If load-bearing walls must be supported, the floor slab must be thickened, as shown in Figure 26.3. If a load is concentrated in a small area, a pier may be placed under the slab to help disperse the weight, as shown in Figure 26.4.

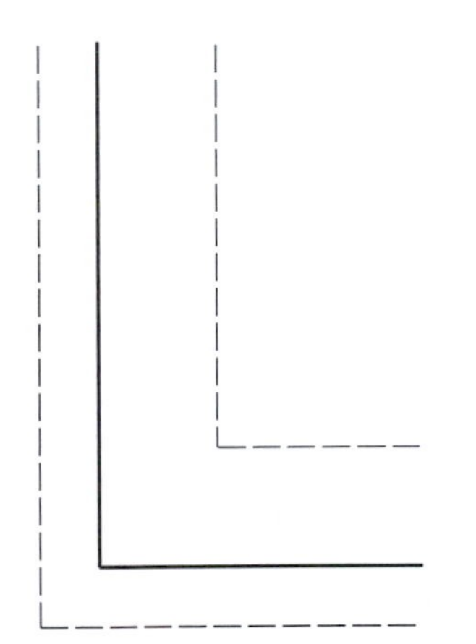

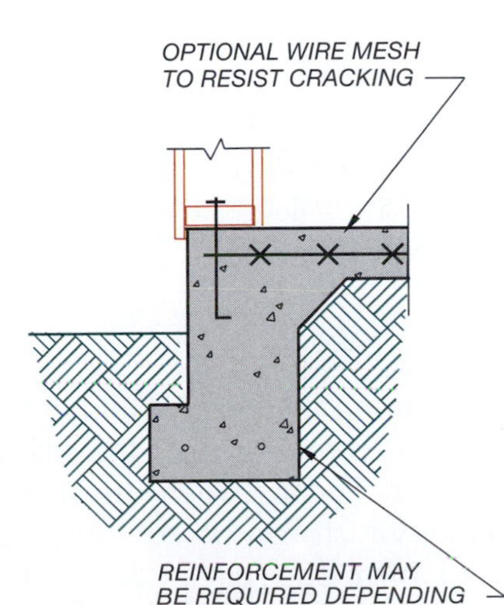

FIGURE 26.1 ■ *The foundation and floor system can often be constructed in one pour, saving time and money.*

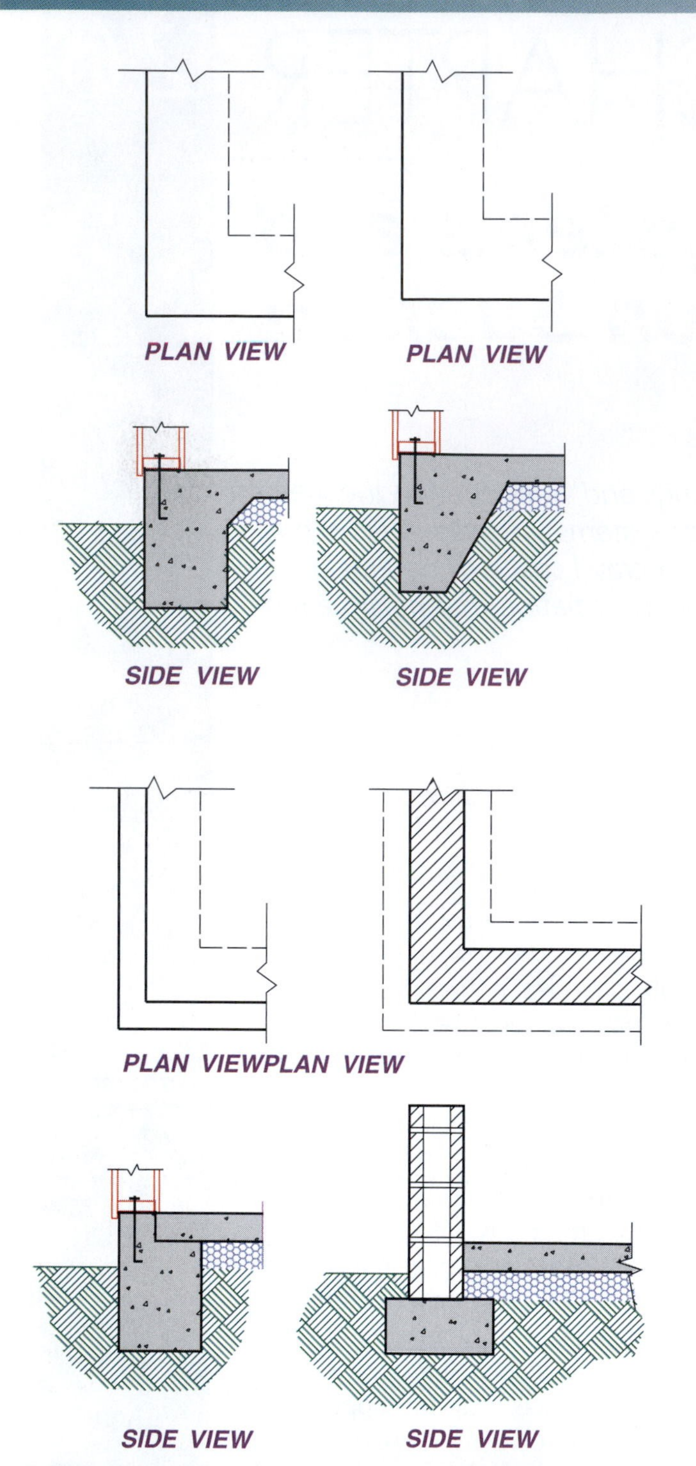

FIGURE 26.2 ■ *Common foundation and slab intersections.*

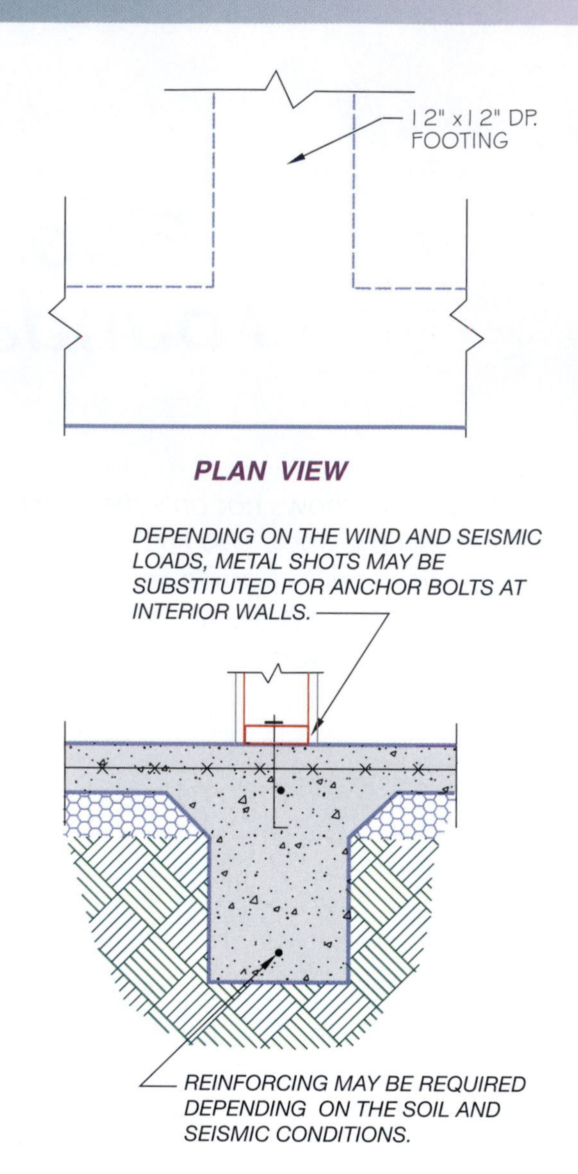

FIGURE 26.3 ■ *A continuous footing is placed under an interior load-bearing wall.*

Slab Joints

Concrete tends to shrink approximately 0.66'' per 100' (16.7 mm per 30 000 mm) as the moisture in the mix hydrates and the concrete hardens. Concrete also continues to expand and shrink throughout its life, depending on the temperatures and the moisture in the supporting soil. This shrinkage can cause the floor slab to crack. To help control possible cracking, three types of joints can be placed in the slab: control, construction, and isolation joints (see Figure 26.5). The location of control and isolation joints must be referenced by the drafter, based on the designer's specifications.

Control Joints

As the slab contracts during the initial drying process, the lower surface of the slab rubs against the soil, creating tensile stress. The friction between the slab and the soil causes cracking, which can be controlled by a ***control*** or ***contraction joint.*** Such a joint does not prevent cracking, but it does control where the cracks will develop in the slab. Control joints are created by cutting the fresh concrete or sawing the concrete within 6 to 8 hours of placement. Joints are usually one-quarter of

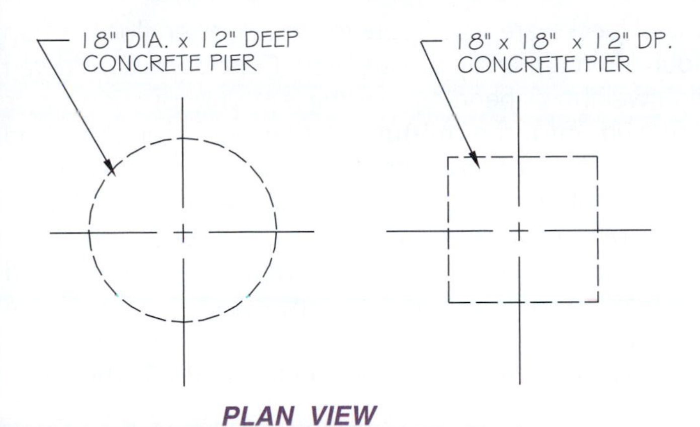

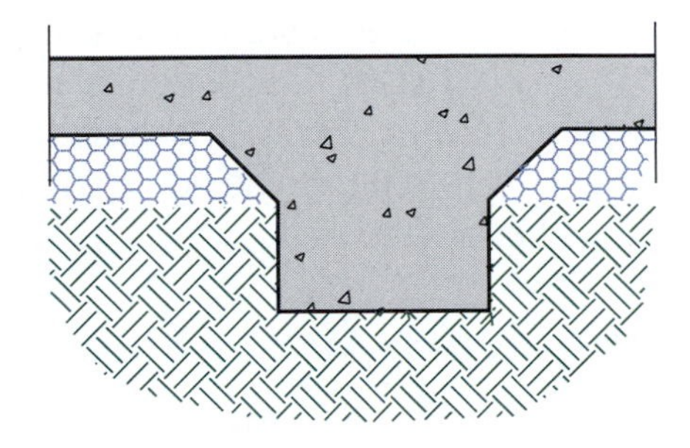

FIGURE 26.4 ■ *A concrete pier is poured under loads concentrated in small areas. Piers are typically round or rectangular, depending on their location. Square piers are used when the exterior footing needs to be reinforced, or if the footing is so large that it must be framed with lumber. Round piers are used when the pier is placed inside the foundation. The load to be supported and the soil bearing capacity determine the size of each pier.*

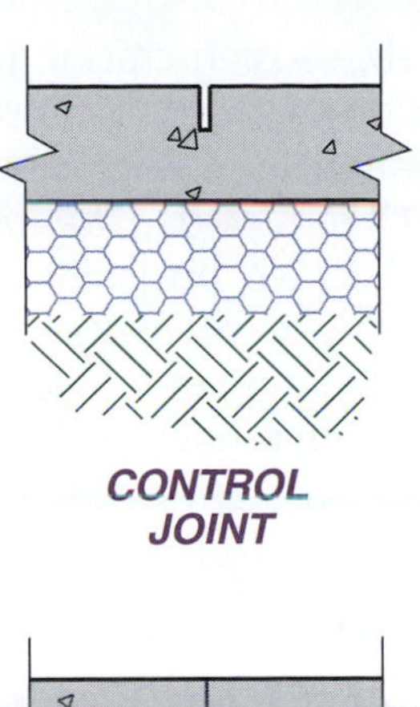

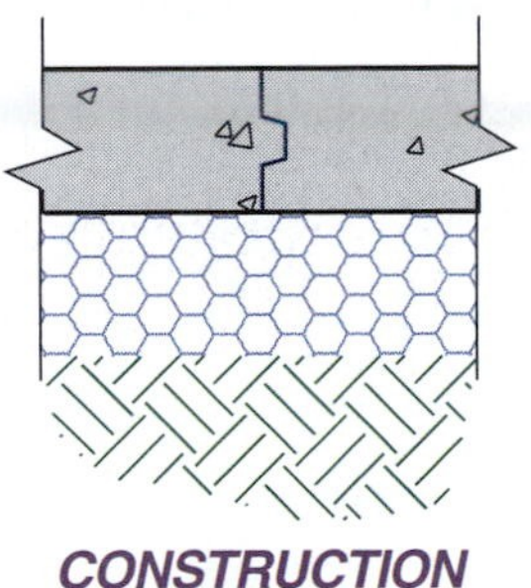

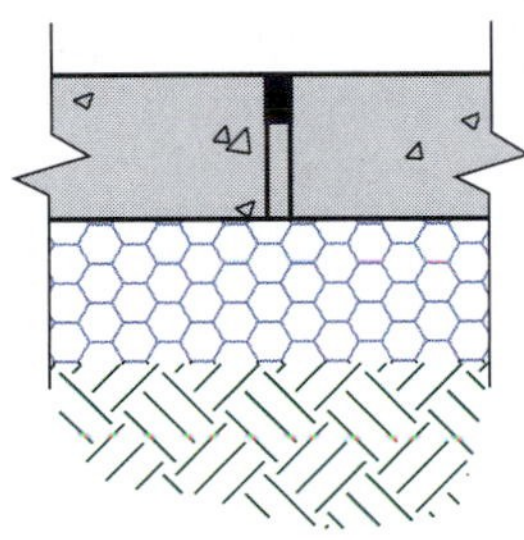

FIGURE 26.5 ■ *Joints are placed in concrete to control cracking. A control joint is placed in the slab to weaken the slab and cause cracking to occur along the joint rather than throughout the slab. When construction must be interrupted, a construction joint is formed to increase bonding with the next day's pour. Isolation joints are provided to keep stress from one structural material from cracking another.*

the slab depth. Because the slab has been weakened, any cracking due to stress will result along the joint. The American Concrete Institute (ACI) suggests that control joints be placed a distance in feet equal to about 2 1/2 times the slab depth in inches. For a 4'' (100 mm) slab, joints would be placed at approximately 10' (3000 mm) intervals. The locations of control joints are usually specified by notes for residential slabs. The spacing and method of placement can be specified in the general notes for the foundation plan.

Construction Joints

When concrete construction must be interrupted, a ***construction joint*** is used to provide a clean surface where work can be resumed. Because a vertical edge of one slab will have no bond to the next slab, a keyed joint is used to provide support between the two slabs. The key is typically formed by placing a beveled strip about one-fifth of the slab thickness and one-tenth of the slab thickness in width to mold the slab. The method used to form the joint can be specified with other foundation notes, but the crew placing the concrete will determine the location.

Isolation Joints

An ***isolation*** or ***expansion joint*** is used to separate a slab from an adjacent slab, wall, or column or some other part of the structure. The joint prevents forces from an adjoining structural member from being transferred into the slab, causing cracking. Such a joint also allows for expansion of the slab caused by moisture or temperature. Isolation joints are typically between 1/4 and 1/2'' (6 and 13 mm) wide. The location of isolation joints should be specified on the foundation plan. Because of the small size of residential slabs, isolation joints are not usually required. Slabs subject to severe freezing conditions often have an isolation joint to separate the slab from the stem wall.

Slab Placement

The slab may be placed above grade, below grade, or at grade level.

Above Ground Slabs

Residential slabs are often placed above grade for hillside construction to provide a suitable floor for a garage. A platform made of wood or steel materials can be used to support a lightweight concrete floor slab. Plywood floor sheathing, covered with 55 lb. building paper protects the wood from moisture in the concrete. Ribbed metal decks are also used for the heavier floors often found in multifamily construction. Concrete is considered lightweight depending on the amount of air that is pumped into the mixture during the manufacturing process. The components of an above-ground concrete floor should be noted on a framing plan but not drawn. The strength of the concrete and its weight should also be noted. The foundation plan shows the columns and footings used to support the increased weight of the floor. An example of a framing plan to support an aboveground concrete slab is shown in Figure 26.6. The foundation plan for the same area is shown in Figure 26.7. The construction process will also require details similar to those in Figure 26.8. Multilevel custom homes and multifamily projects may use precast hollow core slab units to help provide large open areas beneath the slabs. Figures 26.9a and b show concrete panels being lifted into position for a residential addition.

At Grade and Below Grade Slabs

Slabs built below grade are most commonly used in basements. When used at grade, the slab is usually placed just above grade level. The IRC requires the slab

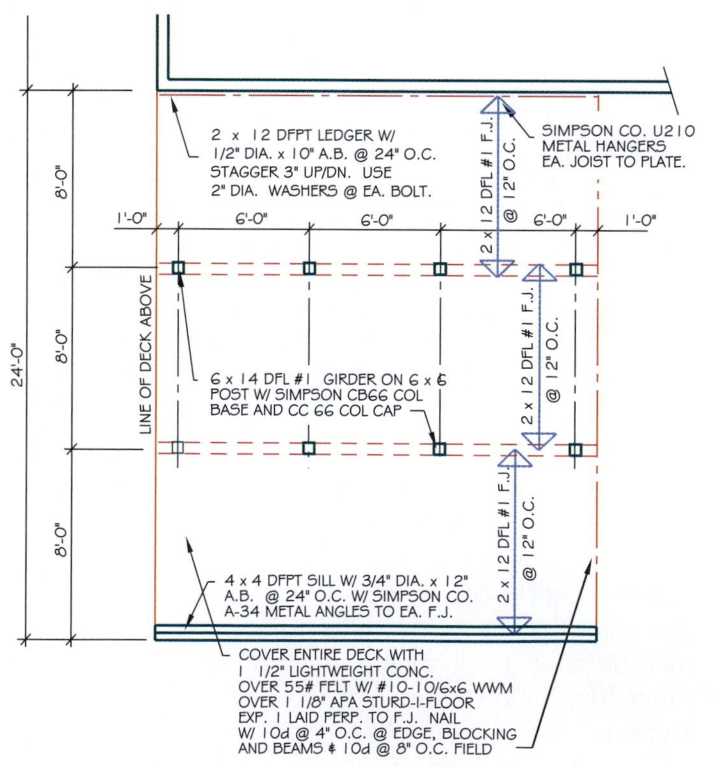

FIGURE 26.6 ■ *An aboveground concrete floor is often used for high end multifamily projects, and for a garage floor and driveway for hillside residential construction. The framing plan shows the framing of the platform used to support the concrete slab.*

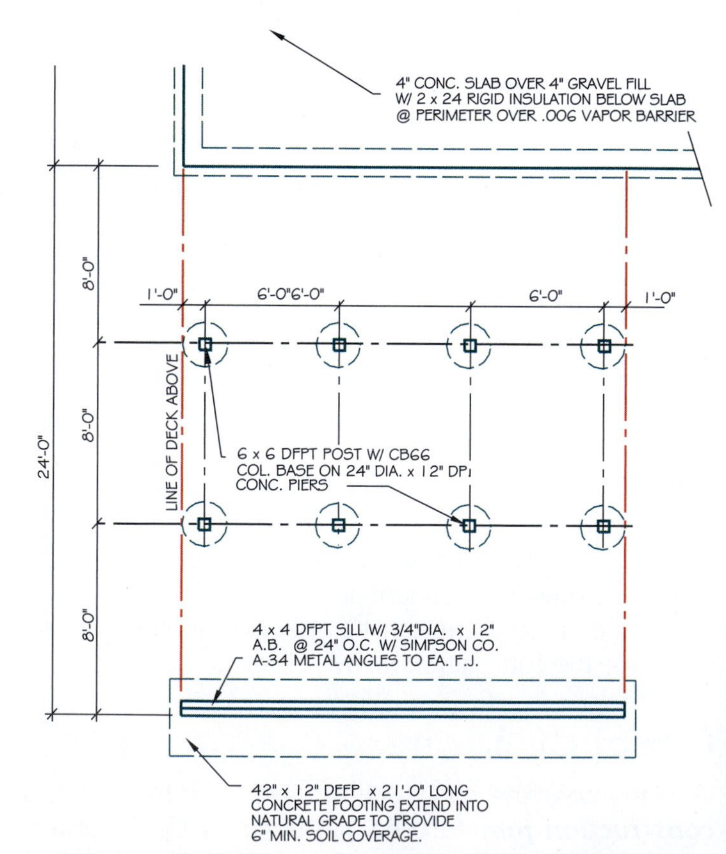

FIGURE 26.7 ■ *The foundation plan for an aboveground concrete slab shows the support piers for the framing platform.*

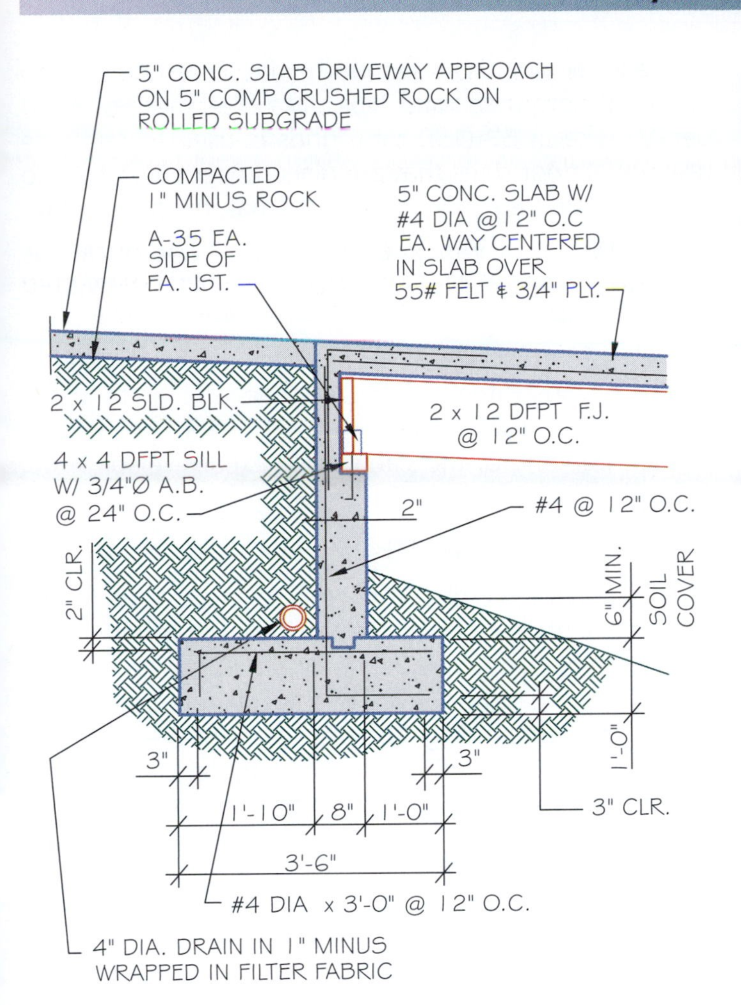

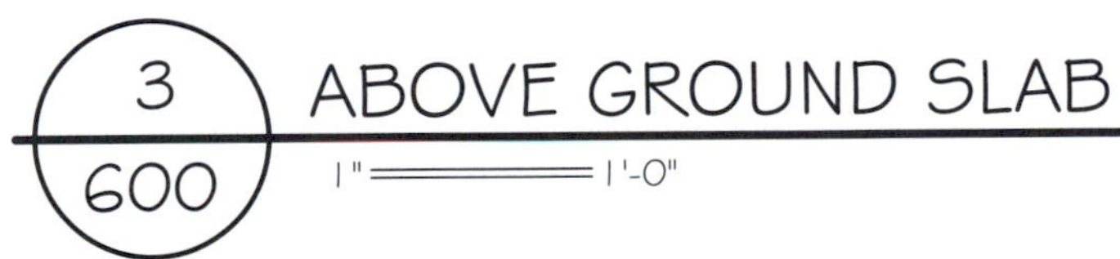

FIGURE 26.8 ■ *In addition to the framing and foundation plans, details are used to clarify the concrete reinforcement. This detail shows how the above ground slab intersects the on-grade slab.*

to be 6'' (150 mm) minimum above grade, but many municipalities require the top of the slab to be 8'' (200 mm) above the finish grade to keep structural wood away from ground moisture.

Slab Preparation

When a slab is built at grade, approximately 8 to 12'' in. (200 to 300 mm) of topsoil and vegetation is removed to provide a stable, level building site. Excavation usually extends about 5' (1500 mm) beyond the building edge to allow for the operation of excavating equipment needed to trench for the footings. Once forms for the footings have been set, fill material can be spread to support the slab. The IRC requires the slab to be placed on a 4'' (100 mm) minimum base of compacted sand, gravel fill, or crushed stone. The availability of materials will dictate the type of fill material used. The fill material provides a level base for the concrete slab and helps eliminate cracking caused by settling of the ground under the slab.

FIGURE 26.9a ■ *Precast hollow core slab units are lifted into position for a residential addition.* Courtesy David P. Scholz, Architect, DPS-labs.

FIGURE 26.9b ■ *Precast hollow core slab units were used to provide a large open space free of columns and beams at the lower floor level.* Courtesy David P. Scholz, Architect, DPS-labs.

The slab is required to be placed over 6-mil polyethylene sheet plastic to protect the floor from ground moisture. When a plastic vapor barrier is to be placed over gravel, a 1'' (25 mm) layer of sand should be specified to cover the gravel fill to avoid tearing the vapor barrier. An alternative is to use 55# rolled roofing in place of the plastic. The fill material and the vapor barrier are not shown on the foundation plan but are specified with a note. A typical note to specify the concrete and fill material might read:

4'' CONC. SLAB OVER .006 VISQUEEN OVER 1'' SAND FILL OVER 4'' COMPACTED GRAVEL FILL.

Slab Reinforcement

When the slab is placed on more than 4'' (100 mm) of uncompacted fill, welded wire fabric should be specified to help the slab resist cracking. Spacing and sizes of wires of welded wire fabric are identified by style and designations. A typical designation specified on a foundation plan might be

6 × 12—W16 × W8,

where

6 = longitudinal wire spacing
12 = transverse wire spacing
16 = longitudinal wire size
8 = transverse wire size

FIGURE 26.10 ■ *Welded wire mesh is often placed in concrete slabs to reduce cracking.* Courtesy Jim Hallmark, VOBB, Verot Oakes Building Blocks.

The letter W indicates smooth wire. D can be used to represent deformed wire. Typically a steel mesh of number 10 wire in 6'' (150 mm) grids is used to reinforce residential floor slabs that are placed over fill. A note to specify reinforced concrete and fill material might read:

4'' CONC. SLAB W/6 × 6 W12 × 12 WWM OVER .006 VISQUEEN OVER 1'' SAND FILL OVER 4'' COMPACTED GRAVEL FILL.

Figure 26.10 shows an example of the mesh used in concrete slabs. Steel reinforcing bars similar to those shown in Figure 26.11 can be added to a floor slab to prevent bending of the slab due to expansive soil.

Steel Rebar Placement In Floor Slabs

Although mesh is placed in a slab to limit cracking, steel reinforcement is placed in the concrete to prevent cracking due to bending. Steel reinforcing bars can be laid in a grid pattern near the surface of the concrete that will be in tension from bending. The placement of the reinforcement in the concrete is important to the effectiveness of the reinforcement. The amount of concrete placed around the steel is referred to as coverage. Proper coverage strengthens the bond between the steel and concrete and also protects the steel from corrosion if the concrete is exposed to chemicals, weather, or water. Proper coverage is also important to protect the steel from damage by fire. If steel is required to reinforce a residential concrete slab, an engineer will typically determine the size, spacing, coverage, and grade of bars to be used. As a general guideline to placing steel, the ACI recommends:

- For concrete cast against and permanently exposed to earth (footings): 3'' (75 mm) minimum coverage.

FIGURE 26.11 ■ *Steel reinforcing is used in place of welded wire mesh when the slab is placed over fill material.* Courtesy Constructionphotographs.com.

- For concrete exposed to weather or earth such as basement walls: 2'' (50 mm) coverage for #6–18 bars and 1 1/2'' (38 mm) coverage for #5 and W31/D31 wire or smaller.
- For concrete not exposed to weather or in contact with ground: 1 1/2'' (38 mm) coverage for slabs, walls, and joists; 3/4'' (19 mm) coverage for #11 bars and smaller; and 1 1/2'' (38 mm) coverage for #14 and #18 bars.

Wire mesh and steel reinforcement is not shown on the foundation plan but is specified by a note similar to that in Figure 26.6.

Post-tensioned Concrete Floor Systems

The methods of reinforcement mentioned thus far assume that the slab will be poured over stable soil. Concrete slabs can often be poured over unstable soil by using a method of reinforcement known as ***post-tensioning.*** This method of construction was originally developed for slabs that were to be poured at ground level and then lifted into place for multilevel structures. Adopted for the use of residential slabs, post-tensioning allows concrete slabs to be poured on grade over expansive soil. The technology has advanced sufficiently so that post-tensioning is now widely used even on stable soils.

For design purposes, a concrete slab can be considered as a wide, shallow beam supported by a concrete foundation at the edge. Although a beam may sag at the center from loads and gravity, a concrete slab can either sag or bow, depending on the soil conditions. Soil at the edge of a slab will be exposed to more moisture than soil near the center of the slab. This differential in moisture content can cause the edges of the slab to heave, creating tension in the bottom portion of the slab and compression in the upper portion. Center lift conditions can result as the soil beneath the interior of the slab becomes wetter and expands, as the perimeter of the slab dries and shrinks, or as a combination of the two factors. As the center of the slab heaves, tension is created in the upper portion of the slab with compression in the lower portion. Because concrete is very poor at resisting stress from tension, the slab must be reinforced with a material such as steel that has high tensile strength. Steel tendons with anchors can be extended through the slab as it is poured. Usually between 3 and 10 days after the concrete has been poured, these tendons are stretched by hydraulic jacks, which place approximately 25,000 lb of force on each tendon. The tendon force is transferred to the concrete slab through anchorage devices at the ends of the tendons. This process creates an internal compressive force throughout the slab, increasing the ability of the slab to resist cracking and heaving. Post-tensioning usually allows for a thinner slab than normally would be required to span over expansive conditions, elimination of other slab reinforcing, and elimination of most slab joints.

Two methods of post-tensioning are typically used for residential slabs: flat slab and ribbed slab. The flat slab method uses steel tendons ranging in diameter from 3/8 to 1/2'' (10 to 13 mm). The maximum spacing of tendons recommended by the Post Tensioning Institute (PTI) are:

- 3/8'' (9 mm) diameter: 5'-0'' (1500 mm) spacing
- 7/16'' (11 mm) diameter: 6'-10'' (1800 mm) spacing
- 1/2'' (13 mm) diameter: 9'-0'' (2700 mm) spacing

The exact spacing and size of tendons must be determined by an engineer based on the loads to be supported and the strength and conditions of the soil. When required, the tendons can be represented on the foundation plan, as shown in Figure 26.12. Details will also need to be provided to indicate how the tendons will be anchored as well as to show the exact locations of the tendons. Figure 26.13 is an example of a tendon detail. In addition to representing and specifying the steel throughout the floor system, the drafter will need to specify the engineer's require-

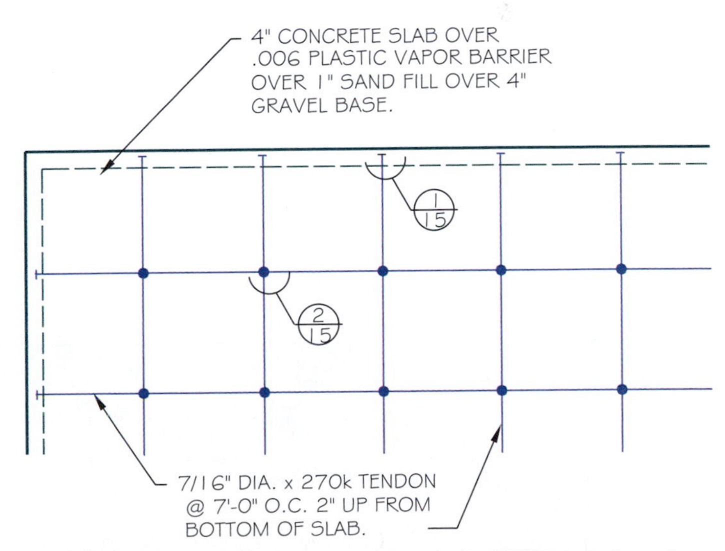

FIGURE 26.12 ■ *When a concrete slab is post-tensioned, the tendons and anchors used to support the floor slab must be represented on the foundation plan.*

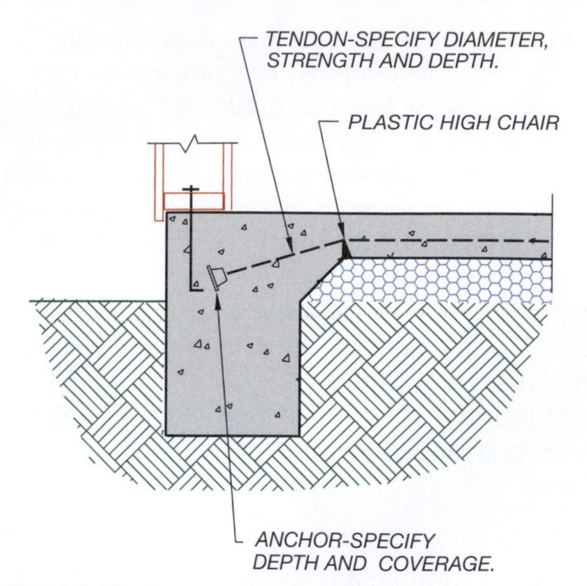

FIGURE 26.13 ■ *Tendon details should be drawn to reflect the design of the engineer.*

FIGURE 26.14 ■ *The preparation for a post-tensioned on-grade floor slab is with the use of concrete ribs placed below the slab. These beams reduce the span of the slab over the soil and provide increased support. The width, depth, and spacing are determined by the engineer based on the strength and condition of the soil and the size of the slab.* Courtesy Jim Hallmark, VOBB, Verot Oakes Building Blocks.

ments for the strength of the concrete at 28 days, the period when the concrete is to be stressed, as well as what strength the concrete should achieve before stressing.

A second method of post-tensioning an on-grade floor slab is with the use of concrete ribs or beams placed below the slab, similar to the foundation in Figure 26.14. These beams reduce the span of the slab over the soil and provide increased support. The width, depth, and spacing are determined by the engineer based on the strength and condition of the soil and the size of the slab. Figure 26.15 shows an example of a beam detail that a drafter would be required to draw to show the reinforcing specified by the engineer. Figure 26.16 is an example of how these beams could be shown on the foundation plan.

Slab Insulation

Depending on the risk of freezing, some municipalities require the concrete slab to be insulated to prevent heat loss. The insulation can be placed under the slab or on the outside edge of the stem wall. When it is placed under the slab, a rigid insulation material at least 2 × 24'' (50 × 600 mm) should be used to insulate the slab. When placed on the exterior side of the stem wall, the insulation should extend past the bottom of the foundation. Care must be taken to protect exposed insulation on the exterior side of the wall. This can usually be done by placing a protective covering such as 1/2'' (13 mm) concrete board over the insulation. Figure 26.17a and b shows common methods of insulating concrete floor slabs. Insulation is not shown on the foundation plan but is represented by a note on the foundation plan and specified in sections and footing details.

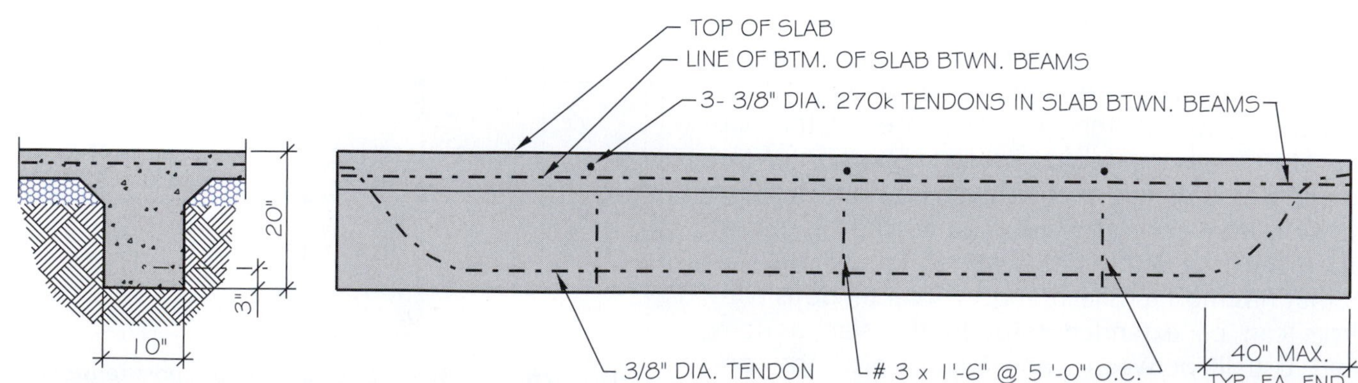

FIGURE 26.15 ■ *Concrete beams can be placed below the floor slab to increase the effect of the slab tensioning. Details must be drawn to indicate how the steel will be placed in the beam, as well as how the beam steel will interact with the slab steel.*

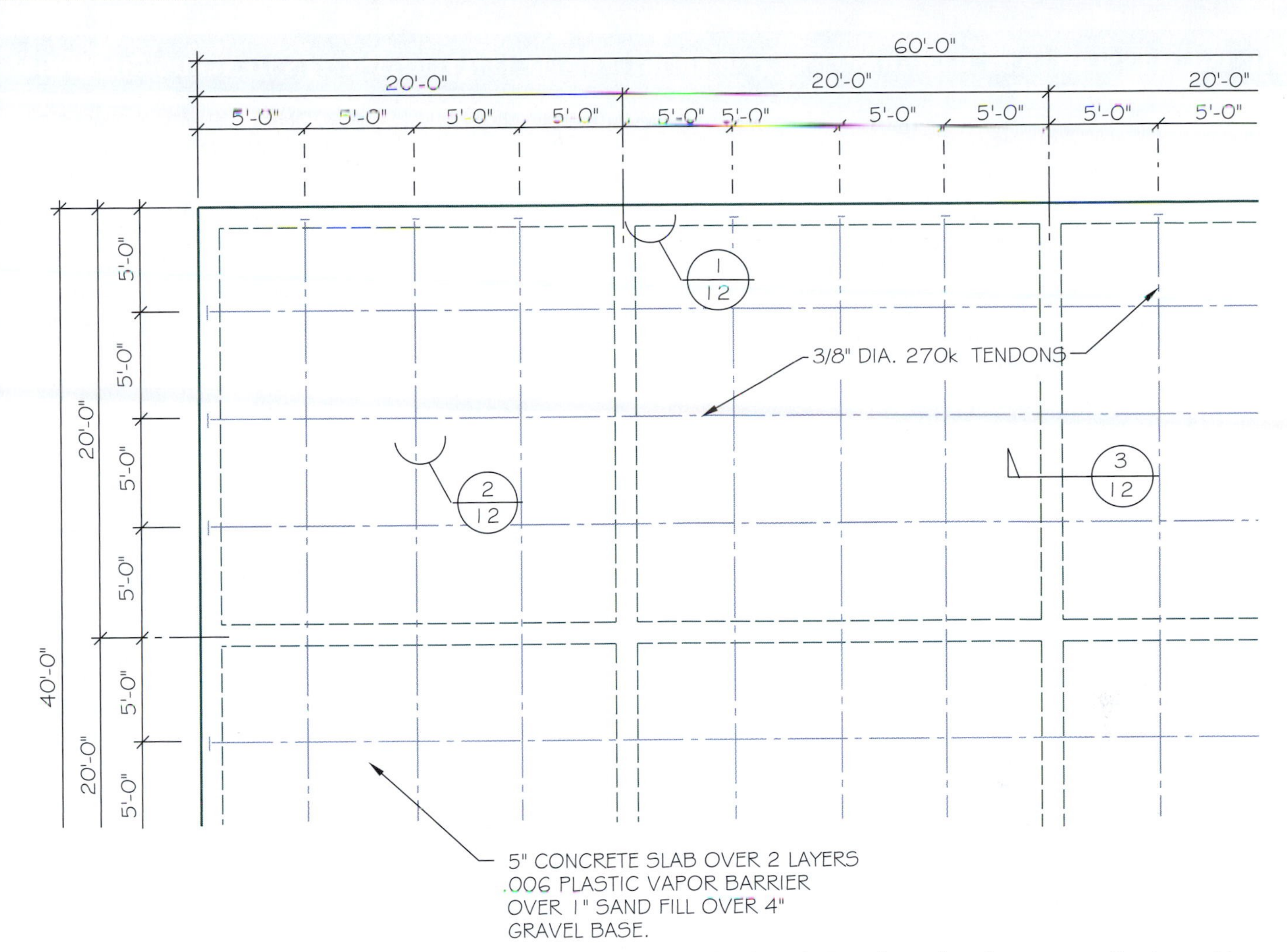

FIGURE 26.16 ■ *Beams located below the slab must be located on the foundation plan using the same methods used to represent an interior footing.*

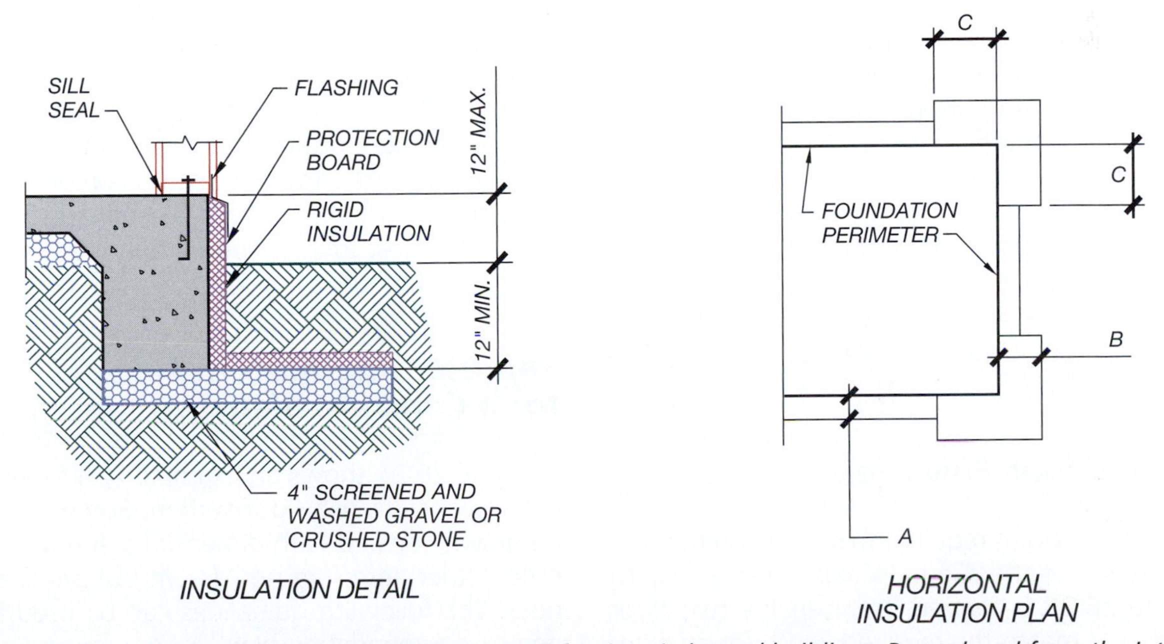

FIGURE 26.17a ■ *Insulation placement for frost-protected footings in heated buildings. Reproduced from the International Residential Code / 2006. Copyright © 2006.* Courtesy International Code Council, Inc.

MINIMUM INSULATION REQUIREMENTS FOR FROST-PROTECTED FOOTINGS IN HEATED BUILDINGS[a]

AIR FREEZING INDEX (°F-DAYS)[b]	VERTICAL INSULATION R-VALUE[c,d]	HORIZONTAL INSULATION R-VALUE[c,e]		HORIZONTAL INSULATION DIMENSIONS PER FIGURE R403.3(1) (INCHES)		
		ALONG WALLS	AT CORNERS	A	B	C
1,500 or less	4.5	NR	NR	NR	NR	NR
2,000	5.6	NR	NR	NR	NR	NR
2,500	6.7	1.7	4.9	12	24	40
3,000	7.8	6.5	8.6	12	24	40
3,500	9.0	8.0	11.2	24	30	60
4,000	10.1	10.5	13.1	24	36	60

For SI: 1 inch = 25.4 mm, °C = [(°F)-32]/1.8.

a. Insulation requirements are for protection against frost damage in heated buildings. Greater values may be required to meet energy conservation standards. Interpolation between values is permissible.

b. See Figure R403.3(2) for Air Freezing Index values.

c. Insulation materials shall provide the stated minimum R-values under long-term exposure to moist, below-ground conditions in freezing climates. The following R-values shall be used to determine insulation thicknesses required for this application: Type II expanded polystyrene—2.4R per inch; Type IV extruded polystyrene—4.5R per inch; Type VI extruded polystyrene—4.5R per inch; Type IX expanded polystyrene—3.2R per inch; Type X extruded polystyrene—4.5R per inch. NR indicates that insulation is not required.

d. Vertical insulation shall be expanded polystyrene insulation or extruded polystyrene insulation.

e. Horizontal insulation shall be extruded polystyrene insulation.

FIGURE 26.17b ■ *Minimum insulation requirements for a concrete slab. Reproduced from the International Residential Code / 2006. Copyright © 2006.* Courtesy International Code Council, Inc.

Plumbing and Heating Requirements

Plumbing and heating ducts must be placed under the slab before the concrete is poured. On residential plans, plumbing is usually not shown on the foundation plan. Generally the skills of the plumbing contractor are relied on for the placement of required utilities. Although piping runs are not shown, terminations such as floor drains are often shown and located on a concrete slab plan. Figure 26.18 shows the placement of plumbing materials prior to pouring the slab. If heating ducts will be placed under the slab, they are usually drawn on the foundation plan, as shown in Figure 26.19.

Changes in Floor Elevation

The floor level is often required to step down to meet the design needs of the client. A stem wall similar to that in Figure 26.20 is formed between the two floor levels and should match the required width for an exterior stem wall. The lower slab is typically thickened to a depth of 8'' (200 mm) to support the change in elevation. Steps between floors greater than 24'' (600 mm) should be designed as a retaining wall and detailed using the methods described in Chapters 25 and 28. Figure 26.21 shows how a lowered slab can be represented. The step often occurs at what will be the edge of a wall when the framing plan is complete. Great care must be taken to coordinate the dimensions of a floor plan with those of the foundation plan, so that walls match the foundation.

Representing Material for a Concrete Slab

Figure 26.21 shows an example of a foundation plan with a concrete slab floor system. Some materials must be drawn, located with dimensions and referenced in a note. Other items are not drawn but are specified in a note. The following guidelines can be used for placing items on a foundation plan.

FIGURE 26.18 ■ *Plumbing must be placed prior to pouring the concrete slab.* Courtesy Dr. Carol Ventura.

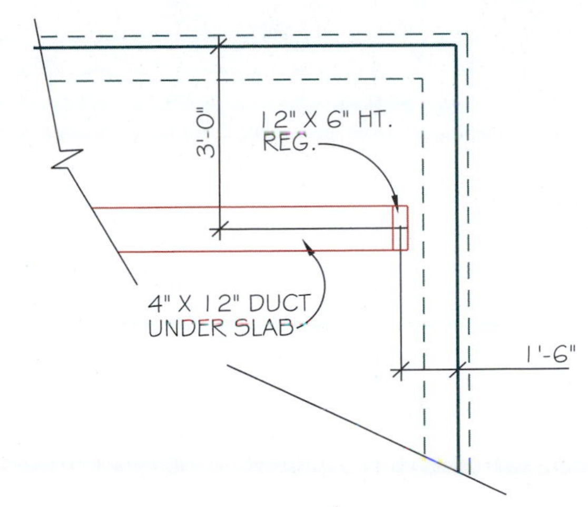

FIGURE 26.19 ■ *HVAC ducts and registers that are to be located in a concrete floor must be shown on the foundation plan.*

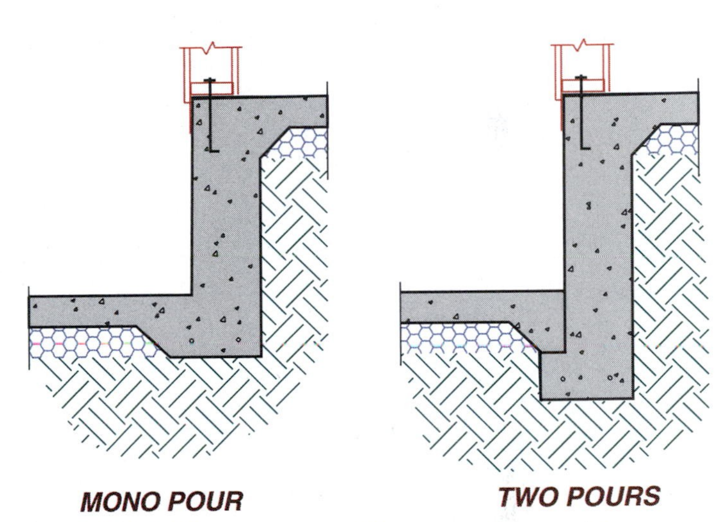

FIGURE 26.20 ■ *A stem wall is created between floor slabs when a change in elevation is required.*

Common Components Shown on a Slab Foundation

Each of the followings materials is shown in Figure 26.21. The following items must be represented and dimensioned on the foundation plan:

Outline of slab
Interior footing locations
Changes in floor level
Floor drains
Exterior footing locations
Ducts for mechanical
Metal anchors
Patio slabs

Common Components Specified by Note Only

The specification for each of the following materials is shown in Figure 26.21. The following items are not drawn but must be referenced by a note on the foundation plan:

Slab thickness and fill material
Wire mesh
Mudsill size
Vapor barriers
Pier sizes
Assumed soil strength
Reinforcing steel
Anchor bolt size and spacing
Insulation
Slab slopes
Concrete strength

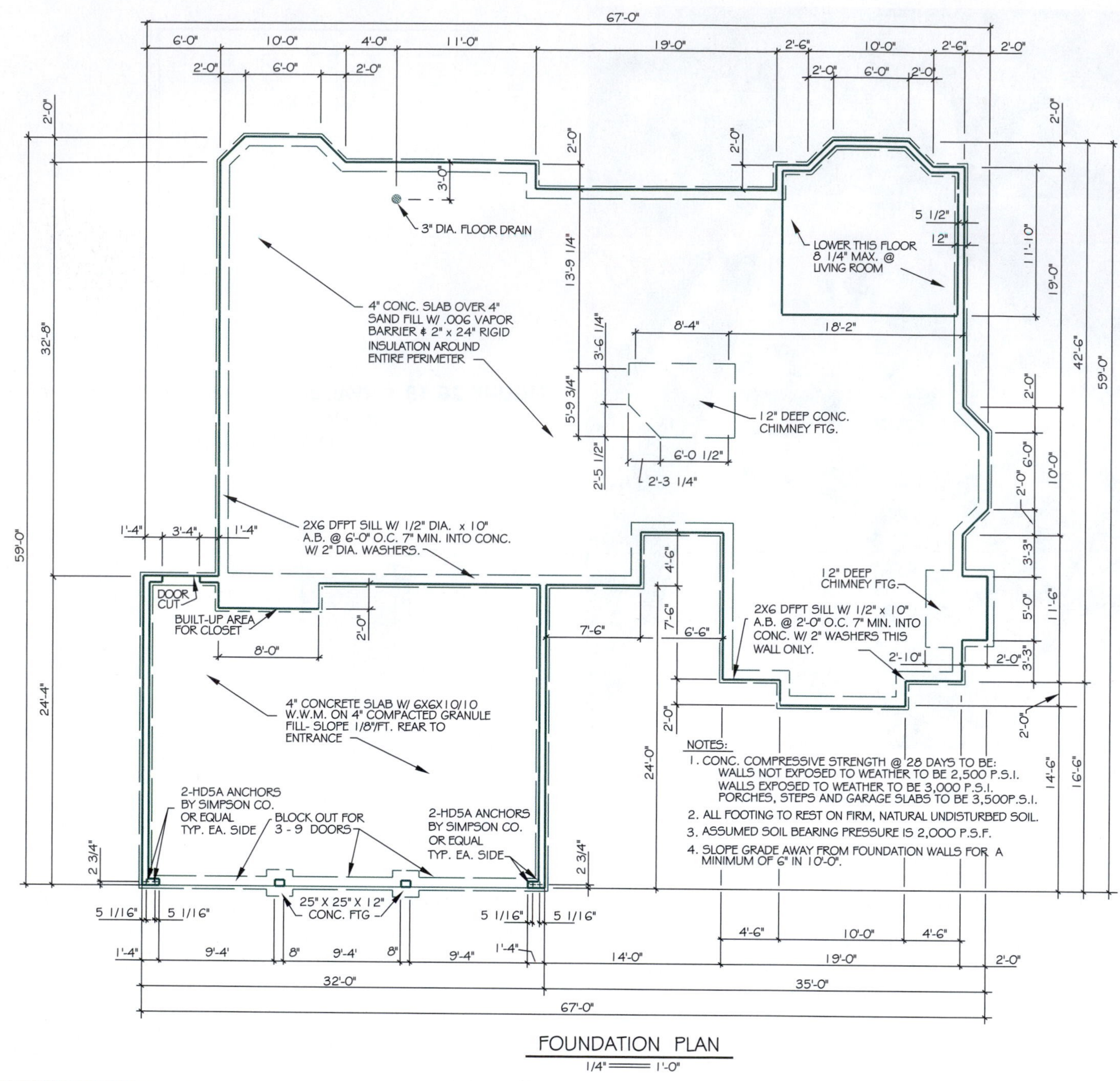

FIGURE 26.21 ■ *A foundation plan for a home with a concrete slab floor system.*

FLOOR SYSTEMS ABOVE A CRAWL SPACE

The crawl space is the area formed between the floor system and the ground. The IRC requires a minimum of 18'' (450 mm) from the bottom of the floor decking to the ground and 12'' (300 mm) from the bottom of girders to the ground. Two common methods of providing a crawl space below the floor are by using floor joists or the post-and-beam system. An introduction to each system is needed to complete the foundation. See Section 6 for a review of each type of floor system as it relates to the entire structure.

Joist Floor Framing

A common method of framing a wood floor is with repetitive wood members called floor joists. Floor systems made with joists are often used on sloping sites because the system requires less interior supports than with a post and beam system. Floor joists are used to span between the foundation walls and are shown in Figure 26.22a. Sawn lumber ranging in size from 2 × 6 through 2 × 12 (50 × 150 to 50 × 300) are used to frame the floor platform. Trusses, steel joists, and joists made from engineered lumber similar to those in Figure 26.22b are also used to frame the floor system. The floor joists are usually placed at 16'' (400 mm) o.c., but the spacing of joists may change depending on the span, the material used, and the load to be supported.

Constructing a Joist Floor

To construct a joist floor system, a pressure-treated sill is bolted to the top of the foundation wall with the anchor bolts that were placed when the foundation was formed. The floor joists can then be nailed to the sill. An alternative used to reduce the distance from the finished floor to the finish grade is to set the floor joists flush with the sill, as in Figure 26.23. This method requires the use of metal hangers to support the joists. With the floor joists in place, plywood floor sheathing is installed to provide a base for the finish floor. The size of the subfloor depends on the spacing of the floor joists and the floor loads that must be supported. The live load to be supported will also affect the thickness of the plywood to be used. The APA has span tables for plywood from 7/16'' through 7/8'' (11 through 22 mm) to meet the various conditions that might be found in a residence. A subfloor can also be made from 1'' (25 mm) material such as 1 × 6 (25 × 150 mm) tongue-and-groove (T&G) lumber, although this method requires more labor. The floor sheathing is not represented on the foundation plan but is specified in a general note. Figure 26.24 shows common methods of drawing floor joists on the foundation plan.

Supporting Floor Joists

When the distance between the foundation walls is too great for the floor joists to span, a girder is used to support the joists. A girder is a horizontal load-bearing member that spans between two or more supports at the foundation level. Girders may be drawn using a sin-

A

B

FIGURE 26.22 ■ *(A) A floor framed with floor joists uses wood members ranging from 2 × 6 through 2 × 12 (50 × 150 to 50 × 300) to span between supports. Solid blocking is placed between the joists at 10'-0'' (3000 mm) o.c. (B) Engineered lumber is a popular material for joist floor systems. Joists are typically placed at 16'' (400 mm) o.c. Depending on the load to be supported, the allowable deflection, and the span, a 12 or 24'' (300 or 600 mm) spacing may also be used.* Courtesy LeRoy Cook.

FIGURE 26.23 ■ *Although floor joists are normally placed above the sill, they can be mounted flush to the sill to reduce the distance from the finish floor to the finish grade. The weight of the joist is transferred to the sill using joist hangers.* Courtesy Jordan Jefferis.

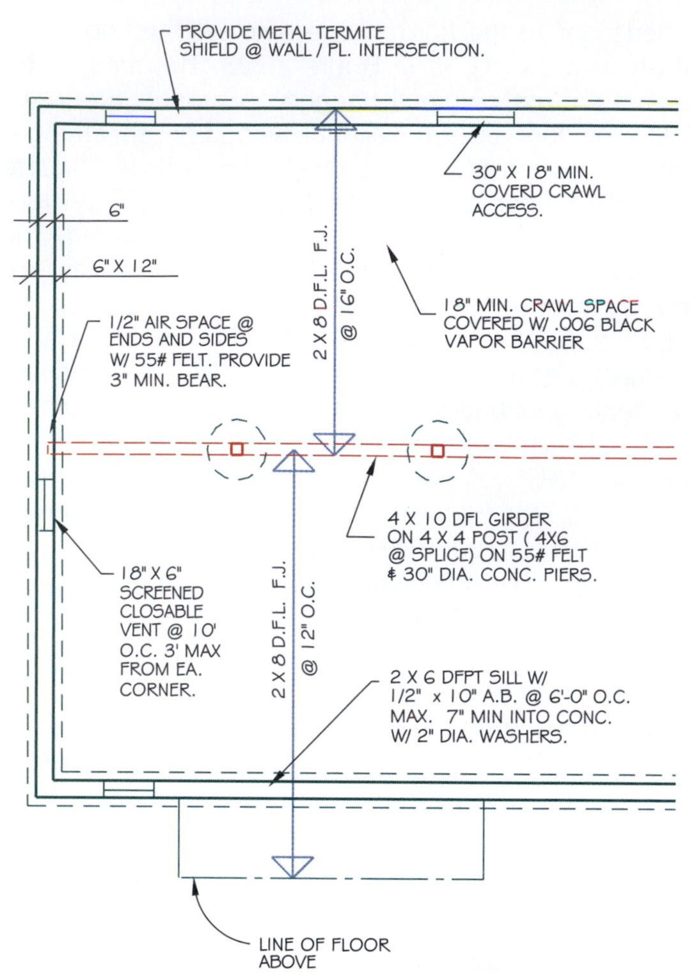

FIGURE 26.24 ■ *Common methods of representing and specifying joist floor components in plan view.*

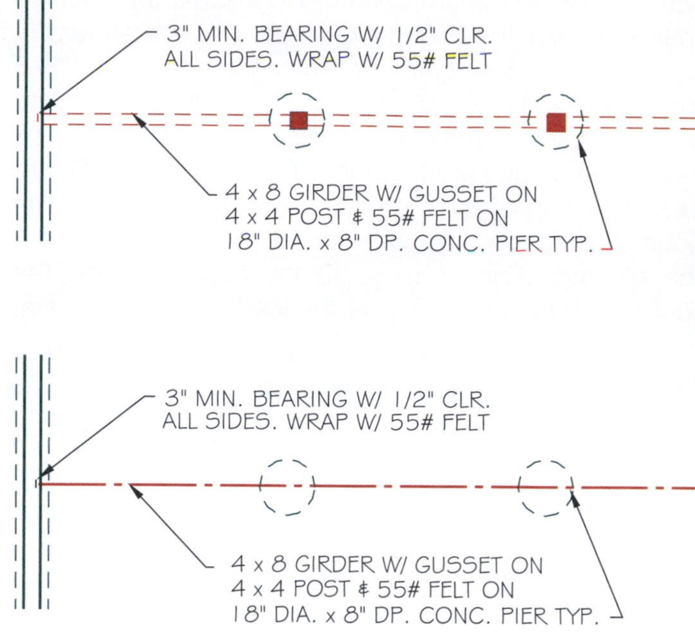

FIGURE 26.25 ■ *Common methods of representing girders, piers, and posts in plan view.*

FIGURE 26.26 ■ *Steel girders and columns have been placed in preparation for the placement of joists for the floor system.* Courtesy W. Lee Roland, Builder (rolandbuilder.com). Brent W. Roland, photographer.

gle bold line or pairs of thin lines similar to those in Figure 26.25. The girder is usually supported in a concrete beam pocket, where it intersects the foundation wall. Depending on the load to be supported and the area you are in, either a wood, laminated wood, engineered wood product, or steel member may be used for the girder and for the support post. No matter what the material is, the girder must be attached to the post. When wood members are used, a gusset attaches the post and girder together. A steel connector can also be used to make wood-to-wood connections or wood-to-steel connections. A steel girder may require no intermediate supports or can be welded to a steel column. Figure 26.26 shows the use of a steel girder to form the floor system over a basement.

A concrete pier is placed under the intermediate girder supports to resist settling. The post must be attached to the pier in some seismic zones. A metal rod inserted into a predrilled hole in the post can be used in areas of low seismic risk. A metal strap or post base may be required where high wind or seismic risk is greater. Figure 26.27 shows common connections required for a joist floor system. Figure 26.24 shows methods of drawing the girders, posts, piers, and beam pocket on the foundation plan. Figure 26.28 shows a portion of the floor framing in place to support a cantilevered window seat. Figure 26.29 shows a complete foundation plan using floor joists to support the floor.

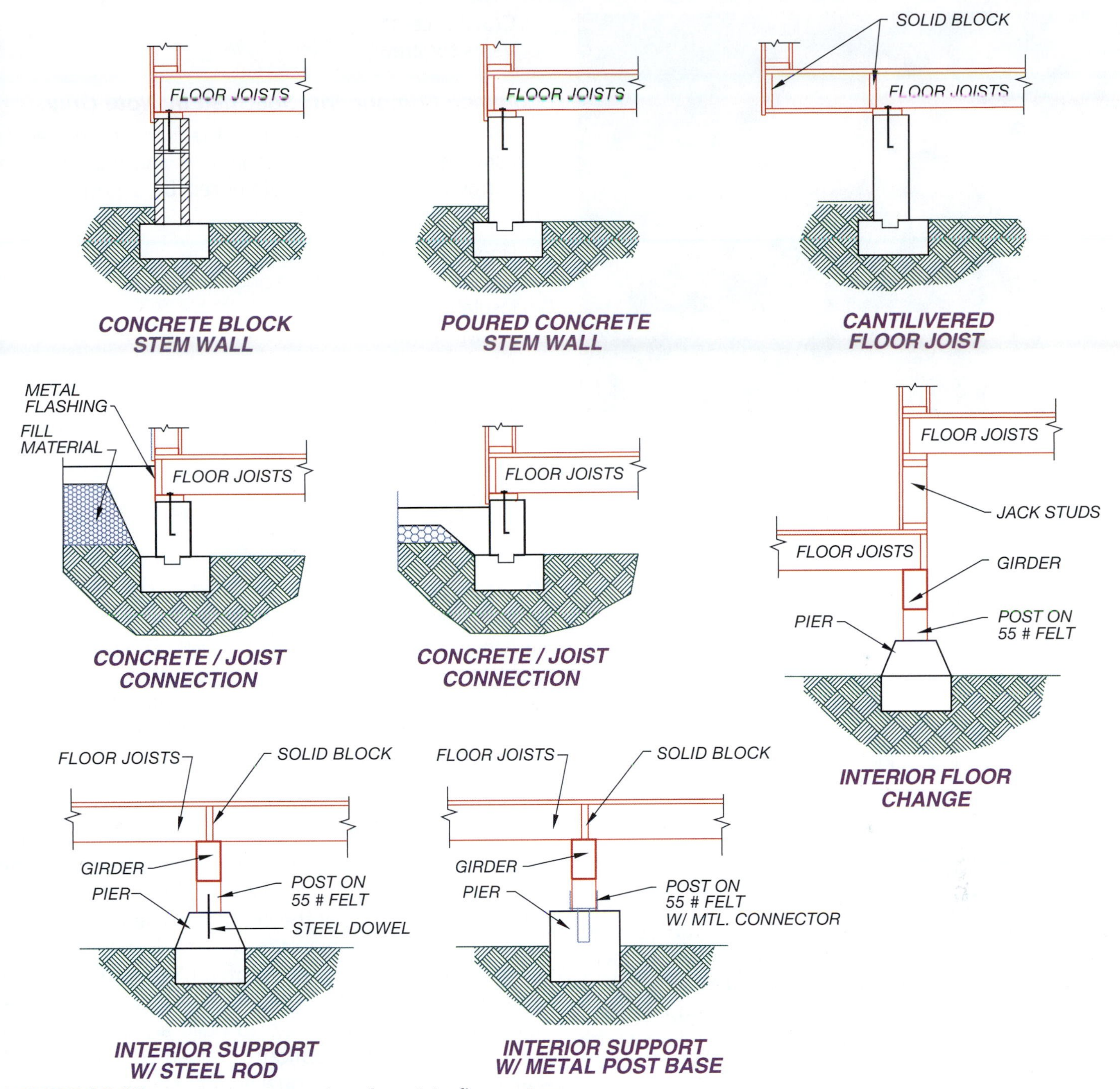

FIGURE 26.27 ▪ *Common connections for a joist floor system.*

FIGURE 26.28 ■ *Placement of floor materials for a cantilevered floor system. Joists are doubled at the edge of the projection to support wall loads.* Courtesy Megan Jefferis.

Representing Materials for a Joist Floor System

Figure 26.29 shows an example of a foundation plan with a concrete slab floor system. Some materials must be drawn, located with dimensions, and referenced in a note. Other items are not drawn but are specified in a note. The following guidelines can be used for placing items on a foundation plan.

Common Components Shown with a Joist Floor System

The following items must be represented on a foundation plan with a joist floor system. Unless marked with an asterisk, all of the components require dimensions to provide location information. The location of objects marked with an asterisk are specified in a note and do not require dimensions. Items that must be drawn include:

Foundation walls
Door openings in foundation walls
Fireplace chimney
Floor joists
Girders (bearing walls and floor support)
Girder pockets
Exterior footings
Metal anchors
Fireplace footings
The outline of floor cantilevers
Interior piers
Changes in floor levels
Crawl access*
Vents for crawl space*

Common Components Specified by Note Only

The specifications for each of the following materials are shown in Figure 26.29. The following materials are not drawn but must be referenced by a note on the foundation plan:

Floor joist size and spacing
Anchor bolt size and spacing
Insulation
Crawl height
Assumed soil strength
Subfloor material
Girder sizes
Mudsill size
Vapor barrier
Concrete strength
Wood type and grade

Post-and-Beam Floor Systems

A post-and-beam floor system is built using a standard foundation system. It is a popular floor framing choice for flat or low-sloped building sites. Rather than having floor joists span between the foundation walls, a series of beams are used to support the subfloor, as shown in Figure 26.30. Once the mudsill is bolted to the foundation wall, the beams are placed so that the top of each beam is flush with the top of the mudsill. The beams are placed at 48'' (1200 mm) o.c., but the spacing can vary depending on the size of the floor decking to be used. The area of the country where the structure is to be built affects how the subfloor is made. Generally plywood with a thickness of 1 1/8'' (28.5 mm) and an APA rating of STURD-I-FLOOR 2-4-1 with an exposure rating of EXP-1 is used to build a post-and-beam subfloor. When the subfloor is glued to the support beams, the strength and quality of the floor is greatly increased because squeaks, bounce, and nail popping are eliminated. An alternative to plywood is to use 2'' (50 mm) material such as 2 × 6 (50 × 150) T&G boards laid perpendicular or diagonally to the girders.

Girder Support

The girders are supported by wooden posts as they span between the foundation walls. Posts are usually placed at 8' (2400 mm) o.c., but spacing can vary depending on the load to be supported and the size of the girder. Each post is supported by a concrete pier similar

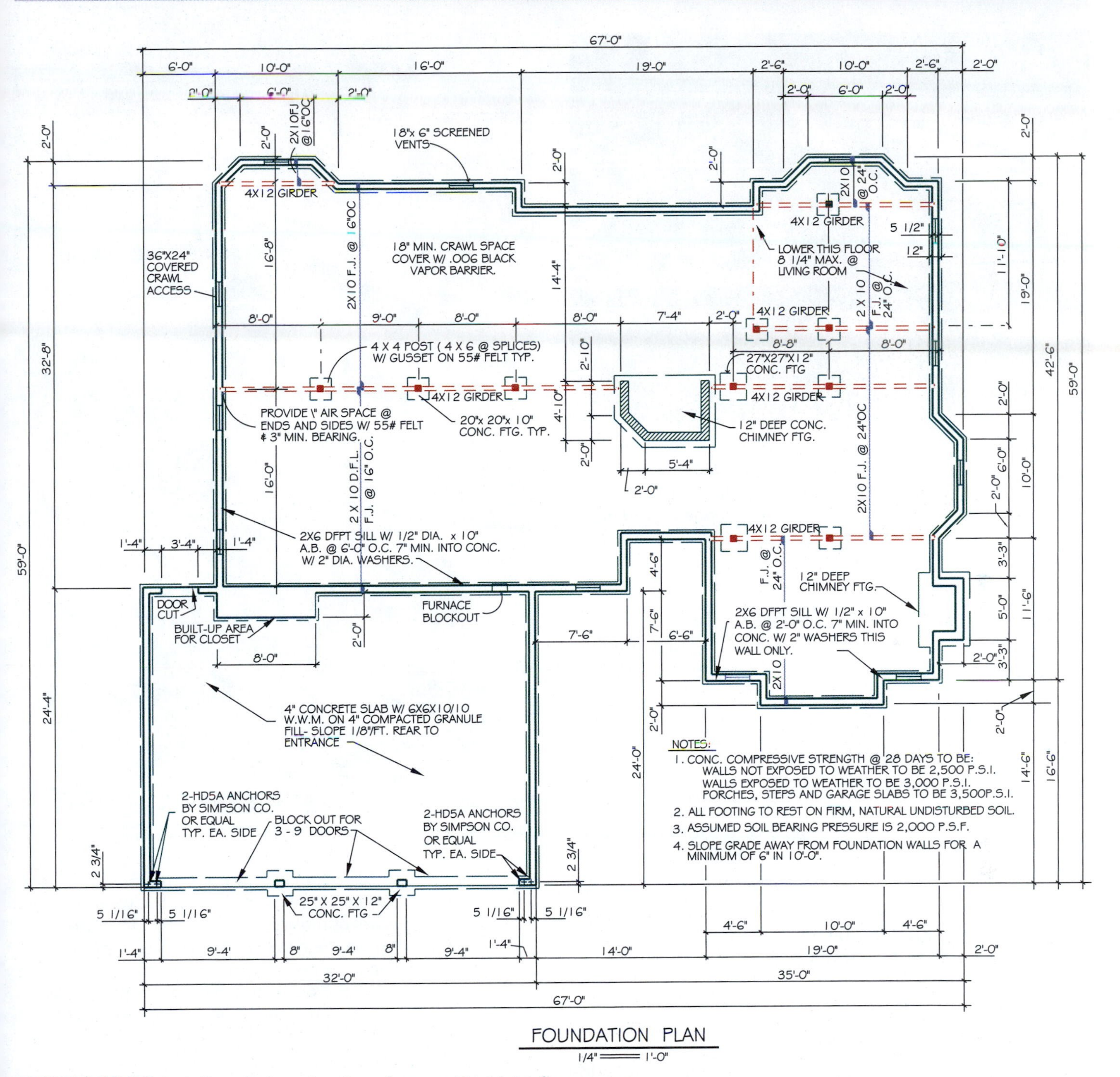

18"x 6" SCREENED VENTS
4X12 GIRDER
36"X24" COVERED CRAWL ACCESS
18" MIN. CRAWL SPACE COVER W/ .006 BLACK VAPOR BARRIER.
2X10 F.J. @ 16" O.C.
LOWER THIS FLOOR 8 1/4" MAX. @ LIVING ROOM
4 X 4 POST (4 X 6 @ SPLICES) W/ GUSSET ON 55# FELT TYP.
27"X27"X12" CONC. FTG
PROVIDE 1" AIR SPACE @ ENDS AND SIDES W/ 55# FELT & 3" MIN. BEARING.
20"x 20"x 10" CONC. FTG. TYP.
12" DEEP CONC. CHIMNEY FTG.
2X10 F.J. @ 24"OC
2X10 D.F.L. F.J. @ 16" O.C.
2X6 DFPT SILL W/ 1/2" DIA. x 10" A.B. @ 6'-0" O.C. 7" MIN. INTO CONC. W/ 2" DIA. WASHERS.
12" DEEP CHIMNEY FTG.
DOOR CUT
BUILT-UP AREA FOR CLOSET
FURNACE BLOCKOUT
2X6 DFPT SILL W/ 1/2" x 10" A.B. @ 2'-0" O.C. 7" MIN. INTO CONC. W/ 2" WASHERS THIS WALL ONLY.
4" CONCRETE SLAB W/ 6X6X10/10 W.W.M. ON 4" COMPACTED GRANULE FILL- SLOPE 1/8"/FT. REAR TO ENTRANCE
NOTES:
1. CONC. COMPRESSIVE STRENGTH @ 28 DAYS TO BE: WALLS NOT EXPOSED TO WEATHER TO BE 2,500 P.S.I. WALLS EXPOSED TO WEATHER TO BE 3,000 P.S.I. PORCHES, STEPS AND GARAGE SLABS TO BE 3,500P.S.I.
2. ALL FOOTING TO REST ON FIRM, NATURAL UNDISTURBED SOIL.
3. ASSUMED SOIL BEARING PRESSURE IS 2,000 P.S.F.
4. SLOPE GRADE AWAY FROM FOUNDATION WALLS FOR A MINIMUM OF 6" IN 10'-0".
2-HD5A ANCHORS BY SIMPSON CO. OR EQUAL TYP. EA. SIDE
BLOCK OUT FOR 3 - 9 DOORS
25" X 25" X 12" CONC. FTG
67'-0"
32'-0"
35'-0"
FOUNDATION PLAN
1/4" = 1'-0"

FIGURE 26.29 ■ *A foundation plan for a home with a joist floor system.*

FIGURE 26.30 ■ *Girders are placed at 48'' (1200 mm) o.c. with supports at 8'-0'' (2400 mm) o.c. Additional girders can be added to provide support for braced wall panels. Blocking is used between girders to provide support for plumbing.* Courtesy Zachary Jefferis.

to that in Figure 26.31. Figure 26.32 shows common components of a post-and-beam floor system. Beams, posts, and piers are drawn on the foundation plan, as shown in Figure 26.24.

Representing Materials for a Post-and-Beam Floor System

Figure 26.33 shows an example of a foundation plan with a post-and-beam floor system. Some materials must be drawn, located with dimensions and referenced in a note. Other items are not drawn but are specified in a note. The following guidelines can be used for placing items on a foundation plan.

Common Components of a Post-and-Beam System

The following items must be represented on a foundation plan using a post-and-beam floor system. Unless marked with an asterisk, all of the components require dimensions to provide location information. The location of objects marked with an asterisk are specified in a note and do not require dimensions. Items that must be drawn include:

Foundation walls
Openings in walls for doors
Fireplace chimney
Girders (beams)
Girder pockets

FIGURE 26.31 ■ *Posts for a wood floor system sit on a shingle or 55# felt placed above a concrete pier. The shingle is placed to stop moisture from being drawn into the post. A vapor barrier keeps moisture from passing from the soil into the crawl space. At the top of the post, a gusset is used to connect the girder to the post. Crossbracing between posts may be required, depending on the height of the post and the seismic zone.* Courtesy Matthew Jefferis.

Crawl access*
Exterior footings
Metal anchors
Fireplace footings
Piers
Changes in floor levels
Vents for crawl space*

Common Components Specified by Note Only

The specification for each of the following materials can be found in Figure 26.33. The following materials

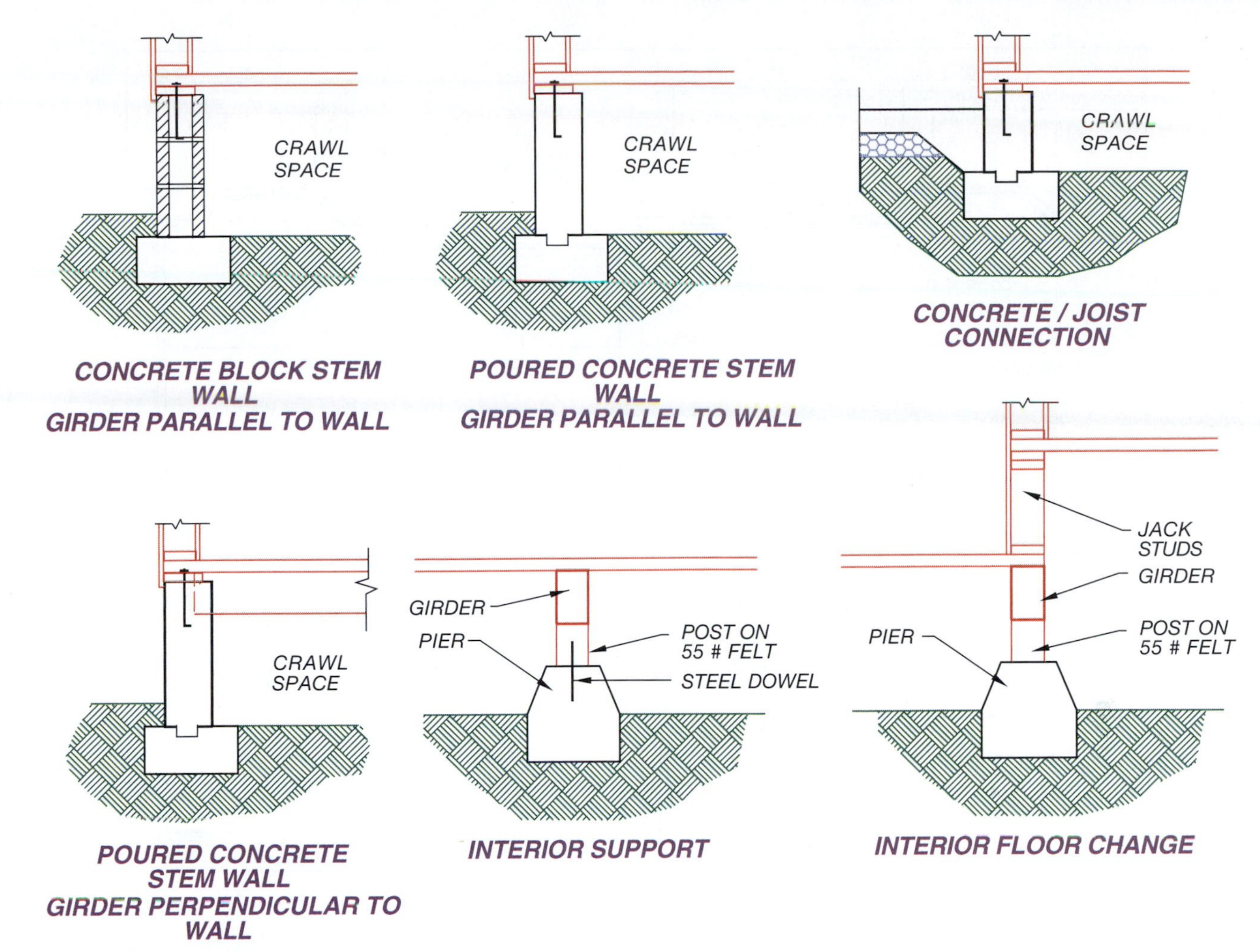

FIGURE 26.32 ■ *Common connections for a post-and-beam floor system.*

are not drawn but must be referenced by a note on the foundation plan:

Anchor bolt spacing
Mudsill size
Insulation
Minimum crawl height
Subfloor material
Girder sizes
Vapor barrier
Concrete strength
Wood type and grade
Assumed soil strength

COMBINED FLOOR METHODS

Floor systems and foundation construction methods may be combined, depending on the building site. This is typically done on partially sloping lots when part of a structure may be constructed with a slab and part of the structure with a joist floor system, as shown in Figure 26.34. The slab is built in an area requiring little or no excavation, and the wood floor system eliminates the need for placing compacted fill to support a concrete slab. A residence with a basement is another example of construction that requires combined floor methods. A home with a partial basement will require the use of a concrete floor in the basement area, with either a joist or post-and-beam floor system over the crawl space. A joist floor for the crawl space is easier to match the floor system that will be used above the basement area. Figure 26.35a shows a residence with a partial basement. The right portion of the plan uses a joist floor system over the crawl area. The left side of the structure has a basement with a concrete slab floor. Figure 26.35b shows a residence with a full basement. The entire basement can be constructed using the methods that were introduced earlier, in the discussion of concrete floors. Figure 26.35c shows a residence with a full basement. The major difference between b and c is that the retaining wall does not totally

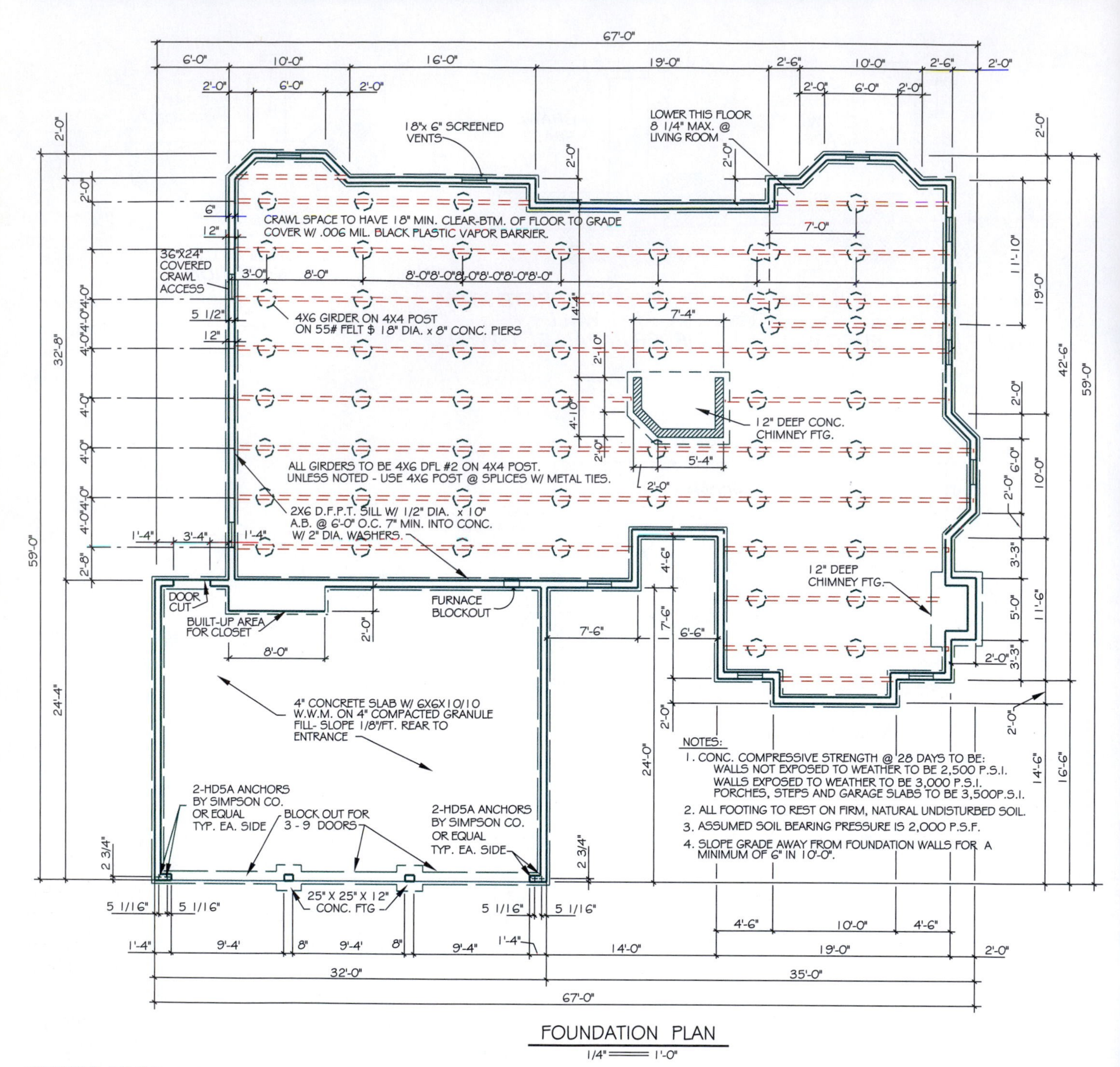

67'-0"
LOWER THIS FLOOR 8 1/4" MAX. @ LIVING ROOM
18"x 6" SCREENED VENTS
CRAWL SPACE TO HAVE 18" MIN. CLEAR-BTM. OF FLOOR TO GRADE COVER W/ .006 MIL. BLACK PLASTIC VAPOR BARRIER.
36"X24" COVERED CRAWL ACCESS
4X6 GIRDER ON 4X4 POST ON 55# FELT $ 18" DIA. x 8" CONC. PIERS
12" DEEP CONC. CHIMNEY FTG.
ALL GIRDERS TO BE 4X6 DFL #2 ON 4X4 POST. UNLESS NOTED - USE 4X6 POST @ SPLICES W/ METAL TIES.
2X6 D.F.P.T. SILL W/ 1/2" DIA. x 10" A.B. @ 6'-0" O.C. 7" MIN. INTO CONC. W/ 2" DIA. WASHERS.
12" DEEP CHIMNEY FTG.
DOOR CUT
BUILT-UP AREA FOR CLOSET
FURNACE BLOCKOUT
4" CONCRETE SLAB W/ 6X6X10/10 W.W.M. ON 4" COMPACTED GRANULE FILL- SLOPE 1/8"/FT. REAR TO ENTRANCE
2-HD5A ANCHORS BY SIMPSON CO. OR EQUAL TYP. EA. SIDE
BLOCK OUT FOR 3 - 9 DOORS
25" X 25" X 12" CONC. FTG
NOTES:
1. CONC. COMPRESSIVE STRENGTH @ 28 DAYS TO BE: WALLS NOT EXPOSED TO WEATHER TO BE 2,500 P.S.I. WALLS EXPOSED TO WEATHER TO BE 3,000 P.S.I. PORCHES, STEPS AND GARAGE SLABS TO BE 3,500P.S.I.
2. ALL FOOTING TO REST ON FIRM, NATURAL UNDISTURBED SOIL.
3. ASSUMED SOIL BEARING PRESSURE IS 2,000 P.S.F.
4. SLOPE GRADE AWAY FROM FOUNDATION WALLS FOR A MINIMUM OF 6" IN 10'-0".
FOUNDATION PLAN
1/4" = 1'-0"

FIGURE 26.33 ■ *A foundation plan for a post-and-beam floor system.*

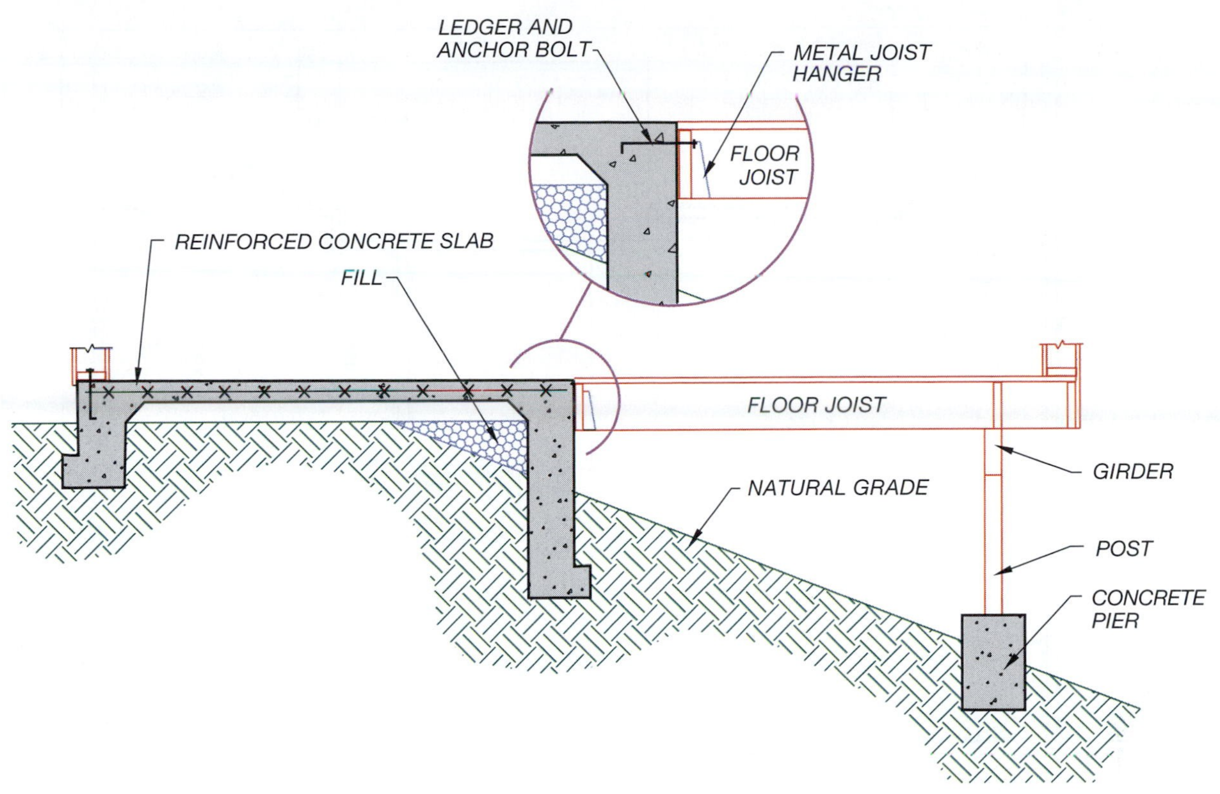

FIGURE 26.34 ■ *Floor systems combining concrete slabs and floor joists are often used on sloping sites to help minimize fill material. A ledger is used to provide anchorage to floor joists where they intersect the concrete slab. Metal joist hangers are used to join the joists to the ledger.*

enclose the basement. This type of basement is ideal for homes built on sloping sites. The basement is built on the low side of the site, allowing some exterior walls to be constructed of wood. Notice on each side of the foundation that a 48'' (1200 mm) high retaining wall has been used. Careful grading allows the 8'-0'' (2400 mm) high retaining wall to be reduced and eliminated for a portion of the basement. Depending on your area, the drawings for the basement may require an engineer's stamp. Verify with your local building department to determine the maximum wall heights allowed without an engineer's stamp. Chapter 28 explores the drawing needed to detail the basement walls.

One component typically used when floor systems are combined is a ledger. A ledger is used to provide support for floor joists and subfloor when they intersect the concrete. Unless felt is placed between the concrete and the ledger, the ledger must be pressure-treated lumber. The ledger can be shown on the foundation plan but is generally specified by note only.

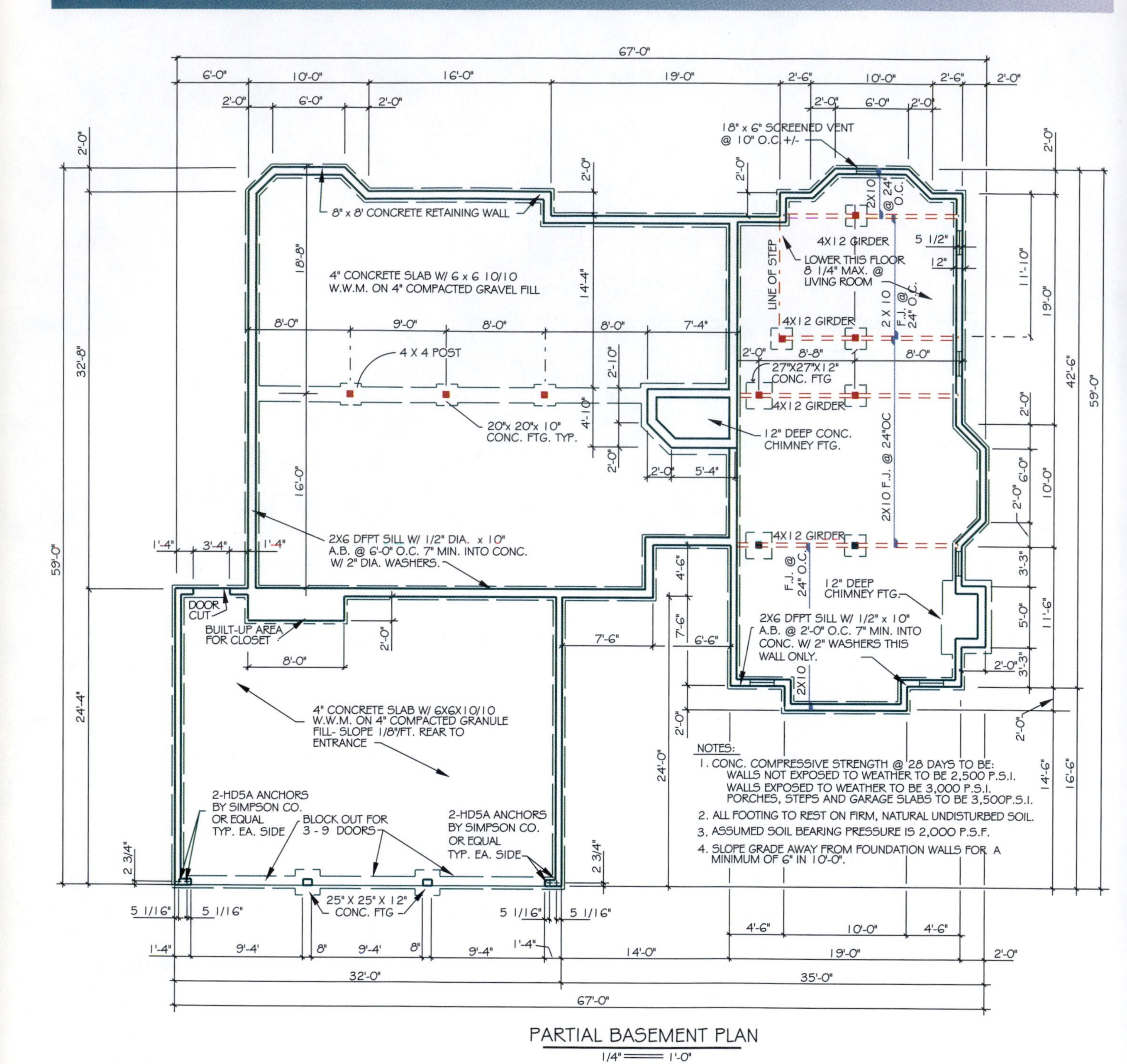

FIGURE 26.35a ■ *A foundation plan for a home with a partial basement. The basement has a concrete floor, and the floor over the crawl space is supported by joists.*

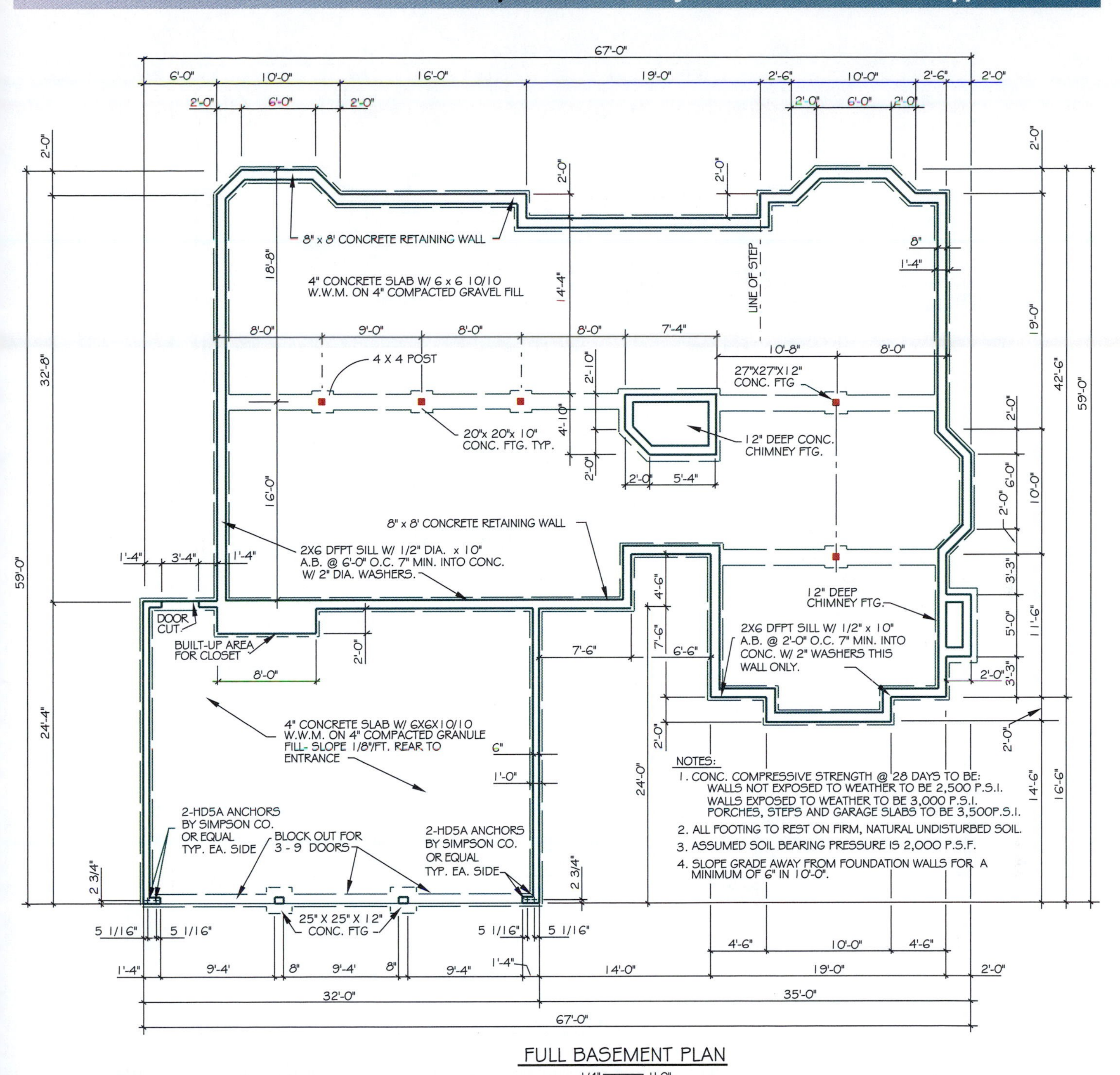

FIGURE 26.35b ■ *A foundation plan for a home with a full basement. Window wells must be added using the guidelines presented in Chapter 7 if habitable space is located in the basement.*

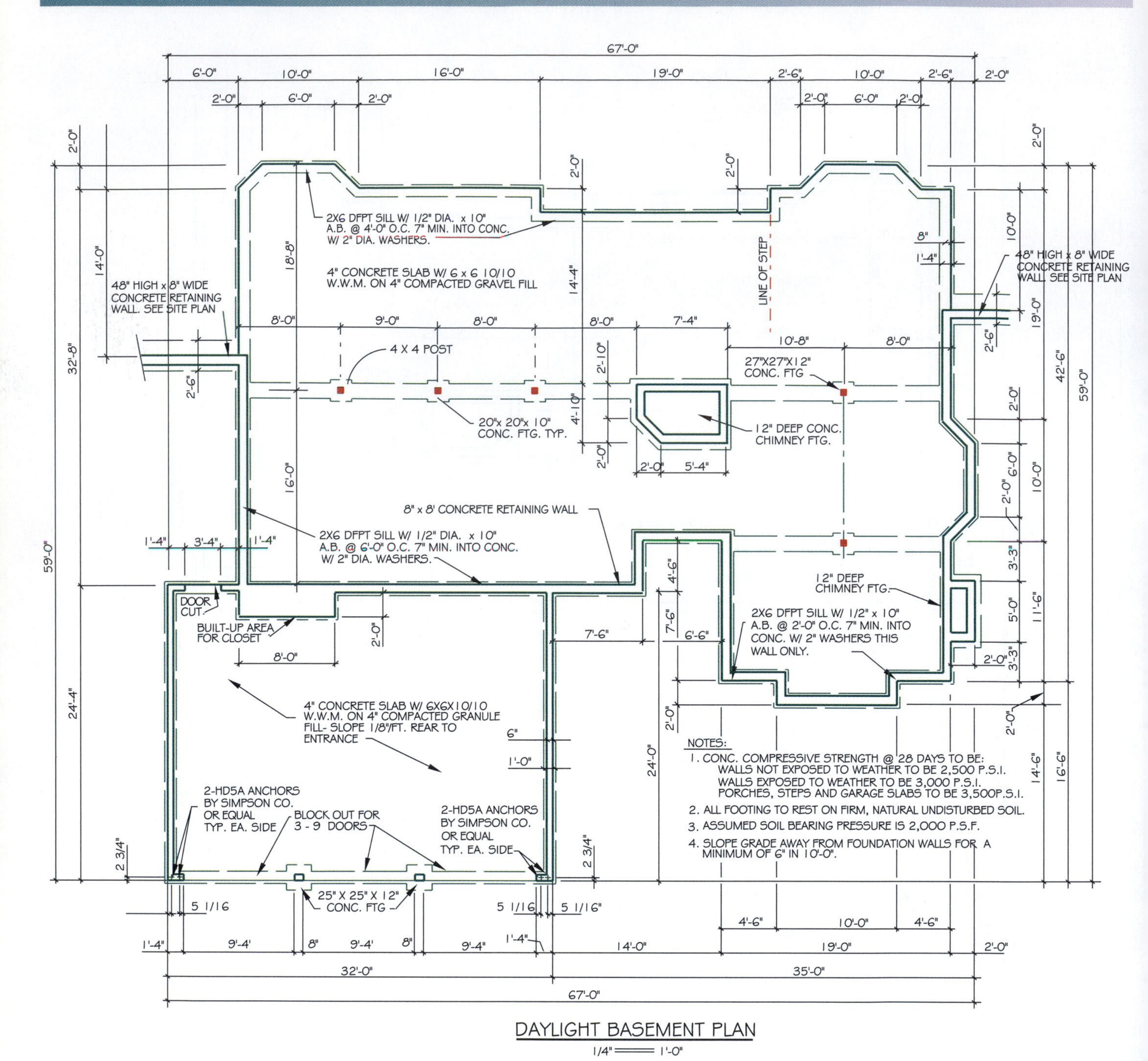

FIGURE 26.35c ■ *A foundation plan for a home with a daylight basement. When a home is built on a sloping site, the low side of the basement can be constructed using slab-on-grade construction methods. The 48'' (1200 mm) retaining walls on each side of the home allow standard wood construction to be used for lower walls.*

CHAPTER 26

Floor Systems and Foundation Support Test

DIRECTIONS

Answer the following questions with short, complete statements. Type your answers using a word processor or neatly print your answers on lined paper.

1. Place your name, the chapter number, and the date at the top of the sheet.
2. Type the question number and provide the answer in the form of a statement that includes part of the question. You do not need to write out the entire question.

Note: The answers to some questions may not be contained in this chapter and will require you to do additional research.

QUESTIONS

26.1. Why is a concrete slab called an on-grade floor system?
26.2. What is the minimum thickness for a residential slab?
26.3. Why are control joints placed in slabs?
26.4. What is the minimum amount of fill required under an on-grade concrete slab?
26.5. What thickness of vapor barrier is to be placed under a slab?
26.6. What are the minimum height limitations required in the crawl area?
26.7. What is the purpose of a girder?
26.8. How are floor joists attached to the foundation wall?
26.9. How are girders supported at the foundation wall?
26.10. What is a common spacing for beams in a post-and-beam floor system?

27 CHAPTER

Foundation Plan Layout

The foundation plan should be created in the same file that contains the floor and framing plans. It should also be plotted at the same scale as the floor and framing plans. Although the floor plan can be traced to obtain overall sizes, this practice can lead to major errors in the foundation plan. If you trace a floor plan that is slightly out of scale, you will reproduce the same errors in the foundation plan. A better method is to draw the foundation plan using the dimensions that are found on the framing plan. If the foundation cannot be drawn using your dimensions, your framing plan may be missing dimensions or may contain errors. Great care needs to be taken with the foundation plan. If the foundation plan is not accurate, changes may be required that affect the entire structure.

This chapter includes guidelines for several types of foundation plans including: concrete slab, joist construction, post-and-beam, and full basements. Each is based on the floor plan that was used for examples in Section 3. The home is a two-story hillside home. The lower floor for each foundation style will always be formed with a concrete floor. It is the floor system over the single level portion of the home that will vary. Before attempting to draw a foundation plan, study the completed plan that follows each example so you will know what the finished drawing should look like.

The foundation plan can be drawn by dividing the work into six stages:

1. *Planning*
2. *Drawing foundation members*
3. *Drawing floor framing members*
4. *Dimensioning*
5. *Annotating*
6. *Evaluating*

DRAWING PLANNING

The process of drawing the foundation plan is very similar to the process used to create each of the previous plan views. Because the drawing will usually be placed in the same file as the other plan views, the template will need only minor adjustments. Template options that must be altered before starting the drawing include adding appropriate layers to contain the foundation information, and adjusting linetypes to represent girders, footings, and piers. A third area of preparation that should be completed before the drawing is started is to track all loads through the structure.

Adding Layers

Before symbols can be placed on the foundation drawing, layers need to be created to separate the concrete and floor information from other information in the drawing file. Layers for the foundation and floor framing information should start with a prefix of *FNDN*. Layers containing foundation information should be named with a modifier listed in Appendix F. Sub names of *ANNO, DIMN,*

FOOT, OUTL, and *SYMB* will always be needed. As with any other drawing, create additional layers as needed to ensure that only layers that will be used are added to the drawing. For foundations with wood floor systems, additional layers might include *BEAM, JOST, PIER,* and *VENT.*

Adjusting Linetypes

As you progress through the drawing, you will use several types of line quality. When drawing the foundation plan, use the following line widths:

- Use thick, continuous lines to represent stem walls, and the outlines of all concrete slabs.

All other objects should be drawn using thin lines using the methods described in Chapter 25 and 26. Linetypes should include:

- Use thin, dashed lines to represent all footings and beams. Lines should be approximately 3/4'' (19 mm) long. Although the line type is the same, be sure and use separate layers for girders and footings.
- Use thin, hidden lines to represent all piers. Lines should be approximately 1/4'' (6 mm) long.

Drawing Considerations

Because other plans have already been completed, much of the information on those drawings can be useful in completing the foundation plan. With the walls of the framing plan displayed, the line representing the outer side of the stem walls can be drawn, using the walls of the framing plan as a guide. Be sure to set OSNAP to ON. Once the outer edge has been placed, freeze the wall layer and continue to create the foundation walls. The OFFSET command should be used to lay out the thickness of the stem walls and footings. Corners can be adjusted by using the FILLET or TRIM command. The CHANGE command can be used to change the lines representing the footings from continuous to hidden. By following the step-by-step instructions for a particular foundation type, the plan can be completed.

Many of the dimensions used on the framing plan can also be used on the foundation plan. A layer such as *BASEDIM* can be used for placing dimensions required by the framing and foundation plans such as the overall and major jogs. General notes can be typed and stored as a WBLOCK and reused on future foundation plans. Many offices store lists of local notes required for a particular type of foundation as a WBLOCK and insert them into a drawing. Once inserted, the notes can be edited and moved to the desired position. When completed, the foundation can be stored separately from the floor plan to make plotting easier.

Determining Loads

The type of framing system used to frame the residence will affect the foundation plan. Homes with roofs framed with trusses will have all of their bearing points located at exterior walls that are parallel to the ridge. Some trusses are three point bearing trusses that require support at the foundation level for the interior wall that supports the additional bearing point. Homes with roofs framed with stick construction will require bearing walls or beams to support the ridge. Each type of framing will affect the supports that need to be represented on the foundation plan.

Supporting a Truss-Framed Roof

Figure 27.1 shows the roof plan for a simple home framed using trusses to support the roof. Each of the exterior walls parallel to the ridge will be bearing walls and need a continuous footing to support their loads. Although the walls perpendicular to the ridge are nonbear-

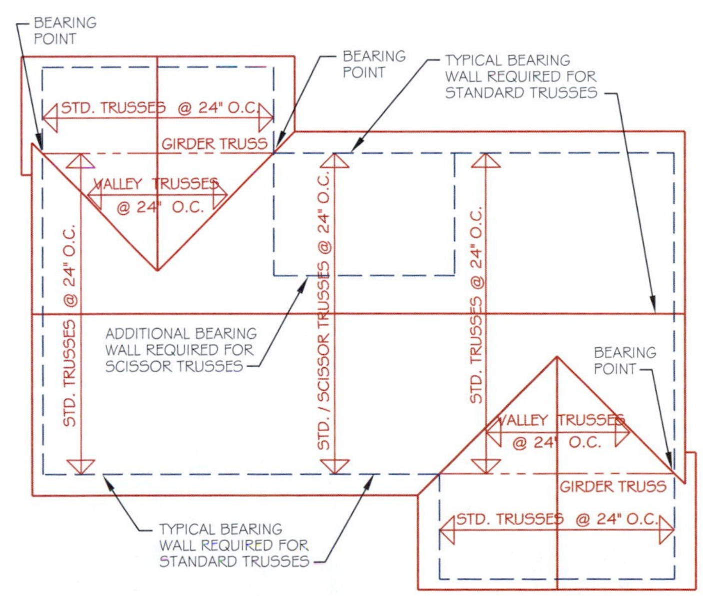

FIGURE 27.1 ▪ *The roof plan for a simple home with a roof framed with trusses. Each of the walls parallel to the ridge are bearing walls and need a continuous footing to support the roof and wall loads. Point loads from girder trusses will also need support at the foundation level.*

ing, the IRC requires that a continuous footing be provided for all exterior walls. Trusses that are partially vaulted may require a third bearing point, as shown in Figure 27.2, where the standard and scissor portions of the truss intersect. The foundation for such a home would resemble the plan in Figure 27.3. In addition to the continuous footings required for a truss roof, individual piers may be required to reinforce the foundation where loads from window supports are located. See Chapter 23 for a review of determining loads and pier sizes.

A second area that needs extra foundation support are the bearing points for a girder truss. Although the company that supplies the trusses will determine the loads on the girder truss, the design team must determine the loads to be supported and distributed through the girder truss into the foundation.

Load Study—Upper

Figure 27.4 shows a portion of the planning study for the home that was started in Chapter 11. The plan views in Figure 11.15, 11.13, and 24.29 provide the needed plans to determine the information needed to plan the foundation. The windows on the upper west side include 9', 7', 6' and 5'. The 9' opening in bedroom 1 has a w = 400 lb (8 + 2) ×40 lb and an R value of 1800 lb (400 lb × 4.5'). Because the foundation will support 2000 lb/ft, no additional support will be needed for this window. This process will need to be repeated for each window on the upper floor. Once the reactions are determined, use the table in Figure 23.26 to determine the required pier size. The other windows will have reactions of:

7' window = R = 2800 lb
6' window = R = 2040 lb
5' window = R = 2000 lb
3.5' window = R = 1400 lb

Any additional loads for interior walls, and other upper bearing walls will also need to be examined. This same process will need to be repeated for the lower floor.

Load Study—Lower

Each of the loads from the upper floor as well as the loads from the main and lower floor framing systems must be transferred into the foundation. As the load path for the west wall is followed to the main floor, each of the loads from the upper windows will transfer through the cantilevered floor joists and into the bearing wall. This is true except for the load from the southwest window in the living room. For this window, loads will transfer through the double floor joists below each window trimmer and into the trimmers for the window in the southwest corner of the family room. The reactions for this window would be:

Upper 5' window = R = 2000 lb
Lower floor weight = R = 1500 lb
(12' × 50 lb) × 2.5'
Total reaction = 3500 lb

Dividing the reaction of 3500 lb by the soil pressure of 2000 lb requires a pier size with an area of 1.75 sq ft. An 18'' square pier will be required to meet the load.

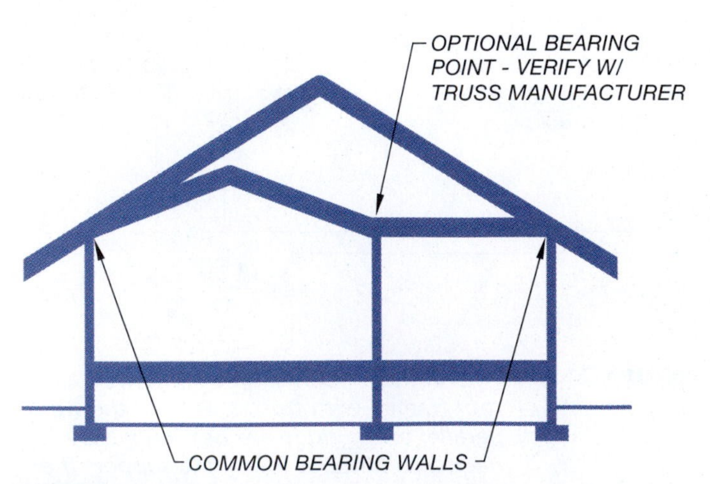

FIGURE 27.2 ■ *Trusses typically have two bearing points. Trusses that are partially vaulted may require a third bearing point where the standard and scissor portions of the truss intersect.*

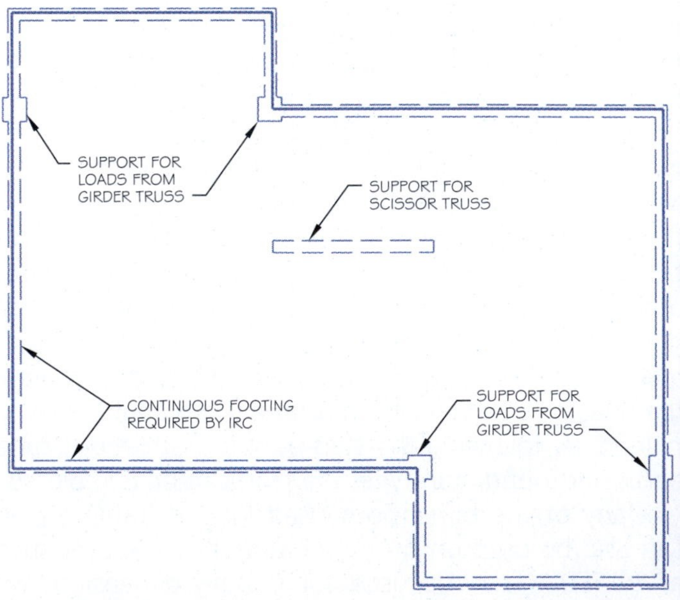

FIGURE 27.3 ■ *The foundation for the home in Figure 27.2 would require continuous footings around the exterior of the foundation, continuous footings under the interior bearing wall that will support the interior bearing of the vaulted trusses, and piers to support each of the point loads from the girder trusses.*

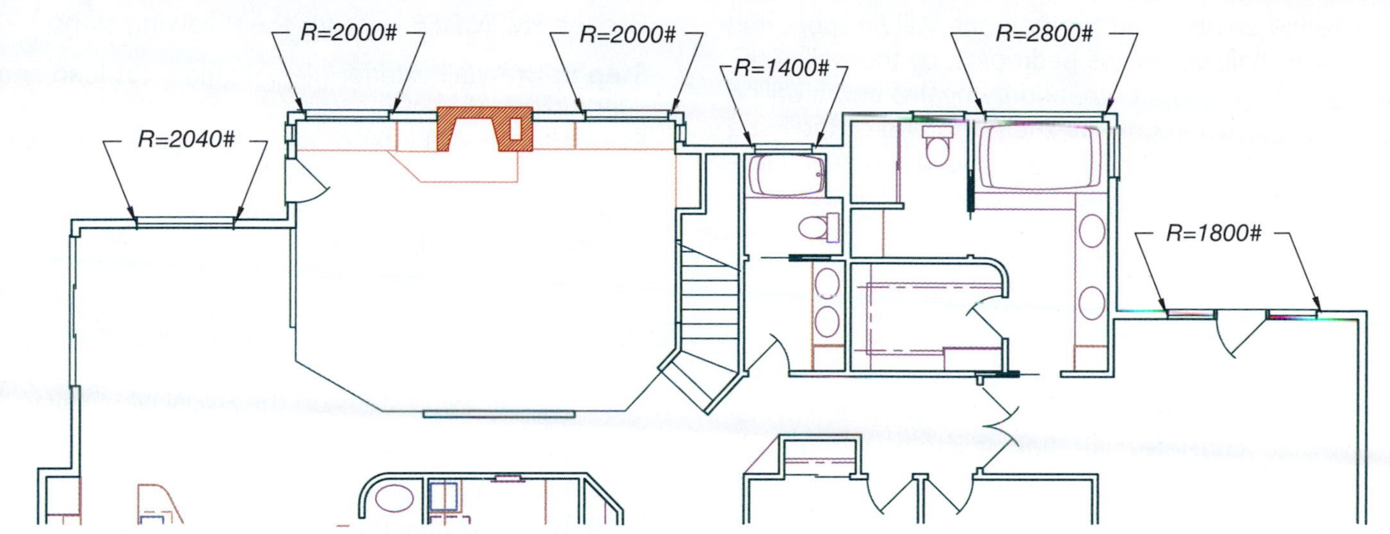

FIGURE 27.4 ■ *A portion of the planning study for the home that was started in Chapter 11. The calculation used to determine the headers for the framing plan provide the needed information to determine the required footings. The reactions from each header calculation will need to be transferred into the soil through the foundation.*

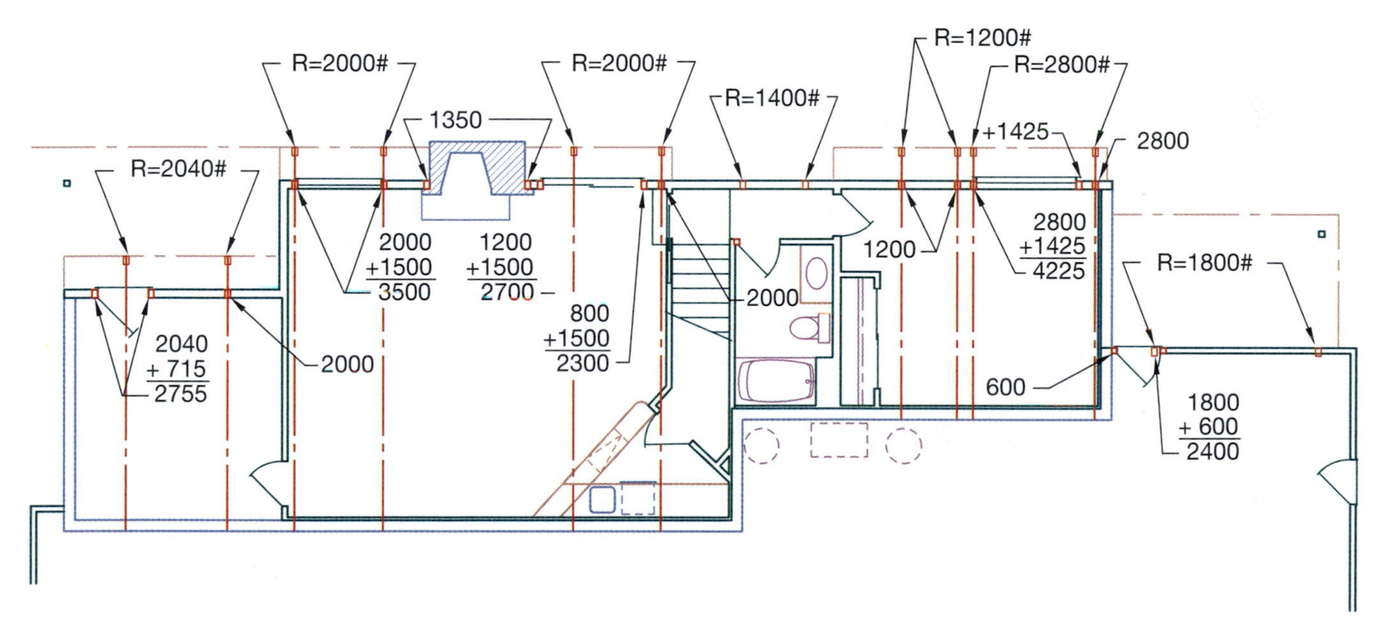

FIGURE 27.5 ■ *The reactions from the upper floor are added to the reactions of the lower floor to determine the total loads to be supported by the foundation. Once all reactions have been located, the foundation plan can be started.*

Each of the remaining openings on the lower floor will need to be examined to determine any required piers. Because the beam sizes were determined as each level of the framing plan was completed, this will mean that you only need to consult your calculations and select the appropriate reaction. This process can be eased if you'll create a nonplotting layer for each level of the framing plan loads. A layer such as *MAIN FRAM LOADS* or *LOW FRAM LOADS* can be used to place markers for each reaction. Once the upper loads have been marked, the layer can be displayed on the lower floor to identify where each of the upper loads will require bearing points. Figure 27.5 shows the loads for the upper floor plan transferred to the lower plan and then down into the foundation.

Supporting a Stick-Framed Roof

The process for planning a foundation to support a stick-framed roof and main floor framing is similar to the process for a truss roof. Each reaction from the beams that were specified on the framing plan must be transferred through the structure and into the foundation plan. Based on the information from Figure 22.5 the loads for a stick-framed roof with three bearing points will typically be half of the loads for the truss roof with two bearing points. Load for the ridge will be supported by the west hall wall by the bedrooms, by the wall at the east side of the living room, and by beams in the ceiling level by the dining room and the north end of the living room (see Figure 24.38). Once each load has been identified, the foundation plan drawing can be started.

CONCRETE SLAB FOUNDATION LAYOUT

The following steps can be used to draw a foundation with a concrete slab floor system and a truss framed roof. Not all steps will be required for every house. Use the *FNDN SLAB* layer for the following steps.

Step 1. Using your framing plan, locate the exterior edge of the slab. The edge of the slab should match the exterior side of the exterior walls on the floor plan.
Step 2. Use the OFFSET command to locate the interior side of the stem wall around the slab at the garage area. See Figure 25.10 for a review of minimum foundation sizes.
Step 3. Block out the locations of doors in the garage stem walls. Trace the outline of the doors from the framing plan. Use the OFFSET command to add an additional 2'' (50 mm) each side. Allow for the door size plus 4'' (100 mm).
Step 4. Use the OFFSET command to add a line to represent the support ledge if brick veneer is to be used. Assume an additional 4'' (100 mm) will be added.
Step 5. Draw the size of the chimney based on measurements from the framing plan. Represent the size of the chimney, not the fireplace and hearth.

Use the *FNDN FOOT* layer for the following steps:

Step 6. Draw the exterior footing width. Use the OFFSET command to offset the exterior edge of the slab as required, and then use the PROPERTIES command to assign the lines to the proper layer and linetypes.

Representing Footings and Interior Piers

Step 7. Draw the fireplace footing so that it extends 6'' (150 mm) minimum beyond the face of the fireplace.
Step 8. Draw all interior footings as required for load bearing walls.

Use the *FNDN PIER* layer for the following steps.

Step 9. Draw all exterior piers required for load and lateral support.
Step 10. Draw all interior piers as required for spot loads.
Step 11. Draw any exterior piers that might be required for decks or porches.

Representing Miscellaneous Materials

Place the following steps on the *FNDN MISC* layer.

Step 12. Draw changes in the floor levels.
Step 13. Draw floor drains.
Step 14. Draw heating registers if required.

You now have drawn all of the information that is required to represent the floor and foundation systems. Your drawing should now resemble the drawing in Figure 27.6.

Placing Dimensions

Follow Steps 15 through 20 to place the required dimensions on the drawing. Place the following information on the *FNDN DIMS* layer unless noted. Your drawing should resemble Figure 27.7 when complete.

Step 15. Thaw the *FRAM DIMS BASE* layer. This will display the overall and major jog dimensions on each side of the foundation.
Step 16. Dimension all door openings in the garage stem wall.
Step 17. Place dimensions to locate the fireplace.
Step 18. Place dimensions for all exterior footings.
Step 19. Place dimensions for all interior footings and piers.
Step 20. Place dimensions to locate all heating and plumbing materials that are below the slab.

Placing Annotation

The final drawing procedure is to specify the materials that are to be used. Place the following information on the *FNDN ANNO* layer. Figure 27.8 is an example of the notes that are required on the foundation. Figure 27.9 shows the same home completed using the prescriptive path of the IRC. Use the following steps to complete the foundation plan.

Step 21. Insert and edit the general notes to describe all concrete and floor system materials.

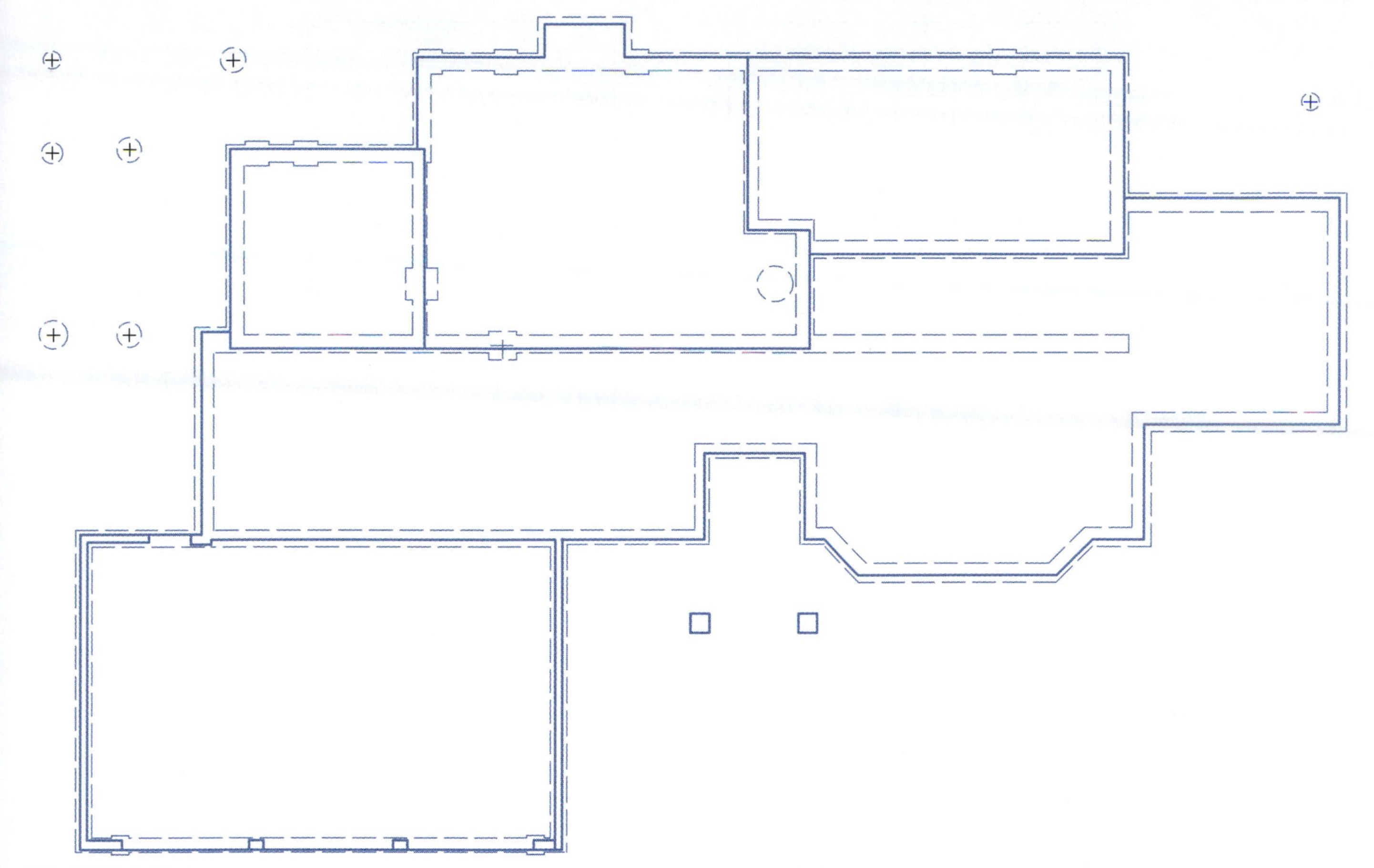

FIGURE 27.6 ■ *The location of stem walls, footings, door locations, and chimney for a concrete slab foundation plan. Any finishing materials such as lower slabs and plumbing, electrical, and HVAC material that will be below the slab must also be shown.*

Step 22. Draw metal connectors required for uplift or lateral loads. This example is designed by an engineer and reflects the use of shear panels.
Step 23. Place all required local notes to describe each material. See the list of required notes in Chapter 26.
Step 24. Place a title and scale under the drawing.
Step 25. Use a print of your plan to evaluate your drawing. Use the checklist found in Appendix D on the CD.

FOUNDATION PLAN WITH JOIST CONSTRUCTION

The following steps can be used to draw a foundation with continuous footings and a joist floor system. Not all steps will be required for every house. This example assumes the use of trusses to frame the roof. Use the *FNDN WALL* layer for the following steps.

Representing Stem Walls

Step 1. Using your framing plan, locate the exterior edge of the slab. The edge of the slab should match the exterior side of the exterior walls on the floor plan.
Step 2. Determine the foundation wall thickness. Use the OFFSET command to locate the interior side of the stem wall. See Figure 25.10 for a review of minimum foundation sizes.
Step 3. Block out the locations of doors in the garage stem walls. Trace the outline of the doors from the framing plan. Use the offset command to add an additional 2'' (50 mm) each side. Allow for the door size plus 4'' (100 mm).
Step 4. Draw a crawl access in the stem wall.
Step 5. Use the OFFSET command to add a line to represent the support ledge if brick veneer is to be used. Assume an additional 4'' (100 mm) will be added.

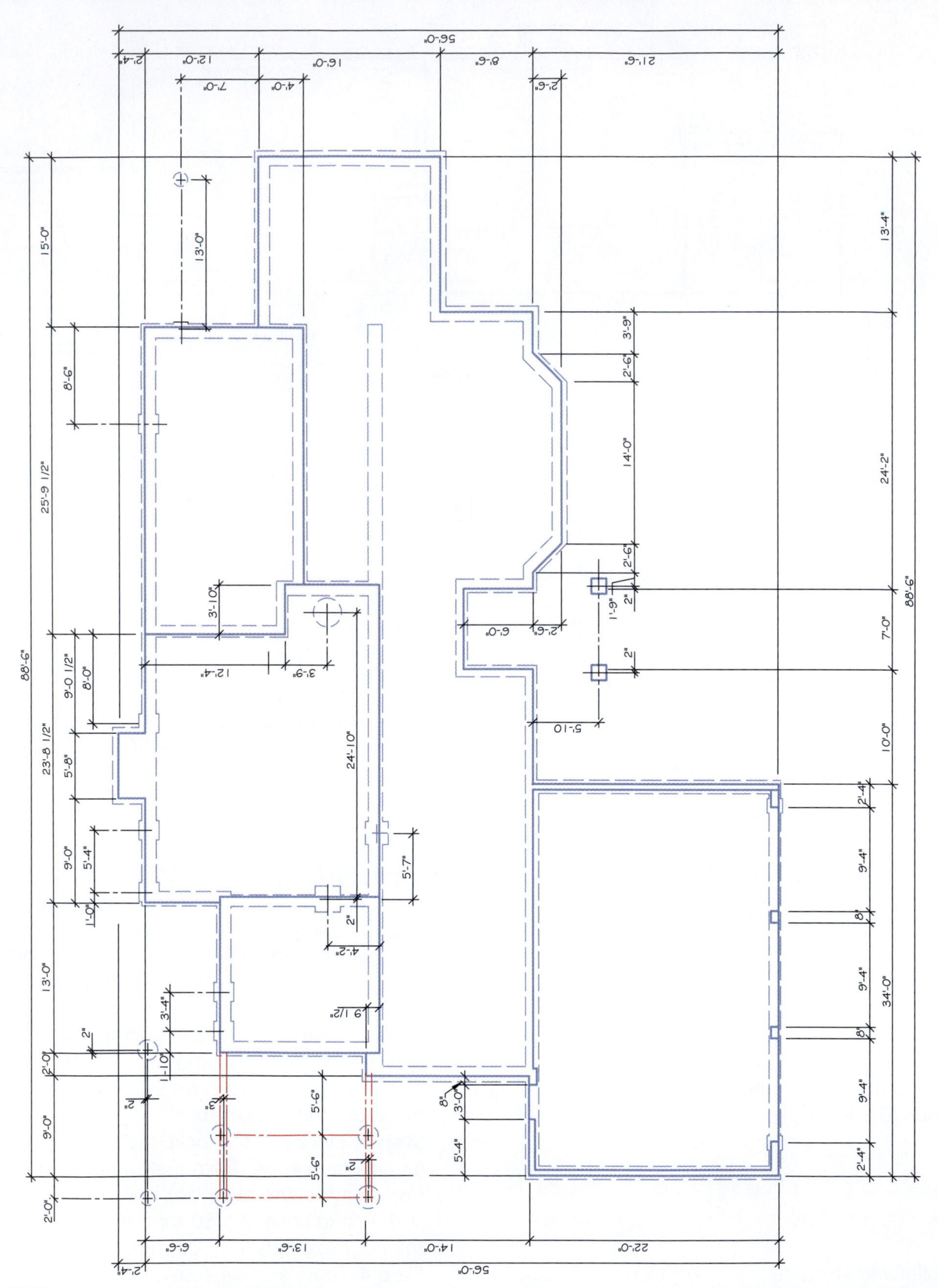

FIGURE 27.7 ■ *Thaw the FRAM DIMS BASE layer that was created on the framing plan to display the overall and major jog dimensions on each side of the foundation.*

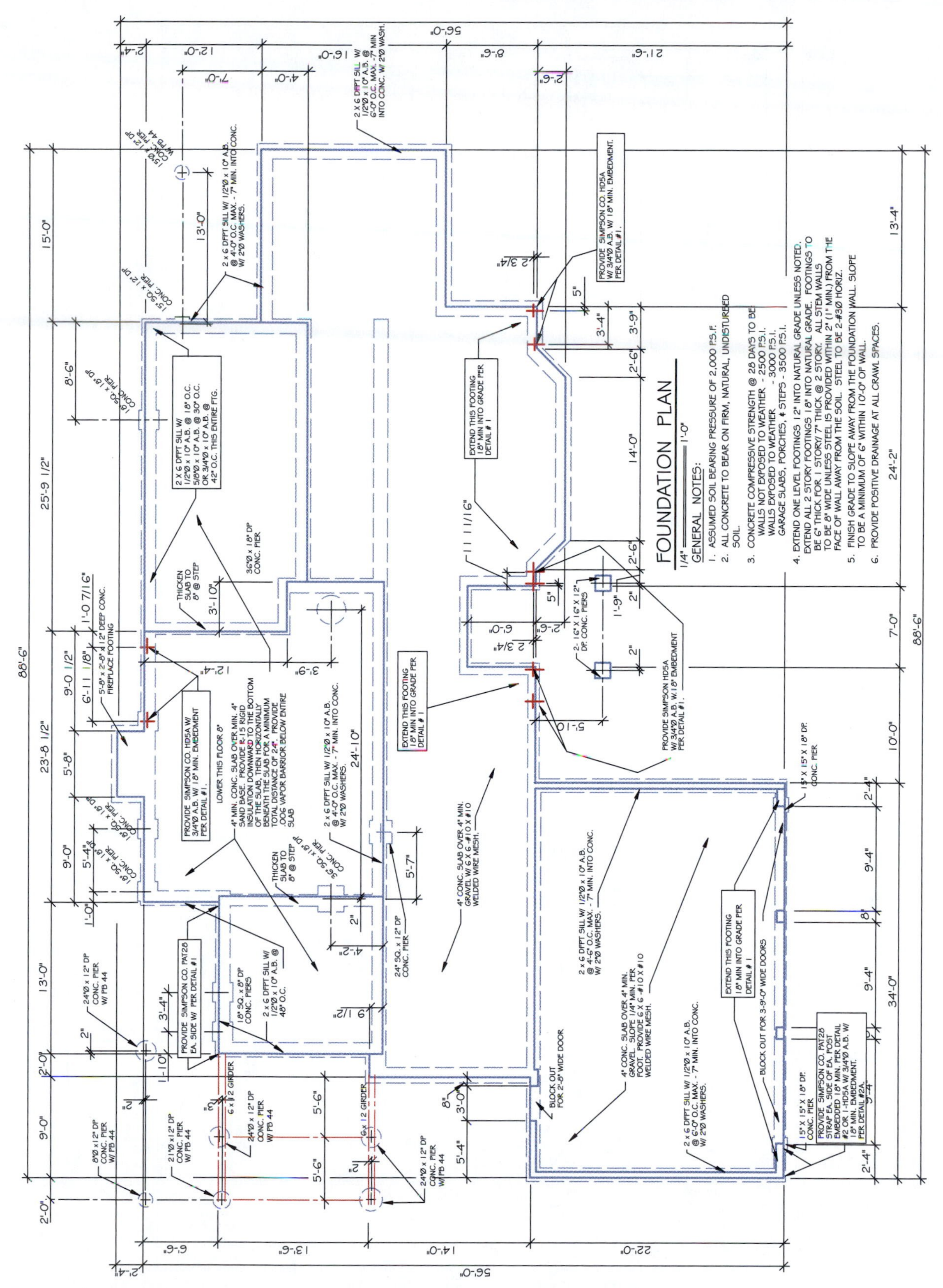

FIGURE 27.8 ■ *The foundation plan for a structure with a concrete slab is finished by adding dimensions to describe concrete features, adding annotation to describe all materials and lateral bracing requirements. Lateral bracing requirements are based on specific engineering design.*

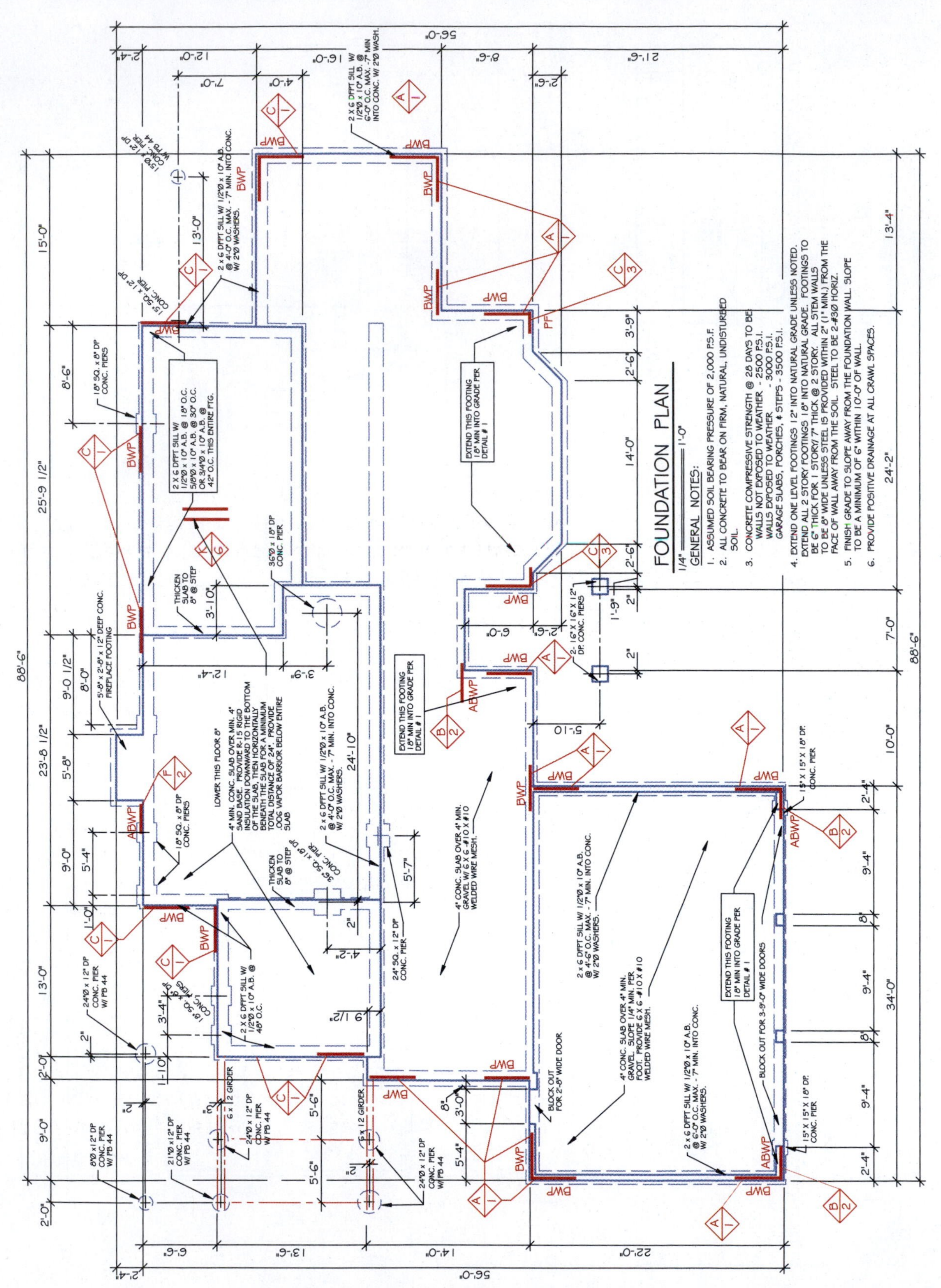
FOUNDATION PLAN
1/4" = 1'-0"
GENERAL NOTES:
1. ASSUMED SOIL BEARING PRESSURE OF 2,000 P.S.F.
2. ALL CONCRETE TO BEAR ON FIRM, NATURAL, UNDISTURBED SOIL.
3. CONCRETE COMPRESSIVE STRENGTH @ 28 DAYS TO BE: WALLS NOT EXPOSED TO WEATHER - 2500 P.S.I. WALLS EXPOSED TO WEATHER - 3000 P.S.I. GARAGE SLABS, PORCHES, & STEPS - 3500 P.S.I.
4. EXTEND ONE LEVEL FOOTINGS 12" INTO NATURAL GRADE UNLESS NOTED. EXTEND ALL 2 STORY FOOTINGS 18" INTO NATURAL GRADE. FOOTINGS TO BE 6" THICK FOR 1 STORY/7" THICK @ 2 STORY. ALL STEM WALLS TO BE 8" WIDE UNLESS STEEL IS PROVIDED WITHIN 2" (1" MIN.) FROM THE FACE OF WALL AWAY FROM THE SOIL. STEEL TO BE 2-#30 HORIZ.
5. FINISH GRADE TO SLOPE AWAY FROM THE FOUNDATION WALL. SLOPE TO BE A MINIMUM OF 6" WITHIN 10'-0" OF WALL.
6. PROVIDE POSITIVE DRAINAGE AT ALL CRAWL SPACES.

FIGURE 27.9 ■ *The foundation plan for a structure with a concrete slab and lateral bracing to meet the prescriptive path of the IRC. The plan is completed by adding dimensions to describe all concrete features, annotation to describe all materials, and lateral bracing requirements.*

Step 6. Draw the size of the chimney based on measurements from the floor plan. Represent the size of the chimney, not the fireplace and hearth.

Representing Footings and Interior Piers

Use the *FNDN FOOT* layer for the following steps:

Step 7. Draw the exterior footing width. Use the OFFSET command to offset the exterior edge of the slab as required, and then use the PROPERTIES command to assign the lines to the proper layer and linetypes.
Step 8. Draw the interior footings below the slab for all bearing walls.
Step 9. Draw the chimney footing so that it extends 6'' (150 mm) minimum beyond the face of the fireplace.

When you are finished with this step, your drawing should resemble Figure 27.10.

Step 10. Use the *FNDN MISC* layer to draw changes in the floor levels.
Step 11. Draw all required girders using thin dashed lines to support floor joists, load-bearing walls, and changes in floor elevation on the *FNDN GIRD* layer.

Use the *FNDN PIER* layer for the following steps:

Step 12. Draw all exterior and interior piers that are required for spot loads.
Step 13. Draw any exterior piers that might be required for decks or porches.
Step 14. Draw the piers to support the girders.

When you are finished with this step, your drawing should resemble Figure 27.11.

See Figure 27.12 for steps 15 through 17.

Representing Miscellaneous Materials

Step 15. Use the *FNDN JOST* layer to draw or insert the arrow symbols to represent the floor joist direction.

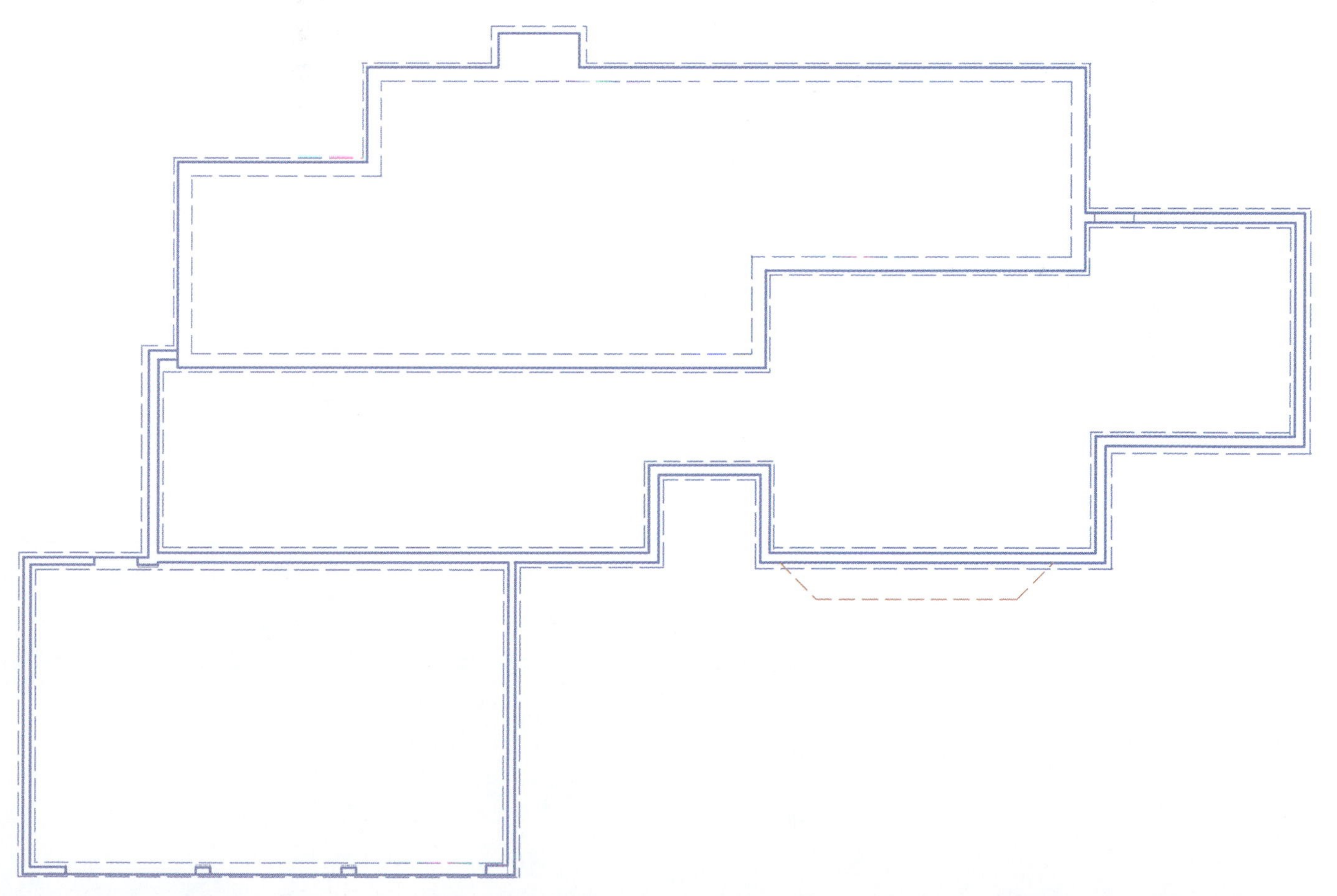

FIGURE 27.10 ■ *The location of stem walls, footings, door locations, and chimney for a foundation plan with a joist floor system. Any finishing materials such as lower slabs and plumbing, electrical, and HVAC material that will be below the slab must also be shown.*

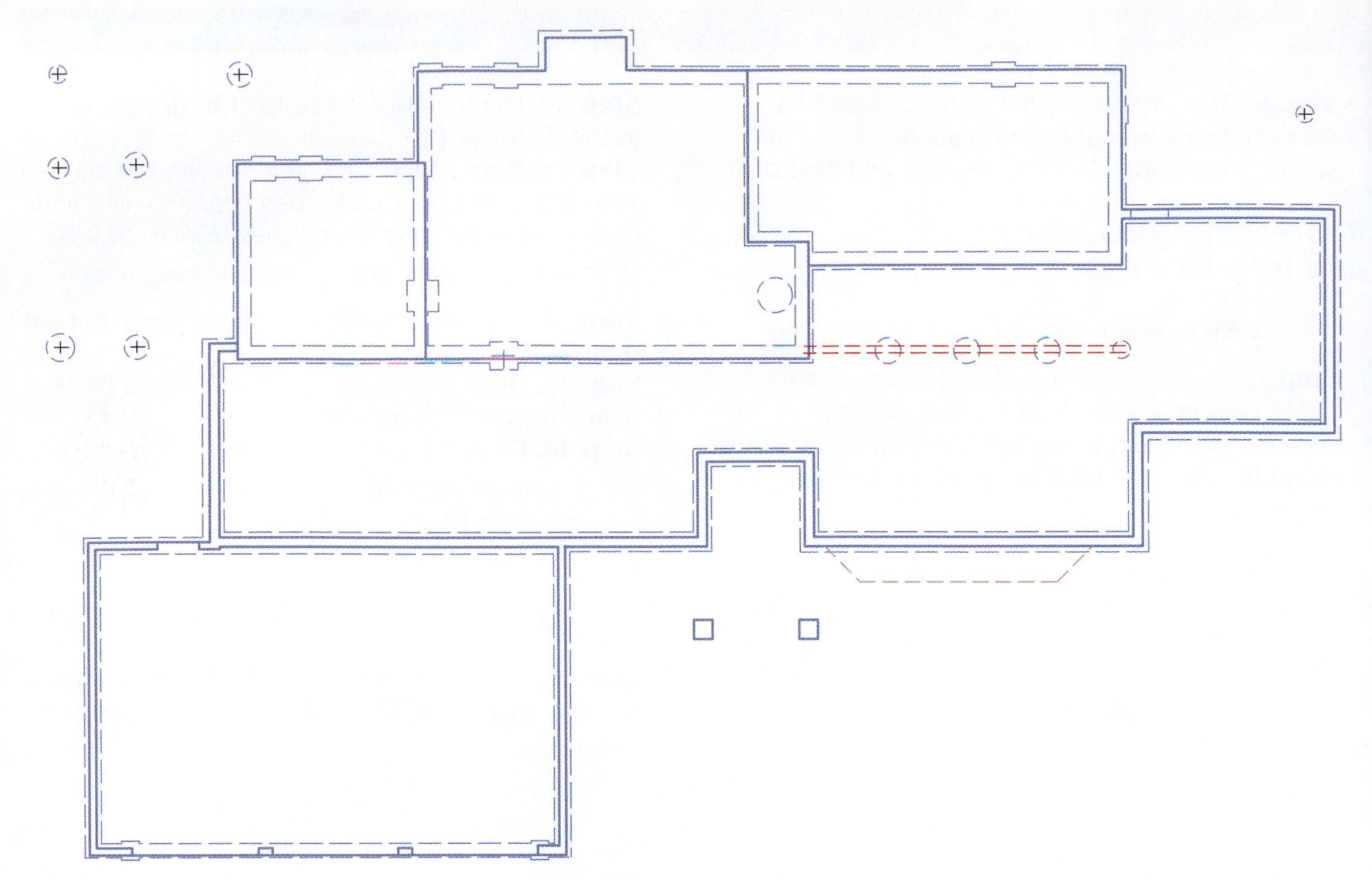

FIGURE 27.11 ■ *Draw the girders, joists, and piers required to support the loads from the upper floor over the crawl space.*

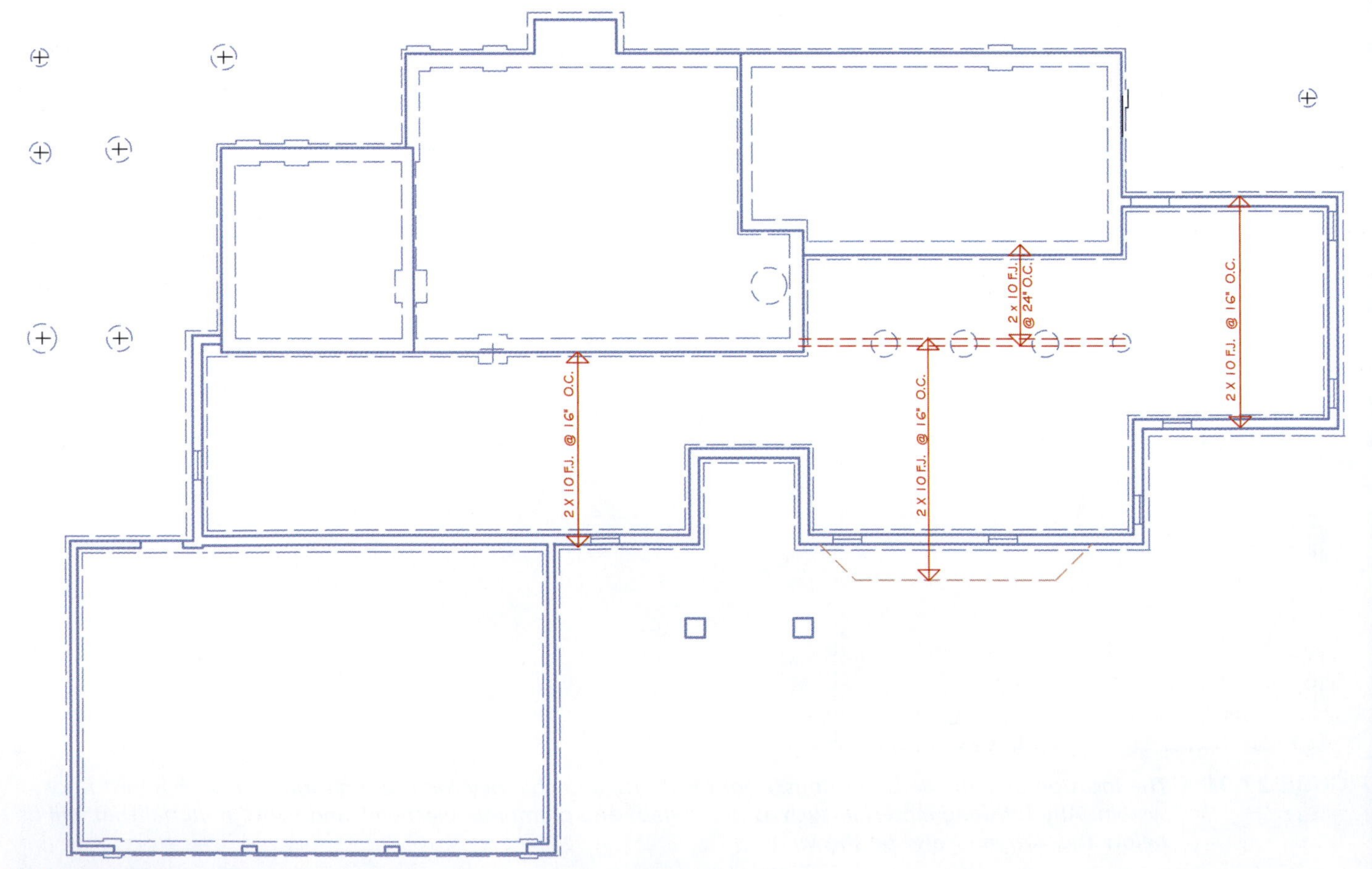

FIGURE 27.12 ■ *Draw or insert all symbols to represent the floor joists over the crawl space.*

Step 16. Use the *FNDN MISC* layer to draw vents in the walls surrounding the crawl space.
Step 17. Use the *FNDN PATT* layer to crosshatch the masonry chimney if concrete block or brick is used.

You have now drawn all of the information that is required to represent the floor and foundation systems. Follow Steps 18 through 25 to dimension on the drawing.

Placing Dimensions

Place the following information on the *FNDN DIMS* layer unless noted. Your drawing should resemble Figure 27.13 when complete.

Step 18. Thaw the *FRAM DIMS BASE* layer. This will display the overall and major jog dimensions on each side of the foundation
Step 19. Dimension all door openings in the garage stem wall.
Step 20. Place dimensions to locate the chimney.
Step 21. Place dimensions to locate all exterior footings.
Step 22. Place dimensions to locate all girders.
Step 23. Place dimensions to locate all piers.
Step 24. Place dimensions to locate all heating and plumbing materials that are below the slab.
Step 25. If required, place dimensions to locate metal connectors.

Placing Annotation

The final drawing procedure is to specify the materials that are to be used. This is done with general and local notes in the same method that was used on the framing plan. Place the following information on the *FNDN ANNO* layer. Figure 27.14 is an example of the notes that are required on the foundation using the prescriptive path of the IRC. Use the following steps to complete the foundation plan:

Step 26. Insert and edit the general notes to describe all concrete and floor system materials.
Step 27. Place all required local notes to describe each material. See the list of required notes in Chapter 26.
Step 28. Place a title and scale under the drawing.
Step 29. This example is designed to meet the prescriptive path of the IRC. Use the *FNDN ANNO LATT* layer and insert all required reinforcements for braced wall lines.
Step 30. Use the *FNDN ANNO LATT SCHD* layer and insert the appropriate lateral bracing schedule. If space is not available for the schedule, place the schedule on another page, as close to the foundation as possible.
Step 31. Use a print of your plan to evaluate your drawing. Use the checklist found in Appendix D on the CD.

STANDARD FOUNDATION WITH POST-AND-BEAM FLOOR SYSTEM

The following steps can be used to draw a foundation with continuous footings and a post-and-beam floor system. This example assumes the use of trusses to frame the roof. Not all steps will be required for every house. When the foundation plan is complete, it should resemble the plan shown in Figure 27.19. Use the *FNDN WALL* layer for the following steps:

Representing Stem Walls

Step 1. Using your framing plan, locate the exterior edge of the slab. The edge of the slab should match the exterior side of the exterior walls on the floor plan.
Step 2. Determine the foundation wall thickness. Use the OFFSET command to locate the interior side of the stem wall. See Figure 25.10 for a review of minimum foundation sizes.
Step 3. Block out the locations of doors in the garage stem walls. Trace the outline of the doors from the framing plan. Use the offset command to add an additional 2'' (50 mm) each side. Allow for the door size plus 4'' (100 mm).
Step 4. Draw a crawl access in the stem wall.
Step 5. Use the OFFSET command to add a line to represent the support ledge if brick veneer is to be used. Assume an additional 4'' (100 mm) will be added.
Step 6. Draw the size of the chimney based on measurements from the floor plan. Represent the size of the chimney, not the fireplace and hearth.

Representing Girders, Footings, and Interior Piers

Use the *FNDN FOOT* layer for the following steps.

Step 7. Draw the exterior footing width. Use the OFFSET command to offset the exterior edge of the

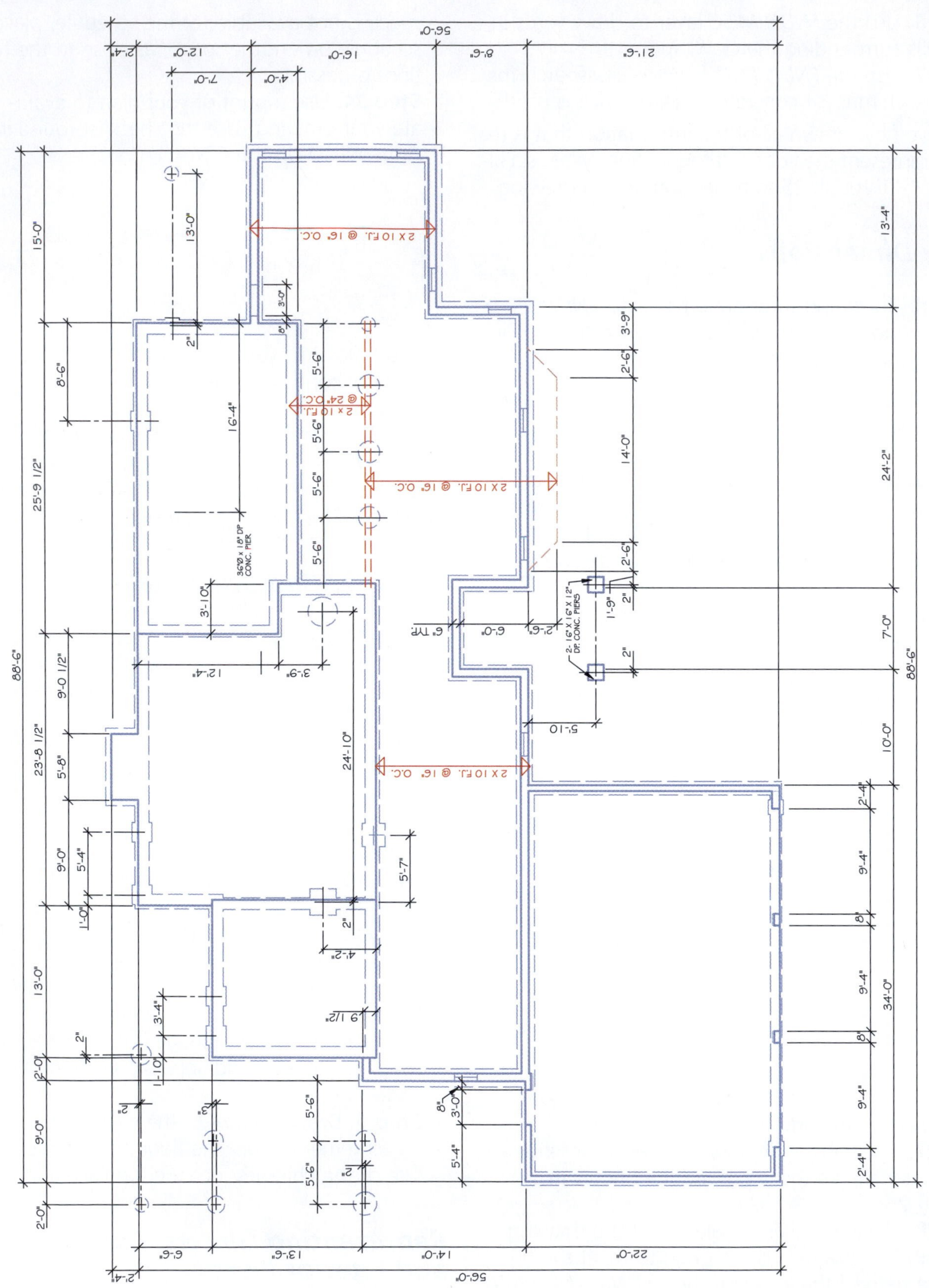

FIGURE 27.13 ■ *Thaw the FRAM DIMS BASE layer that was created on the framing plan to display the overall, and major jog dimensions on each side of the foundation. Add dimensions to describe all concrete features.*

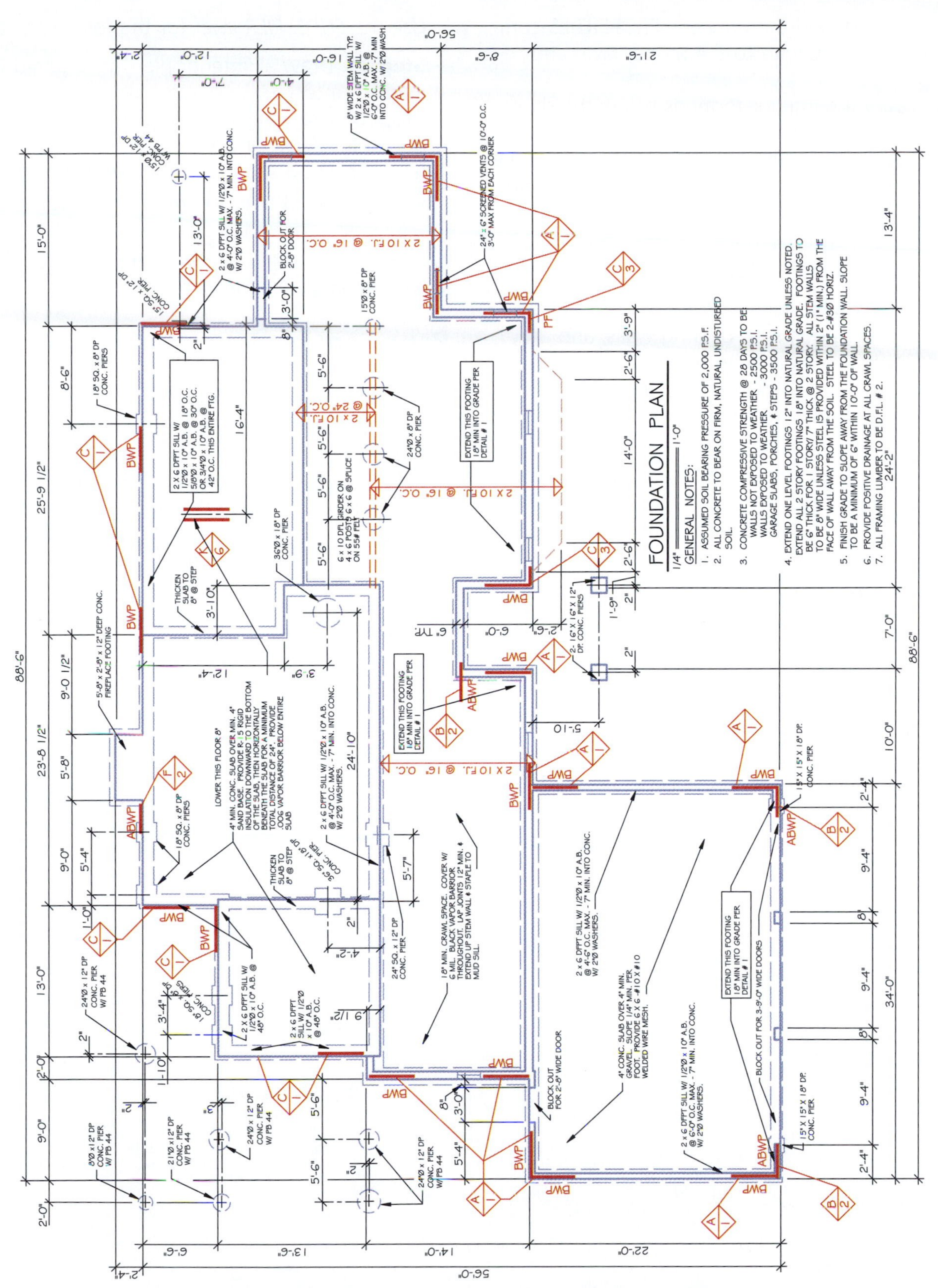
FOUNDATION PLAN
1/4" = 1'-0"
GENERAL NOTES:
1. ASSUMED SOIL BEARING PRESSURE OF 2,000 P.S.F.
2. ALL CONCRETE TO BEAR ON FIRM, NATURAL, UNDISTURBED SOIL.
3. CONCRETE COMPRESSIVE STRENGTH @ 28 DAYS TO BE: WALLS NOT EXPOSED TO WEATHER - 2500 P.S.I. WALLS EXPOSED TO WEATHER - 3000 P.S.I. GARAGE SLABS, PORCHES, & STEPS - 3500 P.S.I.
4. EXTEND ONE LEVEL FOOTINGS 12" INTO NATURAL GRADE UNLESS NOTED. EXTEND ALL 2 STORY FOOTINGS 18" INTO NATURAL GRADE. FOOTINGS TO BE 6" THICK FOR 1 STORY/ 7" THICK @ 2 STORY. ALL STEM WALLS TO BE 8" WIDE UNLESS STEEL IS PROVIDED WITHIN 2" (1" MIN.) FROM THE FACE OF WALL AWAY FROM THE SOIL. STEEL TO BE 2-#30 HORIZ.
5. FINISH GRADE TO SLOPE AWAY FROM THE FOUNDATION WALL. SLOPE TO BE A MINIMUM OF 6" WITHIN 10'-0" OF WALL.
6. PROVIDE POSITIVE DRAINAGE AT ALL CRAWL SPACES.
7. ALL FRAMING LUMBER TO BE D.FL. # 2.

FIGURE 27.14 ■ *The foundation plan for a structure with a joist floor system is finished by adding annotation to describe all materials and lateral bracing requirements to meet the prescriptive path of the IRC.*

slab as required, and then use the PROPERTIES command to assign the lines to the proper layer and linetypes.

Step 8. Draw the interior footings below the slab for all load-bearing walls.

Step 9. Draw the chimney footing so that it extends 6'' (150 mm) minimum beyond the face of the fireplace.

When you are finished with this step, your drawing should resemble Figure 27.15.

Step 10. Draw all required girders using thin dashed lines to support floor, load-bearing walls, and changes in floor elevation on the *FNDN GIRD* layer.

- Start the layout 48'' (1200 mm) from the exterior side of the longest stem wall that is parallel to the ridge.
- Use the offset command and place the balance of girders at 48'' (1200 mm) o.c.
- Add girders to support load-bearing walls or changes in floor elevation that are not supported by the girders placed in the previous step.

Use the *FNDN PIER* layer for the following steps:

Step 11. Draw all interior piers as required for spot loads.

Step 12. Draw any exterior piers that might be required for decks or porches.

Step 13. Draw the piers to support the girders.

- Place piers at 8' (2400 mm) o.c. starting at the longest edge of the stem wall.

When you are finished with this step, your drawing should resemble Figure 27.16.

Representing Miscellaneous Materials

See Figure 27.17 for Steps 14 through 16.

Step 14. Use the *FNDN MISC* layer to draw changes in the floor levels.

Step 15. Use the *FNDN MISC* layer to draw vents in the walls surrounding the crawl space.

Step 16. Use the *FNDN PATT* layer to crosshatch the stem wall and chimney if concrete block or brick is used.

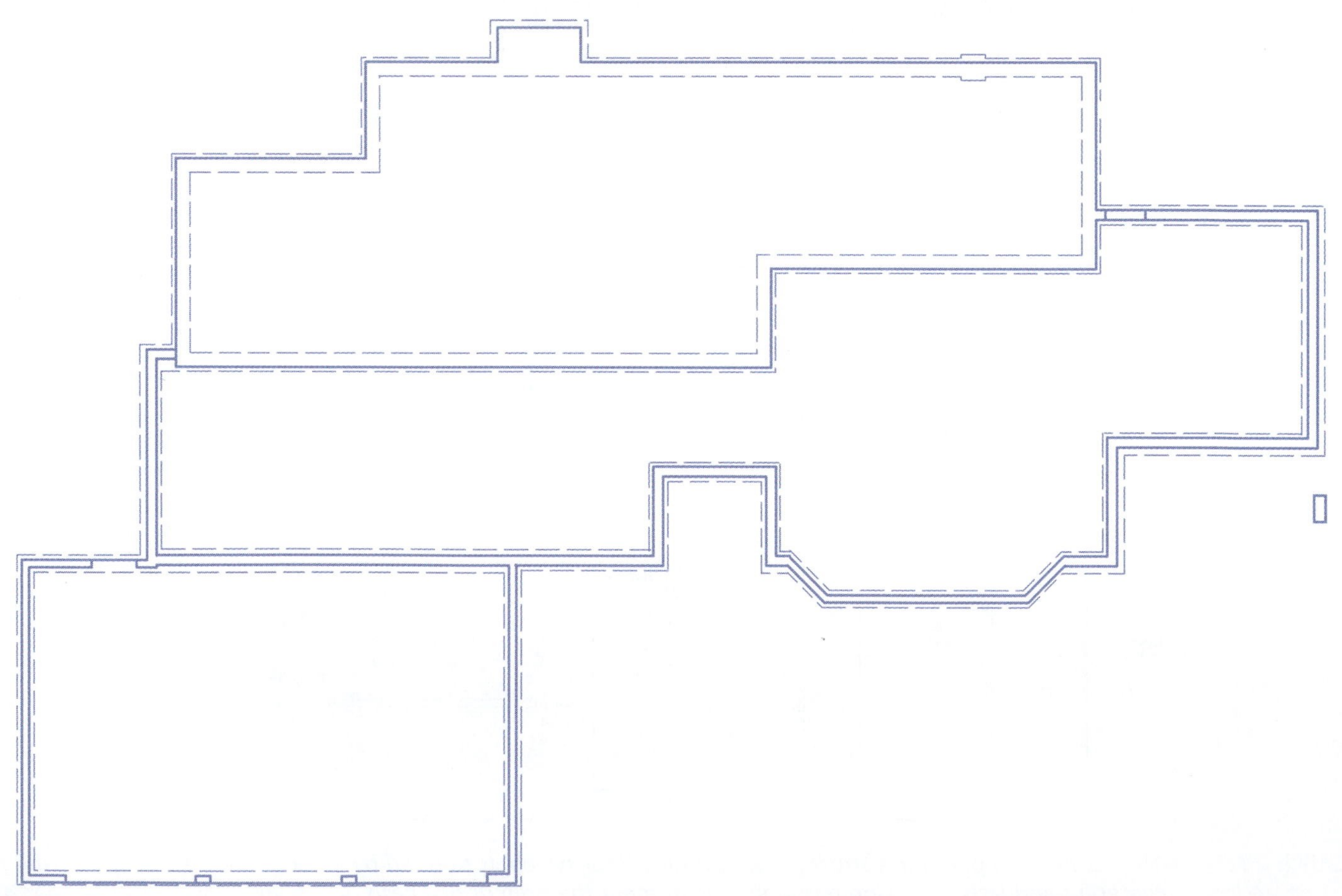

FIGURE 27.15 ■ *The location of stem walls, footings, door locations, and chimney for a foundation plan with a post-and-beam floor system. Any finishing materials such as lower slabs and plumbing, electrical, and HVAC material that will be below the slab must be shown.*

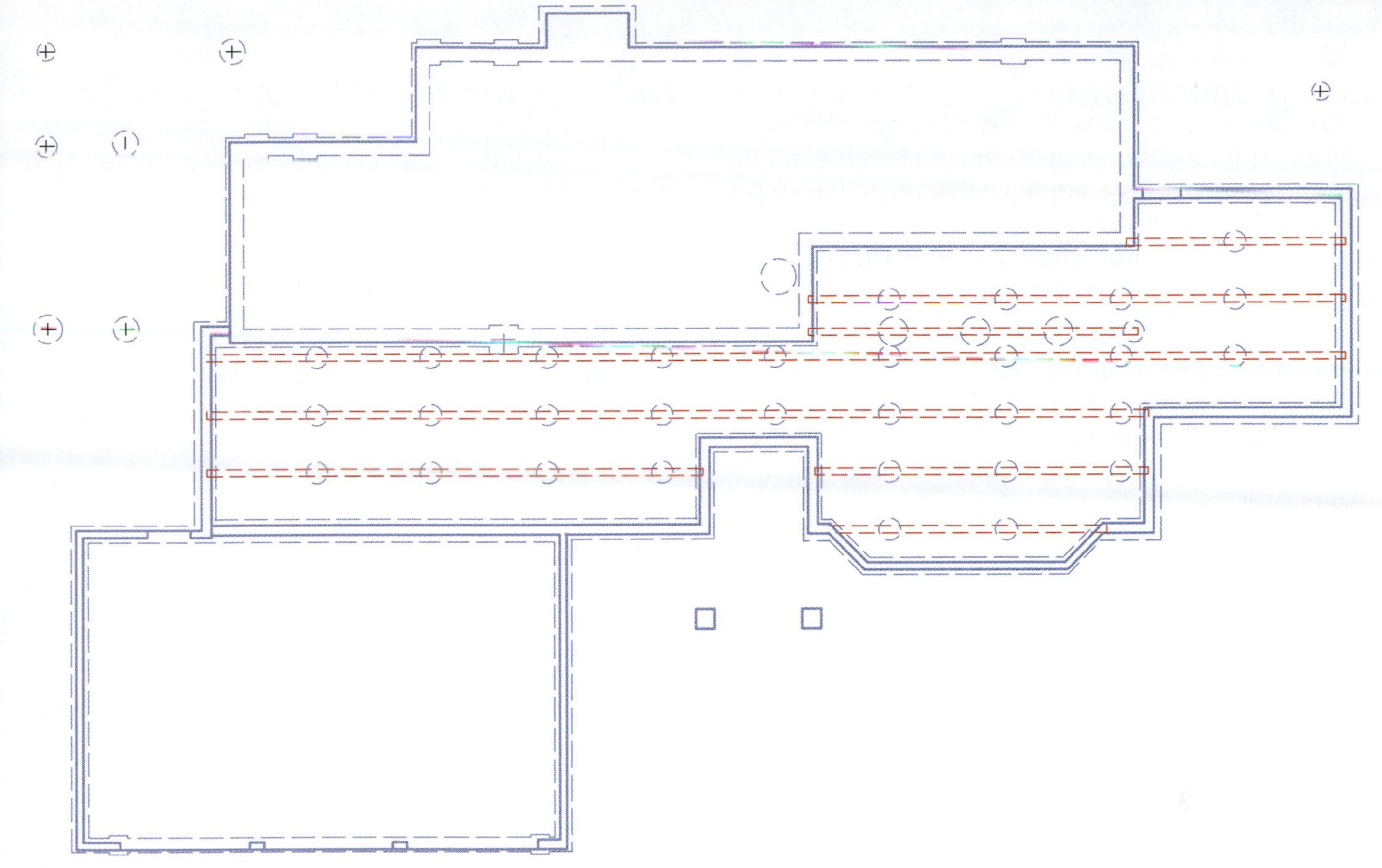

FIGURE 27.16 ■ *Draw the girders and piers required to support the upper floors over the crawl space.*

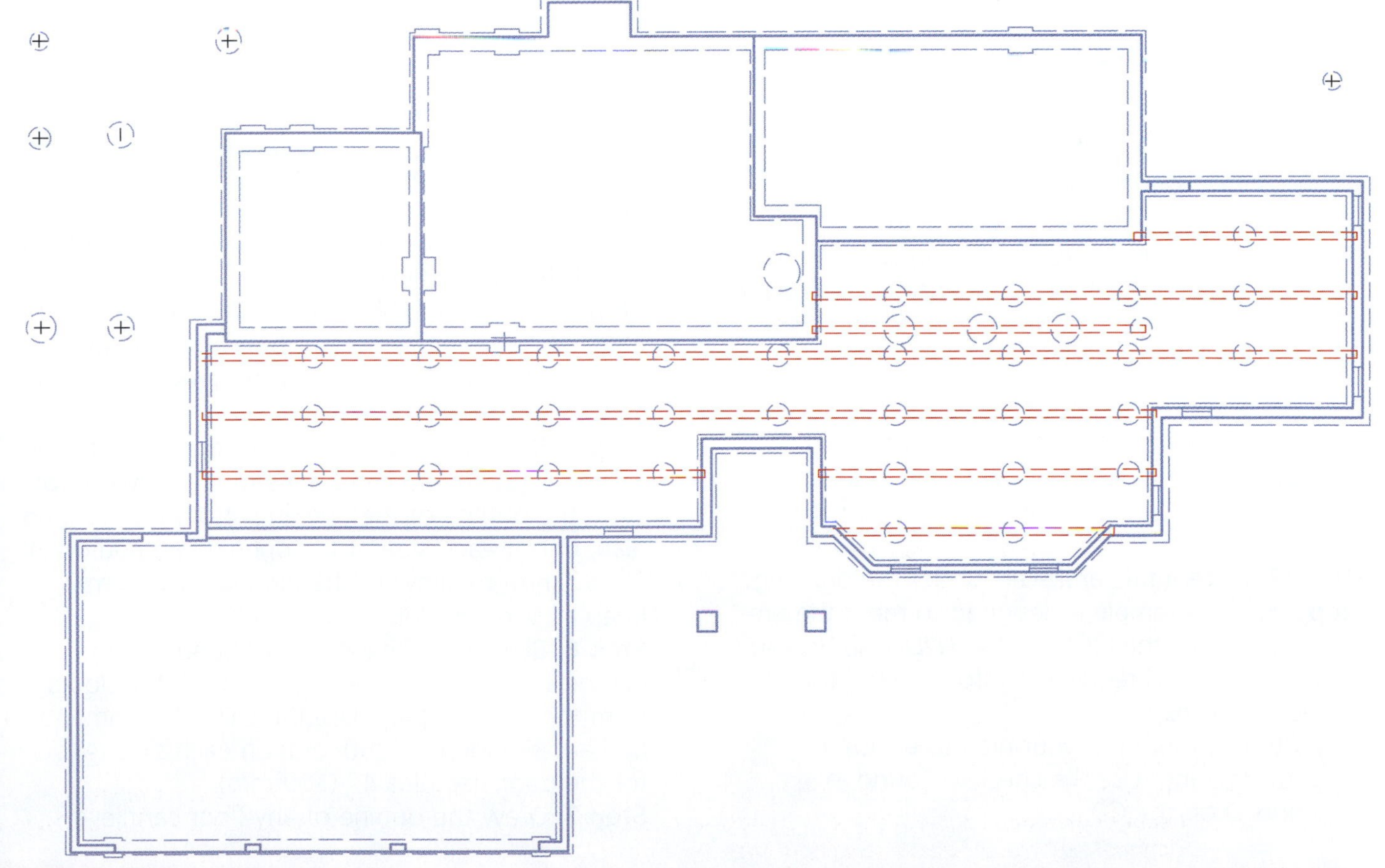

FIGURE 27.17 ■ *Draw any changes in floor elevations and add miscellaneous symbols such as vents and under-slab HVAC and plumbing materials.*

Placing Dimensions

All of the information that is required to represent the floor and foundation systems has now been drawn. Follow steps 17 through 24 to dimension on the drawing. Place the following information on the *FNDN DIMS* layer unless noted. Your drawing should resemble Figure 27.18 when complete.

Step 17. Thaw the *FRAM DIMS BASE* layer. This will display the overall and major jog dimensions on each side of the foundation.
Step 18. Dimension all door openings in the garage stem wall.
Step 19. Place dimensions to locate the chimney.
Step 20. Place dimensions to locate all exterior footings.
Step 21. Place dimensions to locate all girders. Using one dimension of 4'-0'' TYP. is very poor practice. Locate them all!
Step 22. Place dimensions to locate all piers.
Step 23. Place dimensions to locate all heating and plumbing materials that are under the slab.
Step 24. If required, place dimensions to locate metal connectors.

Placing Annotation

The final drawing procedure is to specify all of the materials that are to be used. This is done with general and local notes in the same method that was used on the framing plan. Place the following information on the *FNDN ANNO* layer. Figure 27.19 is an example of the notes that are required on the foundation using the prescriptive path of the IRC. Use the following steps to complete the foundation plan:

Step 25. Insert and edit the general notes to describe all concrete and floor system materials.
Step 26. Place all required local notes to describe each material. See the list of required notes in Chapter 26.
Step 27. Place a title and scale under the drawing.
Step 28. This example is designed to meet the prescriptive path of the IRC. Use the *FNDN ANNO LATT* layer and insert all required reinforcements for braced wall lines.
Step 29. Use a print of your plan to evaluate your drawing. Use the checklist found in Appendix D on the CD.

Full Basement

The foundation plan for a home with a full basement will have similarities to a slab foundation system. Figure 27.23 shows an example of a foundation plan with a basement slab and a joist floor system for the main floor that is not over the basement. This example assumes:

- The use of trusses to frame the roof.
- The use of concrete block units to form all stem walls. Although the goal is to keep as many sizes as possible modular, the overall sizes of this home have not been altered.
- Conformance to the IRC prescriptive path for lateral bracing.
- The use of exterior retaining walls to allow access to the family room.
- Window and door locations in the retaining wall have been altered slightly to conform to the use of CMUs.

Not all steps will be required for every house. Use the *FNDN WALL* layer for the following steps:

Representing Retaining and Stem Walls

Step 1. Using your framing plan, locate the exterior edge of the slab for the main floor slab.
Step 2. Using your framing plan, locate the exterior edge of the exterior walls for the lower floor.
Step 3. Use the OFFSET command to locate the interior side of the basement retaining wall that surrounds the lower floor.
Step 4. Use the OFFSET command to locate the interior side of the stem wall around the slab at the garage area. See Figure 25.10 for a review of minimum foundation sizes.
Step 5. Block out the locations of openings in the retaining walls including access to the crawl space. Trace the outline of the openings from the framing plan. Offset each side of the opening an additional 2'' (50 mm) to allow for framing 4'' (100 mm). Keep all sizes modular if possible.
Step 6. Block out the locations of doors in the garage stem walls. Trace the outline of the doors from the framing plan. Use the OFFSET command to add an additional 2'' (50 mm) on each side. Allow for the door size plus 4'' (100 mm).
Step 7. Draw the outline of any floor cantilevers.

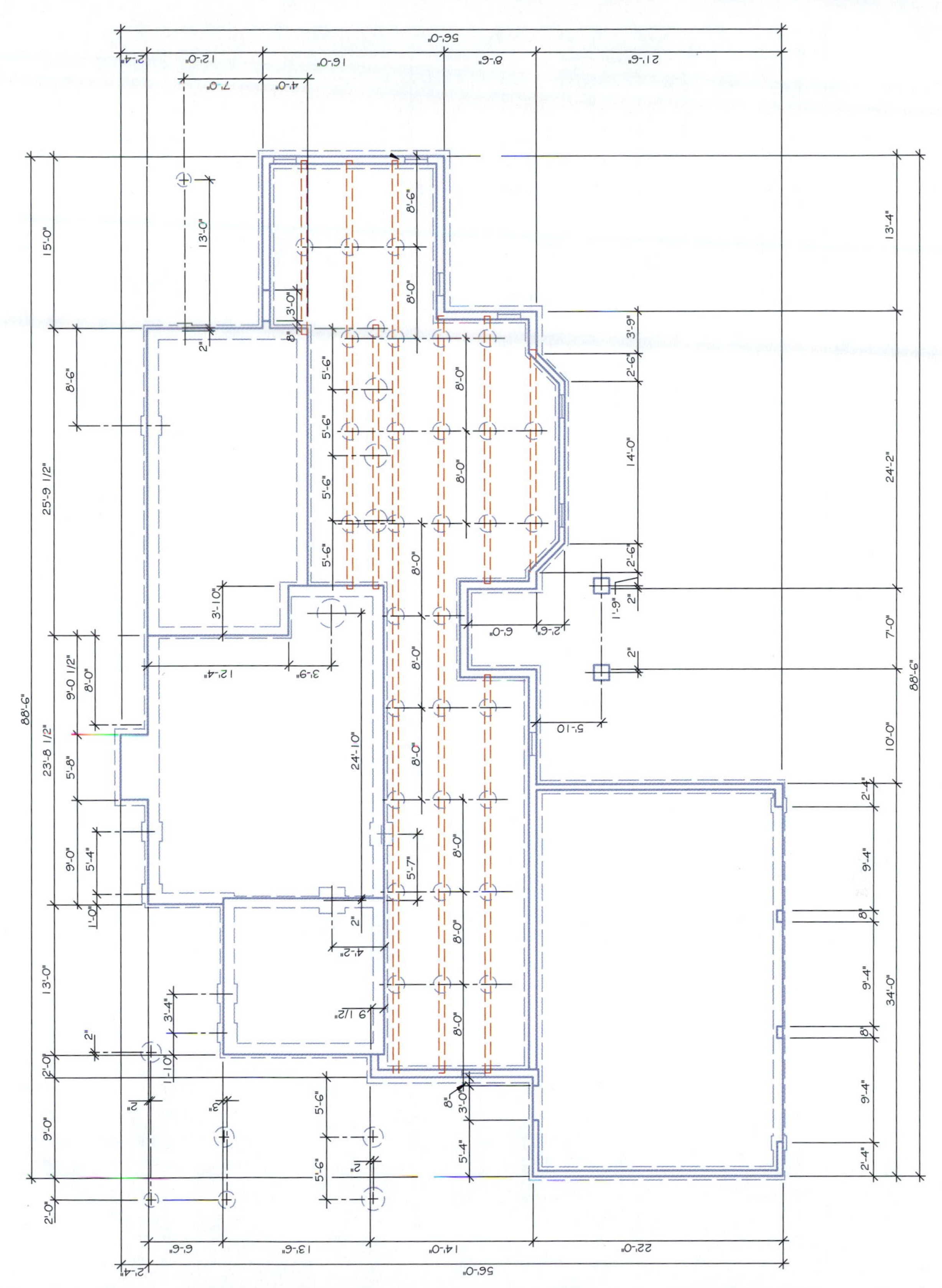

FIGURE 27.18 ■ *Thaw the FRAM DIMS BASE layer that was created on the framing plan to display the overall and major jog dimensions on each side of the foundation. Add dimensions to describe all concrete features.*

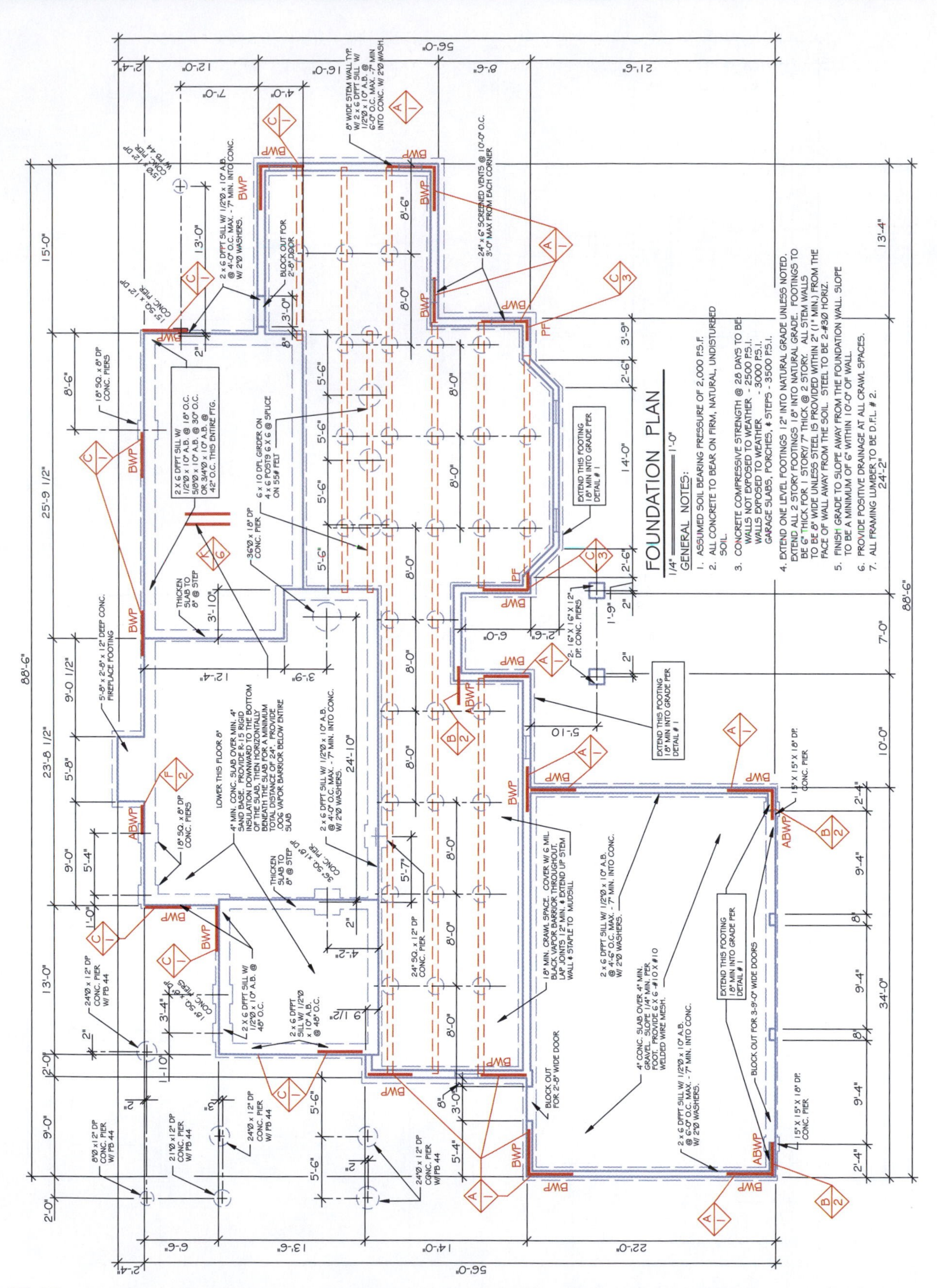
FOUNDATION PLAN
1/4" = 1'-0"
GENERAL NOTES:
1. ASSUMED SOIL BEARING PRESSURE OF 2,000 P.S.F.
2. ALL CONCRETE TO BEAR ON FIRM, NATURAL, UNDISTURBED SOIL.
3. CONCRETE COMPRESSIVE STRENGTH @ 28 DAYS TO BE:
WALLS NOT EXPOSED TO WEATHER - 2500 P.S.I.
WALLS EXPOSED TO WEATHER - 3000 P.S.I.
GARAGE SLABS, PORCHES, & STEPS - 3500 P.S.I.
4. EXTEND ONE LEVEL FOOTINGS 12" INTO NATURAL GRADE UNLESS NOTED. EXTEND ALL 2 STORY FOOTINGS 18" INTO NATURAL GRADE. FOOTINGS TO BE 6" THICK FOR 1 STORY/ 7" THICK @ 2 STORY. ALL STEM WALLS TO BE 8" WIDE UNLESS STEEL IS PROVIDED WITHIN 2" (1" MIN.) FROM THE FACE OF WALL AWAY FROM THE SOIL. STEEL TO BE 2-#30 HORIZ.
5. FINISH GRADE TO SLOPE AWAY FROM THE FOUNDATION WALL. SLOPE TO BE A MINIMUM OF 6" WITHIN 10'-0" OF WALL.
6. PROVIDE POSITIVE DRAINAGE AT ALL CRAWL SPACES.
7. ALL FRAMING LUMBER TO BE D.F.L. # 2.

FIGURE 27.19 ■ *The foundation plan for a structure with a post-and-beam floor system is finished by adding annotation to describe all materials and lateral bracing requirements to meet the prescriptive path of the IRC.*

Step 8. Use the OFFSET command to add a line to represent the support ledge if brick veneer is to be used. Assume that an additional 4'' (100 mm) will be added.
Step 9. Draw the size of the chimney based on measurements from the floor plan. Represent the size of the chimney, not the fireplace and hearth. Your drawing should now resemble Figure 27.20.

Representing Footings and Piers

Use the *FNDN FOOT* layer for the following steps:

Step 10. Draw the exterior footing width. Use the OFFSET command to offset the exterior edge of the slab as required, and then use the PROPERTIES command to assign the lines to the proper layer and linetypes.
Step 11. Draw the chimney footing so that it extends at least 6'' (150 mm) beyond the face of the fireplace.
Step 12. Draw all interior footings as required for load-bearing walls.

Use the *FNDN PIER* layer for the following steps:

Step 13. Draw all interior piers required for spot loads.
Step 14. Draw any exterior piers that might be required for decks or porches.

Representing Miscellaneous Materials

Step 15. Use the *FNDN GIRD* layer to draw all required girders to support floor joists, load-bearing walls, and changes in floor elevation.
Step 16. Use the *FNDN JOST* layer to draw or insert the arrow symbols to represent the floor joist spans.

Place the following steps on the *FNDN MISC* layer:

Step 17. Draw changes in the floor levels.
Step 18. Draw window wells using thin lines. Verify required sizes for emergency egress in Chapter 4.
Step 19. Draw windows using the methods introduced in Chapter 11.
Step 20. Draw vents in the walls surrounding the crawl space.
Step 21. Draw any required floor drains.
Step 22. Draw heating registers if required.
Step 23. Use the *FNDN PATT* layer to crosshatch any walls where concrete blocks are used.

Your drawing should now resemble Figure 27.21.

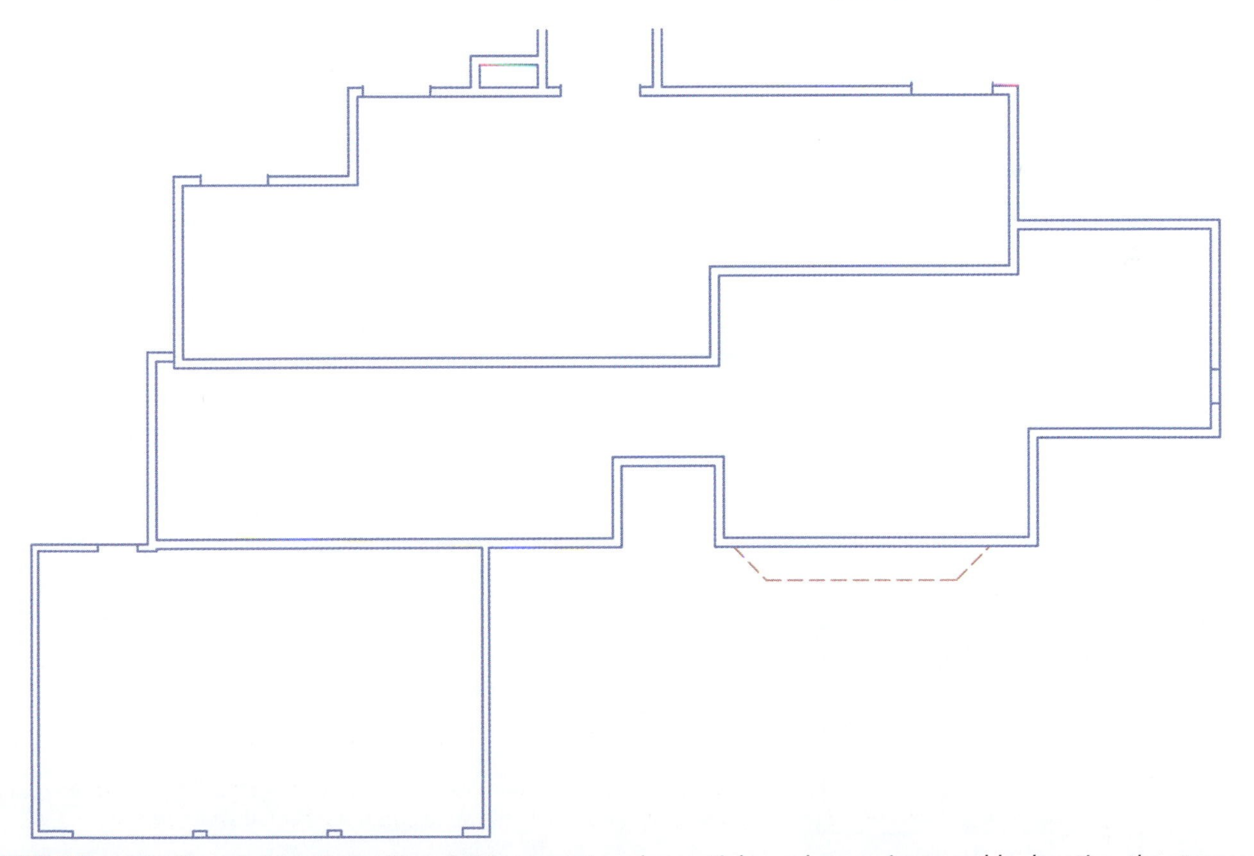

FIGURE 27.20 ■ *The foundation plan with a full basement and a partial crawl space is started by locating the stem walls, retaining walls, and openings in the basement wall.*

Placing Dimensions

Follow steps 24 through 30 to place the required dimensions on the drawing. Place the following information on the *FNDN DIMS* layer unless noted. Your drawing should resemble Figure 27.22 when complete.

Step 24. Thaw the *FRAM DIMS BASE* layer. This will display the overall and major jog dimensions on each side of the foundation.
Step 25. Dimension all openings in the stem and retaining walls. Do not dimension vent openings.
Step 26. Place dimensions to locate the chimney.
Step 27. Place dimensions for all exterior footings.
Step 28. Place dimensions to locate all girders.
Step 29. Place dimensions for all interior footings and piers.
Step 30. Place dimensions to locate all heating and plumbing materials to be placed below the slab.

Placing Annotation

The final drawing procedure is to specify the materials that are to be used. Place the following information on the *FNDN ANNO* layer. Figure 27.23 is an example of the notes that are required on the foundation completed using the prescriptive path of the IRC. Use the following steps to complete the foundation plan.

Step 31. Insert and edit the general notes to describe all concrete and floor system materials.
Step 32. Place all required local notes to describe each material. See the list of required notes in Chapter 26.
Step 33. Place a title and scale under the drawing.
Step 34. This example is designed to meet the prescriptive path of the IRC. Use the *FNDN ANNO LATT* layer and insert all required reinforcements for braced wall lines.
Step 35. Use the *FNDN ANNO LATT SCHD* layer and insert the appropriate lateral bracing schedule. If space is not available for the schedule, place the schedule on another page, as close to the foundation as possible.
Step 36. Use a print of your plan to evaluate your drawing. Use the checklist found in Appendix D on the CD.

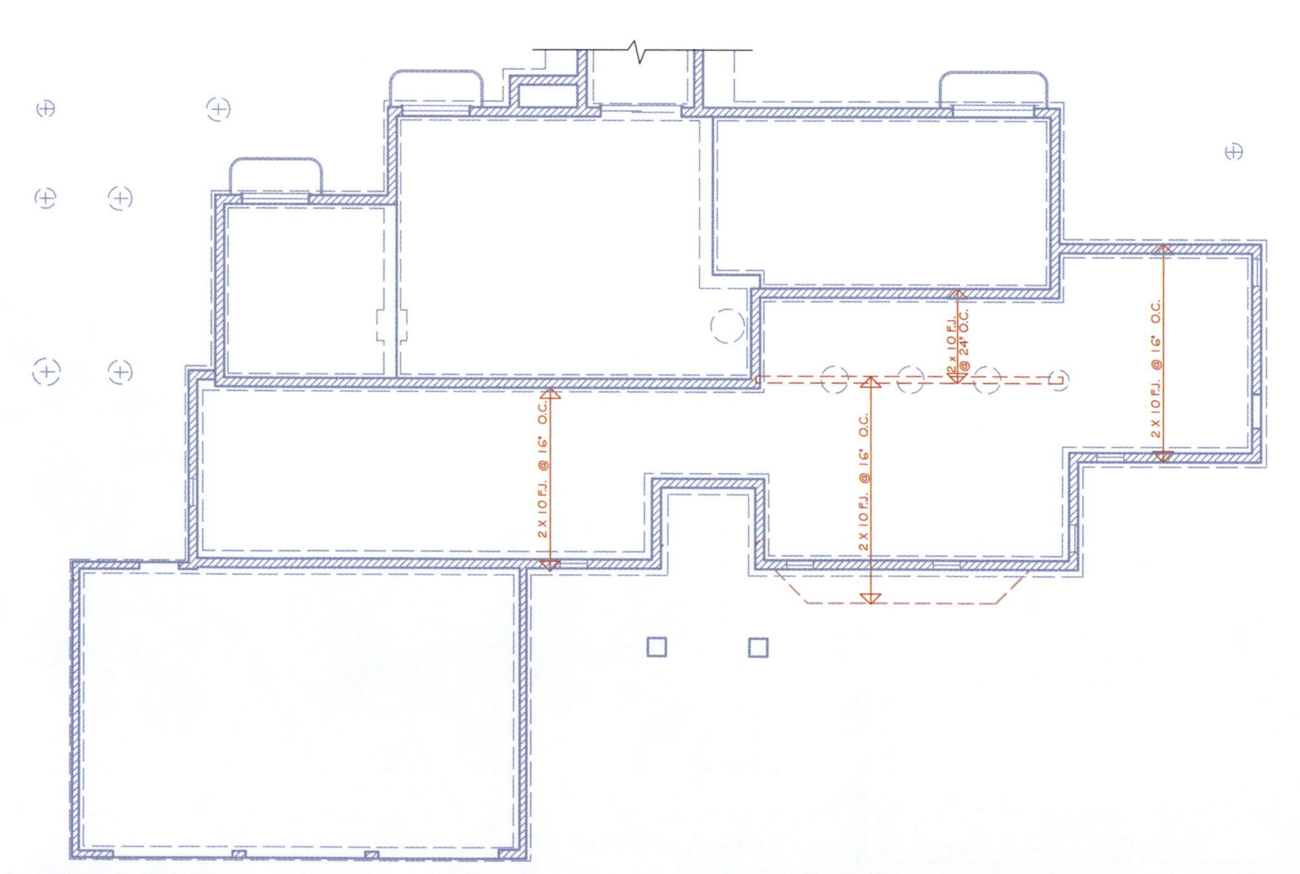

FIGURE 27.21 ■ *Add materials such as window wells, window symbols, girders, and joists to support the floor over the crawl space. Provide a hatch pattern if CMUs are used to build the walls.*

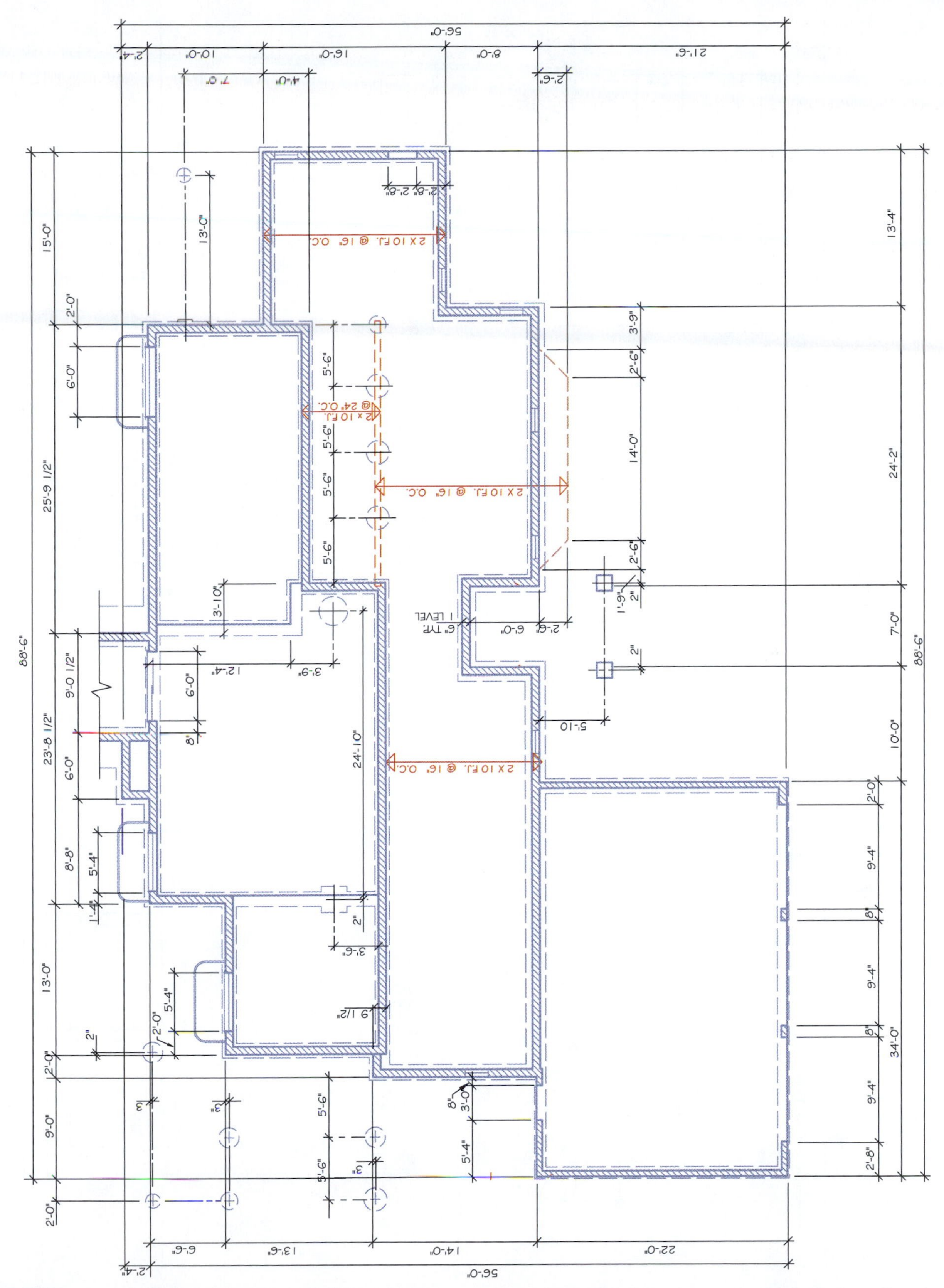

FIGURE 27.22 ■ *Thaw the FRAM DIMS BASE layer that was created on the framing plan to display the overall and major jog dimensions on each side of the foundation. Add dimensions to describe all concrete features.*

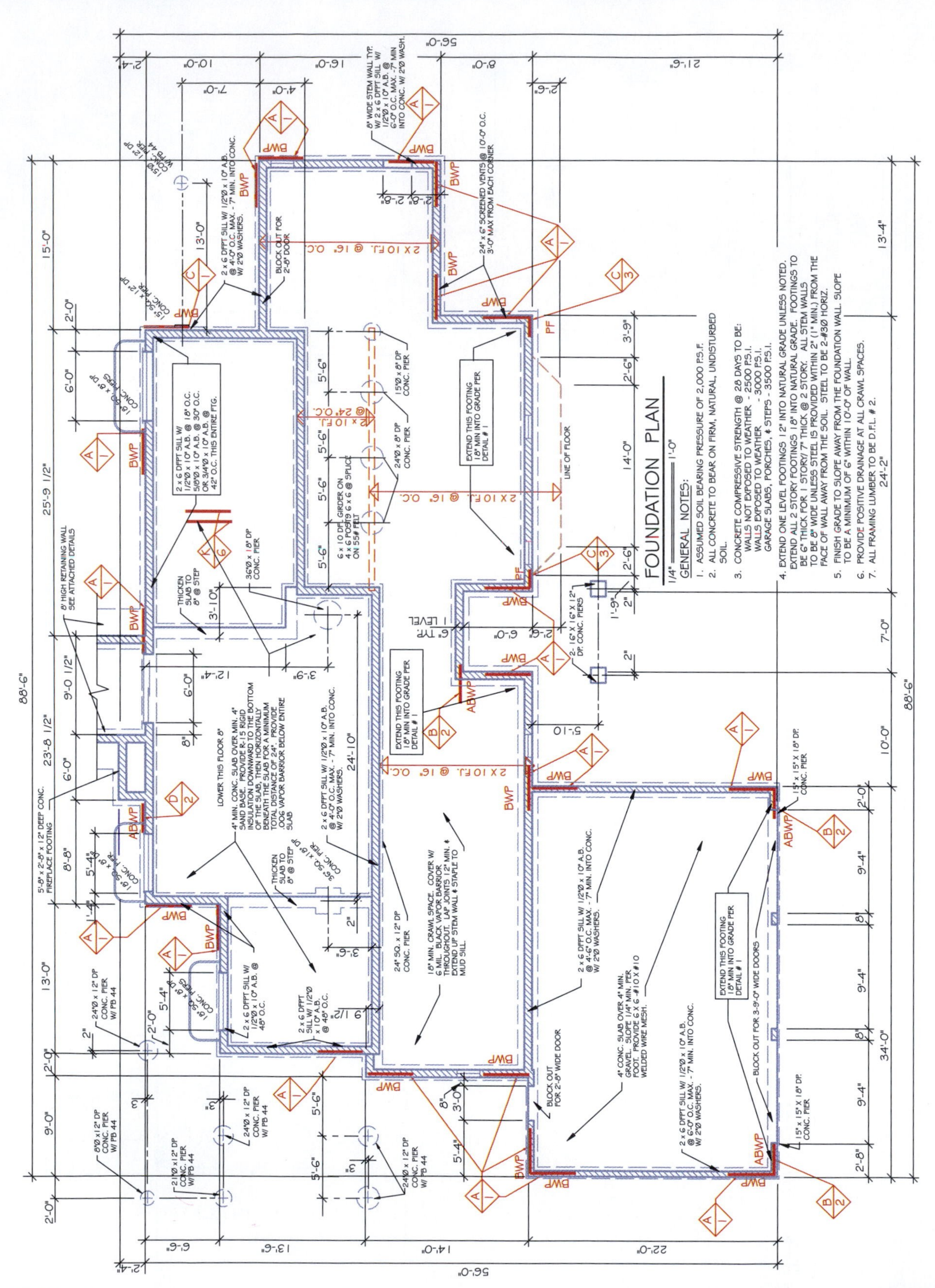
FOUNDATION PLAN
1/4" = 1'-0"
GENERAL NOTES:
1. ASSUMED SOIL BEARING PRESSURE OF 2,000 P.S.F.
2. ALL CONCRETE TO BEAR ON FIRM, NATURAL, UNDISTURBED SOIL.
3. CONCRETE COMPRESSIVE STRENGTH @ 28 DAYS TO BE:
WALLS NOT EXPOSED TO WEATHER - 2500 P.S.I.
WALLS EXPOSED TO WEATHER - 3000 P.S.I.
GARAGE SLABS, PORCHES, & STEPS - 3500 P.S.I.
4. EXTEND ONE LEVEL FOOTINGS 12" INTO NATURAL GRADE UNLESS NOTED. EXTEND ALL 2 STORY FOOTINGS 18" INTO NATURAL GRADE. FOOTINGS TO BE 6" THICK FOR 1 STORY/ 7" THICK @ 2 STORY. ALL STEM WALLS TO BE 8" WIDE UNLESS STEEL IS PROVIDED WITHIN 2" (1" MIN.) FROM THE FACE OF WALL AWAY FROM THE SOIL. STEEL TO BE 2-#3Ø HORIZ.
5. FINISH GRADE TO SLOPE AWAY FROM THE FOUNDATION WALL. SLOPE TO BE A MINIMUM OF 6" WITHIN 10'-0" OF WALL.
6. PROVIDE POSITIVE DRAINAGE AT ALL CRAWL SPACES.
7. ALL FRAMING LUMBER TO BE D.F.L. # 2.

FIGURE 27.23 ■ *The foundation plan for a structure with a full basement and joist floor system is finished by adding annotation to describe all materials and lateral bracing requirements to meet the prescriptive path of the IRC.*

CHAPTER 27

Foundation-Plan Layout Test

DIRECTIONS

Answer the following questions with short, complete statements. Type your answers using a word processor or neatly print your answers on lined paper.

1. Place your name, the chapter number, and the date at the top of the sheet.
2. Type the question number and provide the answer in the form of a statement that includes part of the question. You do not need to write out the entire question.

Note: The answers to some questions may not be contained in this chapter and will require you to do additional research.

QUESTIONS

27.1. At what scale will the foundation be drawn?

27.2. List five items that must be shown on a foundation plan for a concrete slab.

27.3. What general categories of information must be dimensioned on a slab foundation?

27.4. Show how floor joists are represented on a foundation plan.

27.5. How large an opening should be provided in the stem wall for a garage door 8'-0'' wide?

27.6. How much space should be provided for a 3' entry door in a post-and-beam foundation? Explain your answer.

27.7. How are the footings represented on a foundation plan?

27.8. Show two methods of representing girders.

27.9. What type of line quality is typically used to represent beam pockets?

27.10. What are the disadvantages of tracing a print of the floor plan to lay out a foundation plan?

PROBLEMS

Draw a foundation plan that corresponds to the floor plan problem from Chapter 11 that you have drawn. Draw the required foundation plan using the guidelines given in this chapter. Design a floor system that is suitable for the residence and your area of the country.

- Draw the foundation plan using the same type and size of drawing material that you used for the floor plan.
- Use the same scale that was used to draw the floor plan.
- Refer to your framing plan to determine dimensions and position of load-bearing walls.
- Refer to your framing plan to determine the type and location of lateral bracing.
- Refer to the text of this chapter and class lecture notes for complete information.
- When your drawing is complete, turn in a print to your instructor for evaluation.

SECTION 8

Details and Sections

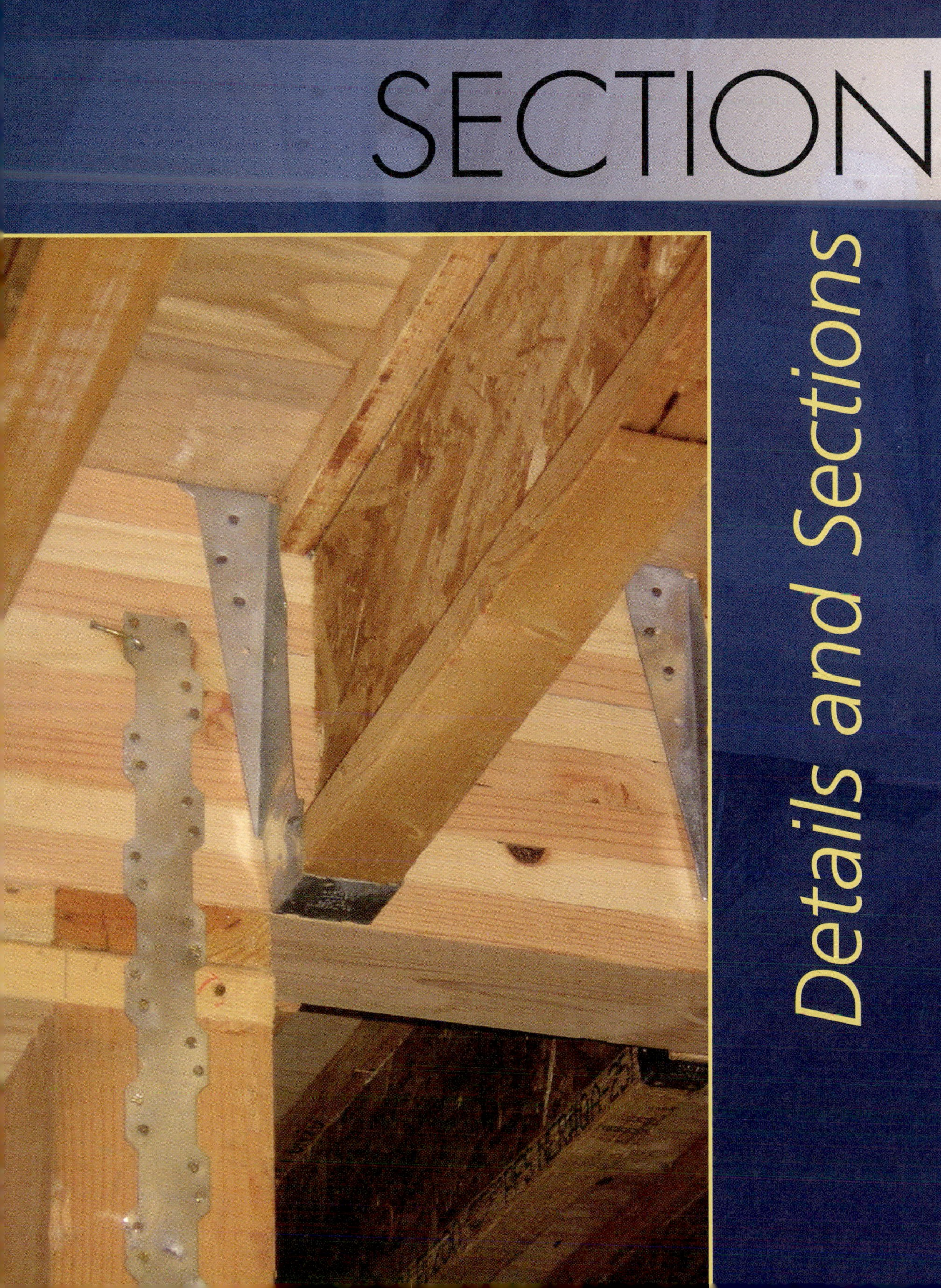

28 CHAPTER

Sectioning Basics

Sections are drawn to show the vertical relationships of the structural materials called for on the floor, framing, roof, and foundation plans. The sections show the methods of construction for the framing crew. Before drawing sections, it is important to understand the different types of sections, their common scales, and the relationship of the cutting plane to the section.

TYPES OF SECTIONS

Three types of sections may be drawn for a set of plans: full sections, partial sections, and ***details.***

Full Sections

For a simple home, only one section might be required to explain fully the types of construction to be used. Each major structural material must be drawn and specified as well as providing the vertical dimensions. A full section can be seen in Figure 28.1a. Notice that the roof is framed with standard/scissor trusses, the exterior walls are framed with 2 × 6 (50 × 150) studs, and the floor system is post and beam with a 7 3/4'' (195 mm) step. Vertical relationships for the roof pitch, wall height, window and door header heights, foundation, and crawl heights are also provided.

An alternative method of drawing full sections is to use the drawing as a reference map. This type of section will have dimensions to explain various heights and room titles to explain the rooms being viewed. Very little text is provided to explain materials, and details are referenced to the section to explain all materials. Figure 28.1b shows a section with minimal information. Notice that information in the background is also represented.

Partial Wall Sections

For a more complex residence, more than one section may be required to specify each major type of construction. Some offices use a combination of partial and full sections to explain the required construction procedures. A partial section will show only typical roof, wall, floor, and foundation information for one typical wall rather than the full structure. A partial section may be drawn at a scale of 3/8'' = 1'-0'', 1/2'' = 1'-0'', or 3/4'' = 1'-0'' (1:32, 1:24, 1:16), depending on office procedure. An example of a partial section is shown in Figure 28.2.

The partial section is typically supplemented with full sections drawn at a scale of 1/4'' = 1'-0'' or 3/8'' = 1'-0'' (1:48 or 1:32). Only material that has not been specified on the partial section is noted on the full sections. The use of partial sections has become popular in many professional offices because of the use of computers. Several different partial sections can be created to reflect major types of construction such as one- or two-level construction, truss or stick roof, concrete slab, post-and-beam or joist foundations, and various wall coverings. These partial sections can be stored in a library and inserted into each set of plans as needed. Partial sections can also be used on complex structures to serve as a reference for details of complicated areas.

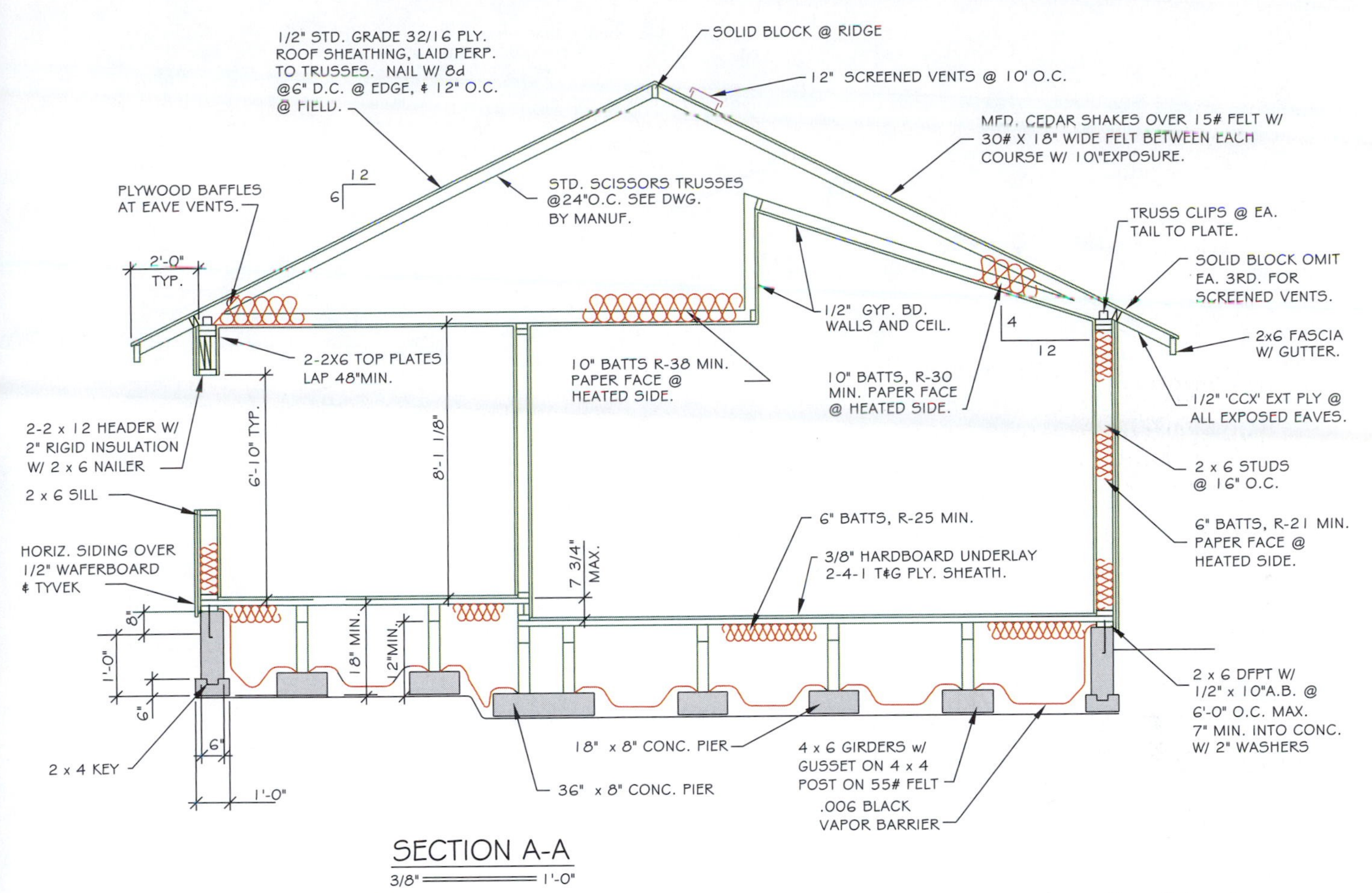

FIGURE 28.1a ■ *A full section shows framing members used in a specific area of a structure and specifies all framing members.*

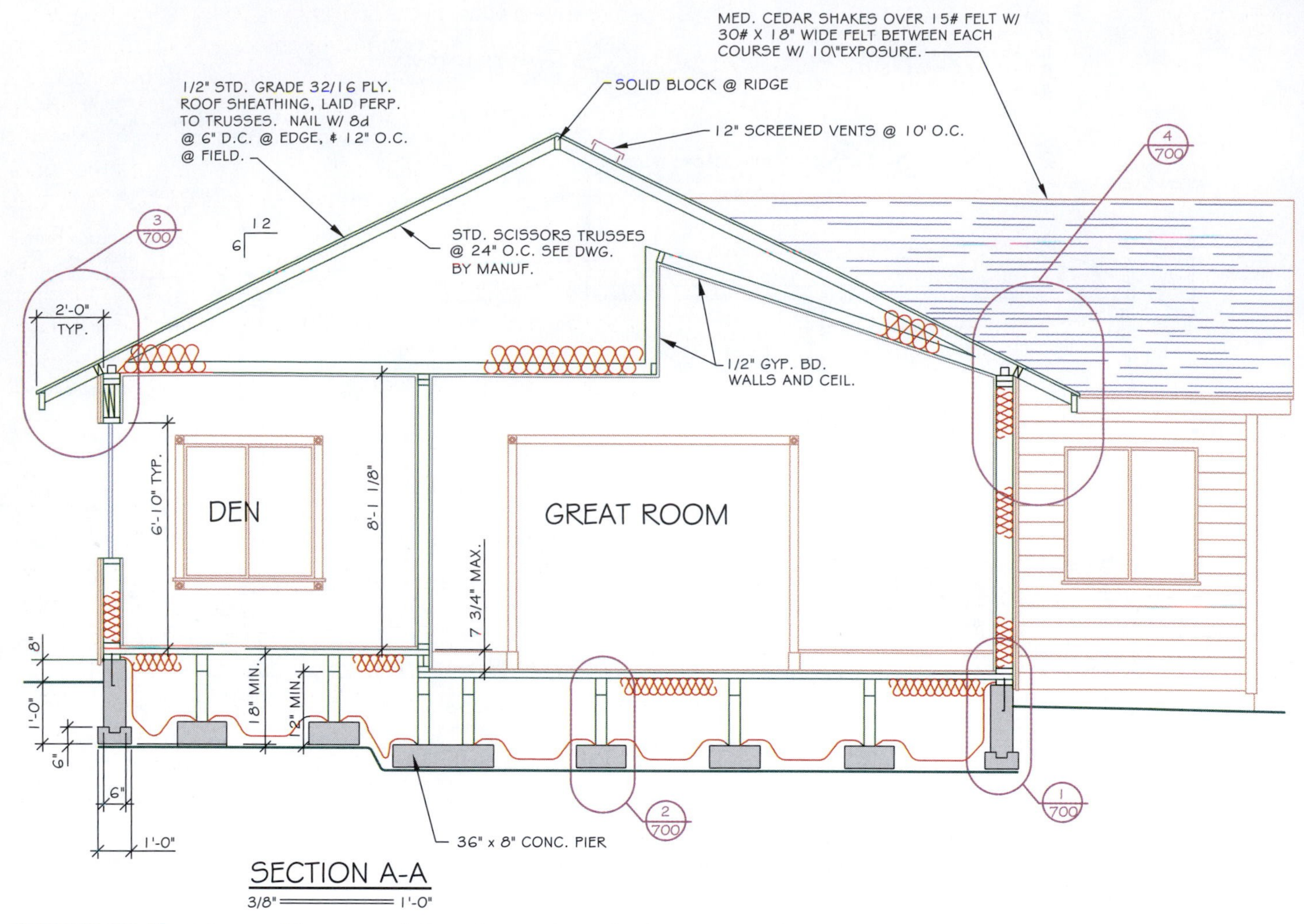

FIGURE 28.1b ■ *Some offices use a full section to show the framing members used in a specific area of a structure as well as materials that lie beyond the cutting plane. Notes are kept to a minimum, with most information placed on details referenced to the section.*

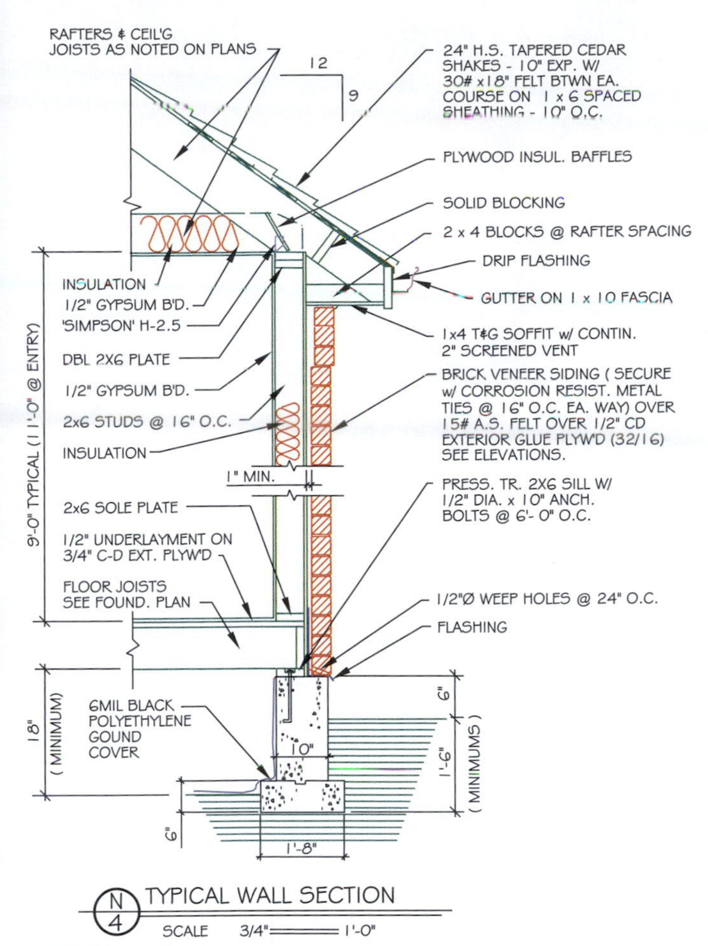

FIGURE 28.2 ■ *A partial section can be used to show the typical roof, wall, floor, and foundation construction materials of a specific structure.*

Details

Details are enlargements of specific areas of a structure and are typically drawn where several components intersect or where small members are required. Figure 28.3 shows an example of a partial section of a hillside residence built on a piling foundation. Figure 28.4 shows two details that relate to the partial section in Figure 28.3. Details are typically drawn at a scale of 1/2'' = 1'-0'' through 3'' = 1'-0'' (1:16 through 1:4), depending on the complexity of the intersection.

Most offices have a library of stock details. These are details of items such as footings that remain the same. Common stock foundations might include:

- 1-level concrete slab
- 1-level concrete slab with masonry veneer
- 1-level concrete slab w/ exterior insulation flush wall
- 1-level concrete slab w/ exterior insulation projected wall

The same details would typically exist for two-level construction, for post-and-beam construction, and for joist construction. Using a computer, the typical detail can be drawn with all required dimensions and all notations added. Once complete, copies can be made for each required detail, each copy can be edited and saved. The first detail may take an hour to complete, the second detail should just take a few minutes of editing time. Figure 28.5 shows a detail created by editing the master detail. By combining the information on the framing plans and the sections and details, the contractor should be able to make accurate estimates of the amount of material required and the cost of completing the project.

The architectural team will be required to draw sections to comply with the building permit application process and to explain to the construction crew each type of construction method to be used on the project. Many municipalities will accept a partial section as sufficient to meet the demands for the building permit. Most architects and designers will provide far more sections to explain each type of construction and changes in shape or size of the structure. Although not specifically required for a building permit, details will also be provided to explain the intersections of structural materials, changes in levels, and the application of materials. Details for the home started in Chapter 11 would include:

- Site-related work
- Construction and reinforcing of footings
- Lateral supports and connections between floors
- The foundation retaining wall
- Interior trim details

PLOTTING SCALES

To make the sections easier to read, sections have become somewhat standardized in several areas, including scales and alignment. Sections are typically drawn at a scale of 3/8'' = 1'-0'' (1:32). Scales of 1/8'' or 1/4'' (1:96 or 1:48) may be used for supplemental sections requiring little detail. A scale of 3/4'' = 1'-0'' (1:16) or larger may be required to draw some construction details. Several factors influence the choice of scale in drawing sections:

1. Size of the drawing sheet to be used
2. Size of the project to be drawn
3. Purpose of the section
4. Placement of the section

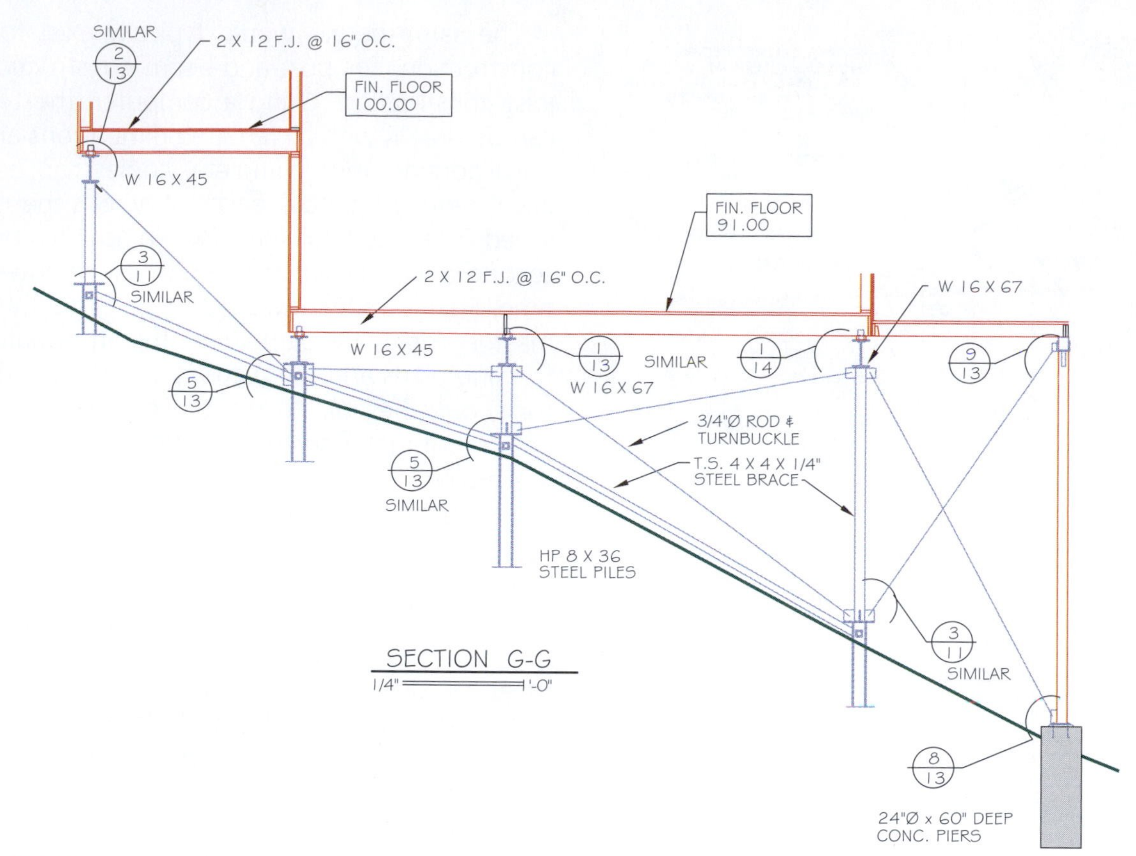

FIGURE 28.3 ■ *This partial section for a hillside residence is used to provide supplemental information about one specific area of the structure.*

Factors 1 and 2 need little discussion. The floor plan determines the size of the project. Once the sheet size is selected for the floor plan, that size should be used throughout the entire project. The placement of the section as it relates to other drawings should have only a minor influence on the scale. It may be practical to put a partial section in a blank corner of a drawing, but don't let space dictate the scale.

The most important factor should be the purpose of the section. If the section is used merely to show the shape of the project, a scale of 1/8'' = 1'-0'' (1:96) would be fine. This type of section is rarely required in residential drawings but is often used in drawing multi-unit residential projects. When it is used for residential projects, this type of section is used as a reference map to locate structural details. See Figure 28.6 for an example of a shape section for a shear wall in a multilevel apartment project.

The primary section is typically drawn at a scale of 3/8'' = 1'-0'' (1:32). This scale provides two benefits, one to the print reader and one to the CAD technician. The main advantage of using this scale is the ease of distinguishing each structural member. At a smaller scale, separate members such as the finished flooring and the rough flooring are difficult to draw and read. Without good clarity, problems could arise at the job site. At 3/8'' (1:32) scale, you'll have a bigger drawing on which to place the notes and dimensions.

A scale of 1/2'' = 1'-0'' (1:24) is not widely used in most offices. This scale is very clear, but sections are so large that a great deal of paper is required to complete the project. Often, if more than one section must be drawn, the primary section is drawn at 3/8'' = 1'-0'' (1:32), and the other sections are drawn at 1/4'' = 1'-0'' (1:48). By combining drawings at these two scales, typical information can be placed on the larger section, and the smaller sections are used to show variations with little detail.

SECTION ALIGNMENT

In drawing sections as with other parts of the plans, the drawing is read from the bottom or right side of the page. The cutting plane on the framing plan shows

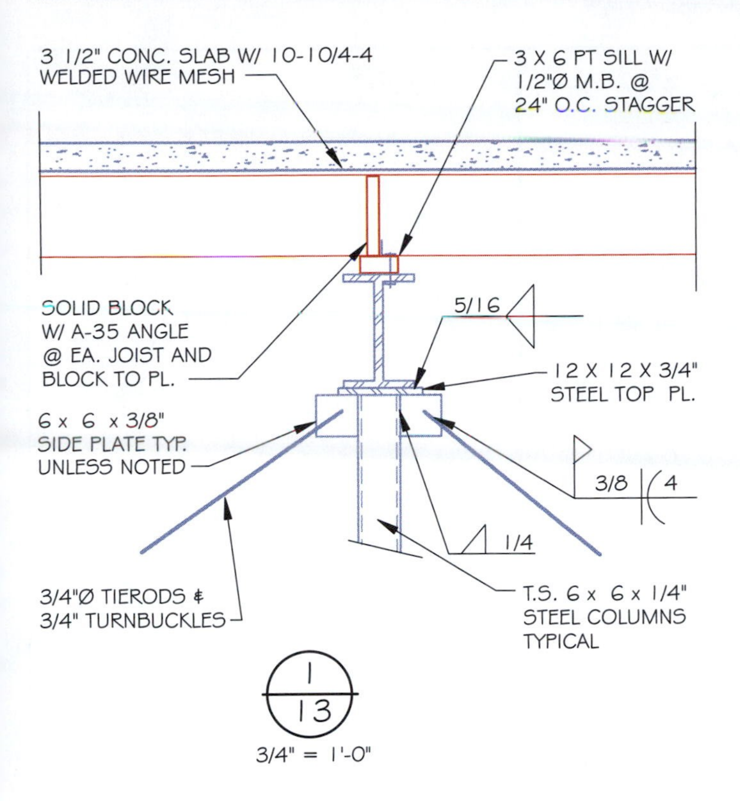

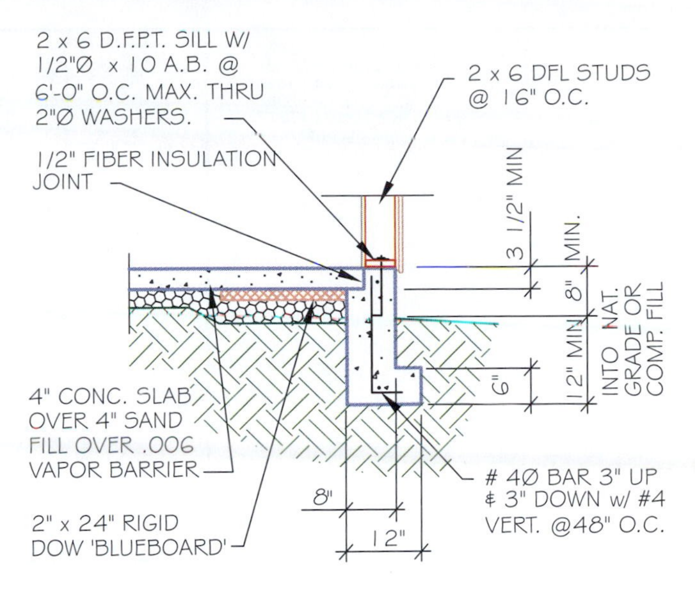

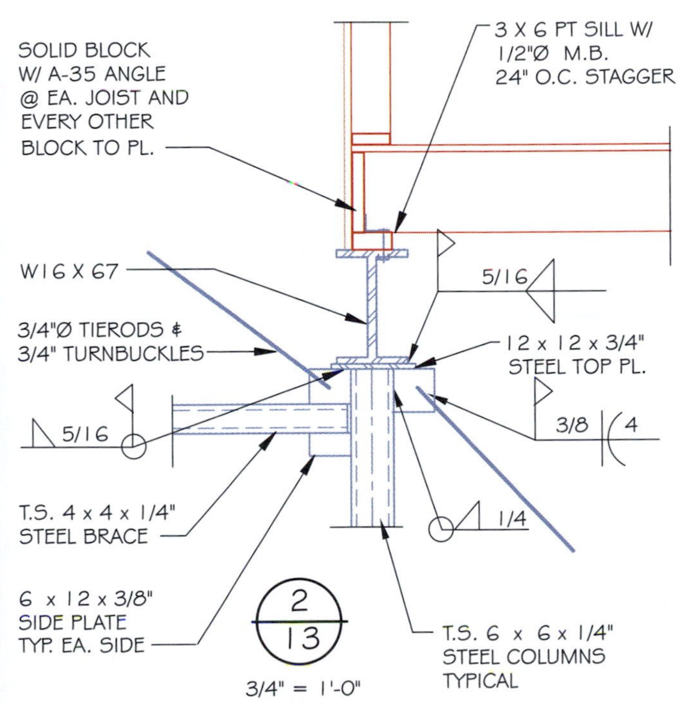

FIGURE 28.4 ■ *Details provide information about a small, complicated area of a structure, such as the intersection of various materials. These details are referenced to the partial section in Figure 28.3.*

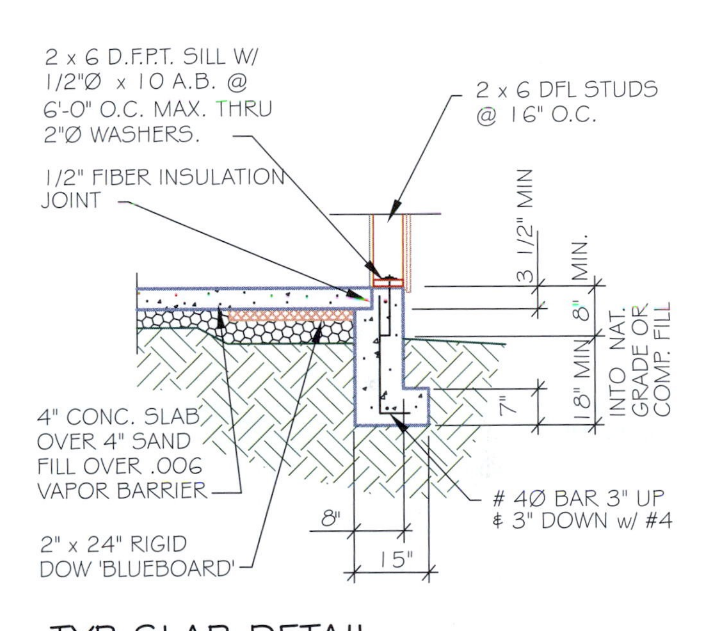

FIGURE 28.5 ■ *Details of common areas of construction such as a footing are drawn and saved in a detail library for future use. The detail at the top was edited to meet the needs of a 1-level project.*

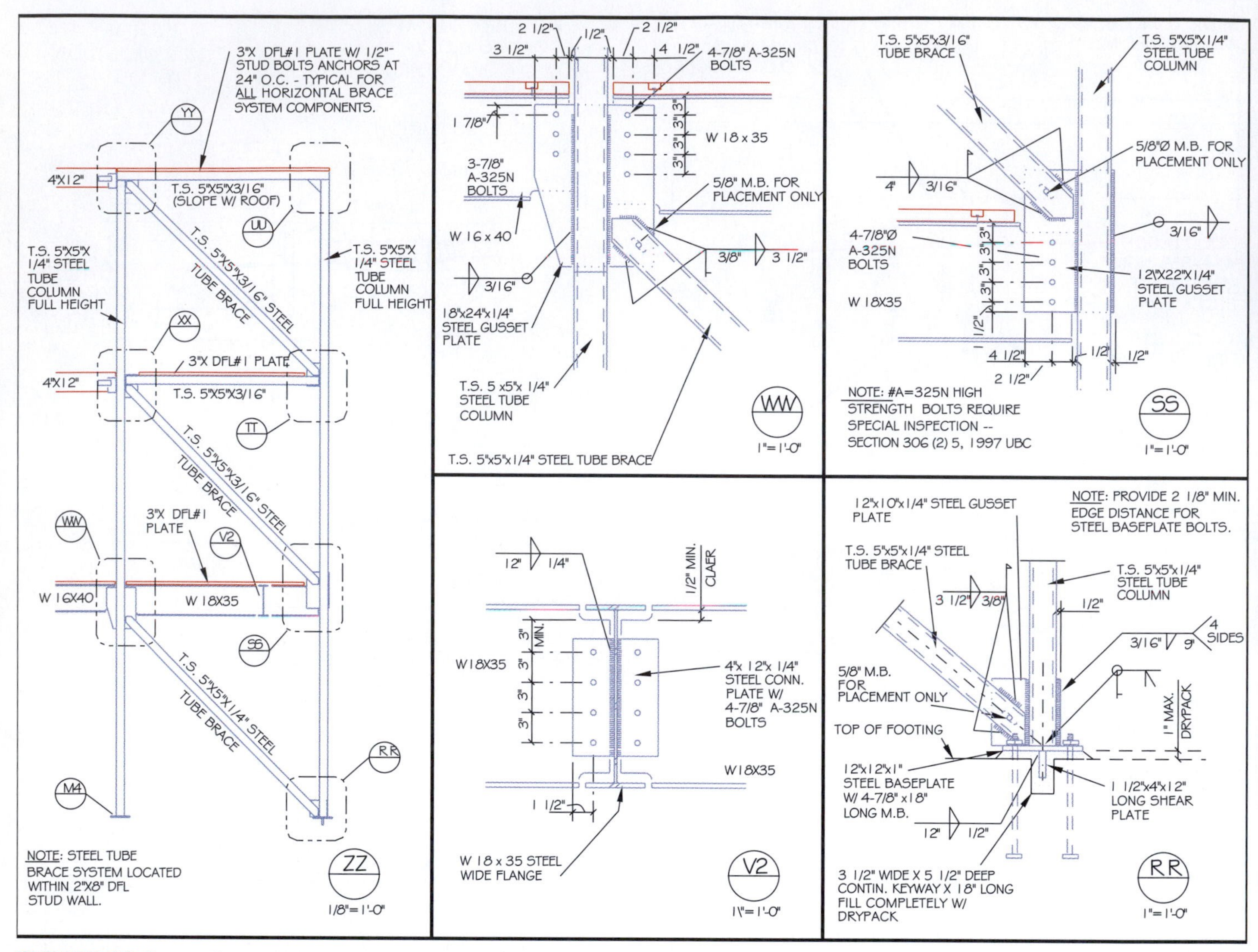

FIGURE 28.6 ■ *A shape section serves as a reference map for detail markers. Typically, it shows no specific details of an intersection.* Courtesy Kenneth D. Smith & Associates, Inc.

which way the section is being viewed. The arrows of the cutting plane should be pointing to the top or left side of the paper, depending on the area of the building being sectioned (see Figure 28.7). Where possible, the cutting plane should extend through the entire structure. The cutting plane can be broken for notes or dimensions to maintain clarity. On complex structures, the cutting plane can be jogged to show material clearly and avoid having to draw a second section. Figure 28.8 shows an example of a framing plan with section markers. Figure 11.18 shows section cutting plane lines for the home that has been used throughout the text.

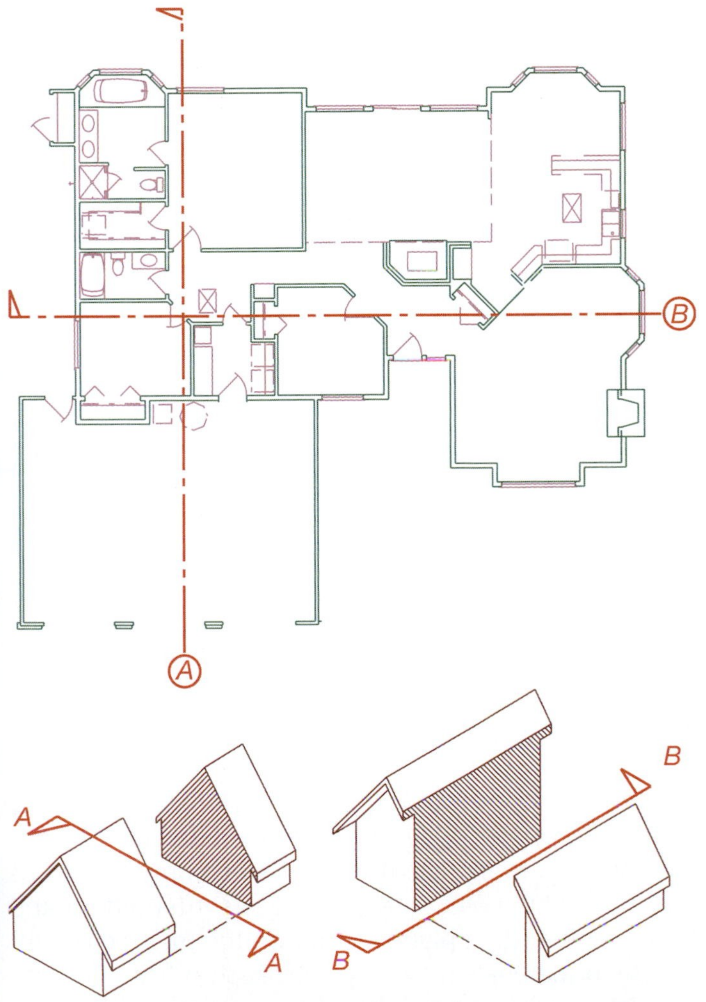

FIGURE 28.7 ■ *Cutting planes on the framing plan show the direction from which the section is to be viewed. Always try to keep the cutting-plane arrows pointing to the left or to the top of the page.*

REPRESENTING AND LOCATING MATERIALS

The type of section to be drawn will dictate the amount of information to be displayed and how the material will be represented. The smaller the plotting scale, the less information will be presented and the fewer number of linetypes will be used. In addition to using larger plotting scales, details require more attention to line contrast and the use of more varied line weights. Careful consideration must also be given to how materials will be represented.

Adding Layers

Before details can be started, layers need to be created to separate information by material and by lineweight and line types. Layers should start with a prefix of *DETL* and information should be named with a modifier listed in Appendix F. Sub-names of *ANNO, DIMN, FOOT, OUTL,* and *SYMB* will always be needed. As with any other drawing, create additional layers as needed to ensure that only layers that will be used are added to the drawing.

Using Line Contrast

Although standards vary, some details will require a minimum of four different lineweights to provide contrast between materials. Unfortunately there is no standard of *"always use this lineweight."* The lineweights used will vary depending on the plotting scale of the detail and the materials that must be represented. The use of .0 (default) for thin lines and .60 lines will serve as a starting point for all details to provide contrast for thin and thick lines. When drawing foundation details to be plotted at a scale of 3/8'' = 1'-0'' (1:32), a thickness of .90 can be used to represent the outline of concrete and a weight of 1.00 to represent the finished grade.

Note:

The goal of any detail is to clearly represent material. Because there is no set standard for lineweight, draw a few lines using varied lineweights and then make a test plot. Line thickness should be thick enough to provide contrast, but not so thick that they bleed into other objects.

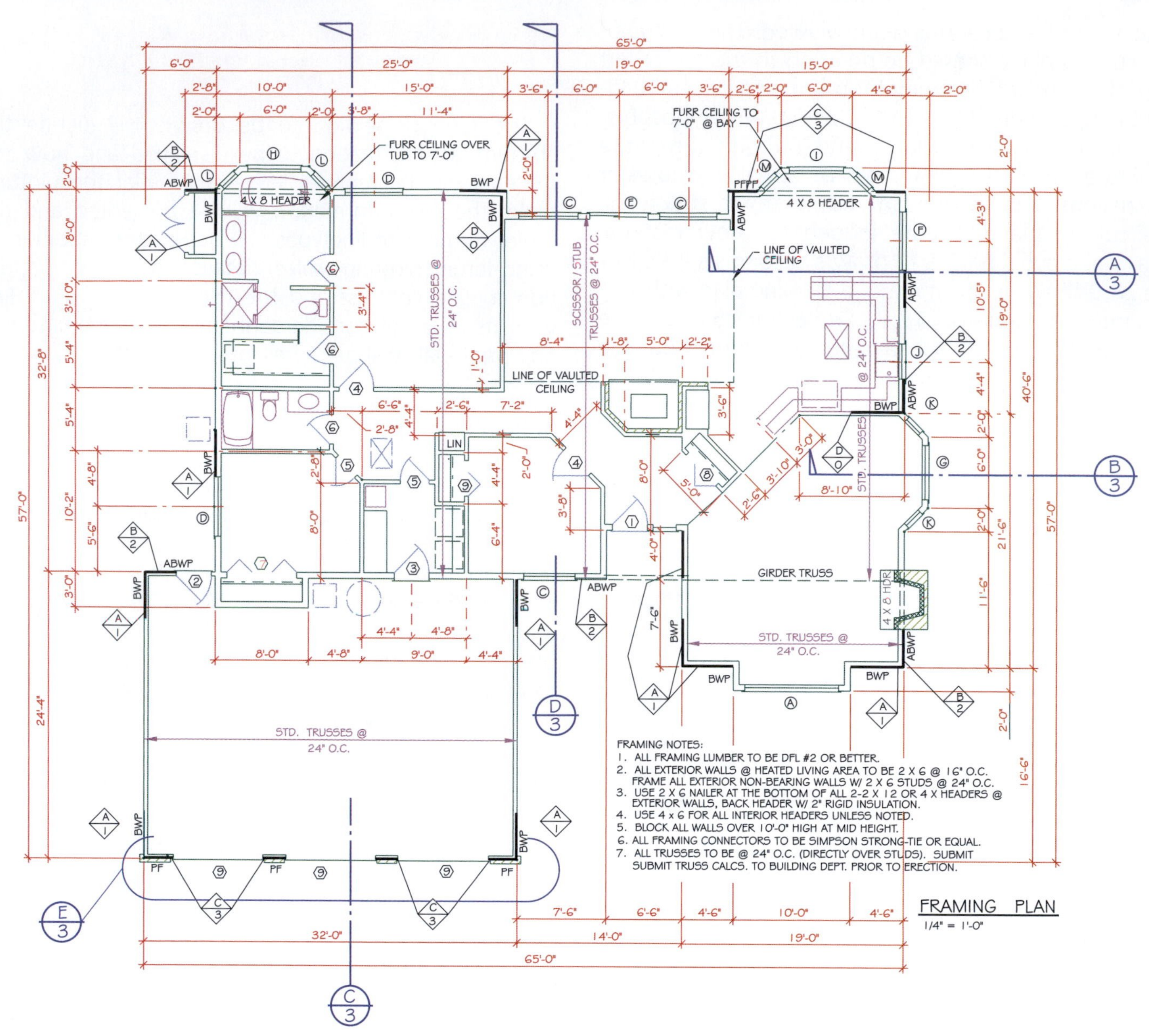
FURR CEILING OVER TUB TO 7'-0"
FURR CEILING TO 7'-0" @ BAY
4 X 8 HEADER
4 X 8 HEADER
STD. TRUSSES @ 24" O.C.
SCISSOR / STUB TRUSSES @ 24" O.C.
LINE OF VAULTED CEILING
LINE OF VAULTED CEILING
GIRDER TRUSS
4 X 8 HDR
STD. TRUSSES @ 24" O.C.
STD. TRUSSES @ 24" O.C.
FRAMING NOTES:
1. ALL FRAMING LUMBER TO BE DFL #2 OR BETTER.
2. ALL EXTERIOR WALLS @ HEATED LIVING AREA TO BE 2 X 6 @ 16" O.C. FRAME ALL EXTERIOR NON-BEARING WALLS W/ 2 X 6 STUDS @ 24" O.C.
3. USE 2 X 6 NAILER AT THE BOTTOM OF ALL 2-2 X 12 OR 4 X HEADERS @ EXTERIOR WALLS, BACK HEADER W/ 2" RIGID INSULATION.
4. USE 4 x 6 FOR ALL INTERIOR HEADERS UNLESS NOTED.
5. BLOCK ALL WALLS OVER 10'-0" HIGH AT MID HEIGHT.
6. ALL FRAMING CONNECTORS TO BE SIMPSON STRONG-TIE OR EQUAL.
7. ALL TRUSSES TO BE @ 24" O.C. (DIRECTLY OVER STUDS). SUBMIT SUBMIT TRUSS CALCS. TO BUILDING DEPT. PRIOR TO ERECTION.
FRAMING PLAN
1/4" = 1'-0"

FIGURE 28.8 ■ *Cutting-plane markers may extend through a structure, broken at text or dimensions to provide clarity, or jogged to eliminate a second section.*

Once the lineweights have been selected, a method will need to be selected on how to assign lineweights. Two common methods to assign lineweights include:

- Assign and name layers such as *THIN, THICK, VERY THICK,* and *MEGA THICK.*
- Assign layers names based on materials such as *WOOD, STEEL, CONCRETE,* or *SOIL.* Lineweights can then be assigned to objects on those layers using the PROPERTIES command. See the guidelines for naming layers on the student CD.

In plotting, gray scale can be used to provide contrast to existing and new materials by assigning gray to existing materials and black to new materials.

Representation of Material

The method used to represent each material will vary depending on the scale that is used. Different methods are also used to represent materials that are continuous and are cut by the cutting plane or are intermittent and beyond the cutting plane. Although the method of representing materials may vary with each office, it is critical that each material be distinguished from other materials to provide drawing clarity. Common materials displayed in section are shown in Figure 28.9. It is also important not to spend more time than necessary detailing materials. If a product is delivered to the site ready to be installed, minimal attention representing the

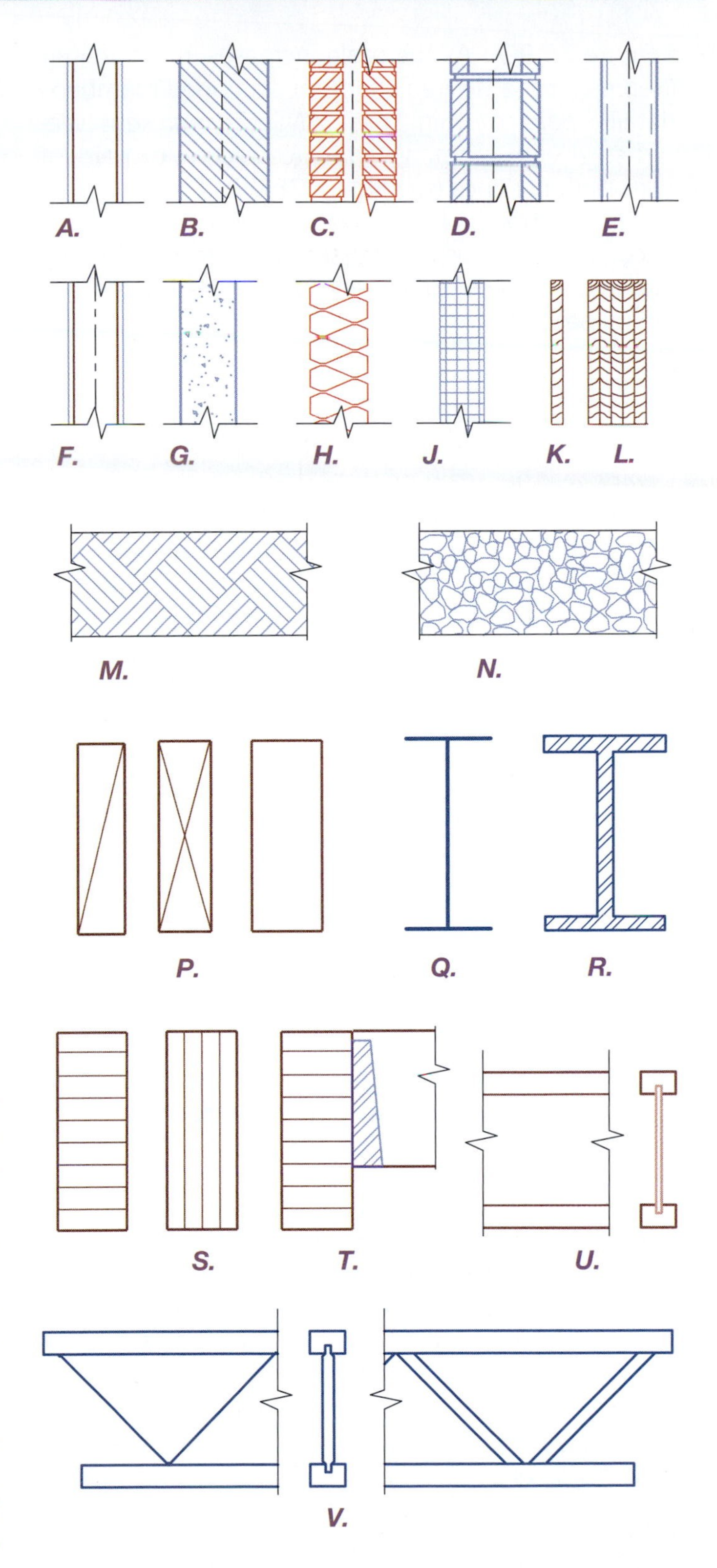

FIGURE 28.9 ■ *Common symbols for representing materials in sections and details. Materials include (A) wood-framed wall; (B) small-scale masonry; (C) double-wythe brick wall; (D) concrete masonry units; (E) steel tubes; (F) steel I or W shapes; (G) poured concrete walls; (H) batt insulation; (J) rigid insulation; (K) small-scale plywood; (L) large-scale plywood; (M) soil; (N) gravel; (P) wood and timber in end view (blocking and two methods of showing continuous members); (Q) small-scale steel shapes in end view; (R) large-scale steel shapes in end view; (S) laminated timbers in end view; (T) wood member supported by a metal hanger in side view on a laminated member in end view; (U) solid-web trusses in side and end view; and (V) open-web trusses in side and end views.*

product is required. If a component must be constructed at the job site, the drawings must provide enough information for all of the different trades depending on the drawings.

Wood, Timber, and Engineered Products

Notice in Figure 28.5 that thin lines represent studs that lie beyond the cutting plane and the plates are represented by thick lines. On small-scale sections, the lumber and timber products can be drawn using their nominal size. Thin materials such as plywood in Figure 28.9K may have to be exaggerated so that they can be clearly represented. Trusses perpendicular to the cutting plane can be represented by thick lines showing the shapes of the truss. When parallel to the cutting plane, the chords and webs can be represented by thin lines similar to those in Figure 28.9V.

In partial sections and details, the actual size of lumber and timber should be represented. In addition to using thick lines to outline members shown in end views, several methods can be used to represent the material. Figures 28.9P and 28.9S show common methods to represent materials such as plates, ledgers, and beams. Plywood, sheet rock, and other finishes are represented by hatch patterns similar to those shown in Figures 28.9A, K, and L.

Steel

The size of the drawing will affect how sectioned steel members are represented. At small scale, a solid, thick line represents the desired shape of steel members (see Figure 28.9Q). As the scale increases, pairs of lines can represent the desired shape of sectioned members. In details, pairs of thin lines *(ANSI32)* represent sectioned steel with a hatch pattern consisting of pairs of parallel diagonal lines (see Figure 28.9R). Thin lines representing the nominal thickness are used to represent steel columns, beams, or trusses that are beyond the cutting plane. Steel trusses are represented much like wood trusses.

Unit Masonry

Methods of representing brick and masonry products vary as the scale of the drawing increases. In small-scale sections, units are typically hatched with diagonal lines and no attempt is made to represent cavities or individual units. As the size of the drawing increases, individual units are represented, as well as cavities within the unit and grouting between the units. Individual hatch patterns are used to differentiate between the masonry unit and the grout. Steel reinforcing can be represented by either a hidden or continuous polyline. Figures 28.9B, C, and D show an example of a wall section representing unit masonry and brick veneer.

Concrete

The edges of poured members are represented by thick lines and a hatch pattern consisting of dots and small triangles (see Figure 28.9G). Because of the complexity of concrete construction, the section depends on many details to show construction of each concrete member.

Glazing

Glass is represented in sections by a single line or pairs of lines depending on the drawing scale. In full and partial sections, glass is generally represented by thin lines, with little attention given to intersections between the glazing and window frames. As the drawing scale is increased, the detail in representing the glass and the frame also increases.

Insulation

The type of insulation will dictate how it is drawn. Batt insulation is generally represented as shown in Figure 28.9H. Depending on the complexity of the section, the insulation may be shown across its entire span or in only one portion of the section. When only a portion of the insulation is drawn, notes must be placed that clearly define the limits of the insulation. Rigid insulation can be represented as shown in Figure 28.9J with the same considerations as in showing batt insulation. As the scale increases, insulation should be shown throughout the entire detail.

Locating Materials with Dimensions

Dimensions are an important element of full, partial, and wall sections. Both vertical and horizontal dimensions may be placed on sections, while partial sections and details generally show only vertical dimensions. On small-scale sections, the use of dimensions depends on the area being represented.

Vertical Dimensions

Vertical heights can be represented by using typical dimension methods or elevation symbols from a known point. Each type of dimension is usually placed on the outside of the section. Dimensions are generally given from the bottom of the sole plate to the top of the top plate for wood frame structures. This dimension also provides the height from the top of the plywood floor to the bottom of the framing member used to frame the next level. A common alternative is to provide a height from the top of the floor sheathing to the top of the next level of floor sheathing. Other common vertical exterior dimensions include:

Steel stud walls: from plate to top of channel
Structural steel: to top of steel member
Masonry units: to top of unit with distance and number of courses provided
Concrete slab: from top of slab or panel

The job captain will generally provide the exact dimensions for inexperienced drafters on a check print. Once the major shapes of the structure have been defined, dimensions should be provided to define openings, floor changes, or protective devices. Openings are located by providing a height from the top of the floor decking or sheathing to the bottom of the header. Changes in floor height and the height of landings are dimensioned in a similar method as changes in height between floor levels. Other common interior dimensions that should be provided include height of railings, partial walls, balconies, planters, and decorative screens. When possible, interior dimensions should be grouped together.

Horizontal Dimensions

The use of horizontal dimensions on full sections varies greatly for each office. With the exception of footing widths, horizontal dimensions are usually not placed on partial sections or details. When provided, horizontal dimensions generally are located from grid lines to the desired member. Exterior wood and concrete members should be referenced to their edge. Interior wood members are referenced to a centerline. Interior concrete members are referenced to an edge. Steel members are referenced to their centers. The distance for roof overhangs and balcony projections also may be placed on sections.

Drawing Symbols

The sections use symbols that match those of the floor, roof, and elevation drawings to reference material. Symbols that might be found on the section include:

Grid markers
Elevation markers
Section markers
Detail markers
Room names and numbers

Examples of each are shown in Figure 28.10. Grid markers should match in both style and reference symbol to those used on other drawings so the sections can be easily matched to other drawings. Elevations that are specified on the floor plan and elevation drawings should also be referenced on the section by use of a datum line or an elevation placed over a leader line.

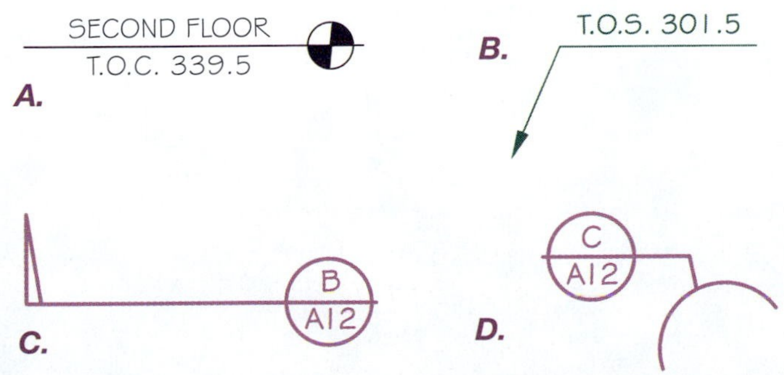

FIGURE 28.10 ■ *Symbols used on sections and details are common to other architectural drawings and include (A) elevation marker; (B) elevation marker; (C) section marker; and (D) detail markers.*

Each section and detail is referenced to other drawings by a section marker, which defines the page the section is drawn on and which section is being viewed. A reference such as 3 over A-7 would indicate that the section is drawing number 3 on page A-7. The smaller the scale used to draw the section, the more likely it is that section and detail markers will be used to reference other drawings to the section. Detail markers are especially prevalent on sections where enlarged views of intersections are provided.

Drawing Notations

Annotation on each type of section is used to specify materials and explain special installation procedures. As with other drawings, notes may be placed as either local or keyed notes. Most offices use local notes with a leader line that connects the note to the material. Local notes should be aligned to be parallel to the drawing to aid the print reader. On full sections, notes need to be placed neatly throughout the entire drawing. Wherever possible, notes should be placed on the exterior of the building. For wall sections, aligned notes can greatly add to drawing neatness.

The smaller the scale, the more generic the notes tend to be on a section. For instance, on a full section, roofing that might be specified as:

300# COMP. ROOF SHINGLES OVER 15# FELT OVER 1/2" OSB

would be referenced by complete notes for the roofing, insulation, and roof sheathing in the roofing details. Related notes should be grouped together within the same area of a section.

3D Details

Throughout this text, information has been presented using orthographic projections. Not because it's the best way, but because it's just the way the construction industry has represented objects for the last 100 years. The use of computers, and specifically 3D drawing programs, has greatly aided the development of 3D drawings for construction details. Figure 28.11a shows the use of a 3D drawing to depict the placement of a timber truss in a vaulted ceiling. Drawings created in 3D offer an excellent view of how major components relate to each other. Figure 28.11b shows a 3D rendering of the truss placement, and Figure 28.11c shows the placement of the truss at the job site.

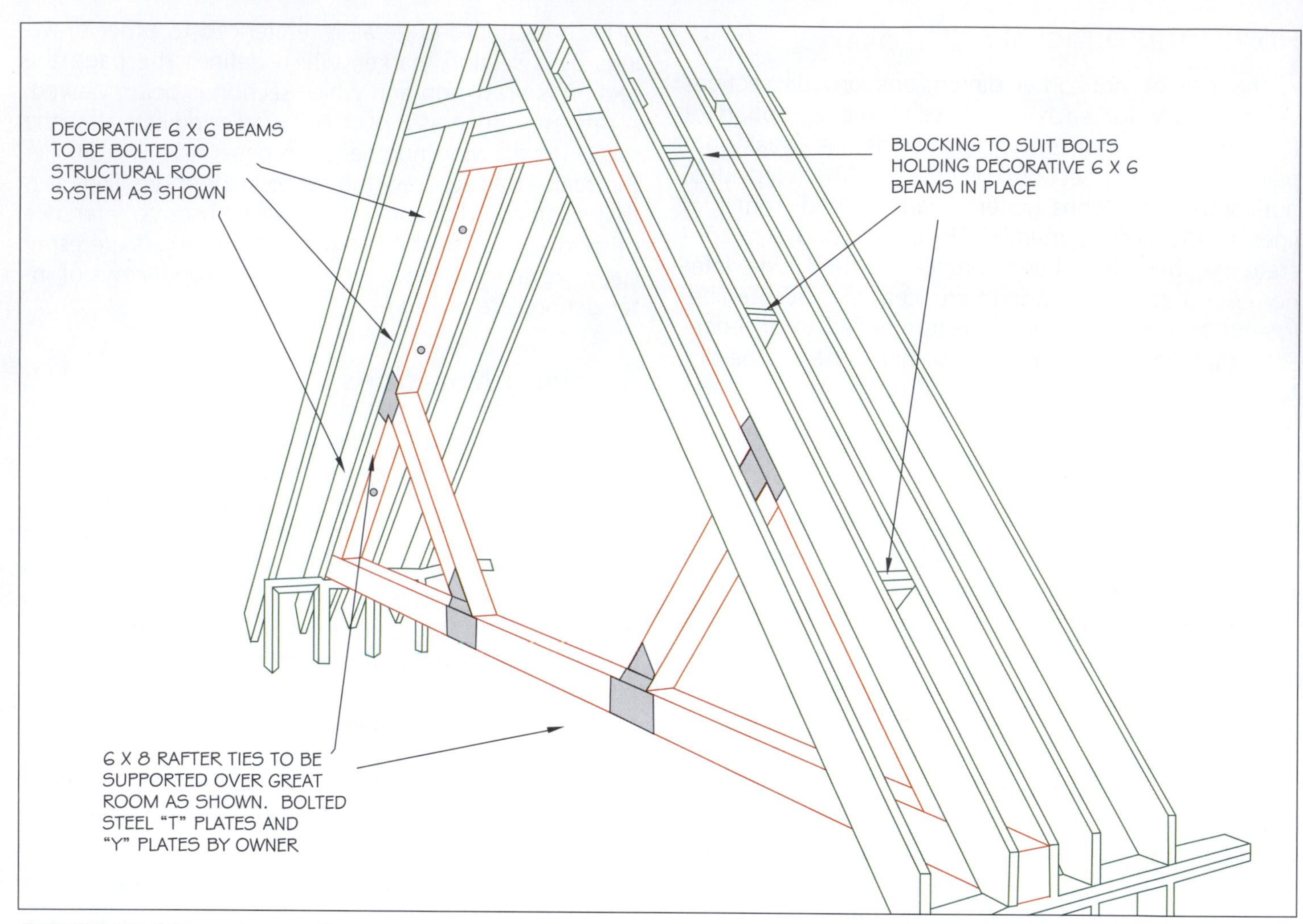
DECORATIVE 6 X 6 BEAMS
TO BE BOLTED TO
STRUCTURAL ROOF
SYSTEM AS SHOWN
BLOCKING TO SUIT BOLTS
HOLDING DECORATIVE 6 X 6
BEAMS IN PLACE
6 X 8 RAFTER TIES TO BE
SUPPORTED OVER GREAT
ROOM AS SHOWN. BOLTED
STEEL "T" PLATES AND
"Y" PLATES BY OWNER

FIGURE 28.11a ■ *The use of computers, and specifically 3D drawing programs, has greatly aided the development of 3D drawings for construction details. This 3D drawing shows the placement of a timber truss in a vaulted ceiling.* Courtesy David P. Scholz, Architect, DPS-labs.

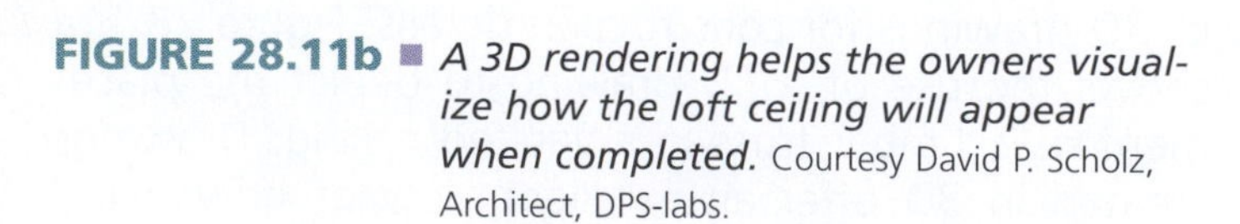
FIGURE 28.11b ■ *A 3D rendering helps the owners visualize how the loft ceiling will appear when completed.* Courtesy David P. Scholz, Architect, DPS-labs.

FIGURE 28.11c ■ *The results of the detail shown in Figure 28.11a.* Courtesy David P. Scholz, Architect, DPS-labs.

CHAPTER 28

Additional Readings

More so than any other drawings, the creation of details will require the use of vendor catalogs. The following websites can be used as a resource to help you keep current with building materials.

ADDRESS	COMPANY OR ORGANIZATION
www.comfast.com	Concrete Fastening Systems
www.strongtie.com	Simpson Strong-Tie
www.trusjoist.com	Trus Joist-A Weyerhaeuser Business

CHAPTER 28

Sectioning Basics Test

DIRECTIONS

Answer the following questions with short, complete statements. Type your answers using a word processor or neatly print your answers on lined paper.

1. Place your name, the chapter number, and the date at the top of the sheet.
2. Type the question number and provide the answer in the form of a statement that includes part of the question. You do not need to write out the entire question.

Note: The answers to some questions may not be contained in this chapter and will require you to do additional research.

QUESTIONS

28.1. What is a full section?
28.2. When could a partial section be used?
28.3. What is a stock detail and when would it be used?
28.4. From which drawings does a drafter get the needed information to draw a section?
28.5. What is the most common scale for drawing full sections?
28.6. What factors influence the scale of a detail?
28.7. What is a cutting plane and how does it relate to a section?
28.8. In which directions should the arrows on a cutting plane be pointing?
28.9. What type of section might be drawn at a scale of 1/8'' = 1'-0''?
28.10. What factors influence the choice of scale for the section?

DRAWING PROBLEMS

The following details will be completed as generic details. They are not drawn for a specific project but will be saved in a library for use on future projects. Once needed, the detail can be edited to meet the needs of a specific project. Use the following minimum sizes and materials to complete the drawings.

Minimum Drawing Standards

Unless noted, draw all details using a minimum scale of 1/2'' = 1'-0''. Bigger is OK; smaller is not.

- NO ISOMETRIC DRAWINGS ARE TO BE DRAWN. Convert isometric drawings to 2D details using the appropriate lineweight. Save each drawing as in individual drawing file. Use the problem number as the file name.
- All text to be Stylus Bt or other architectural style font with 1/8'' text.
- Place leaders with a shoulder using the leader command. Keep arrows approximately the same size as text.
- Text may be placed inside of details where space allows. Maintain 3/4'' clear between the drawing and dimensions or text. Never label material twice when more than 1 view is provided.
- Use dimensions for locations where possible rather than notes.
- Use side-by-side fractions (1/2), not top-over-bottom.
- Provide detail marker w/ 1/4'' text (detail # over page #). Your text should fit neatly in the circle without touching the circle. The detail number should match the problem number.

- Provide a title (1/4'' text) over the scale (at 1/8'' text) for all details. Provide a block reference for all details (detail number from book).
- Draw all material sizes to meet or exceed IRC minimum requirements.
- Assume all floor joists to be 2 × 10 and all I joists to be 9 1/2'' TJI pro 150 joists by Trus-Joist. Assume all trusses to have 2 × 4 cords. Use appropriate Simpson hangers for each type of joist. Specify floor joists with a generic note such as:

Floor Joists—see Foundation Plan for Size and Spacing

- Provide #4 diameter steel 3'' up from bottom of the footing and 3'' down from top of stem wall with 4 vertical bar placed at 48'' o.c. maximum from the footing to the stem wall. (The vertical reinforcement is not required for monolithic pours.)
- Assume all rafters are to be 2 × 6 at 24'' o.c. with 24'' overhangs and 2 × 8 fascias and barge rafters. All ceiling joists are to be 2 × 6 at 16'' o.c. Provide minimum required bearing for rafters at plate connection. Specify rafters and ceiling joists with generic notes such as:
  - 2× RAFTERS—SEE ROOF FRAMING PLAN FOR SIZE AND SPACING
  - 2× CEIL. JSTS—SEE FRAMING PLAN FOR SIZE AND SPACING
- Assume all beams are 4 × 12 DFL #2 unless noted.
- Assume all girders to be 4 × 8 on 4 × 4 post (4 × 6 at splices) on 15''Ø × 8'' deep concrete piers.
- Floor decking over floor joists to be 3/4'' plywood. Decking for post and beam systems to be 1 1/8'' Sturd-I-ply.
- Provide all insulation to meet minimum code standards.
- Use a distance of 3/4'' for the OFFSET command when representing 1/2'' material or smaller.
- Show and specify finishing materials with generic notes such as:
- EXTERIOR SIDING—SEE ELEVATION

Interior Finish—see Finish Schedule

- Show and specify a light weight roofing material over solid roof sheathing.

FOUNDATION PROBLEMS

Unless told otherwise by your instructor, base all foundation sizes on the minimum sizes required by the IRC. See Figure 25.10 for required foundation sizes. Create a drawing template suitable for all foundation details using appropriate layers, lineweight, text, and dimension sizes. Place all annotation and dimensions to explain all construction. Show and specify required insulation.

Concrete Slab Details

28.1. Use the attached sketch and create a detail showing the foundation for a one-level concrete floor system. Show and specify # 10 × 10, 4 × 4 welded wire mesh 2'' down from the top of the slab. Provide steel as required for your seismic zone. Show a wall framed with 2 × 6 studs at 16'' o.c.

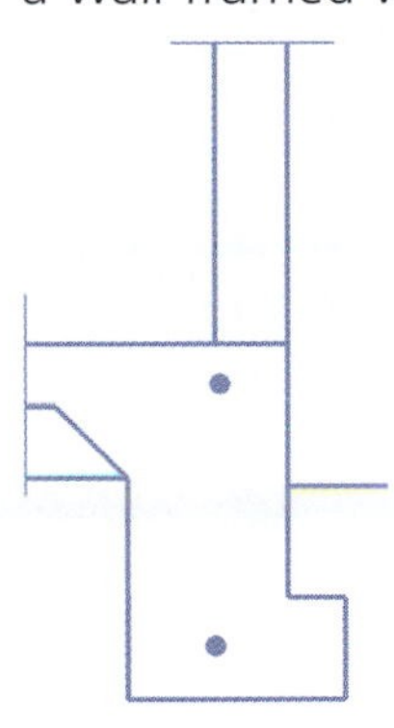

28.2. Use the drawing created in Problem 28.1 to create a detail showing the foundation for a two-level concrete floor system.

28.3. Use Figure 25.14f as a guide and draw a detail showing a monolithic footing for a one-level residence. Assume a 2 × 6 stud wall with Hardiplank siding over 1/2'' OSB underlayment.

28.4. Edit the drawing created in Problem 28.3 to create a detail showing the foundation for a one-level concrete floor system. Provide 2'' rigid insulation on the outer side of the foundation. Cover the insulation with 1/2'' concrete board and use metal flashing to protect the insulation. Show a 6'' wide wood framed wall flush with the edge of the concrete. Make a second copy showing a two-level footing.

28.5. Create a detail showing the foundation for a one-level concrete floor system. Provide 2'' rigid insulation on the outer side of the foundation. Cover the insulation with 1/2'' concrete board and use metal flashing to protect the insulation. Show a 6'' wide wood framed wall with a 2'' projection past the slab edge. Make a second copy showing a two-level footing. Save the drawings for future use.

28.6. Use Figure 25.14d as a guide and draw a detail showing monolithic footing for a one-level residence with a 4'' thick independent slab. Assume a 2 × 4 stud wall with 1'' exterior stucco.

28.7. Use the attached sketch and create a detail showing an interior one-level concrete floor system. Show a 4'' stud wall anchored to the concrete with Ramset type fasteners. Select a strike anchor that will have a 1 1/2'' minimum embedment. Make a second copy showing a two-level footing. Save the drawings for future use.

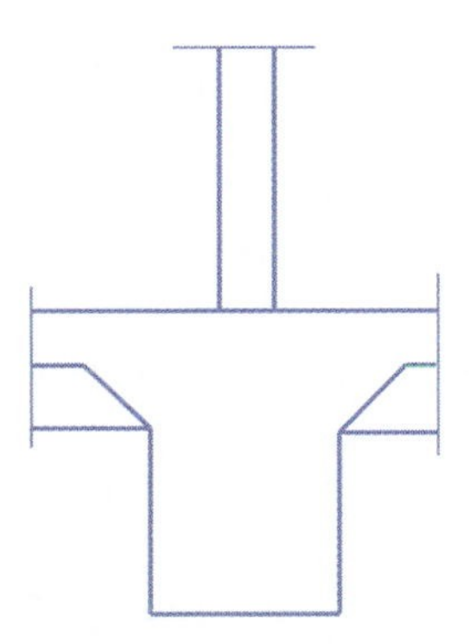

28.8. Use the left portion of Figure 26.19 as a guide and draw a detail of a concrete slab with a 7 3/4'' step. Provide a minimum concrete thickness of 8'' at the step. Provide WWM set 2'' down from the top of the upper slab. Show a 2 × 4 stud wall bolted to the upper slab.

28.9. Use the drawing created in Problem 28.1 to create a detail showing a 1-level concrete footing supporting 4'' wide brick veneer. Extend the footing to be 16'' wide. Edit the detail so that a two-story footing is also created.

Floor Joist Details

28.10. Use the attached sketch and create a detail showing a one-level concrete T footing supporting a joist floor system. Show and specify a 2 × 6 stud wall with double-wall construction and generic finishing materials. Provide steel as required for your seismic zone. Save the drawing for future use.

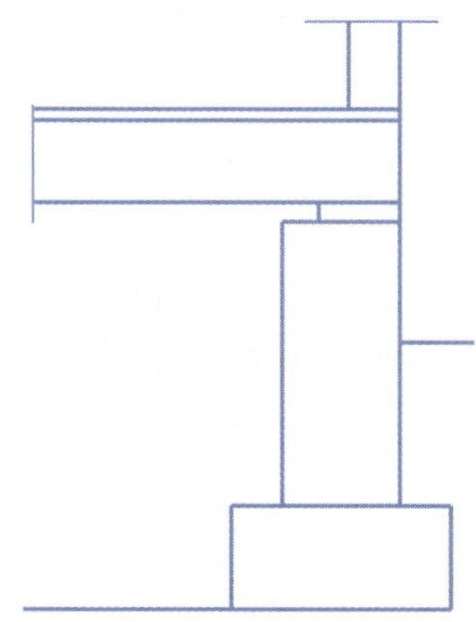

28.11. Use the drawing created in Problem 28.10 to create a detail showing the foundation for a two-level footing for a joist floor system.

28.12. Use the drawing created in Problem 28.10 to create a detail showing a one-level footing for a joist floor system constructed using engineered lumber.

28.13. Create a detail showing a two-level footing for a joist floor system constructed using engineered lumber.

28.14. Create a detail showing a one-level concrete footing with a 6'' wide CMU stem wall supporting a joist floor system. Show and specify a 2 × 4 stud wall with double- wall construction and generic finishing materials. Provide steel as required for your seismic zone.

28.15. Create a detail showing a two-level concrete footing with an 8'' wide CMU stem wall supporting a joist floor system. Show and specify a 2 × 6 stud wall with double- wall construction and generic finishing materials. Provide steel as required for your seismic zone.

28.16. Use Figure 26.27 as a guide to draw a detail showing a joist floor system supported on an interior girder and a concrete pier.

28.17. Use Figure 26.27 as a guide to draw a detail showing a joist floor system with an 18'' step between floor levels. Support each floor with a girder placed below the lower floor level.

28.18. Use Figure 26.27 as a guide to draw a detail showing a joist floor system with a 7 3/4'' maximum step between the floor levels. Support the upper floor joists on the girder and hang the lower joists from the girder. Use appropriate joist hangers for the lower joists.

28.19. Use Figure 26.27 as a guide to draw a detail showing a joist floor system intersecting a 4'' concrete garage slab. Assume the top of the slab is 8'' below the wood floor level. Use 26 gauge flashing to protect the floor framing. Thicken the slab as required at the edge.

28.20. Use Figure 26.27 as a guide to draw a detail showing a joist floor system with an 18'' cantilever measured from the outside edge of the wall to the outside edge of the stem wall. Show a 2 × 6 stud wall with double-wall construction.

28.21. Create a detail showing an engineered joist floor system with an 18'' cantilever measured from the outside edge of the wall to the outside edge of the stem wall. Show a 2 × 6 stud wall with double-wall construction.

28.22. Create a detail showing a two-level concrete footing with an 8'' wide CMU stem wall supporting engineered floor joists. Show and specify a 2 × 6 stud wall with double-wall construction and generic finishing materials. Provide steel as required for your seismic zone. Increase the footing as required to support 4'' brick veneer that extends 48'' above the finish grade.

28.23. Use Figure 26.34 as a guide to draw a detail showing the intersection of a joist floor system and a concrete slab. Extend the concrete footing 24'' into the grade. Use a 3 × 10 DFPT ledger bolted to the concrete with 1/2'' A.B. at 32'' o.c. staggered 3'' up/down. Specify appropriate metal joist hangers. Reinforce the slab with the appropriate WWM and specify steel in the footing and stem wall as per minimum standards presented earlier.

28.24. Use Problem 28.10 as a base to draw a detail showing the floor joists parallel to the stem wall. Show at least three joists.

28.25. Create a detail showing engineered floor joists parallel to the stem wall. Show at least three joists.

28.26. Use Problem 28.10 as a base to draw a detail showing the top of the floor joists set flush with the top of the mudsill. Use appropriate metal hangers to hang the floor joists from the top of the ledger.

28.27. Create a detail showing the top of the engineered floor joists set flush with the top of the mudsill. Use appropriate metal hangers to hang the floor joists from the top of the mudsill.

28.28. Create a detail showing the intersection of a wood deck with a wood floor supported on a concrete stem wall. Frame the floor using sawn lumber. Use a continuous rim joist with solid blocking between the floor joists. Set the top of the deck so that it is 2'' below the finish floor level of the residence. Build the deck using 2 × 4 decking laid flat with a 1/4'' gap supported on 2 × 8 treated floor joists. Hang the floor joists from a 2 × 10 treated ledger. Attach the ledger to the residence using nailing from the standard nailing schedule.

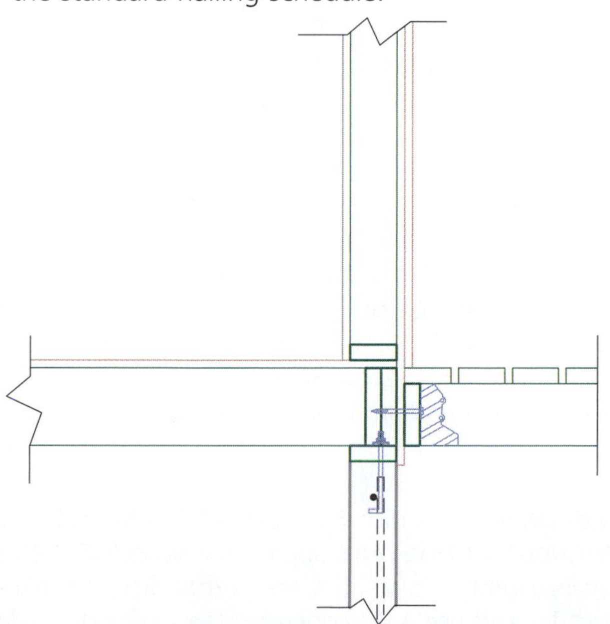

28.29. Create a detail showing the intersection of a wood deck with a wood floor supported on a concrete stem wall. Frame the floor using engineered lumber. Use a continuous rim joist with solid blocking between the floor joists. Set the top of the deck to be 2'' below the finish floor level of the residence. Build the deck using 1 1/2'' lightweight concrete over 30# felt over 3/4'' plywood decking supported on 2 × 10 treated floor joists. Hang the floor joists from a 3 × 10 treated ledger. Attach the ledger to the residence using 3/8''Ø lag bolts set at 24'' o.c. staggered 3'' up/down.

Post-and-Beam Details

28.30. Use the attached sketch and create a detail showing a one-level concrete T footing supporting a post-and-beam floor system. Show the beams parallel to the stem wall. Show and specify a 2 × 6 stud wall with double-wall construction and generic finishing materials. Provide steel as required for your seismic zone.

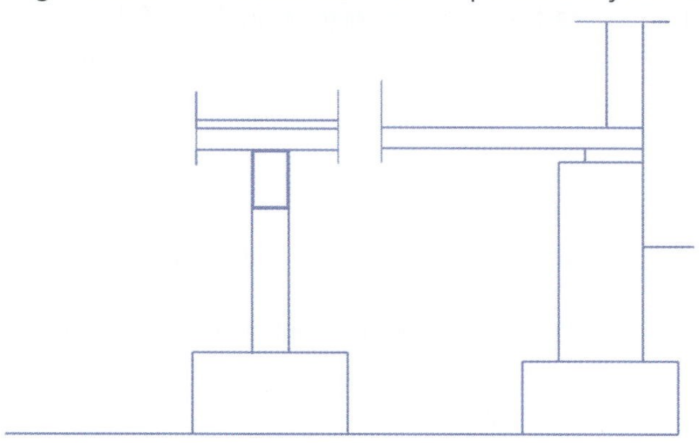

28.31. Create a detail showing a one-level concrete footing with a 6'' CMU stem wall supporting a post-and-beam floor system. Show the beams parallel to the stem wall. Show and specify a 2 × 4 stud wall with single wall construction and generic finishing materials. Provide steel as required for your seismic zone.

28.32 Create a detail showing a two-level concrete T footing supporting a post-and-beam floor system. Show the beams parallel to the stem wall. Show and specify a 2 × 6 stud wall with double-wall construction and generic finishing materials. Provide steel as required for your seismic zone.

28.33. Create a detail showing a two-level concrete footing with an 8'' CMU stem wall supporting a post-and-beam floor system. Show the beams parallel to the stem wall. Show and specify a 2 × 6 stud wall with double-wall construction and generic finishing materials. Provide steel as required for your seismic zone.

28.34. Create a detail showing a one-level concrete T footing supporting a post-and-beam floor system. Show the beams perpendicular to the stem wall. Show and specify a 2 × 6 stud wall with double-wall construction and generic finishing materials. Provide steel as required for your seismic zone.

28.35. Create a detail showing a two-level concrete T footing supporting a post-and-beam floor system. Show the beams perpendicular to the stem wall. Show and specify a 2 × 6 stud wall with double-wall construction and generic finishing materials. Provide steel as required for your seismic zone.

28.36. Use Figure 26.32 as a guide and show a post-and-beam floor with a 7 3/4'' step.

Retaining Walls

28.37. Use the attached drawing as a guide to draw a detail of an 8' high retaining wall made of CMUs to support a floor and upper level wood wall made of 2 × 6 studs. Show the use of sawn lumber joists that are perpendicular to the retaining wall for the upper floor system and use a 4'' concrete slab with no WWM for the lower floor.

Provide a 4'' Ø French drain at the footing in a 12'' wide gravel bed. Reinforce the wall with #5 vertical rebar set 2'' clear of the interior face. Extend a #5 × 18'' L rebar from the footing into the wall at 48'' o.c. Provide (2) #5 continuous rebar 3'' up from the bottom of the footing. Reinforce the wall with #5 horizontal rebar at 16'' o.c. with (2) #5 at mid-height of wall, and (4) #5 at the top of the wall. Provide #5 vertical rebar at 48'' o.c. Attach the mudsill to the wall with 1/2'' Ø A.B. at 24'' o.c. and use a Simpson Company A-34 anchors at each joist to plate connection. Provide two layers of hot asphaltic emulsion on the exterior face of the wall.

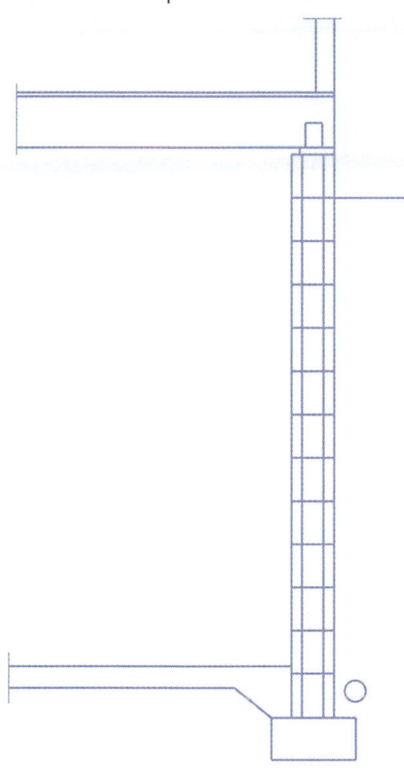

28.38. Use Figure 25.36 as a guide to draw a detail of an 8' high retaining wall made of poured concrete to support a floor and upper level wood wall made of 2 × 6 studs. Show the use of engineered floor joists that are parallel to the retaining wall for the upper floor system and use a 4'' concrete slab with no WWM for the lower floor. Show at least four joist spaces and show solid blocking placed between the joists. Place the blocks at 48'' o.c. and show the blocks extending 48'' out from the wall. Provide a 4''Ø French drain at the footing in a 12'' wide gravel bed. Reinforce the wall with #5 vertical rebar set 2'' clear of the interior face. Extend a #5 × 18'' L rebar from the footing into the wall at 48'' o.c. Provide (2) #5 continuous rebar 3'' up from the bottom of the 16'' × 12'' footing. Provide #5 horizontal rebar at 16'' o.c. with (2) #5 at midheight of wall, and (4) #5 at the top of the wall. Provide #5 vertical rebar at 48'' o.c. Attach the mudsill to the wall with 1/2''Ø A.B. at 24'' o.c. and use a Simpson Company A-34 anchor at 16'' o.c. along the rim joist to the plate. Provide 2 layers of hot asphaltic emulsion on the exterior face of the wall. Cover the exterior wall with 2'' rigid insulation and protect the insulation with 1/2'' concrete board to a depth of 18''. Extend the protection to the mudsill and protect with 26 gauge flashing.

28.39. Use the wall detail drawn in Problem 28.38 as a base to show the intersection of an engineered joist floor system resting on a poured concrete retaining wall. Show a concrete slab intersecting the retaining wall. Show the slab 8'' below the top of the wood floor. Thicken the slab to be 12 × 12 at the edge, and provide a #5Ø 18 × 18 L at 32'' o.c. from the slab to the retaining wall. Provide (2) #5 continuous bars,

3'' up from the bottom of the slab. Provide 26-gauge flashing for any wood exposed to the slab.

28.40. Use the attached sketch to create a detail showing the intersection of a 48'' poured concrete retaining wall and a 2 × 4 stud wall. The height of the wall (48'' max.) is determined from the top of the 4'' concrete floor slab to the line of the finish grade. Extend the wall 6'' above the finish grade. Provide a 30'' wide × 8'' deep footing with (2) #5Ø continuous bars 3'' up from the bottom of the footing. Provide a #5Ø 15'' × 18'' L at 18'' o.c. extending from the footing into the wall. Reinforce the wall with #5Ø at 18'' o.c. 2'' from the tension side of the wall. Provide a 4''Ø drain in an 8 × 30'' gravel bed. Use a 2 × 6 DFPT sill with 1/2'' A.B. placed at 24'' o.c. Show the wood wall supporting an engineered joist floor system with a 12'' cantilever. Cover the exterior wood with 1'' exterior stucco, and cover the wood at the lower level with 5/8'' type X gypsum board.

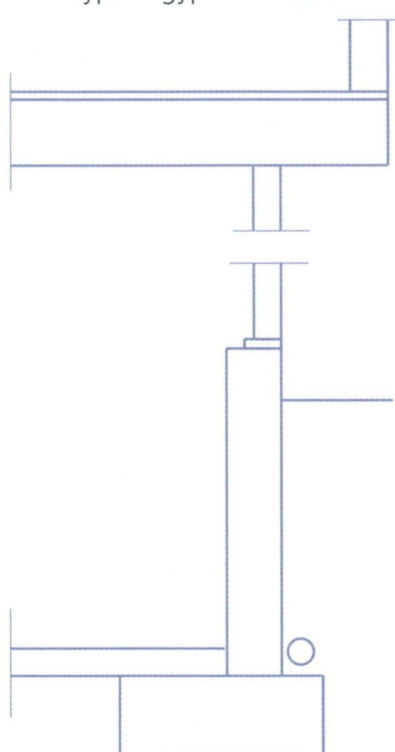

WALL CONSTRUCTION

28.41. Use Figure 20.8 as a guide to draw the header over an opening in a 2 × 6 stud wall with double top plates, a 4 × 8 header, and a 2x nailer. Assume that the top of the top plates is set at 8' and the bottom of the header is set at 6'-8''.

28.42. Use Figure 20.8 as a guide to draw the header over an opening in a 2 × 6 stud wall with double top plates, (2) 2 × 12 headers and a 2 × 6 nailer. Provide 2'' rigid insulation on the interior side of the wall.

28.43. Use Figure 20.7 as a guide to draw a plan view of an exterior corner formed with (3) 2 × 6 studs. Show the exterior side using single-wall construction.

28.44. Use Figure 20.7 as a guide to draw a plan view of an exterior corner formed with (2) 2 × 6 studs. Show the exterior side using double-wall construction.

28.45. Use Figure 20.7 as a guide to draw a plan view of a 2 × 4 interior wall intersection with an exterior wall formed with (2) 2 × 6 studs. Use a flat stud intersection. Show the exterior side using double-wall construction.

28.46. Use Figure 20.7 as a guide to draw a plan view of a 2 × 4 interior wall intersection with an exterior wall formed with (2) 2 × 6 studs. Use a ladder-backed intersection. Show the exterior side using double-wall construction.

28.47. Use Figure 20.7 as a guide to draw a plan view of a 2 × 4 interior wall intersection with an exterior wall formed with (2) 2 × 6 studs. Use a one-stud intersection with the interior finish supported with metal drywall clips placed at 16'' o.c. Show the exterior side using double-wall construction.

28.48. Use Figure 21.25 as a guide to draw an elevation and a section of (2) 2 × 6 king studs and a single trimmer supported on a 4 × 6 post at the lower level. Frame the upper wall with 2 × 6 studs resting on a joist floor system that is supported on a lower wall made of 2 × 6 studs. Tie the trimmers to the post below with a Simpson Company HD-5A connector. (Show just enough framing to explain the connection from the upper to lower floor.)

28.49. Use the information from Figure 24.15 and the rough draft on the student CD to draw an elevation and a section of one side of a portal frame.

28.50. Use the information from Figure 24.15 to draw an elevation and a section of one side of a braced wall panel.

28.51. Use the information from Figure 24.15 to draw an elevation and a section of one side of an alternative braced wall panel.

28.52. Use the information from Figure 24.41 to draw an elevation and a section of one side of the intersection of an upper braced wall panel to a lower braced wall panel.

28.53. Use the information from Figure 24.41 to draw an elevation and a section of one side of the intersection of an upper alternative braced wall panel to a lower alternative braced wall panel.

ROOF-TO-WALL CONNECTIONS

28.54. Use Figure 20.9 as a guide to draw a detail showing the intersection of a standard truss with a 6/12 pitch to a 2 × 6 stud wall.

28.55. Draw a detail showing the intersection of a standard truss with a 6/12 pitch to a 2 × 6 stud wall. Show an enclosed soffit using 1 × 4 T&G cedar. Create a cornice detail using at least three pieces of trim.

28.56. Use Figure 20.21 as a guide to draw a detail showing the intersection of a standard truss with a 6/12 pitch to an 8'' CMU wall. Use a 2 × 6 DFPT sill with 1/2''Ø A.B. at 48'' o.c.

28.57. Use Figure 20.9 as a guide to draw a detail showing the intersection of a standard truss with a 6/12 pitch to a 2 × 6 stud wall. Provide enclosed soffits. Use advanced framing techniques and assume the use of an 8' plate height and a 9' top chord height. Provide solid blocking between the top cords.

28.58. Draw a gable end- wall truss resting on a 2 × 4 stud wall covered with double-wall construction. Assume a 12'' overhang and a 2 × 6 barge rafter.

28.59. Draw the intersection of a scissor truss with a 6/12 pitch to a 2 × 6 stud wall. Show the bottom chord with a 4/12 pitch

28.60. Use Figure 20.5 as a guide to draw a detail showing the intersection of rafters placed at a 6/12 pitch to the top plate of a 2 × 6 stud wall. Assume the use of ceiling joists.

28.61. Draw a detail showing the intersection of rafters placed at a 4/12 pitch to a 2 × 6 top plate on an 8'' CMU wall. Assume the use of ceiling joists.

28.62. Draw a detail showing the intersection of 2 × 12 rafters/ceiling joists set at a 8/12 pitch with the top plate of a 2 × 6 stud wall.

28.63. Draw a detail showing the intersection of 2 × 12 rafters/ceiling joists set at a 7/12 pitch with the top plate of a 2 × 6 stud wall. Use an enclosed eave with a 2 × 6 fascia. Cut the rafter tails as required.

28.64. Draw a detail showing the intersection of 2 × 12 rafters/ceiling joists set at a 7/12 pitch with a 2 × 6 stud wall built with double-wall construction. Support the rafter/ceiling joists with the appropriate metal hangers and with a 2 × 12 ledger that is laid over the OSB sheathing. Provide 26-gauge flashing.

28.65. Draw a detail showing 2 × 12 rafters/ceiling joists that are parallel to a 2 × 6 stud wall built with double-wall construction. Use a 2 × 12 ledger that is laid over the OSB sheathing and flash with 26-gauge flashing.

28.66. Draw a gable end wall showing the rafters and ceiling joists resting on a 2 × 6 stud wall covered with double-wall construction. Assume a 12'' overhang and a 2 × 6 barge rafter.

28.67. Draw a detail showing the intersection of 2 × 12 rafters/ceiling joists set at a .25/12 pitch with the top plate of a 2 × 6 stud wall. Notch the rafter tails to 3 1/2'' and provide a plumb cut. Do not provide a fascia.

28.68. Draw a detail showing the intersection of 2 × 6 rafters with a gable end wall and a flush barge rafter.

28.69. Draw a detail showing two options for the intersection of 2 × 12 rafters/ceiling joists to a 6 × 14 ridge beam. Show one option with the rafters resting on the top of the ridge beam. Provide solid blocking between each rafter and provide screened roof vents for each third rafter. Notch each rafter 1 1/2'' deep × 3'' for airflow. Draw a second option with the rafter/ceiling joists hung from the ridge beam using appropriate metal joist hangers.

PARTIAL WALL SECTIONS

Use a foundation detail and a roof detail to create a partial wall section. Assume the plate to be 8'-1 1/8'' above the finish floor. Edit notes so that wall materials are only specified once per detail.

28.70. Create a partial wall section showing a 1-level concrete slab floor system supporting a standard truss roof.

28.71. Create a partial wall section showing a 2-level concrete slab floor system supporting a truss roof. Show each wall framed with 2 × 6 studs. Show a window 48'' deep in one of the walls. Provide a header framed with (2) 2 × 12s. Show the upper floor framed with engineered joists and provide an 18'' cantilever. Frame the roof using standard trusses.

28.72. Create a partial wall section showing a 1-level T footing supporting sawn floor joists, 2 × 6 studs supporting a standard truss roof.

28.73. Create a partial wall section showing a 1-level T footing supporting engineering floor joists and 2 × 6 studs supporting a scissor truss roof.

28.74. Create a partial wall section showing a 1-level footing with a 6'' wide stem wall made with CMUs supporting engineering floor joists and 2 × 6 studs supporting a standard truss roof.

28.75. Create a partial wall section showing a 1-level T footing supporting a post-and-beam floor system with the girders parallel to the stem wall. Show a 2 × 4 stud wall supporting a standard truss roof.

CHAPTER 29

Section Layout

Two major options available for drawing sections include the use of drawing blocks and completing the section from scratch. Each method will be examined in this chapter. In addition, some architectural CAD programs automatically draw a section from information provided as the floor plan is drawn. As the walls on the floor plan are drawn, heights are established so these dimensions automatically convert to wall heights in the sectional view. This type of parametric design requires that an imaginary cutting plane line be placed through the floor plan in the desired section location. This is followed by computer prompts requesting information such as roof pitch and structural floor thickness. All you do to complete the section is add material symbols, dimensions, and notes. Although this may be your drawing method in the future, the balance of this chapter will assume the use of a traditional 2D drawing program.

ASSEMBLING BLOCKS

When a home will be built using standard construction methods, standard details can be assembled to form a partial wall detail. Standard details can be brought into the drawing from a menu library. Figure 29.1 shows two standard details. Stock details can be merged to form a partial section once the height from the floor to ceiling level is determined. The partial section shown in Figure 29.2 was created by aligning the roof/wall detail directly above the post-and-beam detail. The line representing the top of the floor decking was offset to determine the location of the top of the wall. With the top of the wall located, the roof/wall detail was inserted

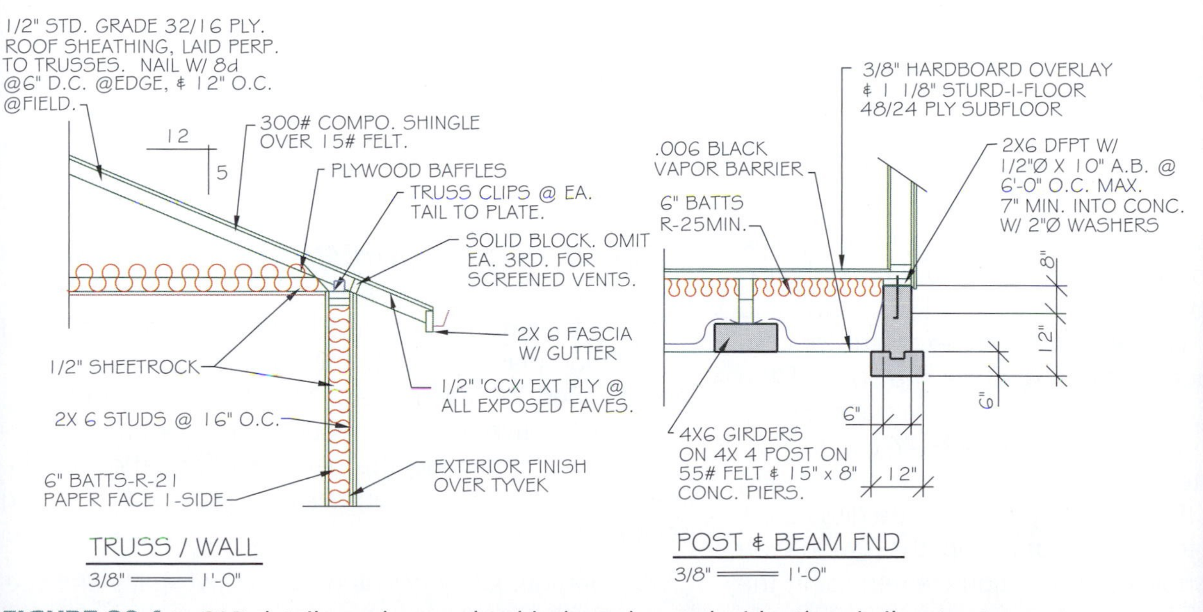

FIGURE 29.1 ■ *CAD details can be saved as blocks and reused with other similar structures.*

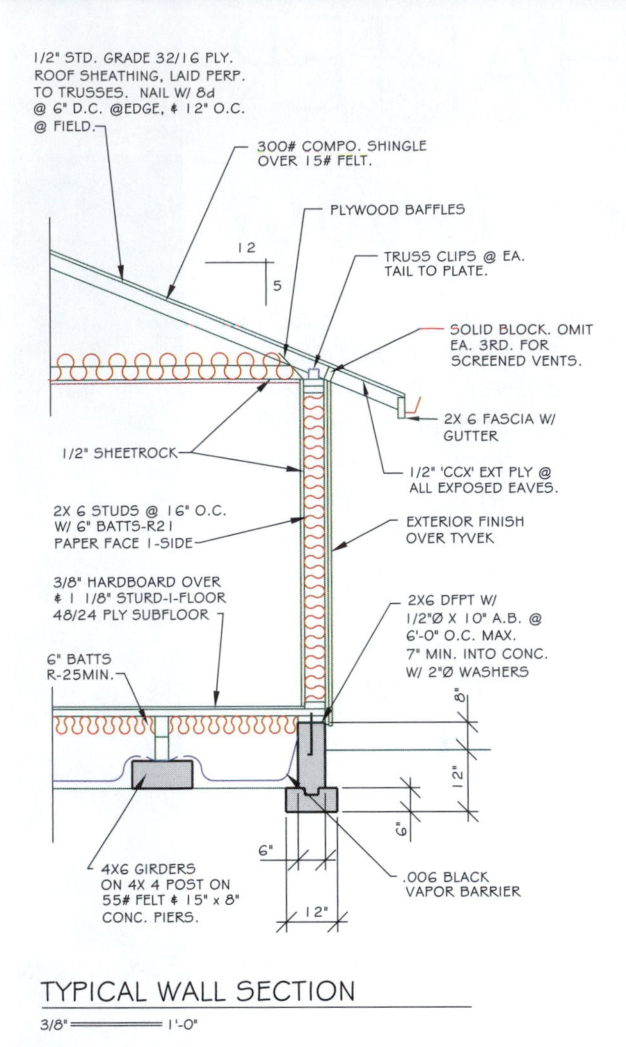

FIGURE 29.2 ■ *Standard details stored in a block library can be joined together to form a partial wall section.*

into the required position. Many building departments will accept a partial section for a building permit for a simple residence.

A partial section can be used to form a full section. Using the dimensions on the framing plan, the line that represents the outside edge of the studs can be offset a distance equal to the width of the house. The partial section can then be mirrored to form the opposite side of the structure, and then lines from one side can be extended to the opposite side. The FILLET, EXTEND, and STRETCH commands of AutoCAD are excellent for connecting the two partial sections. Once the shell of the structure has been drawn, the interior walls and girders can be added using the COPY and ARRAY commands. The location of interior walls can be determined using the dimensions of the framing plan, and the girders are placed according to their location on the foundation plan. Figure 29.3 shows a full section created using the stock details from Figure 29.1.

Note:

Make sure that the MIRRTEXT value is 0 so that all text can be read. You will also need to eliminate some of the text on the drawing. Be sure to keep all related text together.

CREATING SECTIONS WITHOUT BLOCKS

Custom homes often require the use of special building methods that must be reflected in the sections. Sections can be created using seven major stages. By following the step-by-step procedure for each stage, sections can be easily drawn and understood. Read through each stage, and carefully compare the step-by-step instructions with the corresponding illustrations.

This chapter includes guidelines for drawing sections with several types of foundations including a basement with a concrete slab, and the upper level framed with joist construction, post-and-beam, and full basements. Each is based on the floor plan that was used in Chapter 11. The home is a hillside home with a daylight basement. The lower floor for each foundation style will always be formed with a concrete slab. It is the floor system over the single level portion of the home that will vary. Before attempting to draw a section, study the completed plan that follows each example so you will know what the finished drawing should look like.

A section can be drawn by dividing the work into six stages:

1. Evaluating the drawing needs
2. Laying out the overall shape
3. Placing all finishing materials
4. Dimensioning
5. Annotating
6. Evaluating

Stage 1: Evaluating Needs

Using the information on the site, floor, framing, and foundation plans, evaluate the major types of construction needed on the project. For the home started in Chapter 11, five different sections are identified in Figure 11.18. To provide the needed information to the framing crew, five sections will be provided including sections to show:

A—Garage, kitchen/dining over crawl space and the hobby room

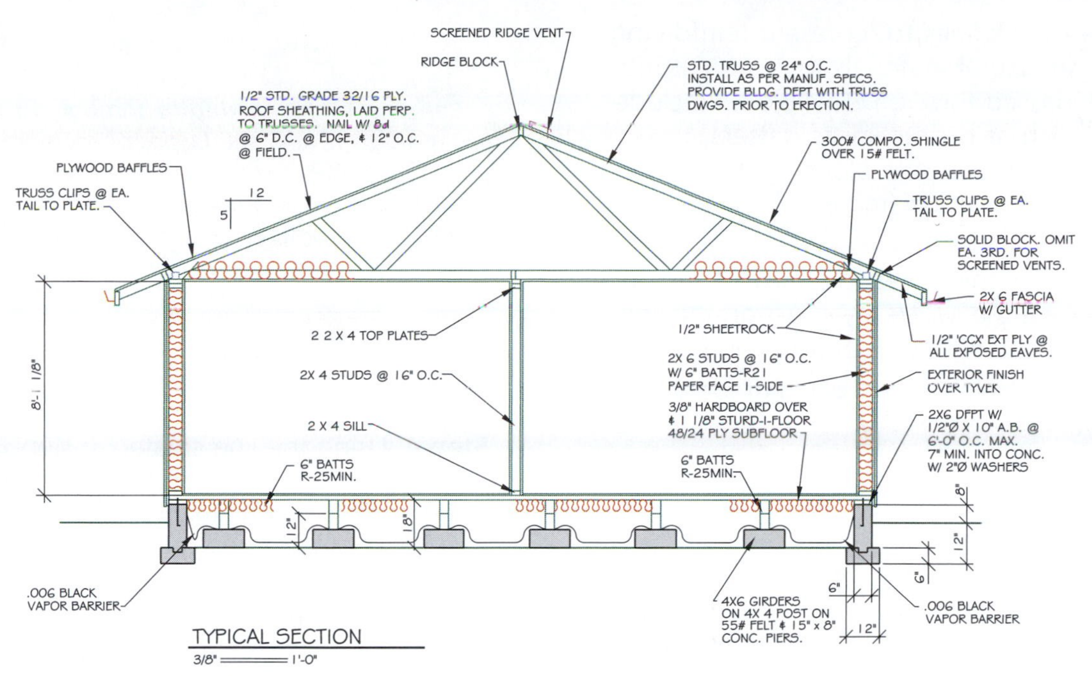

FIGURE 29.3 ■ *Once the width of the structure has been determined, a partial wall section can be used to form a full section using the* MIRROR, FILLET, TRIM, *and* STRETCH *commands.*

B—Utility and vaulted living room over the family room with a cantilever

C—Bedroom, hall, and bath areas over the den with a cantilever

D—Master bedroom with trayed ceiling and deck over the crawl space and deck supports

E—Partial section showing stair and vaulted ceilings (see Chapter 30)

Once the required sections have been determined, the process of drawing the sections for a structure will require minor adjustments to be made to the drawing template. Template options that must be altered before starting the drawing include adding appropriate layers to contain the drawing information, and adjusting linetypes to represent each type of material.

The section that will result from cutting plane AA on the floor plan in Figure 11.18 will be drawn first. The drawing will be drawn at full scale in model space, and plotted using a scale of 3/8'' = 1'-0'' (1:32). This will require:

- An LTSCALE of 16, and a scale factor of 32 for placing text and dimension.
- To produce text with a height of 1/8'' (3 mm) when plotted, use a height of 4'' (100 mm) in model space.

Adding Layers

Before the sections can be started, layers need to be created to separate information by material, by lineweight, and by line types. Layers should start with a prefix of *SECT.* Layers containing detail and section information should be named with a modifier listed in Appendix F. Sub names of *ANNO, DIMN, FOOT, OUTL,* and *SYMB* will always be needed. As with any other drawing, create additional layers as needed to ensure that only layers that will be used are added to the drawing.

Adjusting Linetypes

Although standards will vary, sections will require several different line weights to provide contrast between materials. As with the drawing of details, there are no standards for lineweight. The lineweights that are used will vary depending on the plotting scale and the materials that must be represented. When drawing sections, use the following guidelines for line widths:

- Use .0 (default) for thin lines. These will generally be materials that lie beyond the cutting plane on the *MBND* layer (material beyond section cut).

- Use .60 for thick lines to represent reinforcing steel and structural materials that the cutting plane has passed through. This would include materials such as beams, joists, rafters, top plates, sills, mudsills, and blocking. Place these materials on the *MCUT* (materials cut by sections) layer.
- Use .90 for very thick lines to represent the outline of concrete on the *MCUT* layer. Assign the lineweight using the PROPERTIES command.
- Use 1.00 to represent the finished grade on the *MCUT* layer.

Note:

The goal of any detail is to clearly represent material. Line thickness should be thick enough to provide contrast, but not so thick that it bleeds into other objects. Because there is no set standard for lineweight, draw a few lines using varied lineweights and then make a test plot to compare contrast.

Stage 2: Laying Out the Overall Shape

Use thin lines placed on the *SECT OUTL* layer for this entire stage. To determine the sizes and locations of structural members, refer to the site, grading, floor, framing, and foundation plans. See Figure 29.4 for the layout of section A from Figure 11.18.

Concrete Slab Layout

Step 1. Draw a line to represent the outer edge of one wall of the structure. Use the OFFSET command to represent the width of the building based on the dimensions provided on the framing plan (see Figure 24.37). Use the OFFSET command to represent the following walls:

- The west wall of the garage
- The east wall of the hobby room
- The west wall of the hobby room
- The west edge of the deck
- The line of post to support the deck

Step 2. Draw a line to represent the lower finish floor level.

Step 3. Use the site or grading plan to establish the grade elevations where the section will be cut. Draw lines to represent the natural and finish grades.

Note:

Before completing steps 4 through 10, it may be necessary to review the basics of foundation design in Section 7. All sizes given in the following steps are based on the minimum standards of the IRC (see Figure 25.10) for footing supporting two-level structure (unless otherwise noted). These minimum standards should be compared with local standards.

Step 4. Using the line created in step 2 as a base, use the OFFSET command (4'' [100 mm] down from the top of slab) and create a line to represent the bottom of the 4'' (100 mm) slab.

Step 5. Use the OFFSET command (4'' down from the bottom of the slab) and create a line to represent the bottom of the 4'' (100 mm) gravel. Use the end GRIP and extend this line to also represent the line of the finish grade level.

Step 6. Use the OFFSET command (18'' [450 mm] down from the top of slab) and create a line to represent the bottom of the footing.

Step 7. Use the OFFSET command (7'' [175 mm] up from the bottom of the footing) and create a line to represent the footing thickness.

Step 8. Use the OFFSET command to represent the thickness of the stem wall. Verify this size with local codes. The IRC uses a width of 6''. Many local codes use the traditional width of 8'' (200 mm) for two-level and 10'' (250 mm) for three-level footings.

Step 9. Use the OFFSET command to represent the width of the footings (15''/375 mm) for two stories).

Step 10. Use the OFFSET command to represent the additional footing ledge 4'' (100 mm) wide for brick veneer if required.

Step 11. Block out the interior footings for any load-bearing walls. The drawing should now resemble Figure 29.4. For this home, a retaining wall 48'' (1200 mm) high will be required to support the west wall of the garage. The east wall of the hobby room will be a standard two-story concrete footing. (A perfect opportunity to insert the detail created in Problem 28.2 into the section.)

Wall Layout

With the layout of the foundation and lower floor complete, the lower walls, the main floor system, and the upper walls can be completed. Trim lines as needed

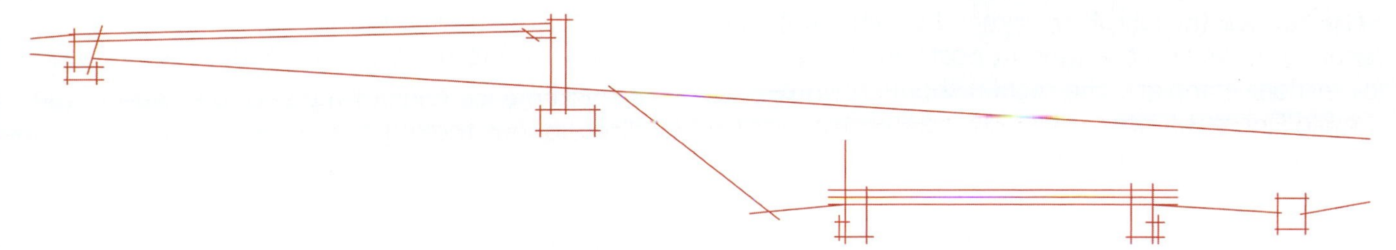

FIGURE 29.4 ■ *The initial steps for the layout of a section for the structure started in Chapter 11. The garage area on the left will have a sloping slab, and the lower floor on the right will have a concrete slab. The center section will be crawl space with no foundation.*

to help with clarity, but remember that this is just the initial layout stage of structural members. Details will be added in the next stage. See Figure 29.6 for the layout of the following steps:

Step 12. Use the OFFSET command (8' [2400 mm] up from the top of slab) and create a line to represent the bottom of the floor joists over the hobby room.

Note:

The walls will be drawn at 8'-0'' (2400 mm) high, but this is not their true height. The true height is the sum of the thickness of two top plates, the studs, and the base plate. The plates are made from 2× (50×) material, which is actually 1 1/2'' (38 mm) thick. The studs are milled in 88 5/8'' (2215) or 92 5/8'' (2315 mm) lengths. Use the longer studs unless your instructor provides other instructions. See Figure 29.5 to determine the true height of the walls. In a two-level house, this would also represent the bottom of the floor joists. In a one-level house, this line will represent the bottom of the trusses or ceiling joists.

Step 13. Use the OFFSET command to represent the interior side of all exterior walls.

Step 14. Locate the interior walls. For this home, the only interior wall is a 6'' (150 mm) wide wall on the lower level at the east side of the hobby room. For other sections, interior walls can be located using the following steps:

- Determine the wall location on the framing plan.
- Use the dimension from the framing plan and the OFFSET command to locate the center of the wall in the section.
- Use the OFFSET command to assign the wall thickness around the centerline.
- Erase the centerline from the wall.

LONG STUD (IN.)		SHORT STUD (IN.)
3	2 Top plates	3
92 5/8	Stud height	88 5/8
+ 1 1/2	Base plate	+ 1 1/2
97 1/8	Total height	93 1/8
8′–1 1/8″	Dimension to appear on section	7′–9 1/8″

FIGURE 29.5 ■ *Determining floor-to-ceiling dimensions. Add the depth of the top plates, the base plate, and the height of the studs. Two common precut stud lengths are given.*

As you complete other sections, after locating the interior walls, make sure that any interior bearing walls line up over a footing. If they do not line up, verify that there are no mistakes in your dimensions, math calculations, or a conflict between the dimensions of the floor and foundation plans.

Step 15. Use the OFFSET command to represent the tops of any required doors and windows. Headers for doors and windows are typically set at 6'-8'' (2000 mm) above the floor. The size of the header will be drawn later.

Step 16. Use the OFFSET command to represent the subsills. Offset the required distance from the bottom of the header to establish the subsill location. To establish the subsill location, verify the window sizes on either the floor plan or window schedule.

Step 17. Use the OFFSET command to represent the bottom of the ceiling joist (8'-0''/2400 mm).

- Show any required 7'-0'' (2100 mm) ceiling soffits for placement of heating ducts in areas such as kitchens, baths, or hallways. On this section, no soffits are required since heating ducts can be run through the crawl space.

Step 18. Use the OFFSET command to represent any required decks and the support posts. With the lower floor complete, the section should now resemble Figure 29.6.

Block out the upper-level walls using the same steps that were used to draw the lower-level walls. See Figure 29.7 for completion of the following steps.

Step 19. Use the OFFSET command to represent the depth (10''/250 mm) of the floor joists. Offset the line that was created in step 12.

Step 20. Use the OFFSET command to represent the interior side of all upper level walls. For this home, all of the walls including the east wall of the garage (for lateral support) will be framed with 6'' (150 mm) material. This home also has a 2'-0'' (600 mm) cantilever of the dining room past the hobby room on the lower level.

Step 21. Use the OFFSET command with a distance of 3/4'' (19 mm) to represent the plywood subfloor.

Step 22. Use the OFFSET command with a distance of 9' (2700 mm) to represent the upper ceiling (2400 mm) above the floor of the residence.

Step 23. From the top of the plywood floor sheathing, draw a line that is 8'' (200 mm) below the finish floor to represent the highest point of the garage slab.

- Slope the line representing the garage floor from a point 8'' (200 mm) below the finish floor of the residence to a point representing the finish grade at the garage door.
- Offset the sloping line of the concrete floor 4'' (100 mm) to represent the bottom of the slab.
- Thicken each edge of the slab to extend to the continuous footing.
- Draw a line below the bottom of the slab to represent the 4'' (100 mm) minimum gravel fill.

Step 24. Use the OFFSET command to represent the interior side of exterior walls.

Step 25. Use the OFFSET command to represent any interior walls.

Step 26. Draw any required windows. The section will now resemble Figure 29.7.

Truss Roof Layout

Use the following steps to complete the layout of a truss roof. The layout of stick-framed roofs is introduced later in this chapter.

Step 27. Use the OFFSET command to place a vertical line to represent the ridge.

Step 28. Use the OFFSET command to locate each overhang.

Step 29. Draw a line to represent the bottom side of the top chord.

- Starting at the intersection of the outer wall and the ceiling line, draw a line at a 6/12 pitch to match the pitch shown in the elevations. A 6/12 pitch can be drawn at an angle of 26 1/2°.
- Because the exact length of the top chord is not known, enter an arbitrary length, such as 12' (3600 mm).

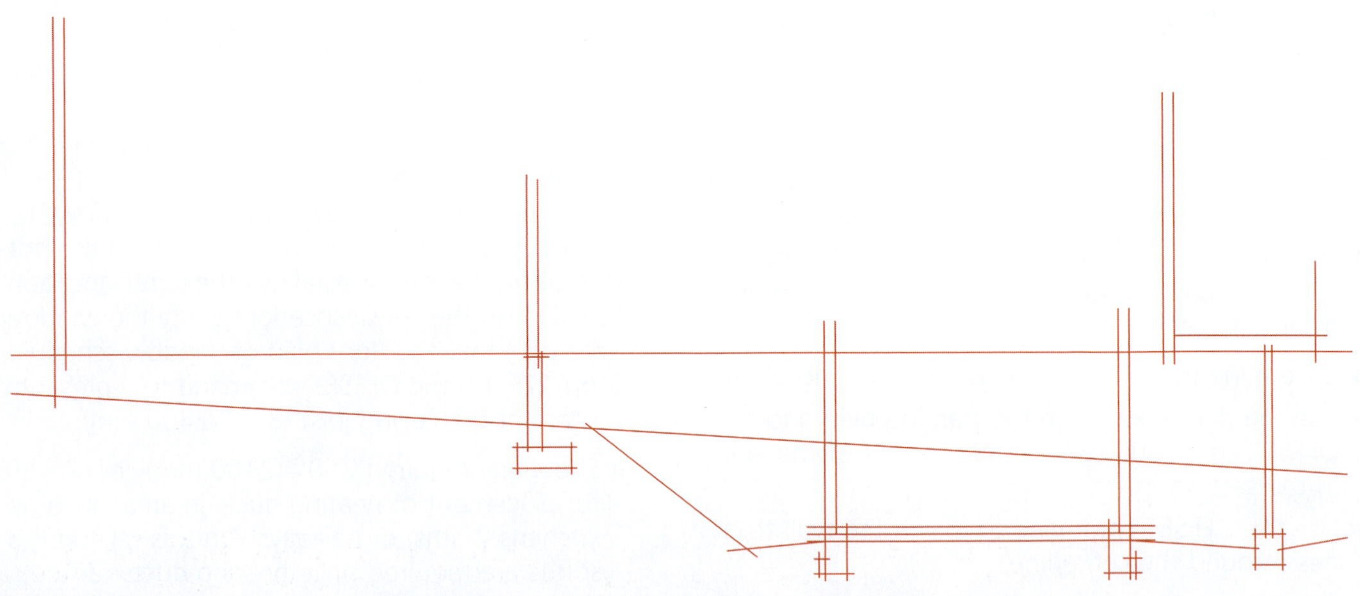

FIGURE 29.6 ■ *Layout of the foundation level and the upper and lower wall locations.*

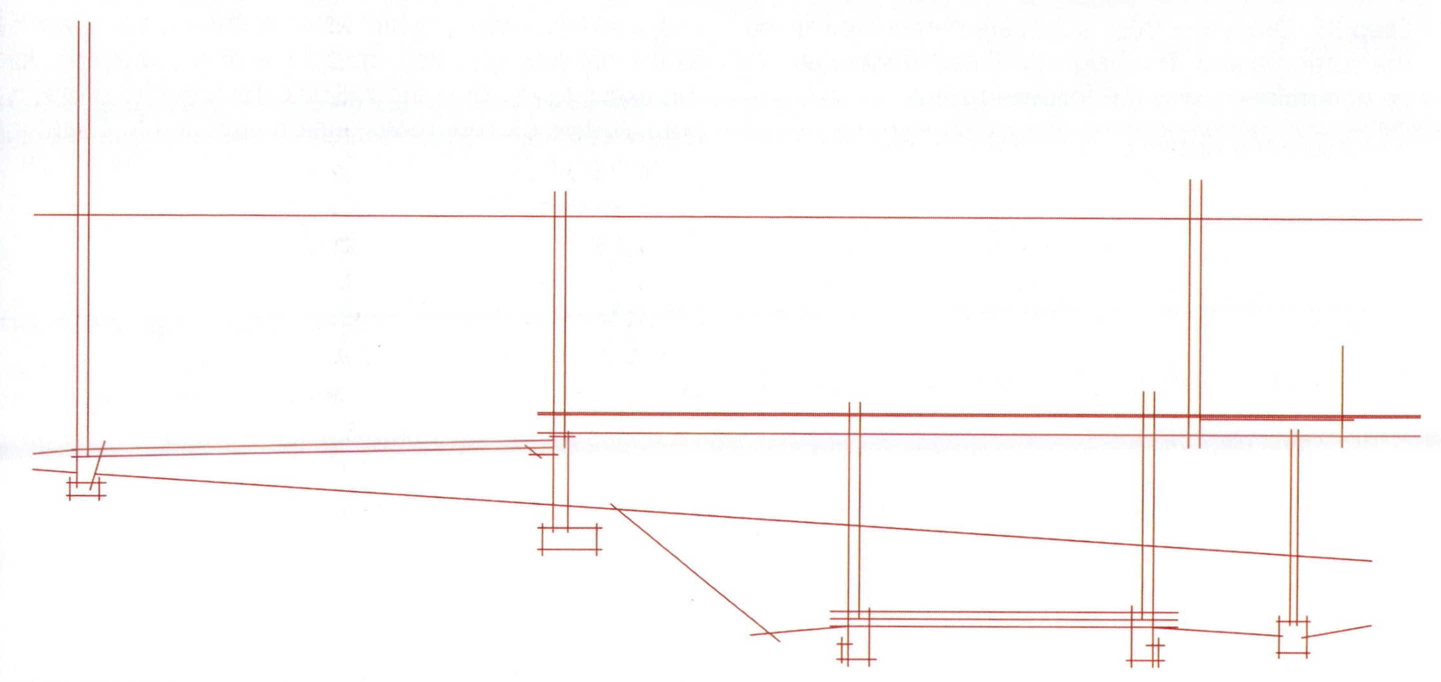

FIGURE 29.7 ■ *Layout of the main floor and the ceiling level.*

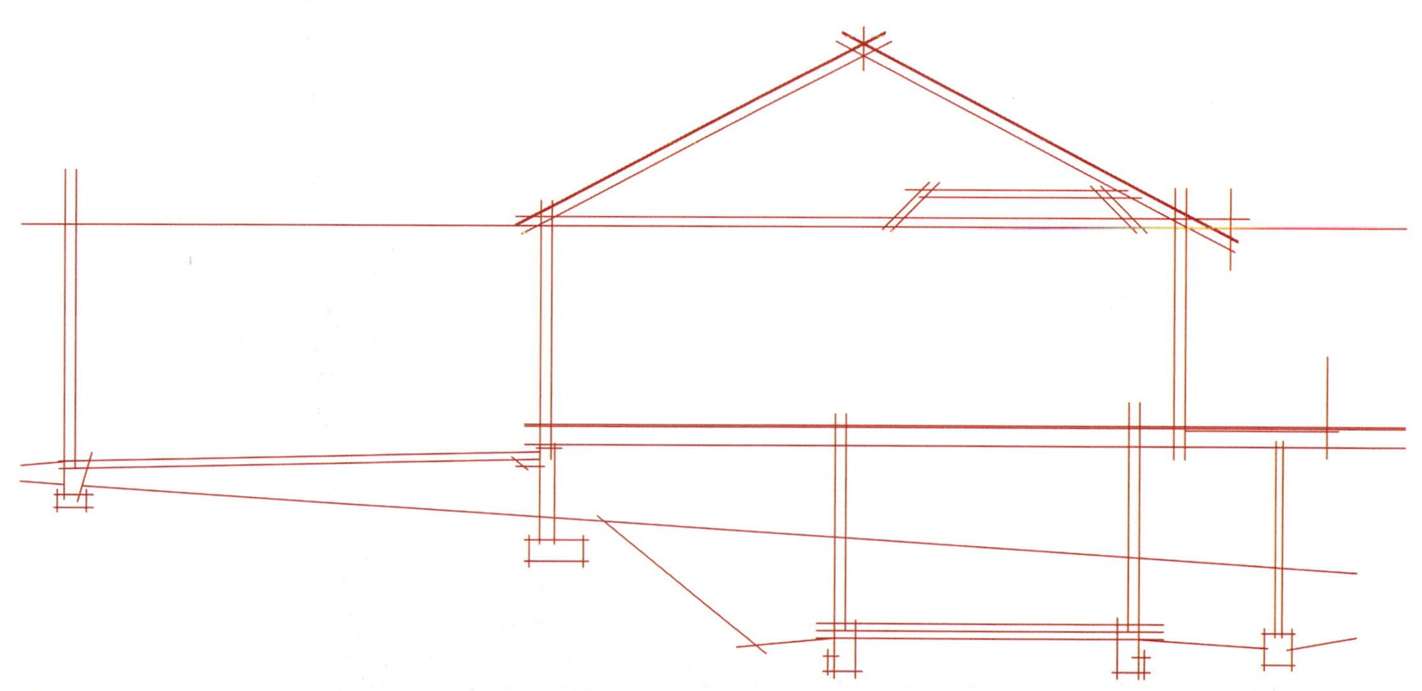

FIGURE 29.8 ■ *Truss roof layout. The truss webs can be omitted, since the truss manufacturer will design and specify the required materials.*

- If the line is too long, it can be trimmed, or, if it is too short, extended. The most important part is to get the angle correct.

Step 30. Use the OFFSET command to represent the top side of the top chord. Assume chords to be 4'' (100 mm) deep.

Step 31. Use the OFFSET command to draw the roof sheathing.

Step 32. Use the OFFSET command to represent the topside of the bottom chord.

Step 33. Locate any soffits, tray, or coffered ceilings specified on the floor plan.

Step 34. Use the MIRROR command to represent the other side of the roof. Your drawing should now resemble Figure 29.8.

Step 35. Draw any trusses that are perpendicular to the cutting plane. The height of these trusses can be determined using the following steps:

- Measure the distance on the framing plan from the supporting wall to the cutting plane.
- Draw this distance on the section. The line representing the distance of the cutting plane to the wall intersects the top chord of the truss (see Figure 29.9).

The completed layout of the trusses is shown in Figure 29.10.

Stage 3: Representing Structural Materials with Finished Quality Lines

Once the entire section has been blocked out, extra lines can be trimmed and details added to the drawing. Use an orderly procedure such as starting at the roof and working down, or start at the foundation and work to the ridge to complete this stage of the drawing. In this example, the drawing will be completed from top to bottom. This procedure was first mentioned in Chapter 23 to ensure that all loads transfer to the foundation. The steps followed to complete the drawing will be divided into roof, upper walls, floor, lower walls, and foundation.

In Chapter 28, continuous members shown in details are traditionally drawn with a diagonal cross (X) placed in the member. Blocking is drawn with one diagonal line (/) through the member. This method of representing sectioned members is time-consuming and may be difficult to read on small-scale sections. Rather than drawing symbols, the members can be drawn much more easily with different line qualities, as shown in Figure 29.11.

As you read through these steps, you will be asked to draw items that have not been laid out. Items such as plates can be drawn once and copied numerous times or

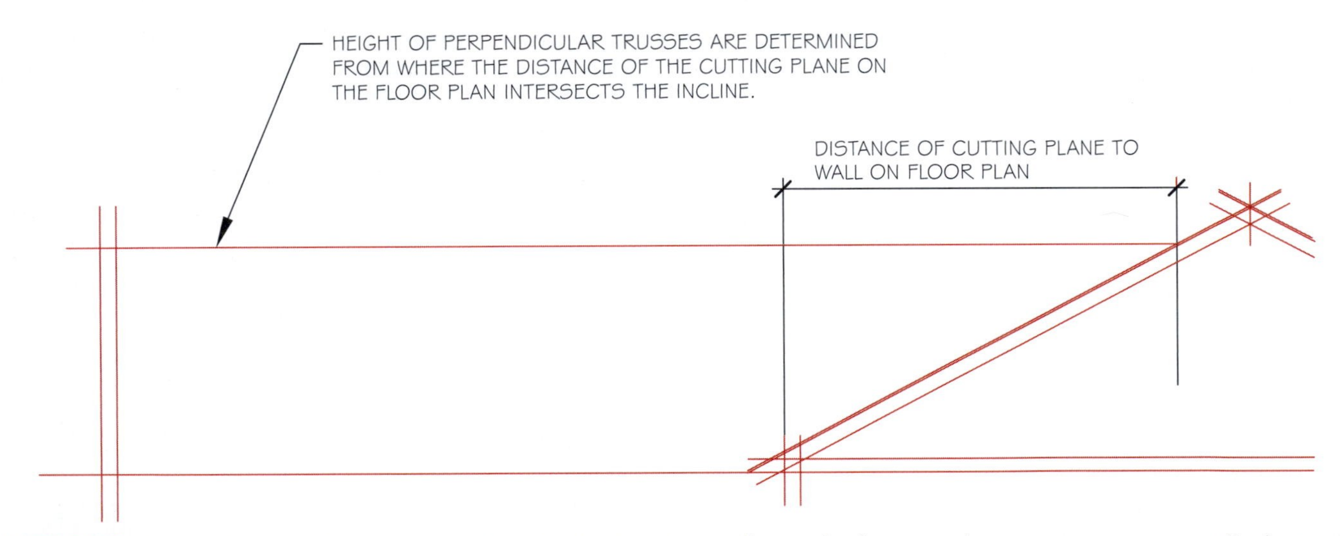

FIGURE 29.9 ■ *The distance of the cutting plane to the bearing wall must be known when trusses are perpendicular to the cutting plane. By examining Figure 11.18, it can be seen that the cutting plane is approximately 17' from the north and south garage walls. Projecting a line into the section where the trusses are parallel to the cutting plane will show how high a truss will be after traveling 17'. Where the line intersects the truss represents the top of the trusses that are perpendicular to the cutting plane.*

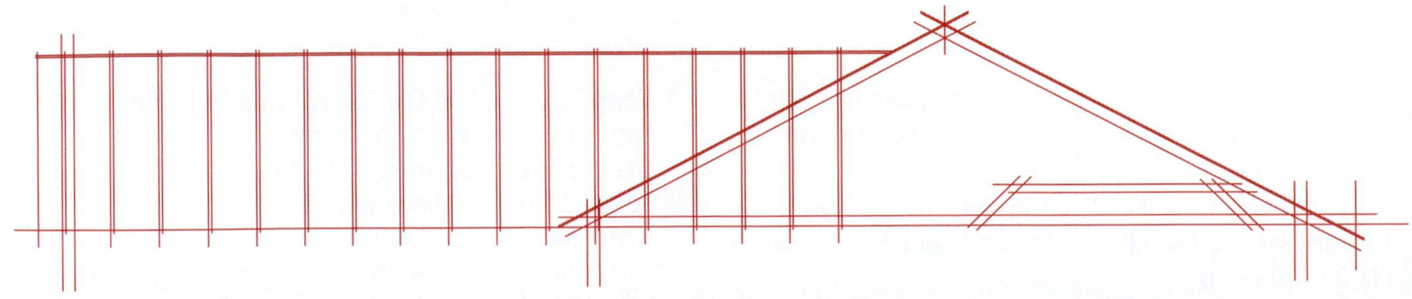

FIGURE 29.10 ■ *Representing trusses that are perpendicular to the cutting plane.*

stored as a block and inserted as necessary. A 2 × 4 (50 × 100) plate can easily be edited to be a 6'' (100 mm) plate, or a 2 × 6 (50 × 150) plate can be copied, rotated, and stretched to make a 2 × 10 (50 × 250) ridge members.

Truss Roof

To gain speed, draw features on one side of the roof and then use the MIRROR command to complete the opposite side. Roof steps 1 through 8 are shown in Figure 29.12. Use the PROPERTIES command as needed to assign the following materials a thin linetype placed on the *SECT MBND* layer unless noted.

Step 1. Use the OFFSET command to represent the roof sheathing with thin lines.
Step 2. Trim the lines representing the top and bottom truss chords as needed and place these lines on the *SECT MBND* layer.

Use the PROPERTIES command as needed to assign the following materials a thick linetype placed on the *SECT MCUT* layer.

Step 3. Draw the eave blocking.
Step 4. Draw the fascia.

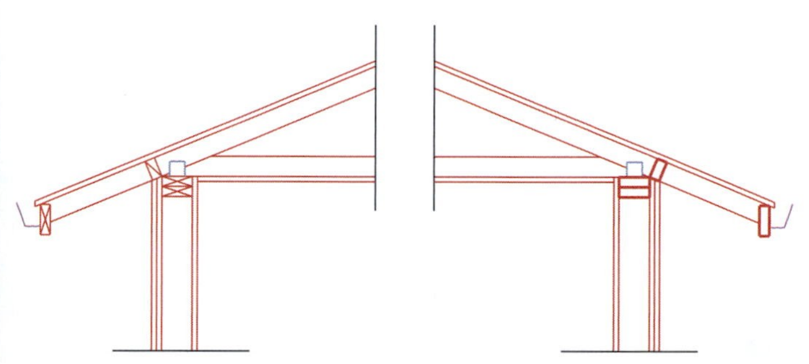

FIGURE 29.11 ■ *Representing structural materials with finished quality lines. Lines to represent the blocking can be omitted for clarity when scales smaller than 3/4'' = 1'-0'' are used.*

Step 5. Use the MIRROR command to place the items from steps 1 through 4 on the opposite side of the roof.
Step 6. Draw the ridge block.
Step 7. Draw any required blocking and gable end-wall bracing.
Step 8. Draw any trusses that are perpendicular to the cutting plane on the *SECT MBND* layer.

- Use thick lines to represent the chords and thin lines to represent the webs.
- Draw one truss and use the ARRAY command to locate the others.
- Use the STRETCH command to adjust the height of the valley trusses.

Your drawing should now resemble Figure 29.12.

Upper Walls and Floor

Use the PROPERTIES command as needed to assign the following materials a thick linetype placed on the *SECT MCUT* layer. See Figure 29.13 for steps 9 through 22.

Step 9. Draw or use the INSERT command to place all double top plates and headers. Stretch the headers to the needed size.
Step 10. Draw or insert all subsills.
Step 11. Draw or insert all bottom plates.
Step 12. Draw all blocks and ledgers as required.

Use the PROPERTIES command as needed to assign the following materials a thin linetype placed on the *SECT MBND* layer unless noted.

Step 13. Draw any required furred ceiling.
Step 14. Draw the plywood subfloor.
Step 15. Draw floor joists.
Step 16. Draw the wall studs.
Step 17. Use the OFFSET command to represent the interior wall and ceiling finishes.
Step 18. Use the OFFSET command to represent the exterior sheathing and siding materials.

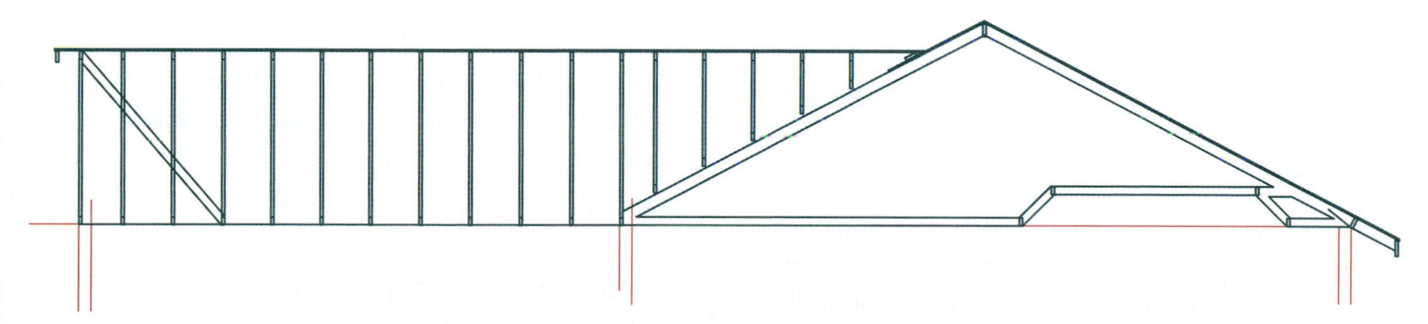

FIGURE 29.12 ■ *All structural members represented with finished linetypes. Repetitive items such as trusses can be drawn once and placed using the* INSERT, COPY, *or* ARRAY *commands.*

Foundation Materials

Trim the following lines as needed and use the PROPERTIES command to assign the needed layers.

Step 19. Trim and place the outline of the concrete on the *SECT MCUT* layer. Assign a very thick line to all concrete.

- Use the HATCH command to place the concrete pattern on the *SECT PATT* layer.

Step 20. Place the finished and natural grade lines on the *SECT MCUT* layer.

- Assign a very thick line to each grade line.
- Use a dashed line to represent the natural grade if a finished grade is also represented.

Step 21. Draw or insert the mudsills, steel rebar if required, and anchor bolts on the *SECT MCUT* layer. Assign bold line types for each item.
Step 22. Draw the 4'' (100 mm) fill material on the *SECT MBND* layer.

- Use the HATCH command to place the gravel pattern on the *SECT PATT* layer.

Note:

This is an excellent time to consider the use of standard details. The blocks created in Chapter 28 as problems 28.2 and 28.40 could be inserted to complete the lower floor and the retaining wall below the interior garage wall.

Representing the Finish Materials

The materials drawn in this stage seal the exterior from the weather and cover the interior side of the framing materials. Start at the roof and work down to the foundation. Place all items in these steps on the *SECT MBND* layer using thin lines for the entire stage unless otherwise noted. See Figure 29.13 for the following steps.

Step 23. Draw hurricane ties approximately 3'' (75 mm) square at the truss/top plate intersection.
Step 24. Draw ridge vents on the back side of the roof that are approximately 3'' high × 12'' long (75 × 300 mm).
Step 25. Draw all window and door glazing.
Step 26. Draw the hardboard underlayment. Note that the plywood subfloor extends under all walls, but the hardboard underlayment does not go under any walls.

Step 27. Draw any required metal connectors such as joist or beam hangers.

Place the following items on the *SECT PATT* layer using thin lines.

Step 28. Draw insulation using the BATT linetype or your own pattern.

- Use batts approximately 10'' (250 mm) deep in flat ceilings.
- Use batts approximately 10'' (250 mm) deep with a 2'' (50 mm) air space above in vaulted ceilings.
- Use 4 or 6'' (100 × 160 mm) batts in exterior wall cavities.
- Use 8'' (200 mm) batts between the floor joists.

Step 29. Draw and crosshatch any required masonry veneer on the *SECT PATT* layer.

Place the following items on the *SECT MCUT* layer using thick lines.

Step 30. Draw baffles at the edge of the insulation. Leave approximately 2'' (50 mm) at the top of the baffle for air flow over the insulation.
Step 31. Draw gutters that are approximately 3 × 3 (75 × 75 mm).

Stage 4: Placing Dimensions

With all materials completely drawn, dimensions must be provided so that the framing crew knows their vertical location. Figure 29.14 shows the needed dimensions. All dimension text should be aligned and 1/8'' (3 mm) high when plotted. Use the DIM command to locate the following dimensions on the *SECT DIMS* layer:

- Floor to ceiling. Place the dimension from the top of the floor sheathing to the bottom of the floor or ceiling joists or trusses.
- Floor to bottom of bedroom windows, 44'' (1100 mm) maximum.
- Finished grade to the top of the slab.
- Depth of footing into the natural grade.
- Height of the footings.
- Width of the footings.
- Width of the stem wall.
- Represent the roof pitch (6/12).
- Eave overhangs.
- Cantilevers.

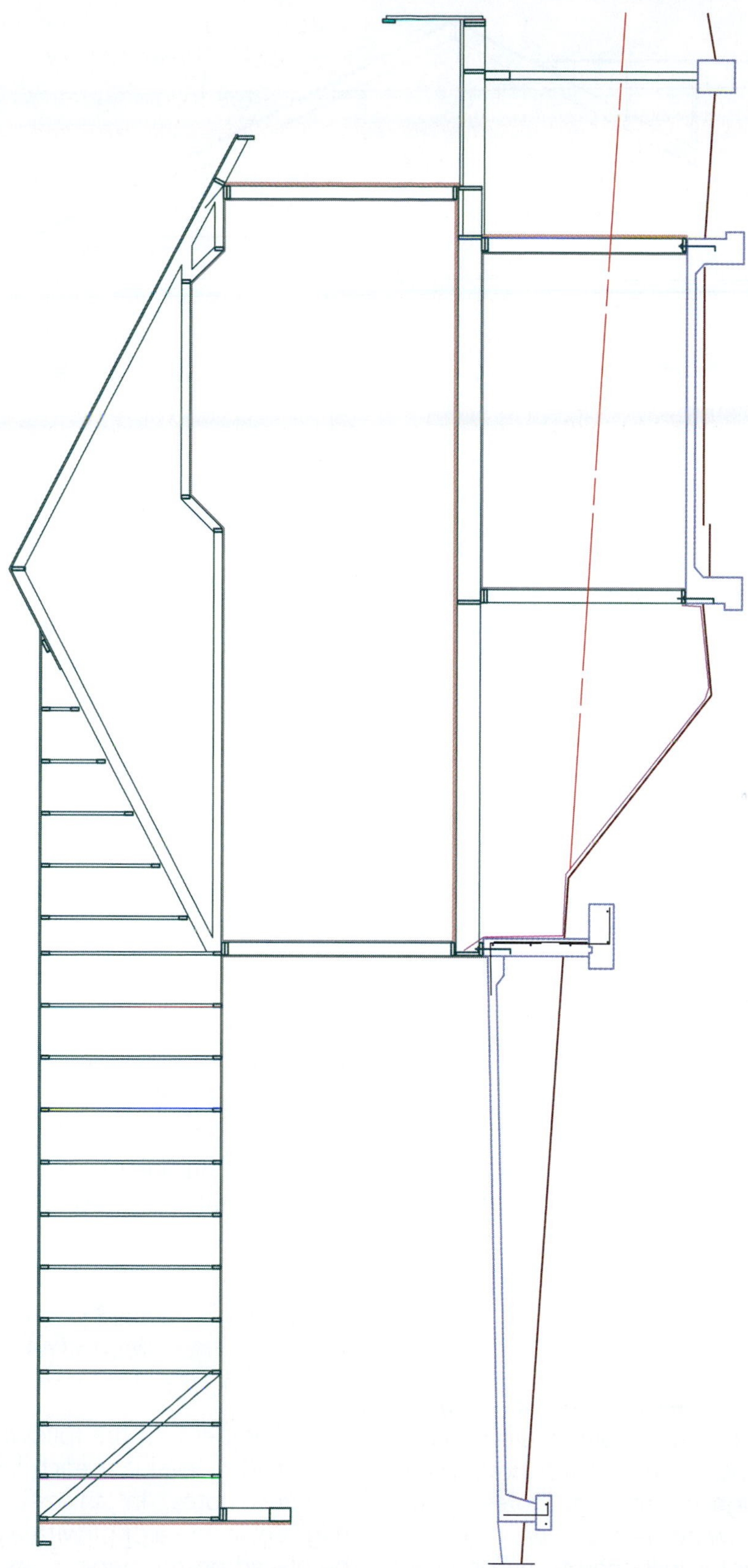

FIGURE 29.13 ■ *All structural materials are now represented. Continuous materials that are cut by the cutting plane should be represented with thick lines. Thin lines are used to represent all materials that lie beyond the cutting plane. Materials such as sheet rock, sheathing, and decking are drawn with thin lines even though the cutting plane passes through these materials.*

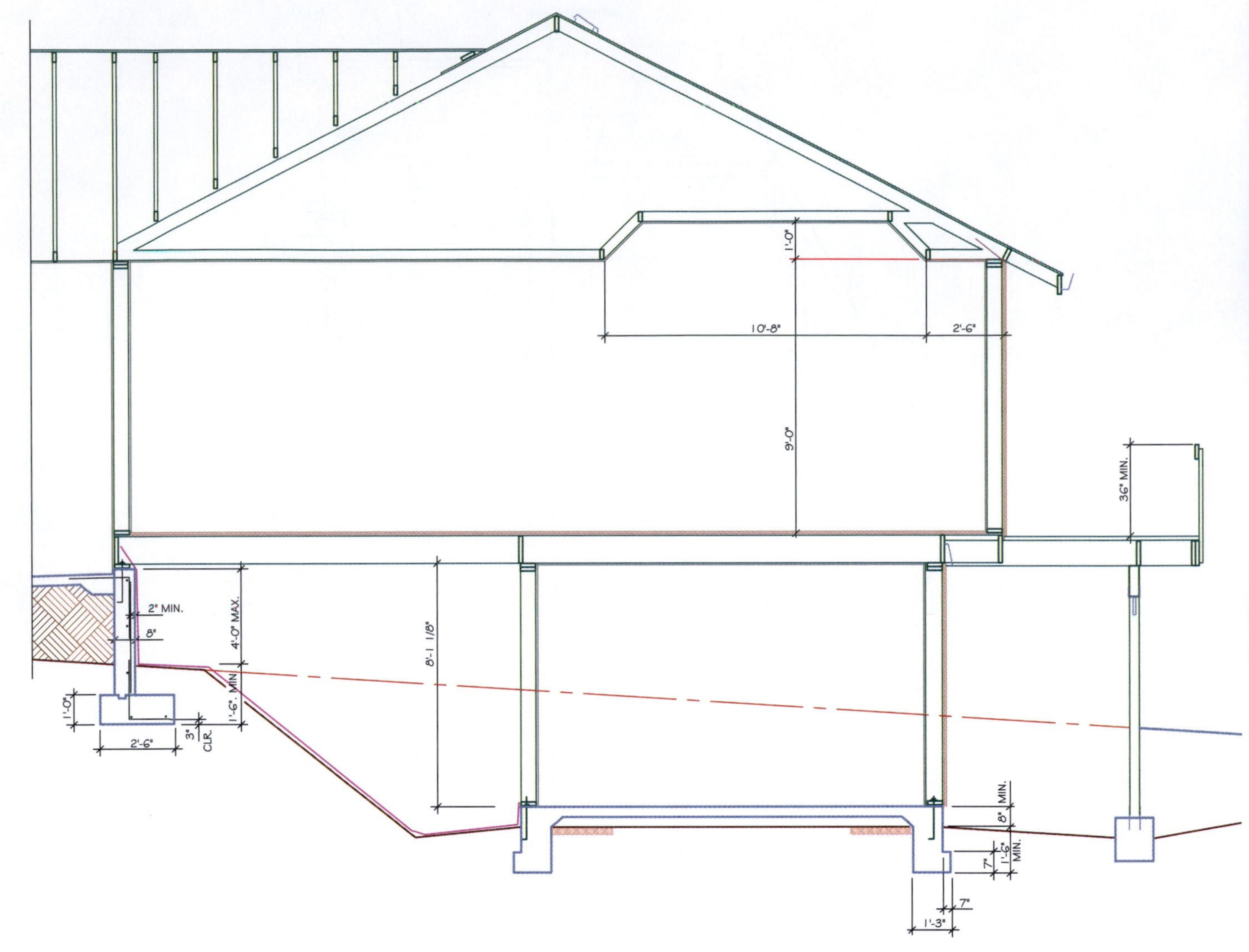

FIGURE 29.14 ■ *Dimensions are required for each section to describe the vertical relationships of structural materials, and to describe minimum clearances that are based on code requirements.*

- Header heights from the rough floor to the bottom of the header.
- Height of brick veneer if necessary.

Stage 5: Adding Annotation

Using the *SECT ANNO* layer, everything that has been drawn and located must now be explained. Offset a line about 1'' (25 mm) from the exterior of the structure to serve as a guideline to align the required notes. Doing so will help make your drawing neat and easy to read. Not all of the following notes will need to go on every section you draw. Typically, the primary section will be fully notated, and then other sections will have only notes to explain varied construction. You will notice as you go through the list of general notes that some are marked (*). For the marked items, there are several options. Ask your instructor for help in determining which grade of material to use.

Note:

Remember to remove the guideline once all text is located.

Remember that the following list is only a guide. You will need to evaluate each section prior to placing the required notes. In an office setting your supervisor might give you a print with all of the notes that need to be placed on the original. As you gain confidence, you will probably be referred to a similar drawing for help in deciding which notes are needed. Eventually you will be able to draw and annotate a section without any guidelines. When complete, your drawing should resemble Figure 29.15. Figures 29.16a and b show the balance of

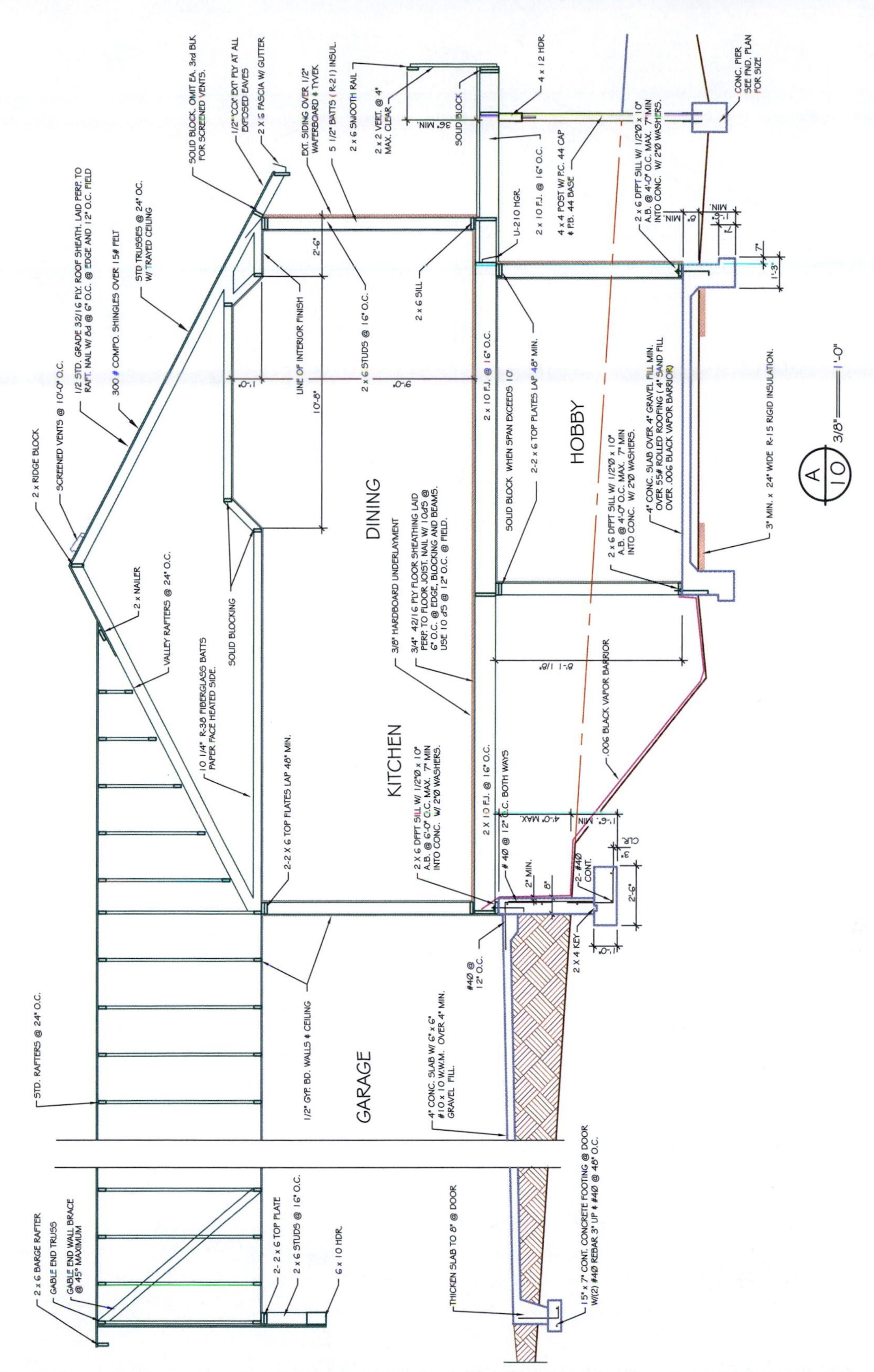

FIGURE 29.15 ■ *Notes are required to describe all structural and finish materials. Some companies use generic notes for structural material such as joists and girders and refer the print reader to other plans for specific sizes. Instead of marking the girder as 6 × 10, a reference such a "Girder—see foundation plan" could be used.*

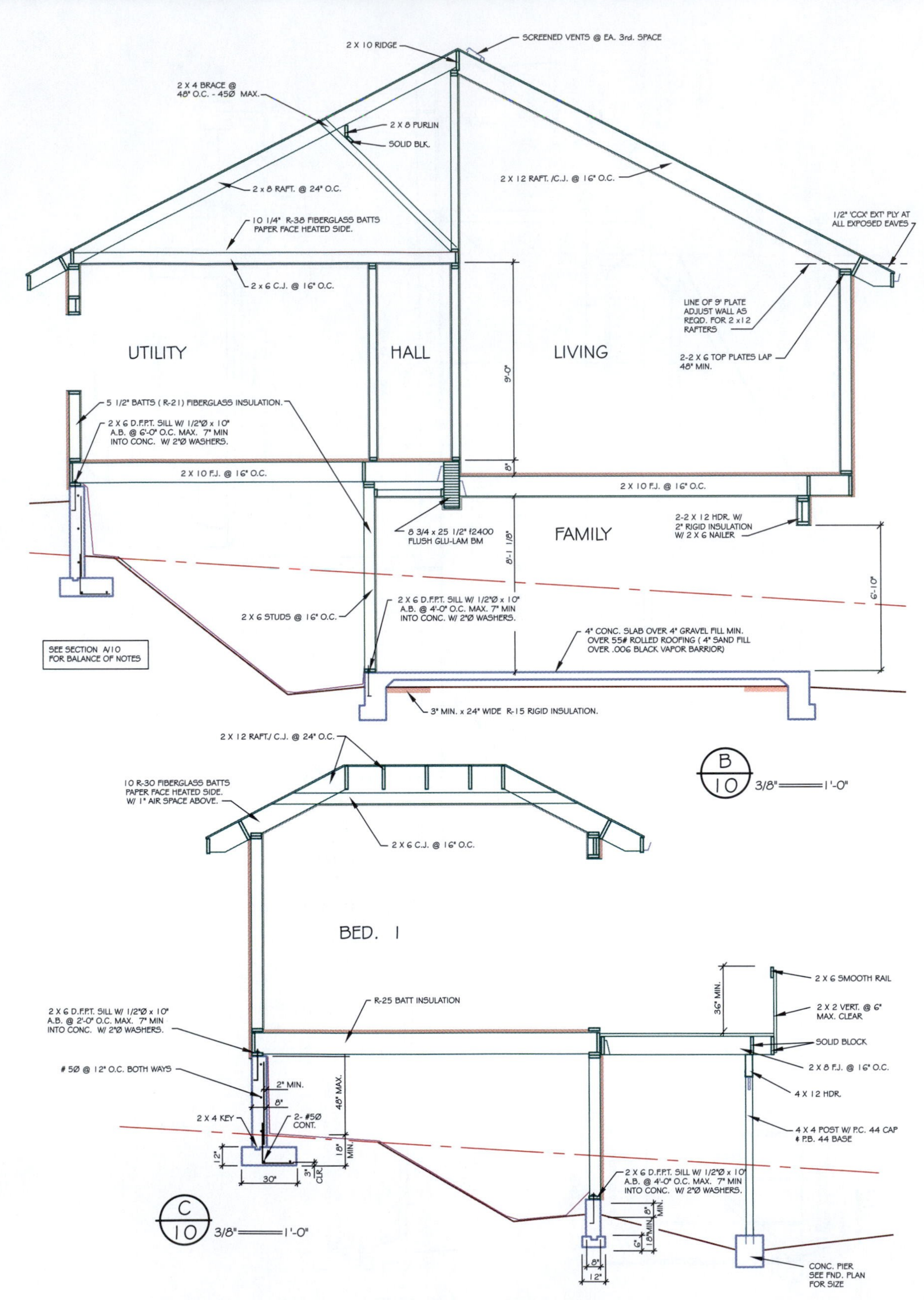
2 X 10 RIDGE
SCREENED VENTS @ EA. 3rd. SPACE
2 X 4 BRACE @ 48" O.C. - 45Ø MAX.
2 X 8 PURLIN
SOLID BLK.
2 X 12 RAFT. /C.J. @ 16" O.C.
2 x 8 RAFT. @ 24" O.C.
10 1/4" R-38 FIBERGLASS BATTS PAPER FACE HEATED SIDE.
1/2" 'CCX' EXT' PLY AT ALL EXPOSED EAVES
2 x 6 C.J. @ 16" O.C.
LINE OF 9' PLATE ADJUST WALL AS REQD. FOR 2 x12 RAFTERS
UTILITY
HALL
LIVING
2-2 X 6 TOP PLATES LAP 48" MIN.
9'-0"
5 1/2" BATTS (R-21) FIBERGLASS INSULATION.
2 X 6 D.F.P.T. SILL W/ 1/2"Ø x 10" A.B. @ 6'-0" O.C. MAX. 7" MIN INTO CONC. W/ 2"Ø WASHERS.
8"
2 X 10 F.J. @ 16" O.C.
2 X 10 F.J. @ 16" O.C.
2-2 X 12 HDR. W/ 2" RIGID INSULATION W/ 2 X 6 NAILER
8 3/4 x 25 1/2" f2400 FLUSH GLU-LAM BM
8'-1 1/8"
FAMILY
6'-10"
2 X 6 STUDS @ 16" O.C.
2 X 6 D.F.P.T. SILL W/ 1/2"Ø x 10" A.B. @ 4'-0" O.C. MAX. 7" MIN INTO CONC. W/ 2"Ø WASHERS.
SEE SECTION A/10 FOR BALANCE OF NOTES
4" CONC. SLAB OVER 4" GRAVEL FILL MIN. OVER 55# ROLLED ROOFING (4" SAND FILL OVER .006 BLACK VAPOR BARRIOR)
3" MIN. x 24" WIDE R-15 RIGID INSULATION.
2 X 12 RAFT./ C.J. @ 24" O.C.
B
10
3/8" = 1'-0"
10 R-30 FIBERGLASS BATTS PAPER FACE HEATED SIDE. W/ 1" AIR SPACE ABOVE.
2 X 6 C.J. @ 16" O.C.
BED. 1
2 X 6 SMOOTH RAIL
36" MIN.
2 X 2 VERT. @ 6" MAX. CLEAR
R-25 BATT INSULATION
2 X 6 D.F.P.T. SILL W/ 1/2"Ø x 10" A.B. @ 2'-0" O.C. MAX. 7" MIN INTO CONC. W/ 2"Ø WASHERS.
SOLID BLOCK
2 X 8 F.J. @ 16" O.C.
5Ø @ 12" O.C. BOTH WAYS
2" MIN.
8"
48" MAX.
4 X 12 HDR.
2 X 4 KEY
2- #5Ø CONT.
4 X 4 POST W/ P.C. 44 CAP & P.B. 44 BASE
12"
18" MIN.
30"
3" CLR.
2 X 6 D.F.P.T. SILL W/ 1/2"Ø x 10" A.B. @ 4'-0" O.C. MAX. 7" MIN INTO CONC. W/ 2"Ø WASHERS.
C
10
3/8" = 1'-0"
8" MIN.
18" MIN.
6"
8"
12"
CONC. PIER SEE FND. PLAN FOR SIZE

FIGURE 29.16a ■ *Sections B and C from the floor plan shown in Figure 11.18.*

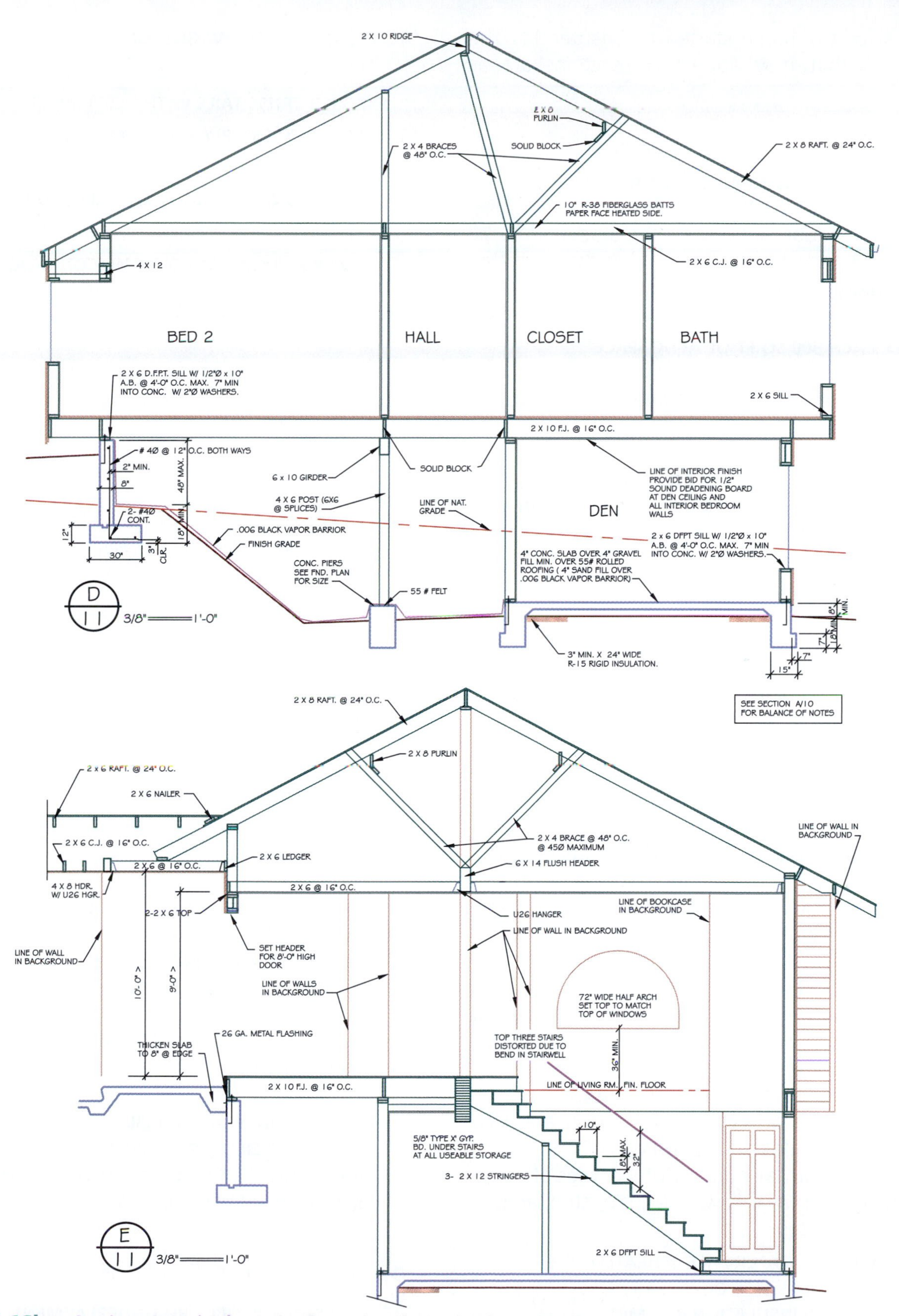

FIGURE 29.16b ■ *Sections D and E from the floor plan shown in Figure 11.18.*

the sections for the home started in Chapter 11. The general notes that may appear on sections are as follows:

Roof

(Fill in the blanks as required to meet local codes.)

See the list of abbreviations on the student CD.

*1. **2 × _ RIDGE BLOCKING**

2. **SCREENED RIDGE VENTS @ 10' o.c. ±. PROVIDE 1 SQ FT PER EACH 300 SQ FT OF ATTIC SPACE.**

*3. **1/2" OSB ROOF SHEATH. LAID PERP. TO TRUSSES. NAIL W/ 8d @ 6" O.C. @ EDGE AND 8d @ 12" O.C. FIELD**

or

1/2" STD. GRADE 32/16 PLY ROOF SHEATH. LAID PERP. TO TRUSSES. NAIL W/ 8d @ 6" O.C. @ EDGE AND 8d @ 12" O.C. FIELD

or

1 × 4 SKIP SHEATH. W/ 3 1/2" SPACING

*4. Choose roofing material to match what has been specified on the exterior elevations such as:

300# (or 235#) COMPO. ROOF SHINGLES OVER 15# FELT

or

MED. CEDAR SHAKES OVER 15# FELT W/ 30# × 18" WIDE FELT BTWN. EA. COURSE W/ 10 1/2" EXPOSURE

or

CONC. ROOF TILES BY (give manufacturer's name, and color weight of tiles and underlayment). **INSTALL AS PER MANUF. SPECS.** (find a list of manufacturers at www.tileroofing.org)

or

METAL ROOFING BY (give the manufacturer's name, type of product, gage, and color). **INSTALL AS PER MANUF. SPECS.** (find a list of manufacturers at www.metalroofing.com)

5. **STD. ROOF TRUSSES @ 24" O.C. SEE DRAW. BY MANUF. SUBMIT TRUSS DRAWINGS TO THE BUILDING DEPARTMENT PRIOR TO ERECTION.**
   - Provide a similar note for other types of trusses such as valley, girder, hip, stub, and scissor trusses.

*6. **_ " BATTS, R-_ PAPER FACE @ HEATED SIDE**

or

_ " BLOWN-IN INSULATION R-_ MIN.

7. **BAFFLES AT EACH EAVE VENT**
8. **SOLID BLOCK—OMIT EA. 3RD FOR SCREENED VENTS**
9. **TRUSS CLIPS @ EA. TAIL TO PLATE**

*10. **_ × _ FASCIA W/ GUTTER**

or

EXPOSED TRUSS TAILS WITH VERTICAL CUT

*11. **1/2" 'CCX' EXT. PLY @ ALL EXPOSED EAVES**

or

1 × 4 T & G DECKING @ ALL EXPOSED EAVES

or

1/2" R.S. PLY SOFFIT COVER W/ CONTINUOUS VENTING

Walls

*1. **2—2 × _ TOP PLATES, LAPPED 48" MIN.**

*2. **2 × _ STUDS @ ____ " O.C.**

*3. **2 × _ SILL**

> **Note:**
>
> *Items 1, 2, & 3 should be listed for one interior and a typical exterior wall.*

*4. **EXT. SIDING OVER 1/2" OSB & TYVEK. SEE ELEVATIONS**

or

EXT. SIDING OVER 3/8" PLY & 15# FELT

*5. **3 1/2" FIBERGLASS BATTS, R-11 MIN.—PAPER FACE ONE SIDE**

or

6" FIBERGLASS BATTS R-19, PAPER FACE ONE SIDE

or

6" FIBERGLASS BATTS R-21

*6. **_ × _ HEADER**

or

(2)—2 × 12 HDRS W/ 2 × 6 NAILER & 2" RIGID INSULATION

7. **LINE OF INTERIOR FINISH**
8. **SOLID BLOCK AT MID HEIGHT FOR ALL WALLS OVER 10' TALL**

*9. **5/8" TYPE 'X' GYP. BD. FROM FLOOR TO BOTTOM OF ROOF SHEATH.**

or

1/2" GYP. BD. WALLS & CEIL.

10. **BRICK VENEER OVER 1" AIRSPACE & TYVEK W/ METAL TIES @ 24" O.C. EA. STUD. PROVIDE 3/16" MIN. WIDE WEEP HOLES @ 32" O.C.**

Upper Floor and Foundation

1. **3/8" MIN. HARDBOARD UNDERLAYMENT**
2. **3/4" PLY. 42/16 FLOOR SHEATH. LAID PERP. TO FLOOR JOISTS. NAIL W/ 10d @ 6" O.C. EDGE, BLOCKING, & BEAMS. USE 10d @ 12" O.C. @ FIELD**

3. **SOLID BLOCK @ 10' O.C. MAX**
4. **__" BATTS—R-__ MIN. (or r-21)**
5. **2 × 6 P.T. SILL W/1/2" Ø × 10" A.B. @ 6'–0" O.C. MAX.—7" MIN. INTO CONC. W/2" Ø WASHERS. PROVIDE 2 BOLTS PER SILL MIN.**

*6. For slabs over natural grade:

4" CONC. SLAB OVER 4" GRAVEL, 6 MIL VAPOR BARRIER, & 2" SAND FILL.

or

4" CONC. SLAB OVER 4" GRAVEL FILL OVER 55# FELT

For slabs over fill material:

4" CONC. SLAB W/ #10 × 10 / 4 × 4 WWM OVER 4" GRAVEL OVER 6 MIL VAPOR BARRIER, & 2" SAND FILL

7. **1/2" CONCRETE BOARD OVER 2" RIGID INSULATION W/ 26 GA. FLASHING**
8. Assign a title to each room in the section.
9. Give the section a title, such as section 'AA'.
10. Give the drawing a scale such as 3/8" = 1'-0".
11. List general notes near the title or near the title block.

***ALL FRAMING LUMBER TO BE DFL # 2 OR BETTER.** (You may need to substitute a different type of wood.)

See sections 'BB' & 'CC' for balance of notes.

Stage 6: Evaluating Your Work

Don't assume that because you did the work and it took a long time, the drawing is complete. Make a plot and evaluate your work for accuracy and quality. Don't just compare it with someone else's drawing. Use the checklist from the student CD and make sure your section matches the material and location that you have specified on the floor, framing, roof, and foundation plans. Your best chance of finding your own mistakes is to get away from your drawing for an hour or two before checking it.

SECTION ALTERNATIVES

The previous portion of this chapter introduced the layout and completion of section A from the home started in Chapter 11. Those methods will not meet the needs of all homes because of variations at the job site, the contractor's personal preference for framing, or the area of the country in which the house is to be built. This portion of the chapter introduces other common construction methods to meet a variety of needs. Unless noted, all sizes given in the following steps are based on the minimum standards of the IRC and should be compared with local standards. Types of sections to be covered in this chapter include:

- Joist foundations
- Post-and-beam foundations
- Full basements
- Trussed framed roofs with a vaulted ceiling
- Conventional roofing with a flat ceiling
- Stick-framed vaulted ceilings

The drawings that will be provided in this portion of the chapter do not match the home from Chapter 11. The following assumptions will be made for each example:

- The home will be a two-story home.
- A load-bearing wall will be placed 12' from the right edge of the home.
- 2 × 10 (50 ×250) floor joists will be used.

Floor Joists/Foundation

Use thin lines placed on the *SECT OUTL* layer for this entire stage. Refer to the site, grading, floor, framing, and foundation plans to determine the sizes and locations of structural members. See Figure 29.17 for steps 1 to 14 showing the layout of a section with a foundation with a joist floor system.

Step 1. Draw a line to represent one edge of the structure. Use the OFFSET command to represent the width of the building based on the dimensions provided on the framing plan.
Step 2. Draw a line to represent the finish floor level.

Note:

Before completing the following steps, it may be necessary to review the basics of foundation design in Section 7. All sizes given in the following steps are based on the minimum standards of the IRC (see Figure 25.10) for footing supporting two-levels (unless otherwise noted) and should be compared with local standards.

Step 3. Using the line created in step 2, use the OFFSET command (10" [250 mm] down from the top of joists) and create a line to represent the bottom of the floor joists.
Step 4. Use the OFFSET command (2" [50 mm] down from the bottom of the joists) and create a line to represent the top of the mudsill.

Step 5. Use the OFFSET command (2'' [50 mm] down from the bottom of the mudsill) and create a line to represent the top of the concrete stem wall.
Step 6. Use the OFFSET command (8'' [200 mm] down from the top of the stem wall) and create a line to represent the finished grade.

- If a site or grading plan is available, locate the natural and finished grades.

Step 7. Use the OFFSET command (18'' [450 mm] down from the finish grade) and create a line to represent the bottom of the footing.
Step 8. Use the OFFSET command (7''/175 mm up from the bottom of the footing) and create a line to represent the footing thickness.
Step 9. Use the OFFSET command to represent the thickness of the stem wall. Verify this size with local codes.

- The IRC allows a width of 6'' (150 mm).
- Many local codes use the traditional width of 7'' (175 mm) for two-level and 8'' (200 mm) for three-level footings.

Step 10. Use the OFFSET command to represent the width of the footing (15'' [375 mm] for two stories). Center the footing on the stem wall.
Step 11. Use the OFFSET command to represent the 4'' (100 mm) wide ledge for brick veneer if required.
Step 12. Using the line created in step 2, use the OFFSET command to represent the 3/4'' (20 mm)-thick plywood subfloor.
Step 13. Lay out required girders beneath the load-bearing walls or as needed on the basis of the joist spans.
Step 14. Lay out the required concrete piers beneath load-bearing walls or as needed on the basis of the joist spans.

Your drawing should now resemble Figure 29.17. Use the following steps to convert the rough layout to the appropriate layers with the correct linetypes.

Representing Structural Materials

Trim the following lines as needed and use the PROPERTIES command to assign the needed layers. See Figure 29.18 for the following steps:

Step 15. Trim and place the outline of the concrete footings, stem walls, and piers on the *SECT MCUT* layer. Assign a very thick line to all concrete.

- Use the HATCH command to place the concrete pattern on the *SECT PATT* layer.

Step 16. Place the finished and natural grade lines on the *SECT MCUT* layer. Assign a very thick line to each grade line and use a dashed line to represent the natural grade.

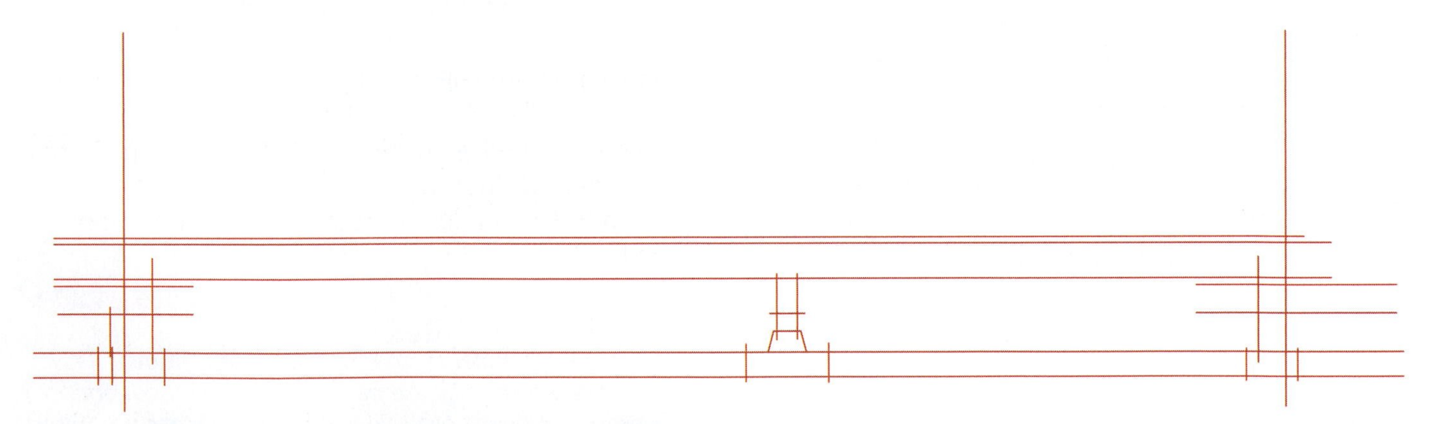

FIGURE 29.17 ■ *The layout for a section with a joist floor system.*

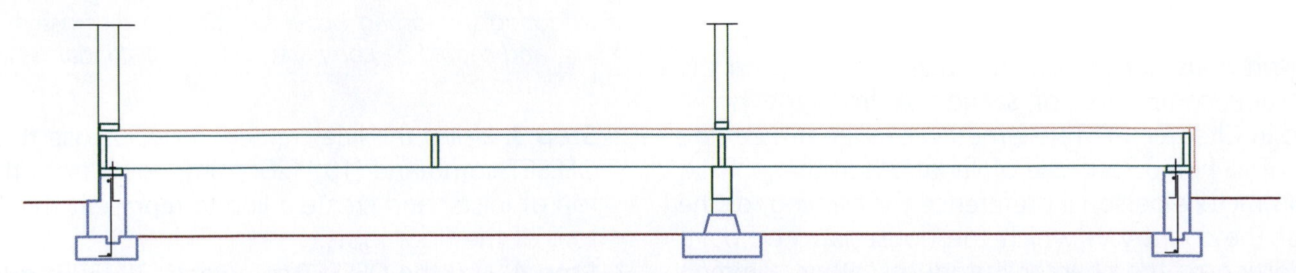

FIGURE 29.18 ■ *A section showing the structural members of a joist floor and foundation system drawn with finished-quality lines. Varied line weights should be used to represent continuous materials that the cutting plane passes through. Thin lines should be used to represent materials that lie beyond the cutting plane.*

Step 17. Draw or insert the mudsills, steel rebar if required, and anchor bolts on the *SECT MCUT* layer. Assign a bold linetype for each item.
Step 18. Draw or insert the girders on the *SECT MCUT* layer. Assign a bold linetype for each item.
Step 19. Trim the lines representing the floor sheathing and place the lines on the *SECT MBND* layer.
Step 20. Trim the support post as required and place the lines on the *SECT MBND* layer. Assign thin lines to represent the post.

- Draw a gusset to connect each girder to the post.

All structural members are now drawn. Your drawing should resemble Figure 29.18.

Representing Finishing Materials

Use Figure 29.19 as a guide to add the finishing materials. Use thin lines to represent the following materials. Place each item on the *SECT MBND* layer.

Note:

A portion of the walls have been drawn to indicate how the floor and wall intersection will be drawn.

Step 21. Draw the hardboard underlayment. Note that the subfloor extends under all walls, but the hardboard underlayment does not go under any walls.
Step 22. Use the BATT linetype to draw batt floor insulation 8'' (200 mm) deep between the floor joists. Place the insulation on the *SECT PATT* layer.
Step 23. Use the SKETCH command to draw the vapor barrier. Leave a small space between the ground and the barrier.
Step 24. Draw any required metal connectors, straps, or flashing.
Step 25. Draw and crosshatch any required brick veneer on the *SECT PATT* layer using thin lines. Your drawing should now resemble Figure 29.19.

Placing Dimensions

Figure 29.20 shows the needed dimensions. All lettering should be aligned. Use the DIM command to locate the following dimensions on the *SECT DIMS* layer:

- Finished grade to the top of stem wall
- Depth of footing into the natural grade
- Height of the footing
- Width of the footing
- Width of the stem wall
- Depth of the crawl space (18'' [450 mm] minimum from grade to the bottom of the joists)
- Bottom of girders to the grade. (12'' [300 mm] minimum)

Floor and Foundation Annotation

Using the *SECT ANNO* layer, letter the floor and foundation notes as shown in Figure 29.20. Typical notes should include the following:

1. **3/8'' MIN. HARDBOARD UNDERLAYMENT**
2. **3/4'' PLY. 42/16 FLOOR SHEATH. LAID PERP. TO FLOOR JOISTS. NAIL W/ 10d @ 6'' O.C. EDGE, BLOCKING, & BEAMS. USE 10d @ 12'' O.C. @ FIELD.**
3. **SOLID BLOCK @ 10'' O.C. MAX.**
4. **__'' BATTS—R-__ MIN. or (R-21).**

*5. **__ × __ GIRDERS**

or

__ × GIRDER SEE FOUNDATION PLAN FOR SIZE & SPACING

*6. **4 × 4 POST, (4 × 6 @ SPLICES), W/ GUSSET TO GIRDER ON 55# FELT ON**
__ × __ DEEP CONC. PIERS.

7. **006 BLACK VAPOR BARRIER**

or

55# ROLLED ROOFING

*8. **2 × __ P.T. SILL W/ 1/2'' Ø A.B. @ 6'-0''. O.C. MAX.— 7'' MIN. INTO CONC.**
W/-2'' Ø WASHERS. PROVIDE 2 BOLTS PER SILL MIN.

9. Assign each room a room title.

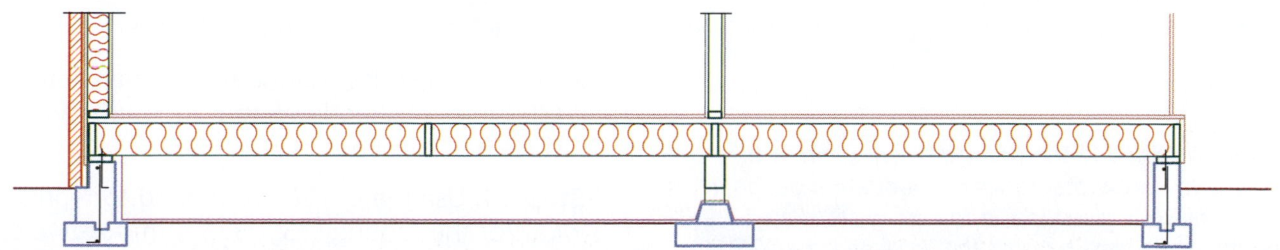

FIGURE 29.19 ■ *Finishing materials for a joist floor and foundation system include the floor overlay, insulation, metal connectors, and vapor barrier.*

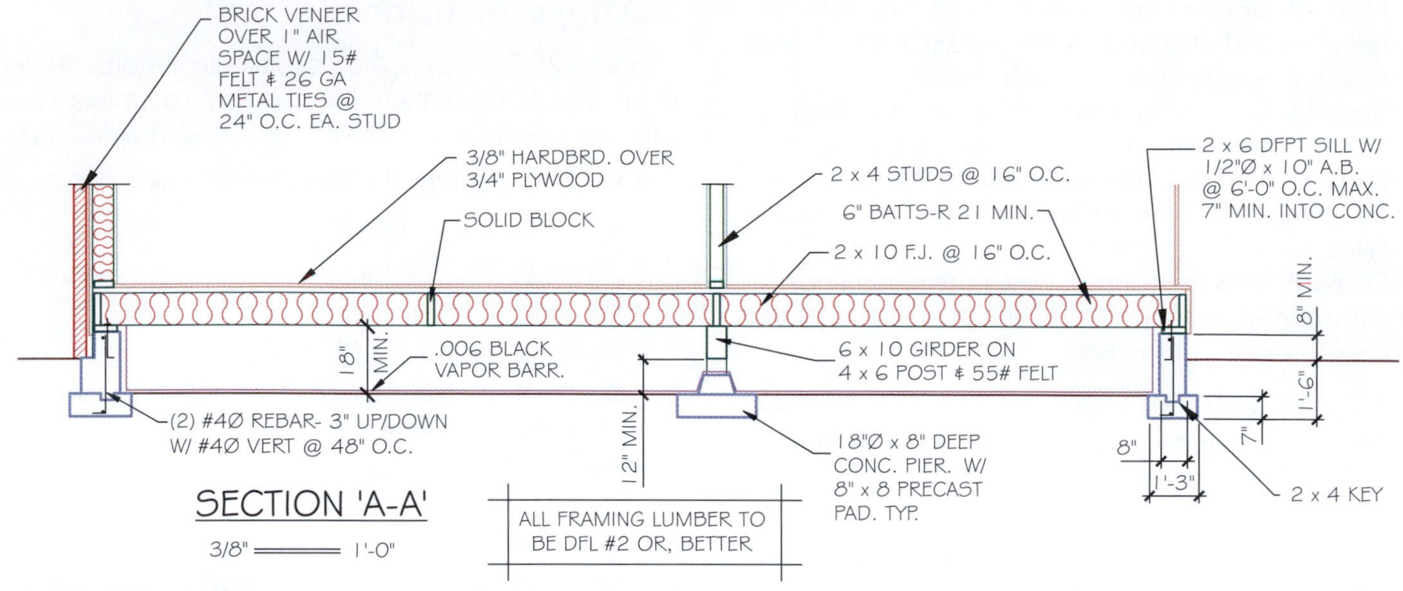

FIGURE 29.20 ■ *Dimensions are required to describe minimum clearances. Annotation is required to describe all structural and finish materials that are not specified in other details or sections.*

10. Give the section a title such as section 'BB'.
11. Give the drawing a scale such as 3/8'' = 1'-0''.
12. List general notes near the title or near the title block.

    *__ALL FRAMING LUMBER TO BE DFL #2 OR BETTER.__ (Instead of DFL you may need to substitute a different type of wood.)

The completed drawings can be seen in Figure 29.20.

Post-and-Beam Foundations

Use thin lines placed on the *SECT OUTL* layer for this entire stage. Refer to the site, grading, floor, framing, and foundation plans to determine the size and locations of structural members. See Figure 29.21 for the following steps showing the layout of a post-and-beam floor system.

Step 1. Draw a line to represent one edge of the structure. Use the OFFSET command to represent the width of the building based on the dimensions provided on the framing plan.

Step 2. Draw a line to represent the finish floor level.

Note:

Before completing the following steps, it may be necessary to review the basics of foundation design in Section 7. All sizes given in the following steps are based on the minimum standards of the IRC (see Figure 25.10) for footing supporting two levels (unless otherwise noted) and should be compared with local standards.

Step 3. Using the line created in step 2, use the OFFSET command (2'' [50 mm] down from the top of the floor) and create a line to represent the bottom of the floor decking.

Step 4. Use the OFFSET command (2'' [50 mm] down from the bottom of the decking) and create a line to represent the top of the mudsill.

Step 5. Use the OFFSET command (2'' [50 mm] down from the bottom of the mudsill) and create a line to represent the top of the concrete stem wall.

Step 6. Use the OFFSET command (8'' [200 mm] down from the top of the stem wall) and create a line to represent the finished grade.

Step 7. Use the OFFSET command (18'' [450 mm] down from the finish grade) and create a line to represent the bottom of the footing.

Step 8. Use the OFFSET command (7'' [175 mm] up from the bottom of the footing) and create a line to represent the footing thickness.

Step 9. Use the OFFSET command to represent the thickness of the stem wall. Verify this size with local codes.

- The IRC allows a width of 6'' (150 mm).
- Many local codes use the traditional width of 8'' (200 mm) for two-level and 10'' (250 mm) for three-level footings.

Step 10. Use the OFFSET command to represent the width of the footing (15''/ 375 mm) for two stories. Center the footing below the stem wall.

Step 11. Use the OFFSET command to represent the 4'' (100 mm) ledge for brick veneer if required.

Step 12. Using the line created in step 2, use the OFFSET command to represent the 2'' (50 mm) floor decking.

Step 13. Lay out required girders beneath the load-bearing wall.

Step 14. Lay out the balance of the girders based on their location on the foundation plan.

Step 15. Lay out required concrete piers beneath the girders based on the sizes specified on the foundation plan. Your drawing should now resemble Figure 29.21.

Finished Line Quality—Structural Materials

Use the following steps to convert the rough layout to the appropriate layers with the correct linetypes. Trim the following lines as needed and use the PROPERTIES command to assign the needed layers. See Figure 29.22 for the following steps:

Step 16. Trim and place the outline of the concrete footings, stem walls and piers on the *SECT MCUT* layer. Assign a very thick line to all concrete.

- Use the HATCH command to place the concrete pattern on the *SECT PATT* layer.

Step 17. Place the finished and natural grade lines on the *SECT MCUT* layer. Assign a very thick line to each grade line and use a dashed line to represent the natural grade.

Step 18. Draw or insert the mudsills, steel rebar if required, and anchor bolts on the *SECT MCUT* layer. Assign a bold line type for each item.

Step 19. Draw or insert the girders on the *SECT MCUT* layer. Assign bold line types for each item.

Step 20. Trim the lines representing the decking and place the lines on the *SECT MBND* layer.

Step 21. Trim the support post as required and place the lines on the *SECT MBND* layer. Assign thin lines to represent the post and draw a gusset to connect each girder to the post.

All structural members are now drawn. Your drawing should resemble Figure 29.22.

Finished-Quality Lines—Finishing Materials

Use Figure 29.23 as a guide to add the finishing materials. Use thin lines to represent the following materials. Place each item on the *SECT MBND* layer.

Note:

A portion of the walls have be drawn to indicate how the floor and wall intersection will be drawn.

Step 22. Draw the hardboard underlayment. Note that the plywood decking extends under all walls, but the hardboard underlayment does not go under any walls.

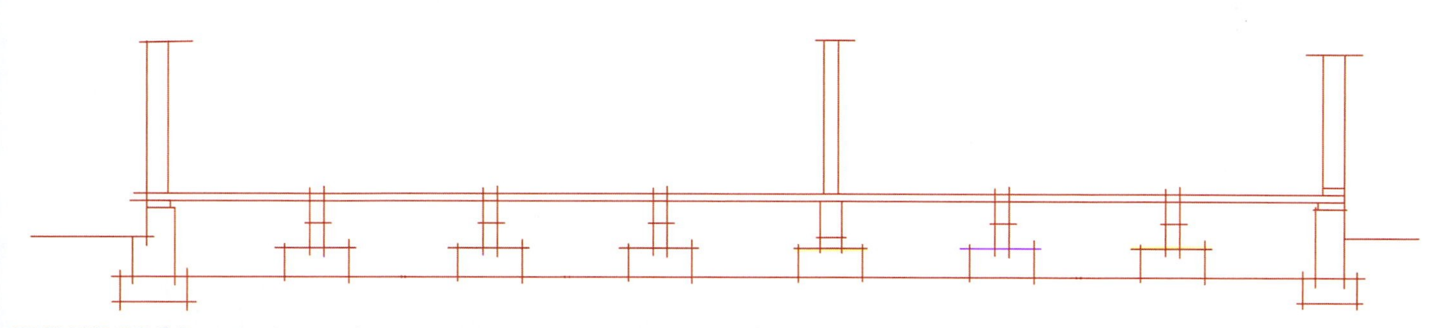

FIGURE 29.21 ■ *The layout for a section with a post-and-beam floor system.*

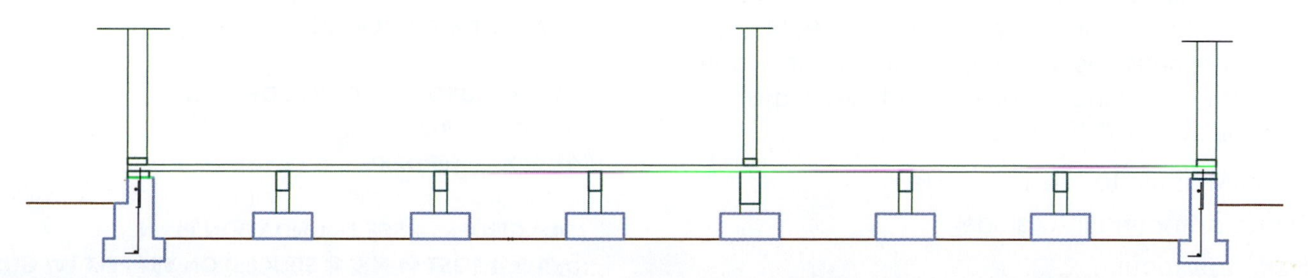

FIGURE 29.22 ■ *Structural materials for a post-and-beam floor system are represented using lines with varied line weight.*

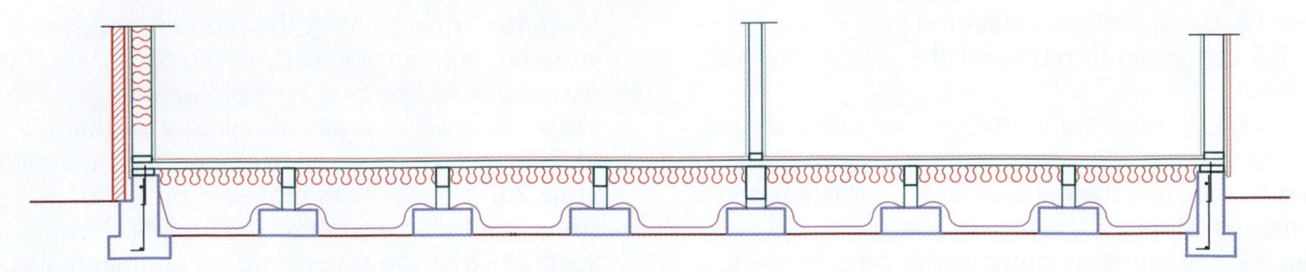

FIGURE 29.23 ■ *Materials such as the floor overlay, insulation, metal connectors, and vapor barrier are added on the SECT PATT and SECT MBND layers.*

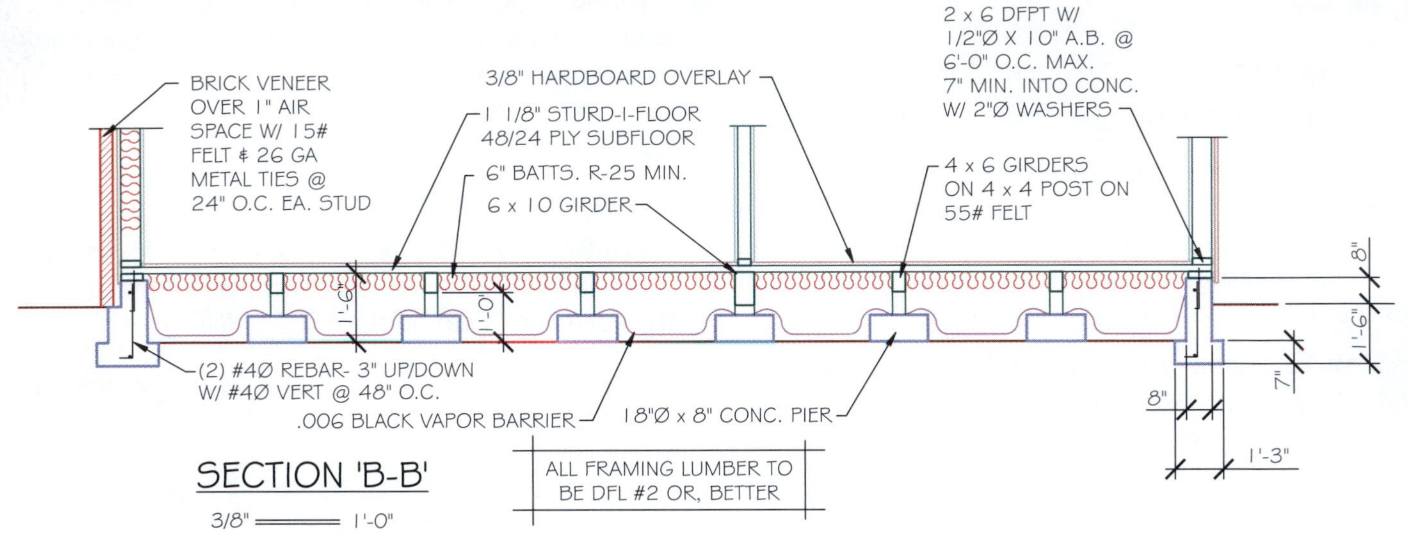

FIGURE 29.24 ■ *Representing the dimensions and annotation that are required to describe the minimum clearances and all structural and finish materials.*

Step 23. Use the BATT linetype to draw 8'' (200 mm) batts floor insulation between the girders. Place the insulation on the *SECT PATT* layer.
Step 24. Use the SKETCH command to draw the vapor barrier. Leave a small space between the ground and the barrier.
Step 25. Draw any required metal connectors, straps, or flashing.
Step 26. Draw and crosshatch any required brick veneer on the *SECT PATT* layer using thin lines.

Placing Dimensions

Dimension all structural materials so that the framing crew knows their vertical location. Figure 29.24 shows the needed dimensions. All lettering should be aligned. Use the DIM command to locate the following dimensions on the *SECT DIMS* layer:

- Finished grade to top of stem wall
- Depth of footing into grade
- Height of footing
- Width of footing
- Width of stem wall
- Depth of crawl space 18'' (450 mm) below the floor decking
- Bottom of girders to grade. 12'' (300 mm) below the girders

Annotation

Using the *SECT ANNO* layer, letter the floor and foundation notes as shown in Figure 29.24. Typical notes should include the following:

1. **3/8'' hardboard overlay**
*2. **2 × 6 T & G DECKING**

or

1 1/8 STURD.I.PLY FLOOR DECKING

3. **6 BATTS R-21 MIN.**
*4. __ × __ **GIRDERS**

or

4× GIRDERS—SEE FOUNDATION PLAN

5. **4 × 4 POST (4 × 6 @ SPLICES) ON 55# FELT W/ GUSSET.**

6. ___ Ø × ___ **DEEP CONC. PIERS**
*7. **.006 BLACK VAPOR BARRIER**
or
55# ROLLED ROOFING
8. **2 × 6 P.T. SILL W/ 1/2" Ø × 10" A.B. @ 6'–0" O.C. MAX.—7" MIN. INTO CONC. THRU 2" Ø WASHERS. PROVIDE 2 BOLTS MIN. PER SILL.**
9. Assign each room a room title.
10. Give the section a title such as section 'BB'.
11. Give the drawing a scale such as 3/8" = 1'-0".
12. List general notes near the title or near the title block.
***ALL FRAMING LUMBER TO BE DFL #2 OR BETTER.** (Instead of DFL you may need to substitute a different type of wood.)

Full Basements with a Concrete Slab

Drawing a section showing a full basement will have similarities to the layout of a section showing a concrete slab. Depending on the seismic zone where the home will be built, an engineer's drawing may be required for the retaining walls. If so, you will need to follow the engineers' design standards similar to those shown in Figure 29.25. Understanding engineers' design standards, or calculations (calcs), can be very frustrating. Calcs are generally divided into the areas of the item to be designed, the formulas to determine the size, and the solution.

You do not need to understand the formulas that are used, but you must be able to convert the solution into a drawing. The solution in Figure 29.25 is a series of notes that are listed under the heading USE. For an entry-level drafter, the engineer will usually provide a sketch similar to the drawing in Figure 29.26 to explain the calculations. By using the written calculations and the sketch, a drafter can make a drawing similar to Figure 29.27.

Using the steps shown in Figure 29.28 as a guide, draw a home with a basement. Use thin lines placed on the *SECT OUTL* layer for this entire stage. Refer to the site, grading, floor, framing, and foundation plans to determine the sizes and locations of structural members.

Step 1. Draw a line to represent one edge of the structure. Use the OFFSET command to represent the width of the building based on the dimensions provided on the framing plan.
Step 2. Draw a line to represent the lower finished floor level.
Step 3. Using the line created in step 2 as a base, use the OFFSET command (4" [100 mm] down from the top of slab) and create a line to represent the bottom of the 4" (100 mm) slab.
Step 4. Use the OFFSET command (4" [100 mm] down from the bottom of the slab) and create a line to represent the bottom of the 4" (100 mm) gravel. This line will also represent the top for the footing.
Step 5. Use the OFFSET command (8" [200 mm] down from the bottom of the gravel) and create a line to represent the bottom of the footing. Footings are typically 8" (200 mm) deep for a one-story building and 12" (300 mm) deep for two or more stories.
Step 6. Use the OFFSET command to represent the thickness of the retaining wall. (Assume 8"/200 mm based on Figure 29.25).
Step 7. Use the OFFSET command to represent the width of the footings (16"/400 mm). Center the footing under the retaining wall.
Step 8. Use the OFFSET command to represent the additional 4" (100 mm) footing ledge for brick veneer if required. An alternative to the ledge method is to form the wall 4" (100 mm) out from the face of the exterior wall to support the brick. See Figure 29.29.
Step 9. Block out the interior footings for any load-bearing walls. The drawing should now resemble Figure 29.4. For this home, a 48" (1200 mm) high retaining wall will be required to support the west wall of the garage. The east wall of the hobby room will be a standard two-story concrete footing. (A perfect opportunity to insert the detail created in Problem 28.2 into the section.)
Step 10. Using the line created in step 2 as a base, use the OFFSET command (8' [2400 mm] up from the top of slab and create a line to represent the bottom of the upper-level floor).
Step 11. Using the line created in step 10 as a base, use the OFFSET command 2" (50 mm) down from the bottom of the floor and create a line to represent the top of the retaining wall.
Step 12. Using the line created in step 10 as a base, use the OFFSET command 10" (250 mm) up to create a line to represent the top of the floor joists.
Step 13. Use the OFFSET command to represent the 3/4" (20 mm) plywood subfloor.
Step 14. Using the line created in step 11 as a base, use the OFFSET command 8" (200 mm) down to create a line to represent the finish grade.
Step 15. Locate and draw the footings for interior bearing walls.
Step 16. Use the OFFSET command to locate the steel reinforcing for the wall per Figure 29.25. The

8' HIGH BEAM TYPE BLOCK RETAINING WALL

KENNETH D. SMITH - ARCHITECT
El Cajon, Ca. 92020

DESIGN TYPE- 8' BEAM TYPE BLOCK RETAINING WALL
M=(0.1283) (960) (8) = 985'#
TRY #5Ø 16" O.C. VERT. PLACED 2" FROM
INSIDE FACE OF 8" BLOCK

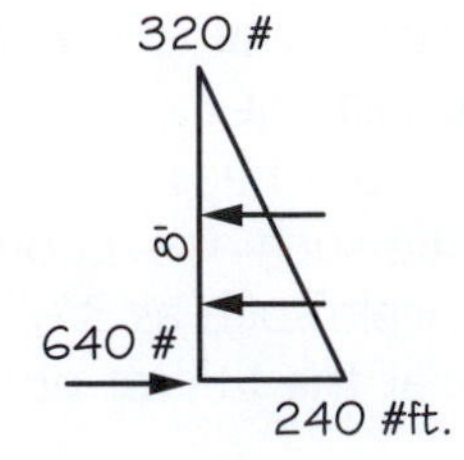

$np = \frac{(43)(0.31)}{(16)(5.62)} = J=861 \quad 2/KJ=558$

$f_m = \frac{(1.33)(12)(985)(558)}{(16)(562)^2} = 174.5$ p.s.i.

$f_s = \frac{(12)(1.33)(985)}{(0.31)(.861)(5.62)} = 10{,}500$ p.s.i. $\quad U = (11.0)\frac{16}{1.963} = 88.5$ p.s.i.

$V = \frac{(1.33)(640)}{(16)(.861)(5.62)} = 11.0$ p.s.i. $\quad ht = \frac{96}{8} = 12$

$f_a = (135)(.94) = 127$ p.s.i.

ALLOW FOR ROOF, 2nd FLOOR, & 1 st. FLOOR LOADING OF WALL FOR f_a

$f_a = \frac{(40)(40)+(8)(32)+(2)(8)(16)+(4)(137)+(8)(63)+(8)(63)=(92)(5)}{(7.62)(12)}$

$f_a = 29.4$ p.s.i.

MIN. HORIZ. STEEL= (0.007)(8)(96)=0.54in.^{2} $\quad \frac{174.5}{225} + \frac{29.4}{127} = \boxed{1.00}$ o.k.

USE:
- # 5Ø VERT. @ 16" O.C. PLACED 2" FROM FACE OF BLOCK AWAY FROM SOIL.
- USE 8" GRADE 'A' CONC. BLOCK- SOLID GROUT ALL STEEL CELLS.
- INTERMEDIATE GRADE DEFORMED BARS LAP 40 DIA. @ SPLICES.
- DBL. ALL VERT. STEEL BESIDE OPENINGS & WALL ENDS & MATCH ALL DOWEL STEEL OUT OF FTGS. FOR SIZE & POSITION.
- USE (2)- #4 Ø UNDER & OVER ALL OPENINGS (UNLESS NOTED OTHERWISE) & EXTEND 24" BEYOND OPENINGS. PLACE 1 1/2" - 2" UP FROM BTM. OF LINTEL IN INVERTED BOND BEAM BLOCK OR EQUAL & GROUT LINTEL CELLS SOLID.
- USE (4)- #4 Ø CONT @ TOP OF WALL IN 8" X 16" BLOCK BOND BM. & (2)- #4 Ø CONT. IN BOND BLOCK @ MIDHEIGHT (4' MAX.)
- GROUT AS PER SECTION 609 & MORTAR AS PER SECTION 607 TYPE 'S' 2003 IRC W/ CENTERING BRACKET @ TOP & BTM.
- PROVIDE CLEANOUT HOLES @ BTM. OF STEEL CELLS WHERE WALLS ARE GROUTED IN MORE THAN 4' LIFTS.
- AT TOP OF WALLS, USE 4" X 3" X 1/4" X 3" LONG (OR EQUAL) ∠ W/ 3/4" Ø X 10" A.B. THRU 3" LEG & PLATE INTO WALL & 3/4" Ø BOLT THRU 4" LEG @ JOIST (OR BLOCK WHERE ⊥ TO WALL W/ 2" Ø WASHERS AGAINST WOOD WHERE JOIST ∥ WALL BLOCK OUT 4' @ 32" O.C. & USE 10d @ 4" O.C. FOR SUBFLOOR.
- WATERPROOF ENTIRE WALL & USE 4" Ø DRAIN TILE & GRAVEL @ BTM. OF WALL.
- DO NOT BACK FILL UNTIL FLOOR IS IN PLACE.

FIGURE 29.25 ■ *A drafter is often required to work from engineers' calculations when drawing retaining walls. Calculations usually show the math formulas to solve a problem and the written solution to the problem. It is the written solution to the problem that the drafter must make into a drawing. Each of the materials in the "Use" portion of the calculations must be shown and specified in the section.*

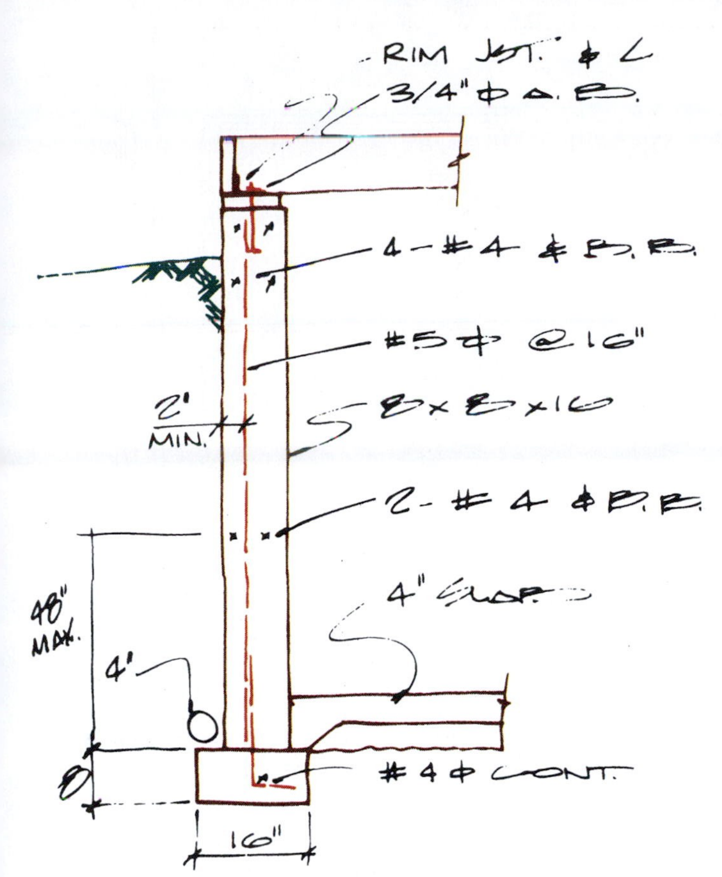

FIGURE 29.26 ■ *A sketch is often given to the drafter to help explain the written calculations. Many of the written calculations appear on the sketch but are not written in proper form.*

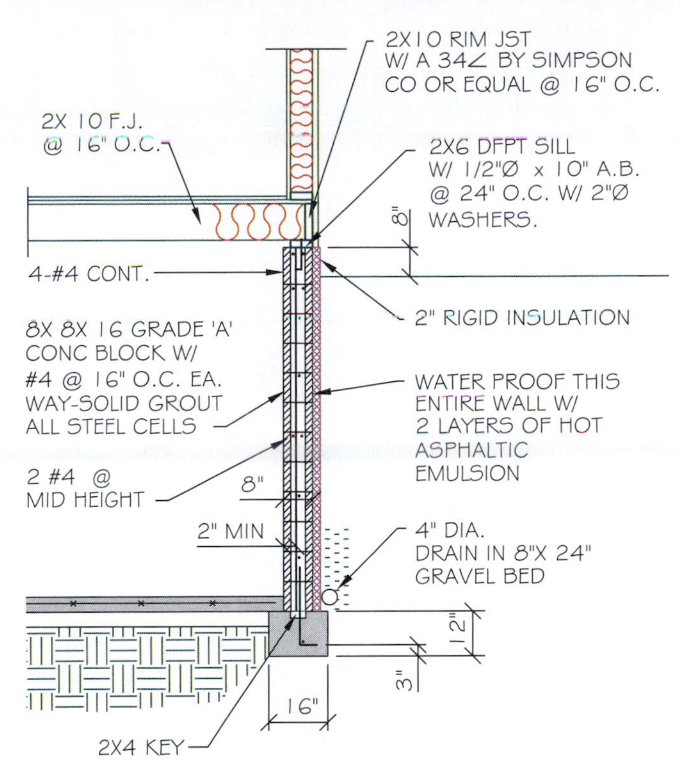

FIGURE 29.27a ■ *A section of an 8' (2400 mm) concrete block retaining wall based on the specification shown in Figures 29.25 and 29.26.*

size and location of the steel will vary greatly, depending on the seismic area and how the wall is to be loaded. One typical placement pattern puts the vertical steel 2'' (50 mm) from the inside (tension) face and the horizontal steel 18'' (450 mm) o.c. starting at the top of the slab. If CMUs are to be used, horizontal bars are usually placed at 16'' (400 mm) o.c. The layout of the section will now resemble Figure 29.28.

Representing Materials

Trim the following lines as needed and use the PROPERTIES command to assign the needed layers. See Figure 29.30 for the following steps:

Step 17. Trim and place the outline of the concrete footings, slab, and retaining walls on the *SECT MCUT* layer. Assign a very thick line to all concrete.

- Use the HATCH command to place the concrete pattern on the *SECT PATT* layer.

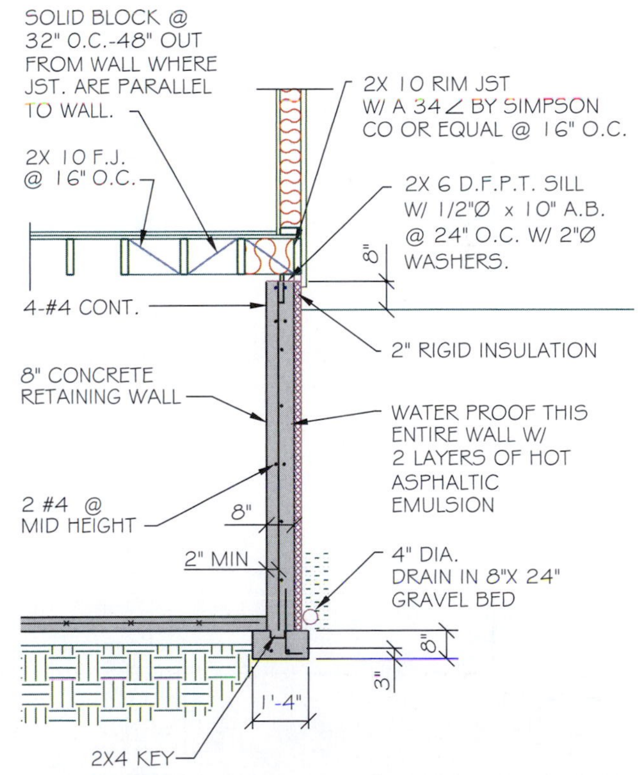

FIGURE 29.27b ■ *A section of an 8' (2400 mm) poured concrete retaining wall with joist parallel to the retaining wall. When the floor joists are parallel to the wall, blocking must be placed between the floor joists to help resist the soil loads.*

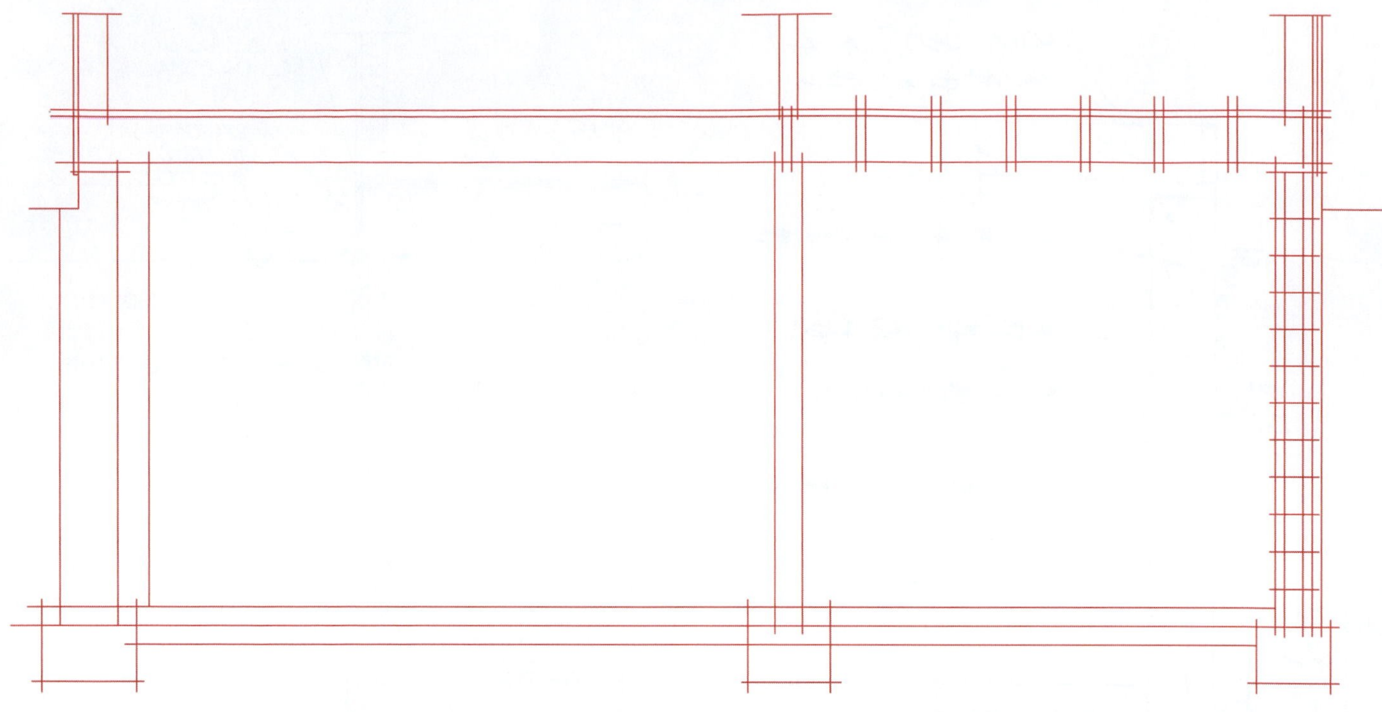

FIGURE 29.28 ■ *The initial layout of a full basement with the left wall showing a poured concrete wall. The right wall is drawn showing 8 × 8 × 16'' (200 × 200 × 400 mm) concrete blocks. The two wall systems are not used together but are shown here to illustrate construction methods.*

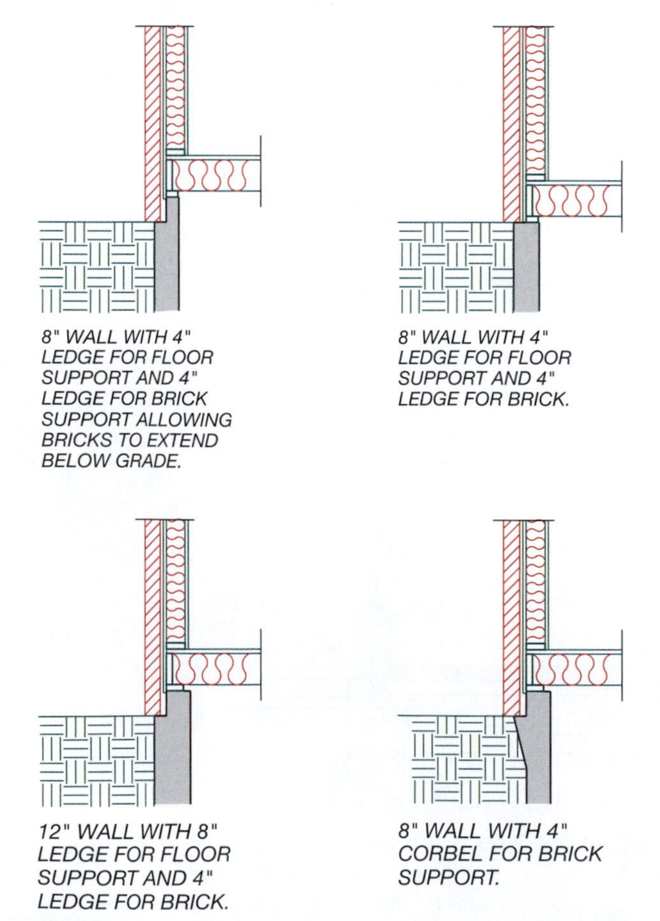

8" WALL WITH 4" LEDGE FOR FLOOR SUPPORT AND 4" LEDGE FOR BRICK SUPPORT ALLOWING BRICKS TO EXTEND BELOW GRADE.
8" WALL WITH 4" LEDGE FOR FLOOR SUPPORT AND 4" LEDGE FOR BRICK.
12" WALL WITH 8" LEDGE FOR FLOOR SUPPORT AND 4" LEDGE FOR BRICK.
8" WALL WITH 4" CORBEL FOR BRICK SUPPORT.

FIGURE 29.29 ■ *Common methods for supporting brick veneer.*

- If concrete blocks are used, use the *SECT PATT* layer to draw the division lines of the blocks and crosshatch the blocks with thin lines.

Step 18. Place the finished and natural grade lines on the *SECT MCUT* layer.

- Assign a very thick line to each grade line.
- Use a dashed line to represent the natural grade if a finished grade is also represented.

Step 19. Draw or insert the mudsills, steel rebar, and anchor bolts on the *SECT MCUT* layer. Assign a bold line type for each item.

Step 20. Draw blocking between the floor joists that are parallel to the wall on the *SECT MBND* layer. Assign a thin line type for each item.

Step 21. Trim the lines representing the floor sheathing and place the lines on the *SECT MBND* layer.

Step 22. Trim the lines representing the floor joists.

- Represent joists parallel to the wall with thick lines on the *SECT MCUT* layer.
- Represent the joists perpendicular to wall with thin lines on the *SECT MBND* layer.

All structural materials have now been represented. The drawing should resemble Figure 29.30. Use the following steps to represent the finishing materials. See Figure 29.31 for the completion of all materials.

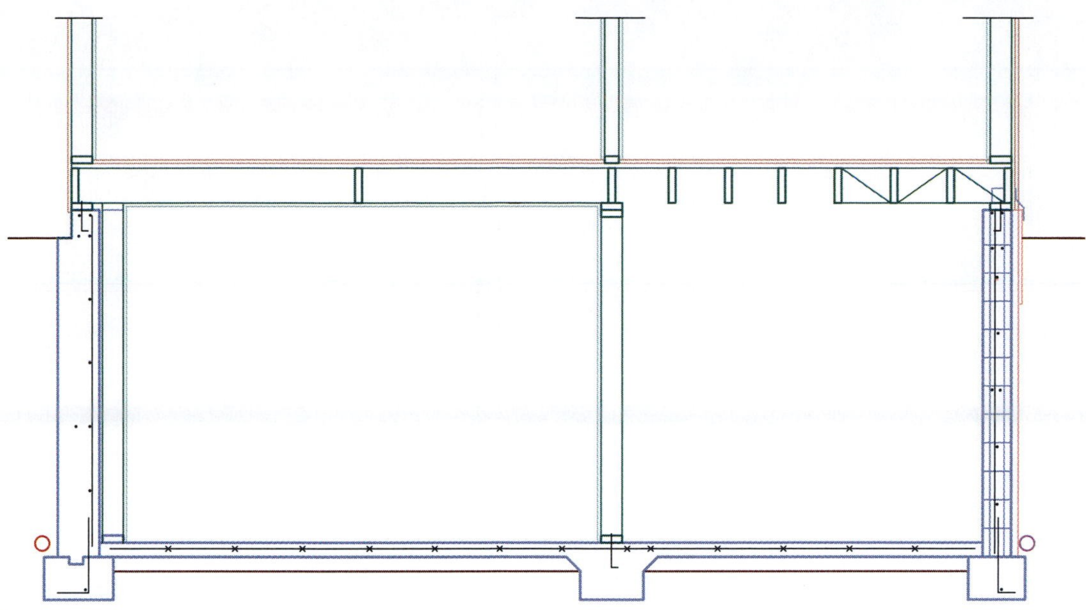

FIGURE 29.30 ■ *Structural materials drawn using the proper lineweights.*

Step 23. Use the OFFSET command to draw any required furring strips on the interior side of the retaining wall.
Step 24. Use the OFFSET command to draw any required insulation and concrete board on the exterior side of the retaining wall with thin lines on the *SECT MBND* layer.

- Place hatch patterns for the insulation on the *SECT PATT* layer.
- Place the flashing for the insulation/concrete board at the siding on the *SECT MCUT* layer using thick lines.

Step 25. Draw any required floor insulation with thin lines using the BATT linetype on the *SECT MBND* layer. Place the insulation pattern on the *SECT PATT* layer.

Drawing the following items on the *SECT MBND* layer using thin lines.

Step 26. Draw the hardboard underlayment. Note that the subfloor extends under all walls, but the hardboard underlayment does not go under any walls.
Step 27. Draw any required metal connectors, straps, or flashing.
Step 28. Draw welded wire mesh in any slabs placed over fill.
Step 29. Draw the 4'' (100 mm)-diameter drain at the intersection of the wall and the footing.
Step 30. Draw the 8 × 24'' (200 × 600 mm) gravel bed at the base of the wall.
Step 31. Draw any fill material.

- Use the HATCH command to place the gravel pattern on the *SECT PATT* layer.

Step 32. Draw and crosshatch any required brick veneer on the *SECT PATT* layer using thin lines. Your drawing should now resemble Figure 29.31.

Placing Dimensions

Dimension all structural materials so that the framing crew knows their vertical location. Figure 29.32 shows the needed dimensions. All lettering should be aligned. Use the DIM command to locate the following dimensions on the *SECT DIMS* layer:

- Finished grade to top of the retaining wall
- Floor to ceiling height
- Height from floor to finish grade—8' (2400 mm) maximum retaining
- Height of the footing
- Width of the footing
- Width of the retaining wall
- The location of the steel to the edge of the wall (2''/50 mm clear)
- Distance from the steel to the bottom of the footing (3''/75 mm clear)

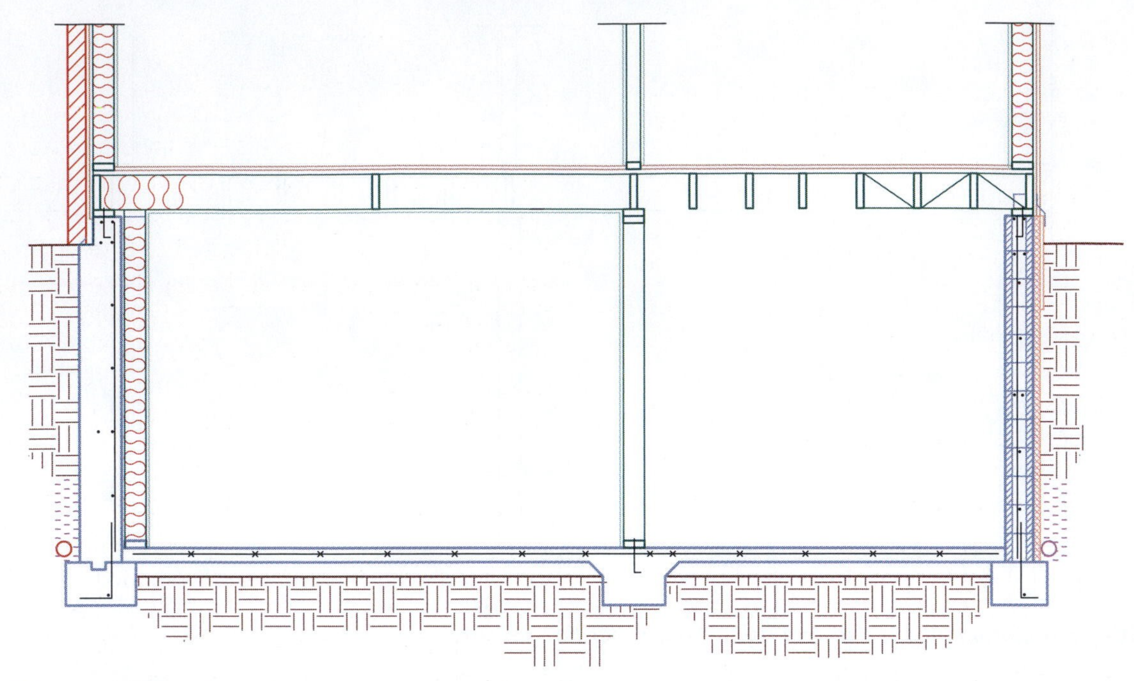

FIGURE 29.31 ■ *Representation of all finish materials for a basement system.*

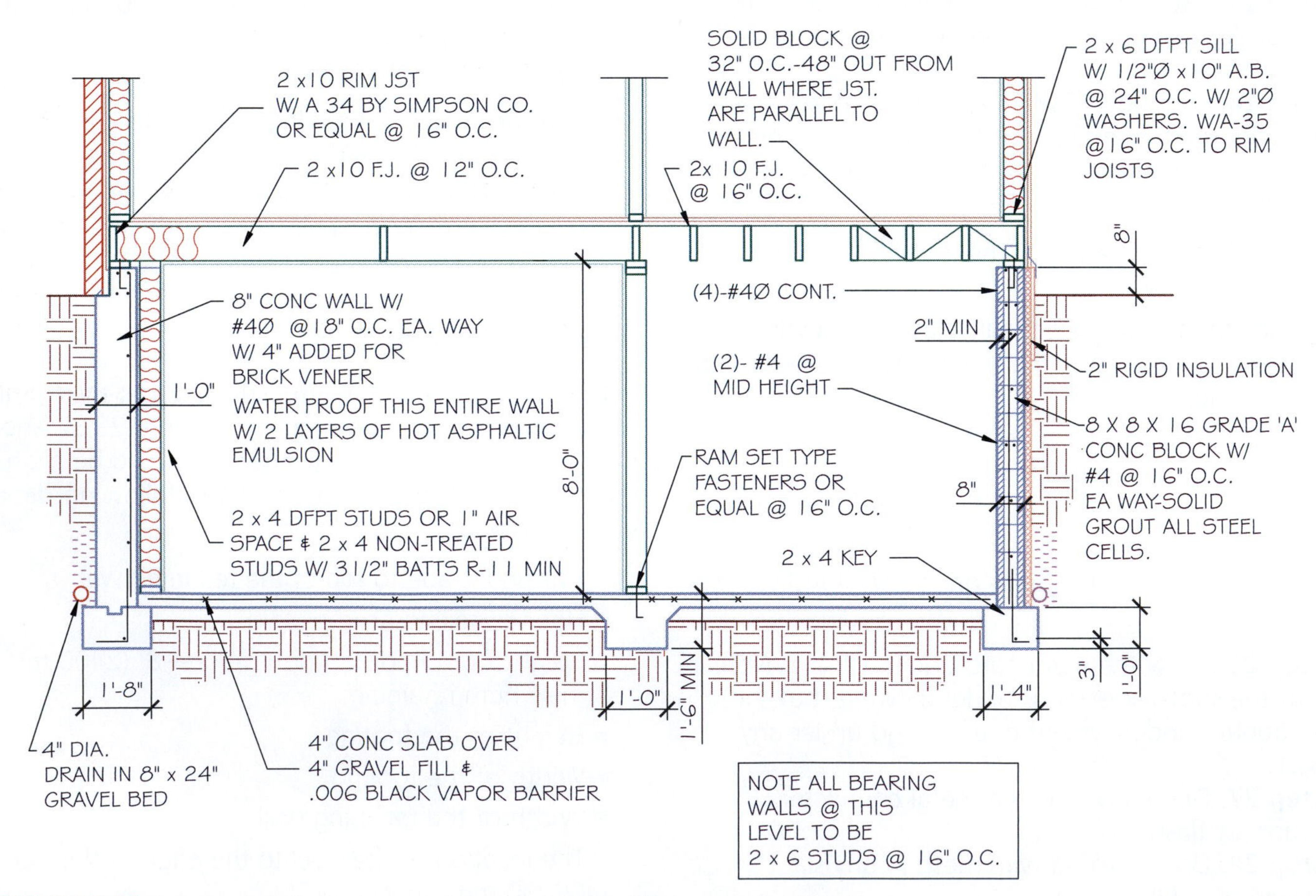

FIGURE 29.32 ■ *Required dimensions and annotation for a basement section.*

Annotation

Step 33. Letter the section using the materials specified in Figures 29.25 and 29.26. Remember that these notes are only guidelines and may vary slightly because of local standards. When complete, the drawing should resemble Figure 29.32. Place the required annotation on the *SECT ANNO* layer. Typical notes should include the following.

1. **2 × __ FLOOR JOIST @ ___" o.c.**
2. **SOLID BLOCK @ 48" O.C.—48" OUT FROM WALL WHERE JOISTS ARE PARALLEL TO WALL**
3. **2 × __ RIM JOIST W/A-34 ANCHORS BY SIMPSON CO. OR EQUAL @ 16" O.C.**
4. **2 × 6 P.T. SILL W/ 1/2" Ø × 10" A.B. @ 24" O.C. W/2" Ø WASHERS**
5. **8" CONC. WALL W/ #5 Ø @ 18" O.C. EA. WAY.**

   or

   8 × 8 × 16 GRADE "A" CONC. BLOCKS W/ #5Ø @ 16" O.C. EA. WAY—SOLID GROUT ALL STEEL CELLS
6. **WATERPROOF THIS WALL W/ 2 LAYERS OF HOT ASPHALTIC EMULSION**
7. **1/2" CONCRETE BOARD OVER 2" RIGID INSULATION W/ 26 GA. FLASHING**
8. **4" Ø DRAIN IN 8" × 24" MIN. GRAVEL BED**
9. **2 × 4 KEY**
10. **4" CONC. SLAB OVER 4" GRAVEL FILL, 1" SAND FILL & 55# FELT**

    or

    4" CONC. SLAB OVER 4" GRAVEL FILL AND 6 MIL VAPOR BARRIER, & 2" SAND FILL

    or

    4" CONC. SLAB W/ 6" × 6"—#10 × #10 WWM OVER 4" GRAVEL FILL, 6 MIL VAPOR BARRIER, & 2" SAND FILL

Trussed Framed Roofs with a Vaulted Ceiling

The following example will show the layout of a partially vaulted ceiling. The truss will be constructed using an exterior pitch of 6/12 and an interior pitch of 4/12 will be used on the left side of the roof. An interior wall will be located 12' from the right exterior wall. Use thin lines placed on the *SECT OUTL* layer for this entire stage. To determine the sizes and locations of structural members, refer to the floor and framing plans. The following roof steps are shown in Figure 29.33.

Step 1. Draw a line to represent the outer edge of one wall of the structure. Use the OFFSET command to represent the width of the building based on the dimensions provided on the framing plan.
Step 2. Draw a horizontal line to represent the ceiling of the structure.
Step 3. Use the OFFSET command to represent the interior side of each exterior wall.
Step 4. Use the OFFSET command to represent the location of any interior walls.

- Locate the center of the wall based on the dimensions from the framing plan.
- Use the offset command to locate the edges of the wall.

Step 5. Use the OFFSET command to represent any required 7'-0" (2100 mm) ceiling soffits for placement of heating ducts in areas such as kitchens, baths, or hallways. On this section, no soffits are required, since heating ducts can be run through the crawl space.
Step 6. Use the OFFSET command to place a vertical line to represent the ridge.

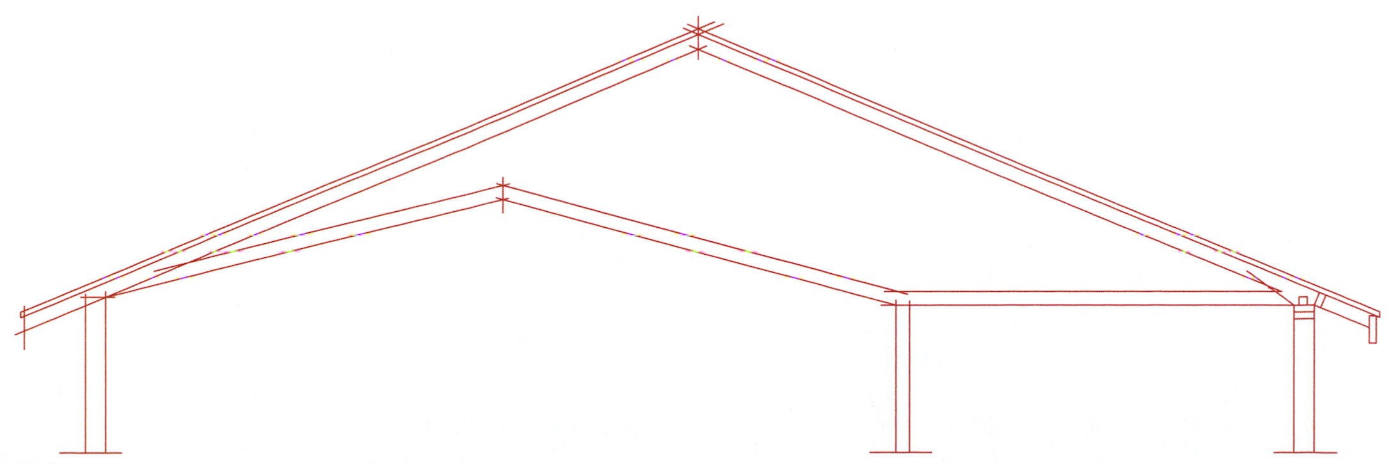

FIGURE 29.33 ■ *The layout of a standard/scissor truss section can be completed using the* OFFSET, TRIM, FILLET, *and* STRETCH *commands.*

Step 7. Use the OFFSET command to locate each overhang.
Step 8. Draw a line to represent the bottom side of the top cord.

- Starting at the intersection of the outer wall and the ceiling line, draw a line at a 6/12 pitch (26 1/2°). Because the exact length of the top chord is not known, enter an arbitrary length such as 12' (3600 mm).
- If the line is too long, it can be trimmed, or, if it is too short, extended. The most important part is to get the angle correct.

Step 9. Use the OFFSET command to represent the topside of the top chord. Assume chords to be 4'' (100 mm) deep.
Step 10. Use the OFFSET command to draw the roof sheathing.
Step 11. Draw a line to represent the topside of the bottom chord.

- For the right side of the section, the ceiling will be flat.
- For the left side of the section, start at the interior side of the exterior wall, and project a line at the required angle (18 1/2°).
- Project the right side of the vaulted ceiling from the left side of the interior wall (−18 1/2°).
- Draw a line to represent the inclined chord for the vaulted ceilings.
    - Keep the bottom chord a minimum of two pitches less than the top chord.

Step 12. Use the MIRROR command to represent the top chords and the exterior roof components for the other side of the roof.
Step 13. Draw any trusses that are perpendicular to the cutting plane. The height of these trusses can be determined by following the example in Figure 29.9.

The completed layout can be seen in Figure 29.33.

Representing Structural Members with Finished Quality Lines

Use the PROPERTIES command as needed to assign the following materials a thin linetype placed on the *SECT MBND* layer unless noted. See Figure 29.34 for an example of how the section should appear.

Step 1. Use the OFFSET command to represent the roof sheathing with thin lines.
Step 2. Trim the lines representing the top and bottom truss chords as needed.

Use the PROPERTIES command as needed to assign the following materials a thick linetype placed on the *SECT MCUT* layer.

Step 3. Draw the eave blocking.
Step 4. Draw the fascia.
Step 5. Use the MIRROR command to place the items from steps 1 through 4 on the opposite side of the section.
Step 6. Draw the ridge block.
Step 7. Draw any required blocking.
Step 8. Draw any trusses that are perpendicular to the cutting plane on the *SECT MBND* layer.

- Use thick lines to represent the chords and thin lines to represent the webs.
    - Note: Many professionals do not draw the webs and rely on the drawings provided by the truss manufacturer to detail the construction of the trusses.
- Draw one truss and use the ARRAY command to locate the others.
- Use The STRETCH command to adjust the height of the valley trusses.

Your drawing should now resemble Figure 29.34.

Representing the Finish Materials

The materials drawn in this stage seal the exterior from the weather and cover the interior side of the framing materials. Start at the roof and work down to the foundation. Place all items in these steps on the *SECT MBND* layer using thin lines for the entire stage unless otherwise noted. See Figure 29.35 for the following steps.

Step 9. Draw hurricane ties approximately 3'' (75 mm) square at the truss/ top plate intersection for each bearing wall.
Step 10. Draw ridge vents on the back side of the roof that are approximately 3'' high × 12'' long (75 × 300 mm).
Step 11. Draw any required metal connectors such as joist or beam hangers.

Place the following items on the *SECT PATT* layer using thin lines.

Step 12. Draw insulation using the BATT linetype or your own pattern.

- Use approximately 10'' (250 mm) deep batts in flat ceilings.

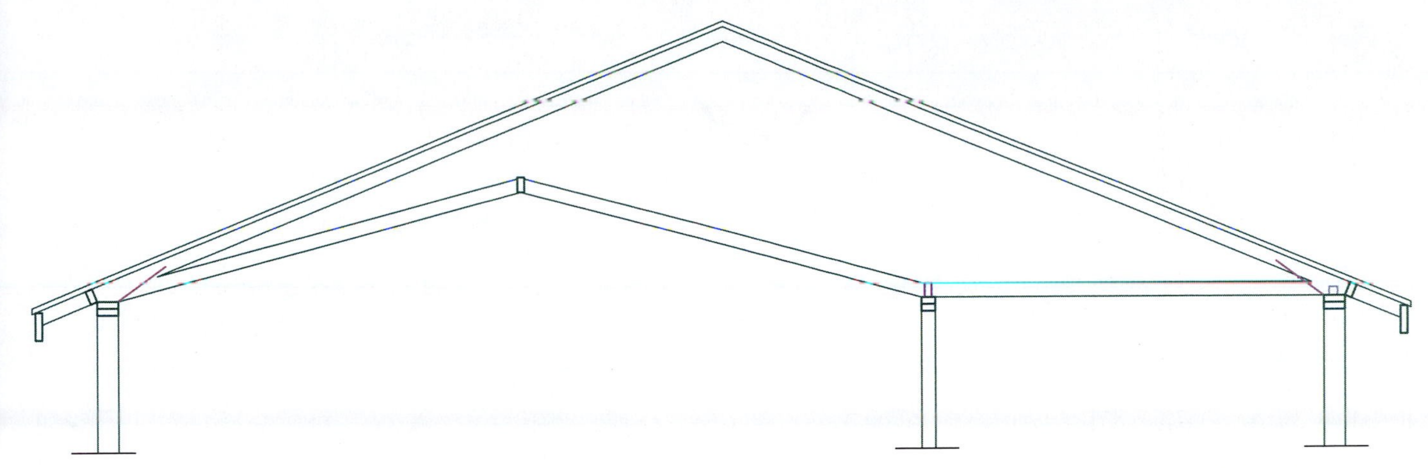

FIGURE 29.34 ■ *A section showing the structural members of a standard/scissor truss system drawn with finished-quality lines. Varied lineweights should be used to represent continuous materials that the cutting plane passes through. Thin lines should be used to represent materials that lie beyond the cutting plane.*

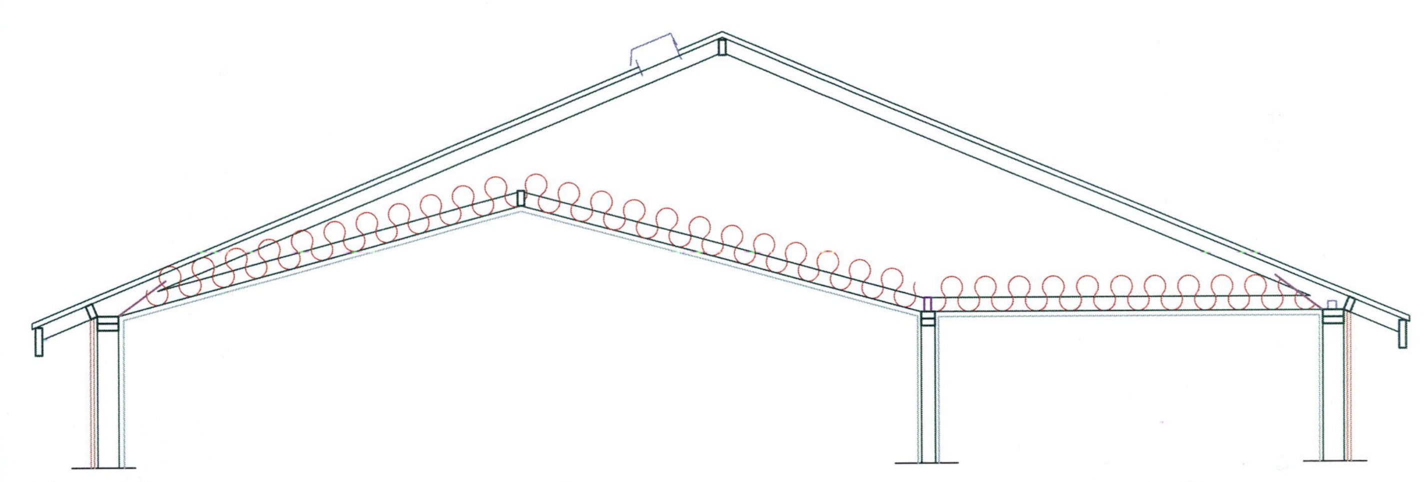

FIGURE 29.35 ■ *Representation of the finish materials include the insulation, vents, truss clips, and baffles.*

- Use approximately 10'' (250 mm) deep batts with 2'' (50 mm) air space above in vaulted ceilings.
- Use 4 or 6'' (100 × 150 mm) batts in exterior wall cavities.

Place the following items on the *SECT MCUT* layer using thick lines.

Step 13. Draw baffles at the edge of the insulation. Leave approximately 2'' (50 mm) at the top of the baffle for air flow over the insulation.

Step 14. Draw gutters that are approximately 3 × 3 (75 × 75).

Dimensioning

Provide dimensions so that the framing crew knows the vertical location of each material. Figure 29.36 shows the needed dimensions. All dimension text should be aligned and 1/8'' (3 mm) high when plotted. Use the DIM command to locate the following dimensions on the *SECT DIMS* layer:

- Eave overhangs
- Exterior roof pitch
- Roof pitch of the vaulted ceiling

Annotation

Label the materials in the section using the following notes. Remember that these notes are only guidelines and may vary slightly because of local standards. When complete, the drawing should resemble Figure 29.36. Typical notes should include the following:

1. **STD./SCISSOR TRUSSES @ 24'' O.C. SEE DRAW. BY MANUF. SUBMIT TRUSS DRAWINGS TO THE BUILDING DEPARTMENT PRIOR TO ERECTION**

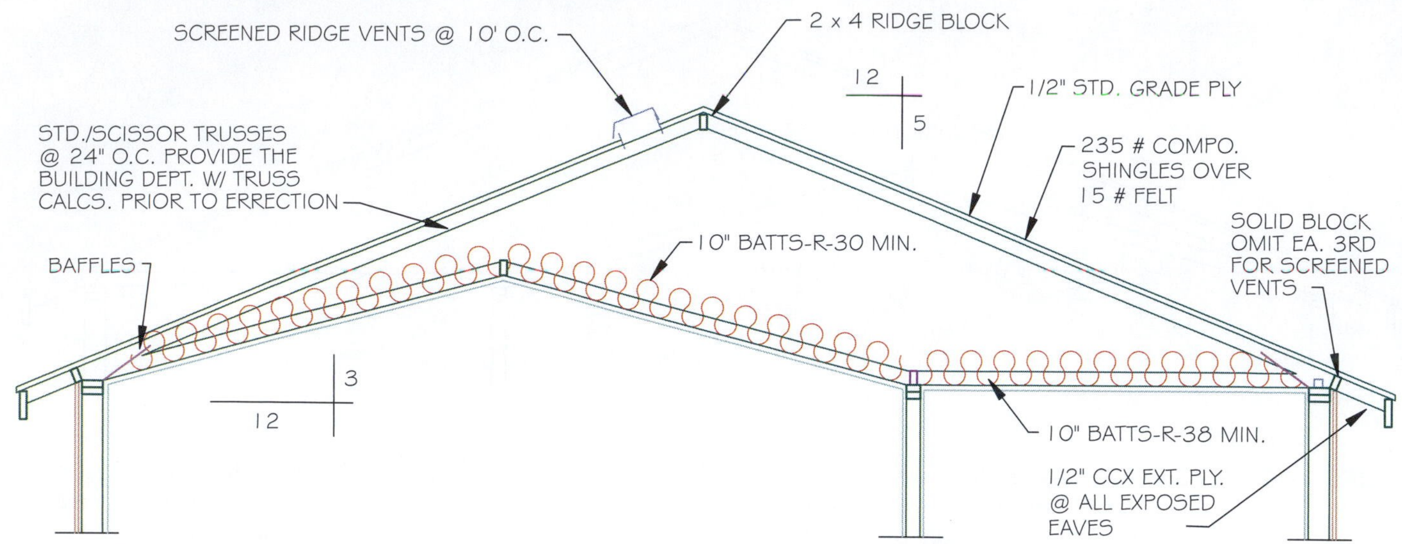

FIGURE 29.36 ■ *Representing the dimensions and annotation required to describe minimum clearances and all structural and finish materials.*

Provide a similar note for other types of trusses such as valley, girder, hip, stub, and scissor trusses.

*2. For flat ceiling: **10 BATTS, R-38 PAPER FACE @ HEATED SIDE**

or

__" BLOWN-IN INSULATION R-__ MIN.

For vaulted ceiling: **10" BATTS, R-30 MIN.**

3. **BAFFLES AT EAVE VENTS**
4. **SOLID BLOCK—OMIT EA. 3RD FOR SCREENED VENTS**
5. **TRUSS CLIPS @ EA. TAIL TO PLATE**
6. **__ × __ FASCIA W/ GUTTER**

*7. **1/2" 'CCX' EXT. PLY @ ALL EXPOSED EAVES**

or

1 × 4 T & G DECKING @ ALL EXPOSED EAVES

or

1/2" R.S. PLY SOFFIT COVER W/ CONTINUOUS VENTING

Conventional Roof Framing

The following example will show the layout of a roof framed with conventional framing methods. The roof will be constructed using a pitch of 6/12. An interior wall will be located 12' from the right exterior wall. Use thin lines placed on the *SECT OUTL* layer for this entire stage. To determine the sizes and locations of structural members, refer to the floor and framing plans. The following roof steps are shown in Figure 29.37.

Step 1. Draw a line to represent the outer edge of one wall of the structure. Use the OFFSET command to represent the width of the building based on the dimensions provided on the framing plan.
Step 2. Draw a horizontal line to represent the ceiling of the structure.

Step 3. Use the OFFSET command to represent the interior side of each exterior wall.
Step 4. Use the OFFSET command to represent the location of any interior walls.

- Locate the center of the wall based on the dimensions from the framing plan.
- Use the offset command to locate the edges of the wall.

Step 5. Use the OFFSET command to represent any required 7'-0'' (2100 mm) ceiling soffits for placement of heating ducts in areas such as kitchens, baths, or hallways.
Step 6. Use the OFFSET command to place a vertical line to represent the ridge.
Step 7. Use the OFFSET command to locate each overhang.
Step 8. Draw a line to represent the bottom side of the rafters.

- Starting at a point 3'' (75 mm) in from the outer edge of the exterior wall and the ceiling line, draw a line at a 6/12 pitch (26 1/2°). Because the exact length of the rafter is not known, enter an arbitrary length such as 12' (3600 mm).
- If the line is too long, it can be trimmed, or, if it is too short, extended. The most important part is to get the angle correct.

Step 9. Use the OFFSET command to represent the top side of the rafter. Base the thickness on the minimum standards that were presented at the start of the chapter.
Step 10. Use the OFFSET command to draw the roof sheathing.

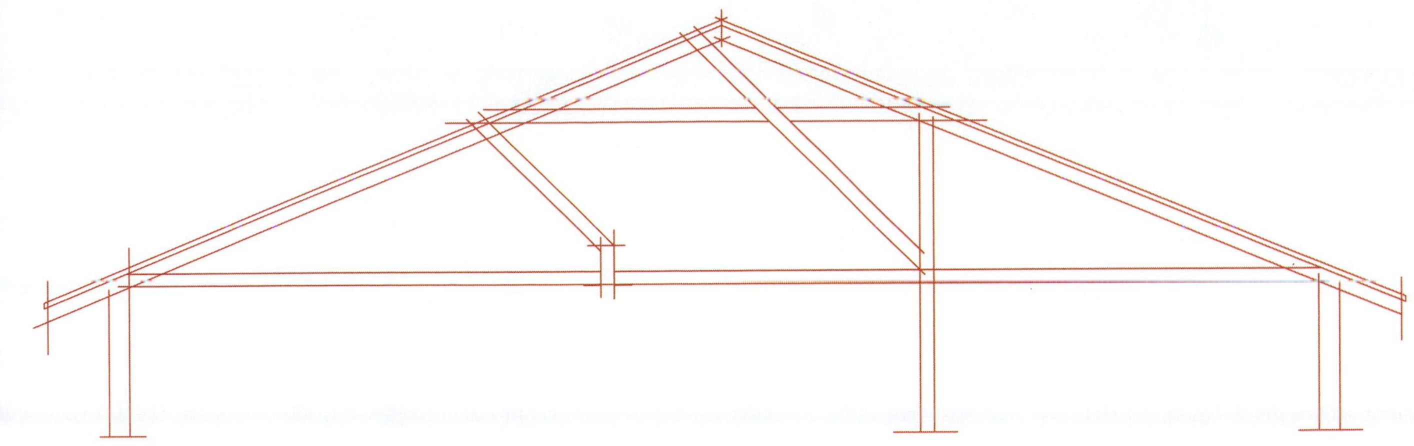

FIGURE 29.37 ■ *The layout of a conventionally framed roof section can be completed using the* OFFSET, TRIM, FILLET, *and* STRETCH *commands.*

Step 11. Use the OFFSET command to draw the top of the ceiling joists.
Step 12. Use the MIRROR command to represent the roof components on the other side of the section.
Step 13. Draw any trusses that are perpendicular to the cutting plane.
Step 14. Draw the ridge brace and support.
Step 15. Draw the purlin braces and supports.
Step 16. Draw the strong backs.
Step 17. Draw the purlins.
Step 18. Draw the collar ties. The depth of the collar ties should match or exceed the depth of the rafters.

The completed layout can be seen in Figure 29.37.

Representing Materials with Finished Quality Lines

To gain speed, draw features on one side of the roof and then use the MIRROR command to complete the opposite side of the section. See Figure 29.38 for the following steps. Use the PROPERTIES command as needed to assign the following materials a thin linetype placed on the *SECT MBND* layer.

Step 1. Use the OFFSET command to represent the roof sheathing with thin lines.
Step 2. Trim the lines representing the top and bottom of the rafters as needed and place these lines on the *SECT MBND* layer.

Use the PROPERTIES command as needed to assign the following materials a thick linetype placed on the *SECT MCUT* layer.

Step 3. Draw the eave blocking.
Step 4. Draw the fascia.
Step 5. Use the MIRROR command to place the items from steps 1 through 4 on the opposite side of the section.
Step 6. Draw any rafters that are perpendicular to the cutting plane on the *SECT MBND* layer.

- Use thick lines to represent the rafter and the ceiling joists where they intersect the cutting plane.
- Use thin lines to represent the rafters where it is beyond the cutting plan.
- Use the ARRAY command to locate the others.
- Use The STRETCH command to adjust the height of the rafters.

Use the PROPERTIES command as needed to assign the following materials a thick linetype placed on the *SECT MCUT* layer.

Step 7. Draw the ridge board so that it is 2'' (50 mm) deeper than the rafters.
Step 8. Draw any required blocking.
Step 9. Draw the strong backs.
Step 10. Draw all plates.
Step 11. Draw all purlins.

Use the PROPERTIES command as needed to assign the following materials a thin linetype placed on the *SECT MBND* layer.

Step 12. Draw all purlin and ridge braces.
Step 13. Draw the collar ties.

Your drawing should now resemble Figure 29.38.

Representing the Finish Materials

The materials drawn in this stage seal the exterior from the weather and cover the interior side of the framing materials. Start at the roof and work down to the foundation. Place all items in these steps on the *SECT MBND* layer using thin lines for the entire stage

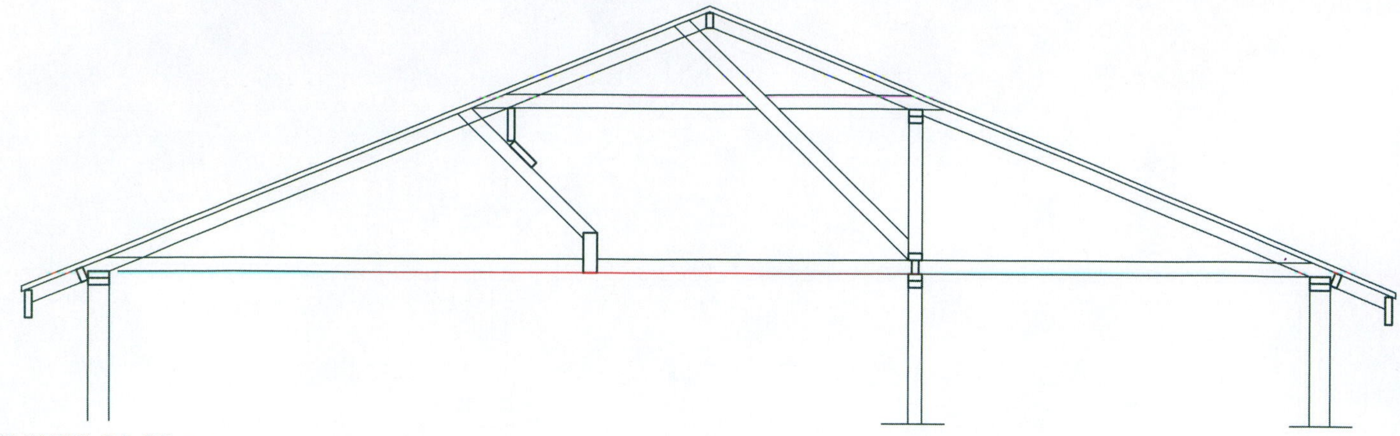

FIGURE 29.38 ■ *Representation of the interior supports for a conventionally framed roof. The strong back (step 9) can be used in place of a bearing wall or a flush header.*

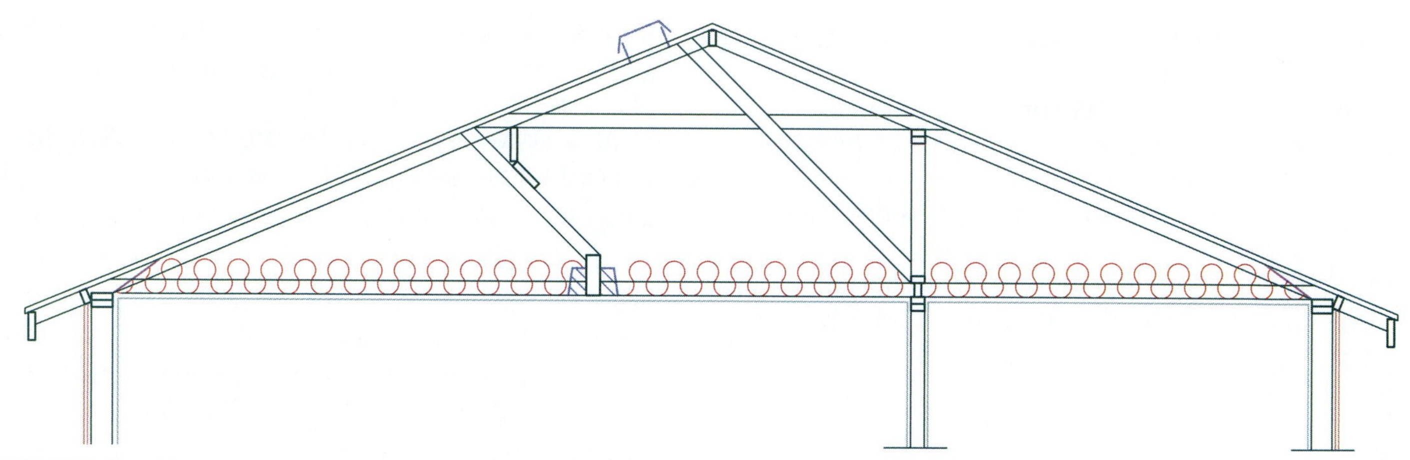

FIGURE 29.39 ■ *The completed representation of the finishing materials. The purlin, purlin brace, and collar ties may or may not be required, depending on the width of the structure and the rafter material and size.*

unless otherwise noted. See Figure 29.39 for the following steps.

Step 14. Draw the finished ceiling and wall materials.
Step 15. Draw ridge vents on the back side of the roof that are approximately 3'' high × 12'' long (75 × 300 mm).
Step 16. Draw any required metal connectors such as joist or beam hangers.

Place the following items on the *SECT PATT* layer using thin lines.

Step 17. Draw insulation using the BATT linetype or your own pattern.

- Use batts approximately 10'' (250 mm) deep in flat ceilings.
- Use 4 or 6'' (100 × 160 mm) batts in exterior wall cavities.

Place the following items on the *SECT MCUT* layer using thick lines.

Step 18. Draw baffles at the edge of the insulation. Leave approximately 2'' (50 mm) at the top of the baffle for air flow over the insulation.
Step 19. Draw gutters that are approximately 3 × 3 (75 × 75).

Dimensions

Provide dimensions so that the framing crew knows the vertical location of each material. Figure 29.40 shows the needed dimensions. All dimension text should be aligned and 1/8'' (3 mm) high when plotted. Use the DIM command to locate the following dimensions on the *SECT DIMS* layer:

- Eave overhangs

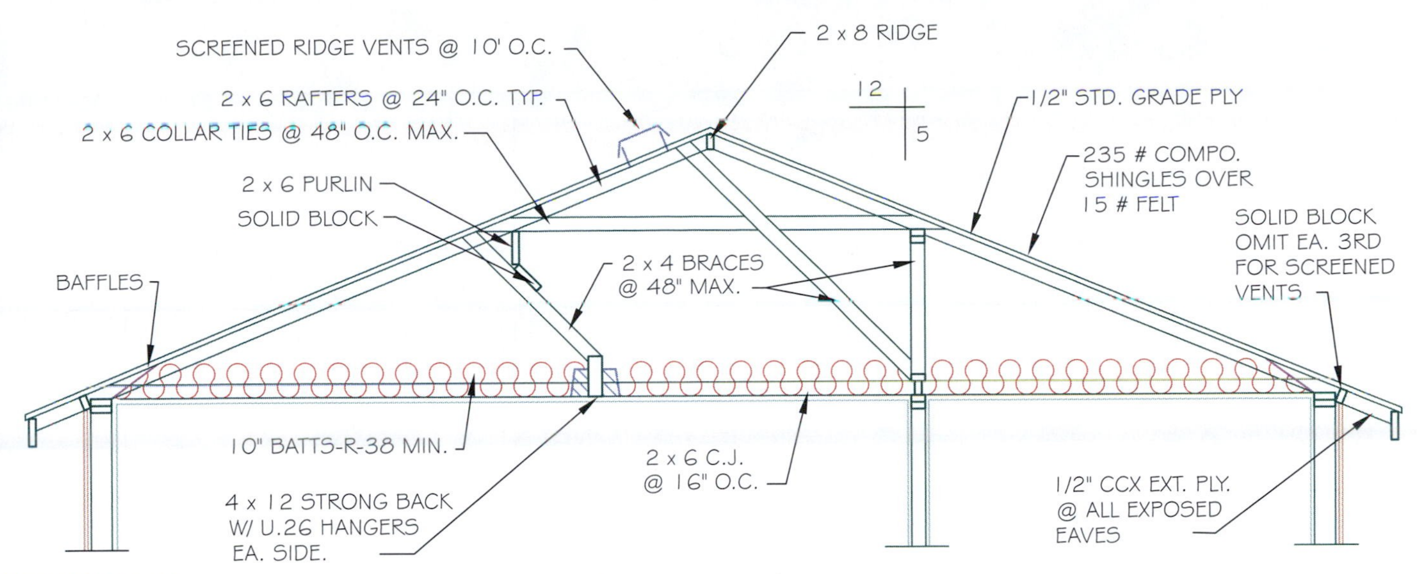

FIGURE 29.40 ■ *Notes are required to describe all framing materials.*

- Exterior roof pitch
- Roof pitch of the vaulted ceiling

Notes

In addition to the notes presented in the first section of this chapter, see Figure 29.40 for the notes needed to describe the interior supports. Roof notes should include the following materials on the *SECT ANNO* layer:

1. **2 × __ RIDGE BOARD**
2. **2 × __ RAFTERS @ ___ O.C.**
3. **2 × __ COLLAR TIES @ 48'' O.C. MAX.**
4. **2 × __ PURLIN**
5. **2 × 4 BRACE @ 48'' O.C. MAX.**

Stick-Framed Vaulted Ceilings

The following example will show the layout of roof framed with conventional framing methods with a partial vaulted ceiling. If the entire roof level is to have vaulted ceilings, the layout process is very similar to the procedure for the layout of a stick roof with flat ceilings. Usually though, part of the roof has standard 2× (50×) rafters with flat ceilings and a vaulted ceiling using 2 × 12 (50 × 300) rafter/ceiling joists. In order to make the two roofs align on the outside, the plate must be lowered in the area having the vaulted ceilings.

For this example, the roof will be constructed using a pitch of 6/12. An interior wall will be located 12' from the right exterior wall. The right side of the structure will have standard construction and the left side of the residence will have vaulted ceilings. A 6 × 14 ridge will be used.

Use thin lines placed on the *SECT OUTL* layer for this entire stage. To determine the sizes and locations of structural members, refer to the floor and framing plans. The following roof steps are shown in Figure 29.41.

Step 1. Draw a line to represent the outer edge of one wall of the structure. Use the OFFSET command to represent the width of the building based on the dimensions provided on the framing plan.
Step 2. Draw a horizontal line to represent the ceiling of the structure.
Step 3. Use the OFFSET command to represent the interior side of each exterior wall.
Step 4. Use the OFFSET command to represent the location of any interior walls.

- Locate the center of the wall based on the dimensions on the framing plan.
- Use the offset command to locate the edges of the wall.

Step 5. Use the OFFSET command to represent any required 7'-0'' (2100 mm) ceiling soffits for placement of heating ducts in areas such as kitchens, baths, or hallways.
Step 6. Use the OFFSET command to place a vertical line to represent the ridge.

- Offset as required to represent each edge of the beam.

Step 7. Use the OFFSET command to locate each overhang.
Step 8. Draw a line to represent the bottom side of the rafters.

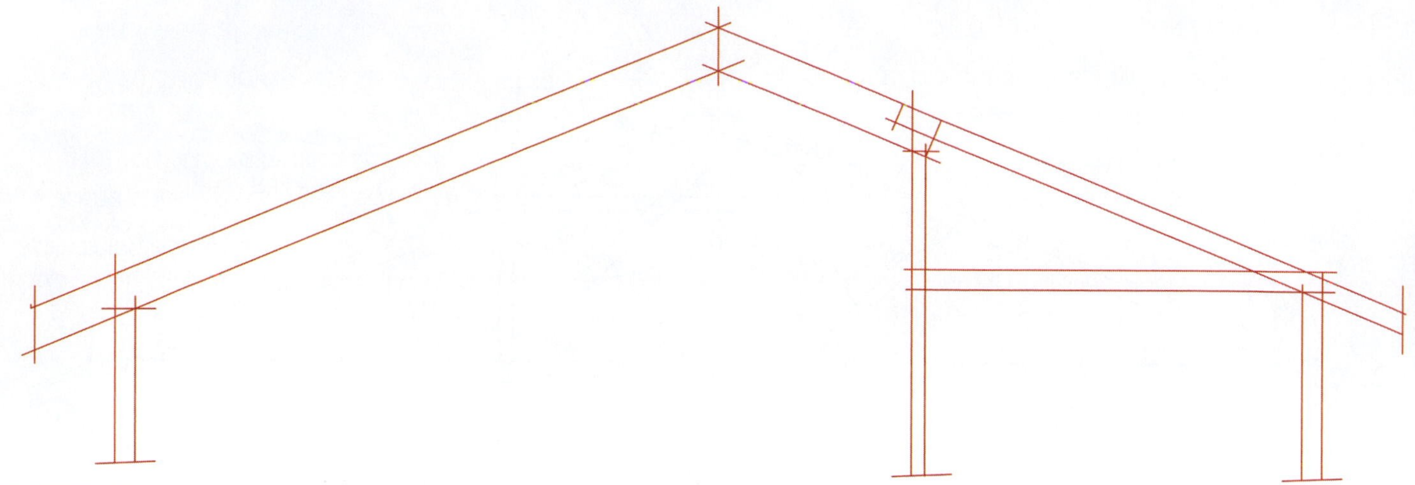

FIGURE 29.41 ■ *Layout of a stick-framed vaulted ceiling.*

- Use 2 × 12 (50 × 300 mm) rafters on the left side of the home. Use rafter sizes from the minimum standards presented earlier in this chapter for the right half of the home.
- Starting at a point 3'' (75 mm) in from the outer edge of the exterior wall and the ceiling line, draw a line at a 6/12 pitch (26 1/2°). Because the exact length of the rafter is not known, enter an arbitrary length such as 12' (3600 mm).
- If the line is too long, it can be trimmed, or, if it is too short, extended. The most important part is to get the angle correct.

Step 9. Use the OFFSET command to represent the top side of the rafter. Base the thickness on the minimum standards that were presented at the start of the chapter.
Step 10. Use the OFFSET command to draw the roof sheathing.
Step 11. Use the OFFSET command to draw the top of the ceiling joists where needed.
Step 12. Use the MIRROR command to represent the other side of the roof components.
Step 13. Draw any rafters or trusses that are perpendicular to the cutting plane.
Step 14. Draw the ridge brace and support.
Step 15. Draw the purlin braces and supports if required.
Step 16. Draw the strong backs if required.
Step 17. Draw the purlins if required.
Step 18. Draw the collar ties if required. The depth of the collar ties should match or exceed the depth of the rafters.

The completed layout can be seen in Figure 29.41.

Representing Materials with Finished Quality Lines

Use the PROPERTIES command as needed to assign the following materials a thin linetype placed on the *SECT MBND* layer. See Figure 29.42. for an example of how the section should appear.

Step 1. Use the OFFSET command to represent the roof sheathing with thin lines.
Step 2. Trim the lines representing the top and bottom of the rafters as needed.

Use the PROPERTIES command as needed to assign the following materials a thick linetype placed on the *SECT MCUT* layer.

Step 3. Draw the ridge beam.
Step 4. Draw the eave blocking.
Step 5. Draw the fascia.
Step 6. Draw or insert all plates.
Step 7. Draw any required blocking.
Step 8. Use the MIRROR command to place the items from steps 1 through 7 on the opposite side of the roof.
Step 9. Draw any rafters that are perpendicular to the cutting plane on the *SECT MBND* layer.

- Use thick lines to represent the rafter and the ceiling joists where they intersects the cutting plane.
- Use thin lines to represent the rafter where it is beyond the cutting plane.
- Use the ARRAY command to locate other required rafters.
- Use the STRETCH command to adjust the height of the rafters.

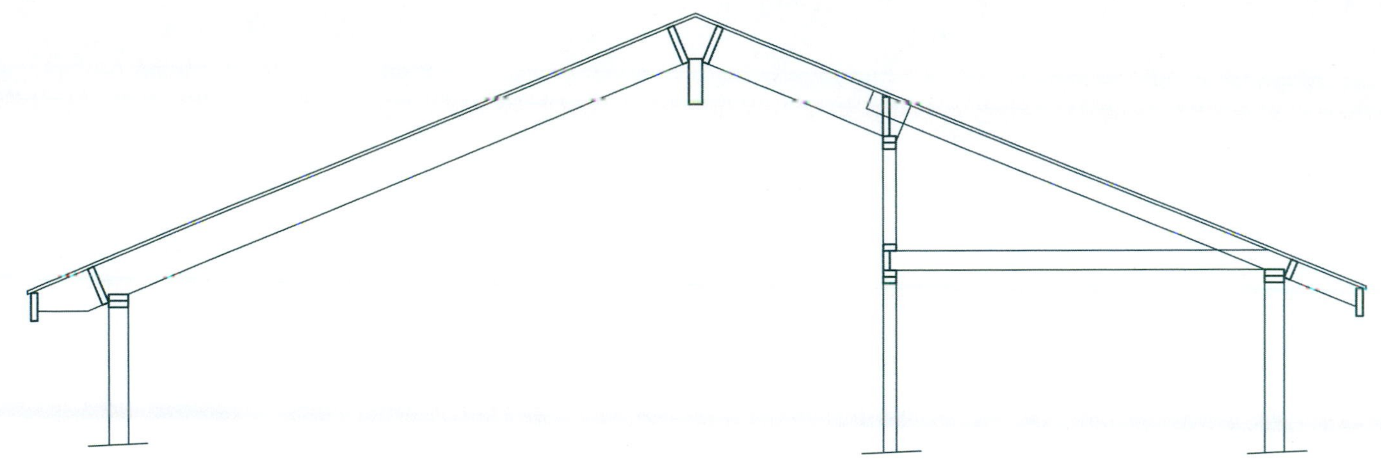

FIGURE 29.42 ■ *Representing the structural materials for a vaulted roof requires varied line weights.*

Use the PROPERTIES command as needed to assign the following materials a thick linetype placed on the *SECT MCUT* layer.

Step 10. Draw or insert the strong backs if required.
Step 11. Draw or insert all required purlins.

Use the PROPERTIES command as needed to assign the following materials a thin linetype placed on the *SECT MBND* layer.

Step 12. Draw all required purlin braces.
Step 13. Draw all required collar ties.

All structural materials have now been drawn and should now resemble Figure 29.42

Representing the Finish Materials

The materials drawn in this stage seal the exterior from the weather and cover the interior side of the framing materials. Start at the roof and work down to the foundation. Place all items in these steps on the *SECT MBND* layer using thin lines for the entire stage unless otherwise noted. See Figure 29.43 for the following steps.

Step 14. Draw the finished ceiling and wall materials.
Step 15. Draw ridge vents that are approximately 3'' high × 12'' long (75 × 300 mm).
Step 16. Draw any required metal connectors such as joist or beam hangers.

Place the following items on the *SECT PATT* layer using thin lines.

Step 17. Draw insulation using the BATT linetype or your own pattern.

- Use approximately 10'' (250 mm)-deep batts in flat ceilings.
- Use 4 or 6'' (100 × 150 mm) batts in exterior wall cavities.
- Place the following items on the *SECT MCUT* layer using thick lines.

Step 18. Draw baffles at the edge of the insulation. Leave approximately 2'' (50 mm) at the top of the baffle for air flow over the insulation.
Step 19. Draw gutters that are approximately 3 × 3 (75 × 75).

Your drawing should now resemble Figure 29.43.

Dimensioning

Provide dimensions so that the framing crew knows the vertical location of each material. Figure 29.44 shows the needed dimensions. All dimension text should be aligned and 1/8'' (3 mm) high when plotted. Use the DIM command to locate the following dimensions on the *SECT DIMS* layer:

- Eave overhangs
- Exterior roof pitch
- Roof pitch of the vaulted ceiling
- Plate height supporting the vaulted ceilings. List this dimension as **7'-6'' ± VERIFY AT JOB SITE.** By telling the framer to verify the height at the job site, you are alerting the framer to a problem. This is a problem that can be solved much more easily at the job site than at a computer station.

Annotation

See the notes presented earlier in this chapter for a complete listing of notes that apply to the roof. In addition to those notes, see Figure 29.43 for the notes needed to describe the roof completely. Place any re-

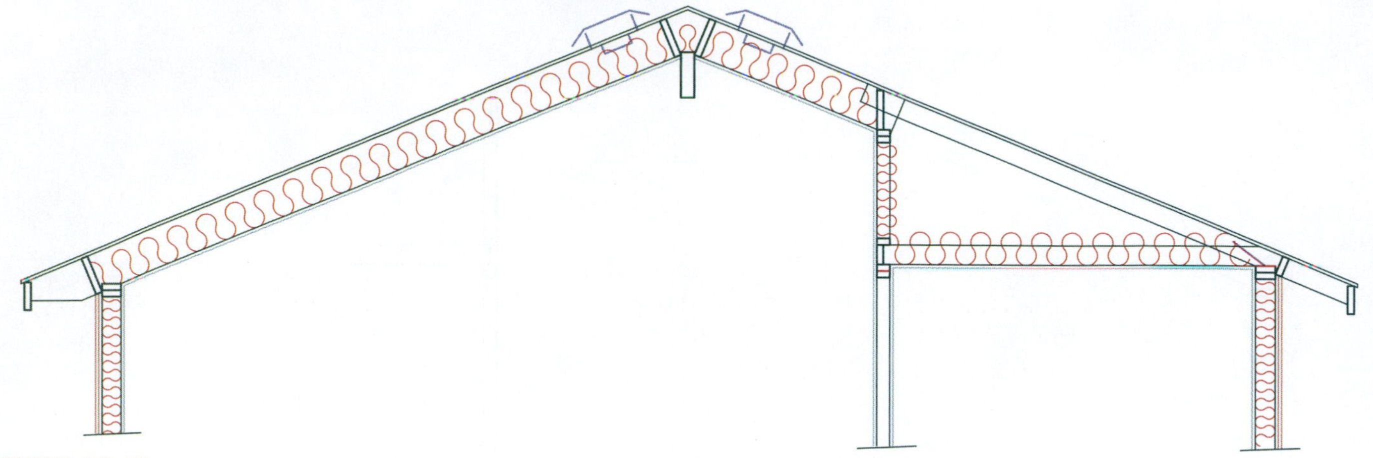

FIGURE 29.43 ■ *Representing the finishing materials for a vaulted roof; this includes drawing the ridge vents, insulation, and metal hangers.*

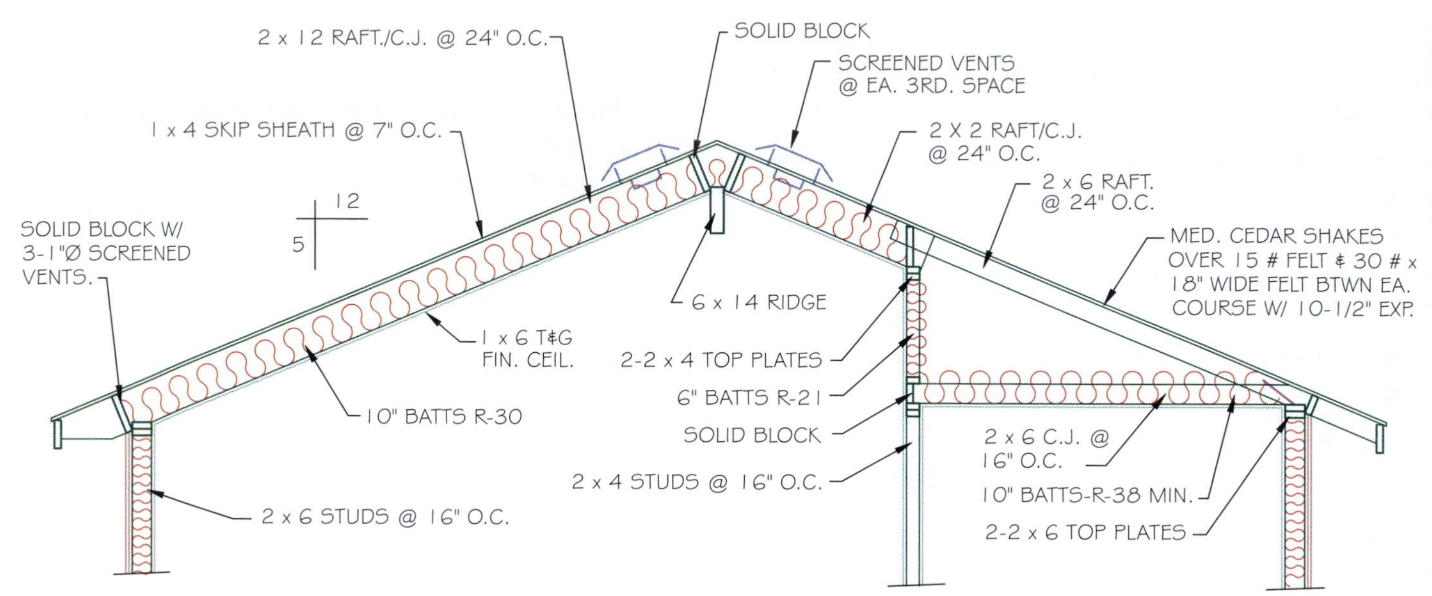

FIGURE 29.44 ■ *Annotation for a vaulted ceiling.*

quired notes on the *SECT ANNO* layer. The notes should include the following:

1. **__ × __ DFL# __ RIDGE BEAM**
2. **2 × 12 RAFT./C.J. @ __ '' O.C.**
3. **PROVIDE A 1 1/2'' × 3'' NOTCH AT EA. RAFT. FOR AIR FLOW @ RIDGE VENTS**
4. **U-210 JST. HGR. BY SIMPSON CO. OR EQUAL**
5. **SOLID BLOCK W/ 3–1'' DIA SCREENED VENTS @ EA. SPACE**
6. **NOTCH RAFT. TAILS AS REQD.**
7. **LINE OF INTERIOR FINISH**
8. **10" BATTS—R-30 MIN. W/ 2'' AIR SPACE ABOVE**

CHAPTER 29

Section Layout Test

DIRECTIONS

Answer the following questions with short, complete statements. Type your answers using a word processor or neatly print your answers on lined paper.

1. Place your name, the chapter number, and the date at the top of the sheet.
2. Type the question number and provide the answer in the form of a statement that includes part of the question. You do not need to write out the entire question.

Note: The answers to some questions may not be contained in this chapter and will require you to do additional research.

QUESTIONS

29.1. Define the following terms:
- a. Rafter
- b. Truss
- c. Ceiling joist
- d. Collar tie
- e. Jack stud
- f. Rim joist
- g. Chord
- h. Sheathing

29.2. On a blank sheet of paper, sketch a section view of a conventionally framed roof showing all interior supports.

29.3. Give the typical sizes of the following materials:
- a. Mudsill
- b. Stud height
- c. Roof sheathing
- d. Wall sheathing
- e. Floor decking for a post and beam floor
- f. Underlayment
- g. Floor decking for a joist floor system
- h. Support piers for a post-and-beam floor system
- i. Bird blocking width
- j. Rafter/ceiling joist size

29.4. What is the thickness of the vapor barrier used under a crawl space?

29.5. The top of the concrete slab must be ____'' above the finished grade.

29.6. Basement walls are typically ____'' wide.

29.7. Describe a common blocking pattern that is used to support a joist floor at the top of a concrete retaining wall.

29.8. Sketch and label the framing for a stick-framed floor showing typical exterior and interior supports.

29.9. Use the drawing on the student CD and identify each part of the drawing for this question.

29.10. Use the drawing on the student CD and identify each part of the drawing for this question.

PROBLEMS

Use your site, framing, roof, and foundation plans to evaluate the construction methods used and determine which sections and partial sections need to be drawn. Lay out a freehand sketch of the sections you will need, showing all wall locations, floor supports, and roof construction. Include sections showing stair, retaining walls, cantilevers, and masonry fireplace construction if required. Use a drawing template that will match the floor plan to lay out and draw all needed sections. Choose a scale based on the reading material in this section and use the examples given in each chapter to help you complete your drawings.

CHAPTER 30

Stair Construction and Layout

More than a means of traveling from floor to floor, stairs similar to those in Figure 30.1 can provide an elegant focal point to a home. Stairs were introduced with floor plans in Section 3. Minimal information was provided in that section—only enough so that you could draw stairs on floor plans. This chapter introduces methods for drawings stairs in section. Commands such as ARRAY, OFFSET, TRIM, and FILLET can be used to quickly reproduce repetitive elements of the stair. These commands are introduced as the step-by-step instructions are given. Steps are provided to aid in the planning, layout, and drawing of straight-run, open, and L- and U-shaped stairways.

This chapter assumes the use of AutoCAD to draw the stair section. The section is even easier to draw with a parametric CAD system. Parametric systems require the starting point of the stairs, the type of stair construction to be used, rise and run, and stair width and direction to be given before the program automatically calculates the rise of each step. The program asks you to provide handrails with or without balusters and gives you several options for handrail ends. After the required information has been provided, the stair is drawn automatically.

STAIR TERMINOLOGY

There are several basic terms you will need to be familiar with in working with stairs. Each is shown in Figure 30.2:

- ***Run:*** Horizontal distance from end to end of the stairs.
- ***Rise:*** Vertical distance from top to bottom of the stairs.
- ***Tread:*** Horizontal step of the stairs. It is usually made from 1'' (25 mm) material on enclosed stairs and 2'' (50 mm) material on open stairs. Tread width is the measurement from the face of the riser to the nosing. The nosing is the portion of the tread that extends past the riser.
- ***Riser:*** Vertical backing between the treads. It is usually made from 1'' (25 mm) material for enclosed stairs and is not used on open stairs.
- ***Stringer*** or ***stair jack:*** Support for the treads. A 2 × 12 (50 × 300) notched stringer is typically used for enclosed stairs. For an open stair, a 4 × 14 (100 × 350) is common, but sizes vary greatly. Figure 30.3 shows stringers, risers, and treads.
- ***Kick block*** or ***kicker:*** Used to keep the bottom of the stringer from sliding on the floor when downward pressure is applied to the stringer.
- ***Headroom:*** Vertical distance measured from the tread nosing to a wall or floor above the stairs. Building codes specify a minimum size.
- ***Handrail:*** Railing that you slide your hand along as you walk down the stairs.

FIGURE 30.1 ■ *In addition to being a key component in the traffic flow of a home, stairs can be used to add elegance.* Courtesy Jordan Jefferis.

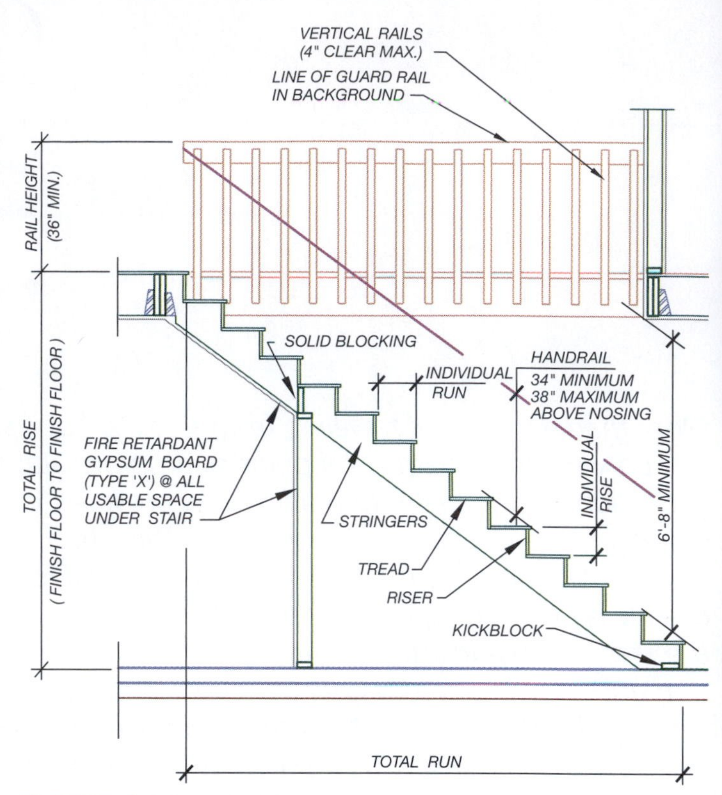

FIGURE 30.2 ■ *Common terms used to describe stairs.*

- ***Guardrail:*** Railing placed around an opening in the floor for the stairs.

Gypsum (GYP.) board 1/2'' (13 mm) thick is required by the IRC for enclosing all usable storage space under the stairs. Figure 30.4 shows common stair dimensions based on the IRC.

DETERMINING RISE AND RUN

The IRC dictates the maximum rise of the stairs. To determine the actual rise, the total height from floor to floor must be known. Review Chapter 29. Total rise can be found by adding the floor-to-ceiling height, the depth of the floor joists, and the depth of the floor covering. The total rise can then be divided by the maximum allowable rise to determine the number of steps required, as shown in Figure 30.5.Once the required rise is determined, this information should be stored in your memory for future reference. Of the residential stairs you will lay out in your career, probably 99% will have the same rise. So with a standard 8'-0'' (2400 mm) ceiling, you will always need 14 risers.

Once the number of risers is known, the required number of treads can be found easily, since there will always be one less tread than the number of risers. Thus, a typical stair for a house with 8'-0'' (2400 mm) ceilings will have 14 risers and 13 treads. If each tread is 10 1/2'' (267 mm) wide, the total run can be found by multiplying 10 1/2'' (267 mm) (the width) by 13 (the number of treads required). With this basic information, you are ready to draw the stairs. The layout for a straight stairway is described first.

STRAIGHT STAIR LAYOUT

The straight-run stair goes from one floor to another in one straight run. An example of a straight-run stair is shown in Figure 30.6. Figure 30.7 shows the stair section for the same house with the basement option. Each stair

FIGURE 30.3 ■ *Stairs with the lower stringers, treads, and risers in place, ready for the upper portion of the run.* Courtesy Benigno Molina-Manriquez.

STAIR TYPE	IRC
Straight stair	
Max. rise	7 3/4″ (195 mm)
Min. run	10″ (250 mm)
Min. headroom	6′–8″ (2000 mm)
Min. tread width	3′–0″ (900 mm)
Handrail height	34″ (850 mm) min. 38″ (950 mm) max.
Guardrail height	36″ (900 mm) min.
Winders	
Min. tread depth	6″ (150 mm) min. 10″ (250 mm) @ 12″ (300 mm)
Spiral	
Min. width	26″ (650 mm)
Min. tread depth	7 1/2″ (190 mm) @ 12″ (300 mm)
Max. rise	9 1/2″ (240 mm)
Min. headroom	6′–6″ (1950 mm)

FIGURE 30.4 ■ *Basic stair dimensions required by the IRC.*

Step 1. Determine the total rise in inches.

3/4	19	plywood
9 1/4	235	floor joist
3	76	top plates
92 5/8	2353	studs
1 1/2	38	bottom plate
107.125″	2721 mm	total rise

Step 2. Find the number of risers required. Divide the total rise of 107.125″ (2721 mm) by the maximum individual riser height of 7 3/4″ (197 mm).

$$7\,3/4\overline{)107.125} = 13.8 \qquad 197\overline{)2721} = 13.8$$

Because you cannot have .8 risers, the number will be rounded up to 14 risers.

Step 3. Find the number of treads required.
Number of treads equal Rise − run.
14 − 1 = 13 treads required.

Step 4. Multiply the length of each tread by the number of treads to find the total run.

FIGURE 30.5 ■ *Determining the rise and run needed for a flight of stairs.*

can be completed using the four steps of layout, drawing construction, adding dimensions, and annotation.

Stair Planning

The objects drawn during this stage will not be displayed on the completed drawings. Objects can be frozen or placed on nonplotting layers. Use thin continuous lines to place the following steps on the *DETL OUTL* layer unless noted. See Figure 30.8 for steps 1 through 4.

Step 1. Draw walls that may be near the stairs.
Step 2. Draw lines to represent each floor level.
Step 3. Draw a line to represent one end of the stairs. If no dimensions are available on the framing plans, scale your drawing.
Step 4. Determine the required risers and treads. Use the OFFSET command to represent the total run of the stairs.

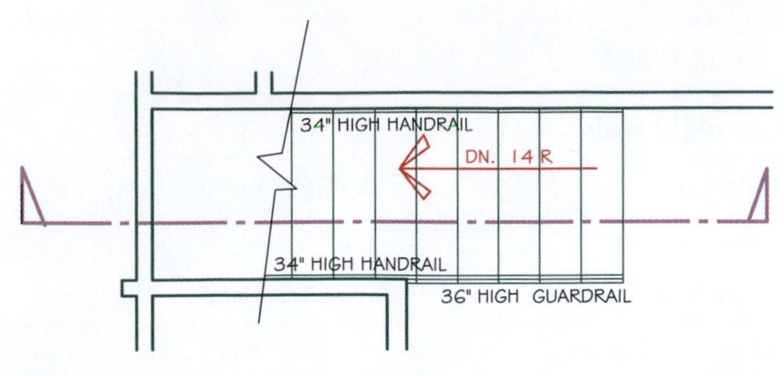

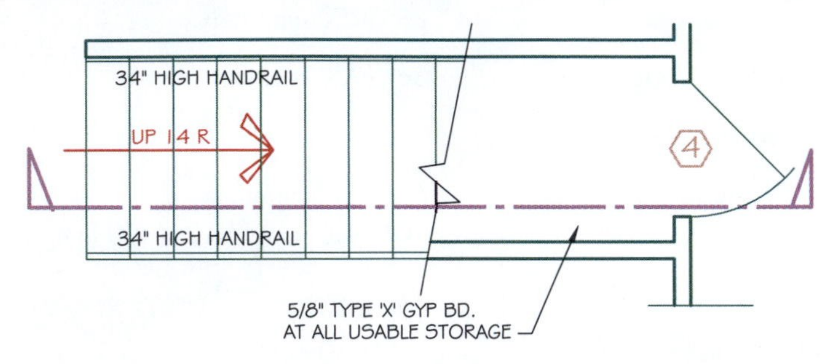

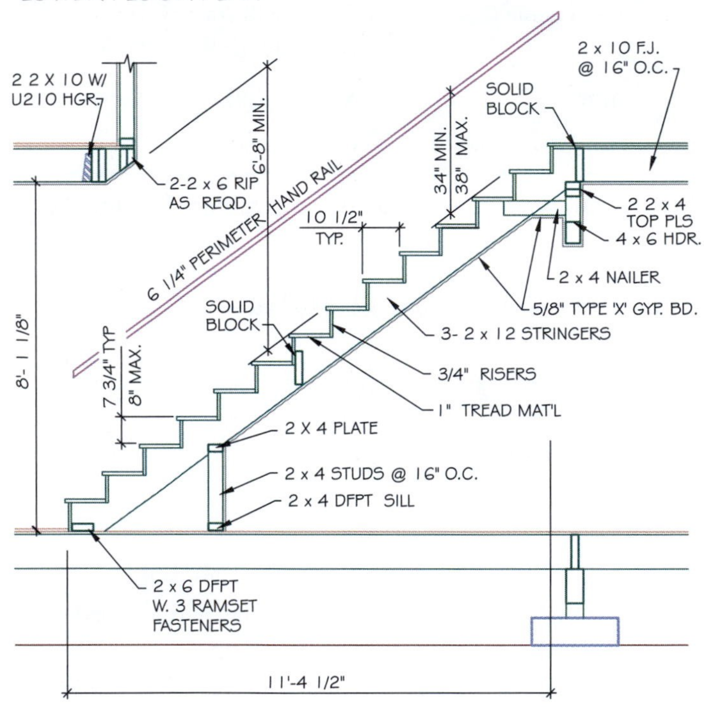

FIGURE 30.6 ■ *The plan and section views for a straight-run enclosed stair.*

Step 5. Draw a line from floor to floor and use the DIVIDE command to determine the required risers. See Figure 30.9.
Step 6. Draw a line from end to end of the stairs and use the DIVIDE command to determine the required treads. See Figure 30.9.
Step 7. Draw a line that passes through the nose of each step. Use the OFFSET command to represent the bottom of the stringer.
Step 8. Use the FILLET, CHAMFER, and OFFSET commands to represent the outline of the treads and risers, as shown in Figure 30.10.

Stair Construction

Once the location of the basic materials have been represented, the objects that will be used to complete the drawing can be represented. Use the *DETL WOOD* layer to draw the stairs unless otherwise noted. See Figure 30.11 for steps 9 through 12.

Step 9. Use the grid that was just created, and draw a line to represent the location of each tread and riser.
Step 10. Use the OFFSET command and the line created in Step 7 to represent the bottom side of the stringer. Assume a depth of 12'' (300 mm).
Step 11. Draw the upper stringer support or support wall where the stringer intersects the floor. Use thick lines to accent any structural wood that the cutting plane has passed through. Place thick lines on the *DETL MCUT* layer.
Step 12. Use the *DETL MBND* layer and draw metal hangers if there is no support wall.

See Figure 30.12 for steps 13 through 18. Place each item on the *DETL MBND* layer unless otherwise noted.

Step 13. Draw any intermediate support walls.
Step 14. Draw the gypsum board in all usable storage below the stairs.
Step 15. Draw any floors or walls that are over the stairs.
Step 16. Draw the handrail.
Step 17. Draw the kick block with bold lines on the *DETL MCUT* layer.
Step 18. Draw solid blocking on the *DETL MCUT* layer.

Adding Dimensions

Use the *DETL DIMS* layer to locate all of the following material. See Figure 30.13 for steps 19 through 24.

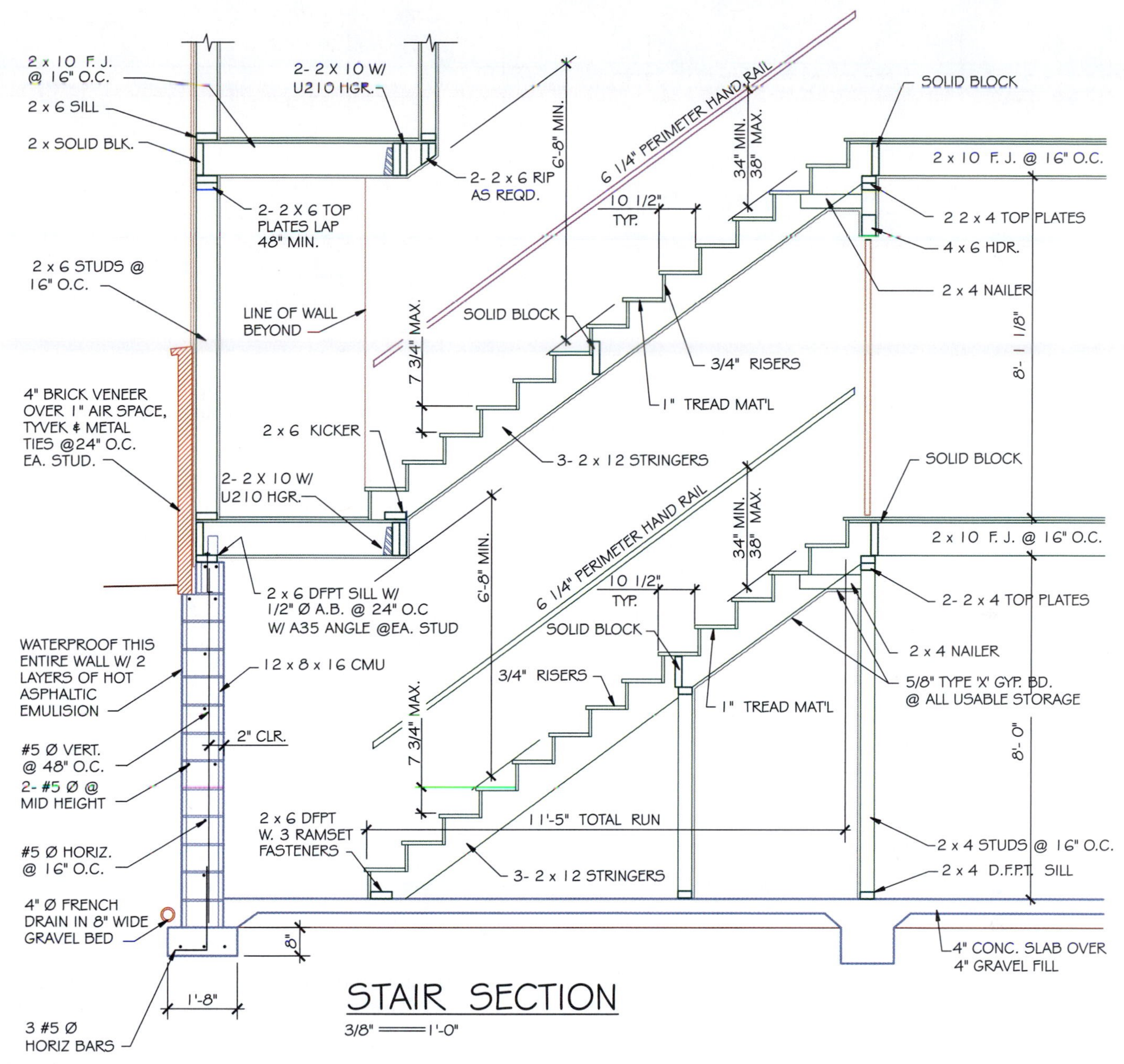

FIGURE 30.7 ■ *Straight-run stairs for a multilevel home.*

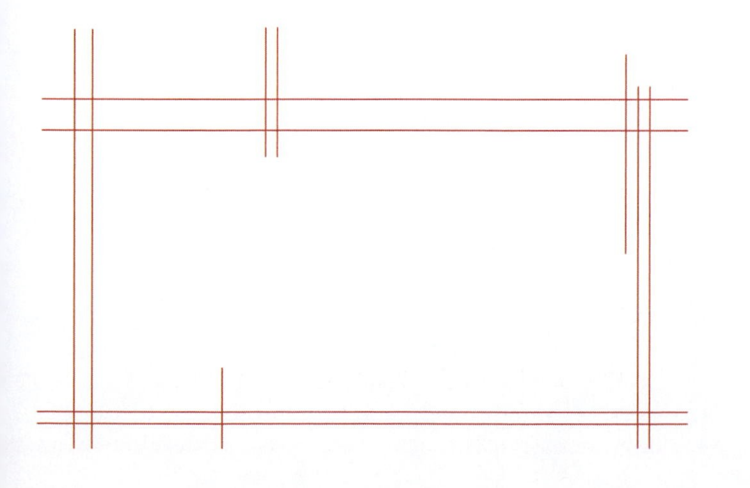

FIGURE 30.8 ■ *Draw lines to represent the walls, floor, and each end of the stairs using the* DETL OUTL *layer.*

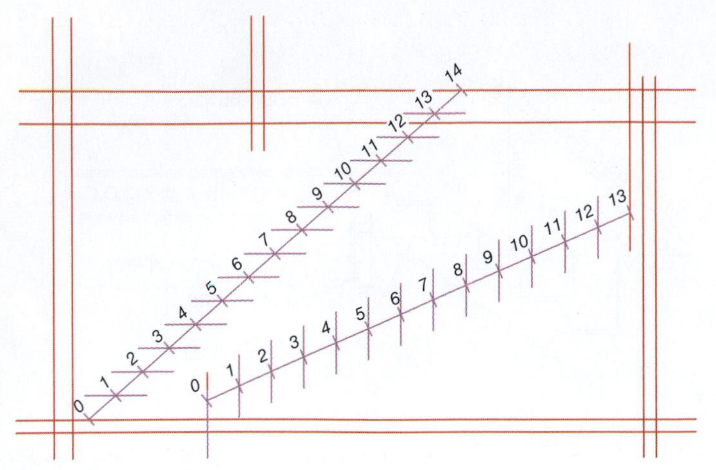

FIGURE 30.9 ■ *Determining the risers and treads.*

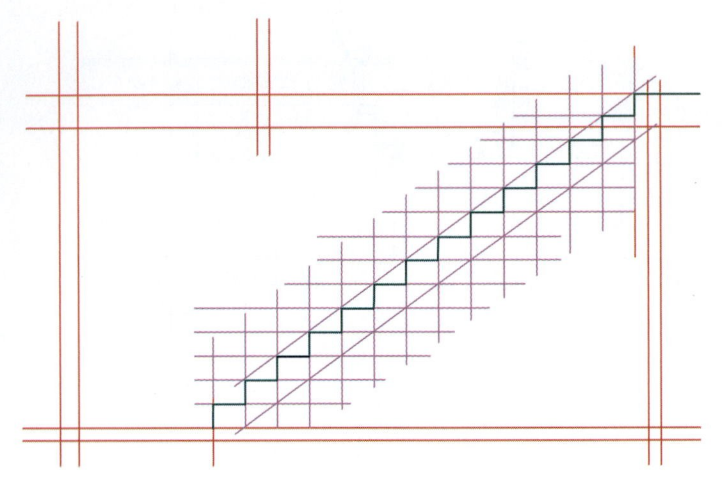

FIGURE 30.10 ■ *Outline the treads, stringer, and risers on the DETL OUTL layer.*

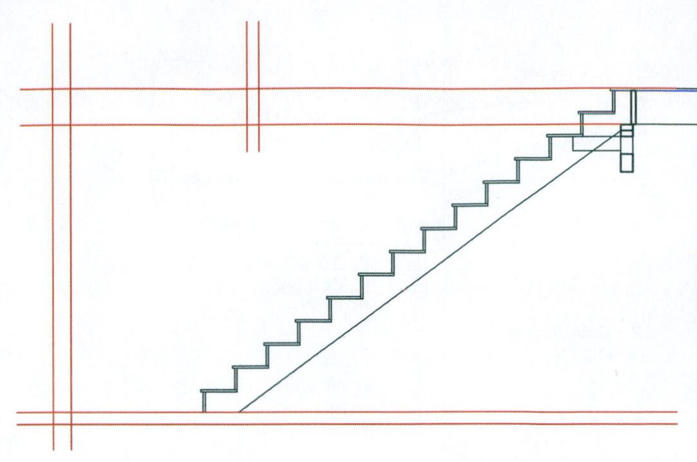

FIGURE 30.11 ■ *Representing the treads, risers, and the stringer with the layout grid frozen.*

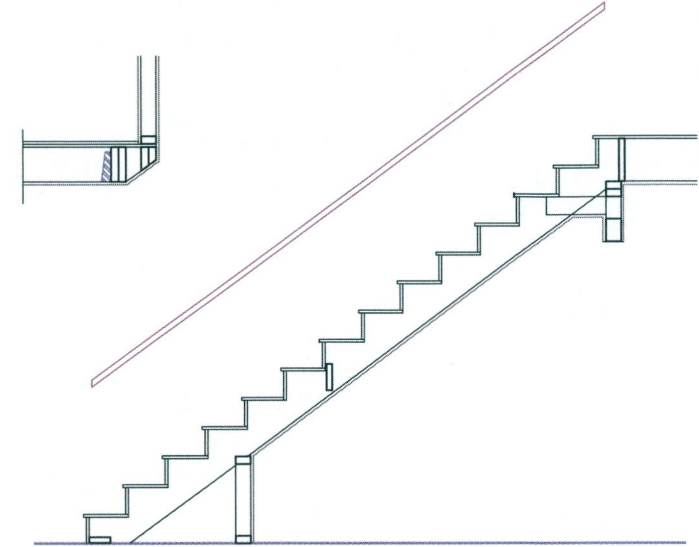

FIGURE 30.12 ■ *Add materials such as the kick block, support walls, blocking, fire-rated gypsum board, and handrails.*

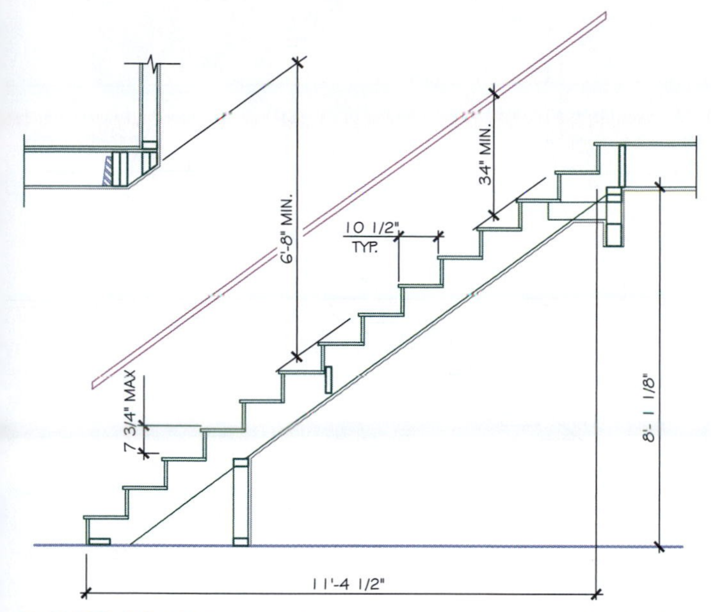

FIGURE 30.13 ■ *Stair dimensions are placed on the DETL DIMS layer to show the individual and total rise and run, the location of the handrail, and the height between floors.*

Place the required leader and dimension lines to locate the following dimensions:

Step 19. Total rise
Step 20. Total run
Step 21. Rise
Step 22. Run
Step 23. Headroom
Step 24. Handrail

Adding Annotation

See Figure 30.14 for typical notes placed on stair sections using the *DETL ANNO* layer. Verify local variations with your instructor. Common materials that must be represented include:

- The size and type of the stringer, tread material, and riser material
- The upper floor framing members
- The kick-block size and method of attachment to the floor

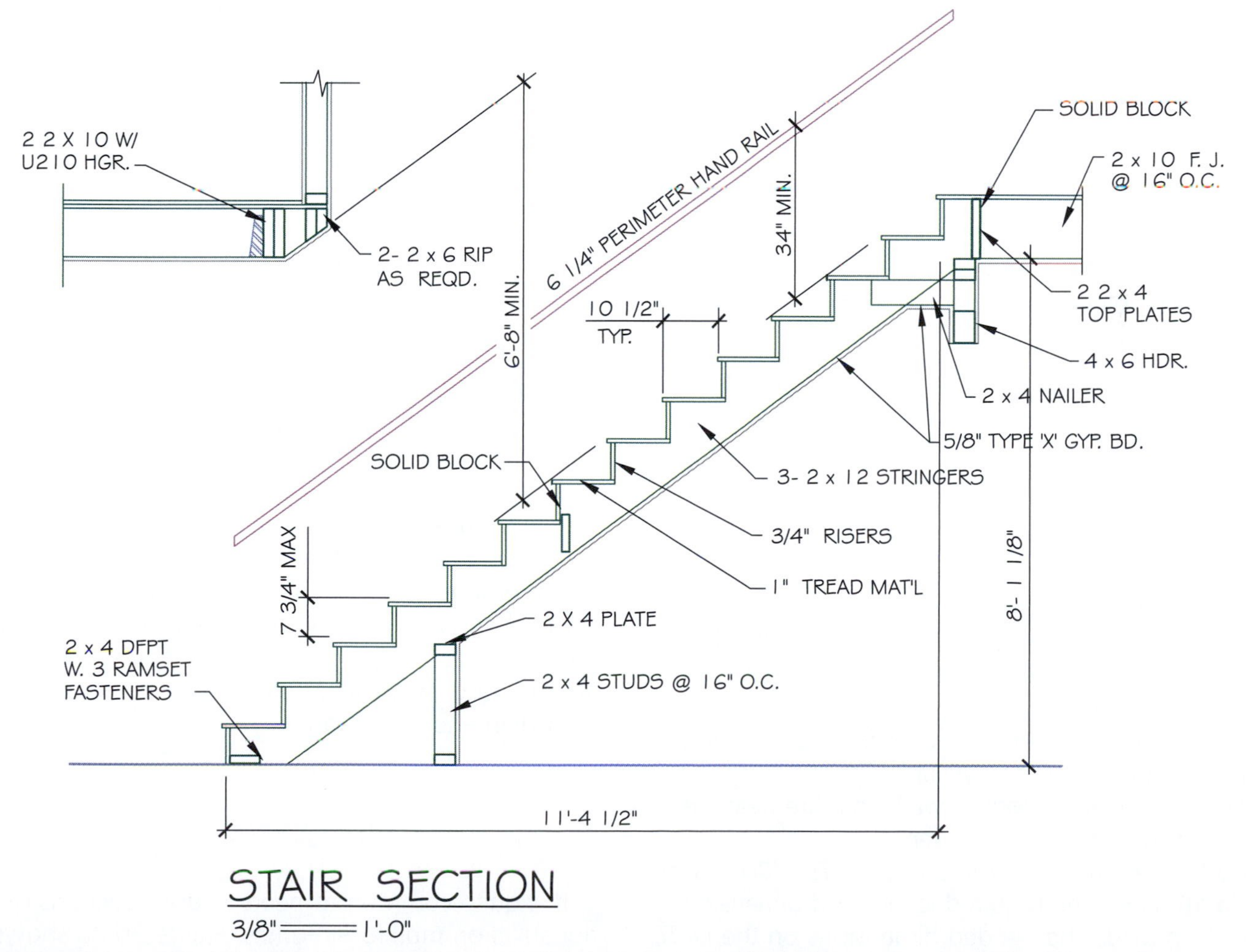

FIGURE 30.14 ■ *The stair drawing is completed by adding notes on the DETL ANNO layer to describe all materials.*

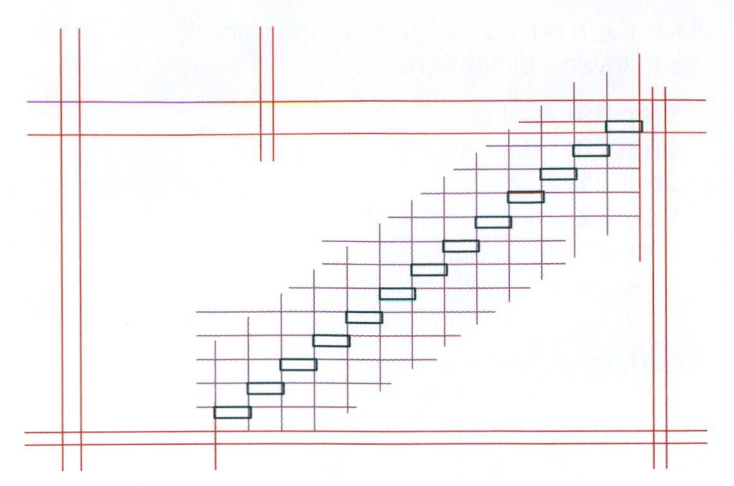

FIGURE 30.15 ■ *Layout of an open tread stair.*

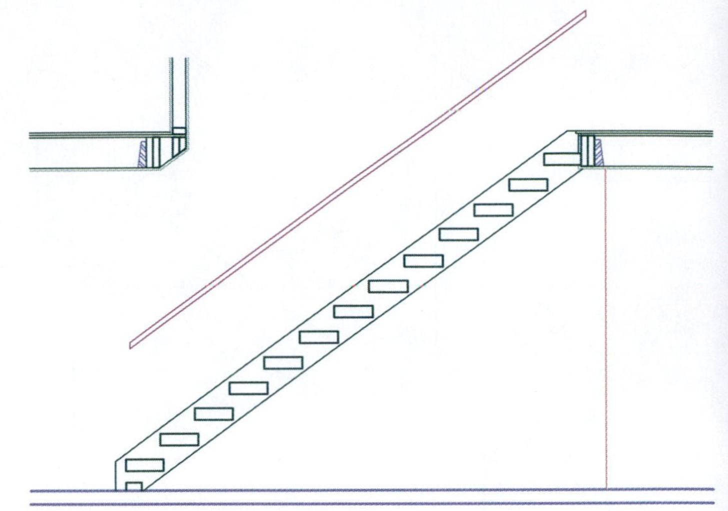

FIGURE 30.16 ■ *Structural material for an open tread stair.*

- Any stair blocking
- The handrail
- Fire protection below the stairs

OPEN STAIRWAY LAYOUT

An open stairway is similar to a straight enclosed stairway. It goes from one level to the next in a straight run. The major difference is that with the open stair, there are no risers between the treads. This allows for viewing from one floor to the next, creating an open feeling. Use the *DETL WOOD* layer to draw the following materials unless otherwise noted. See Figure 30.15 for steps 1, 2, and 3.

Step 1. Start the drawing by following steps 1 through 6 of the enclosed stair layout.
Step 2. Draw the 3 × 12'' (75 × 300 mm) treads.
STEP 3. Draw the 14'' (360 mm)-deep stringer centered on the treads.

See Figure 30.16 for steps 4 through 9.

Step 4. Draw the treads on the *DETL MCUT* layer.
Step 5. Draw the stringer on the *DETL MBND* layer.
Step 6. Draw the upper stringer supports on the *DETL MCUT* layer.
Step 7. Draw the metal hangers for the floor and stringer on the *DETL MBND* layer.
Step 8. Draw any floors or walls that are near the stairs on the *DETL MBND* layer.
Step 9. Draw the handrail on the *DETL MBND* layer.
Step 10. Place the required leader and dimension lines to provide the needed dimensions on the *DETL DIMS* layer. See steps 19 through 25 of the enclosed stair layout for a guide to the needed dimensions. See Figure 30.17.
Step 11. Place the required notes on the section on the *DETL MBND* layer. Use Figure 30.17 as a guide. Have your instructor specify local variations.

L- AND U-SHAPED STAIRS

The U-shaped stair is often used in residential design. Rather than going up a whole flight of steps in a straight run, this stair layout introduces a landing. The landing is usually located at the midpoint of the run, but it can be offset, depending on the amount of room allowed for stairs on the floor plan. The stairs may be either open or enclosed, depending on the location. The layout is similar to the layout of the straight-run stair but requires a little more planning in the layout stage because of the landing. Lay out the distance from the start of the stairs to the landing based on the floor plan measurements. Then proceed using a method similar to that used to draw the straight-run stair. See Figure 30.18 for help in laying out the section. Figure 30.19 shows what a U-shaped stair looks like on the floor plan and in section.

EXTERIOR STAIRS

It is quite common to need to draw sections of exterior stairs on multilevel homes. Figure 30.20 shows two different types of exterior stairs. Although there are

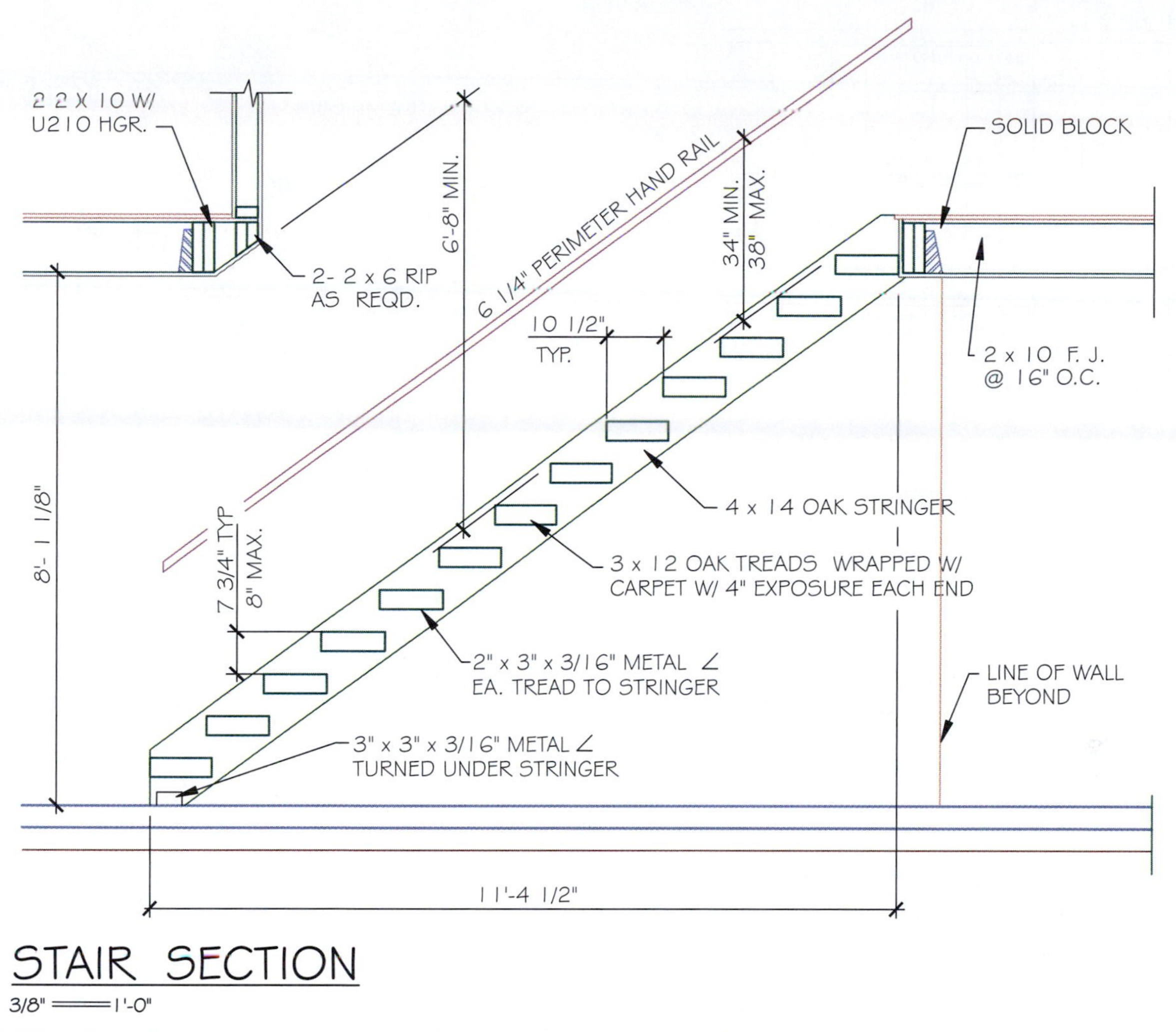

FIGURE 30.17 ■ *Completed open tread stair with dimensions and notes added.*

many variations, these two options are common. Both can be laid out by following the procedure for straight-run stairs.

There are some major differences in the finishing materials. Notice that there is no riser on the wood stairs, and the tread is thicker than the tread of an interior step. Usually the same material used on the deck is used for treads. In many parts of the country, a nonskid material should also be called for to cover the treads. The concrete stair can also be laid out by following the procedure for straight-run stairs. Once the risers and run have been marked off, the riser can be drawn. Notice that the riser is drawn on a slight angle. It can be drawn at about 10° and not labeled. This is something the flatwork crew will determine at the job site, depending on their forms.

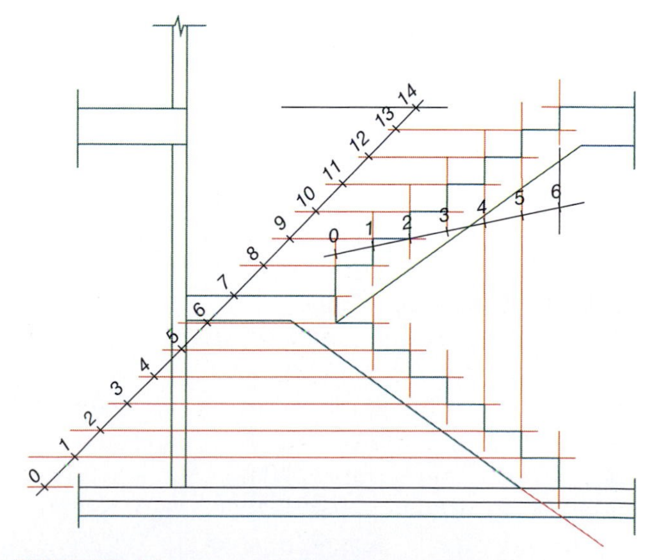

FIGURE 30.18 ■ *Layout of U-shaped stair runs.*

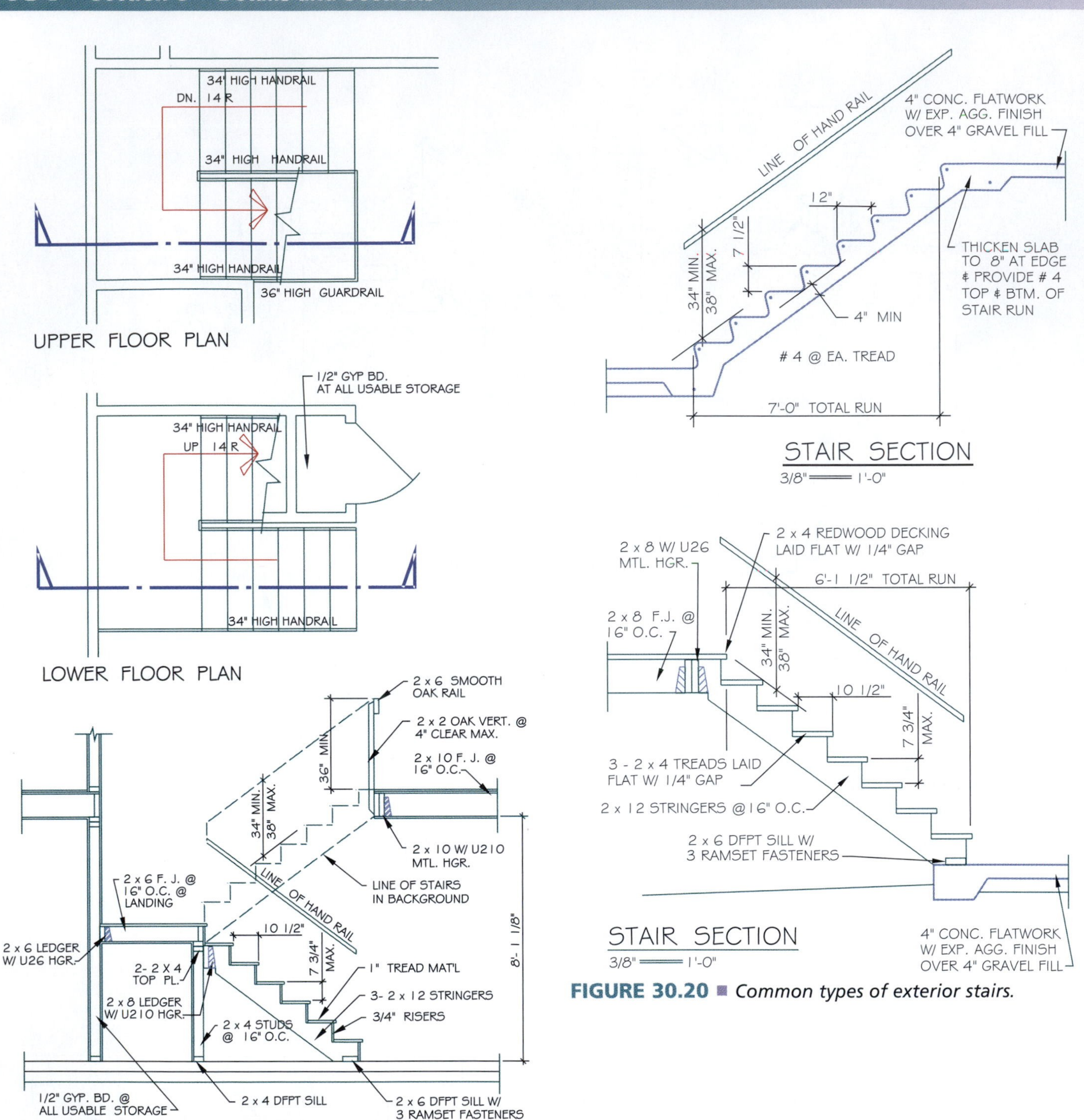
34" HIGH HANDRAIL
DN. 14 R
34" HIGH HANDRAIL
34" HIGH HANDRAIL
36" HIGH GUARDRAIL
UPPER FLOOR PLAN
1/2" GYP BD. AT ALL USABLE STORAGE
34" HIGH HANDRAIL
UP 14 R
34" HIGH HANDRAIL
LOWER FLOOR PLAN
2 x 6 SMOOTH OAK RAIL
2 x 2 OAK VERT. @ 4" CLEAR MAX.
2 x 10 F. J. @ 16" O.C.
36" MIN.
34" MIN. 38" MAX.
2 x 10 W/ U210 MTL. HGR.
LINE OF STAIRS IN BACKGROUND
LINE OF HAND RAIL
2 x 6 F. J. @ 16" O.C. @ LANDING
10 1/2"
7 3/4" MAX.
8'-1 1/8"
2 x 6 LEDGER W/ U26 HGR.
2- 2 X 4 TOP PL.
1" TREAD MAT'L
2 x 8 LEDGER W/ U210 HGR.
3- 2 x 12 STRINGERS
3/4" RISERS
2 x 4 STUDS @ 16" O.C.
1/2" GYP. BD. @ ALL USABLE STORAGE
2 x 4 DFPT SILL
2 x 6 DFPT SILL W/ 3 RAMSET FASTENERS
STAIR SECTION
3/8" = 1'-0"
LINE OF HAND RAIL
4" CONC. FLATWORK W/ EXP. AGG. FINISH OVER 4" GRAVEL FILL
12"
7 1/2"
34" MIN. 38" MAX.
THICKEN SLAB TO 8" AT EDGE & PROVIDE # 4 TOP & BTM. OF STAIR RUN
4" MIN
4 @ EA. TREAD
7'-0" TOTAL RUN
STAIR SECTION
3/8" = 1'-0"
2 x 8 W/ U26 MTL. HGR.
2 x 4 REDWOOD DECKING LAID FLAT W/ 1/4" GAP
6'-1 1/2" TOTAL RUN
2 x 8 F.J. @ 16" O.C.
34" MIN. 38" MAX.
LINE OF HAND RAIL
10 1/2"
7 3/4" MAX.
3 - 2 x 4 TREADS LAID FLAT W/ 1/4" GAP
2 x 12 STRINGERS @ 16" O.C.
2 x 6 DFPT SILL W/ 3 RAMSET FASTENERS
STAIR SECTION
3/8" = 1'-0"
4" CONC. FLATWORK W/ EXP. AGG. FINISH OVER 4" GRAVEL FILL

FIGURE 30.19 ■ *U-shaped stair.*

FIGURE 30.20 ■ *Common types of exterior stairs.*

CHAPTER 30

Additional Readings

The following websites can be used as a resource to help you keep current with changes in stair materials.

ADDRESS	COMPANY OR ORGANIZATION
www.arcways.com	Arcways, Inc.
www.stairwaysinc.com	Stairways, Inc.
www.southernstaircase.com	Southern Staircase
www.theironshop.com	The Iron Shop
www.yorkspiralstair.com	York Spiral Stair

CHAPTER 30

Stair Construction and Layout Test

DIRECTIONS

Answer the following questions with short, complete statements. Type your answers using a word processor or neatly print your answers on lined paper.

1. Place your name, the chapter number, and the date at the top of the sheet.
2. Type the question number and provide the answer in the form of a statement that includes part of the question. You do not need to write out the entire question.

Note: The answers to some questions may not be contained in this chapter and will require you to do additional research.

QUESTIONS

30.1. What is a tread?

30.2. What is the minimum headroom required for a residential stair?

30.3. What is the maximum individual rise of a step?

30.4. What member is used to support the stairs?

30.5. What is the maximum spacing allowed between the verticals of a railing?

30.6. Describe the difference between a handrail and a guardrail.

30.7. How many risers are required if the height between floors is 10' (3000 mm)?

30.8. Sketch a section for three common stair types.

30.9. If a run of 10'' (250 mm) is to be used, what will be the total run when the distance between floors is 9' (2700 mm)?

30.10. What is a common size for treads in an open-tread layout?

CHAPTER 31

Fireplace Construction and Layout

The type and location of the fireplace must be considered when the floor plan is drawn. As the sections are being drawn, consideration must be given to the construction of the fireplace and chimney. The most common construction materials for a fireplace and chimney are masonry and metal. The metal, or zero-clearance, fireplace is manufactured and does not require a section to explain its construction. A metal fireplace is surrounded by wood walls that can be represented on the floor or framing plans. Figure 31.1 shows a metal fireplace installed in a wood-framed chase. The interior can be covered in masonry, stone, tile, or some other noncombustible material, as in Figure 31.2. Notice in Figure 31.1 that a metal chimney has been provided. Metal fireplaces that do not require a chimney are also available. Figure 31.3 shows a gas-burning fireplace that is vented through the wall.

Three common types of metal fireplaces include direct-vent, top-vent, and vent-free fireplaces. Direct-vent *fireplaces vent out the back of the appliance and through the house wall to outside air, as shown in Figure 31.3.* Top-vent *fireplaces can be used as a fireplace insert or as a freestanding unit as seen in Figure 31.1. When used as an insert, the unit is installed in an existing masonry fireplace and vents through the existing fireplace chimney. When installed as a fireplace, a metal chimney is provided and both are enclosed with a wood-framed chase.*

FIGURE 31.1 ■ *A metal fireplace insert encased with standard wood framing. The vertical duct on the left side supplies combustion air.* Courtesy Zachary Jefferis.

FIGURE 31.2 ■ *Metal fireplace units can be covered in masonry, stone, tile, or some other noncombustible material.* Courtesy LeRoy Cook.

A vent-free fireplace has no exhaust vent. A gas line is connected to the unit and the flame burns inside the fireplace without a vent. Not all states allow the use of vent-free gas fireplaces because of conflicts with building codes that require airtight, energy-efficient construction. Although exterior air is required to be provided for factory-built fireplaces, the IRC allows the use of vent-free units if the room is mechanically ventilated and controlled, so that the indoor air pressure is either positive or neutral. Many vent-free fireplaces are equipped with an oxygen-depletion sensor that terminates the gas supply to the fire if it senses a lack of oxygen in the room.

In working with factory-built fireplaces, several items must be specified. The IRC requires the chimney for direct- and top-vent fireplaces to conform to UL standards. Factory-built fireplaces producing gases with a temperature above 1000°F (538°C) at the chimney entrance must meet UL 959. Although a section is not required, each of these chimney specifications should be provided on the floor plan or interior elevations, along with instructions to provide and install all materials per the manufacturer's specifications.

FIGURE 31.3 ▪ *The exterior view of a self-venting gas fireplace; it requires a vent similar to that of a dryer.* Courtesy LeRoy Cook.

FIREPLACE TERMS

Fireplace construction has its own vocabulary that must be understood in order to draw a section. Figure 31.4 shows the major components of a fireplace and chimney.

The size of the ***fireplace opening*** should be given much consideration, since it is important for the appearance and operation of the fireplace. If the opening is too small, the fireplace will not produce sufficient heat. If the opening is too large, the fire could make a room too hot. Common design guidelines suggest that the opening be approximately 1/30 of the room area for small rooms and 1/65 of the room area for large rooms. Figure 31.5 shows suggested fireplace opening sizes relative to room size. The ideal dimensions for a single-face fireplace have been determined to be 36'' (900 mm) wide and 26'' (660 mm) high. These dimensions may be varied slightly to meet the size of the brick or to fit other special dimensions of the room.

The ***hearth*** is the floor of the fireplace and consists of inner and outer parts. The inner hearth is the floor of the firebox. This hearth is made of fire-resistant brick and holds the burning fuel. The structural portion of the hearth must be made of 4'' (100 mm) minimum reinforced masonry or concrete, but the portion that extends in front of the fireplace can be at least 2'' (50 mm) thick. An ***ash dump*** is usually located in the inner hearth of a masonry fireplace. The ash dump is an opening in the inner hearth that allows the ashes to be dumped. The ash dump must be located so that the ash removal will not create a hazard to combustible materials. The ash dump is covered with a small metal plate, which can be removed to provide access to the ash pit. The ***ash pit*** is the space below the fireplace where the ashes can be stored.

The outer hearth may be made of any noncombustible material. The material is usually selected to

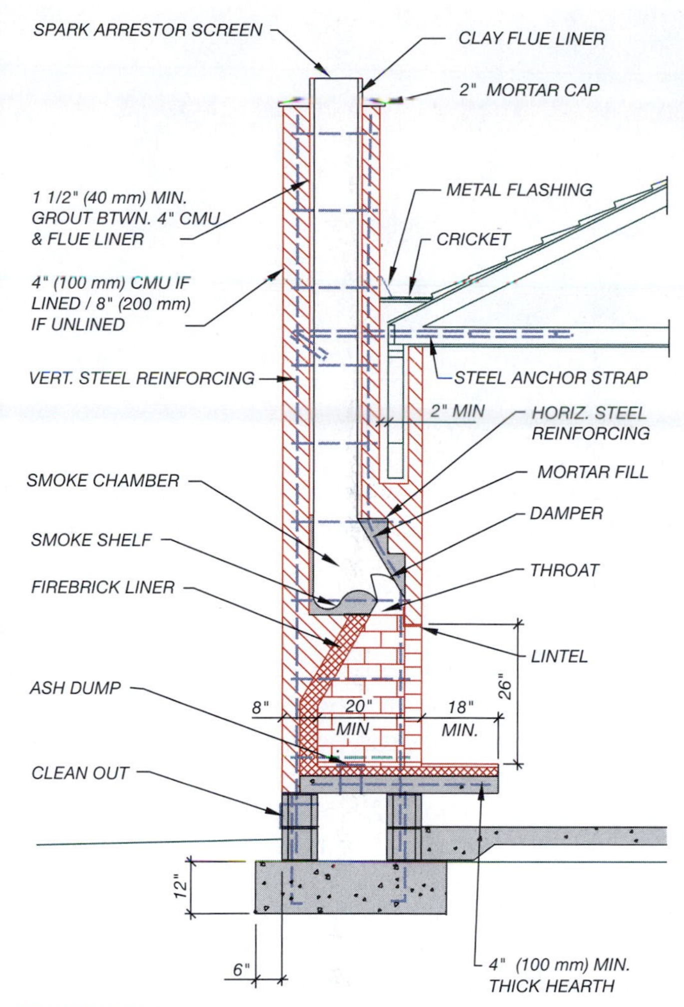

FIGURE 31.4 ■ *Parts of a masonry fireplace and chimney.*

blend with other interior design features and may be brick, tile, marble, or stone. The outer hearth protects the combustible floor around the fireplace. Minimum requirements for the outer hearth projection based on the IRC include:

- The hearth must extend 16'' (400 mm) in front of the fireplace opening and 8'' (200 mm) beyond each side of the fireplace.
- If the size of the fireplace opening is greater than 6 sq ft (0.557 m^2), the hearth extension must be increased to 20'' (500 mm) minimum with at least 12'' (300 mm) side extensions.

The fireplace opening is the front of the ***firebox.*** The opening may be on one, two, or three faces of the fireplace (see Figure 31.6a and b). The firebox is where the combustion of the fuel occurs. Its sides should be slanted slightly to radiate heat into the room. The rear wall should be inclined to provide an upward draft into the upper part of the fireplace and chimney. The firebox is usually constructed of fire-resistant brick set in fire-resistant mortar. Figure 31.7 shows minimum wall thickness for the firebox. The minimum depth of the firebox is required to be 20'' (500 mm). The firebox depth should be proportional to the size of the fireplace opening. Providing a proper depth ensures that smoke will not discolor the front face (breast) of the fireplace. With an opening of 36 × 26'' (900 × 660 mm), a depth of 20'' (500 mm) should be provided for a single-face fireplace. Figure

SUGGESTED WIDTH OF FIREPLACE OPENINGS APPROPRIATE TO ROOM SIZE					
SIZE OF ROOM		IF IN SHORT WALL		IF IN LONG WALL	
(FT)	(MM)	(IN.)	(MM)	(IN.)	(MM)
10 × 14	(3048 × 4267)	24	(610)	24–32	(610–813)
12 × 16	(3658 × 4877)	28–36	(711–914)	32–36	(813–914)
12 × 20	(3658 × 6096)	32–36	(813–914)	36–40	(914–1016)
12 × 24	(3658 × 7315)	32–36	(813–914)	36–48	(914–1219)
14 × 28	(4267 × 8534)	32–40	(813–1016)	40–48	(1016–1219)
16 × 30	(4877 × 9144)	36–40	(914–1016)	48–60	(1219–1524)
20 × 36	(6096 × 10 973)	40–48	(1016–1219)	48–72	(1219–1829)

FIGURE 31.5 ■ *The size of the fireplace should be proportioned to the size of the room. This will give both a pleasing appearance and a fireplace that will not overheat the room.*

FIGURE 31.6a ■ *A two-sided fireplace.* Courtesy Heatilator, Inc.

FIGURE 31.6b ■ *A three-sided fireplace.* Courtesy Majestic.

31.8 lists recommended fireplace opening-to-depth proportions.

Above the fireplace opening is a lintel. The lintel is a reinforced masonry or steel angle that supports the fireplace face. The minimum required bearing length for the lintel at each end of the fireplace is 4'' (100 mm). The throat of a fireplace is the opening at the top of the firebox that opens into the chimney. The throat must be at least 8'' (200 mm) above the fireplace opening and must be a minimum of 4'' (100 mm) in depth. The cross-sectional area of the passageway above the firebox—including the throat, ***damper,*** and ***smoke chamber***—can't be less than the cross-sectional area of the flue. The throat should be closable when the fireplace is not in use. This is done by installing a damper. The damper must be made of ferrous metal that extends the full width of the throat and is used to prevent heat from escaping up the chimney when the fireplace is not in use. When fuel is being burned in the firebox, the damper can be opened from within the room that contains the fireplace to allow smoke from the firebox into the smoke chamber of the chimney.

The smoke chamber acts as a funnel between the firebox and the chimney. The shape of the smoke chamber should be symmetrical, so that the chimney draft pulls evenly and creates an even fire in the firebox. The smoke chamber should be centered under the flue in the chimney and directly above the firebox. The overall size of the smoke chamber is regulated by the IRC. The inside height of the chamber can't be greater than the inside width of the fireplace opening. The walls of the smoke chamber must be made of solid masonry units, stone, concrete, or hollow masonry units that have been filled with grout. The material is corbelled to form the tapered shape, but the IRC requires that the interior surface of the masonry units can't be exposed to the smoke chamber. The walls of the smoke chamber can't be greater than 30° when formed from corbelled masonry units, but they can be inclined up to 45° if prefabricated linings are used. If the walls of the chamber are made with a lining of 2'' (50 mm) firebrick or of 5/8'' (15.9 mm) vitrified clay, the four chamber walls must have a minimum thickness of 6'' (150 mm), including the liner. If no lining is provided, the four chamber walls must each be a minimum of 8'' (200 mm) thick. A ***smoke shelf*** is located at the bottom of the smoke chamber behind the damper. The smoke shelf prevents downdrafts from the ***chimney*** from entering the firebox.

The chimney is the upper extension of the fireplace and is built to carry off the smoke from the fire. The main components of the chimney are the flue, lining, anchors, cap, and spark arrester. The wall thickness of the chimney will be determined by the type of flue construction. Masonry chimneys are not allowed to change in size or shape within 6'' (150 mm) above or below any combustible floor, ceiling, or roof component penetrated by the chimney. Figure 31.7 shows the minimum wall thickness for chimneys.

The ***flue*** is the opening inside the chimney that allows smoke and combustion gases to pass from the firebox away from the structure. A flue may be constructed of normal masonry products or may be covered with a flue liner. The size of the flue must be proportional to the size of the firebox opening and the number of open faces of the fireplace. A flue that is too small will not allow the fire to burn well and will cause smoke to exit through the front of the firebox. A flue that is too large for the firebox will cause too great a

GENERAL CODE REQUIREMENTS		
ITEM	LETTER	UNIFORM BUILDING CODE* 1997
Hearth Slab Thickness	A	4″
Hearth Slab Width (Each side of opening)	B	8″ Fireplace opg. < 6 sq. ft. 12″ Fireplace opg. ≥ 6 sq. ft.
Hearth Slab Length (Front of opening)	C	16″ Fireplace opg. < 6 sq. ft. 20″ Fireplace opg. ≥ 6 sq. ft.
Hearth Slab Reinforcing	D	Reinforced to carry its own weight and all imposed loads
Thickness of Wall or Firebox	E	10″ common brick or 8″ where a fireback lining is used Jts. in fireback 1/4″ max.
Distance from Top of Opening to Throat	F	6″
Smoke Chamber Edge of Shelf Rear Wall—Thickness Front & Side Wall—Thickness	G	 6″ 8″
Chimney Vertical Reinforcing	H	Four #4 full length bars for chimney up to 40″ wide. Add two #4 bars for each additional 40″ or fraction of width or each additional flue.
Horizontal Reinforcing	J	1/4″ ties at 18″ and two ties at each bend in vertical steel
Bond Beams	K	No specified requirements
Fireplace Lintel	L	Noncombustible material
Walls with Flue Lining	M	Solid masonry units or hollow masonry units grouted solid with at least 4″ nominal thickness
Walls with Unlined Flue	N	8″ solid masonry
Distance Between Adjacent Flues	O	4″ including flue liner
Effective Flue Area (Based on Area of Fireplace Opening)	P	Verify with local code
Clearances Wood Frame Combustible Material Above Roof	R	1″ when outside of wall or 1/2″ gypsum board 2″ when entirely within structure 6″ min. to fireplace opening Combustible material within 12″ (305 mm) of the fireplace opening can't extend more than 1/8″ for each inch (3/25 mm) distance from the opening. 2′ at 10′
Anchorage Strap Number Embedment into Chimney Fasten to Bolts	S	 3/16″ × 1″ 2 12″ hooked around outer bar w/6″ ext. 4 joists Two 1/2″ dia.
Footing Thickness Width	T	 12″ min. 6″ each side of fireplace wall
Outside Air Intake	U	Optional
Glass Screen Door		Optional

FIGURE 31.7a ■ *General code requirements for fireplace and chimney construction. The letters in the second column will be helpful in locating specific items in Figure 31.7b.*

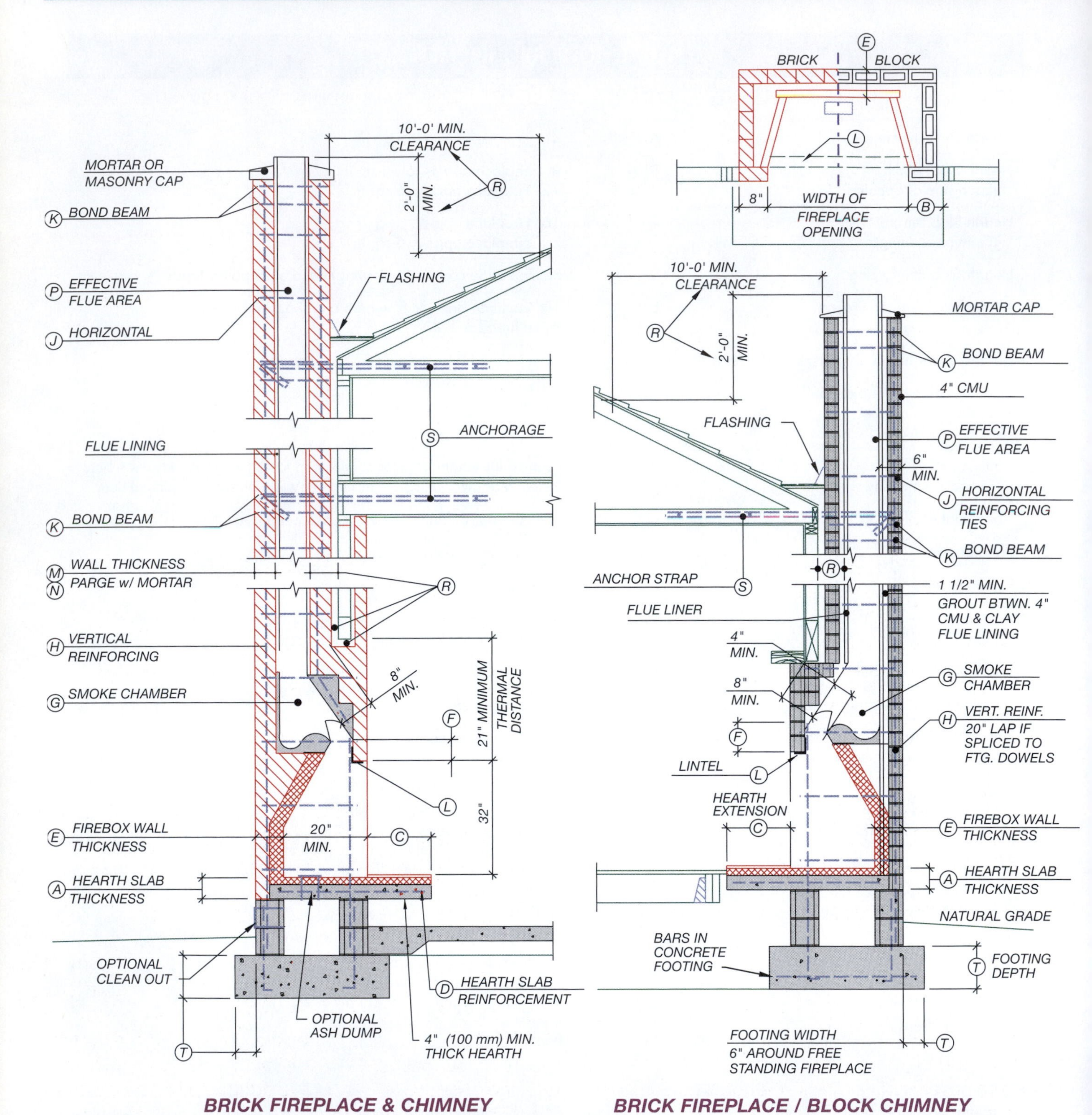

FIGURE 31.7b ■ *Brick fireplace and chimney components. The circled letters refer to item references in Figure 31.7a.*

FIREPLACE TYPE	WIDTH OF OPENING (W)		HEIGHT OF OPENING (H)		DEPTH OF OPENING (D)	
	(IN.)	(MM)	(IN.)	(MM)	(IN.)	(MM)
Single face	28	700	24	600	20	500
	30	760	24	600	20	500
	30	760	26	660	20	500
	36	900	26	660	20	500
	36	900	28	700	22	560
	40	1000	28	700	22	560
	48	1200	32	800	25	635
Two faces adjacent "L" or corner type	34	865	27	685	23	585
	39	990	27	685	23	585
	46	1170	27	685	23	585
	52	1320	30	760	27	635
Two faces* opposite look-through	32	800	21	530	30	760
	35	890	21	530	30	760
	42	1070	21	530	30	760
	48	1200	21	530	34	865
Three face* 2 long, 1 short 3-way opening	39	990	21	530	30	760
	46	1170	21	530	30	760
	52	1320	21	530	34	865
Three face* 1 long, 2 short 3-way opening	43	1090	27	685	23	585
	50	1270	27	685	23	585
	56	1420	30	760	27	686

*Fireplaces open on more than front and one end are **not** recommended.

FIGURE 31.8 ■ *Guide for fireplace opening-to-depth proportions.*

draft through the house as the fire draws its combustion air. Flue sizes are generally required by code to be either 1/8 or 1/10 the size of the fireplace opening. Figure 31.9 shows recommended areas for residential fireplaces.

A ***chimney liner*** is usually built of fire clay or terra cotta. The liner is built into the chimney to provide a smooth surface to the flue wall and to reduce the width of the chimney wall. The smooth surface of the liner will help reduce the buildup of soot, which could cause a chimney fire. The ***chimney cap*** is the sloping surface on the top of the chimney (see Figure 31.10). The slope prevents rain from collecting on the top of the chimney. The flue normally projects 2 to 3'' (50 to 75 mm) above the cap so that water will not run down the flue. The ***chimney hood*** is a masonry or metal covering that may be placed over the flue for protection from the elements. The masonry cap is built so that openings allow for the prevailing wind to blow through the hood and create a draft in the flue. The metal hood can be rotated by wind pressure to keep the opening of the hood downwind and thus prevent rain or snow from entering the flue. A *spark arrester* is a screen placed at the top of the flue inside the hood to prevent combustibles from leaving the flue.

Chimney Reinforcement

In areas subject to seismic damage, a minimum of four vertical reinforcing bars 1/2'' (13 mm) in diameter (#4) must be used in the chimney, extending from the foundation up to the top of the chimney. These vertical bars are supported at 18'' (450 mm) intervals with 1/4'' (6 mm) horizontal rebar. Two additional #4 bars are required for each additional 40'' (1000 mm) of width or fraction thereof or for each additional flue added to the chimney. Two #4 rebar are also installed when the vertical steel is bent for a change in chimney width. In addition to the reinforcement in the chimney, the chimney must be attached to the structure. This is done with steel anchors that connect the fireplace to the framing members at each floor and ceiling level that is more than 6' (1800 mm) above the ground. Two steel straps

TYPE OF FIREPLACE	WIDTH OF OPENING W IN.	HEIGHT OF OPENING H IN.	DEPTH OF OPENING D IN.	AREA OF FIRE-PLACE OPENING FOR FLUE DETERMINATION SQ IN.	FLUE SIZE REQUIRED AT 1/10 AREA OF FIREPLACE OPENING	FLUE SIZE REQUIRED AT 1/8 AREA OF FIREPLACE OPENING
	28	24	20	672	8 1/2 × 13	8 1/2 × 17
	30	24	20	720	8 1/2 × 17	13″ round
	30	26	20	780	8 1/2 × 17	10 × 18
	36	26	20	936	10 × 18	13 × 17
	36	28	22	1008	10 × 18	10 × 21
	40	28	22	1120	10 × 18	10 × 21
	48	32	25	1536	13 × 21	10 × 21
	60	32	25	1920	17 × 21	21 × 21
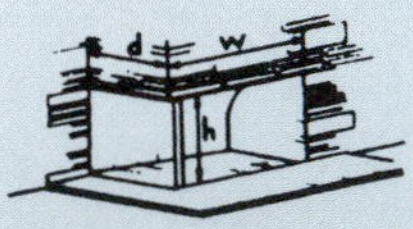	34	27	23	1107	10 × 18	10 × 21
	39	27	23	1223	10 × 21	13 × 21
	46	27	23	1388	10 × 21	13 × 21
	52	30	27	1884	13 × 21	17 × 21
	64	30	27	2085	17 × 21	21 × 21
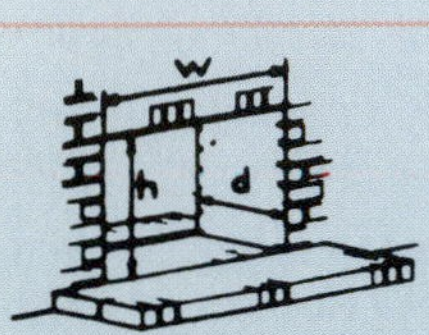	32	21	30	1344	13 × 17 or 10 × 21	17 × 17 or 13 × 21
	35	21	30	1470	17 × 17 or 13 × 21	17 × 21
	42	21	30	1764	17 × 21	17 × 21
	48	21	34	2016	17 × 21	21 × 21
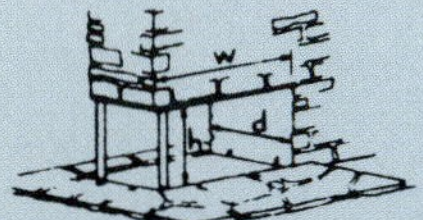	39	21	30	1638	13 × 21 or 17 × 17	17 × 21
	46	21	30	1932	17 × 21	21 × 21
	52	21	34	2184	17 × 21	21 × 21
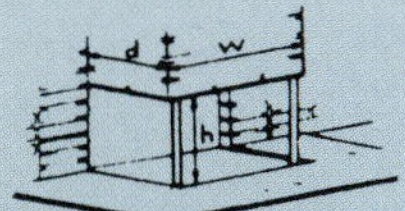	43	27	23	1782	13 × 21 or 17 × 17	17 × 21
	50	27	23	1971	17 × 21	17 × 21
	56	30	27	2490	21 × 21	21 × 21 or 2—10 × 21 ▲
	68	30	27	2850	13 × 21 & ▲ 10 × 21	2—13 × 21 or 2—17 × 17 or ▲ 10 × 18 & 17 × 21

▲ *Rather than using two flue liners in chimney, the flue is often left unlined with 8" masonry walls. Unlined flues must have a minimum area of 1/8 fireplace opening.*

FIGURE 31.9 ■ *Recommended flue areas for residential fireplaces.*

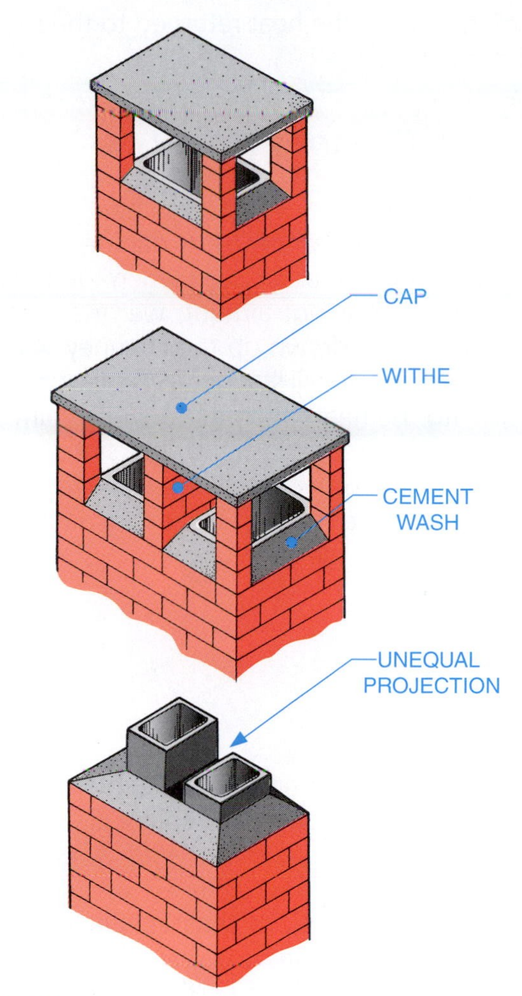

FIGURE 31.10 ■ *Chimney caps are often provided to control the airflow of the chimney.*

3/16 × 1'' (5 × 25 mm) must be embedded at least 12'' (25 mm) into the chimney; they must be hooked around the outer steel rebar and extend 6'' (150 mm) beyond the bend. The straps must attach to four floor or ceiling joists with two 1/2'' (12.7 mm) bolts. These straps can be omitted if the chimney is built completely within the exterior walls.

ENERGY-EFFICIENT FIREPLACES

In addition to adding charm and elegance to a room, a fireplace also needs to be functional. No matter what the material is, consideration must be given to the air that will be used by the fireplace. Three points to consider are:

- How will air be supplied to the fire
- How to use the heated air created by the fire
- How to minimize heated air escaping through the chimney

Providing Air Supply

The construction methods introduced in Chapters 14 and 20 have greatly reduced the amount of air that enters the structure. When a fireplace is added to a relatively airtight residence, the fireplace can cool a room if proper precautions are not taken. Air must be provided to the fireplace for combustion, and air is expelled up the chimney after combustion. As the air is expelled after combustion, new air must be drawn into the fire. If the air for combustion is taken from the room being heated, a draft will be created. To reduce drafts and maximize the heat produced by the fire, combustion air must be taken from an outside source or from spaces in the residence that are ventilated with outside air, such as non–mechanically ventilated crawl or attic spaces. The restrictions by the IRC specify that the exterior air intake can't be located in a basement or garage, and the air intake can't be placed at an elevation higher than the firebox. When the floor plan was completed, a note was placed on the plan reading:

PROVIDE A SCREENED, CLOSABLE VENT WITHIN 24'' OF THE FIREBOX.

Providing the note on the floor plan will help ensure that a draft is not created. As the fireplace section is drawn, the vent for air intake should be indicated. The vent should be placed on an exterior wall or in the back of the fireplace in a position high enough above the ground that it will not be blocked by snow. Combustion kits are available that can be installed in the duct to heat the outside air before it enters the firebox. Another alternative to drawing air from the outside is to draw combustion air from the cold air return ducts of the heating system. The HVAC contractor can calculate the needed size and provide a duct that takes the necessary cool air from inside the home to supply the fireplace.

Using the Heated Air Efficiently

Once the fire has heated the air in the firebox, that air must be used efficiently to heat the structure. Chapter 10 introduced passive solar heating and the use of a masonry mass to store heat. A masonry fireplace and chimney will provide mass to absorb, store, and radiate the heat back into the room long after the fire is out. When a fireplace is built on an outer wall, the mass of the fireplace is often outside the structure in order to increase the usable floor space. With the chimney outside the structure, the interior face of the fireplace is the only surface that radiates heat back into the room. The amount of heat radiated into a room is increased as more of the fireplace and chimney mass is located inside the structure. Placing the fireplace and chimney totally within a structure will allow all surfaces of the masonry to radiate heat into the structure.

Altering the shape of the firebox will also increase the amount of air radiated back to the room. Figure 31.5 shows common sizes of fireplace openings. Increasing the width and height of the opening and decreasing the depth of the fireplace will increase the amount of heat in the room. This altered firebox shape is based on the Rumford design that was popular in the late 1800s. It has been reinvented and has become popular in many areas and is now permitted by the IRC. Increasing the length of the side walls and decreasing the length of the rear walls will also increase the heat returned to the room. Figure 31.11 shows the difference between a Rumford-type firebox and traditional designs. A third method of increasing the heat returned to the room is a mechanical blower.

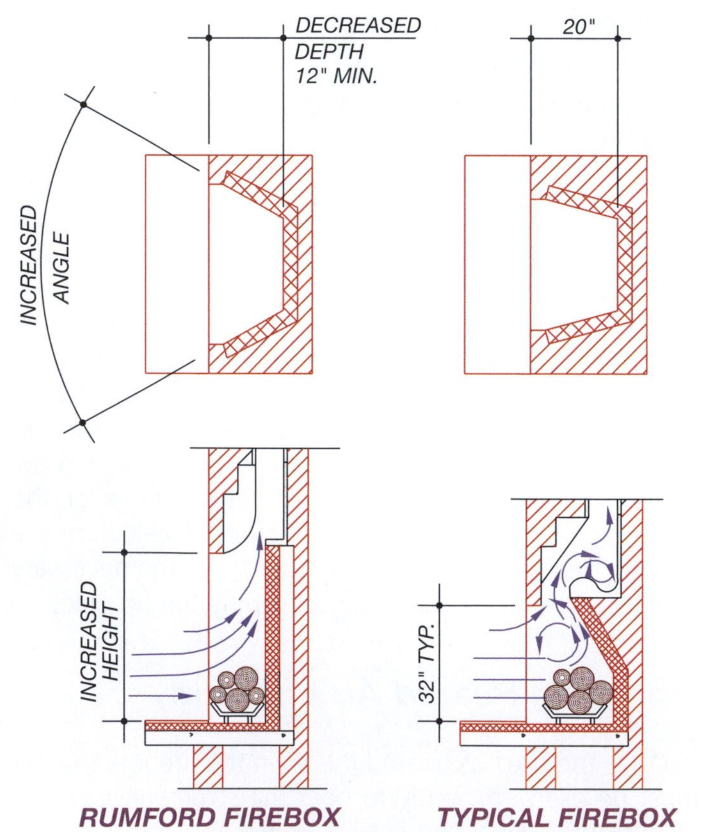

FIGURE 31.11 ■ *Comparison of a Rumford-style firebox and traditional construction. The Rumford style, popular in the mid-1800s, has been reinvented to increase the amount of heated air returned to the room containing the fireplace.*

Decreasing Chimney Heat Loss

The traditional means of reducing heat loss up the chimney has been a damper. The damper is opened while the fire is burning and closed when the fire is out. Because the damper is not airtight, warm air from the room will rise and be drawn up the chimney. As the air in the chimney cools, it will start a reverse current, drawing cold air into the heated room. Placing a chimney on an outer wall will increase this tendency, since the chimney will be cooler. In very cold climates, the reverse convection current can be reduced by adding an additional damper at the top of the chimney.

DRAWING A FIREPLACE SECTION

The fireplace section can be a valuable part of the working drawings. Although a fireplace drawing is required by most building departments when a home has a masonry fireplace and chimney, you may not be required to draw a fireplace section. Because of the similarities of fireplace design, several different fireplace drawings are often kept on file as stock details at many offices. These typical sections are saved in the symbols library and inserted into the working drawings when necessary. A stock detail can be seen in Figure 31.12.

If a fireplace section is required, a print of the floor plans will be needed to help determine wall locations and the size of the fireplace.

Fireplace Layout

Use the *DETL OUTL* layer to draw the following steps. See Figure 31.13 for steps 1 through 4.

Step 1. Draw two lines to represent the width of the fireplace. Assume a depth of 20'' (500 mm) for the firebox and 8'' (200 mm) for the rear wall.

Step 2. Draw all of the walls, floors, and ceilings that are near the fireplace. Be sure to maintain the required minimum distance from the masonry to the wood as determined by the code in your area.

Step 3. Draw the foundation for the fireplace. The footing must be 6'' (150 mm) wider on each side than the fireplace and 12'' (300 mm) deep.

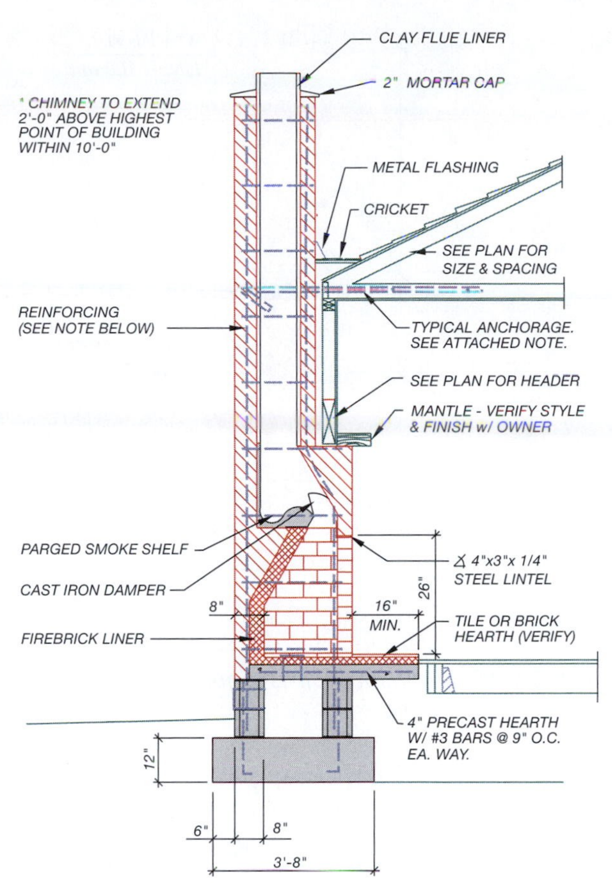

FIGURE 31.12 ■ *Typical sections are saved in the symbols library and inserted when necessary into the drawing set.*

Step 4. Draw the hearth. The hearth will vary in thickness depending on the type of floor material used. If the fireplace section is being drawn for a house with a wood floor, the hearth will require a concrete slab of least 4'' (100 mm) projecting from the fireplace base to support the finished hearth.

Use the *DETL OUTL* layer to draw the following steps. See Figure 31.14 for steps 5 through 8.

Step 5. Draw the firebox. Assume that a 36 × 26 × 20'' (900 × 660 × 500 mm) firebox will be used.

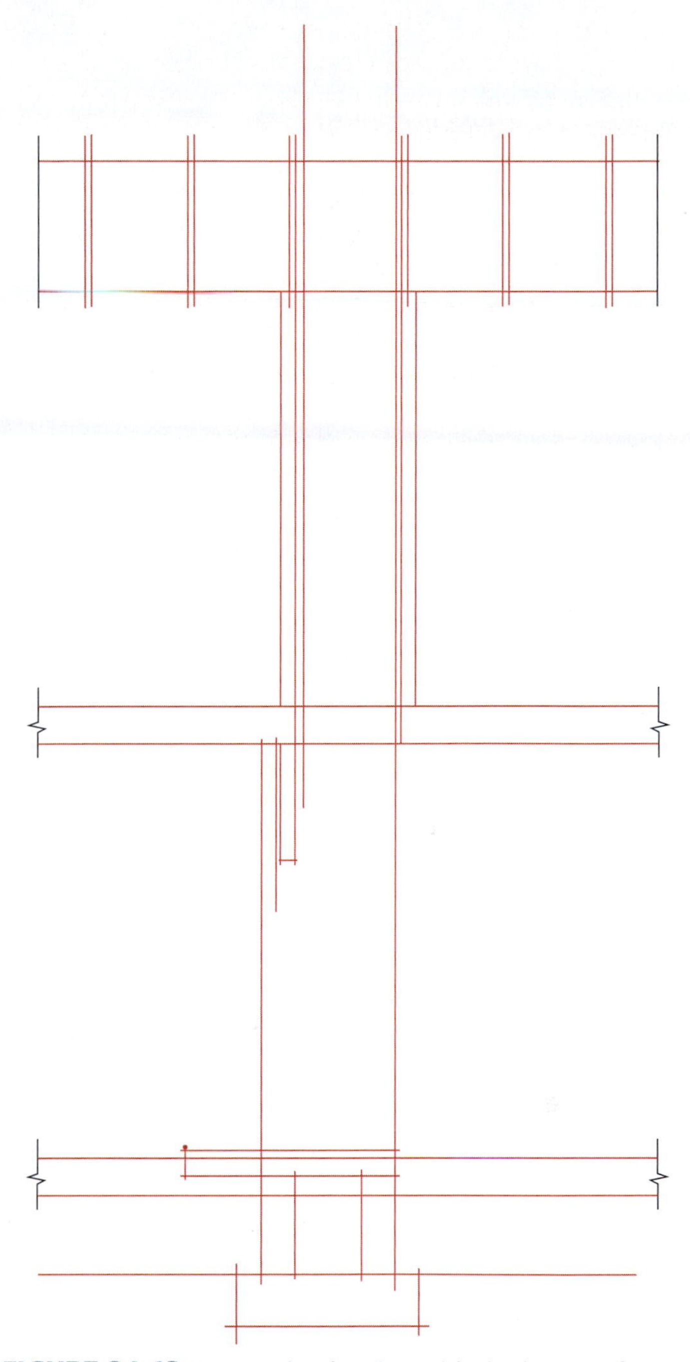

FIGURE 31.13 ■ *Start the drawing with the layout of major items, such as the chimney, footing, and major framing members. Place these members on the DETL OUTL layer.*

See Figure 31.15 for guidelines for representing the firebox.

Step 6. Determine the size of flue required. A 36 × 26'' opening has an area of 936 sq in. By examining Figure 31.16 you can see that a flue with an area of 91 sq in. is required. The flue walls can be constructed using either 8'' (200 mm) of masonry or 4'' (100 mm) of masonry and a clay flue

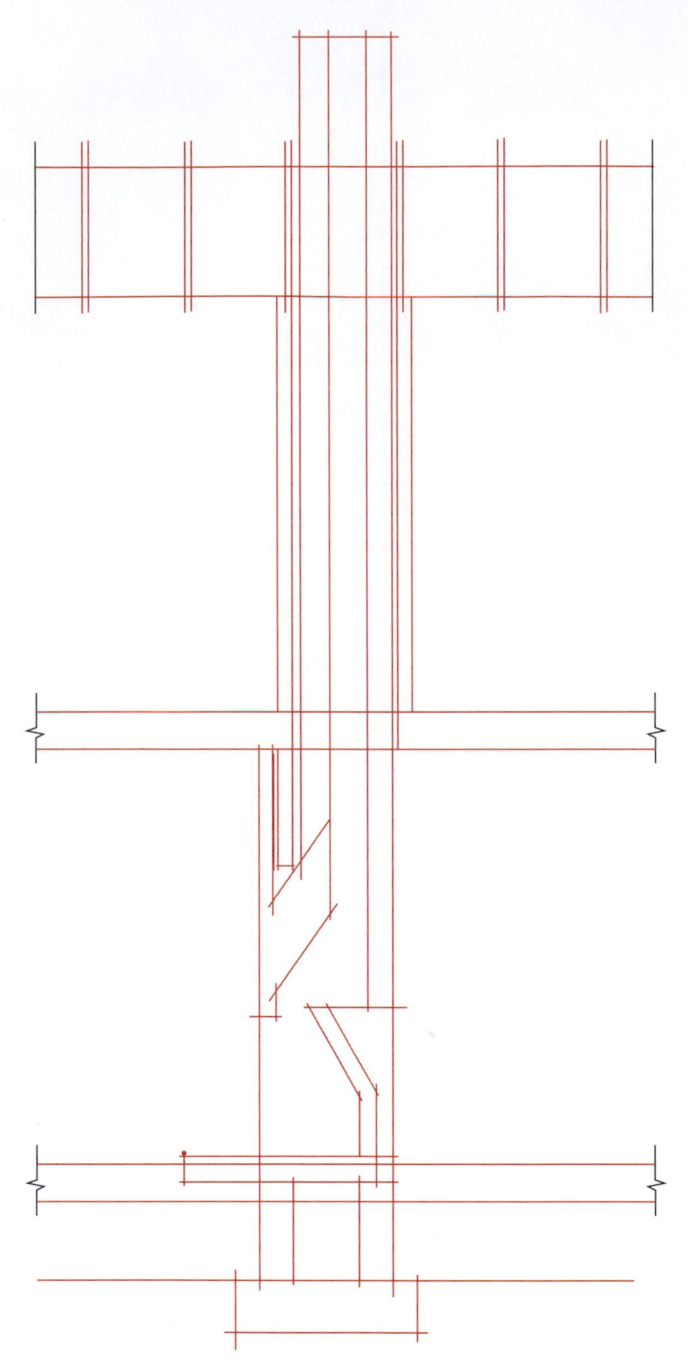

FIGURE 31.14 ■ *Layout of the firebox and flue using the DETL OUTL layer.*

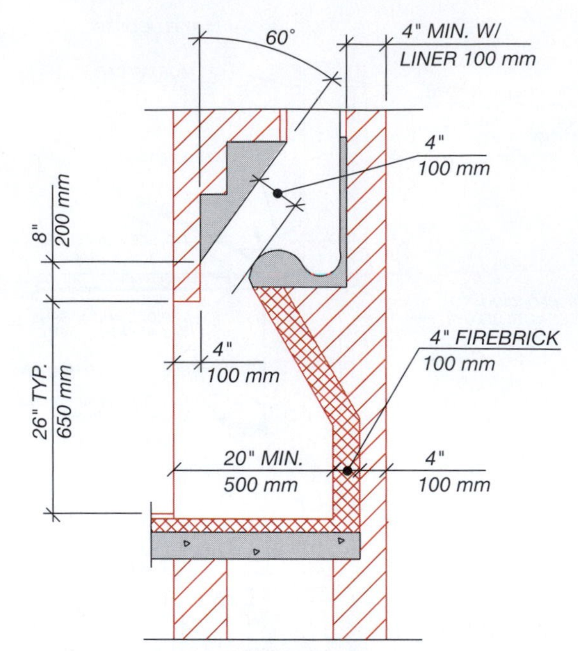

FIGURE 31.15 ■ *Minimum sizes required by the IRC for the firebox.*

liner. If a liner is used, a 13'' (330-mm) round liner is the minimum size required.

Step 7. Draw the flue with a liner. Draw the interior face of the liner. The thickness will be shown later.

Step 8. Determine the height of the chimney. The IRC requires the chimney to project 24'' (600 mm) minimum above any construction within 10' (3000 mm) of the chimney.

Representing Finishing Materials

Refer to Figure 31.17 for steps 9 and 10.

Step 9. Use the *DETL MBND* layer to represent all materials that extend beyond the cutting plane.

Step 10. Use the *DETL MCUT* layer to represent the outline of all framing materials that the cutting plane has passed through.

See Figure 31.18 for steps 11 through 13.

Step 11. Use the *DETL MCUT* layer to highlight the masonry, flue liner, and hearth.

Place steps 12 and 13 on the *DETL PATT* layer.

Step 12. Use the ANSI31 pattern to crosshatch the masonry.

Step 13. Use the ANSI37 pattern to crosshatch the firebrick.

See Figure 31.19 for steps 14 through 16. Place each item created in these steps on the *DETL MCUT* layer.

Step 14. Draw all reinforcing steel with bold lines. Check local building codes to determine what steel will be required.

Step 15. Draw the lintel with very bold lines.

Step 16. Draw the damper with a very bold line on the *DETL MCUT* layer.

See Figure 31.20 for steps 17 and 18.

Step 17. Dimension the drawing. Items that must be dimensioned include the following:

a. Height of the chimney above the roof

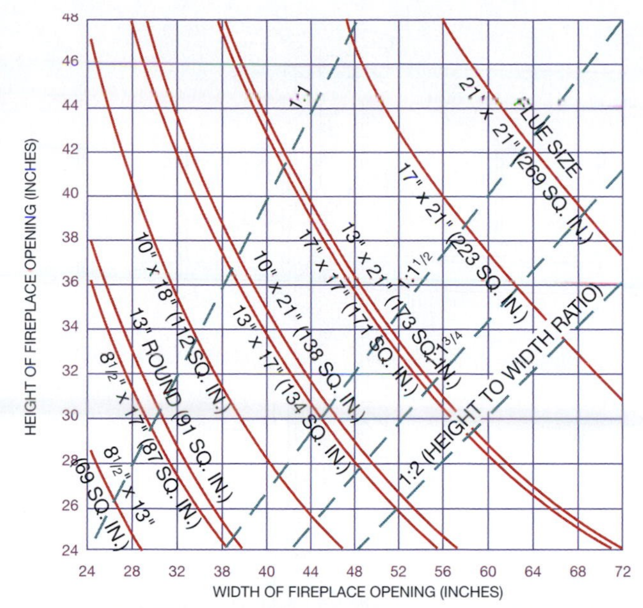

GRAPH TO DETERMINE THE PROPER FLUE SIZE FOR A SINGLE-FACE FIREPLACE, BY MATCHING THE WIDTH AND HEIGHT OF THE FIREPLACE OPENING, THE MINIMUM FLUE SIZE CAN BE DETERMINED. ONCE THE MINIMUM FLUE SIZE HAS BEEN DETERMINED, THE CHIMNEY SIZE CAN BE DETERMINED.

Nominal Dimension of Flue Lining	Actual Outside Dimensions of Flue Lining	Effective Flue Area	Max. Area of Fireplace Opening	Minimum Outside Dimension of Chimney
8½" Round	8½" Round	39 sq. in.	390 sq. in.	17" x 17"
8½" x 13" oval	8½" x 12¾"	69 sq. in.	690 sq. in.	17" x 21"
8½" x 17" oval	8½" x 16¾"	87 sq. in.	870 sq. in.	17" x 25"
13" Round	12¾" Round	91 sq. in.	1092 sq. in.	21" x 21"
10" x 18" oval	10" x 17¾"	112 sq. in.	1120 sq. in.	19" x 26"
10" x 21" oval	10" x 21"	138 sq. in.	1380 sq. in.	19" x 30"
13" x 17" oval	12¾" x 16¾"	134 sq. in.	1340 sq. in.	21" x 25"
13" x 21" oval	12¾" x 21"	173 sq. in.	1730 sq. in.	21" x 30"
17" x 17" oval	16¾" x 16¾"	171 sq. in.	1710 sq. in.	25" x 25"
17" x 21" oval	16¾" x 21"	223 sq. in.	2230 sq. in.	25" x 30"
21" x 21" oval	21" x 21"	269 sq. in.	2690 sq. in.	30" x 30"

FIGURE 31.16 ■ *Common flue and chimney sizes based on the area of the fireplace opening.*

b. Width of hearth
c. Width and height of firebox
d. Width and depth of the chimney footing
e. Wood-to-masonry clearance

Step 18. Place annotation on the drawing using Figure 31.20 as a guide. Notes will vary, depending on the code that you follow and the structural material that has been specified on the floor, framing, and foundation plans.

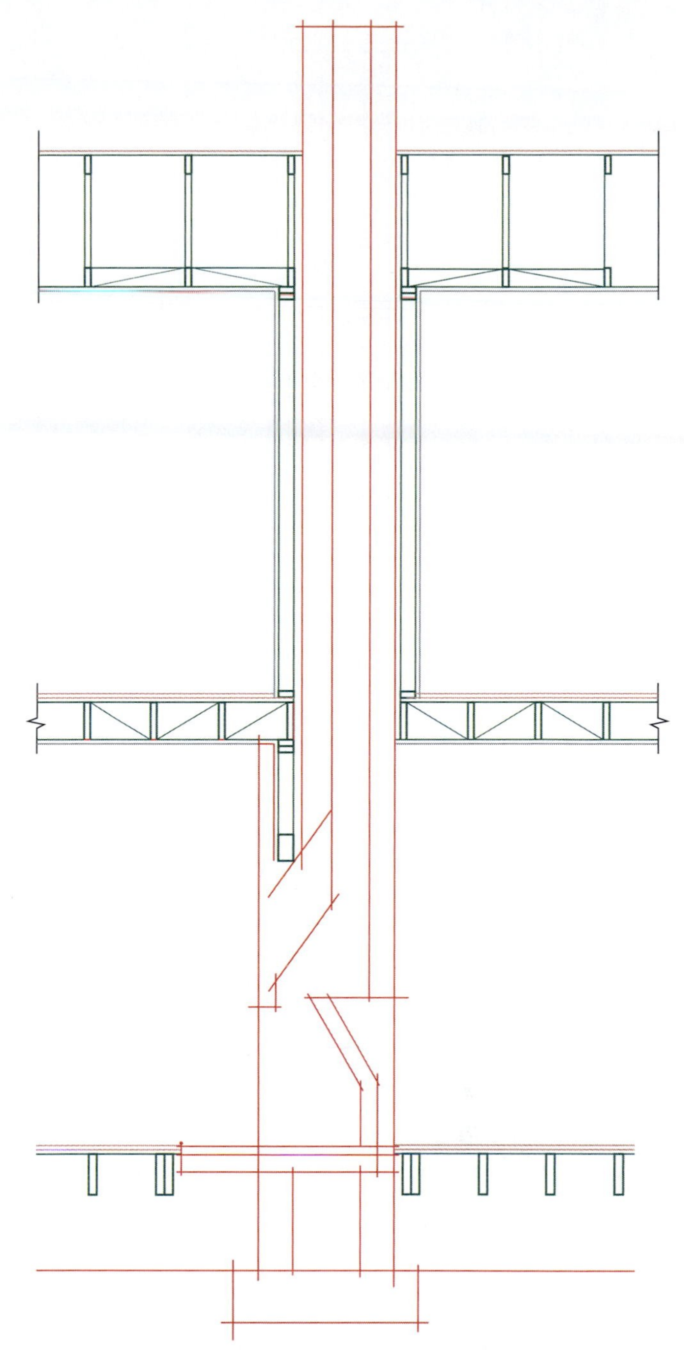

FIGURE 31.17 ■ *Draw all framing members using the DETL MCUT and DETL MBND layers.*

FIREPLACE ELEVATIONS

The fireplace elevation is a drawing of the fireplace and the materials used on the interior of the home. Fireplace elevations must show the size of the firebox and the material that will be used to decorate the face of the fireplace. See Figure 31.21 and use steps 1 through 8 to complete the initial layout of the fireplace elevation. Use the *I ELEV OUTL* layer to draw the following items.

Step 1. Draw the size of the wall that will contain the fireplace.
Step 2. Draw the width of the chimney.
Step 3. Draw the height and width of the hearth if a raised hearth is to be used.
Step 4. Draw the fireplace opening.

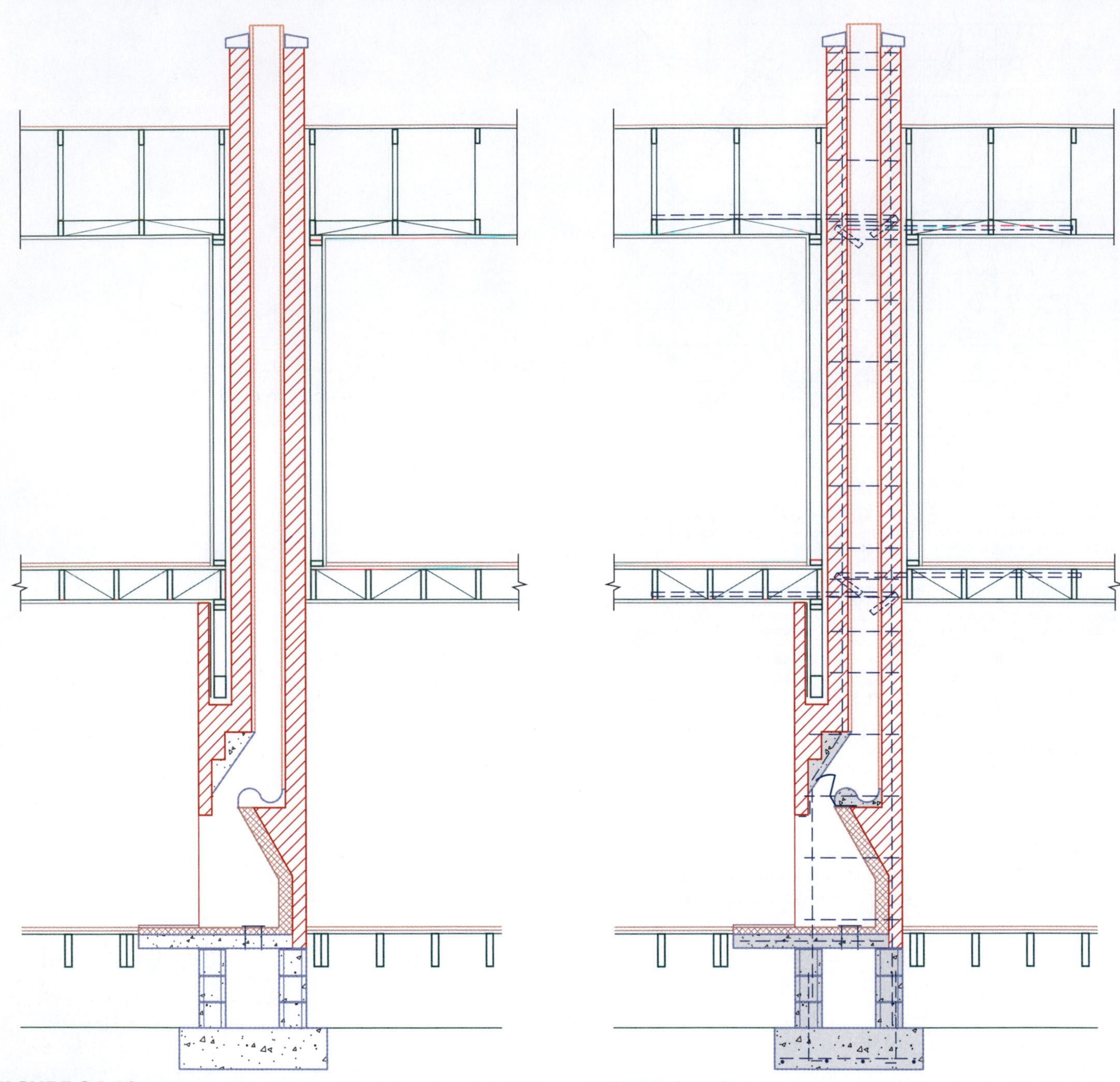

FIGURE 31.18 ■ *Draw all masonry materials using the DETL PATT and DETL MBND layers.*

FIGURE 31.19 ■ *Draw all reinforcing steel with thick lines on the DETL MCUT layer.*

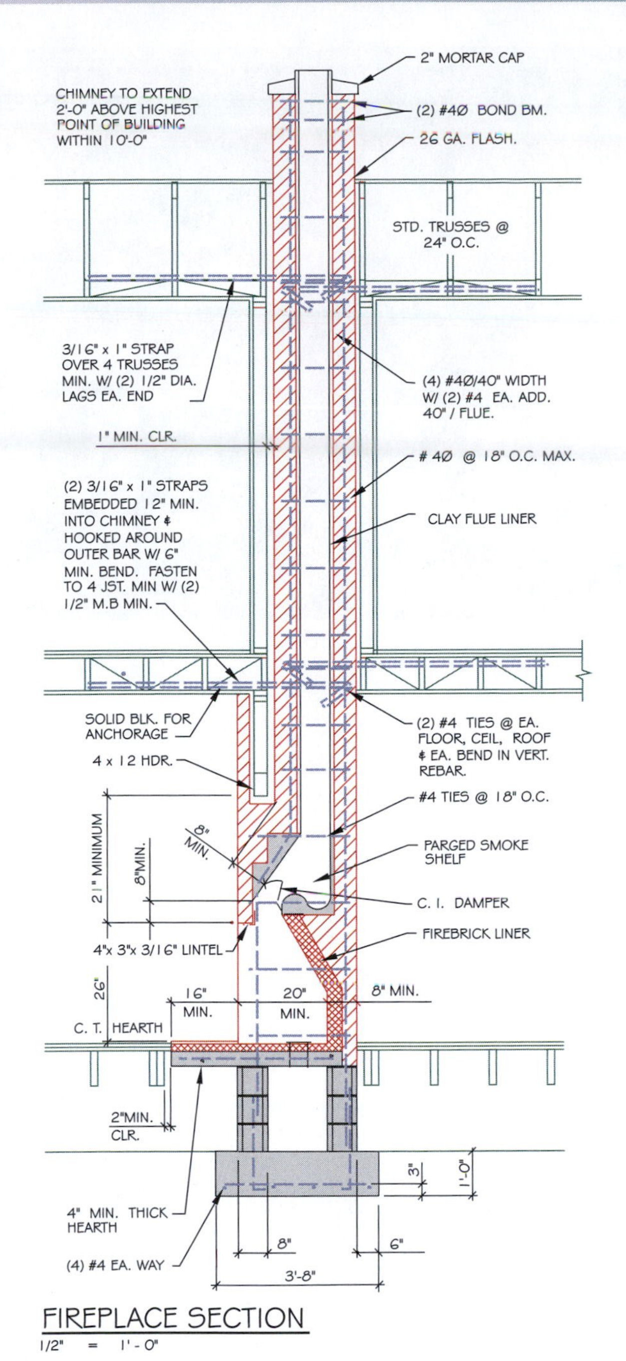

FIGURE 31.20 ■ *Complete the drawing by placing the required notes and dimensions on the DETL ANNO and DETL DIMS layers.*

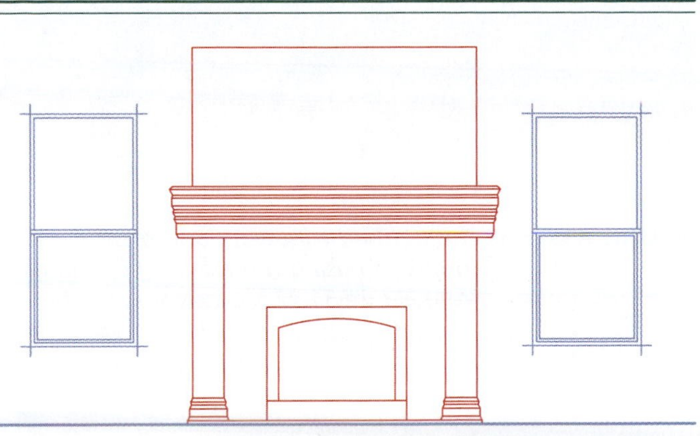

FIGURE 31.21 ■ *Start the fireplace elevation layout by drawing the outline of each major item on the I ELEV OUTL layer.*

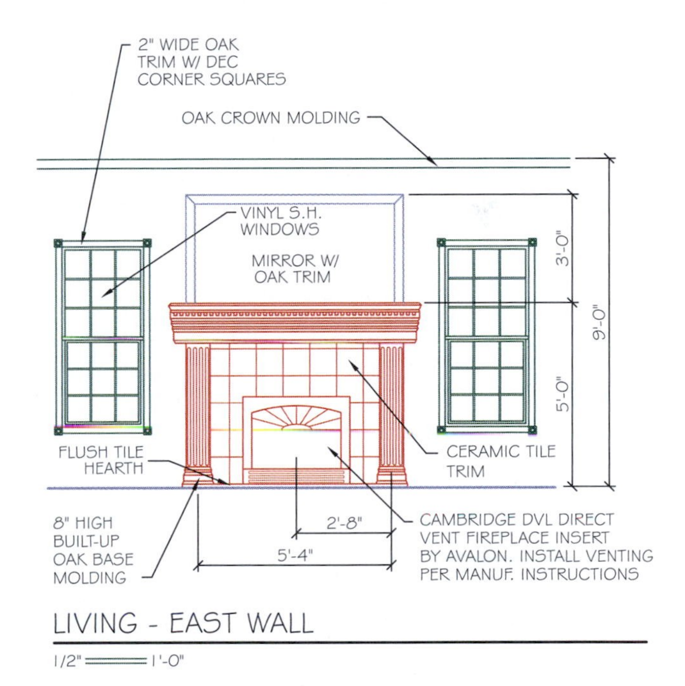

FIGURE 31.22 ■ *The completed fireplace elevation with all materials represented, specified, and dimensioned. The drawing can be inserted into the working drawings with other cabinet drawings or near other floor-related information, depending on available space.*

Step 5. Use the *I ELEV CASE* layer to draw all required crown moldings.
Step 6. Use the *I ELEV FNSH* layer to draw the mantel and all required trim.
Step 7. Use the *I ELEV PATT* layer to draw any required brick or tile patterns.
Step 8. Use the *I ELEV FNSH* layer to draw any required doors or windows that are beside the fireplace.

Once the finishing materials have been drawn, dimensions and annotation can be placed on the elevation as in Figure 31.22. Place dimensions on the *I ELEV DIMS* layer to describe:

1. Room height
2. Hearth height
3. Fireplace opening height and width
4. Mantel height

Place any required notes on the *I ELEV ANNO* layer. Notes will need to be placed on the drawing to describe all materials. The completed project is shown in Figure 31.23.

FIGURE 31.23 ■ *The finished trim work and mantle were completed using the working drawings from Figure 31.22.* Courtesy Megan Jefferis.

CHAPTER 31

Additional Readings

The following websites can be used as a resource to help you keep current with changes in fireplace materials.

ADDRESS	COMPANY OR ORGANIZATION
www.arcat.com/arcatcos	ARCAT, Inc (building product information)
www.bia.org	Brick Industry Association
www.rumford.com	Buckley Rumford Company (fireplaces)
www.centralfireplace.com	Central Fireplace (gas fireplace)
www.fireplace-stove-insert.com	Directory of Fireplace Products
www.eleganceinstone.com	Elegance in Stone (fireplace surrounds)
www.maconline.org	Masonry Advisory Council
www.vermontcastings.com	Vermont Castings/Majestic Fireplaces

CHAPTER 31

Fireplace Construction and Layout Test

DIRECTIONS

Answer the following questions with short, complete statements. Type your answers using a word processor or neatly print your answers on lined paper.

1. Place your name, the chapter number, and the date at the top of the sheet.
2. Type the question number and provide the answer in the form of a statement that includes part of the question. You do not need to write out the entire question.

Note: The answers to some questions may not be contained in this chapter and will require you to do additional research.

QUESTIONS

31.1. What purpose does a damper serve?

31.2. What parts does the throat connect?

31.3. What is the most common size for a fireplace opening?

31.4. What is the required flue area for a fireplace opening of 44 × 26'' (1100 × 650 mm)?

31.5. What flue size should be used for a single-face fireplace opening of 1340 sq in. (864,514 mm^2)?

31.6. How is masonry shown in cross section?

31.7. Why is a fireplace elevation drawn?

31.8. Why is fireplace information often placed in stock details?

31.9. Where should chimney anchors be placed?

31.10. How far should the hearth extend in front of the fireplace with an opening of 7 sq ft (0.65 m^2)?

SECTION 9

Supplemental Drawings

32 CHAPTER

Renovations, Remodeling, and Additions

You will often be confronted by a client who needs to alter an existing structure. The changes may come in the form of alterations, renovations, remodeling, or an addition.

DETERMINING THE TYPE OF PROJECT

Nonstructural changes such as removing or replacing cabinets are an example of alterations. The IRC does not require a building permit for nonstructural changes. The code specifically excludes such projects as painting, papering, paneling, tiling, carpeting, cabinets, countertops, or other finishing projects from needing a permit. Even though a permit is not required, drawings will be useful to aid in communication between the owner and the required contractors who will complete the work.

Renovations

A renovation usually involves removing and replacing non-structural materials such as cabinets and may include minor electrical or mechanical repairs. You do not need a permit to do the following minor repairs and maintenance on a one- or two-family dwelling:

- Paint buildings that are not historic landmarks.
- Blow insulation into existing homes.
- Install storm windows.
- Install window awnings not more than 54'' deep (and not in a design zone) that are supported by an exterior wall and do not project beyond the property line.
- Replace interior wall, floor, or ceiling coverings, such as wallboard or sheet vinyl.
- Put up shelving and cabinets.
- Install gutters and downspouts. (A plumbing permit may still be required for storm water disposal.)
- Replace or repair siding on a wall that is 3' or more from a property line.
- Replace or repair roofing if there is no replacement of sheathing. (A maximum of three layers of roofing is allowed.)
- Replace doors or windows if the existing openings aren't widened.
- Build a fence up to 6' high.
- Pave a walkway.
- Build a patio or deck that is not more than 30'' above grade.

The IRC allows such projects without requiring a building permit, but drawings are still helpful to clearly communicate the work to be completed.

Remodels

A remodel that involves moving, adding, or removing walls requires a building permit. Most building departments require a building permit even if the remodel consists of nonbearing walls contained within the limits of the existing structure. An addition is defined by the IRC as any change in the size of an existing structure caused by increasing the existing floor area or by increasing or altering the height of a structure. A permit is required to construct, enlarge, alter, move, or demolish any one- or two-family dwelling or similar structure. Examples of projects that require a permit include:

- Adding a room.
- Building, demolishing, or moving a carport, garage, or shed of more than 200 sq ft.
- Finishing an attic, garage, or basement to make additional living space.
- Cutting a new window or door opening or widening existing openings.
- Moving, removing, or adding walls.
- Applying roofing when all of the old roofing is removed and new sheathing is installed.
- Building a stairway.
- Building a retaining wall more than 4' high.
- Building a deck more than 30'' above grade.
- Putting up a fence more than 6' high.
- Moving more than 50 yd^3 of earth or any amount of cut or fill on sites affected by waterways or slope hazards.

PERMIT REQUIREMENTS

Building departments require plans that clearly define the location, nature, and extent of the work to be completed. Drawings must contain sufficient detail to allow construction to conform to the existing code. ***Building permits*** required for a remodel or addition includes the structural, electrical, plumbing, and mechanical permits. Some municipalities require the permits to be obtained by the property owner or by the licensed contractor who will complete the work. They do not allow the project designer to obtain the permits, but a member of the design team can prepare the necessary paperwork. If it is legal in your area for the designer to obtain the permits, working with the building department can be a valuable learning experience that will improve your drawings.

Structural Permits

To obtain a structural permit, drawings must be submitted that include:

- The address and ***legal description*** of the property.
- A description of the work proposed.
- The owner's name, address, and phone number. If a contractor is to do the work, provide the contractor's name, address, phone number, and state license number.

Three sets of plans that clearly show all work on the building and where the building sits on the property are typically required. Exact requirements must be determined with the local building department. Plans normally required for a residential addition include a site plan, floor plan, framing plan, foundation plan, exterior elevations, and cross sections showing construction details. In addition to these drawings, written specifications for major equipment and materials, energy documentation, structural calculations, and required fire-protection equipment must be provided. Requirements for each of these drawings are discussed later in this chapter.

Electrical Permits

An electrical permit is required if the following work will be preformed:

- To install or alter any permanent wiring or electrical device
- To run any additional wiring, put in an electrical outlet or light fixture, install a receptacle for a garage-door opener, or convert from fuse box to circuit breakers
- To install or alter low-voltage systems such as security alarms

The local building department responsible for your area issues permits for electrical work. Drawings are usually not necessary to get a permit to do residential electrical work, but you will need to know the structure's square footage, the panel's amperage, and the number of circuits to be added to complete the necessary electrical forms. Fees charged for the permit are based on these figures.

Plumbing Permits

A plumbing permit is required to do the following work:

- To replace water heaters and to alter piping inside a wall or ceiling or beneath a floor
- To do emergency repair, alteration, or replacement of freeze-damaged or leaking concealed piping if new piping exceeds 3'

- To remodel or add to a one- or two-family dwelling when existing plumbing is to be relocated. This includes the installation of building sewers, water service, and rain drains outside the building.

The local building department responsible for your area issues permits for plumbing. Drawings are usually not necessary to get a permit to do residential plumbing work, but you need to know the number of fixtures being added. Fees charged for the permit are based on the size and complexity of the plumbing work being done. Some municipalities base the plumbing fees on the number of fixtures being added or the number of feet of water and sewer lines or the number of rain drains to be added. Some municipalities require a one-line diagram showing pipe sizes for fresh- and wastewater lines. These drawings generally are provided by the plumbing contractor and not supplied by the drafter.

Mechanical Permits

A mechanical permit is required to perform the following work:

- Installing or changing any part of a heating or cooling system that must be vented into any kind of chimney, including unvented decorative appliances.
- Installing a woodstove, fireplace insert, pellet stove, or related venting.
- Installing, altering, or repairing gas piping between the meter and an appliance (indoors or outdoors).
- Installing bath fans, dryer exhausts, kitchen range exhausts, and appliances that are required to be vented.

Plans generally are not required to get a permit to do mechanical work on a dwelling, but you will be expected to briefly describe the work proposed. For example, the sort of appliance to be installed, whether you will be installing a new vent or new ductwork. If new gas piping is to be installed, the number of outlets (future gas appliances) to be installed must also be specified.

Fire Protection

Depending on the municipality governing the construction site, a sprinkler system may be required when a large addition or renovation is undertaken. Municipalities under the jurisdiction of the NFPA (National Fire Protection Association) may require fire protection above what the IRC requires. Homes larger than 3500 sq ft may be required to have sprinkler systems installed to protect them. The designer will need to verify with the building and fire departments regarding specific requirements. Typically NFPA requirements are based on the proximity of existing fire protection, access to the site, distance to the nearest fire hydrant, and size of the structure. For the house considered in this chapter, sprinklers would have been required based on the home size but were waived because two fire hydrants are adjacent to the property.

DRAWING CONSIDERATIONS FOR AN ADDITION

When a home is remodeled or space is added, the same types of drawings required for new construction must be submitted to the building department. The minimum drawings required to obtain a building permit would include site, floor, foundation plans, sections, and exterior elevations. Electrical, framing, roof plans, and interior elevations may also be required depending on the complexity of the structure. The major difference between the drawings required for new construction and those for a remodel or addition is the need to clearly distinguish between types of construction. Another major difference is not in the drawings but in the project's starting point.

Defining Construction Types

It is extremely important that major types of construction be defined on each of the drawings explaining the alterations. Major types of construction to be defined include, new material, existing material to remain, existing material to be removed, future work, items to be relocated, and material not in the contract. The drawing where the information is placed will affect how it is represented. Figure 32.1 shows an example of a floor plan with several types of construction. Because the use of symbols to represent each type of construction will vary with each office, a legend must be provided to clearly distinguish each material used. Common symbols are discussed as each drawing type is introduced. A common method to represent materials is by the use of color. Using a plot style of gray scale, new materials can be represented in black and existing materials in gray, using

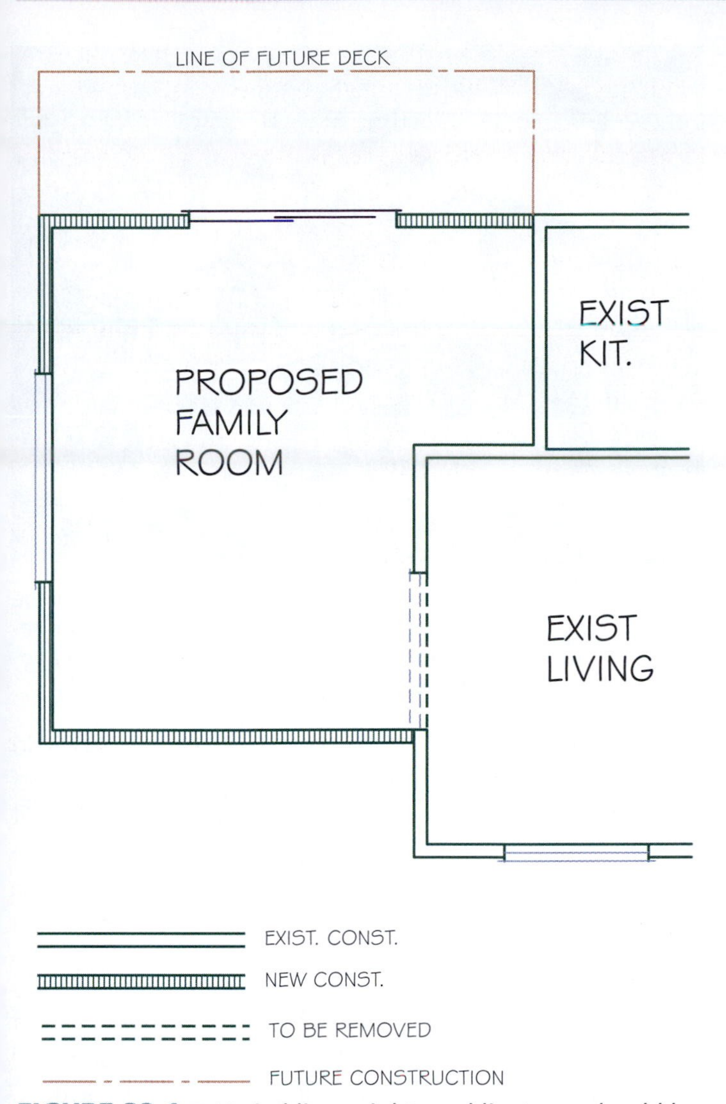

FIGURE 32.1 ■ *Varied lineweights and linetypes should be used to distinguish between types of construction.*

AutoCAD color numbers 8 or 9. Dashed lines are typically used to represent materials that are to be removed. Black dashed lines can be used to represent structural materials to be removed and gray dashed lines to represent nonstructural materials to be removed.

Defining a Project

The initial problem to overcome for a remodel or addition is to determine the extent of the existing structure to be affected by the proposed project. To complete the site plan, dimensions for each exterior surface will need to be determined so the footprint of the structure can be drawn on the site plan. In completing other drawings, only the existing rooms affected by the new construction must be represented. Before any dimensions can be considered, the scope of the job site must be considered.

Basic Site Information

The street and legal description of the property are major pieces of information needed to start a remodel or addition. The client will provide the street address of the job site during the initial contact. He or she may also be able to be able to provide a legal description. See Chapter 8 for a review of the different types of legal descriptions. If the owner is unable to provide the legal description, it can be obtained from a local title company. Obtain the name of a title company in your area from the Yellow Pages or the Internet. The customer service department of the title company can provide you with three valuable pieces of information about the proposed job site for free. These include a map with a legal description, a deed showing current ownership, a printout describing the property, and a record of the tax history. You should have the map in your possession prior to making the initial site visit.

The site map will provide the size of the proposed job site and help you to determine the property lines. If you know that a site is 50' (15 000 mm) wide and you determine that the structure is 38' (11 400 mm) wide, you know that 12' (3600 mm) of side yard is available. If you find you have 14' (4200 mm) at the jobsite, you know you either have the wrong lot on the map or you've made a mistake in your measurements. A second piece of information that should be obtained before visiting the site is gained from the zoning department governing the site. Once you have the street address, the setbacks for construction can be determined. This information may also be available by visiting the zoning department web page of the appropriate municipality. Information that should be requested when making contact with the zoning department includes:

- The zoning of the proposed job site
- The types of uses (occupancy) allowed for this property
- Building department requirements for construction
- The size of the front, side, and rear yards
- Size limitations for existing and new structures
- Height limitations
- Whether certain materials such as overhangs, decks, or masonry chimneys extend into the setbacks

- The minimum drawings the municipality requires to obtain a permit

This information usually can be obtained from the website of the governing zoning department.

Once this information is obtained, you're ready to make the initial site visitation.

Initial Site Visitation

As you prepare to visit the site, consider the needed materials, the measurements that must be made, and the measuring methods to be used.

Materials

The following basic equipment should be taken with you as you visit the job site:

- Measuring tapes, camera, clipboard
- Writing equipment: pens (felt/ball, multicolor) pencil
- Flashlight, ladder/step stool and a compass

A 100' (30 000 mm) and a 25' (7500 mm) tape will be useful in making measurements. The 100' (30 000 mm) tape will be used to measure overall sizes of the structure, setbacks, and the width of the street from curb to curb. It should be a flexible tape that will not be damaged if run over by a car. The short tape should be wide enough so it can be extended about 10' (3000 mm) in a vertical direction without bending. The shorter tape will be useful to take most interior dimensions and to determine the vertical dimensions on the outside of the structure. Just as important as the tools is the measuring system that will be used. The best results will be achieved if you are consistent in how measurements are written in your notes. Dimensions should all be written as feet and inches (14'-6'') or in inches (174''). Be consistent as you record measurements, and do not mix feet and inch measurements and inch only measurements. At the risk of seeming anal, you must be consistent in recording distances so that you'll understand your notes after they've sat for a few days.

Taking a camera with you will allow you to take photos of the job site, the surrounding view, each side of the existing structure, and important interior features. Take photos that locate every major feature on the exterior of the structure. Photos should be taken of all surfaces of the structure, even if the surface will not be affected by the new construction. A photo of an unaltered side may be useful to determine angles or size of a feature on an affected side. Figure 32.2 shows the exterior of the structure to be altered in this chapter. A clipboard is a useful tool to provide support while sketching and placing field notes. In adverse weather, a piece of plastic to protect your sketches will also be useful. Writing equipment should include a combination of pens and pencils. Working in multiple colors or mediums will help keep the structure separate from the dimensions and notes explaining the structure. Depending on the weather, ballpoint pens may be better than felt pens. The ink from most felt pens runs if exposed to water. A flashlight is a useful tool if measurements need to be taken in the attic or crawl space. The ladder will provide safe access into the attic.

FIGURE 32.2 ■ *The northeast elevation of this structure will be remodeled. The deck and porch roof will be removed to provide space for a new kitchen and family room.* Courtesy Matthew Jefferis.

Gathering Information

Begin gathering information about the project before you even get out of your car. Take notes about the existing roof shape and major materials such as roofing or siding. It will also be helpful to approximate the roof pitch. Even though you will take measurements, making a note to yourself may help when you are back at a workstation trying to guess what you've written. After you talk to the client and walk around the job site, obtain the following information.

- Sketch the building footprint on a site plan.
- Use a compass to determine north and record it on your site sketch.
- Indicate material on each side at property line. (Trees, fences, hedge, open space)

  F R LS R
- Determine the distance from the structure to existing fences. A fence may not be exactly on a prop-

erty line, but it is a good guide to locating the property edge.

- Locate sewer cleanouts and plumbing stacks on the plan.

Obtain the following dimensions to complete specific drawings. Express dimensions as feet and inch measurements (12'-0''), or as inch units (144''), depending on your preference for entering units at a computer.

Site

Gather information to locate each major feature that may be required by a building department to obtain a building permit including:

- Determine the overall length of each side of the structure. This information is required to draw the building footprint. These sizes are determined by taking total length measurements of each surface or by taking measurements from one opening to the next.
- Determine the driveway location relative to the structure and the street. Note the driveway material. Measure the length and the width of the driveway.
- Determine the sidewalk locations. For homes on acreage, this may be only an estimate or may not apply. For city lots, measure from curb to curb to determine the street width. Measure from the face of the curb to the exterior edge of the sidewalk, then measure the sidewalk width. For many municipalities, the property line is within a few inches of the interior edge of the sidewalk.
- Determine each side yard distance to the structure. On paper you'll dimension from the structure to a line that does not exist at the job site. Look for a fence or a row of shrubs that may indicate the property edge. If you're working on a city lot, measuring to a neighboring structure and then splitting the distance will give you an approximate location. If the house is near the side yard setback, but the location can't be determined, draw the home at the property line and list the minimum required distance. On rural property, an approximation such as 300' ± is adequate.
- Determine the distance from any other structure to the proposed structure. Locate other structures on the site such as a detached garage, barn, or pump house.
- Determine the size of other structures on the property. Determine the overall dimensions of other buildings to locate the footprint of each building on the site plan.
- Determine the location and diameter of trees or other obstructions near where the alteration will occur. As shown in Figure 32.3, locate a tree by taking measurements from three different locations and obtaining the trunk diameter. Measure the distances from known locations on the exterior of the structure. At your workstation, the three locations become the center point and the lengths become the radius. The arcs will intersect at the location of the tree.
- Determine the locations of water and sewer lines or the septic tank and drain field. Sewer is usually provided in the street for most city sites. Usually a sewer cleanout is located outside of a bath or kitchen to indicate the location of the sewer. Where public sewers are not available, a cleanout can often be found in a rear or side yard near a bath or kitchen. If there is a septic tank, a recessed area can often be found about 10' from the cleanout. The property owner typically will have the map that was required to obtain the original building permit. Such a map is useful to approximate the location of the tank and the drainage field. Keep in mind that most building departments need only the approximate location of the septic system. It is generally the contractor's

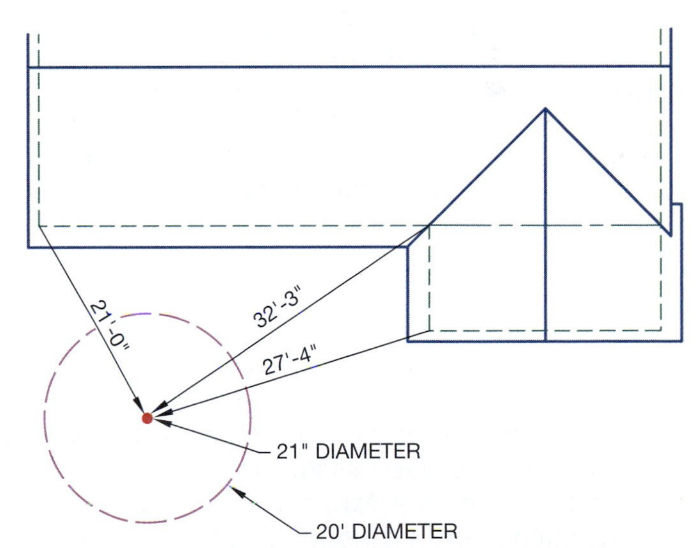

FIGURE 32.3 ■ *A tree can be located by taking three different dimensions from known locations and obtaining the diameter of the trunk. Once back at your workstation, you make the three locations the center point and each distance becomes a radius. The circles will intersect at the location of the tree.*

responsibility during the initial excavation to verify that the new addition will not be built over, or within 10' of a septic tank.

If public water is provided, the main supply line can be located by the location of the water meter. For rural sites, the well location must also be considered in relation to the septic system. Most municipalities require at least 100' (30 000 mm) between the well and the septic system.

Determine the location of all exterior openings in the portion of the structure to be remodeled. The sizes of all exterior openings in the affected rooms must be located. Even if a door or window will be removed, it should be located. Attach a tape measure on one end of a structure and then determine the distance to each edge of an opening to locate all doors and windows on the drawings. Although the size of the trim needs to be known when the elevations are drawn, it is not needed to locate the windows on the floor plan. In addition to locating the openings on the outside of the structure, also locate them from interior walls. This dual measuring serves as a method to cross check dimensions and increase the accuracy of your drawings.

Note:

The siding, corner trim, and wall sheathing add inches that distort your measurements. Taking measurements at the foundation eliminates this accumulation.

The location of the ridge and how it relates to windows or doors should also be noted on a sketch. While measuring the overall size of the structure, take a measurement to locate the ridge. Since you know that it should be in the center of the two exterior walls, this measurement will help determine the overall size and to complete the roof plan.

Elevations

Once the measurements to locate the footprint of the house have been taken, information can be gathered to complete the elevations. Before the elevations for the addition can be drawn, the elevations of the existing structure must be drawn. Information to create the exterior elevations of the existing structure is gathered from the interior and exterior of the structure.

Exterior Measurements. Make a sketch of each surface of the house that will be affected by the new addition and place the following information on the sketch.

- Determine the finished floor location by measuring the distance from the doorsill to the finished grade. Because the grade will not be level, this measurement should be taken at each exterior door. Be sure to count the number of steps and their height.
- Determine the vertical location of all openings. This requires measuring the height of all openings and also the height of the sill above the bottom of the siding. The actual height of the window above the floor can easily be determined when measuring the interior. Comparing interior dimensions with exterior dimensions will accurately locate the bottom of the siding to the floor line. While measuring windows, don't forget to make note of the window type.
- Determine the siding reveal. If horizontal siding is used, measure the exposure of several pieces of siding. If you measure four pieces of siding and determine exposures of 6, 6.125, 5 7/8 and 6'', use the average height of 6'' to estimate vertical distances that are too high to measure with your tape measure.
- Determine the height of exterior walls.
- Measure from the bottom of the siding to the top of the wall at the eave.
- Measure from the top of the windows to the top of the wall at the eave. Count the total number of pieces of horizontal siding from the bottom to the top of the wall at the eave.
- Determine the ridge height. Measure this distance from the top of the wall to the ridge. If a window is located in a gable end, measure the height of the window, the height of the window above the lower wall, and the distance from the top of the window to the bottom of the roof at the ridge.
- Determine the height of trim above ground level. Determine the size and location of any exterior trim.
- Determine the overhang size. Measure all overhangs and note the size of the rafter or truss tails. Note if the eaves are covered with plywood or individual pieces of wood such as 1 × 4s. If the eaves are enclosed make note of the venting method.

Interior Measurements. When you move inside the structure to gather information, two key pieces of information are needed to complete the exterior elevations. Key information includes:

- The height of finished floor to ceiling
- The height of windows above finished floor

Although each of these measurements has been recorded in measuring outside, recording these measurements from the inside will provide a means to check other measurements.

Floor

The floor plan is created from both external and internal measurements. Start the drawing with the external measurements taken to determine the overall footprint. The internal measurements serve as a check when placing the openings in exterior walls and provide information needed to locate the internal walls. It is important to remember that when you work with new construction, you typically use walls 6 and 4'' (150 and 100 mm) wide. When you draw an existing floor plan, these sizes do not apply. Measure at an opening to determine the exact width of exterior and interior walls.

Interior Measurements. Key information to be determined inside the home related to the floor plan includes:

- The overall size of each room. The measurements needed to draw the size of rooms can be determined by laying the tape measure on the floor, by hooking the beginning end under door or window trim, or by having a friend hold the other end of the tape. If help is available, have your assistant read the dimension to you so that you can write the dimension on the sketch. Say the dimension back to your assistant as you write it down to ensure that it has been accurately recorded. Although only the rooms affected by the addition or remodel need to be drawn, it is best to measure all of the rooms. If the scope of the job is altered, an additional trip to the job site will not be required.
- Locate all doors and windows. Work through the house room-by-room and locate all door and window openings. Note the size of each opening as well as the location in the room.
- Indicate all outlets, switches, and heating vents. Note the location of each electrical fixture on your sketch, but there is no need to have exact measurements. Noting that a plug is under the left side of a window allows you to locate the plug if you need to update the electrical drawing. Lighting fixtures for each room as well as the location of each switch that controls the fixtures for the room will also aid in drawing the electrical plan. For homes heated by forced air, note the location and the size of all heat registers and the cold air returns.
- Locate and determine the size of the existing crawl access and attic access.

Sections

The sections are drawn using the measurements that were taken for the site, floor, and elevations. The only two measurements that need to be taken to complete the sections will be the height of the attic and the height from the bottom of the floor to the ground in the crawl space. The height of the attic is determined from the top of the ceiling joists to the bottom of the ridge board. On newer homes, you'll be determining the height of the ridge by measuring the height between the chords. Be sure to note the following information while you are in the attic:

- Rafter (truss chord) size/spacing
- Ceiling joist (truss chord) size/spacing
- Height from top of ceiling joists to the bottom of the ridge (chord to chord)

When measuring the crawl space, determine the following information:

- Construction type (floor joists or post and beam)
  - If floor joists are used note the size and spacing
- Crawl height from decking to ground/beam to ground
- Placement of access and ventilation
- Girder size/locations/direction

DRAWING THE EXISTING STRUCTURE

Once the initial measurements are obtained, these sketches can be used to create the preliminary working drawings. Drawings to be started include the site plan, floor plan, foundation plan, elevations, sections, and details. Throughout this chapter, the development of plans for a family room/kitchen/pantry will be explored. The project consists of removing the existing kitchen and nook areas and expanding the proposed kitchen and providing a new walk-in pantry in the spaces occupied by the current kitchen and family room. A new nook and family room will be added to the north face of the home.

How the drawings will be created must be considered before starting the project. Just as with new construction, drawings are usually assembled by placing all plan views in one drawing file or by using externally referenced drawings. Since the drawings will be created by one office and not require the help of consultants, layers will be used on this project to separate each drawing.

Drawing the Site Plan

The site plan for a remodel or an addition is similar to a site plan for new construction. The plan must show the size of the job site, the street, easements, setbacks, and the footprint of the existing structure, including decks. Once the existing structure is drawn, the space for the addition can be determined. For the addition used in this chapter, the home is in a rural setting and is not located near any setbacks. The existing home is near the top edge of a steep slope that will dictate the size of the addition.

Start the site plan using an existing drawing template or create a template based on the information in Chapter 9. In addition to the layers required for a new site plan, create layers for the following information:

SITE ANNO EXIST	(thin black lines)
SITE DIMS EXST	(thin black lines)
SITE OUTL EXST	(thick gray continuous lines)
SITE PROP	(thin black phantom lines)
SITE PROP BEAR	(thin black lines)
SITE UTIL	(gray, varied linetypes and weights)

Common items to be represented on a site plan include:

- Draw the site using the sizes specified on the map provided by the owner or title company.
- Provide a north arrow.
- Site orientation to the street, and show access to the residence including walks, driveway, front door, and garage.
- Clearly indicate the centerline of the street or access road.
- Locate all setbacks and obstructions to the proposed alterations.
- Show the footprint of the existing structure in bold lines.
- Show existing grade elevations at the property corners and at the corners of the existing structure that will be affected by the proposed project.
- Show existing landscaping that may impact the proposed project.
- Show existing fencing, decorative walls, and decking that will be impacted by the project.
- Show deck walkways, pools, spa, and fountains that may impact the project.
- Use the guidelines in Chapter 9 to complete the initial layout of the site plan. Figure 32.4 shows the existing features for the proposed project. This drawing will serve as a starting point as the project develops and will become the base for the final site plan to be submitted to obtain a building permit.

Drawing the Floor Plan

The complexity of the project determines how the floor plan will be drawn. On a simple project, the existing structure, material to be removed, and new material can be shown on the same floor plan using methods similar to those shown in Figure 32.5. If all information is combined in one floor plan, the print reader will have quick access to all needed information. On complex alterations, separate plans are often used to separate the material to be removed from the new work to be completed. No matter the method used to display the materials, the drawings are started in the same way. On the project for this chapter, the designer placed the existing material on a separate plan from the floor plan showing the new construction.

Start the existing floor plan by using the dimensions determined at the job site. Use the drawing file that contains the site plan and create new layers to contain the floor material. Use the guidelines in Chapter 11 and create new layers for the floor information. Create layers for the following information:

FLOR CABS EXST	(thin gray continuous lines)
FLOR DOOR EXST	(thin gray continuous lines)
FLOR WALL EXST	(thick gray continuous lines)

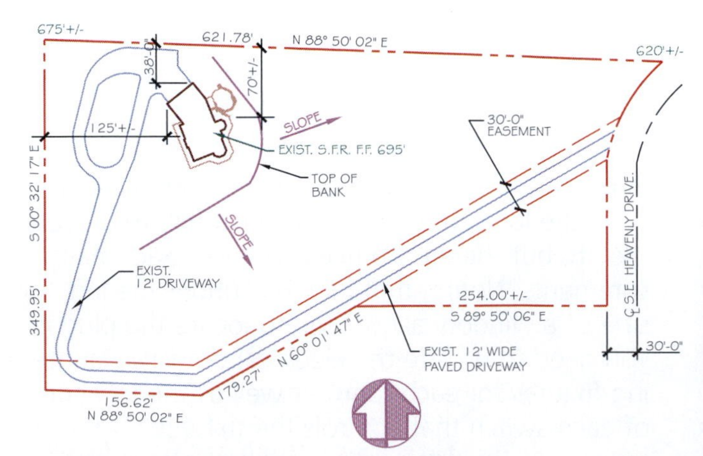

FIGURE 32.4 ■ *Using the guidelines from Chapter 9 and field notes, the initial layout of the site plan can be started. This drawing will provide a base for the design process.*

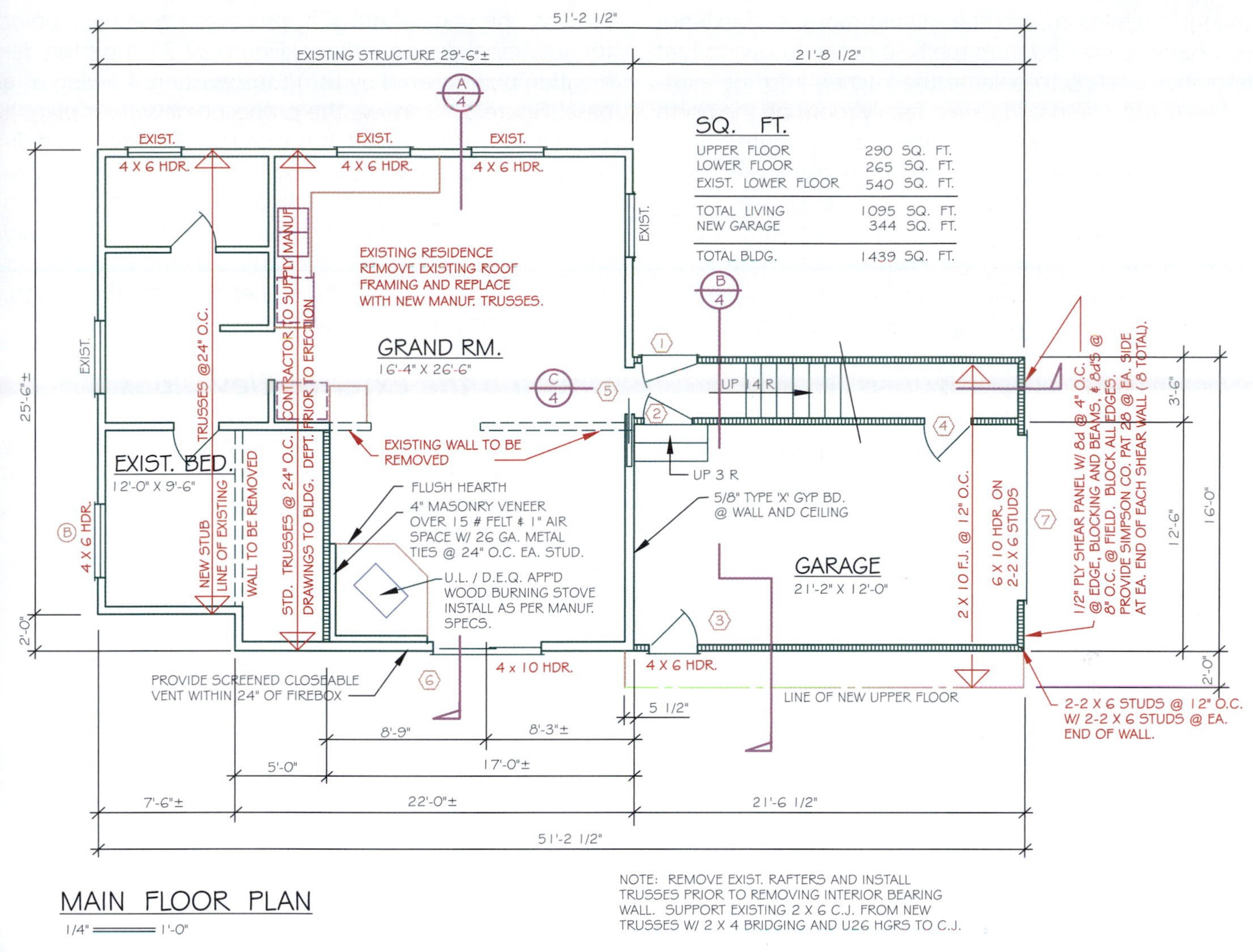

FIGURE 32.5 ■ *The proposed floor plan for a garage addition and a new upper floor over an existing one-level residence. The complexity of the project will determine how the floor plan will be drawn. On a simple project, the existing structure, material to be removed, and new material can be shown on the same floor plan using methods presented in Chapter 10. Combining all information in one floor plan gives the print reader quick access to all needed information.*

FLOR WALL REMVE	(thick gray hidden lines)
FLOR WALL PATT	(thin gray lines)
FLOR GLAZ EXST	(thin gray continuous lines)

By using the footprint of the structure on the site plan as a base, the existing walls can be drawn based on the field notes. Each of the rooms touched by the addition need to be drawn. Figure 32.6 shows the drawing of the

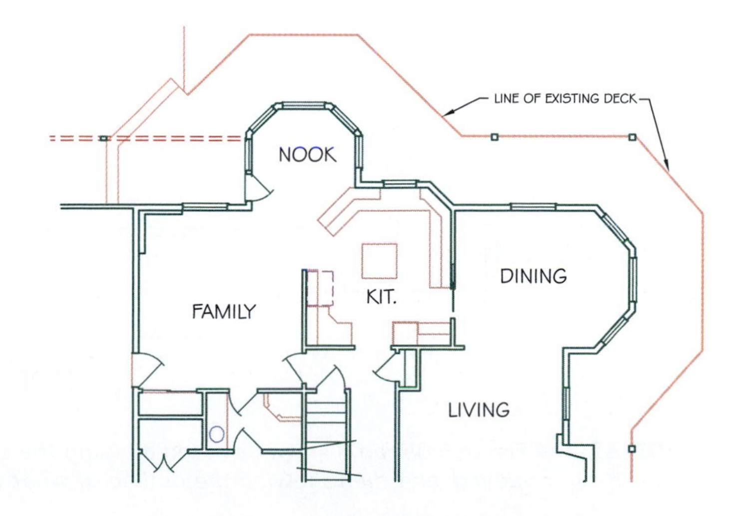

FIGURE 32.6 ■ *The existing walls based on the field notes. Each of the rooms that will be touched by the addition will need to be drawn. Knowing that the owner intends to enlarge the kitchen and add a new family room on the north side of the home allows the rooms on the south side of the home to be ignored.*

existing family room, kitchen, dining room, and exterior deck that was created from the field notes. Knowing that the owner intends to enlarge the kitchen into the existing family room and add a new family room on the north side of the home allows the rooms on the south side of the home to be ignored. Because the plans will document the removal and addition of the kitchen, large amounts of detailing are required to document the project. To clarify information for each of the construction crews, a separate demolition and floor plan will be provided. Separate plans will also be provided to document the electrical and framing information.

Once the base floor plan is created, it can be altered to show material that will remain or be removed. As you begin the new plans, it is wise to keep an unaltered copy of the existing floor plan for future reference. If the drawing suffers major changes during the design process, the original drawing provides a reference point for additional design options. Figure 32.7 shows the demolition plan created by using the existing drawing as a base. Figure 32.8 shows the proposed new floor plan. It was created using the same drawing with the demolition layers frozen and the new construction layers displayed. Notice that only minimal information is placed on both plans. The drawings are still in the preliminary stage, and subject to many changes as the client works with the designer. Once the owner confirms the design, the drawings can be completed.

Drawing the Exterior Elevations

Start the existing elevations by drawing the view that shows the best contour of the existing roof. Use the guidelines that were presented when elevations

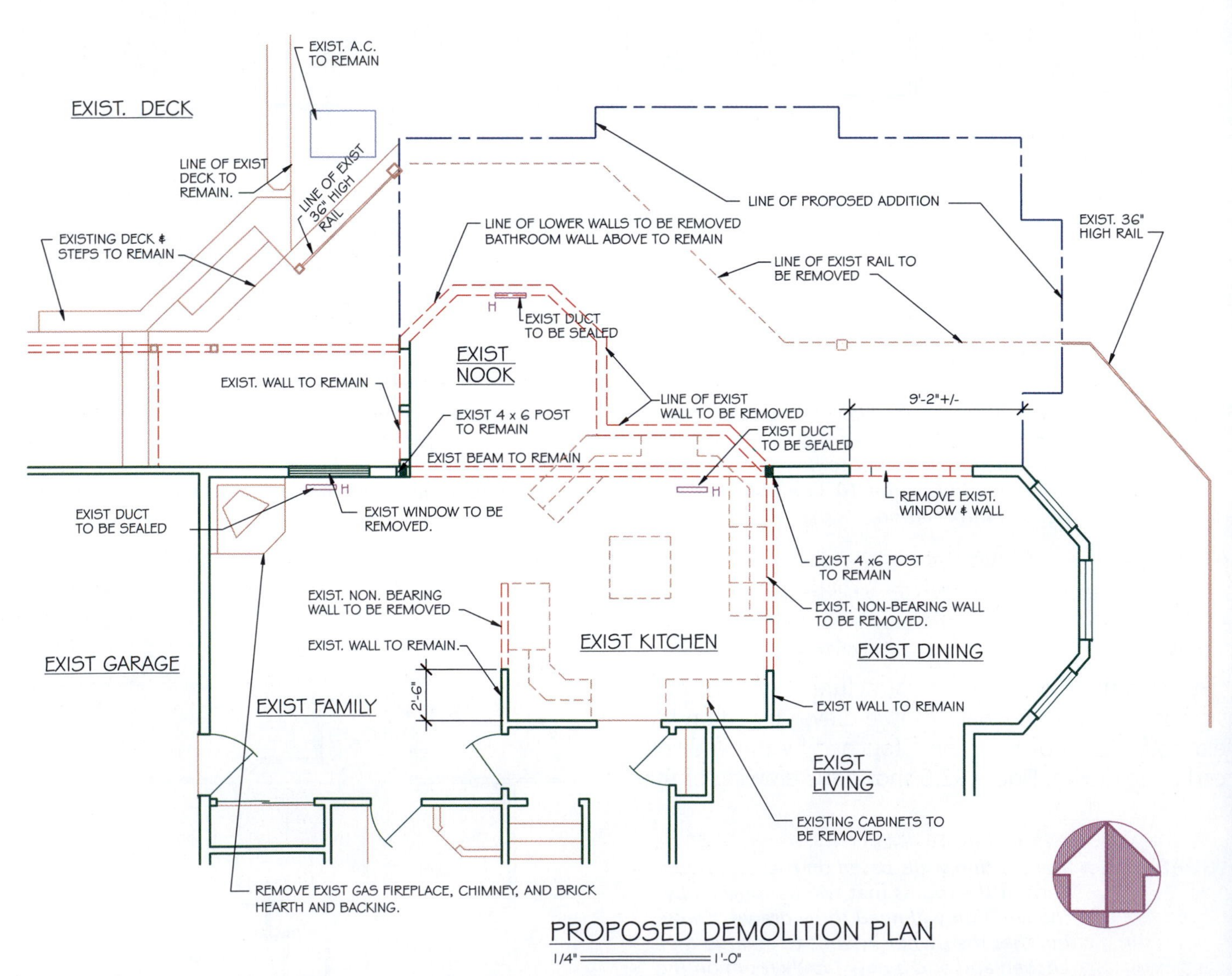

FIGURE 32.7 ■ *The demolition plan was created by using the existing drawing as a base, indicating material that will be removed, and then showing the outline of what will be added.*

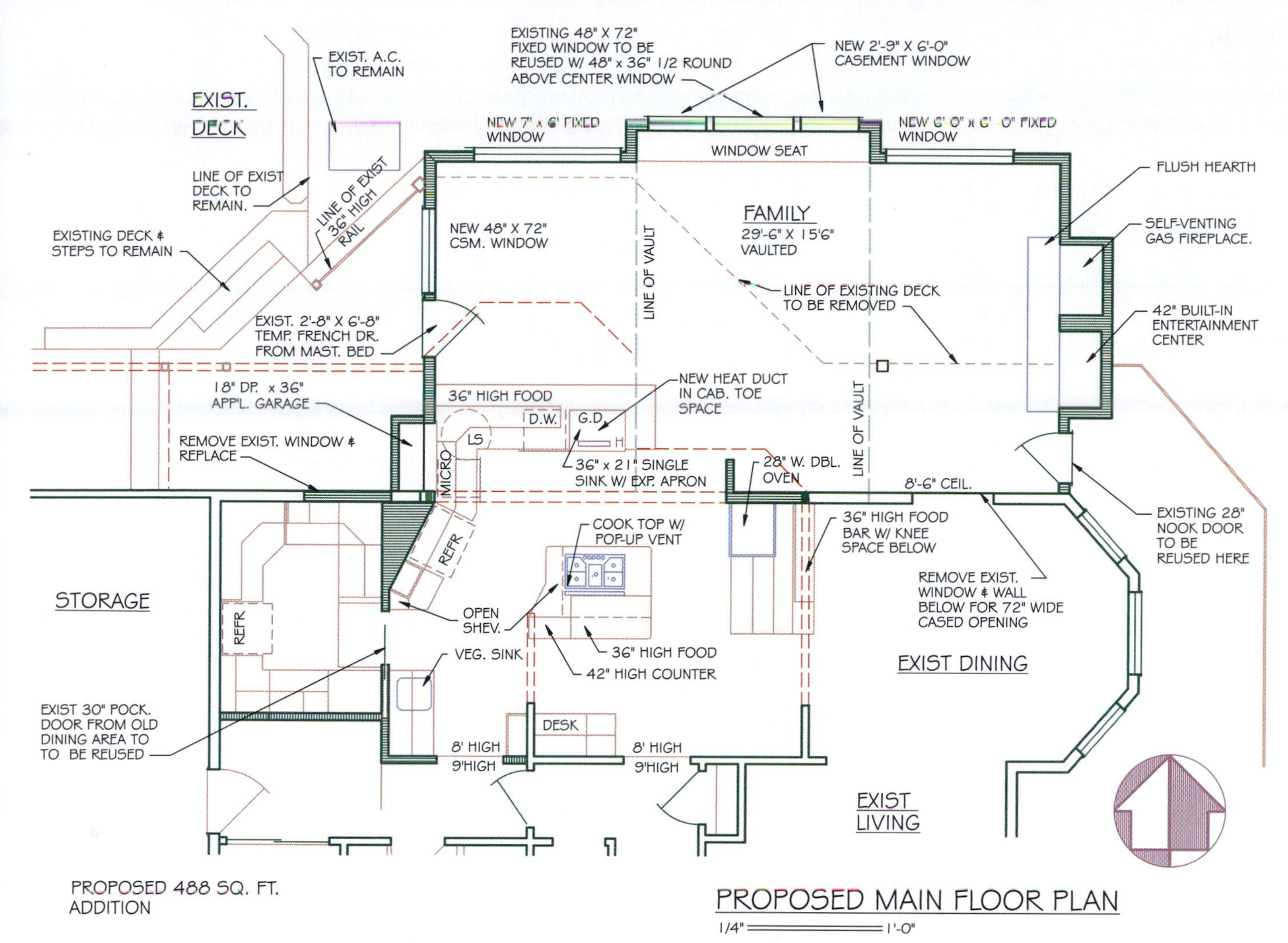

FIGURE 32.8 ■ *The proposed new floor plan was created using the same drawing shown in Figure 32.7 with the demolition layers frozen and the new construction layers displayed. Only a minimum of information has been placed on the plan because it is still in the preliminary stage and requires the owner's approval.*

were explored. Drawing the existing elevations is similar to drawing the exterior elevations for new construction with two exceptions. Knowing that some of the structure will be either removed or covered by the new addition, don't spend time representing siding, roofing, or other exterior details. A second difference is the relationship of the structure to the soil elevations. Keep the drawings simple and show only the following features:

- Shapes of major features of the existing structure
- Existing openings such as doors, windows, and skylights
- Existing chimneys
- Existing porches, decks, and railings
- The slope of the finished grade

Use the guidelines in Chapter 17 to begin the initial layout of the existing elevations. Because these elevations will eventually be used to show the intersections of existing and new material, it will be helpful to establish separate layers to represent the existing and new materials. To represent the existing materials, use layer titles such as:

ELEV ANNO EXST	(thin black lines)
ELEV DIMS EXST	(thin black lines)
ELEV FND EXST	(thin gray dashed lines)
ELEV OUTL EXST	(thin gray lines)
ELEV PATT EXST	(thin gray lines)
ELEV WALL EXST	(thin gray continuous lines)

Figure 32.9 shows the three exterior elevations that will be affected by the proposed project.

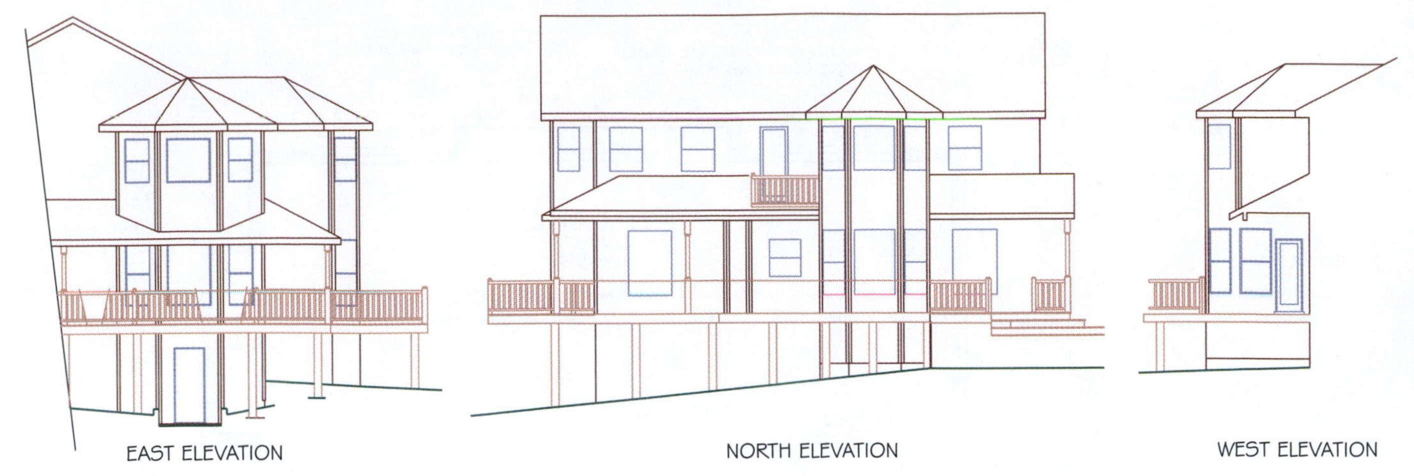

FIGURE 32.9 ■ *The existing exterior elevations will serve as a base for the design of the proposed project.*

Drawing the Foundation Plan

The foundation plan is used to show the intersection of the existing and new foundation and floor systems. The shape of the foundation is drawn using the outline of the existing floor plan. Use the guidelines presented in Chapter 27 to start the layout of the existing materials. If the loads over the existing foundation remain unchanged, you have to show only the general pattern and type of the existing floor system. If new loads are to be imposed over the existing construction, new support eventually will need to be represented on the foundation plan. For now, concentrate on representing the existing materials. Use the sketches from the job site and the guidelines presented in Chapter 26 to layout the existing floor system and foundation material. The drawing should show all materials that will be affected by the new addition. This would include:

- Existing footings and stem walls
- Existing floor system members
- Existing crawl access
- Existing vents
- Existing material that needs to be removed

Create the foundation plan in the drawing file that contains the site as well as the floor plans. Use layers to separate the foundation material from the other plan views and to separate the existing and new foundation materials. Use layer titles such as:

FNDN ANNO EXST	(black lines)
FNDN BEAM EXST	(thin gray lines, hidden)
FNDN DIMS EXST	(black lines)
FNDN FOOT EXST	(thin gray dashed lines)
FNDN JSTS EXST	(thin gray continuous lines)
FNDN WALL EXST	(thick gray lines)

Figure 32.10 shows an example of the foundation plan used to show existing construction.

Drawing Sections

A section of the existing structure is useful for planning intersections between new and existing materials. Use the field notes from the site visit and draw a section of the existing structure. Pass the cutting plane through an area of the existing structure that will be affected by the proposed addition. Since the existing framing is generally not exposed during the site visitation, represent materials you expect to find. Structures built after the 1940s most likely are made using western platform construction. Homes built prior to the 1940s are usually built using balloon framing. Review previous chapters for a review of each type of construction. Existing roof construction can be confirmed by looking in the attic to determine the sizes of rafters and ceiling joists and whether trusses were used. The floor construction can be confirmed by entering the crawl space. The preliminary section should show the following materials:

- Existing foundation and relationship to the finish grade
- Existing floor framing
- Existing wall framing
- Existing roof and ceiling construction

Use the guidelines from Chapter 29 to begin the initial layout of the sections. Because these drawings will eventually be used to show the intersections of existing and new material, it is helpful to establish separate lay-

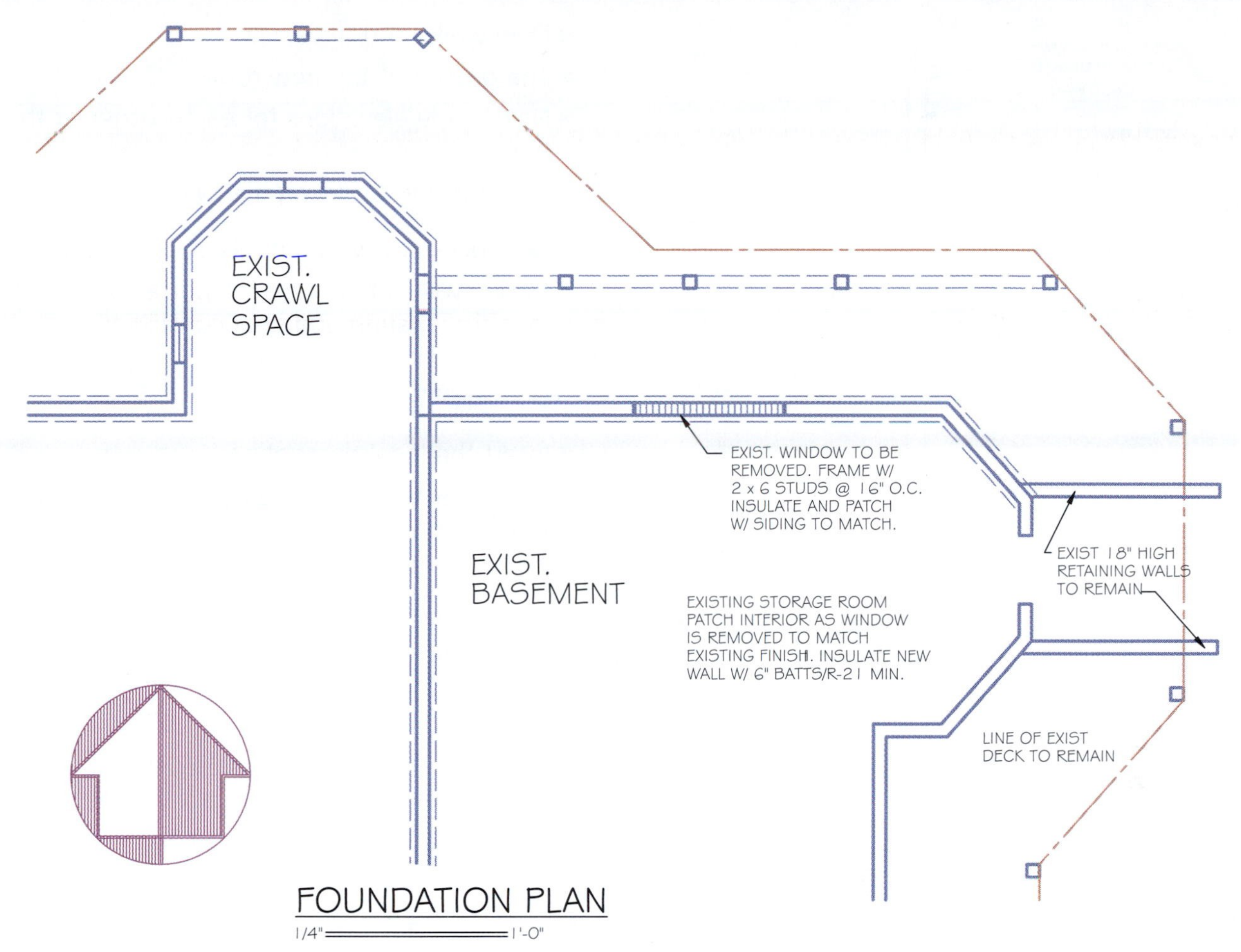

FIGURE 32.10 ■ *Enough of the existing foundation plan must be drawn to show any new loads that must be supported, as well as how new loads will be supported.*

ers to represent existing and new materials. Use layer titles such as:

SECT EXST ANNO	(thin black lines)
SECT EXST DIMS	(thin black lines)
SECT EXST PATT	(thin gray lines)
SECT EXST THIN	(thin gray lines; show material behind the cutting plane)
SECT EXST THCK	(thick gray lines; material cut by the cutting plane)

Figure 32.11 shows one of the sections used to show existing construction. On this project, the following sections were required:

- One will show how the addition will tie into the two-story portion of the house.
- The second section shows how the new project will tie into the structure where the lower floor is removed and the second level remains.
- The third section is a longitudinal section of the new construction and is not started at this point.

REPRESENTING THE PROPOSED CHANGES

Once the existing materials have been represented, the new construction is represented on each of the drawings. The new material is completed using methods similar to those introduced in the previous chapters. The biggest difference in working on an addition or renovation is that the new material on each drawing must be clearly defined. Because gray was used to represent the existing material on each of the drawings, new material should be plotted using black lines. Materials are drawn using any color if a black pen is assigned to the color for plotting purposes. See the AutoCAD Help menu for adjusting plot styles.

Completing the Site Plan

The site plan must show the size of the job site as well as the street, easements, setbacks, and footprint of the existing structure, including decks. The footprint

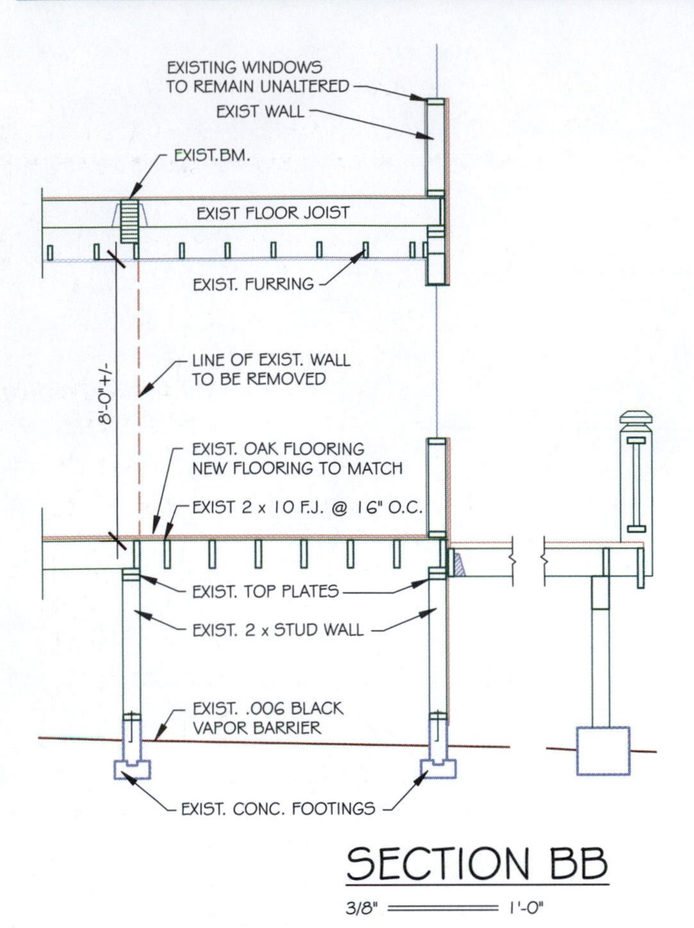

FIGURE 32.11 ■ *Sections must be drawn to show how new construction will tie in with existing construction.*

of the new construction must be clearly indicated in a manner that clearly distinguishes between old and new construction. Although a complete topographic plan is rarely required, the slope of the site must be indicated to ensure proper drainage. Site plans for rural settings may also require the location of wells, septic tanks, and utilities to be represented. The drawing can be completed by using the existing site plan as a base. Notice the modifier *NEWW* (new work) has been added to many of the layer titles. Material to be included can be controlled by using layer titles such as:

SITE ANNO NEWW	(thin black lines)
SITE DIMS NEWW	(thin black lines)
SITE MATL NEWW	(thin black lines)
SITE OUTL NEWW	(thick black continuous lines)
SITE PATT NEWW	(thin black lines)

In addition to the materials shown on the existing site plan, the following items should be represented:

Drawing

- The footprint of new structure in bold lines
- New walks and driveways
- The outline of the new roof
- New grade elevations for each corner of the addition
- New fencing or decorative walls
- New deck walkways, pools, spa, fountains

Provide two different hatch patterns to distinguish between the existing and new construction.

Required Text

- Insert the legal description.
- Label the existing residence as an existing single-family residence (SFR).
- Label the addition as PROPOSED ADDITION.
- Provide the square footage of the existing and the new construction.
- List the finish floor elevation of the proposed construction.
- Label the street, sidewalks, driveways, decks, fencing, grade elevations, concrete flatwork, stairs, patios, walkways, decks, and rails.

Dimensions

Provide dimensions to specify the:

- Lot size
- All setbacks
- Easements
- Centerline of road
- Location of the existing residence and the new construction to each property line

Figure 32.12 shows the completed site plan for the residential addition used throughout this chapter.

Completing the Floor-Related Plans

The floor plan for the project in this chapter shows all new materials. Material to be removed is shown on the demolition plan, and electrical information is shown on a separate electrical plan. Cabinet notes, reference symbols, and framing information are each shown on separate plans. When the floor plan is complete, drawing symbols and notes should clearly define all new materials. A legend should be included or a note provided to explain symbols that define existing and new walls. Other than clearly defining the existing and new por-

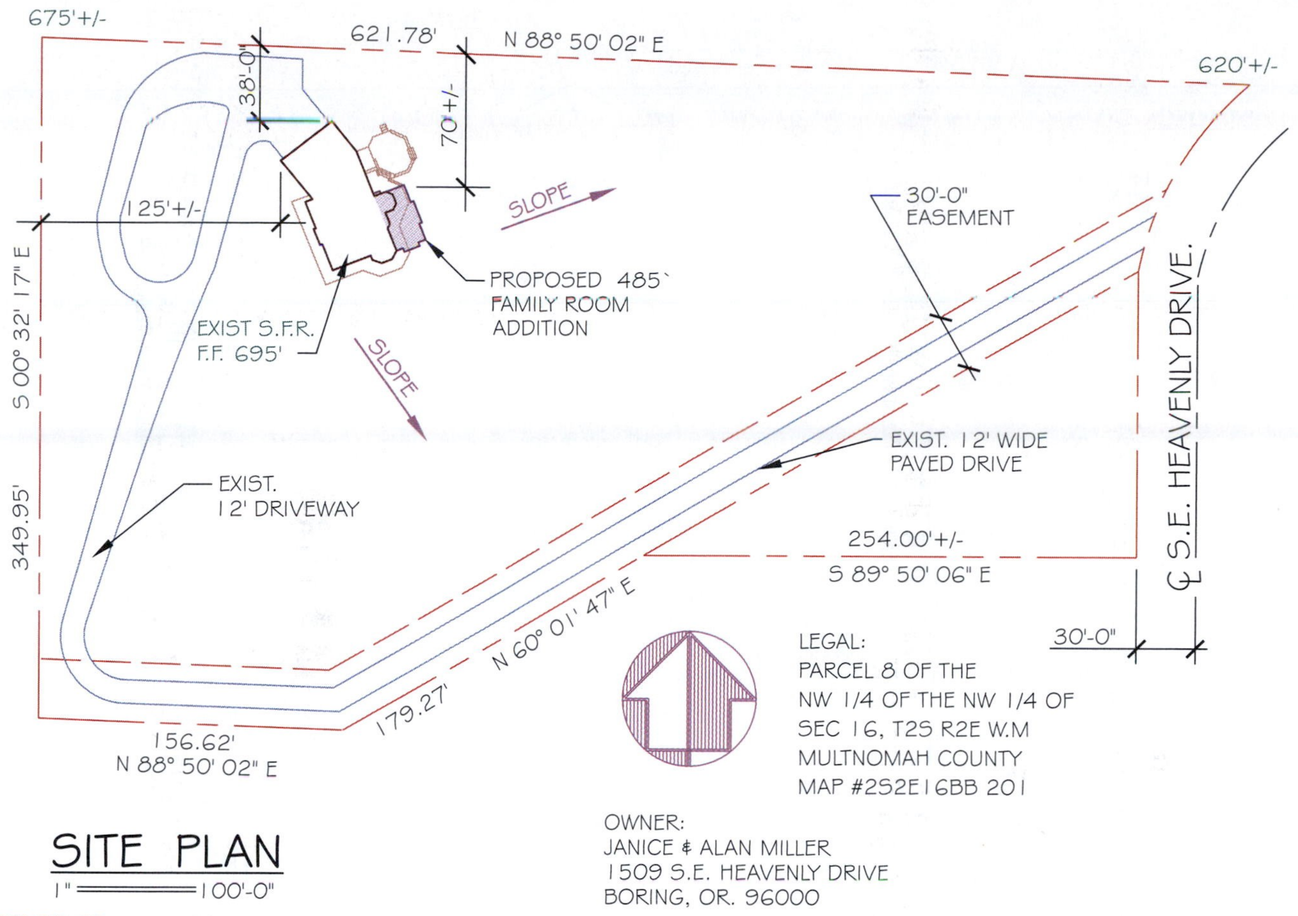

FIGURE 32.12 ■ *In addition to showing material typically represented on a site plan, the plan must clearly represent and distinguish between existing and new materials.*

tions of a project, the floor plan is completed using the guidelines presented in Chapter 11.

The drawing can be completed using the existing floor plan as a base. Material to be included can be controlled by using layer titles such as:

FLOR ANNO NEWW	(thin black lines)
FLOR CABS NEWW	(thin black continuous lines)
FLOR CABS SYMB	(thin black continuous lines)
FLOR DOOR NEWW	(thin black continuous lines)
FLOR WALL EXIST	(thick lines, gray, continuous)
FLOR WALL NEWW	(thick black continuous lines)
FLOR WALL PATT NEWW	(thin black lines)
FLOR GLAZ NEWW	(thin black continuous lines)

Figure 32.13 shows the completed floor plan for the project. Notice that existing walls are drawn with gray lines and new walls have been hatched with a series of thin parallel lines. By using the floor plan with the demolition plan, the owner, contractor, and building officials should have a clear understanding of the new project. Because the floor plan relates to information that will be placed on several different plan views, be sure to:

- Provide a north arrow to reference the floor plan to other plan views.
- Display all plan views in the same orientation. Don't rotate a plan to crowd in additional information.
- Always show the same amount of the existing structure to help the print reader switch between plan views.

Completing the Demolition Plan

The demolition plan shows the existing walls and materials to be removed to prepare the site for the new project. On the project used in this chapter, part of the existing deck, walls, windows, doors, plumbing, and HVAC will be removed or altered in some manner. Existing electrical service will also be affected, but it will be indicated on the electrical plan. Use the existing floor

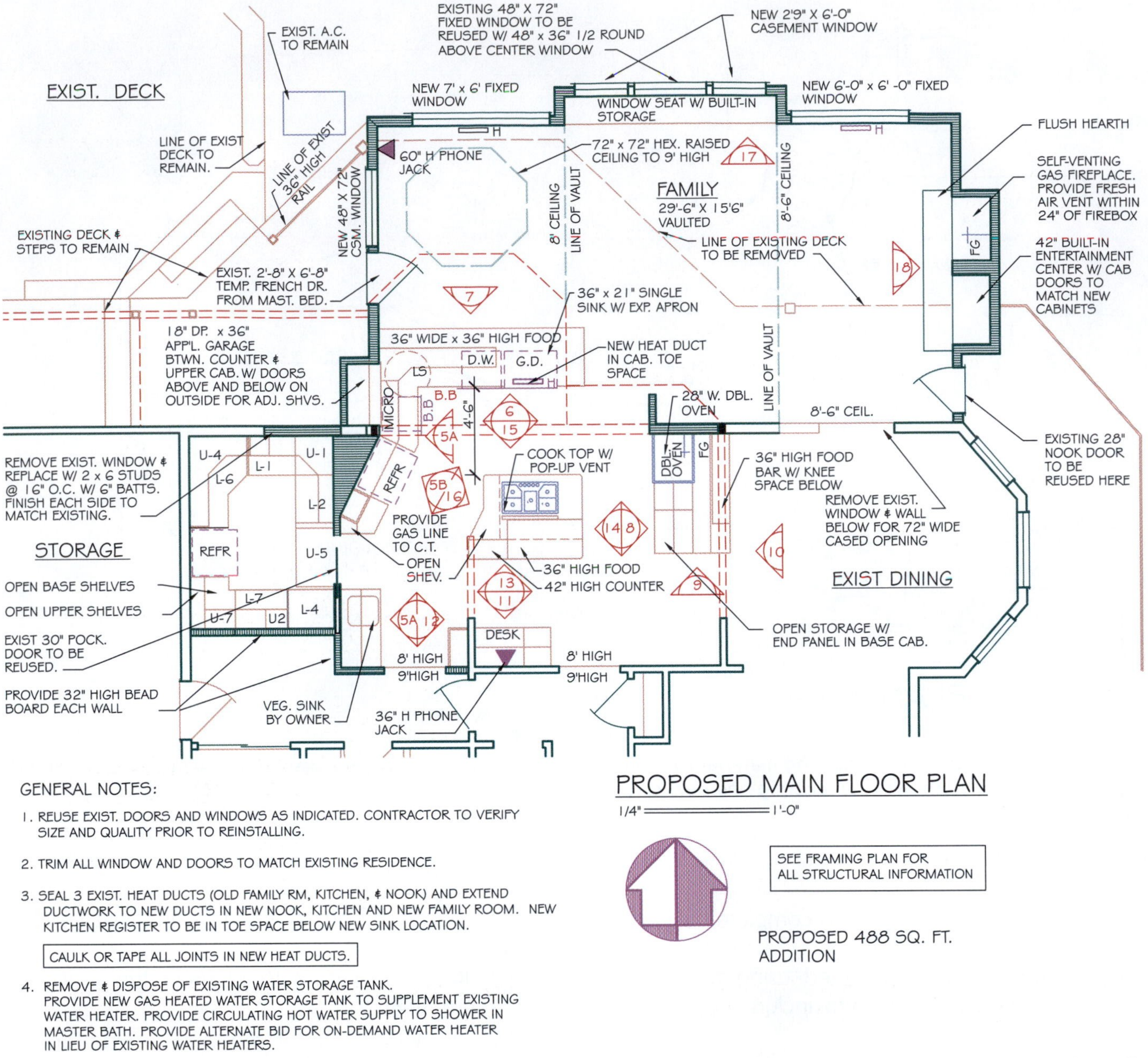

EXIST. DECK
EXIST. A.C. TO REMAIN
EXISTING 48" X 72" FIXED WINDOW TO BE REUSED W/ 48" x 36" 1/2 ROUND ABOVE CENTER WINDOW
NEW 2'9" X 6'-0" CASEMENT WINDOW
NEW 7' x 6' FIXED WINDOW
WINDOW SEAT W/ BUILT-IN STORAGE
NEW 6'-0" x 6' -0" FIXED WINDOW
FLUSH HEARTH
LINE OF EXIST DECK TO REMAIN.
LINE OF EXIST 36" HIGH RAIL
60" H PHONE JACK
72" x 72" HEX. RAISED CEILING TO 9' HIGH
SELF-VENTING GAS FIREPLACE. PROVIDE FRESH AIR VENT WITHIN 24" OF FIREBOX
NEW 48" X 72" CSM. WINDOW
8' CEILING
LINE OF VAULT
FAMILY
29'-6" X 15'6" VAULTED
8'-6" CEILING
EXISTING DECK & STEPS TO REMAIN
LINE OF EXISTING DECK TO BE REMOVED
42" BUILT-IN ENTERTAINMENT CENTER W/ CAB DOORS TO MATCH NEW CABINETS
EXIST. 2'-8" X 6'-8" TEMP. FRENCH DR. FROM MAST. BED.
36" x 21" SINGLE SINK W/ EXP. APRON
18" DP. x 36" APP'L. GARAGE BTWN. COUNTER & UPPER CAB. W/ DOORS ABOVE AND BELOW ON OUTSIDE FOR ADJ. SHVS.
36" WIDE x 36" HIGH FOOD
D.W.
G.D.
NEW HEAT DUCT IN CAB. TOE SPACE
LINE OF VAULT
28" W. DBL. OVEN
8'-6" CEIL.
EXISTING 28" NOOK DOOR TO BE REUSED HERE
REMOVE EXIST. WINDOW & REPLACE W/ 2 x 6 STUDS @ 16" O.C. W/ 6" BATTS. FINISH EACH SIDE TO MATCH EXISTING.
COOK TOP W/ POP-UP VENT
36" HIGH FOOD BAR W/ KNEE SPACE BELOW
REMOVE EXIST. WINDOW & WALL BELOW FOR 72" WIDE CASED OPENING
STORAGE
PROVIDE GAS LINE TO C.T.
OPEN SHEV.
36" HIGH FOOD
42" HIGH COUNTER
EXIST DINING
OPEN BASE SHELVES
OPEN UPPER SHELVES
EXIST 30" POCK. DOOR TO BE REUSED.
OPEN STORAGE W/ END PANEL IN BASE CAB.
DESK
8' HIGH
9'HIGH
PROVIDE 32" HIGH BEAD BOARD EACH WALL
VEG. SINK BY OWNER
36" H PHONE JACK

GENERAL NOTES:

1. REUSE EXIST. DOORS AND WINDOWS AS INDICATED. CONTRACTOR TO VERIFY SIZE AND QUALITY PRIOR TO REINSTALLING.

2. TRIM ALL WINDOW AND DOORS TO MATCH EXISTING RESIDENCE.

3. SEAL 3 EXIST. HEAT DUCTS (OLD FAMILY RM, KITCHEN, & NOOK) AND EXTEND DUCTWORK TO NEW DUCTS IN NEW NOOK, KITCHEN AND NEW FAMILY ROOM. NEW KITCHEN REGISTER TO BE IN TOE SPACE BELOW NEW SINK LOCATION.

CAULK OR TAPE ALL JOINTS IN NEW HEAT DUCTS.

4. REMOVE & DISPOSE OF EXISTING WATER STORAGE TANK. PROVIDE NEW GAS HEATED WATER STORAGE TANK TO SUPPLEMENT EXISTING WATER HEATER. PROVIDE CIRCULATING HOT WATER SUPPLY TO SHOWER IN MASTER BATH. PROVIDE ALTERNATE BID FOR ON-DEMAND WATER HEATER IN LIEU OF EXISTING WATER HEATERS.

PROPOSED MAIN FLOOR PLAN
1/4" = 1'-0"

SEE FRAMING PLAN FOR ALL STRUCTURAL INFORMATION

PROPOSED 488 SQ. FT. ADDITION

FIGURE 32.13 ■ *The floor plan must show portions of the existing plan that will be affected by the addition as well as all new construction.*

plan as a base for the demolition plan. Material to be included can be controlled by using layer titles such as:

FLOR DEMO ANNO	(thin black lines)
FLOR DEMO CABS	(thin gray dashed lines)
FLOR DEMO DIMS	(thin black lines)
FLOR DEMO MISC REMV	(thin gray dashed lines)
FLOR DEMO WALL REMV	(thick gray dashed lines)
FLOR DEMO WALL REMN	(thick gray continuous lines)

Materials that should be represented on a demolition plan include:

- Existing walls, doors, windows, and cabinets to remain
- Existing walls, doors, windows, and cabinets to be removed
  - Indicate material to be removed and reused
  - Indicate material to be removed and discarded
- Notes to specify all materials to be altered

Notice in Figure 32.14 that continuous black lines represent existing materials that will remain; materials to be removed are indicated by dashed gray lines. The outline of the deck to be removed is shown with dashed lines, and a phantom line has been used to indicate the boundary of the proposed project. The results of the demolition plan can be seen in Figures 32.15a and b.

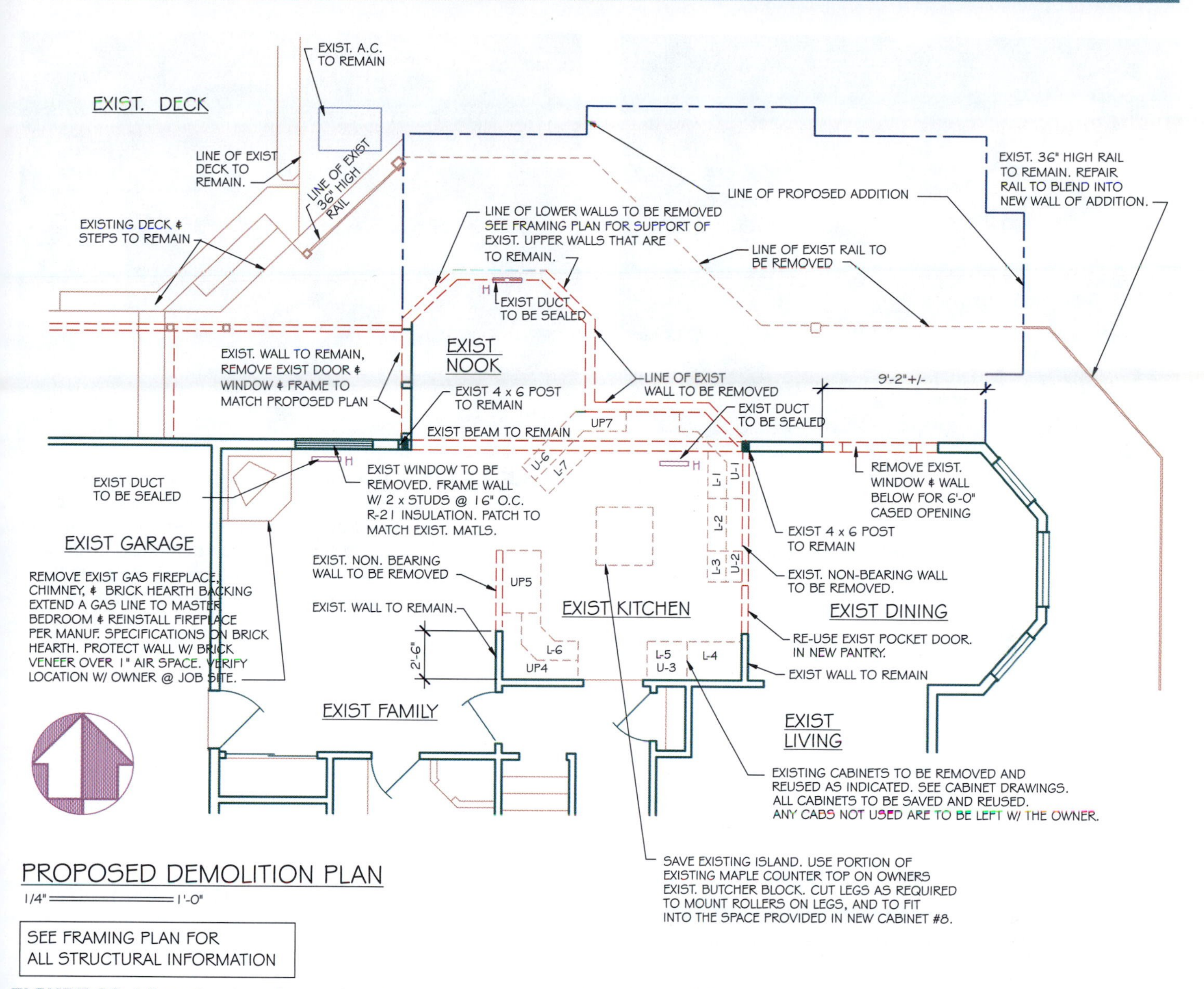

FIGURE 32.14 ■ *The demolition plan shows the existing walls and materials to be removed to prepare the site for the new project. On a simple project, this information can be placed on the floor plan.*

Completing the Electrical Plan

The electrical plan for a renovation, addition, or remodel will be similar to the plan required for new construction introduced in Chapter 12. Use the completed floor plan as a base and freeze all the floor-related materials except for the walls, windows, doors, cabinets, appliances, and plumbing symbols. New fixtures, switches, and plugs can be indicated on the plan. The plan should also show existing switches, fixtures, or plugs that need to be removed or altered in any way. A note may be sufficient to explain what is to be removed or altered. Existing fixtures, plugs, and switches that do not need to be altered do not need to be represented on the plan. Existing and new circuits do not need to be indicated on the drawings. The electrician will design the circuits that will support the indicated fixtures and plugs. In addition to the layers used to describe the existing and new features of the floor plan, materials to be included can be controlled by using layer titles such as:

ELEC ANNO	(thin black lines)
ELEC SCHD	(thin black continuous lines)
ELEC SYMB	(thin black continuous lines)
ELEC WIRE	(thin black dashed lines)

Materials that should be represented on the electrical plan include:

- Existing and new walls, doors, windows, and cabinets

FIGURE 32.15a ■ *The results of the demolition plan with the deck removed and the wall prepared for removal. The wall will not be removed until the new shell is in place.* Courtesy David Jefferis.

FIGURE 32.15b ■ *The kitchen cabinets have been removed (see outline in floor) and the existing wall between the kitchen and the family room is prepped for demolition. All existing electrical and plumbing must be rerouted to allow the wall to be removed.*

- New electrical fixtures including lights, plugs, and switches
- Notes to explain all new electrical work to be preformed by the electrician

Figure 32.16 shows the electrical plan for the proposed addition and some of the resulting work.

Completing the Framing Plan

The framing plan for a remodel is similar to the framing plan required for new construction that was introduced in Chapter 24. Use the completed floor plan as a base and freeze all of the floor-related materials except for the walls, windows, doors, cabinets, appliances, and plumbing symbols. As with other plan views, this drawing must clearly represent new features, existing materials, and material to be removed. The drawing must also clearly show all new and existing framing members. Be sure to distinguish between walls to be removed and new beams and headers. Careful use of layers and color is needed since beams and removed materials are generally represented by hidden or dashed lines on a framing plan. In addition to the layers used to describe the existing and new features of the floor plan, materials to be included can be controlled by using layer titles such as:

FRAM ANNO	(thin black lines)
FRAM BEAM	(thin black dashed lines)
FRAM DIMS	(thin black lines)
FRAM JSTS (or TRUS)	(thin black continuous lines)
FRAM LATL	(thin black continuous lines)
FRAM LATL SCHD	(thick black continuous lines)
FRAM SECT PLAN	(thin black phantom lines)

Materials that should be represented on the framing plan include:

- Existing and new walls, doors, windows, and cabinets
- Framing members, including beams and joists (or trusses)
- Framing fasteners and connectors
- Lateral bracing requirements
- Dimensions to locate all walls, openings, and beams
- Notes to describe all new framing materials and procedures to be completed

Figure 32.17a shows the framing plan for the proposed addition. Notice that existing walls are represented by gray lines in order to provide a stark contrast with new materials. Figure 32.17b shows the results of the information placed on the framing plan

Completing the Elevations

The exterior elevations for a remodel are similar to the elevations required for new construction introduced in Chapter 18. Use the drawings of the existing home as a base and a copy of the new floor plan to project all

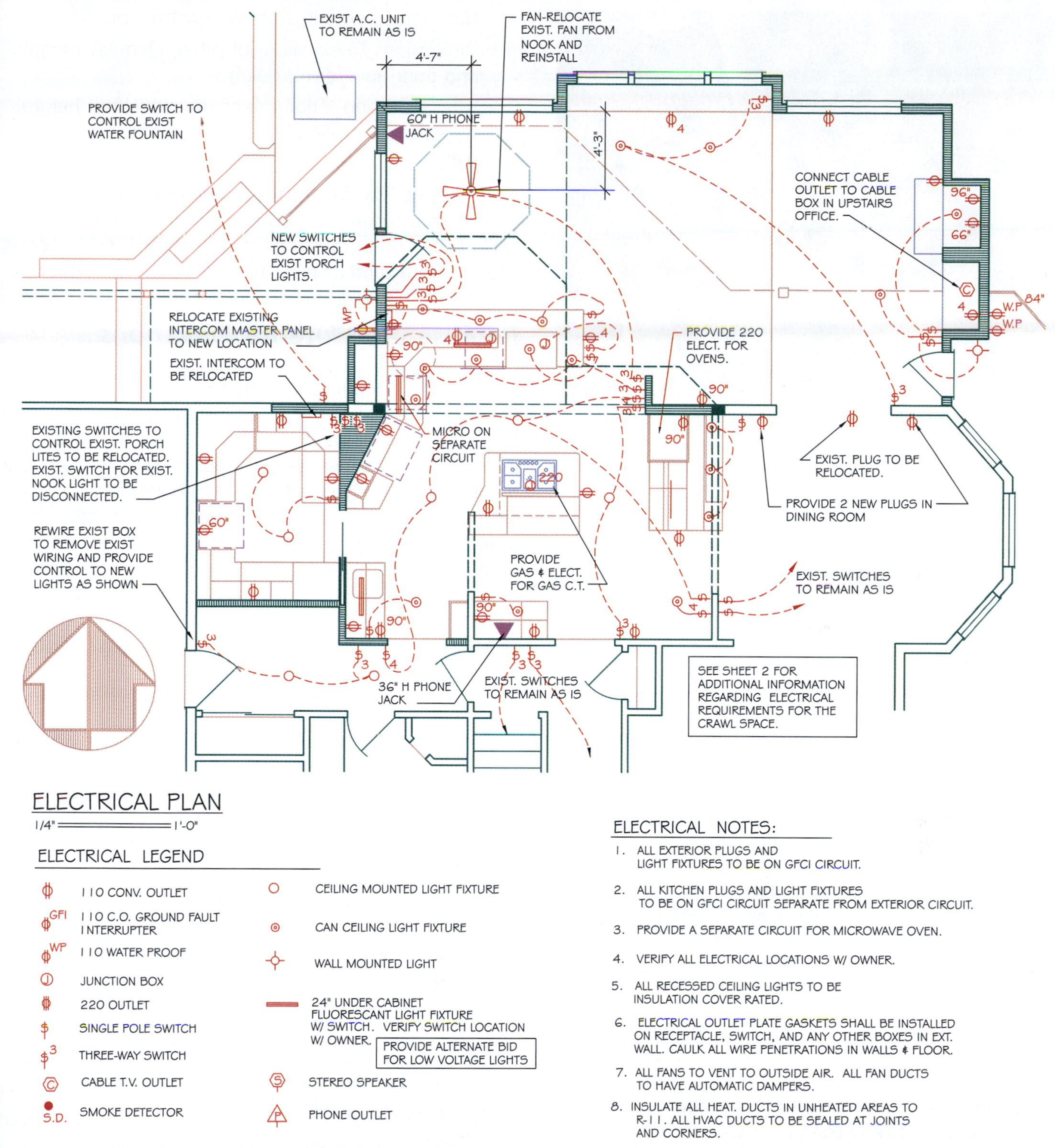

EXIST A.C. UNIT TO REMAIN AS IS
FAN-RELOCATE EXIST. FAN FROM NOOK AND REINSTALL
4'-7"
4'-3"
PROVIDE SWITCH TO CONTROL EXIST WATER FOUNTAIN
60" H PHONE JACK
CONNECT CABLE OUTLET TO CABLE BOX IN UPSTAIRS OFFICE.
NEW SWITCHES TO CONTROL EXIST PORCH LIGHTS.
RELOCATE EXISTING INTERCOM MASTER PANEL TO NEW LOCATION
PROVIDE 220 ELECT. FOR OVENS.
EXIST. INTERCOM TO BE RELOCATED
EXISTING SWITCHES TO CONTROL EXIST. PORCH LITES TO BE RELOCATED. EXIST. SWITCH FOR EXIST. NOOK LIGHT TO BE DISCONNECTED.
MICRO ON SEPARATE CIRCUIT
EXIST. PLUG TO BE RELOCATED.
PROVIDE 2 NEW PLUGS IN DINING ROOM
REWIRE EXIST BOX TO REMOVE EXIST WIRING AND PROVIDE CONTROL TO NEW LIGHTS AS SHOWN
PROVIDE GAS & ELECT. FOR GAS C.T.
EXIST. SWITCHES TO REMAIN AS IS
36" H PHONE JACK
EXIST. SWITCHES TO REMAIN AS IS
SEE SHEET 2 FOR ADDITIONAL INFORMATION REGARDING ELECTRICAL REQUIREMENTS FOR THE CRAWL SPACE.
ELECTRICAL PLAN
1/4" = 1'-0"
ELECTRICAL LEGEND
110 CONV. OUTLET
110 C.O. GROUND FAULT INTERRUPTER
110 WATER PROOF
JUNCTION BOX
220 OUTLET
SINGLE POLE SWITCH
THREE-WAY SWITCH
CABLE T.V. OUTLET
SMOKE DETECTOR
CEILING MOUNTED LIGHT FIXTURE
CAN CEILING LIGHT FIXTURE
WALL MOUNTED LIGHT
24" UNDER CABINET FLUORESCANT LIGHT FIXTURE W/ SWITCH. VERIFY SWITCH LOCATION W/ OWNER.
PROVIDE ALTERNATE BID FOR LOW VOLTAGE LIGHTS
STEREO SPEAKER
PHONE OUTLET
ELECTRICAL NOTES:
1. ALL EXTERIOR PLUGS AND LIGHT FIXTURES TO BE ON GFCI CIRCUIT.
2. ALL KITCHEN PLUGS AND LIGHT FIXTURES TO BE ON GFCI CIRCUIT SEPARATE FROM EXTERIOR CIRCUIT.
3. PROVIDE A SEPARATE CIRCUIT FOR MICROWAVE OVEN.
4. VERIFY ALL ELECTRICAL LOCATIONS W/ OWNER.
5. ALL RECESSED CEILING LIGHTS TO BE INSULATION COVER RATED.
6. ELECTRICAL OUTLET PLATE GASKETS SHALL BE INSTALLED ON RECEPTACLE, SWITCH, AND ANY OTHER BOXES IN EXT. WALL. CAULK ALL WIRE PENETRATIONS IN WALLS & FLOOR.
7. ALL FANS TO VENT TO OUTSIDE AIR. ALL FAN DUCTS TO HAVE AUTOMATIC DAMPERS.
8. INSULATE ALL HEAT. DUCTS IN UNHEATED AREAS TO R-11. ALL HVAC DUCTS TO BE SEALED AT JOINTS AND CORNERS.

FIGURE 32.16a ■ *The electrical plan must show all new fixtures and switches as well as material to be moved.*

FIGURE 32.16b ■ ***Electricians use the electrical plan to determine how new features will be located.*** Courtesy Michael Jefferis.

new features onto the existing drawing. Create a block of the floor plan that includes all features affecting the exterior of the project. Freeze all layers containing interior material. Once the block is created, insert the block into the drawing containing the existing exterior. Insert and rotate the block as described in Chapter 18 to aid projection of the new material. In addition to the layers describing the existing material, create layers to describe new materials using layer titles such as:

ELEV NEWW	(thin black continuous lines)
ELEV ANNO NEWW	(thin black lines)
ELEV DIMS NEWW	(thin black lines)
ELEV FINH	(thin black continuous lines)
ELEV FNDN NEWW	(thin black dashed lines)
ELEV LATL	(thin black continuous lines)
ELEV OUTL NEWW	(thin black lines)
ELEV PATT NEWW	(thin black lines)
ELEV SIDG NEWW	(thin black continuous lines)
ELEV WALL NEWW	(thin black continuous lines)

Materials that should be represented on the exterior elevations include:

- Existing walls, openings, and roof that are impacted by the new construction
- The outline of all new construction including roof, walls, and foundation
- All new openings including skylights, windows, and doors
- All new decks, required supports, and railings
- Notation to explain demolition, repairs, and remaining material
- Notation to explain all new construction
- Dimensions to explain roof pitch, chimney heights, and floor and ceiling heights
- Lateral bracing requirements (optional but helpful to represent material specified on the framing plan)

Figure 32.18a shows the completed elevations for this project. Figure 32.18b shows the completed project. As with other drawings, the existing construction is shown in gray and all new materials are shown in black.

Representing Interior Elevations

Draw the interior elevations for an addition, renovation, or remodel using the methods presented in Chapter 18 to draw the elevations for new construction. Review the step-by-step instructions to complete the drawings. The only difference from new construction is that some materials are to be specified as existing construction. Existing walls or openings may be represented with gray lines; but since the majority of the interior elevations are showing only new cabinets, existing materials are usually not important. Materials to be shown on the interior elevations are controlled by using layer titles such as:

I ELEV ANNO	(thin black lines)
I ELEV DIMS	(thin black lines)
I ELEV SYMB	(thin black continuous lines)
I ELEV THCK	(thick black continuous lines)
I ELEV THIN	(thin black continuous lines)

Materials that should be represented on the interior elevations include:

- Outline of new and existing walls, ceilings, and floors
- All new and existing openings such as doors and windows
- All appliances and fixtures
- All cabinet doors, drawers, and open shelves
- Outlines of all cabinets
- All counters and backsplashes
- Notation to explain all materials, appliances, and fixtures
- Dimensions to explain all the heights and widths of cabinets and appliances

The interior elevations need to be referenced to the floor plan to show the viewing plane for each elevation. The elevation reference symbols are seen in Figure 32.13. Figure 32.19a shows a portion of the completed

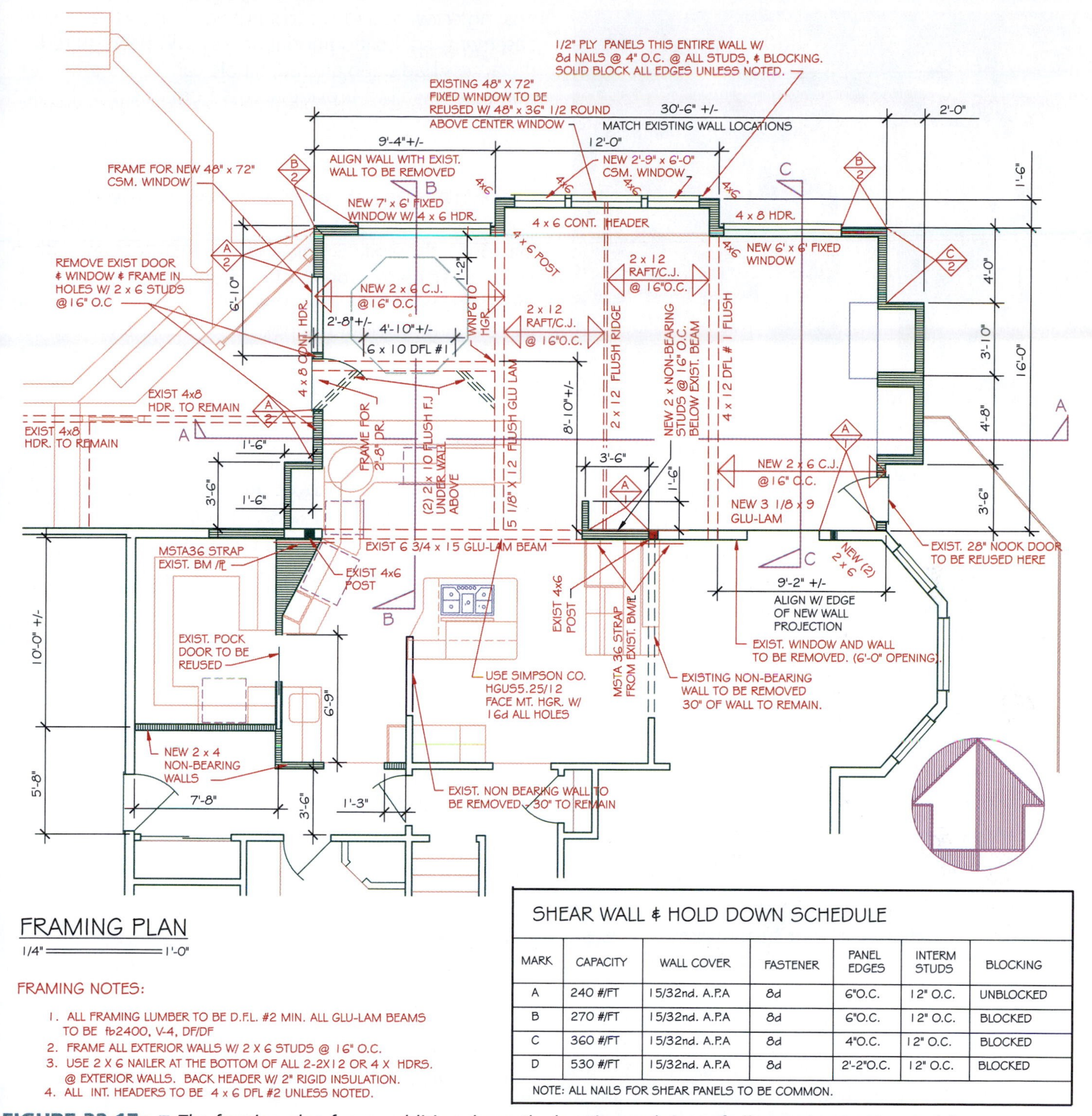

FRAMING NOTES:

1. ALL FRAMING LUMBER TO BE D.F.L. #2 MIN. ALL GLU-LAM BEAMS TO BE fb2400, V-4, DF/DF
2. FRAME ALL EXTERIOR WALLS W/ 2 X 6 STUDS @ 16" O.C.
3. USE 2 X 6 NAILER AT THE BOTTOM OF ALL 2-2X12 OR 4 X HDRS. @ EXTERIOR WALLS. BACK HEADER W/ 2" RIGID INSULATION.
4. ALL INT. HEADERS TO BE 4 x 6 DFL #2 UNLESS NOTED.

SHEAR WALL & HOLD DOWN SCHEDULE

MARK	CAPACITY	WALL COVER	FASTENER	PANEL EDGES	INTERM STUDS	BLOCKING
A	240 #/FT	15/32nd. A.P.A	8d	6"O.C.	12" O.C.	UNBLOCKED
B	270 #/FT	15/32nd. A.P.A	8d	6"O.C.	12" O.C.	BLOCKED
C	360 #/FT	15/32nd. A.P.A	8d	4"O.C.	12" O.C.	BLOCKED
D	530 #/FT	15/32nd. A.P.A	8d	2'-2"O.C.	12" O.C.	BLOCKED

NOTE: ALL NAILS FOR SHEAR PANELS TO BE COMMON.

FIGURE 32.17a ■ *The framing plan for an addition shows the location and sizes of all new structural materials.*

FIGURE 32.17b ■ *Based on the information on the framing plan, the floor, walls and roof framing is placed. Compare this work shown in this photo with the home in Figure 32.15a.* Courtesy Sara Jefferis.

interior elevations for the kitchen renovation/family room addition. Figure 32.19b shows a portion of the completed cabinets. Figures 19.25, 19.26, and 19.27 show the drawings for the food bar that extends into the existing dining room. Unlike other drawings for this project, all existing and new materials are shown in black. Because the existing kitchen cabinets are still in good condition, they will be removed and reused in the new pantry that will occupy the old family room.

Completing the Foundation Plan

The foundation plan is used to show the intersection of the new foundation and floor systems with the existing structure. The drawing is completed using the guidelines presented in Chapter 27 and the existing foundation drawing. A key consideration for an addition is how to join the new concrete to the existing concrete. Steel reinforcing bars are often used to tie the new stem wall to the existing concrete. Holes are drilled into the existing concrete and held in place by epoxy cement. Other considerations for an addition include crawl space access and ventilation. Although both are considered when drawing a new foundation, each must now be reconsidered so that new construction will not block airflow or access to the existing crawl space. On this particular project, doors at each end of the addition will provide access. New vents will be provided along the perimeter of the addition and existing vents will remain to provide airflow to the existing crawl space. The existing crawl access for this project is from the existing basement, so no additional access will be required. If the access had been altered, a hole can be cut in the existing stem wall to provide access from one crawl space to the other.

Although not required for this project, loads from new construction are often imposed over existing floor systems. If a load is transferred into an existing wood floor system, support must be provided to transfer the load into the soil. By placing a concrete pier with a wood post below the existing floor, the load is transferred from the floor to the soil. An alternative is to place a girder below the new load that spans between existing supports. If a new load is applied to an existing concrete slab, a portion of the slab will need to be removed to allow a concrete pier to be placed below the slab. The slab can then be patched to match the existing floor level. A similar situation occurs when a new load is imposed over an existing stem wall and footing. If the load is greater than the assumed design load for the soil or the concrete, a new footing will need to be provided under the existing footing. A new footing is not required on this project because the engineer determined the loads to be applied to the existing retaining wall could safely be distributed into the existing footing.

The foundation plan should be drawn in the file containing the existing foundation plan. Create additional layers to contain all the new material that must be represented. Using Chapter 27 as a guide, create layers for the foundation and floor material such as:

FNDN ANNO NEWW	(thin lines, black)
FNDN BEAM NEWW	(thin hidden black lines)
FNDN DIMS NEWW	(black lines)
FNDN FOOT NEWW	(thin dashed black lines)
FNDN JSTS JSTS	(thin continuous black lines)
FNDN LATL	(thin continuous black lines)
FNDN LATL SCHD	(thick black continuous lines)
FNDN WALL NEWW	(thick black lines)

Materials that should be represented on the foundation plan include:

Existing stem walls and footings
Existing material that will be removed
New footings and stem walls
New floor system members
New crawl access
New vents
New hold down and connectors required for lateral bracing
Dimensions to locate the edges of all new concrete and connections for lateral bracing

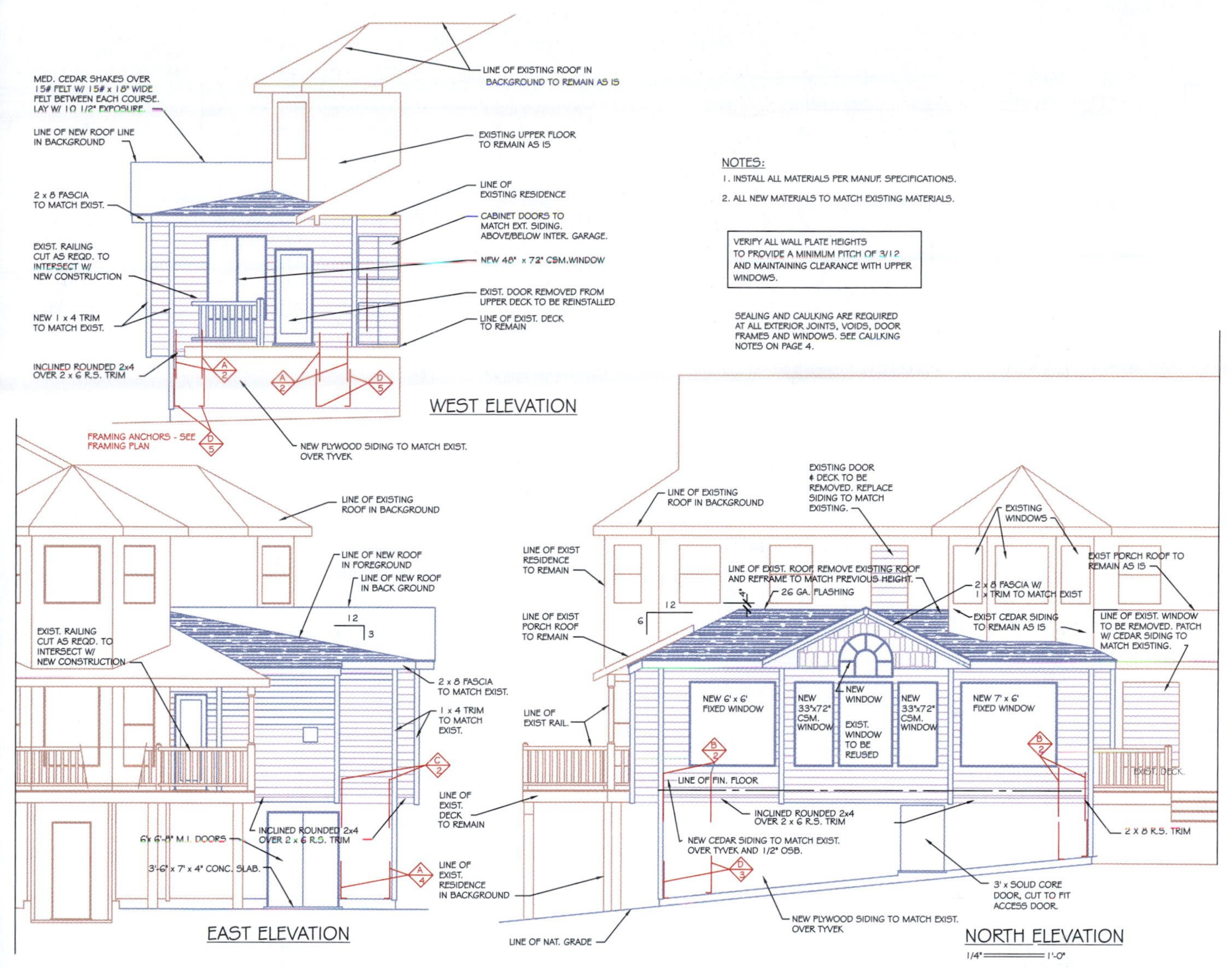
WEST ELEVATION
EAST ELEVATION
NORTH ELEVATION
NOTES:
1. INSTALL ALL MATERIALS PER MANUF. SPECIFICATIONS.
2. ALL NEW MATERIALS TO MATCH EXISTING MATERIALS.
VERIFY ALL WALL PLATE HEIGHTS TO PROVIDE A MINIMUM PITCH OF 3/12 AND MAINTAINING CLEARANCE WITH UPPER WINDOWS.
SEALING AND CAULKING ARE REQUIRED AT ALL EXTERIOR JOINTS, VOIDS, DOOR FRAMES AND WINDOWS. SEE CAULKING NOTES ON PAGE 4.
LINE OF EXISTING ROOF IN BACKGROUND TO REMAIN AS IS
EXISTING UPPER FLOOR TO REMAIN AS IS
LINE OF EXISTING RESIDENCE
NEW PLYWOOD SIDING TO MATCH EXIST. OVER TYVEK
FRAMING ANCHORS - SEE FRAMING PLAN
LINE OF NAT. GRADE

FIGURE 32.18a ■ *The elevations will show each side of the addition and how the new construction will blend with the existing material.*

FIGURE 32.18b ■ *The completed exterior based on the exterior elevations.*

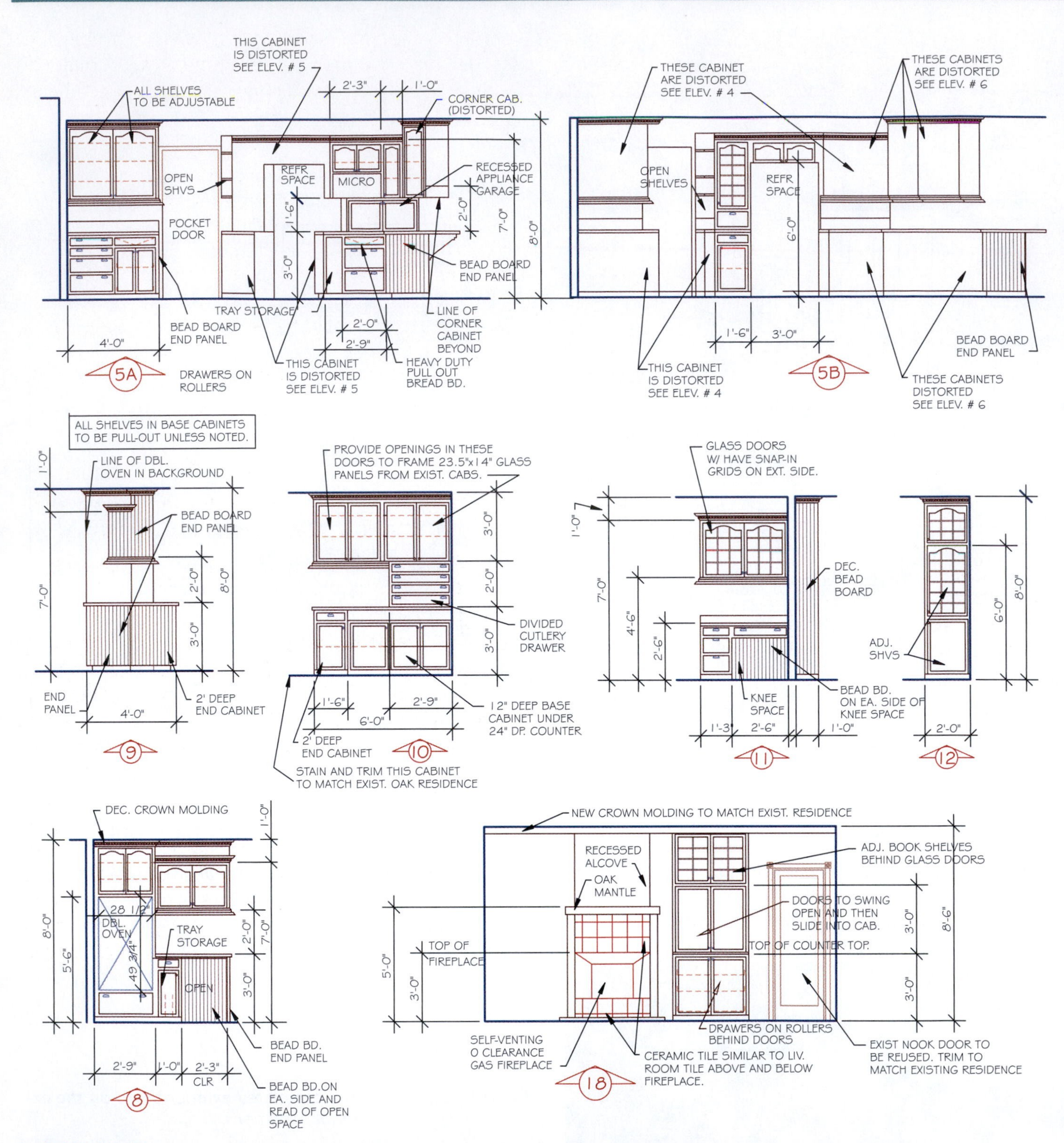

FIGURE 32.19a ■ *The interior elevations will be used by each trade to complete its work.*

FIGURE 32.19b ■ *The completed cabinets based on the interior elevations.* Courtesy Tereasa Jefferis.

Dimensions to locate all new beams, floor cantilevers, and spot piers and footings
Notes to describe all existing materials
Notes to describe all new materials and conditions provided at the foundation level

Figure 32.20a shows the foundation plan for the addition and how it will tie into the existing foundation. Figure 32.20b shows the stem walls and footings that were placed based on the foundation plan.

Completing the Sections and Details

The sections that were used to show the existing structure will be used to show how the new project will merge with existing structure. The guidelines presented in Chapter 29 for drawing sections for new construction can be used to represent new materials for an addition. Represent existing materials that will be removed by dashed lines. Use gray lines to draw existing materials that will remain. In representing new material on the existing sections, use layer titles such as:

SECT ANNO NEWW	(thin black lines)
SECT DIMS NEWW	(thin black lines)
SECT PATT NEWW	(thin gray lines)
SECT THIN NEWW	(thin gray lines to show material behind the cutting plane)
SECT THCK NEWW	(thick gray lines to show material cut by the cutting plane)

Materials that should be represented on the sections include:

- Typical construction of the existing structure
- Typical construction of the new walls, floor, ceiling, and roof
- Intersections of the new and old floor
- Intersections of the new and old roof construction
- Notes to specify all existing and new materials
- Dimensions to specify the heights of all existing and new constructions

As the existing structure was shown in section, it is mentioned that three sections are required for this project. Section AA (Figure 32.21a) shows how the walls surrounding the nook will be removed and how the vaulted ceiling ties into the flat ceilings. This section also shows how the new floor ties into the existing floor system and how the partial floor slab will be placed. Section BB (Figure 32.21b) shows how the new walls and ceilings tie into the existing construction at the nook and second floor bathroom. Section CC (not shown) shows how the new floor, ceiling, and roof will intersect with the existing structure. This section also shows the new beam that will support the upper floor once the north wall of the existing dining room is removed.

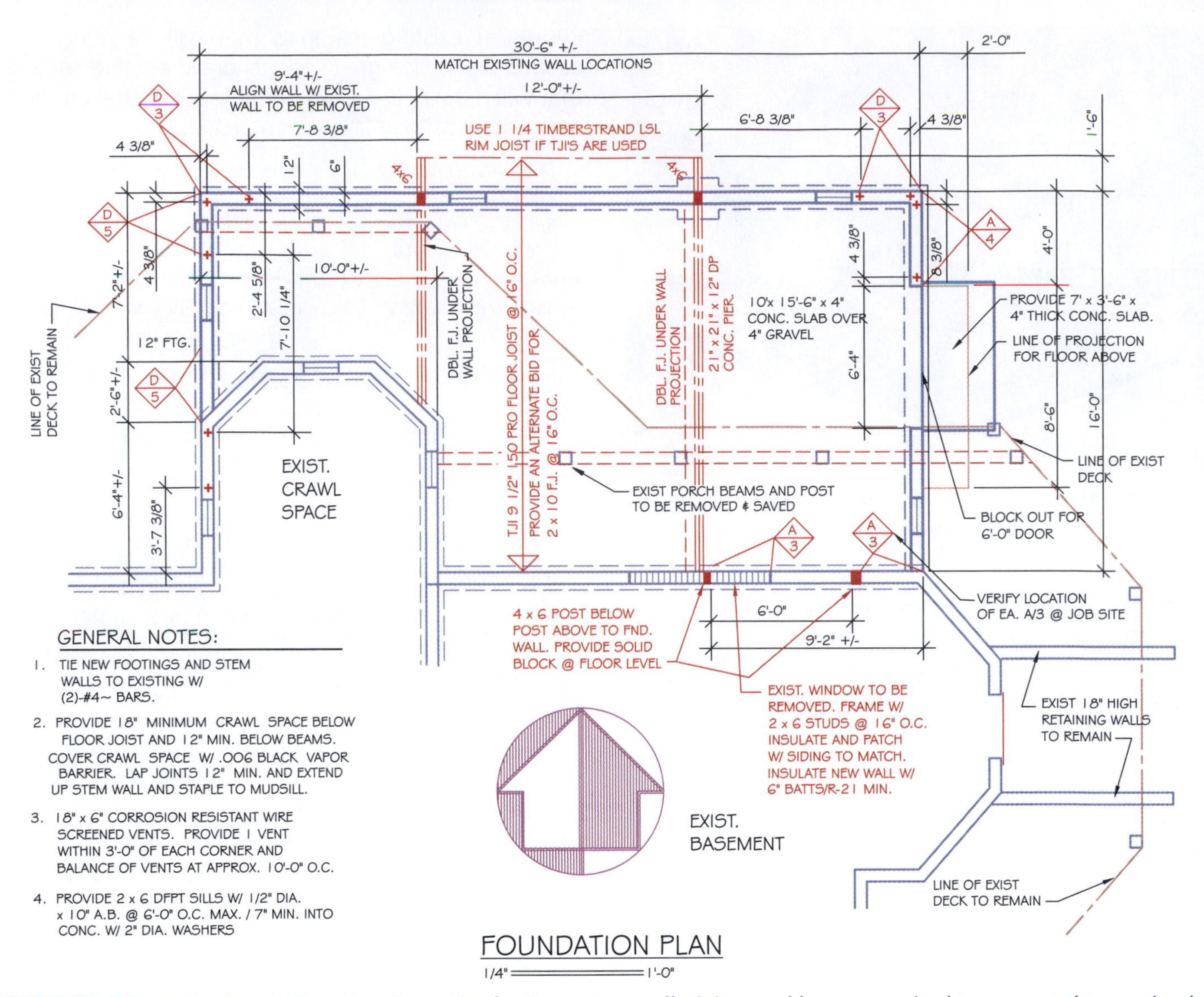

FIGURE 32.20a ■ *The foundation plan shows the footings, stem walls, joists, and beams required to support the new loads.*

FIGURE 32.20b ■ *The completed concrete work based on the foundation plan.*

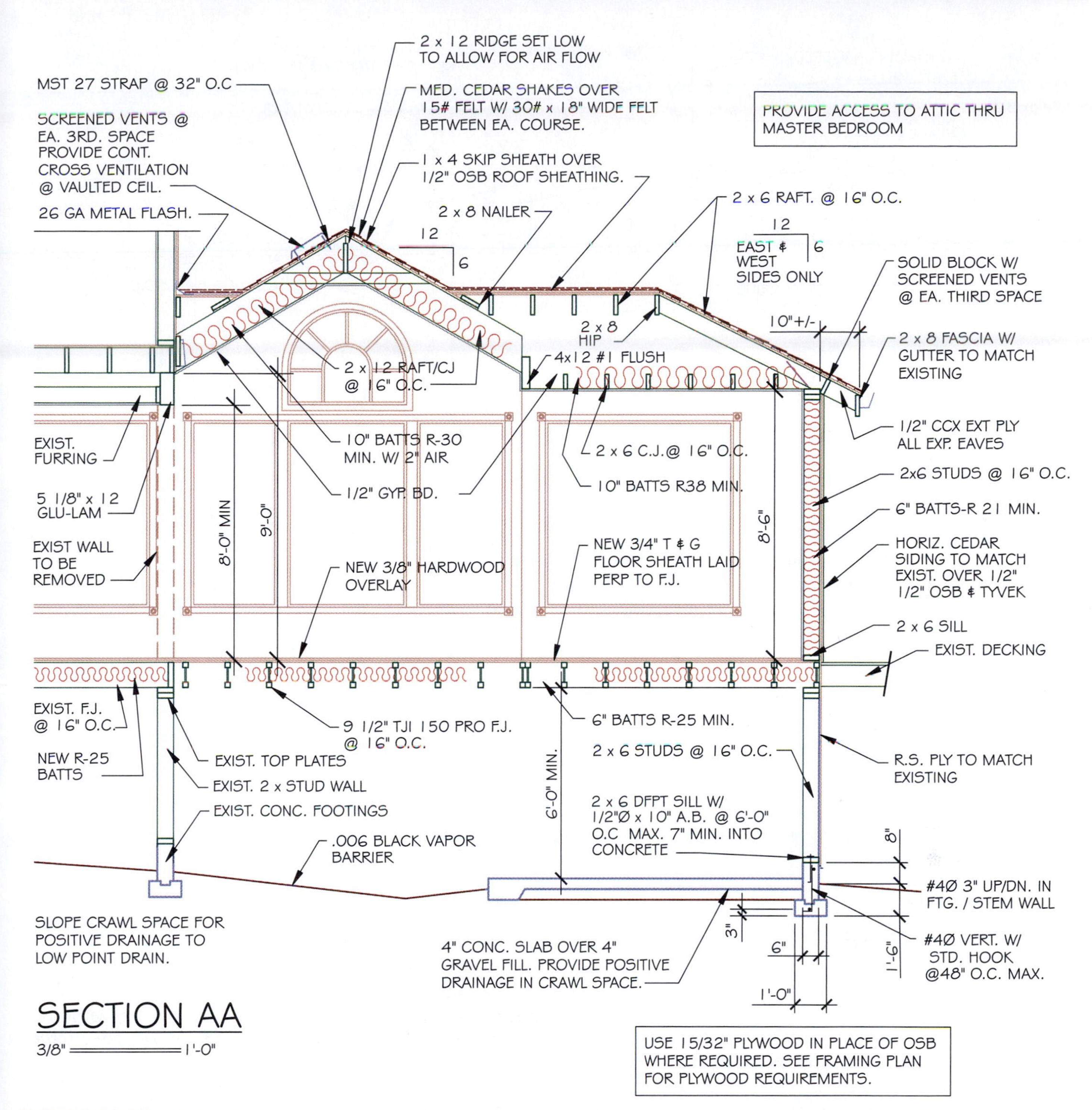
2 x 12 RIDGE SET LOW
TO ALLOW FOR AIR FLOW
MST 27 STRAP @ 32" O.C
MED. CEDAR SHAKES OVER
15# FELT W/ 30# x 18" WIDE FELT
BETWEEN EA. COURSE.
PROVIDE ACCESS TO ATTIC THRU
MASTER BEDROOM
SCREENED VENTS @
EA. 3RD. SPACE
PROVIDE CONT.
CROSS VENTILATION
@ VAULTED CEIL.
1 x 4 SKIP SHEATH OVER
1/2" OSB ROOF SHEATHING.
2 x 6 RAFT. @ 16" O.C.
26 GA METAL FLASH.
2 x 8 NAILER
12
6
12
EAST & WEST SIDES ONLY
6
SOLID BLOCK W/
SCREENED VENTS
@ EA. THIRD SPACE
2 x 8
HIP
10"+/-
2 x 8 FASCIA W/
GUTTER TO MATCH
EXISTING
4x12 #1 FLUSH
2 x 12 RAFT/CJ
@ 16" O.C.
1/2" CCX EXT PLY
ALL EXP. EAVES
EXIST.
FURRING
10" BATTS R-30
MIN. W/ 2" AIR
2 x 6 C.J.@ 16" O.C.
2x6 STUDS @ 16" O.C.
10" BATTS R38 MIN.
5 1/8" x 12
GLU-LAM
1/2" GYP. BD.
6" BATTS-R 21 MIN.
9'-0"
8'-0" MIN
8'-6"
EXIST WALL
TO BE
REMOVED
NEW 3/4" T & G
FLOOR SHEATH LAID
PERP TO F.J.
HORIZ. CEDAR
SIDING TO MATCH
EXIST. OVER 1/2"
1/2" OSB & TYVEK
NEW 3/8" HARDWOOD
OVERLAY
2 x 6 SILL
EXIST. DECKING
EXIST. F.J.
@ 16" O.C.
9 1/2" TJI 150 PRO F.J.
@ 16" O.C.
6" BATTS R-25 MIN.
NEW R-25
BATTS
EXIST. TOP PLATES
2 x 6 STUDS @ 16" O.C.
R.S. PLY TO MATCH
EXISTING
EXIST. 2 x STUD WALL
EXIST. CONC. FOOTINGS
6'-0" MIN.
2 x 6 DFPT SILL W/
1/2"Ø x 10" A.B. @ 6'-0"
O.C MAX. 7" MIN. INTO
CONCRETE
8"
.006 BLACK VAPOR
BARRIER
#4Ø 3" UP/DN. IN
FTG. / STEM WALL
SLOPE CRAWL SPACE FOR
POSITIVE DRAINAGE TO
LOW POINT DRAIN.
3"
4" CONC. SLAB OVER 4"
GRAVEL FILL. PROVIDE POSITIVE
DRAINAGE IN CRAWL SPACE.
6"
1'-6"
#4Ø VERT. W/
STD. HOOK
@48" O.C. MAX.
1'-0"
SECTION AA
3/8" = 1'-0"
USE 15/32" PLYWOOD IN PLACE OF OSB
WHERE REQUIRED. SEE FRAMING PLAN
FOR PLYWOOD REQUIREMENTS.

FIGURE 32.21a ■ *A section through the new family room and existing upper floor bathroom looking north.*

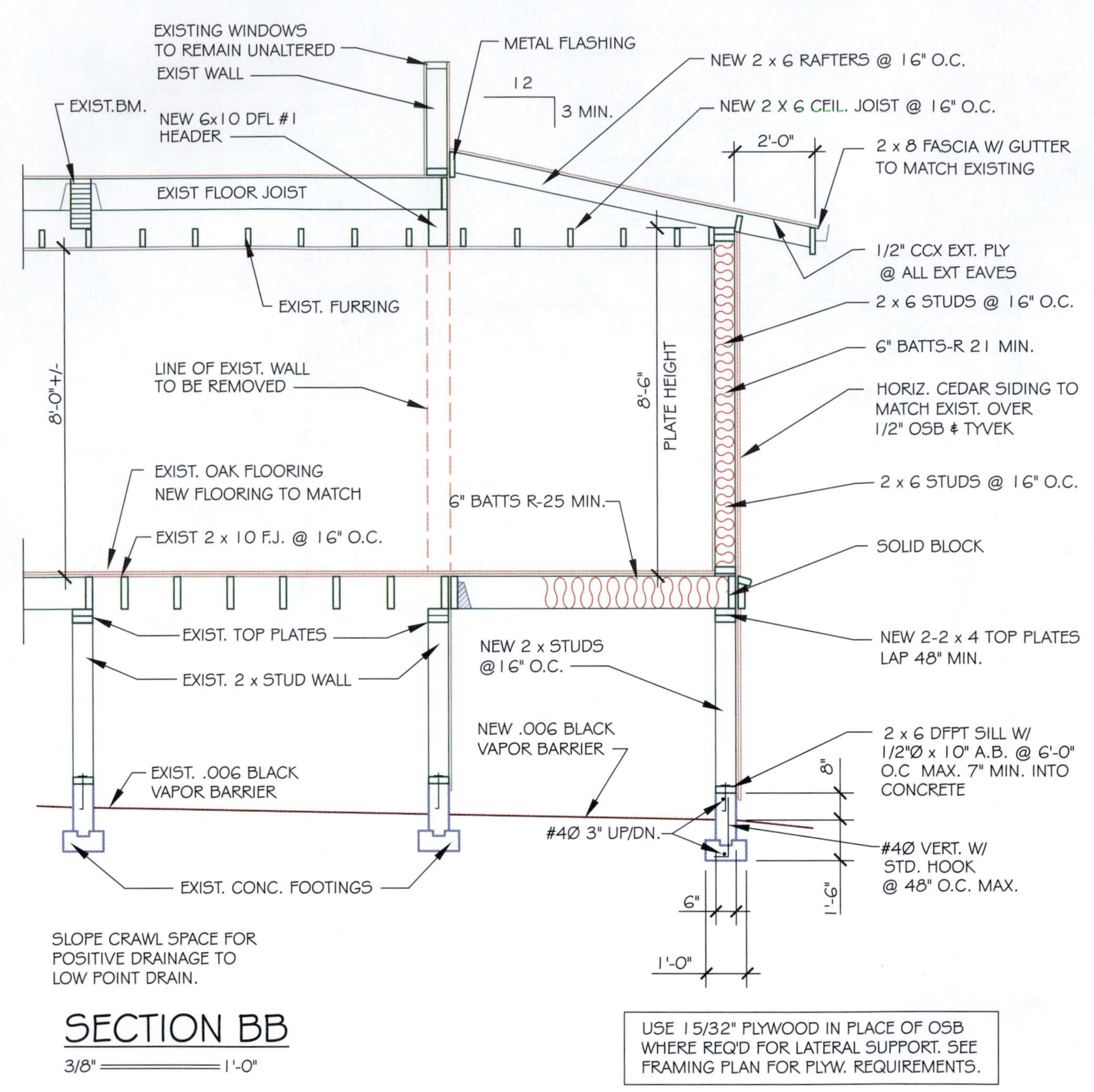

FIGURE 32.21b ■ *Section B shows the supports for the existing upper floor and how the new addition will blend with the existing materials. Additional sections were also provided to show how the new addition will tie into the existing basement and upper bedroom window.*

CHAPTER 32

Additional Readings

ADDRESS	COMPANY OR ORGANIZATION
www.home-improvement-guides.com	Home Improvement Guides
www.interiordec.about.com	About Interior Decorating
www.iprovement.net	Improvement Net
www.letsrenovate.com	Home Remodeling Center
www.nahb.org	National Association of Home Builders
www.nari.org	National Association of the Remodeling Industry
www.nkba.org	National Kitchen and Bath Association
www.renovatorsplace.com	Renovators Place
www2.remodeling.hw.net	Remodeling Online
www.remodelers.com	National Association of Remodelers
www.roomadditions.com	Room Additions
www.updaterenovate.com	Update Renovate

CHAPTER 32

Renovations, Remodeling, and Additions Test

DIRECTIONS

Answer the following questions with short, complete statements. Type your answers using a word processor or neatly print your answers on lined paper.

1. Letter your name, the chapter number, and the date at the top of the sheet.
2. Letter the question number and provide the answer in the form of a statement that includes part of the question. You do not need to write out the entire question.

Note: The answers to some questions may not be contained in this chapter and will require you to do additional research.

QUESTIONS

32.1. What are the major differences between a renovation, a remodel, and an addition?

32.2. Which of the three types of projects presented in this chapter require drawings to obtain a building permit?

32.3. What permits are required for a remodel in the municipality where you live?

32.4. What drawings are required to obtain a building permit for the structural portion of a remodel in the municipality where you live?

32.5. How can you obtain a site map of a prospective remodeling project in your area?

32.6. List three pieces of equipment that are useful in measuring a job site.

32.7. List five common limitations that may influence a remodeling project.

32.8. Explain how to measure the setbacks at a job site.

32.9. Explain how the height of an existing structure can be determined.

32.10. How can the existing materials be determined in drawing a section of the existing structure?

DRAWING PROBLEMS

Use a floor plan for any project from Chapter 11 and the corresponding site plan from Chapter 9 to complete the following project.

32.1. Redesign the existing kitchen/family room to be suitable for a family that likes to entertain. Provide seating at a food bar for four to six people. Assume that the existing roof is framed with trusses so that interior walls may be altered, but keep the footprint of the house as is (as the owners have requested). Complete the drawings that would be required by the building department that governs your area.

32.2. Use the elevations from any project in Chapter 18 and design two different exterior renovations.

32.3. Design a single-level addition that will include the removal and redesign of a new kitchen and the addition of a bedroom suite, bathroom, and family room that blends in with the structure from Chapter 11. The existing kitchen and bedroom space can be reused as needed to meet the needs of the new owner. The kitchen should include an island with seating for four. The family room should include a new masonry fireplace and chimney.

32.4. Design an addition consisting of a den and a master bedroom suite and master bath that will be added above the existing main floor living area. Determine a suitable area to place a stair to access the upper floor.

32.5. Design an addition of a family room, bathroom, and one bedroom that will be sited in a daylight basement below the existing main floor living area. Determine a suitable area to place a stair to access the lower floor. Assume that the roof over the main level is a truss roof. Blend all materials to match.

ABBREVIATIONS

Drafters and designers use many abbreviations to conserve space. Using standard abbreviations ensures that drawings are interpreted accurately.

Here are common guidelines for proper use:

1. A period is used only when the abbreviation may be confused with a word.
2. Several words use the same abbreviation; use is defined by the location.

access panel	ap
acoustic	ac
acoustic plaster	ac pl
actual	act
addition	add
adhesive	adh
adjustable	adj
aggregate	aggr
air conditioning	a.c.
alternate	alt
alternating current	ac
aluminum	alum
amount	amt
ampere	amp
anchor bolt	a.b.
angle	∠
approved	appd
approximate	approx
architectural	arch
area	a
asbestos	asb.
asphalt	asph
asphaltic concrete	asph conc
at	@
automatic	auto
avenue	ave
average	avg
balcony	balc
basement	basm
bathtub	bt.
batten	batt.
beam	bm
bearing	br
benchmark	B.M.
bending moment	M
better	btr
between	btwn
beveled	bev
bidet	bdt
block	blk
blocking	blkg
blower	blo
board	bd.
board feet	bd ft
both sides	b.s.
both ways	b.w.
bottom	btm
bottom of footing	b.f.
boulevard	blvd
brass	br
brick	brk
British thermal unit	Btu
bronze	brz
broom closet	bc
building	bldg
building line	BL
built-in	blt-in
buzzer	buz
by	×
cabinet	cab
cast iron	c.i.
catalog	cat
catch basin	C.B.
caulking	calk
ceiling	clg
ceiling diffuser	c.d.
ceiling joist	c.j. or ceil. jst
cement	cem
center	ctr
center line	CL
center to center	c/c
centimeter	cm
ceramic	cer
chamfer	cham
channel	c
check	chk
cinder block	cin blk
circuit	cir
circuit breaker	cir bkr
class	cl
cleanout	c.o.
clear	clr
cold water	c.w.
column	col
combination	comb
common	com
composition	comp
computer-aided drafting	CAD

concrete	conc
concrete masonry unit	cmu
conduit	cnd
construction	const
continuous	cont
contractor	contr
control joint	c.j.
copper	cop
corridor	corr
corrugate	corr
countersink	csk
courses	c.
cubic	cu.
cubic feet	cu ft
cubic feet per minute	cfm
cubic inch	cu in
cubic yard	cu yd
damp proofing	dp
damper	dpr
dead load	dl
decibel	db
decking	dk
deflection	d.
degree	° or deg
design	dsgn
detail	det.
diagonal	diag
diameter	Ø or dia
diffuser	dif
dimension	dimen
dining room	dr
dishwasher	d/w
disposal	disp
ditto	'' or do
division	div
door	dr
double	dbl
double hung	dh
Douglas fir	df
down	dn
downspout	d.s.
drain	d
drawing	dwg
dryer	D
drywall	D.W.
each	ea
each face	E.F.
each way	E.W.
elbow	el
electrical	elect
elevation	elev
enamel	enam
engineer	engr
entrance	ent.
Equal	eq
Equipment	equip
estimate	est
excavate	exc
exhaust	exh
existing	exist
expansion joint	exp jt
exposed	expo
extension	extn
exterior	ext.
fabricate	fab
face brick	f.b.
face of studs	f.o.s.
Fahrenheit	F
feet / foot	' or ft
feet per minute	fpm
finished	fin
finished floor	fin fl
finished grade	fin gr
finished opening	f.o.
firebrick	fbrk
fire hydrant	F.H.
fireproof	f.p.
fixture	fix
flammable	flam
flashing	fl
flexible	flex
floor	flr
floor drain	f.d.
floor joist	fl jst
floor sink	f.s.
fluorescent	fluor
folding	fldg
foot	(') ft
footcandle	fc
footing	ftg
foot pounds	ft lb
forced air unit	FAU
foundation	fnd
front	fnt
full size	fs
furnace	furn
furred ceiling	fc
future	fut
gage	ga
gallon	gal
galvanized	galv
galvanized iron	g.i.
garage	gar
gas	g
girder	gird
glass	gl
glue laminated	glu-lam
grade	gr
grating	grtg
gravel	gvl
grille	gr
ground	gnd
ground fault circuit interrupter	GFCI or GFI
grout	gt
gypsum	gyp
gypsum board	gyp bd
hardboard	hdb
hardware	hdw
hardwood	hdwd
head	hd.
header	hdr
heater	htr
heating	htg
heating/ventilating/ air conditioning	hvac
height	ht
hemlock	hem
hemlock-fir	hem-fir
hollow core	h.c.
horizontal	horiz
horsepower	h.p.
hose bibb	h.b.
hot water	h.w.
hot water heater	h.w.h.
hundred	c
illuminate	illum.
incandescent	incan.
inch	'' or in.
inch pounds	in. lb.
incinerator	incin.
inflammable	infl.
inside diameter	i.d.
inside face	i.f.
inspection	insp
install	inst
insulate	ins
insulation	insul
interior	int.
iron	i
jamb	jmb.
joint	jt.
joist	jst.
junction	jct.

junction box	J-box
kiln dried	k.d.
kilowatt	kW
kilowatt hour	kWh
Kip (1,000 lb)	K
kitchen	kit.
knockout	k.o.
laboratory	lab
laminated	lam
landing	ldg
laundry	lau
lavatory	lav
length	lgth
level	lev
light	lt
linear feet	lin ft
linen closet	lin
linoleum	lino
live load	LL
living room	Liv
long	lg
louver	lv
machine bolt	m.b.
manhole	m.h.
manufacturer	manuf
marble	mrb
masonry	mas
material	matl
maximum	max
mechanical	mech
medicine cabinet	m.c.
medium	med
membrane	memb
metal	mtl
meter	m
mile	mi
minimum	min
minute	(') min
mirror	mirr
miscellaneous	misc
mixture	mix
model	mod
modular	mod
molding	mldg
mullion	mull
natural	nat
natural grade	nat. gr.
noise reduction coefficient	n.r.c.
nominal	nom
not applicable	n.a.
not in contract	N.I.C.
not to scale	N.T.S.
number	# or no
obscure	obs
on center	O.C.
opening	opg
opposite	opp
ounce	oz
outside diameter	O.D.
outside face	O.F.
overhead	ovhd
painted	ptd
pair	pr
panel	pnl
parallel	// or par.
part	pt
partition	part
pavement	pvmt
penny	d
perforate	perf
perimeter	per
permanent	perm
perpendicular	$\perp$ or perp
pi (3.1416 . . .)	π
plaster	pls.
plasterboard	pls. bd.
plastic	plas
plate	pl
platform	plat
plumbing	plmb
plywood	ply
polyethylene	poly
polyvinyl chloride	pvc
position	pos
pound	# or lb
pounds per square foot	psf
pounds per square inch	psi
prefabricated	prefab
preferred	pfd
preliminary	prelim
pressure treated	p.t.
property	prop
pull chain	p.c.
pushbutton	p.b.
quality	qty
quantity	qty
radiator	rad
radius	r or rad
range	r
receptacle	recp
recessed	rec
redwood	rdwd
reference	ref
refrigerator	refr
register	reg
reinforcing	reinf
reinforcing bar	rebar
required	reqd
return	ret
revision	rev
ridge	rdg
riser	ris
roof drain	R.D.
roofing	rfg
room	rm
rough	rgh
rough opening	r.o.
round	Ø or rd
safety	saf
schedule	sch
screen	scrn
screw	scr
second	sec
section	sect
select	sel
select structural	sel. st.
self-closing	s.c.
service	serv
sewer	sew
sheathing	shtg
sheet	sht
shower	sh
siding	sdg
sill cock	s.c.
similar	sim
single hung	s.h.
soil pipe	s.p.
solid block	sol. blk.
solid core	S.C.
Southern pine	SP
specifications	specs
spruce-pine-fir	SPF
square	sq
square feet	sq ft.
square inch	sq in.
stainless steel	sst.
standard	std
standpipe	st. p.
steel	stl
stirrup	stir
stock	stk
storage	sto
storm drain	S.D.

street	st
structural	str
structural clay tile	S.C.T.
substitute	sub
supply	sup
surface	sur
surface four sides	S4S
surface two sides	S2S
suspended ceiling	susp clg
switch	sw
symbol	sym
synthetic	syn
system	sys
tangent	tan
tee	T
telephone	tel
television	tv
temperature	temp
terra-cotta	t.c.
terrazzo	tz
thermostat	thrm
thickness	thk
thousand	m
thousand board feet	MBF
threshold	thr
through	thru
toilet	tol
tongue and groove	T & G
top of wall	t.o.w.
total	tot
tread	tr
tubing	tub
typical	typ
unfinished	unfin
utility	util.
V-joint	v-jt
valve	v
vanity	van
vapor barrier	v.b.
vapor proof	v prf
ventilation	vent
vent pipe	vp
vent stack	v.s.
vent through roof	v.t.r.
vertical	vert
vertical grain	vert gr
vinyl	vin
vinyl asbestos tile	v.a.t.
vinyl base	v.b.
vinyl tile	v.t.
vitreous	vit
vitreous clay tile	v.c.t.
volt	v
volume	vol
wainscot	wsct
wall vent	w.v.
washing machine	wm
waste stack	w.s.
water closet	w.c.
water heater	w.h.
waterproof	w.p.
watt	W
weather stripping	ws
weatherproof	wp
weep hole	wh
weight	wt
welded wire fabric	wwf
welded wire mesh	wwm
white pine	wp
wide flange	W
width	w
window	wdw
with	w/
without	w/o
wood	wd
wrought iron	w.i.
yard	yd
yellow pine	yp
zinc	zn

GLOSSARY

Accessibility Ability to go in, out, and through a building and its rooms with ease regardless of disability.

Accessible route The walking surface from the exterior access through the residence that is required to connect all spaces of the dwelling unit. If only one route is provided, it cannot pass through a bathroom, closet, or similar space.

Acoustics The science of sound and sound control.

Active solar system A system that uses mechanical devices to absorb, store, and use solar heat.

Adobe Heavy clay used in many southwestern states to make sun-dried bricks.

Aggregate Stone, gravel, cinder, or slag used as one of the components of concrete.

Air duct A pipe, typically made of sheet metal or flexible foil-covered fiberglass, that carries air from a source such as a furnace or air conditioner to a room within a structure.

Air trap A U-shaped pipe placed in wastewater lines to prevent backflow of sewer gas.

Air-dried lumber Lumber that has been stored in yards or sheds for a period of time after cutting. Building codes typically assume a 19% moisture content for determining joist and beam sizes of air-dried lumber.

Alcove A small room adjoining a larger room, often separated by an archway.

Aligned text Text that is rotated so that it can be read when looking from the right side of the drawing page.

Alternative braced wall panel (ABWP) A method of bracing a braced wall line that uses panels with a minimum length of 2'-8'' (800 mm) to resist lateral loads.

Ampere (amp) A measure of electrical current.

Anchor A metal tie or strap used to tie building members to each other.

Anchor bolt An L-shaped threaded bolt used to fasten wood structural members to masonry or concrete.

Angle iron A structural piece of steel shaped to form a 90° angle.

Apron The inside trim board placed below a windowsill. The term is also used to describe a curb around a driveway or parking area.

Architect A licensed professional who is responsible for the design of a structure and the way the building relates to the environment.

Architect's scale A tool used to measure distance with divisions based on 1/16'' for full scale. Other divisions are based on common fractions such as 1/8'' = 1'-0'', 1/4'' = 1'-0'', and 3/4'' = 1'-0''.

Areaway A subsurface enclosure to admit light and air to a basement. Sometimes called a window well.

Asbestos A mineral that does not burn or conduct heat; it is usually used for roofing material.

Ash dump An opening in the hearth where ashes can be dumped.

Ash pit An area in the bottom of the firebox of a fireplace to collect ash.

Ashlar masonry Squared masonry units laid with a horizontal bed joint.

Asphalt An insoluble material used for making floor tile and for waterproofing walls and roofs.

Asphalt shingle Roof shingles made of asphalt-saturated felt and covered with mineral granules.

Asphaltic concrete A mixture of asphalt and aggregate that is used for driveways.

Assessed value The value assigned by governmental agencies to determine the taxes to be assessed on structures and land.

Atrium An inside courtyard of a structure that may be either open at the top or covered with a roof.

Attic The area formed between the ceiling joists and rafters.

Awning window A window that is hinged along the top edge.

Backfill Earth, gravel, or sand placed in the trench around the footing and stem wall after the foundation has cured.

Baffle A shield, usually made of scrap material, used to keep insulation from plugging eave vents. Also used to describe wind- or sound-deadening devices. The term is also used to describe the feeling some students have after taking a test for which they were unprepared.

Balance A principle of design dealing with the relationship between the various areas of a structure as they relate to an imaginary centerline.

Balcony An aboveground deck that projects from a wall or building with no additional supports.

Balloon framing (eastern framing) A construction method that has vertical wall members that extend uninterrupted from the foundation to the roof.

Balusters One of a series of closely spaced ornamental vertical supports for a railing.

Balustrade A low ornamental railing used on the roofs of Georgian and Federal style homes to surround a flattened central area of a low-pitched hipped roof, forming what is often referred to as a "widow's walk" (from which sea captains' wives were supposed to have watched for their husbands' ships). Balustrades on Federal style houses are usually found above the exterior walls.

Band joist (rim joist) A joist set at the edge of the structure that runs parallel to the other joists.

Banister A handrail beside a stairway.

Barge rafter The inclined trim that hangs from the projecting edge of a roof rake.

Base cabinets Cabinets that sit on the floor.

Base course The lowest course in brick or concrete masonry unit construction.

Baseboard The finish trim where the wall and floor intersect.

Baseboard heater An electric heater that extends along the floor.

Baseline A reference line in mapping.

Basement A level of a structure that is built either entirely below grade level (full basement) or partially below grade level (daylight basement).

Basement wall The portion of a wall that is partially or totally below grade and encloses a basement.

Bath, full A bath with a lavatory, toilet, and tub or a combination tub-and-shower unit.

Bath, half A bathroom with a lavatory and water closet (a toilet).

Bath, suite A bath placed with the master bedroom that typically includes the features of a full bath with separate adjoining areas for a toilet, tub or spa, and shower.

Bath, three-quarter A bathroom that includes a shower, toilet and lavatory, but no tub.

Batt Insulation, usually made of fiberglass, to be used between framing members.

Batten A board used to hide the seams when other boards are joined together.

Battlement A parapet wall with open spaces once intended for shooting.

Bay window A window placed in a projection of an exterior wall that extends all the way down to the foundation. In plan view, the wall projection may be rectangular, polygonal, or curved. See also oriel window.

Beam A horizontal structural member that is used to support roof or wall loads (often called a header).

Beamed ceiling A ceiling with support beams that are exposed to view.

Bearing plate A support member, often a steel plate, used to spread weight over a larger area.

Bearing wall A wall that supports vertical loads in addition to its own weight.

Benchmark A reference point used by surveyors to establish grades and construction heights.

Bending One of three major forces acting on a beam. It is the tendency of a beam to bend or sag between its supports.

Bending moment A measure of the forces that cause a beam to break by bending. Represented by "M."

Beveled siding Siding that has a tapered thickness.

Bibb An outdoor faucet that is threaded so that a hose may be attached. (Represented by H.B. on floor plans.)

Bill of material A part of a set of plans that lists all of the material needed.

Bird block (eave blocking) A block placed between rafters or trusses to maintain a uniform spacing and to keep animals out of the attic.

Bird's mouth A notch cut into a rafter to provide a bearing surface where the rafter intersects the top plate.

Blind nailing Driving nails in such a way that the heads are concealed from view.

Blocking (bridging) Framing members, typically of wood, placed between joists, rafters, or studs to provide rigidity.

Board and batten A type of siding using vertical boards with small wood strips (battens) used to cover the joints of the boards.

Board foot The amount of wood contained in a piece of lumber 1'' thick by 12'' wide by 12'' long (25 × 300 × 300 mm).

Bolt, anchor An L-shaped bolt used to connect wood members to concrete.

Bolt, carriage A bolt with a rounded head and a threaded shaft used for connecting steel and other metal members as well as timber connections.

Bolt, machine A bolt with a hexagonal head and a threaded shaft used to attach steel to steel, steel to wood, or wood to wood.

Bond The mortar joint between two masonry units or a pattern in which masonry units are arranged. Also used to describe the top secret spy (007) Bond, James.

Bond beam A reinforced concrete beam used to strengthen masonry walls.

Bottom chord The lower, usually horizontal, member of a truss used to support the ceiling material.

Box beam A hollow built-up structural unit.

Boxed eave Rafter or truss tails that are enclosed by either plywood or 1× material.

Boxed soffit See Eave, boxed.

Braced wall line Each exterior surface of a residence.

Braced wall panel A method of reinforcing a braced wall line using 48'' (1200 mm) wide panels to resist lateral loads.

Branch lines Water feeder lines that branch off the main line to supply fresh water to fixture groups in the home.

Breaker An electrical safety switch that automatically opens the circuit when excessive amperage occurs in the circuit.

Breezeway A covered walkway with open sides between two different parts of a structure.

Bridging Cross blocking between horizontal members used to add stiffness. Also called blocking.

Btu British thermal unit used to measure heat. Each Btu is the amount of heat required to raise the temperature of 1 lb (0.454 kg) of water 1° Fahrenheit. The measurement also assumes that the heating is done at a constant pressure of one atmosphere (air pressure at sea level).

Bubble drawings Freehand sketches used to determine room locations and relationships.

Building code Legal requirements designed to protect the public by providing guidelines for structural, electrical, plumbing, and mechanical areas of a structure. (See International Residential Code.)

Building envelope The portion of a building that encloses the treated environment, including the walls, ceiling or roof, and floor.

Building line An imaginary line determined by zoning departments to specify the limits where a structure may be built (also known as a setback).

Building paper A waterproofed paper used to prevent the passage of air and water into a structure.

Building permit A permit to build a structure issued by a governmental agency after the plans for the structure have been examined and the structure is found to comply with all building code requirements.

Built-up beam A beam built of smaller members that are bolted or nailed together.

Built-up roof A roof composed of three or more layers of felt, asphalt, pitch, or coal tar.

Bullnose Rounded edges of cabinet trim.

Butt joint The junction where two members meet in a square-cut joint; end to end or edge to edge.

Buttress A projection from a wall often located below roof beams to provide support to the roof loads and to keep long walls in the vertical position.

Cabinet work The interior finish woodwork of a structure, especially cabinetry.

CAD technician A term used to describe a person doing work on a computer that had once been done manually by drafters.

Camber A curve built into a laminated beam to increase the load that can be supported.

Cant strip A small built-up area between two intersecting roof shapes to divert water.

Cantilever Construction material that extends past its supports.

Carport A covered automobile parking structure that is not fully enclosed.

Carriage The horizontal part of a stair stringer that supports the tread.

Carriage bolt A bolt used for connecting wood to steel or other metal members.

Casement window A hinged window that swings outward around a vertical axis.

Casing The metal, plastic, or wood trim around a door or a window.

Catch basin An underground reservoir for water drained from a roof before it flows to a storm drain.

Cathedral window A window with an upper edge that is parallel to the roof pitch.

Caulking A soft, waterproof material used to seal seams and cracks in construction.

Cavity wall A masonry wall formed with two wythes with an air-space between each face.

CC&Rs (covenants, conditions, and restrictions) The name given to an agreement between two or more property owners that is recorded on the deed of each property. It consists of a covenant, which is a binding agreement; a condition, which is a statement of what is required as part of an agreement; and a restriction, which is a principle limiting the extent of something.

Ceiling joists The horizontal members of the roof used to resist the outward spread of the rafters and to provide a surface for installing the finished ceiling.

Cement A powder of alumina, silica, lime, iron oxide, and magnesia pulverized and used as an ingredient in mortar and concrete.

Central heating A heating system that delivers heat throughout a structure from a single source.

Cesspool An underground catch basin for the collection and dispersal of sewage.

Chair rail A molding placed horizontally on the wall at the height where chair backs would otherwise damage the wall.

Chamfer A beveled edge formed by removing the sharp corner of a piece of material.

Channel A standard form of structural steel with three sides at right angles to each other forming the letter C.

Chase A recessed area formed between structural members that conceals electrical, mechanical, or plumbing materials.

Check Lengthwise cracks in a board caused by natural drying.

Check valve A valve in a pipe that permits flow in only one direction.

Chimney An upright structure connected to a fireplace or furnace that passes smoke and gases to outside air.

Chimney cap The sloping surface on the top of the chimney.

Chimney hood A covering placed over the flue to keep the elements from entering the flue.

Chimney liner A fire clay or terra-cotta liner built into a chimney to provide a smooth surface to the chimney flue.

Chord The upper and lower members of a truss that are supported by the web.

Cinder block A block made of cinder and cement used in construction.

Circuit The various conductors, connections, and devices found in the path of electrical flow from the source through the components and back to the source.

Circuit breaker A safety device that opens and closes an electrical circuit.

Civil engineer A licensed professional who is responsible for the design and supervision of the land drawings such as a topography map, grading plans, street design, and other land-related improvements.

Civil engineer's scale A tool used to measure distance with divisions based on one-tenth of an inch for full scale or for 1'' = 10' or 100'. Other scales include 20, 30, 40, 50, and 60 divisions per inch.

Clapboard A tapered board used for siding that overlaps the board below it.

Cleanout A fitting with a removable plug that is placed in plumbing drainage lines to allow access for cleaning out the pipe.

Clearance A clear space between building materials to allow for air-flow or access.

Clerestory A window or group of windows that are placed above the normal window height, often with the top edge parallel to the rake of the roof.

Code A performance-based description of the desired results with wide latitude allowed to achieve the results.

Coffered ceiling A ceiling formed using beams and trim to create a pattern of recessed panels or grid-like compartments in a ceiling.

Collar ties Horizontal ties placed between rafters near the ridge to help resist the tendency of the rafters to separate.

Colonial A style of architecture and furniture adapted from the American colonial period.

Column A vertical structural support, usually made of steel.

Common rafter See rafter, common.

Common wall The partition that divides two different dwelling units.

Complex beam Beam with a non-uniform load at any point on it and has supports that are not located at its end.

Compression A force that crushes or compacts.

Concentrated load A load centralized in a small area. The weight supported by a post results in a concentrated load.

Concrete A building material made from cement, sand, gravel, and water.

Concrete blocks Blocks of concrete that are precast. The standard size is 8 × 8 × 16''.

Condensation The formation of water on a surface when warm air comes in contact with a cold surface.

Conditional use A use for a property that is not allowed outright by zoning regulations, but may be allowed on a case-by-case basis, if certain conditions are met.

Conduction The process of transferring heat energy between molecules within an object or between the molecules of two or more objects that are in contact.

Conductor Any material that permits the flow of electricity. The term also applies to a drainpipe that diverts water from the roof (a downspout).

Conduit A bendable metal, fiber pipe, or tube used to enclose one or more electrical wires.

Coniferous Cone-bearing trees such as cedars, cypresses, douglas-firs, firs, junipers, larches, pines, redwoods, and spruces that have year-round leaves that are long, thin, and needle-like.

Construction joint A joint used when a concrete pour must be interrupted that is used to provide a clean surface when work is resumed.

Continuous beam A single beam that is supported by more than two supports.

Contour lines Lines placed on a topography map or grading plan to represent all points with a specific elevation.

Contours A line that represents land formations.

Contraction joint An expansion joint in a masonry wall or slab formed to control where cracking will occur. See Control joint.

Contractor The manager of a construction project, or one specific phase of it.

Control joint (contraction joint) An expansion joint in a masonry wall formed by raking mortar from the vertical joint.

Control-point survey A survey method that establishes elevations that is recorded on a map.

Convection The transfer of heat from a heated surface to a fluid moving over the heated surface, or is transferred by molecules in a fluid from one heated molecule to another.

Convenience outlet An electrical receptacle that allows current to be drawn for an appliance.

Coping A masonry cap placed on top of a block or brick wall to protect it from water penetration.

Corbel A ledge formed in a wall by building out successive courses of masonry.

Cornice The part of the roof that extends out from the wall. Sometimes referred to as the eave.

Counter flash A metal flashing used under normal flashing to provide a waterproof seam.

Course A continuous row of building material such as shingles, stone, or brick.

Court An exterior space that is at grade level, enclosed on three or more sides by walls or a building that is open and unobstructed to the sky.

Cove lighting Lighting concealed behind a cornice or other ceiling features that directs the light upward.

Crawl space The area between the floor joists and the ground.

Cricket A diverter built to direct water away from an area of a roof where it would otherwise collect, as behind a chimney.

Cripple A wall stud that is cut at less than full length (also referred to as a jack stud).

Cross bracing Boards fastened diagonally between structural members, such as floor joists, to provide rigidity.

Csi Uniform drawing system.

Cul-de-sac A dead-end street with no outlet that provides a circular turnaround.

Culvert Underground passageways for water, usually part of a drainage system.

Cupola A short windowed tower or dome typically located in the center of a flat or low-sloped roof on traditional homes to provide light or ventilation.

Cure The process of concrete drying to its maximum design strength, usually taking 28 days.

Cut material Soil that is removed in order to lower the original ground elevation.

Damper A movable plate that controls the amount of draft for a woodstove, fireplace, or furnace.

Datum A reference point for starting a survey.

Daylight The point represented on a grading plan that represents the intersection between cut and fill.

Dead load The weight of building materials or other immovable objects in a structure.

Deadening board A material used to control the transmission of sound.

Deciduous Broad-leafed trees that seasonally shed their leafs.

Deck An exterior floor supported on at least two opposing sides by adjoining structures, posts, or piers.

Decking A wood material used to form the floor or roof, typically used in 1 and 2'' thicknesses.

Deflection The tendency of a structural member to bend under a load and as a result of gravity.

Demand factor An assumption based on NEC standards that assumes that not all lights will be on at the same time, so the lighting load can be reduced.

Density The number of people allowed to live in a specific area of land or to work in a specific area of a structure.

Dentil Molding made from a series of closely spaced rectangular blocks. It is typically found below the cornice along the roofline of a building, but it can be used as a decorative band anywhere on a structure.

Designer Although the definition varies from state to state, the term is used by those who have had formal training and passed a competency test to design and draw residences but is not licensed as an architect.

Details Enlargements of specific areas of a structure that are drawn where several components intersect or where small members are required.

Diaphragm A rigid plate that acts much like a beam and can be found in the roof level, walls, or floor system.

Diffusers The outlets that supply treated air from the HVAC system into a room.

Dimension line A line that extends between two extension lines to show the length of a specific feature.

Dimension lumber Lumber ranging in thickness from 2 to 4'' (50 to 100 mm) and having a moisture content of less than 19%.

Distribution panel Panel where the conductor from the meter base is connected to individual circuit breakers, that are connected to separate circuits for distribution to various locations throughout the structure.

Diverter A metal strip used to divert water.

Dormer A structure that projects from a sloping roof to form another roofed area. This new area is typically used to provide a surface to install a window.

Double glazing Glazing in a door or window that is constructed from two layers of glazing.

Double hung A type of window that allows the upper and lower halves to slide past each other, thus providing an opening at the top and bottom of the window.

Double-acting door A door that swings in two directions.

Double-wall construction A method of construction used in cold climates that places the exterior finishing material over sheathing placed over a water-resistant membrane placed over the wall studs.

Downspout A pipe that carries rainwater from the gutters of the roof to the ground.

Drafter The person who uses the proper line properties, dimension, and text styles to create the drawings and details for another person's creations.

Drain A collector for a pipe that carries waste water from each plumbing fixture to the waste line of the building drainage system.

Drainage grate A metal cover that lets water flow into a catch basin without allowing anyone to fall in. Water flows through the grate and into a catch basin, after which it is funneled into pipes that connect to public storm sewers.

Dressed lumber Lumber that has been surfaced by a planing machine to give the wood a smooth finish.

Dry rot Caused by fungi, a type of wood decay that leaves the wood a soft powder.

Dry well A shallow well used to disperse water from the gutter system.

Drywall An interior wall covering installed in large sheets made of gypsum board.

Ducts Pipes, typically made of sheet metal, used to conduct hot or cold air of the HVAC system.

Duplex outlet A standard electrical convenience outlet with two receptacles.

Dutch door A door divided horizontally in the center so that each half of it may be opened separately.

Dutch hip A type of roof shape that combines features of a gable and a hip roof.

Dynamic loads The loads imposed on a structure from a sudden gust of wind or from an earthquake.

Easement A right to make limited use of another's real property that is recorded on the deed and survives any sale of the property for use as a public right of way, such as a utility easement that grants access to private land to place or maintain a utility.

Eave The lower part of the roof that projects from the wall (also see cornice).

Eave, boxed An eave with a covering applied directly to the bottom side of the rafter or truss tails.

Effluent Treated sewage from a sewage treatment plant or septic tank.

Egress A term used in building codes to describe access.

Elastic limit The extent to which a material can be bent and still return to its original shape.

Elbow An L-shaped plumbing pipe.

Electrical engineers Licensed professionals responsible for the design of lighting and communication systems. They supervise the design and installation of specific lighting fixtures, communication services, surround-sound systems, security features, and requirements for computer networking.

Elevation The height of a specific point in relation to another point. The exterior views of a structure.

Eminent domain The right of a government to condemn private property so that it may be obtained for public use.

Enamel A paint that produces a hard, glossy, smooth finish.

Engineered lumber Structural components made by turning small pieces of wood into framing members such as joists, studs, and rafters.

Engineers Licensed professionals who apply mathematical and scientific principles to the design and construction of structures. They include structural, electrical, mechanical, and civil engineers.

Entourage The surroundings of a rendered building consisting of ground cover, trees, people, and automobiles.

Equity The value of real estate in excess of the balance owed on the mortgage.

Ergonomics The study of human space and movement needs as they relate to a given work area, such as a kitchen.

Excavation The removal of soil for construction purposes.

Expansion joint A joint installed in concrete construction to reduce cracking and to provide workable areas.

Extension lines Thin lines showing the extent of a dimension.

Fabrication Work done on a structure away from the job site.

Facade The exterior covering of a structure.

Face brick Brick used on the visible surface to cover other masonry products.

Face grain The pattern in the visible veneer of plywood.

Fanlight A semicircular or semielliptical nonopening transom window with a horizontal sill; this is placed above a door or another window typically found above the main entry door of houses in the Federal style.

Fascia A horizontal board nailed to the ends of rafters or trusses to conceal those ends.

Federal Housing Administration (FHA) A governmental agency that insures home loans made by private lending institutions.

Felt A tar-impregnated paper used for water protection under roofing and siding materials. Sometimes used under concrete slabs for moisture resistance.

Fenestration Windows or doors located in the building envelope.

Fiber bending stress (F_b) The measurement of structural members used to determine their stiffness.

Fiberboard Fibrous wood products that have been pressed into a sheet. Typically used for the interior construction of cabinets and for a covering for the subfloor.

Field weld A weld that is performed at the job site.

Fill Material used to raise an area for construction. Typically gravel or sand is used to provide a raised, level building area.

Filled insulation Insulation material that is blown or poured into place in attics and walls.

Fillet weld A weld between two surfaces that butt at 90° to each other, with the weld filling the inside corner.

Finished grade The shape of the ground once all excavation and movement of earth has been completed.

Finished lumber Wood that has been milled with a smooth finish suitable for use as trim and other finish work.

Finished size Sometimes called the dressed size, the finished size represents the actual size of lumber after all milling operations and is typically about 1/2'' (13 mm) smaller than the nominal size, which is the size of lumber before planing.

Nominal Size (inches)	Finished Size (inches)
1	3/4
2	1 1/2
4	3 1/2
6	5 1/2
8	7 1/2
10	9 1/2
12	11 1/2
14	13 1/2

Fire cut An angular cut on the end of a joist or rafter that is supported by masonry. The cut allows the wood member to fall away from the wall without damaging a masonry wall when wood is damaged by fire.

Fire door A door used between different types of construction that has been rated as being able to withstand fire for a certain amount of time.

Fire rating A rating given to building materials to specify the amount of time the material can resist damage caused by fire.

Fire wall A wall constructed of materials resulting in a specified time that the wall can resist fire before structural damage will occur.

Firebox The combustion chamber of the fireplace where the fire occurs.

Firebrick A refractory brick capable of withstanding high temperatures and used for lining fireplaces and furnaces.

Fireplace insert A metal fireplace inserted into a masonry fireplace to control drafts and increase heat production. The unit must be vented using the existing chimney.

Fireplace opening The open area between the side and top faces of the fireplace.

Fireproofing Any material used to cover structural materials to increase their fire rating.

Fire-stop Blocking placed between studs or other structural members to resist the spread of fire.

Fitting A standard pipe or tubing joint—such as a tee, elbow, or reducer—used to join two or more pipes.

Fixed window A window designed without hinges so it cannot be opened.

Flagstone Flat stones used for floor and wall coverings.

Flashing Metal used to prevent the leakage of water through surface intersections.

Flat roof A roof with a minimal roof pitch, usually about 1/4'' per 12'' (6 per 25 mm).

Flitch beam A built-up beam consisting of steel plates bolted between wood members.

Floor joists Repetitive horizontal structural members of the floor framing system that are used to span between the stem wall or girders to provide support to the subfloor.

Floor plan Architectural drawing of a room or building as seen from above.

Floor plug A 120-V convenience outlet located in the floor.

Flue A passage inside of the chimney to conduct smoke and gases away from a firebox to outside air.

Flue liner A terra-cotta pipe used to provide a smooth flue surface so that unburned materials will not cling to the flue.

Footing The lowest member of a foundation system used to spread the loads of a structure across supporting soil.

Forced-air heating system A system that blows treated air through ducts using a fan located in the heating or cooling device.

Formal balance Symmetrical arrangement of space so that one side of the structure or room matches the opposite side.

Foundation The system used to support a building's loads consisting of the stem walls, footings, and piers. The term is used in many areas to refer to the footing.

Frame The structural skeleton of a building.

French doors Exterior or interior doors that have glass panels and swing into a room.

Frieze A horizontal band decorated with designs or carvings that runs above doorways or windows or below the cornice.

Frost line The average depth to which soil will freeze.

Furring Wood strips attached to structural members that are used to provide a level surface for finishing materials when different sized structural members are used.

Gable A type of roof with two sloping surfaces that intersect at the ridge of the structure.

Gable end wall The triangular wall that is formed at each end of a gable roof between the top plate of the wall and the rafters.

Galvanized Steel products that have had zinc applied to the exterior surface to provide protection from rusting.

Gambrel A type of roof formed with two planes on each side of the ridge. The lower pitch is steeper than the upper portion of the roof.

General notes Notes that apply to the overall drawing or to a specific group of items in a drawing.

Geothermal system Heating and cooling system that uses the constant, moderate temperature of the ground to provide space heating and cooling, or domestic hot water, by placing a heat exchanger in the ground, or in wells, lakes, rivers, or streams.

Girder A horizontal support beam used at the foundation level to support the floor joists. In a post and beam system, a girder is used to support the floor decking.

Glazing All areas that let in natural light, including windows, clerestories, skylights, glass doors, glass block walls, and glass portions of doors.

Glued-laminated beam (glu-lam) A structural member made up of layers of lumber that are glued together.

Grade The designation of the quality of a manufactured piece of wood.

Grading The moving of soil to effect the elevation of land at a construction site. Grading plan A drawing used to show the finished soil configuration of the building site.

Gravel stop A metal strip used to retain gravel at the edge of built-up roofs.

Gravity Uniform force that affects all structures due to the gravitational force from the earth.

Gravity heat system A heating system that allows warm air to rise naturally without a fan.

Gray water Wastewater from a shower or bath, laundry water, and rain runoff collected from roof gutters that is then recycled.

Green board A type of water-resistant gypsum board designed to be used in a high-moisture area, such as behind a shower enclosure.

Green lumber Lumber that has not been kiln-dried and still contains moisture.

Ground An electrical connection to the earth by means of a rod.

Ground fault circuit interrupter (gfci or GFI) A 120-V convenience outlet with a built-in circuit breaker that must be used within 60'' of any water source.

Grout A mixture of cement, sand, and water used to fill joints in masonry and tile construction.

Guardrail A horizontal protective railing used around stairwells, balconies, and changes of floor elevation greater than 30''.

Gusset A metal or wood plate used to strengthen the intersection of structural members.

Gutter A device mounted on the eave for the collection of rainwater from the roof to downspouts.

Gypsum board An interior finishing material that is installed in large shapes; it is made of gypsum and fiberglass covered with paper.

Habitable space Areas used for sleeping, living, cooking, or dining.

Half bath See Bath, half.

Half-timber A frame construction method where spaces between wood members are filled with masonry.

Hanger A metal support bracket used to attach two structural members.

Hard conversions See metric conversions.

Hardboard Sheet material formed of compressed wood fibers.

Head The upper portion of a door or window frame.

Header A horizontal structural member used to support other structural members over openings, such as doors and windows.

Header course A horizontal masonry course with the end of each masonry unit exposed.

Headroom The vertical clearance over a stairway.

Hearth The fire-resistant floor extending in front and to the side of the firebox.

Heartwood The inner core of a tree trunk.

Heat pump A unit designed to produce forced air for heating and cooling.

Hip A traditional roof shape formed by four or more inserting planes. The term is also used to describe an exterior edge formed by two sloping roof surfaces.

Hip roof A roof shape with four sloping sides.

Hopper window A window that is hinged at the bottom and swings inward.

Horizon line A line drawn parallel to the ground line that represents the intersection of ground and sky.

Horizontal shear (F_v) One of three major forces acting on a beam, it is the tendency of the fibers of a beam to slide past each other in horizontal direction.

Hose bibb A water outlet that is threaded to receive a hose.

Hue A term used to represent what you typically think of as the color.

Humidifier A mechanical device that controls the amount of moisture inside of a structure.

Hurricane ties Metal connectors used to connect roof members to wall members to resist uplift.

Hydroelectric power Electricity generated by the conversion of the energy created by falling water.

I beam The generic term for a wide-flange or American standard steel beam with a cross section in the shape of the letter I.

Illustrator A person with a background in art and architecture who produces drawings showing a proposed structure realistically.

Indirect lighting Mechanical or artificial lighting that is reflected off a surface.

Infiltration The flow of air through building intersections.

Informal balance Nonsymmetrical placement achieved by placing shapes of different sizes in various positions around the imaginary centerline.

Insulated concrete form (ICF) An energy-efficient wall framing system made by placing poured concrete in polystyrene forms that are left in place to create a superinsulated wall.

Insulation Material used to restrict the flow of heat, cold, or sound from one surface to another.

Intensity The brightness or strength of a specific color.

Interior decorator A person who decorates the interiors of buildings with the aim of making rooms more attractive, comfortable, and functional.

Interior designer A person who works with the structural designer to optimize and harmonize the interior design of structures in regard to how a space will be used, the amount of light that will be required, acoustics, seating, storage, and work areas.

International building code (IBC) A national building code for multifamily, commercial, and other large structures involving public occupancy.

International Residential Code (IRC) A national building code for one- and two-family dwellings.

Interpolation A combination of rounding off and guessing based on known points. If you know that two points have a change of elevation of 12'' (25 mm), you can identify a point half way between these two points and assign it an elevation representing a 6'' (13-mm) difference in height.

Irrigation plan A drawing usually completed by a technician working for landscape architect that shows how landscaping will be maintained.

Isolation joint See expansion joint.

Isometric drawings Drawings that appear to be three-dimensional by showing three surfaces of an object in one view.

Jack rafter A rafter cut shorter than the other rafters to allow for an opening in the roof.

Jack stud (cripple) A wall member cut shorter than other studs to allow for an opening such as a window.

Jalousie A type of window made of thin horizontal panels that can be rotated between the open and closed position.

Jamb The vertical members of a door or window frame.

Joist A horizontal structural member used in repetitive patterns to support floor and ceiling loads.

Junction box A box that protects electrical wiring splices in conductors or joints in runs.

Kick block (kicker) A block used to keep the bottom of the stringer from sliding on the floor when downward pressure is applied to the stringer.

Kiln dried Lumber dried in a kiln or oven. Kiln-dried lumber has a lower moisture content than air-dried lumber.

King stud A full-length stud placed at the end of a header and beside the trimmer stud supporting the header.

Kip A term used in some engineering formulas to represent 1000 lb.

Knee wall A wall of less than full height.

Knot A branch or limb of a tree that is cut through in the process of manufacturing lumber.

Lag screw A screw used for wood-to-wood or wood-to-steel connections.

Lally column A vertical steel column used to support floor or foundation loads.

Laminated Several layers of material that have been glued together under pressure.

Landing A platform between two flights of stairs.

Landscape plan A drawing that shows the location, type, size, and quantity of all vegetation required for the project as well as hard scaping such as patios, walkways, fountains, pools, sports courts, and other landscaping features.

Lateral Sideways motion in a structure caused by wind or seismic forces. The term is also used to describe the pipe that connects the construction site to the public sewer pipe.

Lateral loads A load resulting from wind or earthquakes that pushes a structure sideways.

Lath Wood or sheet metal strips that are attached to the structural frame to support plaster.

Lattice A grille made by crisscrossing strips of material.

Lavatory A bathroom sink or a room that is equipped with a washbasin.

Leach lines Waste drain lines placed in an absorbent soil to disperse liquid material from a septic system.

Ledger A horizontal member that is attached to the side of wall members to provide support for rafters or joists.

Legal description The description used to describe a parcel of land for legal purposes, such as the recording of a deed of ownership.

Legends A type of drawing used to explain symbols that have been used on a specific drawing project.

Lintel A horizontal steel member used to provide support for masonry over an opening.

Lintel block A long rectangular stone block that spans a door or window opening to support the weight of the structure above the opening.

Lisp A programming language used to customize CADD software.

Live load The load from all movable objects within a structure including loads from furniture and people. External loads from snow and wind are also considered live-loaded.

Load path The route used to transfer the roof loads into the walls, then into the floor, and then to the foundation.

Load-bearing wall A support wall that holds floor or roof loads in addition to its own weight.

Local notes Drawing notes that refer to a specific material or area of a project.

Lookout Bracing between the wall and subfascia or end cap to which the soffit is attached. Also a beam used to support eave loads.

Louver An opening with horizontal slats to allow for ventilation.

Low-e glass Low-emission glass that has a transparent coating on its surface to acts as a thermal mirror.

Main The water supply line that extends from the water meter into the home to deliver potable water.

Manifold A distribution center between the main, branch, and riser lines.

Mansard A four-sided, steep-sloped roof used to enclose the upper level of a structure.

Mantel A decorative shelf above the opening of a fireplace.

Market value The amount for which property can be sold.

Masonry The use of brick, stone, or concrete blocks to construct a wall.

Master format A list of numbers and titles created to organize information into a standard sequence relating to construction requirements, products, and activities.

Mechanical engineer A licensed professional who is responsible for the sizing and layout of heating, ventilation, and air-conditioning (HVAC) systems and plans how treated air will be routed throughout the project.

Mesh A metal reinforcing material placed in concrete slabs and masonry walls to help resist cracking.

Metal tie A manufactured piece of metal for joining two structural members together.

Metal wall ties Corrugated metal strips used to bond brick veneer to its support wall.

Metric measurement hard conversions Made by using a mathematical formula to change a value of one system (e.g., 1'') to the equivalent value in another system (e.g., 25.4 mm). A 6'' distance would be 152 mm (6 × 25.4).

Metric measurement soft conversions Conversions made by using a mathematical formula to change a value from one system (e.g., 1'') to a rounded value in another system (e.g., 25 mm). A 6'' distance would be 150 mm.

Metric scale A tool used to measure distance, with the millimeter used as the basic unit of measurement.

Millwork Finished woodwork manufactured in a milling plant for use as window and door frames, mantels, moldings, and stairway components.

Mineral wool An insulating material made of fibrous foam.

Modular cabinet Prefabricated cabinets that are constructed in specific sizes called modules. Modular cabinets are usually available in 3'' (75 mm) widths.

Module A standardized unit of measurement.

Modulus of elasticity (E) The degree of stiffness of a beam.

Moisture barrier Typically a plastic material used to restrict moisture vapor from penetrating into a structure.

Molding Decorative strips, usually made of wood, used to conceal the seam in other finishing materials.

Moment The tendency of a force to rotate around a certain point.

Monolithic Concrete construction created in one pour. Monument A point established by the U.S. Geological Society (USGS) that is marked by a steel rod or a benchmark. A monument is referred to as the true point of beginning in a metes-and-bounds legal description.

Mortar A combination of cement, sand, and water used to bond masonry units together.

Moving loads Loads that are not stationary, such as those produced by automobiles and construction equipment.

Mudroom A room or utility entrance where soiled clothing can be removed before entering the main portion of the residence.

Mudsill (base plate) The horizontal wood member that rests on concrete to support other wood members.

Mullion A horizontal or vertical divider between sections of a window.

Muntin A horizontal or vertical divider within a section of a window.

Nailer A wood member bolted to concrete or steel members to provide a nailing surface for attaching other wood members.

National CAD standard CAD guidelines established by the National Institute of Building Sciences to ensure uniform drawings.

National Council for Interior Design Qualification (NCIDQ) The board that regulates the standards to become a professional interior designer.

National Kitchen & Bath Association (NKBA) An organization created to ensure quality among kitchen and bath designers.

Natural grade Soil in its unaltered state.

Nested joists The practice of placing a steel joist around another joist so that the strength of the joist is doubled.

Net size Final size of wood after planing.

Neutral axis The axis formed where the forces of compression and tension in a beam reach equilibrium.

Newel The end post of a stair railing.

Nominal size An approximate size achieved by rounding the actual material size to the nearest larger whole number.

Nonbearing wall A wall that supports no loads other than its own.

Nonferrous metal Metal, such as copper or brass, that contains no iron.

Non-habitable spaces Areas including closets, pantries, bath or toilet rooms, hallways, utility rooms, storage spaces, garages, darkrooms, and other similar spaces.

Nosing The rounded front edge of a tread that extends past the riser.

Obscure glass Glass that is not transparent.

On center (o.c.) A measurement taken from the center of one member to the center of another member.

Oriel window A window placed in a projection of an exterior wall that does not extend all the way to the foundation. In plan view, the wall projection may be rectangular, polygonal, or curved. (See bay window.)

Orientation The locating of a structure on property based on the location of the sun, prevailing winds, view, and noise.

Oriented strand board (OSB) A sheet of material made of layers of wood chips laminated together with glue under extreme pressure. The standard size is a 4 × 8' sheet; it is typically used for the same applications as plywood.

Outlet An electrical receptacle that allows for current to be drawn from the system.

Outrigger A support for roof sheathing and the fascia that extends past the wall line perpendicular to the rafters.

Overhang The horizontal measurement of the distance the roof projects from a wall.

Palladian window A large window, divided into three parts typical of classical architectural styles, such as Georgian and Federal. The arched center section is larger than the two rectangular side sections. A typical location for a Palladian window is above the front door.

Parapet A portion of wall that extends above the edge of the roof.

Parging A thin coat of plaster used to smooth a masonry surface.

Parquet flooring Wood flooring laid to form patterns.

Partition An interior wall.

Party wall A wall dividing two adjoining spaces such as apartments or offices.

Passive solar system System that uses natural architectural means to radiate, store, and radiate solar heat.

Patio A ground-level exterior entertainment area made of concrete, stone, brick, or treated wood.

Pediment A low-pitched triangular gable based on the Greek Revival style of architecture; it is placed over a door or window or on the front of a building.

Penny The length of a nail represented by the lowercase letter d; "10d" is read as "ten penny."

Percolation test A test used to determine whether the soil can accommodate a septic system.

Permit plans Plans required by a government agency for permission to build a structure.

Perspective A drawing method that provides the illusion of depth by the use of vanishing points.

Pictorial drawing A drawing that shows an object in a three-dimensional format.

Picture plane The plane onto that the view of the object is projected.

Pier A concrete or masonry foundation support.

Pilaster A reinforcing column built into or against a masonry wall.

Piling A vertical foundation support driven into the ground to provide support on stable soil or rock.

Pitch A description of roof angle comparing the vertical rise to the horizontal run.

Plank Lumber that is 1 1/2 to 2 1/2'' thick.

Plaster A mix of sand, cement, and water used to cover walls and ceilings.

Plat A map of an area of land showing the boundaries of individual lots.

Plate Horizontal pieces of wood used at the top and bottom of a wall to keep the studs in position.

Platform framing (western platform) A building construction method where each floor acts as a platform in the framing.

Plenum An enclosed air space for transporting air from the HVAC system. The air pressure in the plenum is greater than the pressure in the structure, thus causing air to flow from the furnace through the plenum and into the residence.

Plot A parcel of land. Also used to refer to the process of making a paper copy of a CAD drawing.

Plumb True vertical.

Plywood Wood composed of three or more layers bonded with glue, with the grain of each layer placed at 90° to the next.

Pocket door A door that slides into a pocket built into the wall.

Point of beginning Fixed location on a plot of land where the survey begins.

Porch A roofed entrance to a structure that is open to the air and supported on one side by the structure with the remaining sides supported by columns or arches.

Portal frame (PF) A method of reinforcing a braced wall line that uses two panels with a minimum width of 22 1/2'' (570 mm). The panels are connected by a header to resist lateral loads.

Porte cochere A roofed area covering a driveway at the entrance to a structure that is large enough for vehicles to pass through providing shelter while entering or leaving a vehicle.

Portico A roofed area that leads to the main entrance to structure that is open to the air on one or more sides and supported on one side by the structure and on the others by columns or arches. Porticos are common on Federal, Early Classical Revival, and Greek Revival homes. Portland cement A hydraulic cement made of silica, lime, and aluminum; it has become the most common cement used in the construction industry because of its strength.

Post A vertical wood structural member usually 4 × 4 or larger.

Post-tensioning A process of reinforcing concrete slabs that are poured over unstable soil using reinforcement placed in tension.

Power-driven steel studs Smooth bolt-like fasteners driven by a powder-actuated fastening tool to attach wood members to concrete or steel members.

Precast A concrete component that has been cast in a location other than where it will be used.

Prefabricated units Buildings or components that are built away from the job site and transported there, ready to be used.

Preliminary drawings Drawings made from bubble drawings used to plan basic design concepts of a structure; such drawings comprise the site, the floor, the front elevation, and a typical cross section.

Presentation drawing Drawings used to convey basic design concepts.

Prestressed concrete A concrete component placed in compression as it is cast to help resist deflection.

Prevailing winds Winds identified by the direction from which they most frequently blow in a given area of the country.

Profile Vertical section of the surface of the ground and/or of underlying earth that is taken along any desired fixed line.

Proportion A pleasing relationship related to both size and balance. Rectangles using the proportions 2:3, 3:5, and 5:8 have long been considered to be pleasing.

Punch out A hole in the web of a steel framing member allowing for the installation of plumbing, electrical, and other trade components.

Purlin A horizontal roof member that is laid perpendicular to rafters to help limit deflection.

Purlin brace A support member that extends from the purlin down to a load-bearing wall or header.

Pyramidal roof A hipped roof that lacks a ridge. The four isosceles-triangular planes of the roof meet at a common apex resembling a pyramid. Low-slope pyramidal roofs are common on Greek Revival houses.

Quad A courtyard surrounded by the walls of buildings.

Quarry tile An unglazed machine-made tile.

Quarter round Wood molding that has the profile of one-quarter of a circle.

Quatrefoil window A round window common in Moorish, Gothic, and Mission style architecture. The window is composed of four equal lobes, like a four-petaled flower.

Quoins Heavy blocks of stone, designed to reinforce masonry wall, found at the corners of a brick building. They are often found on Georgian and some Federal and Greek Revival houses. In a brick house, the quoins usually consist of granite blocks, but they may also be formed from bricks and painted in the trim color. Wood quoins are made to imitate stone and are strictly ornamental features.

Rabbet A rectangular groove cut on the edge of a board.

Radiant barrier Material made of aluminum foil with backing that is used to stop heat from radiating through an attic.

Radiant heating Heat emitted from a particular material such as brick, electric coils, or a hot water pipe without use of air movement.

Radon A naturally occurring radioactive gas, usually found in a basement, that breaks down into carcinogenic compounds when it is inhaled over a long period of time.

Rafter The inclined structural member of a roof system designed to support roof loads.

Rafter, common A rafter that extends the full length between the support wall and the ridge board.

Rafter, hip A rafter used at the intersection of two roof planes that forms a sloping ridge (an exterior roof corner).

Rafter, jack A rafter that spans from the supporting wall to a hip or valley rafter so that the jack rafter is not the full length of a common rafter.

Rafter, valley A rafter placed at the intersection of two roof planes, so that it forms an interior roof intersection.

Rafter/ceiling joist An inclined structural member that supports both the ceiling and the roof materials.

Rake A roof extension projecting over an end wall that follows the slope of the roof.

Rake joint A recessed mortar joint.

Reaction The upward forces acting at the supports of a beam.

Rebar Reinforcing steel used to strengthen concrete.

Reference bubble A symbol used to designate the origins of details and sections.

Register An opening in a duct for the supply of heated or cooled air.

Reinforced concrete Concrete that has steel rebar placed in it to resist tension.

Relative humidity The amount of water vapor in the atmosphere compared to the maximum possible amount at the same temperature.

Remodel A construction project that involves moving, adding, or removing structural members.

Renovation The altering or removal and replacement of nonstructural materials such as cabinets as well as the making of minor electrical or mechanical repairs.

Retaining wall A wall, usually made of masonry, that is designed to resist soil loads.

R-factor A unit of thermal resistance applied to the insulating value of a specific building material.

Rheostat An electrical control device used to regulate the current reaching a light fixture. A dimmer switch.

Rhythm A principle of design related to the repetitive elements that provide order and uniformity.

Ribbon A structural wood member framed into studs to support joists or rafters.

Ridge The uppermost area of two intersecting roof planes.

Ridge beam A beam located at the ridge used to support the roof framing members. A beam can also be used as a decorative member at an interior ridge.

Ridge blocks Blocks that are placed between trusses at the ridge to maintain a uniform spacing and provide a nailing surface for the roof sheathing.

Ridge board The horizontal member at the ridge that runs perpendicular to the rafters. The rafters are aligned against the ridge to resist the downward force of the rafters.

Ridge brace A support member used to transfer the weight from the ridge board to a bearing wall or beam. The brace is typically spaced at 48'' o.c. and may not exceed a 45° angle from vertical.

Rim joist A joist at the perimeter of a structure that runs parallel to the other floor joist. Sometimes referred to as a band or header.

Rim track A metal track installed over the ends of steel floor joists to support the ends and close off the space between them.

Rise The amount of vertical distance between one tread and another. Also refers to the angle of the roof based on 12 horizontal units.

Riser The vertical member of stairs between the treads. The term is also used to refer to a water supply pipe that extends vertically one or more stories to carry water to fixtures.

Roll roofing Roofing material of fiber or asphalt that is shipped in rolls.

Roof drain A receptacle for removal of roof water.

Rough floor The subfloor, usually plywood, that serves as a base for the finished floor.

Rough lumber Lumber that has not been surfaced but has been trimmed on all four sides.

Rough opening The unfinished opening provided between framing members to allow for the placement of doors, windows, or skylights.

Rough-in To add framing for a feature such as a door or window that will be installed at a future date. The opening is covered with the finish materials until the door or window is needed. The term is also used to describe the HVAC, plumbing or electrical work that is done in the joist or stud space prior to adding the finish material.

Rowlock A pattern for laying masonry units so that the ends of the units are exposed.

Run The horizontal distance of a set of steps or the measurement describing the depth of one step. Also used to describe the horizontal measurement from the outside edge of the wall to the centerline of the ridge.

R-value Measurement of thermal resistance used to indicate the effectiveness of insulation.

Saddle A small gable-shaped roof used to divert water from behind a chimney.

Sanitary sewer A sewer line that carries sewage without any storm, surface, or groundwater.

Sash An individual frame around a window.

Sawn lumber Lumber that has been cut from logs without being altered by engineering.

Scab A short member that overlaps the butt joint of two other members used to fasten those members.

Scale A ratio that is used to reduce or enlarge the size of a drawing for plotting. Scale is also used to refer to a measuring tool.

Schedule A written list of similar components such as windows and doors.

Sconce A wall-mounted light fixture that provides indirect lighting by directing light either up or down, depending on the shape of the sconce.

Scratch coat The first coat of stucco that is scratched to provide a good bonding surface for the second coat.

Seasoning The process of removing moisture from green lumber by either air (natural) or kiln drying.

Section A type of drawing showing an object as if it had been cut through to show interior construction.

Seismic forces Earthquake-related forces.

Self-drilling screws A fastener with a drilling point that is able to penetrate heavy-gauge metal.

Septic system A sewage disposal system consisting of a storage tank and an absorption field. Sewage is stored in the septic tank and the liquid waste is dispersed into the drainage field.

Septic tank A tank where sewage is decomposed by bacteria and dispersed by drain tiles.

Service connection The wires that run to a structure from a power pole or transformer.

Setback The minimum distance required between the structure and the property line.

Shake A hand-split wooden roof shingle.

Shear The stress that occurs when two forces from opposite directions are acting on the same member. Shearing stress tends to cut a member just as scissors cut paper.

Shear panel A wall panel designed by an engineer to resist wind or seismic forces.

Shear walls Walls designed by an engineer to resist lateral loads.

Sheathing A thin covering that is usually made of OSB or plywood with a thickness of between 3/8 and 3/4'' (9.5 and 19 mm); it is placed over walls, floors, and roofs to serve as a backing for the finish materials.

Shim A piece of material used to fill a space between two surfaces.

Shiplap A siding pattern of overlapping rabbeted edges.

Sidelight A window or a series of small fixed panes arranged vertically and found on either side of the main entry door of many Federal and Greek Revival style homes.

Sill A horizontal wood member placed at the bottom of walls and openings in walls.

Simple beam Beam with a uniform load evenly distributed over its entire length and supported at each end.

Single-wall construction A construction method used in temperate climates that places the exterior finishing material directly over a water-resistant membrane and the wall studs.

Site orientation Placement of a structure on a property with certain environmental and physical factors taken into consideration.

Skip sheathing Material such as 1 × 4s (25 × 100s) that are laid perpendicular to the rafters to support roofing such as wood shingles and concrete or clay tiles.

Sky window A opening in the wall and roof that combines features of a skylight and a window using glazing on a wall that extends to meet glazing on a portion of the roof.

Skylight An opening in the roof to allow light and ventilation; it is usually covered with glass or plastic.

Slab A concrete floor system typically poured at ground level.

Sleepers Strips of wood placed over a concrete slab in order to attach other wood members.

Smoke chamber The portion of the chimney located directly over the firebox that acts as a funnel between the firebox and the chimney.

Smoke shelf A shelf located at the bottom of the smoke chamber to prevent down-drafts from the chimney from entering the firebox.

Soffit A lowered ceiling, typically found in kitchens, halls, and bathrooms to allow for recessed lighting or HVAC ducts. The term is also used to describe an enclosed area below the overhangs that is used to protect the rafter or truss tails from the elements.

Softwood Wood that comes from cone-bearing trees.

Soil line Disposal lines in the waste water system.

Soil stack The main vertical wastewater pipe.

Soil vent A vent that runs up the wall and vents out the roof, allowing vapor to escape and ventilating the system.

Solar heat Heat that comes from energy generated from sunlight.

Solar panel Solar cells that convert sunlight into electricity.

Solarium Glassed-in porch on the south side of a house that is directly exposed to the sun's rays.

Soldier A masonry unit laid on end with its narrow surface exposed.

Sole plate The plate placed at the bottom of a wall.

Spackle The covering of sheetrock joints with joint compound.

Span The horizontal distance between two supporting members.

Spark arrester A screen placed at the top of the flue to prevent combustibles from leaving the flue.

Specification writer A person who provides written specifications to supplement the working drawings regarding the quality of materials to be used and labor to be supplied.

Specifications A written statement that describe the characteristics of a particular aspect of a project, such as a description of materials or equipment, construction systems, standards, and workmanship.

Splice Two similar members that are joined together in a straight line, usually by nailing or bolting.

Split-level A house that has two levels, one about half a level above the other.

Spot grade An contour elevation determined by a survey team of a specific location at the job site.

Square An area of roofing covering 100 sq ft.

Stack A vertical drain lines that carry waste from the home to the sewer main.

Stack line The vertical drain line that carries waste from the home to the sewer main.

Stair jack See stringer.

Stairwell The opening in the floor where a stair will be framed.

Station point The position of the observer's eye in a perspective drawing.

Stations Numbers given to the horizontal lines of a grid survey.

Stick framing Framing one member at a time on the job site instead of raising prefabricated units.

Stile A vertical member of a cabinet, door, or decorative panel.

Stirrup A U-shaped metal bracket used to support wood beams.

Stock Common sizes of building materials.

Stock plans Houses with several different options designed to appeal to a wide variety of people.

Stop A wooden strip used to hold a window in place.

Storm sewer A municipal drainage system used to dispose of groundwater, rainwater, surface water, or other nonpolluting waste separately from sewage.

Stress A live or dead load acting on a structural member. Stress results as the fibers of a beam resist an external force.

Stressed-skin panel A hollow built-up member typically used as a beam.

Stretcher A course of masonry laid horizontally with the end of the unit exposed.

Stringer (stair jack) The inclined support member of a stair that supports the risers and treads.

Strong back A beam used to support ridge and ceiling loads. It is placed above the ceiling joists when perpendicular to the joists and between the joist when it is parallel to the joists.

Stucco A type of plaster made from Portland cement, sand, water, and a coloring agent that is applied to exterior walls.

Studs The vertical repetitive framing members of a wall that are usually 2 × 4 (50 × 100) or 2 × 6 (50 × 150) in size.

Subfloor The flooring surface laid on the floor joists which serves as a base layer for the finished floor.

Subsill A sill located between the trimmers and bottom side of a window opening. It provides a nailing surface for interior and exterior materials.

Sump A recessed area in a basement floor that collects water so that it can be removed by a pump.

Sun tunnel A type of sky light that delivers light into a room using a reflective, flexible tunnel to connect the roof opening with an opening in the ceiling. A diffuser is mounted at the ceiling end of the tunnel to disperse the light.

Surfaced lumber Lumber that has been smoothed on at least one side.

Survey map Map of a property showing its size, boundaries, and topography.

Swale A recessed area formed in the ground to help divert ground water away from a structure.

Tamp To compact soil or concrete.

Temporary loads Loads that must be supported for a limited time.

Tensile strength The resistance of a material or beam to the tendency to stretch.

Tension Forces that cause a material to stretch or pull apart.

Termite shield A strip of sheet metal used at the intersection of concrete and wood surfaces near ground level to prevent termites from entering the wood.

Terra cotta Hard-baked clay typically used as a liner for chimneys.

Terrain The shape and character of a given stretch of land.

Thermal break Material used to prevent or reduce the direct transmission of heat or cold between two surfaces.

Thermal conductivity The rate of heat flow through 1 sq ft (0.0929 m^2) of homogeneous material 1'' (25 mm) thick with a temperature difference of 1°F (−17.2°C) between the opposite sides of the material.

Thermal conductor A material suitable for transmitting heat.

Thermal resistance Represented by the letter R, resistance measures the ability of a material to resist the flow of heat.

Thermostat A mechanical device for controlling the output of HVAC units.

Threshold The beveled member directly under a door.

Throat The narrow opening to the chimney that is just above the firebox. The throat of a chimney is where the damper is placed.

Timber Lumber with a cross-sectional size of 4 × 6 or larger.

Toenail Nails driven into a member at an angle.

Tongue and groove A joint where the edge of one member fits into a groove in the next member.

Top plate A horizontal structural member located on top of the studs used to hold the wall together.

Topographical survey The measurement of a property that provides the contour of the land surface, with the grades measured in relation to sea level.

Topography Physical description of land surface showing its variation in elevation and location of features such as rivers, lakes, or towns.

Township A 6-sq-mi parcel of land referenced to a baselines or meridian.

Track A U-shaped member used for applications such as top and bottom track for walls and rim track for floor joists.

Traffic flow The route that people follow as they move from one area of a residence to another, using hallways or portions of a room.

Transom Originally used to denote a horizontal crossbar in a window. It later came to mean a window positioned above such a crossbar. Today the term is most commonly used for a shallow, rectangular window located immediately above a door.

Trap A U-shaped vented fitting below plumbing fixtures that provides a liquid seal to prevent the emission of sewer gases without affecting the flow of sewage or waste water.

Tray ceiling A ceiling constructed with the sides angling at approximately 45° or curving to a higher flat ceiling so that it resembles an inverted tray.

Tread The horizontal member of a stair on which the foot is placed.

Tributary width The accumulation of loads directed to a structural member. It is always half the distance between the beam to be designed and the next bearing point.

Trimmer A stud of less than full height that is used to support a header. The term is also used to describe a joist or rafters used to frame an opening in a floor, ceiling, or roof.

Truss A prefabricated or job-built construction member formed of triangular shapes used to support a roof or floor loads over long spans.

Truss clips See hurricane ties. Used to tie the roof framing members to the wall.

Ultimate strength The unit stress within a member just before it breaks.

Underlayment Thin material, usually plywood, waferboard, or hardboard 3/8 or 1/2'' (9 or 13 mm) thick, used to provide a smooth, impact-resistant surface on which to install the finished flooring.

UniFormat An arrangement of construction information based on physical parts of a facility called systems and assemblies.

Unit stress The maximum permissible stress a structural member can resist without failing.

Unity A principle of design related to the common design or decorating pattern that ties a structure together.

Uplift The tendency of structural members to move upward due to wind or seismic pressure.

Valley The internal corner formed between two intersecting roof structures.

Value The darkening or lightening of a hue.

Valve A fitting that is used to control the flow of fluid or gas to a fixture or appliance.

Vapor barrier Material used to block the flow of water vapor into a structure. Typically 6-mil (0.006'') black plastic.

Varge rafter See barge rafter.

Variance A legal request by a property owner to allow a modification from a standard or requirement of the zoning code.

Vault An inclined ceiling area.

Veneer A thin outer covering or non-load–bearing masonry face material.

Vent pipes Pipes that allow air into the waste lines to facilitate drainage. Each plumbing fixture is connected to a vent stack.

Vent stack A vertical pipe of a plumbing system used to equalize pressure within the system and vent sewer gases.

Ventilation The process of supplying and removing air from a structure.

Vertical shear A stress acting on a beam, that causes a beam to drop between its supports.

Vestibule A small entrance or lobby.

Vicinity map A map that shows major roads and landmarks surrounding a specific area.

Virtual reality A computer-simulated world that appears to be real.

Volt A unit of measurement of electrical force or potential that makes electricity flow through an electrical wire. For a specific load, the higher the voltage, the more electricity will flow.

Wainscot Paneling applied to the lower portion of a wall.

Wallboard Large, flat sheets of gypsum, typically 3/8, 1/2, or 5/8'' (9.5, 13, or 16 mm) thick, used to finish interior walls.

Warp Variation from true shape.

Water closet A toilet.

Waterproof Material or a type of construction that prevents the absorption of water.

Watt A unit of power measurement based on the potential and the current. The amount of power to be used per fixture is be determined by multiplying the current (amps) by volts (potential) to determine the watts.

Weather strip A fabric or plastic material placed along the edges of doors, windows, and skylights to reduce air infiltration.

Web stiffener Additional material that is attached to the web of a steel W- or C-section member to strengthen the member against web crippling.

Webs Interior members of the truss that span between the top and bottom chord.

Weep hole An opening in the bottom course of a masonry to allow for drainage.

Weld A method of providing a rigid connection between two or more pieces of steel.

Wetland Lowland areas such as marshes, swamps, and bogs that in their normal condition are saturated with moisture and therefore provide the natural habitat for certain wildlife.

Widow's walk A deck above the highest level of the roof from which sea captains' wives were supposed to have watched for their husbands' ships.

Work triangle The relationship between the work areas formed by drawing lines from the centers of the storage, preparation, and cleaning areas. This triangle outlines the main traffic area required to prepare a meal with food taken from the refrigerator, cleaned at the sink, and cooked at the microwave or stove, and then returning leftovers to the refrigerator.

Working drawings The drawings that will be used to obtain a permit and build a structure.

Wythe A single-unit thickness of a masonry wall.

Zoning An ordinance that regulates the location, size, and type of a structure in a building zone.

Zoning regulations Limits to the uses of property and the structures that may be built by controlling the use of land, lot sizes, types of structure permitted, building heights, setbacks, and density (the ratio of land area to improvement area).

Acronyms

AAC	Autoclaved aerated concrete
ACH	Air exchange per hour
ACI	American Concrete Institute
ADA	Americans with Disabilities Act
AFT	Advanced framing techniques
AFPA	American Forest and Paper Association
AFT	Advanced framing techniques
AIA	American Institute of Architects
AIBD	American Institute of Building Designers
AISC	American Institute of Steel Construction
AITC	American Institute of Timber Construction
ANSI	American National Standards Institute
APA	The Engineered Wood Association (Formally the American Plywood Association)
ASAI	American Society of Architectural Illustrators
ASCE	American Society of Civil Engineers
ASHRAE	American Society of Heating, Refrigerating, and Air Conditioning
ASID	American Society of Interior Designers
ASLA	American Society of Landscape Architects
ASLD	American Society of Landscape Architects
ASTM	American Society for Testing and Materials
AKBD	Associate of Kitchen & Bath Designer
BOCA	Building Officials and Code Administrators International, Inc.
CAD	Computer-aided design
CADD	Computer-aided drafting and design
CC&R	Covenants, Conditions, and Restrictions
CSA	Canadian Standards Association
CMKBD	Certified Master Kitchen & Bathroom Designer
CMU	Concrete masonry units
CSI	The Construction Specifications Institute

DOE	Department of Energy's Energy Network
EPA	U.S. Environmental Protection Agency
EPS	Expanded polystyrene Foam
FHA	Federal Housing Authority
FSC	Forest Stewardship Council
HGTV	Home & Garden Television
HDF	High density fiberboard
HUD	Department of Housing and Urban Development
IBC	International Building Code
ICC	International Code Council
ICBO	International Conference of Building Officials
IRC	International Residential Code
LSL	Laminated strand lumber
LVL	Laminated veneer lumber
LRFD	Load reduction factor design standards
MBI	Modular Building Institute
MDB	Medium density fiberboard
MDL	Mechanically evaluated lumber
NAHB	National Association of Home Builders
NBS	National Bureau of Standards
NCIDQ	National Council for Interior Design Qualification
NCMA	National Concrete Masonry Association
NDS	National Design Specifications
NFPA	National Fire Protection Association
NIBS	National Institute of Building Sciences
NIST	National Institute of Standards and Technology
NKBA	National Kitchen & Bath Association
NMHC	National Multi-Housing Council
NSPE	National Society of Professional Engineers
OSB	Oriented Strand Board
OSHA	Occupational Safety & Health Administration
PSL	Parallel strand lumber
PTI	Post Tensioning Institute
PVC	Polyvinyl chloride
SBC	Standard Building Code
SBCCI	Southern Building Code Congress International, Inc.
SFPA	Southern Forest Products Association
SIP	Structural insulated panels
SPIB	Southern Pine Inspection Bureau
TLC	The Learning Channel
T&G	Tongue and groove
UBC	Uniform Building Code
UL	Underwriters Laboratories Inc.
USMA	U.S. Metric Association
VOC	Volatile organic compound
WWPA	Western Wood Products Association
WBD	We be done

INDEX